Biology 12

Biology 12

Authors

Maurice Di Giuseppe
Toronto Catholic District School Board

Angela Vavitsas
Toronto District School Board

Dr. Bob Ritter
Edmonton Catholic School Board

Douglas Fraser
District School Board Ontario North East

Anu Arora
Peel District School Board

Beth Lisser
Peel District School Board

Program Consultant

Maurice Di Giuseppe
Toronto Catholic District School Board

THOMSON

NELSON

Australia   Canada   Mexico   Singapore   Spain   United Kingdom   United States

THOMSON
★
NELSON

Authors
Maurice Di Giuseppe, Angela Vavitsas,
Dr. Bob Ritter, Douglas Fraser,
Anu Arora, Beth Lisser

Contributing Writers
Michael Pidgeon
Barry LeDrew

Director of Publishing
David Steele

Publisher
Kevin Martindale

Program Manager
Lisa Dimson

Project Editor
Lee Ensor

Developmental Editors
Vicki Austin
Barb Every
Lee Geller
Susan Quirk
Tony Rodrigues

Editorial Assistant
Lisa Kafun

Senior Managing Editor
Nicola Balfour

Senior Production Editor
Joanne Close

Copy Editor
Susan Till

Proofreader
Dawn Hunter

On-line Quiz Editor
Karim Dharssi

Production Coordinator
Sharon Latta Paterson

Creative Director
Angela Cluer

Art Director
Ken Phipps

Art Management
Kyle Gell
Allan Moon
Suzanne Peden

Illustrators
AMID Studios
Greg Banning
Andrew Breithaupt
Steven Corrigan
Deborah Crowle
Kyle Gell design
Imagineering
Irma Ikonen

Kathy Karakasidis
Margo Davies LeClair
Dave Mazierski
Dave McKay
Marie Price
Dino Pulera
Bart Vallecoccia
Cynthia Watada

Design and Composition Team
AMID Studios
Kyle Gell design

Interior Design
Kyle Gell
Allan Moon

Cover Design
Ken Phipps

Cover Image
Nancy Kedersha/Science Photo
Library

Photo Research and Permissions
Robyn Craig

Printer
Transcontinental Printing Inc.

**National Library of Canada
Cataloguing in Publication Data**

Main entry under title:
Nelson biology 12 / Maurice
Di Giuseppe ... [et al.].

Includes index.
ISBN 0-17-625987-2

1. Biology. I. Di Giuseppe, Maurice
II. Title: Biology 12. III. Title:
Biology twelve. IV. Title: Nelson
biology twelve.

QH308.7.N454 2002   570
C2002-901578-2

Reviewers

▸ CONTENTS

▶ Unit 4 Evolution

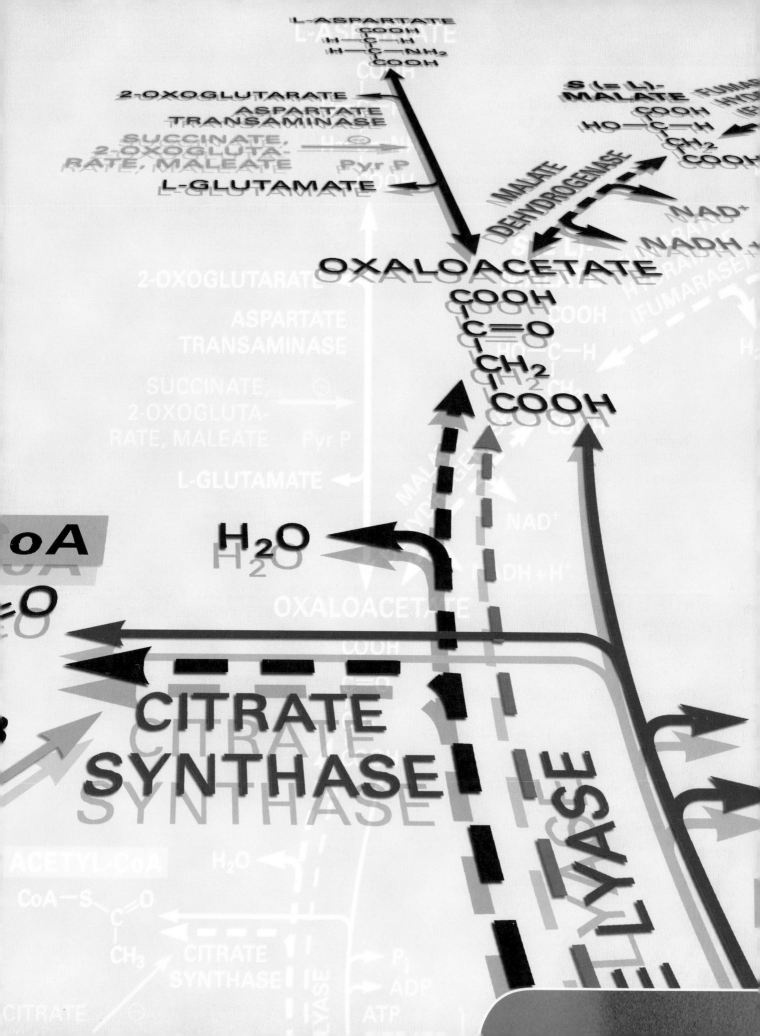

Metabolic Processes

Life is not easy to define. It has been said that living organisms are more than the sum of their parts. While philosophers continue to debate the mysteries of life, biotechnologists, physicians, ecologists, and others who work directly with living things require a scientific definition to guide their actions and decisions in their work. Extensive research over the past 150 years has confirmed that the cell is the basic unit of life and that new cells arise only by the reproduction of already existing cells.

At the most basic level, the cell is a highly organized assemblage of atoms and molecules, programmed by genetic instructions in DNA to efficiently carry out the chemical reactions that produce the telltale signs of life. These signs include reproduction, growth and differentiation, response to internal and external stimuli, and evolution by natural selection. You will study each of these life-sustaining processes in the units that follow, but first you must become familiar with the atoms, molecules, and reactions that form the foundation on which life is built.

Living organisms are characterized by a balance between chemical reactions that degrade molecules (catabolism) and those that build them up (anabolism). These reactions—called metabolism—also govern the energy economy of the cell or organism. In this unit, you will learn about metabolic processes and the chemical pathways in which they occur.

▶ Overall Expectations

In this unit, you will be able to

- describe the structure and function of the macromolecules necessary for the normal metabolic functions of all living things, and the role of enzymes in maintaining normal metabolic functions;

- investigate the transformation of energy in the cell, including photosynthesis and cellular respiration, and the chemical and physical properties of biological molecules;

- explain ways in which knowledge of the metabolic processes of living systems can contribute to technological development and affect community processes and personal choices in everyday life.

▶ Prerequisites

Concepts

- cell theory
- differences between animal and plant cells
- atomic theory and chemical bonds
- carbohydrates, lipids, proteins, and nucleic acids
- enzymes
- phospholipid bilayer membrane
- diffusion, osmosis, facilitated diffusion, active transport
- endocytosis and exocytosis
- energy flow in photosynthesis and cellular respiration
- ATP and cellular energy

Skills

- select and use suitable instruments to precisely measure length and mass
- use chemical reagents, laboratory glassware, and heating apparatus in an appropriate and safe manner
- calculate the concentration of a solute in an aqueous solution

Knowledge and Understanding

1. Identify the cell components shown in **Figure 1**, and describe a function or characteristic of each component.

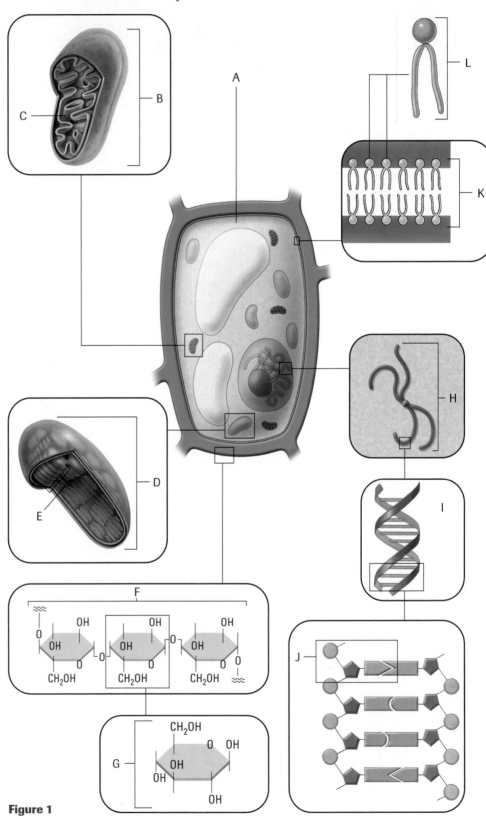

Figure 1

2. Match the term in column A with the corresponding image, formula, or equation in column B.

A.

1. diffusion

2. ATP

3. protein

4. osmosis

5. polysaccharide

6. photosynthesis

7. endocytosis

8. lipids

9. cellular respiration

B.

(a)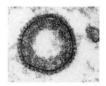

(b)

(c) $6CO_2 + 12H_2O + light \longrightarrow C_6H_{12}O_6 + 6O_2 + 6H_2O$

(d)

(e)

(f)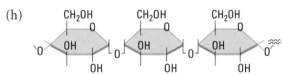

(g) $C_6H_{12}O_6 + 6O_2 + 36ADP + 36Pi \longrightarrow CO_2 + 6H_2O + 36ATP$

(h)

(i)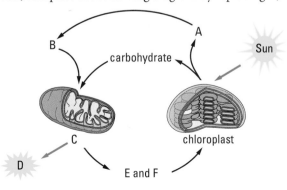

selectively permeable membrane   before   after

3. In your notebook, complete the following diagram by replacing A, B, C, D, E, and F with the correct labels.

A

B

carbohydrate

Sun

chloroplast

C

D

E and F

1

The Chemical Basis of Life

In this chapter, you will be able to

- define terms related to metabolic processes;
- apply the laws of thermodynamics to respiration and photosynthesis;
- explain how the functional groups within biological molecules contribute to the molecule's function;
- describe the chemical structure, mechanisms, and dynamics of enzymes;
- describe the reactions of hydrolysis, condensation, reduction and oxidation, and neutralization;
- describe how molecules function within energy transformations in the cell;
- investigate the structures of biological molecules by using computer-generated images and molecular models;
- design and carry out an experiment on acids and bases, metabolic rate, membrane transport, and enzyme activity;
- describe technological applications of enzyme activity;
- explain the relevance of metabolic processes in your personal life.

It is amazing to realize that many millions of complex chemical reactions take place within our bodies every moment of our lives. What is even more remarkable is that they occur without us being consciously aware of them. Supplied with the necessary raw materials, our cells carry out the functions of life automatically. We don't know the molecular details until we purposefully conduct investigations to find out.

For the past two hundred years, biologists have studied cells and organisms at every level and have discovered that chemical reactions are the basis of virtually every biological process. Although we have come to understand much about the chemistry of biology, we still have a long way to go to solve the remaining mysteries. Nevertheless, we have gathered much evidence for the chemical basis of life.

Most animals die in several weeks without food, a few days without water, and a few minutes without oxygen. Many species of archaebacteria will die in the presence of oxygen. Without a daily intake of magnesium, body muscles become sore and weak, the human brain fails to work properly, and plants turn yellow and die. The common herbicide 2,4-D stimulates weeds to grow in such an uncoordinated fashion that they literally grow themselves to death, and the anti-cancer drug, Taxol, stops cell growth by disrupting spindle fibre function during mitosis. These and other effects confirm that living organisms are composed of chemicals that react with each other and with the great variety of substances in the environment. Chemical reactions transform tadpoles into frogs and caterpillars into butterflies; they make leaves change colour in the fall; they make us feel hungry when we need food. When we are scared, chemical reactions in our muscles allow us to run away from danger. Chemistry drives life.

REFLECT on your learning

1. What property of water explains each of the following observations?
   (a) Sugar, salt, alcohol, and acids dissolve well in water.
   (b) Wet clothes take a long time to dry at room temperature.
2. What does the expression "like dissolves like" mean?
3. Why are carbon-based molecules so common in living organisms?
4. Why can cows survive on grass but humans cannot?
5. Peeled potatoes turn brown. Why does a quick dip in scalding water stop the browning reaction?

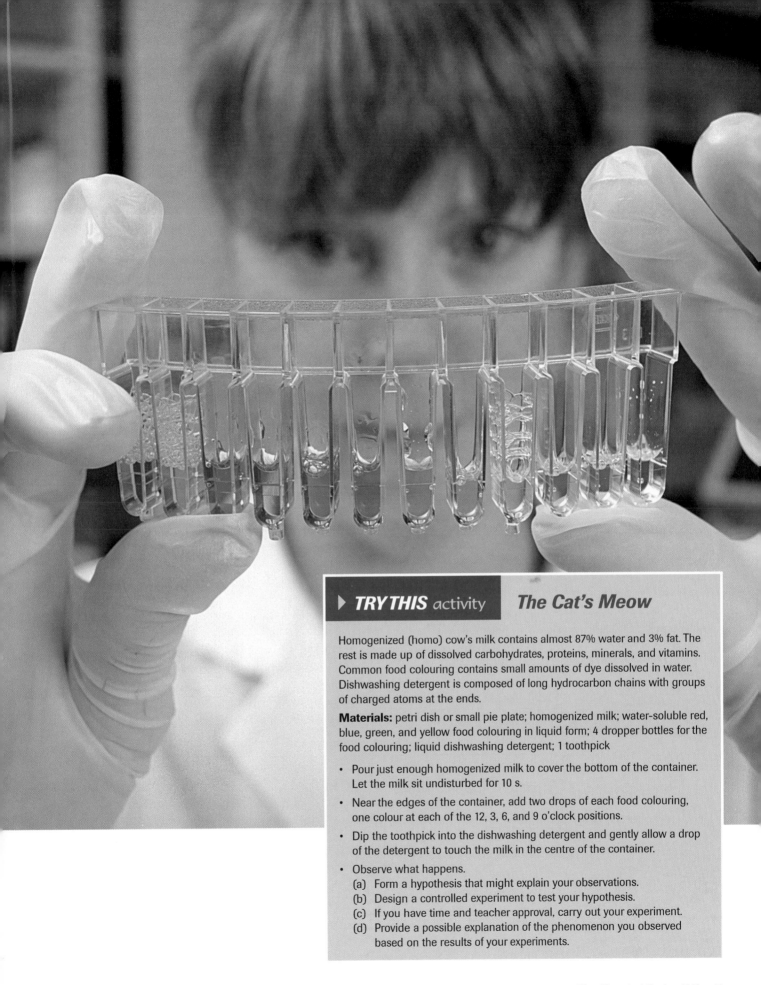

▶ *TRY THIS* activity   *The Cat's Meow*

Homogenized (homo) cow's milk contains almost 87% water and 3% fat. The rest is made up of dissolved carbohydrates, proteins, minerals, and vitamins. Common food colouring contains small amounts of dye dissolved in water. Dishwashing detergent is composed of long hydrocarbon chains with groups of charged atoms at the ends.

Materials: petri dish or small pie plate; homogenized milk; water-soluble red, blue, green, and yellow food colouring in liquid form; 4 dropper bottles for the food colouring; liquid dishwashing detergent; 1 toothpick

- Pour just enough homogenized milk to cover the bottom of the container. Let the milk sit undisturbed for 10 s.
- Near the edges of the container, add two drops of each food colouring, one colour at each of the 12, 3, 6, and 9 o'clock positions.
- Dip the toothpick into the dishwashing detergent and gently allow a drop of the detergent to touch the milk in the centre of the container.
- Observe what happens.
  (a) Form a hypothesis that might explain your observations.
  (b) Design a controlled experiment to test your hypothesis.
  (c) If you have time and teacher approval, carry out your experiment.
  (d) Provide a possible explanation of the phenomenon you observed based on the results of your experiments.

mass number, number of
protons and neutrons

$$^A_Z X$$

atomic number,
number of protons

Figure 1
Symbol representing an individual
atom of an element

Living things are composed of matter. Matter has mass, occupies space, and is found in a bewildering variety of forms. According to the Bohr–Rutherford model, the atom is composed of an extremely small nucleus containing positively charged protons (p^+) and neutral neutrons (n^0) surrounded by tiny negatively charged electrons (e^-). The atomic number, Z, is equal to the number of protons in the nucleus and also the number of electrons in a neutral atom. The mass number, A, equals the sum of the protons and neutrons in the nucleus. The elements are arranged in order of increasing atomic number in the periodic table. Carbon is element six on the table. It has six protons and six neutrons in its nucleus and six electrons surrounding the nucleus. Thus, for carbon, Z = 6 and A = 12. It is useful to symbolize individual atoms with a shorthand notation that places the atomic number at the bottom left corner of the element's chemical symbol and the mass number at the top left corner (**Figure 1**). An atom of sulfur may be represented as $^{32}_{16}S$. This atom may also be symbolized as S-32 or sulfur-32.

Isotopes

The masses of the elements given in the periodic table are not whole number values because, in nature, most elements contain atoms with different numbers of neutrons. We call these atoms isotopes. Isotopes are atoms of an element with the same atomic number but a different mass number. Since the atomic number is the same, the number of protons and electrons are the same. A difference in the number of neutrons in the nucleus distinguishes isotopes from one another.

The element carbon, C, a major constituent of living organisms, consists of three isotopes: $^{12}_6C$, $^{13}_6C$, and $^{14}_6C$. Carbon-12 accounts for about 99% of the carbon atoms in nature. Carbon-13 makes up most of the rest, with carbon-14 present in trace amounts. In a series of investigations led by Ernest Rutherford (1871–1937) and Canadian scientist Harriet Brooks (1876–1933), it was discovered that the nucleus of some isotopes spontaneously breaks apart or decays. Isotopes that can decompose in this way are called **radioisotopes** and are said to be radioactive. Radioactivity results in the formation of different elements, the release of a number of subatomic particles, and radiation. Carbon-12 and carbon-13 are stable isotopes of carbon, but carbon-14 is radioactive. It spontaneously decays into nitrogen-14. Hydrogen is composed of three isotopes: ^1_1H (protium), ^2_1H (deuterium), and radioactive ^3_1H (tritium). **Table 1** summarizes the basic characteristics of the three isotopes of carbon and hydrogen.

radioisotopes radioactive atoms of an element that spontaneously decay into smaller atoms, subatomic particles, and energy

Table 1 Three Isotopes of Carbon and Hydrogen

Name, symbol	Atomic number (Z)	Mass number (A)	Protons	Neutrons	Relative abundance	Structural stability
carbon-12, $^{12}_6C$	6	12	6	6	98.9%	stable
carbon-13, $^{13}_6C$	6	13	6	7	1.1%	stable
carbon-14, $^{14}_6C$	6	14	6	8	trace	radioactive
hydrogen-1, ^1_1H	1	1	1	0	99.8%	stable
hydrogen-2, ^2_1H	1	2	1	1	0.2%	stable
hydrogen-3, ^3_1H	1	3	1	2	trace	radioactive

Every radioactive isotope has a characteristic property called its **half-life**. The half-life of a radioisotope is the time it takes for one half of the atoms in a sample to decay. The half-life of different radioisotopes varies considerably, but the rate of decay of a particular isotope is constant. Radioisotopes are both useful and dangerous. Two useful applications of radioisotopes are radiometric dating and radioactive tracers.

Carbon-14 makes its way into plants as they absorb a mixture of radioactive and non-radioactive carbon dioxide from the air or water for photosynthesis. It enters the bodies of other organisms through the food chain. When the organism is alive, the ratio of carbon-12 to carbon-14 is the same as it is in the atmosphere. When an organism dies, it stops absorbing carbon from the environment. The amount of carbon-12 remains constant, but the amount of carbon-14 decreases in a predictable way because it is radioactive. Measuring the ratio of carbon-12 to carbon-14 in a dead or fossilized organism allows scientists to calculate the time that has elapsed since the organism's death. This process is known as carbon-14 dating, one type of radiometric dating.

Cells generally use radioactive atoms in the same way that they use nonradioactive isotopes of the same element. However, since radioisotopes emit radiation as they decay, their location may be readily detected. **Radioactive tracers** are radioisotopes used to follow chemicals through chemical reactions and to trace their path as they move through the cells and bodies of organisms. Radioactive isotopes have found many applications in biological, chemical, and medical research.

Radiolabelled molecules (molecules containing specific radioisotopes) are synthesized to investigate a variety of reaction mechanisms and biochemical processes. Melvin Calvin, a pioneer in photosynthesis research, used carbon-14-labelled molecules to determine the sequence of reactions in photosynthesis (Chapter 3). Radioisotopes are used in the study of many biochemical reactions (covered in Units 1 and 3) and in DNA sequencing procedures (covered in Unit 2). Since most compounds of biological importance contain carbon and hydrogen, carbon-14 and hydrogen-3 (tritium) are commonly used as tracers in biological research.

Besides being research tools, radioisotopes are used in the relatively new field of nuclear medicine for diagnosis and treatment (**Table 2**). For example, the thyroid gland produces hormones that have great influence over growth and metabolism. This gland is located in front of the windpipe and is the only organ of the body that actively absorbs iodine. If a patient's symptoms point to abnormal levels of thyroid hormone output, the physician may administer a small amount of radioactive iodine-131 and then use a photographic device to scan the thyroid gland. The resulting images help to identify the possible causes of the condition. **Figure 2** shows three images obtained using this technique.

half-life the time it takes for one half of the nuclei in a radioactive sample to decay

radioactive tracers radioisotopes that are used to follow chemicals through chemical reactions and trace their path as they move through the cells and bodies of organisms

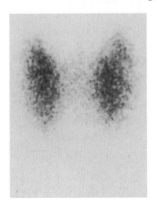
normal

enlarged

cancerous

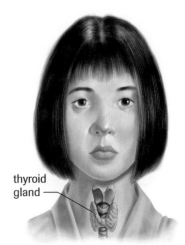
thyroid gland

Figure 2
Location of thyroid gland and scans of the thyroid gland from three patients

The Chemical Basis of Life **9**

Table 2 Radioisotopes Used in Nuclear Medicine

Radioisotope	Medical uses	Half-life
technetium-99 (the most commonly used radioisotope in medicine)	to view the skeleton and heart muscle but also the brain, thyroid, lungs, liver, spleen, kidney, gall bladder, bone marrow, and salivary glands	6.02 h
iodine-125	to evaluate the filtration rate of the kidneys and to determine bone density measurements	42 d
iodine-131	to view and treat thyroid, liver, and kidney diseases and various cancers	8.0 d
phosphorus-32	to treat polycythemia vera (excess red blood cells)	14.3 d
strontium-89	to relieve the pain of secondary cancers lodged in the bone	46.7 h
indium-111	to study the brain, the colon, and sites of infection	2.8 d
fluorine-18	to image tumours and localized infections	110 min

Figure 3
Scientist wearing a dosimeter badge

Although very useful in biological research and medicine, radiation from decaying radioisotopes poses a hazard to living organisms. High-energy radiation penetrates tissues and damages cellular molecules, and regular exposure can lead to radiation sickness, genetic mutations, and, at high enough doses, cell death. Because of this, living organisms must be shielded from sources of high-energy radiation. Researchers and technicians who regularly work with radioactive materials wear radiation-sensitive badges called dosimeters to monitor their daily exposure (**Figure 3**).

▶ *Practice*

Understanding Concepts

1. Determine the name of the element, the atomic number, and the number of neutrons in the nucleus of an atom that has 19 protons and a mass number of 39.

2. Write the shorthand notation for the element containing
   (a) 24 protons
   (b) 15 electrons, and 17 neutrons

3. Which subatomic particle of the nucleus is present in the same amount in all isotopes of a particular element?

4. Iodine-131 has a half-life of eight days. What mass of iodine-131 would remain from a 20-g sample after 32 days?

5. Why is carbon-14 useful in radioactive dating, but not as useful in nuclear medicine? The half-life of C-14 is 5730 years.

Applying Inquiry Skills

6. In humans, dietary calcium helps maintain strong bones and teeth. Calcium is absorbed into the bloodstream by the cells that line the small intestine. Bone cells called osteoblasts absorb calcium from the blood and use it to form the mineralized part of bone called bone matrix. Osteoporosis is a condition in which bones become thin and brittle because of a decrease in the density of bone matrix. Explain how radioisotopes could be used to determine whether intestinal cells or osteoblasts lack the ability to process calcium in a person with osteoporosis.

Making Connections

7. Comment on this statement: "The health benefits gained from the use of radioactive pharmaceuticals in this generation threaten the health of generations to come."

Answers

1. K, Z=19, number of neutrons=20
2. (a) $^{52}_{24}$Cr
   (b) $^{32}_{15}$P
4. 1.25 g

Chemical Bonding

Electrons are not arranged haphazardly about the nucleus of an atom. They occupy positions at various distances from the nucleus, depending on their energy content. These positions are called energy levels because as the electron moves to new positions farther from the nucleus, its potential energy increases. The energy levels are called $n=1$, $n=2$, $n=3$, and so forth ($n=1$ is closest to the nucleus). Early experiments in atomic structure conducted by Ernest Rutherford (1911), Niels Bohr (1913), and others seemed to indicate that electrons move in the space about the nucleus in circular orbits like planets revolve around the sun.

However, more recently we have learned that electrons are so small that it is impossible to know exactly where they are and how they are moving at any given time. Today, using statistical analyses, scientists can determine locations around the nucleus where electrons are most likely to be found. These volumes of space are called **orbitals**. An orbital can accommodate no more than two electrons. The energy levels around the nucleus of an atom have different orbitals; however, similarly shaped orbitals may be found at each level. An electron in the first energy level, $n=1$, most likely occupies a spherical orbital called the $1s$ orbital, as shown in **Figure 4(a)** (electron orbitals). When two electrons occupy an orbital, they pair up, forming a more stable arrangement than does an electron by itself. Although electrons occupy positions within three-dimensional spaces about the nucleus, it is still convenient to illustrate them as dots on shells surrounding the nucleus, as illustrated in **Figure 4** (electron-shell diagrams). In the second energy level, $n=2$, electrons may occupy a spherical orbital called the $2s$ orbital (larger than the $1s$ orbital), or any one of three dumbbell-shaped $2p$ orbitals, as depicted in **Figure 4(b)**. Since only two electrons can occupy the same orbital, energy level $n=1$ may contain up to two electrons (one pair), and energy level $n=2$ may contain up to eight electrons (four pairs). Higher energy levels contain more electrons with higher energy values, and the orbitals are called $3s$, $3p$, $3d$, and so on. The arrangement of electrons in the orbitals is called the atom's electron configuration. **Table 3** (page 12) lists the number of electrons that may occupy orbitals in energy levels 1 to 3.

orbitals volumes of space around the nucleus where electrons are most likely to be found

	(a) 1s orbital	**(b)** 2s and 2p orbitals		**(c)** Neon ($_{10}$Ne): 1s, 2s, and 2p
electron orbitals	1s orbital (2e⁻)	2s orbital (2e⁻)	2p orbitals 2p$_z$ orbital (2e⁻) 2p$_y$ orbital (2e⁻) 2p$_x$ orbital (2e⁻)	
electron-shell diagrams				

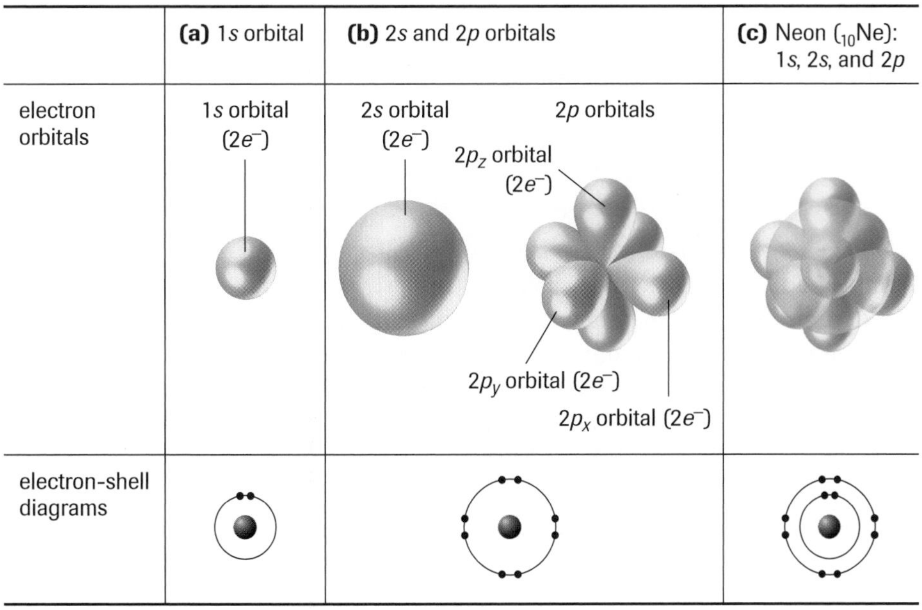

Figure 4
Each orbital can hold up to two electrons.

(a) The $1s$ orbital is in the energy level closest to the nucleus, $n=1$.

(b) The $2s$ orbital is in energy level $n=2$ and is spherical like the $1s$ orbital. Energy level $n=2$ also contains three dumbbell-shaped $2p$ orbitals oriented at right angles to one another along the x-axis ($2p_x$), along the y-axis ($2p_y$), and along the z-axis ($2p_z$).

(c) The noble gas neon has 10 electrons, two in a $1s$ orbital, two in a $2s$ orbital, and one pair in each of the three $2p$ orbitals. It is an unreactive element because it has full outer s and p orbitals.

Table 3 Number of Electrons in Various Energy Levels

Energy level (n)	Orbital types	Maximum number of electrons allowed
1	1s	2
2	2s, 2p	8
3	3s, 3p, 3d	18

valence electrons electrons located in outermost s and p orbitals that determine an atom's chemical behaviour

The outermost s and p orbitals are called valence orbitals, and the electrons in these orbitals are called **valence electrons**. The chemical behaviour of an atom is determined by the number and arrangement of its valence electrons. A special stability is associated with having full outermost s and p orbitals. Helium (Z=2) contains two valence electrons in a 1s orbital. Since there are no p orbitals in energy level n=1, the 1s orbital is the only valence orbital. Helium is inert (nonreactive) because the 1s orbital is full with a pair of electrons. Neon (Z=10), shown in **Figure 4(c)** on page 11 is inert because its outermost s and p orbitals, the 2s and 2p orbitals, are full with four pairs of electrons (one pair in the 2s orbital and one pair in each of the three 2p orbitals). These elements are called noble gases because they do not attempt to gain, lose, or share electrons with other atoms; they are stable and do not normally participate in chemical reactions. All other elements attempt to gain, lose, or share valence electrons to achieve the same stable electron configuration as a noble gas. These interactions are the root cause of chemical reactions and are responsible for the formation of chemical bonds between atoms.

It is convenient to draw diagrams of the elements showing only the valence electrons. Such diagrams are called Lewis dot diagrams (**Figures 5 and 6**).

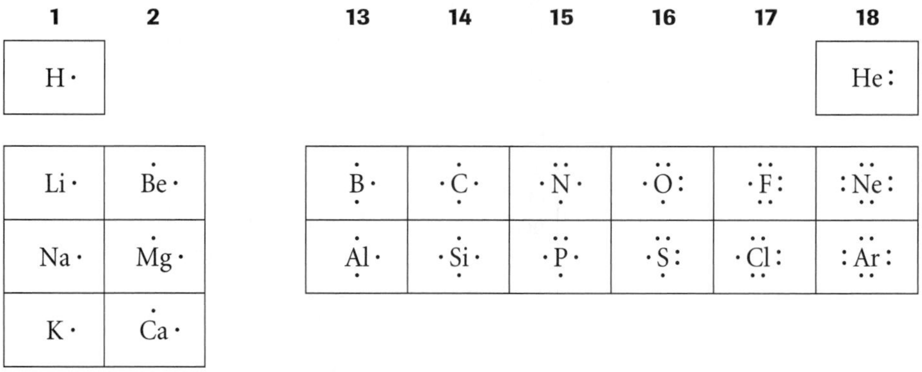

1	2		13	14	15	16	17	18
H·								He:
Li·	Be·		B·	·C·	·N·	·O:	·F:	:Ne:
Na·	Mg·		Al·	·Si·	·P·	·S:	·Cl:	:Ar:
K·	Ca·							

Figure 5
Lewis dot diagrams of the first 20 elements

The periodic table is an arrangement of the elements in order of increasing atomic number. Vertical columns on the table are called groups or families; horizontal rows are called periods. Elements in the same group contain the same number of valence electrons. Thus, group 1 elements all contain one valence electron, group 2 elements contain two valence electrons, and group 17 elements contain seven valence electrons. The noble gases constitute group 18 and contain eight valence electrons (a stable octet).

LEARNING TIP

Lewis Dot Diagrams
When drawing Lewis dot diagrams, first place electrons one at a time at the 12, 3, 6, and 9 o'clock positions, as required. Then, add more electrons by pairing them one at a time in the same order as you placed the first four (**Figure 6**).

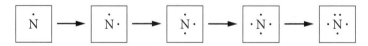

Figure 6
The progressive placement of electrons in Lewis dot diagrams

Compounds are stable combinations of atoms of different elements held together by chemical bonds. Two of the most common bonds are the ionic bond and the covalent bond. When atoms lose electrons, they become positively charged ions called cations. When atoms gain electrons, they become negatively charged ions called anions. An ionic bond is a force of attraction between cations and anions. An ionic compound, such as sodium chloride, $NaCl_{(s)}$, is also called a salt or an ionic solid. A covalent bond is formed when two atoms share one or more pairs of valence electrons. In a hydrogen molecule, two hydrogen atoms share a pair of electrons (one from each atom) and, therefore, possess a single covalent bond. Some molecules contain double and triple covalent bonds. For example, oxygen molecules, $O_{2(g)}$, are composed of two oxygen atoms held together with a double covalent bond (two shared pairs of electrons). Nitrogen molecules, $N_{2(g)}$, are composed of two nitrogen atoms with a triple covalent bond between them (three shared pairs of electrons).

Covalent bonds are stronger than ionic bonds. Ionic and covalent bonds are generally called **intramolecular forces of attraction** (Because they hold the atoms of a molecule or ions of an ionic solid together). Compounds may be described by using chemical formulas and Lewis diagrams (**Table 4**).

intramolecular forces of attraction the covalent bond that holds the atoms of a molecule together, and the ionic bonds that hold ions together in a salt

Table 4 Ionic and Covalent Substances

Bond type	Constituent entities	Force of attraction	Chemical formula	Lewis diagram[1]	Name
ionic	metal cations[2] and nonmetal anions[3]	electrostatic attraction between oppositely charged ions	$NaCl_{(s)}$	$Na+[Cl^-]$	sodium chloride
			$MgF_{2(s)}$	$[F^-]Mg^{2+}[F^-]$	magnesium fluoride
covalent	neutral atoms	electrostatic attraction between nuclei and valence electrons of neutral atoms	$H_2O_{(l)}$		water
			$CO_{2(g)}$		carbon dioxide
			$C_6H_{12}O_{6(aq)}$		glucose

[1] For covalent molecules, each horizontal or vertical line represents a pair of electrons.

[2] From groups 1, 2, or 3 of the periodic table

[3] From groups 16 or 17 of the periodic table

Electronegativity and the Polarity of Covalent Bonds

electronegativity a measure of an atom's ability to attract a shared electron pair when it is participating in a covalent bond

All the atoms in the periodic table have been assigned an **electronegativity** number (E_n) on the basis of experimental results. The larger the electronegativity number, the stronger the atom attracts the electrons of a covalent bond. **Figure 7** shows a periodic table that indicates electronegativity numbers. Note that the noble gases are not assigned an E_n because they do not participate in chemical bonding ($E_n = 0$ for noble gases). When an electron pair is unequally shared, the atom that attracts the pair more strongly takes on a partial negative charge ($\delta-$) and the atom that attracts more weakly takes on a partial positive charge ($\delta+$). The lowercase Greek letter delta (δ) denotes a partial charge. The charge is partial because the electron pair spends time around both atoms, but spends more time around one atom than it does around the other. This difference in electron attraction forms a polar covalent bond.

1	2	3	4	5	6	7	8	9	10	11	12	13	14	15	16	17	18
H 2.1																	He –
Li 1.0	Be 1.5											B 2.0	C 2.5	N 3.1	O 3.5	F 4.1	Na –
Na 1.0	Mg 1.3											Al 1.5	Si 1.8	P 2.1	S 2.4	Cl 2.9	Ar –
K 0.9	Ca 1.1	Sc 1.2	Ti 1.3	V 1.5	Cr 1.6	Mn 1.6	Fe 1.7	Co 1.7	Ni 1.8	Cu 1.8	Zn 1.7	Ga 1.8	Ge 2.0	As 2.2	Se 2.5	Br 2.8	Kr –
Rb 0.9	Sr 1.0	Y 1.1	Zr 1.2	Nb 1.3	Mo 1.3	Tc 1.4	Ru 1.4	Rh 1.5	Pd 1.4	Ag 1.4	Cd 1.5	In 1.5	Sn 1.7	Sb 1.8	Te 2.0	I 2.2	Xe –
Cs 0.9	Ba 0.9	La 1.1	Hf 1.2	Ta 1.4	W 1.4	Re 1.5	Os 1.5	Ir 1.6	Pt 1.5	Au 1.4	Hg 1.5	Tl 1.5	Pb 1.6	Bi 1.7	Po 1.8	At 2.0	Rn –
Fr 0.9	Ra 0.9	Ac 1.0															

Figure 7
Table of electronegativities

The electronegativity difference (ΔE_n) is the difference in electronegativity number between two atoms participating in a covalent bond. This difference will be zero only if the electronegativity number of two atoms is the same. In this case, the electron pair will be shared equally, and a nonpolar covalent bond is formed. If the electronegativity difference is greater than zero but less than 1.7, the bond is polar covalent. The polarity of the bond increases as the ΔE_n approaches 1.7. When the ΔE_n is equal to or greater than 1.7, the bond is considered ionic (**Figure 8**). Thus, the ionic bond may be considered an extreme form of polar covalency. You will notice that the electronegativity differences between atoms in groups 1 or 2, and atoms in groups 16 or 17, are generally greater than or equal to 1.7. Atoms from these two sets of groups generally form ionic compounds.

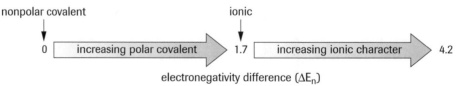

Figure 8
The relationship between electronegativity and bonding

Molecular Shape

A molecule's biological function is determined by the types of bonds between its atoms, and by its overall shape and polarity. The types of atoms determine the types of bonds, and the orientation of bonding electron pairs determines molecular shape. When atoms react to form covalent bonds, their valence electron orbitals undergo a process called **hybridization** that changes the orientation of the valence electrons (**Figure 9**).

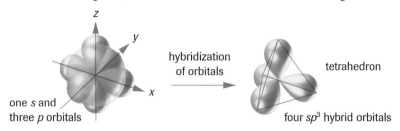

hybridization a modification of the valence orbitals that changes the orientation of the valence electrons

Figure 9
Hybridization of one *s* orbital and three *p* orbitals forms a new orbital system (*sp*³ hybrid) with four orbitals that point toward the verticies of a regular tetrahedron whose internal angles are all 109.5°. As usual, only two valence electrons can occupy each orbital.

Hybridization is a complex process. The Valence Shell Electron Pair Repulsion (VSEPR) theory, developed by Canadian chemist Ronald J. Gillespie (1939–), is a useful and relatively simple way of predicting molecular shape. This theory states that, since electrons are all negatively charged, valence electron pairs repel one another and will move as far apart as possible. The majority of molecules of biological importance possess up to four valence electron pairs around a central atom (an exception is phosphorous, which has five pairs). **Table 5** illustrates the resulting shape of molecules containing one to four valence electron pairs. Note that nonbonding pairs occupy more space than do bonding pairs and will repel and compress the bond angles of the bonding pairs. Also note that the orbital model and the VSEPR model predict similar molecular shapes.

Table 5 Molecular Shapes

Name and formula (shape)	Orbital model	VSEPR model	Bonding valence electron pairs	Nonbonding valence electron pairs	Ball-and-stick diagram	Space-filling diagram
methane, CH_4 (tetrahedral)			4	0		
ammonia, NH_3 (pyramidal)			3	1		
water, H_2O (angular)			2	2		
hydrogen chloride, HCl (linear)			1	3		

Molecular Polarity

Covalent bonds may be polar or nonpolar. However, the polarity of a molecule as a whole is dependent on bond polarity *and* molecular shape. **Figure 10(a)** shows that symmetrical molecular structures produce nonpolar molecules (whether the bonds are polar or not). Asymmetrical molecular shapes produce nonpolar molecules if the bonds are all nonpolar, and they produce polar molecules if at least one bond is polar, as **Figure 10(b)** illustrates.

(a)

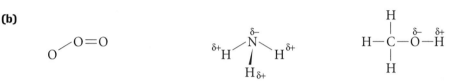

carbon tetrachloride, $CCl_{(l)}$

- tetrahedral shape
- symmetrical arrangement of polar covalent bonds
- nonpolar molecule

(b)

ozone, $O_{3(g)}$
- asymmetrical arrangement of nonpolar bonds
- nonpolar molecule

ammonia, $NH_{3(g)}$
- asymmetrical arrangement of polar bonds
- polar molecules

methanol, $CH_3OH_{(l)}$

Figure 10
(a) Symmetrical molecules whose bonds are all polar are nonpolar.
(b) Asymmetrical molecules are nonpolar if all bonds are non-polar, and they are polar if at least one bond is polar.

▶ Practice

Understanding Concepts

8. Hydrogen sulfide, $H_2S_{(g)}$, also known as "sewer gas" is released when bacteria break down sulfur-containing waste.
   (a) Draw a Lewis diagram for hydrogen sulfide.
   (b) Determine whether the S–H bonds in hydrogen sulfide are nonpolar covalent, polar covalent, or ionic.
   (c) Use the VSEPR theory to determine the shape of the hydrogen sulfide molecule.
   (d) Determine whether the hydrogen sulfide molecule is polar or nonpolar. Explain.

9. How can a molecule with polar covalent bonds be nonpolar?

Water

Life as we know it could not exist without water. Approximately two thirds of Earth's surface is covered in water and about two thirds of any organism is composed of water. Bacterial cells are about 70% water by mass. Herbaceous plants, such as celery, are about 90% water and woody plants, such as trees, are about 50% water. Water keeps plant cells rigid and is required for photosynthesis. In humans, lungs are about 90% water, and water makes up about 70% of the brain. Fat tissue is about 25% water, and even bones contain more than 20% water by mass. Water helps control body temperature, keeps eyeballs moist, and lubricates joints. It also acts as a shock absorber, protecting your brain and spinal cord from bruising when your head and back are jarred. Water is so common that its extraordinary properties are easily overlooked.

Water is a colourless, tasteless, odourless substance that can exist as a solid, liquid, or gas in the temperature ranges normally found near Earth's surface. Its physical and chemical properties are a direct result of its simple composition and structure. Water, H_2O, comprises two hydrogen atoms attached to one oxygen atom. Water's polar covalent bonds and asymmetrical structure create a highly polar molecule (**Figure 11**).

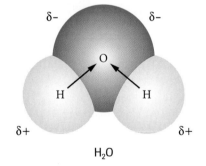

Figure 11
Oxygen, being more electronegative than hydrogen, attracts the electron pair in the bond toward itself, as indicated by the arrows. This unequal sharing of electrons gives the oxygen side of the molecule a slightly negative charge and the hydrogen side a slightly positive charge.

The polarity of water allows it to form chemical bonds with other molecules and ions, including other water molecules. Bonds between molecules are called **intermolecular bonds**. These are the bonds that determine the physical state of molecular substances at a given temperature and pressure. They are broken when solids melt into liquids and liquids evaporate into gases. Intermolecular bonds are weaker than the intramolecular covalent and ionic bonds that hold atoms and ions together in molecules and ionic solids. There are three types of intermolecular bonds: London forces (London dispersion forces), dipole–dipole forces, and hydrogen bonds (H-bonds). London dispersion forces are the weakest, and exist between all atoms and molecules. London forces constitute the only intermolecular forces of attraction between noble gas atoms and between nonpolar molecules. These bonds are formed by the temporary unequal distribution of electrons as they randomly move about the nuclei of atoms. The unequal electron clouds allow the electrons of one neutral atom to attract the nucleus of neighbouring atoms. These bonds are very weak between single atoms, such as helium, and small nonpolar molecules, such as methane, $CH_{4(g)}$ (in natural gas) **Figure 12(a)**. That is why these materials are gases at room temperature. London forces become more significant between large nonpolar molecules, such as octane, $C_8H_{18(l)}$ (in gasoline), a liquid at room temperature, because of the cumulative effect of many London forces between their atoms.

Dipole–dipole forces, illustrated in **Figure 12(b)**, hold polar molecules to one another. In this case, the partially positive side of one polar molecule attracts the partially negative side of adjacent polar molecules. These intermolecular forces of attraction are stronger than London forces.

intermolecular bonds chemical bonds between molecules

(a)

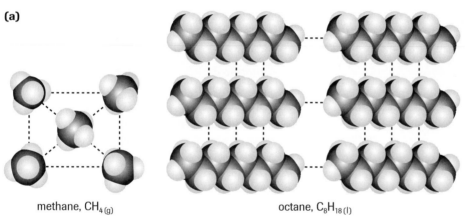

methane, $CH_{4\,(g)}$ octane, $C_8H_{18\,(l)}$

(b)

dipole–dipole forces between hydrogen chloride (HCl) molecules

Figure 12
(a) London forces are weak forces of attraction between all atoms and molecules. They are the only intermolecular forces that hold nonpolar molecules to one another. Small nonpolar molecules, such as methane, are gases at room temperature because of the relative weakness of the London forces between the molecules. Larger nonpolar molecules, such as octane, are liquids at room temperature because of the cumulative effect of many London forces between their atoms.
(b) Dipole–dipole forces hold polar molecules together. These intermolecular forces of attraction are stronger than London forces.

Hydrogen bonds are the strongest intermolecular forces of attraction. They are, in fact, especially strong dipole–dipole forces that only form between an electropositive H of one polar molecule and an electronegative N, O, or F of a neighbouring polar molecule (**Figure 13**, page 18). Water molecules hold onto each other by H-bonds. Hydrogen bonds and an angular shape give water its many unique properties (**Figure 14**, page 18).

London dispersion forces, dipole–dipole attractions, and hydrogen bonds are collectively called **van der Waals forces**.

van der Waals forces intermolecular forces of attraction including London forces, dipole–dipole forces, and hydrogen bonds

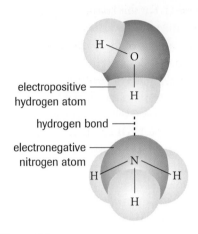

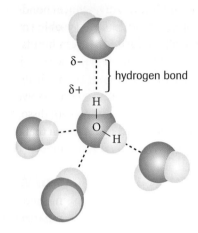

Figure 13
A hydrogen bond joins a water molecule (H₂O) to an ammonia molecule (NH₃).

Figure 14
Water molecules form hydrogen bonds with each other. Each molecule forms hydrogen bonds with up to four neighbouring molecules.

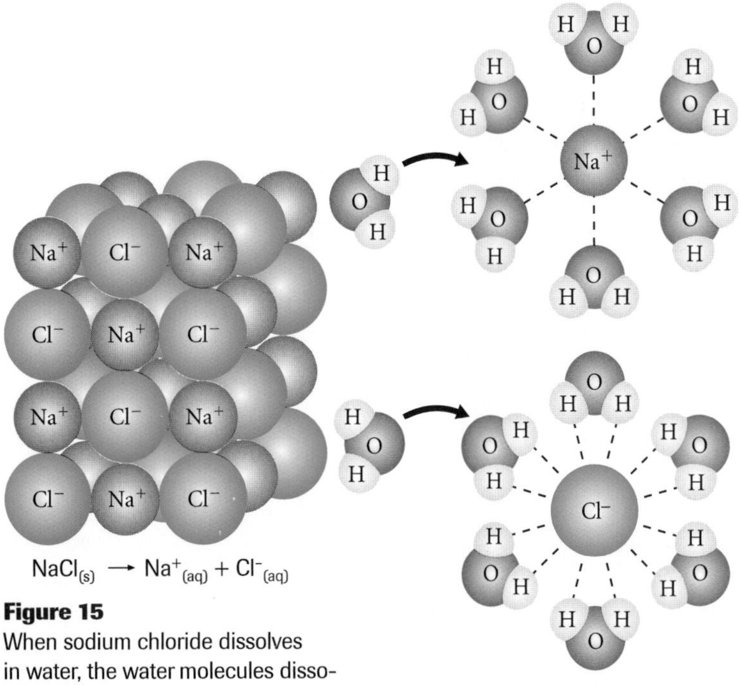

$$NaCl_{(s)} \longrightarrow Na^+_{(aq)} + Cl^-_{(aq)}$$

Figure 15
When sodium chloride dissolves in water, the water molecules dissociate the positive sodium ions and the negative chloride ions. The electronegative oxygen side of the water molecules pulls on the positive sodium ions and then surrounds the dissociated ions. The electropositive hydrogen side of the water molecules dissociates and then surrounds the negative chloride ions.

miscible describes liquids that dissolve into one another

immiscible describes liquids that form separate layers instead of dissolving

Water: The Versatile Solvent

More substances dissolve in water than in any other liquid. Practically all chemicals of biological interest occur in a water medium. The reason for water's excellent dissolving ability relates to its polarity—water molecules provide partial positive and negative charges to which other polar molecules or ions can attach. **Figure 15** illustrates the dissolving of sodium chloride in water. When ionic solids dissolve, the anions and cations dissociate from one another. Dissociation involves the breaking of ionic bonds.

Materials do not have to be ionic to dissolve in water. Polar molecules generally dissolve readily in water. Polar covalent substances, such as sugars and alcohols, dissolve easily in water. Any substance that dissolves in another substance is referred to as soluble. Water has been called the universal solvent because so many different substances dissolve in it. However, some things are easier to dissolve in water than others, and some materials do not dissolve in water to any appreciable degree. Substances that dissolve very little, such as iron and chalk, are called insoluble. The terms **miscible** and **immiscible** are used to describe whether liquids dissolve in each other or not. Ethanol (the alcohol in alcoholic beverages) and ethylene glycol (the alcohol in antifreeze) are miscible with water. Gasoline and oil are immiscible with water but are miscible with each other. The subscript (aq), meaning "aqueous," is used to identify a molecule or ion that is dissolved in water. The formula $C_6H_{12}O_{6(s)}$ represents glucose in solid form; the formula $C_6H_{12}O_{6(aq)}$ symbolizes an aqueous solution of glucose.

Small nonpolar molecules, such as oxygen (O_2) and carbon dioxide (CO_2), cannot form hydrogen bonds with water and, thus, are only slightly soluble. That is why a soluble protein carrier molecule, such as hemoglobin, is needed to transport oxygen, and to a lesser degree, carbon dioxide, within the circulatory systems of animals. Large nonpolar molecules, such as fats and oils, also do not form hydrogen bonds with water. When placed in water, they are excluded from associating with water molecules because the water molecules form hydrogen bonds with one another. For this reason, nonpolar molecules are said to be **hydrophobic** (meaning "water fearing"). Conversely, polar molecules that can form hydrogen bonds with water are described as **hydrophilic** ("water loving"). Oil floats on water because it is less dense than water, but it fails to dissolve in water because it is nonpolar and hydrophobic. Although unable to form solutions with water, nonpolar substances do dissolve in other nonpolar materials. Oil and gasoline are nonpolar substances that do not dissolve in water but dissolve well into each other. "Like dissolves like" is a generalization that describes the fact that polar substances dissolve in other polar substances and nonpolar materials dissolve in other nonpolar materials.

hydrophobic having an aversion to water; the tendency of nonpolar molecules to exclude water

hydrophilic having an affinity for water; the tendency of polar and ionic substances to dissolve in water

The Unique Properties of Water

Its angular shape and hydrogen-bonding characteristics give water a number of extraordinary properties. **Table 6** summarizes some of water's unique characteristics.

Table 6   The Unique Properties of Water

Descriptive characteristic	Property	Explanation	Effect	Example
Water clings.	cohesion	Water molecules form hydrogen bonds with one another.	high surface tension	A water strider (*Gerris* species) walks on the surface of a pond.
	adhesion	Water molecules form hydrogen bonds with other polar materials.	capillary action	Capillary action causes water to creep up a narrow glass tube and paper.
Water absorbs lots of heat.	high specific heat capacity[1]	Hydrogen bonding causes water to absorb a large amount of heat before its temperature increases appreciably and also causes it to lose large amounts of heat before its temperature decreases significantly.	temperature moderation	High heat capacity helps organisms maintain a constant body temperature.
	high specific heat of vaporization[2]	Hydrogen bonding causes liquid water to absorb a large amount of heat to become a vapour (gas).	evaporative cooling	Many organisms, including humans, dissipate body heat by evaporation of water from surfaces, such as skin (by sweating) and tongue (by panting).
Solid water is less dense than liquid water.	highest density at 4°C	As water molecules cool below 0°C, they form a crystalline lattice (freezing). The hydrogen bonds between the V-shaped molecules spread the molecules apart, reducing the density below that of liquid water.	Ice floats on liquid water.	Fish and other aquatic organisms are able to survive in winter.

1 Heat energy, like all other forms of energy, is measured in joules (J). Specific heat capacity is a measure of the extent to which a substance resists changes in temperature when it absorbs or releases heat. The specific heat capacity of water is 4.18 J/(g°C), approximately twice that of most organic compounds.

2 The amount of heat energy required to convert 1 g of liquid water into a vapour is 2.4 kJ, nearly twice as much heat energy as is needed to vapourize 1 g of ethanol.

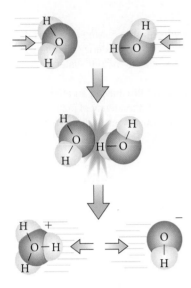

Figure 16
The autoionization of water

Acids, Bases, and Buffers

Pure water never contains only H_2O molecules. At 25°C, two H_2O molecules in every 550 million react with each other. One H_2O molecule transfers an H^+ ion to the other H_2O molecule. This produces an OH^- (hydroxide) ion and an H_3O^+ (hydronium) ion. The process is called the autoionization of water and is summarized in **Figure 16**.

The hydronium ion has been characterized as a substance that gives rise to the properties of acidic solutions—sour taste, the ability to conduct electricity, and the ability to turn blue litmus red. Acids are defined as substances that increase the concentration of $H_3O^+_{(aq)}$ when dissolved in water and that contain at least one ionizable hydrogen atom in their chemical structure. The following equation describes the reaction of hydrogen chloride with water to produce hydrochloric acid:

$$HCl_{(g)} + H_2O_{(l)} \longrightarrow H_3O^+_{(aq)} + Cl^-_{(aq)}$$

The hydroxide ion has the properties of a base—bitter taste, slippery feel, and the ability to conduct electricity and to change the colour of red litmus to blue. Bases are substances that increase the concentration of $OH^-_{(aq)}$ ions in solution. This is accomplished in one of two ways. Ionic bases containing OH^- ions, such as sodium hydroxide (NaOH), dissociate in water to produce OH^- ions directly.

$$NaOH_{(s)} \longrightarrow Na^+_{(aq)} + OH^-_{(aq)}$$

Some bases combine with H^+ ions directly. Ammonia (NH_3), a product of decomposed plant and animal matter, is such a base. Instead of releasing OH^- ions into solution directly, ammonia combines with an $H^+_{(aq)}$ ion, from $H_2O_{(l)}$, to produce an ammonium ion, $NH_4^+_{(aq)}$, and a hydroxide ion, $OH^-_{(aq)}$.

$$NH_{3(g)} + H_2O_{(l)} \longrightarrow NH_4^+_{(aq)} + OH^-_{(aq)}$$

Since pure water contains equal numbers of hydronium and hydroxide ions, it is neutral. When equal amounts of hydronium and hydroxide ions react, water is formed.

$$H_3O^+_{(aq)} + OH^-_{(aq)} \longrightarrow 2H_2O_{(l)} \quad \text{or} \quad H^+_{(aq)} + OH^-_{(aq)} \longrightarrow H_2O_{(l)}$$

neutralization reaction the reaction of an acid and a base to produce water and a salt

Since water is neutral, this reaction is called a **neutralization reaction**, and occurs whenever an acid is mixed with a base. In these reactions, water and a salt are produced.

$$\underset{\text{acid}}{HCl_{(aq)}} + \underset{\text{base}}{NaOH_{(aq)}} \longrightarrow \underset{\text{water}}{H_2O_{(l)}} + \underset{\text{salt}}{NaCl_{(aq)}}$$

The acidity of an aqueous solution may be expressed in terms of hydronium ion concentration, symbolized as $[H_3O^+_{(aq)}]$. The concentration of a solute in aqueous solution is measured in moles of the solute per litre of solution and is symbolized as mol/L, the mole being the amount of any substance that contains 6.02×10^{23} entities of that substance. A hydronium ion concentration of 1.0 mol/L means that the solution contains 1.0 mol of $H_3O^+_{(aq)}$ (6.02×10^{23} $H_3O^+_{(aq)}$ ions) per litre of solution. In pure water at 25°C, the $[H_3O^+_{(aq)}]$ is 1.0×10^{-7} mol/L. It is common in acid–base chemistry for concentrations to be very small values. The pH scale was devised as a more convenient way to express the concentration of $H_3O^+_{(aq)}$. The pH of an aqueous solution is equal to the negative logarithm of the hydronium ion concentration, $-\log_{10}[H_3O^+_{(aq)}]$. Thus, the pH of pure water is 7 since pH $= -\log_{10}(10^{-7}$ mol/L$) = 7$. Any aqueous solution with pH $= 7$ will be neutral, indicating a balance of hydronium and hydroxide ions. Solutions whose pH is less than 7 are acidic and solutions with pH greater than 7 are basic (**Figure 17**).

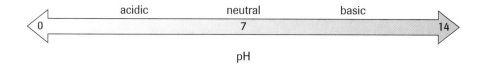

Figure 17
A solution whose pH is 7 is neutral.
Solutions with pH > 7 are basic.
Solutions with pH < 7 are acidic.

Strong and Weak Acids and Bases

Acids and bases may be classified as strong or weak, according to the degree to which they ionize when dissolved in water. **Strong acids**, such as $HCl_{(aq)}$, and **strong bases**, such as $NaOH_{(s)}$, ionize completely when dissolved in water.

$$HCl_{(aq)} + H_2O_{(l)} \xrightarrow{100\%} H_3O^+_{(aq)} + Cl^-_{(aq)}$$

$$NaOH_{(aq)} \xrightarrow{100\%} Na^+_{(aq)} + OH^-_{(aq)}$$

strong acids acids that ionize completely in aqueous solution

strong bases bases that ionize completely in aqueous solution

Weak acids, such as acetic acid, $CH_3COOH_{(aq)}$, and **weak bases**, such as $NH_{3(aq)}$, ionize partially in water.

$$CH_3COOH_{(aq)} + H_2O^{(l)} \underset{}{\overset{1.3\%}{\rightleftharpoons}} H_3O^+_{(aq)} + CH_3COO^-_{(aq)}$$

$$NH_{3(aq)} + H_2O^{(l)} \underset{}{\overset{10\%}{\rightleftharpoons}} NH_4^+_{(aq)} + OH^-_{(aq)}$$

weak acids acids that partially ionize in aqueous solution

weak bases bases that partially ionize in aqueous solution

The equations representing the reactions of weak acids and bases contain double arrows to indicate that these reactions may proceed in both directions—they are reversible reactions. When first placed in water, acetic acid molecules react with water molecules to form hydronium and acetate ions; the forward reaction is favoured. As the concentration of the ions increases, the reverse reaction occurs more frequently. When approximately 1.3% of the acetic acid molecules have ionized, the rates of the forward and reverse reactions become equal and the solution is said to be in a state of **equilibrium**. Once equilibrium is reached, the forward and backward reactions continue to occur at equal rates, and the concentrations of all entities in the solution remain constant. Most organic acids and bases are weak and reach equilibrium.

equilibrium a condition in which opposing reactions occur at equal rates

Conjugate Acids and Bases

According to the Brønsted–Lowry concept, reversible acid–base reactions involve the transfer of a proton. An acid is a proton donor; a base is a proton acceptor. In an acetic acid solution, the forward reaction involves a proton transfer from acetic acid to water. Acetic acid, CH_3COOH, acts as the acid; water, H_2O, acts as the base. In the reverse reaction, the hydronium ion, H_3O^+, acts as the acid, and the acetate ion, CH_3COO^-, acts as the base.

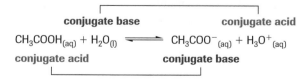

Acetic acid, acetate and hydronium ions, and water, are conjugate acid–base pairs.

$$\underbrace{CH_3COOH_{(aq)}}_{\text{conjugate acid}} + H_2O_{(l)} \rightleftharpoons \underbrace{CH_3COO^-_{(aq)}}_{\text{conjugate base}} + H_3O^+_{(aq)}$$

The Chemical Basis of Life **21**

Acid–Base Buffers

The components of living cells and the internal environments of multicellular organisms are sensitive to pH levels. Most cellular processes operate best at pH 7. Chemical reactions within cells normally produce acids and bases that have the potential to seriously disrupt function. Many of the foods we eat are acidic. Absorbing these acids could affect the pH balance of blood. Living cells use **buffers** to resist significant changes in pH. In living organisms, buffers usually consist of conjugate acid–base pairs in equilibrium. The most important buffer in human extracellular fluid and blood is the carbonic acid (acid)–bicarbonate (base) buffer (**Figure 18**). The components of this buffer are produced when carbon dioxide reacts with water molecules in body fluids. In the formation of the carbonic acid–bicarbonate buffer, carbon dioxide and water react to form carbonic acid, $H_2CO_{3(aq)}$. Carbonic acid then ionizes to form bicarbonate ions $HCO_3^-{}_{(aq)}$ and $H^+{}_{(aq)}$ ions.

buffers chemical systems containing a substance that can donate H^+ ions when they are required and containing a substance that can remove H^+ ions when there are too many in a solution

Figure 18
The formation of the carbonic acid–bicarbonate buffer. Note the double arrows between reactants and products. These are reversible reactions in equilibrium.

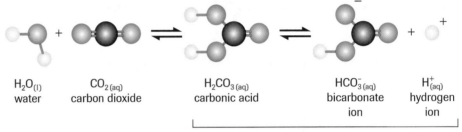

$H_2O_{(l)}$	$CO_{2(aq)}$	$H_2CO_{3(aq)}$	$HCO_3^-{}_{(aq)}$	$H^+_{(aq)}$
water	carbon dioxide	carbonic acid	bicarbonate ion	hydrogen ion

carbonic acid/bicarbonate buffer

When $H^+{}_{(aq)}$ ions enter the bloodstream, after someone eats a vinaigrette salad, for example, $HCO_3^-{}_{(aq)}$ ions react with the $H^+{}_{(aq)}$ ions to produce $H_2CO_{3(aq)}$. Similarly, if a base enters the blood and removes $H^+{}_{(aq)}$ ions, $H_2CO_{3(aq)}$ ionizes to replace missing $H^+{}_{(aq)}$ ions in the blood. Together, the carbonic acid molecules (the conjugate acid) and bicarbonate ions (the conjugate base) help maintain the pH of blood at about 7.4. An increase or decrease in blood pH of 0.2 to 0.4 pH units is fatal if not treated immediately. A pH below 7.35 is described as acidosis and a pH above 7.45 is described as alkalosis.

Proteins can also act as buffers. Hemoglobin helps maintain a constant pH within red blood cells. Proteins are able to buffer solutions because some of the amino acids in their structure may be acidic and others may be basic (discussed further in Section 1.2). Together, the amino acids can remove excess $H_3O^+{}_{(aq)}$ or $OH^-{}_{(aq)}$ from a solution.

DID YOU KNOW ?

Hyperventilation Improves Running Performance
Sprinters take advantage of the carbonic acid–bicarbonate buffer system in blood by hyperventilating shortly before a race. This increases the pH of the blood slightly, which allows better control of the short-term buildup of lactic acid during the sprint.

INVESTIGATION 1.1.1

Buffers in Living Systems (p. 78)
The vast majority of foods we eat are acidic. Fruit, vegetables, salad dressings, wine, and pop all add $H_3O^+{}_{(aq)}$ ions to your blood. People also enjoy alkaline foods and beverages like shrimp, corn, and tonic water. Most cellular proteins, especially enzymes, work best at near-neutral pH, their activity dropping significantly when pH levels deviate from the ideal. How well does a buffer work to maintain a constant pH in cells? In Investigation 1.1.1, you will investigate the properties of acids, bases, and buffers.

SUMMARY · Chemical Fundamentals

- The atomic number of an element is equal to the number of protons in the nucleus. The mass number of an atom is determined by the sum of the protons and neutrons in the nucleus.

- Isotopes are atoms of an element with the same atomic number but a different mass number. They differ from one another by the number of neutrons in their nuclei. Radioisotopes spontaneously break apart.

- Electronegativity is a measure of the ability of an atom to attract the pair of electrons in a covalent bond.

- VSEPR theory is a method for predicting molecular shape based on the mutual repulsion of electron pairs in a molecule (**Figure 19**).

- When $\Delta E_n \geq 1.7$, the bond is ionic. When $0 < \Delta E_n < 1.7$, the bond is covalent.

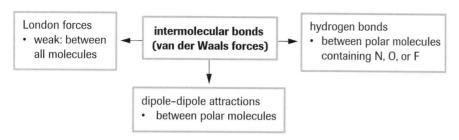

Figure 19

- Hydrogen bonding accounts for the following properties of water: high solubility of polar and ionic substances, cohesion, adhesion, high surface tension, capillarity, high specific heat capacity, and high specific heat of vaporization.

- Acids are substances that increase the concentration of $H_3O^+_{(aq)}$ when dissolved in water. Bases are substances that increase the concentration of OH^- ions in solution.

- A buffer is a chemical system containing a substance that can donate H^+ ions when they are required and containing a substance that can remove H^+ ions when there are too many in a solution.

Section 1.1 Questions

Understanding Concepts

1. List the number of protons, neutrons, electrons, and valence electrons in the atoms of each of the following elements:
   (a) sulfur   (b) calcium   (c) nitrogen

2. Why do the elements with atomic numbers 8, 16, and 34 belong to the same family on the periodic table?

3. How many neutrons are found in an atom of cobalt-60?

4. What type of intermolecular forces of attraction must be overcome to melt each of the following solids?
   (a) ice, $H_2O_{(s)}$ (b) iodine, $I_{2(s)}$

5. State the principle on which the VSEPR theory is based.

6. Use electronegativity values and VSEPR theory to determine whether carbon tetrachloride, CCl_4, and ammonia, NH_3, are polar or nonpolar molecules.

7. Why is table salt, $NaCl_{(s)}$, soluble in ethanol, $CH_3OH_{(l)}$, but not soluble in gasoline, $C_8H_{18(l)}$?

8. What is the difference between a weak acid and a dilute solution of a strong acid?

9. Identify the two conjugate acid–base pairs in the following acid–base equilibrium.

   $HCOOH_{(aq)} + CN^-_{(aq)} \rightleftharpoons HCOO^-_{(aq)} + HCN_{(aq)}$

10. Describe the components of a buffer and the role each plays in helping maintain a constant pH.

11. Determine the pH of each of the following solutions, and state whether each is acidic, basic, or neutral:

    (a) tonic water: $[H_3O^+_{(aq)}] = 3.1 \times 10^{-9}$ mol/L
    (b) wine: $[H_3O^+_{(aq)}] = 2.5 \times 10^{-3}$ mol/L

12. Distinguish between ionic and polar covalent bonds.

13. Distinguish between table sugar dissolving in water and table salt dissolving in water.

14. Identify each of the following substances as either hydrophilic or hydrophobic:
    (a) $C_6H_{6(l)}$ (b) $C_2H_5OH_{(l)}$ (c) $CCl_{4(l)}$

15. What property of water accounts for each of the following observations?
    (a) A steel sewing needle floats on water but a large steel nail sinks.
    (b) Dogs pant on a hot summer day.
    (c) Water creeps up the walls in a flooded room.
    (d) Hands are usually washed in water.

Applying Inquiry Skills

16. Water and varsol (a paint thinner) are immiscible liquids. Describe an experiment using these two materials that would determine whether polystyrene foam is polar or nonpolar.

Making Connections

17. (a) Describe three technological uses of radioisotopes.
    (b) Provide three possible reasons for the high costs generally associated with these uses of radioisotopes.

With the notable exception of water, virtually all chemicals of life are carbon based. Compounds that contain carbon (other than carbon dioxide and a few other exceptions) are called organic compounds, and the chemistry of carbon compounds is called organic chemistry. The name *organic* originates from the belief once held by scientists that organic compounds could only be produced by living organisms. It was thought that a vital force was needed to produce organic compounds. Nonliving matter was thought to lack this vital force and was classified as *inorganic*. In 1828, Frederick Wohler, a German chemist, prepared the organic compound urea (normally found in urine) by heating ammonium cyanate, an inorganic substance. Until then, urea could only be obtained from living things.

Synthetic organic chemistry has grown to become the largest branch of chemistry and the foundation of the petrochemical and pharmaceutical industries. More than 9 million organic chemicals are known, and many of these have been synthesized in the laboratory. The expanding biomaterials industry develops synthetic substances that integrate well with living tissues and withstand the wear and tear associated with normal body function. Biomedical engineers select suitable materials to design and produce replacement body parts. Synthetic materials must possess properties similar to the natural parts they replace and must survive the body's tissue-rejection reactions. For example, plastics, useful organic materials produced from crude oil; Teflon, a slippery material commonly used as a nonstick surface on kitchen utensils; and Dacron, a type of polyester used extensively in clothing, are used to make artificial blood vessels, heart valves, and other replacement body parts (**Figure 1**).

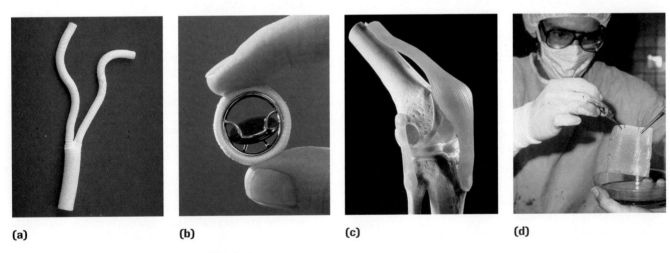

(a)　　**(b)**　　**(c)**　　**(d)**

Figure 1
Synthetic plastics are used to construct replacement body parts, such as artificial blood vessels **(a)**, heart valves **(b)**, and knee joints **(c)**. Silicone has been used to produce artificial skin for burn victims **(d)**.

Advances in synthetic organic chemistry notwithstanding, the major classes of organic chemicals of biological importance continue to be produced only in living cells. These include carbohydrates, lipids, proteins, and nucleic acids. Cells use these compounds for structure, energy, communication, genetic information, and metabolic processing.

Carbon

Carbon is a small, relatively light element with four single valence electrons oriented toward the vertices of a regular tetrahedron. It can form up to four stable covalent bonds with other atoms. Carbon atoms attach to each other to form straight and branched chains and ring structures of various sizes and complexity that act as the backbones of biological molecules. When attached to four hydrogen atoms, carbon forms a tetrahedral methane molecule. Although the carbon–hydrogen bond is slightly polar ($\Delta E_n = 0.4$), the symmetrical geometry of the methane molecule causes it to be nonpolar. Molecules containing only carbon and hydrogen are known as hydrocarbons. Larger hydrocarbons are also nonpolar because of the symmetrical arrangement of their bonds. **Figure 2** shows models of pentane (C_5H_{12}). Other elements such as hydrogen, oxygen, sulfur, and phosphorus may attach to the carbon backbone to form reactive clusters of atoms called **functional groups**. **Table 1** lists the more common functional groups found in biomolecules.

functional groups reactive clusters of atoms attached to the carbon backbone of organic molecules

Table 1 Functional Groups in Biomolecules

Group	Chemical formula	Structural formula	Ball-and-stick model	Found in
hydroxyl	$-OH$	$-OH$		alcohols (e.g., ethanol)
carboxyl	$-COOH$			acids (e.g., vinegar)
amino	$-NH_2$			bases (e.g., ammonia)
sulfhydryl	$-SH$	$-S-H$		rubber
phosphate	$-PO_4$			ATP
carbonyl	$-COH$			aldehydes (e.g., formaldehyde)
	$-CO-$			ketones (e.g., acetone)

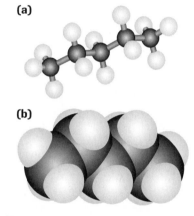

(a)

(b)

Figure 2
(a) A ball-and-stick model of pentane. The bond angles are all tetrahedral at 109.5°.
(b) A space-filling model of pentane. Note the symmetrical structure.

Looking at the structure of each functional group, you will notice that the various atoms always form the same number of covalent bonds with adjacent atoms. The number of covalent bonds an atom can form with neighbouring atoms is called its **bonding capacity**. Hydrogen has a bonding capacity of one, oxygen and sulfur two, nitrogen three, carbon four, and phosphorous five. The hydrocarbon portions of biological molecules are composed of strong, nonpolar, carbon–hydrogen bonds (**Table 2**).

Table 2 The Bonding Capacity of Common Elements

Element	Bonding capacity	Examples		
H	1	O — H | H H_2O (water)	H H \ / C = C / \ H H C_2H_4 (ethane)	
O	2	O = O O_2 (oxygen)	H | H — C — O — H | H CH_3OH (methanol)	
S	2	S — H | H H_2S (hydrogen sulfide)	H | H — C — S — H | H CH_3SH (methanethiol)	
N	3	$N \equiv N$ N_2 (nitrogen)	$H — C \equiv N$ HCN (hydrogen cyanide)	H — N — H | H NH_3 (ammonia)
C	4	H | H — C — H | H CH_4 (methane)	O = C = O CO_2 (carbon dioxide)	
P	5	Cl Cl \ | P — Cl / | Cl Cl PCl_5 (phosphorus pentachloride)	O || HO — P — OH | OH H_3PO_4 (hydrogen phosphate)	

Functional groups possess certain chemical properties that they impart to the molecules to which they are attached. These groups are more reactive than the hydrocarbon portions of biological molecules. Most of the reactions that occur in living organisms involve functional group interactions. The hydroxyl group, $-OH$, and the carboxyl group, $-COOH$, are polar because of the electronegative oxygen atom they contain. Sugars and alcohols are highly soluble in (polar) water because they contain polar hydroxyl groups.

The carboxyl group, $-COOH$, makes a molecule acidic, whereas the amino group, $-NH_2$, makes it basic. Compounds containing the carboxyl group are organic acids called carboxylic acids. Those containing amino groups are organic bases called amines. Amino acids, the basic building blocks of proteins, contain an amino group and a carboxyl group. A number of carboxylic acids are important intermediates in the process of cellular respiration. The phosphate group is found in molecules, such as adenosine triphosphate (ATP) which is used to transfer energy between organic molecules in cells.

> **Practice**

Understanding Concepts

1. Identify and name the functional groups in the molecule in **Figure 3**.

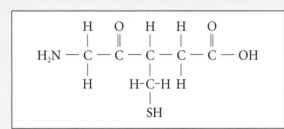

Figure 3

Biological Macromolecules

Many biologically important molecules are **macromolecules**, large molecules sometimes composed of a great number of repeating subunits. There are essentially four major classes of macromolecules in living organisms: carbohydrates, lipids, proteins, and nucleic acids (**Figure 4**).

macromolecules large molecules sometimes composed of a large number of repeating subunits

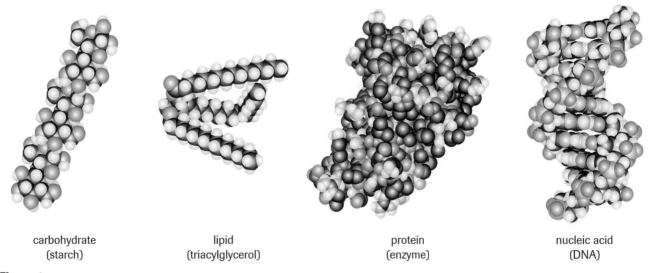

| carbohydrate | lipid | protein | nucleic acid |
| (starch) | (triacylglycerol) | (enzyme) | (DNA) |

Figure 4
Space-filling models of the four classes of biological molecules: carbohydrates, lipids, proteins, and nucleic acids

The Chemical Basis of Life **27**

Complex carbohydrates, proteins, and nucleic acids are polymers, which are composed of long chains of smaller subunits. The most common lipids, triacylglycerols (also called triglycerides) and phospholipids, are not polymers but relatively large molecules composed of several smaller parts. The basic subunits of the four classes of biological macromolecules are listed in **Table 3**.

Table 3 Subunits of Biological Macromolecules

Macromolecule	Subunit
complex carbohydrate (starch)	simple sugar (glucose)
lipid (triacylglycerol)	glycerol and fatty acids
protein	amino acids
nucleic acid (DNA or RNA)	nucleotides

condensation reaction (dehydration synthesis) a reaction that creates a covalent bond between two interacting subunits, linking them to each other

anabolic reactions reactions that produce large molecules from smaller subunits

Although the various macromolecules contain different subunits, they are assembled and disassembled in the same way. A **condensation reaction**, also called a **dehydration synthesis reaction**, creates a covalent bond between two interacting subunits, linking them to each other. This reaction always involves the removal of a hydrogen atom, H, from the functional group of one subunit and an −OH group from the other subunit's functional group. As shown in **Figure 5(a)**, the −OH group and the H come together to form a water molecule, H_2O (from which the terms condensation reaction and dehydration synthesis were derived). This reaction results in a covalent bond between the subunits. Condensation reactions absorb energy in the overall process. They are called **anabolic reactions** because they result in the construction of large molecules from smaller subunits.

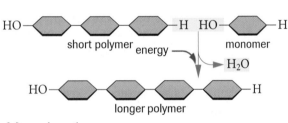

(a) condensation

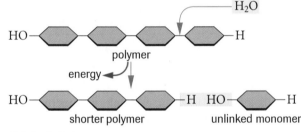

(b) hydrolysis

Figure 5
(a) A condensation reaction joins monomers into polymers. An −OH group from one subunit and an H atom from another join to form a water molecule, resulting in a covalent bond between the subunits. This is an energy-requiring process.
(b) A hydrolysis reaction uses an −OH group and an H atom from a water molecule to break a bond holding subunits together. Energy is released in this process.

catabolic reactions reactions that break macromolecules into constituent individual subunits

hydrolysis reaction a catabolic reaction in which a water molecule is used to break a covalent bond holding subunits together

In some cases, cells must break macromolecules into their individual subunits. These reactions, called **catabolic reactions**, break nutrient macromolecules down in the process of digestion. Catabolic reactions occur primarily by a process essentially the reverse of condensation reactions. In this process, a water molecule is used to break a covalent bond holding subunits together. The water molecule provides an H atom to one subunit and an −OH group to the other; hence, the name given to this process is **hydrolysis reaction** (*hydro*, meaning "water"; *lysis*, meaning "to break apart). Hydrolytic reactions release energy in the overall process, as illustrated in **Figure 5(b)**.

Condensation and hydrolysis reactions require the assistance of special protein molecules, called enzymes, to take place efficiently. Enzymes are biological catalysts, and catalysts are chemicals that speed up chemical reactions without becoming consumed in the process. Enzymes possess special structures that recognize the covalent bonds that must be created or broken. Enzymes will be discussed further in section 1.4.

Carbohydrates

Carbohydrates are among the most common organic materials on Earth. Millions of tonnes are produced by plants and algae every year through the process of photosynthesis. Carbohydrates are used by organisms as sources of energy, as building materials, and as cell surface markers for cell-to-cell identification and communication. Carbohydrates contain carbon, hydrogen, and oxygen atoms in a 1:2:1 ratio. Their empirical formula (the chemical formula that shows the ratio of atoms in the molecule in lowest terms) is $(CH_2O)_n$, where n represents the number of carbon atoms. Carbohydrates may be classified into three groups: monosaccharides, oligosaccharides, and polysaccharides.

The simplest carbohydrates are the simple sugars or monosaccharides (*mono*, meaning "single"; *saccharide* from *saccharum*, meaning "sugar"). The term *saccharide* and the suffix *-ose* both refer to sugars. Simple sugars contain a single chain of carbon atoms to which hydroxyl groups are attached. Monosaccharides may be distinguished by the carbonyl group they possess—aldehyde or ketone—and the number of atoms in their carbon backbone. A sugar with five carbons is called a pentose, one with six carbons, a hexose. The two simplest monosaccharides are dihydroxyacetone (a triose with a ketone group) and glyceraldehyde (a triose with an aldehyde group) (**Figure 6**).

Figure 6
Monosaccharides are distinguished from one another by the carbonyl group they possess (beige), the length of their carbon chains, and the spatial arrangement of their atoms. Note the difference in orientation of hydroxyl groups between glucose and galactose (blue).

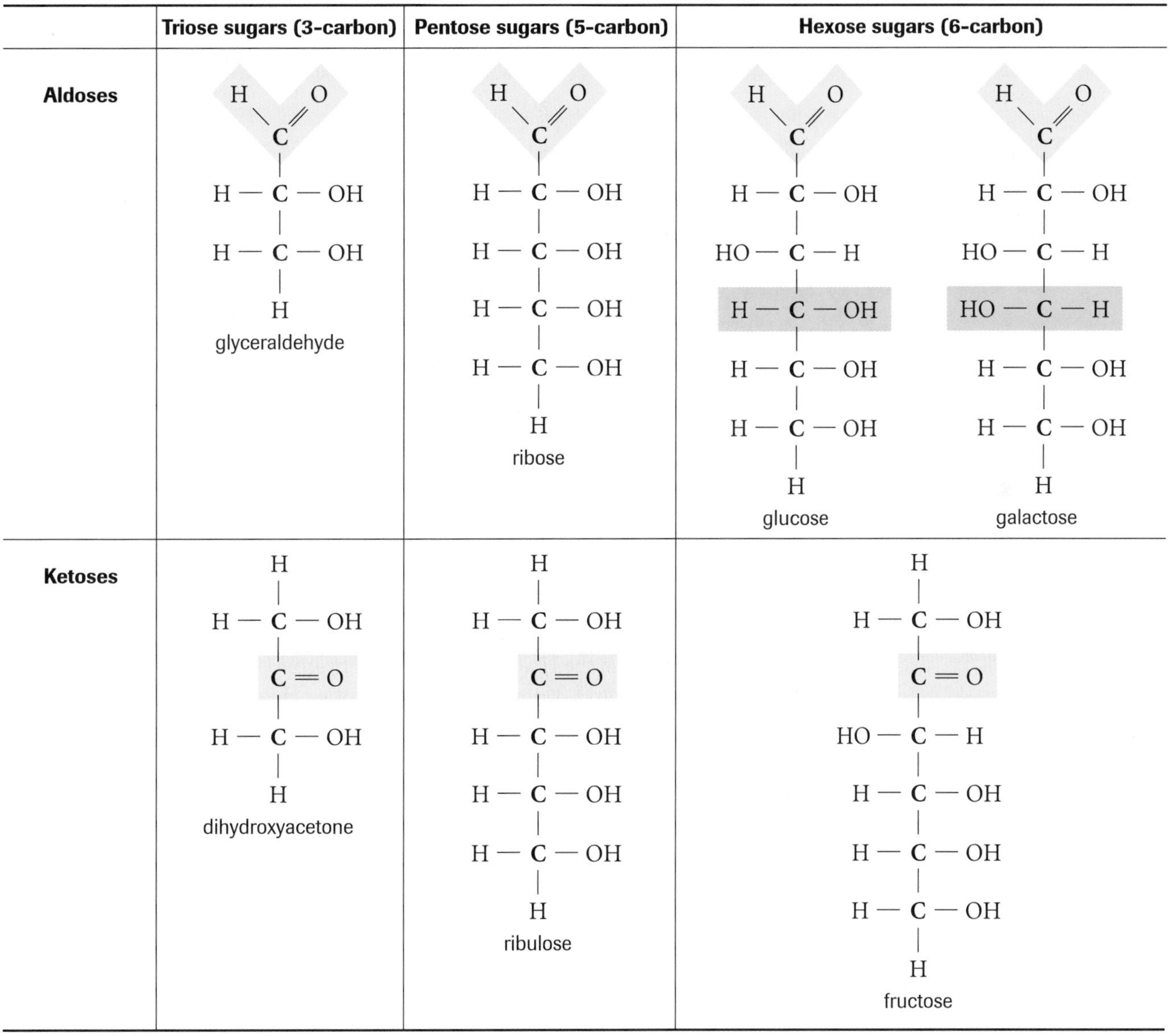

In addition to the two triose sugars, **Figure 6** illustrates two common pentose sugars, ribose (a component of RNA) and ribulose (used in photosynthesis), and three common hexoses, glucose (a source of energy for virtually all cells), galactose (a component of lactose, milk sugar), and fructose (fruit sugar). Another distinguishing feature of simple sugars is the spatial arrangement of their atoms. Molecules with the same chemical formula but with a different arrangement of atoms are called **isomers**. Glucose ($C_6H_{12}O_6$), galactose ($C_6H_{12}O_6$), and fructose ($C_6H_{12}O_6$) are isomers, since they possess the same number and types of atoms, but a different arrangement of those atoms (**Figure 6**). Isomers possess different shapes and different physical and chemical properties.

Monosaccharides with five or more carbons are linear molecules in the dry state, but readily form ring structures when dissolved in water. **Figure 7** shows a glucose molecule forming a six-membered ring. When this occurs, there is a 50% chance that the hydroxyl group at carbon 1 will end up below the plane of the ring. If so, the resulting molecule is called α-glucose. If the hydroxyl group at carbon 1 ends up above the plane of the ring, then β-glucose is formed.

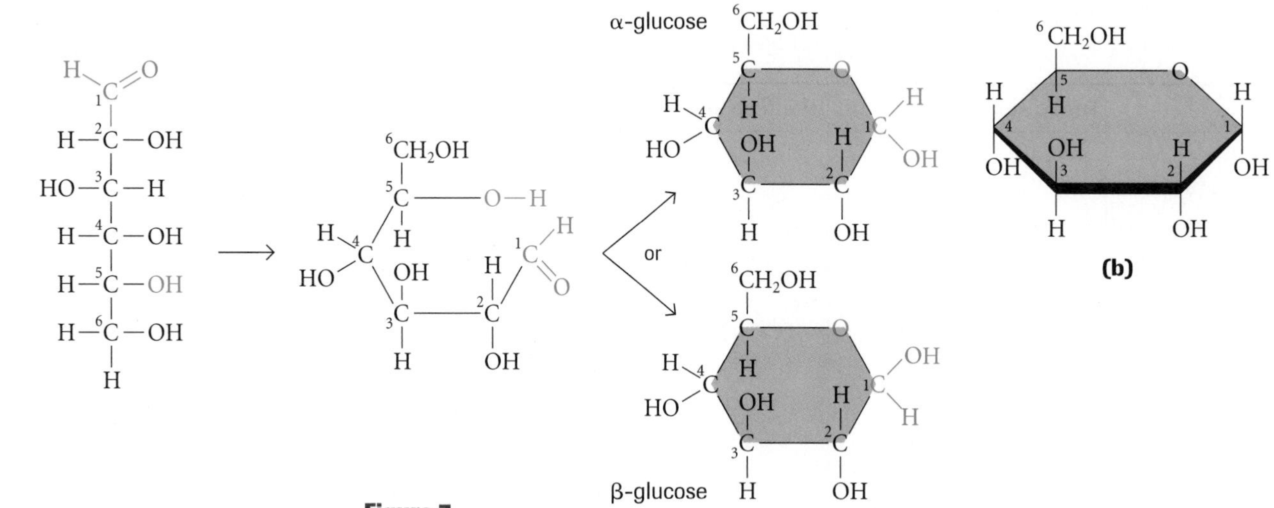

isomers molecules with the same chemical formula but with a different arrangement of atoms

Figure 7

(a) When glucose dissolves in water, the hydroxyl group on carbon 5 reacts with the aldehyde group at carbon 1 to form a closed ring. If the hydroxyl group at carbon 1 lies below the plane of the ring, the resulting molecule is called α-glucose; if the hydroxyl group lies above the plane of the ring, then it is called β-glucose.

(b) In the abbreviated formula for glucose, carbon symbols are omitted and constituent groups are projected above and below the plane of the ring.

oligosaccharides sugars containing several simple sugars attached to one another

glycosidic linkages covalent bonds holding monosaccharides to one another that are formed by condensation reactions in which the H atom comes from a hydroxyl group of one sugar and the −OH group comes from a hydroxyl group of the other

Oligosaccharides are sugars containing two or three simple sugars attached to one another by covalent bonds called **glycosidic linkages**. These bonds form by condensation reactions, in which the H atom comes from a hydroxyl group on one sugar and the − OH group comes from a hydroxyl group on the other (**Figure 8**). Important disaccharides include maltose, shown in **Figure 8(a)**, composed of two α-glucose molecules held together by an α 1–4 glycosidic linkage; sucrose, shown in **Figure 8(b)**, composed of α-glucose and α-fructose with an α 1–2 glycosidic linkage between them; and lactose, composed of α-glucose and α-galactose. Maltose is a disaccharide found in grains that is used in the production of beer, sucrose is table sugar, and lactose is the sugar normally found in milk. Sucrose is used by many plants to transport glucose and other simple sugars from one part of the plant to another. It is found in high concentrations in sugar cane, sugar beets, and sugar maple trees. The glycosidic linkage between the glucose molecules in maltose is called an α 1–4 glycosidic linkage because it forms between the hydroxyl group of carbon 1 on one glucose molecule and the hydroxyl group of carbon 4 of the adjacent molecule, as illustrated in **Figure 8(a)**.

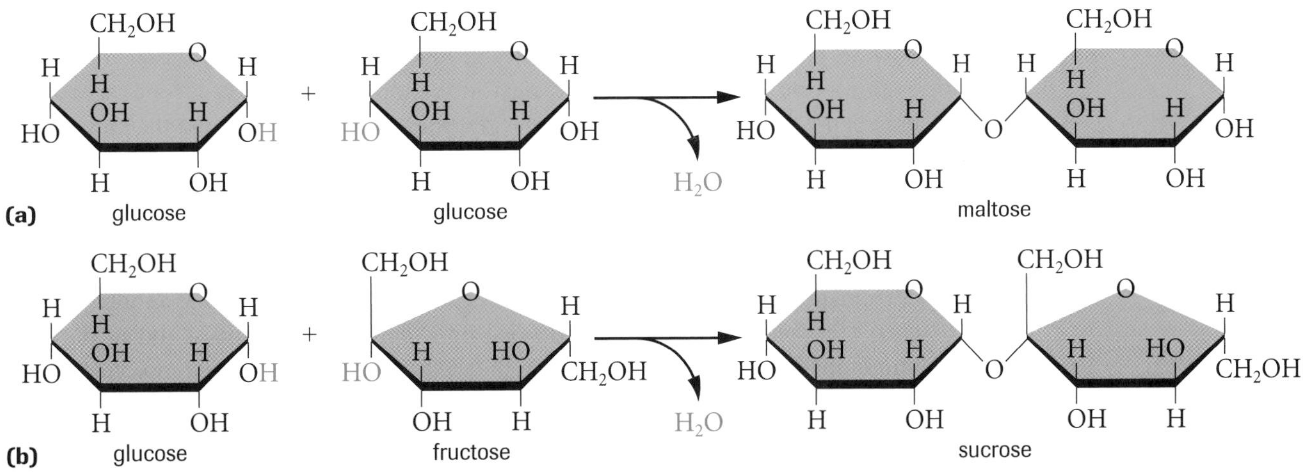

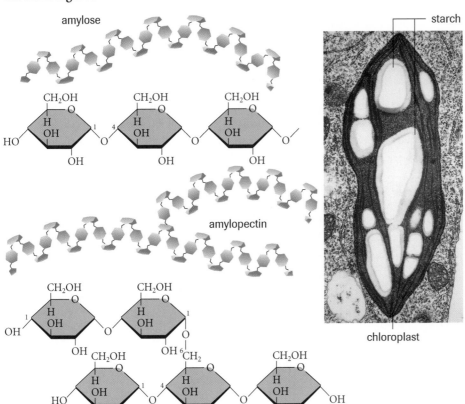

(a) glucose + glucose → maltose (H_2O)

(b) glucose + fructose → sucrose (H_2O)

Figure 8

(a) Maltose, also known as malt sugar, is formed by the condensation of two α-glucose molecules, resulting in an α 1–4 glycosidic linkage.

(b) Sucrose, table sugar, is composed of α-glucose and α-fructose with an α 1–2 glycosidic linkage between them.

Polysaccharides (complex carbohydrates) are monosaccharide polymers composed of several hundred to several thousand monosaccharide subunits held together by glycosidic linkages. Some polysaccharides are straight chains, whereas others are branched. Polysaccharides serve two important functions in living cells: energy storage and structural support. Starch and glycogen are examples of storage polysaccharides; cellulose and chitin are structural polysaccharides.

Plants store the Sun's energy in the form of glucose and other carbohydrates through the process of photosynthesis. In most cases, plants produce much more glucose than they need for their immediate energy requirements. Enzymes in the chloroplasts link excess glucose molecules to one another, forming starch, the main energy storage molecule in plants. Starch is a mixture of two different polysaccharides, amylose and amylopectin, shown in **Figure 9**.

amylose

amylopectin

starch

chloroplast

Figure 9

Amylose is an unbranched α-glucose polymer held together by α 1–4 glycosidic linkages. Amylopectin is a branched α-glucose polymer composed of a main chain with glucose molecules attached by α 1–4 glycosidic bonds and branch points formed by α 1–6 glycosidic linkages. Starch is a combination of amylose and amylopectin. Starch grains can be seen in chloroplasts of cells.

Potato starch is 20% amylose and 80% amylopectin, by mass. Amylose is a straight chain polymer of α-glucose with α 1–4 glycosidic linkages. Amylopectin is a branched-chain α-glucose polymer with α 1–4 linkages in the main chain and α 1–6 linkages at the branch points. The angles at which the glycosidic linkages form cause the polymers to twist into coils that make them insoluble in water. Plants store starch as insoluble granules within chloroplasts, amyloplasts, and other plastids. Some plants store large amounts of starch in specialized structures, such as tubers (potatoes), taproots (carrots and turnip), and the fruits of grains (wheat, corn, and rice).

Humans and other heterotrophs use the plant's stockpile of energy as a source of food energy for themselves. Animals possess enzymes that hydrolyze amylose and amylopectin into individual glucose molecules and then extract energy from these molecules in the process of cellular respiration. Excess glucose molecules are linked to one another to form glycogen, an energy storage molecule in animals. Shown in **Figure 10**, glycogen is similar to amylopectin in structure with α 1–4 main chain linkages and α 1–6 branch-point linkages; however, there are more branches in glycogen. Humans and other animals store small amounts of glycogen in muscle and liver cells. Enzymes in these tissues hydrolyze glycogen into single glucose molecules for energy during bouts of physical exercise. However, glycogen stores are small and are depleted in about one day if not replenished.

Figure 10
Glycogen is structurally similar to amylopectin with α 1–4 glycosidic linkages between α-glucose monomers in the main chain and α 1–6 linkages at the branch points. Glycogen is more highly branched than amylopectin. Notice the helical structure of these three polysaccharides. Glycogen granules can be seen in the cytoplasm of animal cells.

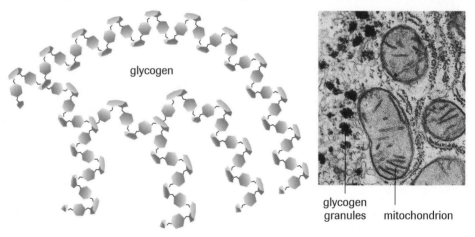

glycogen

glycogen granules   mitochondrion

Cellulose, the primary structural polysaccharide of plants, is a major component of plant cell walls. Worldwide, plants produce approximately 200 billion kilograms of cellulose a year, making it the most abundant organic substance on Earth. Cellulose is a straight-chain polymer of β-glucose held together by β 1–4 glycosidic linkages (**Figure 11**). Hydroxyl groups at the

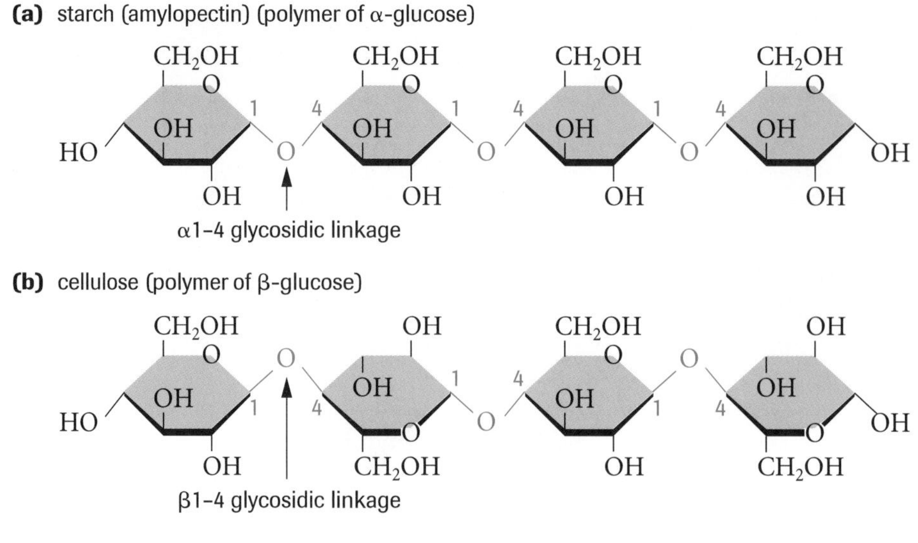

(a) starch (amylopectin) (polymer of α-glucose)

α1–4 glycosidic linkage

(b) cellulose (polymer of β-glucose)

β1–4 glycosidic linkage

Figure 11
Comparing glycosidic linkages in starch and cellulose. Note the alternating orientation of the β-glucose subunits.

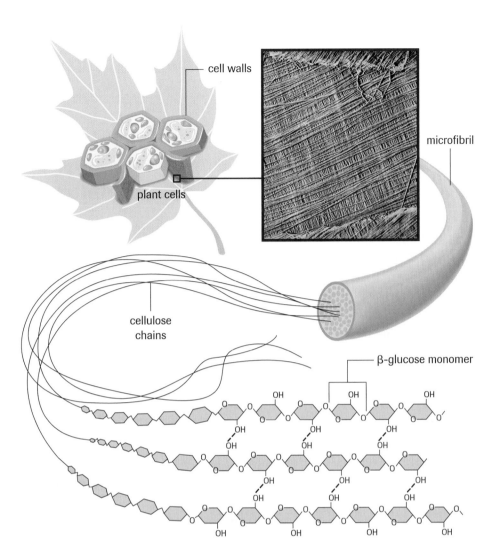

cell walls

microfibril

plant cells

cellulose
chains

β-glucose monomer

OH OH OH

OH OH OH

OH OH OH

OH OH OH

OH OH OH

OH OH OH OH

Figure 12
Cellulose fibres are composed of
many microfibrils, which are, in
turn, composed of many cellulose
molecules held together by
hydrogen bonds.

1 and 4 positions in β-glucose cause every other monomer to be inverted for the glycosidic linkage to form. Unlike amylose and amylopectin, cellulose molecules are neither coiled nor branched. Their straight shape allows the hydroxyl groups of parallel molecules to form many hydrogen bonds, producing tight bundles called microfibrils (**Figure 12**).

Many cellulose microfibrils intertwine to form tough, insoluble cellulose fibres that plants use to build their cell walls. Humans have made use of the tough physical properties of cellulose by using it in wood for lumber and paper, and in cotton and linen for clothing. Digestive enzymes in humans are able to break the glycosidic linkages in starch, but humans are unable to digest the linkages between the β-glucose subunits in cellulose. Certain animals, such as cows, sheep, and rabbits, contain symbiotic bacteria and protists in their digestive tracts that produce enzymes that break the β-glucose linkages in cellulose. This allows the animal to digest the cellulose in grasses and other vegetable matter. Although humans are not able to use cellulose as a source of energy, the cellulose in fresh fruits, vegetables, and grains is a healthy part of a well-balanced diet. Cellulose fibres, called roughage, pass through our digestive systems undigested and gently scrape the walls of the large intestine. This stimulates intestinal cells to secrete mucus, which lubricates feces and aids in the elimination of solid waste. It also helps keep feces moist by binding water in the large intestine.

The hard exoskeleton of insects and of crustaceans, such as crabs and lobsters, and the cell wall of many fungi, is made of chitin, a cellulose-like polymer of N-acetylglucosamine, shown in **Figure 13(a)** on page 34. This monomer is a glucose molecule to which a nitrogen-containing group is attached at the second carbon position. Chitin is the

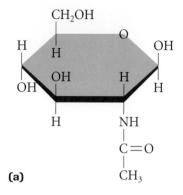

(a)

(b)

(c)

Figure 13
(a) N-acetylglucosamine is the repeating subunit in chitin.
(b) The mantis shrimp *Odontodactylus scyllarus* contains a chitin exoskeleton.
(c) The mushroom *Amanita muscaria* contains chitin in its cell walls.

second most abundant organic material found in nature. Many organisms, such as insects, crustaceans, and mushrooms, have brightly coloured exoskeletons made of pigmented chitin, as illustrated in **Figure 13(b)**. For many of them, coloration is a key to survival, used for defence, concealment, and recognition.

Chitin's physical and chemical properties make it useful in medical applications, such as contact lenses and the tough, yet biodegradable, stitches that decompose as a wound heals.

▶ *Practice*

Understanding Concepts

2. Name two functional groups found in the straight-chain form of glucose.

3. Which of the monosaccharides in **Figure 14** are isomers?

4. What are the differences between a ketotriose and an aldopentose? Provide an example of each.

5. How does the structure of a monosaccharide change when it is placed in an aqueous solution?

6. (a) Why are glycogen and starch called storage polysaccharides?

(b) Why are cellulose and chitin called structural polysaccharides?

7. Distinguish between the glycosidic bonds in amylose and cellulose.

8. Why can humans not obtain energy by eating grass?

Making Connections

9. What are some practical uses for discarded crab and lobster shells?

10. What is "dietary roughage" and why is it good for you?

(a) H₂C=O structure with chain:
H—C—OH
HO—C—H
H—C—OH
H—C—OH
H—C—OH
H

(b) H₂C=O structure with chain:
H—C—OH
H—C—OH
H—C—OH
H—C—OH
H

(c) H₂C=O structure with chain:
H—C—OH
H—C—OH
H

(d) H₂C=O structure with chain:
H—C—OH
HO—C—H
HO—C—H
H—C—OH
H—C—OH
H

Figure 14

Lipids

Lipids are hydrophobic molecules composed of carbon, hydrogen, and oxygen. They are a varied group of molecules that contain fewer polar O–H bonds and more non-polar C–H bonds than do carbohydrates. Thus, they are insoluble in water but soluble in other nonpolar substances. Some are large molecules consisting of long hydrocarbon chains attached to smaller polar subunits, while others are compact molecules consisting of several hydrocarbon rings. Organisms use lipids for storing energy, building membranes and other cell parts, and as chemical signalling molecules. Lipids can be divided into four families: fats, phospholipids, steroids, and waxes.

Fats are the most common energy-storing molecules in living organisms. A gram of fat stores about 38 kJ (9 kilocalories) of chemical energy per gram, more than twice the 17 kJ (4 kilocalories) of chemical energy stored in a gram of carbohydrate or protein. The calorie is a non-SI unit for energy that is commonly used in the food industry and in food science. One calorie (cal) is equal to 4.18 J. One kilocalorie, symbolized 1 kcal, 1 Calorie, or 1 Cal on food labels, is equivalent to 4.18 kJ.

Animals convert excess carbohydrate into fat and store the fat molecules as droplets in the cells of adipose (fat) tissue. A layer of fat under the skin acts as thermal insulation in many animals, such as penguins (**Figure 15**). Plants also store energy in the form of fat. The most common fats in plants and animals are the **triacylglycerols (triglycerides)**. These contain three fatty acids attached to a single molecule of glycerol (**Figure 16**).

triacylglycerols (triglycerides) lipids containing three fatty acids attached to a single molecule of glycerol

Figure 15
A layer of fat under the skin acts as thermal insulation in penguins.

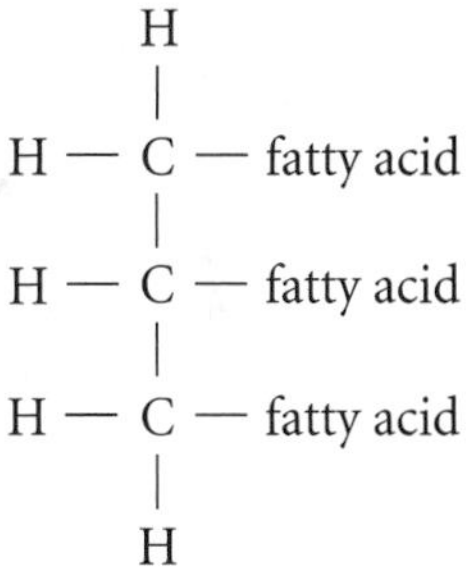

Figure 16
A triacylglycerol

Glycerol is a three-carbon alcohol containing a hydroxyl group attached to each carbon. Fatty acids are long hydrocarbon chains containing a single carboxyl group (−COOH) at one end. The hydrocarbon chains are usually even numbered and 16 or 18 carbon atoms long. Saturated fatty acids have only single bonds between carbon atoms, as shown in **Figure 17(a)** on page 36. Unsaturated fatty acids, such as oleic acid, have one or more carbon–carbon double bonds, as shown in **Figure 17(b)** on page 36. Saturated fatty acids are saturated because they contain the maximum number of hydrogen atoms possible attached to the carbon skeleton. Unsaturated fatty acids have fewer than the maximum number of hydrogens possible. The double bonds between carbons are formed by removing hydrogen atoms from the molecule.

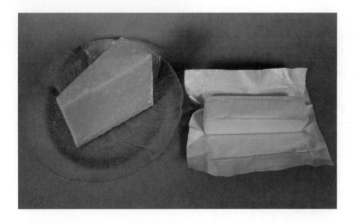

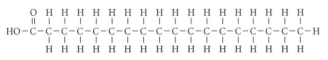

(a)   stearic acid

(b)   oleic acid

single bonds between carbon atoms;
straight fatty acid chains fit close together

double bonds between carbon atoms;
bent fatty acid chains do not fit close together

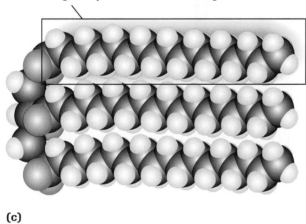

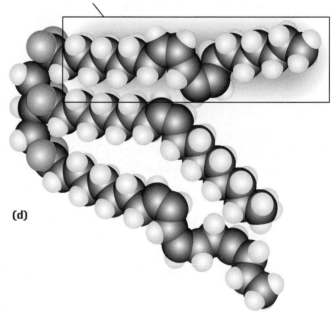

(c)

(d)

Figure 17

(a) Stearic acid is a saturated fatty acid containing 18 carbon atoms and no carbon–carbon double bonds.

(b) Oleic acid is an unsaturated fatty acid also containing 18 carbon atoms but with a carbon–carbon double bond between carbon 9 and 10.

(c) a triacylglycerol containing only saturated fatty acids

(d) a triacylglycerol containing unsaturated and polyunsaturated fatty acids

Fatty acids with many carbon–carbon double bonds are called polyunsaturated fatty acids. Animal fats, such as butter and lard, are mostly composed of triacylglycerols containing only saturated fatty acids, shown in **Figure 17(c)**. The straight hydrocarbon chains of the fatty acids fit closely together, allowing for many van der Waals attractions along their length. This results in a solid consistency at room temperature. Shown in **Figure 17(d)**, the unsaturated and polyunsaturated fatty acids commonly found in the triacylglycerols of plant oils, such as olive oil, corn oil, and peanut oil, are bent at the double bonds. These rigid kinks in the fatty acid tails reduce the number of van der Waals attractions that can form along their lengths, causing oils to be liquids at room temperature. An industrial process called hydrogenation adds hydrogen atoms to the double bonds in unsaturated triacylglycerols of liquid fats, such as corn oil or canola oil, converting them into a semisolid material, such as margarine or shortening.

When glycerol reacts with fatty acids, a condensation reaction takes place between a hydroxyl group of glycerol and the carboxyl group of a fatty acid. The resulting bond is called an **ester linkage**, and the process is known as esterification (**Figure 19**).

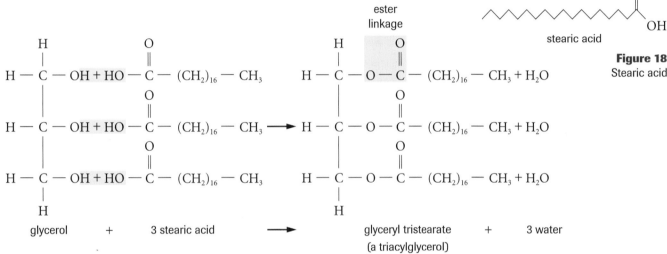

Figure 19
Esterification of glycerol and stearic acid (note the abbreviated structural formulas for the fatty acids)

The membranes of cells are mostly composed of fats called phospholipids. Phospholipids are composed of a glycerol molecule to which is attached two fatty acids and a highly polar phosphate group. The phosphate group can be thought of as a polar head and the fatty acids as long, nonpolar tails (**Figure 20**, page 38). The polar head is hydrophilic and the nonpolar tails are hydrophobic. When added to water, phospholipids form spheres, called **micelles**, shown in **Figure 21(a)** on page 38. The hydrophilic heads dissolve in the water and their hydrophobic tails mix with one another in the centre of the sphere.

Cell membranes normally separate two water compartments, such as extracellular fluid and the cell's cytoplasm. A double layer of phospholipids allows for heads to mix with water in both compartments and tails to mix with one another in the centre of the bilayer, as illustrated in **Figure 21(b)** on page 38. Water and other polar and ionic materials cannot pass through the bilayer because of the highly nonpolar centre. Functional cell membranes contain proteins and hydrophilic pores that form channels through which charged materials can pass.

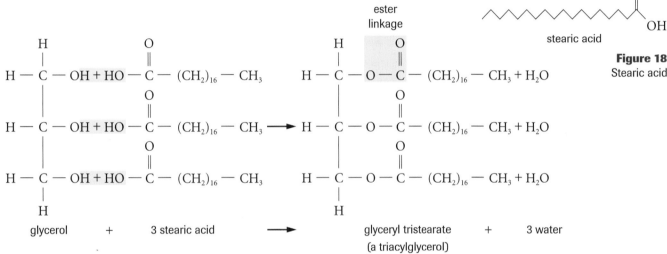

DID YOU *KNOW* ?

Simplified Structural Formulas
Organic chemists and biochemists rarely write out all the atoms in organic molecules. There are usually so many of them that they make the overall shape of the molecule difficult to see. In general, a short-form notation called a wireframe, or Kekule, structure is used. These diagrams do not contain carbon and hydrogen atom symbols. Instead a carbon atom is assumed to exist at every free end or bend in the wire frame. **Figure 18** is a wireframe diagram of stearic acid.

stearic acid

Figure 18
Stearic acid

ester linkage a functional group linkage formed by the condensation of a carboxyl group and a hydroxyl group

micelles spheres formed when phospholipids are added to water

The Chemical Basis of Life **37**

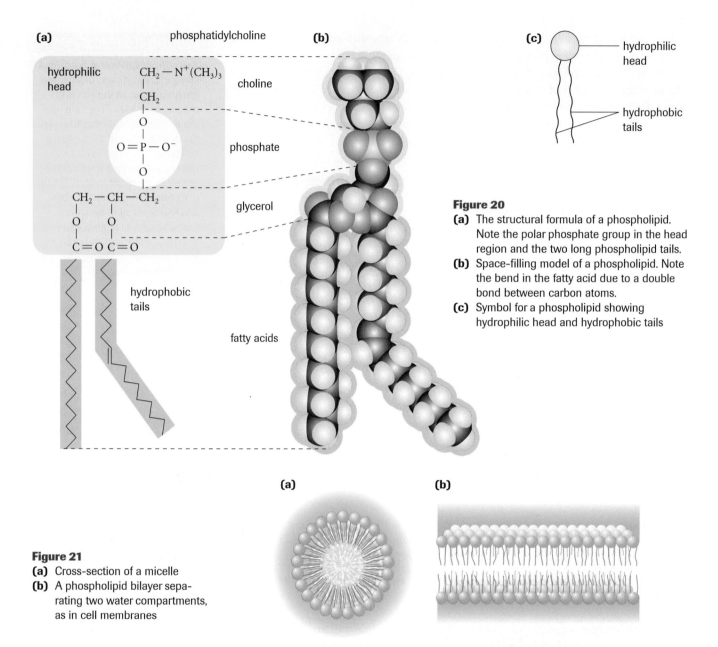

(a)

phosphatidylcholine

hydrophilic head

CH$_2$—N$^+$(CH$_3$)$_3$

CH$_2$

O

O=P—O$^-$

O

CH$_2$—CH—CH$_2$

O   O

C=O   C=O

hydrophobic tails

(b)

choline

phosphate

glycerol

fatty acids

(c)

hydrophilic head

hydrophobic tails

Figure 20

(a) The structural formula of a phospholipid. Note the polar phosphate group in the head region and the two long phospholipid tails.

(b) Space-filling model of a phospholipid. Note the bend in the fatty acid due to a double bond between carbon atoms.

(c) Symbol for a phospholipid showing hydrophilic head and hydrophobic tails

Figure 21

(a) Cross-section of a micelle

(b) A phospholipid bilayer separating two water compartments, as in cell membranes

(a)

(b)

Membranes often contain another class of lipid called sterols. Sterols (also called steroids) are compact hydrophobic molecules containing four fused hydrocarbon rings and several different functional groups. Cholesterol is an important steroid component of cell membranes. High concentrations of cholesterol in the bloodstream and a diet rich in saturated fats have been linked to the development of atherosclerosis, a condition in which fatty deposits, called plaque, form on the inner lining of blood vessels, blocking the flow of blood to tissues (**Figure 22**). Deprived of oxygen and nutrients, the cells of the tissues die. If the tissue happens to be heart muscle, a heart attack may occur. If blood flow to the brain is blocked, the person may suffer a stroke, possibly leading to paralysis and loss of bodily functions.

Cells convert cholesterol into a number of compounds, such as vitamin D (needed for healthy bones and teeth) and bile salts (needed for the digestion of fats in the small intestines). Other steroids include the sex hormones—testosterone, estrogens, and progesterone—that control the development of sex traits and gametes (**Figure 23**).

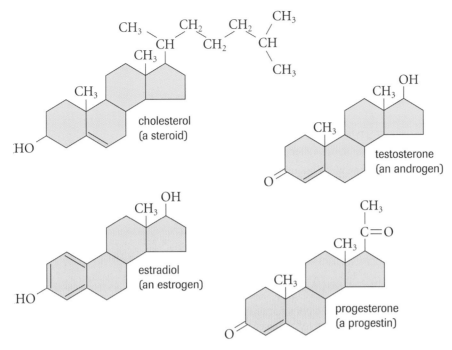

Figure 22
High cholesterol levels and a diet rich in saturated fats may lead to the development of atherosclerosis and coronary heart disease.

Figure 23
Cholesterol and the sex hormones progesterone, estradiol, and testosterone are members of the sterol family of lipids. Notice their structural similarity.

Waxes are lipids containing long-chain fatty acids linked to alcohols or carbon rings. They are hydrophobic molecules with a firm, pliable consistency. This property makes waxes ideally suited to form waterproof coatings on various plant and animal parts. A type of wax called cutin is produced by the epidermal cells of plants, forming a water-resistant coating on the surfaces of stems, leaves, and fruit (**Figure 24**). This helps the plant conserve water and also acts as a barrier to infection. Birds secrete a waxy material that helps keep their feathers dry, and bees produce beeswax to construct honeycombs (**Figure 25**).

Figure 24
Cutin forms a water-resistant coat on cherries.

Figure 25
Honeycombs are constructed of beeswax.

Understanding Concepts

11. What type of functional group linkage holds a fatty acid to glycerol in a triacylglycerol?

12. Distinguish between a saturated and an unsaturated fatty acid in terms of chemical structure and overall shape.

13. Why do polyunsaturated fatty acids tend to be liquids (oils) at room temperature?

14. Draw a Kekule structure for palmitic acid, a 16-carbon fatty acid.

15. (a) What is the shiny material on the surface of an apple (**Figure 26**)?
    (b) What is its purpose?
    (c) How do the material's properties suit its function?

Figure 26

16. (a) How do steroids differ structurally from other lipids?
    (b) List two biological functions of steroids.

Making Connections

17. (a) What process is used to produce vegetable oil margarine?
    (b) Why do producers usually add yellow food colouring to margarine?

18. Why do physicians recommend that people not eat diets rich in cholesterol and/or saturated fats?

Proteins

What do gelatin desserts, hair, antibodies, spider webs, blood clots, egg whites, tofu, and fingernails all have in common? They are all made of protein. Proteins are the most diverse molecules in living organisms and among the most important (**Figure 27**).

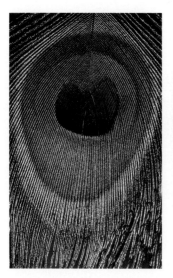

Figure 26
Proteins are used by living organisms to produce a variety of structures. Humans also use proteins to produce useful products.

The genetic information in DNA codes specifically for the production of proteins and nothing else. Proteins accomplish more tasks than any other group of biological molecules, and they make up more than 50% of the dry mass of most cells. They are used as structural building blocks and, as functional molecules, are involved in almost everything that cells do. Cells contain thousands of different proteins, each performing a specific task. As with carbohydrates and lipids, their three-dimensional structure is directly related to their function.

Proteins called enzymes act as biological catalysts, making chemical reactions proceed at a speed that sustains life. Proteins called immunoglobulins protect animals against foreign microbes and cancer cells. A variety of protein molecules help to transport materials through cell membranes and through the bodies of plants and animals. Hemoglobin shuttles oxygen from place to place in mammals, and protein carriers help move sucrose through phloem tissue in plants. Keratin, the most common protein in vertebrates, is a tough, structural protein found in hair and fingernails. Fibrin is the protein that helps blood clot, and collagen forms the protein component of bones, skin, ligaments, and tendons.

Proteins are amino acid polymers folded into specific three-dimensional shapes. A protein's structural characteristics determine its function.

An amino acid is an organic molecule possessing a central carbon atom to which are attached an amino group, a carboxyl group, a hydrogen atom, and a variable group of atoms called a side chain, usually symbolized by the letter R (**Figure 28**). There are 20 different R groups commonly found in living organisms and, therefore, 20 different amino acids (**Figure 29**, pages 42–43).

Glycine (gly) is the simplest amino acid, with a single hydrogen atom as its R group. Notice that the amino and carboxyl groups are ionized. Amino acids are amphiprotic, which means that they possess both acidic (carboxyl) and basic (amino) functional groups. When dissolved in water, the carboxyl group donates an H^+ ion to the amino group, causing the carboxyl group to become negatively charged and the amino group to possess an extra hydrogen and a net positive charge. Amino acids may be polar (hydrophilic), nonpolar (hydrophobic), or charged (acidic or basic), depending on the nature of their side chains. Proline (pro) is the only amino acid whose side chain forms a covalent bond with its own amino group. Acidic amino acids possess a carboxyl group on their side chains, and basic amino acids contain amino groups on their side chains.

Proteins consist of one or more amino acid polymers that have twisted and coiled into a specific shape. The final shape, or **conformation**, of the polymer is determined by the sequence of amino acids it contains. An amino acid polymer is called a polypeptide. Polypeptides are constructed in the cytoplasm of cells through a complex process called protein synthesis (studied in Chapter 5). Multiple copies of each of the 20 amino acids are dissolved in the cytoplasm of all cells. These are constructed by the cell from simpler compounds or obtained in the diet from proteins, in food, that the organism absorbs and digests. Of the 20 amino acids found in the food humans normally eat, 8 are considered to be **essential amino acids**. This means that the body cannot synthesize them from simpler compounds; they must be obtained from the diet. The 8 essential amino acids are tryptophan, methionine, valine, threonine, phenylalanine, leucine, isoleucine, and lysine. In protein synthesis, genetic information contained in DNA (sometimes RNA), directs ribosomes, RNA, and enzymes to join amino acids to one another in a particular sequence. The bonds that hold the amino acids together are called **peptide bonds**. Peptide bonds are formed by a condensation reaction between the amino group of one amino acid and the carboxyl group of an adjacent amino acid (**Figure 30**, page 43). This type of functional group linkage is called an amide bond.

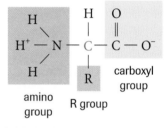

Figure 28
Generalized structural formula for amino acids

conformation the three-dimensional shape of a protein determined by the sequence of amino acids it contains

essential amino acids amino acids that the body cannot synthesize from simpler compounds; they must be obtained from the diet

peptide bonds the amide linkage that holds amino acids together in polypeptides

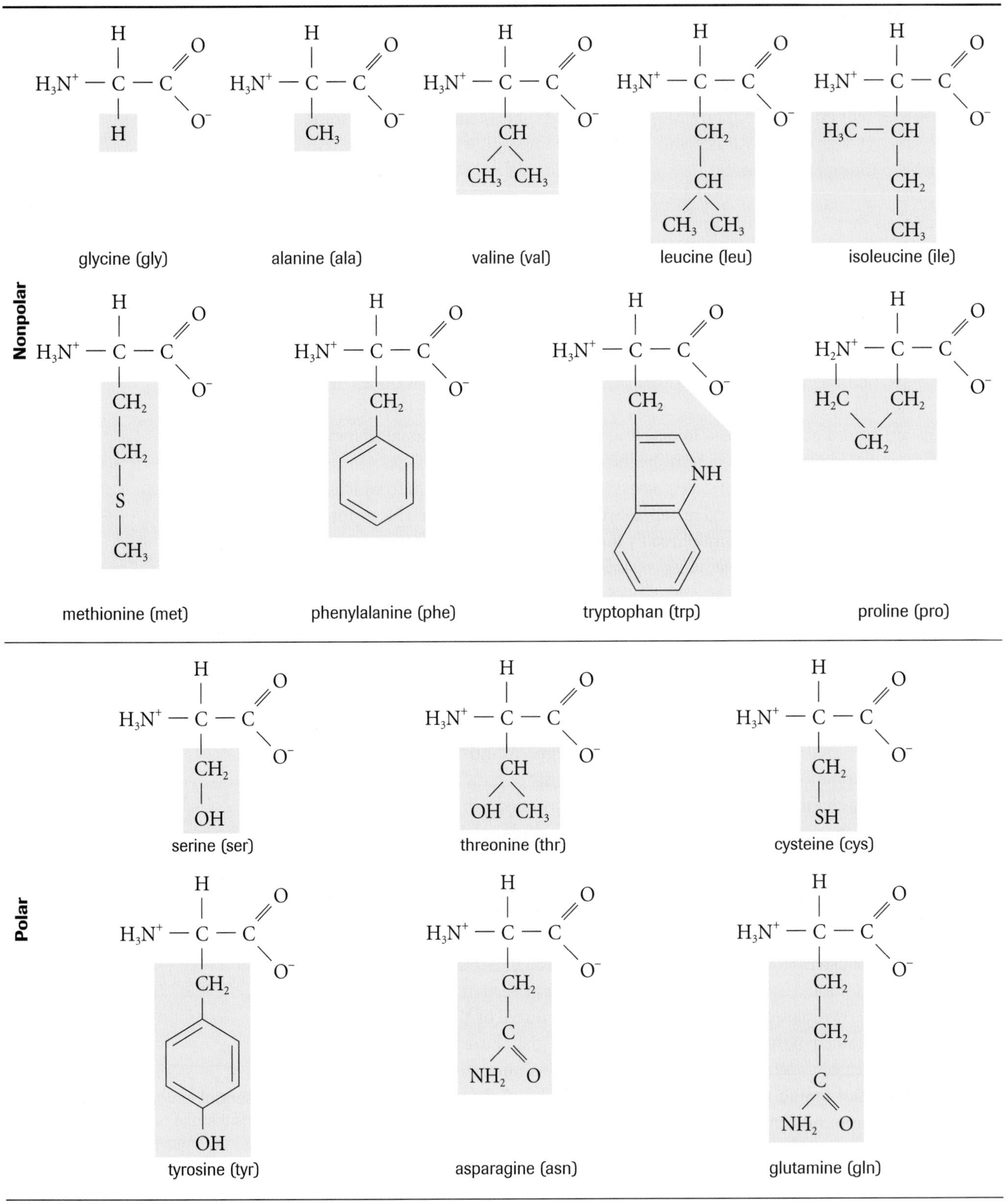

Figure 29
The 20 amino acids grouped according to their properties. The side chains (R groups) are highlighted. The amino and carboxyl groups are shown in their ionized form as they would be when dissolved in water at pH 7, the pH inside a living cell. Each amino acid is named and the abbreviation of the name is shown in parentheses.

Figure 29 *continued*

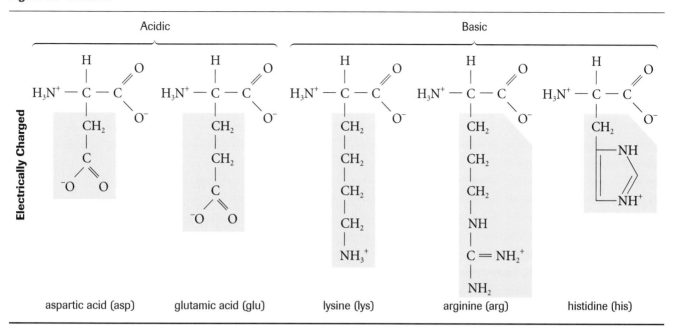

Electrically Charged

| Acidic | | Basic | | |

aspartic acid (asp)   glutamic acid (glu)   lysine (lys)   arginine (arg)   histidine (his)

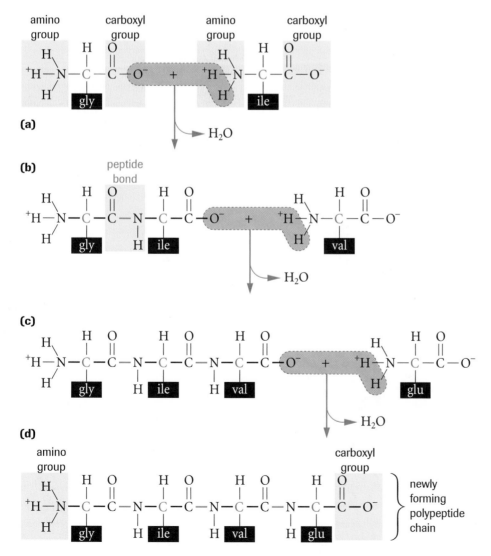

(a)

(b)

(c)

(d)

Figure 30

(a) The first two amino acids shown are glycine (gly) and isoleucine (ile). They are at the start of the sequence for one of the two polypeptide chains that make up the protein insulin in cattle. A water molecule forms as the by-product of the condensation reaction.

(b) Through the condensation reaction, the isoleucine becomes joined to the glycine by a peptide bond. Another condensation reaction takes place between isoleucine and valine (val) and water forms.

(c) A condensation reaction occurs between valine and glutamate (glu).

(d) Glutamate is the fourth amino acid in this growing polypeptide chain. DNA specifies the order in which the different amino acids follow one another. The process is called protein synthesis.

amino terminus the free amino group at one end of a polypeptide

carboxyl terminus the free carboxyl group at one end of a polypeptide

globular proteins protein molecules composed of one or more polypeptide chains that take on a rounded, spherical shape

The polypeptide chain will always have an amino group at one end, called the **amino terminus** (A-terminus), and a carboxyl group at the other, called the **carboxyl terminus** (C-terminus). Polypeptides range in length from a few amino acids to more than a thousand. All copies of a particular polypeptide have the same sequence of amino acids, this sequence being coded for by a particular gene in a DNA molecule. The sequence of amino acids determines a polypeptide's three-dimensional conformation and, thus, its functional characteristics. Many structural proteins are roughly linear in shape and arranged to form strands or sheets. **Globular proteins** are composed of one or more polypeptide chains that take on a rounded, spherical shape. Many enzymes and other functional proteins are globular.

Globular proteins may be described in terms of four levels of structure: primary, secondary, tertiary, and quaternary. The first three structures—primary, secondary, and tertiary—apply to the individual polypeptide chains of a protein, while the fourth, quaternary structure, describes the interactions that occur between polypeptide strands in proteins composed of two or more polypeptides (**Figure 31**).

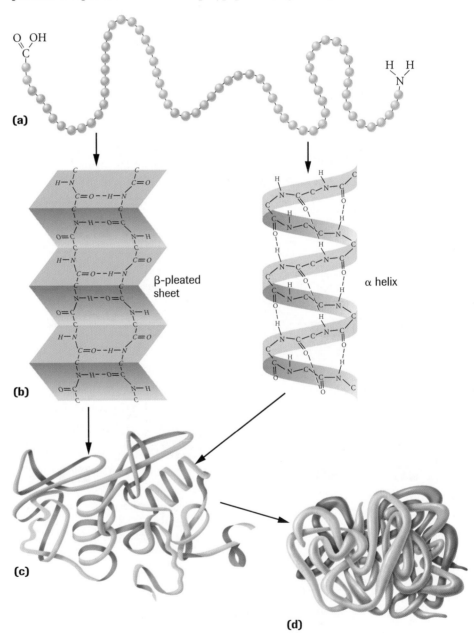

Figure 31
(a) The primary structure of a protein is the sequence of amino acids in the polypeptide strand.
(b) Hydrogen bonds that form with nearby amino acids coil and fold the polypeptide into α helices and β-pleated sheets; these constitute the polypeptide's secondary structure.
(c) The polypeptide folds further to form its tertiary structure. These folds are stabilized by R group–R group interactions.
(d) The clustering of two or more polypeptides in tertiary structure generates the quaternary structure of a protein.

Primary structure, shown in **Figure 31(a)**, is the unique sequence of amino acids in a polypeptide chain. Each of the amino acids in a polypeptide is referred to as a **residue**. Fredrick Sanger, a British biochemist, was the first to determine the primary structure of a protein. He received the Nobel Prize in 1958 for determining the sequence of the 51 amino acids in the protein hormone insulin. When Sanger announced that he wanted to determine the sequence of amino acids in a naturally occurring protein, his professor said that he was mad! Sanger received another Nobel Prize in 1980 for his work on gene structure.

Primary structure is determined by the nucleotide sequence in a gene in DNA. During protein synthesis (discussed further in Chapter 5), the nucleotide sequence governs the order in which individual amino acids are linked to form the polypeptide chains that fold into functional protein molecules. With 20 different amino acids to choose from, a polypeptide containing 200 amino acids could be arranged into 20^{200} different sequences—more than all the atoms in the known universe! The final conformation of a protein is very sensitive to primary structure. Changing the sequence by one amino acid could alter the three-dimensional shape to the point that the protein loses function or is rendered useless. This happens in sickle cell anemia, an inherited condition in which a single amino acid substitution changes the conformation of a polypeptide in hemoglobin, causing red blood cells to become deformed. The deformed cells clog blood vessels and impede blood flow to the tissues.

During protein synthesis, amino acids are added to the growing chain one at a time. As the chain grows, it coils and folds at various locations along its length. These coils and folds are collectively referred to as the polypeptide's **secondary structure**. In certain regions, a hydrogen bond forms between the electronegative oxygen of the carboxyl group (C=O) of one peptide bond and the electropositive hydrogen of an amino group (N–H) four peptide bonds away. These H bonds, repeated over a certain length of the chain, shape portions of the polypeptide into a coil called an **α helix**, shown in **Figure 31(b)**. Fibrous proteins, such as α-keratin, the protein in hair, are almost completely shaped into an α helix. Many globular proteins, such as lysozyme, an enzyme that acts as a natural disinfectant in saliva, sweat, and tears, have regions of α helix separated by nonhelical regions called **β-pleated sheets**. This type of secondary structure occurs when two parts of the polypeptide chain lie parallel to one another. In this case, hydrogen bonds form between the oxygen atoms of carboxyl groups on one strand and the hydrogen atoms of amino groups on an adjacent strand, as shown in **Figure 31(b)**.

Some fibrous proteins, such as the protein in the silk that spiders use to construct their webs, contain large amounts of β-pleated sheets (**Figure 32**). When building their webs, spiders secrete silk fibres in liquid form. These solidify when exposed to air. The hydrogen bonds that form in the many regions of β-pleated sheets make the silk stronger than steel (spider silk can be stretched about 40% before it breaks; steel can only be stretched about 8%).

primary structure the unique sequence of amino acids in a polypeptide chain

residue an amino acid subunit of a polypeptide

secondary structure coils and folds in a polypeptide caused by hydrogen bonds between hydrogen and oxygen atoms near the peptide bonds

α helix a type of polypeptide secondary structure characterized by a tight coil that is stabilized by hydrogen bonds

β-pleated sheet polypeptide secondary structures that form between parallel stretches of a polypeptide and are stabilized by hydrogen bonds

Figure 32
Hydrogen bonding within proteins in spider silk contributes to its strength.

Strong forces of attraction and repulsion between the polypeptide and its environment force it to undergo additional folding into a **tertiary structure**, illustrated in **Figure 31(c)** on page 45. Amino acids containing polar R groups, such as serine, tyrosine, and glutamine, are attracted to water, while those containing nonpolar R groups are excluded. As a result, amino acids with hydrophobic groups, such as valine and phenylalanine, congregate in the interior of the folded polypeptide, away from water. Tertiary structure is stabilized by a number of different R-group interactions. Hydrogen bonds between certain polar side chains, ionic bonds between oppositely charged side chains, and van der Waals forces between nonpolar R groups, all help to keep the polypeptide folded into tertiary structure. Proline, the only amino acid in which the R group is attached to the amino group, forms a natural kink where it occurs in an α helix or a β-pleated sheet. This also helps shape tertiary structure. Whenever the sulfur-containing R groups of two cysteine residues are brought close to one another, a covalent bond called a **disulfide bridge** (-S-S-) may form. Disulfide bridges are strong stabilizers of tertiary structure (**Figure 33**).

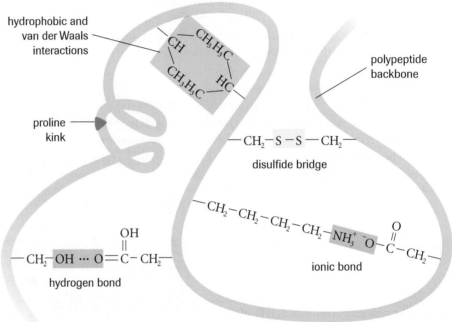

Figure 33
Hydrophobic interactions, hydrogen bonds, ionic bonds, and disulfide bridges help to hold polypeptides in tertiary structure.

In some cases, two or more polypeptide subunits must come together to form a functional protein. Such proteins are said to be in **quaternary structure**, shown in **Figure 31(d)** on page 45. Collagen (the tough fibrous protein found in skin, bones, tendons, and ligaments), keratin (the fibrous protein in hair) (**Figure 34**), and hemoglobin, the globular protein that transports oxygen in red blood cells (**Figure 35**) are examples of proteins in quaternary structure. Hemoglobin is composed of four polypeptides in tertiary structure: two identical α-chains and two identical β-chains. Each subunit has a nonproteinaceous heme group containing an iron atom that binds oxygen.

It is evident that a polypeptide's final shape, its tertiary structure, is determined by its primary structure. However, chemical and physical environmental factors also help to determine the final shape of the molecule. Within a cell, a protein is constructed in an aqueous environment at a neutral pH and a particular temperature. If a protein produced in this environment is placed in different environmental conditions, it may unravel. A change in the three-dimensional shape of a protein caused by changes in temperature, pH, ionic concentration, or other environmental factors is called denaturation.

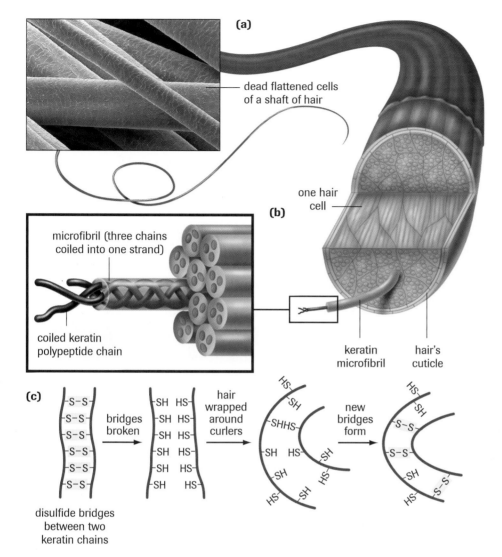

(a)

dead flattened cells
of a shaft of hair

one hair
cell

(b)

microfibril (three chains
coiled into one strand)

coiled keratin
polypeptide chain

keratin
microfibril

hair's
cuticle

(c)

bridges
broken

hair
wrapped
around
curlers

new
bridges
form

disulfide bridges
between two
keratin chains

Figure 34
(a), **(b)** Hair cells develop from modified skin cells. These cells synthesize polypeptide chains of the protein keratin. Disulfide bridges link three chains together as fine fibres, which get bundled into larger, cable-like fibres. The larger fibres almost fill the cells, which eventually die off. Dead, flattened cells form a tubelike cuticle around the developing hair shaft.

(c) For a permanent wave, hair is exposed to chemicals that break disulfide bridges. When hairs wrap around curlers, the polypeptide chains of the hairs are held in new positions. Exposing the hair to a chemical that encourages the formation of disulfide bridges causes new disulfide bridges to form. The bonds now form between different sulfur-bearing amino acids than before. The displaced bonding locks the hair into curled positions.

A denatured protein cannot carry out its biological functions. Various chemicals and heat disrupt the hydrogen bonds, ionic bonds, disulfide bridges, and hydrophobic interactions that help hold a polypeptide in tertiary structure. A protein denatured by heat or chemicals will usually return to its normal shape when the denaturing agent is removed, as long as its primary structure remains intact. If the denaturing agent breaks the peptide bonds that hold the amino acids together, the protein will be destroyed.

Protein denaturation can be both dangerous and useful. Protein enzymes function best within a narrow range of temperature, pH, and salt concentration. Enzymes in thermophiles (archaebacteria that normally live in water at temperatures around 100°C) can not survive at room temperature because their proteins and enzymes would denature. Gastrin, a digestive enzyme in the stomach, works best at pH 2 and is denatured in the small intestine, where the pH increases to about 10. Prolonged fever above 39°C could denature critical enzymes in the brain, leading to seizures and possibly death. Curing meats with high concentrations of salt or sugar, and pickling meat or vegetables in vinegar, preserves the food by denaturing the enzymes in bacteria that would otherwise spoil the food. The practice of blanching fruits and vegetables by quickly dipping them in boiling water denatures enzymes that would otherwise cause them to turn brown when exposed to air. Heat can be used to denature the proteins in hair, allowing people to temporarily straighten curly hair or curl straight hair.

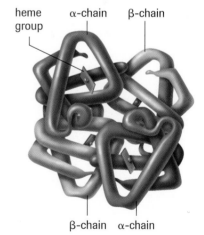

heme
group

α-chain

β-chain

β-chain

α-chain

Figure 35
Hemoglobin, a protein that transports oxygen in blood, consists of four polypeptide chains, two α-chains, and two β-chains. The α-chain and β-chain differ slightly in their amino acid sequence.

In general, small proteins are able to refold into tertiary structure once the denaturing agent is removed; however, large globular proteins fail to refold spontaneously because many intricate folds are needed to reestablish tertiary structure. In fact, recent research into how proteins fold when they are created in protein synthesis has revealed that special proteins, called **chaperone proteins**, aid the growing polypeptide to fold into tertiary structure. Some forms of cystic fibrosis, a genetic disease characterized by the inability of certain proteins to move ions across cell membranes, are caused by a deformation in the proteins' three-dimensional structure. Protein sequencing studies have shown that the affected proteins possess the same primary structure as normal proteins of the same type, causing researchers to speculate that the proteins fail to fold because of a deficiency in specific chaperone proteins.

Humans consume animal flesh (meat) as a source of protein. However, the fibrous proteins contained in muscle cells, the most common type of meat, makes raw meat difficult to chew. Heating denatures the fibrous protein molecules and partially decomposes them. In many cases, the primary purpose of cooking is to denature the protein component of food. Foods are usually a rich mixture of nutrients including carbohydrates, lipids, proteins, nucleic acids, vitamins, and minerals.

Case Study Sequencing a Polypeptide

Protein sequencing allows scientists to determine the primary structure of proteins and enzymes. To obtain the sequence, first the protein is hydrolyzed into its individual amino acids using concentrated acid. Then, the number and type of each amino acid in the polypeptide is identified. In practice, it is very difficult to isolate one molecule of the protein in solution. Usually a pure sample containing thousands of molecules of the same protein is used. For example, a sample containing the following hexapeptide is hydrolyzed:

H_2N–glycine-alanine-valine-phenylalanine-threonine–alanine-COOH

The hydrolysis reaction separates glycine, alanine, and threonine units in a 1:2:1 molecular ratio. However, information regarding the sequence is lost as the individual molecules randomly mix in the solution.

The amino-terminal residue of a polypeptide (glycine in this case) can be identified by first attaching a "flag" molecule to this residue while the peptide is intact. Then, the modified residue from the polypeptide is removed by hydrolyzing the entire polymer and isolating and purifying the modified amino-terminal amino acid (**Figure 36**).

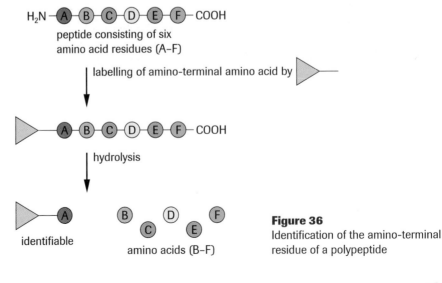

Figure 36
Identification of the amino-terminal residue of a polypeptide

Unfortunately, the need to hydrolyze the amino-terminal amino acid (to isolate, purify, and identify it) hydrolyzes the other peptide bonds as well, destroying the rest of the sequence. In 1947, Pehr Victor Edman, working at the Rockefeller Institute in Princeton, New Jersey, discovered that a substance called phenylisothiocyanate (PITC) (**Figure 37**) could bind to the amino terminus and weaken the adjacent peptide bond. When the polypeptide is subjected to trifluoroacetic acid (**Figure 38**), the modified amino-terminal amino acid breaks off, leaving the rest of the polypeptide intact. The modified amino-terminal amino acid, now called a phenylthiohydantoin (PTH) derivative, is separated from the rest of the polypeptide, purified, and identified. The procedure is repeated with the rest of the polypeptide, resulting in the determination of the peptide sequence (**Figure 39**).

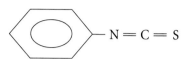

Figure 37
Phenylisothiocyanate

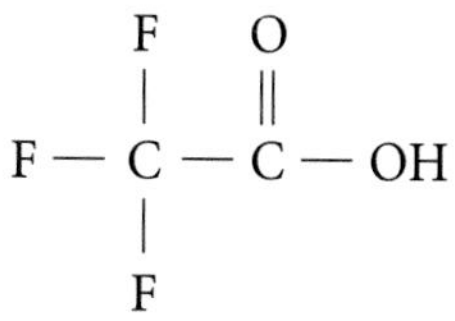

Figure 38
Trifluoroacetic acid

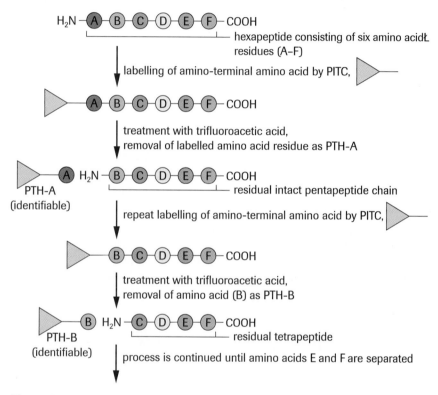

Figure 39
The Edman sequential degradation of a polypeptide

This process has been fully automated and can be used with extremely small quantities of polypeptide or protein in the picomole (10^{-12} mol) range. These procedures are frequently referred to as "wet" methods because they are done in solution. Although precise, the method begins to fail after about 20 to 40 residues have been determined, so that it is necessary to break proteins into smaller pieces before beginning the process. The Edman degradation is then used to sequence the smaller fragments of the original chain. Although the amino acid sequences of the fragments are obtained, the order in which these smaller segments occur in the original protein is lost. To determine the overall sequence of original peptides, the original polypeptide must be cleaved by enzymes that are known to hydrolyze a particular peptide bond. Using two such enzymes (each cleaving the original polypeptide into different pieces), performing an Edman degradation on each fragment, and comparing the results allows the sequence of the original polypeptide to be determined (**Figure 40**, page 50). Today, automated sequencers carry out complete Edman degradations on polypeptides containing tens of amino acids in less than one day.

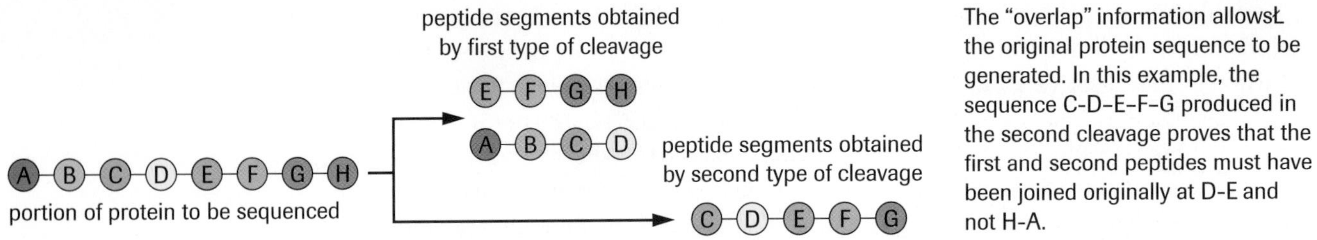

peptide segments obtained
by first type of cleavage

portion of protein to be sequenced

peptide segments obtained
by second type of cleavage

The "overlap" information allows
the original protein sequence to be
generated. In this example, the
sequence C–D–E–F–G produced in
the second cleavage proves that the
first and second peptides must have
been joined originally at D-E and
not H-A.

Figure 40
Protein sequencing by producing
overlapping peptide fragments
(mapping)

▶ *Case Study Questions*

Understanding Concepts

1. Why is protein sequencing useful?

Making Connections

Conduct library and/or Internet research to answer questions 2 to 5.

GO   www.science.nelson.com

2. What was the first protein to be fully sequenced? Who conducted the experiment and when was it accomplished? What substance was used to identify the amino-terminal amino acid? What problems were encountered in the process?

3. What "wet" methods could be used to identify the PTH derivative produced in the Edman process?

4. Identify another method used to sequence a protein. Describe the method.

5. A number of organizations on the Internet offer protein sequencing and protein synthesis services. Search the Internet for these organizations. What types of organizations usually do this sort of work? Describe a typical pricing structure.

▶ *Practice*

Understanding Concepts

19. What two functional groups do all amino acids have in common?

20. (a) Write an equation showing the formation of a peptide bond between serine and aspartic acid.
    (b) Identify the amino terminus, the carboxyl terminus, and the peptide bond.

21. Define primary structure and describe how it is specified by genes.

22. (a) Name the two types of secondary structure.
    (b) What type of intermolecular bond stabilizes secondary structure?

23. How does the amino acid proline affect tertiary structure?

24. Distinguish between a polypeptide and a protein.

25. Define *protein denaturation*.

26. Why do proteins have more diverse functional roles in a cell than do carbohydrates?

Applying Inquiry Skills

27. (a) Egg whites are composed almost entirely of the protein albumin. In a fresh egg, the albumin is in the form of a yellowish liquid, as illustrated in **Figure 41(a)**. **Figure 41(b)** shows that, when heated, egg whites become white, solid, and opaque. Predict the physical characteristics of fresh egg whites that have been placed in vinegar.
    (b) Develop a hypothesis to explain why egg whites will not regain their original characteristics when removed from heat or vinegar.

▶ *Practice* continued

Figure 41
(a) Raw egg whites are yellowish and liquid.
(b) Heated egg whites are white and solid.

Making Connections

28. Aside from prescribing medications, how might a hospital treat a person who has had a high fever for several days? Why?

29. (a) How are fruits and vegetables blanched?
    (b) What does blanching do to the food?
    (c) Why would a food-processing company blanch certain fruits and vegetables before packaging them in transparent jars?

▶ *EXPLORE* an issue

Take a Stand: Functional Foods and Nutriceuticals

Decision-Making Skills

○ Define the Issue   ● Analyze the Issue   ● Research
● Defend the Position   ● Identify Alternatives   ● Evaluate

If it is not exactly a food and it is not quite a drug, what is it? It is a nutriceutical, also known as a functional food. These are dietary products that claim to provide health benefits beyond the nutritional value of a food. The term combines the words *nutrient* and *pharmaceutical*, and refers to foodstuffs that claim to prevent or control disease and improve physical or mental performance. You may have already used products with extract of *Gingko biloba* in hopes of improving your memory or with St. John's Wort for a personality adjustment. *Echinacea* has become a popular cold remedy and Straight A Lollipops claim to help children think straight.

Recent studies have shown that red wine and chocolate contain large quantities of natural antioxidants called flavonoids. Antioxidants are substances that deactivate dangerous byproducts of metabolism, called free radicals. Free radicals have been found to cause damage to membranes and DNA, and are thought to contribute to the development of cancer, heart disease, and stroke.

All sorts of functional foods are appearing on store shelves, and that has prompted calls for government regulation in Canada and around the world. These products are presently being produced and sold without any government regulation.

Wine and chocolate producers would love to be able to adorn their labels with health claims, but dieticians worry that consuming these and other products might actually aggravate health problems. For example, *Gingko biloba* acts as a blood thinner. People who are presently taking blood-thinning agents

to control a medical condition could experience an undesired effect by ingesting *Gingko biloba*. Nutriceutical promoters claim that there is so little of the active ingredient in functional foods that the products are rarely dangerous. Consumers and health officials alike want package labels to list the amount of active ingredients in the products, and Health Canada says that regulations should be in place shortly.

Some activists in this area claim that labelling is not enough. They insist that nutriceuticals must undergo clinical trials similar to those that potential drugs must go through before being sold to the public. Detractors say that this would increase the price of these beneficial foods, making them less accessible to people who need or want them.

Statement
Nutriceuticals should be regulated as drugs under the *Canada Food and Drug Act*.

- In your group, research the issue. Search for information in newspapers, periodicals, CD-ROMs, and on the Internet.

   www.science.nelson.com

- Identify organizations, and government agencies that have addressed the issue.
- Identify the perspectives of opposing positions.
  (a) Write a position paper summarizing your opinion.

Nucleic Acids

Nucleic acids are informational macromolecules. They are used by all organisms to store hereditary information that determines structural and functional characteristics. They are the only molecules in existence that can produce identical copies of themselves, and in so doing, allow organisms to reproduce. The instructions for creating an organism are stored in code along coiled chains of deoxyribonucleic acid, referred to as DNA.

Another nucleic acid, ribonucleic acid, RNA, reads the information in DNA and transports it to the protein-building apparatus of the cell. Here, the instructions direct the production of structural and functional proteins. DNA and RNA are nucleotide polymers. A nucleotide subunit in both molecules contains a nitrogenous base, a five-carbon (pentose) sugar, and a phosphate group, as shown in **Figure 42(a)**.

DNA contains the sugar deoxyribose, whereas RNA contains ribose. As **Figure 42(b)** shows, the only difference between these two monosaccharides is the lack of an oxygen atom at carbon 2 in deoxyribose, which accounts for its name. There are five types of organic bases found in nucleic acids: adenine (A), guanine (G), cytosine (C), thymine (T), and uracil, (U). **Figure 42(c)** illustrates that cytosine, thymine, and uracil are single-ringed pyrimidines, while adenine and guanine are larger double-ringed purines.

DNA has nucleotides containing the bases A, G, C, and T, while RNA contains A, G, and C, and U instead of T. Notice the structural similarity between U and T. When a DNA or RNA molecule is built, specific enzymes link a series of nucleotides to one another forming a polymer called a **strand**. The enzymes facilitate the formation of covalent bonds between the phosphate group of one nucleotide and the hydroxyl group attached to the number 3 carbon of the sugar on the adjacent nucleotide. A phosphodiester bond forms as the result of a condensation reaction between two −OH groups. RNA molecules coil into a helix, but remain single stranded.

DNA molecules are also helical in structure, but they are composed of two DNA strands wound around each other like the inside and outside rails of a circular staircase, as illustrated in **Figure 42(d)**. The two strands in DNA are held together by hydrogen bonds between nitrogenous bases on adjacent strands. There is a strict rule regarding base pairing. Adenine (A) always bonds to thymine (T) with two hydrogen bonds; guanine (G) always bonds to cytosine (C) with three hydrogen bonds. Each strand of the DNA molecule has a free phosphate group at one end and a free sugar (deoxyribose) at the other end. Hydrogen bonds will form properly only if one strand is upside down compared with the other strand. This means that the free phosphate end of one strand lines up with the free sugar end of the adjacent strand. Because of this, the two strands are said to run **antiparallel**. Every nucleotide pair is composed of a purine (double ring) facing a pyrimidine (single ring).

Nucleotides are not only used in the construction of DNA and RNA. They are also important intermediates in a cell's energy transformations. The nucleotide **adenosine triphosphate (ATP)** is used to drive virtually all the energy-requiring reactions in a cell. Nucleotide derivatives, such as nicotinamide adenine dinucleotide (NAD^+) and flavin adenine dinucleotide (FAD), are used in the production of ATP. $NADP^+$, a nucleotide similar to NAD^+, is used as a coenzyme in photosynthesis (Chapter 3) and cyclic adenosine monophosphate (cAMP) is used as a "second messenger" in various hormone interactions (Chapter 8).

All organisms pass on to their offspring their unique sequence of nucleotides in DNA. According to the theory of evolution, organisms possessing similar features are more closely related than those sharing fewer features. Thus, humans seem to be more closely related to apes than they are to crickets. But are we more related to gorillas than we are to gibbons? A smaller number of anatomical differences may produce inconclusive results. New studies comparing the sequence of nucleotides in DNA are beginning to yield more

strand a single nucleotide polymer

antiparallel describes two adjacent nucleotide polymers running in opposite directions relative to one another

adenosine triphosphate (ATP) a nucleotide derivative that acts as the primary energy-transferring molecule in living organisms

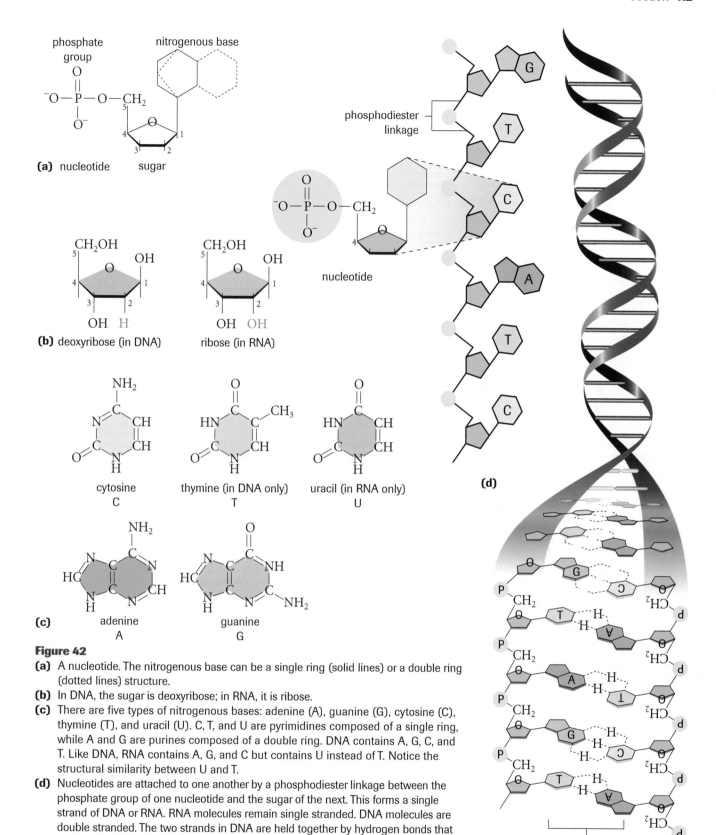

Figure 42

(a) A nucleotide. The nitrogenous base can be a single ring (solid lines) or a double ring (dotted lines) structure.

(b) In DNA, the sugar is deoxyribose; in RNA, it is ribose.

(c) There are five types of nitrogenous bases: adenine (A), guanine (G), cytosine (C), thymine (T), and uracil (U). C, T, and U are pyrimidines composed of a single ring, while A and G are purines composed of a double ring. DNA contains A, G, C, and T. Like DNA, RNA contains A, G, and C but contains U instead of T. Notice the structural similarity between U and T.

(d) Nucleotides are attached to one another by a phosphodiester linkage between the phosphate group of one nucleotide and the sugar of the next. This forms a single strand of DNA or RNA. RNA molecules remain single stranded. DNA molecules are double stranded. The two strands in DNA are held together by hydrogen bonds that form between a nitrogenous base on one strand and a complementary base on the adjacent strand. Notice how the two strands run antiparallel—one strand being upside down compared with the other strand—and how every nucleotide pair is composed of a purine (double ring) facing a pyrimidine (single ring). Also note that thymine (T) is always attached to adenine (A) by two hydrogen bonds, and cytosine (C) to guanine (G) by three hydrogen bonds.

Biological Macromolecules in 3-D (p. 80)

Structural formulas written on paper simply do not convey a realistic view of the three-dimensional shapes of molecules and the reaction mechanisms involved in chemical reactions. What do these molecules look like? In Activity 1.2.1, you will manipulate ball-and-spring models of biomolecules and see interactive computer-generated graphics of key macromolecules.

convincing evidence of evolutionary kinship. It is hypothesized that more closely related organisms contain more closely related sequences of nucleotides in their DNA molecules. In nucleic acid hybridization studies, researchers force the DNA molecules from two different organisms to unwind and separate into individual strands by breaking the hydrogen bonds between them. They allow the separated strands from one organism to mix with the separated strands of the other organism. As the strands mingle, complementary bases will attach to one another with hydrogen bonds, forming hybrid DNA molecules. Heating the hybrid molecules and measuring the amount of heat needed to break the hydrogen bonds and separate the strands provides a measure of the similarity in nucleotide sequence between the strands. More heat will be required to separate the hybrid DNA molecules of more closely related species because these combinations will produce more base pairs. **Table 4** summarizes the main organic compounds in living things.

Table 4 Summary of the Main Organic Compounds in Living Organisms

Category	Main subcategories	Some examples and their functions
Carbohydrates contain an aldehyde or a ketone group and one or more hydroxyl groups.	monosaccharides (simple sugars)	glucose: energy source
	oligosaccharides (several monosaccharides attached by glycosidic linkages)	sucrose (a dissaccharide): most common form of sugar transported through plants
	polysaccharides (complex carbohydrates composed of many monosaccharides attached by glycosidic linkages)	starch, glycogen: energy storage cellulose: structural roles, plant cell walls
Lipids are largely hydrocarbon; generally, they do not dissolve in water but dissolve in nonpolar substances, such as other lipids.	*Lipids with fatty acids*	
	triglycerides: one, two, and three fatty acids attached to a glycerol molecule	fats (e.g., butter), oils (e.g., corn oil): energy storage
	phospholipids: phosphate group, one other polar group, and (often) two fatty acids attached to a glycerol molecule	phosphatidycholine: key component of cell membranes
	waxes: long-chain fatty acids attached to an alcohol portion	waxes in cutin: water conservation at above-ground surfaces of plants
	Lipids with no fatty acids	
	sterols: four carbon rings; with varying number, position, and types of functional groups	cholesterol: component of animal cell membranes; precursor of many steroids/hormones and vitamin D
Proteins are composed of one or more polypeptide chains, each with as many as several thousand covalently linked amino acids.	*Fibrous proteins*	
	long strands or sheets of polypeptide chains; often tough, water insoluble	keratin: structural element of hair and nails collagen: structural element of bone and cartilage
	Globular proteins	
	one or more polypeptide chains folded and linked into globular shapes; many roles in cell activities	enzymes: increase rates of reactions hemoglobin: oxygen transport insulin: controls glucose metabolism antibodies: tissue defence
Nucleic acids (and nucleotides) are chains of units (or individual units) that each consist of a five-carbon sugar, phosphate, and a nitrogen-containing base.	adenosine phosphates	ATP: major energy carrier cAMP: messenger in hormone regulation
	nucleotide coenzymes	NAD^+, $NADP^+$, FAD^+: transport of protons (H^+), and electrons from one reaction site to another
	nucleic acids: chains of thousands to millions of nucleotides	DNA, RNA: storage, transmission, and translation of genetic information

SUMMARY *The Chemicals of Life*

- **Figure 43** shows the divisions of biomolecules.

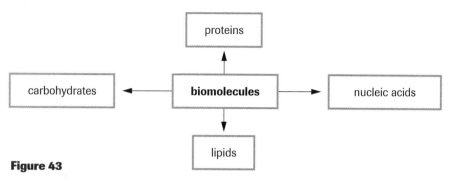

Figure 43

Divisions of biomolecules

- A condensation reaction links the subunits of macromolecules together. A water molecule is formed in the process. Hydrolysis reactions use water molecules to break macromolecules into their constituent subunits.

- Carbohydrates contain carbon, hydrogen, and oxygen atoms in a 1:2:1 ratio. Monosaccharides are simple sugars. Oligosaccharides are monosaccharide polymers containing two or three simple sugars. Polysaccharides are polymers composed of several hundred to several thousand monosaccharide subunits held together by glycosidic linkages.

- Lipids (fats) are hydrophobic molecules composed of carbon, hydrogen, and oxygen. The most common fats are the triacylglycerols (triglycerides) composed of glycerol and three fatty acids. Fatty acids are long hydrocarbon chains containing a single carboxyl group ($-COOH$) at one end. Cholesterol is an important sterol component of cell membranes. Other steroids include the sex hormones testosterone, estrogens, and progesterone. Phospholipids are key components of biological membranes.

- Proteins are amino acid polymers folded into specific three-dimensional shapes. An amino acid is an organic molecule possessing a central carbon atom to which are attached an amino group, a carboxyl group, a hydrogen atom, and a variable group of atoms called a side chain, usually symbolized by the letter R.

- **Table 5** shows the four structural levels of proteins.

Table 5 Structural Levels of Proteins

Structural name	Structural description	Bonding characteristics
primary	sequence of amino acids	covalent peptide bonds
secondary	α helix	hydrogen bonds
	β-pleated sheet	hydrogen bonds
tertiary	complex folding	R-group–R-group interactions (H-bonds, ionic bonds, hydrophobic interactions), proline kinks, and disulfide bridges
quaternary	several polypeptides in tertiary structure interacting with one another	R-group–R-group interactions between globular polypeptide chains

- A change in the three-dimensional shape of a protein caused by changes in temperature, pH, ionic concentration, or other environmental factor is called denaturation. A denatured protein cannot carry out its biological functions.
- Nucleic acids are informational macromolecules composed of nucleotide polymers. DNA and RNA are important nucleic acids.

Understanding Concepts

1. Why are hydrocarbons all nonpolar molecules?

2. List the elements found in all carbohydrates, indicating the atomic ratio in which they are found.

3. (a) Define *functional group*.
   (b) What advantage is conferred to biological molecules by having functional groups?

4. Why are monosaccharides more soluble in water than are triacylglycerols?

5. In your notebook, copy and complete **Table 6**.

Table 6 The Functions of Carbohydrates

Carbohydrate (or derivative)	Main function
glucose	
	energy storage in plant cells
chitin	
	plant cell wall component
glycogen	

6. Describe the difference between a condensation reaction and a hydrolysis reaction.

7. (a) Define *bonding capacity*.
   (b) What is the valence of carbon? nitrogen? oxygen?
   (c) Why is hydrogen never a central atom in a complex organic molecule?

8. Distinguish between oligosaccharides and polysaccharides and provide two examples of each.

9. Why are glucose and galactose isomers?

10. Why do animals use lipids instead of carbohydrates as energy-storing molecules?

11. (a) How many fatty acids are attached to a glycerol molecule in a triacylglycerol? in a phospholipid?
    (b) What two functional groups react when a fatty acid bonds to a glycerol molecule?

12. Detergent molecules are phospholipidlike molecules with highly polar heads and nonpolar tails. Why is it a good idea to wear rubber gloves when washing dishes using detergents?

13. Distinguish between a polypeptide and a protein.

14. Copy the following amino acid formula (**Figure 44**) into your notebook.

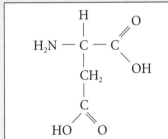

Figure 44

(a) Draw a circle around the amino group and a square around the carboxyl group.
(b) What functional group is found in the R-group side chain?
(c) Is this amino acid polar, nonpolar, acidic, or basic?

15. What are essential amino acids?

16. State two similarities and two differences between an α helix and a β-pleated sheet.

17. (a) List the four types of bonds that stabilize a protein's tertiary structure.
    (b) Which of these bonds is the strongest?

18. Describe two differences between RNA and DNA nucleotides.

19. State the rule for base pairing in DNA, indicating the number of hydrogen bonds that forms in each case.

Applying Inquiry Skills

20. Aspartame, also known as NutraSweet, is an artificial sweetener that was discovered by James Schlatter in 1965. The chemical name for aspartame is aspartyl-phenylalanine-1-methyl ester. This molecule is a dipeptide with a methyl group attached. It contains the amino acids phenylalanine and aspartic acid. Many people fear that consuming large quantities of aspartame may be bad for your health. A series of laboratory studies showed that aspartame is metabolized in the body to its components: aspartic acid, phenylalanine, and methanol. Aspartic acid and phenylalanine are naturally occurring amino acids and are found in all proteins. Methanol is well known to be poisonous in large quantity. **Table 7** shows the amount of methanol you get from the same volume of a number of common beverages.

▶ **Section 1.2** *Questions* *continued*

Table 7 Quantity of Methanol in Common Beverages

Beverage	Quantity of methanol (g/100mL)
orange juice	0.018
apple juice	0.021
diet soft drink containing aspartame	0.024
grape juice	0.046
tomato juice	0.085

(a) What conclusions can you draw from these results?
(b) What further studies would you suggest before advising on the safety of aspartame consumption?

21. A group of researchers at the University of Washington in Seattle have found that they can produce functional proteins with only five of the twenty amino acids.
   (a) Using concepts learned in this section regarding primary, secondary, and tertiary structure, predict which five amino acids the scientists used.
   (b) Conduct Internet research to explore the details of this study and to evaluate your prediction. When evaluating your prediction, do not focus on the specific amino acids you chose. Instead compare the assortment of R-groups in your list and theirs.

 www.science.nelson.com

Making Connections

22. (a) What does "like dissolves like" mean?
   (b) Provide an example of this generalization from everyday life.

23. Chitin is used to produce a certain type of surgical thread.
   (a) What physical and chemical properties make it an ideal material for this purpose?
   (b) Propose another product that chitin could be used for in everyday life.
   (c) Why should products made of chitin be relatively inexpensive to mass produce?

24. (a) What carbohydrate is found in dietary roughage?
   (b) Describe a lunch that would provide a good amount of this material.

25. Soybeans have been called the "miracle crop" partially because of the carbohydrate-rich meal obtained from the seed, but also due to the versatility of the soybean oil. Conduct Internet research to answer the following questions about soybean-based products.
   (a) Describe four nonfood uses of soybeans.
   (b) For each of the uses described in (a), explain why the soybean alternative is preferable to the conventional material.

 www.science.nelson.com

26. Carnuba wax is a natural wax that is used in the cosmetic, pharmaceutical, cleansing, and food industries. Conduct library and Internet research to answer the following questions about carnuba wax.
   (a) What is the most common source of carnuba wax?
   (b) Describe the physical properties of carnuba wax that gives the material its useful characteristics.
   (c) Describe five different common uses of carnuba wax.

 www.science.nelson.com

1.3 An Introduction to Metabolism

energy the ability to do work

Living organisms are in a constant struggle for survival. They labour to build their bodies from available raw materials, protect those structures from the destructive forces in their environment, and reproduce to exist over long periods of time. These activities can only be accomplished by doing work. Living organisms must continually capture, store, and use **energy** (defined as the ability to do work) to carry out the functions of life. Without energy, work cannot be done, and life comes to an end. Although we can see manifestations of life, such as movement, growth, and reproduction, on a human scale, organisms, in fact, do all their work at the molecular level. Through an elaborate series of highly controlled chemical reactions, cells manage the materials and energy they use to keep themselves alive. In some cases, they need to break larger compounds, such as amylose, into glucose, and glucose into carbon dioxide and water. As mentioned in section 1.2, reactions that result in the breakdown of complex substances are called catabolic reactions. In other cases, cells build complex substances from simpler subunits, such as DNA from nucleotides; these are the anabolic reactions. At any given instant, we may consider the biological component of a living organism as the product of all the anabolic and catabolic processes occurring within its body. We call the sum of all anabolic and catabolic processes in a cell or organism **metabolism**.

metabolism the sum of all anabolic and catabolic processes in a cell or organism

kinetic energy energy possessed by moving objects

Work is done when one object applies a force on another object and changes its position or state of motion. You do work when you kick a ball or ride a bike or even think a thought. The energy that organisms use to do work comes in many forms, such as light, sound, and electricity. At a more fundamental level, all forms of energy can be classified as kinetic energy or potential energy. **Kinetic energy** is the energy possessed by moving objects, such as raindrops falling through the air, the heart muscle contracting during a heartbeat, and chromosomes moving along spindle fibres during mitosis. Kinetic energy comes in many forms, including thermal energy or heat (the random motion of particles), mechanical energy (the coordinated motion of particles), electromagnetic energy (the motion of light), and electrical energy (the motion of charged particles). **Potential energy** is stored energy. An object possesses potential energy because of its position within an attractive or repulsive force field. As with kinetic energy, potential energy also comes in several different forms, such as gravitational potential energy (the attraction between two objects) and chemical potential energy (the attraction of electrons to protons in a chemical bond). A diver about to dive from a platform has potential energy because of the force of gravity between him and Earth and the distance between the two. The diver did not always have this potential energy. He gained it because his muscles did **work** to lift him to his present height. When work is done, energy is transferred from one body or place to another. When he dives, he will fall. He will gain kinetic energy as his speed increases and lose potential energy as the distance between him and the water decreases. During the descent, potential energy is converted into kinetic energy. Energy is not created or destroyed in the process; it is only converted from one form into another (**Figure 1**). This example illustrates one of the fundamental laws of nature called the first law of thermodynamics.

potential energy energy stored by virtue of an object's position within an attractive or repulsive force field

work the transfer of energy from one body or place to another

The First Law of Thermodynamics

The total amount of energy in the universe is constant. Energy cannot be created or destroyed but only converted from one form into another. If an object or process gains an amount of energy, it does so at the expense of a loss in energy somewhere else in the universe.

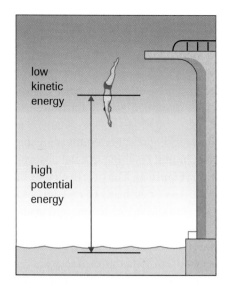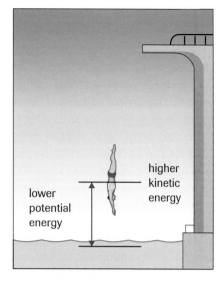

Figure 1
As the diver falls, his gravitational potential energy is converted into kinetic energy.

Nature does not always provide energy in a readily usable form. In most cases, organisms obtain energy in one form and convert it into another before it can be used. Plants, for example, capture light energy in photosynthesis and convert it into chemical potential energy in glucose and other carbohydrate molecules. Through the reactions of cellular respiration, the chemical energy in glucose is passed on to ATP, the primary energy-transferring compound in living cells, and to other energy-transferring molecules. The ATP molecules, in turn, may be used to activate protein carriers that transport ions through a cell membrane against a concentration gradient thus building up another form of energy, chemical potential energy. Chemical potential energy is an important form of energy in living systems.

Molecules possess stability because of the chemical bonds between their atoms. When atoms form covalent bonds, they achieve a greater stability by attaining a stable valence electron configuration. However, some chemical bonds are more stable than others. **Bond energy** is a measure of the stability of a covalent bond. It is measured in kilojoules (kJ) and is equal to the minimum energy required to break one mole of bonds between two types of atoms. It is also equal to the amount of energy released when a bond is formed. **Table 1** lists the average bond energies of the most common types of chemical bonds found in biologically important molecules.

We assume that the energy needed to break a bond is equivalent to the relative stability of that bond. A bond that requires 799 kJ/mol to break, such as the C=O bond, is almost twice as stable as one that breaks with the application of 391 kJ/mol, such as the N–H bond. In a chemical reaction, the bonds between reactant molecules must be broken and the bonds between product molecules must form. Energy is absorbed when reactant bonds break and energy is released when product bonds form.

A **potential energy diagram** shows the changes in potential energy that take place during a chemical reaction (**Figure 2**, page 60). In this diagram, the reactants are placed at a potential energy level corresponding to the relative stability of the bonds between their atoms. The amount of energy needed to strain and break the reactants' bonds, called the **activation energy**, is equal to the difference between the potential energy level of the transition state and the potential energy of the reactants. If the activation

bond energy the minimum energy required to break one mole of bonds between two species of atoms; a measure of the stability of a chemical bond

Table 1 Average Bond Energies

Bond type	Average bond energy (kJ/mol)
H–H	436
C–H	411
O–H	459
N–H	391
C–C	346
C–O	359
C=O	799
O=O	494

potential energy diagram a diagram showing the changes in potential energy that take place during a chemical reaction

activation energy the difference between the potential energy level of the transition state and the potential energy of reacting molecules

transition state in a chemical reaction, a temporary condition in which the bonds within reactants are breaking and the bonds between products are forming

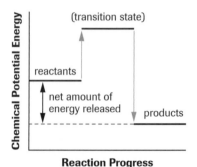

Figure 2
Changes in potential energy during an exothermic reaction

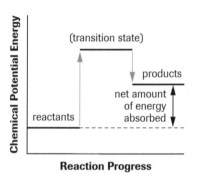

Figure 3
Changes in potential energy during an endothermic reaction

Figure 4
Exothermic reactions in *Mycena spp.*, a bioluminescent fungus, release light.

energy is provided, the reactants will reach the **transition state**. This is a temporary condition in which the bonds within reactants are breaking and bonds between products are forming. Energy is released when product bonds form. If the bonds in products are more stable than those in the reactants, more energy is released during bond formation than was absorbed during bond breaking. This will result in a net energy output. This type of reaction is called an exothermic reaction. **Figure 2** illustrates the energy level diagram for an exothermic reaction.

In some cases, the amount of energy absorbed in breaking reactant bonds is greater than the energy released in the formation of product bonds, resulting in a net absorption of energy. This is called an endothermic reaction (**Figure 3**). The overall change in energy that occurs in a chemical reaction is called the heat or enthalpy of reaction and is symbolized by ΔH. The value of ΔH is positive for endothermic reactions and negative for exothermic reactions.

A common exothermic reaction in living organisms is combustion. The energy change that occurs in combustion is commonly called the heat of combustion, or $\Delta H_{combustion}$. When organic compounds undergo combustion they react with oxygen and produce carbon dioxide and water. The following is the equation for the combustion of glucose:

$$C_6H_{12}O_{6(s)} + 6O_{2(g)} \longrightarrow 6CO_{2(g)} + 6H_2O_{(l)} \qquad \Delta H_{combustion} = \frac{kJ/mol\ C_6H_{12}O_{6(s)}}{2870\ kJ/mol}$$

▶ **TRY THIS** activity　　**The Combustion Accountant**

Using the bond energy values in **Table 1** on page 59 and the concept that energy is absorbed when reactant bonds break and released when product bonds form, calculate the $\Delta H_{combustion}$ of one mole of glucose.

Materials: balance sheet or computer spreadsheet, calculator or computer

(a) Using structural formulas that show all the bonds between atoms, write a balanced chemical equation for the combustion of glucose.

(b) Enter the bond energy values into the appropriate cells of the balance sheet or spreadsheet.

(c) Calculate the heat of combustion.

(d) Compare your calculated value to an accepted value by calculating a percentage difference using the following equation:

$$\% \text{ difference} = \frac{\text{difference between values}}{\text{accepted value}} \times 100\%$$

(e) Account for any percentage difference.

Many endothermic and exothermic reactions take place in the cells of living organisms. These reactions absorb and release energy in many different forms. Photosynthesis reactions in plants absorb light energy, while fireflies and bioluminescent fungi give off light (**Figure 4**). However, the most common form of energy absorbed and released by chemical reactions in living things is thermal energy as heat.

Given the required activation energy, it seems reasonable that an exothermic reaction would proceed spontaneously since the products are more stable than the reactants (products possess less potential energy than reactants). But why would an endothermic reaction proceed to completion when the products are less stable than the reactants? Studies in the field of thermodynamics have determined that energy is not the only factor that determines whether a chemical or physical change will occur spontaneously (a spontaneous change occurs without continuous outside assistance). A physical property called entropy must also be taken into consideration. **Entropy** is a measure of the randomness or disorder in energy or in a collection of objects. Entropy increases when disorder increases. Greater disorder is achieved when the arrangement of a collection of objects becomes more randomly assorted. Allowing a package of playing cards to fall through the air and flutter to the ground constitutes an increase in entropy (**Figure 5**). The cards are more randomly assorted by the time they reach the ground than they were in the package.

entropy a measure of the randomness or disorder in a collection of objects or energy; symbolized by *S*

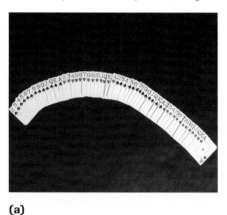
(a)

(b)

(c)

Figure 5
(a) A new package of playing cards is highly ordered.
(b) Throwing cards into the air increases entropy.
(c) The entropy of the cards on the floor is higher than the entropy of the cards in the package.

In chemical reactions, entropy increases when

- solid reactants become liquid or gaseous products;
- liquid reactants become gaseous products;
- fewer moles of reactant molecules form a greater number of moles of product molecules;
- complex molecules react to form simpler molecules (polymers into monomers, or glucose into carbon dioxide and water);
- solutes move from an area of high concentration to an area of lower concentration until they are uniformly distributed in the given volume (diffusion).

The universe favours an increase in entropy. Thus, there are two factors that need to be taken into consideration when determining whether a given chemical or physical change will occur spontaneously: energy and entropy. Exothermic reactions (favoured) that involve an increase in entropy (favoured) occur spontaneously since both changes are favourable. Endothermic reactions (not favoured) involving a decrease in entropy (not favoured) do not occur spontaneously since neither change is favoured. What about cases in which the energy change is exothermic (favoured) and the entropy decreases (not favoured) and others in which the energy change is endothermic (not favoured) but entropy increases (favoured)? The spontaneity of these reactions will depend on the temperature in which they occur. **Table 2** summarizes these concepts.

Table 2 Spontaneous and Nonspontaneous Changes

	Exothermic reaction (favoured)	**Endothermic reaction** (not favoured)
increase in entropy (favoured)	spontaneous at all temperatures	not spontaneous at low temperatures; spontaneous at high temperatures
decrease in entropy (not favoured)	not spontaneous at high temperatures; spontaneous at low temperatures	not spontaneous at all temperatures; these changes proceed only with a net input of energy

Free Energy

In the late 19th century, Josiah Willard Gibbs, an American physicist, discovered a relationship between the energy change, entropy change, and the temperature of a reaction; this relationship predicts whether a reaction will proceed spontaneously or not. Gibbs distinguished between energy and free energy. He defined **free energy** (also called **Gibbs free energy**) as energy that can do useful work.

free energy (Gibbs free energy)
energy that can do useful work

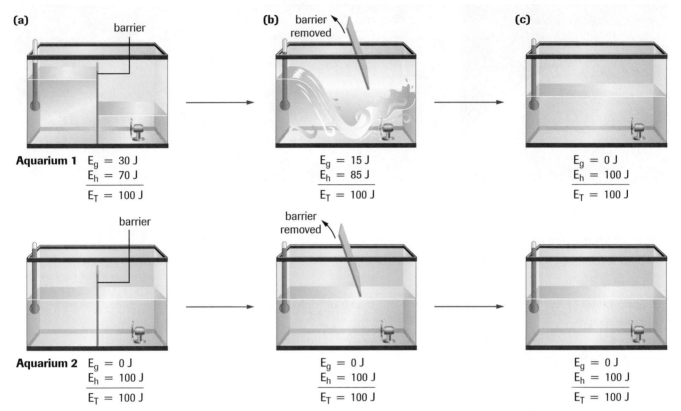

(a)

barrier

Aquarium 1
$E_g = 30$ J
$E_h = 70$ J
_____
$E_T = 100$ J

(b) barrier removed

$E_g = 15$ J
$E_h = 85$ J
_____
$E_T = 100$ J

(c)

$E_g = 0$ J
$E_h = 100$ J
_____
$E_T = 100$ J

barrier

Aquarium 2
$E_g = 0$ J
$E_h = 100$ J
_____
$E_T = 100$ J

barrier removed

$E_g = 0$ J
$E_h = 100$ J
_____
$E_T = 100$ J

$E_g = 0$ J
$E_h = 100$ J
_____
$E_T = 100$ J

Figure 6

Aquariums 1 and 2 ultimately achieve the same outcome. All energy is in the form of thermal energy (the random movement of particles). Key to symbols: E_g = gravitational potential energy; E_h = thermal energy; E_T = total energy

Figure 6 illustrates the difference between energy and free energy. In this hypothetical situation, two aquariums of the same size contain the same amount of water. The water temperature in aquarium 1 is lower than that of aquarium 2, such that aquarium 1 contains 30 J of thermal energy, E_h, less than aquarium 2, as shown in **Figure 6(a)**. Both aquariums contain 100 J of total energy, E_T, since aquarium 1 possesses 30 J of gravitational potential energy because of a reservoir of water that is raised to a higher level and held there by a removable barrier. Aquarium 2 also has a removable barrier, but the water level is the same on both sides of the barrier. A fan is placed in both aquariums, as shown. When the barrier is removed in aquarium 1, as illustrated in **Figure 6(b)**, water falls and does work as it turns the blades of the fan. Water moves from a more ordered state in part (a) (lower entropy, higher free energy) to a less ordered state in part (b) (higher entropy, lower free energy), resulting in the release of free energy as work is done on the blades of the fan. However, no work can be done on the fan in aquarium 2 when the barrier is lifted. All of the energy in this aquarium is thermal. The molecules of water in aquarium 2 are in constant motion, and hit the blades of the fan in every direction with approximately equal frequency and strength. Thus, the blades cannot be moved in any one direction and work cannot be done. In this example, the water in aquarium 1 could do work because it contained free energy, but the water in aquarium 2 could not do work because it contained no free energy, even though the total amount of energy in both aquariums is equal.

In aquarium 1, the positions and energy of the water molecules are more disordered after the barrier is removed and the water falls onto the blades of the fan. The change is spontaneous because there is a decrease in Gibbs free energy—the component of total energy that can do useful work. A change in Gibbs free energy is symbolized as ΔG and is equal to $G_{final} - G_{initial}$. Changes are spontaneous if ΔG is negative (G_{final} is less than $G_{initial}$). A change with a negative ΔG in one direction has an equivalent positive ΔG in the reverse direction. Thus, a reaction that is spontaneous in one direction is nonspontaneous in the reverse direction. In the aquarium example, the high water reservoir in aquarium 1 cannot be reestablished spontaneously (water does not move against gravity on its own). Work must be done to move the water to a higher level (greater order and more free energy). The energy needed to do this work can only be obtained from a change that releases free energy by becoming more disordered (negative ΔG). Thus, some of the energy released in aquarium 1 can be used to pump water. However notice that the temperature of aquarium 1 increases as the water falls. In general, when energy is changed from one form to another, some is lost as heat.

These results are not restricted to this one case, but apply to all changes that occur in the universe as a whole. Thus, all changes either directly or indirectly result in an increase in the entropy (overall disorder) of the universe. This is called the second law of thermodynamics (also known as the law of entropy).

The Second Law of Thermodynamics

The entropy of the universe increases with any change that occurs. Mathematically, $\Delta S_{universe} > 0$.

Living organisms seem to violate the second law of thermodynamics. Anabolic processes within cells build highly ordered structures such as proteins and DNA from a random assortment of amino acids and nucleotides dissolved in cell fluids. Cells assemble highly organized membranes as they grow and intricate spindle fibre systems when they divide. On a larger scale, organisms organize the world around them by building highly ordered structures, such as nests, webs, homes, and automobiles. These are all changes involving an increase in Gibbs free energy. Each of these activities seems to violate the second law of thermodynamics by apparently causing the universe to become a little more ordered. However, studies show that, in each and every case, the apparent order created by anabolic processes is accompanied by an even greater disorder caused by energy-yielding catabolic processes. By coupling free-energy-yielding catabolic processes with energy-requiring anabolic processes, living things build up their bodies and the world around them. They do this at the expense of the entropy of the universe as a whole.

To illustrate, consider a boy lifting a potato chip in an effort to place it in his mouth. A potato chip would not normally do this spontaneously. Moving the chip to the mouth involves an increase in energy (higher gravitational potential energy) and a decrease in entropy (a more ordered state), and thus an overall increase in Gibbs free energy. This change will not occur without doing work to contract the muscles of the arm. Free energy is needed to do the work.

The person obtains the necessary free energy by converting highly ordered nutrient molecules (such as glucose) into more disordered carbon dioxide, water molecules, and other waste products, through the catabolic reactions of digestion and cellular respiration. In the end, the amount of disorder created by the conversion of glucose into carbon dioxide and water is greater than the amount of order created by moving the potato chip to the mouth, resulting in a net amount of disorder in the universe as a whole.

Living organisms obey the second law of thermodynamics. They create order out of chaos in a local area of the universe at the expense of creating a greater amount of disorder in the universe as a whole. Tracing the source of free energy, we notice that the glucose molecule was originally produced in a plant cell by using the free energy of the Sun (in the form of light energy) to assemble carbon dioxide and water into glucose. This process involves a positive change in Gibbs free energy. The apparent order created by the photosynthetic process is accompanied by an even greater disorder of the Sun as it converts hydrogen into helium and releases a host of subatomic particles, including the light energy used as a source of free energy by plants. Photosynthesis creates order on Earth at the expense of a greater disorder on the Sun. Over time, the Sun will become completely disordered and will stop shining. In fact, all of the sources of free energy in the universe will eventually become completely disordered. The second law of thermodynamics predicts that the universe will experience a final "heat death" in which all particles and energy move randomly about, unable to do useful work. Life will come to an end, stars and suns will stop shining, and all concentrations of free energy will eventually disintegrate. All the energy that there ever was will still be there, except that it will be uniformly distributed throughout the space of the universe, unable to apply an effective push or a pull on anything. Chaos will reign supreme.

The terms *endothermic* and *exothermic* are used to describe the change in total energy that takes place in a chemical reaction. We can now describe the energy change in terms of a change in Gibb's free energy. An **exergonic reaction** is spontaneous and is accompanied by a decrease in Gibbs free energy (ΔG is negative). In these reactions, the value of ΔG provides a measure of the amount of free energy released by the reaction. An **endergonic reaction** is not spontaneous and is accompanied by an increase in Gibbs free energy (ΔG is positive). The value of ΔG provides a measure of the amount of free energy needed to drive the reaction.

Figure 7 summarizes the relationships between free energy, spontaneous change, and work. In **Figure 7(a)**, the diver releases free energy as he falls. The system of Earth and diver becomes more disordered as a result, and the released free energy does work on the surface of the water when he lands. In **Figure 7(b)**, an accumulation of molecules or ions on one side of a selectively permeable membrane represents a more ordered, high free-energy system. The particles will spontaneously diffuse to the other side, becoming more disordered as they become more uniformly distributed, releasing free energy in the process. In **Figure 7(c)**, molecules, such as glucose and other highly ordered compounds, possess free energy that is released in catabolic (breakdown) reactions, such as cellular respiration. The free energy released can be used by cells to do useful cellular work.

exergonic reaction a chemical reaction in which the energy of the products is less than the energy of the reactants; chemists call it an exothermic reaction

endergonic reaction a chemical reaction in which the energy of the products is more than the energy of the reactants; chemists call it an endothermic reaction

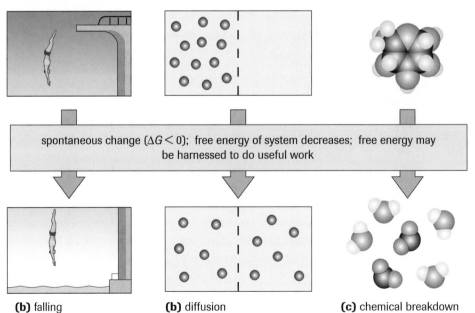

spontaneous change ($\Delta G < 0$); free energy of system decreases; free energy may be harnessed to do useful work

(b) falling   **(b)** diffusion   **(c)** chemical breakdown

Figure 7
The relationships between spontaneous change free energy, and work

In the overall process of cellular respiration, 2870 kJ of free energy per mole of glucose processed is made available to the cell for performing cellular work:

$$C_6H_{12}O_6 + 6O_2 \longrightarrow 6CO_2 + 6H_2O \qquad \Delta G = -2870 \text{ kJ/mol } C_6H_{12}O_6 \text{ used}$$

Conversely, in the process of photosynthesis, 2870 kJ of work is done by light energy for every mole of glucose formed in the overall process (this amount of free energy is absorbed in the overall process):

$$6CO_2 + 6H_2O \longrightarrow C_6H_{12}O_6 + 6O_2 \qquad \Delta G = +2870 \text{ kJ/mol } C_6H_{12}O_6 \text{ formed}$$

Metabolic Reactions Are Reversible

The reactions of metabolism are enzyme catalyzed and are all reversible. When a reversible reaction reaches equilibrium, its ΔG value equals zero and its free energy content is zero. A cell whose reversible reactions have reached equilibrium is a dead cell. Living organisms avoid this situation by preventing the accumulation of solutes in solution. Most of the reactions of metabolism are a series of chain reactions in which the product of one reaction is the reactant of another reaction. The final product of the chain is usually removed from the solution by excretion into the external environment or precipitating it out of solution as a solid. At the end of photosynthesis, glucose molecules are polymerized into insoluble starch granules, and in cellular respiration, the carbon dioxide and water products are expelled into the atmosphere as waste.

Adenosine Triphosphate

Adenosine triphosphate (ATP) is the primary source of free energy in living cells. This compound contains the purine nitrogenous base adenine attached to the five-carbon sugar ribose, which is bound to a chain of three phosphate groups, as illustrated in **Figure 8(a)**. When the cell requires free energy to drive an endergonic reaction, an enzyme called ATPase catalyzes the hydrolysis of the terminal phosphate of an ATP molecule, resulting in a molecule of adenosine diphosphate, ADP, a molecule of inorganic phosphate, P_i, and the release of 31 kJ/mol of free energy, as depicted in **Figure 8(b)**. This reaction may be represented by the following chemical equation:

$$ATP + H_2O \longrightarrow ADP + P_i \qquad \Delta G = -31 \text{ kJ/mol}$$

It should be noted that the hydrolysis of the terminal phosphate of ATP yields an energy value of 31 kJ/mol under standard conditions in a laboratory. In a living cell, this value is closer to 54 kJ/mol.

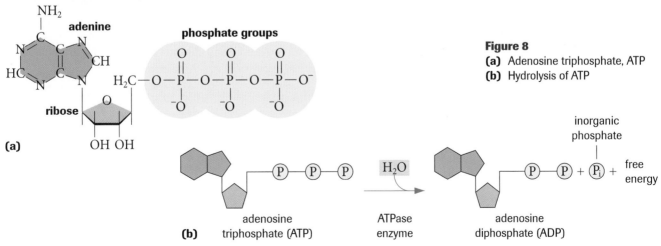

(a)

Figure 8
(a) Adenosine triphosphate, ATP
(b) Hydrolysis of ATP

The Chemical Basis of Life **65**

The hydrolysis of the terminal phosphate of ATP is an exergonic process resulting in a more stable product, ADP + P$_i$. The relative instability of the terminal phosphate bond is caused by the concentration of negative charge in the triphosphate tail of the ATP molecule. The free energy release of approximately 54 kJ/mol makes this reaction a relatively good source of energy for driving endergonic processes in the cell. In most cases, the free energy is not released as heat since this would increase the temperature of the cell to dangerous levels. Instead of releasing energy directly, the hydrolysis of ATP is usually coupled with an endergonic process that attaches the inorganic phosphate group to another molecule directly associated with the work the cell needs to do (**Figure 9**). Attaching a phosphate group to an organic molecule, such as ADP, is called **phosphorylation**, which causes the molecule to become more reactive. In active transport, ATP phosphorylates protein carriers, which change their shape and move ions against their concentration gradient as a result.

An active cell requires large amounts of ATP to power all the endergonic reactions that must occur. A single working muscle cell uses about 600 million ATP molecules per minute—an enormous amount. An active human body consumes its own mass in ATP per day. These molecules do not have to be provided from an external source because ADP and P$_i$ are condensed into ATP in the process of cellular respiration. **Figure 10** shows how the cell recycles ATP.

phosphorylation the process of attaching a phosphate group to an organic molecule

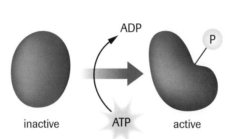

Figure 9
ATP may drive endergonic reactions.

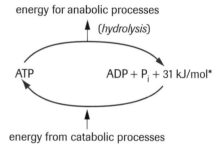

Figure 10
Cellular recycling of ATP
(31kJ/mol is the theoretical ΔG value)

oxidation a chemical reaction in which an atom loses one or more electrons

reduction a chemical reaction in which an atom gains one or more electrons

redox reaction a chemical reaction involving the transfer of one or more electrons from one atom to another; a reaction in which oxidation and reduction occur

reducing agent the substance that loses an electron in a redox reaction; the substance that causes the reduced atom to become reduced

oxidizing agent the substance that gains an electron in a redox reaction; the substance that causes the oxidized atom to become oxidized

Redox Reactions

Many chemical reactions involve the transfer of one or more electrons from one reactant to another. The process of losing electrons is called **oxidation** and the process of gaining electrons is called **reduction**. An electron transfer between two substances always involves an oxidation and a reduction, thus the name **redox reaction**. The substance that gains the electron is reduced and the substance that provides the electron is called the **reducing agent**. The substance that loses the electron is oxidized and the substance that takes the electron is called the **oxidizing agent**. The reaction between sodium and chlorine that produces sodium chloride is an example of a simple redox reaction. Here an electron is transferred from sodium to chlorine. Sodium is oxidized and chlorine is reduced (**Figure 11**).

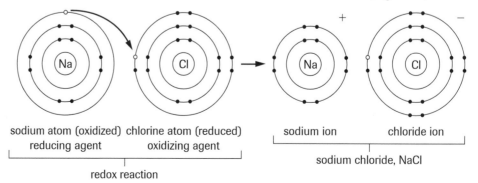

Figure 11
The redox reaction between sodium and chlorine produces sodium chloride

In some cases, a series of redox reactions occur in which the product of one redox reaction is the reactant of the next in the series. In this case, a substance that was reduced in the first reaction becomes oxidized in the next reaction. This can happen over a number of reactions, resulting in the transport of electrons through a series of increasingly stronger electron carriers. This can be symbolized using a coupled redox equation (**Figure 12**). Since the electron moves to successively stronger electron acceptors, free energy is released in each step of the process. This is the basis for the electron transport chains in photosynthesis and cellular respiration.

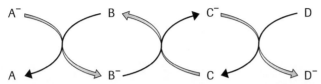

relative strength of oxidizing agents B < C < D (D is the strongest oxidizing agent)

Figure 12
A coupled redox equation

Reduction and oxidation may also be achieved by the partial transfer of electrons from one atom to another. In this case, electrons in a covalent bond move closer to a more electronegative atom, constituting the effective loss of electrons. Combustion reactions are a good example of this type of redox reaction. In the combustion of methane, $CH_{4(g)}$, oxygen acts as the oxidizing agent (**Figure 13**). The shift in electron position from a less electronegative element (carbon and hydrogen) to a more electronegative element (oxygen) results in a decrease in the potential energy of the electrons and a release of free energy. Thus, the combustion of methane is an exergonic process that releases large amounts of free energy. Overall, the controlled combustion of glucose in cellular respiration is an exergonic redox reaction similar to the combustion of methane. This process will be discussed further in Chapter 2.

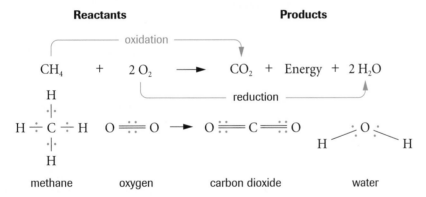

Figure 13
In the combustion of methane, electrons move from carbon and hydrogen toward oxygen, a more electronegative atom. Carbon is oxidized and oxygen is reduced in the process.

SUMMARY   *An Introduction to Metabolism*

- Free energy is the ability to do useful work. Kinetic energy is the energy possessed by moving objects. Potential energy is the energy stored by virtue of an object's position within an attractive or repulsive force field.

- Anabolic reactions build complex molecules and structures from simpler subunits, whereas catabolic reactions break down complex structures into their constituent subunits.

- Metabolism is the sum of all anabolic and catabolic processes in a cell or organism.

- The first law of thermodynamics states that the total amount of energy in the universe is constant. Energy cannot be created or destroyed but only converted from one form to another. The second law of thermodynamics states that all changes, either directly or indirectly, increase the entropy of the universe.

- Reversible reactions are reactions that may proceed in the forward and reverse directions.

- Oxidation is a chemical reaction in which an atom loses one or more electrons. Reduction is a chemical reaction in which an atom gains one or more electrons. A redox reaction (also known as oxidation–reduction) is a chemical reaction involving the transfer of one or more electrons from one atom to another.

▶ Section 1.3 Questions

Understanding Concepts

1. Distinguish between energy and work.

2. Identify each of the following activities as either anabolic or catabolic:
   (a) protein synthesis
   (b) digestion
   (c) DNA synthesis
   (d) photosynthesis
   (e) cellular respiration

3. Define *metabolism*.

4. If energy cannot be destroyed, why do organisms have to continually harness energy?

5. (a) Draw a labelled potential energy diagram for a hypothetical exergonic chemical reaction.
   (b) Distinguish between the relative stability of reactant and product bonds in an exergonic reaction.

6. Determine whether the following changes cause an increase or a decrease in the entropy of the system involved. Explain your reasoning.
   (a) an arm is raised

   (b) protein is digested into amino acids in the duodenum
   (c) chromosomes move along spindle fibres
   (d) oxygen diffuses into capillaries in alveoli in the lungs
   (e) a cell divides

7. Define *Gibbs free energy*.

8. (a) What is meant by the heat death of the universe?
   (b) How is this situation related to the death of an organism today?

9. (a) What is the value of ΔG for a system at equilibrium?
   (b) How does this situation relate to the saying, "Old biochemists never die, they just fail to react"?

10. Why is the terminal phosphate bond in ATP relatively unstable?

Making Connections

11. (a) How does the second law of thermodynamics relate to the production of pollution in our world?
    (b) Based on your answer to part (a), what can we do about pollution?

Enzymes are protein catalysts. A catalyst is a substance that speeds up a chemical reaction without being consumed in the process. In catalyzed reactions, the reactants are converted into products faster than they would be without the catalyst, and, at the end of the process, the catalyst is regenerated intact, ready to catalyze the same reaction again. Chemical change takes place when the bonds between reactant molecules break, and a rearrangement of atoms creates the bonds between the atoms of product molecules. For this to occur, reactant molecules must collide with enough force and with the correct geometric orientation for bond breaking to occur. Only under these conditions can the transition state be reached and product molecules formed. As mentioned earlier, all reactions, whether endergonic or exergonic, possess an activation energy (E_A) barrier that must be overcome for reaction to occur (**Figure 1**). Heat provides the activation energy for many reactions. The spark from a spark plug provides the activation energy for the combustion of gasoline in an internal combustion engine. Without the spark, the gasoline–oxygen collisions are ineffective and do not allow the reactants to reach the transition state.

Although an increase in temperature increases the rate of most reactions, proteins are denatured at high temperatures and lose their function. This could be devastating for the cell. Therefore, living cells cannot rely on high levels of heat as a source of activation energy. Catalysts allow reactions to proceed at suitable rates at moderate temperatures by reducing the activation energy (E_A) barrier (**Figure 1**).

(a)

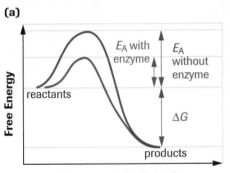

(b)

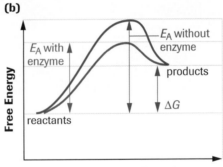

Figure 1
A catalyst speeds up endergonic and exergonic reactions by lowering the activation energy, but it does not change the value of ΔG.
(a) The effect of a catalyst on an exergonic reaction
(b) The effect of a catalyst on an endergonic reaction

The catalyst does not affect the free energy change (ΔG) of a reaction. It cannot change an endergonic reaction into an exergonic reaction; it can only decrease the potential energy level of the transition state and, thus, allow a greater proportion of colliding reactants to reach the transition state and become products. Catalysts can only speed up reactions that would normally take place anyway. Since a catalyst works by reducing the energy of activation, it speeds up forward and reverse reactions equally. Thus, it cannot affect the position of equilibrium, only the speed in which equilibrium is reached.

The **substrate** is the reactant that an enzyme acts on when it catalyzes a chemical reaction. The substrate binds to a particular site on the enzyme to which it is attracted. Enzymes are proteins in tertiary or quaternary structure with complex conformations. They are very specific for the substrate to which they bind. In most cases, they will not bind isomers of their substrate. The names of enzymes usually end in -*ase*. Thus, amylase catalyzes the breakdown of amylose into maltose subunits, and maltase catalyzes

substrate the reactant that an enzyme acts on when it catalyzes a chemical reaction

the breakdown of maltose into individual glucose molecules. An enzyme-catalyzed reaction is usually written with the name of the enzyme over the arrow, as follows:

$$\text{amylose} + H_2O \underset{\text{amylase}}{\rightleftharpoons} \text{maltose}$$

$$\text{maltose} + H_2O \underset{\text{maltase}}{\rightleftharpoons} \alpha\text{-glucose}$$

Notice that reactions are reversible.

A Model of Enzyme Activity

In an enzyme-catalyzed reaction, the substrate binds to a very small portion of the enzyme. The location where the substrate binds to the enzyme is called the **active site**, and is usually a pocket or groove in the three-dimensional structure of the protein (**Figure 2**). The substrate and the active site must possess compatible shapes for binding to occur. As the substrate enters the active site, its functional groups come close to the functional groups of a number of amino acids. The interactions between these chemical groups cause the protein to change its shape, thereby better accommodating the substrate. This is known as the **induced-fit model** of enzyme–substrate interaction. The attachment of the substrate to the enzyme's active site creates the **enzyme–substrate complex**.

active site the location where the substrate binds to an enzyme

induced-fit model a model of enzyme activity that describes an enzyme as a dynamic protein molecule that changes shape to better accommodate the substrate

enzyme–substrate complex an enzyme with its substrate attached to the active site

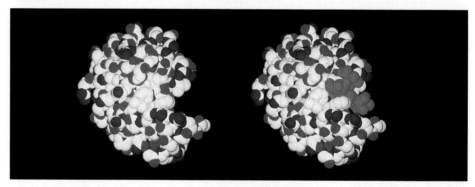

Figure 2
This photo illustrates the binding of a substrate to the active site of an enzyme. The action induces a conformational change in the enzyme's structure that prepares the substrate for reaction.

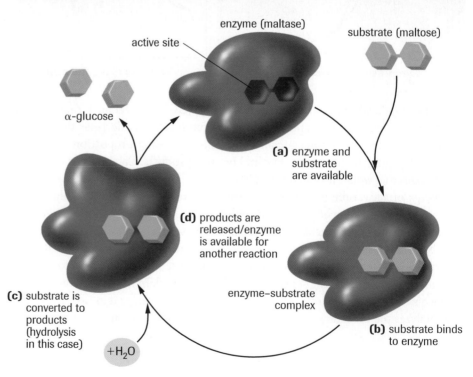

Figure 3
The enzyme maltase catalyzes the hydrolysis of maltose into two separate glucose molecules by breaking the α 1–4 glycosidic linkage.

In **Figure 3**, the enzyme maltase catalyzes the hydrolysis of maltose into two separate glucose molecules by breaking the α1–4 glycosidic linkage. **Figure 3(a)** shows the active site ready to receive the substrate. In (**b**), the substrate enters the active site and forms weak bonds while the enzyme's conformation changes to better accommodate the substrate (induced fit). In (**c**), the glycosidic linkage between the two glucose molecules is broken (using a water molecule) while the substrate is in the active site. In (**d**), breaking the glycosidic linkage changes the conformation of the protein slightly; it loses affinity for the product molecules and releases them. The active site now becomes available for another maltose to attach. The recycling of enzyme molecules causes cells to catalyze many reactions with relatively small numbers of enzymes.

> ▶ **TRY THIS** activity
>
> ## Synthesizing a Paper Clip Polymer with a Paper Enzyme
>
> **Materials:** coloured paper clips, 5-cm x 21.5-cm strip of paper

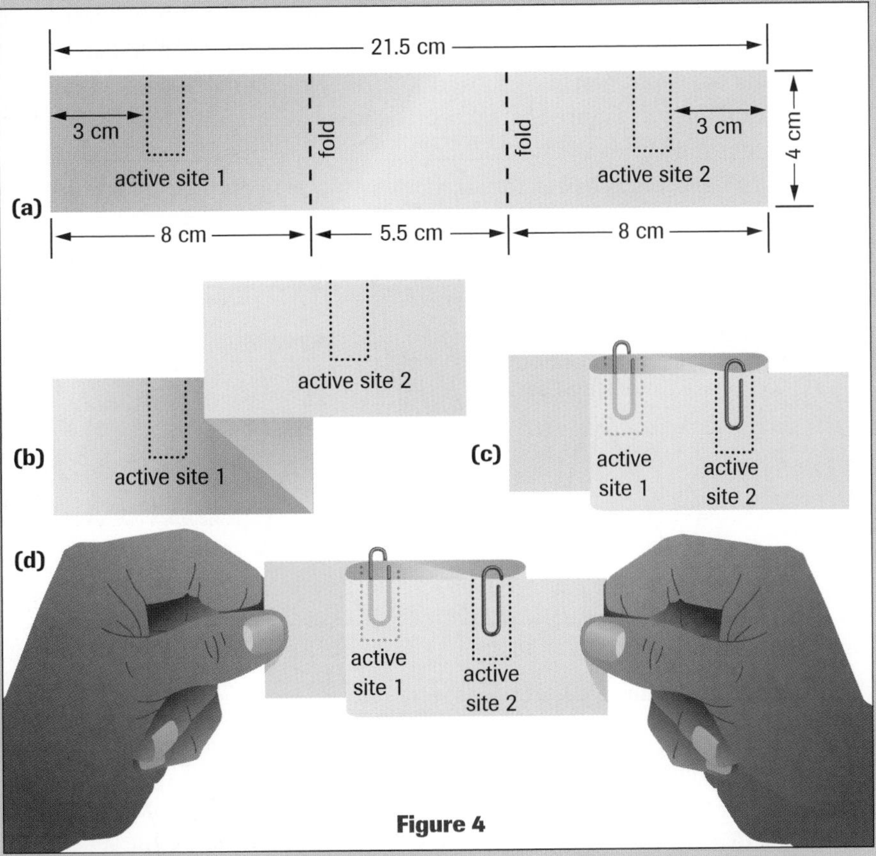

Figure 4

- Prepare the strip of paper as in **Figure 4(a)**.
- Fold the strip of paper as in **Figure 4(b)**.
- Place paper clip substrate 1 on active site 1, spanning back two layers of paper enzyme. Place paper clip substrate 2 on active site 2, spanning front two layers of paper enzyme, as in **Figure 4(c)**.
- Briskly pull the two tabs apart to activate the paper enzyme, as illustrated in **Figure 4(d)**.
  (a) Explain how the action of the paper enzyme relates to a real enzyme-catalyzed condensation reaction.
  (b) Try to produce a "triclipide" or a "tetraclipide" with one pull of the tabs.
  (c) Create a different enzyme simulation, perhaps a simulation of an enzyme-catalyzed hydrolytic process.

INVESTIGATION 1.4.1

Factors Affecting the Rate of Enzyme Activity (p. 82)

Enzymes are protein catalysts that, in most cases, catalyze only one chemical reaction. The ability of an enzyme to bind to its substrate and effectively catalyze a reaction is largely determined by its three-dimensional structure and certain environmental conditions. What are the conditions that optimize enzyme activity and how do changes affect enzyme function? In Investigation 1.4.1, you will design controlled experiments to examine the rate of a common enzyme's activity under various conditions.

Enzymes, such as maltase, decrease the energy of activation by stretching and bending chemical bonds that normally break in the reaction. Heat energy provided by the cell's internal environment and the action of the enzyme brings the substrate molecule to the transition state. In other cases, the active site may provide an acidic environment in an otherwise neutral part of the cell. Acidic R groups may be prevalent in the area of the active site. This may provide the low pH environment needed for certain reactions to take place.

Enzyme-catalyzed reactions can be saturated. There are a limited number of specific enzyme molecules in a cell at any one time. Since it takes some time for a catalyst to catalyze a particular reaction, the speed at which a catalyzed reaction proceeds cannot increase indefinitely by increasing the concentration of the substrate.

Temperature and pH affect enzyme activity. As with all other reactions, enzyme-catalyzed reactions increase in speed with an increase in temperature, as illustrated in **Figure 5(a)**. However, as the temperature increases beyond a critical point, thermal agitation begins to disrupt protein structure, resulting in denaturation and loss of enzyme function. Every enzyme has an optimal temperature at which it works best. Enzyme activity decreases above and below the optimal temperature. Most human enzymes work best at around 37°C, normal body temperature. There are enzymes in certain species of archaebacteria that work best at or above 100°C. Enzymes also have an optimal pH in which they work best, as **Figure 5(b)** shows. The digestive enzyme pepsin works best in the acidic environment of the stomach, pH 2. The digestive enzyme trypsin has an optimal pH of 8 and works best in the alkaline environment of the small intestine.

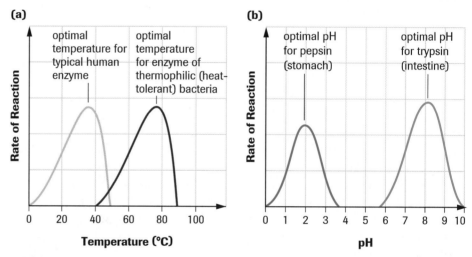

(a)   **(b)**

Figure 5
Enzymes have an optimal temperature **(a)** and optimal pH levels **(b)**.

cofactors nonprotein components, such as dissolved ions, that are needed for some enzymes to function

coenzymes organic nonprotein cofactors that are needed for some enzymes to function

Some enzymes require either nonprotein **cofactors**, such as inorganic substances or organic **coenzymes**, before they can work properly. These may bind to the active site with covalent bonds, or they may bind weakly with the substrate. Cofactors include zinc ions (Zn^{2+}) and manganese ions (Mn^{2+}). Coenzymes include the derivatives of many vitamins. One of the most important coenzymes is nicotinamide adenine dinucleotide (NAD^+), a derivative of vitamin B_3 (niacin). This substance acts as an electron carrier in cellular respiration. A similar compound called nicotinamide adenine dinucleotide phosphate ($NADP^+$) performs a similar role in photosynthesis. Many coenzymes shuttle molecules from one enzyme to another. Their activities will be discussed further in Chapter 2.

Enzyme Inhibition

A variety of substances inhibit enzyme activity. **Competitive inhibitors** are so similar to the enzyme's substrate that they are able to enter the enzyme's active site and block the normal substrate from binding, as shown in **Figures 6(a)** and **(b)**. This process is reversible and can be overcome by increasing the concentration of the enzyme's substrate, allowing it to compete favourably with the inhibitor. **Noncompetitive inhibitors** differ from the competitive types. They do not compete with an enzyme's substrate for the active site. Instead they attach to another site on the enzyme, causing a change in the enzyme's shape, as illustrated in **Figure 6(c)**. This changes the active site in such a way that it loses affinity for its substrate. Alternatively, the inhibitor may affect those parts of the active site that perform the work of catalysis, resulting in a loss of enzyme activity. An example is DDT, a poison that inhibits enzymes of the nervous system. However, not all enzyme inhibitors are poisons; some are used by the cell to control enzyme activity.

competitive inhibitors substances that compete with the substrate for an enzyme's active site

noncompetitive inhibitors substances that attach to a binding site on an enzyme other than the active site, causing a change in the enzyme's shape and a loss of affinity for its substrate

(a)

(b)

(c)

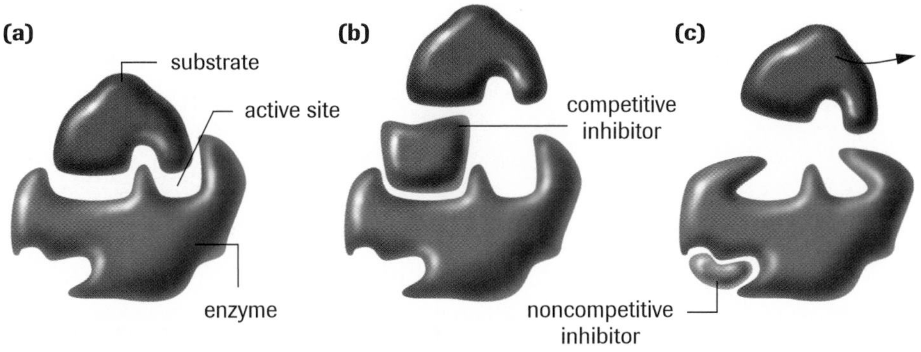

Figure 6
(a) A substrate normally binds to the active site of an enzyme.
(b) A competitive inhibitor competes with the substrate for the active site.
(c) A noncompetitive inhibitor binds to a site other than the active site and changes the shape of the enzyme to the point that it loses its affinity for the substrate.

Allosteric Regulation

Cells must control enzyme activity to coordinate cellular activities. They may accomplish this in two ways: by restricting the production of a particular enzyme, or by inhibiting the action of an enzyme that has already been produced. Some enzymes possess receptor sites, called **allosteric sites**, that are some distance away from the active site. Substances that bind to the allosteric sites may inhibit or stimulate an enzyme's activity. Allosterically controlled enzymes are usually composed of proteins in quaternary structure having several subunits, each with an active site. Binding an **activator** to an allosteric site stabilizes the protein conformation that keeps all of the active sites available to their substrates (**Figure 7**). Binding an **allosteric inhibitor** stabilizes the inactive form of the enzyme. Noncompetitive inhibitors attach to the allosteric sites of certain enzymes. The binding of an activator or inhibitor to one allosteric site will affect the activity of all the active sites on the enzyme. Allosteric regulators attach to their sites using weak bonds.

allosteric sites receptor sites, some distance from the active site of certain enzymes, that bind substances that may inhibit or stimulate an enzyme's activity

activator a substance that binds to an allosteric site on an enzyme and stabilizes the protein conformation that keeps all the active sites available to their substrates

allosteric inhibitor a substance that binds to an allosteric site on an enzyme and stabilizes the inactive form of the enzyme

(a) active form   inactive form   **(b)** active form stabilized by an allosteric activator molecule   inactive form stabilized by an allosteric inhibitor molecule

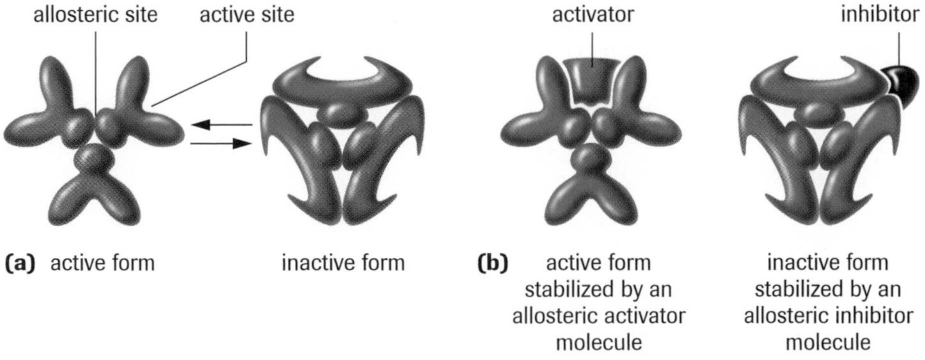

Figure 7
(a) Changes in the shape of allosteric enzymes
(b) Allosteric regulation

Feedback Inhibition

feedback inhibition a method of metabolic control in which a product formed later in a sequence of reactions allosterically inhibits an enzyme that catalyzes a reaction occuring earlier in the process

Feedback inhibition is a method used by cells to control metabolic pathways involving a series of sequential reactions, each catalyzed by a specific enzyme. In this case, a product formed later in the sequence of reaction steps allosterically inhibits an enzyme that catalyzes a reaction occuring earlier in the process (**Figure 8**). This effectively reduces the production of the inhibitor, which is, at the same time, a product of the process. The inhibitor binds to the allosteric site of the enzyme using weak bonds. As the product is used up over time, its concentration decreases. This causes the enzyme to be in the active form more often and the production of the inhibitor product increases. As this occurs, inhibition of the enzyme increases again and the production of the inhibitor is reduced once again. Thus, the amount of product is kept tightly controlled by the feedback inhibition process. Feedback inhibition is one of the most common control mechanisms used in metabolism.

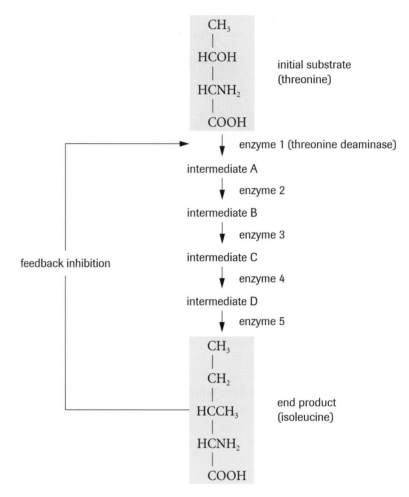

Figure 8
The production of the amino acid isoleucine is regulated by feedback inhibition. In this case, the end product, isoleucine, allosterically inhibits enzyme 1, threonine deaminase.

Finally, cells control metabolic processes by restricting the location of enzymes and enzyme complexes to certain locations within the cell. In some cases, enzymes are incorporated into specific membranes or fluid-filled spaces in the cell. Some of the enzymes for cellular respiration are inside the mitochondria, whereas others are dissolved in the cell's cytoplasm. Thus, the rate of the inner mitrochondrial reactions can be controlled by restricting the movement into the mitochondrion of intermediates that are formed in the cytoplasm.

Commercial and Industrial Uses of Enzymes

Many enzymes are used in industrial processes and consumer products. One of the largest industrial users of enzymes is the starch-processing industry. Starch from corn, wheat, and other grains is a plentiful carbohydrate resource, consisting of amylose and amylopectin. Starch can be converted to glucose syrups by hydrolysis, and the syrups find many applications as sweeteners in foods, such as candy, biscuits, jams and jellies, and pharmaceuticals like cough syrups, tonics, and vitamin preparations. Hydrolysis can be accomplished without enzymes, but this method can generate distasteful flavours and unsightly colours. Enzymatic hydrolysis offers a cleaner and more efficient hydrolysis. In a common process, α-amylase, produced by the bacterium *Bacillus licheniformis*, hydrolyzes starch to maltose. Then the enzyme glucoamylase, produced by moulds such as *Aspergillus* and *Rhizopus*, is added to hydrolyze maltose into individual glucose molecules.

$$
\text{amylose and amylopectin} \xrightarrow[\textit{B. licheiforms}]{\text{amylase}} \text{maltose} \xrightarrow[\textit{Aspergillis} \text{ or } \textit{Rhizopus} \text{ species}]{\text{glucoamylase (maltase)}} \alpha\text{-glucose}
$$

The most frequently used enzymes in the dairy industry are the protein-hydrolyzing enzymes, called proteases. They are most commonly used to coagulate milk for the manufacture of cheese. The cheese industry often uses the term *rennet* to refer to an enzyme preparation that coagulates milk. Historically, rennet was obtained from the stomach of calves. The primary enzyme obtained from this source is chymosin, which normally aids calves in the digestion of milk proteins. Since around 1990, genetic engineering has made it possible to produce chymosin from microbial sources. The gene for chymosin was removed from the DNA of a calf stomach cell and inserted into the DNA of certain bacteria and yeast cells (genetic engineering is discussed further in Chapter 6). The microbes reproduce rapidly in large vats containing a nutrient broth and secrete exact copies of calf chymosin. When making cheese, bacteria are added to milk to begin the curdling process. The bacteria feed on lactose in the milk and produce lactic acid as a waste product. The lactic acid lowers the pH and begins to denature milk proteins. Chymosin is then added to hydrolyze casein, a protein that makes up about 85% of the total protein in milk, causing it to coagulate into semisolid cheese curd. The curd shrinks and separates from a greenish liquid called whey. Whey is removed from the mixture and the curd is further processed according to the type of cheese being produced.

Many dairy products are made from cow's milk. Lactose, a disaccharide composed of glucose and galactose, accounts for 40% of the solids in cow's milk. Lactose is removed from some milk products because many people are intolerant to it and suffer gastrointestinal distress when exposed to it in their diet. Lactose-intolerant individuals may easily absorb and metabolize glucose and galactose. Lactose is hydrolyzed into glucose and galactose by the enzyme α-galactosidase (lactase), which may be obtained from several microorganisms, including *Aspergillus niger* and *Saccharomyces lactis*.

$$
\text{lactose (milk sugar)} \xrightarrow[\textit{Aspergillis niger} \text{ or } \textit{Saccharomyces lactis}]{\text{lactase}} \text{glucose + galactose}
$$

In addition to causing illness in people who are intolerant, lactose causes severe problems in cheese making. Large quantities of whey are made as a byproduct of cheese production. Whey contains the lactose present in the original milk. Problems arise because lactose is not particularly soluble and tends to crystallize out of solution. This leads to an undesirable grainy texture in some cheeses and ice creams. The enzyme α-galactosidase produced by *S. lactis* is used to solve this problem because it is active at 4°C, the temperature at which milk is normally stored. This means that the enzyme can be added to the milk during the overnight cold storage period used in dairies.

Fat-hydrolyzing enzymes called lipases are used by the dairy industry to develop characteristic cheeses, especially in Italian cheeses and other varieties with strong flavours. Although mozzarella receives most of its flavour from the proteases, lipases produce the strong-flavoured cheeses like Romano. These enzymes hydrolyze the fat in milk to produce free fatty acids. The type and concentration of fatty acids give the cheeses their distinct flavours.

In addition to the food industry, enzymes are also used in the cleaning industry. Dirt commonly found on clothing includes proteins, starch, and lipids. Although it is possible to remove these stains with soaps and detergents, enzymes allow stains to be removed at lower temperatures and with less mechanical agitation in a washing machine. Enzymes are also more effective than nonbiological cleaning agents at removing stains such as blood, grass, milk, and perspiration. The cleaning industry commonly adds proteases and amylases to detergents to help remove protein and carbohydrate stains from clothing and other fabrics. Additional uses of enzymes are outlined in **Table 1**.

Table 1 Additional Uses of Enzymes

Product/Process	Effects of enzymes
ethanol	enzymes convert starch into sugars, which are then converted into ethanol
animal feed	degradation of feed components for improved feed utilization and nutrient digestion
baking	modification of flour for improved baking properties
brewing	faster maturation of beer, chill-proofing, removal of carbohydrates for light beers
dairy	enzymes used in cheese making, removal/conversion of lactose in milk
detergent	active biological component of washing powders or liquids; breakdown of starch and fatty stains by proteases, amylases, and lipases; colour brightening and softening of cotton garments by cellulases
leather	soaking of hides and skins; unhairing, batting, and defatting
protein	improvement of nutritional and functional properties of animal and vegetable proteins; development of flavour bases from proteins
pulp and paper	pulp bleaching, viscosity control in starch-based coatings, de-inking for recycling programs
starch	production of dextrose, fructose, and special syrups for the baking and soft-drink industries
textiles	stone washing of denim (in combination with pumice stones), bio-polishing and softening of cotton, bleaching, cleanup, removal of starch from woven materials
wine and juice	degradation of pectin for clarification and increase in juice yields

SUMMARY   *Enzymes*

- An enzyme is a biological protein catalyst. The substrate is the reactant that an enzyme acts on when it catalyzes a chemical reaction. The active site is the location where the substrate binds to an enzyme.

- The induced-fit model of enzyme–substrate interaction describes a protein as a dynamic molecule that changes its shape to better accommodate the substrate.

- A variety of substances inhibit enzyme activity. Competitive inhibitors enter the enzyme's active site and block the normal substrate from binding.

- Noncompetitive inhibitors attach to another site on the enzyme, causing a change in the enzyme's shape. This changes the active site in such a way that it loses affinity for its substrate.

- Some enzymes possess receptor sites called allosteric sites that are some distance from the active site. Substances that bind to the allosteric sites may inhibit or stimulate an enzyme's activity.

- Feedback inhibition is a method used by cells to control metabolic pathways in which a product formed later in the sequence of reaction steps allosterically inhibits an enzyme that catalyzes a reaction occuring earlier in the process.

- Cells also control metabolic processes by restricting the location of enzymes and enzyme complexes to certain locations within the cell.

- Enzymes are used in many commercial and industrial processes such as cheese-making, cleaning, brewing, and paper-making.

▶ *Section 1.4* Questions

Understanding Concepts

1. Define *catalyst*.

2. Draw a labelled free-energy diagram to illustrate the effect of an enzyme on the activation energy of a hypothetical reaction. (Assume it is an exergonic reaction.)

3. What is meant by the statement, "an enzyme cannot affect the free-energy change of a reaction"?

4. Describe the induced-fit model of enzyme activity. Use diagrams if necessary.

5. How does an enzyme lower the activation energy of a biochemical reaction?

6. How do competitive enzyme inhibition and noncompetitive enzyme inhibition differ?

7. Enzymes in the testicles of males are responsible for both sperm and hormone (testosterone) production. Some of these enzymes have an optimal temperature of 33°C (4°C lower than normal body temperature). If this temperature is elevated or lowered, sperm and testosterone production are adversely affected.
   (a) What anatomical features help the testicles maintain this temperature?
   (b) Describe two lifestyle choices that could affect sperm and hormone production.

8. What happens to an enzyme after it has catalyzed a reaction?

Applying Inquiry Skills

9. The browning reaction that occurs when the flesh of fresh fruit comes into contact with air is caused by the naturally occurring enzyme polyphenol oxidase. Suggest a hypothesis for controlling the formation of the brown product and design a controlled experiment to test your hypothesis.

Making Connections

10. Papain and bromelain are the two most commonly used enzymes in commercial meat tenderizers. Conduct library or Internet research to answer the following questions regarding these two enzymes.
    (a) What are meat tenderizers? What are they used for?
    (b) What type of enzyme are papain and bromelain? (What is their substrate?)
    (c) What is the source of the papain and bromelain found in commercial meat tenderizer preparations?
    (d) What is the Milk Clot Assay (MSA), and how is it used in the meat processing industry?
    (e) The antemortern method of tenderizing meat involves the physical injection of solution of papain or bromelain into the living animal. The enzyme tenderizes muscle tissue while the animal is alive. Discuss this method with fellow classmates and write a brief position statement on the ethics of this procedure.

 www.science.nelson.com

11. Recent advances in wound treatment include the use of enzymatic debridement preparations. Conduct library or Internet research to answer the following questions regarding novel debridement procedures.
    (a) What is debridement?
    (b) Why is debridement performed?
    (c) What debridement methods are available to the physician treating a wound?
    (d) What types of enzymes are used in chemical debridement procedures?
    (e) What are the benefits and drawbacks of enzymatic debridement?

 www.science.nelson.com

⚗ INVESTIGATION 1.1.1

Inquiry Skills

○ Questioning	○ Planning	● Analyzing
● Hypothesizing	● Conducting	● Evaluating
● Predicting	● Recording	● Communicating

Buffers in Living Systems

Buffers are aqueous solutions that resist a change in pH when small amounts of acid ($H_3O^+_{(aq)}$ ions), or base ($OH^-_{(aq)}$ ions) are added. In general, an ideal buffer consists of an equilibrium mixture of a weak acid and its anion in approximately equal concentrations. A typical example is the acetic acid–acetate ion buffer:

$$HC_2H_3O_{2\,(aq)} + H_2O_{(l)} \rightleftharpoons H_3O^+_{(aq)} + C_2H_3O_2^-_{(aq)}$$
acetic acid   water   hydronium ion   acetate ion

In this case, the solution contains $HC_2H_3O_2$ molecules and $C_2H_3O_2^-$ ions in equilibrium.

If the base $OH^-_{(aq)}$ is added to the buffer solution, the $OH^-_{(aq)}$ ions will react with the $HC_2H_3O_{2(aq)}$ on the left side of the equilibrium equation to produce $C_2H_3O_2^-_{(aq)}$ and $H_2O_{(l)}$:

$$OH^-_{(aq)} + HC_2H_3O_{2(aq)} \longrightarrow C_2H_3O_2^-_{(aq)} + H_2O_{(l)}$$
(from added base) (from buffer)

If acid $H_3O^+_{(aq)}$ is added, the $H_3O^+_{(aq)}$ will react with $C_2H_3O_2^-_{(aq)}$ on the right side to produce $HC_2H_3O_{2(aq)}$ and $H_2O_{(l)}$.

$$H_3O^+_{(aq)} + C_2H_3O_2^-_{(aq)} \longrightarrow HC_2H_3O_{2(aq)} + H_2O_{(l)}$$
(from added acid) (from buffer)

The carbonic acid–bicarbonate buffer plays an important role in maintaining the blood's pH at or near 7.43. The buffer is created by carbon dioxide, $CO_{2(g)}$, dissolving into aqueous blood plasma. The following equilibrium exists in blood:

$$CO_2(g) + H_2O(l) \rightleftharpoons H_2CO_{3(aq)} + H_2O_{(l)}$$
$$H_3O^+_{(aq)} + HCO_3^-_{(aq)}$$

Excess acid, $H_3O^+_{(aq)}$, reacts with $HCO_3^-_{(aq)}$ to produce $H_2CO_{3(aq)}$ and $H_2O_{(l)}$. Excess base, $OH^-_{(aq)}$, reacts with $H_2CO_{3(aq)}$ to produce $H_2O_{(l)}$ and $HCO_3^-_{(aq)}$.

In this investigation, you will examine the resistance of an acetic acid–acetate buffer to changes in its pH (buffering capacity). In addition, you will analyze the conditions of alkalosis and acidosis in a simulated blood plasma containing a carbonic acid–bicarbonate buffer.

Questions

What is the difference in buffering capacity between a buffer comprising 0.1 mol/L acetic acid and 0.1 mol/L acetate, and an acetic acid solution?

What happens to the pH of a carbonic acid–bicarbonate buffer when $CO_{2(g)}$ is added?

Prediction

(a) Make a prediction regarding the difference in buffering capacity between the acetic acid–acetate buffer and the acetic acid solution.

(b) Make a prediction regarding the effect on the pH of a carbonic acid–bicarbonate buffer when $CO_{2(g)}$ is added to the solution.

Experimental Design
Part I

Increasing amounts of a 0.1 mol/L sodium hydroxide ($NaOH_{(aq)}$) solution are added to a buffer comprising 0.1 mol/L acetic acid and 0.1 mol/L acetate. The procedure is repeated with distilled water in the spot plate. The number of drops of $NaOH_{(aq)}$ needed to neutralize the two solutions is compared to determine whether there is a difference in buffering capacity.

Part II

Expired air containing $CO_{2(g)}$ is blown through a straw into a buffer comprising 0.1 mol/L carbonic acid and 0.1 mol/L bicarbonate to determine its effect on the pH of the solution.

Materials
safety goggles
24-well spot plate
distilled water
buffer of 0.1 mol/L acetic acid and 0.1 mol/L acetate (in dropper bottle)
universal pH indicator and colour chart or pH meter
0.01 mol/L sodium hydroxide, $NaOH_{(aq)}$ (in dropper bottle)
0.05 mol/L acetic acid, $CH_3COOH_{(aq)}$ (in dropper bottle)
simulated blood plasma–buffer of 0.1 mol/L carbonic acid and 0.1 mol/L bicarbonate (in dropper bottle)
20 mm × 150 mm test tube
10-mL graduated cylinder
wrapped drinking straw

Sodium hydroxide can cause blindness. Wear goggles and use the chemical with care.

Procedure

Part I

1. Put on your safety goggles.

2. Wash the spot plate with distilled water and then dry it.

3. Using a dropper, place one drop of acetic acid–acetate buffer into 12 wells of the clean spot plate.

4. Place one drop of universal indicator into each well.

5. Place one drop of 0.01 mol/L sodium hydroxide, $NaOH_{(aq)}$, into the first well, two drops into the second well, etc.

(c) For the well that became neutral (pH = 7), note the number of drops of sodium hydroxide that were added.

6. Place one drop of 0.05 mol/L acetic acid into 12 wells.

7. Place one drop of universal indicator into each well.

8. Place one drop of 0.01 mol/L sodium hydroxide $(NaOH_{(aq)})$ into the first well, two drops into the second well, etc.

(d) For the well that became neutral (pH = 7), note the number of drops of sodium hydroxide that were added.

(e) Compare the value obtained in (c) with that obtained in (d).

9. Discard the contents of the spot plate in a sink with plenty of running water.

Part II

10. Pour about 10 mL of simulated blood plasma (buffer of 0.1 mol/L carbonic acid and 0.1 mol/L bicarbonate) into a 20 mm × 150 mm test tube. Set up two test tubes. Use one as a control.

11. Add one or two drops of universal indicator.

(f) Record the pH of the solution.

(g) Determine the acid–base condition of the plasma (is it normal, acidic, basic?).

12. Using a drinking straw, gently blow air into the solution until a distinct colour change occurs. Record the new pH of the solution.

(h) Determine the metabolic state of the plasma (is its acid–base balance in a normal state, a state of alkylosis, or a state of acidosis?).

13. Continue blowing into the simulated plasma solution with the straw until another distinct colour change is observed. Record the new pH of the solution.

(i) Determine the metabolic state of the plasma (is its acid–base balance in a normal state, a state of alkylosis, or a state of acidosis?).

14. Discard the contents of the test tube in a sink with plenty of running water. Throw the straw into the garbage. Wash your hands with soap and water.

Analysis

(j) Summarize your results for Parts I and II in the form of tables or charts.

(k) Analyze your results for trends and patterns.

Evaluation

(l) In your report, evaluate your predictions, taking into account possible sources of error. Draw reasonable conclusions.

(m) Describe how you could improve the experimental methods used in this activity.

(n) Suggest other experiments you could perform to extend your knowledge of buffers.

Biological Macromolecules in 3-D

In this activity, you will build molecular models of biologically important molecules using ball-and-spring molecular model-building kits and then you will view these and other molecules of interest using computer-generated graphics.

Procedure

Part I: Building Biomolecule Models with Ball-and-Spring Model-Building Kits

1. On a large uncluttered table, open the molecular model kit and identify the colours representing carbon (C), hydrogen (H), oxygen (O), nitrogen (N), and sulfur (S). Separate the short springs (single bonds) from the long springs (double bonds).

2. Build a molecule of methane, CH_4.

(a) Describe the molecule's overall shape, structural symmetry, and bond angles.

(b) Is CH_4 polar or nonpolar?

3. Replace one H atom with a C atom and fill any open bonding positions on the C atom with H atoms.

(c) In your notebook, draw a structural formula, write a chemical formula, and name this compound.

(d) Describe the molecule's overall shape, structural symmetry, and bond angles.

(e) Is this molecule polar or nonpolar?

4. Repeat step 3 until you have created a straight-chained hydrocarbon containing six carbon atoms and 14 hydrogen atoms, which is hexane, C_6H_{14}.

5. Build another hexane molecule and place it alongside the first one you built.

(f) What types of intermolecular forces help hold hexane molecules to each other?

(g) What aspect of the molecules' geometry makes these bonds effective?

6. Convert one of the terminal methylene groups ($-CH_3$) *on each molecule* into a carboxyl group:

(h) What type of biomolecule have you created?

7. Orient the two molecules so that portions of the molecules that would normally form intermolecular bonds lie next to each other.

(i) What intermolecular forces help to hold these molecules together?

8. Remove an H atom from the third C and the fourth C of one molecule. Replace these atoms with a long spring, forming a double bond between these carbon atoms.

(j) Describe the change in the overall shape of the molecule caused by the introduction of a carbon–carbon double bond.

9. Keeping the carboxyl groups as close to one another as possible, orient the molecules parallel to each other.

(k) How does the presence of the carbon–carbon double bond influence the effectiveness of the intermolecular attractions?

(l) What term describes the presence of a double bond in the hydrocarbon portion of this molecule?

10. Construct eight water molecules and place them around the organic molecules oriented according to the type of intermolecular bonds they would form.

(m) What is the name of the intermolecular bonds that hold water molecules to these organic molecules?

11. Remove the carbon–carbon double bond and return the molecule to its original condition.

12. Using the two 6-carbon molecules, add, remove, and rearrange atoms on both molecules to construct two straight-chain glucose molecules. Use **Figure 6** in section 1.2 (page 29) as a guide.

(n) What two functional groups do these molecules contain?

13. Attach a small piece of tape to the carbon atoms that are double bonded to an oxygen atom at one end of each molecule. These are designated carbon-1 atoms.

14. Using **Figure 7** in section 1.2 (page 30) as a guide, carefully change each straight-chained glucose molecule into an α-glucose ring. You may need to reposition the hydroxyl groups to ensure that they are arranged above and below the plane of the ring according to the diagrams in **Figure 7**.

15. Orient the two α-glucose rings so that the hydroxyl group on carbon-1 of one ring faces the hydroxyl group of carbon-4 on the other.

16. Construct an α 1–4 glycosidic linkage by simulating a condensation reaction between the two molecules.

(o) What is the name of the resulting molecule?

(p) What other compound forms as a result of the condensation reaction?

17. Disassemble all molecules and return the balls and springs to their respective compartments.

18. Using **Figure 29** in section 1.2 (page 42) as a guide, construct the amino acid glycine. Note that the figure shows amino acids in their ionized form. You will construct the molecules in nonionized form (**Figure 1**).

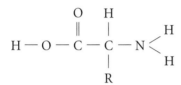

Figure 1

(q) Name the functional groups in the molecule.

(r) Identify the R-group side chain. Is this group charged, polar, or nonpolar?

19. Construct a cysteine molecule.

(s) What side-chain R group does cysteine contain?

20. Form a peptide bond between the two amino acid molecules.

(t) What functional groups were used to create the peptide bond?

(u) In your notebook, draw an image of the peptide bond and the four atoms that surround it, showing the covalent bonds that hold this structure together.

(v) What compound was formed in the condensation process other than the dipeptide?

21. Disassemble all molecules and return the balls and springs to their respective compartments.

Part II: Viewing Biomolecules Using Computer-Generated Graphics: Molecules of Life

1. Log onto your computer and follow instructions from your teacher to access the *Molecules of Life* computer modelling program.

 www.science.nelson.com

2. Read the *Molecules of Life—Introduction*, which describes the software and the mouse actions that allow you to view and manipulate the computer generated images of molecules. When you have finished reading the instructions and practicing molecule manipulations, move on to the *Exploring Molecules* section.

3. Complete each of the molecule sections as listed and answer the questions on the computer as well as the analysis questions below.

Analysis

4. *Molecular structure*

(a) What characteristics are typical of molecules that are gases at room temperature?

(b) Compare and contrast the characteristics of soluble and insoluble molecules.

5. *Carbohydrates*

(c) Describe the general molecular characteristics of mono and disaccharides. What features account for their high solubility in water?

(d) What element forms the bridge between the monomers that make up maltose and sucrose? What molecule was made during the condensation reactions that formed these disaccharides?

(e) Compare and contrast the overall shapes of cellulose, amylose, and amylopectin. How do their differences in structure account for their different biological functions?

(f) What features of cellulose molecules account for both the strength and the water absorbing qualities of paper towels?

(g) What type of intermolecular force dominates the interactions of carbohydrates and water?

6. *Lipids*

(h) Describe the general characteristics of all fatty acids.

(i) Compare the shape and molecular structure of saturated and unsaturated fatty acids. Provide examples to support your description.

(j) What functional groups are involved in the formation of a triacylglycerol from glycerol and fatty acids?

(k) What is the fundamental difference between a triacylglycerol and a phospholipid? How does this difference account for the hydrophobic and hydrophilic nature of the phospholipid molecule?

(l) What characteristics of wax make it well suited to the role of nature's main waterproofing agent?

7. *Amino Acids and Proteins*

(m) Compare and contrast the variety of amino acids with that of monosaccharides and fatty acids. Which group exhibits the greatest structural variability?

(n) How does the molecular structure of urea suit its biological function?

(o) Describe the appearance and arrangement of amino acids in α-helix and β-pleated sheet regions of polypeptides. What type of intermolecular forces are involved in maintaining protein secondary structure?

(p) Describe and give examples of proteins that have a variety of tertiary structures. Relate the tertiary structure of these proteins to their biological functions.

(q) What special role is played by the amino acid cysteine? How do disulphide bridges influence protein shape and stability?

(r) Do all proteins have quaternary structure? Support your answer with examples.

8. *Nucleotides*

(s) Describe the structure of ATP. What is the primary function of ATP within living organisms?

(t) What bases are building blocks of nucleotides within RNA and DNA, respectively.

(u) List some of the structural similarities and differences between transfer RNA and DNA.

(v) How are the bases paired within the DNA double helix? What type of bonding is responsible for the forces maintaining these base pairings and the double helix?

INVESTIGATION 1.4.1

Factors Affecting the Rate of Enzyme Activity

Inquiry Skills

○ Questioning ● Planning ● Analyzing
● Hypothesizing ● Conducting ● Evaluating
● Predicting ● Recording ● Communicating

Enzymes are functional proteins that catalyze specific reactions in living organisms. Most are proteins in tertiary or quaternary structure whose overall shape affects their performance. In this investigation, you will design and conduct controlled experiments to examine how different environmental conditions affect enzyme activity.

You will use catalase as the enzyme under study. This enzyme is found in the cells of most organisms and may be obtained easily from beef or chicken liver, potatoes, and yeast. Catalase catalyzes the decomposition of hydrogen peroxide, $H_2O_{2(aq)}$, into water, $H_2O_{(l)}$, and oxygen, $O_{2(g)}$, according to the following equation:

$$2H_2O_{2(aq)} \xrightarrow{\text{catalase}_{(aq)}} 2H_2O_{(l)} + O_{2(g)}$$

Catalase is the enzyme; hydrogen peroxide is the substrate. Hydrogen peroxide decomposes spontaneously under normal conditions. The rate of reaction is greatly increased by the action of catalase.

Question

How do changes in substrate concentration, enzyme concentration, and environmental factors, such as temperature and pH, affect the rate of enzyme activity?

Hypothesis

In Parts I, II, III, and IV of this investigation, you will examine how a change in temperature, pH, substrate concentration, and enzyme concentration affect the rate of catalase activity. For Part V, make a hypothesis regarding another factor that may affect this enzyme's activity.

Prediction

(a) Part I: Predict the effect that a change in temperature will have on the rate of catalase activity.

(b) Part II: Predict the effect that a change in pH will have on the rate of catalase activity.

(c) Part III: Predict the effect that a change in substrate concentration will have on the rate of catalase activity.

(d) Part IV: Predict the effect that a change in enzyme concentration will have on the rate of catalase activity.

(e) Part V: Based on your hypothesis, predict the effect that a change in a variable of your choice will have on the rate of catalase activity.

Experimental Design

In this investigation, you will design the experiments, the controls, and the materials and methods that you will use to record and display your observations; however, you will use

⚗ INVESTIGATION 1.4.1 *continued*

the provided assay technique to measure the rate of enzyme activity. An assay is an analysis performed to determine the presence, absence, quantity, or rate of change of a substance or process.

The following assay will be used: A filter paper disc coated with the enzyme is dropped into a vial of substrate (hydrogen peroxide). As the hydrogen peroxide breaks down into water and oxygen gas, the bubbles of oxygen collect underneath the filter and make it rise to the surface of the hydrogen peroxide. The time it takes for the filter to rise is an indication of the rate of enzyme activity.

The materials list that follows includes the enzyme, the substrate, and only those materials necessary to conduct the assay.

Materials

catalase (liver extract)
hydrogen peroxide (3% solution)
forceps
identical filter paper discs (cut with a hole punch)
vials
marking pencils
stopwatch or timer

 Hydrogen peroxide is corrosive and may cause burns to the respiratory tract, skin, and eyes. Do not inhale vapours. Wear safety goggles and gloves.

Procedure

Part I

Design a procedure to determine the effect of temperature on the rate of catalase activity. Write a list of the materials you need to perform your experiment. Indicate the precautions you would follow when using the required equipment.

Part II

Design a procedure to determine the effect of pH on the rate of catalase activity. Write a list of the materials you need to perform your experiment. Indicate the precautions you would follow when using the required equipment.

Part III

Design a procedure to determine the effect of substrate concentration, $[H_2O_{2(aq)}]$, on the rate of catalase activity. Write a list of the materials you need to perform your experiment. Indicate the precautions you would follow when using the required equipment.

Part IV

Design a procedure to determine the effect of enzyme concentration, catalase, on the rate of catalase activity. Write a list of the materials you need to perform your experiment. Indicate the precautions you would follow when using the required equipment.

Part V

Design a procedure to determine the effect that a variable of your choice has on the rate of catalase activity. Write a list of the materials you need to perform your experiment. Indicate the precautions you would follow when using the required equipment.

When your teacher has approved your procedure and the materials you will use, carry out the procedure with your lab group. Be sure to include a diagram of your experimental setup and all safety precautions.

Analysis

(f) For the variables you tested, summarize your results in tables and draw suitable graphs.

(g) Analyze your results for trends and patterns.

Evaluation

(h) In your report, evaluate your predictions, taking into account possible sources of error. Draw reasonable conclusions.

(i) Describe how you could improve your experimental methods and the assay technique.

(j) Suggest other experiments you could perform to extend your knowledge of enzyme activity.

Key Expectations

- Explain the relevance, in your personal life and the life of the community, of the study of cell biology and related technologies. (1.1, 1.2)

- Identify and describe the four main types of biochemical reactions: redox, hydrolysis, condensation, and neutralization. (1.1, 1.2, 1.3)

- Formulate operational definitions of the terms related to metabolic processes. (1.1, 1.2, 1.3, 1.4)

- Design and carry out an experiment related to a cell process, controlling the major variables and adapting or extending procedures where required. (1.1, 1.4)

- Describe technological applications of enzyme activity in the food and pharmaceutical industries. (1.2)

- Identify the functional groups within biological molecules and explain how they contribute to the function of each molecule. (1.2)

- Investigate the structures of biological molecules and functional groups using computer-generated, three-dimensional images and/or by building molecular models. (1.2)

- Describe how various molecules function within energy transformations in the cell. (1.2, 1.3)

- Investigate and explain the relationship between metabolism and the structure of biomolecules, using problem-solving techniques. (1.2, 1.3, 1.4)

- Describe the chemical structure, mechanisms, and dynamics of enzymes in cellular metabolism. (1.2, 1.4)

- Apply the laws of thermodynamics to the transfer of energy in the cell, particularly with respect to cellular respiration and photosynthesis. (1.3)

- Relate knowledge of metabolism to the fields of chemical thermodynamics and physical energy. (1.3)

Key Terms

activation energy

activator

active site

adenosine triphosphate (ATP)

allosteric inhibitor

allosteric sites

alpha (α) helix

amino terminus

anabolic reactions

antiparallel

β-pleated sheet

bond energy

bonding capacity

buffers

carboxyl terminus

catabolic reactions

chaperone proteins

coenzymes

cofactors

competitive inhibitors

condensation reaction (dehydration synthesis)

conformation

disulfide bridge

electronegativity

endergonic reaction

energy

entropy

enzyme–substrate complex

equilibrium

essential amino acids

ester linkage

exergonic reaction

feedback inhibition

free energy (Gibbs free energy)

functional groups

globular proteins

glycosidic linkages

half-life

hybridization

hydrolysis reaction

hydrophilic

hydrophobic

immiscible

induced-fit model

intermolecular bonds

intramolecular forces of attraction

isomers

kinetic energy

macromolecules

metabolism

micelles

miscible

neutralization reaction

noncompetitive inhibitors

oligosaccharides

orbitals

oxidation

oxidizing agent

peptide bonds

phosphorylation

potential energy

potential energy diagram

primary structure

quaternary structure

radioactive tracers

radioisotopes

redox reaction

reducing agent

reduction

residue

secondary structure

strand

strong acids

strong bases

substrate

tertiary structure

transition state

triacylglycerols (triglycerides)

valence electrons

van der Waals forces

Valence Shell Electron Pair Repulsion (VSEPR) theory

weak acids

weak bases

work

▸ *MAKE* a summary

Construct a large concept map beginning with the term biological macromolecule in the centre of a blank sheet of paper. Include concepts relating to the chemical structures and biological functions of the various molecules discussed in this chapter.

In your notebook, indicate whether each statement is true or false. Rewrite a false statement to make it true.

1. Radioisotopes are used to date prehistoric life forms.

2. Glucose forms a six-sided ring when dissolved in water.

3. The pH of vinegar is lower than the pH of pure water.

4. Glycerol is hydrophobic.

5. All polypeptides contain an amino terminus and a carboxyl terminus.

6. Fibrous proteins are rich in α helix.

7. Chaperone proteins help assemble DNA molecules.

8. The amount of adenine equals the amount of cytosine in a molecule of DNA.

9. Competitive inhibitors attach to the active site of an enzyme.

10. In feedback inhibition, an enzyme's substrate attaches to an allosteric site.

In your notebook, record the letter of choice that best answers the question.

11. Which of the following properties of water helps move water to the tops of trees?
    (a) high specific heat of vaporization
    (b) adhesion
    (c) high solubility
    (d) high surface tension
    (e) lower density as a solid than as a liquid

12. Which of the following is not a type of intermolecular force of attraction?
    (a) covalent bond
    (b) London dispersion force
    (c) dipole–dipole interaction
    (d) hydrogen bond
    (e) hydrophobic interaction

13. Which of the following does a buffer comprise?
    (a) a strong acid only
    (b) a weak base only
    (c) a strong acid and a strong base
    (d) a weak acid and a weak base
    (e) an alcohol and a base

14. Which of the following polysaccharides contains more α1–6 glycosidic linkages?
    (a) amylose   (d) chitin
    (b) glycogen   (e) amylopectin
    (c) cellulose

15. How many fatty acids are in a phospholipid molecule?
    (a) 1   (c) 3   (e) 5
    (b) 2   (d) 4

16. Which of the following bonds holds together a protein's primary structure?
    (a) hydrogen bonds   (d) covalent bonds
    (b) ionic bonds   (e) dipole–dipole interactions
    (c) London forces

17. Which of the following is not a component of DNA?
    (a) deoxyribose   (d) adenine
    (b) uracil   (e) thymine
    (c) phosphate

18. The backbone of a DNA molecule is held together by which of following?
    (a) glycosidic linkages
    (b) peptide linkages
    (c) phosphodiester linkages
    (d) ester linkages
    (e) disulfide bridges

19. Which of the following is a statement of the second law of thermodynamics?
    (a) Energy is conserved in the universe.
    (b) Entropy is conserved in the universe.
    (c) Living organisms are exempt from this law.
    (d) The entropy of the universe always increases.
    (e) The energy of the universe always increases.

20. How does an enzyme speed up a chemical reaction?
    (a) by providing additional heat energy
    (b) by reducing the potential energy of reactant molecules
    (c) by lowering the activation energy of the reaction
    (d) by changing the position of equilibrium
    (e) by becoming consumed in the reaction

Understanding Concepts

1. Complete **Table 1**.

Table 1 Functional Groups in Biomolecules

Functional group name	Functional group formula	Biological molecule
aldehyde		glucose
	C = O	fructose
phosphate		nucleotide
sulfhydryl	— S — H	
	O‖ — C — OH	fatty acid
amino		

2. What are two structural similarities and one structural difference between amylose and amylopectin?

3. Distinguish between α-glucose and β-glucose.

4. (a) Why are humans not able to digest cellulose?
   (b) What animals are able to digest cellulose? Why?

5. Why can cellulose form strong fibres but starch molecules cannot?

6. How does the cliché, "Too much of a good thing may be bad for you" apply to cholesterol?

7. What is the difference between a saturated fatty acid and an unsaturated fatty acid?

8. (a) What chemical process converts vegetable oil into margarine?
   (b) What does this process do to the fatty acid molecules in vegetable oil?

9. List two sterols other than cholesterol and describe a function for each.

10. What physical properties of waxes make them ideal as coatings on animal and plant body surfaces?

11. Proteins perform more functions than either carbohydrates or lipids.
    (a) What structural characteristics enable proteins to be so varied?
    (b) List three major functions of proteins.

12. When an organism is deprived of food for a long time, it catabolizes its bodily stores of fats, carbohydrates, proteins, and nucleic acids for energy. Why are fats and carbohydrates used for this purpose before proteins and nucleic acids?

13. Copy the structural formula of **Figure 1** into your notebook.

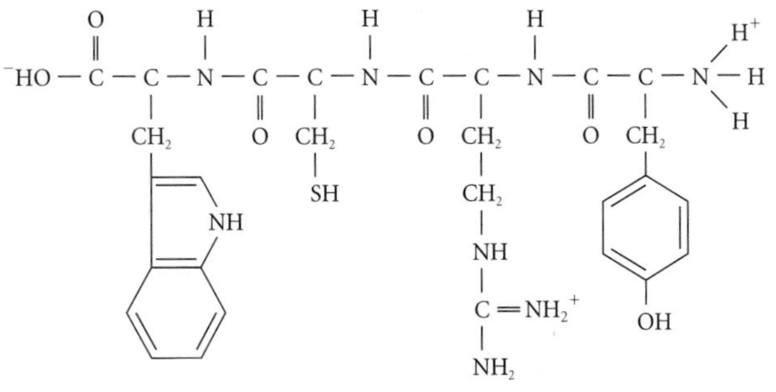

Figure 1

(a) Draw arrows pointing to the peptide bonds in the molecule.
(b) Draw a circle around the carboxyl terminus and a square around the amino terminus.
(c) Use the table of amino acids in **Figure 29** on page 42 of section 1.2 to identify each amino acid in the molecule.
(d) Using the abbreviations, list the amino acids, in order, from left to right, beginning with the carboxyl terminus and ending with the amino terminus. Use dashes to represent the peptide bonds.

14. Distinguish between globular proteins and fibrous proteins. Provide an example of each.

15. Define *quaternary structure*. Provide two examples of proteins that are only functional in quaternary structure.

16. What do chaperone proteins do?

17. Describe two differences between RNA and DNA nucleotides.

18. Why does DNA possess hydrogen bonds but RNA does not?

19. Identify each of the following as a source of kinetic energy or potential energy:
    (a) blood flow
    (b) ATP
    (c) concentration gradient
    (d) muscle contraction
    (e) glucose

20. How do living organisms drive endergonic processes?

21. Explain why heterotrophs are thermodynamically dependent on autotrophs.

22. (a) Draw a labelled free-energy diagram for the hydrolysis of the terminal phosphate of ATP.
    (b) Is this an endergonic or an exergonic process?
    (c) How is the energy usually dissipated within a living cell?

23. Distinguish between oxidation and reduction.

24. Why can the removal of an electron from an atom and the addition of an oxygen to an atom both be considered oxidations?

25. Explain why enzymes function best at an optimum
    (a) pH   (b) temperature

Applying Inquiry Skills

26. A biologist investigating enzyme function plotted the activity of a particular enzyme (*y*-axis) versus pH (*x*-axis) (**Figure 2**). Experiment A was performed at 30°C, experiment B at 10°C, and experiment C at 100°C.

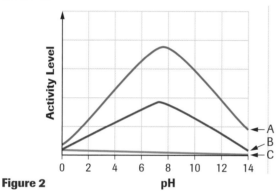

Figure 2

(a) Account for the changes in lines A and B.
(b) Why is there no hump in the middle of line C?
(c) After conducting experiment C, the investigator cools the reaction vessel to 30°C and repeats the experiment. Draw an enzyme activity graph for this experiment based on your predictions of the results.

27. The basic principle behind the permanent-wave process for hair involves breaking the existing disulfide bonds between cysteine residues in the keratin molecules and then reforming new disulfide bonds after the hair fibres have been rearranged by the hair stylist. A reducing agent is first applied to the hair to break the disulfide bridges. The hair is then arranged into the desired shape and an oxidizing agent applied to form new disulfide bonds between a different set of cysteine residues.
    (a) What type of chemical bonds are being broken in this process?
    (b) What type of protein structure is being changed in this process?
    (c) Why must gloves be worn when applying the reducing agent to the hair?
    (d) Suggest a less expensive method for changing the shape of your hair.

Making Connections

28. List three everyday products that are largely made of cellulose.

29. Restaurants prefer to use saturated fats, such as shortening or lard, to fry food because they can be heated to higher temperatures than unsaturated fats, such as vegetable oil, before they boil and evaporate quickly. This cuts down on cooking time and reduces costs.
    (a) Why do saturated fats have a higher boiling point than unsaturated fats?
    (b) A high intake of saturated fats in the diet has been linked to heart disease and stroke. Should restaurants be forced to use sources of unsaturated fats for cooking? Discuss.

30. Medical isotopes are usually produced in nuclear reactors far away from the hospitals and clinics where they are used. Describe some of the problems that may be associated with the distribution, storage, and disposal of these materials and their waste products.

31. The "hot" ingredient in peppers of the genus *Capsicum* is a compound called capsaicin. Conduct library or Internet research to answer the following questions about capsaicin.
    (a) Draw the structural formula for capsaicin.
    (b) Explain how capsaicin's chemical structure allows the compound to produce a persistent burning sensation.
    (c) How is the burning sensation produced by capsaicin and that produced by heat similar?
    (d) What part of a pepper contains the highest concentration of capsaicin?
    (e) What is the Scoville Test for capsaicin? Which peppers have the lowest test score? Which have the highest? What is the test score for pure capsaicin?
    (f) Why is drinking water an ineffective way of relieving the burning sensation caused by hot peppers? What beverage works better? Why?
    (g) Describe three useful nonfood applications of capsaicin

 www.science.nelson.com

2

Cellular Respiration

▶ **In this chapter, you will be able to**

- describe the chemical structure and mechanisms of enzymes in cellular respiration;

- describe how such molecules as glucose, ATP, pyruvic acid, NADH, and oxygen function in the energy transformations in a cell;

- explain the role of mitochondria in cellular respiration;

- compare the processes of aerobic and anaerobic respiration;

- formulate operational definitions of the terms related to metabolic processes;

- investigate and explain the relationship between metabolism and the structure of biomolecules;

- design and carry out a controlled experiment related to cellular respiration;

- interpret qualitative and quantitative observations, gathered through investigation, of the products of cellular respiration;

- relate the process of cellular respiration to the laws of thermodynamics;

- explain the relevance of the study of metabolism to your personal life.

On a warm summer day in 1974, 8-year-old Sarah suddenly felt pins and needles in the muscles of her legs as she walked. Within a year, she couldn't walk without experiencing muscle pain, shortness of breath, and a racing heart. By the age of 16, Sarah was attending school in a wheelchair. Tests showed that her blood was highly acidic. Her doctors were puzzled. After extensive study, a team of geneticists, biochemists, and physicians finally determined the cause of the problem. Sarah's symptoms were due to a substance common to all of the cells of her body: lactic acid.

When Sarah engaged in light physical activity, her muscles generated as much lactic acid as would normally be produced during prolonged strenuous exercise. The lactic acid buildup was responsible for the muscle pain, but why was it accumulating abnormally? Sarah's problem was narrowed down to her energy production system. But where? The trouble could lie within one of the three components of cellular respiration: glycolysis, the Krebs cycle, or the electron transport system.

Biochemists traced Sarah's problem to the electron transport system in her mitochondria. This chain of proteins allows for the efficient production of ATP, and in Sarah's case, one of the proteins, cytochrome a_3, was defective. Her muscles could not make sufficient amounts of ATP and, instead, produced excess lactic acid. Observations showed that Sarah's muscle cells had produced huge numbers of mitochondria to make up for the low energy yield.

Researchers found a substitute for the faulty protein: a simple mixture of vitamin C and a modified form of vitamin K. Sarah began feeling better almost instantly. By the age of 20, she no longer used a wheelchair and was completely free of symptoms. The mixture did not cure Sarah's metabolic defect; it simply filled the gap caused by the defective protein. Without a team of investigators who understood cellular respiration, she would have been denied the opportunity to lead a normal life.

💡 REFLECT on your learning

1. (a) What do organisms do with the oxygen they absorb from the air?
   (b) What is the source of carbon in the carbon dioxide excreted by these organisms?
   (c) Why is carbon dioxide excreted?

2. (a) Why do bakers add yeast to flour and water when making bread?
   (b) When yeast is added to grape juice at room temperature, vigorous bubbling occurs. What gas produces the bubbles?
   (c) After a while, the bubbling stops. Why does it stop?
   (d) What type of beverage is produced by this process?
   (e) What is the name of this process?

3. (a) After a long, hard run, your muscles feel sore and stiff. What is the cause of these symptoms?
   (b) Why do you pant at the end of the run?

▶ TRY THIS activity   *Clothespins and Muscle Fatigue*

Automobiles and machines must be supplied with gasoline or electricity as a source of energy before they can move. Your muscles require energy in the form of ATP to contract. Muscles can produce ATP by using oxygen (aerobic respiration) or not using it (anaerobic respiration). Anaerobic respiration in muscle cells produces lactic acid. When muscles do a lot of work quickly, the buildup of lactic acid reduces their ability to contract until exhaustion eventually sets in and contraction stops altogether. This is called muscle fatigue.

Materials: clothespin, timer

- Hold a clothespin in the thumb and index finger of your dominant hand.

- Count the number of times you can open and close the clothespin in a 20-s period while holding the other fingers of the hand straight out. Make sure to squeeze quickly and completely to get the maximum number of squeezes for each trial.

- Repeat this process for nine more 20-s periods, recording the result for each trial in a suitable table. Do not rest your fingers between trials.

- Repeat the procedure for the nondominant hand.
  - (a) Prepare a suitable graph of the data you collected.
  - (b) What happened to your strength as you progressed through each trial?
  - (c) Describe how your hand and fingers felt during the end of your trials.
  - (d) What factors might cause you to get more squeezes (to have less fatigue)?
  - (e) Were your results different for the dominant and the nondominant hand? Explain why they would be different.
  - (f) Your muscles would probably recover after 10 min of rest to operate at the original squeeze rate. Explain why.

photoautotrophs organisms that can build all the organic compounds required for life from simple inorganic materials, using light in the process

heterotrophs organisms that feed on other organisms to obtain chemical energy

chemoautotrophs organisms that can build all the organic compounds required for life from simple inorganic materials without using light energy

aerobic cellular respiration harvesting energy from organic compounds using oxygen

Energy enables life. Over the past 3.5 billion years, living organisms have evolved into structures that efficiently harness free energy from their environment, convert it into usable forms, and use it to power the endergonic processes of life. Through photosynthesis, **photoautotrophs** (Greek for "light-using self-feeders"), such as green plants and photosynthetic microorganisms, transform light energy into the chemical potential energy in glucose and other carbohydrates. They are the dominant autotrophs on Earth and are self-sufficient for their energy needs. All other living things are **heterotrophs** (Greek for "other feeders"), which rely on autotrophs for energy. The vast majority of organisms are heterotrophs, including all animals and fungi, and most protists and bacteria. Most of these organisms obtain energy and building materials by ingesting the bodies of autotrophs or, through a food chain, the bodies or remnants of fellow heterotrophs that ultimately relied on autotrophs for sustenance. Virtually everything a heterotroph eats was once alive. A small number of microorganisms, such as archaebacteria, "eat" inorganic matter and are able to extract energy from such things as iron and sulfur-containing compounds, much like a car battery that extracts energy from sulfuric acid, $H_2SO_{4(aq)}$. This is the way the first organisms on Earth got their energy because there was not enough organic material to use and photosynthesis had not yet developed. These organisms are called **chemoautotrophs** (Greek for "chemical self-feeders") and are usually found in extreme environments, such as volcanoes, sulfur springs, and salt flats.

With the exception of the chemoautotrophs, all organisms use glucose, $C_6H_{12}O_6$, as a primary source of energy. Through a series of enzyme-controlled redox reactions, organisms break the covalent bonds in this molecule and rearrange them into new and more stable configurations. As with all exergonic processes, the greater stability of the covalent bonds in the products results in the release of free energy. Cells have developed several different mechanisms to extract the energy they need from available nutrients. In each case, redox reactions are involved. In one method, the redox reactions result in the transfer of electrons from glucose to oxygen. Glucose is oxidized to carbon dioxide and oxygen is reduced to water. The overall reaction is as follows:

$$C_6H_{12}O_{6(aq)} + 6O_{2(g)} \longrightarrow 6CO_{2(g)} + 6H_2O_{(l)} + \text{energy via heat and ATP}$$

This process is known as **aerobic cellular respiration**. The term *aerobic* means that oxygen is used in the process; however, do not confuse the term *respiration* with the physical process of breathing air into the lungs. Cellular respiration refers to the chemical process summarized in the preceding equation. In fact, aerobic respiration is accomplished through a series of about 20 reactions in which the product of one reaction is the reactant of the next, with each step catalyzed by a specific enzyme. The previous equation merely represents a summary of the process. It only shows the initial reactants and final products, giving no information about the step-by-step changes that occur in between. Nevertheless, cellular respiration is no different from the combustion of a hydrocarbon fuel, such as propane in a gas barbecue or gasoline in a car engine. All are redox reactions that use oxygen as the oxidizing agent, and all are exergonic processes that release free energy as they occur. ⚗

Nutrients, such as glucose and fatty acids, along with hydrocarbon fuels, like propane and gasoline, contain a number of carbon–hydrogen (C–H) bonds. Oxygen oxidizes these bonds in two ways. Carbon–hydrogen bonds are relatively nonpolar ($\Delta E_n = 0.4$), the

⚗ **INVESTIGATION 2.1.1**

Oxygen Consumption in Germinating and Nongerminating Pea Seeds (p. 126)
Plant seeds contain living embryos that require energy to carry out the functions of life. When they germinate, they experience high rates of growth and cell division. What happens to a plant seed's rate of energy metabolism when it germinates and starts to grow? Investigation 2.1.1 provides you with an opportunity to conduct controlled experiments on the relationship between growth and the rate of metabolic activity.

electron pairs being almost equally shared by the two atoms. The overall equation for aerobic respiration tells us that 12 hydrogen atoms break away from glucose and attach to six oxygen atoms from the six oxygen (O_2) molecules to become six water (H_2O) molecules.

$$C_6H_{12}O_{6(aq)} + 6O_{2(g)} \longrightarrow 6CO_{2(g)} + 6H_2O_{(l)}$$

This is called oxidation because hydrogen atoms carry electrons away from carbon atoms in glucose to oxygen atoms. When the hydrogen atoms form covalent bonds with oxygen, the shared electron pairs occupy positions closer to the oxygen nuclei than they did when they were part of the glucose molecule. This happens because oxygen is much more electronegative than carbon. An electron occupying an energy level closer to a nucleus is equivalent to a skydiver occupying a position closer to the ground—both possess less free energy. Therefore, as electrons (in hydrogen atoms) move from less electronegative carbon atoms in glucose to highly electronegative oxygen atoms, they lose potential energy and achieve a less ordered state. (It is less ordered for electrons to occupy an energy level closer to a nucleus, just as it is for a skydiver to be closer to Earth's surface). The decrease in potential energy coupled with an increase in entropy causes a decrease in free energy and an overall exergonic process.

However, this explanation only accounts for half of the combustion process. The other half involves the attachment of the remaining oxygen atoms to carbon atoms, forming the six carbon dioxide molecules on the product side of the overall equation.

$$C_6H_{12}O_{6(aq)} + 6O_{2(g)} \longrightarrow 6CO_{2(g)} + 6H_2O_{(l)}$$

This change also constitutes an oxidation because, once bound to carbon atoms, the highly electronegative oxygen atoms draw the shared electron pairs to themselves. This results in a change that is considered equivalent to carbon losing electrons. As with the transfer of hydrogen atoms to oxygen, the attachment of oxygen atoms to carbon places the electrons in a more stable configuration that results in the release of free energy. Overall, the aerobic oxidation of glucose involves the movement of valence electrons from a higher free energy state in glucose to a lower free energy state in carbon dioxide and water (**Figure 1**).

The overall change includes a decrease in potential energy and an increase in entropy. This is a "downhill" process that yields 2870 kJ of free energy per mole of glucose (approximately 180 g).

$$C_6H_{12}O_{6(aq)} + 6O_{2(g)} \longrightarrow 6CO_{2(g)} + 6H_2O_{(l)} \qquad \Delta G = -2870 \text{ kJ/mol glucose}$$

(The standard free energy change value of -2870 kJ/mol glucose is obtained under standard laboratory conditions of 25°C and 101.3 kPa. When measured in the non-standard conditions found within a cell, the value is closer to -3012 kJ/mol glucose. The standard value will be used in calculations throughout this book.)

When glucose is burned in a test tube, carbon dioxide and water are also formed, and all the energy is given off as heat and light (a flame) (**Figure 2**, page 92). In a living cell, the free energy released from the combustion of glucose would also dissipate as heat and light, but cells have evolved methods to trap some of the energy (about 34% of it) by moving the positions of electrons in certain molecules to higher free energy states (such as in ATP). These molecules then become readily available sources of free energy to power endergonic processes throughout the cell.

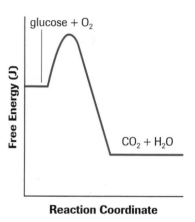

Figure 1
Free energy diagram for the combustion of glucose

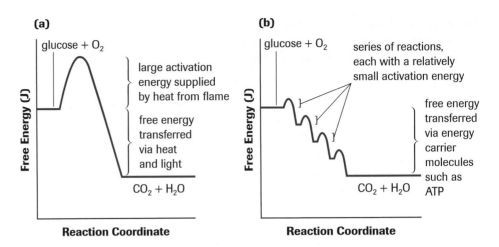

Figure 2
Energy level diagrams for combustion of glucose
(a) Combustion of glucose in a test tube
(b) Combustion of glucose by cellular respiration

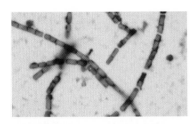

(a)

(b)

Figure 3
(a) Gas gangrene is caused by the anaerobic bacterium *Clostridium perfringens.*
(b) The facultative anaerobe *Salmonella enteritidis* is one cause of food poisoning.

obligate anaerobes organisms that cannot live in the presence of oxygen and obtain energy by oxidizing inorganic substances

obligate aerobes organisms that obtain energy by oxidizing organic substances using oxygen

facultative anaerobes organisms that obtain energy by oxidizing inorganic substances with or without oxygen

As with all endergonic or exergonic reactions, reactants do not react with one another unless the necessary activation energy is provided. In aerobic respiration, the reactants, glucose and oxygen, are stable covalent compounds. When they come in contact, the oxygen atoms in the O_2 molecules attract the electrons of the H–C bonds in glucose. However, oxygen is not a strong enough oxidizer to strip the electrons from these bonds just by bumping into a glucose molecule at room temperature (25°C), or even at normal body temperature (37°C). If it could, life would not exist. With so much oxygen in the air (21%), all glucose molecules—and, for that matter, virtually all organic molecules—would be very quickly converted into carbon dioxide, water, and energy. Activation energy prevents spontaneous combustion and allows living things to control the oxidation process. The amount of activation energy needed for the combustion of glucose in air is relatively high. In fact, you must provide the heat of a flame to begin the combustion in a test tube (**Figure 2(a)**).

Clearly this method would not do in a living cell. That is where enzymes come into the picture. Specific enzymes catalyze every step in the aerobic respiration process. They lower the activation energy and allow the reactions to occur at a rate that is consistent with the needs of the cell, a process illustrated in **Figure 2(b)**. Enzyme activity regulates the entire process. The available free energy is transferred to a number of energy-carrier molecules, including ATP. The control of cellular respiration is discussed in more detail in section 2.2.

Oxygen is not the only possible electron acceptor in the oxidation of glucose in a cell. In fact, some microorganisms use NO_2, SO_4, CO_2, and even Fe^{3+} as final electron acceptors. These organisms are known as **obligate anaerobes** and include such notable bacteria (and their associated disorders) as *Clostridium tetani* (tetanus), *Clostridium botulinum* (food poisoning), and *Clostridium perfringens* (gas gangrene), shown in **Figure 3(a)**. These organisms must live in environments that contain no oxygen. In fact, they die when even small amounts of oxygen are present. Conversely, most animals, plants, protists, fungi, and bacteria are **obligate aerobes**. They require oxygen as the final electron acceptor and cannot survive without it. Organisms that can tolerate aerobic and anaerobic conditions are called **facultative anaerobes**. Most of these are bacteria and include *Escherichia coli* (dysentery), *Vibrio cholerae* (cholera), and *Salmonella enteritidis* (food poisoning), shown in **Figure 3(b)**.

Cellular Respiration: The Big Picture

- Photoautotrophs produce glucose by photosynthesis, whereas heterotrophs obtain energy by ingesting the bodies or remnants of autotrophs or fellow heterotrophs.

- In aerobic respiration, oxygen is used to oxidize glucose molecules. This is a combustion reaction, the products of which are carbon dioxide, water, and energy in the form of heat and ATP:

$$C_6H_{12}O_{6(aq)} + 6O_{2(g)} \longrightarrow 6CO_{2(g)} + 6H_2O_{(l)} + \text{energy via heat and ATP}$$

- Free energy is released in the combustion of glucose. Electrons move from a higher free energy state in the C–H bonds of glucose and O–O bonds of oxygen to a more stable configuration in the C–O bonds of carbon dioxide and the O–H bonds of water.

- The activation energy of combustion and the use of specific enzymes allow the cell to control the reactions of cellular respiration.

- Obligate anaerobes must live in environments that contain no oxygen. Obligate aerobes require oxygen as the final electron acceptor and cannot survive without it. Organisms that can tolerate aerobic and anaerobic conditions are called facultative anaerobes.

▶ *Section 2.1* **Questions**

Understanding Concepts

1. (a) What is the source of the carbon atoms in the $CO_{2(g)}$ that humans exhale?
   (b) At the end of cellular respiration, where are the hydrogen atoms of glucose?
   (c) Use the overall equation of cellular respiration and the definition of oxidation to explain how glucose is oxidized in the respiration process.

2. (a) Define the terms *autotroph* and *heterotroph*.
   (b) Identify each of the organisms in **Figure 4** as a heterotroph or an autotroph.

3. (a) Explain why the free energy diagram in **Figure 5** cannot represent the process of cellular respiration.
   (b) Sketch a free energy diagram that accurately represents the cellular respiration process.

4. Distinguish between obligate anaerobes and facultative anaerobes, and provide an example of each.

Making Connections

5. *Clostridium* species include bacteria that produce serious infections in humans.
   (a) Are *Clostridium* species obligate aerobes, obligate anaerobes, or facultative anaerobes? Explain.
   (b) What waste products of energy metabolism do *Clostridium* species excrete?
   (c) Describe three infections in humans caused by *Clostridium* species.

(a)　　　　　**(c)**

(b)　　　　　**(d)**　　　　　**(e)**

Figure 4
(a) dry yeast; **(b)** fern; **(c)** lichen; **(d)** amoeba; **(e)** *Anabaena*

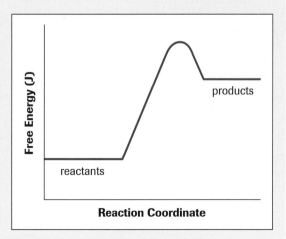

products

Free Energy (J)

reactants

Reaction Coordinate

Figure 5

The sequence of chemical reactions and energy changes that occurs in aerobic respiration may seem complex, but keep in mind the overall equation:

$$C_6H_{12}O_{6(aq)} + 6O_{2(g)} \longrightarrow 6CO_{2(g)} + 6H_2O_{(l)}$$

It provides you with a reminder of the three overall goals of the process:

1. to break the bonds between the six carbon atoms of glucose, resulting in six carbon dioxide molecules
2. to move hydrogen atom electrons from glucose to oxygen, forming six water molecules
3. to trap as much of the free energy released in the process as possible in the form of ATP

The entire process occurs in four stages and in three different places within the cell:

Stage 1: Glycolysis—a 10-step process occurring in the cytoplasm

Stage 2: Pyruvate oxidation—a one-step process occurring in the mitochondrial matrix

Stage 3: The Krebs cycle (also called the tricarboxylic acid cycle, the TCA cycle, or the citric acid cycle)—an eight-step cyclical process occurring in the mitochondrial matrix

Stage 4: Electron transport and chemiosmosis (oxidative phosphorylation)—a multistep process occurring in the inner mitochondrial membrane

Figure 1 shows the four stages of respiration in graphic form and indicates their locations within the cell. This illustration reappears at certain points within the chapter to remind you of the stage and location of a particular set of reactions.

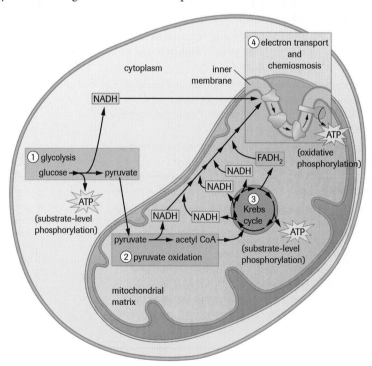

Figure 1
In eukaryotic cells, glycolysis occurs in the cytoplasm (pink), pyruvate oxidation and the Krebs cycle occur in the mitochondrial matrix (blue and purple respectively), and the electron transport chain is embedded in the inner mitochondrial membrane (green).

The ultimate goal of cellular respiration is to extract energy from nutrient molecules and store it in a form that the cell can use for its many and varied energy-requiring activities. In cellular respiration, the primary energy transfer is from glucose to ATP.

Energy Transfer

Before we describe the series of chemical reactions in glycolysis, the first stage of cellular respiration, we will outline the mechanisms used by cells to convert chemical potential energy from one form into another. The first law of thermodynamics tells us that energy cannot be created or destroyed—it can only be changed from one form into another or transferred from one object to another. Many chemical changes take place during cellular respiration, but only some of them transfer significant amounts of energy from one compound to another. The ultimate goal is to capture as much of the available free energy as possible in the form of ATP. This goal is accomplished through two distinctly different energy-transfer mechanisms called **substrate-level phosphorylation** and **oxidative phosphorylation**.

Substrate-Level Phosphorylation

In substrate-level phosphorylation, ATP is formed directly in an enzyme-catalyzed reaction. In the process, a phosphate-containing compound transfers a phosphate group directly to ADP, forming ATP. During the process, approximately 31 kJ/mol of potential energy is also transferred (**Figure 2**). (The value of 31 kJ/mol is determined under standard laboratory conditions and is used in all calculations in this book. In a living cell, the value is closer to 50 kJ/mol.) For each glucose molecule processed, four ATP molecules are generated this way in glycolysis and two in the Krebs cycle (**Figure 3**, yellow starbursts).

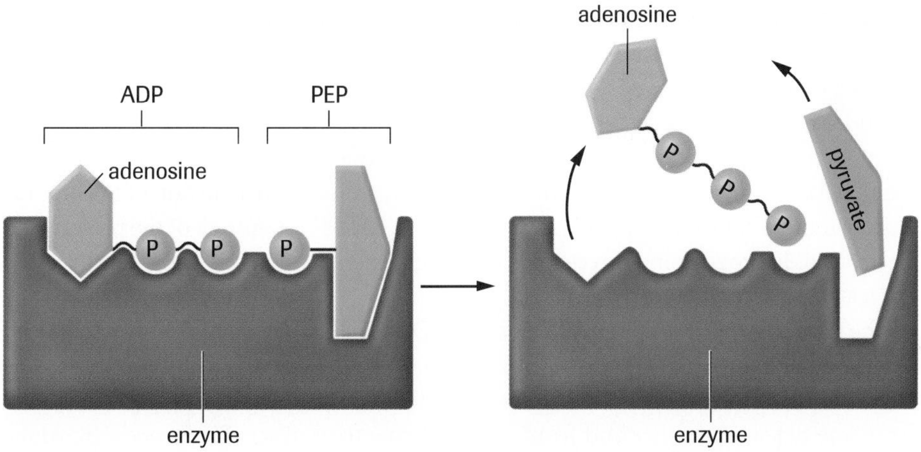

substrate-level phosphorylation mechanism forming ATP directly in an enzyme-catalyzed reaction

oxidative phosphorylation mechanism forming ATP indirectly through a series of enzyme-catalyzed redox reactions involving oxygen as the final electron acceptor

nicotinamide adenine dinucleotide, NAD$^+$ coenzyme used to shuttle electrons to the first component of the electron transport chain in the mitochondrial inner membrane

Figure 2
In this example of substrate-level phosphorylation, a phosphate-containing molecule called phosphoenolpyruvate (PEP) transfers its phosphate group to ADP, forming ATP. In the process, 31 kJ/mol of free energy is also transferred. This particular reaction occurs in glycolysis.

Oxidative Phosphorylation

In oxidative phosphorylation, ATP is formed indirectly. This process is oxidative because it involves a number of sequential redox reactions, with oxygen being the final electron acceptor. It is a more complex process than substrate-level phosphorylation and yields many more ATP molecules for each glucose molecule processed.

Oxidative phosphorylation begins when a compound called **nicotinamide adenine dinucleotide, NAD$^+$**, removes two hydrogen atoms (two protons and two electrons) from

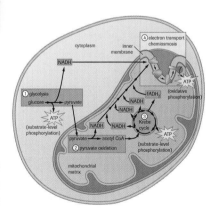

Figure 3
Substrate-level phosphorylation in cellular respiration

DID YOU KNOW?

Coenzymes and Vitamins

NAD^+ is a derivative of vitamin B_3 (niacin or niacinamide) (**Figure 4**). Vitamins are organic compounds that are required in very small amounts in the diet. Many vitamins, especially B vitamins like niacin, function as coenzymes in energy metabolism.

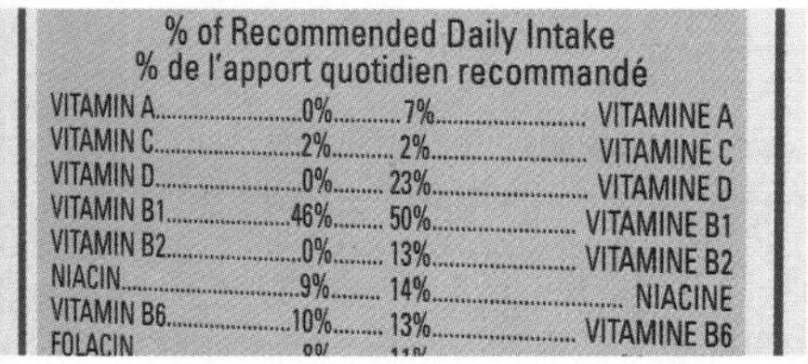

Figure 4
NAD^+ is derived from vitamin B_3 (niacin).

NAD^+: oxidized form of nicotimamide adenine dinucleotide

NADH: reduced form of nicotimamide adenine dinucleotide

Figure 5
NAD^+ is reduced to NADH + H^+ (NADH) by the transfer of two hydrogen atoms from a portion of the original glucose molecule.

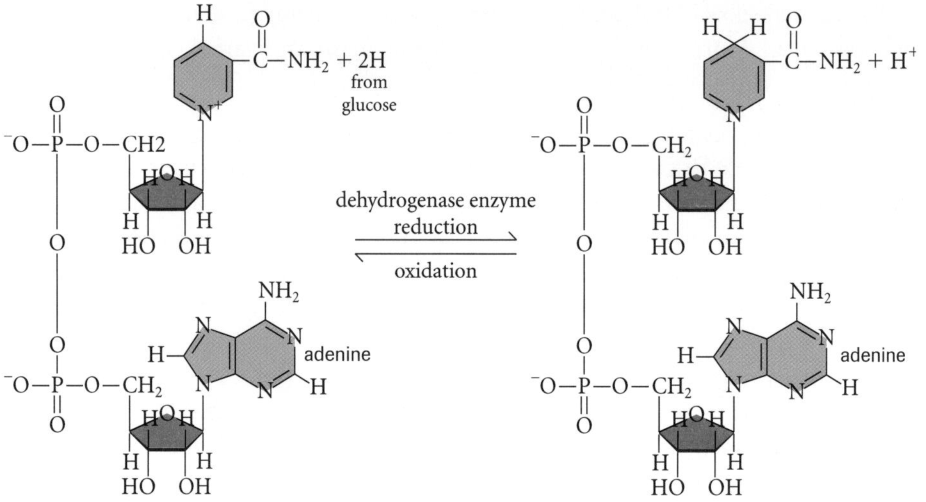

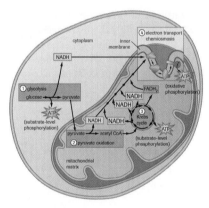

Figure 6
NADH is formed in glycolysis, pyruvate oxidation, and three steps of the Krebs cycle (yellow boxes).

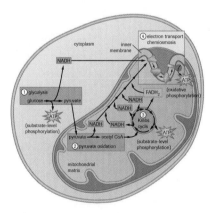

Figure 7
$FADH_2$ is formed in one step of the Krebs cycle (blue box).

a portion of the original glucose molecule. In the process, two electrons and one proton attach to the NAD^+, reducing it to NADH, while the remaining proton dissolves into the surrounding solution as $H^+_{(aq)}$. A dehydrogenase enzyme catalyzes this reaction. Thus, NAD^+ is the oxidized form of nicotinamide adenine dinucleotide and NADH + H^+ (shortened to NADH) is the reduced form (**Figure 5**). NAD^+ reduction occurs in one reaction of glycolysis (stage 1), during the pyruvate oxidation step (stage 2), and in three reactions of the Krebs cycle (stage 3) (**Figure 6**).

Another coenzyme called flavin adenine dinucleotide, FAD, performs a function similar to NAD^+. FAD is also reduced by two hydrogen atoms from a portion of the original glucose molecule. Its reduced form is symbolized as $FADH_2$ because all of the protons and electrons of hydrogen bind directly to the molecule. FAD is reduced to $FADH_2$ in one of the reactions of the Krebs cycle (**Figure 7**).

The reductions of NAD^+ to NADH and FAD to $FADH_2$ are energy-harvesting reactions that will eventually transfer most of their free energy to ATP molecules. The reduced coenzymes act as mobile energy carriers within the cell, moving free energy from one place to another and from one molecule to another. The process by which a cell transfers free energy from NADH and $FADH_2$ to ATP has not yet been discussed. It involves stage 4 of the cellular respiration process (electron transport and chemiosmosis) and requires the use of free oxygen molecules. All the reduced coenzymes are formed in the first three stages (glycolysis, pyruvate oxidation, and the Krebs cycle); therefore, the chemical reactions in these stages will be outlined before the mechanisms by which the reduced coenzymes power ATP synthesis are described. In the discussion, particular attention will be paid to the reactions in which NADH and $FADH_2$ are formed. The reactions in which ATP is produced directly by substrate-level phosphorylation will also be noted.

Stage 1: Glycolysis

The first 10 reactions of cellular respiration are collectively called **glycolysis** (Greek for "sugar splitting") (**Figure 8**). Starting with glucose, a 6-carbon sugar, glycolysis produces two 3-carbon pyruvate (pyruvic acid), molecules (**Figure 9**). The carbon backbone of glucose is essentially split in half. All the reactions in glycolysis occur in the cytoplasm, each step catalyzed by a specific enzyme. Glycolysis is an anaerobic process; it does not require oxygen.

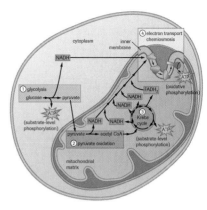

Figure 8
Glycolysis in cellular respiration (pink box)

glycolysis a process for harnessing energy in which a glucose molecule is broken into two pyruvate molecules in the cytoplasm of a cell

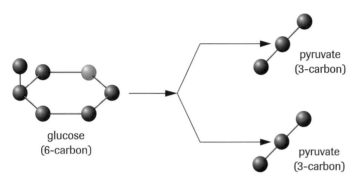

Figure 9
In a series of reactions called glycolysis, a 6-carbon glucose molecule is split into two 3-carbon pyruvate molecules.

Figure 11 (page 98) outlines the 10 reactions of the glycolytic pathway. As you study them, note the following:

- Two ATP molecules are used in the first five steps of the process, one in step 1 and one in step 3. These reactions "prime" the glucose molecule by adding phosphate groups to its structure, which prepares the molecule for cleavage in steps 4 and 5 and a return on the energy investment in the last five steps.

- Fructose 1,6-bisphosphate is split into dihydroxyacetone phosphate (DHAP) and glyceraldehyde 3-phosphate (G3P) in steps 4 and 5. The isomerase enzyme in this step immediately converts DHAP into G3P, resulting in two molecules of G3P. Steps 6 through 10 happen exactly the same way for each of the G3P molecules.

- In step 6, two NADH molecules are produced, one from each of the two G3P molecules processed.

- Two ATP molecules are produced by substrate-level phosphorylation in step 7, one ATP for each of the 1,3-bisphosphoglycerate (BPG) molecules processed.

- In step 10, two more ATP molecules are formed by substrate-level phosphorylation as two molecules of phosphoenolpyruvate (PEP) are converted into two pyruvate molecules.

The overall chemical equation for glycolysis is the following:

glucose + 2ADP + 2P_i + 2NAD$^+$ $\longrightarrow$ 2 pyruvate + 2ATP + 2(NADH + H$^+$)

The following is the energy yield for glycolysis:

4 ATP produced
2 ATP used
___
2 ATP produced net (may be used by the cell immediately)
2 NADH produced (may be further processed by some cells to obtain more ATP)

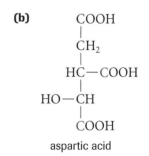

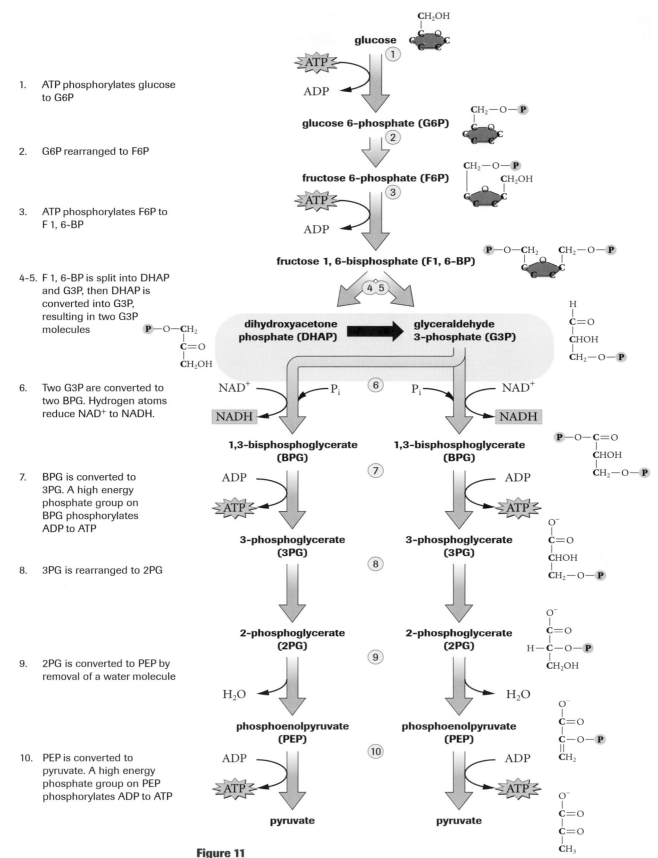

1. ATP phosphorylates glucose to G6P

2. G6P rearranged to F6P

3. ATP phosphorylates F6P to F 1, 6-BP

4-5. F 1, 6-BP is split into DHAP and G3P, then DHAP is converted into G3P, resulting in two G3P molecules

6. Two G3P are converted to two BPG. Hydrogen atoms reduce NAD^+ to NADH.

7. BPG is converted to 3PG. A high energy phosphate group on BPG phosphorylates ADP to ATP

8. 3PG is rearranged to 2PG

9. 2PG is converted to PEP by removal of a water molecule

10. PEP is converted to pyruvate. A high energy phosphate group on PEP phosphorylates ADP to ATP

Figure 11

Glycolysis converts a glucose molecule into two pyruvate molecules. Some of the free energy released is captured in the form of ATP and NADH.

The energy conversion efficiency of glycolysis (per mole glucose processed) is calculated as follows:

2 mol ATP $\times$ 31 kJ/mol ATP = 62 kJ

total free energy in 1 mol of glucose = 2870 kJ

energy conversion efficiency = $\dfrac{62\ kJ}{2870\ kJ} \times 100\% = 2.2\%$

Glycolysis alone is not a highly efficient energy-harnessing mechanism. It transfers only about 2.2% of the free energy available in 1 mol of glucose to ATP. Some of the energy is released as heat during the process, but the vast majority is still trapped in the two pyruvate and two NADH molecules. The 2.2% conversion efficiency value applies to glycolysis only; it does not take into consideration the possibility of obtaining additional ATP by further processing pyruvate and NADH in the reactions of aerobic respiration (stages 2, 3, and 4).

Glycolysis is thought to be the earliest form of energy metabolism. The first cells to emerge on Earth probably used this process to harness energy and, today, the simplest organisms continue to use it for all their energy needs. Glycolysis yields two ATP molecules from each glucose molecule processed. This may be sufficient energy for the needs of certain microorganisms, but it is not enough to satisfy the energy needs of most multicellular organisms. Nevertheless, all organisms, large and small, multicellular or not, carry out glycolysis either as their only source of ATP or as the first part of a more elaborate and more productive energy-yielding process, such as aerobic respiration. In addition to glycolysis (stage 1), three more processes are associated with aerobic respiration: pyruvate oxidation (stage 2), the Krebs cycle (stage 3), and electron transport and chemiosmosis (stage 4). In eukaryotes, these processes all occur in the cell's mitochondria and require oxygen.

Mitochondria

Mitochondria (singular: mitochondrion) are round or sausage-shaped organelles that are usually scattered throughout a cell's cytoplasm. These vital organelles specialize in the production of large quantities of ATP, the main energy-carrying molecule in living cells. The process that produces ATP in mitochondria cannot proceed without free oxygen. Three stages of aerobic cellular respiration take place within mitochondria: pyruvate oxidation, the Krebs cycle, and electron transport and chemiosmosis. Only **eukaryotic cells** contain mitochondria. **Prokaryotic cells** carry out all the stages of cellular respiration within the cytoplasm.

Mitochondria possess a double membrane (referred to as an envelope) composed of a smooth outer membrane and a highly folded inner membrane (**Figure 12**, page 100). The folds of the inner membrane are called **cristae** (singular: crista). The outer membrane plays a role similar to that of the cell membrane, but the inner membrane performs many functions associated with cellular respiration. It has numerous substances, such as proteins and enzymes, attached to its inner surface or embedded in its phospholipid bilayer that participate in the reactions of respiration. The inner membrane also creates two compartments within the mitochondrion. The mitochondrial **matrix** is a protein-rich liquid that fills the innermost space of a mitochondrion, and a fluid-filled **intermembrane space** lies between the inner and outer membrane. Both these compartments play a critical role in aerobic respiration.

Mitochondria have their own DNA (symbolized as mtDNA), RNA, and ribosomes. These components allow them to reproduce. Many of the features of mitochondrial DNA resemble those found in prokaryotes, such as bacteria. For example, mtDNA is

mitochondria eukaryotic cell organelle in which aerobic cellular respiration occurs

eukaryotic cells cells possessing a cell nucleus and other membrane-bound organelles

prokaryotic cells cells possessing no intracellular membrane-bound organelles or nucleus

cristae the folds of the inner mitochondrial membrane

matrix the fluid that fills the interior space of the mitochondrion

intermembrane space the fluid-filled space between the inner and outer mitochondrial membranes

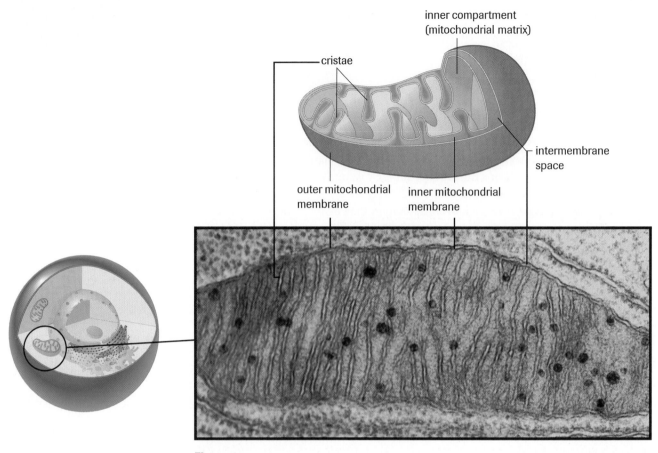

Figure 12
Sketch and transmission electron micrograph (thin section) of a typical mitochondrion

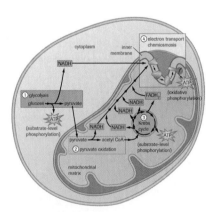

Figure 13
Pyruvate oxidation in cellular respiration

Figure 14
Pyruvate oxidation results in three changes to pyruvate:
1. A CO_2 portion is removed.
2. NAD^+ is reduced by two H atoms, obtained from food.
3. Coenzyme A is attached to the remaining acetic acid portion (acetyl group).

circular like that of bacteria. This observation has strengthened the theory (called the *endosymbiosis hypothesis*) that mitochondria are the evolutionary descendants of prokaryotes that established a symbiotic relationship with the ancestors of eukaryotic cells (discussed further in section 5.7 of Chapter 5).

Stage 2: Pyruvate Oxidation

The two pyruvate molecules formed in glycolysis are transported through the two mitochondrial membranes into the matrix (**Figure 13**). There, a multienzyme complex catalyzes the following three changes (**Figure 14**):

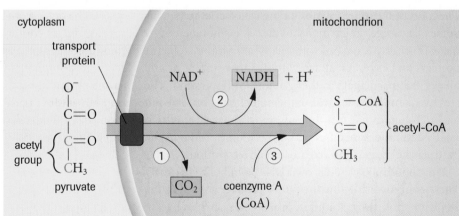

1. A low-energy carboxyl group is removed as CO_2. This is a decarboxylation reaction catalyzed by the enzyme pyruvate decarboxylase.

2. The remaining two-carbon portion is oxidized by NAD^+. In the process, NAD^+ gains two hydrogen atoms (two protons and two electrons) from organic molecules of food, and the remaining two-carbon compound becomes an acetic acid (acetate) group. This reaction transfers potential energy to NAD^+. It is a redox reaction—pyruvate is oxidized, and NAD^+ is reduced.

3. A sulfur-containing compound called coenzyme A (CoA) is attached to the acetate component, forming acetyl-CoA. The carbon–sulfur bond that holds the acetyl group to coenzyme A is unstable. This prepares the two-carbon acetyl portion of this molecule for further oxidation in the Krebs cycle. CoA is a derivative of vitamin B_5, also known as pantothenic acid.

The following is the overall equation for this process. (Remember that glycolysis produces two pyruvate molecules from one glucose molecule.)

$$2\,\text{pyruvate} + 2\,NAD^+ + 2\,\text{CoA} \longrightarrow 2\,\text{acetyl-CoA} + 2\,\text{NADH} + 2H^+ + 2\,CO_2$$

The two molecules of acetyl-CoA enter the Krebs cycle where additional free energy transfers occur. The two molecules of NADH proceed to stage 4 (electron transport and chemiosmosis) to produce ATP by oxidative phosphorylation. The two CO_2 molecules produced during pyruvate oxidation diffuse out of the mitochondrion and then out of the cell as a low-energy waste product. The two H^+ ions remain dissolved in the matrix.

Acetyl-CoA is a central molecule in energy metabolism. Almost all molecules that are catabolized for energy are converted into acetyl-CoA, including proteins, lipids, and carbohydrates. Acetyl-CoA is multifunctional; it can be used to produce fat or ATP. If the body needs energy, acetyl-CoA enters the Krebs cycle, ultimately transferring most of its free energy to ATP. If the body does not need energy, acetyl-CoA is channelled into an anabolic pathway that synthesizes lipids as a way of storing large amounts of energy as fat. The pathway taken by acetyl-CoA depends on the levels of ATP in the cell. If ATP levels are low, acetyl-CoA goes into the Krebs cycle to increase ATP production. If ATP levels are high, acetyl-CoA goes on to produce lipids. This explains why animals accumulate fat when they consume more food than their bodies require to satisfy their energy needs. All nutrients, whether protein, lipid, or carbohydrate, are converted to acetyl-CoA and then channelled toward fat production or ATP production, depending on the organism's immediate energy needs.

Krebs cycle a cyclic series of reactions that transfers energy from organic molecules to ATP, NADH, and $FADH_2$ and removes carbon atoms as CO_2

Stage 3: The Krebs Cycle

In 1937, Sir Hans Krebs (1900–81), a biochemist working at the University of Sheffield in England, discovered the series of metabolic reactions that became known as the **Krebs cycle** (**Figure 15**). He received the 1953 Nobel Prize in Physiology or Medicine for this important discovery. Fritz Albert Lipmann (1899–1986) shared the Nobel Prize with Krebs for his discovery of coenzyme A and the key role it plays in metabolism.

The Krebs cycle is an eight-step process, each step catalyzed by a specific enzyme. It is a cyclic process because oxaloacetate, the product of step 8, is the reactant in step 1 (**Figure 16**, page 102). Key features of the Krebs cycle are outlined in **Table 1** (page 103).

The following is the overall chemical equation for the Krebs cycle:

oxaloacetate + acetyl-CoA + ADP + P_i + 3NAD^+ + FAD $\longrightarrow$

CoA + ATP + 3NADH + 3H^+ + $FADH_2$ + 2CO_2 + oxaloacetate

(Oxaloacetate is shown as reactant and product to indicate that the process is cyclic.)

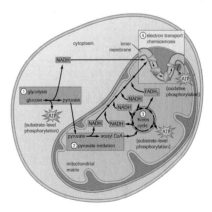

Figure 15
The Krebs cycle in cellular respiration

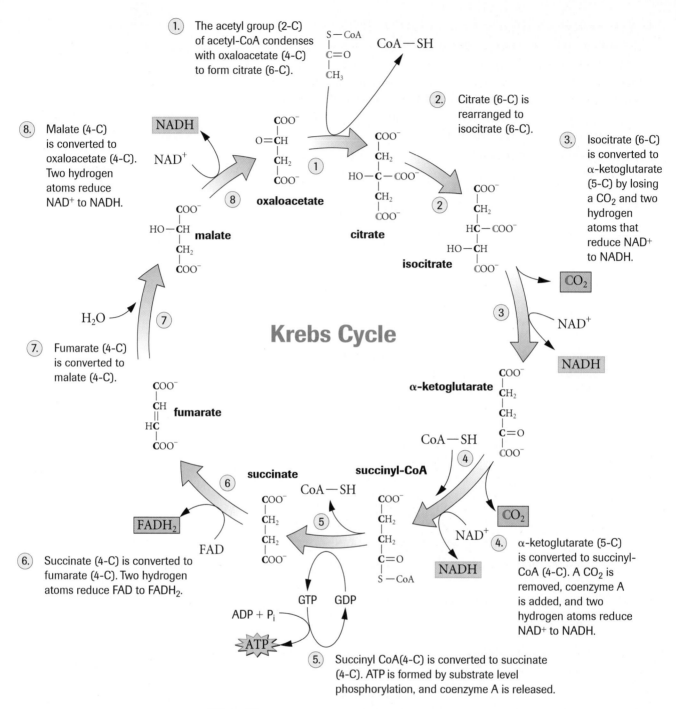

1. The acetyl group (2-C) of acetyl-CoA condenses with oxaloacetate (4-C) to form citrate (6-C).

2. Citrate (6-C) is rearranged to isocitrate (6-C).

3. Isocitrate (6-C) is converted to α-ketoglutarate (5-C) by losing a CO_2 and two hydrogen atoms that reduce NAD^+ to NADH.

8. Malate (4-C) is converted to oxaloacetate (4-C). Two hydrogen atoms reduce NAD^+ to NADH.

7. Fumarate (4-C) is converted to malate (4-C).

6. Succinate (4-C) is converted to fumarate (4-C). Two hydrogen atoms reduce FAD to $FADH_2$.

5. Succinyl CoA(4-C) is converted to succinate (4-C). ATP is formed by substrate level phosphorylation, and coenzyme A is released.

4. α-ketoglutarate (5-C) is converted to succinyl-CoA (4-C). A CO_2 is removed, coenzyme A is added, and two hydrogen atoms reduce NAD^+ to NADH.

Krebs Cycle

Figure 16
The Krebs cycle begins when acetyl-CoA condenses with oxaloacetate to form citrate. In one turn of the cycle, the last two carbon atoms of the original glucose molecule are removed as CO_2, and free energy is transferred to ATP, NADH, and $FADH_2$.

By the end of the Krebs cycle, the original glucose molecule is entirely consumed. The six carbon atoms leave the process as six low-energy CO_2 molecules, which are released by the cell as waste. All that is preserved of the original glucose molecule is most of its energy, which is stored in the form of four ATP molecules (two from glycolysis and two from the Krebs cycle) and 12 reduced coenzymes (two NADH from glycolysis, two NADH from the pyruvate oxidation step, six NADH from the Krebs cycle, and two

Table 1 Key Features of the Krebs Cycle

- Since two molecules of acetyl-CoA are formed from one molecule of glucose, the Krebs cycle occurs twice for each molecule of glucose processed (**Figure 17**).

- Acetyl-CoA enters the cycle at step 1. It reacts with an existing molecule of oxaloacetate (OAA) to produce a molecule of citrate. This is why the cycle is sometimes called the citric acid cycle. Note that this reaction converts a four-carbon compound (OAA) into a six-carbon compound (citrate) by the addition of the two-carbon acetyl group of acetyl-CoA. This releases CoA, which can be used to process another pyruvate molecule in the pyruvate oxidation step. Thus, CoA is recycled. Also notice that oxaloacetate contains two carboxyl groups and citrate has three carboxyl groups. Because of this, the cycle is sometimes referred to as the tricarboxylic acid cycle (TCA cycle).

- Energy is harvested in steps 3, 4, 5, 6, and 8.

- In steps 3, 4, and 8, NAD^+ is reduced to NADH.

- In step 5, ATP is formed by substrate-level phosphorylation. In this reaction, a phosphate group from the matrix displaces CoA from succinyl-CoA. The phosphate group is then transferred to guanosine diphosphate (GDP), forming guanosine triphosphate (GTP). Next, the phosphate group condenses with ADP, forming ATP. Overall, free energy is transferred from succinyl-CoA to ATP by a form of substrate-level phosphorylation.

- Energy is harvested in step 6. However, this reaction is not exergonic enough to reduce NAD^+ to NADH. Instead, a molecule of FAD is reduced to $FADH_2$, storing free energy in this form. This step is closely linked to the electron transport chain in the inner mitochondrial membrane.

- The last four carbon atoms of the original glucose molecule leave as fully oxidized CO_2 molecules in steps 3 and 4 (remember that it takes two turns of the Krebs cycle to process one glucose molecule). Like the two CO_2 molecules produced in the pyruvate oxidation steps, these four carbon atoms diffuse out of the mitochondrion and eventually out of the cell as low energy metabolic waste.

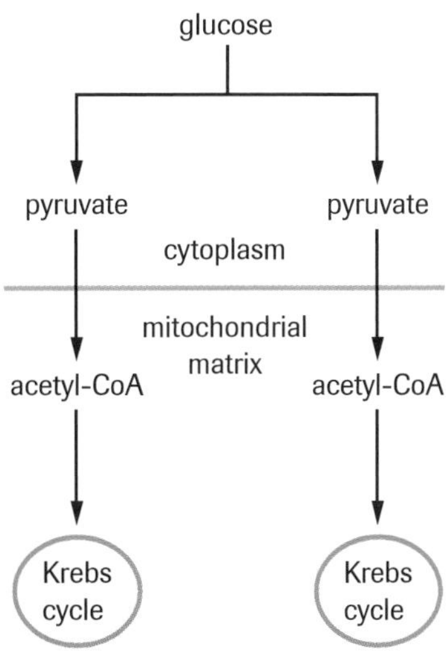

glucose

pyruvate   pyruvate

cytoplasm

mitochondrial matrix

acetyl-CoA   acetyl-CoA

Krebs cycle   Krebs cycle

Figure 17

$FADH_2$ from the Krebs cycle). Most of the free energy stored in NADH and $FADH_2$ molecules will eventually be transferred to ATP in the next (and last) stage of cellular respiration—an elaborate series of processes called electron transport and chemiosmosis.

However, before we move to stage 4, let's take a close look at the six carbon atoms of the original glucose molecule and see what happens to them as they move through the reactions of the first three stages of the process. The following summarizes the fate of glucose's carbon atoms:

electron transport chain (ETC) a series of membrane-associated protein complexes and cytochromes that transfer energy to an electrochemical gradient by pumping H^+ ions into an intermembrane space

$$\underset{\substack{\text{glucose}}}{\text{CCCCCC}} \xrightarrow{\text{glycolysis}} \underset{\substack{\text{2 pyruvate}}}{\text{CCC} + \text{CCC}} \xrightarrow[\text{oxidation}]{\text{pyruvate}} \underset{\substack{\text{2 acetyl-CoA} + 2CO_2}}{\text{CC} + \text{CC} + CO_2 + CO_2} \xrightarrow{\text{Krebs cycle}} \underset{\substack{\text{4CO}_2}}{CO_2 + CO_2 + CO_2 + CO_2}$$

Notice that by the end of the Krebs cycle, all six carbon atoms of glucose have been oxidized to CO_2 and released from the cell as metabolic waste. All that is left of the original glucose molecule is some of its free energy in the form of ATP and the reduced coenzymes, NADH and $FADH_2$. The reduced coenzymes now go on to stage 4 of the process, electron transport and chemiosmosis, where much of their free energy will be transferred to ATP.

Stage 4: Electron Transport and Chemiosmosis

NADH and $FADH_2$ eventually transfer the hydrogen atom electrons they carry to a series of compounds, mainly proteins, which are associated with the inner mitochondrial membrane, called the **electron transport chain (ETC)** (**Figure 18**). The components of the ETC are arranged in order of increasing electronegativity, with the weakest attractor of electrons (NADH dehydrogenase) at the beginning of the chain and the strongest (cytochrome oxidase) at the end. Each component is alternately reduced (by

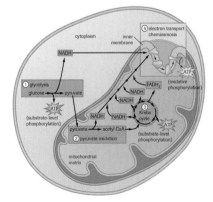

Figure 18
The electron transport chain and chemiosmosis in cellular respiration

gaining two electrons from the component before it in the chain) and oxidized (by losing the two electrons to the component after it in the chain); the electrons shuttle through the ETC like a baton handed from runner to runner in a relay race. As the electrons move from molecule to molecule in the ETC, they occupy ever more stable positions relative to the nuclei of the atoms they associate with. The free energy released in the process is used to move protons (H^+ ions) from the mitochondrial matrix. They move through three proton pumps, one in each of three membrane-associated protein complexes, into the fluid-filled intermembrane space. By the time the two electrons reach the last component of the ETC, they are in a very stable position. A highly electronegative substance is required to oxidize this last protein. Oxygen, one of the most electronegative substances on Earth, is used to do this. It strips the two electrons from the final protein complex in the chain and, together with two protons from the matrix, forms water. As such, oxygen acts as the final electron acceptor in the electron transport process. The actual components of the ETC are (in order of increasing electronegativity) NADH dehydrogenase, ubiquinone (Q), the cytochrome b-c_1 complex, cytochrome c, and the cytochrome oxidase complex.

Figure 19 illustrates how the components of the ETC are arranged in the membrane and shows the path taken by electrons (long red arrow) and protons (H^+ ions) (short blue arrows). The process begins with NADH giving up its two electrons to the first protein complex in the ETC, NADH dehydrogenase. The mobile electron carriers, Q and cytochrome c, shuttle the electrons from one protein complex to the next until they reach the final protein complex in the chain, the cytochrome oxidase complex. Finally, the enzyme cytochrome oxidase, which is part of this complex, catalyzes the reaction between the electrons, protons, and molecular oxygen to form water.

The electron transport process is highly exergonic. As mentioned earlier, the free energy lost by the electron pair during electron transport is used to pump three protons into the intermembrane space. This mechanism converts one form of energy into another—the chemical potential energy of electron position is converted to electrochemical potential energy of a proton gradient that forms across the inner mitochondrial membrane.

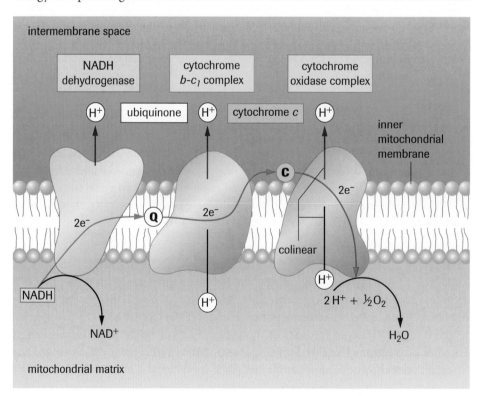

Figure 19
Electron transport chain (ETC)

Electrochemical potential energy is the type of stored energy possessed by a charged battery. It is caused by an accumulation of charged objects (ions, protons, electrons, etc.) on one side of an insulator. (The nature of the proton gradient that forms across the inner mitochondrial membrane will be described in more detail later in this section.) The transfer of a pair of electrons from NADH to oxygen through the electron transport chain is an exergonic process with a free energy change (ΔG) of -222 kJ/mol NADH. Much of this energy becomes stored in the electrochemical gradient and will be used to power ATP synthesis in the next part of the process, called chemiosmosis.

Before chemiosmosis is described, it is important to distinguish between NADH and $FADH_2$ in terms of their relationship with the electron transport system. NADH and $FADH_2$ do not transfer their electrons to the electron transport chain in the same way. NADH passes its electrons on to the first protein complex, NADH dehydrogenase, and $FADH_2$ transfers its electrons to Q, the second component of the chain (**Figure 20**). Thus, the free energy released by the oxidation of $FADH_2$ is used to pump two protons into the intermembrane space, while NADH oxidation pumps three. The result is that two ATP are formed per $FADH_2$ and three ATP molecules are formed per NADH.

Also, a distinction must be made between the NADH molecules produced in glycolysis and those produced in the pyruvate oxidation step and Krebs cycle. NADH produced by glycolysis in the cytoplasm (cytosolic NADH) may diffuse through the outer mitochondrial membrane into the intermembrane space, but not through the inner membrane into the matrix. Since the inner membrane is impermeable to NADH, it has two shuttle systems that pass electrons from cytosolic NADH in the intermembrane space to the matrix. The first and most common shuttle, called the glycerol-phosphate shuttle, transfers the electrons from cystolic NADH to FAD to produce $FADH_2$. Like $FADH_2$ produced in the Kreb's cycle, it transfers its electrons to Q, resulting in the synthesis of two ATP molecules by chemiosmosis. The second shuttle, called the aspartate shuttle, transfers electrons to NAD^+ instead of FAD, forming NADH, and then three ATP molecules. However, in this chapter we will assume that the transfer is made to FAD, using the glycerol-phosphate shuttle.

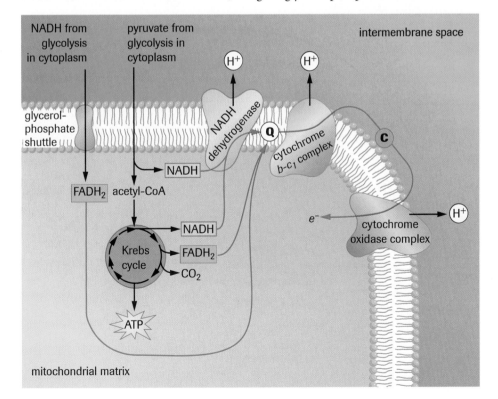

Figure 20
Reduced coenzymes give up their electrons to different components of the electron transport chain.

The many folds of the inner membrane increase surface area and allow multiple copies of the ETC to be located throughout the mitochondrion. At any given time, there are a limited number of NAD^+ and FAD molecules in a cell and, therefore, they must be recycled. Once NADH and $FADH_2$ give up their electrons to NADH dehydrogenase and Q, respectively, the resulting oxidized compounds pick up more hydrogen atoms in glycolysis, pyruvate oxidation, or the Krebs cycle.

Chemiosmosis and Oxidative ATP Synthesis

The protons that accumulate in the intermembrane space of the mitochondrion during electron transport create an **electrochemical gradient** that stores free energy. This gradient has two components: an electrical component caused by a higher positive charge in the intermembrane space than in the matrix, and a chemical component caused by a higher concentration of protons in the intermembrane space than in the matrix. The intermembrane space essentially becomes an H^+ reservoir because the inner mitochondrial membrane is virtually impermeable to protons. The electrochemical gradient creates a potential difference (voltage) across the inner mitochondrial membrane similar to that in a chemical cell or battery. Unable to diffuse through the phospholipid bilayer, the protons are forced to pass through special proton channels associated with the enzyme ATP synthase (ATPase). The free energy stored in the electrochemical gradient produces a **proton-motive force (PMF)** that moves protons through an ATPase complex. As protons move through the ATPase complex, the free energy of the electrochemical gradient is reduced. This energy drives the synthesis of ATP from ADP and inorganic phosphate (P_i) in the matrix (**Figure 21**).

Thus, some of the free energy lost by the electrochemical gradient is harvested as chemical potential energy in ATP. This mechanism of ATP generation was first worked out by Peter Mitchell in 1961, whose theory started a revolution in the way scientists thought about bioenergetics. However, it took him a long time to convince an initially hostile scientific community that his ideas were reasonable. He received the Nobel Prize

electrochemical gradient a concentration gradient created by pumping ions into a space surrounded by a membrane that is impermeable to the ions

proton-motive force (PMF) a force that moves protons through an ATPase complex on account of the free energy stored in the form of an electrochemical gradient of protons across a biological membrane

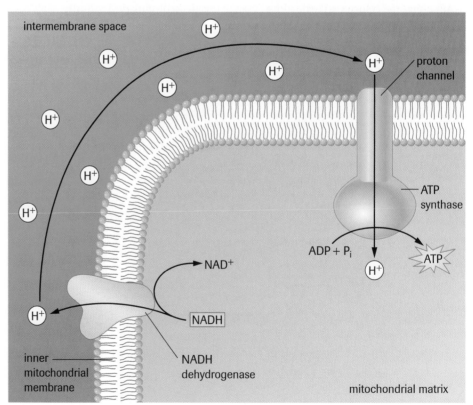

Figure 21
One molecule of ATP is synthesized from ADP and P_i as an H^+ ion passes through the ATPase complex into the mitochondrial matrix from the H^+ reservoir in the intermembrane space.

in Chemistry in 1978 for "his contribution to the understanding of biological energy transfer through the formulation of the chemiosmotic theory." Mitchell called the process **chemiosmosis** because the energy that drives the synthesis of ATP comes from the "osmosis" of protons through a membrane from one compartment into another. Although it is a misnomer according to today's definition of osmosis, the term *chemiosmosis* continues to be used to describe this process.

After they are formed by chemiosmosis, the ATP molecules are transported through both mitochondrial membranes by facilitated diffusion into the cytoplasm, where they are used to drive endergonic processes, such as movement, active transport, and synthesis reactions throughout the cell.

Electron transport followed by chemiosmosis is the last stage of the oxidative phosphorylation process that began with the reduction of NAD^+ and FAD with hydrogen atoms from the original glucose molecule. But how is chemiosmosis linked to electron transport in the ETC? The electron transport chain gets its electrons from the hydrogen atoms that NADH obtained from glucose. Some of the free energy released by the electrons as they travel through the chain is harnessed by pumping protons into the H^+ reservoir and creating the electrochemical gradient. The energy of the gradient is reduced as the protons pass through the ATPase complex back into the mitochondrial matrix. Although some of the energy dissipates as heat, much of it is captured by condensing ADP and P_i into ATP at the ATPase complex.

The continual production of ATP by this method is dependent on the establishment and maintenance of an H^+ reservoir. This condition requires the continual movement of electrons through the ETC, which, in turn, is dependent on the availability of oxygen to act as the final electron acceptor. This explains why animals have lungs and fish have gills. Oxygen is needed to keep the electrons flowing through the ETC. Electrons are "pulled down" the chain in an energy-yielding "fall," similar to gravity pulling a skydiver down toward the centre of Earth. The energy released in the fall keeps protons moving into the H^+ reservoir so that they can "fall back" into the matrix and drive the synthesis of ATP.

Without food (glucose), there will be no electrons in the first place. This is one of the reasons why heterotrophs must continually eat and why photoautotrophs must continually photosynthesize. If oxygen is not available, or its supply is interrupted, electrons cannot continue to flow through the ETC because there is no substance available to act as the final electron acceptor. Although there are many chemicals in a cell, none is electronegative enough to oxidize the last protein in the chain. Without oxygen to free up that last protein, the chain soon becomes clogged with stationary electrons. H^+ ions cannot be pumped into the intermembrane space, and those that are there soon move into the matrix until protons are equally distributed across the inner membrane of the mitochondrion. If this happens, chemiosmosis stops, and ATP synthesis grinds to a halt. At the other end of the chain, NADH and $FADH_2$ are no longer able to give up their electrons (as hydrogen atoms) to proteins in the ETC because these proteins cannot get rid of the electrons that they are already holding on to. All of the available NAD^+ and FAD remain in reduced form as NADH and $FADH_2$, unable to remove any more hydrogen atoms from glucose.

As you can see, the three stages of oxidative phosphorylation (pyruvate oxidation, the Krebs cycle, and electron transport and chemiosmosis) are all linked to one another and are all dependent on glycolysis for the production of pyruvate. It is said that ATP synthesis by chemiosmosis is *coupled* with electron transport, and both of these are dependent on the availability of electrons (from food such as glucose) and oxygen (for its ability to act as a final electron acceptor).

An overview of the mechanism of oxidative phosphorylation is presented in **Figure 22** (page 108). The illustration shows that NADH shuttles electrons from glucose to the ETC, where they lose potential energy as they are transported from protein to protein in

chemiosmosis a process for synthesizing ATP using the energy of an electrochemical gradient and the ATP synthase enzyme

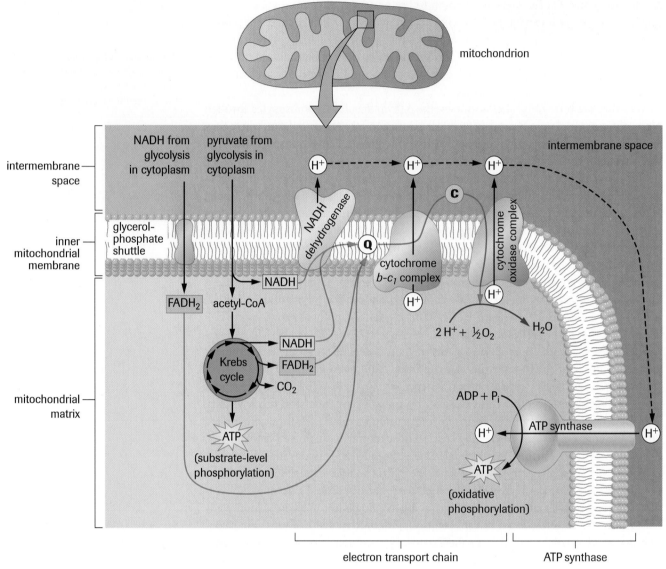

Figure 22
An overview of oxidative phosphorylation

the chain. Some of the energy is harnessed in the form of an electrochemical gradient caused by the accumulation of protons in the intermembrane space of the mitochondrion. The energy of the gradient is used to power ATP synthesis in the matrix as protons diffuse through ATPase complexes back into the matrix. The continual movement of electrons through the ETC is dependent on the availability of molecular oxygen to act as a final electron acceptor. The formation of water using electrons from the ETC and protons from the matrix is catalyzed by cytochrome oxidase, the last protein in the chain.

In oxidative phosphorylation, electrons flow "downhill" (**Figure 23**).

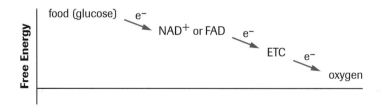

Figure 23
The exergonic ("downhill") flow of electrons in aerobic respiration

The Energetics of Oxidative Phosphorylation

Water can be formed in a test tube by combining hydrogen, $H_{2(g)}$, and oxygen, $O_{2(g)}$, according to the following equation:

$$H_{2(g)} + \frac{1}{2}O_{2(g)} \longrightarrow H_2O_{(l)}$$

This reaction is highly exergonic. In fact, it is explosive! A large amount of energy is released rapidly in the test tube reaction because bonding electrons quickly move much closer to a nucleus in water than they were in the hydrogen and oxygen molecules, as illustrated in **Figure 24(a)**. **Figure 24(b)** shows that the same reaction occurs at the end of the electron transport chain, where an oxygen atom ($\frac{1}{2}O_2$) combines with two electrons and two protons (two H atoms) to form water. As you can see, the same reaction may occur by two totally different mechanisms.

In cellular respiration, the source of hydrogen is glucose. The electron transport chain separates the electrons of the hydrogen atoms from their protons. The protons dissolve in the mitochondrial matrix and the electrons move through the ETC, occupying more stable configurations as they move to ever more electronegative components. Energy is released at each step. Most of the energy dissipates as heat, but a significant amount is harnessed as ATP is synthesized by chemiosmosis. The electrons, already in a much more stable state at the end of the ETC, gain a little more stability when they are captured by oxygen and reunited with protons in the matrix to form water. This removes low-energy electrons from the ETC, and makes room for other electrons coming down the chain.

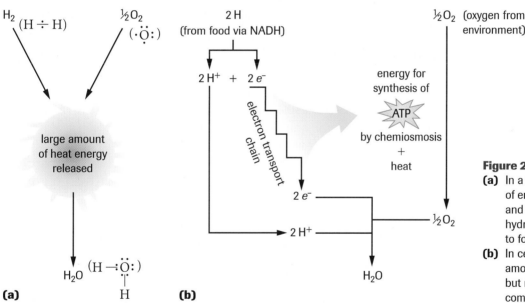

(a)   **(b)**

Figure 24
(a) In a test tube, a large amount of energy in the form of heat and light is released when hydrogen and oxygen react to form water.
(b) In cellular respiration, the same amount of energy is released, but not all at once and not completely as heat and light.

The Aerobic Respiration Energy Balance Sheet

How much energy was transferred from glucose to ATP in the entire aerobic respiration process? We may calculate two values in answer to this question: a theoretical value and an actual value. Although the actual value gives a more realistic total, it too varies according to the type of cell and various environmental conditions. **Figure 25** (page 110) summarizes the theoretical yield of 36 ATP and its sources.

The actual ATP yield is less than 36 for the following two reasons:

1. The inner mitochondrial membrane is not completely impermeable to H^+ ions. Thus, some H^+ ions leak through the phospholipid bilayer of the membrane, reducing the number that go through the ATPase complex to produce ATP.

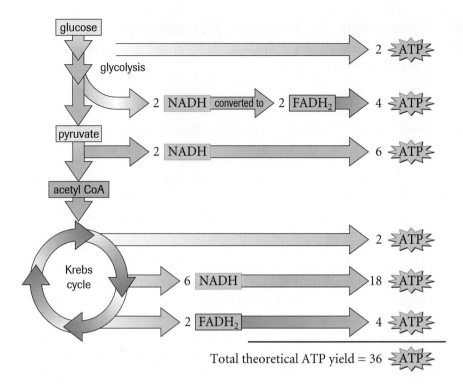

Figure 25
Theoretical coenzyme and ATP yield from the aerobic respiration of one glucose molecule

2. Some of the protons in the H$^+$ reservoir are used by the cell for other energy-requiring activities. The result is a reduction in the number of ATP molecules produced by chemiosmosis.

It has been estimated that the equivalent of 2.5 ATP molecules (not 3) are realistically produced for every NADH and approximately 1.5 ATP molecules (not 2) are produced for each FADH$_2$. This difference reduces the number of ATP produced from NADH to 25 (10 NADH $\times$ 2.5 ATP/NADH) and the number produced from FADH$_2$ to 3 (2 FADH$_2$ $\times$ 1.5 ATP/FADH$_2$), resulting in an actual yield of 30 ATP per glucose molecule (or 30 mol ATP per mol glucose) molecules from one glucose molecule.

Efficiency of Energy Conversion for Aerobic Respiration

Aerobic respiration is a much more efficient energy conversion mechanism than glycolysis. Aerobic respiration captures approximately 32% of the available free energy in glucose. Using an actual yield of 30 ATP per glucose molecule (or 30 mol ATP per mol glucose), the efficiency may be calculated as follows:

efficiency = 30 mol ATP $\times$ 31 kJ/mol ATP / 2870 kJ $\times$ 100% = 32% (rounded)

The greater ability of aerobic respiration to harness energy from nutrients makes multicellular life possible.

The efficiency of energy conversion for an automobile engine is estimated to be approximately 25%. In both cases, the remainder of the energy is given off as heat.

Metabolic Rate

metabolic rate the amount of energy consumed by an organism in a given time

An organism's **metabolic rate** is the amount of energy consumed by the organism in a given time. This value is also a measure of the overall rate at which the energy-yielding reactions of cellular respiration take place. Metabolic rate will increase when work is done, but even when at rest, an organism uses energy to keep cells alive. In the case of the human body, energy is used for breathing, maintaining body temperature, contracting muscles, and maintaining brain function. The minimum amount of energy needed to keep an organism alive is called the **basal metabolic rate (BMR)**. The BMR accounts for about 60% to 70% of the energy a human body uses in a day. In general, BMR is measured in units of kilojoules per square metre of body surface per hour: $kJ/m^2/h$.

The BMR does not remain constant over time but changes with growth, development, and age. A newborn baby's BMR is approximately 100 $kJ/m^2/h$. As the baby grows, the BMR increases, reaching a maximum of about 220 $kJ/m^2/h$ by the end of the first year. After that, as age increases, the BMR gradually decreases (**Figure 26**).

BMR varies not only with age, but also according to gender and health. In general, a healthy adult male has a BMR of about 167 $kJ/m^2/h$, and a healthy woman has a BMR of approximately 150 $kJ/m^2/h$. The BMR may be estimated experimentally by measuring the amount of thermal energy lost by a person's body over a given time. This is done while a person is lying at rest in a human calorimeter, such as the Benzinger calorimeter.

The calorimeter is an insulated vessel containing a known mass of water and a thermometer. As the person lies perfectly still in the calorimeter, thermal energy expended by the body is transferred to the water, causing its temperature to rise. The amount of thermal energy released is calculated on the basis that 4.2 J of energy is required to raise the temperature of 1.0 g of water by 1°C. Since the thermal energy expended by the body comes from the oxidation of food, the calculated energy value is proportional to the person's BMR. To complete the BMR calculation, the person's surface area must be measured. This measurement is made by encasing the person's entire body in a wax mould, and then flattening out the mould and measuring its surface area. As you might imagine, this is a very demanding process for the human subject. Thankfully, such measurements have led to formulas that can estimate a person's surface area without subjecting the individual to the wax mould. The following equation estimates human body surface area (BSA) in square metres (m^2):

$$BSA = m^{0.425} \times h^{0.725} \times 0.007\,184$$

(m = body mass in kilograms, h = height in centimetres)

basal metabolic rate (BMR) the minimum amount of energy on which an organism can survive

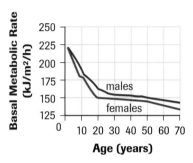

Figure 26
Changes in the human BMR with age

nomograms graphical methods for determining the value of an unknown quantity when the values of other quantities that it is mathematically related to are known

---

Calculating Human Body Surface Area (BSA)

SAMPLE problem

Example
(a) Calculate the body surface area of an adult male who is 1.75 m tall and has a mass of 85.0 kg.
(b) Calculate the person's BMR if he dissipates 322 kJ of thermal energy in 1.0 hours while in a Benzinger calorimeter.

Solution
(a) h = 1.75 m = 175 cm

m = 85.0 kg

$BSA = m^{0.425} \times h^{0.725} \times 0.007\,184$

$= 85.0^{0.425} \times 175^{0.725} \times 0.007\,184$

$= 6.46 \times 42.3 \times 0.007\,184$

$= 2.01\ m^2$

The man's body surface area is approximately 2.01 m^2.

(b) $BMR = 322kJ \div 2.01\ m^2 \div 1.0\ h$

$= 160\ kJ/m^2/h$

To make things even easier, **nomograms** have been developed that allow the BSA to be determined graphically (**Figure 27**, page 112).

Mass (kg)

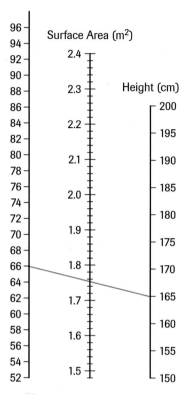

Figure 27
Line up a ruler from the body mass in kg to the body height in cm. The point where the ruler crosses the surface area line gives the body's surface area in m², as shown by the red line for a person who is 165 cm tall and has a mass of 66 kg (1.755 m²).

Answer

2. (a) BSA = 1.89 m²
   (b) BMR = 155 kJ/m²/h

△ INVESTIGATION 2.2.1

Energy Consumption During Exercise (p. 128)
Cellular respiration provides energy in the form of ATP for all the energy-requiring activities of an organism. How many teaspoons of table sugar does a human body have to respire to carry out a specific function? Investigation 2.2.1 provides you with an opportunity to determine the number of teaspoons of table sugar respired when exercising.

The BMR is measured when a person is lying at rest—even the slightest movement will result in a value higher than the basal value. BMR represents a base line for the measurement of metabolic activity. It is not applicable to everyday life. A person's metabolic rate changes according to activity level. The metabolic rate of an active person may be measured in terms of energy expenditure over a given period time. **Table 2** lists the average energy expenditures for different kinds of activities for men and women in kilojoules per minute. △

Table 2 Average Energy Expenditures for Different Types of Human Activities

Activity	Average Energy expediture (kJ/min)	
	Woman	*Man*
sleeping	3.8	4.2
sitting	5.0	5.8
light work	15.0	17.0
bicycling (20km/h)	37.0	41.0
heavy work	57.0	63.0

As with BMR, metabolic rates also decrease with age. This is partly because, as we get older, the body becomes more efficient at doing the same tasks over and over again. It is also because in general, people become less physically active as they get older. As level of activity declines, the amount of muscle tissue decreases, and energy requirements diminish.

> ▶ **Practice**

Understanding Concepts

1. Why is body surface area used in the determination of basal metabolic rate?
2. (a) Use the BSA equation to determine the body surface area of a teenager whose mass is 78 kg and whose height is 1.70 m. Use the nomogram in **Figure 27** to check your answer.
   (b) Calculate the teenager's BMR as she dissipates 439 kJ of thermal energy in 1.5 hours while in a Benzinger calorimeter.

Controlling Aerobic Respiration

The reactions of aerobic respiration are regulated by various feedback inhibition and product activation loops (**Figure 28**). A major control point is the third reaction of glycolysis, which is catalyzed by the allosteric enzyme phosphofructokinase. The final product of respiration, ATP, inhibits phosphofructokinase, while ADP stimulates its activity. Therefore, less ATP will be produced when ATP levels are high and ADP levels are low, and vice versa. If citrate, the first product of the Krebs cycle, accumulates in the mitochondria, some will pass into the cytoplasm and inhibit phosphofructokinase, thereby slowing down glycolysis. As citrate is used in the Krebs cycle, or in other metabolic processes, its concentration will decrease; phosphofructokinase inhibition will be reduced, and the rate of glycolysis (and the rest of the respiratory process) will increase.

Another important control mechanism involves NADH. A high concentration of NADH in a cell indicates that the electron transport chains are full of electrons and ATP production is high. In this case, NADH allosterically inhibits pyruvate decarboxylase and reduces the amount of acetyl-CoA that is fed into the Krebs cycle, restricting the amounts of NADH produced.

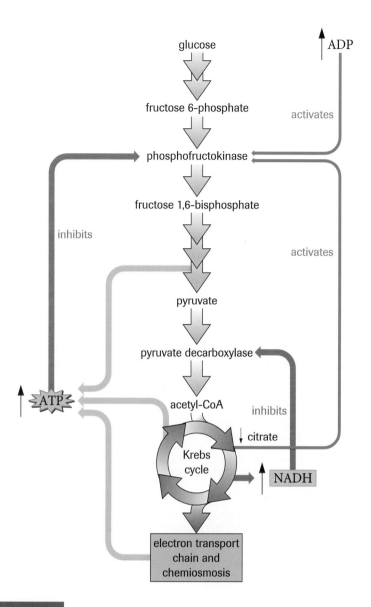

Figure 28
Aerobic respiration is regulated by the relative concentrations of ATP and ADP and the concentrations of citrate and NADH.

SUMMARY *Cellular Respiration: The Details*

- Cellular respiration begins in the cytoplasm and, in the case of aerobic respiration, is completed in mitochondria in eukaryotes; it takes place in the cytoplasm of prokaryotes.

- Mitochondria possess two membranes: a smooth outer membrane and a highly folded inner membrane that contains many proteins used in cellular respiration.

- ATP may be formed by substrate-level phosphorylation or oxidative phosphorylation. Substrate-level phosphorylation does not require oxygen; oxidative phosphorylation does. In substrate-level phosphorylation, a phosphate group is attached to ADP in an enzyme-catalyzed reaction. Oxidative phosphorylation is made up of redox reactions involving NAD^+, FAD, an electron transport chain, the inner mitochondrial membrane, ATPase, and oxygen as the final electron acceptor.

- Glycolysis occurs in the cytoplasm. It produces two three-carbon pyruvate molecules from a six-carbon glucose molecule. Glycolysis produces two ATP (net) and two NADH.

- Pyruvate oxidation occurs in the mitochondria. In the process, a CO_2 portion is cleaved from pyruvate and removed from the cell as waste. The remaining two-carbon acetyl group attaches to coenzyme A to produce acetyl-CoA. In this reaction, two NADH and two CO_2 are formed (one for each of the two pyruvate molecules).

- The Krebs cycle occurs in the mitochondrial matrix. It begins when acetyl-CoA reacts with oxaloacetate to produce citrate. The two carbon atoms introduced by acetyl-CoA are removed as two CO_2, one ATP molecule is produced by substrate-level phosphorylation, one $FADH_2$ and three NADH are produced, and the final step regenerates oxaloacetate.

- The electron transport chain, associated with the inner mitochondrial membrane, transports electrons through a series of redox reactions that release the free energy used to pump protons into the mitochondrial intermembrane space, creating an electrochemical gradient that is a source of free energy.

- In chemiosmosis, protons move through ATPase complexes embedded in the inner membrane, releasing free energy that drives the synthesis of ATP.

- Oxygen is the final acceptor of electrons that pass through the electron transport chain. If oxygen is not available, the Krebs cycle, electron transport, and chemiosmosis come to a halt.

- **Table 3** summarizes the inputs (reactants) and outputs (products) of cellular respiration. Compare the reactants and products in the table with the figures that illustrate the various stages of the process.

Table 3 Summary of Cellular Respiration

	Glycolysis (per glucose)	Pyruvate oxidation (per glucose)	Krebs cycle (per glucose)	Electron transport and chemiosmosis (per glucose, theoretical yield)
Location	cytoplasm	mitochondrial matrix	mitochondrial matrix and inner membrane	inner mitochondrial membrane and intermembrane space
Reactants	glucose 2 ATP 2 NAD^+ 4 ADP 2 P_i	2 pyruvate 2 NAD^+ 2 CoA	2 acetyl-CoA 2 oxaloacetate 6 NAD^+ 2 ADP 2 P_i 2 FAD	6 NADH (Krebs) 2 NADH (pyruvate oxidation) 2 $FADH_2$ (Krebs) 2 $FADH_2$ (from 2 cytosolic NADH) 32 ADP 32 P_i 6 O_2 12 H^+
Products	2 pyruvate 4 ATP 2 NADH 2 H^+ 2 ADP	2 acetyl-CoA 2 NADH 2 H^+ 2 CO_2	2 CoA 4 CO_2 2 oxaloacetate 6 NADH 6 H^+ 2 $FADH_2$ 2 ATP	8 NAD^+ 4 FAD^+ 24 H^+ 32 ATP 6 H_2O
ATP required	2	none	none	none
ATP produced	4	none	2	32
Net ATP produced	2	none	2	32

Total ATP produced by aerobic respiration of 1 glucose molecule = 36

- The theoretical yield of ATP in aerobic respiration is 36; however, the actual yield is about 30 ATP because some free energy is lost by the permeability of the inner mitochondrial membrane to protons and by the use of some energy for other endergonic reactions. The efficiency of energy conversion is approximately 32%.

- Phosphofructokinase, the enzyme that catalyzes step 3 of glycolysis, controls cellular respiration. It is activated by ADP and citrate and inhibited by ATP. NADH inhibits pyruvate decarboxylase, an enzyme that catalyzes the conversion of pyruvate to acetyl-CoA.

- An organism's metabolic rate is the amount of energy consumed in a given time, and a measure of the overall rate at which the energy-yielding reactions of cellular respiration take place.

▶ *Section 2.2* *Questions*

Understanding Concepts

1. What raw materials are needed for a cell to produce a molecule of ATP by substrate-level phosphorylation?

2. (a) In eukaryotic cells, where does glycolysis occur?
   (b) What does *glycolysis* mean?

3. Which stores more potential energy: one molecule of glucose or two molecules of pyruvic acid? Explain.

4. (a) List the final products of glycolysis.
   (b) What two products of glycolysis may be transported into mitochondria for further processing?

5. Compare substrate-level phosphorylation and oxidative phosphorylation.

6. How do ADP and ATP differ in structure? in free energy content?

7. Arrange the following types of cells in order of increasing number of mitochondria in the cytoplasm: nerve cell, skin cell, fat cell, heart muscle cell. Provide a rationale for your sequence.

8. Describe two functions that mitochondrial membranes serve in energy metabolism.

9. (a) Why is every reaction of cellular respiration catalyzed by a specific enzyme?
   (b) What would happen to an organism that lacked the gene for hexokinase, the enzyme that catalyzes the first reaction in glycolysis?

10. Describe the function of NAD$^+$ and FAD in cellular respiration.

11. What are the final products of aerobic cellular respiration?

12. Why is aerobic respiration a more efficient energy-extracting process than glycolysis alone?

13. As a result of glycolysis, pyruvate oxidation, and the Krebs cycle, only a small portion of the energy of glucose has been converted to ATP. In what form is the rest of the usable energy found at this stage of the process?

14. (a) What part of a glucose molecule provides electrons in cellular respiration?

(b) Describe how electron transport complexes set up a proton gradient in response to electron flow.
(c) How is the energy used to drive the synthesis of ATP?
(d) What is the name of this process?
(e) Who discovered this mechanism?

15. (a) Distinguish between an electron carrier and a terminal electron acceptor.
   (b) What is the final electron acceptor in aerobic respiration?

16. Explain how the following overall equation for cellular respiration is misleading:

$$C_6H_{12}O_6 + 6O_2 \longrightarrow 6CO_2 + 6H_2O$$

17. Explain why CO_2 does not serve as a source of free energy in living systems.

18. (a) Distinguish between metabolic rate and basal metabolic rate.
   (b) Explain how and why metabolic rate changes as we grow older.

Applying Inquiry Skills

19. (a) How could a pH meter be used to support Peter Mitchell's chemiosmotic theory?
   (b) What other common laboratory apparatus could be used to test the theory? Briefly describe a procedure that uses the instrument.
   (c) Strong detergents disrupt and rupture phospholipid bilayers. When a suspension of mitochondria in a test tube is treated with detergents, electron transport is detected, but ATP is not produced. Explain how this supports the theory of chemiosmosis.

20. (a) Use the nomogram in **Figure 29** on page 116 to determine the surface area of a teacher who is 180 cm tall and has a mass of 80 kg.
   (b) If the basal energy requirement for this individual is 160 kJ/m^2/h, calculate the total energy content of the food the teacher must consume to function at rest for 24 h.
   (c) Predict the BMR of a student whose height is 165 cm and whose mass is 90 kg.
   (d) Determine the solution for (c) and evaluate your prediction. ▶

Mass (kg)

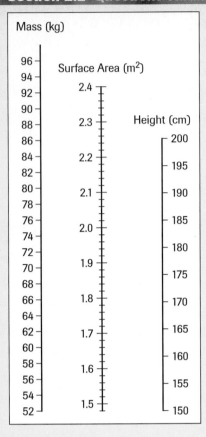

Figure 29

Making Connections

21. (a) Why must mitochondria be able to reproduce?
    (b) At conception, only the nucleus of a sperm cell enters the egg cell. Considering this, how are all of the mitochondria in a fully grown individual genetically related?
    (c) One class of metabolic disorders is caused by defective mitochondria. Research at the library or on the Internet to find out about two mitochondrial diseases. Describe the symptoms of the disorders, the part(s) of the mitochondrion that are defective, and treatments that are available for the conditions.

 www.science.nelson.com

22. Several vitamins, especially the B vitamins, play key roles in energy metabolism. A number of vitamin B deficiencies, if left unchecked, can cause serious illness. Conduct library and/or Internet research to answer the following questions:
    (a) What is meant by the term vitamin B complex? List the names of the vitamins that are part of this group.
    (b) Why are these vitamins called water-soluble vitamins? What other water-soluble vitamins are there?

(c) Select two B vitamins and briefly describe
    (i) their function in energy metabolism;
    (ii) functions not directly associated with energy metabolism;
    (iii) good natural sources;
    (iv) deficiency symptoms.

 www.science.nelson.com

23. Hummingbirds (**Figure 30**) have the highest metabolic rate of any animal.

Figure 30
The male rufus hummingbird, like many diminutive animals, must consume relatively large quantities of food to maintain body temperature and meet its body's demands for energy.

(a) Conduct library or Internet research to complete the comparison chart shown in **Table 4**.

Table 4

Function	Hummingbird	Human
Resting heart rate (beats/min)		
Breathing rate (breaths/min)		
Fastest speed (km/h)		
Average lifespan (yrs.)		

(b) Determine the mass of hamburger meat that a human would have to consume in one day to obtain as much energy as a hummingbird uses in one day.

 www.science.nelson.com

Carbohydrates are the first nutrients most organisms catabolize for energy. In some cases, living things must be able to metabolize other energy-rich nutrients to obtain energy in times of starvation. Most organisms possess metabolic pathways that, when necessary, metabolize proteins, lipids, and nucleic acids. In each case, the larger molecules are first digested into their component parts, which the cell may reassemble into macromolecules for its own use. Otherwise, they may be metabolized for energy by feeding into various parts of glycolysis or the Krebs cycle (**Figure 1**).

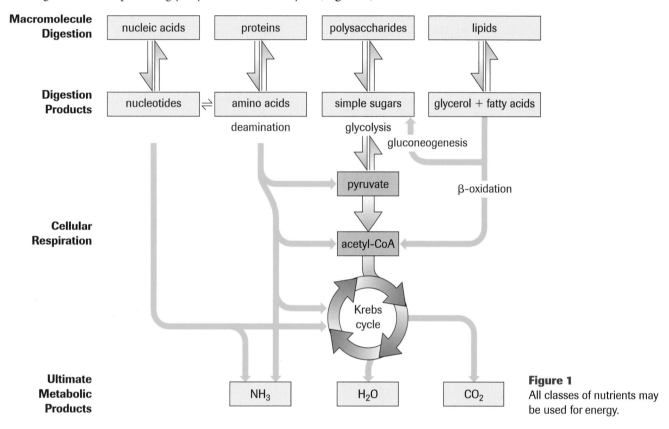

Figure 1
All classes of nutrients may be used for energy.

Protein Catabolism

Under normal conditions, proteins are first digested into individual amino acids, which are absorbed and used to produce the cell's own proteins. In the first stage of protein metabolism, amino groups are removed from the amino acids in a process called **deamination**. This process converts the amino group of the amino acids into ammonia, NH_3, a waste product in animals and a useful byproduct in plants. Other chemical reactions convert the remaining portions of the amino acids into various components of glycolysis or the Krebs cycle. The point of entry into these metabolic pathways depends on the identity of the amino acid. For example, leucine is converted into acetyl-CoA, alanine is converted into pyruvate, and proline is converted into α-ketoglutarate. These metabolic intermediates then feed into their respective catabolic pathways and the amino acid's energy is transferred to ATP.

deamination the first step in protein catabolism, involving the removal of the amino group of an amino acid as ammonia

Lipid Catabolism

Triacylglycerols are first digested into glycerol and fatty acids. The glycerol portion may be converted into glucose in a process called gluconeogenesis, or it may be changed into dihydroxyacetone phosphate (DHAP) and then 3-phosphoglycerate (G3P) and fed into the glycolytic pathway for energy production. Fatty acids are transported into the matrix of mitochondria where they undergo a process called **β-oxidation**. In this catabolic process, a number of enzymes sequentially remove two-carbon acetyl groups from the fatty acids beginning at the carboxyl end of the chains. The acetyl groups are combined with coenzyme A molecules to form acetyl-CoA, which in turn feed into the Krebs cycle.

A 12-carbon fatty acid like lauric acid is cleaved five times to produce six (two-carbon) acetyl-CoA molecules. Every cleavage reaction uses one ATP and produces one NADH and one $FADH_2$. Therefore, five cleavages use five ATP molecules and produce five NADH and five $FADH_2$ molecules. The overall equation for the oxidation of the 12-carbon fatty acid is:

$$\text{Lauric acid} + 5\text{CoA} + 5\text{FAD} + 5\text{NAD}^+ + 5H_2O \longrightarrow 6\text{acetyl-CoA} + 5FADH_2 + 5NADH^+ + 5H^+$$

The reduced coenzymes eventually produce a total of 20 ATP molecules by oxidative phosphorylation (assuming 2.5 ATP per NADH and 1.5 ATP per $FADH_2$). Subtracting the five ATP molecules used in the cleavage steps results in a net yield of 15 ATP molecules. The six acetyl-CoA molecules enter the Krebs cycle and produce six ATP by substrate-level phosphorylation, 18 NADH, and six $FADH_2$ molecules. The reduced coenzymes will produce a total of 54 ATP by oxidative phosphorylation (assuming 2.5 ATP per NADH and 1.5 ATP per $FADH_2$). The total yield is 15 ATP + 54 ATP + 6 ATP = 75 ATP. Two glucose molecules containing an equivalent number of carbon atoms produce only 60 ATP molecules, 20% fewer than the lipid produces. In general, carbohydrates are denser than fats. Thus, fats produce a little more than twice the amount of energy than an equivalent mass of carbohydrate. Carbohydrates supply approximately 16 kJ/g, whereas fats provide approximately 38 kJ/g. The energy yields of glucose and a 12-carbon fatty acid (laurate) are compared in **Table 1**.

Table 1 Energy Yields of Glucose and Laurate

	12 Carbon Atoms from Glucose (two 6-C glucose molecules)	12 Carbon Atoms from Laurate (one 12−C fatty acid molecule)
ATP produced	76 (2 × 38)	97
ATP used	4 (2 × 2)	5
Theoretical net ATP yield	72 (2 × 36)	92
Actual net ATP yield	60 (2 × 30)	75
Energy yield (kJ/g)	16	38

Anaerobic Pathways

Glycolysis allows organisms to obtain energy from nutrients in the absence of oxygen. However, step 6 (G3P to BPB) of the glycolytic pathway reduces NAD^+ to NADH. As mentioned earlier, cells possess a limited supply of NAD^+. If glycolysis continues without a mechanism to oxidize NADH back into NAD^+, step 6 will be blocked and glycolysis will come to a halt. In oxidative respiration, the ETC oxidizes NADH to NAD^+, allowing glycolysis to continue. Organisms have evolved several ways of recycling NAD^+ and

β-oxidation the sequential removal of acetyl groups in the catabolism of fatty acids

allowing glycolysis to continue when oxygen is not available. One method involves transferring the hydrogen atoms of NADH to certain organic molecules instead of the electron transport chain. This process is called **fermentation**. Bacteria have evolved dozens of different forms of fermentation, but eukaryotes primarily use two methods: **ethanol fermentation** and **lactate (lactic acid) fermentation**.

Ethanol Fermentation

In ethanol fermentation, NADH passes its hydrogen atoms to acetaldehyde, a compound formed when a carbon dioxide molecule is removed from pyruvate by the enzyme pyruvate decarboxylase, as shown in **Figure 2**. This forms ethanol, the alcohol used in alcoholic beverages. This process allows NAD^+ to be recycled and glycolysis to continue. The two ATP molecules produced satisfy the organism's energy needs, and the ethanol and carbon dioxide are released as waste products.

Humans have learned ways of making use of these metabolic wastes. Ethanol fermentation carried out by yeast (a variety of single-celled fungi) is of great historical, economic, and cultural importance. Breads and pastries, wine, beer, liquor, and soy sauce are all products of fermentation (**Figure 3**).

Bread is leavened by mixing live yeast cells with starches (in flour) and water. The yeast cells ferment the glucose from the starch and release carbon dioxide and ethanol. Small bubbles of carbon dioxide gas cause the bread to rise (or leaven) and the ethanol evaporates away when the bread is baked. In beer making and winemaking, yeast cells ferment the sugars found in carbohydrate-rich fruit juices, such as grape juice. The mixture bubbles as the yeast cells release carbon dioxide gas and ethanol during fermentation. In winemaking, fermentation ends when the concentration of ethanol reaches approximately 12%. At this point, the yeast cells die as a result of ethanol accumulation and the product is ready to be consumed as a beverage. Flooded plants undergo ethanol fermentation in the roots and may die if oxygen is not returned to the roots. This is why it is important not to overwater houseplants.

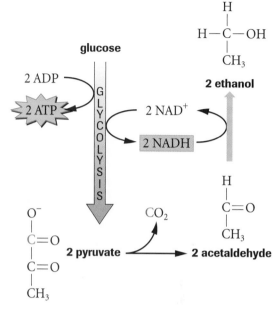

Alcohol fermentation in yeast

Figure 2
Ethanol fermentation creates ethanol and carbon dioxide from glucose. In the process, NADH is oxidized to NAD^+, allowing glycolysis to continue.

fermentation a process in which the hydrogen atoms of NADH are transferred to organic compounds other than an electron transport chain

ethanol fermentation a form of fermentation occurring in yeast in which NADH passes its hydrogen atoms to acetaldehyde, generating carbon dioxide, ethanol, and NAD^+

lactate (lactic acid) fermentation a form of fermentation occurring in animal cells in which NADH transfers its hydrogen atoms to pyruvate, regenerating NAD^+ and lactate

Figure 3
Ethanol fermentation is used in the production of baked goods and products such as wine, beer, and soy sauce.

Take a Stand: Fetal Alcohol Syndrome (FAS)

Fetal alcohol syndrome (FAS) is a serious health problem that develops in some unborn babies when the mother drinks alcohol (ethanol) during pregnancy. The ethanol enters the woman's bloodstream and circulates to the fetus by crossing the placenta. Normal cell development in the brain and other body organs of the fetus is negatively affected. Babies with FAS tend to be born with a lower-than-normal birth weight and body length. They usually have smaller-than-normal heads, deformed facial features, abnormal joints and limbs, coordination problems, learning difficulties, and shorter-than-normal memories. As they mature, people with FAS often have difficulty in school, in caring for themselves and their children, and in recognizing inappropriate sexual behaviour. They also may experience related psychological and social problems.

A safe amount of drinking during pregnancy has not been determined because no one metabolizes alcohol in the same way. For this reason, all major authorities suggest that women should not drink at all during pregnancy. Unfortunately, women sometimes do not reduce or stop drinking until a pregnancy is confirmed. By then, the embryo or fetus may have gone through several weeks of critical development, a period during which exposure to alcohol can be very damaging. However, recent studies suggest that persistent heavy drinking or drinking a large amount of alcohol at any one time may be more dangerous to

the fetus than drinking small amounts more frequently. Heavy drinkers are those who have two or more drinks per day. The Royal College of Obstetricians and Gynecologists recently conducted a large study that included 400 000 women, all of whom had consumed alcohol during pregnancy. Not a single case of FAS occurred and no adverse effects on children were found when consumption was under 8.5 drinks over the course of a week. A recent analysis of seven major medical research studies involving more than 130 000 pregnancies suggests that consuming 2–14 drinks over the course of a week does not increase the risk of giving birth to a child with either malformations or FAS.

Statement

Women should not drink even small amounts of alcoholic beverages while pregnant.

- Consider the information provided and add your own ideas.
- Conduct research to learn more about the issue.

 www.science.nelson.com

- In a group, discuss the ideas.
- Communicate your opinions and arguments in a position paper or in an organized small group discussion.

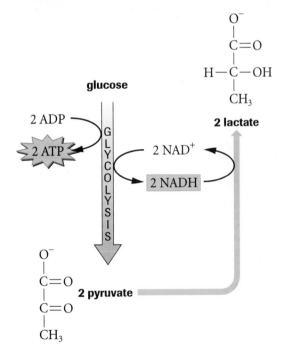

Lactic acid fermentation in muscle cells

Figure 4
Lactate fermentation produces lactic acid from glucose. In the process, NADH is oxidized to NAD$^+$, allowing glycolysis to continue.

Lactate (Lactic Acid) Fermentation

Under normal conditions, animals such as humans catabolize glucose by aerobic respiration. However, during strenuous exercise, muscle cells respire glucose faster than oxygen can be supplied. Under such conditions, oxidative respiration slows down and lactate fermentation, shown in **Figure 4**, begins. In lactate fermentation, NADH produced in glycolysis transfers its hydrogen atoms to pyruvate in the cytoplasm of the cell, regenerating NAD$^+$ and allowing glycolysis to continue. This results in a change of pyruvate into lactate. The accumulation of lactate molecules in muscle tissue causes stiffness, soreness, and fatigue. Lactate is transported through the bloodstream from the muscles to the liver. When vigorous exercise ceases, lactate is oxidized back to pyruvate, which then goes through the Krebs cycle and oxidative phosphorylation. The extra oxygen required to catabolize lactate to CO_2 and H_2O (through the aerobic pathway) is referred to as **oxygen debt**. Panting after bouts of strenuous exercise is the body's way of "paying" the oxygen debt (**Figure 5**).

Exercise Physiology: VO$_2$ max and the Lactate Threshold

Exercise physiology is a branch of biology that deals with the body's biological responses to exercise. Scientists in this field try to answer such questions as "Why do muscles become sore and fatigued after a bout of strenuous exercise? How can athletes train to control fatigue and maximize the amount of oxygen that enters their bloodstream? Why does exercise deplete the body of its water reserves and how can athletes avoid dehydration?" Exercise physiologists search for solutions to practical problems faced by individuals who engage in sports and athletic activities. The most common problem faced by athletes is a shortage of energy. Therefore, particular emphasis is placed on the study of aerobic and anaerobic metabolism and its relationship to cardiopulmonary fitness, also known as aerobic fitness. **Aerobic fitness** is a measure of the ability of the heart, lungs, and bloodstream to supply oxygen to the cells of the body (especially the muscle cells) during physical activity. Aerobic fitness is one of the factors used by physiologists to judge a person's overall physical fitness. Other factors include muscular strength, muscular endurance, flexibility, and body composition (the ratio of fat to bone to muscle).

Since muscle cells need energy from ATP to contract, it is assumed that ATP production (by aerobic respiration) will be increased if more oxygen is absorbed and used by the cells of the body (especially muscle cells) in a given period of time. Exercise physiologists measure a value called the **maximum oxygen consumption, ($VO_{2\ max}$)**, as a measure of a body's capacity to generate the energy required for physical activity. $VO_{2\ max}$ measures the maximum volume of oxygen, in millilitres, that the cells of the body can remove from the bloodstream in one minute per kilogram of body mass while the body experiences maximal exertion. $VO_{2\ max}$ values are typically expressed in mL/kg/min, and are measured directly by a maximal exercise test, also known as a treadmill exercise test. During the test, the person or animal is forced to move faster and faster on a treadmill while expired air is collected and measured by a computer (**Figure 6**). The entire test usually lasts between 10 and 15 min. Needless to say, the test is not pleasant since one must achieve a rather painful state of maximal exertion. Indirect methods of estimating the value of $VO_{2\ max}$ have been developed that require much less physical strain.

Figure 5
Marathon runners are fatigued after a race because of the accumulation of lactate in their muscles. Panting provides the oxygen needed to respire the excess lactate.

oxygen debt the extra oxygen required to catabolize lactate to CO_2 and H_2O

aerobic fitness a measure of the ability of the heart, lungs, and bloodstream to supply oxygen to the cells of the body during physical exercise

maximum oxygen consumption, $VO_{2\ max}$ the maximum volume of oxygen, in millilitres, that the cells of the body can remove from the bloodstream in one minute per kilogram of body mass while the body experiences maximal exertion

Figure 6
A maximal exertion test being conducted in a human performance lab. The apparatus in this photo is used to make precise measurements of $VO_{2\ max}$.

DID YOU KNOW ?

Death and Rigor Mortis

Cellular death occurs when cellular respiration and metabolism cease. When all cells are dead, the body is dead. However, when an organism dies, all cells of the body do not die simultaneously. In many countries a person is considered legally dead only when the cells of the brain stem no longer function. There are two things that happen soon after death: one is a gradual drop in body temperature, and the other is stiffening of the muscles, known as rigor mortis. Rigor mortis is not caused by the drop in body temperature, but by the fermentation of glucose in muscle cells, leading to high levels of lactic acid. The lactic acid causes muscle tissue to become rigid. Rigor mortis sets in much sooner if death occurs immediately following strenuous activity, such as running.

ACTIVITY 2.3.1

Estimating VO$_{2 \text{ max}}$ (p. 131)

A maximal exertion test involves a lot of effort on the part of the patient and the use of expensive, high-tech equipment. Activity 2.3.1 will allow you to estimate your VO$_{2 \text{ max}}$ using a method that requires little equipment and a 1.6-km walk.

DID YOU KNOW ?

Record Values for VO$_{2 \text{ max}}$

The highest VO$_{2 \text{ max}}$ value recorded for a male is from a champion Norwegian cross-country skier who had a VO$_{2 \text{ max}}$ of 94 mL/kg/min. The highest value recorded for a female is 74 mL/kg/min from a Russian cross-country skier. In contrast, poorly conditioned adults may have values below 20mL/kg/min!

In general, individuals with higher VO$_{2 \text{ max}}$ values may be considered more aerobically fit than individuals with lower values. **Tables 2** and **3** show the range of VO$_{2 \text{ max}}$ values in females and in males, according to age.

Table 2 VO$_{2 \text{ max}}$ for Females (mL/kg/min)

Age	Very poor	Poor	Fair	Good	Excellent	Superior
13–19	< 25.0	25.0–30.9	31.0–34.9	35.0–38.9	39.0–41.9	> 41.9
20–29	< 23.6	23.6–28.9	29.0–32.9	33.0–36.9	37.0–41.0	> 41.0
30–39	< 22.8	22.8–26.9	27.0–31.4	31.5–35.6	35.7–40.0	> 40.0
40–49	< 21.0	21.0–24.4	24.5–28.9	29.0–32.8	32.9–36.9	> 36.9
50–59	< 20.2	20.2–22.7	22.8–26.9	27.0–31.4	31.5–35.7	> 35.7
60+	< 17.5	17.5–20.1	20.2–24.4	34.5–30.2	30.3–31.4	> 34.1

Table 3 VO$_{2 \text{ max}}$ for Males (mL/kg/min)

Age	Very poor	Poor	Fair	Good	Excellent	Superior
13–19	< 35.0	35.0–38.3	38.4–45.1	45.2–50.9	51.0–55.9	> 55.9
20–29	< 33.0	33.0–36.4	36.5–42.4	42.5–46.4	46.5–52.4	> 52.4
30–39	< 31.5	31.5–35.4	35.5–40.9	41.0–44.9	45.0–49.4	> 49.4
40–49	< 30.2	30.2–33.5	33.6–38.9	39.0–43.7	43.8–48.0	> 48.0
50–59	< 26.1	26.1–30.9	31.0–35.7	35.8–40.9	41.0–45.3	> 45.3
60+	< 20.5	20.5–26.0	26.1–32.2	32.3–36.4	36.5–44.2	> 44.2

VO$_{2 \text{ max}}$ values vary between 20 mL/kg/min and 90 mL/kg/min. The average value for a typical North American is about 35 mL/kg/min, while elite endurance athletes reach values of 70 mL/kg/min. **Figure 7** illustrates average VO$_{2 \text{ max}}$ values for the athletes of various sports.

VO$_{2 \text{ max}}$ values may be increased with exercise and training, but genetic variation helps to explain why everyone cannot train to be an elite athlete. Exercising harder, more frequently, and for longer durations will increase VO$_{2 \text{ max}}$ values to a degree. However, VO$_{2 \text{ max}}$ values also decrease with age. In any case, there is not always a direct correla-

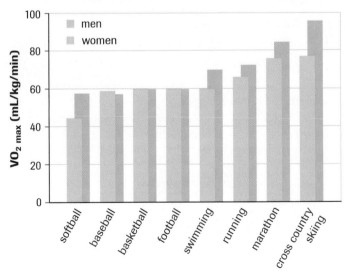

Maximal Oxygen Uptake Values (VO$_{2 \text{ max}}$) for Popular Sports

Figure 7
VO$_{2 \text{ max}}$ values for athletes in popular sports

tion between $VO_{2\ max}$ values and overall athletic performance. Although it is true that elite athletes have $VO_{2\ max}$ values that are higher than the population mean, factors such as mental attitude, running efficiency, and the amount of lactate produced during exercise greatly influence overall performance.

Since oxygen cannot reach all the body's mitochondria all the time, lactate fermentation occurs continuously as you exercise. However, as exercise intensity increases, lactate production increases. The lactate threshold (LT) is the value of exercise intensity at which blood lactate concentration begins to increase sharply (**Figure 8**). Exercising at or below this intensity may be sustainable for hours, but exercising beyond the lactate threshold may limit the duration of the exercise because of increased pain, muscle stiffness, and fatigue.

In general, athletic training improves blood circulation and increases the efficiency of oxygen delivery to the cells of the body. The result is a decrease in lactate production at any given exercise intensity level and an increase in the lactate threshold. With a higher lactate threshold, the person will be able to sustain greater exercise intensities and improved athletic performance. One measure of performance is the percentage of $VO_{2\ max}$ at which the LT is reached. Untrained individuals usually reach the LT at about 60% of $VO_{2\ max}$. Elite endurance athletes typically reach their lactate thresholds at or above 80% of $VO_{2\ max}$.

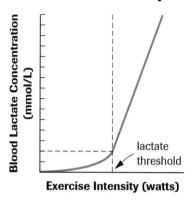

Blood Lactate Concentration vs Exercise Intensity

Blood Lactate Concentration (mmol/L)

lactate threshold

Exercise Intensity (watts)

Figure 8
The lactate threshold

SUMMARY *Related Pathways*

- Lipids, proteins, and nucleic acids are sometimes catabolized for their energy content. Proteins undergo deamination, a process in which the amino group is removed from amino acids, converted into ammonia, and excreted. In β-oxidation, fatty acids are sequentially degraded into two-carbon acetyl portions that are converted into acetyl-CoA and respired through the Krebs cycle, electron transport, and chemiosmosis.

- When oxygen is not available, eukaryotes still carry out glycolysis by transferring the hydrogen atoms in NADH to pyruvate or acetaldehyde. The NAD^+ molecules formed allow glycolysis to continue. Lactate fermentation forms lactic acid. In ethanol fermentation, a molecule of CO_2 is removed from pyruvate, forming a molecule of acetaldehyde. The acetaldehyde is converted to ethanol by attaching hydrogen from NADH.

- Lactate fermentation occurs in animal muscle cells during strenuous exercise. Ethanol fermentation occurs in yeast cells and is used in wine, beer, and bread making.

- The maximum oxygen uptake, or $VO_{2\ max}$, is the maximum volume of oxygen that the cells of the body can remove from the bloodstream in one minute per kilogram of body mass while the body experiences maximal exertion. The lactate threshold (LT) is the value of exercise intensity at which blood lactate concentration begins to increase sharply.

Understanding Concepts

1. What happens to acetyl-CoA if a cell already has sufficient quantities of ATP?

2. List two differences between aerobic respiration and fermentation.

3. A student regularly runs 3 km each afternoon at a slow, leisurely pace. One day, she runs 1 km as fast as she can. Afterward, she is winded and feels pain in her chest and leg muscles. What is responsible for her symptoms?

4. Name a nonalcoholic final product of alcoholic fermentation, other than ATP.

5. In addition to ATP, name the other final products of both types of fermentation.

6. (a) How many molecules of ethanol are produced by the fermentation of one molecule of glucose?
   (b) How many molecules of carbon dioxide are produced during the fermentation of one molecule of glucose?
   (c) How much oxygen is used during the fermentation of one glucose molecule?

7. Name an organism in which alcoholic fermentation takes place.

8. (a) What happens to lactic acid after it is formed in a muscle cell?
   (b) Explain oxygen debt.

9. How does a human feel the presence of lactic acid in the tissues of the body?

10. Define *maximum oxygen consumption, $VO_{2\ max}$*.

11. How do $VO_{2\ max}$ values vary with age? How do they vary with increases in body mass? Explain.

12. Why are $VO_{2\ max}$ values not perfectly correlated with overall athletic performance?

13. (a) Determine the value of the lactate threshold from **Figure 9**.
    (b) What does this value mean?

Making Connections

14. When Henry Ford built the first Model T in 1908, he expected it to run on pure ethanol produced by fermenting corn. From 1920 to 1924, the Standard Oil Company in Baltimore produced and sold a mixture of ethanol and gasoline called gasohol. However, high corn prices and transportation difficulties terminated the project.
    (a) Research gasohol on the Internet or at the library. List three advantages and three disadvantages of gasohol production in Canada.
    (b) Comment on the viability of a gasohol industry in Canada.

 www.science.nelson.com

15. Marathon runners have discovered that taking walk breaks during a race may get them to the finish line faster than running all the way.
    (a) Using concepts discussed in this section, explain why walking breaks may be beneficial to a marathon runner.
    (b) What is "carbohydrate loading"? How is it done? Why do some atheletes practise carbohydrate loading? List some of the adverse effects of carbohydrate loading.

 www.science.nelson.com

16. Alcoholic beverages, such as wine and beer, have been produced by humans since the earliest days of agriculture. How do you suppose the process of fermentation was first discovered?

17. Conduct library and/or Internet research to answer the following questions:
    (a) How do long-distance runners make use of the lactate threshold in their training?
    (b) What is blood doping? What are the perceived metabolic benefits of this practice? What are some of the dangers associated with blood doping?

 www.science.nelson.com

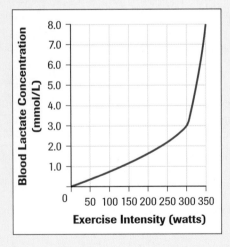

Figure 9

Molecular Biologist

A molecular biologist must have at least a four-year bachelor of science degree in molecular biology to work in this field, and a masters or doctorate to conduct high-level research. A strong high-school background in biology is essential to be admitted to these programs. Molecular biologists study plants, animals, and humans to understand the basis of disease. They use highly sophisticated equipment that separates molecules electrically so that they can probe the basis of disease at the molecular or genetic level. Molecular biologists carry out research in university laboratories, government research centres, and in private research facilities.

Biochemist

Candidates for four-year bachelor of science programs specializing in biochemistry must have a strong background in biology, chemistry, and math. Biochemists work in all areas of the pharmaceutical industry, from developing products like vaccines, to quality control and sales and marketing. Biochemists develop products that involves organic molecules using spectrometers, amino acid analyzers, and computers. These scientists work extensively in both private and public sectors. Biochemists are also highly prized as high-school science teachers because of their knowledge of both chemistry and biology.

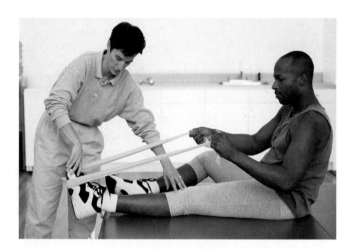

Kinesiologist

Kinesiologists work in the field of human movement, helping people to rehabilitate from physical injuries. To be accepted into a kinesiology program, which is a four-year bachelor of science program specializing in this area, candidates must possess high marks in biology, chemistry, math, and physics. After graduation, they must become a member of the Ontario Kinesiologist Association. Kinesiologists work in a variety of settings, including hospitals, clinics, fitness centres, and crown corporations such as the Workplace Safety and Insurance Board. Some large corporations, such as General Motors, hire kinesiologists to advise on ways to improve workers' safety and efficiency on assembly lines.

▶ Practice

Making Connections

1. Identify several careers that require knowledge about metabolic processes. Select a career that interests you from your list. What university programs lead to that career? Which universities offer degree programs in this area? Which high-school subjects are required to enter the program?

2. How are the concepts you have learned in this unit used in the career?

3. Investigate and describe responsibilities and duties involved in this career. What appeals to you about the career? What do you find less attractive?

4. Survey the newspaper or conduct an Internet search to identify career opportunities in this area.

 www.science.nelson.com

INVESTIGATION 2.1.1

Oxygen Consumption in Germinating and Nongerminating Pea Seeds

Inquiry Skills

- ○ Questioning
- ● Planning
- ● Analyzing
- ○ Hypothesizing
- ● Conducting
- ● Evaluating
- ● Predicting
- ● Recording
- ● Communicating

Aerobic cellular respiration involves the release of energy from organic compounds by a series of enzyme-catalyzed chemical reactions in the mitochondria of each cell. The following equation summarizes the process:

$$C_6H_{12}O_6 + 6O_2 \longrightarrow 6\,CO_2 + 6\,H_2O + 2870\text{ kJ}$$

The overall rate of cellular respiration could be determined by measuring the utilization of glucose, the consumption of O_2 gas, the production of CO_2 gas, or the release of energy.

Before they germinate, pea seeds contain embryos that are alive, but dormant. When the necessary conditions are met, germination occurs. During germination, the embryo grows and cells begin to divide rapidly. In this investigation, the relative volume of O_2 consumed by germinating and nongerminating (dry) peas will be measured. You will use an apparatus called a respirometer to measure oxygen production (**Figure 1**). In addition, you will measure the rate of respiration of peas at two different temperatures. The setup and part of the procedure is provided, but you will be required to design the procedure for using the respirometer to investigate how cellular respiration is affected by different temperatures and how it is affected by using dry pea seeds.

It is important to remember that gases and other fluids flow from regions of high pressure to regions of lower pressure. The CO_2 produced during cellular respiration will be removed by reaction with potassium hydroxide, $KOH_{(s)}$, to form solid potassium carbonate, $K_2CO_{3(s)}$, according to the following equation:

$$CO_{2(g)} + 2\,KOH_{(s)} \longrightarrow K_2CO_{3(s)} + H_2O_{(l)}$$

Cellular respiration uses oxygen gas and produces carbon dioxide gas. Since carbon dioxide is being removed, any change in the volume of the gas in the respirometer will be directly related to the amount of oxygen consumed. In the experimental apparatus, if water temperature and volume remain constant, the water will move toward the region of lower pressure. During respiration, oxygen will be consumed. Its volume will be reduced because the carbon dioxide produced is being converted to a solid. The net result is a decrease in gas volume

within the tube, and a related decrease in pressure in the tube. A respirometer containing glass beads alone will be used to detect any changes in volume due to atmospheric pressure changes or temperature changes. The amount of oxygen consumed will be measured over a period of time.

Question

How much oxygen is consumed by germinating and nongerminating pea seeds?

Prediction

(a) Predict whether germinating or nongerminating pea seeds will consume more oxygen in a given period of time and predict the effect a change in temperature will have on the rate of oxygen consumption.

Materials

safety goggles	2 two-hole test-tube
laboratory apron	stoppers
pea seeds	tape
water	2 millimetre rulers
paper towels	petroleum jelly
2 large test tubes	2 pieces of rubber tubing
nonabsorbent cotton	2 pinch clamps
laboratory scoop or forceps	2 utility clamps
potassium hydroxide, KOH,	utility stand
pellets	medicine dropper
2 straight glass tubes	food colouring
2 bent glass tubes	

Procedure

1. Place some pea seeds in water 24 h before the investigation. Keep some of the pea seeds dry. The wet seeds should be kept in damp paper towels until they germinate. Check them daily to make sure they do not become mouldy.

2. Place 10 seeds in a test tube and place a layer of cotton above the seeds. Using forceps or a scoop, add approximately 30 KOH pellets to the top of the cotton.

INVESTIGATION 2.1.1 *continued*

✋ **KOH is strongly alkaline. Avoid any contact with your skin. Wash with cold water for several minutes if you get any KOH on your skin. KOH could cause blindness if it goes in the eyes. If it comes in contact with the eyes, wash with water for 15 min, and seek medical help immediately.**

3. Using straight and bent glass tubing, and the two-hole test-tube stopper, set up a respirometer, as shown in **Figure 1**.

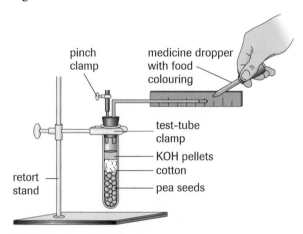

Figure 1
Respirometer

4. Use tape to attach a millimeter ruler to the end of the glass tubing, and seal all stopper openings with petroleum jelly.

5. With a test-tube clamp, attach the respirometer to a utility stand. Attach a pinch clamp to the rubber tubing.

6. Use a medicine dropper to add a few drops of food colouring to the bent glass tubing. Close the pinch clamp.

(b) In which direction does the food colouring begin to move?

(c) Why is the ruler used?

(d) Why is KOH needed?

7. Use the respirometer to investigate how cellular respiration is affected by different temperatures and by using germinating versus nongerminating pea seeds.

(e) Identify the independent and dependent variables.

(f) Suggest appropriate controls for your experiment.

(g) Prepare a complete list of materials that you will need.

(h) Write a procedure for your experiments.

8. Submit your procedure and materials list to your teacher for approval and then carry out the experiments.

Analysis

(i) Explain how the respirometer works.

(j) Why does the test tube have to be completely sealed around the stopper?

(k) Explain why water moves into the bent tube of the respirometer.

(l) A student places a respirometer in cold water and notes that the droplet of food colouring in the glass tube moves toward the respirometer. Another respirometer is placed in warm water and the student notes that the food colouring moves away from the respirometer. Why does the food colouring move in opposite directions? Explain why allowing the respirometer to stabilize before closing the pinch clamp would avoid this problem.

(m) Draw a graph of the dye movement by plotting the distance along the y-axis and the time along the x-axis.

(n) Draw conclusions from your experimental results. Make sure you support your conclusions with observations.

Evaluation

(o) Evaluate your predictions on the basis of your observations.

(p) Identify sources of error and suggest improvements to the experimental design.

Energy Consumption During Exercise

Inquiry Skills

○ Questioning   ○ Planning   ● Analyzing
○ Hypothesizing   ● Conducting   ● Evaluating
● Predicting   ● Recording   ● Communicating

In this investigation, you will compare your power during aerobic and anaerobic exercise and determine the amount of glucose that you consume.

Question

How does the power of aerobic exercise compare to the power of anaerobic exercise, and how much glucose does exercise consume?

Prediction

(a) Predict how many times more powerful aerobic exercise is than anaerobic exercise.

(b) Predict how many teaspoons of table sugar you will "burn" exercising your arm to exhaustion.

Materials

dumbbell or mass (2.5 kg or 80% of the heaviest mass you can lift once)
metre stick

Procedure

1. Place the mass or dumbbell on a desk.

2. Grip the mass or dumbbell with your hand and place your arm flat on the desk with arm, hand, and dumbbell resting on the desktop.

3. Place a metre stick on the edge of the desk so that there is no movement. Have a partner hold a metre stick securely on the desk (**Figure 1**).

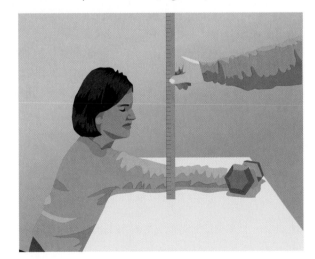

Figure 1
Setting up the apparatus

4. Lift the dumbbell by bending your arm at the elbow one time so that your partner can measure the vertical distance the dumbbell covers when it reaches its highest point. Measure the lowest point on one edge of the dumbbell before you lift it and the same edge of the dumbbell after the lift (**Figure 2**).

Figure 2

5. While you perform step 6, your partner will
   - note the time elapsed at the end of Stage 1 (Aerobic Exercise; exercise in which no muscular discomfort is felt) and again at the end of Stage 2 (Anaerobic Exercise; exercise in which you feel muscle soreness)
   - count the number of lifts in each stage
   - make sure you cover the full distance on each lift of the dumbbell

6. Lift the dumbbell repeatedly and quickly until your arm feels sore. Tell your partner to note the end of Stage 1 (Aerobic Exercise). Continue to lift the dumbbell repeatedly and quickly until you can no longer lift it anymore. Tell your partner to note the end of Stage 2.

INVESTIGATION 2.2.1 *continued*

Observations

Complete the following measurements and calculations.

(c) Dumbbell mass = _____ kg

(d) Vertical distance dumbbell moves = _____ m

(e) Time elapsed by the end of Stage 1 = _____ s
= Stage 1 total time

(f) Number of lifts in Stage 1 = _____

(g) Total time elapsed by the end of Stage 2 =
_____ s (subtract Stage 1 time from total
time to determine Stage 2 time, _____ s)

(h) Number of lifts in Stage 2 = _____

Calculations

1. To lift the dumbbell, you must overcome the gravitational force on the dumbbell. The gravitational force on the dumbbell equals the dumbbell's mass multiplied by the acceleration due to gravity, 9.81 m/s^2.

   Force needed to lift the dumbbell
   = mass of the dumbbell $\times$ 9.81 m/s^2 (F = ma)

(i) Force = ____ kg $\times$ 9.81 m/s^2
= _____ kg m/s^2

2. Each time you lift the dumbbell, you must give it gravitational potential energy (E_g). This energy equals *mgh* (mass $\times$ acceleration due to gravity $\times$ vertical distance).

(j) E_g = _____ kg $\times$ 9.81 m/s^2 $\times$ _____ m
= _____ J

   We will assume that each time you lower the dumbbell, gravity does an equal but opposite amount of work on the dumbbell.

3. You can determine the amount of work done per lift, because work is equal to the change in potential energy.

(k) Work = ΔE_g
= _____ J / lift

4. To find the total work done by your arm, multiply the amount of work needed to raise the dumbbell once by the number of times the dumbbell was lifted in each stage.

(l) Total work done during Stage 1
= _____ J / lift $\times$ ____ lifts
= _____ J

(m) Total work done during Stage 2
= _____ J / lift $\times$ ____ lifts
= _____ J

5. Power indicates the rate at which you do work. It is determined by dividing total work by total time ($P = W/t$). One watt is equivalent to one joule per second. You will determine your power during aerobic exercise (Stage 1) and anaerobic exercise (Stage 2).

(n) Power during Stage 1 = _____ J $\div$ _____ s
= _____ W

(o) Power during Stage 2 = _____ J $\div$ _____ s
= _____ W

6. In cellular respiration, one mole of ATP allows a muscle to perform 2870 kJ of work

(p) Moles ATP used during exercise (Stages 1 and 2)
= _____ J total work $\times$ (1 kJ / 1000 J) $\times$
(1 mol ATP / 2870 kJ)
= _____ mol ATP

7. On average, cellular respiration produces 30 moles of ATP molecules per mole of glucose respired.

(q) moles of glucose used
= _____ mol ATP $\times$ 1 mol glucose/30 mol ATP
= _____ mol glucose

8. Each mole of sucrose (table sugar) is approximately equivalent to 2 moles of glucose. Determine the number of moles of sucrose used:

(r) _____ mol glucose used $\times$ (1 mol sucrose / 2 mol glucose)
= _____ mol sucrose used

9. The chemical formula of sucrose is $C_{12}H_{22}O_{11}$. The molar mass of sucrose is 342 g/ mol. Determine the number of grams of sucrose used:

(s) _____ mol sucrose used $\times$ 342 g/mol
= _____ g sucrose

10. There are approximately 4 g of sucrose in a teaspoon of table sugar. Determine the number of teaspoons of sugar you may have used for the total exercise performed.

(t) _____ g sucrose $\times$ (1 teaspoon sugar / 4 g sucrose)
= _____ teaspoons of table sugar

Analysis

(u) After lifting the dumbbell, did you feel hot? Explain your answer using the concepts of cellular respiration in this chapter.

(v) Could you tell when most of your muscles went into anaerobic respiration? What evidence was there for this?

Evaluation

(w) How many times more powerful were you during aerobic exercise than during anaerobic exercise? How did this compare with your predictions?

(x) List sources of error and suggest some improvements to the experimental design.

ACTIVITY 2.3.1

Estimating VO₂ max

The Rockport Fitness Walking Test is a standard test used to estimate the value of VO₂ max. In this test, you will walk a distance of 1.6 km on level ground as quickly as possible, without running. You will measure the time it takes for you to complete the walk and your heart rate after the walk. These measurements, and your age, gender, and mass, are substituted into an equation that yields an estimate of your VO₂ max. The variables, symbols, and units are presented in **Table 1**.

Table 1

Variable	Symbol	Unit
age	a	years
gender	g	male or female
mass	m	kg
walk time	t	min
heart rate	r	beats per min

Table 2

m (kg)	a (years)	g (male = 1, female = 0)	t (min)	r (beats/min)

Materials

bathroom scale
digital stopwatch
calculator

Procedure

1. Copy **Table 2** into your notebook. Use the bathroom scale to measure your mass in kilograms and record the value in the m column in the table.

2. Walk a measured 1.6-km distance as fast as you can with a partner. Have your partner measure the time it takes for you to complete the walk (in seconds) using a digital stopwatch. Your partner should walk with you, carrying the stopwatch, so that he or she can measure your heart rate at the end of the walk. Record your walk time in column **t** in **Table 2**.

3. Have your partner measure your heart rate at the end of the walk by taking your radial pulse for 1 min (the pulse rate must be taken within 15 s of completing the walk). Record your heart rate in column **r** in **Table 2**.

4. If you are a female, enter the number 0 in column **g**, and if you are a male, record the number 1 in this column.

5. Record your age (in years) in column **a**. You may use a fractional value.

Analysis

(a) Substitute the values of the five variables (in the chart) into the following equation to calculate your estimated VO₂ max:

$VO_{2\ max}$ (mL/kg/min)
$= 132.853 - 0.1696\ (m) - 0.3877\ (a) + 6.3150\ (g) - 3.2649\ (t) - 0.1565\ (r)$

Evaluation

(b) Compare your estimated VO₂ max to the values in **Tables 2** and **3** in section 2.3 (p. 122). Interpret your findings.

(c) Explain why this activity only provides an estimate of your VO₂ max. List some of the sources of error and provide suggestions for reducing the size of the error.

(d) Why do you think males are given a value of 1 and females given a value of 0 for the gender factor in the VO₂ max equation?

Key Expectations

- Design and carry out an experiment related to a cell process, controlling the major variables and adapting or extending procedures where required. (2.1)

- Compare matter and energy transformations associated with the processes of aerobic and anaerobic cellular respiration. (2.1, 2.2, 2.3)

- Describe how such molecules as glucose, ATP, pyruvic acid, NADH, and oxygen function within energy transformations in the cell, and explain the roles of such cell components as mitochondria and enzymes in the processes of cellular respiration. (2.1, 2.2, 2.3)

- Describe the chemical structure, mechanisms, and dynamics of enzymes in cellular metabolism. (2.1, 2.2, 2.3)

- Explain the relevance, in your personal lives and in the community, of the study of cell biology and related technologies. (2.1, 2.2, 2.3)

- Formulate operational definitions of the terms related to metabolic processes. (2.1, 2.2, 2.3)

- Investigate and explain the relationship between metabolism and the structure of biomolecules, using problem-solving techniques. (2.1, 2.2, 2.3)

- Interpret qualitative and quantitative observations, gathered through investigation, of the products of cellular respiration. (2.2)

- Relate knowledge gained from current studies of metabolism to learning in the fields of chemical thermodynamics and physical energy. (2.2)

Key Terms

aerobic cellular respiration

aerobic fitness

basal metabolic rate (BMR)

β-oxidation

chemiosmosis

chemoautotrophs

cristae

deamination

electrochemical gradient

electron transport chain (ETC)

ethanol fermentation

eukaryotic cells

facultative anaerobes

fermentation

glycolysis

heterotrophs

intermembrane space

Krebs cycle

lactate (lactic acid) fermentation

matrix

maximum oxygen consumption, VO_2 max

metabolic rate

mitochondria

nicotinamide adenine dinucleotide, NAD^+

nomograms

obligate aerobes

obligate anaerobes

oxidative phosphorylation

oxygen debt

photoautotrophs

prokaryotic cells

proton-motive force (PMF)

substrate-level phosphorylation

▶ *MAKE* a summary

Draw a large, well-labelled poster summarizing the four stages of cellular respiration on a letter-size sheet of paper. Have the area of the sheet represent the cytoplasm of an animal cell. Draw a very large mitochondrion covering at least one half of the area. Add coloured cartoons representing each stage of the process and place them in their respective locations. Use arrows to indicate the movement of intermediate molecules. Show the ATP yield from each stage and the overall ATP yield from the entire process.

In your notebook, indicate whether each statement is true or false. Rewrite a false statement to make it true.

1. During cellular respiration, the oxygen we inhale ends up in the carbon dioxide we exhale.

2. Overall, glycolysis is an endergonic process.

3. Electrons combine with oxygen and protons at the end of the electron transport chain.

4. The following reaction takes place in animal cells but not in plant cells:

$$C_6H_{12}O_6 + 6O_2 \longrightarrow 6CO_2 + 6H_2O$$

5. Pyruvate is a three-carbon compound that is converted into acetyl-CoA in the mitochondrial matrix before entering the Krebs cycle.

6. NADH is reduced by the first protein complex of the electron transport chain.

7. If a poison blocks the flow of protons through ATP synthase, the Krebs cycle will continue, but glycolysis will not.

8. Only the amino group of amino acids can be used for energy.

9. In oxidative phosphorylation, 2 NADH and 1 $FADH_2$ theoretically produce 8 ATP.

10. Fermentation oxidizes NADH to NAD^+ so that glycolysis can continue.

11. Maximum oxygen consumption, $VO_{2\,max}$, increases as people grow older.

In your notebook, record the letter of the choice that best answers the question.

12. The complete oxidation of glucose to carbon dioxide and water has a ΔG of -2870 kJ/mol glucose. Aerobic respiration captures about 930 kJ of this energy. What happens to the rest of the energy?
    (a) It is used for the chemiosmotic pumping of protons.
    (b) It is lost as heat and an increase in entropy.
    (c) It is used in substrate-level phosphorylation.
    (d) It is used to synthesize NADH.
    (e) It is used in active transport.

13. In glycolysis, glucose must first be activated. Which of the following does this require?
    (a) two molecules of O_2
    (b) two molecules of H_2O
    (c) two molecules of ATP
    (d) two molecules of NAD
    (e) two molecules of ADP

14. Why is anaerobic respiration less efficient than aerobic respiration?
    (a) Fewer molecules of ATP are produced per glucose processed.
    (b) Glucose is not completely oxidized to CO_2 and H_2O.
    (c) The potential energy of NADH is not transferred to ATP.
    (d) The electron transport system is not used.
    (e) all of the above

15. Which of the following is a product of lactate fermentation?
    (a) acetaldehyde
    (b) NAD^+
    (c) ethanol
    (d) carbon dioxide
    (e) FADH

16. When the supply of acetyl-CoA exceeds the demand for energy, some of the acetyl-CoA can be diverted toward the synthesis of which one of the following?
    (a) ethanol   (d) FADH
    (b) fatty acids   (e) NADH
    (c) ATP

17. Cyanide poisoning occurs because cyanide binds cytochrome oxidase, preventing it from transferring electrons to the final electron acceptor of the mitochondrial electron transport chain. What is that acceptor molecule?
    (a) O_2 (d) CO_2
    (b) ATP   (e) H_2O
    (c) NAD^+

18. If the ATP level of a cell becomes too high, ATP can bind to phosphofructokinase and block the formation of fructose 1,6-bisphosphate. This is an example of which of the following?
    (a) chemiosmosis
    (b) substrate-level phosphorylation
    (c) a reduction reaction
    (d) feedback inhibition
    (e) enzyme activation

19. Which of the following is associated with increased levels of aerobic fitness?
    (a) a low $VO_{2\,max}$
    (b) a high $VO_{2\,max}$
    (c) a low lactate threshold
    (d) a low $VO_{2\,max}$ and a low lactate threshold
    (e) a high $VO_{2\,max}$ and a low lactate threshold

Understanding Concepts

1. Draw a concept map, linking the terms with arrows and a connecting word or short phrase: *glycolysis, aerobic respiration, anaerobic respiration, Krebs cycle, electron transport chain, fermentation, glucose, lipids, proteins, ATP, plant cells, animal cells, mitochondrial matrix, intermembrane space, chemiosmosis, NADH, NAD$^+$, H$^+$ ions, lactic acid, ethanol.*

2. Name a reaction or pathway of cellular respiration that occurs in each of the following locations:
   (a) cytoplasm
   (b) mitochondrial matrix
   (c) inner mitochondrial membrane

3. Describe how each of the following compounds participates in energy metabolism:
   (a) ubiquinone (Q)
   (b) FADH$_2$
   (c) pyruvic acid
   (d) ATP synthase

4. (a) How many ATP molecules are produced for each NADH molecule and each FADH$_2$ that enters the electron transport system?
   (b) Why do electrons carried by NADH result in the synthesis of more ATP molecules than those carried by FADH$_2$?

5. (a) What is the net gain in ATP when one glucose molecule undergoes aerobic cellular respiration?
   (b) What is the net gain in ATP when one glucose molecule undergoes alcoholic fermentation?

6. Draw a labelled diagram of a mitochondrion, identifying key structures and indicating the position of the H$^+$ reservoir.

7. Name the four stages of aerobic cellular respiration, indicating where each stage occurs in a cell.

8. What happens to NAD$^+$ during glycolysis and the Krebs cycle?

9. Why is FADH$_2$, instead of NADH, formed in one step of the Krebs cycle?

10. (a) What is an electrochemical gradient?
    (b) How is an electrochemical gradient created during electron transport in cellular respiration?
    (c) How is the energy of the electrochemical gradient used to power ATP synthesis?

11. Oxygen is toxic or unavailable to some cells. How do yeast cells produce ATP from glucose?

12. Compare the oxidized form of NAD with its reduced form.

13. Under what conditions do muscle cells produce lactate from pyruvate?

14. How many carbon atoms are present in one ethanol molecule? Compare this figure with that of pyruvate. If there is a difference, explain why.

15. Where in a cell is lactic acid produced?

16. How many acetyl-CoA molecules are produced from one glucose molecule during aerobic respiration?

17. (a) In what part of the mitochondrion does the Krebs cycle occur?
    (b) Where in a mitochondrion are the enzymes for the electron transport chain located?

18. What happens to an animal cell that cannot respire or ferment glucose?

19. Approximately what percentage of the energy stored in glucose is captured and stored in ATP during aerobic respiration?

20. Identify the labelled components of the mitochondrion on the electron micrograph in **Figure 1**.

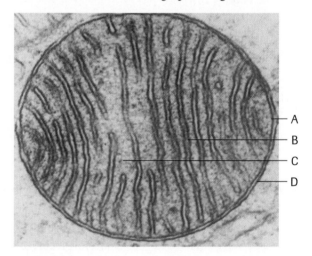

Figure 1

21. Dinitrophenol, DNP, was prescribed in low doses in the 1940s to help obese patients lose weight. Its use for this purpose was discontinued when several users died. DNP makes the inner membrane of the mitochondrion leaky to protons. Explain why this drug would eventually kill a person.

22. During a heart attack, blood flowing to the heart muscle is interrupted by blockage of a coronary artery. How would you expect the metabolism in the heart to change?

23. Explain why it is essential that muscle cells convert pyruvate into lactic acid during strenuous exercise,

even though the cell obtains very little energy in this process and the accumulation of lactic acid causes muscle fatigue and pain.

24. Contrast basal metabolic rate, BMR, and maximum oxygen consumption, $VO_{2\ max}$.

25. Why will a person with a higher $VO_{2\ max}$ feel less tired than a person with a lower $VO_{2\ max}$ (after doing the same amount of physical exercise)?

Applying Inquiry Skills

26. A geneticist gives you two test tubes containing two types of yeast cells that are the same in every way except that one can carry out only aerobic respiration and the other can carry out only anaerobic respiration. The tubes are labelled A and B, and they look the same. Yeast from tube A grows rapidly, whereas yeast from tube B grows slowly. Which tube contains cells capable of performing only aerobic respiration? How did you make your choice? Devise an experiment to verify your results.

27. A newborn is diagnosed with severe neurological abnormalities. Urinalysis reveals abnormally large amounts of α-ketogutarate in the urine. Propose a hypothesis to explain the test result

Making Connections

28. The soybean is the seed of the soybean plant. Soy foods have been a staple part of the Chinese diet for over 4000 years. Research soybean food products on the Internet.
    (a) Name five different commercially available soybean food products.
    (b) Which of the soy-based foods are produced using a fermentation process?
    (c) Name the microorganisms used in the fermentation processes you described in (b).
    (d) Outline the process for making one fermented soybean product.

 www.science.nelson.com

29. *Lactobacillus* bacteria are used to sour milk in the production of yogurt.
    (a) What form of anaerobic respiration do *Lactobacillus* bacteria undergo?
    (b) Why does this form of metabolism increase the shelf life of foods?

Extension

30. (a) A triacylglycerol molecule contains three molecules of palmitic acid (palmitate). Assuming complete catabolism, calculate the actual number of ATPs produced by β-oxidation of the three molecules of palmitic acid (palmitate).
    (b) Calculate the theoretical energy yield from the β-oxidation of three moles of palmitate.
    (c) How many grams of palmitate would have to be consumed to obtain the amount of energy calculated in (b)? (assume 100% metals)
    (d) Use the data in **Table 2** on page 112 to calculate the number of minutes a woman would have to ride a bicycle (at 20 km/h) to expend the energy gained by metabolizing three moles of palmitate.

31. Drinking alcoholic beverages is legal for individuals in many countries once they reach a certain age. Laws restricting the sale and consumption of alcoholic beverages vary from one jurisdiction to another.
    (a) Perform Internet or library research to determine the legal drinking age in various jurisdictions in Canada, the United States, Britain, and one other European country.
    (b) Attempt to determine why these jurisdictions have chosen these particular ages.
    (c) Research the toxicity of ethanol in alcoholic beverages, such as beer, wine, and whisky.
    (d) Write a short position statement on the proposition that human consumption of alcoholic beverages should be prohibited completely.

 www.science.nelson.com

Photosynthesis

In this chapter, you will be able to

- describe the chemical structure, mechanisms, and dynamics of enzymes and other components of photosynthesis;

- describe how such molecules as glucose, chlorophyll, CO_2, ATP, $NADP^+$, and H_2O function in photosynthesis, and explain the role of chloroplasts;

- compare the matter and energy transformations associated with the process of photosynthesis;

- use problem-solving techniques to investigate and explain relationships between metabolism and the structure of biomolecules;

- design and carry out an experiment related to photosynthesis, controlling the major variables and adapting or extending procedures where required;

- determine the similarities and differences between mitochondria and chloroplasts, and compare the processes of cellular respiration and photosynthesis.

Living organisms are powered by the Sun. Plants, algae, some protists, and certain bacteria capture about 5% of the Sun's energy and, through the process of photosynthesis, make it available to almost every other living thing. You know plants use the Sun's energy because they wither and die if kept in the dark, but animals and mushrooms do not appear to need light energy for survival; raccoons and bats, for example, seem to do well in both darkness and in light. How then do we make sense of the opening statement? How can sunlight be the source of energy for virtually every living thing?

Photosynthetic organisms—plants and cyanobacteria, for example—are largely self-sufficient. These photoautotrophs absorb carbon dioxide, water, and radiant energy directly from the environment and, through photosynthesis, transform these components into sugars and oxygen. Then, through aerobic respiration, they use the oxygen they produce to transfer energy from glucose to ATP.

Heterotrophs, such as raccoons, mushrooms, and bats, cannot synthesize glucose by photosynthesis. So how are they solar powered? They obtain nutrients from other living organisms in their environment. A herbivore, such as a rabbit, consumes plants only. In this case, there is a fairly direct link between the energy of light and the energy of a hopping rabbit. The path is not so direct in a carnivore, such as a lion. It obtains energy-rich nutrients by consuming the flesh of other animals.

Almost every food chain begins with a photosynthetic organism that transforms the energy of light into the chemical potential energy in sugars, such as glucose. Thus, lions obtain energy by eating wildebeests, wildebeests eat grass, and grass obtains energy from sunlight. With the exception of chemoautotrophs, the Sun powers virtually all living organisms.

REFLECT on your learning

1. (a) Distinguish between cyanobacteria and photosynthetic protists.
   (b) Which of these two groups of organisms contains chloroplasts?

2. Describe the endosymbiotic theory.

3. (a) Write the overall equation for photosynthesis.
   (b) Do the O_2 molecules produced in photosynthesis come from CO_2 or H_2O or both?
   (c) What is the purpose of water in photosynthesis?

4. Why are deciduous leaves green in the summer and yellow in the fall?

5. What does *carbon fixation* mean?

6. The process of photosynthesis requires energy in the form of ATP. How do plants obtain ATP for photosynthesis?

7. How are the processes of photosynthesis and cellular respiration dependent on each other?

▶ TRY THIS activity   *Do Plants "Exhale"?*

We know that plants and animals take in oxygen and give off carbon dioxide as they undergo cellular respiration. However, plants also photosynthesize. Are any gases released by plants as a result of photosynthesis?

Materials: two 20 mm × 200 mm test tubes; 500-mL beaker; potted houseplant with runners, such as a spider plant, or a water plant (e.g., *Elodea*); wood splint; secured Bunsen burner

- Fill a 500-mL beaker with 400 mL of water.

- Fill a test tube with water and, without spilling, turn it upside down into the water in the beaker. If an air bubble remains in the test tube, repeat the procedure until there is no bubble or until the bubble is as small as possible.

- Place the other test tube in the beaker repeating the steps above.

- Carefully place a spider-plant runner or sprig of a water plant into one of the test tubes, as shown in **Figure 1**, and leave the other test tube filled with water only.

Figure 1
A spider-plant runner can be used to determine whether plants give off a gas.

- Leave the apparatus in bright sunlight or under a spotlight until there is almost no water left in the tube containing the plant. Observe the test tubes every 15 min over several hours.

- Identify the gas in the test tube using the glowing splint test.
  - (a) What happened to the glowing splint when it was lowered into the test tube?
  - (b) What gas collected in the test tube?
  - (c) How do you know that the gas came from the plant?

chlorophyll the light-absorbing green-coloured pigment that begins the process of photosynthesis

Photosynthesis is carried out by a number of different organisms, including plants, algae, some protists, and cyanobacteria (**Figure 1**). These organisms all contain the green-coloured pigment called **chlorophyll**. Chlorophyll absorbs light energy and begins the process of photosynthesis. Several types of chlorophyll are found in photosynthetic organisms; chlorophyll *a* (blue-green) and chlorophyll *b* (yellow-green) are two common forms. Both molecules are composed of a porphyrin ring attached to a long hydrocarbon tail (**Figure 2**). Porphyrin rings are not only found in chlorophyll molecules, but they also form a part of the oxygen-carrying component of hemoglobin and the cytochromes found in the electron transport chains in mitochondria.

INVESTIGATION 3.1.1

Light and Photosynthesis (p. 183)
Plants seem to need light to stay healthy and alive. What do plants use light for? In Investigation 3.1.1, you will analyze leaves from plants that have been exposed to Sun and from those that have been "starved" of light, and you will test for the presence of starch.

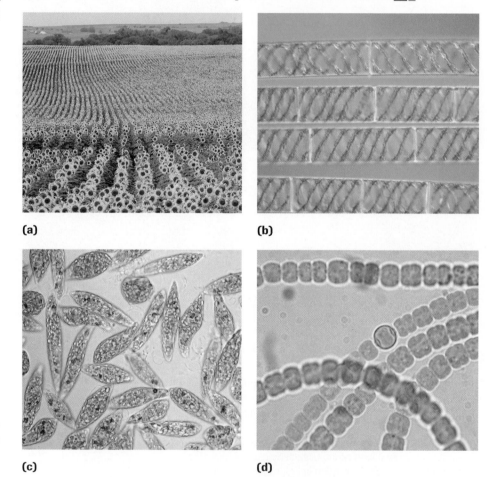

(a) **(b)**

(c) **(d)**

Figure 1
(a) Sunflowers
(b) *Spirogyra*, a photosynthetic alga
(c) *Euglena gracilis*, a photosynthetic protist
(d) *Anabaena*, a cyanobacterium

porphyrin the light-absorbing portion of the chlorophyll molecule, containing a magnesium atom surrounded by a hydrocarbon ring

The **porphyrin** portion of chlorophyll contains a magnesium atom at its centre surrounded by a hydrocarbon ring with alternating single and double bonds. Delocalized electrons in the alternating single-double bonds in the ring absorb light energy and begin the photosynthetic process. Chlorophylls *a* and *b* differ in only one respect: chlorophyll *a* contains a methyl group ($-CH_3$) at position $-R$ in the molecule and chlorophyll *b* contains an aldehyde group ($-COH$) at that position (**Figure 2**). The difference in functional group affects the type of light energy that the molecules can absorb. The hydrocarbon tail (also called the phytol tail or phytol chain) anchors the chlorophyll molecule in a membrane by associating with the hydrophobic regions of the phospholipid bilayer. All photosynthetic organisms use chlorophyll *a* as the primary light-absorbing pigment.

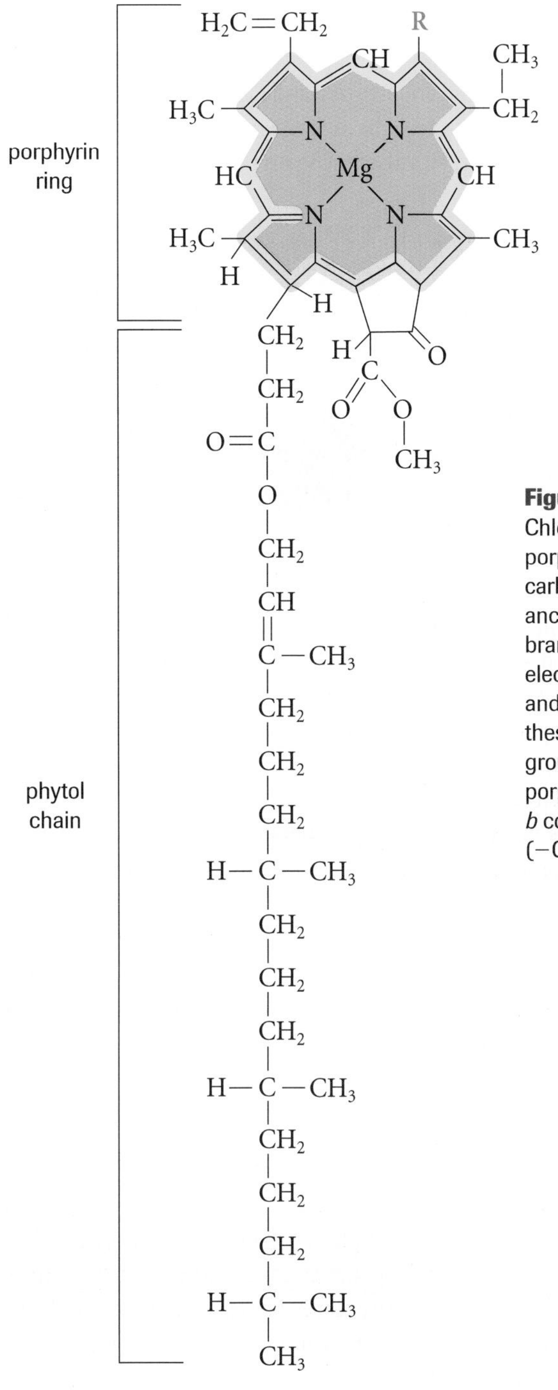

porphyrin ring

phytol chain

Figure 2
Chlorophyll molecules contain a porphyrin ring and long hydrocarbon tail. The hydrophobic tail anchors the molecule into a membrane. The porphyrin ring contains electrons that absorb light energy and begin the process of photosynthesis. Chlorophyll *a* has a methyl group ($-CH_3$) at position $-R$ in the porphyrin ring, whereas chlorophyll *b* contains an aldehyde group ($-CHO$) in that position.

DID YOU *KNOW* ?

Chlorophyll's Root
The *chloro* in chlorophyll does not mean that the molecule contains chlorine. *Chloros* is a Greek prefix that means "yellow-green." The element chlorine gets its name because it is a yellow-green gas at room temperature (**Figure 3**).

Figure 3
Chlorine gas

Prokaryotic Autotrophs: Cyanobacteria

Cyanobacteria, formerly known as blue-green algae, make up the largest group of photosynthesizing prokaryotes. These organisms are unicellular, but may grow in colonies large enough to see. Cyanobacteria live in many different environments including oceans, freshwater lakes and rivers, rocks, and soil. They grow rapidly in nutrient-rich water and are known to cause cyanobacterial blooms (**Figure 4**, page 140). Cyanobacterial blooms discolour water and may be toxic to fish, birds, humans, and other mammals. Although not usually poisonous, dense blooms of the cyanobacterium *Microcystis aeruginosa* can pose an environmental hazard. These organisms produce a toxin called microcystin that in humans may cause symptoms such as headache, vomiting, diarrhea,

cyanobacteria the largest group of photosynthesizing prokaryotes

and itchy skin. Small animals may become sick and die from contact with high concentrations of the toxin. Cyanobacterial blooms usually develop in water that is rich in nitrates and phosphates commonly found in fertilizer and detergent runoff from homes, farms, and industry. On rocks, cyanobacteria associate with fungi to produce lichens (**Figure 5**). Cyanobacteria have even been found growing on polar bears, imparting a green tinge to their fur.

(a)

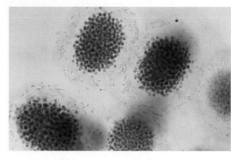
(b)

Figure 4
(a) An algal bloom caused by the cyanobacterium *Microcystis*
(b) *Microcystis sp.*

Figure 5
Flavopunctelia soredica, a lichen composed of cyanobacteria living in close association with a fungus

Having evolved between 2.5 and 3.4 billion years ago, cyanobacteria were probably the first organisms to use sunlight in the production of organic compounds from water and carbon dioxide. Being the first cells to produce oxygen on a large scale, they paved the way for heterotrophic life on Earth. Cyanobacteria are closely related to the chloroplasts of higher plants. The endosymbiotic theory proposes that an ancestor of cyanobacteria was engulfed by an ancestor of today's eukaryotic cells, an association that was mutually beneficial. The cyanobacterium was protected from a harsh external environment, while the eukaryotic host obtained food molecules produced by the engulfed photosynthetic bacterium. It is believed that this association eventually gave rise to plant cells. Like chloroplasts, cyanobacteria contain chlorophyll *a* and carry out photosynthesis. Unlike plant cell chloroplasts, they contain chlorophyll *d*, photosynthetic pigments called phycobilins, and typical prokaryotic cell walls. As with all prokaryotes, cyanobacteria lack membrane-bound organelles like nuclei and mitochondria. Instead, infoldings of the cell membrane are used as sites of photosynthesis and respiration.

Eukaryotic Autotrophs: Algae, Photosynthetic Protists, and Plants

Unlike cyanobacteria, algae, some protists, and plant cells contain chlorophyll within the photosynthetic membranes of discrete organelles called **chloroplasts** (**Figure 6**). Because they contain chlorophyll, chloroplasts give leaves, stems, and unripened fruit their characteristic green colour. Since chloroplasts are found only in these parts, no other structures in a plant are able to photosynthesize. Leaves, in particular, are the primary photosynthetic organs of most plants.

Figure 6
(a) The green alga *Spirogyra* contains a single spiral chloroplast per cell.
(b) onion leaf chloroplasts

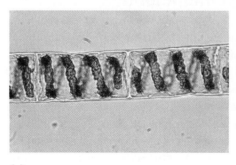
(a)

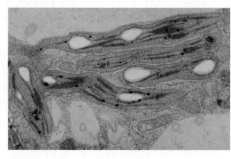
(b)

The overall process of photosynthesis is summarized in the following chemical equation:

$$6CO_{2(g)} + 6H_2O_{(l)} + \text{light energy} \xrightarrow{\text{chlorophyll}} C_6H_{12}O_{6(aq)} + 6O_{2(g)}$$

carbon dioxide   water   glucose   oxygen

This equation represents the overall process for the production of glucose, one of the most common sugars formed by photosynthesis. However, since a variety of other simple sugars may be formed by this process, we may write the equation in the following general form (using the empirical formula for simple sugars):

$$CO_{2(g)} + H_2O_{(l)} + \text{light energy} \xrightarrow{\text{chlorophyll}} [CH_2O]_{(aq)} + O_{2(g)}$$

The overall process is essentially the reverse of cellular respiration. To undergo photosynthesis, a plant cell must contain chlorophyll, and it must be able to obtain carbon dioxide, water, and light energy from its environment. Plants, in fact, possess specialized structures that allow them to accomplish these tasks very efficiently.

Leaves: The Photosynthetic Organs of Plants

Leaves are generally thin and broad, as in maple leaves, shown in **Figure 7(a)**, or thin and narrow, as in pine needles, shown in **Figure 7(b)**. In either case, their structure and arrangement on stems and branches maximizes the surface area exposed to sunlight and limits the distance that gases, such as CO_2, need to travel to reach the chloroplasts. An aerial view reveals a dense "leaf mosaic" that effectively intercepts light that penetrates Earth's atmosphere (**Figure 8**).

epidermis layer the transparent, colourless layer of cells below the cuticle of a leaf, stem, or root

mesophyll layers the photosynthetic cells that form the bulk of a plant leaf

guard cells photosynthetic epidermal cells of a leaf or stem that form and regulate the size of an opening called a stoma

(a)   **(b)**

Figure 7
(a) Maple leaves are thin and broad. **(b)** Pine leaves are thin and narrow.

Figure 8
A mixed forest showing a leaf mosaic that effectively captures sunlight

The primary function of leaves is photosynthesis. The surface of a leaf is uniformly coated with a water-resistant, waxy **cuticle**, illustrated in **Figure 9** (page 142), that protects the leaf from excessive absorption of light and evaporation of water.

The transparent, colourless **epidermis layer** allows light to pass through to the mesophyll cells where most of the photosynthesis takes place. Chloroplasts are most abundant in the cells of the spongy and palisade **mesophyll layers**. **Guard cells** create microscopic openings called **stomata** (singular: *stoma*) that regulate the exchange of carbon dioxide and oxygen with the atmosphere, and also allow water vapour to escape by **transpiration**. The exchange of gases and transpiration of water are regulated by the

stomata openings on the surface of a leaf that allow for the exchange of gases between air spaces in the leaf interior and the atmosphere

transpiration the loss of water vapour from plant tissues, primarily through stomata

vascular bundles a system of tubes and cells that transport water and minerals from the roots to the leaf cells and carry carbohydrates from the leaves to other parts of a plant, including the roots

size of the kidney-shaped guard cells that form stomata (**Figure 10**). **Vascular bundles**, also called veins, transport water and minerals from roots to leaves and carry carbohydrates produced in photosynthesis from leaves to roots and other parts of the plant.

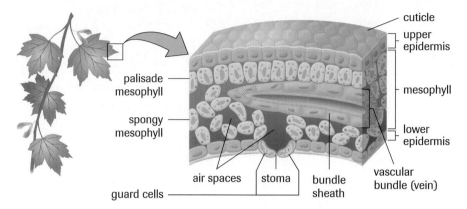

Figure 9
Leaves are the primary organs of photosynthesis in plants. Cross-section of a leaf showing the cuticle, upper and lower epidermis, palisade and spongy mesophyll cells, guard cells and stoma, and a vascular bundle (vein)

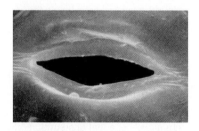

Figure 10
A plant leaf stoma. Two guard cells adjust the size of the opening (the stoma) to regulate the exchange of gases between the leaf and the environment.

Transpiration and Photosynthesis

On a hot summer day, an average-size tree may lose more than 200 L of water through stomata by transpiration. Although stomata cover only 1% to 2% of a leaf's epidermal surface area, they may be responsible for more than 85% of the water lost by a plant. The waxy cuticle and the preponderance of stomata on the surface of the leaf control the amount of water lost by transpiration.

Transpiration assists the photosynthetic process in two ways. First, it creates a "transpiration pull" that helps move water, minerals, and other substances from roots, where they are absorbed, to leaves where they are used. Second, it produces an evaporative cooling effect that prevents leaves from heating to temperatures that could inhibit or even denature the enzymes that catalyze the reactions of photosynthesis. In addition to transpiration, stomata allow carbon dioxide to diffuse into the air spaces within the leaf's mesophyll layers, where most of the photosynthetic cells are located. Plants regulate the size of stomatal openings in response to various environmental conditions in an effort to maximize CO_2 intake and limit water loss. In general, conditions that promote transpiration—such as sunny, warm, dry, windy weather—cause guard cells to reduce the size of stomata.

Opening and Closing Stomata

Guard cells control the size of a stoma by changing their shape in response to changes in environmental conditions. In general, stomata open when guard cells are turgid (swollen) and close when guard cells are flaccid (limp) (**Figure 11**).

The size of a guard cell changes when water moves, by osmosis, into or out of the cell. The direction of osmosis follows the diffusion of potassium ions, $K^+_{(aq)}$, across the guard cell's plasma membrane (cell membrane). Studies show that while potassium ions diffuse passively through the membranes, their movement is coupled with the active

> ▶ *TRY THIS* activity

Opening and Closing Stomata

Stomata may be observed indirectly by making a nail polish film impression of the surface of a leaf and observing the film under a microscope.
Materials: green leaves, clear nail polish, clear adhesive tape, microscope slide, microscope

- Coat a small portion of the underside of a leaf with nail polish and let it dry for about 10 min.
- Once the polish is dry, place a piece of clear adhesive tape over the nail polish film and then carefully remove it.
- Place the tape on a microscope slide.
- Put the slide on the stage of the microscope and observe the stomata under low, medium, and high power.
- Repeat the procedure, this time putting polish on the top surface of the leaf. Compare the abundance of stomata found on the surface with the number found on the underside of the leaf.
- Repeat the procedure with leaves that have been subjected to various environmental conditions (different temperatures, different humidity levels, etc.), and compare the sizes of the stomatal openings.

transport of H^+ ions through membrane-associated proton pumps. Therefore, changes in the size of stomata are dependent on the availability of ATP, one of the products of photosynthesis. When $K^+_{(aq)}$ ions move into guard cells from neighbouring epidermal cells, water follows by osmosis, and guard cells swell. The thicker cell walls forming the stomatal perimeter, along with a series of radial cellulose microfibrils and terminal attachments, cause the cells to buckle outward when they swell up, increasing the size of the stoma, as illustrated in **Figure 11(a)**. When $K^+_{(aq)}$ ions move out of guard cells, water follows by osmosis; the guard cells sag, and the stoma closes, as shown in **Figure 11(b)**.

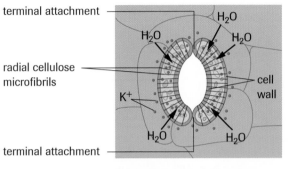 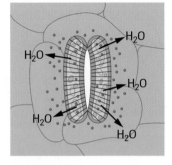

(a) cells turgid-stoma open   **(b)** cells flaccid-stoma closed

Figure 11
(a) When $K^+_{(aq)}$ ions diffuse into guard cells, water moves in by osmosis, guard cells swell, and the stoma opens.
(b) When $K^+_{(aq)}$ ions diffuse out of guard cells, water moves out by osmosis, guard cells become flaccid, and the stoma closes.

In general, stomata are open in the daytime and closed at night. In the morning, when the Sun comes out, light energy (blue light) activates specific receptors in guard cell membranes (blue-light receptors), stimulating proton pumps that drive protons out of the cells. The resulting electrochemical gradient causes K^+ ions to move into the cells, and water to follow by osmosis. The guard cells swell and the stomata open. At the same time, mesophyll cells begin photosynthesizing and using up CO_2 that has accumulated in the air spaces overnight. Reduction in CO_2 concentration in the spaces surrounding guard cells also causes osmosis into the cells, resulting in stomatal opening. Recent studies indicate that stomatal closing at the end of the day generally parallels a decrease in the sucrose content of guard cells (**Figure 12**). Therefore, it appears that the movement of $K^+_{(aq)}$ ions into guard cells, and a decrease in CO_2 concentrations around guard cells, stimulate stomatal opening in the morning, and a reduction in sucrose concentration in guard cells in the evening controls stomatal closing.

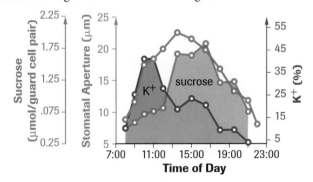

Figure 12
Daily changes in stomatal size (stomatal aperture) and the potassium and sucrose content of guard cells during the day in bean plants (*Vicia faba*)

Chloroplasts

Chloroplasts are the photosynthesis factories of plants and algae. A typical plant cell chloroplast is approximately 3 μm to 8 μm in length and 2 μm to 3 μm in diameter, as shown in **Figure 13** (page 144). Chloroplasts have two limiting membranes, an outer membrane and an inner membrane. These membranes enclose an interior space filled with a protein-rich semiliquid material called **stroma**. Within the stroma, a system of membrane-bound sacs called **thylakoids** stack on top of one another to form characteristic columns

stroma the protein-rich semiliquid material in the interior of a chloroplast

thylakoids a system of interconnected flattened membrane sacs forming a separate compartment within the stroma of a chloroplast

grana (singular: *granum*) a stack of thylakoids

lamellae (singular: *lamella*) unstacked thylakoids between grana

thylakoid membrane the photosynthetic membrane within a chloroplast that contains light-gathering pigment molecules and electron transport chains

thylakoid lumen the fluid-filled space inside a thylakoid

called **grana**. A typical chloroplast has approximately 60 grana, each consisting of 30 to 50 thylakoids. Adjacent grana are connected to one another by unstacked thylakoids called **lamellae**. Photosynthesis occurs partly within the stroma and partly within the **thylakoid membrane**, which contains light-gathering pigment molecules and the electron transport chains that are essential to the process. Thylakoid membranes enclose an interior (water-filled) thylakoid space called the **thylakoid lumen**. The structure of the thylakoid system within the chloroplast greatly increases the surface area of the thylakoid membrane and, therefore, significantly amplifies the efficiency of photosynthesis. Chloroplasts, like mitochondria, contain their own DNA and ribosomes, and the are able to replicate by fission. Starch grains and lipid droplets may also be found in chloroplasts.

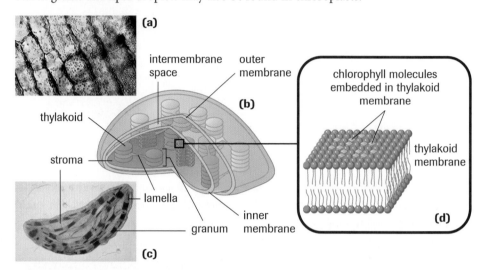

Figure 13
(a) Chloroplasts within plant cells
(b) An artist's representation of a chloroplast, showing key components
(c) An electron micrograph of a chloroplast
(d) Chlorophyll molecules in the thylakoid membrane

▶ *EXPLORE* an issue

Take a Stand: Can Plants Cure Sick Building Syndrome?

Jarmit Singh began a new job at Summit Insurance Co. Ltd. on January 4, 2001. The company was located in a brand-new climate-controlled high-rise in the business district of a large metropolitan area. Jarmit's workstation was on the thirteenth floor in the middle of a large open space with a number of perimeter windows that did not open to the outside. At the end of the first week at his new job, Jarmit developed coldlike symptoms and a persistent headache that his doctor diagnosed as a sinus infection. These symptoms lasted three weeks. In early March, he had another painful sinus infection that lasted six weeks. He made six additional visits to the doctor that year—twice to report severe muscle cramping in his legs and chest while at work, once seeking help for debilitating fatigue and shortness of breath, and three other times complaining of blurred vision, dizziness, and nausea. Through conversations with fellow workers, Jarmit discovered that some of them were experiencing many of the same symptoms. Before he started his new job, Jarmit was a healthy 26-year-old man who did not smoke or drink alcohol, and he maintained a regular exercise program. After consulting a number of medical specialists who claimed that his symptoms were psychosomatic, he was interviewed by an environmental health doctor who suspected that he was suffering from sick building syndrome (SBS), also known as building-related illness

Decision-Making Skills
● Define the Issue   ● Analyze the Issue   ● Research
● Defend the Position   ● Identify Alternatives   ○ Evaluate

(BRI). A friend who heard of Jarmit's diagnosis advised him to bring spider plants to the office since he had read a newspaper article that suggested spider plants help "cure" buildings of SBS.

Statement
Plants may be used as a treatment for sick building syndrome.

- Conduct library and/or Internet research to investigate
  (i) the indicators of SBS;
  (ii) the possible causes of SBS;
  (iii) procedures used to identify SBS;
  (iv) the effectiveness of the use of plants in SBS treatment;
  (v) the types of plants that some have found useful in the treatment of SBS.

 www.science.nelson.com

(a) As a plant biologist and consultant to the Canadian Labour Congress, you will be attending an upcoming World Health Organization (WHO) conference on Environmental Pollution in the Workplace. For the conference, you must prepare a brochure or slide presentation on the effectiveness of the use of plants in the treatment of SBS.

Photosynthetic Organisms

- Photosynthesis is carried out by plants, algae, photosynthetic protists, and cyanobacteria.

- Electrons in the porphyrin ring of chlorophyll absorb light energy and begin the process of photosynthesis.

- The endosymbiotic theory proposes that an ancestor of cyanobacteria was engulfed by an ancestor of today's eukaryotic cell and gave rise to plant cells.

- Algal and plant cells contain chloroplasts. They are most abundant in the meso-phyll and guard cells of plant leaves.

- The following is an overall equation for photosynthesis:

$$6CO_{2(g)} + 6H_2O_{(l)} + \text{light energy} \xrightarrow{\text{chlorophyll}} [CH_2O]_{(aq)} + 6O_{2(g)}$$

- Stomata are open in the daytime and closed at night. Light-activated proton pumps in guard cell membranes cause potassium ions to move from neighbouring epi-dermal cells into guard cells. As a result, water moves by osmosis into guard cells, causing them to swell. Increasing turgor pressure within guard cells causes their membranes to buckle and stomata to open. As the concentration of sucrose in guard cells decreases in the evening, water moves out of the cells and stomata close.

- Chloroplasts have an outer membrane and an inner membrane. The interior space contains a semiliquid material called stroma with a system of membrane-bound sacs called thylakoids, some of which are stacked on top of one another to form grana. Thylakoid membranes contain chlorophyll molecules and electron transport chains.

▶ *Section 3.1* Questions

Understanding Concepts

1. (a) A wolf ate a rabbit that earlier chewed on lettuce. Which organisms are autotrophs? Which are heterotrophs?
   (b) How do all of these organisms rely on sunlight?

2. (a) What do all photosynthetic organisms have in common?
   (b) Where are chlorophyll molecules located in cyanobac-teria?

3. (a) In the endosymbiotic theory, why is the association between prehistoric cyanobacteria and the ancestor of today's eukaryotic cells considered mutually beneficial?
   (b) What evidence is there in today's plant cells sug-gesting that they may have been created by an endosymbiotic relationship between a cyanobacterium and a eukaryotic cell?

4. (a) Why are broad leaves generally thin?
   (b) If all the chloroplasts were removed from the cells of an intact leaf, what colour would you expect the remains to be? Explain.
   (c) How do stomata aid photosynthesis?
   (d) How do proton pumps help regulate the size of stomata?

(e) Why do stomata generally open in the day and close at night?

5. (a) Label parts A, B, C, and D of the chloroplast in **Figure 14**.
   (b) Distinguish between the structure of a chloroplast and the structure of a mitochondrion.

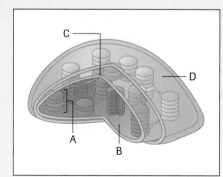

Figure 14
A chloroplast

6. How is the overall equation of photosynthesis not like the reverse of the overall equation of cellular respiration?

7. (a) Name the molecule in **Figure 16**.
   (b) Label parts H and P of the molecule in **Figure 16**.
   (c) Copy the diagram of the phospholipid bilayer in **Figure 15** into your notebook, and place the molecule in position A and B, as it would be oriented in an intact thylakoid membrane.

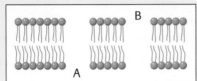

Figure 15
A phospholipid bilayer

Applying Inquiry Skills

8. A student would like to study how environmental factors affect the opening and closing of stomata.
   (a) State a question for such an investigation.
   (b) State a hypothesis.
   (c) Describe an experimental design.

Making Connections

9. (a) What are cyanobacterial blooms?
   (b) How do cyanobacterial blooms adversely affect humans and other animals?
   (c) What adverse effects can they have on the environment?
   (d) Suggest one or two remedies for the problem.

10. Carnivorous plants like the Venus fly-trap (*Dionaea* species) and the pitcher plant (*Cephalotus* species) obtain nutrients by digesting small insects. Conduct library and/or Internet research to determine
   (a) whether these plants are autotrophs or heterotrophs;
   (b) why these plants digest animals.

 www.science.nelson.com

11. Many products containing chlorophyll are sold in pharmacies and health food stores.
   (a) Identify and list three different products containing chlorophyll you can buy at a pharmacy or other store.
   (b) Describe the suggested uses for each of these products.
   (c) Conduct library and/or Internet research to determine the characteristics of chlorophyll that make it suitable for use in these products. If you do not find suitable reasons, form your own hypotheses.

 www.science.nelson.com

Molecule

Figure 16

In plants and algae, all the reactions of photosynthesis take place within chloroplasts and can be broken down into the following three distinct, but connected, stages:

Stage 1: *Capturing light energy*

Stage 2: *Using captured light energy to make ATP and reduced NADP$^+$ (NADPH)*
Nicotinamide adenine dinucleotide phosphate, NADP$^+$, is the energy-shuttling coenzyme used in photosynthesis. Like NAD$^+$ in cellular respiration, NADP$^+$ is reduced by hydrogen atoms to NADPH + H$^+$ (shortened to NADPH for convenience).

Stage 3: *Using the free energy of ATP and the reducing power of NADPH to synthesize organic compounds, such as glucose, from CO$_2$*

The first two stages involve a series of reactions that are directly energized by light, called the **light reactions**. These reactions require chlorophyll and occur on the thylakoid membranes in chloroplasts. They absorb the light energy that is eventually transferred to carbohydrate molecules in the last stage (stage 3) of the process. The reactions of the third stage result in **carbon fixation**—the incorporation of the carbon of CO$_{2(g)}$ into organic compounds, such as glucose. These reactions are endergonic, requiring the energy of ATP and the reducing power of NADPH. Carbon fixation takes place in the stroma by means of a cyclic sequence of enzyme-catalyzed reactions called the **Calvin cycle** or the photosynthetic carbon reduction cycle (**Figure 1**). (A smaller version of **Figure 1** will appear at locations in this chapter where the three stages of photosynthesis are described in more detail. This will help illustrate how each stage fits into the overall process.)

It was thought for many years that the reactions of the Calvin cycle could occur in the absence of light, and they used to be called the "dark reactions" of photosynthesis. However, recent studies have shown that many of the enzymes that catalyze these reactions are activated by light and are either inactive or exhibit low activity in the dark. In addition, the Calvin cycle requires ATP and NADPH formed in the light reactions.

light reactions the first set of reactions of photosynthesis in which light energy excites electrons in chlorophyll molecules, powers chemiosmotic ATP synthesis, and results in the reduction of NADP$^+$ to NADPH

carbon fixation the process of incorporating CO$_2$ into carbohydrate molecules

Calvin cycle a cyclic set of reactions occurring in the stroma of chloroplasts that fixes the carbon of CO$_2$ into carbohydrate molecules and recycles coenzymes

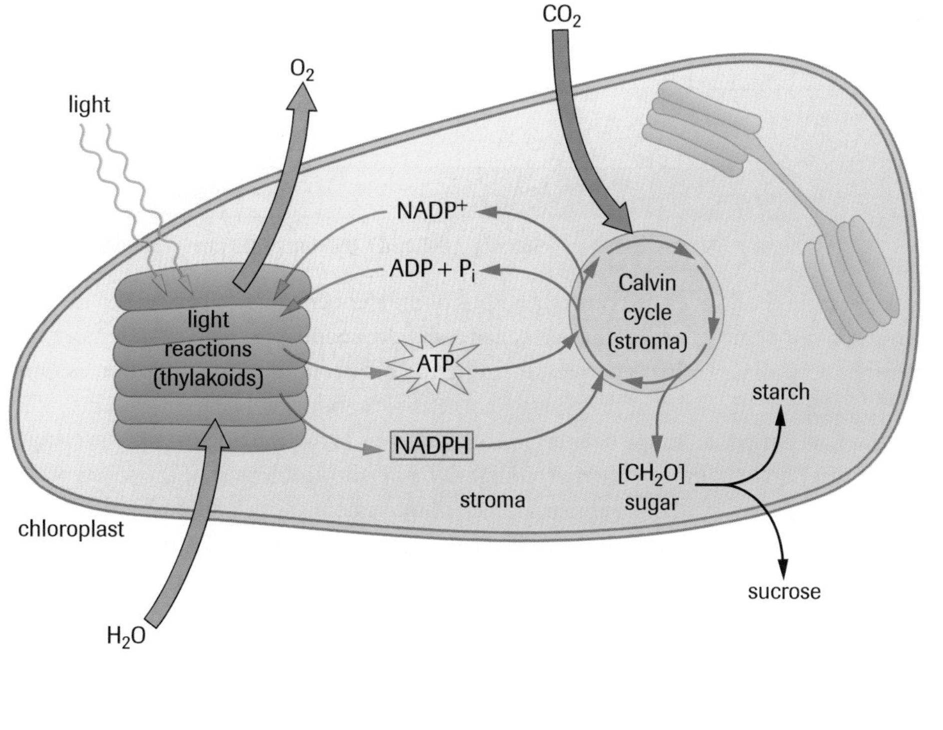

Figure 1
An overview of photosynthesis. The light reactions of photosynthesis occur in the thylakoid membranes of chloroplasts and transfer the energy of light to ATP and NADPH. The Calvin cycle takes place in the stroma and uses NADPH to reduce CO$_2$ to organic compounds, such as glucose and other carbohydrates.

Light

Approximately 1.3 kW/m^2 of light energy reaches Earth's surface from the Sun. Only about 5% of this energy is transferred to carbohydrates by a photosynthesizing leaf (**Figure 2**).

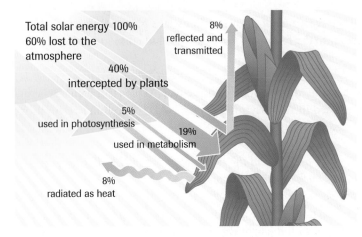

Total solar energy 100%
60% lost to the atmosphere

8% reflected and transmitted

40% intercepted by plants

5% used in photosynthesis

19% used in metabolism

8% radiated as heat

Figure 2
Only 5% of the solar energy incident on a leaf is transferred to carbohydrates.

electromagnetic (EM) radiation
a form of energy that travels at 3×10^8 m/s in wave packets called photons that include visible light

photons packets of EM radiation (also known as quanta)

Light, or **electromagnetic (EM) radiation**, is a form of energy that travels at 3×10^8 m/s in the form of wave packets called **photons** (or quanta), as shown in **Figure 3(a)**. Photons are characterized by a wavelength that is inversely proportional to their energy. Therefore, photons with short wavelengths have high energy and those with long wavelengths have low energy.

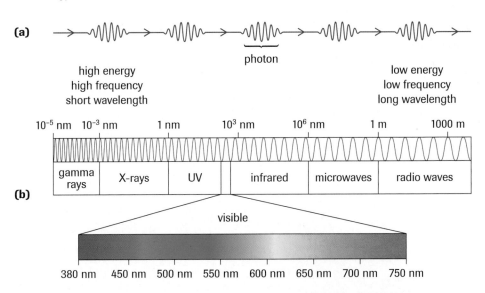

(a)

photon

high energy
high frequency
short wavelength

low energy
low frequency
long wavelength

10^{-5} nm	10^{-3} nm	1 nm	10^3 nm	10^6 nm	1 m	1000 m

(b)

gamma rays	X-rays	UV	infrared	microwaves	radio waves

visible

380 nm   450 nm   500 nm   550 nm   600 nm   650 nm   700 nm   750 nm

Figure 3
(a) Light is a form of electromag-netic radiation that travels as wave packets called photons.
(b) Photons possess a wavelength that is inversely proportional to their energy. Photons with long wavelengths have low energy and photons with short wave-lengths have high energy.

spectroscope an instrument that separates different wavelengths into an electromagnetic spectrum

electromagnetic spectrum pho-tons separated according to wave-length

photosystems clusters of photo-synthetic pigments embedded in the thylakoid membranes of chloroplasts that absorb light energy

Light from the Sun is a mixture of photons of different energies. When passed through a transparent prism in an instrument called a **spectroscope**, photons separate from one another according to their energies, forming the **electromagnetic spectrum**, as illus-trated in **Figure 3(b)**. Most of the photons are invisible to humans, being either in the radio, infrared, or ultraviolet range of energies, but a narrow band, ranging from a wavelength of 380 nm (violet light) to 750 nm (red light), forms the visible part of the spectrum.

Clusters of photosynthetic pigments, called **photosystems**, which are embedded in thy-lakoid membranes, absorb photons of particular wavelengths and, through the light reactions, transfer their energy to ADP, P$_i$, and NADP$^+$, forming ATP and NADPH. The electrons (in the form of H atoms) that are needed to reduce NADP$^+$ to NADPH are sup-plied by water molecules that enter the thylakoids from the stroma. ATP and NADPH are synthesized in the stroma, where the carbon fixation reactions occur.

A Brief History of Photosynthesis Research

In ancient times, it was thought that plants obtained all their food from soil through their roots. In the early 1600s, J.B. Van Helmont, a Belgian physician, set out to determine whether this was true. He planted a willow tree in a pot of soil after having determined the mass of the tree and the soil. Adding only water, he noticed that, after five years, the tree's mass had increased by 74.4 kg, but the soil's mass had decreased by only around 60 g (**Figure 4**). Having proven that the soil was not responsible for the tree's increase in mass, he incorrectly concluded that the absorption of water was responsible.

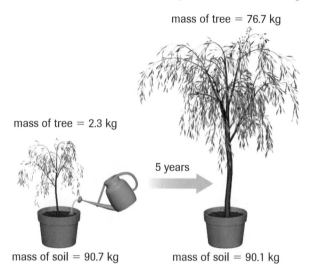

mass of tree = 76.7 kg

mass of tree = 2.3 kg

5 years

mass of soil = 90.7 kg   mass of soil = 90.1 kg

Figure 4
In the early 1600s, Jean Baptiste Van Helmont found that a willow tree grew by 74.4 kg without a comparable decrease in the soil's mass.

In 1771, Joseph Priestley (**Figure 5**), an English scientist, discovered quite by accident that gases in the air play a role in photosynthesis. After burning a candle in a closed container until it went out, he placed a living mint plant in the container. After 10 days, he found that the candle could burn once again. In this way, he showed that plants release a gas into the air that supports combustion (**Figure 6**).

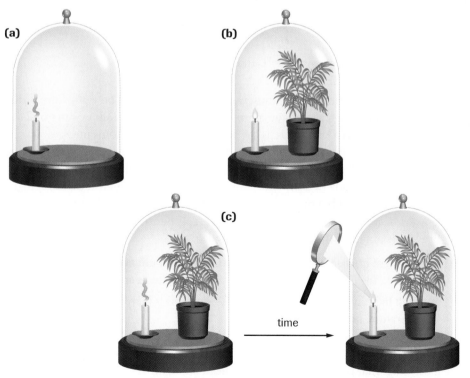

(a)

(b)

(c)

time

Figure 5
Joseph Priestley (1733–1804)

Figure 6
Priestley's experiment
(a) The candle, as Priestley put it, "injures" the air in the bell jar and goes out.
(b) A live plant is placed in a bell jar with a lit candle.
(c) The candle "injures" the air and goes out. After several days, the candle can be lit once again. The plant "restores" the air.

In 1796, Jan Ingenhousz, a Dutch medical doctor, confirmed Priestley's findings and identified the gas released by plants as oxygen. He was the first to realize that sunlight is essential to the process of photosynthesis. He was also the first to note that carbon dioxide was the source of carbon in plants. However, he mistakenly assumed that the oxygen given off in the process was produced from carbon dioxide. Ingenhousz wrote, "Sunshine splits apart the carbon dioxide that a plant has absorbed from the air; the plant throws out at that time the oxygen alone and keeps the carbon to itself as nourishment."

In the early 1930s, a graduate student at Stanford University, C.B. Van Niel (**Figure 7**), working with purple sulfur bacteria, showed that the oxygen in photosynthesis is produced by splitting water, not carbon dioxide. Purple sulfur bacteria absorb hydrogen sulfide, $H_2S_{(g)}$, instead of water, $H_2O_{(l)}$, and they produce pure elemental sulfur as a waste product of photosynthesis instead of oxygen. Van Niel proposed the following equation for photosynthesis in these organisms:

$$CO_{2(g)} + 2H_2S_{(g)} + light \longrightarrow [CH_2O]_{(aq)} + H_2O_{(l)} + 2S_{(s)}$$

carbon dioxide   hydrogen sulfide   carbohydrate   water   sulfur

In green plants, the $H_2S_{(g)}$ is replaced by $H_2O_{(l)}$, and the $2S_{(s)}$ replaced with $O_{2(g)}$.

In 1938, S.M. Ruben, M. Kamen, and their coworkers confirmed Van Niel's findings in an experiment using the heavy isotope of oxygen, ^{18}O, and an alga called *Chlorella*. In the experiment, Ruben and Kamen grew *Chlorella* in water containing a form of "heavy" water, $H_2{}^{18}O$, instead of normal water, $H_2{}^{16}O$. Using an instrument called a mass spectrometer that separates and detects molecules by their mass, they were able to detect ^{18}O in the oxygen molecules released by the organism.

$$CO_{2(g)} + 2H_2{}^{18}O_{(l)} + light \longrightarrow [CH_2O]_{(aq)} + H_2O_{(l)} + {}^{18}O_{2(g)}$$

If, however, *Chlorella* was grown in normal water containing "heavy" carbon dioxide, $C^{18}O_2$, instead of normal carbon dioxide, $C^{16}O_2$, the oxygen given off by the organisms did not contain ^{18}O. This confirmed that the oxygen formed in photosynthesis comes from water, not from carbon dioxide.

Further research in carbohydrate chemistry resulted in the present overall equation (here, the source and destination of the various atoms are indicated):

$$6CO_2 + 12H_2O + light\ energy \longrightarrow C_6H_{12}O_6 + 6H_2O + 6O_2$$

Notice that, in this summary equation, water molecules appear on both sides of the arrow. In fact, water is a reactant and a product in the process. We show water on both sides to indicate that H_2O, not CO_2, is the source of the oxygen atoms in O_2.

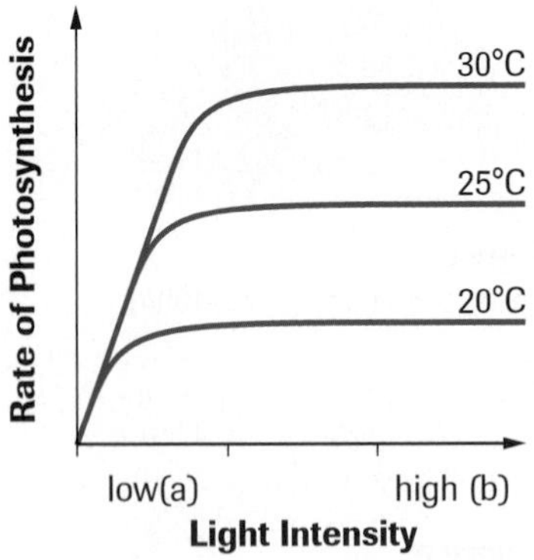

Although certain groups of scientists worked out the role that chemicals play in the process of photosynthesis, others explored the function of light energy in the process. A key figure in the study of light and photosynthesis was the English plant biologist F.F. Blackman. In 1905, Blackman measured the effect that changes in light intensity, CO_2 concentration, and temperature have on the rate of photosynthesis in green plants (**Figure 8**). He obtained the following two results:

1. At low light intensities, the rate of photosynthesis could be increased by increasing the light intensity, but not by increasing temperature.

2. At high light intensities, the rate of photosynthesis is increased by increasing temperature, not light intensity.

Using these results, Blackman concluded that photosynthesis takes place in two stages, an initial light-dependent (photochemical) stage and a second light-independent (biochemical) stage that is primarily affected by heat, not light.

Figure 7
C.B. Van Niel (1897–1985)

Figure 8
(a) At low light intensities, the rate of photosynthesis increases as the light intensity increases, regardless of the temperature.

(b) At high light intensities, the rate of photosynthesis increases with increasing temperature, not with increasing light intensity.

In further experiments, Blackman showed that the rate of photosynthesis is sensitive to the concentration of carbon dioxide. While controlling the temperature, Blackman subjected plants to air containing different concentrations of carbon dioxide. In these investigations, he found that the overall rate of photosynthesis decreased when the concentration of carbon dioxide was lowered (**Figure 9**).

Through the efforts of these and many other scientists, we now know that photosynthesis does, in fact, involve two types of reactions: the light reactions and carbon fixation. The light reactions only take place when light is available, and they are not affected significantly by changes in temperature. These reactions use light and water, and produce NADPH and ATP. We know that the reactions of carbon fixation are dependent on the light reactions for a supply of NADPH and ATP, and require light energy to activate no fewer than five enzymes used in the process. Nevertheless, the overall rate of carbon fixation varies with temperature, not with intensity of light.

After Ingenhousz proved that it was the green part of a plant that carried out photosynthesis, scientists began to purify and experiment with the green material they called chlorophyll. In 1882, the German botanist T.W. Engelmann performed an ingenious experiment in which he used the alga *Spirogyra* to determine whether all colours of the visible spectrum carried out photosynthesis equally well. Engelmann placed a triangular glass prism between the light source and the stage of a microscope. This setup caused the white light to spread into the colours of the spectrum as it passed through a slide on which he placed a sample of *Spirogyra*. He chose *Spirogyra* because it is a filamentous microorganism that possesses a long, spiral chloroplast throughout its length.

Engelmann carefully aligned the *Spirogyra* filament so that it was exposed to different colours (wavelengths) of light across its length. To determine whether photosynthesis occurred throughout the filament's length, he added aerobic bacteria to the slide, realizing that these would accumulate wherever oxygen was being produced. After a period of time, he found that the bacteria accumulated in areas where the filament was exposed to red and blue-violet light, with very few bacteria gathering in the area illuminated with green light (**Figure 10**).

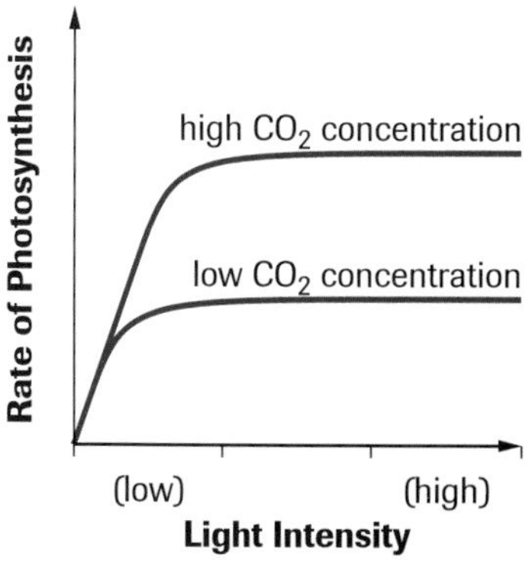

Figure 9
The overall rate of photosynthesis decreases when the availability of carbon dioxide is limited.

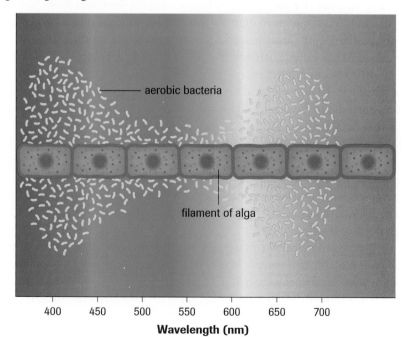

Figure 10
Engelmann's experiment. T.W. Engelmann used *Spirogyra* and aerobic bacteria to determine whether different colours of visible light carry out photosynthesis equally well. He found that the bacteria accumulated in areas where the chloroplast of *Spirogyra* was exposed to red and blue-violet light, indicating that these colours best support photosynthesis, and thereby oxygen production. This is called the action spectrum for photosynthesis.

action spectrum a graph illustrating the effectiveness with which different wavelengths of light promote photosynthesis

absorption spectrum a graph illustrating the wavelengths of light absorbed by a pigment

β-carotene a carotenoid pigment found in photosystems that also forms vitamin A in animals

The **action spectrum** of photosynthesis illustrates the effectiveness with which different wavelengths of light promote photosynthesis. Today, using a sophisticated instrument called a spectrophotometer (see Lab Exercise 3.3.1), the **absorption spectrum** of pigments, such as chlorophyll *a* and chlorophyll *b*, can be determined with accuracy, as **Figure 11** shows.

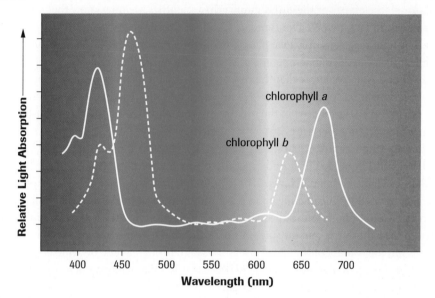

Figure 11
The absorption spectrum of chlorophyll *a* and chlorophyll *b*. Notice how the action spectrum of photosynthesis obtained by Engelmann parallels the absorption spectrum of chlorophyll obtained by spectrophotometric studies.

Chlorophyll and Accessory Pigments

Chlorophylls *a* and *b* absorb photons with energies in the blue-violet and red regions of the spectrum and reflect or transmit those with wavelengths between about 500 nm and 600 nm that our eyes see as green light. This is why most photosynthesizing organisms look green in white light. Comparison of the absorption and action spectra of a plant shows a close correspondence between the two, indicating that most of the wavelengths of light absorbed by chlorophylls are used in photosynthesis.

Chlorophyll *a* is the only pigment that can transfer the energy of light to the carbon fixation reactions of photosynthesis. Chlorophyll *b* acts as an accessory pigment, absorbing photons that chlorophyll *a* absorbs poorly, or not at all. Other compounds called carotenoids also act as accessory pigments. Carotenoids, such as **β-carotene** (**Figure 12**), typically possess two hydrocarbon rings connected by a hydrocarbon chain containing alternating single and double bonds. Like the bonding system in the porphyrin ring of chlorophyll, electrons in the chain are able to absorb light energy in the range from 400 nm to 500 nm (blue-violet).

Since carotenoids reflect or transmit photons in the yellow-to-red range of the spectrum, they appear yellow-orange to the human eye. Some carotenoids play an energy-absorbing role rather than a photosynthetic one. These carotenoids absorb light, which would otherwise damage chlorophyll, and then they lose the energy as heat. Similar compounds are thought to protect the human eye from excessive photon damage. β-carotene, in particular, plays an important role in vertebrate vision. Enzymes split β-carotene into identical halves, producing two molecules of vitamin A. Vitamin A is then oxidized to retinal, a compound needed to make rhodopsin, a photopigment in the rod cells of the eye that helps vertebrates see in low-light conditions. Carrots, a vegetable rich in β-carotene, help maintain good vision.

β-carotene

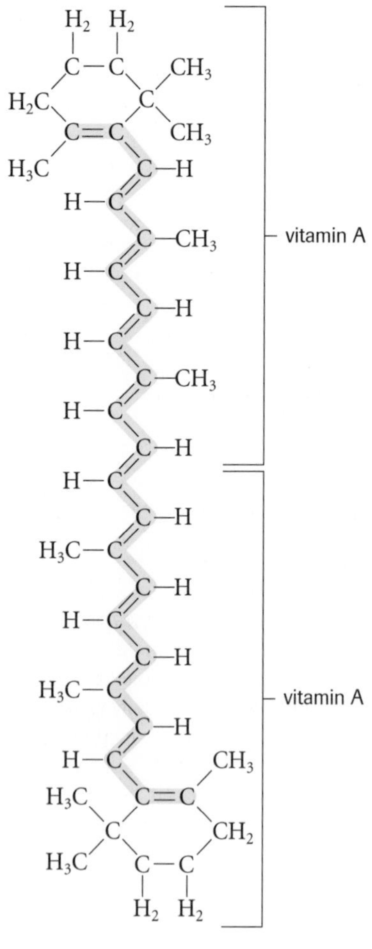

Figure 12
β-carotene is an accessory pigment in photosynthesis and a precursor of vitamin A in vertebrates.

In spring and summer, most leaves appear green because of the high concentration of chlorophyll in the chloroplasts of leaf cells. A number of accessory pigments, including **xanthophylls** (yellow) and carotenoids (yellow-orange), are interspersed in the thylakoid membranes—their colours overwhelmed by the green light reflected by chlorophyll, as shown in **Figure 13(a)**. Other accessory pigments called **anthocyanins** (red, violet, and blue) are primarily located in plant cell vacuoles (not chloroplasts) and are synthesized in the fall in ripening tomatoes or turning maple leaves, for example. With the onset of cooler autumn temperatures, plants stop producing chlorophyll molecules and disassemble those already in the leaves. This causes the yellow, red, and brown colours of autumn leaves to become visible, as **Figure 13(b)** shows.

xanthophylls pigments in chloroplast membranes that give rise to the yellow colour in autumn leaves

anthocyanins pigments in vacuoles that give rise to the red colour in autumn leaves

photosynthetically active radiation (PAR) wavelengths of light between 400 nm and 700 nm that support photosynthesis

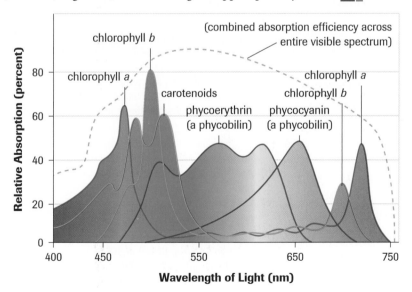

(a)   **(b)**

Figure 13
(a) Summer leaves contain chlorophyll and accessory pigments, but the green colour of chlorophyll overwhelms the colours reflected by the accessory pigments.
(b) Autumn leaves contain less chlorophyll, causing the colours of the accessory and other pigments to become visible.

When the absorption spectra of chlorophylls *a* and *b* are combined with those of the accessory pigments, the absorption range covers almost the entire visible spectrum (**Figure 14**). The wavelengths from 400 nm to 700 nm are called **photosynthetically active radiation (PAR)** because, in general, these wavelengths support photosynthesis.

INVESTIGATION 3.2.1

Identifying Plant Pigments by Chromatography (p. 184)
Look at **Figure 15** with unaided eyes and determine its colour. Now look at the figure with a magnifying glass. What colours do you see? The spring and summer leaves of deciduous trees appear green in colour. Do green leaves contain only green pigments, or is there a mixture of pigments with the green variety predominating? Investigation 3.2.1 will help you decide.

Figure 15

Figure 14
Light is absorbed from the entire visible spectrum when the accessory pigments are combined with chlorophylls *a* and *b*.

SUMMARY Light Energy and Photosynthetic Pigments

- Photosynthesis takes place within chloroplasts in three stages:

    Stage 1: Capturing light energy

    Stage 2: Synthesizing ATP and NADPH

    Stage 3: The Calvin cycle (carbon fixation)

- Stages 1 and 2 are the light reactions; carbon dioxide is fixed into carbohydrates in stage 3.

- The following equation summarizes the overall process of photosynthesis:

$$6CO_2 + 12H_2O + \text{light energy} \longrightarrow C_6H_{12}O_6 + 6H_2O + 6O_2$$

    carbon dioxide   water   glucose   water oxygen

- Light is a form of energy that travels in wave packets called photons.

- Photosystems, embedded in the thylakoid membranes, absorb photons of particular wavelengths and, through the light reactions, transfer their energy to ADP and $NADP^+$, forming ATP and NADPH.

- The light reactions produce NADPH and ATP. The reactions of carbon fixation depend on the light reactions for a supply of NADPH and ATP, and require light to activate certain enzymes.

- Chlorophyll *a* is the only pigment that can transfer the energy of light to the carbon fixation reactions of photosynthesis. Chlorophyll *b* and the carotenoids act as accessory pigments, transferring their energy to chlorophyll *a*.

- In spring and summer, leaves appear green because of the high concentration of chlorophyll. Xanthophyll (yellow), carotenoids (yellow-orange), and anthocyanins (red, violet, and blue) are overwhelmed by the green light reflected by chlorophyll. In autumn, plants stop producing chlorophyll, causing the yellow, red, and brown colours of autumn leaves to show.

▶ Section 3.2 Questions

Understanding Concepts

1. Write a balanced equation to represent the overall reaction of photosynthesis.

2. What are the products of the light reactions of photosynthesis?

3. (a) Define *light*.
   (b) What is a photon?
   (c) How are the wavelength and energy of a photon related?
   (d) Which possesses a higher energy value: red light or green light? Explain.

4. Create a photosynthesis timeline that includes dates, key individuals' names and nationalities, and a brief description of the individuals' contributions to our understanding of the photosynthetic process. Make sure your timeline includes discoveries made within the past 25 years, and at least one discovery made since 1990. Conduct library and/or Internet research to obtain this information.

5. The following statements were made by Priestley regarding his experiment. Restate them in terms that relate to our present knowledge of chemistry and photosynthesis:
   (a) "The lit candle injures the air in the bell jar."
   (b) "The plant restores the air in the bell jar."

6. (a) Distinguish between the action spectrum and the absorption spectrum of chlorophyll *a*.
   (b) Are the two spectra similar to one another? Why or why not?
   (c) Do the action and absorption spectra have to be the same? Explain.

7. Describe two roles that β-carotene may play in plants and one function it performs in humans.

8. (a) What pigments are present in green leaves?
   (b) Explain why yellow-coloured pigments are visible in autumn leaves but not in summer leaves.
   (c) What is photosynthetically active radiation (PAR)?
   (d) If green plants absorb all the wavelengths in PAR, why do they appear green?

▶ **Section 3.2** *Questions* continued

9. (a) What is the difference between a heavy isotope of an element and a radioactive isotope of the same element?
   (b) What did Ruben and Kamen discover about photosynthesis by using a heavy isotope of oxygen?
   (c) What instrument did they use to detect the presence of "heavy" water?

Applying Inquiry Skills

10. (a) Describe Van Helmont's experiment.
    (b) Based on the description given in the text, was this a controlled experiment? If so, describe the control(s) Van Helmont applied to his experiment. If not, what control(s) were missing?

11. The data in **Table 1** were obtained by extracting the pigments from spinach leaves and placing them in an instrument called a spectrophotometer that measures the amount of light (of different wavelengths) absorbed by pigments.

Table 1 Absorption Spectrum of Spinach Leaf Pigments

Wavelength (nm)	Absorbance (%)	Wavelength (nm)	Absorbance (%)
400	0.42	560	0.12
420	0.68	580	0.15
440	0.60	600	0.17
460	0.58	620	0.25
480	0.83	640	0.40
500	0.23	660	0.32
520	0.11	670	0.56
540	0.12	680	0.24

(a) Construct a line graph of the data with percent absorbance along the vertical axis and wavelength along the horizontal axis. Indicate the colours of the visible spectrum corresponding to the wavelengths along the horizontal axis.
(b) Which colours of an intact spinach leaf would be least visible? Why?
(c) What colour is absorbed the least by the pigment extract? Why?

(d) Compare this graph to the absorption spectrum in **Figure 11** on page 152. Which pigment is most likely responsible for the peak at 670 nm?
(e) Why are there no peaks in the range of 500 nm–620 nm?
(f) Which pigments are primarily responsible for absorption in the range of 400 nm–480 nm?

Making Connections

12. Conduct library and/or Internet research to answer the following questions regarding leaves:
    (a) What are bracts?
    (b) What colour changes do bracts of poinsettia (*Euphorbia pulcherrima*) plants (**Figure 16**) undergo? At what time of year does this change normally occur?

Figure 16
Poinsettia plant

(c) Decribe the history of the poinsettia as a popular Christmas plant.

 www.science.nelson.com

13. Many lightbulb manufacturers produce fluorescent tubes labelled as "growlights" that they claim emit "full-spectrum light that imitates sunlight." Conduct library and/or Internet research to determine whether fluorescent tubes labelled as "growlights" are more effective sources of artificial light for growing plants indoors than tubes without this label.

 www.science.nelson.com

Photosynthesis is divided into two sequential processes: the light reactions (stages 1 and 2) and carbon fixation (stage 3).

The Light Reactions

The light reactions begin when photons strike a photosynthetic membrane. This process may be divided into three parts:

1. Photoexcitation: absorption of a photon by an electron of chlorophyll

2. Electron transport: transfer of the excited electron through a series of membrane-bound electron carriers, resulting in the pumping of a proton through the photosynthetic membrane, which creates an H^+ reservoir and eventually reduces an electron acceptor

3. Chemiosmosis: the movement of protons through ATPase complexes to drive the phosphorylation of ADP to ATP

The relationship between the light reactions of photosynthesis and the rest of the process is illustrated in **Figure 1**.

Photoexcitation

Before a photon of light strikes them, a chlorophyll molecule's electrons are at the lowest possible potential energy level, called the **ground state**. During its interaction with a photon, the electron gains energy and is raised to a higher potential energy level. This process is called **excitation**. A ball thrown into the air eventually reaches an unstable position at the top of its flight and, in an instant, returns back to its original lower potential energy level. Likewise, an excited electron is unstable and returns to its ground state in about one billionth of a second. As it returns, the loss in potential energy appears as heat and light. This emission of light by this process is called **fluorescence**. Chlorophyll, like other pigments, emits a photon of light when one of its excited electrons returns to its ground state. However, in most cases, this only happens when the molecule is separated from the photosynthetic membrane in which it is normally embedded. A solution of chlorophyll fluoresces red when exposed to bright white light (**Figure 2**).

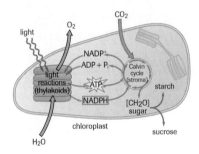

Figure 1
The light reactions of photosynthesis

ground state the lowest possible potential energy level of an atom's electron

excitation the absorption of energy by an atom's electron

fluorescence the release of energy as light as an atom's electron returns to its ground state

Figure 2
(a) A solution containing isolated chlorophyll molecules fluoresces red when exposed to bright white light.
(b) The red light is produced when excited electrons move back to ground state.

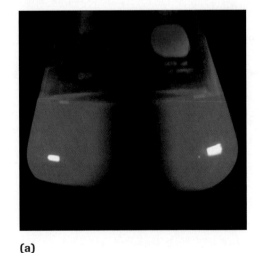

(a)

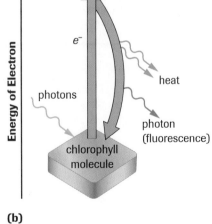

(b)

Most chlorophyll molecules do not fluoresce when associated with a photosynthetic membrane because the excited electron is captured by a special molecule called the **primary electron acceptor**. This is a redox reaction—the chlorophyll is oxidized and the primary acceptor is reduced.

primary electron acceptor a compound embedded in the thylakoid membrane that is reduced by an excited chlorophyll electron

> ▶ **TRY THIS** activity **Chlorophyll Fluorescence**

Materials: chlorophyll extract, 250-mL beaker or 20 mm × 200 mm test tube, bright white light (e.g., overhead-projector beam), green plant leaf

- Place 250 mL of chlorophyll extract into a beaker or 20 mm × 200 mm test tube.
- Hold the solution in a beam of bright white light and observe the colour of the chlorophyll from different angles.

(a) Describe your observations.

- Place the green leaf in the beam of light and observe the colour of the leaf from different angles.

(b) Describe your observations.

(c) Did the chlorophyll solution change colour when exposed to bright white light? If so, why?

(d) Explain any differences in your observations from (a) and (b).

Photosystems

In a functioning chloroplast, light is not absorbed by independent pigment molecules. Instead, light is absorbed by chlorophyll or accessory pigment molecules that are associated with proteins in clusters called photosystems.

A photosystem consists of an **antenna complex** and a **reaction centre** (**Figure 3**). The antenna complex is composed of a number of chlorophyll molecules and accessory pigments set in a protein matrix and embedded in the thylakoid membrane. An antenna pigment absorbs a photon and transfers the energy from pigment to pigment until it reaches a chlorophyll *a* molecule in an area called the reaction centre. An electron of this chlorophyll *a* molecule absorbs the transferred energy and is raised to a high energy level. A redox reaction transfers the excited electron from this chlorophyll molecule to a primary electron acceptor, leaving the chlorophyll in an oxidized state. Isolated chlorophyll molecules fluoresce because there is no primary electron acceptor to receive the photoexcited electron. Thus, photosystems act as the primary light-harvesting units of chloroplasts.

antenna complex a web of chlorophyll molecules embedded in the thylakoid membrane that transfers energy to the reaction centre

reaction centre a transmembrane protein complex containing chlorophyll *a* whose electrons absorb light energy and begin the process of photosynthesis

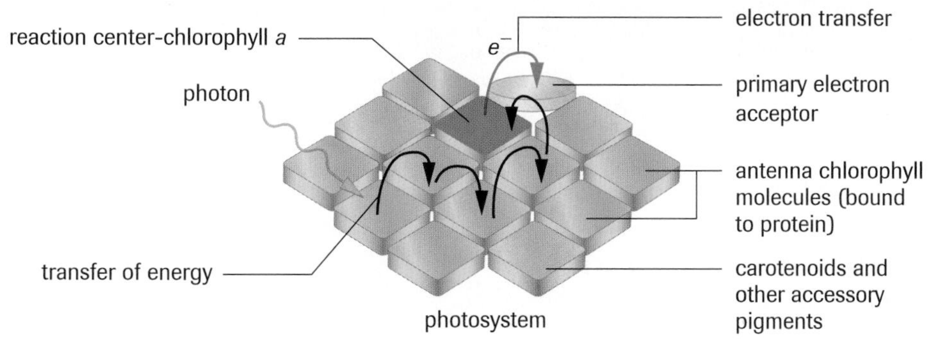

Figure 3
A photosystem is a complex of chlorophyll molecules, accessory pigments, and proteins embedded in a thylakoid membrane. An antenna complex, consisting of several hundred pigment molecules, absorbs photon energy and transmits it from molecule to molecule (black arrows). This continues until it causes the oxidation of a chlorophyll molecule in the reaction centre and reduces a primary electron carrier (red arrow). The reaction centre chlorophyll is always a chlorophyll *a* molecule.

Chloroplast thylakoid membranes contain two types of photosystems called **photosystem I** and **photosystem II**. The chlorophyll *a* molecule in the reaction centre of photosystem I is called P700 because its absorption spectrum peaks at a wavelength of 700 nm (red light). The chlorophyll *a* molecule in the reaction centre of photosystem II is called P680 because it is best at absorbing photons with a wavelength of 680 nm (red light). The P700 and P680 chlorophyll *a* molecules are identical. Their absorption spectra peak at slightly different wavelengths because of the effects of the proteins they are associated with in the reaction centre.

Noncyclic Electron Flow and Chemiosmosis

Plants use photosystems I and II to produce ATP and NADPH. This two-part process is called **noncyclic electron flow** (**Figure 4**). Photosynthesis begins when a photon strikes photosystem II and excites an electron of chlorophyll P680. The excited electron is captured by the primary electron acceptor, pheophytin, and through a series of redox reactions, transferred to plastoquinone, PQ. A **Z protein**, associated with photosystem II, splits water into oxygen, hydrogen ions (protons), and electrons. One of these electrons are used to replace the missing electron in chlorophyll P680. Oxygen leaves the cell and the protons remain in the thylakoid space. The electron passes through a proton pump called the "Q cycle" involving various components of photosystem II, plastiquinone, and

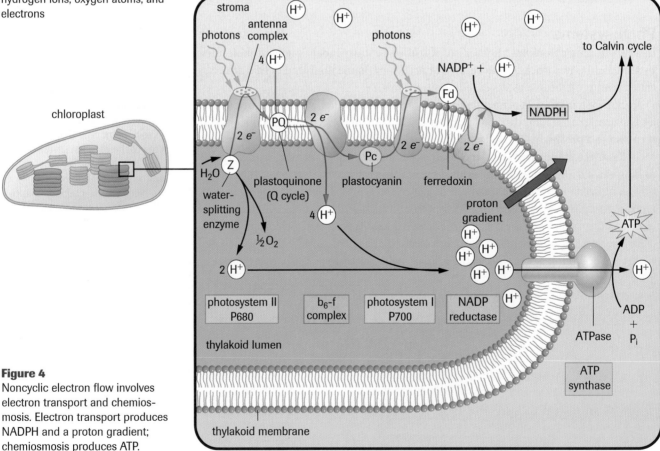

Figure 4
Noncyclic electron flow involves electron transport and chemiosmosis. Electron transport produces NADPH and a proton gradient; chemiosmosis produces ATP.

components of the b_6–f cytochrome complex. The components of the Q cycle transport protons from the stroma into the thylakoid lumen, thus creating an H^+ gradient for chemiosmosis. Then the electron passes through other components of an electron transport chain similar to that in cellular respiration, eventually replacing an electron that is lost by photosystem I when it is struck by a photon. The electron from photosystem I is passed to ferredoxin, then to the enzyme NADP reductase, which uses the electron and H^+ ions from the stroma to reduce $NADP^+$ to NADPH. Protons that accumulate in the thylakoid lumen form an electrochemical gradient that drives the phosphorylation of ADP to ATP as protons move through the ATP synthase complex from the thylakoid lumen into the stroma. Note that two electrons are required to reduce $NADP^+$ to NADPH; thus, the entire process happens twice for each $NADP^+$ reduced to NADPH.

The process is *noncyclic* because once an electron is lost by a reaction centre chlorophyll molecule within a photosystem, it does not return to that system, but instead ends up in NADPH. The following steps summarize noncyclic electron flow:

1. A photon strikes photosystem II and excites an electron of chlorophyll P680. The excited electron is captured by a primary electron acceptor called pheophytin. Through a series of redox reactions, the electron is transferred to an electron carrier called plastoquinone (PQ) and then to an electron transport chain similar to that in cellular respiration. This process occurs twice, causing two electrons to pass through the electron transport chain.

2. A Z protein, associated with photosystem II and facing the thylakoid lumen, splits water into oxygen, hydrogen ions, and electrons. Two of these electrons are used to replace the missing electrons in chlorophyll P680. Oxygen leaves the chloroplast as a byproduct, and the protons remain in the thylakoid space, adding to the H^+ gradient that powers chemiosmosis.

3. The electrons that leave photosystem II pass through the Q cycle, which transports protons from the stroma into the thylakoid lumen, thus creating an H^+ gradient for chemiosmosis. Four protons are translocated into the thylakoid lumen for each pair of electrons that passes through the transport chain. The electrons then move through plastocyanin, P_c, and other components of the electron transport chain, eventually replacing the two electrons that were lost by photosystem I when it was struck by photons.

4. The electrons from photosystem I pass through another electron transport chain containing an iron-containing protein called ferredoxin (Fd). They then move to the enzyme NADP reductase that uses the two electrons and H^+ ions from the stroma to reduce $NADP^+$ to NADPH.

5. Protons that accumulate in the thylakoid lumen contribute to an electrochemical gradient that drives the phosphorylation of ADP to ATP. As protons move through the ATPase complex from the thylakoid lumen into the stroma, ATP is formed. Since light is required for the establishment of the proton gradient, this process is called **photophosphorylation**. The ratio of protons translocated to ATP formed by chemiosmosis has recently been found to be four H^+ per ATP.

The energy changes to electrons as they go through noncyclic electron flow are illustrated in the Z diagram in **Figure 5** on page 160.

photophosphorylation the light-dependent formation of ATP by chemiosmosis in photosynthesis

Figure 5

This energy diagram of noncyclic electron flow shows the pathways and the energy changes that photosystem electrons undergo in noncyclic electron flow. Noncyclic electron flow generates a proton gradient for chemiosmotic ATP synthesis, and NADPH.

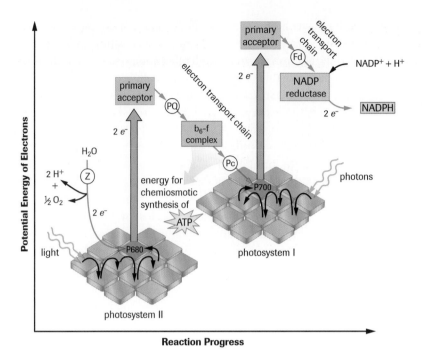

Figure 6

Cyclic electron flow. A photon ejects an electron from photosystem I. The electron is passed to ferredoxin, then through the Q cycle and cytochrome chain and back to chlorophyll P700. This generates a proton gradient for chemiosmotic ATP synthesis, but does not generate NADPH.

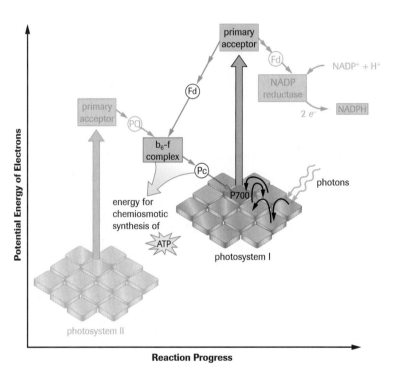

cyclic electron flow flow of photon-energized electrons from photosystem I, through an electron transport chain that produces ATP by chemiosmosis, but no NADPH

Cyclic Electron Flow

In some cases, excited electrons take a cyclic pathway called **cyclic electron flow** that uses photosystem I only (**Figure 6**). In this pathway, a photon ejects an electron from chlorophyll P700 of photosystem I. The electron is passed to ferredoxin, Fd, and then goes through the Q cycle, the b_6-f complex (cytochrome chain), and back to chlorophyll P700. This cyclic pathway generates a proton gradient for chemiosmotic ATP synthesis, but does not release electrons to generate NADPH. Without NADPH, the reactions of carbon fixation cannot occur because electrons in the hydrogen atoms of NADPH are needed to reduce carbon dioxide.

Although the light reactions involve a series of complex processes, the overall goal that is achieved is simple: the energy of light is transferred to ATP and NADPH. Both of these substances play a critical role in the next phase of photosynthesis: carbon fixation.

The Calvin Cycle

The reactions that convert carbon dioxide into carbohydrate molecules occur in the stroma of chloroplasts (**Figure 7**). The process occurs by a cyclic series of reactions. These reactions are similar to the Krebs cycle in that some of the starting material is regenerated in the process. The details of the cycle were determined in the early 1960s by Melvin Calvin (**Figure 8**) and his associates at the University of California at Berkeley. Calvin received the Nobel Prize in Chemistry in 1961. The reactions are called the Calvin cycle in his honour.

The Calvin cycle can be divided into three phases: carbon fixation, reduction reactions, and ribulose 1,5-bisphosphate (RuBP) regeneration (**Figure 9**).

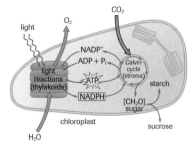

Figure 7
Stage 3 (the Calvin cycle) of photosynthesis (blue circle)

Figure 8
Melvin Calvin (1911–97)

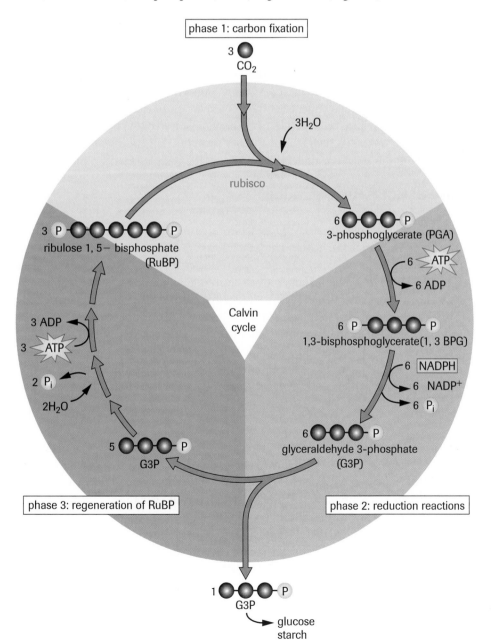

Figure 9
The Calvin cycle can be divided into three phases: carbon fixation, reduction reactions, and RuBP regeneration.

Phase 1: Carbon Fixation

Carbon fixation occurs in the first reaction of the Calvin cycle. Here, CO_2 adds to an already existing 5-carbon molecule, ribulose 1, 5-bisphosphate, RuBP, to form an unstable 6-carbon intermediate. This intermediate instantly splits into two 3-carbon molecules called 3-phosphoglycerate (PGA). If this reaction occurs three times, three molecules of CO_2 (three carbons) react with three molecules of RuBP (3×5 carbons = 15 carbons) to produce six molecules of PGA (6×3 carbons = 18 carbons). Since the first compound to be produced contains three carbon atoms, the Calvin cycle is also known as **C₃ photosynthesis**. Most plants are called C₃ plants because they undergo this form of carbon fixation.

These reactions are catalyzed by the enzyme **ribulose bisphosphate carboxylase/oxygenase (rubisco)**. Rubisco is a very large enzyme that works very slowly. Typical plant enzymes process about 1000 substrate molecules per second. Rubisco averages about three. Thus, many copies of the enzyme are needed to complete the reactions efficiently. It has been estimated that rubisco makes up about half of all the protein in a typical leaf, causing it to be the most abundant protein on Earth. This reaction is exergonic, owing to the high level of chemical potential energy in RuBP in relation to PGA.

Phase 2: Reduction Reactions

In this phase, each of the six molecules of PGA is phosphorylated by an ATP to form six molecules of 1,3-bisphosphoglycerate (1, 3-BPG). Next, a pair of electrons from each of six NADPH molecules reduces six molecules of 1,3-bisphosphoglycerate to six molecules of glyceraldehyde 3-phosphate (G3P or PGAL), a sugar. These reactions are essentially the reverse of glycolysis. At this stage, one molecule of G3P exits the cycle as a final product.

Phase 3: RuBP Regeneration

In a series of enzyme-catalyzed reactions, the remaining five molecules of G3P (5×3 carbons = 15 carbons) are rearranged to regenerate three molecules of RuBP (3×5 carbons = 15 carbons) to complete the cycle. Three molecules of ATP are used in the process. With RuBP regenerated, the cycle may fix more CO_2. As the cycle continues, the G3P molecules that leave are used to synthesize larger sugars, such as glucose and other carbohydrates.

The Calvin cycle illustrated in **Figure 9** (page 161) shows the steps for fixing not one but three carbon dioxide molecules. Three CO_2 molecules must be fixed before one 3-carbon G3P molecule can be removed from the cycle in order to maintain the pool of intermediate molecules needed to sustain the cycle. This means that six turns of the cycle fix enough CO_2 to produce the equivalent of one 6-carbon glucose molecule.

The overall equation for the Calvin cycle (per G3P produced) is as follows:

$$3 \text{ RuBP} + 3CO_2 + 9ATP + 6NADPH + 5H_2O \longrightarrow 9ADP + 8P_i + 6NADP^+ + G3P + 3 \text{ RuBP}$$

Thus, for the net synthesis of one G3P molecule, the Calvin cycle uses nine molecules of ATP and six molecules of NADPH. The light reactions are a source of ATP and NADPH. For every four electrons (from 2 H_2O) transferred in the light reactions, twelve protons are added to the H^+ reservoir in the thylakoid lumen, resulting in the production of three molecules of ATP. Thus, three ATP and two NADPH are produced in the light reactions, and three ATP and two NADPH are used in the Calvin cycle for every G3P produced.

Photosynthesis seems to be self-sufficient for its ATP and NADPH needs. However, the H^+/ATP ratio is an approximation, and any mismatch in the production and use of ATP may be compensated for by the production of additional ATP by cyclic electron flow. The relative rates of cyclic and noncyclic electron flow are regulated by the NADPH to $NADP^+$ ratio in the stroma. When the ratio is high, (e.g., during high light intensities) cyclic electron flow is favoured because the availablility of $NADP^+$ will be low, slowing down noncyclic flow.

Case Study *Discovering Metabolic Pathways* ▾

One of the main problems in biology is to identify the chemical reactions that take place in a cell and to determine the individual steps that make up the process. This is a very difficult task, primarily because the cell is so small, but also because there are about 1000 different reactions taking place at the same time, many of them repeated hundreds of times throughout the cell. In addition, the cell is chock full of thousands of different chemicals. How does one make sense of this chemical complexity?

Scientists have devised some ingenious methods for deciphering the complex chemical pathways of metabolism. One method involves the use of radioisotopes. This approach, called radioisotope labelling, proved ideal for determining the series of reactions that occur in the Calvin Cycle.

In the early 1960s, Melvin Calvin and his associates at the University of California at Berkeley conducted a series of elegant experiments aimed at determining the sequence of reactions involved in the carbon fixation reactions of photosynthesis. Calvin and his colleagues used suspensions of unicellular green algae called *Chlorella* to trace the path of carbon in photosynthesis (**Figure 10**).

The investigators first exposed suspensions of the algae to constant conditions of light and CO_2 to establish a steady rate of photosynthesis. Then they added a solution containing $^{14}CO_2$, which is carbon dioxide in which the normal carbon, ^{12}C, is replaced by the radioisotope, ^{14}C. The radioactive carbon is called a tracer, and the carbon dioxide is described as being labelled or tagged. Algal cells absorb CO_2 and use it in the process of photosynthesis. The cells do not distinguish between radioactive carbon and nonradioactive carbon, and so use $^{14}CO_2$ as they would $^{12}CO_2$.

Calvin and his colleagues allowed the algae in his experiments to undergo photosynthesis using the radioactively labelled $^{14}CO_2$ for various periods of time, and then killed the cells by plunging the suspension into boiling alcohol. The compounds into which the radioactive isotope had become incorporated were then identified by grinding up the dead cells, purifying the liquid extract, and separating the chemicals in the mixture by two-dimensional chromatography (**Figure 11**).

Chromatography is a way of separating the components of a mixture. In this technique, a drop of the mixture is placed in one corner of a square of filter paper. As illustrated in **Figure 11(a)**, one edge of the paper is immersed in a solvent. The solvent moves up the paper by capillary action and carries the substances in the mixture with it, as shown in **Figure 11(b)**. Each compound moves through the paper fibres at a rate that reflects its molecular size and its solubility in the solvent. Once the mixture has separated in one direction, the process is repeated at right angles to the first, usually using a different solvent. **Figure 11(c)** shows the final product: a two-dimensional chromatogram.

Figure 10
Calvin placed suspensions of *Chlorella* in flattened glass "lollipop" containers in his experiments in photosynthesis.

(a)
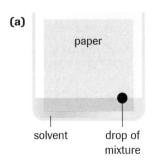
solvent   drop of
               mixture

some hours later
(b)
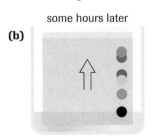
Turn paper 90° clockwise
and use a different solvent.

some hours later
(c)
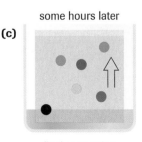
final separation

Figure 11
Identification of compounds by two-dimensional chromatography

In some cases, a piece of the paper containing a spot of the chemical can be cut from the chromatogram and analyzed directly. If the mixture contains radioactively labelled compounds, as in the case of Calvin's experiments, they can be located by placing the chromatogram over a sheet of photographic film (in a darkroom). The radiation emitted by the radioactive compounds will expose the film. Dark spots will be seen on the developed film where radioactive chemicals were located on the chromatogram. The resulting image is called an autoradiogram.

Figure 12 shows two autoradiograms of the type that were produced in Calvin's experiments on photosynthesis. The dark spots show the radioactive compounds produced after 5 s **Figure 12(a)** and 30 s **Figure 12(b)** of photosynthesis. At 5 s, most of the radioactivity is found in the compound 3-phosphoglycerate (PGA). At 30 s, glucose-6-phosphate (glucose-P) and fructose-6-phosphate (fructose-P) have been formed, and a number of amino acids have been created. The small circles at the lower right-hand corners mark the spots where the cell extract was first applied.

Figure 12
Autoradiograms of the type produced in Calvin's experiments on photosynthesis
(a) Radioactive compounds produced after 5 s of photosynthesis
(b) radioactive compounds produced after 30 s of photosynthesis

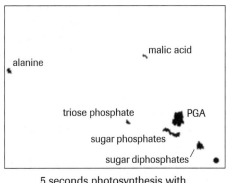

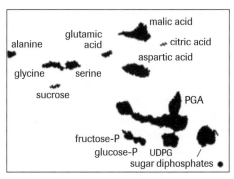

(a) 5 seconds photosynthesis with *Clorella* in $^{14}CO_2$

(b) 30 seconds photosynthesis with *Clorella* in $^{14}CO_2$

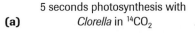

▶ *Case Study* *Questions*

Applying Inquiry Skills

1. Describe the step-by-step procedure Calvin may have used with the lollipop apparatus and autoradiography to determine the order in which compounds are formed in the carbon fixation reactions.

2. Why can the identity of a compound not be determined by cutting out the spot on an autoradiogram and subjecting it to chemical analysis?

3. How could the identity of a compound on an autoradiogram be determined? Suggest two methods.

Making Connections

4. Suggest other possible uses for this technique in science, medicine, or law enforcement.

G3P: A Key Intermediate in Photosynthesis

Higher plants generally fix more carbon dioxide into sugars than they require for their immediate needs. Glyceraldehyde 3-phosphate (G3P), the final product of the Calvin cycle, is used to regenerate RuBP to keep the cycle going and acts as the raw material in the production of other carbohydrates that plant cells use for various structural and functional purposes. A key compound of cellular metabolism is glucose. Glucose may be synthesized from G3P by a series of enzyme-catalyzed reactions in the chloroplast stroma or in the cytoplasm. When optimal conditions allow photosynthesis to produce more glucose than is immediately required, enzymes polymerize glucose into amylose and amylopectin, and store them as starch granules within chloroplasts.

Starch granules may be seen in an electron micrograph of a chloroplast (**Figure 13**). In general, about one third of the G3P not used to regenerate RuBP in the Calvin cycle is converted to starch. Starch is a storage polysaccharide that plants can hydrolyze into individual glucose molecules when environmental conditions, such as a lack of light, inhibit photosynthesis. The other two thirds of the G3P produced is moved out of the chloroplast into the cytoplasm where it is converted into sucrose for transport to other cells of the plant.

Plants possess many nonphotosynthetic tissues, such as roots. The cells of these parts need a supply of carbohydrates to meet their needs. Plants translocate (transport) sucrose from where it is formed in leaf mesophyll cells, through phloem cells in vascular bundles, to all other cells of the plant. Plant cells may hydrolyze sucrose into glucose and fructose, and respire these for energy, or convert them into amino acids and lipids. Alternatively, the glucose portion of sucrose may be polymerized into starch, cellulose, or other polysaccharides. **Figure 14** outlines the possible paths of G3P and other triose phosphates after they are produced by photosynthesis in the chloroplasts of leaf mesophyll cells. 🔬▮

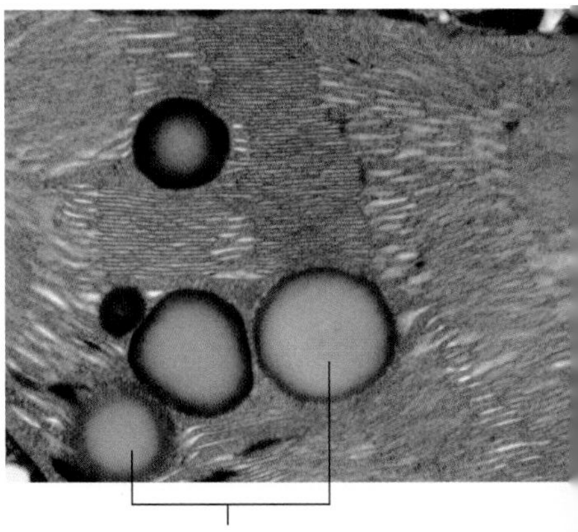

starch granules

Figure 13
Starch granules in a chloroplast

🔬 **INVESTIGATION 3.3.1**

Factors Affecting the Rate of Photosynthesis (p. 188)
Photosynthesis involves light reactions and biochemical reactions that do not directly require light energy. Plants live in a variety of environments on Earth. Do changes in environmental conditions affect the rate of photosynthesis? In Investigation 3.3.1, you will design experiments to measure the rate of photosynthesis in various conditions of light intensity, temperature, CO_2 concentration, and other factors.

DID YOU KNOW ?

Enough Oxygen in 11 Days
One hectare of corn can convert up to 10 000 kg of carbon from CO_2 into 25 000 kg of sugar in one growing season. It also produces enough oxygen per day in midsummer to meet the respiratory needs of about 325 people. This means that the 1 million or so hectares of corn grown in Ontario produce enough oxygen to meet the annual respiratory needs of Ontario's 10 million residents in about 11 summer days!

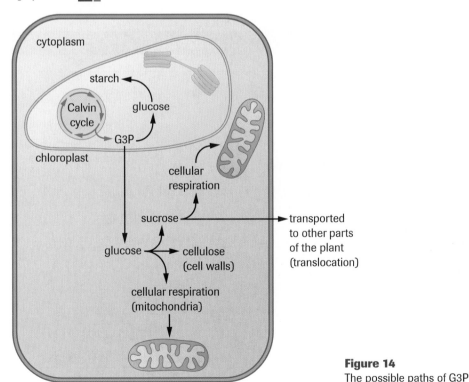

Figure 14
The possible paths of G3P

Photosynthesis: The Details

- The light reactions begin when photons strike a photosynthetic membrane.
- Chlorophyll emits red light when its excited electrons return to their ground state. This happens when the molecule is separated from a photosynthetic membrane, but, to some degree, also occurs in the chloroplasts of living cells. A solution of chlorophyll fluoresces red when exposed to bright white light.
- A photosystem contains chlorophyll molecules, accessory pigments, and proteins embedded in a thylakoid membrane. The antenna complex of a photosystem is composed of a number of chlorophyll molecules and accessory pigments set in a protein matrix and embedded in the thylakoid membrane.
- An antenna pigment absorbs a photon and transfers the energy from pigment to pigment until it reaches a chlorophyll *a* molecule in an area called the reaction centre. An electron of the reaction centre chlorophyll molecule absorbs the transferred energy and is raised to a high energy level. A redox reaction transfers the excited electron to the primary electron acceptor.
- Chloroplast membranes contain two types of photosystems. Photosystem I is called P700 because its absorption spectrum peaks at a wavelength of 700 nm (red light). The chlorophyll *a* molecule in the reaction centre of photosystem II is called P680 because it is best at absorbing photons with a wavelength of 680 nm (red light).
- G3P is the primary end product of the Calvin cycle. It may be converted into glucose and polymerized into starch within the stroma of chloroplasts, or it may be transported into the cytoplasm and used to produce glucose and sucrose. Sucrose is the main carbohydrate transported from mesophyll cells of the leaf to other cells of the plant.

▶ Section 3.3 Questions

Understanding Concepts

1. Define the following terms: *ground state, excitation*, and *fluorescence*.

2. (a) What is the primary function of photosynthesis?
   (b) Where in the chloroplast do the light reactions occur?
   (c) What are the products of the light reactions, if you assume they involve noncyclic electron flow?
   (d) In what phase of photosynthesis are the products of the light reactions used?

3. (a) Name the gas released as a byproduct of the light reactions of photosynthesis.
   (b) Name the molecule that is the source of this gas.

4. (a) How many electrons must be removed from photosystem II to reduce one molecule of NADP⁺?
   (b) Name the series of electron carriers that transport electrons from photosystem II to photosystem I.
   (c) What happens to the free energy that the electrons lose in this process?

5. **Figure 5** on page 160 illustrates the energy changes that take place in noncyclic electron flow. Draw your own diagram or write a story that describes these energy changes in your words.

6. Identify A, B, C, D, and E in **Figure 15**.

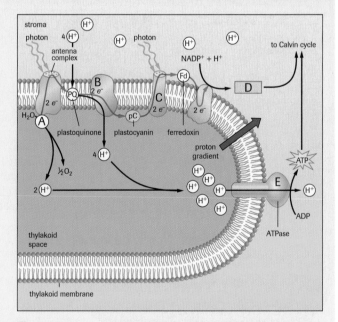

Figure 15

7. The herbicide 3-(3,4-dichlorophenyl)-1,1,-dimethylurea (DCMU) blocks the transport of electrons from photo-system II to the cytochrome b_6-f complex.
    (a) How will this affect a chloroplast's ability to produce ATP? to produce NADPH?
    (b) Why is DCMU an effective herbicide?

8. (a) Distinguish between cyclic and noncyclic electron flow in terms of
    (i) evolution of O_2;
    (ii) production of NADPH;
    (iii) production of ATP;
    (iv) enabling the Calvin cycle to fix CO_2.
    (b) How does cyclic electron flow support photosynthesis?

9. (a) What is the name of the enzyme that catalyzes the carbon fixation reaction of the Calvin cycle?
    (b) What are the two substrates of this enzyme when it acts as a carboxylase?
    (c) What is the name of the product of the first reaction of the Calvin cycle?
    (d) Where in the chloroplast does this reaction occur?

10. (a) Why does a suspension of isolated chloroplasts not synthesize G3P in the dark, given CO_2 and H_2O?
    (b) What would have to be added to the test tube for photosynthesis to occur?

11. What is the name of the three-carbon carbohydrate that is a final product of the Calvin cycle? What are the possible fates of this compound?

12. How many molecules of CO_2 must enter the Calvin cycle for a plant to ultimately produce a sugar, such as sucrose, that contains 12 carbon atoms? How many ATP molecules will be used? How many NADPH molecules will be used? How many photons must be absorbed (assume noncyclic electron flow only)?

13. Which contains more free energy: three molecules of CO_2 or one molecule of phosphoglyceraldehyde? Explain.

Applying Inquiry Skills

14. In an experiment, a bean plant is illuminated with green light and another bean plant of similar size is illuminated with equally intense blue light. If all other conditions are controlled, how will the photosynthetic rates of the two plants most probably compare?

15. (a) Describe a procedure for showing that the chlorophyll in the leaves of a spinach plant fluoresces when exposed to bright white light.
    (b) Will chlorophyll display fluorescence if exposed to bright green light? Explain.

Making Connections

16. Recent advances in remote sensing have made detection of plant health possible on a large scale. Using satellite images, spectral analysis, and other sensing technologies, farmers may now detect problems in large fields of crops before they are identified at ground level. Conduct library and/or Internet research about spectral remote sensing as applied to plants to answer the following questions:
    (a) What characteristic(s) of plants do remote sensing systems detect to provide information regarding a crop's overall health?
    (b) Why would a farmer spend money to have crops tested by these methods? What advantages are gained by the procedure?

 www.science.nelson.com

Alternative Mechanisms of Carbon Fixation

Terrestrial plants absorb CO_2 for photosynthesis through stomata that also allow for transpiration. When the weather is hot and dry, guard cells decrease the size of stomata in an effort to conserve water. As a result, CO_2 concentration in the leaves' air spaces declines. Since the cells continue cellular respiration, oxygen levels increase. Oxygen competes with carbon dioxide for rubisco's active site.

Rubisco is an enzyme that may catalyze two reactions: the addition of carbon dioxide to RuBP (the first reaction of the Calvin cycle, a carboxylation reaction) or the addition of oxygen to RuBP (an oxidation reaction). When oxygen is more plentiful than carbon dioxide, oxygen binds more often, resulting in the oxidation of RuBP, instead of its carboxylation to PGA. This process is called **photorespiration** because it occurs in light.

Photorespiration decreases the production of carbohydrates by photosynthesis, since it removes PGA molecules from the Calvin cycle. When photorespiration occurs, one 3-carbon PGA molecule and one 2-carbon glycolate molecule are formed from one 5-carbon RuBP, instead of the two PGA molecules that are formed when carbon dioxide reacts. Glycolate is partially converted to CO_2 but three of four carbons in two glycolate molecules are subsequently returned to the Calvin Cycle as G3P. In fact, C_3 plants (those that use the Calvin cycle to fix carbon), such as soybeans, Kentucky bluegrass, and sunflowers, lose from one quarter to one half of the carbon they fix to photorespiration. Under normal conditions, the rate of carbon fixation is four times that of oxidation. Thus, about 20% of fixed carbon is lost to photorespiration. In tropical climates, the persistent warm temperatures exacerbate this problem in C_3 plants. Since the optimum temperature for photorespiration (30°C–47°C) is much higher than that for photosynthesis (15°C–25°C), photorespiration rates may still be increasing when that of C_3 photosynthesis is starting to decline (**Figure 1**). Hot, dry, bright days produce the conditions that facilitate photorespiration.

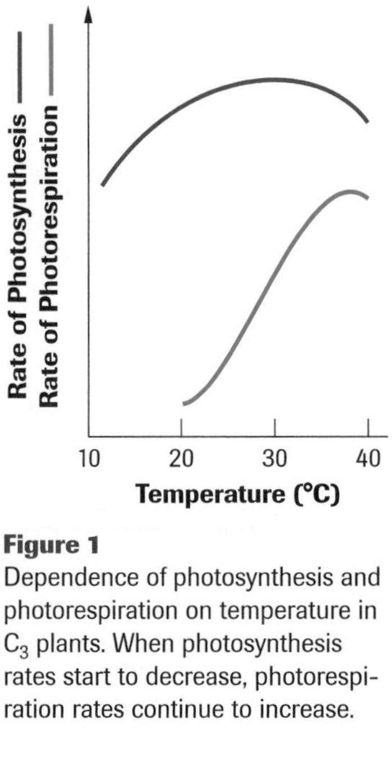

Why do plants undergo photorespiration? One hypothesis maintains that this curious characteristic of rubisco's activity is an evolutionary remnant of an earlier mechanism, better suited to an early Earth whose atmosphere was rich in carbon dioxide and poor in oxygen. When photosynthesis evolved in early prokaryotes, rubisco functioned well as a mechanism of fixing carbon dioxide. Meanwhile, as the byproduct of photosynthesis, oxygen began to accumulate in the atmosphere. As a result, these early autotrophs made life possible for eukaryotic heterotrophic organisms. Unfortunately, rubisco turned out to have an oxygenase activity that was associated with the same active site as the one that fixed carbon dioxide. Over evolutionary time, plants did not evolve a modified rubisco or a new fixing enzyme that would bind carbon dioxide and not oxygen. However, some plant species have evolved alternative mechanisms of carbon fixation that concentrate CO_2 at the site where rubisco is found, thus effectively suppressing the rate of photorespiration. Two important strategies are C_4 photosynthesis and crassulacean acid metabolism (CAM).

photorespiration oxidation of RuBP by rubisco and oxygen in light to form glycolate, which subsequently releases carbon dioxide

Figure 1
Dependence of photosynthesis and photorespiration on temperature in C_3 plants. When photosynthesis rates start to decrease, photorespiration rates continue to increase.

C_4 photosynthesis a photosynthetic pathway of carbon fixation that reduces the amount of photorespiration that takes place by continually pumping CO_2 molecules (via malate) from mesophyll cells into bundle-sheath cells, where rubisco brings them into the C_3 Calvin cycle.

C_4 Plants

Several thousand species of plants—including sugar cane and corn, both members of the grass family, and broad-leaved sorghum—undergo a form of photosynthesis called **C_4 photosynthesis**. In this process, an enzyme called phosphenolpyruvate carboxylase (PEP carboxylase) first catalyzes the addition of a CO_2 molecule to a three-carbon molecule called phosphenolpyruvate (PEP), forming the four-carbon molecule oxaloacetate (OAA). This is why the process is called C_4 photosynthesis or the C_4 pathway.

C_4 plants possess a unique leaf anatomy that facilitates this form of photosynthesis, as shown in **Figure 2(a)**. The leaves of these plants contain two types of photosynthetic cells: bundle-sheath cells surrounding a vein, and mesophyll cells that are located around the bundle-sheath cells. No mesophyll cell of a C_4 plant is more than two or three cells away from a bundle-sheath cell.

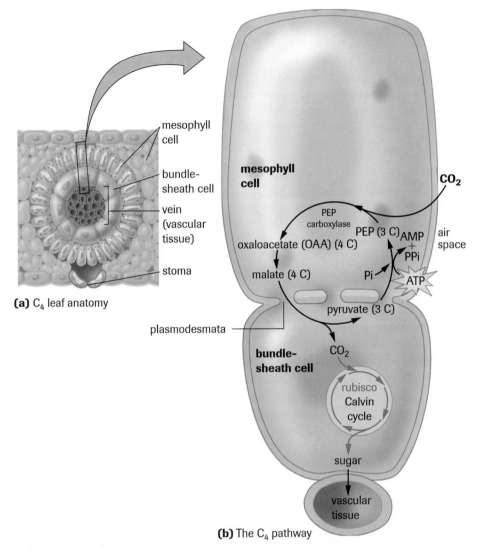

(a) C_4 leaf anatomy

(b) The C_4 pathway

Figure 2
C_4 plant anatomy and reactions
(a) C_4 plant leaves contain two types of photosynthetic cells: bundle-sheath cells surrounding a vein and mesophyll cells that are located around the bundle-sheath cells.
(b) The enzyme PEP carboxylase fixes CO_2 in mesophyll cells by catalyzing the attachment of CO_2 to PEP to form the four-carbon acid OAA, which is then converted into malate. This four-carbon acid enters bundle-sheath cells through cell–cell connections called plasmodesmata. Here, the malate molecule is decarboxylated, releasing carbon dioxide prior to a second fixation by rubisco where the carbon dioxide enters the Calvin cycle. One ATP is used in the process of regenerating PEP.

LEARNING TIP

Notice that ATP is cleaved into AMP (adenosine monophosphate) and PP_i (double-phosphate or pyrophosphate) instead of ADP and P_i in the endergonic step that converts pyruvate into PEP in the C_4 pathway. This transfers approximately the same amount of energy as an ATP → ADP + P_i reaction ($\Delta G = -31$ kJ/mol in both cases).

plasmodesmata membrane-lined channels between plant cells that allow for the movement of some substances from cell to cell

In the C_4 pathway, the enzyme PEP carboxylase fixes CO_2 by catalyzing the attachment of CO_2 to three-carbon PEP, forming four-carbon OAA, as **Figure 2(b)** illustrates. This rection occurs in the cytoplasm, not the chloroplast. OAA is converted to malate (OAA and malate are also found in the Krebs cycle). Malate, a four-carbon acid, diffuses from the mesophyll cells into bundle-sheath cells through cell–cell connections called **plasmodesmata**. Here, a carbon dioxide portion is removed from the malate molecule (decarboxylation). This reaction converts 4-carbon malate into 3-carbon pyruvate, which diffuses back into the mesophyll cell where it is converted into PEP (the compound we started with). This reaction is endergonic and uses one ATP. The CO_2 that was removed from malate in the bundle-sheath cell enters the C_3 Calvin cycle in a second fixation reaction, this time catalyzed by rubisco in the bundle sheath cell.

crassulacean acid metabolism (CAM) a photosynthetic mechanism in which stomata open at night so that plants can take in CO_2 and incorporate it into organic acids, and close during the day to allow the organic acids to release CO_2 molecules that enter the C_3 Calvin cycle to be fixed into carbohydrates

This method of carbon fixation reduces the amount of photorespiration that takes place by continually pumping CO_2 molecules from the mesophyll cells to the bundle-sheath cells (via malate), where rubisco brings them into the Calvin cycle. The concentration of CO_2 in bundle-sheath cells is kept high so that CO_2 outcompetes O_2 in binding to rubisco. Photorespiration is minimized and sugar production is maximized. However, it costs the plant two ATP molecules to transport a CO_2 molecule into a bundle-sheath cell. Since six CO_2 molecules are processed by the Calvin cycle to produce the equivalent of one glucose molecule, the cell must expend the energy of 12 additional ATP molecules to fix six CO_2 molecules. In C_3 photosynthesis, 18 molecules of ATP are used to produce one glucose molecule, whereas, in C_4 plants, 30 ATP molecules are used—almost twice as many. Nevertheless, the process is advantageous in hot tropical climates where photorespiration would otherwise convert more than half of the glucose produced back to CO_2.

CAM Plants

Water-storing plants (known as succulents), such as cacti and pineapples, and members of the *Crassulacea* family (including the jade plant), open their stomata at night and close them during the day—the reverse of other plants (**Figure 3**).

Closing stomata during the day helps conserve water but prevents CO_2 from entering the leaves. In the dark, when stomata open, the plants take in CO_2 and incorporate it into C_4 organic acids using the enzyme PEP carboxylase. This form of carbon fixation is called **crassulacean acid metabolism (CAM)** because it was first discovered in the crassulacean family of plants.

The organic acids produced by CAM plants at night are stored in vacuoles until morning. At this time, stomata close and the organic acids release CO_2 molecules that enter the C_3 Calvin cycle to be fixed into carbohydrates. In C_4 plants, the first part of carbon fixation and the Calvin cycle occur in separate compartments of the leaf. In CAM plants, the two steps occur in the same compartments, but at different times of the day (**Figure 3**).

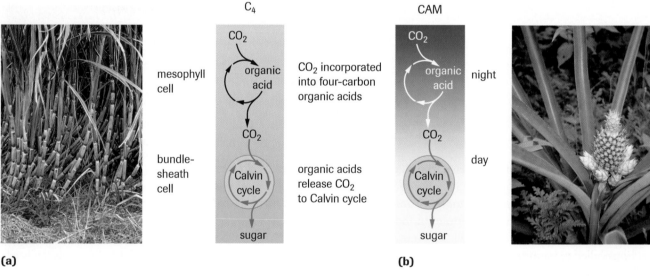

(a)   (b)

Figure 3
Comparing C_4 and CAM processes
(a) In C_4 plants, such as sugar cane, CO_2 is first incorporated into C_4 organic acids in mesophyll cells. The C_4 organic acid enters bundle-sheath cells through cell-cell connections and then releases carbon dioxide, which is fixed via the Calvin cycle.
(b) In CAM plants, such as pineapples, carbon fixation into organic acids occurs at night, and the Calvin cycle occurs in the day.

The C_4 and CAM pathways represent evolutionary solutions to the problem of maintaining photosynthesis when stomata close on sunny, hot, dry days. Both methods initially produce organic acids that eventually transfer CO_2 to the C_3 Calvin cycle. In C_4 plants, these two processes occur in two different types of cells that are connected by plasmodesmata (spatial separation). In CAM plants, the two processes occur in the same compartment, but at different times (temporal separation): carbon fixation into organic acids during the night and the Calvin cycle during the day. The C_4 pathway uses almost twice as much ATP as the C_3 pathway to produce glucose, but without this mechanism, photorespiration would reduce the glucose yield to less than half that produced by the C_3 pathway.

DID YOU KNOW ?

Houseplants

If you grow cacti and succulents at home, they will only show CAM photosynthesis if they are kept in desertlike dry conditions. If you overwater these plants, they cease opening their stomata at night and become conventional C_3 plants, opening their stomata during the day and using a single carbon dioxide fixation.

SUMMARY   *Alternative Methods of Carbon Fixation*

- Alternate methods of carbon fixation are summarized in **Figure 4.**, under normal atmospheric conditions (low CO_2/high O_2)

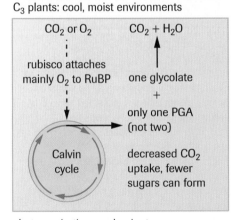

C_3 plants: cool, moist environments

C_4 plants: hot, dry environments

CAM plants: hot, dry, and desert environments

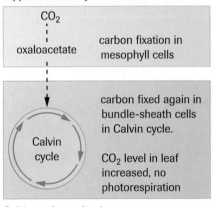

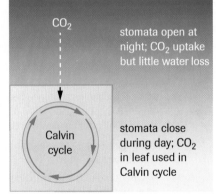

photorespiration predominates

Calvin cycle predominates; no photorespiration

Calvin cycle predominates; no photorespiration

Figure 4
C_3, C_4, and CAM photosynthesis

- A typical C_4 leaf cross section is illustrated in **Figure 5**.

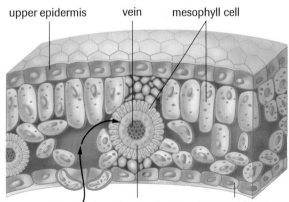

CO_2 moves through stoma, into air spaces in leaf   bundle-sheath cell   lower epidermis

Figure 5
Typical C_4 leaf (cross section)

Understanding Concepts

1. (a) Define *photorespiration*.
   (b) What gas can compete with CO_2 for the binding site of the enzyme rubisco?
   (c) Under normal conditions, what proportion of fixed carbon is affected by photorespiration in C_3 plants?
   (d) Compare the end products of photosynthesis and photorespiration.

2. How does temperature affect the relative amounts of photosynthesis and photorespiration that occur in C_3 plants?

3. (a) Label A, B, C, D, and E in **Figure 5**.
   (b) What type of cell–cell connection do malate and pyruvate go through to move from one cell into the other?

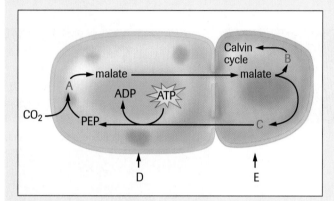

Figure 5

4. (a) What is the main difference between the ideal environments of C_4 plants and CAM plants?
   (b) Name two C_4 plants and two CAM plants.

5. What might have been responsible for the evolution of photorespiration?

6. (a) At what time of the day would you expect to find the most malate in CAM plants?
   (b) When would you find the least amount of malate in CAM plants?
   (c) Why do plants that use CAM photosynthetic pathways close their stomata during the day?

(d) During the cool of evening, CAM plants open their stomata. What gas is preferentially absorbed at this time?
(e) Explain how this gas is stored for daytime use.

Applying Inquiry Skills

7. (a) Write a letter to the Canadian Chemical Association supporting the nomination of rubisco as Enzyme of the Year.
   (b) Write a letter to the Canadian Chemical Association explaining why rubisco does not deserve to be Enzyme of the Year.

8. Write a hypothesis about the acid content of two different CAM plants, one of which has been in sunlight for 24 h previous to testing and one that has been kept in the dark for 24 h.

Making Connections

9. If you were a researcher looking for new plant species that exhibited CAM photosynthesis, what biome or biomes would you explore? Why?

10. If the greenhouse effect persists and gets worse, which photosynthetic pathway (C_3, C_4, or CAM) would benefit the environment the most? Explain.

11. Conduct library and/or Internet research to answer the following questions regarding weeds:
    (a) What is a weed?
    (b) Of the ten most aggressive weeds in the world, eight of them are C_4 plants. Identify and name three common North American C_4 plants that are classified as weeds.
    (c) Some of the world's most productive crops are C_4 plants. Identify and name three important North American crops that are C_4 plants.
    (d) How will escalation of the greenhouse effect (a result of increased atmospheric carbon dioxide concentrations) affect the competition between C_3 and C_4 plants in terms of crop productivity and weed control?

 www.science.nelson.com

Photosynthetic organisms continually exchange materials with their immediate environment. Thus, the rate of photosynthesis may be determined by measuring the rate at which a reactant is used or the rate at which a product is formed. The balanced overall equation for photosynthesis indicates that carbon dioxide and water are consumed and sugars and oxygen are evolved in the process:

$$6CO_{2(g)} + 6H_2O_{(l)} + light \longrightarrow [CH_2O] + 6O_{2(g)}$$

Since the carbohydrate product remains within the cell, it is usually more practical, in a laboratory setting, to determine the rate of photosynthesis by measuring the rate at which a plant absorbs carbon dioxide or the rate at which it evolves oxygen gas. However, carbon dioxide and oxygen participate in two other metabolic processes within a plant: cellular respiration (in all plants) and photorespiration (primarily in C_3 plants). Both of these pathways consume oxygen and evolve carbon dioxide. Like photosynthesis, photorespiration is light dependent, but cellular respiration is independent of light and continues to occur in the dark (**Figure 1**).

Therefore, when carbon dioxide uptake or oxygen gas evolution are being measured, it is necessary to measure *net* gas exchange between a plant and its environment, according to the following equations:

net CO_2 uptake = photosynthetic CO_2 uptake − photorespiratory CO_2 evolution − respiratory CO_2 evolution

net O_2 evolution = photosynthetic O_2 evolution − photorespiratory O_2 uptake − respiratory O_2 uptake

Carbon dioxide uptake is typically measured in units of millimoles of carbon dioxide used (assimilated) per unit surface area (of leaf) per second (mmol $CO_2/m^2/s$). Similarly, oxygen evolution is measured in units of millimoles of oxygen (evolved) per unit surface area (of leaf) per second (mmol $O_2/m^2/s$).

Light Intensity and the Rate of Photosynthesis

Irradiance is measured as light intensity per unit area of leaf (Light intensity is measured in candelas, abbreviated cd). The graph in **Figure 2** illustrates the effect of increasing irradiance on the rate of photosynthesis in C_3 plants, measured as the rate of net CO_2 uptake. The data used to generate this graph (known as a light-response curve) were collected using a plant kept at a constant temperature of 25°C, and a constant ambient CO_2 concentration of 336 parts per million (336 ppm, or 0.0336%). Notice that when the irradiance is zero (darkness), the rate of net CO_2 uptake is a negative value. A negative uptake means that there is a net amount of CO_2 given off by the plant. In the dark, photosynthesis is not occurring, so a net amount of CO_2 is evolved by the process of cellular respiration. As irradiance increases, photosynthesis begins, and the rate of photosynthetic CO_2 uptake increases until it equals CO_2 release by mitochondrial respiration. The point at which the rate of photosynthetic CO_2 uptake exactly equals the rate of respiratory CO_2 evolution is called the **light-compensation point**. The rate of net CO_2 uptake is zero at the light-compensation point.

In the light

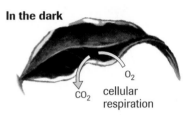

In the dark

Figure 1
Photorespiration requires light, but cellular respiration does not.

light-compensation point the point on a light-response curve at which the rate of photosynthetic CO_2 uptake exactly equals the rate of respiratory CO_2 evolution

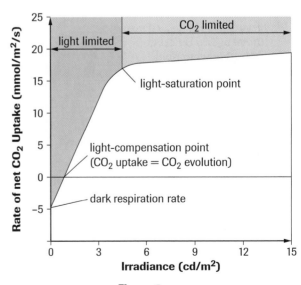

Figure 2
A typical light-response curve of photosynthesis in C_3 plants

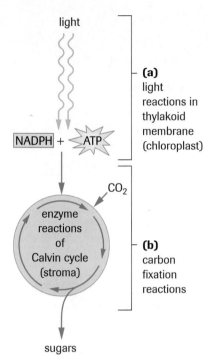

light

NADPH + ATP

(a)
light
reactions in
thylakoid
membrane
(chloroplast)

CO_2

enzyme
reactions
of
Calvin cycle
(stroma)

(b)
carbon
fixation
reactions

sugars

Figure 3
(a) The light reactions of photosynthesis
(b) the carbon fixation reactions of photosynthesis

light-saturation point the irradiance level at which the carbon fixation reactions reach a maximum overall rate

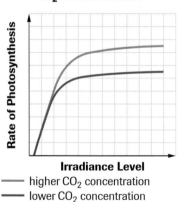

Photosynthesis Light Response Curve at Two Different CO_2 Concentrations

Rate of Photosynthesis

Irradiance Level
— higher CO_2 concentration
— lower CO_2 concentration

Figure 4
The effect of CO_2 concentration on the rate of photosynthesis

Beyond the light-compensation point, the rate of photosynthesis increases in direct proportion with increases in irradiance—the relationship between the rate of photosynthesis and irradiance is linear (**Figure 2**, page 173). The rate of photosynthesis is said to be *light limited* in this range. The reason for this relationship is related to the fact that photosynthesis occurs in two stages: the light reactions and the carbon fixation reactions. The light reactions transfer the energy of light to NADPH and ATP, as shown in **Figure 3(a)**. The carbon fixation reactions use the reducing power of NADPH and the energy of ATP to fix carbon dioxide into carbohydrates, as illustrated in **Figure 3(b)**. As you know, the carbon fixation reactions (the Calvin cycle) are catalyzed by enzymes.

In all enzyme-catalyzed reactions, the rate of reaction increases as the substrate concentration increases, until the enzymes are working at full capacity. At this point, the enzymes are saturated with substrate and cannot convert substrate to product any faster—the reaction has reached its maximum rate. Increasing substrate concentration beyond the saturation point will not increase the reaction rate any further. In the light-limited range of the light-response curve, the carbon fixation reactions have not yet reached their maximum rate (the enzymes are not yet saturated with substrate). Increasing irradiance results in an increase in the production of NADPH and ATP by the light reactions and an increase in the rate of carbon fixation. This results in an increase in the overall rate of photosynthesis. Therefore, in this region of the graph, the light reactions (not the carbon fixation reactions) limit the overall rate of photosynthesis. It is evident that as irradiance increases, we will eventually reach a point at which the reactions of the Calvin cycle reach a maximum overall rate (at a particular temperature). The irradiance level at which this occurs is called the **light-saturation point** (**Figure 2**). Increasing the production of NADPH and ATP in the light reactions with higher intensities of light cannot increase the overall rate of photosynthesis any more because the enzymes of the Calvin cycle are operating at full capacity and cannot convert substrate to product any faster. Thus, the overall rate of photosynthesis reaches a constant value. This fact is illustrated by the plateau in the light-response curve. The plateau is called the *CO_2-limited range* because, under normal conditions, carbon dioxide availability, not irradiance, limits the overall rate of photosynthesis.

The concentration of CO_2 in the atmosphere is approximately 0.037%. Burning fossil fuels and deforestation have led to a general increase in the concentration of CO_2 in the atmosphere. In the last one hundred years, the concentration of atmospheric CO_2 has increased approximately 25% (from about 0.029% to 0.037%). Many scientists fear that elevated CO_2 concentrations contribute to global warming through the greenhouse effect, which may, in time, change global climate patterns. Some plant biologists claim that elevated atmospheric temperatures and rising CO_2 levels may increase photosynthesis rates and possibly lead to higher C_3 crop yields in the future. **Figure 4** illustrates the effect of a higher CO_2 concentration on the rate of photosynthesis.

Temperature and the Rate of Photosynthesis

The light reactions of photosynthesis are a series of redox reactions occurring in the thylakoid membrane. The rate of these reactions is affected by changes in the irradiance level, but is not significantly influenced by changes in temperature. However, the overall rate of the Calvin cycle is affected by changes in temperature because these reactions are catalyzed by enzymes. Enzymes possess optimal temperature conditions in which their activity is maximal. As shown in **Figure 5**, an increase in temperature affects the overall rate of photosynthesis. Between 10°C and 30°C, the rate of photosynthesis increases as the temperature goes up, primarily because of an increase in the overall rate of the enzyme-catalyzed reactions of the Calvin cycle. However, as temperatures approach

40°C and beyond, enzyme activity is affected by changes in enzyme structure (enzymes become denatured), resulting in lower rates. In most cases, photosynthesis will come to a halt in temperatures that cause enzymes to denature.

Oxygen Concentration and the Rate of Photosynthesis

Approximately 21% of the atmosphere is composed of oxygen. Autotrophs produce oxygen as a byproduct of photosynthesis and are responsible for having created virtually all of the oxygen on Earth. As you know, high oxygen levels have an inhibitory effect on photosynthesis because of competition between O_2 and CO_2 for the active sites on rubisco. As shown in **Figure 6**, overall photosynthesis rates are lower at higher oxygen concentrations. When carbon dioxide binds, it is fixed into carbohydrates by photosynthesis; when oxygen binds, photorespiration occurs, and the rate of photosynthesis is reduced.

Photosynthetic Efficiency

The **photosynthetic efficiency** (also called quantum yield) of a plant is defined as the net amount of carbon dioxide uptake per unit of light energy (photons) absorbed. This value may be determined by calculating the initial slope of the light-response curve (the straight line portion at low light intensities). **Figure 7(a)** illustrates the light-response curve for a typical C_3 plant at 10°C and 25°C. Notice that the initial slope of the curve is greater at the lower temperature. As mentioned earlier in this chapter, in C_3 plants, such as sunflowers and wheat, the rate of photorespiration increases more rapidly than does the rate of photosynthesis as temperature increases. Therefore, in most C_3 plants, the net CO_2 uptake per unit of light absorbed decreases as temperature increases, as line A in **Figure 7(b)** shows. This occurs because the rate of CO_2 output by photorespiration increases more than the rate of CO_2 utilization by photosynthesis.

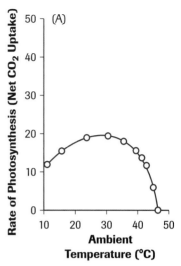

Figure 5
The relationship between ambient temperature and the rate of photosynthesis

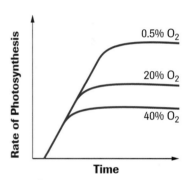

Figure 6
The relationship between the rate of photosynthesis and ambient oxygen concentration

photosynthetic efficiency of a plant is defined as the net amount of carbon dioxide uptake per unit of light energy absorbed; also called quantum yield

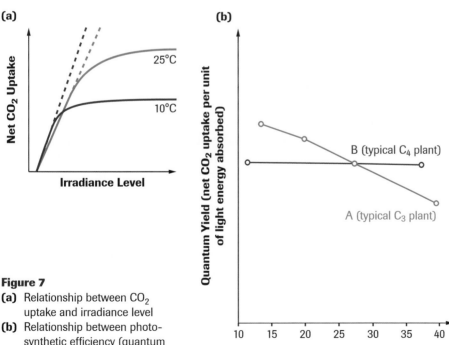

Figure 7
(a) Relationship between CO_2 uptake and irradiance level
(b) Relationship between photosynthetic efficiency (quantum yield) and leaf temperature

Plants that fix carbon dioxide by the C_4 pathway generally exhibit higher photosynthetic rates at higher temperatures than do C_3 plants. This is largely due to the fact that, although photorespiration rates increase with increasing temperatures in C_3 plants (resulting in up to 50% CO_2 loss), the loss of CO_2 by photorespiration is virtually nil in C_4 plants. Thus, in C_4 plants, the net CO_2 uptake per unit of light absorbed remains constant as temperature increases, as line B in **Figure 7(b)** on page 175 illustrates. The graph shows that C_4 plants are more efficient at fixing carbon into carbohydrates at higher temperatures, and C_3 plants are more efficient photosynthesizers at lower temperatures. This is why corn, a C_4 plant, grows best during hot summers. Some of the most aggressive weeds in North America are C_4 plants. As **Figure 8** shows, during hot, dry summers, weeds, such as crabgrass and purslane (C_4 plants), outcompete typical C_3 grasses, such as Kentucky bluegrass, and grow rapidly on lawns.

(a) **(b)**

Figure 8
(a) Crabgrass
(b) Kentucky bluegrass

Sun Plants Versus Shade Plants

You have probably noticed that some plants are better adapted for growth in direct sunlight, whereas others prefer shade or indirect light. This is noticeable by simply walking down an east/west road in an urban residential neighbourhood and noting the difference in vegetation in the front yards on the north and south sides. This is also true of the leaves on some trees, especially on very tall trees. Leaves that develop under the shade of other leaves are structurally and metabolically different from those that are exposed to the Sun's direct rays.

Shade plants, shown in **Figure 9(a)**, typically possess leaves that are thinner, broader, and greener (meaning they contain more chlorophyll) than the leaves of a sun plant, shown in **Figure 9(b)**. As a result, shade plants are, in general, more efficient at harvesting light at low intensities. As **Figure 10** shows, the shade plant has higher photosynthetic efficiency in comparison to the sun plant, which displays a higher light-saturation point and maximum rate of photosynthesis.

In general, sun-adapted plants have higher light-compensation points than do those that grow best in the shade. The light-compensation points for shade plants are lower because these plants have lower respiration rates than do sun plants. It appears that a low respiration rate is an adaptation that allows shade plants to survive in light-limited conditions.

(a)

(b)

Figure 9
(a) a shade plant; **(b)** a sun plant

Figure 10
Light-response curves of sun and shade plants

![Rate of Photosynthesis vs Irradiance Level graph showing curve A (sun plant) and curve B (shade plant)]

- - - photosynthetic efficiency A (sun plant)

- - - photosynthetic efficiency B (shade plant)

| SUMMARY | *Photosynthesis and the Environment* |

- The light-compensation point is the point on a light-response curve at which the rate of photosynthetic CO_2 uptake exactly equals the rate of respiratory CO_2 evolution.

- In a light-response curve, the rate of photosynthesis is light limited between the light-compensation point and the light-saturation point. In this range, providing more light causes the rate of photosynthesis to increase.

- The light-saturation point is the irradiance level at which the carbon fixation reactions reach a maximum overall rate.

- The CO_2-limited phase of the light-response curve is a plateau because carbon dioxide availability, not irradiance level, limits the overall rate of photosynthesis.

- An increase in temperature has virtually no effect on the overall rate of photosynthesis within the light-limited phase, but causes a marked increase in rate within the CO_2-limited phase.

- In general, C_3 photosynthesis rates are lower at higher oxygen concentrations.

- Photosynthetic efficiency (also called quantum yield) of a plant is the net amount of carbon dioxide uptake per unit of light energy absorbed.

- Plants that fix carbon dioxide by the C_4 pathway generally exhibit higher photosynthetic rates at higher temperatures than do C_3 plants. In general, photorespiration rates increase with increasing temperatures in C_3 plants, whereas there is virtually no loss of CO_2 by photorespiration in C_4 plants.

Table 1 Sun Plants versus Shade Plants

	Shade plants	**Sun plants**
light-compensation point	lower	higher
light-saturation point	lower	higher
maximum rate of photosynthesis	lower	higher
concentration of chlorophyll	higher	lower
photosynthetic efficiency	higher	lower

Understanding Concepts

1. What process limits the rate of photosynthesis at low light levels?

2. Why do C_4 plants generally exhibit higher photosynthetic rates at higher temperatures than do C_3 plants?

Applying Inquiry Skills

3. Why is the amount of oxygen evolved per unit time (mmol $O_2/m^2/s$) not a good measure of the rate of photosynthesis of a plant?

4. Two students conducted an experiment to study the rate of photosynthesis in a tomato plant. One student designed an apparatus that measures the rate of water absorption; the other student used a setup that measures oxygen uptake. Will both systems yield reliable evidence of the rate of photosynthesis? Explain.

5. (a) Identify point A on the graph in **Figure 11**.
   (b) What is the significance of point A?
   (c) Identify point B on the graph in **Figure 11**.
   (d) What is the significance of point B?

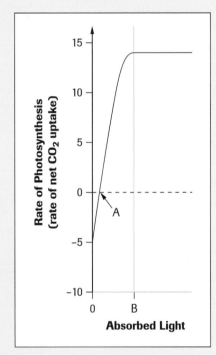

Figure 11
Light-response curve

6. A tobacco plant is first grown outdoors where the carbon dioxide concentration is measured to be 336 ppm, then transferred to an indoor chamber containing a carbon dioxide concentration of 450 ppm. **Figure 12** shows the light-response curve for this plant while growing outdoors. Copy the graph into your notebook and draw on it the light-response curve for the tobacco plant while growing in the chamber. Assume the temperature is constant.

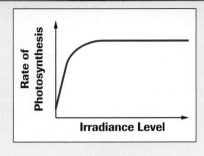

Figure 12
Light-response curve for tobacco plant

7. Which of the light-response curves in **Figure 13** shows the highest photosynthetic efficiency? Explain.

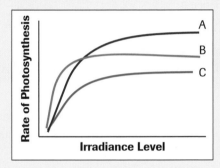

Figure 13
Light-response curves

8. **Figure 14** is a plot of the rate of photosynthesis versus ambient CO_2 concentration for a C_3 plant and a C_4 plant.
   (a) Explain the difference in the two CO_2 compensation points.
   (b) Why does the rate of C_4 plant photosynthesis rise faster than that of C_3 photosynthesis as CO_2 concentration increases?
   (c) What do these curves tell you about the relative success of C_3 and C_4 plants in an environment where CO_2 concentrations are rising?

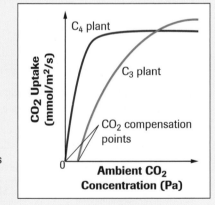

Figure 14
Rate of photosynthesis versus ambient CO_2 concentration

Making Connections

9. How will increasing CO_2 concentrations and rising temperatures affect the ratio of C_3 to C_4 plants on Earth? Do you think this is positive or negative?

Photosynthesis and cellular respiration are closely related to one another. In plants and other autotrophs, both processes may occur within individual cells. This is possible because plants contain mitochondria and chloroplasts. Animals and other heterotrophs undergo cellular respiration, but not photosynthesis. Nevertheless, heterotrophs require the products of photosynthesis to carry out cellular respiration. Photosynthesis uses the products of cellular respiration, and cellular respiration uses the products of photosynthesis (**Figure 1**).

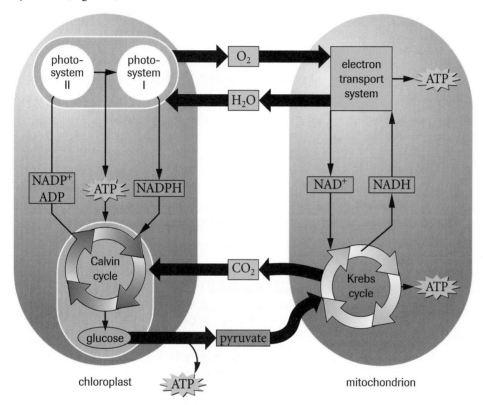

Figure 1
The food/energy cycle. Photosynthesis uses the products of cellular respiration and cellular respiration uses the products of photosynthesis.

Photosynthesis and cellular respiration are related in several other ways:

- The Calvin cycle includes reactions that are similar to reactions in cellular respiration but that occur in reverse.

- The proteins, quinones and cytochromes of the electron transport chains in photosynthetic membranes and in respiratory membranes are similar in structure and, in some cases, are exactly the same.

- Both processes use chemiosmosis to transform energy from one form to another. In mitochondria, H^+ ions are pumped from the matrix into the intermembrane space, with ATP synthesis occurring in the matrix. In chloroplasts, H^+ ions are pumped from the stroma into the thylakoid lumen, with ATP synthesis occurring in the stroma (**Figure 2**, page 180).

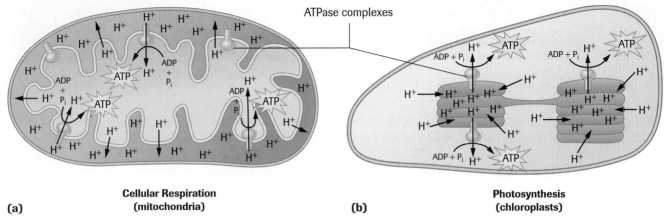

ATPase complexes

Cellular Respiration
(mitochondria)

(a)

Photosynthesis
(chloroplasts)

(b)

Figure 2
Chemiosmosis in chloroplasts and mitochondria
(a) During cellular respiration in mitochondria, H^+ ions move from the matrix into the inter-membrane space. ATP is synthesized in the matrix as H^+ ions move through the ATPase complexes embedded in the inner mitochondrial membrane.
(b) During the light reactions of photosynthesis, H^+ ions move from the stroma into the thy-lakoid lumen. ATP synthesis occurs in the stroma as the H^+ ions move through ATPase complexes embedded in the thylakoid membranes.

The relationships that exist between photosynthesis and cellular respiration reveal a dependency between autotrophs and heterotrophs. Heterotrophs produce the majority of the CO_2 in the environment that autotrophs use in photosynthesis, and autotrophs produce the majority of the O_2 in the environment that heterotrophs and autotrophs use in cellular respiration.

▶ **EXPLORE** an issue

Decision-Making Skills

○ Define the Issue ● Analyze the Issue ● Research
● Defend the Position ● Identify Alternatives ○ Evaluate

Take a Stand: Tropical Rain-Forest Depletion: Is There Cause for Concern?

The destruction of the tropical rain forests, and the resulting extinction of species, is one of the greatest global threats. Many scientists, environmentalists, and concerned citizens believe that the social and economic benefits of preserving the rain forests far exceed all current benefits of destroying them for their commercial resources. Those in favour of rain-forest preservation claim that deforestation causes loss of biodiversity, destruction of forest-based societies, and disruption in climate. The rain forests are often referred to as "the lungs of Earth." However, this is not exactly true. Rain forests are a climax vegetation, where no more carbon can be fixed. If the rain forests were destroyed, however, the fixed carbon would be converted into carbon dioxide and contribute significantly to the greenhouse effect and global warming. The tropical rain forests contain an amount of carbon equal to almost half of the carbon in the atmosphere. Individuals and groups fighting to preserve the forests claim that action must be taken quickly or the rain forests of the world will be destroyed by 2025.

Others disagree. They do not consider deforestation to be a serious problem. They believe that flooding, soil erosion, and the loss of biodiversity are not symptoms of a problem, but natural events occurring in the ongoing evolution of life on Earth. For some, there exists a powerful incentive to clear wood to make room for agricultural land and industrialization. They feel that developing countries have the same right to change their land now as developed countries had in the past. They view interference by developed countries as an infringement on the fundamental rights of sovereign nations to self-determination.

Statement
The governments of developing countries have the right to make room for agriculture and industrialization by clearing tropical rain forests.

• Add your own ideas to the above-mentioned points.
• Find information to help you learn more about the issue.

 www.science.nelson.com

• In a group, discuss the ideas.

(a) Communicate your opinion and arguments through a position paper, an organized small-group discussion, or a formal debate.

 **Comparing Photosynthesis and Cellular Respiration**

- The comparison of photosynthesis and cellular respiration is summarized in **Tables 1** to **5**.

Table 1 Comparison of the Overall Reactions

	Respiration	Photosynthesis
reactants	organic molecules (e.g., glucose)	$CO_2 + H_2O$
products	$CO_2 + H_2O$	organic molecules (e.g., glucose)
energy	released	stored

Table 2 Electrons

	Respiration	Photosynthesis
source	organic molecules (e.g., glucose)	water
carrier(s)	NAD^+, FAD	$NADP^+$

Table 3 Electron Transport System

	Respiration	Photosynthesis
energy profile		
electron source	NADH and $FADH_2$	water
electron sink	oxygen	NADPH
products	ATP	ATP and NADPH

Table 4 ATP Synthesis

	Respiration	Photosynthesis
H^+ ions pumped by ETC	yes	yes
ATP synthesis by chemiosmosis	yes	yes
membrane-embedded ATPase complex	yes	yes

Table 5 Organelle Structure and Function

	Mitochondrion (cristae)	Chloroplast (thylakoid)
inner membrane functions	electron transport H^+ ion transport ATP synthesis	electron transport H^+ ion transport ATP synthesis
contains DNA, ribosomes, etc. for replication	yes	yes
location of H^+ reservoir	intermembrane space	thylakoid lumen
location of ATP synthesis	matrix	stroma

Understanding Concepts

1. Are photosynthesis and respiration exact opposites? Explain.

2. Describe the potential energy change in an electron as it is transported from water to NADPH in light reactions.

3. (a) What would happen to humans and most other living organisms on Earth if photosynthesis stopped?
   (b) Why would this happen?

4. (a) Draw a labelled diagram of a chloroplast next to a mitochondrion within the cytoplasm of a plant cell.
   (b) On your diagram, Indicate the part of the organelle in which the following activities occur:
   (i) glycolysis
   (ii) Krebs cycle
   (iii) H^+ reservoir
   (iv) ATP synthesis
   (v) electron transport
   (vi) light reactions
   (vii) Calvin cycle
   (viii) pyruvate oxidation

5. Explain why the energy profile for photosynthesis is a zig-zag line while the line for respiration is straight.

6. How does a tropical fish aquarium containing fish and plants demonstrate the relationships between photosynthesis and cellular respiration?

Making Connections

7. The diagram on the Unit 1 opening pages is part of a much larger metabolic pathways chart that outlines almost all known metabolic reactions. Conduct Internet research to see the complete chart and to answer the following questions.

 www.science.nelson.com

   (a) Identify all of the major metabolic pathways you have studied in this unit, and note the interactions among them.
   (b) Why do scientists produce such charts? How are they useful?

⚗ INVESTIGATION 3.1.1

Light and Photosynthesis

Inquiry Skills

○ Questioning	○ Planning	● Analyzing
● Hypothesizing	● Conducting	● Evaluating
○ Predicting	● Recording	● Communicating

Plants seem to need light to stay alive and healthy. Is light used to produce nutrient molecules, such as starch? In this investigation, you will analyze leaves from plants that have been partially exposed to light and partially "starved" of light, and you will test for the presence of starch.

Question

Do plants require light to produce starch?

Hypothesis

(a) Construct a hypothesis regarding the need for light to produce starch.

Materials

safety goggles	two 20 mm × 200 mm
laboratory apron	test tubes
black paper or photographic	wax pencil
negative	ethanol
scissors	test-tube rack
plant that has been kept	2 glass, or plastic, petri
in the dark for 48 h	dishes
paper clips	large test-tube holder
250-mL beaker	Lugol's iodine solution
hot plate	in dropper bottle

Procedure

1. Put on your safety goggles and lab apron.

2. Obtain two pieces of opaque black paper or photographic negative and cut out a shape of your choice in the centre, making a stencil. Dispose of the cutout centre. Leave the other piece intact.

3. Remove the plant from the darkroom. Attach the stencil to the top (Sun-facing side) of a leaf and the other piece of paper to the bottom of the same leaf. Secure the pieces of paper to the leaf using two paperclips, as in **Figure 1**.

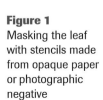

Figure 1
Masking the leaf with stencils made from opaque paper or photographic negative

4. Place the plant on a sunny windowsill or under a bright light source for one or two days. Make sure that the covered leaf faces the Sun or light.

5. After the time period, place 100 mL of water into the 250-mL beaker and heat to a boil on the hot plate.

6. Label a 20 mm × 200 mm test tube "light," and another test tube "without light."

7. Pour 50 mL of ethanol into each of the two labelled test tubes and place the tubes in a test-tube rack.

✋ **Ethanol is highly flammable. Make sure that any flame on your desk or near it is turned off before use.**

8. Remove the black paper or photographic negative from the leaf and place the leaf into the test tube labelled "without light."

9. Remove another leaf from the plant and place it into the test tube labelled "light."

10. Put the test tubes into the beaker containing boiling water and leave the tubes in the beaker until the leaves have no more colour.

11. Label one petri dish "light" and the other petri dish "without light."

12. Use the test-tube holder to remove the test tubes from the boiling water bath.

13. Pour the ethanol into a container designated by your teacher, then remove the leaves from the test tubes and place them on their corresponding petri dishes.

14. Wash the leaves gently under running tap water.

15. Put a few drops of iodine solution onto each leaf and observe the results.

16. Discard the plant leaf in the regular garbage and dispose of other materials according to instructions provided by your teacher.

Analysis

(b) Reread the Question and answer it.

(c) Why were plants kept in the dark for 48 h?

(d) Why were the leaves placed into boiling ethanol?

Evaluation

(e) In your report, evaluate your hypothesis, taking into account possible sources of error

(f) Describe how you could improve the experimental procedure.

(g) Suggest other experiments you could perform to extend your knowledge of light and the production of carbohydrates such as starch.

⚗ **INVESTIGATION 3.2.1**

Inquiry Skills		
○ Questioning	○ Planning	● Analyzing
● Hypothesizing	● Conducting	● Evaluating
● Predicting	● Recording	● Communicating

Identifying Plant Pigments by Chromatography

Coloured images in books and magazines may be deceiving at times. On close inspection with a magnifying glass, what appears as a bright red maple leaf in the middle of the Canadian flag is really composed of thousands of minute red, green, and blue dots. Since there are many more red dots than the other two colours, the red colour overwhelms the blue and green, resulting in a largely red image.

Plant leaves look green. Are they composed of a single pigment that reflects green light only, or are there other substances in the leaf whose colours are overwhelmed by the green? In this investigation, you will separate a green leaf extract to determine what types of pigments are in the leaves of green plants.

Question

What types of pigment are found in green leaf extracts from two different species of plants?

Hypothesis

Note the differences in colour between different plant species.

(a) Why are plant leaves different colours?

Prediction

(b) Predict whether there will be one or more pigments in the green leaf extracts from the two plants.

(c) Predict whether the pigments in both plants will be the same or different.

Materials

safety goggles
laboratory apron
leaves from two different plants
   (e.g., spinach and green cabbage)

mortar and pestle
20 mL of acetone

 Acetone is flammable. Do not use near open flame and use only in a fume hood. Avoid breathing vapours.

2 microcentrifuge tubes
microcentrifuge
wax pencil
two 150-mL test tubes
chromatography solvent
   (90% petroleum ether:10% acetone, v/v)

 Petroleum ether is highly flammable. Use only in a fume hood. Do not use petroleum ether in a room with an open flame.

2 chromatography tubes with cork stopper and hook
scissors
filter paper
sand for grinding
capillary tube for spotting

Procedure
Part I: Extraction of Pigments

1. Put on your safety goggles and lab apron.

2. Grind one leaf, or a portion of a leaf, from each plant separately with a mortar and pestle along with 3 mL of acetone. Make sure you clean the mortar and pestle with 1 mL of acetone between plants. Do not cross-contaminate the specimens in any way. Also, do not overgrind or you will end up with a pulpy mess that absorbs all of the solvent.

INVESTIGATION 3.2.1 *continued*

3. Transfer 1.5 mL of each sample to labelled microcentrifuge tubes. Spin the samples for 1 min at high speed.

4. Pour the supernatant (liquid portion of the mixture) into two clean, labelled 150-mL test tubes, leaving behind the solid pellets of cellular debris.

Part II: Chromatography

(T) **Perform the rest of the investigation under a fume hood.**

5. Place 3.0 mL of the chromatography solvent into a chromatography tube with a cork stopper. Keep the tube stoppered while you prepare the filter paper.

6. For each of the two pigment samples (supernatants from the test tubes), cut a 12-cm strip of filter paper. Handle the paper by the edges to avoid putting skin oils on it. Trim the filter paper strip as shown in **Figure 1**.

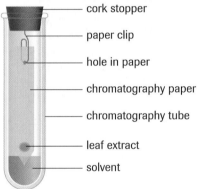

Figure 1
Chromatography setup

7. Apply a spot of one of the pigment extracts 3 cm from the bottom of a strip of filter paper. Repeat with the other pigment extract. To apply enough pigment so that the results will be visible, allow the paper to dry completely and then load more of the same sample onto the paper in the same place. Repeat this 10–20 times, allowing the spot to dry completely between applications. The spot should remain small, about 2 mm in diameter.

8. Hang each piece of filter paper on a hook attached to a cork stopper and replace the stoppers in the chromatography tubes (**Figure 1**). Make sure the extract spots do not touch the solution and that the tubes are tightly closed. Observe the apparatus for 15 min–30 min.

9. Before the solvent reaches the top of the paper, remove the chromatography strip from the tube and replace the stopper in the tube. With a pencil, draw a line along the uppermost point of the solvent on the paper before it dries and becomes invisible. Also mark the top of each pigment band and note its colour before it fades.

(d) Measure the distance from the middle of the original extract spot to each pigment band line and from the extract spot to the solvent line. Calculate the R_f values of each pigment according to the following equation:

$$R_f = \frac{\text{distance of pigment band}}{\text{distance of solvent front}} \times 100\%$$

10. You may dry and keep your chromatogram for future reference. Dispose of the chromatography solution according to your teacher's instructions. Wash your hands with soap and water.

Analysis

(e) Consult a standard biology or biochemistry resource book or the Internet to find standard R_f values for various plant pigment molecules separated by paper chromatography using this solvent mixture. Identify the pigments separated by chromatography by comparing R_f values.

(f) Draw a diagram of each of the chromatography strips side by side on a single sheet of white paper using coloured pencils to represent the separated pigments.

(g) Compare the variety of pigments identified in each plant species.

Evaluation

(h) Account for any pigment tracks that could not be identified by comparing them with standard R_f values. Consult standard reference sources or the Internet.

(i) Evaluate your hypothesis and predictions on the basis of your observations and analyses.

(j) Identify sources of error and suggest improvements to the experimental design.

The Hill Reaction

Inquiry Skills

○ Questioning	○ Planning	● Analyzing
○ Hypothesizing	○ Conducting	● Evaluating
○ Predicting	○ Recording	○ Communicating

In this lab exercise, you will analyze one part of a famous experiment first performed by British plant physiologist Robert Hill around 1939. Hill used the results of his experiment to demonstrate that the consumption of CO_2 (a process called carbon fixation) occurs in a separate set of reactions than those in which light energy transfers electrons down the electron transport chain (the light reactions). In photosynthesizing plant cells, electrons that leave the electron transport chain normally reduce $NADP^+$ to NAPDH. In the Hill reaction, an oxidized dye called 2,6 dichlorophenol-indophenol (DCPIP) is added to a suspension of isolated chloroplasts in a test tube. When exposed to light, the chloroplasts begin the light reactions, sending electrons through the electron transport chains in the chloroplast thylakoid membranes. DCPIP intercepts some of these electrons, changing from blue to colourless. DCPIP is blue when oxidized and colourless when reduced:

$$\text{oxidized DCPIP} + \text{electrons} \longrightarrow \text{reduced DCPIP}$$
$$\text{(blue)} \qquad\qquad\qquad\qquad \text{(colourless)}$$

The loss in blue colour may be detected and measured by a laboratory instrument called a spectrophotometer.

A spectrophotometer is a research instrument that measures the relative amount of light absorbed (or transmitted) by a coloured solution (**Figure 1**). Inside the spectrophotometer, a prism disperses white light into its various colours (wavelengths), and a movable slit allows individual wavelengths of light to pass through a sample of the solution (**Figure 2**). The sample is kept in a special colourless test tube called a cuvette. The cuvette is placed in a chamber that positions the solution directly in the path of the beam of light. The transmitted light strikes a photoelectric cell that converts the light energy into electric current. The amount of current is measured by a galvanometer and displayed on a meter as percent transmittance or percent absorbance.

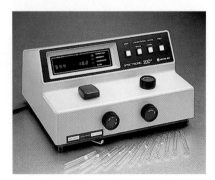

Figure 1
A Spectronic 20 spectrophotometer

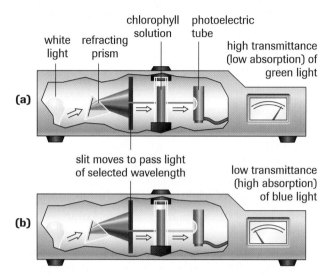

Figure 2
How a spectrophotometer works

Questions

Are chloroplasts necessary for reduction of DCPIP?
Is light necessary for reduction of DCPIP?

Procedure

1. The spectrophotometer was turned on and the wavelength set to 600 nm.
2. Five clean cuvettes were labelled and prepared according to **Table 1**.

 DCPIP is a hazardous chemical. Safety gloves were worn when pipetting, and waste was discarded in an appropriately labelled container.

Table 1 Experimental Mixtures

Cuvettes	Chloroplasts	Phosphate buffer	Water	DCPIP
1 (blank)	none	3 mL	1.5 mL	none
2	0.5 mL	3 mL	1.5 mL	none
3	none	3 mL	1.0 mL	1.0 mL
4	0.5 mL	3 mL	0.5 mL	1.0 mL
5 (dark)	0.5 mL	3 mL	0.5 mL	1.0 mL

3. Each cuvette was gently inverted to mix the solutions.

![] LAB EXERCISE 3.3.1 *continued*

4. Cuvette 5 was completely covered in a piece of aluminum foil and placed in a rack along with the other four cuvettes. The cuvettes were stored in a light-tight box until it was time to take measurements.

5. Cuvette 1 was placed in the spectrophotometer and the percent transmittance was adjusted to 0. Cuvette 1 served as the "blank" (the standard against which transmittance in the other cuvettes was measured).

6. The rack of cuvettes 1–5 was removed from the light-tight box and placed exactly 15 cm away from a bright spotlight. The light was turned on and all the cuvettes were exposed to it for 2 min and then immediately returned to the dark box.

7. Cuvettes were removed from the box one at a time, gently inverted to mix the contents, wiped with tissue, and placed in the spectrophotometer. The percent transmittance value of each was measured and recorded in **Table 2**, and each cuvette was returned to

Table 2 Results of the Hill Reaction

Cuvette	Time (min)	Transmittance at 600 nm (%)
1	2	100.0
	4	100.0
	6	100.0
	8	100.0
	10	100.0
2	2	28.5
	4	29.2
	6	28.0
	8	28.3
	10	28.0
3	2	31.0
	4	30.5
	6	31.0
	8	30.7
	10	30.9
4	2	30.7
	4	45.0
	6	56.2
	8	61.3
	10	64.8
5	2	31.2
	4	32.3
	6	32.7
	8	34.0
	10	31.5

the light-tight box. Cuvette 5 was removed from its aluminum foil wrapper only when taking transmittance measurements in the spectrophotometer.

8. After the percent transmittance of all five cuvettes was measured, they were placed in front of the spotlight for another 2 min, then again measured for transmittance in the spectrophotometer. This procedure was repeated for a total of five measurements each.

9. After the fifth measurement, the contents of the five cuvettes were properly discarded into a liquid waste bottle.

Observations

(a) Graph the results by plotting best-fit lines of the percent transmittance of each cuvette versus exposure time (in minutes).

Analysis

(b) Answer the first question from the Question section and describe the experimental evidence that supports your answer.

(c) Answer the second question and describe the experimental evidence that supports your answer.

(d) Why was a spectrophotometer used in this investigation?

(e) Why was the spectrophotometer set to measure percent transmittance at 600 nm (orange-red light)?

(f) Why were all cuvettes kept in the dark except for the 2 min time period in which the transmittance measurement was taken with the spectrophotometer?

(g) Why were the cuvettes gently inverted to mix the contents in step 3? in step 7? Why was vigorous shaking avoided?

(h) Why were cuvettes wiped clean before transmittance was measured in step 7?

(i) Why were all samples placed exactly 15 cm from the light source and exposed to light from the floodlight for 2 min *all at the same* time? In your answer, explain why cuvette 5 (wrapped in aluminum foil) was also exposed to the light along with the other four cuvettes.

(j) What does the evidence in this investigation tell you about the process of photosynthesis?

Evaluation

(k) Describe sources of error in this investigation and suggest changes that would help reduce or eliminate experimental error.

Factors Affecting the Rate of Photosynthesis

Inquiry Skills

- Questioning
- ○ Hypothesizing
- Predicting
- Planning
- Conducting
- Recording
- Analyzing
- Evaluating
- Communicating

Photosynthesis involves light reactions and biochemical reactions that do not directly require light. Plants live in a variety of environments on Earth. Do changes in environmental conditions affect the rate of photosynthesis? In this investigation, you will measure the rate of photosynthesis in various conditions by quantifying the amount of oxygen being released from a photosynthesizing solution.

Questions

1. How do changes in light intensity, temperature, and CO_2 concentration affect the rate of photosynthesis?

2. Develop your own question regarding the effect of an environmental condition of your choice on the rate of photosynthesis.

Prediction

(a) Predict the effect that changes in each of the following environmental conditions will have on the rate of photosynthesis:
  (i)   light intensity
  (ii) temperature
  (iii) CO_2 concentration
  (iv) another environmental condition of your choice

Experimental Design

The procedure outlined in this investigation provides a method for measuring the rate of photosynthesis of plants submersed in an aqueous sodium bicarbonate buffer (pH 7). Sodium bicarbonate is used as a source of $CO_{2(aq)}$. You will use this procedure to measure the rate of photosynthesis in four experiments that you will design and perform. In each case, you must conduct controlled experiments that allow you to make reasonable comparisons.

(b) Design three experimental procedures to determine
  (i)   the effect of varying light intensity on the rate of photosynthesis;
  (ii) the effect of varying temperature on the rate of photosynthesis;
  (iii) the effect of varying concentration of dissolved carbon dioxide on the rate of photosynthesis.

Have your teacher approve each experimental procedure, then carry out the experiments. Use the procedure outlined in this investigation to measure the rate of photosynthesis in each case. Record all observations and measurements in a suitable table format.

(c) Design an experimental procedure to test the prediction you made in (a) (iv). Obtain teacher approval, then carry out the experiment. Record all observations and measurements in suitable table format.

Materials

safety goggles
laboratory apron
500-mL conical flask or
  large test tube
plants (terrestrial plants or
  water plants)
sodium bicarbonate buffer,
  pH 7
rubber stopper with glass
  tubing
rubber tubing

50-mL burette
distilled water
500-mL beaker
utility stand and clamp
rubber bulb
ice
sodium bicarbonate
thermometer
light intensity meter
other materials and equipment as necessary
200-W light bulb

Procedure

1. Put on your safety goggles and lab apron.

2. Fill the 500-mL conical flask with plant material.

3. Add enough sodium bicarbonate buffer to submerse the plant material.

4. Put the stopper with glass tubing onto the mouth of the conical flask. Make sure that the glass tubing does not touch the contents of the flask.

5. Place 400 mL of water into a 500-mL beaker. Fill the burette with water to the top, then invert it in the beaker and secure it with a clamp to the utility stand (**Figure 1**).

INVESTIGATION 3.3.1 *continued*

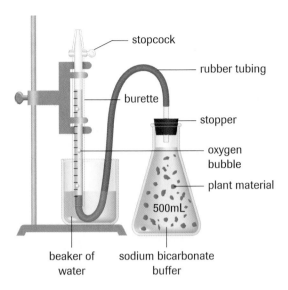

stopcock

rubber tubing

burette

stopper

oxygen bubble

plant material

500mL

sodium bicarbonate buffer

beaker of water

Figure 1
Gas collection apparatus setup

6. Use rubber tubing to connect the open end of the glass top in the stopper of the flask to the bottom of the burette. Be sure there is space for the water to escape when the gas bubbles up the burette.

7. Open the burette stopcock carefully and allow the water level to drop to the 50-mL level on the burette.

8. Subject the system to conditions according to your design. Allow several minutes for the system to stabilize.

9. Follow the rate of photosynthesis by either counting the number of bubbles over 1-min spans or measuring the amount of water displaced over 5-min spans. Measure for a total of 10 min for each condition. If you get no bubbles, check if the meniscus in the tubing is moving. If it is, then your burette valve is leaking. Your teacher will help you correct this problem.

Analysis

(d) In tables, summarize the results for the variables you tested. Draw suitable graphs using your data.

(e) Analyze your results for trends and patterns. Answer the questions in the Questions section above.

Evaluation

(f) In your report, evaluate your predictions, taking into account possible sources of error. Draw reasonable conclusions.

(g) Describe how you could improve your experimental methods and the assay technique.

(h) Suggest other experiments you could perform to extend your knowledge of photosynthetic activity.

(i) The experiments you conducted in this investigation were carried out with plants submersed in water. Design a procedure for carrying out the same types of experiments in air instead of water. Draw a labelled diagram, like Figure 1, to illustrate your experimental design.

(j) What environmental conditions affecting the rate of photosynthesis can be tested in an air environment that could not be tested in a water environment?

Synthesis

(k) Suggest a procedure you may use to identify the type of gas produced by the plant material in these experiments.

Key Expectations

- Using problem-solving techniques, investigate and explain the relationship between metabolism and the structure of biomolecules. (3.1, 3.2)

- Compare the matter and energy transformations associated with the process of photosynthesis. (3.1, 3.2, 3.3, 3.4, 3.5)

- Describe how such molecules as glucose, chlorophyll, CO_2, ATP, $NADP^+$, and water function in photosynthesis, and explain the role of chloroplasts. (3.1, 3.2, 3.3, 3.4, 3.5)

- Describe the chemical structure, mechanisms, and dynamics of enzymes and other components of photosynthesis. (3.1, 3.2, 3.3, 3.4, 3.5)

- Determine the similarities and differences between mitochondria and chloroplasts. (3.1, 3.6)

- Design and carry out an experiment related to a cell process, controlling the major variables and adapting or extending procedures where required. (3.3)

- Compare the processes of cellular respiration and photosynthesis. (3.6)

Key Terms

absorption spectrum

action spectrum

antenna complex

anthocyanins

β-carotene

C_3 photosynthesis

C_4 photosynthesis

Calvin cycle

carbon fixation

chlorophyll

chloroplasts

crassulacean acid metabolism (CAM)

cuticle

cyanobacteria

cyclic electron flow

electromagnetic (EM) radiation

electromagnetic spectrum

epidermis layer

excitation

fluorescence

grana

ground state

guard cells

lamellae

light-compensation point

light reactions

light-saturation point

mesophyll layers

noncyclic electron flow

photons

photophosphorylation

photorespiration

photosynthetic efficiency

photosynthetically active radiation (PAR)

photosystem I

photosystem II

photosystems

plasmodesmata

porphyrin

primary electron acceptor

reaction centre

ribulose bisphosphate carboxylase/ oxygenase (rubisco)

spectroscope

stomata

stroma

thylakoid lumen

thylakoid membrane

thylakoids

transpiration

vascular bundles

xanthophylls

Z protein

▶ **MAKE** a summary

On a letter-size sheet of paper, draw a large, well-labelled poster summarizing the three stages of photosynthesis. Consider the area of the sheet to represent the cytoplasm of a C_3 plant cell. Draw a very large chloroplast covering at least one half of the area. Place different cartoons/drawings representing each stage of the process in the appropriate locations. Use arrows to indicate the movement of intermediate molecules.

In your notebook, indicate whether the statement is true or false. Rewrite a false statement to make it true.

1. In green plants, photosystems I and II are required for the synthesis of NADPH.

2. The first stable product of photosynthesis in C_3 plants is glyceraldehyde 3-phosphate.

3. In normal aerobic environments and saturating levels of light, an increase in CO_2 concentration from 0.03% to 0.06% will result in an increase in the rate of CO_2 fixation in C_3 plants.

4. Photosynthesis is a redox process; that is, H_2O is reduced and CO_2 is oxidized.

5. The enzymes required for CO_2 fixation in plants are located only in thylakoid membranes.

6. RuBP can be called a carbon dioxide acceptor.

7. ATP synthesis in chloroplasts occurs in the stroma.

8. Cyclic electron flow produces ATP and NADPH.

9. C_4 photosynthesis is so named because it produces a four-carbon organic acid as the first stable product of carbon fixation.

10. Chloroplasts use CO_2 to produce glucose; mitochondria use glucose to produce CO_2.

In your notebook, record the letter of the choice that best answers each question.

11. Which of the following organisms would not be an autotroph?
    (a) cyanobacterium
    (b) bee
    (c) cactus
    (d) palm tree
    (e) *Spirogyra*

12. Photosynthesis occurs mainly in which of the following?
    (a) epidermal cells
    (b) the cuticle
    (c) mesophyll cells
    (d) vascular bundle cells
    (e) all of the above

13. Which of the following are raw materials for photosynthesis?
    (a) oxygen and water
    (b) carbon dioxide and water
    (c) glucose and oxygen
    (d) glucose and carbon dioxide
    (e) oxygen and carbon dioxide

14. Which word equation summarizes photosynthesis?
    (a) water + starch $\longrightarrow$ glucose + glucose + glucose
    (b) water + carbon dioxide $\longrightarrow$ oxygen + glucose + water
    (c) glucose + oxygen $\longrightarrow$ water + carbon dioxide
    (d) glucose + fructose $\longrightarrow$ sucrose + water
    (e) carbon dioxide + glucose $\longrightarrow$ water + oxygen

15. When does the process of splitting water to release hydrogen ions, electrons, and oxygen occur?
    (a) during the light reactions
    (b) during the Calvin cycle
    (c) during photorespiration
    (d) during carbon fixation
    (e) during (b) and (d)

16. When does the process of incorporating the carbon of carbon dioxide into carbohydrates occur?
    (a) during the light reactions
    (b) during the Calvin cycle
    (c) during photorespiration
    (d) during carbon fixation
    (e) during (b) and (d)

17. What is the term for an individual flattened membrane-bound sac in the chloroplast?
    (a) granum
    (b) stroma
    (c) thylakoid
    (d) crista
    (e) matrix

18. Which of the following best describes what occurs at light levels below the light-compensation point?
    (a) Photorespiration causes a net evolution of CO_2.
    (b) The Calvin cycle is not operating.
    (c) Photorespiration causes a net uptake of O_2.
    (d) The light reactions are operating in complete darkness.
    (e) both (a) and (c)

Understanding Concepts

1. Which has more energy, short wavelengths or long wavelengths of electromagnetic radiation?

2. What range of wavelengths do plants use in photosynthesis?

3. Which part of photosynthesis uses water? What happens to the water?

4. Why are there various pigments that form the antenna complex in photosystems?

5. What unique role do the pigments of the photochemical reaction centre of photosystems play?

6. (a) Describe the absorption spectrum of chlorophyll *a* versus that of chlorophyll *b*.
   (b) How do the absorption spectra in (a) compare to the action spectrum of photosynthesis?
   (c) Explain why plants do not contain a variety of pigments that would absorb all visible wavelengths, so that they could use the Sun's energy more efficiently.

7. (a) Draw a diagram of a chloroplast (**Figure 1**) into your notebook and label where the following substances would be found: newly formed NADPH, ATP produced by chemiosmosis, rubisco, newly formed G3P, newly formed O_2, photosystem I, photosystem II, and a starch granule.
   (b) Locate the components that are directly affected by photons and colour them green.

Figure 1
A chloroplast

8. When electrons are removed from water, protons are liberated. Does this occur in the stroma or inside the thylakoid? Can protons move directly across the phospholipid bilayer? Why or why not?

9. Some old biology textbooks called the carbon fixation reactions of photosynthesis the "dark reactions."
   (a) Why did they use this term?
   (b) Why is this misleading?

10. (a) Under what conditions will G3P be converted into starch within the chloroplast instead of being transported into the cytoplasm?

(b) If transported into the cytoplasm, what may happen to G3P?

11. What happens to the carbon dioxide used in the chemical reactions of the Calvin cycle?

12. (a) What role does rubisco play in the photosynthetic process?
    (b) What evidence suggests that rubisco is the most abundant protein on Earth?

13. (a) Why are C_4 plants more efficient photosynthesizers than C_3 plants?
    (b) If C_3 photosynthesis is less efficient than C_4 photosynthesis, why have all plants not evolved into C_4 plants over time?

14. Why do many cacti lack woody stems and, instead, are fleshy green throughout?

15. Which of the lines in **Figure 2** is the light-response curve for a typical sun plant? Explain.

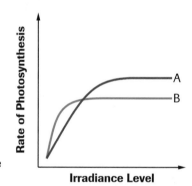

Figure 2
Light-response curve for typical sun plant

16. (a) Describe two structural similarities and two structural differences between chloroplasts and mitochondria.
    (b) Describe two functional similarities and two functional differences between chloroplasts and mitochondria.

Applying Inquiry Skills

17. Why does the light-response curve not pass through the origin?

18. In many terrestrial plants, especially woody species, stomata occur only in the lower epidermis of the leaves. In others, stomata occur in both the upper and lower epidermis, but are usually more abundant in the lower epidermis. The distribution of the stomata on the leaves of some plants is shown in **Table 1**.
    (a) Draw suitable graphs of the data in **Table 1**.
    (b) What is the significance of this pattern of stomatal distribution?

Table 1 Average Number of Stomata per cm^2 of Leaf Area

Species	Upper epidermis	Lower epidermis
apple	0	29 400
Coleus	0	46 100
geranium	1900	5900
pea	10 100	21 600
sunflower	8500	15 600

19 Line B on the graph in **Figure 3** is the light-response curve for a plant growing in an environment in which the CO_2 concentration is 360 ppm. Which of the lines is the light-response curve for the plant growing in a chamber with a CO_2 concentration of 450 ppm?

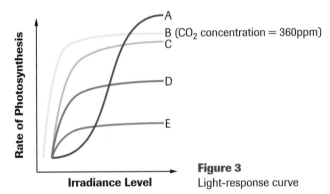

Figure 3
Light-response curve

Making Connections

20. Conduct library and/or Internet research to study the historical development of the endosymbiotic theory of the origins of the chloroplast. Write a brief essay describing the development of the theory and the social and professional challenges faced by its developer(s). In your essay, describe rival theories and evidence in support of, and against, endosymbiosis.

 www.science.nelson.com

21. Recently, some delegates at international conferences on global warming have proposed that planting more trees can offset the carbon dioxide emissions that produce the greenhouse effect and global warming.
    (a) Identify two major greenhouse gases.
    (b) Why are these gases called greenhouse gases?
    (c) What is the greenhouse effect and how does it contribute to global warming?
    (d) Write a position paper on the contention that the strategy of planting more trees is an effective alternative to reducing greenhouse gas emissions.

 www.science.nelson.com

22. Several biotechnology research companies are experimenting with the possibility of producing plastic from plants, called "green plastic." One procedure turns sugar from corn and other plants into polylactide (PLA), a plastic similar to polyethylene terephthalate (PET), which is a petrochemical plastic used in pop bottles and clothing fibres. Conduct Internet research to complete the following tasks:
    (a) Describe one or two other green plastics and their potential uses.
    (b) Compare the costs and benefits of producing green plastics on a large scale with producing conventional oil-based plastics.

 www.science.nelson.com

23. Biomass, plant matter such as trees, grasses, and agricultural crops, may be used as a solid fuel or converted into liquid or gaseous forms for the production of electric power, heat, or chemicals for use in vehicles. Conduct library and/or Internet research to answer the following questions:
    (a) How is electricity generated from biofuel?
    (b) What proportion of electric power production in Canada comes from biomass energy? What is the potential for increasing the amount of electricity produced by biomass?
    (c) Compare the costs and benefits of producing automobile fuel from biomass with producing fuel from petroleum.

 www.science.nelson.com

24. **Figure 5** on page 160 shows a model that illustrates the energy changes that occur in the light reactions of photosynthesis. Create a pictorial model that you would use to explain the process to a student who has never studied photosynthesis.

Extension

25. (a) If a test tube containing an aqueous suspension of chloroplasts is kept in the dark, what substance(s) would you have to add to the suspension for G3P to be formed by photosynthesis? Explain.
    (b) Given the missing substances you added in (a), explain why there is a sudden increase in ATP synthesis when the pH of the external solution is raised to 8.0.

Unit 1
Metabolic Proceses

▶ **Criteria**

Process

- Make and evaluate hypotheses and predictions.
- Design appropriate experimental procedures.
- Choose and safely use laboratory materials.
- Analyze results, using quantitative methods when appropriate.
- Evaluate experimental procedures and suggest improvements.

Product

- Prepare a formal lab report or project display board to communicate the Predictions, Experimental Design, Procedures, Observations, Analysis, and Evaluation of your investigation.
- Demonstrate an understanding of the concepts presented in this unit.
- Use terms, symbols, and SI measurements correctly.

Student Aquarist

An aquarium is a dynamic ecosystem in which the resident organisms perform the functions of life, and respond to changes that directly or indirectly affect those functions. Changes in water temperature, pH, light conditions (intensity, frequency, duration), and the concentrations of dissolved solutes, significantly affect the health and viability of the living organisms.

In Chapter 3 you learned that the principal metabolic processes of cellular respiration and photosynthesis are highly interdependent. To remain healthy, an ecosystem must provide the necessary metabolic inputs and simultaneously absorb outputs to ensure a state of dynamic equilibrium (a metabolic balance). One of the ways in which this balance is achieved is by cycling nutrients like oxygen and carbon dioxide. Through photosynthesis, autotrophs help maintain an oxygen balance for the benefit of heterotrophs and through cellular respiration, heterotrophs help maintain a carbon dioxide balance that primarily benefits autotrophs. However, before they reach a state of balance, newly created ecosystems may be dangerous places for organisms to live. Too much oxygen or too little light may adversely affect the organism's health.

Your task is to set up a simple freshwater ecosystem and then monitor and manipulate biotic and abiotic factors that affect the metabolic health of the aquarium plants and animals.

Before beginning the activity, conduct library or Internet research to obtain information regarding simple aquarium construction, normal (healthy) levels for each of the abiotic components you are analyzing, and techniques that aquarists use to change the concentrations of ammonia, oxygen, and carbon dioxide in aquarium water. Determine the reasonable limits of the biological loading of the aquarium (the amounts of plants and animals introduced into the aquarium with respect to the volume of water), and establish a definition for, and indicators of, metabolic health, as it applies to the living organisms in your aquarium.

 www.science.nelson.com

Question

How do temperature, light conditions, pH, dissolved ammonia, dissolved oxygen, and dissolved carbon dioxide affect the metabolic health of plants, fish, and snails in a freshwater ecosystem?

Hypotheses/Predictions

(a) When a change in an environmental condition is warranted, make testable hypotheses regarding the changes, and predict the outcome of the change. Record hypotheses and predictions in your logbook.

Experimental Design

Maintain your aquarium in ambient conditions for the first week of the activity. After that, you may change conditions by performing controlled experiments of your own design. Controlled changes to any of the monitored biotic or abiotic components are made to enhance the general health and viability of the living organisms in the ecosystem. Remember to change one variable at a time, and allow the ecosystem to adjust to the change for several days before changing another variable.

(b) Report the design and procedures of your experiments to your teacher for approval before proceeding, and make sure you observe and follow appropriate safety procedures.

Materials

10-L aquarium tank or fish bowl

tap water

thermometer

wide-range pH paper and colour chart or
  pH test kit

aquarium gravel

water quality test kits for $[O_{2(aq)}]$,
  $[CO_{2(aq)}]$, and $[NH_{3(aq)}]$

water plants (*Elodea* or *Anacharis*)

1 or 2 guppies

1 small- to medium-sized snail

aquarium heater (if required)

fish food (if required)

other materials as required

Procedure

1. Run the experiments and provide a detailed description of the procedures in your logbook.

2. Once you have completed the study, write a formal lab report, or construct a project board display and submit it to your teacher for assessment.

Observations

(c) Measure water temperature (°C), light duration (hours per day), light intensity (optional), light frequency profile (incandescent, fluorescent, etc.), pH, $[O_{2(aq)}]$, $[CO_{2(aq)}]$, and $[NH_{3(aq)}]$ on a daily basis. Record all qualitative and quantitative data in your logbook in suitable formats such as tables, charts, etc.

Analysis

(d) Answer the Question. Your analysis should attempt to uncover the possible causes and effects of any or all of the following conditions: changes in pH, $[O_{2(aq)}]$, $[CO_{2(aq)}]$, or $[NH_{3(aq)}]$; algal blooms; fish, snail, or plant illness or death; changes in the physical properties (e.g., colour, clarity, viscosity) of the water in the aquarium. Relate your analysis to the processes of photosynthesis and cellular respiration whenever possible.

Evaluation

(e) Evaluate your hypotheses and predictions on the basis of the evidence gathered.

(f) Describe sources of experimental error and suggest improvements that would help reduce or eliminate error in your experiments.

Synthesis

Conduct library or Internet research to help you answer the following questions.

 www.science.nelson.com

(g) Why do aquariums usually have filtration systems? How do these systems help maintain healthy water conditions for plants and animals?

(h) What is the source of dissolved ammonia in an aquarium ecosystem? What happens to the ammonia over time?

(i) What are the major sources of dissolved oxygen and carbon dioxide in a freshwater aquarium? How does the ecosystem maintain a CO_2/O_2 equilibrium?

(j) Can an aquarium ecosystem ever be self-sustaining? Explain.

Understanding Concepts

1. Determine whether each of the compounds in **Figure 1** is polar or nonpolar and explain why.

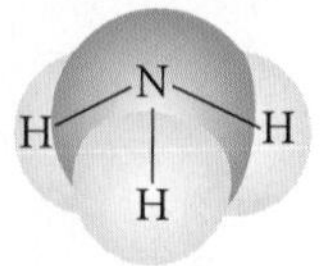

(a) ammonia, NH_3

(b) carbon dioxide, CO_2

(c) pentane, C_5H_{12}

Figure 1

2. Distinguish between the molecular structure of glycogen and amylopectin.

3. Hydrogen bonds are commonly used to hold components of biological macromolecules together. Describe how hydrogen bonds are used in each of the following:
   (a) DNA
   (b) β-pleated sheet
   (c) cellulose fibre

4. Distinguish between a competitive and a noncompetitive inhibitor.

5. Where do animals store most of the glycogen in their bodies? Why?

6. What is the main function of nucleic acids? In your answer, describe the roles of DNA and RNA.

7. Adenosine monophosphate (AMP) is a nucleotide primarily used in RNA synthesis. What is the primary purpose of adenosine triphosphate (ATP)?

8. Jeremy comes home to a messy house. He begins to tidy up by sweeping the floor, organizing his bookcases, making the bed, and gathering his dirty clothes into a basket.
   (a) Has the entropy of the room increased or decreased due to these actions? Explain.
   (b) Has the entropy of the universe increased or decreased due to Jeremy's actions? Explain.
   (c) How could Jeremy have prevented the outcome in (b)?
   (d) Did Jeremy's actions violate the second law of thermodynamics? Explain.
   (e) Draw a free-energy diagram to represent the overall process.

9. Which graph in **Figure 2** represents the relationship between enzyme activity (y-axis) and substrate concentration (x-axis)?

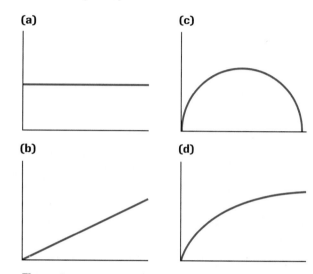

Figure 2

10. Compare the process of feedback inhibition to the function of a furnace thermostat.

11. What is the main difference between an allosteric regulator and a noncompetitive inhibitor?

12. In the following list, identify the carbohydrate, the fatty acid, the amino acid, and the polypeptide. Explain your answers.
    (a) $NH_3 - CHR - COO$
    (b) $C_6H_{12}O_6$
    (c) $(glycine)_{20}$
    (d) $CH_3(CH_2)_{16}COOH$

13. Distinguish among anabolism, catabolism, and metabolism.

14. List four electron carriers found in mitochondria.

15. Aerobic breakdown of sugar yields 2870 kJ of energy per mole, whereas anaerobic breakdown of sugar yields no more than 210 kJ per mole. Explain this difference in terms of the chemistry of cellular respiration.

16. Fatty acids cannot be catabolized by the Krebs cycle. What must happen to fatty acids before organisms can use the energy stored in them?

17. (a) In active muscle tissue, what happens when the supply of oxygen is not adequate for the demands of oxidative phosphorylation?
    (b) Why does deep breathing continue even after strenous exercise (e.g., running) has stopped?

18. Explain how guard cells regulate gas exchange in a leaf.

19. What is the range of wavelengths of light that is visible to humans? Which wavelengths are absorbed by plants?

20. (a) Where are the pigments of photosystems located in plants?
    (b) Where are the pigments of photosystems located in photosynthetic bacteria?

21. Explain why chloroplast thylakoids must be saclike structures.

22. (a) What type of gradient is established to provide the energy for the synthesis of ATP during photosynthesis?
    (b) Describe the difference in pH between the thylakoid interior and the chloroplast stroma during light conditions. How does the pH difference change in the dark?

23. Summarize what happens during the light reactions of photosynthesis.

24. (a) Where in the chloroplast do the light reactions and Calvin cycle reactions of photosynthesis occur?
    (b) In which of these two sets of reactions does the conversion of light energy to chemical potential energy occur?
    (c) What energy-carrying intermediates are shared between these two photosynthetic processes?

25. Draw a labelled diagram illustrating the chemiosmotic mechanism of ATP synthesis in chloroplasts. Write a caption below the diagram that describes the process.

26. Why do plants produce far more sugar than they need? Why do they not stop photosynthesis once immediate needs are met?

27. Why might rubisco be considered the most important enzyme on Earth?

28. Compare the general equations of photosynthesis and aerobic respiration. List the similarities and differences.

29. Write a short essay on the interdependence between plants and animals.

Applying Inquiry Skills

30. From what you know about cohesion, propose a hypothesis to explain why water forms droplets.

31. A student is studying an enzyme-catalyzed chemical reaction. In the reaction, H^+ ions are formed. She found that after a specific time, the reaction stopped, even though there was still a lot of product available in the mixture.
    (a) Propose a hypothesis to explain why the reaction stopped.
    (b) Based on your hypothesis, suggest a solution and explain how it might solve the problem.

32. On a health-food website you read that tablets containing "natural" vitamin C extracted from rose hips are better for you than synthetic vitamin C tablets.
    (a) Given your knowledge of organic compounds, what is your response?
    (b) Design an experiment to test whether synthetic and "natural" vitamin C differ in quality.
    (c) Conduct Internet research to determine the scientific consensus on this debate and write a brief report summarizing your research.

 www.science.nelson.com

33. The maximum efficiency of glucose oxidation is 32%. Assuming that all electrons accepted by NAD or FAD are delivered to oxygen, calculate the proportions of this efficiency that are contributed by glycolysis, pyruvate oxidation, and the Krebs cycle.

34. A 30-year-old, 75-kg runner completed a marathon in 2 h and 35 min. His oxygen consumption at 15-min intervals is shown in **Table 1** (page 198).
    (a) Construct a graph to display these results.
    (b) What is his resting $VO_{2\ max}$?
    (c) Calculate his oxygen consumption while resting and during his highest $VO_{2\ max}$.
    (d) Explain what is happening during each phase of the graph.
    (e) Use your knowledge of oxygen consumption during exercise to explain the oxygen consumption after the race is finished.

Table 1

Time (min)	VO$_{2\ max}$ mL/kg/min
0	15
15	40
30	70
45	90
60	90
75	75
90	70
105	65
120	65
135	65
150	65
165	50
180	45
195	40
210	35
225	30
240	25
255	20
270	15
290	15

35. Only about 5% of the solar radiation that hits a leaf is converted into useful stored energy. This is a relatively low photosynthetic efficiency.
    (a) Explain why so much energy is not available to the leaf.
    (b) If you were a biological engineer, what would you do to create a plant with higher photosynthetic efficiency?

Making Connections

36. You accidentally ruined a wool sweater because you dried it in the hot cycle of the clothes dryer. Wool is rich in keratin, a strand-like, fibrous protein, held together in a helical shape by many hydrogen bonds. Visualize three such strands coiled together, with many disulfide bridges in the end regions holding the three strands together like a rope.
    (a) Explain why the sweater permanently shrinks when it is dried in the dryer.
    (b) What should you do to prevent this from happening to a new wool sweater?

37. Cyanide, a toxic organic compound, binds to an enzyme that is a component of the electron transport system and causes cyanide poisoning. Binding prevents the enzyme from donating electrons to a nearby acceptor molecule in the system.
    (a) What effect will this have on ATP production? Why?
    (b) What effect will this have on a person's health?

38. Barbiturates are one of several classes of drugs that act by preventing the transfer of electrons through the electron transport chain in mitochondria.
    (a) Why are these drugs sometimes called "downers"?
    (b) Why is the uncontrolled use of barbiturates dangerous?

39. Exercise programs are generally aimed at increasing the efficiency of cellular respiration in the cells of certain tissues. Conduct library and/or Internet research to determine how exercise programs promote increased ATP production.

 www.science.nelson.com

40. Write a letter to the Canadian Society of Plant Physiologists nominating chlorophyll *a* as Molecule of the Year. Describe as many merits of the molecule as you can.

41. Conduct library and Internet research to identify a concept or theory regarding the process of photosynthesis that has changed dramatically in the past 200 years. Identify the key scientist(s) involved in the changes and briefly describe the investigation(s) that led to the new concept or theory.

 www.science.nelson.com

42. **Figure 3** shows the changes that have occurred in global atmospheric carbon dioxide concentration over the past 1000 years. Studies show that current concentrations are increasing at about 1 ppm per year.

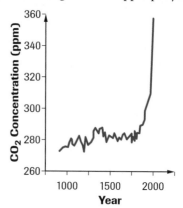

Figure 3

(a) Account for the steep increase in CO_2 concentrations that occurred between the 1700s and the present.

(b) What is primarily responsible for the present 1 ppm per year increase in CO_2 concentration?

(c) What effect might higher atmospheric carbon dioxide concentrations have on:
   (i) C_3 and C_4 plant photosynthesis?
   (ii) stomatal openings?
   (iii) the ratio of C_3 to C_4 plants in the world?

43. You are a biochemist studying the properties of the photosynthetic enzyme, rubisco. You dissolve the enzyme in a solution that contains only carbon dioxide, magnesium ions, and an adequate supply of hydrogen ions.

(a) What must you add to this mixture to be able to convert the substrate carbon dioxide to phosphoglycerate? Explain.

(b) Does this reaction need light to proceed? Why or why not?

Extension

44. A plant cell is compared to a city. Everything that occurs within a city, or that moves in or out of a city, has a comparable activity in the functioning cell. Consider the structure of the cell and the processes of photosynthesis and cellular respiration. Write a brief essay comparing a plant cell to a city focussing on the processes of photosynthesis and cellular respiration.

45. Given C_3, C_4, and CAM plants to choose from, in what plant and tissue would you find
   (a) chloroplasts that produce ATP but not NADPH?

(b) chloroplasts that carry out the complete set of light reactions, but do not make G3P?

(c) cells that produce C_4 acid only in the light?

(d) cells that produce C_4 acids only in the dark?

 www.science.nelson.com

46. (a) Why do C_4 plants possess a lower light-compensation point and a higher light-saturation point than C_3 plants?

(b) Draw light-response curves to illustrate the concepts described in (a).

47. All living organisms depend directly or indirectly on light. Photosynthetic plants must have light to survive. Humans can survive without light but are affected directly and indirectly by light. Just as animals change their behaviour in response to seasonal changes, so do humans. A disorder known as SAD (seasonal affective disorder) has been identified which accounts for significant changes in human behaviour as the seasons change. In everyday language, SAD is often referred to as "the winter blues." It has been estimated that around 2% of the population in northern latitudes suffer symptoms ranging from sleep problems, overeating, and depression. Conduct library and/or Internet research to complete the following:

(a) Describe the symptoms of SAD.

(b) What causes SAD? The explanation should include a complete description of the scientific concepts on which it is based.

(c) What cures or treatments are available?

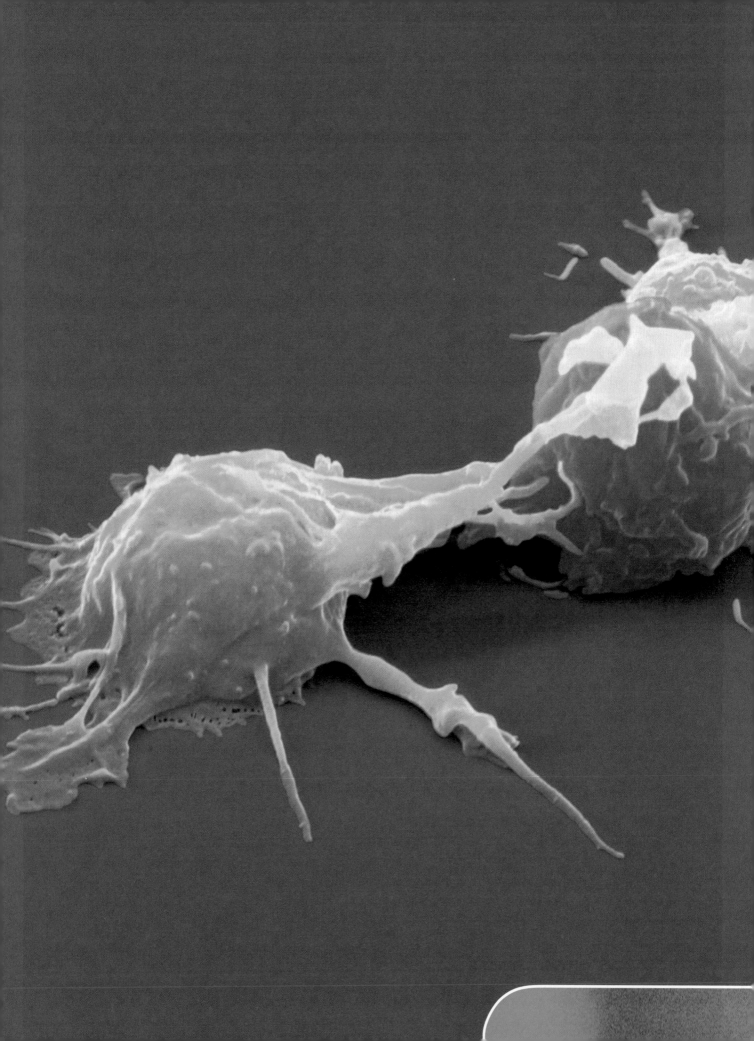

Molecular Genetics

A key difference between cancer cells and normal cells is that cancer cells express proteins called tumour antigens on their surface. The human body recognizes these antigens as foreign and mounts an immune response. The response, however, is usually insufficient to stop the progression of cancer.

A revolutionary treatment being tested is DNA vaccines. They may be used to introduce the immune system to the antigens found in cancer cells. If the body produces antibodies against cancer, it will have a natural, built-in, therapeutic mechanism to control cancer growth.

The development of DNA vaccines would not be possible without concurrent advancement in biotechnology. The genes responsible for the expression of tumour-surface proteins must be isolated and inserted into host cells, which are then injected into the patient. The immune system produces antibodies in response to the proteins, despite the lack of cancer cells.

Preliminary studies in mice are promising. Human trials have yet to be performed with cancer patients. Nonetheless, engineered DNA vaccines may someday replace traditional cancer treatments.

▶ Overall Expectations

In this unit, you will be able to

- understand the concepts of gene and gene expression, and explain the roles that DNA, RNA, and chromosomes play in cellular metabolism, growth, and division, demonstrating an awareness of the universal nature of the genetic code;

- explain, through laboratory activities and conceptual models, processes within the cell nucleus;

- describe some of the theoretical, philosophical, and ethical issues and implications surrounding knowledge gained by genetic research and its technological applications.

ARE YOU READY?

Knowledge and Understanding

1. Prepare a table with the headings Eukaryotic Cell and Prokaryotic Cell, and list the characteristics that enable you to differentiate between the two cell types.

2. Explain how meiosis differs from mitosis.

3. Copy **Figure 1** into your notebook and fill in the missing labels.

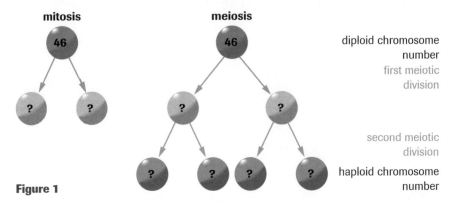

Figure 1

4. Copy the passage below into your notebook. Fill in the blanks with the following words: ultraviolet radiation, mitosis, deoxyribose sugar, interphase, cytosine, mutation, genetic disorder, adenine, nucleus, genes, thymine, phosphate group, chromosomes, chemical agents, nitrogenous base, natural, guanine.

DNA is a polymer made up of units called nucleotides. Each nucleotide is composed of three chemicals: _____, a _____, and a _____. The nitrogenous bases are _____, _____, _____, and _____. In eukaryotes, DNA is located in the _____ of the cell and is organized into _____. DNA has coding regions known as _____ that determine the phenotypical characteristics of an organism. An alteration in the DNA sequence is known as a _____. Such alterations may be caused by _____, _____, or _____ causes (**Figures 2**, **3**, and **4**). Mistakes can be made during the process of _____. During _____, copies of chromosomes are made. If DNA replication does not proceed accurately, it may result in a _____, a positive evolutionary advantage, or it may have no effect at all.

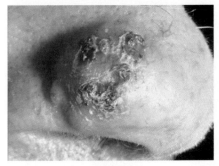

Figure 2
Exposing the skin to long periods of UV radiation is a major risk factor for the development of skin cancer because the probability of a DNA mutation is increased.

Figure 3
Smoking exposes lungs to tar and nicotine, two DNA-damaging agents. Both have been directly linked to the development of lung cancer, a genetic disorder arising from mutations in the DNA sequence.

Figure 4
As we age, the process of copying our DNA becomes less accurate. At the same time, the probability of genetic change on exposure to mutagens increases. Sometimes these errors result in mutations that lead to disease.

5. Compare covalent and hydrogen bonding. Identify which of the following molecules can hydrogen bond with other molecules: CO_2, H_2O, CH_4, HF, $CaCl_2$.

Making Connections

6. Read the following statements and state whether you agree, somewhat agree, are neutral, somewhat disagree, or disagree. Provide reasoning for your opinion.
   (a) Institutions that discover specific DNA sequences should be allowed to patent those sequences.
   (b) Information gained from the Human Genome Project should only be used in medical applications.
   (c) Parents have the right to screen their potential offspring for any genetic disorders or characteristics that they feel are deleterious to their offspring's success in society.
   (d) The government should control all the research relating to the Human Genome Project.
   (e) Human cloning should be allowed to proceed.
   (f) Life insurance companies ask about your family history of heart disease, diabetes, and other medical conditions. Your habits (e.g., smoking) and your age are used to determine the value and cost of your policy. In addition, you must complete a medical physical examination and submit to an HIV test. Insurance companies in the future should have access to your genetic profile.
   (g) Your genetic makeup predetermines your habits, characteristics, and predispositions. The environment plays a small role in who you are.
   (h) Genetically modified foods are having a negative impact on the environment. More testing should be required before they are allowed onto the market.
   (i) Genetically modified foods should be fully labelled in supermarkets.
   (j) Everyone should have access to biotechnological products, no matter the "seriousness" of the need. For example, short people should have access to human growth hormone even though the product was designed to treat dwarfism (**Figure 5**).
   (k) Biotechnology has had a positive influence in the fields of forensics, medicine, and agriculture.
   (l) Medical genetic research will eventually cure all disease.

Figure 5
Both mice are the same age, but the mouse on the left had the gene for human growth hormone inserted into it.

4

DNA: The Molecular Basis of Life

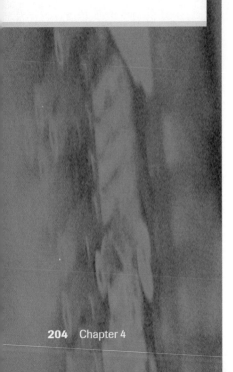

▶ **In this chapter, you will be able to**

- describe the experiments leading to the acceptance that DNA carries the hereditary information for all living things;

- describe in detail the structure of DNA and the experimental evidence that led scientists to deduce its structure;

- explain the current model of DNA replication and a method of repair used to correct errors;

- understand how to extract and characterize a visible mass of DNA by using the properties of DNA and other cellular components;

- discuss the nature of scientific research from social, academic, political, and competitive viewpoints.

About 130 years ago, a Swiss biochemist, Friedrich Miescher, extracted a viscous white substance from the nucleus of cells. The substance was slightly acidic and rich in phosphorous and nitrogen. Another 100 years passed before this substance was confirmed as the hereditary material of all life and its structure deduced by James Watson and Francis Crick. Today we are fully aware of the DNA sequence of numerous organisms, including humans.

In 1990, the Human Genome Project officially commenced. Only 11 years later, two teams announced that they had deciphered the human DNA sequence. The team of J. Craig Venter and his colleagues at Celera Genomics, funded by private investors, worked feverishly for three years to be first in the race to sequence the genome. Eric Lander at the Whitehead Institute's Center for Genome Research in Massachusetts headed the other team, which was publicly funded. The two groups arrived simultaneously at the sequence of adenine (A), thymine (T), guanine (G), and cytosine (C) in humans.

Sequencing the human genome has given scientists more insight into the genetic differences between humans and other organisms. Surprisingly, only about 42 000 protein-encoding genes are present in the human genome, a far lower number than the 100 000 originally predicted. This suggests that the remaining DNA sequence, which does not code for genes, may contribute in other ways to the complexity of higher organism development. Other surprises include the number of repeat sequences we possess and the number of protein families that we share with worms, flies, and plants. Some human genes are clustered, but more than half are dispersed among large tracts of noncoding DNA.

The Human Genome Project promises great hope in the world of medicine. The completion of the project would not have been possible without the concurrent advancement of technology. Supercomputers and sequencers worked 24 hours a day decoding the 3 billion base pairs that make up the human genome. An Atlanta company plans to use a computer that is capable of 7.5 trillion calculations per second to analyze cancer patients' individual genetic profiles so that the most effective treatments for their cancer can be found. The possibilities are endless. Yet, controversy also surrounds the project. Issues such as ownership of genetic material, patents, and the extent of genetic screening have been raised.

Scientists continue to try to understand the molecular and chemical basis of life. Starting with Miescher's original discovery in 1869 to the success of the Human Genome Project, scientists have unravelled a very small, yet significant, amount of genetic information pertaining to one aspect of the phenomenon of life.

💡 REFLECT on your learning ▼

1. Differentiate between DNA and proteins. What varying cellular roles do they play?
2. Describe the physical and chemical characteristics of DNA.
3. What is the significance of DNA replication in your body?
4. Write a short overview, in paragraph form, of the process of DNA replication.

GCGATGATGGCCA

CGCTACTACCGGT

GACAAAAGGTTTCGC

CTGTTTTCCAAAGC

TGTTTA

The Size of the Genome

All organisms, no matter how simple they may seem to us, require DNA in each cell to encode the instructions necessary to live and reproduce. The total DNA of an organism is referred to as the genome. In bacteria, the genomic DNA is circular, accounts for 2% to 3% of the cell's mass, and occupies about 10% of its volume. In this activity, you will make a model of an *Escherichia coli* cell that will be 10 000 times bigger than actual size. You will also gain an appreciation for how compactly DNA is packed within a cell.

Materials: 2-cm gelatin capsule, 10 m of white thread, 10 m of coloured thread

- Try to construct the bacterium by placing the long lengths of thread inside the gelatin capsule. Good luck! It's not easy!
  (a) Why does it take two lengths of thread to represent the chromosome?
  (b) Is the thread that you tried to place in the capsule too thick to represent the actual thickness of the DNA? (What percentage of bacterial cell volume does your thread fill, and what is the actual volume that the DNA occupies in the bacteria?)
  (c) If the human genome is 1000 times bigger than the *E. coli* genome, how many metres of thread would it take to represent the human genome?

deoxyribonucleic acid (DNA)
a double-stranded polymer of nucleotides (each consisting of a deoxyribose sugar, a phosphate, and four nitrogenous bases) that carries the genetic information of an organism

The 21st century holds great promise in the fields of medicine, agriculture, and the environment. Through well-executed and collaborative research, scientists around the globe are uncovering the secrets that govern the continuity of life. Currently, genetics and biotechnology are the largest areas of research. Yet, it was not very long ago that scientists became aware of the existence of **deoxyribonucleic acid (DNA)** and the role it plays as the hereditary information of living things.

The Discovery of DNA

Although the structure of DNA was not determined until the 1950s, scientists 130 years ago were investigating a phosphorous-rich substance found in the nuclei of cells. In 1869, Friedrich Miescher, a Swiss biochemist (**Figure 1**), investigated the chemical composition of DNA using pus cells. Pus cells were readily available from infections because antiseptic techniques were not in widespread use in medicine. Miescher discovered that the nuclei of cells contain large quantities of a substance that does not act like a protein. At the time, proteins were thought to be the hereditary material because they were widely prevalent, complex chemical substances that were known to carry out biological functions. He called this substance *nuclein* because it is predominantly found in the nucleus of cells. Nuclein was renamed deoxyribonucleic acid (DNA) when its chemical composition was determined. The debate as to whether DNA or proteins were the hereditary material continued into the 20th century.

Figure 1
Friedrich Miescher first investigated nuclein, now known as DNA.

🔬 INVESTIGATION 4.1.1

Isolation and Quantification of DNA (p. 224)
Today, DNA can be easily extracted from plant or animal cells in a manner similar to that used by Miescher more than 130 years ago. In Investigation 4.1.1, you will isolate DNA and quantify how much DNA you have extracted by using a spectrophotometer.

The Location of Hereditary Information

Scientists were interested in determining the location of the hereditary information in the cell. In the 1930s, Joachim Hammerling, a Danish biologist, conducted a series of experiments using *Acetabularia*, a one-celled green alga, to address this question. *Acetabularia* proved to be an ideal model organism for Hammerling's experiment because an individual alga consists of three parts in a single cell: a distinct "foot" that contains the nucleus, a stalk, and a cap region. In addition, *Acetabularia* reaches an average length of 5 cm, making it very easy to work with. In his experiment, Hammerling removed the cap from some cells and the foot from others. He observed that the cells whose caps had been excised were able to regenerate new caps. Cells whose feet had been excised were not able to regenerate at all (**Figure 2**). This led Hammerling to hypothesize that the hereditary information driving regeneration is found within the foot of *Acetabularia* and possibly in the nucleus.

To further test his hypothesis, Hammerling continued experimentation using two different species of *Acetabularia*, which differed in the shape of their caps. *A. mediterranea* possesses a disk-shaped cap, whereas *A. crenulata* has a branched, flowerlike cap. Hammerling grafted a stalk of *A. crenulata* to the foot of *A. mediterranea*. The subsequent cap that grew was intermediary in its characteristics between the two species. He then excised this cap and found that the new cap that formed was again an *A. mediterranea* cap. Hammerling concluded that the instructions for specifying the type of cap and for directing the cellular substances to build it are found in the nucleus in the foot of the cell. The first cap was an intermediate type because the substances that determine cap type were still present in the transplanted stalk. When the second cap formed, the substances in the transplant were used up, and the cap was under the control of instructions from the new nucleus. Many years later, experimental results clearly indicated that DNA, not protein, is indeed the molecular material responsible for the transmission of hereditary information.

Experiment 1

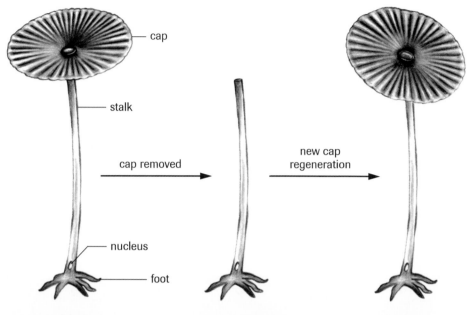

Experiment 2

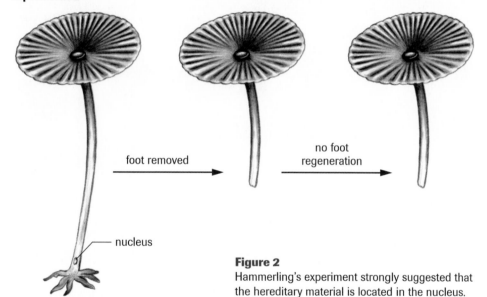

Figure 2
Hammerling's experiment strongly suggested that
the hereditary material is located in the nucleus.

The Transforming Principle

Although it was clear that the nucleus contains the hereditary information, the exact
chemical nature of the hereditary material was unknown. Scientists suspected that DNA
was indeed the hereditary material, but they were uncertain. The nucleus contains the
chromosomes known to be responsible for inheritance, yet chromosomes consist of
both DNA and proteins.

Frederick Griffith's work in the 1920s began because he was trying to develop a vaccine against pneumonia caused by *Streptococcus pneumoniae*. However, his unexpected
experimental observations led scientists to suspect that protein is not the hereditary
material. He was the first to discover the process of transformation.

LAB EXERCISE 4.1.1

**Evidence of Hereditary Material
(p. 226)**
How did Griffith's attempt to create
a vaccine against pneumonia lead
to the discovery of the process of
transformation? In this lab exercise,
you will analyze Griffith's experimental data. You will also analyze
the result's of later experiments
with *Streptococcus pneumoniae*
conducted by Avery, McCarty, and
MacLeod that identified DNA as
the transforming substance.

DNA: The Molecular Basis of Life **207**

DID YOU KNOW ?

Persistence Pays Off

Canadian-born scientists Oswald Avery and Colin MacLeod spent their early years as scientists in Nova Scotia, where they were born. They met in New York, where, together with American scientist Maclyn McCarty, they painstakingly isolated components of pneumo-cocci (*Streptococcus pneumonifor*) for over a decade before identifying DNA as the transforming principle. You can find more information on this classic experiment in an animation by accessing the Nelson science web site.

GO www.science.nelson.com

bacteriophage any bacteria-infecting virus

isotope different atoms of the same element containing the same number of protons but a different number of neutrons

radioisotopes unstable isotopes that decay spontaneously by emitting radiation

Alfred D. Hershey and Martha Chase

It was not until 1952 that DNA was accepted as the hereditary material. That year, Alfred Hershey and Martha Chase conducted experiments using a virus (**bacteriophage** T2) that infects a bacterial host (**Figure 3**). Bacteriophages (commonly called phages) consist of two components: DNA and a protein coat. A bacteriophage infects a bacterial cell by attaching to the outer surface of the cell, injecting its hereditary information into the cell, and producing thousands of new viruses, which then burst out of the cell, resulting in its death. The results of Hershey and Chase's experiments showed that only the DNA from the bacteriophage, and not the protein coat, enters the bacteria to direct the synthesis of new viral DNA and new viral protein coats.

Proteins contain sulfur but no phosphorous, whereas DNA contains phosphorous but no sulfur. Therefore, to track the location of DNA and proteins, Hershey and Chase tagged the viral proteins with an **isotope** of sulfur, ^{35}S, and the viral DNA with an isotope of phosphorus, ^{32}P. ^{35}S and ^{32}P are **radioisotopes** of sulfur and phosphorus, respectively. They are easy to track in an experiment because radioisotopes are unstable and the radiation that they emit as they decay can be measured.

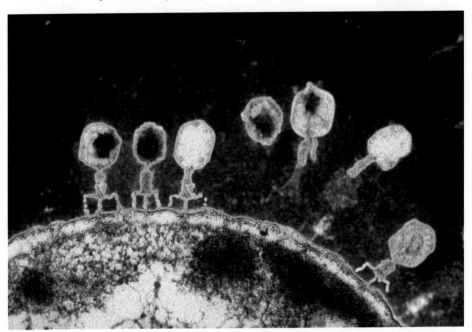

Figure 3

Micrograph of a bacteriophage injecting its DNA into a bacterial cell

Each type of tagged bacteriophage was allowed to infect a separate batch of bacterial host cells and to multiply. The bacterial cells were put into a blender to remove the protein coats of the viruses from the surfaces of the bacteria. The mixtures were then subjected to centrifugation to isolate the individual components (bacteria as a pellet and viral particles in the liquid). The bacterial cells that were exposed to viruses containing radioactively labelled DNA contained ^{32}P. The bacterial cells that were exposed to viruses whose protein coats were radioactively tagged with ^{35}S did not contain any radioactivity; instead, the radioactive ^{35}S was found in the culture medium (**Figure 4**). These experiments illustrate that phosphorus-rich DNA was injected into the bacterial cells. In addition, Hershey and Chase found that the bacteriophages in both experiments reproduced and destroyed the bacterial cells that they had infected. This observation further supported the claim that DNA entering the host bacterial cell carries all the genetic information. Hershey and Chase's experiments ended the debate. DNA was accepted as the hereditary material.

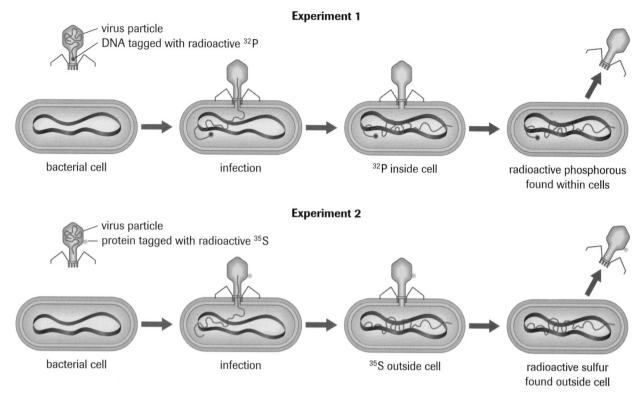

Figure 4
Hershey and Chase's experiment conclusively showed that DNA was the hereditary material.

Understanding Concepts

1. Describe how the experiments of Joachim Hammerling; Frederick Griffith; Oswald Avery, Maclyn McCarty, and Colin MacLeod; and Alfred Hershey and Martha Chase strengthened the hypothesis that DNA is the hereditary material.

2. Explain why Hammerling's experiment cannot be used as conclusive scientific evidence that DNA is the hereditary material.

Applying Inquiry Skills

3. What conclusions, if any, could be made based on the following fictional experimental results? Your conclusions may pertain to the results or the scientific protocol used.
   (a) Hershey and Chase discovered that the bacterial cells that were infected with bacteriophage tagged with radioactive ^{32}P did not contain any radioactive phosphorous. In addition, their experimental results revealed that the bacterial cells that were infected with bacteriophage tagged with radioactive ^{35}S contained radioactive sulfur within their cell walls.
   (b) Hershey and Chase discovered that the bacterial cells that were exposed to radioactive ^{32}P did not contain any radioactive phosphorous. In addition, their experimental results revealed that the bacterial cells that

were infected with bacteriophage tagged with radioactive ^{35}S did not contain any radioactive sulfur either.
   (c) Hammerling observed that after grafting an *A. crenulata* stalk to an *A. mediterranea* foot, the cap that grew was that of the *A. crenulata* species.
   (d) Hammerling observed that, after grafting an *A. crenulata* stalk to an *A. mediterranea* foot, no cap was regenerated.

Making Connections

4. Hammerling chose *Acetabularia* as a model organism for his experiment. Identify some of the characteristics of this green alga that rendered it an ideal organism. Scientists use model organisms in many of their experiments. Identify social, economic, and physical characteristics that would make an organism highly suitable for experimental research. Explain why humans do not make ideal research subjects.

5. Explain why it is important to study both the historic experiments that revealed genetic principles and the principles themselves. Support your reasons, using examples.

6. It can be argued that the repetition of experiments is a waste of time, money, and other valuable resources. Provide arguments that support and dispute this statement. Use examples from the experiments of Griffith and of Avery, McCarty, and MacLeod to strengthen your arguments.

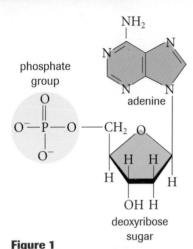

Figure 1
A DNA nucleotide composed of a phosphate group, a deoxyribose sugar, and a nitrogenous base. In this case, the nitrogenous base is adenine.

deoxyribose sugar sugar molecule containing five carbons that has lost the −OH (hydroxyl group) on its 2′ (2 prime) carbon

phosphate group group of four oxygen atoms surrounding a central phosphorus atom found in the backbone of DNA

nitrogenous base an alkaline, cyclic molecule containing nitrogen

nucleotides molecules that consist of a five-carbon sugar (deoxyribose or ribose) with a nitrogenous base attached to their 1′ carbon and a phosphate group attached to their 5′ carbon

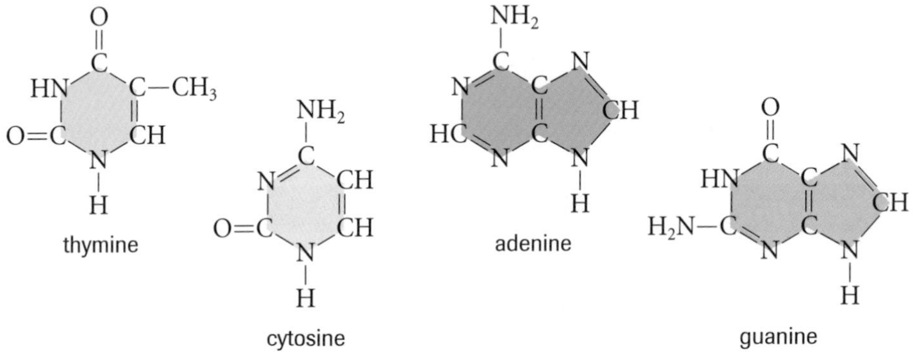

glycosyl bond a bond between a sugar and another organic molecule by way of an intervening nitrogen or oxygen atom

After DNA was identified as the hereditary substance, the next step was to unravel its physical and chemical structure and understand how its structure determines its activity. The collaboration of many scientists helped to put the pieces of the puzzle slowly into place.

The Chemical Composition of DNA

Miescher discovered in the 1860s that nuclein, now known as DNA, is slightly acidic and composed of large amounts of phosphorous and nitrogen. Not until 70 years later were these observations partially explained. In the 1920s, it was determined that DNA comprises three main components: a **deoxyribose sugar**, a **phosphate group** that is negatively charged, and a **nitrogenous base** (**Figure 1**).

The source of variation in DNA is found in the nitrogenous bases. Four nitrogenous bases exist. Adenine (A) and guanine (G) are double-ringed structures known as the purines. Thymine (T) and cytosine (C) are single-ringed structures known as the pyrimidines (**Figure 2**). DNA is composed of many **nucleotides** held together by phosphodiester bonds and is, therefore, a polymer.

Figure 2
The four nitrogenous bases found in DNA

In 1949, Erwin Chargaff analyzed data obtained from chemical analyses of DNA from many different organisms. The data revealed that the proportion of adenine in a DNA molecule is equal to that of thymine, and the proportion of guanine is equal to that of cytosine. He also discovered that the total amount of purines equals the total amount of pyrimidines in an organism's genome. These discoveries were important pieces of information that helped solve the mystery about the structure of DNA.

Structure of Nucleotides

Each of DNA's four nucleotides comprises a deoxyribose sugar (five-carbon cyclic ring structure) attached to a phosphate group and a nitrogenous base (**Figure 3**). In the five-carbon deoxyribose sugar, the first four of the five carbon atoms, together with an oxygen atom, form a five-membered ring. The fifth carbon atom extends out from the ring. The carbon atoms are numbered clockwise, starting with the carbon atom to the immediate right of the oxygen atom. This first carbon atom is designated as 1′ (1 prime), followed by 2′ (2 prime), and so on. A deoxyribose sugar has a hydroxyl group (OH) on the 3′ carbon, and a hydrogen atom (H) on the 2′ carbon. The nitrogenous base is attached to the 1′ carbon of the sugar by a **glycosyl bond** and the phosphate group is attached to the 5′ carbon by an ester bond (**Figure 9** on page 214).

Building a Model for DNA Structure

To understand how this simple polymer could account for such a large number of characteristics in all species, scientists strived to determine its structure. Not only did DNA have to carry hereditary information, it also had to control cellular processes, accommodate individual variation, and be able to replicate itself. Until 1953, scientists had only two clues: the chemical composition of DNA and the relative proportions of the nitrogenous bases to each other. Numerous scientists worldwide were working on the same problems. Linus Pauling was working on the problem in California, Rosalind Franklin and Maurice Wilkins in London, and James Watson and Francis Crick at Cambridge University.

Rosalind Franklin and Maurice Wilkins were each using X-ray diffraction analysis of DNA to determine its structure. In X-ray diffraction, a molecule is bombarded with a beam of X-rays. The X-rays are deflected by the atoms in the molecule, producing a pattern of lighter and darker lines on photographic film. The pattern that is produced is analyzed and the three-dimensional structure is deciphered using mathematics. Although X-ray crystallography was mostly used for nonbiological molecules because they were easily crystallized and isolated, scientists were starting to apply the process to biological molecules.

Maurice Wilkins presented Watson and Crick with an informal look at Rosalind Franklin's diffraction pattern before it was published. The pattern revealed that DNA has the shape of a helix or corkscrew, is about 2 **nm** in diameter, and has a complete helical turn every 3.4 nm. Note that 1 nm (nanometer) is the equivalent of 10^{-9} m. Using Rosalind's information, along with the information available from Chargaff's experiments, Watson and Crick built their famous model of the double-helix structure of DNA in 1953 (**Figure 4**). James Watson, Francis Crick, and Maurice Wilkins were awarded the Nobel Prize in Physiology or Medicine in 1962 for their efforts.

Rosalind Franklin

The value of Rosalind Franklin's contribution to the understanding of the structure of DNA is undisputed. Using her talents in X-ray crystallography, she was able to produce a clear X-ray diffraction pattern that suggested that DNA is helical in nature (**Figure 5**). This evidence was the last piece of the puzzle that Watson and Crick needed to win the scientific race of determining the structure of DNA.

Rosalind Franklin (**Figure 6**, page 212) was born July 25, 1920, to liberal-minded parents, who raised her in the same manner as her male siblings. When her father brought home a carpenter's workbench, she was encouraged to learn to mitre and dovetail. Her machine-shop abilities proved invaluable when setting up her research laboratories. Her formal education took place at St. Paul's Girls School, where she excelled in mathematics, chemistry, and physics. At the age of 15, she decided she would pursue a career in science, despite the lack of female role models in the scientific community in the 1930s. No woman at that point held a major university appointment or was a Fellow of the Royal Society. Franklin was determined and dedicated, two characteristics that contributed to her later success.

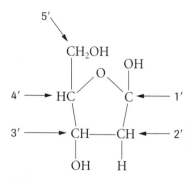

Figure 3
A deoxyribose sugar with numbered carbons

nm nanometer, the equivalent of 10^{-9} m

Figure 4
James Watson and Francis Crick were awarded the Nobel Prize for Physiology or Medicine in 1962 for deducing the structure of DNA.

DID YOU KNOW ?

Elementary, My Dear Crick
Erwin Chargaff visited Watson and Crick in Cambridge in 1952. Crick's lack of knowledge with respect to nitrogenous bases did not impress Chargaff. By the following year, Watson and Crick had constructed their model of DNA. Enjoy Watson and Crick's deductive process in an animation found by accessing the Nelson science web site.

 www.science.nelson.com

LEARNING TIP

Chargaff's Rules
The proportion of A always equals that of T (A = T).
The proportion of G always equals that of C (G = C).
A + G = T + C

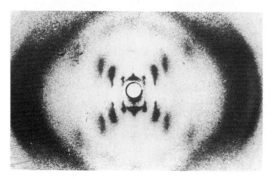

Figure 5
Rosalind Franklin's X-ray diffraction pattern of DNA revealed that it had a helical structure.

Figure 6
Rosalind Franklin's X-ray crystallography was crucial to the determination of the structure of DNA.

Entering Cambridge in 1938, she earned an undergraduate degree in science. On completion, she declined a research scholarship. Instead she accepted a position as a research officer with the British Coal Utilization Research Association (BCURA). Since the association was newly founded, little protocol or hierarchy had been established, making working relations between employees very collaborative. In her four years with the BCURA, she published five papers that examined the microstructure of coal. Her work was highly technical and meticulous, since she used a single helium atom as her unit of measurement. In 1945, she submitted her thesis to Cambridge and was awarded her Ph.D.

Franklin's expertise in X-ray crystallography was nurtured in Paris at the Laboratoire Central des Services Chemiques de l'Etat. From 1947 to 1950, she learned techniques in X-ray diffraction and investigated the structure of graphite. From there, she was invited to work in Professor John Randall's laboratory at King's College in Cambridge in 1951, where she spent the first eight months setting up and fine-tuning the equipment she would require in a basement laboratory. Her challenge was to determine the structure of DNA. Another researcher, Maurice Wilkins, was also working on the problem. Both had their own groups, albeit Franklin's group was much smaller. Wilkins assumed that Franklin was working for him on the project, since he did not have any expertise in X-ray crystallography, but Franklin refused to think of herself as Wilkin's assistant. Between 1950 and 1956 she published 11 papers, only one of which was coauthored. She regarded the determination of the structure of DNA as her problem to solve.

Needless to say, the relationship between Franklin and Wilkins was somewhat hostile, although Wilkins did provide Franklin with crystalline fibres of DNA that he had isolated so that she could analyze them. In 1953, Franklin produced the revealing X-ray diffraction pattern that helped Watson and Crick deduce the helical structure of DNA. Although Franklin had not yet found conclusive evidence of the helical shape of DNA in the pattern, she had publicly predicted in a talk at the university the possibility of a helical structure 14 months earlier based on work that she had completed.

Franklin died of cancer at the age of 37 in 1958. She worked in her basement laboratory until the very end, with the use of a wheelchair. The Nobel Prize awarded for the deduction of the structure of DNA was given to Watson, Crick, and Wilkins in 1962. The Nobel Committee does not award prizes to deceased scientists; it is unclear whether she would have been considered for the prize if she had been alive in 1962.

▶ EXPLORE an issue

Debate: Competition Drives Science

Decision-Making Skills

○ Define the Issue	● Analyze the Issue	● Research
● Defend the Position	● Identify Alternatives	○ Evaluate

Scientists have been described as intelligent, ambitious, and sometimes competitive individuals. Yet, for science to progress, many individuals must work together in a collaborative, communicative atmosphere. Current science demands two conflicting ideologies: competition and collaboration. A fine balance is not necessarily struck between the two. Other factors that come into play are economics, politics, market demand, profit, and patriotism in times of war.

Statement
Competition is the key driving force of science, followed by collaboration.

- Form groups to research this issue. Prepare a position paper in point form that supports or disputes this statement, using a specific example. Some scientists and case studies that may

be used include Robert Oppenheimer's and Phillip Morrison's role in the Manhattan Project; the perception of Linus Pauling as a communist and the denial of a visa for him to visit Watson and Crick in Cambridge; Craig Venter and Eric Lander leading opposing research teams in the Human Genome Project; and Fritz Haber's role in the production of deadly gases during World War I.

- Search for information in periodicals, on CD-ROMS, and on the Internet.

- As a group, present your supported view in a class discussion.

GO www.science.nelson.com

DNA Structure: The Double Helix

According to the model proposed by Watson and Crick, DNA consists of two **antiparallel** strands of nucleotides (**Figure 8**, page 214). The bases of one strand are paired with bases in the other strand facing inward toward each other. The nitrogenous base pairs are arranged above each other, perpendicular to the axis of the molecule. A purine is always bonded to a pyrimidine. Adenine (a purine) is always paired with thymine (a pyrimidine) in the other strand (**Figure 7**), while guanine (a purine) is always paired with cytosine (a pyrimidine) in the other strand. This type of pairing is termed **complementary base pairing**. This structural arrangement was proposed to account for Chargaff's chemical data. An important consequence of the complementary base pairing rule is that if you know the sequence on one strand, you also know the sequence on the complementary strand. Complementary base pairing is fundamental to the storage and transfer of genetic information.

antiparallel parallel but running in opposite directions; the 5′ end of one strand of DNA aligns with the 3′ end of the other strand in a double helix (**Figure 9**, page 214)

complementary base pairing pairing of the nitrogenous base of one strand of DNA with the nitrogenous base of another strand; adenine (A) pairs with thymine (T), and guanine (G) pairs with cytosine (C)

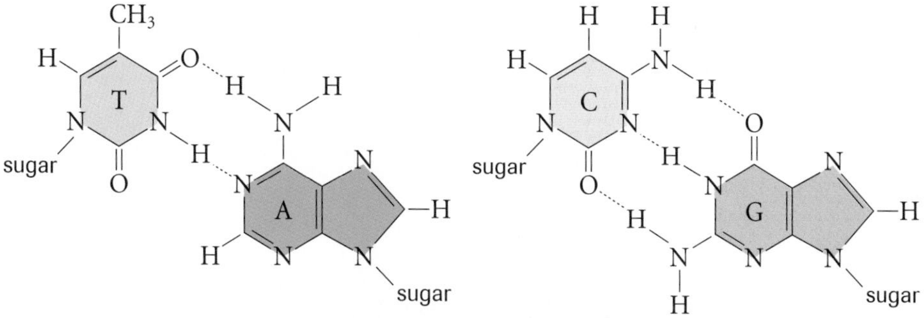

Figure 7
Adenine (a purine) is always hydrogen bonded to thymine (a pyrimidine), while guanine (a purine) is always hydrogen bonded to cytosine (a pyrimidine).

The double-helical structure of DNA as proposed by Watson and Crick is consistent with Rosalind Franklin's and Maurice Wilkins' X-ray diffraction pattern, which indicates that the diameter of DNA is constant at 2 nm. If the two purines bonded together, the DNA molecule would be wider at some points; conversely, if the two pyrimidines bonded together, the molecule would be less than 2 nm in some places. The bases are bonded together by hydrogen bonds, which can only occur in the A–T, G–C arrangement. An individual hydrogen bond is very weak, but large numbers of hydrogen bonds are collectively strong, which explains DNA's high stability as a molecule. Thymine (a pyrimidine) could not bond with guanine (a purine), even though these bases would constitute the desired diameter within the DNA molecule. The molecule would be unstable because of the lack of hydrogen bonding.

Although DNA is found in alternate forms, Watson and Crick's model proposed that the double helix turns in a clockwise direction (**Figure 8**, page 214). This right-handed helix makes one complete turn every 10 nucleotides to account for the 3.4-nm full helical twist. Therefore 0.34 nm represents the distance between adjacent base pairs. Watson and Crick's model placed the deoxyribose sugar and phosphate groups on the perimeters of the molecule; the sugar-phosphate backbone acts as structural support for the bases. The sequence of bases in DNA varies among species.

The two strands of DNA run antiparallel. One strand runs in the 5′ to 3′ direction, while the other strand runs in the 3′ to 5′ direction. The 3′ end terminates with the hydroxyl group of the deoxyribose sugar, and the 5′ end terminates with a phosphate group (**Figure 9**, page 214). Therefore, every DNA molecule has an intrinsic directionality. The sequence of bases is indicated in the following way:

5′–ATGCCGTTA–3′
3′–TACGGCAAT–5′

By convention, only the 5′ to 3′ strand is written, since the complementary strand can easily be deduced from the complementary nature of bases.

Sugar–Phosphate backbone

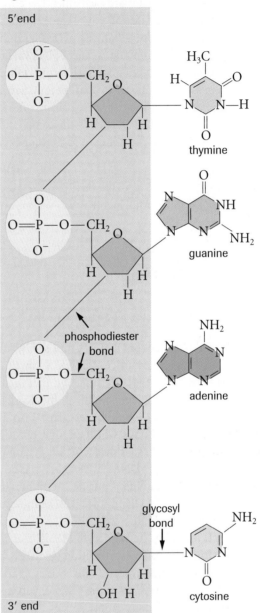

Figure 9
Each DNA strand has a 5′ end where a phosphate group resides, and a 3′ end where the hydroxyl group (−OH) of the deoxyribose sugar is found. The nitrogenous bases are attached to the deoxyribose sugar by a glycosyl bond. The sugar–phosphate backbone is held together by phosphodiester bonds, which are bonds within any molecule that join two parts through a phosphate group. In the case of DNA, a carbon in the pentose sugar of one nucleotide is bound to a carbon in the pentose sugar of the adjacent nucleotide via a phosphate group.

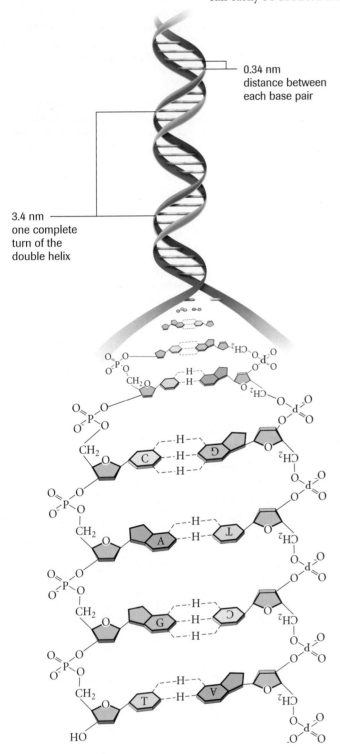

Figure 8
Representation of DNA molecule. Note the hydrogen bonds and the antiparallel strands.

The Directionality of DNA Strands: A Whole-Class Simulation

DNA strands run antiparallel. One strand runs in the 5′ to 3′ direction, while the other strand runs in the 3′ to 5′ direction (**Figure 10**). The antiparallel nature of DNA can be simulated using all students in the class. Each student represents a nucleotide.

- Choose a partner. This person will be your conjugate base.
- Hydrogen bond with your partner using your right hand (shaking hands).
- All sets of partners line up side by side. Each person places their left hand on the right shoulder of the person beside them. This represents a phosphodiester bond.
- At each end of the line of students there will be one free left hand. This represents the phosphate group.

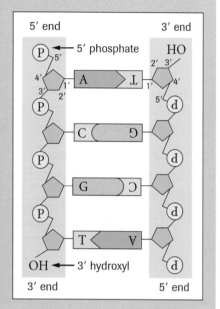

Figure 10
The antiparallel nature of DNA

SUMMARY   DNA Structure

Year	Scientist	Experimental results
Late 1860s	Friedrich Miescher	• isolated nonprotein substance from nucleus of cells; named this substance nuclein
1920s	Frederick Griffith	• experimented using mice and two different strains of pneumococcus bacteria (virulent and nonvirulent); observed that when heat-treated virulent pneumococcus was mixed with nonvirulent pneumococcus and was injected into healthy mice death resulted • discovered the process of transformation
1930s	Joachim Hammerling	• experimented using green alga *Acetabularia;* observed that regeneration of new appendages was driven by the nucleus-containing "foot" of the alga • hypothesized that hereditary information is stored in the nucleus
1944	Oswald Avery, Maclyn McCarty, and Colin MacLeod	• demonstrated that DNA was the transforming principle of pneumococcus bacteria
1949	Erwin Chargaff	• discovered that in the DNA of numerous organisms the amount of adenine is equal to the amount of thymine, and the amount of guanine is equal to that of cytosine
1952	Alfred Hershey and Martha Chase	• used radioactively labelled viruses, infected bacterial cells; observed that the infected bacterial cells contained radioactivity originating from DNA of the virus • suggesting that DNA is hereditary material
1953	Rosalind Franklin	• produced an X-ray diffraction pattern of DNA that suggested it was in the shape of a double helix
1953	James Watson and Francis Crick	• deduced the structure of DNA using information from the work of Chargaff, Franklin, and Maurice Wilkins

Figure 11
An historical overview of DNA structure and function

Understanding Concepts

1. Define the following terms: nucleotide, complementary base pairing, phosphodiester bond, and glycosyl bond.

2. In a DNA molecule, a purine pairs with a pyrimidine. If this is the case, then why can't A–C and G–T pairs form?

3. The following is a segment taken from a strand of DNA: 5′–ATGCCTTA–3′. Write out the complementary strand for this segment. Be sure to show directionality.

4. Summarize the key physical and chemical properties of DNA.

5. Differentiate between a purine and a pyrimidine.

Applying Inquiry Skills

6. A molecule of DNA was analyzed and found to contain 20% thymine. Calculate the percentage of adenine, guanine, and cytosine in this molecule.

7. **Table 1** shows the distribution of nucleotides that originated from analysis of single-stranded DNA and double-stranded DNA. Identify which of the samples A, B, and C are most likely double stranded and which are single stranded, and record your answers in your notebook.

Table 1

Sample	Adenine	Guanine	Thymine	Cytosine
A	10%	40%	10%	40%
B	35%	10%	35%	20%
C	15%	25%	20%	40%

8. In a double helix, there is a complete turn every 3.4 nm, or 10 nucleotides. The human genome contains approximately 3 billion nucleotides. Calculate how long an average human cell's DNA would be if its structure were unwound. Calculate how many complete turns would exist in this molecule.

9. Assume that the DNA molecule in a particular chromosome is 75 mm long. Calculate the number of nucleotide pairs in this molecule has.

Making Connections

10. While trying to determine the structure of the DNA molecule, James Watson and Francis Crick proposed that like bases are bonded to like bases. For example, thymine would be bonded to thymine, cytosine would be bonded to cytosine, and so on. Explain why this proposed model would not fit with Erwin Chargaff's observations. How would this model also contradict Rosalind Franklin's findings in her X-ray crystallography of DNA?

One of the fundamental properties of cells and organisms is their ability to reproduce. All cells are, in principle, capable of giving rise to a new generation of cells by undergoing DNA replication and cell division. During cell division in eukaryotic cells, the replicated genetic material in the nucleus is divided equally between two daughter nuclei in a process known as **mitosis**. This is usually followed by **cytokinesis**, in which the cell is split into two new cells. Mitotic cell division is essential for the growth of tissues during embryonic development and childhood, as well as for tissue regeneration, such as the continuous, daily replacement of thousands of skin cells or the repair of damaged tissue.

It is important that each daughter cell has an exact copy of the parent cell's DNA. When it was discovered by Watson and Crick, the structure of DNA immediately suggested how DNA was able to replicate. The hydrogen bonds between complementary bases could break, allowing the DNA helix to unzip. Each single DNA strand could act as a **template** to build the complementary strand, resulting in two identical DNA molecules, one for each daughter cell.

mitosis division of the nucleus of a eukaryotic cell into two daughter nuclei with identical sets of chromosomes

cytokinesis division of cytoplasm and organelles of a cell into two daughter cells

template a single-stranded DNA sequence that acts as the guiding pattern for producing a complementary DNA strand

DNA Replicates Semiconservatively

Experimental support for the Watson-Crick hypothesis came in 1958 when Matthew Meselson and Franklin Stahl devised a clever experiment that suggested that DNA replication is **semiconservative** (**Figure 1(a)**) and not conservative (**Figure 1(b)**).

semiconservative process of replication in which each DNA molecule is composed of one parent strand and one newly synthesized strand

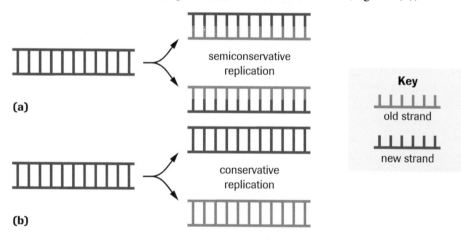

Figure 1
(a) DNA replicates semiconservatively. Each daughter molecule receives one strand from the parent molecule plus one newly synthesized strand.
(b) If DNA were to replicate conservatively, one daughter molecule would be unchanged from the parent molecule and one daughter molecule would consist of newly synthesized strands.

Meselson and Stahl grew *Escherichia coli* bacteria in a nutrient medium that was rich in ^{15}N, a heavy isotope of nitrogen. The bacteria reproduced by the process of cell division for a total of 17 generations. Nitrogen is a component of DNA found in the nitrogenous bases; therefore after 17 generations, both strands of bacterial DNA would be highly labelled with heavy ^{15}N. If the same bacteria are then transferred to a culture medium that contains only a light isotope of nitrogen, ^{14}N, and allowed to reproduce, light nitrogen would be incorporated into the DNA that is subsequently replicated. Hence, after the switch to light nitrogen, the density of nitrogen present in bacterial DNA after each generation would reflect the mode of DNA replication used by *E. coli*. Using this knowledge as a premise for their experimental procedure, Meselson and Stahl devised a new centrifugation method to measure the DNA densities of the bacteria before and after the switch from heavy to light nitrogen.

DID YOU *KNOW*

Room Service, Please
Colleagues in the scientific community heard about the results of Meselson and Stahl's work long before it was published. One colleague, Max Delbruck, eventually locked Meselson and Stahl in a room so that they would finish writing their landmark paper. Watch an animation of Stahl and Meselson's famous experiment by accessing the Nelson science web site.

GO www.science.nelson.com

Meselson and Stahl used density gradient centrifugation to analyze the density of DNA isolated from cells that were grown only in ^{14}N (unlabelled cells) and from cells grown for 17 generations in ^{15}N (labelled cells). They repeated the procedure for samples of cells taken from the two generations following the switch from heavy to light nitrogen. Each sample of DNA extracted from the cells was centrifuged in a solution of cesium chloride, which forms a density gradient when spun for many hours at high speed. The highest density is at the bottom of the centrifuge tubes, while progressively lower densities are toward the top. When a molecule is centrifuged, two forces act on it: the centrifugal force moving it toward the bottom of the tube, and the buoyant force moving it to the top. The movement of the DNA in the gradient stops once the DNA reaches the same density in the gradient as its own. At this point, equilibrium is reached between two opposing forces.

Meselson and Stahl hypothesized that, if the semiconservative mode of replication were used by *E. coli*, all the DNA after one generation after the switch to ^{14}N would consist of one ^{14}N strand and one ^{15}N strand. The isolated DNA should settle as one band of intermediate density after centrifugation. After two generations, half of the DNA would have a density like that after one generation, while the other half would have a density similar to the ^{14}N DNA. Therefore, two different bands would be present in the gradient, one of intermediate and one of light density. On the other hand, if a conservative mode of replication were used, the observations after one generation would be different. The DNA would settle as two different bands after centrifugation, half like the DNA labelled with heavy nitrogen before the switch to light, and half like the unlabelled DNA. After two generations, the same two densities would be present, but in a 3:1 ratio (light:heavy nitrogen).

Meselson and Stahl observed that the DNA isolated from cells before the switch from heavy to light nitrogen settled as one discrete band in a dense region of the cesium density gradient (**Figure 2(a)**). The DNA from the second generation (after one replication) settled as a single band with a density intermediate between DNA containing ^{14}N and DNA containing ^{15}N (**Figure 2(b)**). The DNA from the third generation (after two replications) showed two bands. The bands consisted of an intermediate-density band identical to that in the second generation and a lighter-density band (**Figure 2(c)**). Analysis of the intermediate band showed that it consisted of strands of two different densities. Furthermore, after the bacterial DNA underwent numerous replications, Meselson and Stahl found that the intermediate band never disappeared, suggesting that the initial DNA containing ^{15}N continued to exist and to be replicated, so that a heavy strand (^{15}N) was always paired with a new light strand (^{14}N).

Meselson and Stahl's density gradient experiments supported the idea that individual strands remain intact during DNA replication when they serve as templates for new DNA strands, and after replication, each strand is found paired with a new complementary strand. Therefore, their results were consistent with the hypothesis that DNA replication is semiconservative. Although Meselson and Stahl's conclusion was thought to apply only to bacteria, other investigators soon conducted experiments with eukaryotic cells and found the same semiconservative pattern of DNA replication.

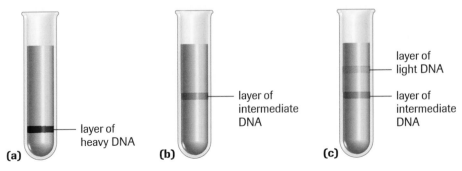

Figure 2
Results of Meselson and Stahl's experiment

(a) — layer of heavy DNA

(b) — layer of intermediate DNA

(c) — layer of light DNA / layer of intermediate DNA

The Process of DNA Replication

Separating the DNA Strands

Scientists have a good understanding of the replication process in prokaryotes from studying the bacterium *Escherichia coli*. Eukaryotic replication is similar to prokaryotic replication but more complex. Replication begins when proteins bind at a specific site on the DNA known as the replication origin. The closed circular DNA of prokaryotes usually has only one such site, whereas the linear DNA of eukaryotes has multiple origins of replication.

In both organisms, the two strands forming a DNA molecule cannot simply be pulled apart because they are held together by hydrogen bonds and are twisted around each other to form a double helix. To expose a template strand, the two parent DNA strands must be unravelled and kept separate. Specific enzymes work together to expose the DNA template strands. **DNA helicase** unwinds the double helix by breaking the hydrogen bonds between the complementary base pairs holding the two DNA strands together. Since the base pairs are complementary and have a natural propensity to **anneal**, the two individual strands are kept apart by **single-stranded binding proteins (SSBs)**. SSBs bind to the exposed DNA single strands and block hydrogen bonding (**Figure 3**). **DNA gyrase** is an enzyme that relieves any tension brought about by the unwinding of the DNA strands during bacterial replication. Gyrase works by cutting both strands of DNA, allowing them to swivel around one another, and then resealing the cut strands. Enzymes from the same family as gyrase have similar functions in eukaryotes.

DNA helicase the enzyme that unwinds double-helical DNA by disrupting hydrogen bonds

anneal the pairing of complementary strands of DNA through hydrogen bonding

single-stranded binding proteins (SSBs) a protein that keeps separated strands of DNA apart

DNA gyrase the bacterial enzyme that relieves the tension produced by the unwinding of DNA during replication

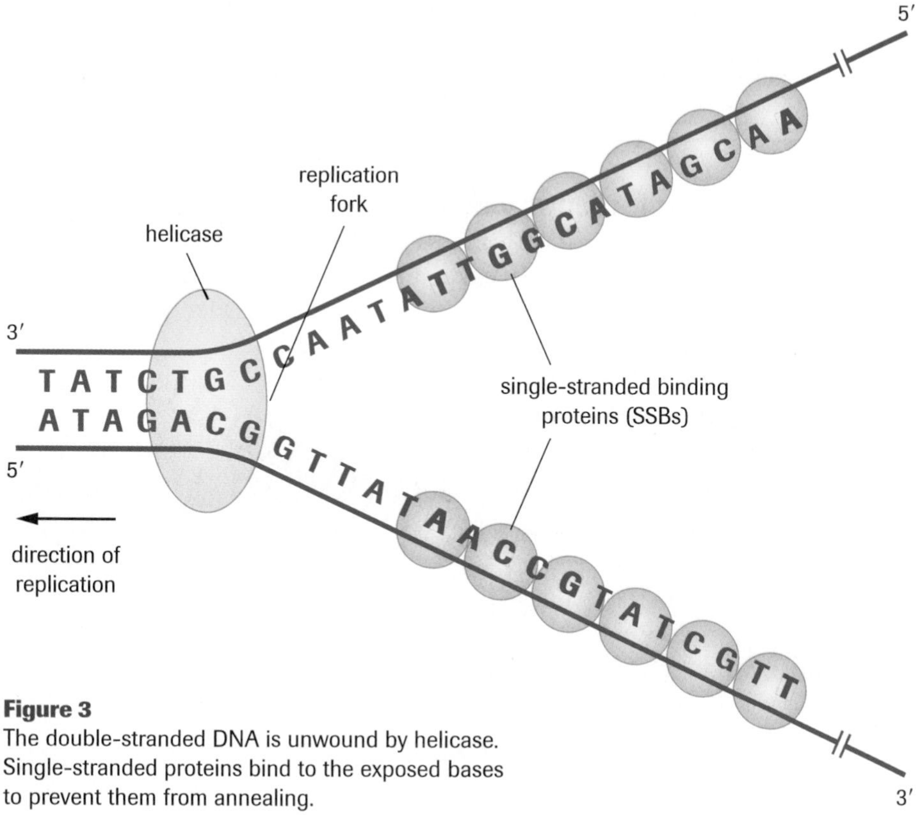

Figure 3
The double-stranded DNA is unwound by helicase. Single-stranded proteins bind to the exposed bases to prevent them from annealing.

Replication begins in two directions from the origin(s) as a region of the DNA is unwound. The DNA cannot be fully unwound because of its large size compared with the size of the cell. For example, the diameter of a cell is approximately 5 μm. The length of DNA in one human chromosome is approximately 1 cm, a 2000-fold difference. Hence, new complementary strands are built as soon as an area of the DNA has been

replication fork the region where the enzymes replicating a DNA molecule are bound to untwisted, single-stranded DNA

replication bubble the region where two replication forks are in close proximity to each other, producing a bubble in the replicating DNA

unwound. As the two strands of DNA are disrupted, the junction where they are still joined is called the **replication fork**. DNA replication proceeds toward the direction of the replication fork on one strand and away from the fork on the other strand. In eukaryotes, more than one replication fork may exist on a DNA molecule at once because of the multiple sites of origin, ensuring the rapid replication of DNA. When two replication forks are quite near each other, a **replication bubble** forms (**Figure 4**).

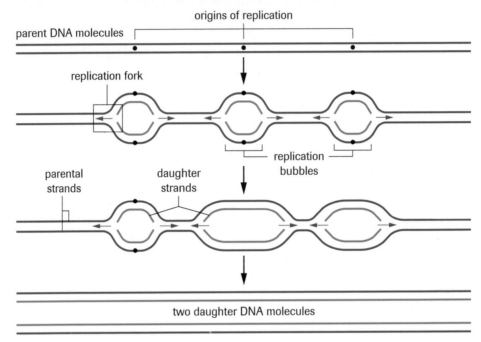

Figure 4
In eukaryotes, DNA replication occurs at more than one site at a time, resulting in hundreds of replication forks across a DNA strand. Eventually the replication bubbles become continuous and the two new double-stranded daughter molecules are completely formed.

Building the Complementary Strands

In prokaryotes, DNA polymerase I, II, and III are the three enzymes known to function in replication and repair. In eukaryotes, five different types of DNA polymerase are at work. The enzyme that builds the complementary strand using the template strand as a guide in prokaryotes is **DNA polymerase III** (**Figure 5**). DNA polymerase III functions only under certain conditions.

DNA polymerase III the enzyme responsible for synthesizing complementary strands of DNA during DNA replication

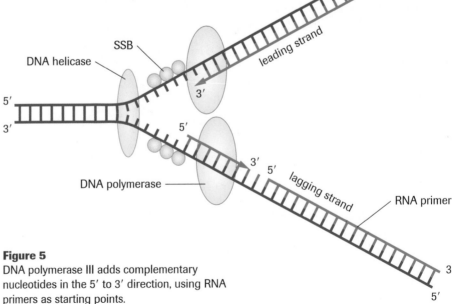

Figure 5
DNA polymerase III adds complementary nucleotides in the 5′ to 3′ direction, using RNA primers as starting points.

First, it synthesizes DNA in the 5′ to 3′ direction and therefore adds free **deoxyribo-nucleoside triphosphates** to a 3′ end of an elongating strand (**Figure 6**). Second, it requires an initial starting 3′ end to commence elongation. Since DNA polymerase III cannot initiate a new complementary strand by itself, an **RNA (ribonucleic acid) primer** of 10 to 60 base pairs of DNA is annealed to the template strand. This primer is synthesized by the enzyme **primase**. The RNA primer marks the initiation sequences as "temporary" for ease of removal later. Once in place, DNA polymerase III can start elongation by adding free deoxyribonucleotide triphosphates to the growing complementary strand.

The free bases in the nucleoplasm used by DNA polymerase III to build complementary strands are deoxyribonucleoside triphosphates. DNA polymerase III uses the energy derived from breaking the bond between the first and second phosphate to drive the dehydration synthesis (condensation reaction) that adds a complementary nucleotide to the elongating strand. The extra two phosphates are recycled by the cell and are used to build more nucleoside triphosphates.

Since DNA is always synthesized in the 5′ to 3′ direction and the template strands run antiparallel, only one strand is able to be built continuously. This strand, which uses the 3′ to 5′ template strand as its guide, is called the **leading strand** and is built toward the replication fork. The other strand is synthesized discontinuously in short fragments in the opposite direction to the replication fork and is known as the **lagging strand**. Primers are continuously added as the replication fork forms along the DNA parent strand, and DNA polymerase III builds in short segments known as **Okazaki fragments** (**Figure 7**), named after the scientist who first discovered them. **DNA polymerase I** removes the RNA primers from the leading strand and fragments of the lagging strand and replaces them with the appropriate deoxyribonucleotides. Since the lagging strand is built in short segments (about 100 to 200 nucleotides in length in eukaryotes and 1000 to 2000 nucleotides in length in prokaryotes), another enzyme, **DNA ligase**, joins the Okazaki fragments into one strand by creation of a phosphodiester bond. As the two new strands of DNA are synthesized, two double-stranded DNA molecules are produced that automatically twist into a helix.

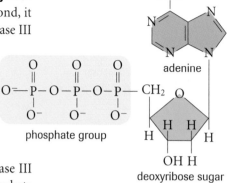

phosphate group

adenine

deoxyribose sugar

Figure 6
Deoxyadenosine triphosphate

deoxyribonucleoside triphosphates molecules composed of a deoxyribose bonded to three phosphate groups and a nitrogenous base

RNA (ribonucleic acid) primer a sequence of 10 to 60 RNA bases that is annealed to a region of single-stranded DNA for the purpose of initiating DNA replication

primase the enzyme that builds RNA primers

leading strand the new strand of DNA that is synthesized continuously during DNA replication

lagging strand the new strand of DNA that is synthesized in short fragments, which are later joined together

Okazaki fragments short fragments of DNA that are a result of the synthesis of the lagging strand during DNA replication

DNA polymerase I an enzyme that removes RNA primers and replaces them with the appropriate deoxyribonucleotides during DNA replication

DNA ligase the enzyme that joins DNA fragments together by catalyzing the formation of a bond between the 3′ hydroxyl group and a 5′phoshate group on the sugar–phospate backbones

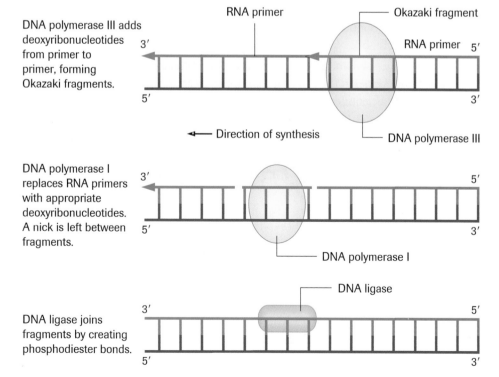

DNA polymerase III adds deoxyribonucleotides from primer to primer, forming Okazaki fragments.

DNA polymerase I replaces RNA primers with appropriate deoxyribonucleotides. A nick is left between fragments.

DNA ligase joins fragments by creating phosphodiester bonds.

Figure 7
Joining of Okazaki fragments by DNA ligase

Ensuring Quality Control of New DNA Strands: DNA Repair

exonuclease an enzyme that cuts outnucleotides at the end of a DNA strand

As complementary sequences are built, DNA polymerase III and DNA polymerase I act as quality control checkers by proofreading the newly synthesized strand. When mistakes occur, either enzyme can function as an **exonuclease**. The enzyme backtracks past the nucleotide on the end of the strand that is incorrectly paired to a nucleotide on the template, excises it, and continues adding nucleotides to the complementary strand. The repair must be made immediately to avoid the mistake from being copied in subsequent replications. Errors missed by proofreading can be corrected by one of several repair mechanisms that operate after the completion of DNA replication.

DID YOU *KNOW* ?

Proofreading for Accuracy

The *Escherichia coli* genome consists of 4.7 million nucleotide pairs. With a replication rate of about 1000 nucleotides per second, the entire genome is replicated in 40 minutes. The proofreading efficiency of DNA polymerase I and polymerase III reduces the error rate to one in 1 billion base pairs, or roughly one error per 1000 cells duplicated! View a complete animation of DNA replication by accessing the Nelson science web site.

GO   www.science.nelson.com

SUMMARY   *DNA Replication and Repair*

1. The enzyme gyrase relieves any tension from the unwinding of the double helix.

2. The enzyme helicase breaks the hydrogen bonds holding the two complementary parent strands together, resulting in an unzipped helix that terminates at the replication fork.

3. Single-stranded binding proteins anneal to the newly exposed template strands, preventing them from reannealing.

4. The enzyme primase lays down RNA primers that will be used by DNA polymerase III as a starting point to build the new complementary strands.

5. DNA polymerase III adds the appropriate deoxyribonucleoside triphosphates to the 3′ end of the new strand using the template strand as a guide. The energy in the phosphate bonds is used to drive the process. The leading strand is built continuously toward the replication fork. A lagging strand composed of short segments of DNA, known as Okazaki fragments, is built discontinuously away from the replication fork.

6. DNA polymerase I excises the RNA primers and replaces them with the appropriate deoxyribonucleotides. DNA ligase joins the gaps in the Okazaki fragments by the creation of a phosphodiester bond.

7. DNA polymerase I and DNA polymerase III proofread by excising incorrectly paired nucleotides at the end of the complementary strand and adding the correct nucleotides.

Figure 8
An overview of DNA replication

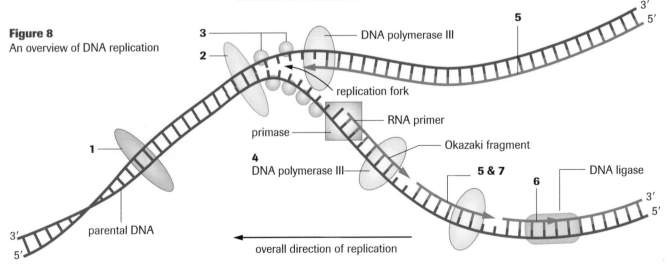

Understanding Concepts

1. Copy **Table 1** into your notebook and fill in the missing information.

Table 1

Enzyme	Function
DNA gyrase	
DNA helicase	
DNA polymerase I	
DNA polymerase III	
DNA ligase	
RNA primase	

2. Compare the mechanisms by which the leading and lagging strands are replicated. Explain why it is necessary for two mechanisms to exist.

3. Matthew Meselson and Franklin Stahl observed a total of three bands (heavy, light, and intermediate) in the density gradient they obtained in their experiment. At which point in their experiment did they observe each type of band? What was the constituent isotope that each band contained?

4. Distinguish between a replication fork and a replication bubble.

5. Explain why it is necessary for eukaryotic DNA to have multiple replication origin sites.

6. Explain why both parental DNA strands may be used to serve as a template for replication. How does DNA's structure allow for this phenomenon to occur?

7. Meselson and Stahl's experiment indicated that DNA replicates semiconservatively. What percentage of DNA double helixes would contain one of the original parent strands after four generations of replication?

Applying Inquiry Skills

8. In a conservative model of replication the original double-stranded molecule is conserved and both strands of the replicated molecule are new. Describe how Meselson and Stahl's results would have differed if DNA replicated conservatively.

Making Connections

9. One of the consequences of DNA polymerase III only being able to elongate and not initiate a complementary strand is that RNA primers need to be added. Since DNA polymerase III only works in the 5′ to 3′ direction, an end of one strand is consistently missing some of its base pairs after replication. To compensate, eukaryotic cells have noncoding sequences of their DNA at the ends called *telomeres*. As people age, telomeres shorten and therefore less protection is afforded against damaging coding DNA. Research the connection between telomeres and aging. Prepare a report of your findings.

 www.science.nelson.com

10. When Watson and Crick built their model of DNA, they immediately proposed a semi-conservative mode of replication. Provide justification for their initial hypothesis.

🔥 INVESTIGATION 4.1.1

Isolation and Quantification of DNA

Inquiry Skills

○ Questioning	○ Planning	● Analyzing
● Hypothesizing	● Conducting	● Evaluating
○ Predicting	● Recording	● Communicating

In Parts 1 and 2 of this investigation you will isolate DNA from onion cells and beef liver. Part 3 verifies the presence of DNA in your extraction using a biological analysis and Part 4 quantifies the amount of DNA using spectrophotometry. Parts 3 and 4 are optional depending on whether your school has the necessary reagents.

Question

How much DNA can be extracted from plant and animal cells using simple laboratory methods?

Hypothesis

(a) Form a hypothesis about the comparative quantity of DNA that will be obtained from each type of cell using similar extraction procedures.

Materials

safety goggles
rubber gloves
fresh beef liver
scissors
mortar and pestle
0.9 % (w/v) solution of
    sodium chloride (NaCl)
three 10-mL graduated
    cylinders
sand (very fine, washed)
cheesecloth
two 50-mL beakers, or
    two large test tubes
10% (w/v) solution of
    sodium dodecyl sulfate
    (SDS)
95% ethanol (chilled)
50-mL graduated cylinder
glass rod

four medium test tubes
4% (w/v) solution of
    sodium chloride (NaCl)
onion
blender (optional)
hot-water bath
ice bath
meat tenderizer solution
    (3 g/50 mL of solution)
diphenylamine solution
25-mL graduated cylinder
Pasteur pipette, or plastic
    graduated eyedropper
distilled water
DNA standard solution
test-tube rack
spectrophotometer
cuvette
facial tissue

Procedure

DNA extraction is the first step in many biotechnological procedures. Cell walls and cell membranes must be disrupted to isolate DNA. In addition, lipids, proteins, and sugars must be separated from nucleic acid. In the following procedure, heat, detergents, salts, and cleaving enzymes are used to minimize contamination from nonnucleic acid molecules and to maximize purification.

Part 1: Extraction of DNA from Beef Liver

1. Obtain a 10-g to 15-g sample of beef liver and place it in the mortar.

2. Using scissors, cut the liver into small pieces.

3. Add 10 mL of 0.9% NaCl solution to the diced liver. Use a 10-mL graduated cylinder to measure out the NaCl. Add a pinch of sand into the mixture to act as an abrasive, and grind the tissue thoroughly for approximately 5 min.

4. Strain the liver cell suspension through several layers of cheesecloth to eliminate any unpulverized liver. Collect the filtrate into a 50-mL beaker.

5. Add 3 mL of 10% SDS solution. If a centrifuge is available, spin the suspension, and remove and save the supernatant. If a centrifuge is not available, mix the suspension thoroughly for 30 s and proceed to step 6.

6. Gently layer twice the volume of cold 95% ethanol on the supernatant as the total volume of the cell suspension–SDS mixture. Use a 50-mL graduated cylinder to measure out the ethanol.

7. Using the glass rod, stir gently and slowly. A white, mucuslike substance will appear at the interface between the solutions. This substance is the DNA–nucleoprotein complex. After the complex has formed, twirl the stirring rod slowly and collect it onto the rod. Record your observations.

8. Place the isolated DNA–nucleoprotein complex into a test tube containing 3 mL of 4% NaCl solution for later use. Use a 10-mL graduated cylinder to measure the 4% NaCl solution.

Part 2: Extraction of DNA from Onion

Onion is used because of its low starch content, which allows for a higher purity DNA extraction.

9. Repeat steps 1–5 using finely chopped onion. Instead of hand chopping the onion, a blender could be substituted, which gives optimum results.

10. Stir the mixture and let it sit for 15 min in a hot water bath at 60°C (Any longer and the DNA starts to break down.)

11. Cool the mixture in an ice-water bath for 5 min, stirring frequently.

⚗ INVESTIGATION 4.1.1 *continued*

12. Add half the volume of meat tenderizer solution as is present in your filtrate and swirl to mix.

13. Repeat steps 6–8.

Part 3: Testing for the Presence of DNA

The presence of DNA may be detected qualitatively with the reagent diphenylamine. Diphenylamine reacts with the purine nucleotides in DNA, producing a characteristic blue colour.

 Diphenylamine solution contains glacial acetic acid. Be very careful not to spill any of the solution on yourself or on any surface. Inform your teacher immediately if any spills occur. Wear safety goggles and rubber gloves when handling this solution.

14. Stir the DNA from the onion and beef liver with their respective glass rods to resuspend them into the 4% NaCl solution.

15. Dispense 15 mL of diphenylamine solution into a 25-mL graduated cylinder. The teacher will direct you to the stock diphenylamine solution, which will have been set up in a burette.

16. Transfer 5 mL of the solution to a 10-mL graduated cylinder with a Pasteur pipette or with a plastic graduated eyedropper.

17. Add 5 mL of diphenylamine solution to the DNA suspension obtained from the onion and from the beef liver.

18. Repeat step 16 and add 5 mL of diphenylamine solution to a test tube containing 3 mL of distilled water (the blank).

19. Repeat step 16 and add 5 mL of diphenylamine solution to a test tube containing 3 mL of DNA standard (the standard).

20. Place all of the test tubes in boiling water for 10 min and record the colour changes. Record your observations.

21. Remove the test tubes from the hot-water bath and place into a test-tube rack. Allow the tubes to cool before proceeding.

Part 4: Quantitative Determination of DNA Concentration Using Spectrophotometry

The principle underlying a spectrophotometric method of analysis involves the interaction of electromagnetic (EM) radiation (light) with matter. When EM radiation strikes an atom, energy in the form of light is absorbed. The remainder of the energy passes through the sample and can be detected.

Hence, the more molecules that are present, the more energy will be absorbed, resulting in a higher absorbance reading. Since the relationship is direct, we can determine the concentration of an unknown by comparing it with a known. In this case, the unknown is the DNA concentration obtained from beef liver extract and from onion extract, and the known is the DNA standard.

22. Set the spectrophotometer to a wavelength of 600 nm.

23. Fill a dry cuvette with the solution that consists of the distilled water and the diphenylamine. This will serve as a blank.

24. Wipe off any fingerprints from the outside of the cuvette by holding the cuvette at the very top and using a facial tissue. Place the blank into the spectrophotometer and set the absorbance to 0.00. This step is analogous to taring the balance. (Your teacher will demonstrate proper use of a spectrophotometer.)

25. Pour the blank solution back into its original test tube and place it in a test-tube rack.

26. Rinse the cuvette with a tiny amount of standard DNA solution (DNA standard and diphenylamine from step 19). Wipe off any fingerprints in the manner described in step 24.

27. Place the DNA standard solution into the spectrophotometer, then record the absorbance.

28. Pour the DNA standard solution into its original test tube and save in case of error.

29. Repeat steps 26–28 with the beef liver extract solution and with the onion extract solution.

Analysis

(b) Propose reasons that the onion cells required heating and the liver cells did not.

(c) DNA was spooled out using a glass rod. How do you account for the "stickiness" of DNA to glass?

(d) Describe the DNA you extracted. If DNA is a rigid structure, why do the DNA strands appear flexible? What features of DNA's structure account for its stiffness? If DNA is rigid, how does it coil tightly into a small space?

(e) Comment on the purity of the DNA extracted.

(f) Compare the amount of DNA extracted from the onion versus that from the liver. Which source of DNA provided more of the molecule? Account for this observation, given your knowledge of cell structure and given differences in the procedure.

(g) What was the purpose of the standard DNA solution? What was the purpose of the blank?

(h) Did the spectrophotometric results correlate with the qualitative observations obtained from the diphenylamine test? Comment.

(i) Calculate the amount of DNA extracted from each source using your standard as a guide.

(j) The liver and onion were chopped very finely. Provide reasoning for this step. If the step was omitted, what effect would this omission have on the results?

(k) SDS is a detergent. Describe how detergents work and explain the role of SDS in the protocol.

(l) How does NaCl contribute to maximum DNA extraction? (Hint: Think about DNA's chemical constituents.) Keep in mind that NaCl is a salt that ionizes in solution.

(m) What is the purpose of adding cold ethanol to each extraction? How does this phenomenon work?

(n) In the extraction of DNA from onion, you added a meat tenderizer solution. The meat tenderizer solution contains an enzyme called papain. What role does papain play in the extraction?

(o) Identify three properties of DNA that are demonstrated by this investigation.

Evaluation

(p) Suggest possible sources of error in this procedure. Indicate the effect of these sources of error on the results.

(q) Using standard scientific format, prepare a written report summarizing your analysis and evaluation of this investigation.

LAB EXERCISE 4.1.1

Evidence of Hereditary Material

Inquiry Skills

○ Questioning	○ Planning	● Analyzing
● Hypothesizing	○ Conducting	● Evaluating
● Predicting	○ Recording	● Communicating

Background Information

In the 1920s, Frederick Griffith, an English medical officer, started experimenting with *Streptococcus pneumoniae*. This bacterium, which causes pneumonia, exists in two forms. One form is surrounded by a polysaccharide coating called a capsule and is known as the S form because it forms smooth colonies on a culture dish. The second harmless form has no coating and is known as the R form because it forms rough colonies on a culture dish (**Figure 1**).

Figure 1   unencapsuled cells (R form)   encapsuled cells (S form)

The following is an abbreviated summary of Griffith's procedures and results:

1. Mouse A was injected with encapsulated cells (S form), while mouse B was injected with unencapsulated cells (R form).

Observation

Mouse A contracted pneumonia and died, while mouse B continued to live. Mouse B was sacrificed, and an autopsy was conducted on both mice. The autopsies revealed living S cells in mouse A's tissues and living R cells in mouse B's tissues.

(a) What conclusions can you derive from the experimental results?

(b) Why might a scientist decide to repeat this experimental procedure on other mice?

2. Encapsulated (S-form) pneumococcal cells were heated, killed, and then injected into mouse C (**Figure 2**).

Observation

Mouse C continued to live. Mouse C was sacrificed and the autopsy revealed that no living S cells were found in the animal's tissues.

(c) What is the significance of this result?

(d) Predict what would have happened to the mouse if the unencapsulated (R-form) cells had been heated and then injected. What would this step have represented in the experimental protocol?

LAB EXERCISE 4.1.1 *continued*

3. The heated encapsulated (S-form) cells were mixed with unencapsulated (R-form) cells. The mixture was grown on a special growth medium. Cells from the culture medium were injected into mouse D (**Figure 2**).

Observation

Mouse D died. An autopsy indicated that the mouse had died of pneumonia; encapsulated (S-form) bacteria and unencapsulated (R-form) bacteria were isolated from the mouse.

(e) Would you have predicted this observation? Explain why or why not.

Analysis

(f) A microscopic examination of the dead and live cell mixture (step 3) revealed cells with and without capsules. What influence did the heat-destroyed cells have on the unencapsulated cells?

(g) Griffith hypothesized that a chemical in the dead, heat-treated, encapsulated cells (step 3) must have altered the living unencapsulated cells and he dubbed this chemical phenomenon *transformation*. In 1944, Oswald Avery, Maclyn McCarty, and Colin MacLeod conducted experiments in test tubes with *Streptococcus pneumoniae* that led them to conclude that DNA is the *transforming principle*, as they called it, and not proteins, as was widely believed. In their experiments, what must have happened to the DNA when the cells divided?

Synthesis

(h) To discover the identity of the transforming principle, Avery and his associates ruptured heat-killed, encapsulated cells to release their contents. RNA, DNA, protein, and purified polysaccharide coats were isolated and were tested for transforming activity. Avery and his associates found that only R cells mixed with purified DNA isolated from dead S cells were transformed to S cells. When R cells were mixed with purified RNA, with the polysaccharide coat, or with protein extracted from dead S cells, only R cell colonies were isolated. Do these results support their hypothesis? Explain.

(i) Predict the experimental results of the following protocols. Support your prediction with a hypotheses.

- Polysaccharide-digesting enzymes are used to digest the encapsulated polysaccharide coat of the heated S form of the bacteria. The treated bacteria are then placed with unencapsulated pneumonia cells, which are then injected into a mouse.

- Heated encapsulated bacteria are treated with DNAase, a DNA-digesting enzyme. The treated bacteria are then mixed with unencapsulated pneumonia cells, which are injected into a mouse.

- All proteins are extracted from the heated encapsulated bacteria. The treated bacteria are then mixed with unencapsulated pneumonia cells, which are injected into a mouse.

(j) Based on the information provided, suggest improvements to the experimental protocols.

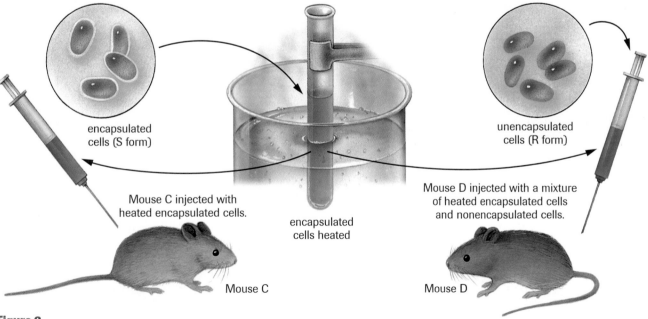

encapsulated cells (S form)

unencapsulated cells (R form)

Mouse C injected with heated encapsulated cells.

encapsulated cells heated

Mouse D injected with a mixture of heated encapsulated cells and nonencapsulated cells.

Mouse C

Mouse D

Figure 2

Key Expectations

- Describe the current model of DNA replication and methods of repair following an error. (4.3)
- Outline contributions of genetic engineers, molecular biologists, and biochemists that have led to the further development of the field of genetics. (4.1, 4.2)
- Investigate and analyze the cell components involved in protein synthesis, using laboratory equipment safely and appropriately. (4.1)
- Explain the roles of evidence and theories in the development of scientific knowledge about genetics. (4.1, 4.2)

Key Terms

anneal

antiparallel

bacteriophage

complementary base pairing

cytokinesis

deoxyribonucleic acid (DNA)

deoxyribonucleoside triphosphates

deoxyribose sugar

DNA gyrase

DNA helicase

DNA ligase

DNA polymerase I

DNA polymerase III

exonuclease

glycosyl bond

isotope

lagging strand

leading strand

mitosis

nitrogenous base

nucleotides

Okazaki fragments

phosphate group

phosphodiester bond

primase

radioactive isotopes

replication bubble

replication fork

RNA (ribonucleic acid) primer

semiconservative

single-stranded binding proteins (SSBs)

template

▶ **MAKE** a summary

In this chapter, you studied the structure of DNA and the mode by which it replicates itself. Create a flow chart that outlines the key steps in DNA replication.

In your notebook, indicate whether each sentence or statement is true or false. Rewrite a false statement to make it true.

1. Adenine and guanine are pyrimidines.

2. The DNA double helix makes a complete turn every 3.4 nm along its length.

3. DNA's backbone is held together by hydrogen bonds while the complementary bases are held together by phosphodiester bonds.

4. If the deoxyribonucleotide sequence in one strand of a short stretch of DNA double helix is 5′–CTGGAT–3′, then the complementary sequence in the opposite strand is 3′–GACCTG–5′.

5. DNA helicase is an enzyme that ensures that the appropriate complementary base pairs have been added to the growing daughter strands during DNA replication.

6. DNA replication is conservative.

7. Joachim Hammerling's experiment using *Acetabularia* revealed that the hereditary information is found in the foot of the alga where the nucleus resides.

8. DNA polymerase III builds a complementary DNA strand during DNA replication in the 3′ to 5′ direction.

9. After one replication of double-stranded DNA, some of the daughter DNA molecules contain no parental material.

10. Erwin Chargaff's rules state that adenine bonds with thymine, and cytosine bonds with guanine.

In your notebook, record the letter of the choice that best completes the statement or answers the question.

11. Which of the following best describes deoxyribonucleic acid (DNA)?
    (a) It consists of a linear backbone of the sugar deoxyribose and phosphates, with nitrogenous bases attached to the sugar residues.
    (b) It consists of a linear backbone of sugar and bases, with a phosphate group attached to each base.
    (c) It consists of a linear backbone of phosphates and bases, with a sugar group attached to each end.
    (d) It is a double helix when present in single-stranded form.
    (e) It consists of a linear backbone of the deoxyribose sugar.

12. Which of the following is associated with only the lagging strand during DNA replication?
    (a) RNA primers   (d) Okazaki fragments
    (b) DNA polymerase I   (e) DNA gyrase
    (c) DNA polymerase III

13. The molecule depicted in **Figure 1** is which one of the following?
    (a) sugar   (d) purine
    (b) nucleotide   (e) pyrimidine
    (c) amino acid

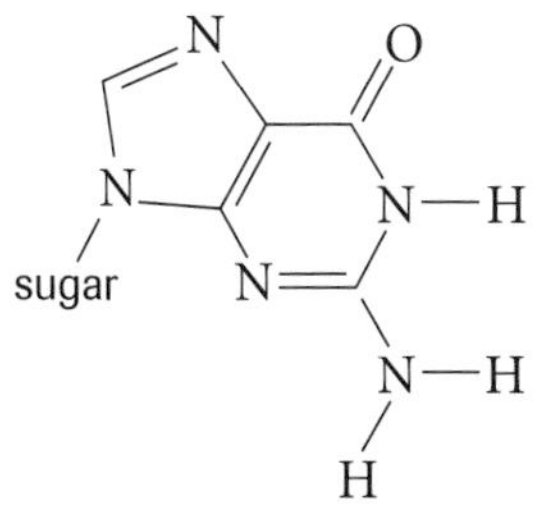

Figure 1

14. The percent composition of adenine in a DNA molecule is found to be 35%. Therefore, which of the following is the percent composition of guanine?
    (a) 35%
    (b) 30%
    (c) 15%
    (d) cannot be determined
    (e) depends on the species

15. DNA replication occurs in which of the following ways?
    (a) in the 5′ to 3′ direction only
    (b) in the 3′ to 5′ direction only
    (c) in both the 5′ to 3′ direction and the 3′ to 5′ direction
    (d) in the 3′ to 5′ direction in the lagging strand
    (e) none of the above

16. Which of the following did Matthew Meselson and Franklin Stahl's experiment illustrate?
    (a) DNA is the hereditary material.
    (b) DNA is found in the nucleus.
    (c) DNA replicates semiconservatively.
    (d) DNA replicates conservatively.
    (e) DNA is composed of a deoxyribose sugar, a phosphate group, and a nitrogenous base.

Understanding Concepts

1. Describe Erwin Chargaff's contribution to the determination of DNA's structure.

2. Explain, using a diagram, why it is not possible for adenine to hydrogen bond with cytosine.

3. Use a diagram to illustrate how the two DNA strands in a double helix run antiparallel. Make sure you label your diagram.

4. How does the fact that DNA replicates semiconservatively decrease the possibility of errors made during DNA replication? Describe another mechanism that minimizes DNA replication error.

5. Numerous enzymes are involved in DNA replication. Outline the role that the following enzymes play: DNA ligase, DNA gyrase, DNA helicase, DNA polymerase I, and DNA polymerase III.

6. What is the complementary strand of AATTGCATA?

7. The sugar found in DNA is a deoxyribose sugar. How would a deoxyribose sugar differ from a ribose sugar? Use a labelled diagram to illustrate your answer.

8. Describe how Rosalind Franklin's scientific findings improved our undertanding of the chemical structure and composition of DNA.

9. DNA polymerase III can only extend an existing DNA strand in the 5′ to 3′ direction. Describe the mechanisms in place that compensate for DNA polymerase III's inability to intitiate a strand and for its stringent directionality.

10. What would likely occur if
    (a) single-stranded binding proteins were malfunctioning during DNA replication?
    (b) primase was inactivated?
    (c) DNA polymerase I was malfunctioning?

11. DNA polymerase III builds complementary DNA strands with free deoxyribonucleoside triphosphates in the nucleoplasm. DNA structure calls for nucleoside monophospates.
    (a) Why are nucleoside triphosphates required?
    (b) What happens to the inorganic phosphate released after the completion of an elongation reaction?

12. Despite the fact that the two DNA strands are held together by weak hydrogen bonds, the DNA molecule is quite stable. Explain how this is possible, given the nature of hydrogen bonding.

13. Explain why DNA can be classified as a polymer.

14. Refer to the diagram of DNA replication (**Figure 1**). Which letter indicates the 3′ ends of molecules, and which indicates the 5′ ends?

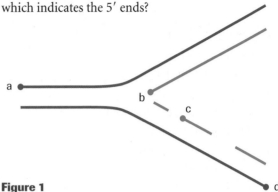

Figure 1

15. If you had a mixture of single-stranded DNA fragments, all four deoxyribonucleotides in their triphosphate form, and DNA polymerase III, what additional component would need to be added to ensure replication? Explain why there would not be any discontinuous replication occurring in the test tube.

16. One strand of a DNA molecule contains the nucleotide proportions 15% adenine (A), 30% thymine (T), 20% guanine (G), and 35% cytosine (C). What proportions of the four base pairs would be expected in the double-stranded form of this DNA?

17. **Table 1** presents the nucleotide base composition of DNA obtained from different organisms.

Table 1

Source	A	T	G	C	$\frac{A+T}{G+C}$	$\frac{A+G}{C+T}$
mouse	29.1	29.0	21.1	21.1	1.38	1.00
corn	25.6	25.3	24.5	24.6	1.04	1.00
frog	26.3	26.4	23.5	23.8	1.11	1.00

All quantities are in %.

Examine the data provided. Given your knowledge of DNA structure and base composition distribution, why is it necessary that the $\frac{A+G}{C+T}$ percentage equal one, and yet not necessary that the $\frac{A+T}{G+C}$ percentage equal one?

18. DNA is the hereditary material of all life. Comment.

Applying Inquiry Skills

19. The DNA chromosome in *E. coli* contains approximately 3 million nucleotides. The average gene contains about 1200 nucleotides. Using this information, calculate the following:

(a) What is the length in metres of the average *E. coli* chromosome? the average *E. coli* gene?

(b) How many complete turns would be found in its double-helical structure?

(c) If *E. coli* were enlarged 10 000 times the normal size, how long (in mm) would its DNA strand be?

(d) If *E. coli* were enlarged 10 000 times the normal size, how long (in mm) would a gene be?

20. The presence of DNA may be detected with the reagent diphenylamine. Diphenylamine reacts with the purine nucleotides in DNA, producing a characteristic blue colour. Using colorimetry, the amount of DNA in a sample may be quantified. Examine **Figure 2**.

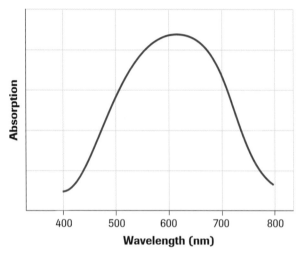

Figure 2
Absorption spectra of DNA reacted with diphenylamine

(a) If you were to choose an optimum wavelength to characterize DNA that has been reacted with diphenylamine, which wavelength would you choose and why, based on experimental results?

(b) Explain why the answer to (a) should not be a wavelength in the blue range of the light spectrum.

(c) Diphenylamine reacts with the purines in DNA. Usually the purine-to-pyrimidine ratio in DNA is 1:1. If the purine proportion of a DNA molecule is 30% and the concentration obtained is 0.5 mg/mL, what is the pyrimidine concentration?

21. To isolate Okazaki fragments, in 1969 Pauling and Hamm observed DNA replication in the presence of a temperature-sensitive strain of *E. coli* whose DNA ligase did not work at temperatures above 40°C, but functioned properly at 25°C. Pauling and Hamm allowed the DNA to replicate at 25°C. Then they transferred the DNA into a medium that contained radioactive nucleotides, while raising the temperature to 40°C.

They repeated the same experiment with a strain of *E. coli* whose DNA ligase that works above 40°C. Use your knowledge of DNA replication, the formation of Okazaki fragments, and the role of DNA ligase, to do the following:

(a) Provide a hypothesis for this experiment.

(b) Explain why it was necessary to repeat the experiment with a strain of *E. coli* whose DNA ligase functioned above 40°C.

(c) Identify, if applicable, where any radioactivity would have been found.

(d) Suggest conclusions that the scientists would have been able to draw from this experiment.

Making Connections

22. Numerous enzymes are involved in DNA replication. DNA polymerase I acts as an exonuclease, as it is able to recognize and excise incorrectly paired nucleotides from the end of a DNA strand. DNA ligase is able to anneal DNA fragments together. Propose possible uses for these enzymes in the field of genetic engineering.

23. The Nobel Prize Committee awards its prizes based on the value of the scientific contribution. It also examines the scientific methods used, including the interpretation and analysis of data. Watson, Crick, and Wilkins were awarded the Nobel Prize in 1962 for determining the structure of DNA. Explain why Chargaff was not included in this group of scientists, despite his important contribution. Should Wilkins have been included? Franklin? Justify your answer. How does Watson and Crick's work meet the parameters set by the Nobel Prize Committee?

Extension

24. Since DNA is held together by hydrogen bonds, heat may be used to separate the two strands. This process is known as denaturation melting. The melting temperature of DNA (T_m equals the temperature at which 50% is denatured) varies depending on the nucleotide base composition (the number of G–C pairs compared to the number of A–T pairs). Also, when exposed to ultraviolet (UV) light, the amount of radiation absorbed varies between denatured DNA and native DNA. This change in absorbance is known as the hyperchromic shift. Research the relationship between melting temperature and base composition, and the relationship between UV absorbance and structure. Propose a hypothesis before you start your research.

chapter

5

Protein Synthesis

Because of the success of the Human Genome Project, scientists have discovered much about the sequence of nucleotides that constitutes the human genome. We now know the order of each nucleotide on the 46 chromosomes that distinguishes us from other organisms. Unfortunately, we know very little about how the sequence governs the development and physiology of a human being. Scientists are interested in using the information gained in the Human Genome Project to determine which genes are "turned on" in a particular cell type during a cellular event and how this activity is related to health and disease.

To initiate protein synthesis in a cell, genes are first transcribed into an RNA complement. The RNA is then translated by ribosomes into protein. If a certain protein is in high demand, numerous RNA transcripts of its gene will be produced. For example, during cell division, the proteins required for this process are in high demand; hence, numerous RNA transcripts of the cell division genes are made. Scientists can compare the region of the genome that is being transcribed in cells undergoing an event like cell division to the same region in cells that are not undergoing the event.

Microarray technology takes advantage of the complementary nature of DNA and RNA to identify the genes in this region. RNA that has been transcribed is isolated and reverse-transcribed into complementary single-stranded DNA (cDNA) that incorporated a fluorescent dye. The cDNA is then placed in contact with glass slides onto which tens of thousands of single-stranded sequences of the human genome have been spotted in an ordered array. The cDNA will hydrogen bond to the sequences to which it is complementary. The tagged sequences can be read by an optical device and interpreted by a computer. In fact, the expression of tens of thousands of genes can be monitored simultaneously in the same experiment, thanks to high-speed computing. The same procedure is applied to control cells that are not undergoing the specific event. Sequences that differ in the amount of cDNA adhesion are identified as genes that are being expressed during the cellular process that is being investigated. Using this technology, we can identify which genes are turned on or off during different cellular events.

REFLECT on your learning

1. Outline the relationship among DNA, RNA, and protein. What role does each play in protein synthesis?

2. Describe an RNA molecule including its physical and chemical structure.

3. Why is it important that genes are turned on only when the proteins that they code for are required in the cell?

4. What is a mutation? Explain why a mutation may be disadvantageous for the well-being of an organism. Provide examples of mutagens that may induce mutations.

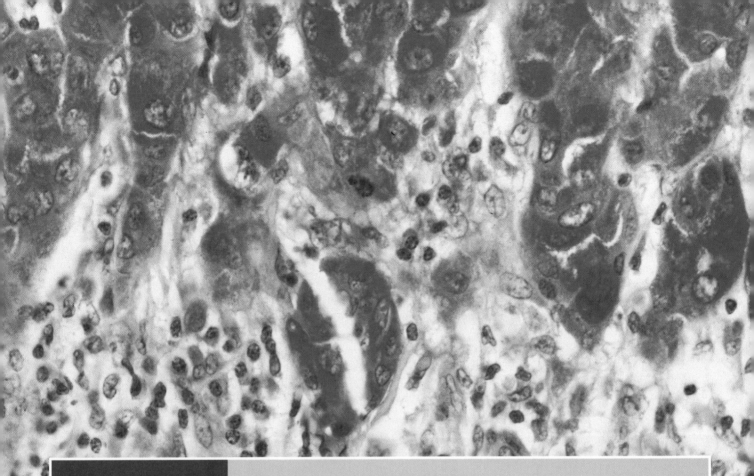

▶ TRY THIS activity *Translating the Code*

The genetic material is composed of nucleotides. The sequence of nucleotides in DNA determines the order in which amino acids will be assembled into proteins. Before DNA can be expressed as protein, it must be transcribed into RNA. In this activity, you will transcribe a sequence of DNA bases into an "RNA" message and then translate the RNA message into a "protein" sentence. Note that in RNA, uracil is complementary to adenine. For example, the DNA sequence AATCGATA would be transcribed into the RNA sequence UUAGCUAU. In effect, thymine is replaced by uracil in RNA.

- Transcribe the following DNA sequences into RNA.
- Translate the RNA into a quotation by using **Table 1**.

Sequence A
AAA TAG TAC TCA ACG TTT CCC GGA TTC ACT TTA GTC ATC
CGG CCT ACA CCC GGC

Sequence B
GTT GAG ATC GAA AAC AAA GAC ATA TTG CCG AAG GAT
GAG ATC CAA GGC CGT

Sequence C
AAA TAT TCA GTA TTT CCC CCA GGT TCA GGG

(a) What part of a protein do the words in sequence A, B, and C represent?

(b) What do the quotations in this activity represent?

(c) What would happen to the message if an error were made in transcribing the DNA?

(d) BONUS! Who are the authors of these quotations?

Table 1 The Code

RNA	Word	RNA	Word	RNA	Word	RNA	Word	RNA	Word	RNA	Word
UUU	the	CUG	right	GUG	do	UGU	expect	GGG	to	GCC	do
UUC	run	AUU	sit	UAU	track	UGC	life	AAA	is	GGC	get
UUG	on	AUC	true	UAG	you	UGA	under	CCC	none	GCA	there
CUU	are	AUA	greatest	CAU	faults	AGU	of	CCU	plant	AAG	trees
CUC	if	AUG	meaning	CAA	even	GGU	be	CCA	conscious	AAC	you'll
CUA	over	GUU	just	CAG	shade	GGA	not	CCG	sit	AAU	whose

genes a sequence of nucleotides in DNA that performs a specific function such as coding for a particular protein

proteins complex molecules composed of one or more polypeptide chains made of amino acids and folded into specific three-dimensional shapes that determine protein function

Gregor Mendel was an Austrian monk whose experiments with garden peas laid the foundation for the science of genetics. His experimental results led to his hypothesis that certain "factors" were responsible for the patterns of inheritance that he observed in his pea plants. The height of the pea plant, the shape and colour of the seed, and the flower colour were determined by the factors that were inherited. Today these factors are known as **genes**, and they direct the production of **proteins** that determine the phenotypical characteristics of organisms. In addition, genes direct the production of other physiologically essential proteins, such as antibodies, and hormones. Proteins drive cellular processes such as metabolism, determine physical characteristics (**Figure 1**), and manifest genetic disorders by their absence or by their presence in an altered form.

Garrod's Hypothesis

The hypothesis that one gene directs the production of one enzyme was first proposed by Archibald Garrod in the early 20th century. Garrod, a British physician, noticed that certain illnesses recur in some families. Garrod analyzed the blood and urine samples of patients with alkaptonuria, a genetic disorder in which the urine appears black because it contains the chemical alkapton. When exposed to air, alkapton darkens in colour. Through biochemical analysis, Garrod confirmed that the chemical alkapton was present in high concentrations in people who had alkaptonuria. He hypothesized that a defective enzyme causes an "inborn error of metabolism" along a reaction pathway (**Figure 2**). Individuals with alkaptonuria have a defective alkapton-metabolizing enzyme, whereas unaffected individuals have an enzyme that breaks down alkapton. Garrod further hypothesized that enzymes are under the control of the hereditary material, and thus an error in the hereditary material resulted in an error in an enzyme.

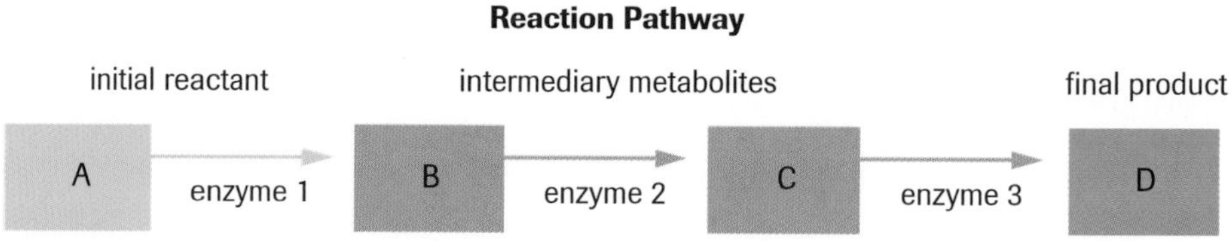

Figure 1
Physical characteristics, such as the colours of the feathers of the peacock, are determined by proteins under gene control.

Reaction Pathway

initial reactant | intermediary metabolites | final product

A → enzyme 1 → B → enzyme 2 → C → enzyme 3 → D

Figure 2
Garrod hypothesized that a defective enzyme causes an "inborn error of metabolism." If there is an accumulation of substance B, then enzyme 2 must be defective; if there is an accumulation of substance C, enzyme 3 is defective. The hereditary material directs the production of enzymes.

George Beadle and Edward Tatum

Thirty-three years after Garrod first hypothesized a relationship between genes and enzymes, George Beadle and Edward Tatum were able to demonstrate experimentally the relationship using the red bread mould *Neurospora crassa*. One strain of *N. crassa* was able to synthesize all the complex amino acids and vitamins it required for optimum growth if it was given a minimal nutrient medium containing simple inorganic salts, sugar, and one of the B vitamins. Mutant (defective) strains of *N. crassa* were created by exposing its spores to X rays or ultraviolet light. Mutant strains were identified among the descendants of these spores when they were unable to grow on minimal medium because they could no longer manufacture one or more of the complex compounds that they required. These strains could grow, however, on a complete nutrient medium. Beadle and Tatum were interested in finding out which **amino acid** or vitamin the mutant strains were

amino acid the monomer unit of a polypeptide chain that is composed of a carboxylic acid, an amino group, and a side group that differentiates it from other amino acids

unable to synthesize. In their experiment, they placed colonies of a mutant strain in vials, each containing minimal medium plus one additional nutrient. Beadle and Tatum observed that one mutant strain only grew on minimal medium containing the amino acid arginine. It was hypothesized that one or more of the enzymes present in the biochemical pathway that synthesized arginine was defective. The biochemical pathway for the synthesis of arginine is shown in **Figure 3**.

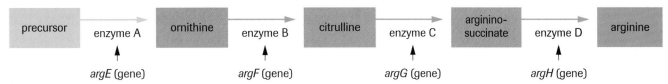

Figure 3
Reaction pathway for the synthesis of arginine in *Neurospora crassa*

Beadle and Tatum tried to determine experimentally the sequence of the enzymes in the pathway and which enzyme was defective by studying which product in the pathway accumulated in the mutant strain. They found that some strains accumulated the precursor and could not grow in the absence of ornithine but could grow in the presence of citrulline and minimal medium or argininosuccinate and minimal medium; therefore, these strains possessed defective enzyme A. Other strains could not grow in the absence of citrulline or ornithine and, therefore, possessed defective enzyme A or B (**Figure 3**). In this case, if the precursor accumulated, the defect was in enzyme A, but if ornithine accumulated, enzyme B was defective. In fact, Beadle and Tatum isolated four distinct mutant strains, each with a defective form of one enzyme in the arginine biochemical pathway. They further showed that a lack of a particular enzyme corresponded to a mutation in a specific gene. From these observations, they were able to conclude that a gene acts by directing the production of only one enzyme. Beadle and Tatum summarized this relationship as the one gene–one enzyme hypothesis.

Vernon Ingram

Today, we know that genes code not only for enzymes, but also for other proteins. Furthermore, some proteins, such as hemoglobin, consist of more than one polypeptide chain, each chain controlled by a different gene. For this reason, Beadle and Tatum's results are commonly known as the one gene–one polypeptide hypothesis. Vernon Ingram demonstrated this relationship while studying the amino acid sequence of hemoglobin from individuals with sickle cell anemia. Hemoglobin comprises two polypeptide chains of one type (α) and two of another type (β). Ingram found that, in sickle cell anemia red blood cells, the amino acid called valine substitutes for only one of the glutamic acid residues in one of the hemoglobin chains (**Figure 4**). This substitution leads to a change

normal hemoglobin β-chain

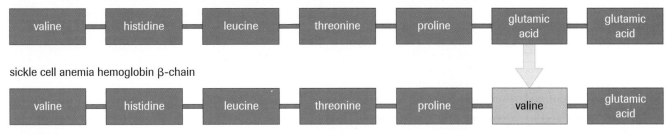

sickle cell anemia hemoglobin β-chain

Figure 4
The first seven amino acids of the hemoglobin β-chain. An amino acid substitution of valine for glutamic acid occurs at the sixth position of the β-globin polypeptides in individuals with sickle cell anemia.

Figure 5

(a) A normal red blood cell; **(b)** a sickle cell anemia red blood cell. The change in protein conformation causes the hemoglobin chains to stick to each other in the shape of rods, resulting in an overall change into sickle-shaped cells.

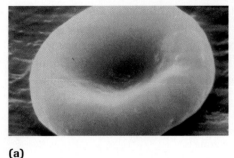
(a)

(b)

in the shape of the red blood cell (**Figure 5**), resulting in sickle cell anemia, a debilitating disease that is sometimes deadly. Ingram's investigation showed that a gene specifies the kind and location of each amino acid in of a given polypeptide chain. The significance of Ingram's findings is that he linked a human hereditary abnormality to a single alteration in the amino acid sequence of a protein. Many hereditary diseases, such as hemophilia and cystic fibrosis, have been traced to alterations in a single gene.

▶ *Section 5.1* *Questions*

Understanding Concepts

1. Describe the relationship between genes and proteins. How might an error in a gene result in a change in protein sequence?

2. Describe Garrod's observational evidence that helped him form his hypothesis. Explain what he meant by the phrase "inborn error of metabolism."

3. What hypothesis did Beadle and Tatum investigate in their experiment? How did the results of their experiment support their hypothesis? Formulate other results that would not have supported their hypothesis.

Applying Inquiry Skills

4. In Beadle and Tatum's experiment using the organism *Neurospora crassa*, why was it important to use organisms that were mutated in one gene? Explain why it was important to grow the mutant strains on minimal media containing only one additional nutrient.

5. The following metabolic pathway exists in a strain of bacteria:

$$A \xrightarrow{\text{enzyme 1}} B \xrightarrow{\text{enzyme 2}} C \xrightarrow{\text{enzyme 3}} D \xrightarrow{\text{enzyme 4}} E$$

with gene 1, gene 2, gene 3, gene 4 controlling enzymes 1–4 respectively.

The bacteria require the product E to survive. The four enzymes required to produce E are controlled by four separate genes (genes 1 to 4). Describe what effect a mutation may have on each gene in terms of enzyme activity, accumulation of substrate (the substance an enzyme acts on), and the growth of the bacteria.

6. A bacterial strain that produces a substance needed for growth is exposed to ultraviolet light, creating three nutritional mutants. The pathway for the substance involves compounds A, B, C, and D, but not necessarily in this sequence. Each mutant has a block in its pathway. The mutants are grown on media containing one of the pathway compounds. The results of the growth are shown in **Table 1**.

Table 1 Patterns of Growth of Different Mutants of Bacteria on Different Media

Mutant	Growth on A	Growth on B	Growth on C	Growth on D
1	growth	no growth	no growth	growth
2	growth	no growth	growth	growth
3	no growth	no growth	no growth	growth

(a) Determine the sequence of compounds in the biochemical pathway.

(b) Identify where in the pathway each mutant has experienced a block.

Making Connections

7. In the isolation of a specific gene, researchers always try to find a family tree that has many members with the respective genetic disorder. Using your knowledge of gene inheritance, explain why this approach produces optimum results.

An organism's genome is housed within the confines of the nucleus. However, proteins are synthesized outside the nucleus, in the cytoplasm, on ribosomes. Since the information for protein synthesis is specified by DNA and DNA is not able to exit the nucleus, a problem exists as to how the blueprint for building proteins will be brought to the ribosomes.

The Central Dogma

DNA is too valuable to the normal functioning of a cell to be allowed to exit the nucleus. If it were allowed to exit and it were distorted, cleaved, or damaged in any way in the caustic environment of the cytoplasm, it would be rendered useless, resulting in the death of the cell, and possibly leading to the death of the organism. In addition, when a protein is required by a cell, it is often required in large amounts. Since there are only two complete copies of DNA in the nucleus of a somatic cell, not many ribosomes could use a specific gene at the same time. All proteins required by the cell would be manufactured at the same rate, a condition that may be disadvantageous to the cell because it would reduce its flexibility in carrying out various activities. Finally, once it had been used by the ribosome, the DNA would have to reenter the nucleus for storage, necessitating a reentry strategy. The complications are endless. The mechanism that has evolved to prevent DNA from exiting the nucleus is pertinent to the survival and long-term normal operation of a cell. Hence, the question remains, how does a ribosome synthesize the protein required if it does not have access to DNA itself?

The answer lies in an intermediate substance known as messenger RNA (ribonucleic acid). DNA is transcribed into a complementary RNA message that is capable of encoding genetic information. Numerous copies of the message can be delivered to the ribosomes in the cytoplasm. The ribosomes then translate the message into polypeptide chains, which are processed into proteins. The entire sequence is described as the central dogma of molecular genetics (**Figure 1**). **Transcription** involves the copying of the information in DNA into messenger RNA, and **translation** involves ribosomes using the messenger RNA as a blueprint to synthesize a protein composed of amino acids. In the more general sense of the terms, when you *transcribe*, you copy from one medium to another, and when you *translate*, you convert into a different language. DNA is transcribed into complementary messenger RNA, and ribosomes translate messenger RNA into specific sequences of amino acids, which are used to build proteins.

Ribonucleic Acid (RNA)

Like DNA, RNA is a carrier of genetic information. However, RNA differs from DNA in many ways. First, RNA contains a ribose sugar rather than a deoxyribose sugar (**Figure 2**, page 238). A ribose sugar has a hydroxyl group on its 2′ carbon. Second, instead of thymine, RNA contains the base uracil. Uracil is similar in structure to thymine, except thymine has a methyl group on its 1′ carbon (**Figure 3**, page 238). Uracil is able to pair with adenine in DNA (**Figure 4**, page 238). The third and final key difference between RNA and DNA is that DNA is double stranded, whereas RNA is single stranded. When a gene is transcribed into messenger RNA, only a single-stranded complementary copy is made. There is no need to waste valuable energy repeating information in double-stranded form. In the complementary copy, uracil is substituted for thymine.

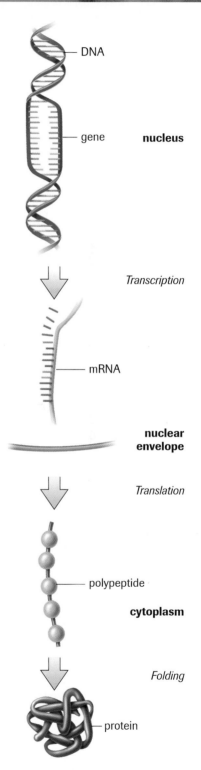

DNA

gene   **nucleus**

Transcription

mRNA

nuclear envelope

Translation

polypeptide

cytoplasm

Folding

protein

Figure 1
The central dogma of molecular genetics

Figure 2

A ribose sugar possesses an −OH group (hydroxyl) on the 2′ carbon. The deoxyribose sugar is missing the −OH group on the 2′ carbon. The *deoxy* part of the name deoxyribose indicates a "loss of oxygen" at position 2.

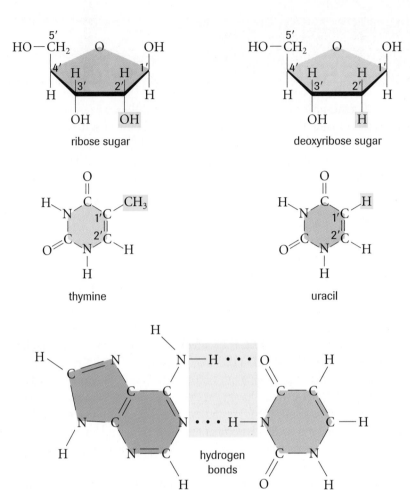

ribose sugar   deoxyribose sugar

Figure 3

Thymine contains a methyl group on the 1′ carbon. Uracil does not.

thymine   uracil

Figure 4

Uracil is able to bind only with adenine because of the specific placement of the two hydrogen bonds.

hydrogen bonds

adenine   uracil

transcription the process in which DNA is used as a template for the production of complementary messenger RNA molecules

translation the process by which a ribosome assembles amino acids in a specific sequence to synthesize a specific polypeptide coded by messenger RNA

messenger RNA (mRNA) the end product of transcription of a gene, mRNA is translated by ribosomes into protein

transfer RNA (tRNA) a form of RNA that is responsible for delivering amino acids to the ribosomes during the process of translation

ribosomal RNA (rRNA) a form of RNA that binds with ribosomal protein to form ribosomes

There are three major classes of RNA molecules: **messenger RNA (mRNA)**, **transfer RNA (tRNA)**, and **ribosomal RNA (rRNA)**. The mRNA, described above, varies in length, depending on the gene that has been transcribed; the longer the gene, the longer the mRNA. The role of tRNA is to transfer the appropriate amino acid to the ribosome to build a protein, as dictated by the mRNA template. The tRNAs are comparatively short in length, averaging 70 to 90 ribonucleotides. The rRNA is a structural component of a ribosome. Along with proteins, it forms the ribosome, which provides the construction site for the assembly of polypeptides.

 **SUMMARY**   *Ribonucleic Acid*

Table 1   Differences Between the Nucleic Acids

Deoxyribonucleic acid	Ribonucleic acid
• contains deoxyribose sugar	• contains ribose sugar
• double stranded	• single stranded
• adenine pairs with thymine guanine pairs with cytosine	• adenine pairs with uracil guanine pairs with cytosine
• resides in the nucleus	• resides both in the nucleus and in the cytoplasm

Table 2 Different Types of Ribonucleic Acid

Types of RNA	Characteristics and key functions
messenger RNA (mRNA)	• varies in length, depending on the gene that has been copied • acts as the intermediary between DNA and the ribosomes • translated into protein by ribosomes • RNA version of the gene encoded by DNA
transfer RNA (tRNA)	• functions as the delivery system of amino acids to ribosomes as they synthesize proteins • very short, only 70–90 base pairs long
ribosomal RNA (rRNA)	• binds with proteins to form the ribosomes • varies in length

Transcription and Translation: An Overview

Transcription can be divided into three sequential processes: initiation, elongation, and termination. During initiation, the **RNA polymerase** binds to the DNA at a specific site, known as the promoter, near the beginning of the gene. Using the DNA as a template, the enzyme RNA polymerase puts together the appropriate ribonucleotides and builds the mRNA transcript, a process known as elongation (**Figure 5**). Shortly after the RNA polymerase passes the end of a gene, it recognizes a signal to stop transcribing, which constitutes termination. The mRNA transcript is then completely released from the DNA and will eventually exit the nucleus.

The second part of the central dogma is the translation of the information contained in mRNA into protein. Translation is also subdivided into three main stages: initiation, elongation, and termination. Initiation occurs when a ribosome recognizes a specific sequence on the mRNA and binds to that site. The ribosome then moves along the mRNA three nucleotides at a time. Each set of three nucleotides codes for an amino acid (**Figure 6**). A tRNA delivers the appropriate amino acid and the polypeptide chain is elongated. Elongation continues until a three-base nucleotide sequence is reached that does not code for an amino acid but is instead a "stop" signal. At this point, the ribosome falls off the mRNA and the polypeptide chain is released.

RNA polymerase enzyme that transcribes DNA into complementary mRNA

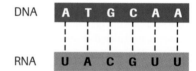

Figure 5
An example of base pairing of RNA with DNA during the process of transcription. Notice that thymine does not exist in RNA but is substituted with the nitrogenous base uracil.

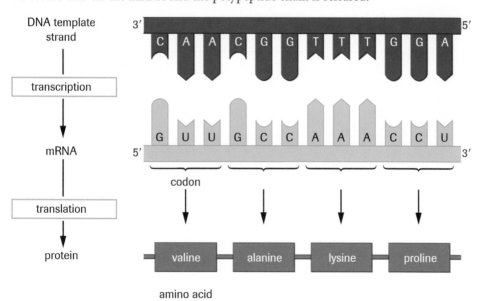

Figure 6
Each set of three nucleotides codes for a specific amino acid.

codon sequence of three bases in DNA or complementary mRNA that serves as a code for a particular amino acid

start codon specific codon (AUG) that signals to the ribosome that the translation commences at that point

stop codons specific codons that signal the end of translation to a ribosome

The Genetic Code

There are 20 amino acids found in proteins but only four different bases in messenger RNA. To code for all 20 amino acids, a sequence of three nucleotides must be used for each amino acid. Each triplet of nucleotides is called a **codon.** If only one nucleotide coded for one amino acid, we would be limited to four amino acids. If two nucleotides (AC, TT, CA, etc.) coded for one amino acid, only 16 amino acid combinations ($4^2 = 16$) would be possible, leaving us with four amino acids for which there is no code. The use of three nucleotides results in 64 different possible combinations ($4^3 = 64$), which easily code for the 20 amino acids that exist (**Figure 7**). **Figure 7** depicts the genetic code. Each triplet of nucleotides codes for a specific amino acid (**Table 3**). More than one codon can code for a single amino acid, indicating a redundancy in the genetic code. For example, UUU, UUC, UCU, and UCC all code for the amino acid phenylalanine (**Figure 7**). Later, you will discover that this redundancy minimizes errors that may lead to serious mutations. One codon serves as the **start codon** (signalling initiation) and others serve as **stop codons** (signalling termination).

1st Base	2nd (middle) Base of a Codon: U	C	A	G	3rd Base
U	UUU Phe UUC Phe UUA Leu UUG Leu	UCU Ser UCC Ser UCA Ser UCG Ser	UAU Tyr UAC Tyr UAA STOP UAG STOP	UGU Cys UGC Cys UGA STOP UGG Trp	U C A G
C	CUU Leu CUC Leu CUA Leu CUG Leu	CCU Pro CCC Pro CCA Pro CCG Pro	CAU His CAC His CAA Gln CAG Gln	CGU Arg CGC Arg CGA Arg CGG Arg	U C A G
A	AUU Ile AUC Ile AUA Ile **AUG Met***	ACU Thr ACC Thr ACA Thr ACG Thr	AAU Asn AAC Asn AAA Lys AAG Lys	AGU Ser AGC Ser AGA Arg AGG Arg	U C A G
G	GUU Val GUC Val GUA Val GUG Val	GCU Ala GCC Ala GCA Ala GCG Ala	GAU Asp GAC Asp GAA Glu GAG Glu	GGU Gly GGC Gly GGA Gly GGG Gly	U C A G

Figure 7
The genetic code comprises 64 triplet nucleotides. Sixty-one of the codons specify 20 different amino acids and three codons signal termination. The start codon, AUG, also codes for methionine (Met) (**Table 3**).

* The codon AUG codes for methionine (Met) and is the (almost) universal start codon. Occasionally GUG or UUG act as start codons.

DID YOU KNOW?

The Universal Language of Life
The genetic code is universal. The same genetic code is used for translation in every organism from bacteria to mammals with only a few isolated exceptions, such as *Paramecium*. The universality of the genetic code is powerful evidence that the evolution of the code happened at such an early stage that all living organisms are descended from a common pool of primitive cells. The universality of the genetic code has important ramifications for genetic engineering. Since bacteria use the same code to build proteins as humans do, we can engineer bacteria to produce useful proteins, such as insulin.

Table 3 Amino Acids and Their Abbreviations

Amino acid	Three-letter abbreviation	Amino acid	Three-letter abbreviation
alanine	Ala	leucine	Leu
arginine	Arg	lysine	Lys
asparagine	Asn	methionine	Met
aspartic acid	Asp	phenylalanine	Phe
cysteine	Cys	proline	Pro
glutamic acid	Glu	serine	Ser
glutamine	Gln	threonine	Thr
glycine	Gly	tryptophan	Trp
histidine	His	tyrosine	Tyr
isoleucine	Ile	valine	Val

▶ *Section 5.2* Questions

Understanding Concepts

1. A cell's DNA is always maintained in the nucleus. Provide the rationale for why this is optimal for a cell's long-term survival.

2. State the central dogma of molecular genetics.

3. The sugar found in DNA is a deoxyribose sugar. How does a deoxyribose sugar differ from a ribose sugar? Use a labelled diagram to illustrate your answer.

4. Describe the role of the following molecules in protein synthesis: ribosomes, mRNA, tRNA, rRNA.

5. Compare DNA and RNA and outline their similarities and differences.

6. Differentiate between transcription and translation in terms of their purpose and location.

7. Transcription and translation can be divided into three main stages: initiation, elongation, and termination. Using point form, list the role and key characteristics of each stage in transcription and translation.

8. The DNA code is read in groups of three nucleotides called codons. Explain why reading the code in pairs of nucleotides is not sufficient.

9. Differentiate between a stop codon and a start codon.

10. The following is the sequence of a fragment of DNA:

    GGATCAGGTCCAGGCAATTTAGCATGCCCCAA

    Transcribe this sequence into mRNA.

11. Using the genetic code, decipher the following mRNA sequence:

    5′ - GGCAUGGGACAUUAUUUUGCCCGUUGUGGUGGGGCGUGA - 3′

Applying Inquiry Skills

12. In a hypothetical situation, 85 amino acids exist and there are still only 4 nucleotides found in nucleic acid. Calculate the minimum number of nucleotides required to code for this large number of amino acids. What is the maximum number of amino acids your answer would code for?

13. The amino acid sequence for a certain peptide is Leu–Tyr–Arg–Trp–Ser. How many nucleotides are necessary in the DNA to code for this peptide?

Making Connections

14. The genetic code is universal. With the exception of a few organisms, all living things use the same code to build proteins. List possible ramifications to the biotechnology industry and to the theory of evolution if this were not the case.

upstream region of DNA adjacent to the start of a gene

promoter sequence of DNA that binds RNA polymerase upstream of a gene

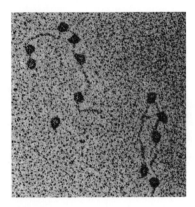

Figure 1
The RNA polymerase (the dark circles) binds to the DNA strand and initiates transcription. Transcription occurs simultaneously at numerous locations along the DNA.

template strand the strand of DNA that the RNA polymerase uses as a guide to build complementary mRNA

coding strand the strand of DNA that is not used for transcription and is identical in sequence to mRNA, except it contains uracil instead of thymine

terminator sequence sequence of bases at the end of a gene that signals the RNA polymerase to stop transcribing

primary transcript mRNA that has to be modified before exiting the nucleus in eukaryotic cells

5′ cap 7-methyl guanosine added to the start of an primary transcript to protect it from digestion in the cytoplasm and to bind it to the ribosome as part of the initiation of translation

Initiation

Transcription commences in both prokaryotes and eukaryotes when the enzyme RNA polymerase binds to the segment of DNA that is to be transcribed and opens the double helix (**Figure 1**). If the RNA polymerase randomly transcribed DNA, the cell would not make the correct polypeptide, nor would the cell make good use of its energy store. The RNA polymerase binds to the DNA molecule **upstream** of the gene to be transcribed. The upstream region is a sequence on one strand of DNA located adjacent to the start of the gene. It indicates where the RNA polymerase should start transcribing and which DNA strand should be transcribed. This upstream region is known as the **promoter** and, in most genes, consists of a characteristic base-pair pattern, one that is high in adenine and thymine bases. The string of A's and T's in the promoter region serves as a recognition site for RNA polymerase. Adenine and thymine share only two hydrogen bonds, as compared with guanine and cytosine, which share three hydrogen bonds. It takes less energy to break two bonds; therefore, the RNA polymerase expends less energy opening up the DNA helix if it possesses a high concentration of adenine and thymine base pairs. The binding site of RNA polymerase only recognizes the promoter region and, therefore, can only bind upstream of a gene-coding region.

Elongation

Once the RNA polymerase binds to the promoter and opens the double helix, it starts building the single-stranded mRNA in the direction of 5′ to 3′. Unlike the case in DNA replication, the RNA polymerase does not require a primer to start building the complementary strand; hence, elongation starts as soon as the RNA polymerase moves to the start of the gene and binds to the promoter. The promoter itself does not get transcribed. The process of elongation of the RNA transcript is similar to that of DNA replication. The RNA polymerase uses only one of the strands of DNA as a template for mRNA synthesis. This chosen DNA strand is called the **template strand**. The strand that is not used for transcription is known as the **coding strand**. The RNA sequence is complementary to the template strand and identical to the coding strand, except that it contains uracil instead of thymine.

Termination

The mRNA strand is synthesized until the end of the gene is reached. RNA polymerase recognizes the end of the gene when it comes across a **terminator sequence**. The terminator sequence differs between prokaryotes and eukaryotes. Generally, when RNA polymerase reaches the terminator sequence, the newly synthesized mRNA disassociates with the DNA template strand. Transcription ceases and the RNA polymerase is free to bind to another promoter region and transcribe another gene. The process of transcription is summarized in **Figure 2**.

Posttranscriptional Modifications

In eukaryotic cells, the mRNA is not ready to leave the nucleus directly following transcription. Modifications need to be made to this **primary transcript**. A **5′ cap** is added to the start of the primary transcript. This 5′ cap consists of 7-methyl guanosine, which forms a modified guanine nucleoside triphosphate (**Figure 3**). The cap protects the mRNA from digestion by nucleases and phosphatases as it exits the nucleus and enters

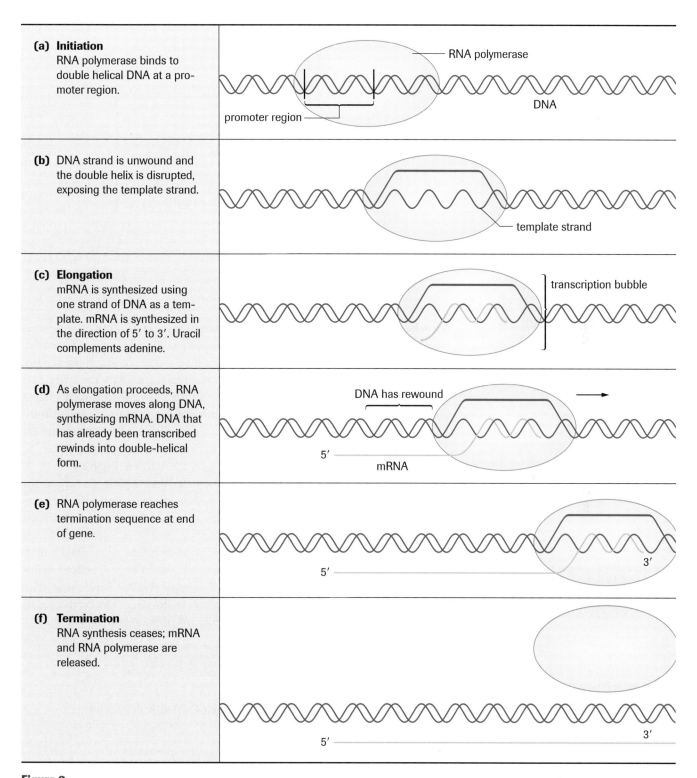

(a) Initiation
RNA polymerase binds to double helical DNA at a promoter region.

RNA polymerase

DNA

promoter region

(b) DNA strand is unwound and the double helix is disrupted, exposing the template strand.

template strand

(c) Elongation
mRNA is synthesized using one strand of DNA as a template. mRNA is synthesized in the direction of 5′ to 3′. Uracil complements adenine.

transcription bubble

(d) As elongation proceeds, RNA polymerase moves along DNA, synthesizing mRNA. DNA that has already been transcribed rewinds into double-helical form.

DNA has rewound

5′

mRNA

(e) RNA polymerase reaches termination sequence at end of gene.

5′

3′

(f) Termination
RNA synthesis ceases; mRNA and RNA polymerase are released.

5′

3′

Figure 2

the cytoplasm of the cell. It also plays a role in the initiation of translation. In addition, a string of approximately 200 adenine ribonucleotides is added to the 3′ end by the enzyme **poly-A polymerase**. The 3′ end is known to contain a **poly-A tail**. This whole process is known as capping and tailing.

Further modifications need to be made. The DNA of a eukaryotic gene comprises coding regions (known as **exons**) and noncoding regions (known as **introns**). The introns

poly-A polymerase enzyme responsible for adding a string of adenine bases to the end of mRNA to protect it from degradation later on

poly-A tail a string of 200 to 300 adenine base pairs at the end of an mRNA transcript

modified guanine nucleoside triphosphate

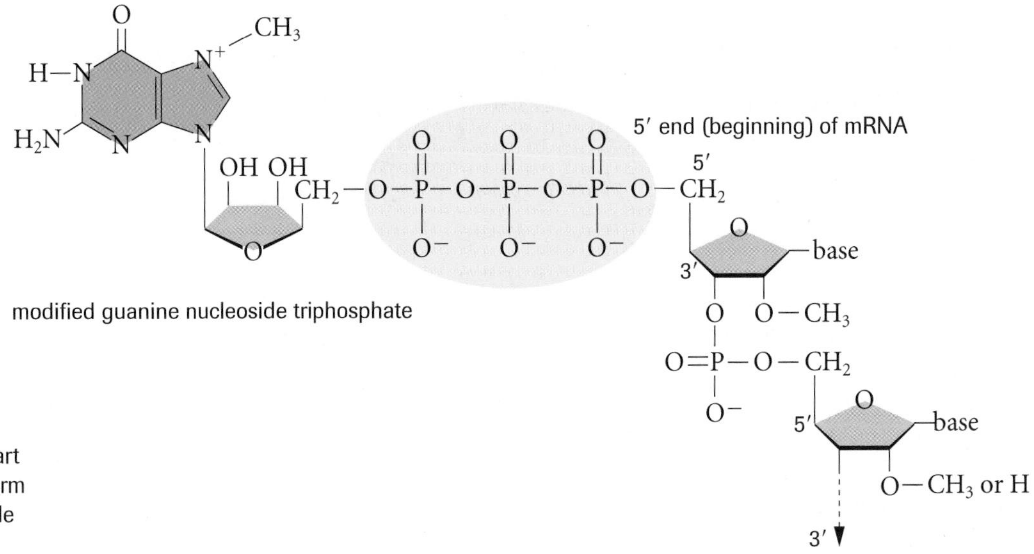

5′ end (beginning) of mRNA

Figure 3
The 5′ cap is added to the start of the primary transcript to form a modified guanine nucleoside triphosphate.

exons segments of DNA that code for part of a specific protein

introns noncoding region of a gene

spliceosomes particles made of RNA and protein that cut introns from mRNA primary transcript and joins together the remaining coding exon regions

mRNA transcript mRNA that has been modified for exit out of the nucleus and into the cytoplasm

are interspersed among the exons; therefore, a primary transcript will contain sections of RNA that do not code for part of the protein. If these noncoding regions are translated, the protein will not fold properly. Improper folding will render the protein dysfunctional and useless to the cell. Before the primary transcript leaves the nucleus, the introns are removed. The introns are removed from the primary transcript by particles made of RNA and proteins known as **spliceosomes**. Spliceosomes cut out the introns and join the remaining exons together so that all the coding regions are now continuous to form the mRNA molecule. The mRNA exits the nucleus, but the spliced-out introns stay within the nucleus, where they are degraded and their nucleotides are recycled (**Figure 4**).

Once the primary transcript has been capped and tailed and the introns excised, the processed transcript is ready to be translated by a ribosome into a protein. It is now called the **mRNA transcript**. Unlike DNA replication, there is no quality control enzyme to ensure that the mRNA transcript is correct. Therefore, more errors are made during transcription than during replication. However, since a single gene is transcribed repeatedly to produce hundreds of transcripts of the same gene, errors are not as detrimental to the cell as those that occur during replication. If an error is made during transcription, the protein is susceptible to degradation once it is synthesized. The correct copies of the mRNA transcript will produce adequate amounts of the required protein.

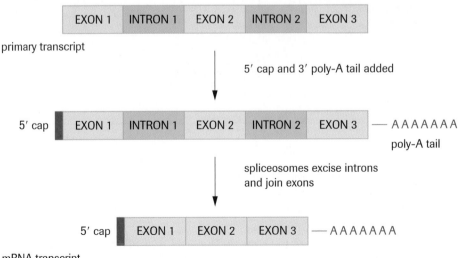

Figure 4
Posttranscriptional modification of the primary transcript. A 5′ cap and poly-A tail are added. Introns are cut out and the remaining exons are put together by phosphodiester bond formation.

Case Study Human Immunodeficiency Virus

Acquired immunodeficiency syndrome, or AIDS, describes a number of disorders associated with infection by the human immunodeficiency virus, or HIV. Two different types of HIV have been identified: HIV-1, discovered in 1981, evolved in chimpanzees before "jumping" species to infect humans; HIV-2, the less prevalent and less virulent form, was described in 1985.

For viruses to survive inside the body, they must be able to evade the defences of the human host. HIV does this in a unique manner—by invading the very cells whose function it is to protect the body from pathogens, or disease-causing agents. Instead of eliminating the virus, this stimulation of host defences actually helps HIV replicate and survive. Although the body does eventually mount an immune response to HIV, the virus is never fully contained, resulting in progressive disease and the development of AIDS. Individuals with HIV or AIDS have damaged immune systems. This condition makes them more susceptible to infections that humans with normal immune systems would be able to fight off. These are known as opportunistic infections. AIDS patients are also at higher risk of developing malignancies (cancers). As of early 2002, there was no cure for AIDS. However, advances in microbiology, genetics, and molecular biology have led to the development of more effective treatment for the disease, and work continues on the development of an HIV vaccine. Prevention of further transmission of HIV and improved treatment of existing cases are difficult challenges presently being tackled.

Unlike the chicken pox and flu viruses, HIV cannot be transmitted through the air, but it is found in human body fluids. It is spread primarily through direct sexual contact and by the introduction of blood or blood components into the bloodstream through blood transfusions or the sharing of needles or syringes for injection drug use. HIV can also be transmitted from infected mothers to their infants during pregnancy, at the time of birth, and through breastfeeding.

The structure of HIV is deceptively simple, consisting of an outer membrane made of proteins and lipids, and an inner coat made of protein that protects an RNA core. RNA encodes the genetic information of the virus. HIV attacks the immune system directly by selectively targeting and infecting helper T cells (**Figure 5**). Helper T cells act as guards against invading pathogens. Thus, HIV destroys the body's own defences, rendering it incapable of defeating other invading organisms.

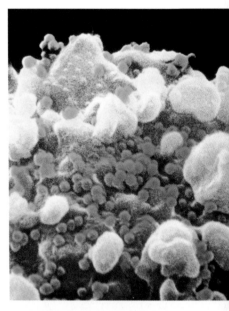

Figure 5
Colour-enhanced scanning electron micrograph of the helper T cell being attacked by HIV particles. The HIV particles appear in green.

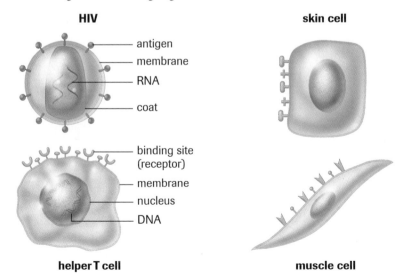

Figure 6
Outer membranes of the helper T cell, skin cell, and muscle cell. Note that the antigens on the HIV membrane are complementary to the binding sites of helper T cells, but not to those of other cells, such as skin or muscle cells.

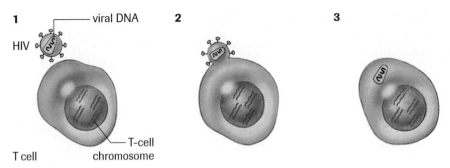

Figure 7
HIV entering the helper T cell

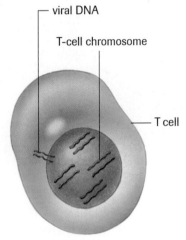

Figure 8
The RNA in HIV is converted into double-stranded DNA.

With infections other than HIV, proteins on the surface of helper T cells recognize foreign substances, which starts a chain of events to produce an immunologic attack and removal of the invading organism. HIV infects helper T cells by locking into specific binding sites on the cell surface, called receptors, much like a key into a lock (**Figure 6**, page 245). Once HIV binds to these sites, the viral and cell membranes fuse. The entire virus enters the cell, where it loses its coat and the RNA core is set free (**Figure 7**).

HIV belongs to a group of viruses known as retroviruses, whose genetic information is composed of RNA instead of DNA. The RNA of retroviruses encodes a special enzyme called reverse transcriptase, which converts the genetic message contained in RNA into a complementary copy of single-stranded DNA. The single-stranded DNA is then converted to double-stranded DNA by the same enzyme (**Figure 8**).

The newly constructed double-stranded viral DNA slips into the nucleus of the infected cell. Here it is spliced into the infected helper T cell's DNA, such that the instructions for HIV proteins are now part of the helper T cell's genome. The virus may remain dormant for many years. When viral DNA is integrated into a host cell's genome it is known as a provirus (**Figure 9**).

Activation of the helper T cells results in transcription of the integrated viral DNA into viral mRNA, which enters the cytoplasm. The transcribed mRNA attaches itself to ribosomes and directs them to produce many copies of viral proteins and enzymes (**Figure 10**). They will form the protective coat for the newly released RNA.

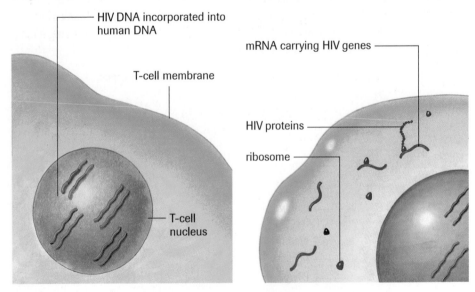

Figure 9
HIV genomic DNA is incorporated into human DNA.

Figure 10
HIV mRNA directs the production of viral proteins and enzymes.

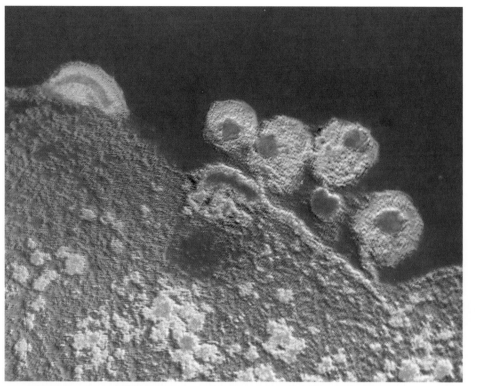

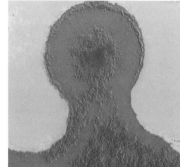

(a)

(b)

(c)

Figure 11
(a) Electron micrograph showing HIV being released from the cell membrane of a helper T cell;
(b) and **(c)** a closer view of the infected helper T cell as the virus becomes coated by the cell
membrane and escapes

Instead of performing T-cell functions, the once-healthy helper T cell is now transformed into an HIV factory. Viral proteins, destined for the HIV membrane, enter the host cell's membrane. Viral RNA and structural proteins migrate to just inside the host cell's membrane, where they pinch off from the cell to form a new viral particle, as illustrated in **Figure 11(a)**. **Figures 11(b)** and **(c)** show that, as the viral particle escapes, it becomes coated by the membrane. The newly released viral particle infects other helper T cells. Release of many viral particles eventually weakens the host T cell and it dies.

One of the challenges of finding a cure for AIDS stems from the ability of HIV to mutate to avoid immune detection and destruction. Examples of such mutations include changes to the proteins on its outer membrane. Killer T cells play an important role in containing HIV replication by recognizing the HIV-infected cells that display remnants of viral protein on their surface. When mutations of viral proteins occur, the killer T cells can no longer recognize HIV-infected cells and the virus escapes immune recognition.

In developed countries, the numbers of new AIDS diagnoses and deaths have been steadily falling. These reductions result from a number of factors, a major one being the development and use of potent anti-HIV drugs. There are several different classes of agents that work by interfering with different stages in the HIV life cycle. Patients usually receive combinations of three or more drugs at one time. Such combinations are known as highly active antiretroviral therapy (HAART). Despite the beneficial effects of HAART, many patients find taking many pills several times a day extremely difficult. As well, the medications are expensive, can have serious or intolerable side effects, and are not curative.

Major breakthroughs in the efforts to limit the spread of AIDS have also come in the form of tests for detecting HIV. Since 1985, blood collected by the Red Cross, and now the Canadian Blood Services, has been screened for the presence of HIV. As a result, the risk of acquiring HIV through a blood transfusion is extremely low.

There are several approaches to HIV prevention. Education about HIV and other sexually transmitted diseases, treatment and clean needle programs for injection drug users, and the use of antiretroviral drugs to prevent transmission of HIV from mother to infant have all proved successful. The development of a safe and effective HIV vaccine remains a priority in AIDS research. Although many of the developed nations of the world have health budgets that allow for HIV/AIDS testing and treatment, less developed countries are struggling to deal with the consequences of having exerted inadequate control over the disease in its early years. Furthermore, since the countries that are most in need cannot afford the costs of HAART and the high expenses associated with HIV, the majority of worldwide infections go untreated. Thus, despite the medical, social, and political advances in HIV management to date, many challenges still need to be addressed before global control of HIV is realized.

▶ *Case Study* Questions

Understanding Concepts

1. Can HIV attach itself to a muscle cell or a skin cell?

2. Explain why you cannot get AIDS by shaking hands. (Use the information that you have gained about binding sites.)

3. Why is the enzyme that converts the RNA in HIV into DNA referred to as reverse transcriptase?

4. What happens to the viral DNA if the helper T cell divides?

5. Explain why it is possible for a human to be infected with HIV and not exhibit any of the symptoms of AIDS.

6. Indicate why people infected with HIV most often die of another infection, such as pneumonia.

7. David Vetter, "the boy in the plastic bubble," suffered from a disorder called severe combined immunodeficiency syndrome. How does this disorder differ from acquired immunodeficiency syndrome? (Severe combined immunodeficiency syndrome is discussed in more detail in Chapter 10, section 10.1.)

8. Why is it so difficult to destroy a virus that mutates frequently?

9. Canadian Blood Services inquires about a person's travels before blood donations are accepted. Explain why this practice can be classified as preventive.

10. Can HIV be transmitted through either food or beverages? Explain your answer.

Making Connections

11. Should health-care workers, such as doctors, dentists, and nurses, be screened for HIV? Justify your answer. What precautions do medical professionals take to protect themselves from the AIDS virus?

12. How can the spread of AIDS be prevented? List numerous policies that might minimize the spread of the disease.

13. Currently, blood tests are available to screen for the AIDS virus. Although these tests are very effective in detecting the virus, they fail if the virus has been contracted recently. Research how the AIDS test works. Explain why the test may fail to detect the virus after a recent exposure to HIV.

 www.science.nelson.com

SUMMARY   *An Overview of Transcription*

1. Initiation of transcription commences when the RNA polymerase binds to the promoter region of the gene to be transcribed. At this point, the DNA is unwound and the double helix is disrupted.

2. RNA polymerase moves past the promoter until it reaches the start sequence of the gene to be transcribed.

3. A complementary RNA strand is synthesized in the direction of 5′ to 3′, using one strand of DNA as a template. This step is known as elongation. The complement of adenine in RNA is uracil.

4. Once the terminator sequence is reached by the RNA polymerase, transcription ceases. The mRNA is separated from the DNA and the RNA polymerase falls off the DNA molecule, constituting termination of transcription. The DNA molecule reforms its double-helical shape.

5. The mRNA at this point is not in final form; posttranscriptional modifications need to be made. 7-methyl guanosine, known as the 5′ cap, is added to the 5′ end. A string of 200 to 300 adenine ribonucleotides are added to the 3′ end by the enzyme poly-A polymerase. This is known as the poly-A tail.

6. Introns (noncoding regions) are cut out of the primary transcript by particles known as spliceosomes. The spliceosomes then rejoin the remaining exons (coding regions).

7. The mRNA transcript is ready to exit the nucleus.

Note that this overview summarizes transcription in eukaryotes. Minor differences for transcription in prokaryotes are discussed in section 5.7.

▶ **Section 5.3** *Questions*

Understanding Concepts

1. Explain the role of the following in transcription: RNA polymerase, poly-A polymerase, and spliceosomes.

2. A short fragment of a particular gene includes the following sequence of nucleotides:

   3′- TACTACGGTAGGTATA - 5′

   Write out the mRNA sequence.

3. A short fragment of a particular gene includes the following sequence of nucleotides:

   3′- GGCATGCACCATAATATCGACCTTCGGCACGG - 5′

   (a) Identify the promoter region. Justify your answer.
   (b) Explain the purpose of the promoter in transcription.

4. Differentiate between introns and exons.

5. In eukaryotes, mRNA must be modified before it is allowed to exit the nucleus.
   (a) Describe the modifications that are made to eukaryotic primary transcript.
   (b) Explain how these modifications ensure that mRNA survives in the cytoplasm and is translated into a functioning protein.

6. Typically, DNA is double stranded and RNA is single stranded. Provide reasons to account for this difference in structure. While formulating your answer, consider the role of DNA and RNA in protein synthesis.

7. In paragraph form, summarize the process of transcription. The areas of initiation, elongation, and termination should each be covered in a separate paragraph.

8. Explain the ramifications to the process of transcription if the following occurs:
   (a) The termination sequence of a gene is removed.
   (b) Poly-A polymerase is inactivated.

   (c) The enzyme that adds the 5′ cap is nonfunctional.
   (d) Spliceosomes excise exons and join the remaining introns together.
   (e) The RNA polymerase fails to recognize the promoter sequence.

9. The process of gene transcription is similar in many respects to the process of DNA replication. In a table, outline the similarities between the two processes.

Applying Inquiry Skills

10. As a graduate student in a university laboratory, you are challenged with the problem of determining whether the mRNA that you have been given is from a eukaryotic cell or a prokaryotic cell. You have been provided with a nucleotide sequencer, which will help determine the DNA sequence for you. Indicate what features in the sequence you will look for to determine whether the mRNA is eukaryotic or prokaryotic.

11. A molecular biologist isolated two DNA fragments of a genetic sequence that were believed to be the promoter region for a certain gene from a bacterial cell. The biologist subjected the DNA fragments to heat and recorded the temperature at which the double strands pulled apart. DNA helix A unwound at 84°C, whereas DNA helix B unwound at 65°C. Which of the two strands is most likely to be the promoter? Explain your answer.

12. A biotechnologist isolated a gene from a human heart cell and spliced it into a bacterium's genome. The protein that was synthesized from this gene was useless because it contained many more amino acids than the protein produced by the heart cell. Explain the biotechnologist's observations. How could the biotechnologist correct this problem?

reading frame one of three possible phases in which to read the bases of a gene in groups of three

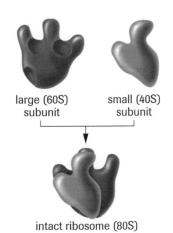

large (60S)   small (40S)
subunit   subunit

intact ribosome (80S)

Figure 1
Ribosomes consist of a large subunit and a small subunit. The subunits are a combination of ribosomal RNA (rRNA) and protein.

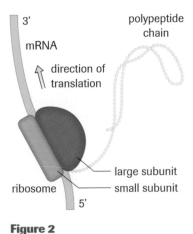

Figure 2
The large and small subunit of a ribosome work together to translate a strand of mRNA into protein. The protein grows as the ribosome moves farther along the mRNA strand.

anticodon group of three complementary bases on tRNA that recognizes and pairs with a codon on the mRNA

The Ribosome

The next step in the process of manufacturing a protein from a DNA blueprint is for the mRNA to exit the nucleus. Once the mRNA enters the cytoplasm, it is ready to be translated. Ribosomes bind to the mRNA, recognizing the 5′ cap in eukaryotes. This process constitutes the initiation of translation of nucleic acid into protein.

In eukaryotes, the ribosome consists of two subunits, a large subunit of 60S and a small subunit of 40S (**Figure 1**). The S refers to the rates at which various components sediment when centrifuged. The rates are a measure of size. The two subunits bind to the mRNA, clamping the mRNA between them. The ribosome moves along the mRNA in the 5′-to-3′ direction, adding a new amino acid to the growing polypeptide chain each time it reads a codon (**Figure 2**). Ribosomes synthesize different proteins by associating with different mRNAs and reading their coding sequences. The phase in which the mRNA is read, in triplets of nucleotides, each of which encodes an amino acid, is known as the **reading frame**. The reading frame can differ, depending on the base pair from which the ribosome starts reading. Consider the following mRNA sequence:

AUGCCAGAUGCCAUCCAAGGCC

It is possible to read the codons as follows:

AUG CCA GAU GCC AUC CAA GGC

Yet, if the ribosome begins reading from the second nucleotide, this will be the reading frame:

UGC CAG AUG CCA UCC AAG GCC

Therefore, it is important that mRNA be positioned correctly.

The Role of Transfer RNA (tRNA)

The ribosome alone cannot synthesize the polypeptide chain. The correct amino acids must be delivered to the polypeptide-building site. The molecule that delivers the amino acids is tRNA. A tRNA is a small, single-stranded nucleic acid whose structure resembles a cloverleaf. At one arm of tRNA, a sequence of three bases (the **anticodon**) recognizes the codon of the mRNA, and the opposite arm carries the corresponding amino acid (**Figure 3**). The recognition by tRNA of mRNA is facilitated by the complementarity of base pairing. For example, if the mRNA has the codon UAU, the complementary base sequence on the anticodon arm of tRNA is AUA.

Therefore, the tRNA carrying the amino acid tyrosine at one end has the anticodon AUA at the other arm. Every tRNA carries only one specific amino acid, which means that at least 20 different tRNAs are required. Recall that there are 64 possible codons. In reality, anywhere from 20 to 64 types of tRNA molecules are available, depending on the organism. The third base in each codon may differ between two codons that code for the same amino acid. For example, UAU and UAC both code for tyrosine. If the tRNA's anticodon is AUA, it can still bind to the codon UAC. This flexibility makes it possible for the correct amino acid to be added in the growing polypeptide chain, despite errors made in the gene sequence of mRNA. The proposal that a tRNA can recognize more than one codon by unusual pairing between the first base of the anticodon and the third base of a codon is known as the wobble hypothesis.

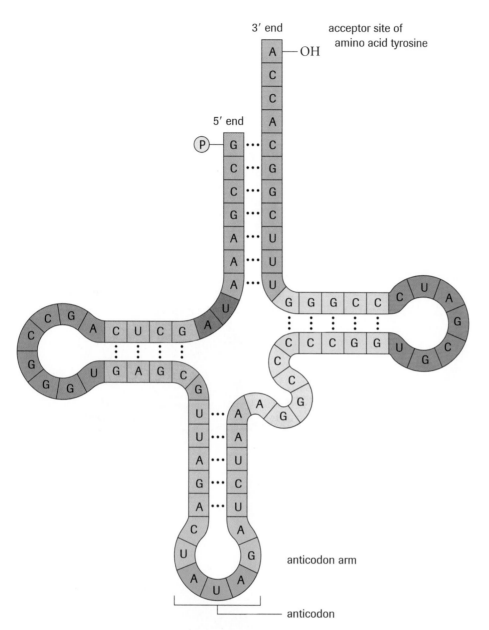

Figure 3
The tRNA molecule has a cloverleaf structure. The molecule folds to form this structure because of hydrogen bonding. The anticodon is located on the anticodon arm and the amino acid is covalently bound to the adenine nucleotide at the 3′ end (acceptor site). Aminoacyl–tRNA synthetase charges the tRNA by adding the appropriate amino acid. In this case, the amino acid that would be added is tyrosine because the anticodon is AUA.

A tRNA molecule containing its corresponding amino acid attached to its acceptor site at the 3′ end is called **aminoacyl-tRNA**. The enzymes responsible for adding the appropriate amino acid to each tRNA are called aminoacyl–tRNA synthetases. There are at least 20 of these enzymes, each of which is specific for a particular amino acid and tRNA.

aminoacyl–tRNA a tRNA molecule with its corresponding amino acid attached to its acceptor site at the 3′ end.

Elongation of the Polypeptide Chain

The first codon that is recognized by the ribosome is the start codon AUG. AUG ensures that the correct reading frame is used by the ribosome. The AUG codon codes for methionine, which means that every protein initially starts with the amino acid methionine. The ribosome has two sites for tRNA: the **A (acceptor) site** and the **P (peptide) site**. As shown in **Figure 4(a)** on page 252, the tRNA that carries methionine enters the P site.

The next tRNA carrying the required amino acid enters the A site, as shown in **Figure 4(b)**. **Figure 4(c)** shows how the methionine amino acid is bonded to the second amino acid. The ribosome shifts over (translocates) one codon. The third amino acid enters

A (acceptor) site site in the ribosome where tRNA brings in an amino acid

P (peptide) site site in the ribosome where peptide bonds are formed between adjoining amino acids on a growing polypeptide chain

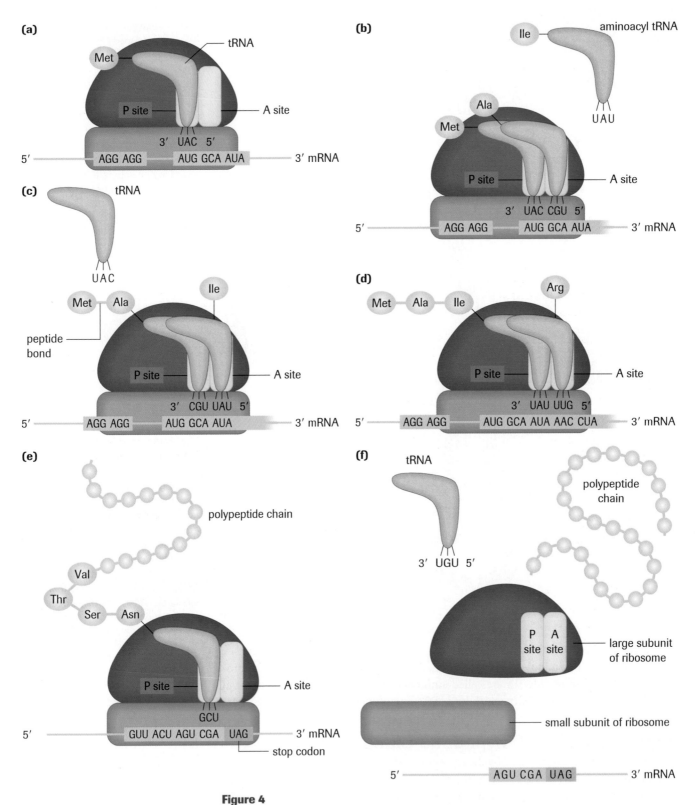

Figure 4

Protein synthesis: **(a)** The first tRNA that is brought into the P site carries methionine because the start codon is AUG. **(b)** The second tRNA enters the A site. **(c)** A peptide bond forms between methionine and alanine. The ribosome translocates one codon over and the next tRNA brings in the appropriate amino acid into the A site. **(d)** The ribosome moves along the mRNA and another amino acid is added to the chain. **(e)** The process is repeated until the ribosome reaches a stop codon for which no tRNA exists. **(f)** A release-factor protein aids in the dismantling of the ribosome-mRNA complex, releasing the polypeptide chain.

the A site, while the second amino acid moves to the P site, as illustrated in **Figure 4(c)**. A peptide bond is formed between the second and third amino acid. The process is analogous to a ticker tape running through a ticker-tape machine, except that the ribosome "machine" moves along the mRNA "ticker tape." This process continues and the growing polypeptide chain trails in the cytoplasm. The tRNAs that have been released (that have exited from the P site) are recycled by aminoacyl–tRNA synthetase adding a corresponding amino acid to them.

Termination of Protein Synthesis

The ribosome will eventually reach a stop codon. The three stop codons are UGA, UAG, and UAA. Since these codons do not code for an amino acid, there are no corresponding tRNAs, as shown in **Figure 4(e)**. A protein known as a **release factor** recognizes that the ribosome has stalled and aids in the release of the polypeptide chain from the ribosome. **Figure 4(f)** shows that the two subunits of the ribosome fall off the mRNA, and translation officially ceases. Sugars or phosphate may be added to some of the amino acid residues in processes known as glycosylation and phosphorylation, respectively. Also, enzymes may cleave the chain at specific places, or some of the amino acids in the chain may be altered in other ways. 🔳 🔳

SUMMARY   *An Overview of Translation*

1. Ribosome subunits (large and small) bind to the 5′ cap of the mRNA transcript, sandwiching the mRNA between them, and translation commences. The large ribosome subunit contains two sites, the A (acceptor) site and the P (peptide) site.

2. The ribosome moves along the mRNA, reading the code in triplets known as codons. Translation does not occur until it reads the triplet AUG, which is the start codon. The start codon corresponds to the amino acid methionine.

3. When the start codon is in the P site, a tRNA delivers the amino acid methionine. tRNA recognizes the codon because of the complementary anticodon.

4. The second codon is now in the A site. The appropriate tRNA delivers the next amino acid in the protein sequence. A peptide bond is formed between methionine and the second amino acid. The ribosome shifts over one codon. The tRNA that delivered methionine is released to obtain another methionine amino acid. The second amino acid tRNA shifts over to the P site. The third amino acid, coded by the third codon, is brought in to the A site by the next tRNA. A peptide bond is formed between the second and third amino acid.

5. The process of elongation continues until a stop codon is read in the A site. The stop codons are UAG, UGA, and UAA. The ribosome stalls.

6. A protein known as the release factor recognizes that the ribosome has stalled and causes the ribosome subunits to disassemble, releasing the mRNA and newly formed protein.

7. The protein is folded and modified and then targeted to the area of the cell where it is required.

Note that this overview summarizes translation in eukaryotes. Differences for translation in prokaryotes are discussed in section 5.7.

DID YOU KNOW?

Methionine Goes Formyl
In eukaryotes, the first amino acid to enter the P site is methionine. In prokaryotes, a slight modification is made: the methionine is tagged with a formyl group. Therefore, in prokaryotes, the first amino acid to enter is formyl–methionine (fMet).

release factor a protein involved in the release of a finished polypeptide chain from the ribosome

INVESTIGATION 5.4.1

Protein Synthesis and Inactivation of Antibiotics (p. 268)
Each protein has a specific function. Its presence or absence in a cell may make the difference between life and death. Bacteria that carry an ampicillin-resistance gene produce a protein that inactivates the antibiotic ampicillin. What happens when they are grown on ampicillin-rich media? Investigation 5.4.1 allows you to observe the effects of the presence and function of a specific gene.

ACTIVITY 5.4.1

**Synthesis of Protein:
A Simulation Activity (p. 269)**
Activity 5.4.1 guides you through the process of transcription and translation for a particular protein. Can you organize the mRNA transcript found within the human genome into a translatable reading frame that codes for a protein?

▶ Section 5.4 Questions

Understanding Concepts

1. Differentiate between the following:
   (a) P site and A site
   (b) codon and anticodon
   (c) start codon and stop codon
   (d) charged tRNA and uncharged tRNA

2. Three types of RNA must be functioning properly to ensure normal protein synthesis. Identify the three types of RNA that are needed and explain the critical role that each one plays during translation.

3. List the possible anticodons for threonine, alanine, and proline.

4. Errors in the process of transcription are occasionally made. Explain why an error in the third base of a triplet may not necessarily result in a mutation. Provide an example to support your answer using the genetic code.

5. Using the wobble hypothesis and the genetic code, calculate the minimum number of tRNAs that a cell can utilize to pair up with 61 codons.

6. The following sequence was isolated from a fragment of mRNA:

   5' - GGC CCA UAG AUG CCA CCG GGA AAA GAC UGA GCC CCG - 3'

   Translate the sequence into protein starting with the start codon.

7. All incoming tRNAs must enter the A site first and then shift to the P site after peptide bond formation. Name the exception to this rule and explain this occurrence.

8. In paragraph form, summarize the process of translation. The areas of initiation, elongation, and termination should each be covered in a separate paragraph.

9. Explain why the process of translation has been appropriately named.

Making Connections

10. Many antibiotics function by inhibiting protein synthesis in bacteria. Choose one of the following antibiotics and use the Internet or periodicals to research how the antibiotic inhibits protein synthesis. Report your findings in paragraph form.
    (a) chloramphenicol
    (b) streptomycin
    (c) tetracycline

    GO www.science.nelson.com

Approximately 42 000 genes exist that code for proteins in humans. Not all proteins are required at all times. Insulin is only required in a cell when the glucose level is high. It would be inefficient for a cell to transcribe and translate the insulin gene when glucose levels are low. Cells have developed methods by which they can control the transcription and translation of genes, depending on their need. Some genes are always needed in a cell and are constantly being transcribed and translated. These are known as **housekeeping genes**. **Transcription factors** turn genes on when required. **Gene regulation** is vital to an organism's survival. Regulation of genes can occur at four levels (**Table 1**). Some genes are regulated at the transcriptional and posttranscriptional level; the extent to which they are transcribed is regulated. Other genes are regulated at the translational and posttranslational level; mRNA that has been transcribed does not necessarily get translated.

housekeeping genes genes that are switched on all the time because they are needed for life functions vital to an organism

transcription factors proteins that switch on genes by binding to DNA and helping the RNA polymerase to bind

gene regulation the turning on or off of specific genes depending on the requirements of an organism

Table 1 Four Levels of Control of Gene Expression in Eukaryotic Cells

Type of control	Description
transcriptional	It regulates which genes are transcribed (DNA to mRNA) or controls the rate at which transcription occurs.
posttranscriptional	The mRNA molecules undergo changes in the nucleus before translation occurs. Introns are removed and exons are spliced together.
translational	It controls how often and how rapidly mRNA transcripts will be translated into proteins. This control affects the length of time it takes for mRNA to be activated and the speed at which cytoplasmic enzymes destroy mRNA.
posttranslational	Before many proteins become functional, they must pass through the cell membrane. A number of control mechanisms affect the rate at which a protein becomes active and the time that it remains functional, including the addition of various chemical groups.

lactose a disaccharide that consists of the sugars glucose and galactose

β-galactosidase the enzyme responsible for the breakdown of lactose into its component sugars, glucose and galactose

The *lac* Operon

Lactose is a disaccharide found in milk or milk products that consists of two sugars: glucose and galactose (**Figure 1**). *Escherichia coli* bacterial cells found on the intestinal lining of mammals can use the energy supplied by lactose for growth. To use the energy of the disaccharide, *E. coli* must split lactose into its two monomer sugars.

β-galactosidase is the enzyme responsible for the degradation of lactose. It would not be economical for *E. coli* to produce the enzyme β-galactosidase when lactose is not present, especially as mammals mature and their milk intake decreases. Using negative regulation to control the transcription and translation of the β-galactosidase gene, bacterial cells only produce the enzyme when necessary. A negative control system blocks a particular activity in the cell. In this case, the production of β-galactosidase is blocked if lactose is not present.

The gene for β-galactosidase is part of an **operon**. An operon is located on the bacterial chromosome and comprises a cluster of structural genes, a promoter, and, between these two regions, a short sequence of bases known as an **operator**. Only prokaryotes use operons to regulate genes and their respective products. The promoters and operators of operons function only as regulatory sequences of DNA; they do not code for structural proteins.

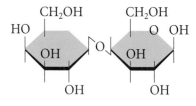

Figure 1
Lactose is a disaccharide of glucose and galactose.

operon a cluster of genes under the control of one promoter and one operator in prokaryotic cells; acts as a simple regulatory loop

operator regulatory sequences of DNA to which a repressor protein binds

lac operon a cluster of genes under the control of one promoter and one operator; the genes collectively code for the enzymes and proteins required for a bacterial cell to use lactose as a nutrient

LacI protein a repressor protein that binds to the *lac* operon operator, preventing RNA polymerase from transcribing the *lac* operon genes

repressor protein regulatory molecules that bind to an operator site and prevent the transcription of an operon

signal molecule a molecule that activates an activator protein or represses a repressor protein

inducer a molecule that binds to a repressor protein and causes a change in conformation, resulting in the repressor protein falling off the operator

DID YOU *KNOW* ❓

Gene Expression Is Regulated
In 1961, French scientists François Jacob and Jacques Monod proposed the operon model of gene regulation in *E. coli*, and the existence of messenger RNA. They determined that the regulation of the β-galactosidase gene was linked to the levels of lactose in the cell. Find out more about their classic experiment in an online animation.

GO   www.science.nelson.com

The ***lac* operon** consists of a cluster of three genes that code for proteins involved in the metabolism of lactose. The three genes are *lacZ*, *lacY*, and *lacA*. The *lacZ* gene encodes the enzyme β-galactosidase; *lacY* encodes β-galactosidase permease, an enzyme that causes lactose to permeate the cell membrane and enter the cell; and *lacA* encodes a transacetylase of unknown function. The **LacI protein** is a **repressor protein** that blocks the transcription of the β-galactosidase gene by binding to the lactose operator and getting in the way of the RNA polymerase, as shown in **Figure 2(a)**. The promoter and operator regions overlap; when the LacI protein binds to the operator, it covers part of the promoter, the binding site for RNA polymerase. To put it simply, a roadblock is put into effect. The roadblock ensures that if no lactose is present in the cell's environment, the *lac* operon genes are not transcribed and then translated into their respective proteins. In this way, the level of lactose is an effector, meaning that it controls the activity of a specific set of genes.

If lactose is introduced into a cell's environment, the roadblock must be removed to ensure that β-galactosidase is manufactured. The presence of lactose itself removes the roadblock, and is therefore known as a **signal molecule** or an **inducer**. Lactose binds to the LacI protein. The binding of lactose to LacI protein changes the conformation of the LacI protein. This change results in the inability of the new complex (lactose and LacI) to stay bound to the operator region of the *lac* operon. The complex falls off the DNA, allowing the RNA polymerase to proceed onward and transcribe the *lac* operon, as illustrated in **Figure 2(b)**.

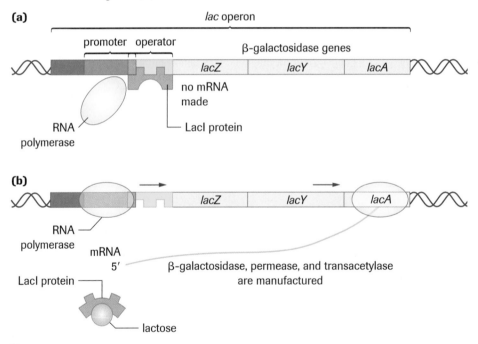

Figure 2
(a) When lactose is not present in the cell environment, the LacI protein binds to the *lac* operator, covering part of the promoter and, thereby, blocking transcription.
(b) Lactose binds to the LacI protein, changing its shape. It can no longer bind to the *lac* operator and transcription proceeds.

The *trp* Operon

Tryptophan is an amino acid that is used by *E. coli* cells for the production of protein. *E. coli* cells located on the intestinal lining of a mammal can absorb tryptophan from the mammal's diet. If found elsewhere, *E. coli* must manufacture its own tryptophan. Once a high concentration of this amino acid is present, the genes for tryptophan production are no longer transcribed.

The ***trp* operon** is another example of coordinated regulation. In contrast to the *lac* operon, whose transcription is induced when lactose is present, the *trp* operon is repressed when high levels of tryptophan are present. So, in this case, the effector is the level of tryptophan. The *trp* operon consists of five genes. These five genes code for five polypeptides that make three enzymes needed to synthesize tryptophan. When tryptophan levels are high, this amino acid binds to the *trp* repressor protein, altering its shape. The *trp* repressor–tryptophan complex can now bind to the *trp* operator. Since tryptophan itself is needed to inactivate the *trp* operon, it is called a **corepressor**. When the level of this amino acid decreases, the shape of the *trp* repressor protein changes because of the lack of the tryptophan corepressor. The *trp* repressor can no longer stay bound to the *trp* operator and it falls off. The RNA polymerase is free to transcribe the *trp* operon genes, resulting in an increase in tryptophan production (**Figure 3**).

***trp* operon** a cluster of genes in a prokaryotic cell under the control of one promoter and one operator; the genes govern the synthesis of the necessary enzymes required to synthesize the amino acid tryptophan

corepressor a molecule (usually the product of an operon) that binds to a repressor to activate it

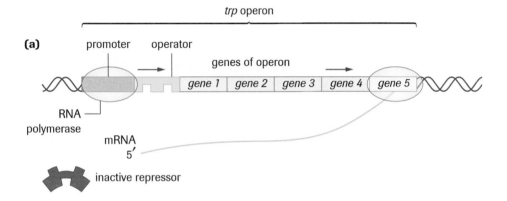

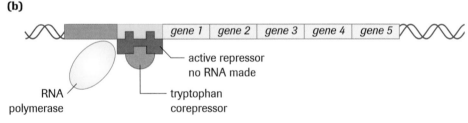

Figure 3
(a) Lack of tryptophan inactivates the repressor and transcription proceeds.
(b) Tryptophan acts as a corepressor and binds to the tryptophan repressor. The complex can now bind to the *trp* operator and transcription is blocked.

 ***Control Mechanisms***

Operons are gene regulation mechanisms found in prokaryotes. The genes code for proteins participating in the same metabolic pathway and are often found adjacent to each other. This strategy allows prokaryotes to regulate the level of synthesis of the end product of the pathway. The *lac* operon is an example of enzyme induction, whereas the *trp* operon is an example of enzyme repression (**Table 2**). Induction and repression are precise regulatory mechanisms that respond to specific substances, called effectors, that control the activity of a specific set of genes. The effector in the *lac* operon is the level of lactose, whereas the effector in the *trp* operon is the level of tryptophan.

Table 2 Comparison of the *lac* Operon and *trp* Operon as a Means of Gene Regulation

lac operon	*trp* operon
• It regulates the production of β-galactosidase and other proteins involved in the metabolism of lactose.	• It regulates the production of the amino acid tryptophan.
• It consists of a cluster of three genes under the control of one promoter and one operator.	• It consists of a cluster of five genes under the control of one promoter and one operator.
• The LacI repressor protein binds to the operator when lactose levels are low.	• The corepressor tryptophan binds to the *trp* repressor protein, and the complex binds to the operator when tryptophan levels are high.
• High levels of lactose induce the operon.	• High levels of tryptophan repress the operon.

▸ Section 5.5 Questions

Understanding Concepts

1. Define the following terms: *operon, operator, corepressor, repressor, housekeeping gene, signal molecule*.

2. Explain why it is to a cell's advantage to have some of its genes under regulation.

3. Describe the state of the *lac* operon system if the level of lactose is low, indicating the activity and state of all major enzymes and proteins found in this system. What changes take place if lactose is suddenly made available as a nutrient to the bacteria?

4. The gene that regulates the production of the LacI protein is found further upstream of the *lac* operon. Explain the potential ramifications to the functioning of the *lac* operon and the production of its products under the following conditions:
   (a) A mutation is found in the LacI protein gene.
   (b) A mutation is found in the second gene of the *lac* operon.
   (c) The RNA polymerase has difficulty binding to the promoter.

5. Describe the state of the *trp* operon system if the level of tryptophan is high, indicating the activity and state of all major enzymes and proteins found in this system. What changes take place if the tryptophan levels fall?

Applying Inquiry Skills

6. A researcher was trying to determine whether two molecules (molecule A and molecule B) were corepressors or inducers in their respective operon systems. Data were collected regarding the levels of protein and the amount of gene transcription for the genes in their respective operons. The data is shown in **Table 3**.

Table 3

	Level of protein	Transcription of gene 1	Transcription of gene 2
Molecule A	high	low	low
	low	high	high
Molecule B	high	high	high
	low	low	low

(a) Determine whether molecule A and molecule B are inducers or corepressors. Justify your answers.
(b) Identify which system resembles the *lac* operon system and which resembles the *trp* operon system.
(c) State a generality about operon systems involving an inducer and about operon systems involving a corepressor.

Mutations are errors made in the DNA sequence that are inherited. These errors may have deleterious side effects, no effect, or positive side effects for an organism. For example, some mutations in the genome lead to the development of diseases that may be inherited. Mutations in the cystic fibrosis transmembrane regulator (CFTR) gene that result in cystic fibrosis can be passed on from parent to child—a negative side effect. However, organisms have evolved as a result of mutations that have been naturally selected. The large size of the human brain arose from a series of mutations—a positive side effect. The side effect of a mutation may not surface immediately, especially in the case of eukaryotes. Diploid organisms have two copies of each gene; hence, if an error is made in one copy, the other copy may compensate. Each of us may have numerous deleterious mutations in our genome. If humans were haploid, much of the world's population would not be alive because of the deleterious mutations that we carry.

mutations changes in the DNA sequence that are inherited

Types of Mutations

Different types of mutations exist and the extent of their effect varies widely. One type of mutation is called a **silent mutation**, which has no effect on the operation of the cell. Usually silent mutations occur in the noncoding regions (introns) of DNA. If these areas are cut out of the primary mRNA transcript during the process of transcription, the mutation never surfaces. Another way a mutation may be silenced is through the redundant nature of the genetic code. For example, the amino acid phenylalanine is coded for by the base sequences UUU and UUC on mRNA. If there is a mistake made during transcription or a mutation occurs in the DNA so that the third base is now a G instead of an A (complementary base C instead of U on mRNA), phenylalanine will still be the amino acid that is translated. The effect of the mutation will go unnoticed.

silent mutation a mutation that does not result in a change in the amino acid coded for and, therefore, does not cause any phenotypic change

Not all mutations are silent. A **missense mutation** arises when a change in the base sequence of DNA alters a codon, leading to a different amino acid being placed in the protein sequence. Sickle cell anemia occurs as a result of a missense mutation. A third type of mutation is a **nonsense mutation**. A nonsense mutation occurs when a change in the DNA sequence causes a stop codon to replace a codon specifying an amino acid. During translation, only the part of the protein that precedes the stop codon is produced, and the fragment may be digested by cell proteases. Nonsense mutations are often lethal to the cell.

missense mutation a mutation that results in the single substitution of one amino acid in the resulting polypeptide

nonsense mutation a mutation that converts a codon for an amino acid into a termination codon

Missense and nonsense mutations arise from the **substitution** of one base pair for another. Another type of mutation is a **deletion**, which occurs when one or more nucleotides are removed from the DNA sequence. For example, if the third base is removed in the sequence AUG GGA UUC AAC GGA AUA, the codon sequence becomes AUG GAU UCA ACG GAA UA. The original sequence codes for the amino acids methionine–glycine–phenylalanine–asparagine–glycine–isoleucine. As a result of the deletion the second sequence codes for the amino acids methionine–asparagine–leucine–threonine–glutamine. The protein structure has been drastically altered and will inevitably result in a defective protein. Such shifts in the reading frame usually result in drastic errors. Another way that a shift in the reading frame can occur is by the **insertion** of a nucleotide. Since the DNA sequence is read in triplets of nucleotides, inserting an extra nucleotide will cause different amino acids to be translated, similar to a deletion mutation.

substitution the replacement of one base in a DNA sequence by another base

deletion the elimination of a base pair or group of base pairs from a DNA sequence

insertion the placement of an extra nucleotide in a DNA sequence

When a mutation changes the reading frame, it is called a **frameshift mutation**. Insertions and deletions can both cause frameshift mutations. A deletion or insertion of

frameshift mutation a mutation that causes the reading frame of codons to change, usually resulting in different amino acids being incorporated into the polypeptide

two nucleotides will also result in a shift of the reading frame; however, a deletion or insertion of three nucleotides does not have this effect. The insertion or deletion of three nucleotides results in the addition or removal of one amino acid. The presence or absence of the amino acid will cause a change in protein conformation, but it may not have as serious an impact as would not building the appropriate protein at all.

Silent, insertion, and deletion mutations may all occur at a certain point in the base sequence and only involve one base pair. Despite the variation in their effect, these mutations all belong to a category known as **point mutations**. Point mutations are specific to one base pair (**Figure 1**).

Another category of mutations involves large segments of DNA and is apparent at the chromosomal level. **Translocation** is characterized by the relocation of groups of base pairs from one part of the genome to another. Usually translocations occur between two nonhomologous chromosomes. A segment of one chromosome breaks and releases a fragment, while the same event takes place with another chromosome. The two fragments exchange places, sometimes disrupting the normal structure of genes. When unrelated gene sequences become contiguous and are transcribed and translated, the result is a fusion protein with a completely altered function, if any. Some types of leukemia are associated with translocations and their respective fusion proteins.

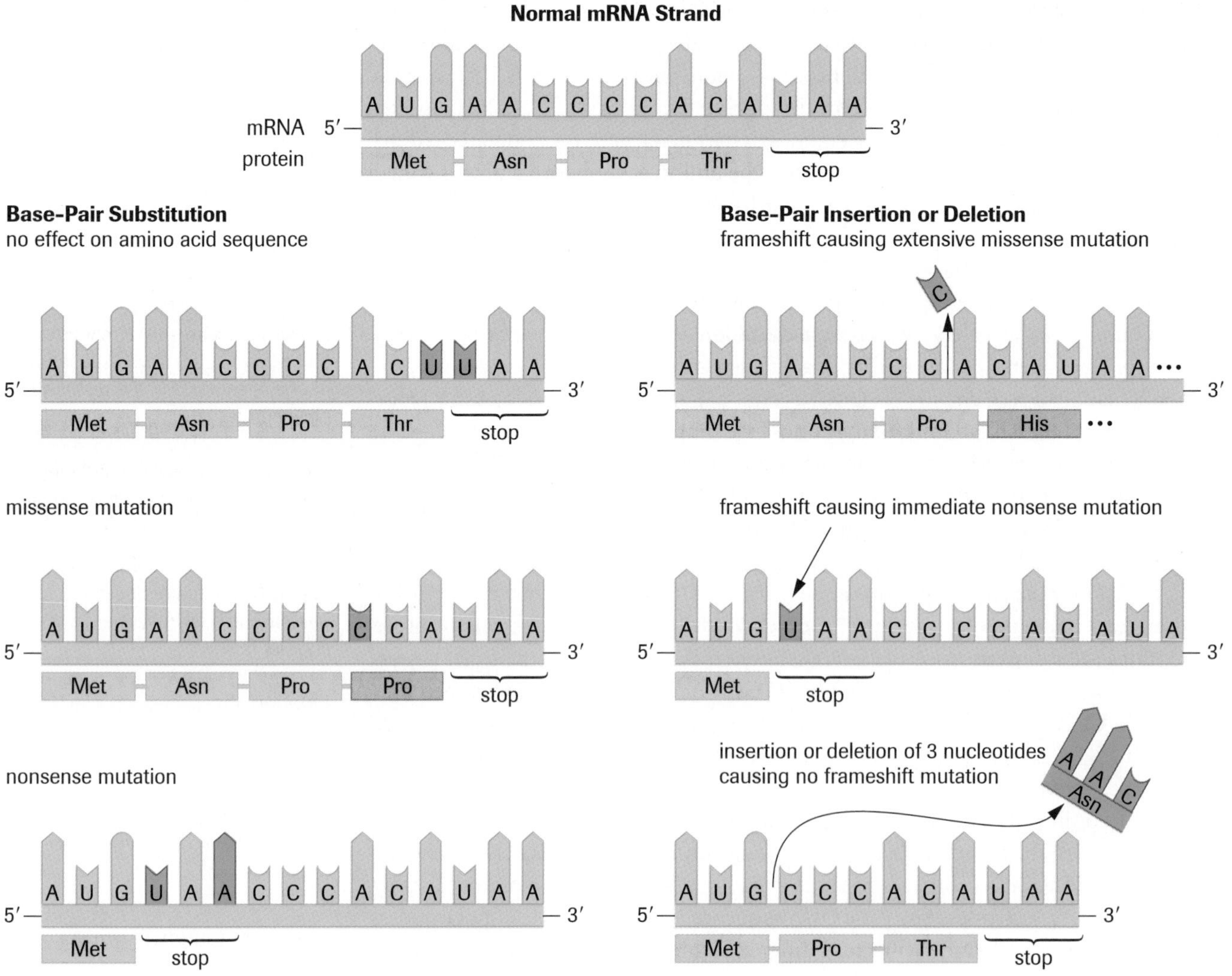

point mutations mutations at a specific base pair in the genome

translocation the transfer of a fragment of DNA from one site in the genome to another location

Figure 1
A summary of different types of mutations that may occur in a DNA sequence, affecting the transcribed RNA sequence.

Some fragments of DNA are consistently on the move. These "jumping genes" or **transposable elements** move from one location to another in the genome. If they happen to fall within a coding region of a gene, they will disrupt the correct transcription of the gene, leaving it inactive. Barbara McClintock, an American genetic botanist, was awarded the Nobel Prize in Physiology or Medicine in 1983 for her discovery of these transposable elements (**Figure 2**).

Finally, an **inversion** is a chromosomal segment that has reversed its orientation in the chromosome. There is no gain or loss of genetic material, but, depending on where the break occurs, a gene may be disrupted or come under other transcriptional control.

Causes of Genetic Mutations

Some mutations are simply caused by error of the genetic machinery and are known as **spontaneous mutations**. DNA polymerase I rereads the duplicated DNA to check for errors, yet on occasion it misses a base or two, which results in a point mutation. Mutations may also arise from exposure to **mutagenic agents** and are said to be **induced mutations**. Some examples of mutagenic agents include ultraviolet (UV) radiation, cosmic rays, X rays, and certain chemicals.

The depletion of the ozone layer allows more damaging UV light to penetrate our troposphere. UV light possesses more energy than visible or infrared light; a high-frequency UV light contains enough energy to cause a point mutation. The rate of skin cancer has been on the rise in Canada for years (**Figure 3**). It has now been predicted that one of every nine Canadians will develop skin cancer at some point in life. Similarly, X rays are also high-frequency, high-energy radiation and have the ability to break the backbone of a DNA molecule. The backbone is annealed by enzymes that repair DNA, but sometimes bases are lost in the process. X-ray technicians always leave the room during the taking of an X ray. If they stayed within the room throughout the day, they would be exposed to seriously dangerous doses of radiation. As a patient, you are only exposed to a single dose once. In addition, only a specific area is exposed, while the remaining area is usually covered with a lead apron. Pregnant women are advised to avoid X rays because the developing fetus may be harmed. Natural radiation also may cause a mutation in a developing fetus. Airline passengers travel above the clouds that filter a large amount of radiation from the sun. Pregnant women are warned not to fly excessively because of the increased exposure to natural radiation.

transposable elements segments of DNA that are replicated as a unit from one location to another on chromosomal DNA

inversion the reversal of a segment of DNA within a chromosome

Figure 2
U.S. genetic botanist Barbara McClintock was awarded a Nobel Prize for her discovery of transposable elements in Indian corn.

spontaneous mutations mutations occurring without chemical change or radiation but as a result of errors made in DNA replication

mutagenic agents agents that can cause a mutation

induced mutations mutations caused by a chemical agent or radiation

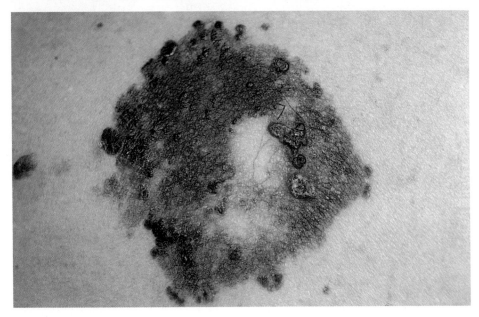

Figure 3
Melanoma is a very aggressive skin cancer that can be induced by prolonged exposure to UV light.

Cancer is considered a genetic disease because it is always a result of a mutation in the genetic sequence. Mutations result in oncogenes, which are mutant versions of genes that control cell growth and division. Cystic fibrosis is another example of a genetic disease. It is caused by any one of at least 60 different types of mutations in the CFTR gene. The severity of cystic fibrosis varies with the location of the mutation. Missense, splicing, nonsense, and frameshift mutations have been characterized in the CFTR gene, all causing cystic fibrosis. The most common mutation is a 3-base-pair deletion discovered in Toronto at the Hospital for Sick Children by Dr. Lap-Chee Tsui and his research team.

Induced mutations may also be caused by exposure to chemicals. Numerous pesticides have been linked to deleterious mutations and have now been banned in Canada and other parts of the world.

Chemicals that resemble DNA can also cause mutations. This capability has been taken advantage of in AIDS therapy. Some of the chemicals in AIDS drugs resemble the nitrogenous bases in DNA and insert themselves into the virus's DNA. When the viral DNA is to be duplicated, the DNA polymerase stops because it does not recognize the substituted chemical as a base. The progression of the disease is slowed.

Some chemicals are more dangerous than others. Ethidium bromide is a chemical commonly used in biotechnology to visualize DNA. If UV light is shone on a fragment of DNA containing ethidium bromide, the DNA will phosphoresce. Biotechnologists use extreme care when handling ethidium bromide because it can bind to DNA and lead to a mutation.

▶ EXPLORE an issue

Debate: Cell Phones and Brain Cancer

Decision-Making Skills

○ Define the Issue ● Analyze the Issue ● Research
● Defend the Position ● Identify Alternatives ○ Evaluate

The use of cell phones has increased exponentially over the past 10 years. More and more people are using them as a means of communication. Despite the convenience, numerous drawbacks and concerns have surfaced about the widespread use of cell phones. One issue that has come to light is the possible relationship between cell phones and cancer (**Figure 4**).

Statement
The electromagnetic radiation associated with the use of cell phones does not contribute to the development of brain cancer.

- In your group, research this issue. Gather information about how cell phones work, the source and effects of electromagnetic radiation from their use, and scientific studies in this area.

- Search for information in periodicals, newspaper articles, and on the Internet.

 www.science.nelson.com

(a) Write a list of points and counterpoints which your group has considered.
(b) Decide whether your group agrees or disagrees with the statement.
(c) Prepare to defend your group's position in a class discussion.

Figure 4
Cell phones may be linked to brain cancer. More research is necessary to determine conclusively whether such a relationship exists.

 **Mutations**

Table 1 Types of Mutations

Category	Type	Result
point mutation	**substitution** AAG CCC GGC AAA AAG ACC GGC AAA	**missense mutation** only one amino acid substituted
	deletion AAG CCC GGC AAA AAC CCG GCA AA ↑	**frameshift mutation** can result in many different amino acids substituted or a stop codon read (nonsense mutation)
	addition/insertion AAG CCC GGC AAA AAG ACC GGG CAA A	
chromosomal	**translocation** chromosome 1 5′ AAATTCG GCACCA 3′ chromosome 2 5′ TAGCCC AAGCGAG 3′ ↓ chromosome 1 5′ TAGCCC GCACCA 3′ chromosome 2 5′ AAATTCG AGCGAG 3′	inactivation of gene if transloca-tion is within a coding segment

	inversion
	
	normal chromosome 5′ AATTGGCCATA ATATGAA AAGCCC 3′
	                    3′ TTAACCGGTAT TATACTT TTCGGG
	
	       ↓
	
	after inversion 5′ AATTGGCCATA TTCATAT AAGCCC
	                     TTAACCGGTAT AAGTATA TTCGGG	

▶ *Section 5.6 Questions*

Understanding Concepts

1. Define the following terms: *mutation, frameshift mutation, point mutation, nonsense mutation, missense mutation.*

2. Explain why mutations, such as nitrogen-based additions, are often much more harmful than nitrogen-base substitutions.

3. Which of two types of mutations, nonsense or missense, would be more harmful to an organism? Justify your answer using your knowledge of protein synthesis.

4. Name three factors that can produce gene mutations.

5. Using the genetic code, list all the codons that can possibly be altered to become a stop codon by substitution by one base.

Applying Inquiry Skills

6. Determine whether or not the following mutations would be harmful to an organism. Translate the mRNA sequence into protein to help you decide. The mutation is indicated in red.

   (a) AUG UUU UUG CCU UAU CAU CGU
       AUG UUU UUG CCU UAC CAU CGU

   (b) AUG UUU UUG CCU UAU CAU CGU
       AUG UUU UUG CCU UAA CAU CGU

   (c) AUG UUU UUG CCU UAU CAU CGU
       AUG UUU CUU GCC UUA UCA UCG U

   (d) AUG UUU UUG CCU UAU CAU CGU
       AUG UUU UUG CCU AUC AUC GU

   (e) AUG UUU UUG CCU UAU CAU CGU
       UGC UAC UAU UCC GUU UUU GUA

7. Which of the following amino acid changes can result from a single base-pair substitution?
   (a) arg to leu   (b) cys to glu
   (c) ser to thr   (d) ile to ser

Making Connections

8. Explain why a food dye that has been identified as a chem-ical mutagen poses greater dangers for a developing fetus than for an adult.

9. List three changes that can be made to your personal lifestyle that would reduce the odds of a mutation taking place.

coupled transcription–translation a phenomenon in which ribosomes of bacteria start translating an mRNA molecule that is still being transcribed

Protein synthesis in prokaryotes and eukaryotes differs. Some of the major differences are as follows:

1. Prokaryotic organisms do not possess a nuclear membrane. Therefore, once transcription by RNA polymerase has begun, translation can begin, even though the full gene has yet to be transcribed. This phenomenon is known as **coupled transcription–translation** (Figure 1).

2. The genes of prokaryotic organisms do not contain any noncoding regions (introns). Some archaebacteria possess introns that distinguish them from eubacterial genomes.

3. In prokaryotes, the ribosome recognizes the start of an mRNA transcript by a unique sequence of purine-rich bases known as the Shine–Dalgarno sequence. In eukaryotes, ribosomes recognize the 5′ cap that has been placed on the mRNA.

4. Ribosomes in eukaryotes are larger than those found in prokaryotes.

5. The tRNA for methionine is the first tRNA to enter the P site of a ribosome during translation in both prokaryotes and eukaryotes. In prokaryotes, the methionine at the start of translation is tagged with a formyl group.

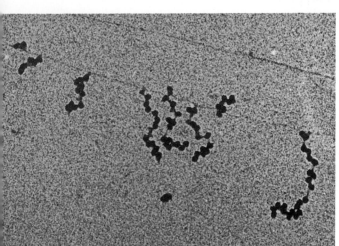

Figure 1
Transcription and translation are coupled in prokaryotes. Since prokaryotes lack a nuclear envelope, mRNA molecules do not need to be completely formed before crossing a nuclear membrane, as eukaryotic mRNA molecules do. Hence, translation can begin immediately.

6. Eukaryotic organisms do not possess operons.

7. The prokaryotic genome is a circular chromosome. The eukaryotic genome is organized into many chromosomes.

Endosymbiotic Relationships Between Organelles and Cells

Since prokaryotes and eukaryotes share numerous similarities in their mechanisms for protein synthesis, the idea that eukaryotic cells may have evolved from an **endosymbiotic** relationship between the prokaryotic ancestors of mitochondria and host prokaryotic cells is not surprising. Before they were enveloped by prokaryotic cells, mitochondria may have been primitive prokaryotic cells with compartmentalized organelles.

Mitochondria in eukaryotic cells resemble prokaryotic cells in the following respects:

1. Mitochondria have circular genomes that are not contained within a nucleus.

2. The sequence of mitochondrial DNA is very similar to the genomes of bacterial cells.

3. Mitochondria divide by the process of **fission** within a eukaryotic cell, similar to bacteria.

4. Mitochondria possess their own system of DNA synthesis, transcription, and translation, indicating that mitochondria may once have been free-living cells.

The genomes of different types of mitochondria from different organisms vary greatly, suggesting evolutionary change over time. Nevertheless, the characteristics of mitochondria within a eukaryotic cell suggest that, long ago, mitochondria were entities of their own.

ACTIVITY 5.7.1

**Protein Synthesis:
A Very Close Look (p. 270)**
The process of transcription and translation in eukaryotes and prokaryotes has been observed with the electron microscope. In Activity 5.7.1, you will examine electron micrographs that show the cellular structures involved in protein synthesis. Can you identify some of the key enzymes and organelles involved in transcription and translation in both prokaryotes and eukaryotes?

Recently a similar relationship has been discovered between a small biflagellate alga, known as a cryptomonad, and a secondary compartmentalized structure found within it, likely derived from a primitive red alga. This compartmentalized eukaryotic structure contains a chloroplast, a plasma membrane, and a small volume of cytoplasm surrounding a tiny nucleus called a **nucleomorph**. The nucleomorph contains 531 genes organized into 3 chromosomes. Most of the genes are concerned with functions such as DNA replication within the nucleomorph itself. A small number of the genes, 30 of them, govern the synthesis of proteins that are involved in maintaining the chloroplast.

The partnership between the cryptomonad and the compartmentalized structure is endosymbiotic. By using energy from sunlight, the chloroplast in the compartmentalized structure synthesizes food that the alga can use. Evolutionary advantage favours an endosymbiotic relationship between a cell and a given internal cellular structure.

endosymbiotic physical and chemical contact between one species and another species living within its body, which is beneficial to at least one of the species

fission asexual reproduction typical of bacteria in which the cell divides into two daughter cells

nucleomorph tiny nucleus containing genomic material found within a eukaryotic endosymbiotic structure originally believed to be derived from primitive red alga

SUMMARY *Eukaryotes and Prokaryotes*

Table 1 Key Differences Between Eukaryotes and Prokaryotes

	Prokaryotes	**Eukaryotes**
Genome	• small and circular • all regions are coding, except for promoters and operators • presence of operons	• large and arranged in chromosomes • consists of coding and noncoding regions • absence of operons
Transcription	• coupled with translation • lack of introns means no excision	• occurs in the nucleus • introns excised by spliceosomes and exons joined together
Translation	• commences with formyl–methionine • ribosome recognizes Shine–Dalgarno sequence on mRNA as binding site • ribosomes are smaller than in eukaryotes	• commences with methionine • ribosome recognizes 5′ cap on mRNA as binding site • occurs in the cytoplasm • ribosomes are larger than prokaryotes

▶ Section 5.7 Questions

Understanding Concepts

1. Define the term *endosymbiosis*.

2. Explain why prokaryotes are able to undergo coupled transcription-translation while eukaryotes cannot.

3. What evidence exists that mitochondria and nucleomorphs may have possibly existed as free-living entities?

Making Connections

4. Using your knowledge of mitochondrial function and structure, and information you find on the Internet and at the library, explain why an endosymbiotic relationship may have evolved between mitochondria and cells. Repeat this exercise for the relationship between cryptomonads and the compartmentalized structure within them that contains a nucleomorph. Present your explanations in paragraph form.

 www.science.nelson.com

5. Eukaryotes are more complex than prokaryotes. Using your knowledge of the structural and molecular differences between the two, explain how this is the case. Incorporate specific examples in your answer and present your explanation in paragraph form.

chromatin complex of DNA and histone proteins located in the nucleus of eukaryotes

histones positively charged proteins that bind to negatively charged DNA in chromosomes

nucleosome a complex of eight histones enveloped by coiled DNA

supercoiling DNA folded into a higher level of coiling than is already present in nucleosomes

variable number tandem repeats (VNTRs) repetitive sequences of DNA that vary among individuals; also known as microsatellites

telomeres long sequences of repetitive, noncoding DNA on the end of chromosomes

centromeres constricted region of chromosome that holds two replicated chromosome strands together

pseudogenes DNA sequences that are homologous with known genes but are never transcribed

LINEs repeated DNA sequences of 5000 to 7000 base pairs in length that alternate with lengths of DNA sequences found in the genomes of higher organisms

SINEs repeated DNA sequences of 300 base pairs in length that alternate with lengths of DNA sequences found in the genomes of higher organisms

---

🔍 ACTIVITY 5.8.1

Comparison of Eukaryotic and Prokaryotic Genomes (p. 271)
Eukaryotic and prokaryotic DNA share numerous common properties, but they also have differences. Activity 5.8.1 leads you to discover what some of these differences are.

In the nucleus, the human genome is organized into chromosomes, which consist of a complex of protein and DNA called **chromatin**. One unbroken double-stranded DNA helix forms each chromosome. If the nucleotides of all the chromosomes in a genome could be stretched out in one long double helix, they would measure 1.8 m in length. An individual nucleus is 5 μm (a micrometre is one millionth of a metre), visible only with the aid of a microscope, so it is only by being intricately intertwined that the DNA can fit in the minute space of a cell's nucleus.

Packed tightly inside the nucleus, the DNA is organized hierarchically. First, every 200 nucleotides, the DNA is coiled around a core group of eight stabilizing proteins, known as **histones**. The positively charged histones are strongly attracted to negatively charged DNA. This complex of histones enveloped by coiled DNA is called a **nucleosome**. Second, a series of nucleosomes coil into chromatin fibres. Lastly, these fibres fold into the final chromatin structure by means of a higher level of coiling, known as **supercoiling** (Figure 1).

In humans, there are 46 chromosomes: 44 are somatic and the other two are sex chromosomes. The chromosomes vary in size and have been numbered 1 to 22, according to size, with chromosome 1 being the largest. Only a fraction of the human genome is known to code for specific proteins. Currently, we estimate that 42 000 genes exist within the very small percentage of DNA coding for proteins. More than 95% of the human genome is noncoding.

Noncoding regions are filled with **variable number tandem repeats (VNTRs)**, also known as microsatellites. These are sequences of base pairs that repeat over and over again (e.g., TAGTAGTAGTAG) and vary among individuals. The length of microsatellites varies, as does their position in the genome. Some microsatellites have been found within genes. Huntington's disease is associated with a repetitive sequence within a gene. Not all microsatellites are detrimental; repetitive DNA is used as a defence mechanism against the shortfalls of DNA replication. For instance, the ends of chromosomes have long sequences of repetitive noncoding DNA, known as **telomeres**, that protect cells from losing valuable genomic material during DNA replication. Telomeric DNA protects chromosomes by binding proteins that stop the ends from being degraded and sticking to other chromosomes. Repetitive DNA sequences are also found in the region of the **centromeres**, which play a role during cell division.

In addition, the genome contains **pseudogenes**. A pseudogene has a nucleotide sequence similar to a functioning gene but does not seem to express any RNA or protein. Two types of pseudogenes are thought to be crippled copies of known functional genes arranged as **LINEs** (**l**ong **i**nterspersed **n**uclear **e**lements) and **SINEs** (**s**hort **i**nterspersed **n**uclear **e**lements). The function of LINEs and SINEs is not clear.

In the last century, we have discovered a great deal about the human genome, but we are still only scratching the surface. In the next chapter, biotechnological techniques are discussed that promise to reveal even more about the structure, function, and organization of our genome, and its relationship to normal physiological development and disease. 🔍

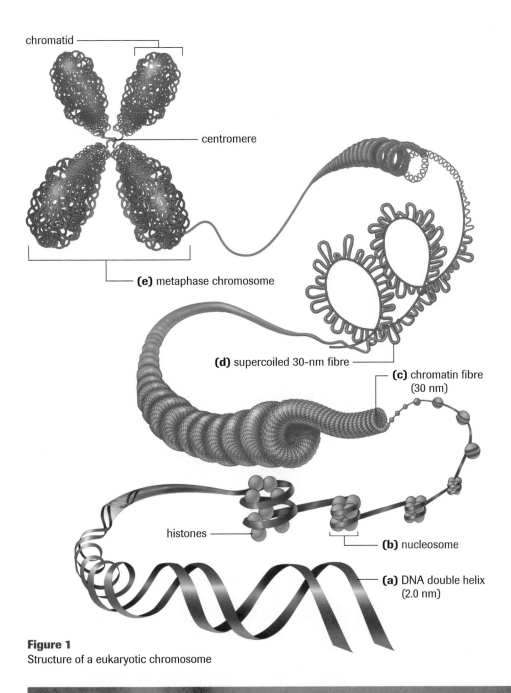

chromatid

centromere

(e) metaphase chromosome

(d) supercoiled 30-nm fibre

(c) chromatin fibre
(30 nm)

histones

(b) nucleosome

(a) DNA double helix
(2.0 nm)

Figure 1
Structure of a eukaryotic chromosome

▶ *Section 5.8* *Questions*

Understanding Concepts

1. Differentiate between the following:
   (a) nucleosome and histone
   (b) telomeres and centromeres
   (c) SINEs and LINEs

2. In a flow chart, illustrate the organization of genetic material within a eukaryotic genome.

Making Connections

3. Describe the ramifications if a SINE were found in
   (a) an exon;   (b) an intron;   (c) a telomeric region.

4. The number of variable number tandem repeats (VNTRs) in a specific area of a chromosome is not constant between two homologous chromosomes. In addition, the number of VNTRs between different individuals also varies. Provide a possible use of this variation in the field of forensics.

5. Repetitive DNA sequences are found in centromeres and telomeres. Explain why this is an advantage. Why would it not be advantageous to have exons in the telomeric and centromeric areas of the chromosome?

INVESTIGATION 5.4.1

Protein Synthesis and Inactivation of Antibiotics

In this investigation, you will examine the effects of ampicillin on two types of bacteria. *E. coli* MM294/pAmp contains a gene insert that directs the synthesis of a protein that inactivates ampicillin, whereas *E. coli* MM294 does not. Ampicillin inhibits bacterial growth by interfering with cell wall biosynthesis.

Question

What effect does the presence of an ampicillin-resistance gene in a bacterium have on its growth on ampicillin-rich media?

Prediction

(a) Based on your knowledge of protein synthesis, make a prediction about the survival of *E. coli* MM294/pAmp and *E. coli* MM294 on ampicillin-rich media.

Materials

apron
safety goggles
gloves
10% bleach
2-LB agar plates
2-LB + ampicillin
  (LB/amp) plates

masking tape
permanent marker
inoculating loop
Bunsen burner
MM294 culture
MM294/pAMP culture
37°C incubator

Wear safety goggles at all times.

Wear gloves when performing the experiment. Disposable latex gloves are best avoided since allergic reactions to latex have been widely reported. Disposable polyethylene, PVC, or neoprene gloves are recommended.

Wipe down all surfaces with 10% bleach before and after the laboratory exercise.

All resulting cultures must be immersed in 10% bleach before disposal to ensure sterilization.

Do not leave a lit Bunsen burner unattended. Refer to the Appendix for a review of the safe use of a Bunsen burner.

Wash your hands thoroughly at the end of the laboratory.

Procedure

1. Put on your safety goggles and gloves, and wipe down your bench with a 10% bleach solution.

2. Obtain 2-LB plates and 2-LB/amp plates from your teacher.

3. Label the bottom of each plate with your name and the date, using a permanent marker.

4. Label the 2-LB plates "− amp" for the *E. coli* MM294 cells. Label the 2-LB/amp plates "+ amp" for the *E. coli* MM294/pAMP cells.

5. Hold your inoculating loop like a pencil and sterilize it in the nonluminous flame of the Bunsen burner until it becomes red hot. Cool the sterilized loop by touching it to the edge of the agar on one of the LB plates.

6. Using the sterilized loop, pick up one colony of *E. coli* MM294 from a start culture plate. Glide the inoculating loop across an LB agar plate, making sure not to gouge the agar (**Figure 1**).

Figure 1
Pattern of streaking on an agar plate

7. Resterilize your loop as directed in step 5.

8. Repeat step 6 with *E. coli* MM294 streaked on an LB/amp plate.

9. Resterilize your loop as directed in step 5.

10. Repeat step 6 with *E. coli* MM294/pAmp streaked on the other LB plate.

11. Resterilize your loop as directed in step 5.

12. Repeat step 6 with *E. coli* MM294/pAmp streaked on the other LB/amp plate.

13. Sterilize and cool your inoculating loop.

14. Place all four streaked plates in a stack and tape them together.

15. Place the streaked plates upside down in the incubator. Alternatively, if you do not have an incubator, place the plates in a warm part of the room for a couple of days.

INVESTIGATION 5.4.1 *continued*

16. Disinfect your laboratory bench using the bleach solution.

17. Wash your hands thoroughly with soap and water.

> Once the experiment has been completed, flood plates with bleach to kill the bacterial colonies that have been cultured. Alternative is to place plates in an autoclave before they are disposed of.

Analysis

(b) Copy **Table 1** in your notebook and use it to record your observations.

Table 1

	MM294/pAmp + pAmp	MM294 − pAmp
LB/amp plate		
LB plate		

(c) Compare your results to your prediction. Explain any possible causes for variation.

(d) What evidence is there to indicate that protein was synthesised by the bacteria?

(e) Why was it important to streak out both types of bacteria on both types of plates?

(f) This experiment contains both positive and negative controls. Identify them. What information do the controls provide in this experiment?

(g) Why was it important to cool the inoculating loop before obtaining a bacterial colony from a stock plate?

(h) Why was it important to resterilize the inoculating loop between transfers of bacteria?

Evaluation

(i) Suggest possible sources of error in this procedure and indicate their effect on the results.

(j) Using standard scientific format, prepare a written report.

Synthesis

(k) *E. coli* strains containing the genetic sequence pAmp are resistant to ampicillin. Research how the ampicillin can be deactivated by β-lactamase, the protein that the ampicillin-resistance gene codes for.

(l) Predict what would happen if there was an error in the genetic sequence that codes for β-lactamase.

ACTIVITY 5.4.1

Synthesis of a Protein: A Simulation Activity

In this activity, you will be provided with the DNA nucleotide sequence that codes for a hypothetical protein. The code will be provided to you in three fragments. You will have to transcribe the code into mRNA, remove an intron segment, and translate the mRNA into the protein. In addition, you will have to identify the beginning fragment, the middle fragment, and the end fragment.

Procedure

1. Copy each of the following sequences onto a separate piece of paper.

Sequence A
TCTTCCCTCCTAAACGTTCAACCGGTTCTTAATCCGC
CGCCAGGGCCCCGCCCCTCAGAAGTTGGT

Sequence B
TCAGACGTTTTTGCCCCGTAACAACTTGTTACAA
CATGGTCATAAACGTCAGAGATGGTCAATCTCTTAAT
GACT

Sequence C
TACAAACATGTAAACACACCCTCAGTGGACCAACT
CCGCAACATAAACCAAACACCGCTCGCGCCGAAAAA
GATATGG

2. Divide the sequences into triplets (codons) by putting a slash between each group of three bases.

3. Transcribe the DNA into mRNA.

4. Identify the middle, end, and beginning sequence. Use your knowledge of start and stop codons to help you figure it out.

5. Remove codons 24 to 66, including codon 66.

6. Translate the mRNA into protein using the genetic code.

Analysis

(a) Which fragment was the beginning fragment? How do you know?

(b) Which fragment was the end fragment? How do you know?

(c) Codons 24 to 66 represent an intron. At what point in the process of protein synthesis are introns removed? What is the name of the enzyme responsible for this excision?

(d) How many amino acids does this protein contain?

(e) Is this genetic sequence eukaryotic or prokaryotic? How do you know?

(f) If you worked backward, starting with the amino acid sequence of the protein, would you obtain the same DNA nucleotide sequence? Why or why not?

(g) Provide the anticodon sequence that would build this protein.

ACTIVITY 5.7.1

Protein Synthesis: A Very Close Look

By studying electron micrographs, scientists have been able to obtain even more valuable information about numerous biochemical cellular processes. In this activity, you will examine electron micrographs that illustrate different aspects of protein synthesis. Using your knowledge of the process of protein synthesis, you will identify organelles and enzymes involved in protein synthesis.

Procedure

1. Examine each electron micrograph (**Figures 1** to **5**).

Analysis

(a) The electron micrograph in **Figure 1** depicts an mRNA strand as it undergoes posttranscriptional modification in a eukaryotic cell. Given your knowledge of this process, identify the enzyme depicted by the large white spot.

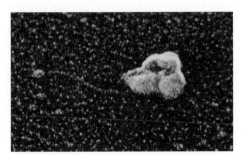

Figure 1

(b) Explain the function of the enzyme in **Figure 1**.

(c) The electron micrograph in **Figure 2** depicts ribosomes translating an mRNA sequence. Identify the ribosomes. Can you distinguish the two subunits of the ribosome?

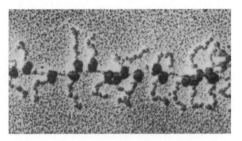

Figure 2

(d) **Figure 3** depicts the process of transcription. What enzyme is represented by the dark spots? Why does more than one enzyme exist on the strand of DNA?

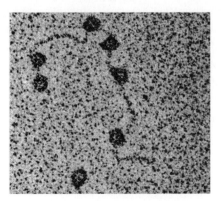

Figure 3

ACTIVITY 5.7.1 *continued*

(e) In the electron micrograph in **Figure 4**, more than one ribosome is translating the mRNA at one time. Identify the ribosomes.

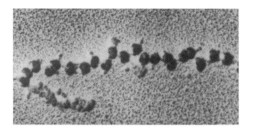

Figure 4

(f) The polysome shown in **Figure 4** exists in prokaryotes and eukaryotes. Why is it more efficient to translate an mRNA strand more than once and simultaneously by many ribosomes?

(g) Identify the ribosomes in **Figure 5**.

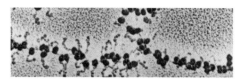

Figure 5

(h) In **Figure 5**, why is the polypeptide chain increasingly longer as you move across the mRNA?

ACTIVITY 5.8.1

Comparison of Eukaryotic and Prokaryotic Genomes

The organization of eukaryotic and prokaryotic genomes differs greatly. In addition, prokaryotic RNA does not undergo posttranscriptional modification. In this activity, you will be provided with an example of a eukaryotic genome and a prokaryotic genome. You will identify different areas of the genomes, including promoter regions, microsatellite regions, telomeric regions, and possible gene regions. Assume that all these regions are found on the one strand provided.

Procedure

1. Copy each of the following sequences onto a separate piece of paper. Assume that the sequences represent an organism's entire genome.

Genome A

5′- GCAGGCCATATAAAATAGCGCCATACTAGATACGGG
CCATATTATTGCATATCCGCCGATTACAGGATTTAATTT
GGGAATTCCCCGATTAACGCGATCGATCGGGCCATATC
GATATGCATCGTAATCCGGTAGATTCACAGGTAG -3′

Genome B

5′- GCATACCCAAATTAATAACGGCGGTAGGCGACTCATT
CTGATATACGCATCGGCATTTACCTACGGCCGGCCGGC
CGGCCGGCCTAGATTTACCGCATTTACCGGCCGCATCG
GATCGGGATTAGCATAATTAAAATGCATCGGCGTAGTAG
GCAATCGGCGCAGCCGAGCCACCTCCCGGAGAATCATC
ATCATCATCATCATCATCATCATCATCATCATACGGAT
AGATCCATTACCATGCGATTTAAAGGCCATTCATGGGC
CCCCGATTTATCCATTTAGGCCGGATTCCATGGATTCAT
TTCCATTTTTCGGCATCATCATCATCATCATCATCATCAT
CATCATCATCATCATCATCAT -3′

Analysis

(a) Based on the size of the genome, which of the sequences would you consider eukaryotic and which prokaryotic? Justify your answer.

(b) The prokaryotic genome contains one ten base pair promoter region, whereas the eukaryotic genome contains two. Find these regions and circle them. How do you know these are the promoter regions?

(c) Which of the two genomes has telomeric DNA? Circle the telomeric region. Where is the telomeric region found and what is its function?

(d) Does the eukaryotic genome contain any microsatellites? Circle the areas if they exist. What sort of sequence is found in a microsatellite region?

(e) Identify the areas that may be sequences that code for a gene in both genomes. Identify the difference between the eukaryotic and prokaryotic sequences. How many genes are found in the eukaryotic genome? (Hint: Look for start codons.)

(f) Using the Internet or library resources, find other examples of prokaryotic and eukaryotic sequences. Compare the sequences for size, gene clustering, and the presence of microsatellites.

 www.science.nelson.com

(g) Why is it important that genomic databases are shared among researchers?

Chapter 5 SUMMARY

Key Expectations

- Compare the structure and function of RNA and DNA, and explain their roles in protein synthesis. (5.2, 5.3, 5.4)

- Explain the steps involved in protein synthesis (transcription and translation) and the control mechanisms for genetic expression, using regulatory proteins (e.g., *lac* operon, *trp* operon). (5.2, 5.3, 5.4, 5.5)

- Illustrate the genetic code by examining/analyzing a segment of DNA. (5.4, 5.8)

- Describe how mutagens, such as radiation and chemicals, can change the genetic material in cells by causing mutations. (5.6)

- Interpret micrographs that demonstrate cellular structures involved in protein synthesis. (5.7)

Key Terms

5′ cap

A (acceptor) site

amino acid

aminoacyl–tRNA

anticodon

β-galactosidase

centromeres

chromatin

coding strand

codon

corepressor

coupled transcription–translation

deletion

endosymbiotic

exons

fission

frameshift mutation

gene regulation

genes

histones

housekeeping genes

induced mutations

inducer

insertion

introns

inversion

*Lac*I protein

lac operon

lactose

LINEs

messenger RNA (mRNA)

missense mutation

mRNA transcript

mutagenic agents

mutations

nonsense mutation

nucleomorph

nucleosome

operator

operon

point mutations

poly-A polymerase

poly-A tail

primary transcript

promoter

proteins

pseudogenes

P (peptide) site

reading frame

release factor

repressor protein

ribosomal RNA (rRNA)

RNA polymerase

signal molecule

silent mutation

SINEs

spliceosomes

spontaneous mutations

start codon

stop codons

substitution

supercoiling

telomeres

template strand

terminator sequence

transcription

transcription factors

transfer RNA (tRNA)

translation

translocation

transposable elements

trp operon

upstream

variable number tandem repeats (VNTRs)

▶ MAKE a summary

In this chapter, you have studied the effects of errors in a given DNA sequence, the regulation of protein synthesis, and the process of protein synthesis from gene to protein. In addition, you have examined the organization of the genome.

1. Starting with the title "The Human Genome," produce a flow chart that illustrates the flow from gene to protein. Your flow chart should include as many key concepts as possible.

In your notebook, indicate whether each statement is true or false. Rewrite a false statement to make it true.

1. In eukaryotes, transcription occurs in the cytoplasm and translation occurs in the nucleus.

2. During transcription, RNA polymerase uses the template strand of DNA as a template to synthesize the complementary mRNA.

3. RNA is comprised of ribose sugar, a phosphate, and a nitrogenous base. The nitrogenous base may be adenine, uracil, guanine, or cytosine.

4. Beadle and Tatum's investigation of metabolism in the organism *Neurospora crassa* led to the one gene–one polypeptide hypothesis.

5. When a bacteria's environment is high in lactose, the LacI protein is bound tightly to the operator of the *lac* operon.

6. A missense mutation results in the substitution of one or more amino acids in the protein chain.

7. Telomeres are examples of positive variable number tandem repeats that protect the chromosome.

8. The role of ribosomal RNA is to deliver amino acids to the ribosome to undergo protein synthesis.

9. The complementary mRNA sequence to the DNA sequence ATGGGCCATAC is UACCCGGUAUG.

In your notebook, record the letter of choice that best completes the statement or answers the question.

10. Which of the following best describes ribonucleic acid (RNA)?
   (a) composed of a linear backbone of sugar deoxyribose and phosphates with amino acids attached to the sugar residues
   (b) identical to DNA except that the base uracil is present instead of the base thymine
   (c) identical to DNA except that the sugar ribose is present instead of the sugar deoxyribose
   (d) identical to DNA except that it is always single stranded, whereas DNA is always found as a double helix containing two strands
   (e) similar to DNA but it contains ribose in place of deoxyribose and uracil instead of thymine

11. What does the central dogma specify?
   (a) A DNA sequence encodes an RNA sequence that encodes protein.
   (b) An RNA sequence encodes protein that encodes DNA.
   (c) RNA plays a major role in gene expression.
   (d) Genes are encoded in the nucleus.
   (e) RNA sequences encode DNA, which encode proteins.

12. The anticodons of arginine, serine, and tyrosine are GCA, UCA, and AUA, respectively. If a segment of DNA codes for these three amino acids in the sequence listed (arginine, serine, tyrosine), which of the following would be the sequence of deoxyribonucleotides of the strand of the DNA molecule that coded for these amino acids?
   (a) GCATCAATA   (d) GCAUCAAUA
   (b) CGUAGUUAU   (e) CGTUCAAUA
   (c) CGATCTAUA

13. Which of the following are the enzymes involved in posttranscriptional modification?
   (a) RNA polymerase and spliceosomes
   (b) RNA polymerase and poly-A polymerase
   (c) poly-A polymerase and spliceosomes
   (d) RNA polymerase, poly-A polymerase, and spliceosomes
   (e) none of the above

14. Base-pair substitutions involving the third base of a codon may not result in an error in the polypeptide. What is partly responsible for this avoidance of error?
   (a) Base-pair substitutions are corrected before transcription begins.
   (b) Base-pair substitutions are restricted to introns, and these regions are later deleted from the mRNA.
   (c) Most tRNAs bind tightly to a codon with only the first two bases of the anticodon.
   (d) A signal-recognition particle corrects coding errors before the mRNA reaches the ribosome.
   (e) Transcribed errors attract spliceosomes which then stimulate splicing and correction.

15. What does the base sequence AAU GGC code for?
   (a) one specific amino acid, followed by a nonsense codon
   (b) two enzymes
   (c) two specific amino acids
   (d) three nucleotide base pairs
   (e) one enzyme and one specific amino acid

16. What does an operon typically consist of?
   (a) an operator, a promoter, and a cluster of genes
   (b) a cluster of genes
   (c) a promoter and a cluster of genes
   (d) an operator and a cluster of genes
   (e) a promoter and an operator

Chapter 5 REVIEW

Understanding Concepts

1. Copy the following in your notebook and fill in the blanks:
   (a) Eukaryotic genomes are divided into two areas. The noncoding regions are known as the _____ _____ and the coding regions are known as _____. Noncoding regions usually contain a large number of _____ _____ that consist of many repetitive sequences. If the repetitive sequence is found in the _____ or in the _____ _____ region of the chromosome, it will likely cause no harm.
   (b) The specific DNA regions where an RNA polymerase attaches and initiates RNA synthesis are called _____. In prokaryotes, many genes may be under the control of the _____ _____ and an _____. The whole system is known as an _____.
   (c) The transfer of genetic information from DNA to RNA is called _____. The enzyme that facilitates the process is called _____ _____ and can only synthesize _____ _____ in the _____ to _____ direction. Synthesis ceases when the _____ is reached.

2. What is the function of each of the following in protein synthesis: tRNA, mRNA, and rRNA?

3. Differentiate between transcription and translation. Use a table to organize your answer.

4. Using the digestion of lactose as an example, explain how prokaryotic cells, such as bacteria, regulate structural genes.

5. Briefly explain the four levels of control in eukaryotic cells:
   (a) transcriptional   (c) translational
   (b) posttranscriptional (d) posttranslational

6. For each pair of mutations, identify which type would be the least harmful to an organism and explain why:
   (a) substitution versus deletion
   (b) an addition in an intron region versus an addition in an exon region
   (c) a substitution in a telomeric region versus a substitution in a promoter site
   (d) inversion versus substitution

7. The following is a sequence of DNA for a hypothetical peptide. Translate this sequence into protein using the genetic code.
   5'- AAGTACAGCAT - 3'
   3'- TTCATGTCGTA - 5'

8. Describe the consequences to protein synthesis if the following were inactivated:
   (a) spliceosomes
   (b) RNA polymerase
   (c) poly-A polymerase
   (d) tRNA
   (e) ribosomes

9. Provide two reasons that explain why coupled transcription-translation is not possible in eukaryotes.

10. The following polypeptide sequence has been obtained from a protein found in eukaryotic cells: met–gly–pro–val–arg.
    (a) List a possible mRNA sequence that may have coded for this protein.
    (b) Explain why more than one sequence may exist.
    (c) How does variability in the mRNA offset mutations? Where does this variability originate?

11. Promoter regions are usually very high in adenine and thymine. Explain how this helps initiate transcription.

12. Explain why mitochondria and nucleomorphs can be considered descendants of a prokaryotic organism.

13. Describe the state of the *trp* operon under the following conditions. Make sure you include the state of RNA polymerase, the tryptophan molecule, and the tryptophan repressor.
    (a) Tryptophan levels are high.
    (b) Tryptophan levels are low.

14. Every codon consists of a triplet of base pairs. Explain why amino acids cannot be coded with just two base pairs.

15. In what ways is the structure of mRNA similar to DNA? How does mRNA differ from DNA?

Applying Inquiry Skills

16. The following metabolic pathway exists in a strain of bacteria. The bacteria require the product E to survive. It is believed that 4 enzymes are required to produce E and are controlled by 4 separate genes (genes 1 to 4).

$$A \xrightarrow{\text{enzyme 1}} B \xrightarrow{\text{enzyme 2}} C \xrightarrow{\text{enzyme 3}} D \xrightarrow{\text{enzyme 4}} E$$

After numerous experiments that mutated each gene separately, the following results were obtained:

- A mutation that affected enzyme 1 resulted in an accumulation of A.
- A mutation that affected enzyme 2 resulted in an accumulation of B.
- A mutation that affected enzyme 3 resulted in an accumulation of B.
- A mutation that affected enzyme 4 resulted in an accumulation of B.

(a) What can you conclude about enzymes 2, 3, and 4 from the results of the experiment?

(b) How many genes are actually present in the coding for this biochemical pathway?

17. Hypothetically, there are 126 amino acids and only 5 nucleotides available. Calculate the minimum number of nucleotides required to code for all 126 amino acids.

18. The following is an mRNA segment:

5′- GGC CAG AAA CAA GAA - 3′

(a) Translate this mRNA into protein.

(b) For each of the amino acids, write out all the tRNA anticodons that could have delivered these specific amino acids to the ribosome. Do not neglect the wobble hypothesis in your answer.

Making Connections

19. After elucidating the structure of DNA, Francis Crick hypothesized the idea known today as the central dogma. He stated that the transfer of information from nucleic acid to nucleic acid, or from nucleic acid to protein, may be possible, but from protein to protein, or protein to nucleic acid, is impossible. Describe the features of the molecules involved in translation and transcription that make Crick's hypothesis true.

20. Recently, the Human Genome Project (HGP) was completed. The HGP has provided us with a blueprint of the base pairs that constitute our genome. Despite this great advancement, we are far away from realizing the numerous medical treatments that will eventually be made available because of or as a result of the project. Currently, scientists are starting to work on the Human Proteome Project, which involves linking genes to proteins that are both functional and dysfunctional. Explain why there would be limited progress in medical research if scientists were

restricted to working with DNA sequences and ignored proteins altogether.

21. Explain what would happen if a drug became attached to a specific repressor protein and prevented the protein from binding to an operator gene.

22. The 2 molecules 5-bromouracil (BU) and 2-aminopurine (AP) (**Figure 1**) are known to cause mutations. Examine the structure of the bases carefully. (Hint: Compare them to the structure of bases found in DNA.) Determine how these molecules can cause a mutation in a given DNA sequence.

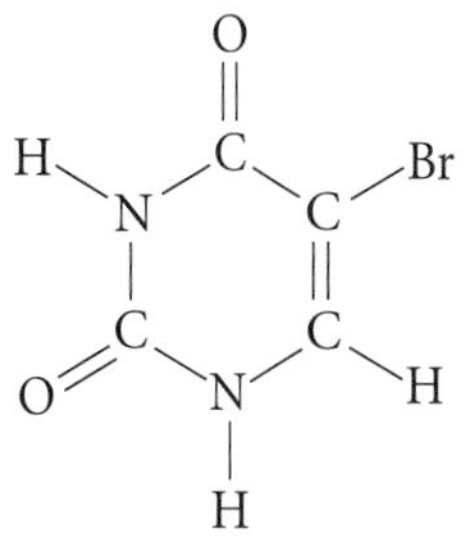

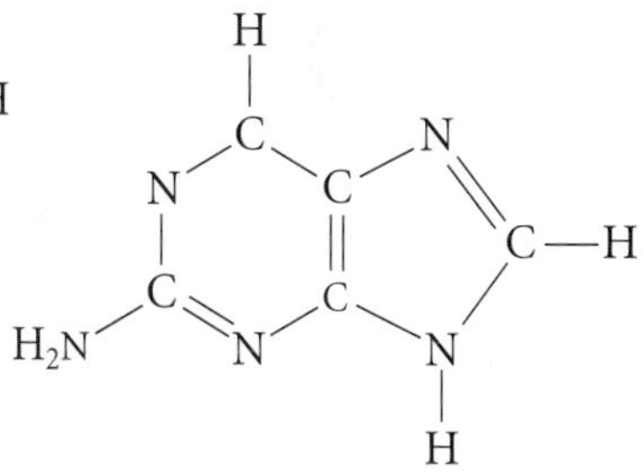

Figure 1
5-bromouracil and 2-aminopurine are known mutagens.

Extensions

23. The mutation that causes sickle cell anemia involves the substitution of the amino acid called valine for another amino acid called glutamic acid. Research the structure of valine and glutamic acid and, with your knowledge of chemistry, propose why this substitution results in a large conformational change for the hemoglobin protein. List other amino acids that could have been substituted instead of valine that may not have caused such serious side effects. List amino acids that are similar to glutamic acid that would probably cause similar side effects.

24. One of the first therapies for HIV was the drug AZT (azidothymidine). Currently, a combination of a variety of drugs is being used to combat the disease. Research these drugs and investigate how they inhibit the expression of the AIDS virus. Report your findings in paragraph form.

6

Biotechnology

In this chapter, you will be able to

- demonstrate an understanding of genetic manipulation, including the processes of recombinant DNA technology, DNA sequencing, and the polymerase chain reaction;

- apply the principles and processes of recombinant DNA technology to construct and test for the presence of a specific gene insert;

- gain an understanding of the industrial, medical, and agricultural applications of recombinant DNA technology;

- describe the functions of the cell components used in genetic engineering, such as restriction endonucleases, methylases, and plasmids;

- outline the contributions of genetic engineers, molecular biologists, and biochemists that have led to further developments in the field of genetics;

- digest DNA using restriction endonucleases and separate the fragments using gel electrophoresis;

- describe the major findings that have arisen from the Human Genome Project;

- describe, and explain the implications of, the principal elements of Canadian regulations on biotechnological products.

The use of living organisms, or substances from living organisms, to develop an agricultural, medicinal, or environmental product or process is not a recent phenomenon. Humans have manipulated biological organisms to obtain a desired product or effect for thousands of years. For example, alcohol fermentation, the conversion of sugar into ethanol by microbes, is an example of biotechnology that has been used by many cultures around the world for many centuries.

A major area of research that has surfaced in the field of biotechnology over the past century is bioremediation. Bioremediation involves the use of living microorganisms to transform undesirable and harmful substances into nontoxic compounds. An example of this type of harmful substance is the oil that was spilled into the environment on March 24, 1989, in South Central Alaska, when the *Exxon Valdez* ran aground. More than 20% of the ship's cargo (approximately 50 million litres of oil) was spilled in about five hours. The oil posed a threat to the welfare of all the plant and animal life in the area. In addition, the poisonous compounds found in the oil would eventually have been consumed by humans via the food chain.

Oil spills can be cleaned up using various methods. The oil may be treated with chemicals, surrounded by physical barriers, or pumped away from the site into storage tanks. When an oil spill of the magnitude of the *Exxon Valdez* occurs, other strategies are also required. Scientists have been aware for many decades that bacteria are able to degrade oil naturally into harmless small molecules. The microbes release enzymes that degrade the oil into carbon dioxide and water. The inset photo shows rocks cleaned by bioremediation.

Currently, research is being conducted into speeding up this natural process. For example, by adding nutrients, such as oxygen, to the spill site, the bacteria will propagate more quickly, and a higher population will be available to break down the oil. Another strategy is to add more oil-digesting bacteria to the spill site. A third approach is to genetically engineer oil-digesting bacteria so that they are more effective consumers of the oil compounds. Also being explored is adding emulsifiers to the oil so that more surface area is available to the microbes, thereby increasing the rate of degradation.

The most effective method of dealing with an oil spill is that of prevention, yet if the unfortunate occurs, the use of living organisms is a safe and viable option.

REFLECT on your learning

1. Define biotechnology. Explain why ethanol, a product of the fermentation of sugar, is considered a biotechnological product.

2. Provide an example of a medical biotechnological application, an agricultural biotechnological application, and a forensic biotechnological application.

3. Compile a list of the obstacles that a biotechnologist may face when trying to insert a foreign gene into a bacterial cell, an animal cell, and a plant cell. Use your knowledge of plant, animal, and bacterial organelles and genome structure to formulate your answer.

▶ TRY THIS activity

DNA Palindromic Sequences

One of the most useful tools in biotechnology is that of restriction enzymes. Restriction enzymes scan double-stranded DNA for specific palindromic sequences and cut the DNA into two fragments at these positions. What is meant by a "palindromic sequence"? A palindrome is something that reads the same forward or backward. An example of a literary palindrome is "Madam I'm Adam." One of the longest palindromes in English is "A man, a plan, a canal, Panama." In DNA, a palindromic sequence occurs when its complementary DNA sequence is the same as the DNA sequence read backward. An example of a palindromic sequence in DNA is GAATTC, since the complementary sequence is CTTAAG.

• In your notebook, copy the following hypothetical sequence of DNA:

ATGAATTCAGGCTAGGATATCCCAAGCTCGAGAATATCCAAGATCTTGGCCACC

(a) Write out the sequence for the complementary strand.
(b) Identify the palindromic sequences in the original fragment.
(c) Using your knowledge of palindromes, write a 6-base-pair sequence that is a palindrome.
(d) For an added challenge, attempt the following:
  (i) Write a palindrome using the English language.
  (ii) The dates October 11, 2001, and October 22, 2001, were palindromes: 10 11 01 and 10 22 01. Using two digits for month, date, and year, determine when the next palindrome will occur.

Carpenters require tools such as hammers, screwdrivers, and saws; surgeons require scalpels, forceps, and stitching needles; and mechanics require hoists, wrenches, and pumps. These individuals use their implements to modify, deconstruct, or build a system that they are working with. Just like any other technicians, molecular biologists use tools to complete a project. The tools in their laboratories may aid them in investigating genetic disorders, altering the genetic makeup of organisms so that they produce useful products such as insulin, or analyzing DNA evidence in a criminal investigation. A key difference between the tools a molecular biologist uses and those of an electrician or a computer engineer, for example, is that, in many instances, the tools of the molecular biologist are living biological organisms or biological molecules. Molecular biological tools are dynamic, and, depending on the environment they are in, they behave accordingly. Using these tools, molecular biologists can cut, join, and replicate DNA. This makes it possible for molecular biologists to treat specific DNA sequences as modules and move them at will from one DNA molecule to another, forming **recombinant DNA**. In recombinant DNA technology, molecular biologists use tools and processes that enable them to analyze and alter genes and their respective proteins. This research has led to exciting new advances in biotechnology.

recombinant DNA fragment of DNA composed of sequences originating from at least two different sources

Restriction Endonucleases

A commonly used tool in molecular biology is **restriction endonucleases**. Restriction endonucleases, otherwise known as restriction enzymes, are molecular scissors that can cut double-stranded DNA at a specific base-pair sequence. Each type of restriction enzyme recognizes a characteristic sequence of nucleotides that is known as its **recognition site**. Molecular biologists can use these enzymes to cut DNA in a predictable and precise manner.

Most recognition sites are four to eight base pairs long and are usually characterized by a complementary palindromic sequence (**Table 1**). For example, the restriction enzyme *Eco*RI binds to the following base-pair sequence: 5′-GAATTC-3′/3′-CTTAAG-5′. It is palindromic because both strands have the same base sequence when read in the 5′ to 3′ direction. *Eco*RI scans a DNA molecule and only stops when it is able to bind to its recognition site. Once bound, it disrupts, via a hydrolysis reaction, the phosphodiester bond between the guanine and adenine nucleotides on each strand. Subsequently, the hydrogen bonds of complementary base pairs in between the cuts are disrupted (**Figure 1**). The result is a cut within a DNA strand, producing two DNA fragments where once there was only one.

restriction endonucleases enzymes that are able to cleave double-stranded DNA into fragments at specific sequences; also known as restriction enzymes

recognition site a specific sequence within double-stranded DNA, usually palindromic and consisting of four to eight nucleotides, that a restriction endonuclease recognizes and cleaves

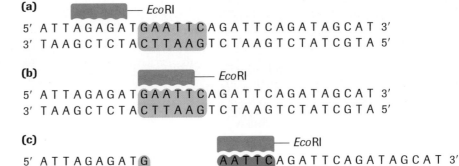

Figure 1
Cleavage of DNA sequence using restriction enzyme *Eco*RI:
(a) *Eco*RI scans the DNA molecule;
(b) *Eco*RI binds to the recognition site;
(c) *Eco*RI disrupts the phosphodiester bonds. Two fragments with complementary ends are produced.

Table 1 List of Restriction Enzymes and Their Respective Recognition Sites

Microorganism of origin	Enzyme	Recognition site	After restriction enzyme digestion
Escherichia coli	*Eco*RI	5′-GˇAATTC-3′ 3′-CTTAAG-5′	5′-G AATTC-3′ 3′-CTTAA G-5′
Serratia marcescens	*Sma*I	5′-CCCˇGGG-3′ 3′-GGGCCC-5′	5′-GGG CCC-3′ 3′-CCC GGG-5′
Arthrobacter luteus	*Alu*I	5′-AGˇCT-3′ 3′-TCGA-5′	5′-AG CT-3′ 3′-TC GA-5′
Streptomyces albus	*Sal*I	5′-GˇTCGAC-3′ 3′-CAGGTG-5′	5′-G TCGAC-3′ 3′-CAGGT G-5′
Haemophilus parainfluenzae	*Hind*III	5′-AˇAGCTT-3′ 3′-TTCGAA-5′	5′-A AGCTT-3′ 3′-TTCGA A-5′

The ends of DNA fragments produced from a cut by different restriction endonucleases differ, depending on where the phosphodiester bonds are broken in the recognition site. In the example in **Table 1**, *Eco*RI produces **sticky ends**; that is, both fragments have DNA nucleotides that are now lacking their respective complementary bases. These overhangs are produced because *Eco*RI cleaves between the guanine and the adenine nucleotide on each strand. Since A and G are at opposite ends of the recognition site on each of the complementary strands, the result is the overhang. Another restriction endonuclease, *Sma*I, produces **blunt ends**, which means that the ends of the DNA molecule fragments are fully base paired (**Table 1**). Since *Sma*I cleaves between the cytosine and guanine nucleotides and these nucleotides are directly opposite each other in their complementary strands, the result is a blunt cut without overhang. Restriction endonucleases that produce sticky ends are a generally more useful tool to molecular biologists than those that produce blunt ends. Sticky-end fragments can be joined more easily to other sticky-end fragments that have been produced by the same restriction endonuclease through complementary base pairing. Although another enzyme is needed to complete the phosphodiester linkage, the complementary bases on each fragment are able to hydrogen bond (anneal), facilitating the initial joining of the fragments.

Restriction endonucleases recognize four-base-pair to eight-base-pair sequences as their recognition site, resulting in a relatively low frequency of cuts in comparison to a two-base-pair recognition site. For example, the probability of finding a six-base-pair sequence, such as that of *Eco*RI, is $4 \times 4 \times 4 \times 4 \times 4 \times 4$, once in every 4096 nucleotides. The probability of finding two specific bases next to each other is 4×4, once in every 16 nucleotides. The decrease in occurrence of longer recognition sites results in fewer cuts. This is important in genetic engineering when a molecular biologist wants to excise a piece of DNA that includes a whole gene. If the cuts are more frequent, the gene itself may be cut and would have to be isolated in several fragments. If the recognition site comprises eight base pairs, its frequency is lower. This may also pose a problem, though, since much larger fragments than originally desired would be produced after digestion. Usually restriction enzymes that cleave at six-base-pair recognition sites result in a frequency of cleavage that can be used for many applications.

sticky ends fragment end of a DNA molecule with short single-stranded overhangs, resulting from cleavage by a restriction enzyme

blunt ends fragment ends of a DNA molecule that are fully base paired, resulting from cleavage by a restriction enzyme

A number of the tools that molecular biologists use to manipulate DNA are biological molecules themselves. Restriction enzymes are an example because they are isolated and purified solely from bacteria. In fact, one of the roles of restriction enzymes in a bacterium is to provide a crude immune system. When a virus (known as a bacteriophage) attacks a bacterial cell, it injects its DNA into the bacterium and leaves its protein coat outside the cell wall. As bacteriophage DNA is being injected into a bacterial cell, the bacteria's restriction endonuclease starts to scan this foreign DNA, looking for recognition sites. Because recognition sites inevitably exist in the foreign DNA, the restriction endonuclease cleaves the bacteriophage DNA into fragments (**Figure 2**). The bacteriophage DNA is inactivated. Fragments cannot be transcribed or translated into anything useful. The bacteria cell's own genome is protected and it can continue functioning as a bacterial cell, as opposed to a bacteriophage production factory.

Figure 2

Foreign DNA that enters bacteria is digested using a restriction enzyme. This prevents the foreign DNA from taking over the cellular machinery available in the bacteria to propagate itself and its products.

(a) A bacteriophage injects its DNA into a host bacterium. The host bacterium recognizes the DNA as foreign.

(b) *Eco*RI digests foreign DNA into small fragments, rendering it useless.

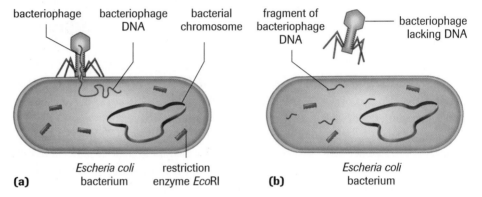

Restriction enzymes are named according to the bacteria from which they originate. For example, the restriction enzyme ***Bam*HI** is named as follows:

B   represents the genus *Bacillus*
am   represents the species *amyloliquefaciens*
H   represents the strain
I   means that it was the first endonuclease isolated from this strain

Following the same pattern, the rationale for the name of the restriction enzyme ***Hind*II** is the following:

H   represents the genus *Haemophilus*
in   represents the species *influenzae*
d   represents the strain Rd
II   means that it was the second endonuclease isolated from this strain

Generally speaking, the first letter is the initial of the genus name of the organism from which the enzyme is isolated. The second and third letters are usually the initial letters of the species name. The fourth letter indicates the strain, while the numerals indicate the order of discovery of that particular enzyme from that strain of bacteria.

The presence of restriction enzymes was first suspected in the early 1960s. Scientists noticed that when DNA from one strain of *E. coli* was inserted into a different strain of *E. coli*, the DNA did not survive but instead was cut into pieces. Two enzymes were soon discovered that could cleave DNA randomly. Hamilton Smith at John Hopkins University confirmed the presence of site-specific restriction enzymes with the accidental discovery of *Hind*II in 1970. In his research, Smith was trying to understand how some bacteria resist invasion by viruses. He found that when foreign DNA was inserted into the host bacteria, it was quickly degraded, yet the bacteria's own DNA remained intact. He hypothesized that an enzyme exists within the host bacterium that is responsible for this degradation. Further research led to the isolation of *Hind*II. Smith shared the Nobel Prize in 1978 for his discovery. Since then, more than 2500 restriction endonucleases with specificity for about 200 different target sites have been isolated from prokaryotic organisms. Today, approximately 200 of these restriction endonucleases are available commercially to molecular biologists for use in laboratory investigations.

> ▶ *Practice*

Understanding Concepts

1. The following sequence of DNA was digested with the restriction endonuclease *Sma*I:

   5′-AATTCGCCCGGGATATTACGGATTATGCATTATCCGCCCGGGATATTTTAGCA-3′

   3′-TTAAGCGGGCCCTATAATGCCTAATACGTAATAGGCGGGCCCTATAAAATCGT-5′

   *Sma*I recognizes the sequence CCCGGG and cuts between the C and the G.
   (a) Identify the location of the cuts.
   (b) How many fragments will be produced if *Sma*I digests this sequence?
   (c) What type of ends does *Sma*I produce?

2. *Hind*III recognizes the sequence AAGCTT and cleaves between the two A's. What type of end is produced by cleavage with *Hind*III?

3. Explain why restriction endonucleases are considered to be molecular tools.

4. Calculate the number of expected cuts in a DNA sequence of 75 000 base pairs by a restriction enzyme that has a six-base-pair recognition site.

5. Identify the palindromic sequences in the following DNA fragment. You will need to write the complementary strand to do so.

   GCGCTAAGGATAGCATTCGAATTCCCAATTAGGATCCTTTAAAGCTTATCC

Methylases

Restriction endonucleases must be able to distinguish between foreign DNA and the genetic material of their own cells; otherwise a bacterium's DNA would be in jeopardy of being cleaved by its own immune system. **Methylases** are specific enzymes found in prokaryotes and eukaryotes. In prokaryotes, they modify the recognition site of a respective restriction endonuclease by placing a **methyl group** on one of the bases, preventing the restriction endonuclease from cutting the DNA into fragments. *Eco*RI methylase, for example, adds a methyl group to the second adenine nucleotide in the *Eco*RI recognition site (**Figure 3**), preventing *Eco*RI from cutting the DNA. When foreign DNA is introduced into the bacterium, it is not methylated, rendering it defenceless against the bacterium's restriction enzymes.

Methylases are important tools for a molecular biologist when working with prokaryotic organisms. They allow the molecular biologist to protect a gene fragment from being cleaved in an undesired location.

DNA Ligase

If genes can be cut out of source DNA, there must also be a way to join them to foreign target DNA. If two fragments of nucleic acids have been generated using the same restriction enzyme, they will naturally be attracted to each other at their complementary sticky ends. Hydrogen bonds will form between the complementary base pairs, yet this is not a stable arrangement. The phosphodiester linkage between the backbones of the double strands must be reformed. DNA ligase is the enzyme used for joining the cut strands of DNA together. Using a condensation reaction, DNA ligase drives out a molecule of water and reforms the phosphodiester bond of the backbone of the DNA. Joining strands with blunt ends using DNA ligase is very inefficient. Molecular biologists use **T4 DNA ligase**, an enzyme that originated from the T4 bacteriophage, to join blunt ends together (**Figure 4**).

DID YOU *KNOW* ?

Eukaryotic Methylation
The purpose of methylases in eukaryotic cells differs from that of prokaryotic cells. Methylases in eukaryotes are connected with the control of transcription. In addition, approximately 2% of mammalian ribosomal RNA is methylated after it is transcribed.

methylases enzymes that add a methyl group to one of the nucleotides found in a restriction endonuclease recognition site, altering its chemical composition

methyl group CH₃

```
         CH₃
          |
     G A A T T C
     C T T A A G
              |
             CH₃
```

Figure 3
At a methylated *Eco*RI site, *Eco*RI restriction enzyme is no longer able to cut.

T4 DNA ligase an enzyme used to join together DNA blunt or sticky ends

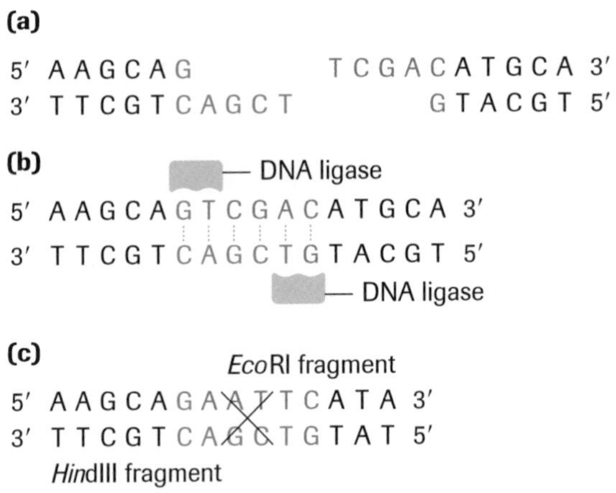
(a)
5′ A A G C A G T C G A C A T G C A 3′
3′ T T C G T C A G C T G T A C G T 5′
(b)
DNA ligase
5′ A A G C A G T C G A C A T G C A 3′
3′ T T C G T C A G C T G T A C G T 5′
DNA ligase
(c)
EcoRI fragment
5′ A A G C A G A A T T C A T A 3′
3′ T T C G T C A G C T G T A T 5′
HindIII fragment

Figure 4
DNA ligase is able to join complementary sticky ends produced by the same restriction enzyme via a condensation reaction.
(a) Complementary sticky ends produced by *Hin*dIII
(b) Hydrogen bonds form between complementary bases. DNA ligase reconstitutes the phosphodiester bond in DNA backbones.
(c) If fragments are not complementary, then hydrogen bonds will not form.

Practice

Understanding Concepts

6. Explain why cut fragments that result in sticky ends are easier to join than blunt-end fragments. Use a diagram to help you explain.
7. Distinguish between DNA ligase and T4 DNA ligase.
8. Annealing together sticky-end fragments that have been produced by the same restriction endonuclease is much more efficient compared with annealing blunt-end fragments produced by the same restriction endonuclease. Explain.
9. Describe the mechanism used by bacteria to protect their chromosomal DNA from damage by restriction endonucleases.
10. State the type of reaction used by DNA ligase to anneal two DNA fragments together. List the products of this reaction.

Gel Electrophoresis

Once the desired gene has been excised from its source DNA, it must be separated from the remaining unwanted fragments. DNA fragments can be separated using the process of **gel electrophoresis**. Gel electrophoresis takes advantage of the chemical and physical properties of DNA (**Figure 5**).

gel electrophoresis separation of charged molecules on the basis of size by sorting through a gel meshwork

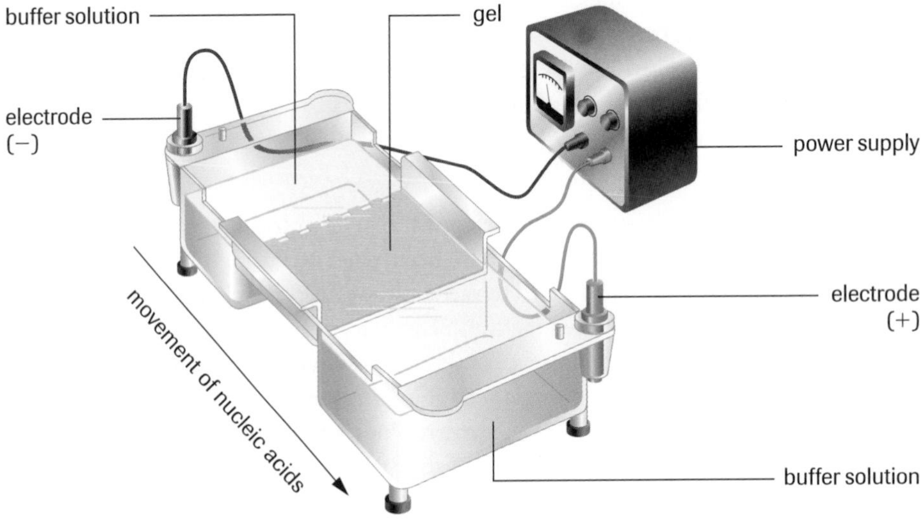
buffer solution
gel
electrode (−)
power supply
electrode (+)
movement of nucleic acids
buffer solution

Figure 5
In a common gel electrophoresis setup, DNA is loaded into wells at one end of the gel and then migrates toward the positive electrode at the opposite end. The rate of migration of fragments varies with size.

DNA is negatively charged. Each nucleotide possesses a constituent phosphate group that carries a net charge of -1. In addition, the molar mass of each nucleotide is relatively consistent. The small difference in molar mass between nucleotides is negligible because of the ratio of purines to pyrimidines in a DNA molecule. These two properties contribute to each nucleotide having the same charge-to-mass ratio. Hence, the only difference between two fragments of DNA that are of differing lengths is the number of nucleotides. This difference can be exploited to separate DNA fragments according to size.

Gel electrophoresis can be envisioned as a molecular sieve. DNA that has been subjected to restriction endonuclease digestion will be cleaved into fragments of different lengths. DNA fragments migrate through the gel at a rate that is inversely proportional to the logarithm of their size. The shorter the fragment is, the faster it will travel because of its ability to navigate through the pores in the gel more easily than a large fragment can. Larger fragments are hampered by their size. An analogous situation would be a group of people running through a forest in which the trees are equidistant from each other. If 30 people are forced to hold hands while they run through the forest, it will take them a very long time to navigate from one end to the other. If people are allowed to run in triplets, they will travel a lot faster. Finally, an individual person will have absolutely no problem getting through. Hence, the longer a nucleotide chain, the longer it takes for the migration.

Gel electrophoresis takes advantage of DNA's negative charge. A solution containing different-size fragments to be separated is placed in a well (**Figure 6**). A well is a depression at one end of the gel. The gel itself is usually a square or rectangular slab and consists of a buffer containing electrolytes and **agarose**, or possibly **polyacrylamide**. The DNA solution containing fragments to be separated is mixed with a loading dye containing glycerol. The loading dye allows visualization of the DNA solution. Glycerol is a heavy molecule that causes the DNA solution to sink down into the well. The gel is loaded while it is submerged in a tray containing an electrolytic solution called the buffer. Using direct current, a negative charge is placed at one end of the gel where the wells are, and a positive charge is placed at the opposite end of the gel. The electrolyte solution conveys the current through the gel. The negatively charged DNA will migrate toward the positively charged electrode, with the shorter fragments migrating faster than the longer fragments, achieving separation. Small molecules found within the loading dye migrate ahead of all the DNA fragments. Since the small molecules can be visualized, the electrical current can be turned off before they reach the end of the gel.

agarose gel-forming polysaccharide found in some types of seaweed that is used to form a gel meshwork for electrophoresis

polyacrylamide artificial polymer used to form a gel meshwork for electrophoresis

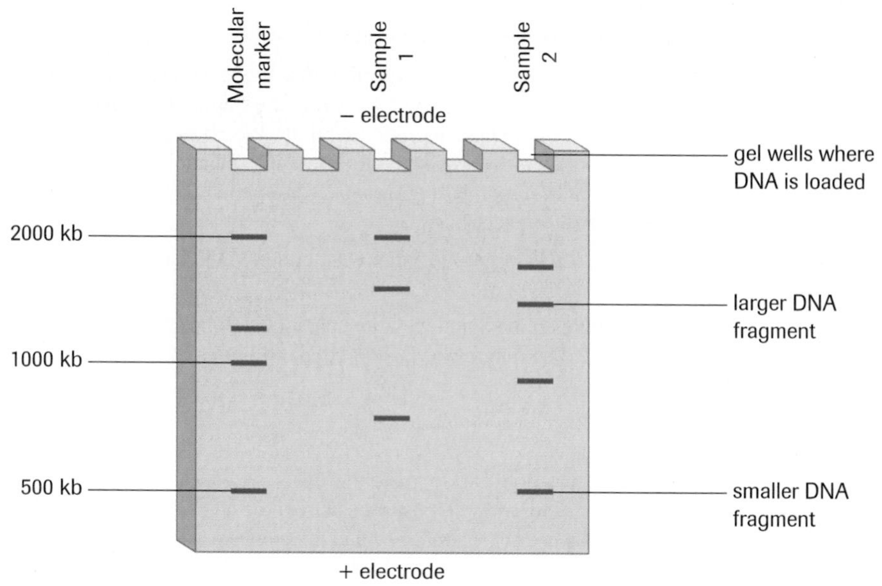

Figure 6
All fragments maintain the same charge-to-mass ratio and experience resistance from the gel. Smaller fragments experience less resistance and travel farther than larger fragments. A molecular marker is run alongside to determine the size of the fragments.

ethidium bromide a carcinogenic, flat molecule that inserts itself among the rungs of the ladder of DNA and fluoresces under UV light

fluoresces glows under UV light because of the excitation of a molecule's electrons

---

🔬 INVESTIGATION 6.1.1

Restriction Enzyme Digestion of Bacteriophage DNA (p. 310)
Using restriction enzymes and gel electrophoresis, molecular biologists are able to excise and isolate target sequences from DNA. How would the banding patterns compare if the same fragment of DNA were digested with different restriction enzymes? In Investigation 6.1.1, you will conduct gel electrophoresis of bacteriophage DNA that has been digested with restriction enzymes.

Figure 7
Example of DNA banding pattern produced by the digestion of DNA using restriction enzymes and then the separation of DNA fragments by gel electrophoresis

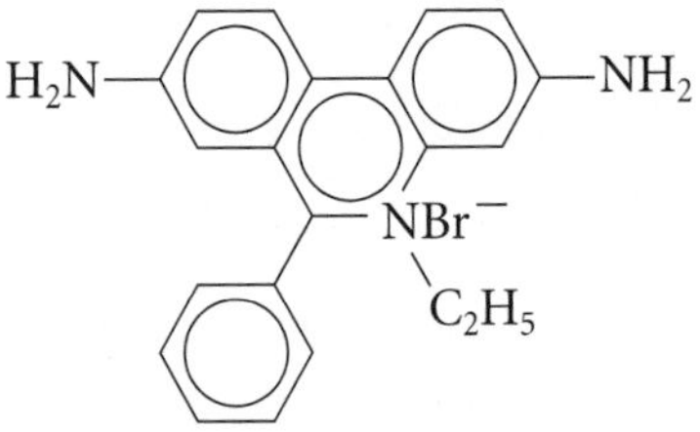

Once gel electrophoresis is complete, the DNA fragments are made visible by staining the gel (**Figure 7**). The set of fragments generated with a particular restriction enzyme produces a banding pattern characteristic for that DNA. The most commonly used stain is **ethidium bromide** (**Figure 8**). Ethidium bromide is a flat molecule that **fluoresces** under ultraviolet (UV) light and is able to insert itself among the rungs of the ladder of DNA. When the gel is subjected to UV light, the bands of DNA are visualized because the ethidium bromide is inserted among the nucleotides. The size of the fragments is then determined using a molecular marker as a standard. The molecular marker, which contains fragments of known size, is run under the same conditions (in the same gel) as the digested DNA. The researcher can then plot the log of the known fragment sizes versus the distance travelled from the wells. The resulting graph can be used to determine the size of the unknown fragments through interpolation. In addition to being able to estimate the size of a desired fragment, a researcher can compare fragment sizes from different sources and can also excise the desired fragment out of the gel. Once the region of the gel containing the desired fragment has been cut out, the region can be purified for further use.

Gel electrophoresis is not limited to the separation of nucleic acids but is also commonly applied to proteins. Proteins are usually run on polyacrylamide gels, which have smaller pores, because proteins are generally smaller in size than nucleic acids. 🔬

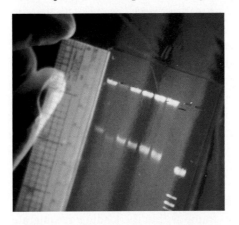

Figure 8
Ethidium bromide molecule

▶ *Practice*

Understanding Concepts

11. Explain why DNA fragments can be separated using an electrical current.

12. The following fragments were obtained after digestion of a sequence using restriction enzymes: 5.3 kb, 2.1 kb, 6.8 kb, 3.6 kb, 1.1 kb, 2.2 kb. Order the fragments, moving from the fastest migration to the slowest migration on a gel. Which two fragments may be difficult to isolate on a gel? Why? Sketch the resulting gel after running the fragments on it. Assume that the fragments were all run in one lane. Label the positions of the positive and negative electrodes.

13. If the concentration of agarose is increased, what will happen to the isolating power of the gel? Explain.

14. (a) Why is ethidium bromide an ideal molecule for staining DNA gels?
    (b) What precautions should be taken when using ethidium bromide?

Applying Inquiry Skills

15. Predict what would happen if
    (a) an electrolyte solution was not used to make the gel;
    (b) the electrodes were reversed, with the positive electrode being closest to the wells;
    (c) the gel constituency was not consistent, resulting in some areas of the gel having a higher agarose concentration and others having a lower agarose concentration.

Plasmids

As discussed, a desired sequence of nucleotides from source DNA can be cut out of the genome, separated, and annealed into another DNA molecule. In some cases, gene fragments are excised with the goal of having the gene expressed as a useful protein. However, for a gene to be expressed, cellular machinery is required. Bacteria often provide the necessary machinery.

The gene for insulin, for example, has been isolated and can now be expressed by bacterial cells. Insulin is a protein produced by the pancreas that controls the levels of sugar in the blood. Using the technology of genetic engineering, the human insulin gene is inserted into bacterial cells that express the gene and build the protein insulin. A large percentage of people with diabetes have benefited from the supply of insulin produced by bacteria. Originally, these patients depended on insulin that was extracted from animals, which can cause allergic reactions.

Bacteria are able to express foreign genes inserted into **plasmids** (**Figure 9**). Plasmids are small, circular, double-stranded DNA molecules lacking a protein coat that naturally exist in the cytoplasm of many strains of bacteria. Plasmids are independent of the chromosome of the bacterial cell and range in size from 1000 to 200 000 base pairs. Using the enzymes and ribosomes that the bacterial cell houses, DNA contained in plasmids can be replicated and expressed.

The bacterial cell benefits from the presence of plasmids. Plasmids often carry genes that express proteins able to confer antibiotic resistance (**Figure 10**). They also protect bacteria by carrying genes for resistance to toxic heavy metals, such as mercury, lead, or cadmium. In addition, some bacteria carry plasmids possessing genes that enable the bacteria to break down herbicides, certain industrial chemicals, or the components of petroleum. The relationship between bacteria and plasmids is endosymbiotic; both the bacteria and the plasmid benefit from the mutual arrangement.

plasmids small circular pieces of DNA that can exit and enter bacterial cells

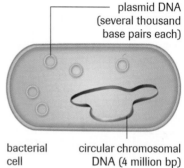

Figure 9
Chromosomal and plasmid DNA coexist in an endosymbiotic relationship.

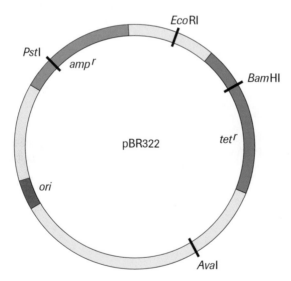

Figure 10
Genetically engineered plasmid pBR322 has two genes that confer antibiotic resistance. One gene confers tetracycline resistance (*tet^r*) and another gene codes for ampicillin resistance (*amp^r*). Plasmids also contain an origin of replication (*ori*), a region that directs the plasmid's replication. Plasmids contain recognition sites for restriction enzymes. A few of the restriction enzyme recognition sites are shown.

DID YOU KNOW ?

Plasmids: Beneficial Guests
Japanese scientists were the first to discover plasmids that carry genes for multiple drug resistance. The bacterium *Shigella*, which causes dysentery, developed resistance to as many as four antibiotics, including tetracycline, streptomycin, chloramphenicol, and the sulfonamides. The multidrug resistance was due to a plasmid within the bacterium that carried genes for resistance and could be passed naturally from bacterium to bacterium.

copy number number of copies of a particular plasmid found in a bacterial cell

🔬 INVESTIGATION 6.1.2

Calcium Chloride Transformation of *E. coli* with Ampicillin-Resistant Plasmid (p. 314)
Plasmids can enter bacterial host cells using the classical calcium chloride transformation method. How efficient is this method in promoting the entry of a foreign plasmid to a host bacterium? In Investigation 6.1.2, you will use the calcium chloride transformation protocol and calculate transformation efficiency.

multiple-cloning site region in plasmid that has been engineered to contain recognition sites of a number of restriction endonucleases

Plasmids also possess a characteristic **copy number**. The higher the copy number, the higher the number of individual plasmids in a host bacterial cell. If more copies of a plasmid exist, more protein will be synthesized because of the larger number of gene copies carried by the plasmid. The number of copies plays a role in the phenotypic manifestation of a gene. For example, the more copies of an antibiotic-resistance gene there are, the higher the resistance to the antibiotic. 🔬

Restriction endonucleases are used to splice a foreign gene into a plasmid. Artificial plasmids have been engineered to contain a unique region that can be cut by many restriction enzymes. The recognition sites of many restriction enzymes have been positioned very close together in this one area and are not found anywhere else on the plasmid's DNA sequence. This site is called the **multiple-cloning site**. Recognition sites are present only once in the plasmid and, therefore, the restriction enzyme can only make one cut in the DNA. It would be of no value to a molecular biologist if the plasmid were cut into many fragments by the same restriction enzyme. One cut results in the circular plasmid becoming linear.

If the foreign gene has been excised using the same restriction enzyme, it will possess the same complementary ends as the linearized plasmid. When placed together, the sticky fragments will anneal. Then the foreign gene will permanently become part of the plasmid after the phosphodiester bonds are reestablished (**Figure 11**). For example, if the desired foreign DNA fragment is excised out of its source DNA with *Eco*RI, then the circular plasmid is cut open at the multiple cloning site by digesting with the same restriction enzyme (in this case, *Eco*RI). The desired fragment and plasmid fragments are brought into proximity by placing them into the same solution, where they anneal

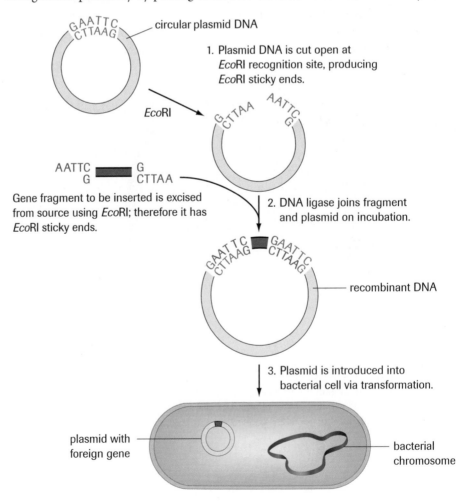

Figure 11
A foreign gene is introduced into a plasmid. The plasmid is now an example of recombinant DNA, which can be introduced into a bacterial cell to produce numerous copies of the gene.

because of the complementary sticky ends produced by digestion with the same enzyme. DNA ligase is then added to re-form the phosphodiester bonds between the fragments, resulting once again in a circular piece of DNA that now carries the foreign gene fragment. The plasmid is now recombinant DNA, a combination of the original plasmid DNA and the foreign DNA. The plasmid may now be introduced into a bacterial cell, where it replicates to form many copies within the cell. The gene has been **cloned** because, as the plasmid replicates, many copies of the recombinant DNA are produced, each of which includes a copy of the original inserted gene.

cloned a fragment of DNA that has been introduced into a foreign cell, resulting in exact copies of the original DNA fragment being made when the foreign cell replicates and divides

> ▶ *Practice*

Understanding Concepts

16. Define *copy number, multiple cloning site, recombinant DNA,* and *plasmid*.

17. Explain why plasmids are able to exit and enter bacterial cells easily compared with bacterial chromosomes.

18. Viruses, like plasmids, can act as DNA vectors. Given your knowledge of viral infection, explain why viruses would also serve as good molecular tools for introducing foreign genes into cells.

ACTIVITY 6.1.1

Constructing a Plasmid Map (p. 316)

To incorporate a gene into a plasmid and produce recombinant DNA, the molecular biologist requires a restriction enzyme map of the plasmid. How is the information that is gained from restriction enzyme digests used to construct plasmid maps? Activity 6.1.1 walks you through the process of plasmid mapping.

Transformation

Because of the small size of plasmids, bacteria are able to take them up under specific conditions. The introduction of DNA from another source is known as **transformation**, and a bacterium that has taken in a foreign plasmid is referred to as being transformed. Hence, plasmids can be used as **vectors** to carry a desired gene into a **host cell**.

If a bacterium readily takes up foreign DNA, it is described as a **competent cell**. Most bacteria are not naturally competent but can be chemically induced in the laboratory to become so with the aid of calcium chloride. Calcium chloride is a salt that ionizes into Ca^{2+} and Cl^-. Bacterial cells are suspended in a solution of calcium chloride at 0°C. The bacterial membrane contains exposed phosphates that are negatively charged. The positively charged calcium ions stabilize the negative charges of the phosphates. In addition, the low temperature "freezes" the cell membrane, which is fluid, making it more rigid.

Now that the cell membrane has been stabilized both physically and chemically, the plasmid DNA is introduced into the solution. The negatively charged phosphates of the DNA are also stabilized by the calcium cations (**Figure 12**). At this point, the entire solution is

transformation introduction of foreign DNA, usually by a plasmid or virus, into a bacterial cell

vectors vehicles by which DNA may be introduced into host cells

host cell a cell that has taken up a foreign plasmid or virus and whose cellular machinery is being used to express the foreign DNA

competent cell a cell that readily takes up foreign DNA

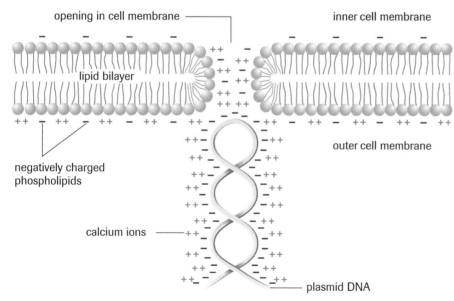

Figure 12
Calcium ions neutralize the negative charge from the phosphate group on the plasmid DNA and on the phospholipids found in the cell membrane, minimizing the repelling effect of like charges. DNA can then enter the bacterial cell more easily.

subjected to a quick heat shock treatment of 42°C that lasts for approximately 90 seconds and creates a draft. The outside environment of the cell is now at a slightly higher temperature than the inside of the cell. The resulting draft sweeps the plasmids into the bacterial cell through pores in its membrane. Finally, the bacterial cells are allowed to recover from the experience. They are incubated in a nutrient media suspension at a temperature of 37°C.

Selective plating is a method that can be used to isolate the cells with recombinant DNA (Investigation 5.4.1 in Chapter 5) (**Figure 13**). The vectors used for cloning carry an antibiotic-resistance gene. Therefore, if the transformation is a success, the bacteria will be able to grow on media that contain the antibiotic. If no growth is observed, the bacteria were not transformed and were eliminated by the antibiotic.

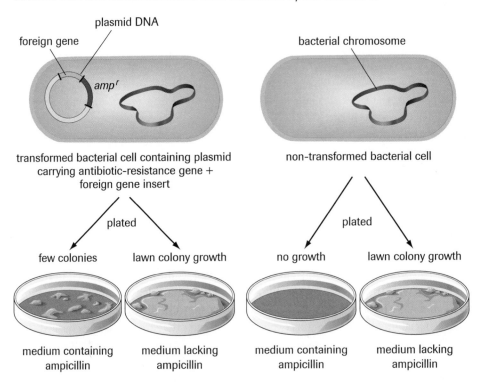

Figure 13
The success of a transformation can be assessed using selective plating. Positive and negative controls would also have to be conducted to increase the validity of the experiment.

It is also necessary to check that the foreign gene actually exists in the transformed bacteria. The bacteria may exhibit antibiotic resistance by taking up the plasmid that has failed to have the foreign gene fragment inserted. Colonies of bacteria are selected that have been transformed. Individual colonies are removed from the plate and allowed to proliferate in liquid media until enough bacterial cells can be harvested to extract a suitable amount of plasmid DNA. The extracted plasmid DNA is subjected to a restriction enzyme digestion to release the cloned fragment from the vector. If the expected pattern of bands is observed on the gel, then that colony does carry a recombinant DNA plasmid.

The calcium chloride method is the classical method of transforming cells. Today, electroporators—chambers that subject the bacteria to an electric shock—are also used. The electric shock loosens the structure of the cell walls and allows foreign DNA to enter. Plant cells provide more of a challenge to molecular biologists because of their rigid cell wall. Modern electrical "gene guns" are used to "shoot" DNA through the cell wall and membrane. The initial projectile is a droplet of water, which is charged and fired down a tube via an electrical discharge. The water droplets hit a carrier sheet that contains the DNA wrapped around gold particles. The force of the impact causes the DNA to be propelled toward the target cells, where it is able to penetrate cell walls and cell membranes.

▶ *Practice*

Understanding Concepts

19. Describe the process of transformation using calcium chloride, a gene gun, and an electroporator.

20. Explain why plasmids are considered vectors.

21. Plasmids usually carry a marker gene, such as antibiotic resistance. Explain how these marker genes may be used to check if an inserted gene has been successfully introduced into a plasmid and a bacterial cell.

Applying Inquiry Skills

22. Predict what would happen in a calcium chloride transformation if
    (a) neither the DNA nor the bacterial cell was subjected to the calcium chloride treatment;
    (b) only the DNA was subjected to the calcium chloride treatment.

SUMMARY   *Biotechnological Tools and Techniques*

Table 2   Key Tools of Molecular Biology

Tool	Use	Example
restriction endonuclease	bacterial enzyme that cleaves DNA sequences at a specific recognition site	*Bam*HI recognition site: 5′-GGATCC-3′ 3′-CCTAGG-5′ DNA sequence before cleavage: 5′-TCAGCGGATCCCAT-3′ 3′-AGTCGCCTAAGCTA-5′ DNA sequence after cleavage with *Bam*HI 5′-TCAGCG GATCCCAT-3′ 3′-AGTCGCCTAG GGTA-5′
methylase	bacterial enzymes that add a methyl group to recognition sites to protect DNA from cleavage by restriction enzyme	Methylase *Bam*HI (M. *Bam*HI) adds methyl group ($-CH_3$) to second guanine nucleotide recognition site: 5′-GGATCC-3′ 3′-CCTAGG-5′ DNA sequence no longer cleaved by *Bam*HI methyl group changes recognition site
DNA ligase	enzyme that joins complementary fragments by reconstituting phosphodiester bond of DNA backbone	DNA fragments before subjection to DNA ligase 5′-ATAGTG-3′ 5′-AATTCGG-3′ 3′-TATCACTTAA-5′ 3′-GCC-5′ DNA fragments after subjection to DNA ligase 5′-ATAGTGAATTCGG-3′ 3′-TATCACTTAAGCC-5′ two fragments are joined

Table 2 *(continued)*

Tool	Use	Example
gel electrophoresis	process by which DNA fragments of different lengths are separated by electrical current, negative charge of DNA, and constant charge-to-mass ratio	DNA fragments of 750, 1250, 1500, and 2000 base pairs run on a gel
plasmid	small circular DNA that has the ability to enter and replicate in bacterial cells and, therefore, can be used as a vector to introduce new genes into a bacterial cell	plasmid containing multiple cloning site, ampicillin-resistant gene, and other restriction enzyme sites

The Major Steps in the Cloning of DNA

Figure 14 demonstrates the following steps in DNA cloning:

1. Generation of DNA fragments using restriction endonucleases
   - Appropriate restriction endonucleases need to be used to ensure that the gene fragment in question is excised completely from the source DNA.
   - More than one restriction endonuclease may be used at one time.

2. Construction of a recombinant DNA molecule
   - The target gene fragment is ligated to a DNA vector (plasmids are one example) and is now recombinant DNA.
   - The vector can replicate autonomously in an appropriate host organism.

3. Introduction into a host cell
   - Bacterial host cells can be manipulated to take up the recombinant DNA using electroporators, gene guns, or classical transformation protocols, such as calcium chloride.
   - Once the bacterium takes up the recombinant DNA, it is referred to as being transformed.

4. Selection
   - Cells that have been successfully transformed with the recombinant DNA must be isolated.
   - The desired cells are usually chemically selected by the presence of a marker (e.g., antibiotic resistance) on the vector.
   - Growth of colonies on media containing the chemical indicates successful transformation of the recombinant DNA vector.
   - Individual colonies are isolated from media containing the chemical and are grown in culture to produce multiple copies (clones) of the incorporated recombinant DNA.

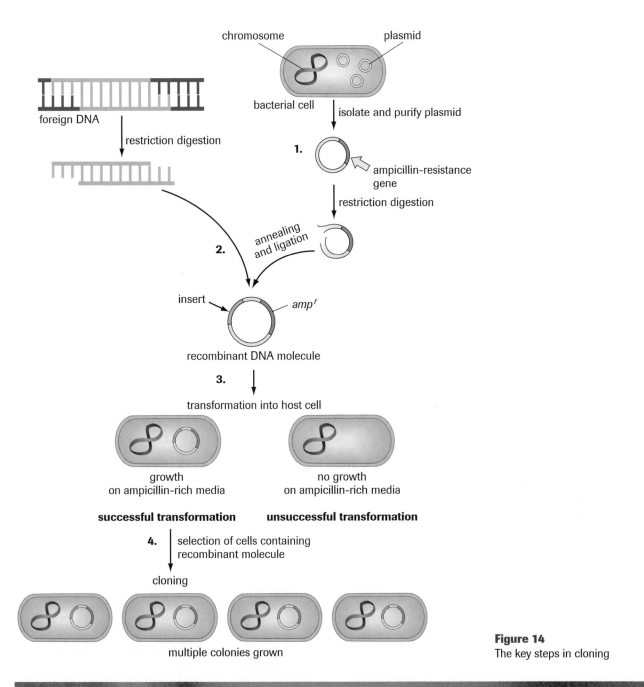

Figure 14
The key steps in cloning

Understanding Concepts

1. Define *restriction endonuclease, methylase,* and *transformation*.

2. Restriction endonucleases are found in many species of bacteria.
   (a) Describe their role and function in a bacterial cell.
   (b) How does the role of restriction endonucleases in nature differ from the role of restriction endonucleases in the laboratory setting?

3. (a) Distinguish between blunt ends and sticky ends.
   (b) Provide a few examples of restriction enzymes that produce blunt ends and of restriction enzymes that produce sticky ends.

(c) Name the bonds that are formed when sticky ends anneal and those that are formed when blunt ends anneal.

4. Explain why different DNA molecules that were cut with the same restriction enzyme carry the same nucleotide sequence at the cut ends. Use a diagram to support your explanation.

5. Define *recognition site*. Using examples to support your answer, depict the palindromic nature of recognition sites.

6. Restriction enzymes cut at recognition sites that are usually six to eight base pairs in length. Provide reasons why a two-base-pair recognition site would be too short ▶

to be useful and a 14-base-pair recognition site may be too long to be useful in the field of genetic engineering.

7. When molecular biologists try to introduce a foreign gene into a plasmid, they are more efficient if the corresponding ends of the plasmid and foreign gene have been produced by the same restriction enzyme and have sticky ends. Provide reasons why this might be the case.

8. Outline the process of gel electrophoresis. List the properties of DNA that are used to achieve band separation.

9. Why is it necessary to stain gels with ethidium bromide after they have been run? How does ethidium bromide help in visualizing the location of DNA on a gel?

10. Why is it important that the multiple cloning site contains a restriction enzyme recognition site that is only found once in the plasmid? Describe what would happen if one of the recognition sites found in the multiple cloning site was also found in another location in the plasmid.

11. Identify the challenges and limitations that the field of genetic engineering would face if the following were true:
    (a) Restriction endonucleases only recognize three-base-pair recognition sites.
    (b) Plasmids do not replicate autonomously from the main chromosome of a bacterial cell.
    (c) Plasmids have very low copy numbers.

12. (a) Explain what is meant when a cell is competent.
    (b) Describe the process that molecular biologists apply to induce competency in cells.
    (c) What role in this process is played by calcium chloride? by heat shock?

13. Why are viruses, as opposed to plasmids, preferred vectors in plant cells?

Applying Inquiry Skills

14. The gel shown in **Figure 15** was run after bacterial DNA was digested using restriction enzymes. In your notebook, indicate on the gel
    (a) where the positive electrode was located;
    (b) where the negative electrode was located;
    (c) the location of the largest band;
    (d) the location of the smallest band;
    (e) the number of cuts that were made on the linear fragment of DNA to produce this number of bands.

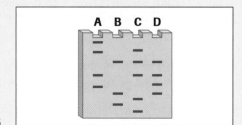

Figure 15

15. An unknown plasmid was digested with *Eco*RI, *Bam*HI, and *Pst*I. The results of the digestions are shown in **Table 1**.

Table 1

*Eco*RI (base pairs)	*Bam*HI (base pairs)	*Pst*I (base pairs)
800	2400	750
1600		850
		800

(a) How many times does each of the enzymes cut this plasmid?
(b) What is the total size of this plasmid?

16. The plasmid shown in **Figure 16** was digested using different restriction enzymes whose sites have been mapped. The plasmid is 7896 base pairs in length and contains a gene that confers antibiotic resistance to tetracycline.

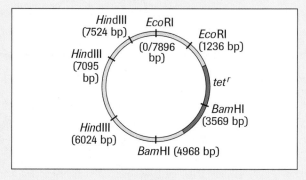

Figure 16

(a) Determine the size and number of fragments that would be produced if the plasmid was digested with *Eco*RI; *Bam*HI; *Hind*III; *Eco*RI and *Hind*III; *Eco*RI, *Hind*III, and *Bam*HI.
(b) Which one of the cuts would inactivate the antibiotic resistance. Why?
(c) Sketch the resulting gel if the DNA were digested with all three of the enzymes.
(d) The DNA was treated with M. *Eco*RI methylase before it was digested with all three enzymes. Sketch the resulting gel.

17. A fragment of DNA is composed of 100 000 base pairs. Calculate the expected frequency of cuts if a restriction enzyme is used that has a recognition site six base pairs long.

18. Calcium chloride is the salt of choice used for the purpose of neutralizing the phosphate groups of both the DNA and the lipids of cell membranes to facilitate transformation. Make a list of other chloride-based salts that would probably work in the same capacity.

Making Connections

19. The transformation of bacterial cells is much more successful than the transformation of eukaryotic animal cells from humans and other higher-order organisms. Using your knowledge of genetic differences between prokaryotes and eukaryotes, provide reasons why this is the case.

The principles of **genetic engineering** were experimentally established in the early 1970s by two American researchers. Stanley Cohen, who was investigating plasmids, collaborated with Herbert Boyer, a researcher who was interested in restriction endonucleases. Cohen and Boyer designed and executed a series of experiments that resulted in a method of selecting, recombining, and introducing new genes into bacteria via plasmid vectors. Their experiments had no commercial applications at the time, but, within a decade, this situation changed. In 1976, Herbert Boyer cofounded Genentech, the first biotechnology company to go public on the stock market. In 1978, somatostatin was the first human hormone produced by this new technology. The techniques developed by Cohen and Boyer are now commonplace in molecular biology laboratories worldwide. Cohen and Boyer's monumental discovery is one of the most important in the history of biomedical research.

Many new products have been developed through genetic engineering, some of which are of medical significance, such as insulin. More than 90% of diabetics rely on the supply of human insulin produced by bacteria. Other products have an agricultural application. Frost damage has been minimized by spraying crops with bacteria that have been engineered to disable them from forming an ice nucleation factor. Ice nucleation factor is a protein found on bacterial coats that allows ice to seed and crystallize.

The growth hormone **somatropin** is also being synthesized in large quantities using the tools of molecular biology. This genetically engineered product is a drug that is identical to the human growth hormone somatotropin. It is used to treat human growth deficiency that results from a genetic mutation causing dwarfism and Turner's syndrome. Since it also helps build muscle, it is used as a treatment for AIDS-associated wasting syndrome. Before the hormone was commercially available as a product of genetic engineering, human growth hormone was obtained from the pituitary glands of cadavers. The treatment was accompanied by risks since the cadaver pituitaries had the potential to carry infectious disease, such as Creutzfeldt–Jakob disease, a fatal brain disease. Growth hormone from animals was not a viable option, because it does not function in humans. Genetic engineers realized that if bacteria could be induced to produce human growth hormone, an abundant and safe supply would become available (**Figure 1**).

Somatropin, which is synthesized by recombinant *E. coli* bacteria, consists of 191 amino acids. A precursor gene that codes for 217 amino acids is found in the human genome. The first 26 amino acids serve as a signal sequence for the human growth hormone and are removed in the endoplasmic reticulum after translation, leaving the remaining 191 amino acids. Genetic engineers excised the human growth hormone gene using restriction enzymes. The cuts were made in such a fashion that the base pairs that coded for the first 26 codons were also removed, because bacteria do not remove signal peptides as eukaryotic cells do. The gene was inserted after the *lac* promoter in a plasmid and introduced into *E. coli* cells. The *E. coli* cells are grown in a medium rich in the chemical isopropyl thiogalactoside (IPTG). IPTG is an inducer of the *lac* promoter that is not consumed by bacterial cells. Therefore, in the presence of IPTG, bacteria are induced to produce somatropin. In the absence of IPTG, somatropin is not produced, which allows biotechnology companies to control the amount of hormone produced. The somatropin that is harvested from bacteria is sold to patients for medical purposes.

genetic engineering altering the sequence of DNA molecules

DID YOU KNOW?

DNA Experiments Suspended
After the first experiments involving recombinant DNA were announced in the early 1970s, scientists around the world recognized the potential power and risks in the new technology. In 1974, Stanford University biochemist Paul Berg drafted a letter calling for a voluntary world moratorium on recombinant DNA work (now known as genetic engineering) until public safety could be assured. More than 140 scientists, legal experts, bioethicists, and members of the media from 17 nations met at the Asilomar Conference in California to establish regulations for the control of recombinant DNA. The regulations were adopted in June 1976, ending the moratorium.

somatropin a drug identical to human growth hormone (somatotropin) used to treat growth deficiency

Recently, controversy has surrounded the availability and use of somatropin. The ability of somatropin to build muscle cells has made it attractive to athletes, a number of whom have been found to be using somatropin (a banned substance in sporting events) during international competitions. Another form of the drug, bovine somatotropin, is used in some American states to boost milk production in cows, but bovine somatotropin has been banned in Canada.

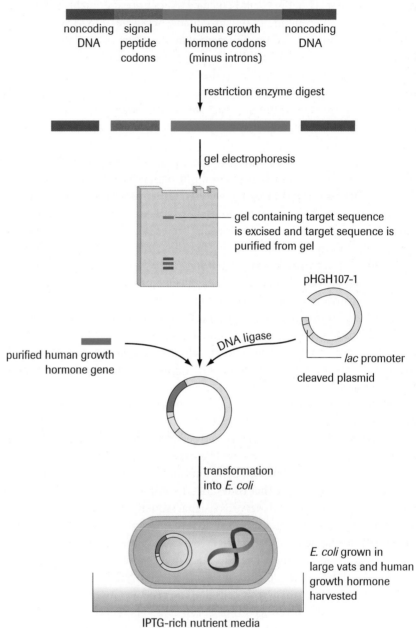

human DNA fragment containing gene for human growth hormone

| noncoding DNA | signal peptide codons | human growth hormone codons (minus introns) | noncoding DNA |

restriction enzyme digest

gel electrophoresis

gel containing target sequence is excised and target sequence is purified from gel

pHGH107-1

purified human growth hormone gene

DNA ligase

lac promoter

cleaved plasmid

transformation into *E. coli*

E. coli grown in large vats and human growth hormone harvested

IPTG-rich nutrient media

Figure 1
Overview of protocol used to insert human growth hormone gene into bacteria

EXPLORE *an issue*

Take a Stand: Genetic Engineering Guidelines and Regulations

In 1975, a group of more than 100 scientists convened at a meeting in Asilomar, California to develop experimental guidelines for recombinant DNA technology. Currently, recombinant DNA technology is progressing at an unforeseen pace.

Statement

International guidelines and parameters overseeing public and private recombinant DNA technology need to be established and reviewed annually.

- Using the Internet or other resources, research the guidelines that were set in place approximately 30 years ago in Asilomar, California. In a group, discuss these regulations and assess whether they are appropriate currently.

GO ➤ www.science.nelson.com

(a) Consider the statement as a group. Compile a list of points and counterpoints using supporting evidence from your research.

(b) Decide whether your group agrees or disagrees with the statement.

(c) Prepare to defend your group's position in a class discussion.

(d) As a group, propose regulations that should be established with respect to medical and agricultural genetic engineering. Provide a rationale for your set of regulations, focusing on issues related to economics, society, and the environment. Place your regulations in chart form with the associated rationale beside them.

(e) As a group, present your regulations to the class.

▶ *Section 6.2* **Questions**

Understanding Concepts

1. Provide reasons why the insertion of a gene into a foreign organism would be considered a form of engineering.

2. Explain why it is necessary to remove the DNA sequence coding for the 26 amino acid signal sequence from the somatotropin gene before it is inserted into a bacterial plasmid.

3. Explain how the technique of recombinant DNA is used to manufacture the protein somatotropin. Support your explanation using diagrams.

Applying Inquiry Skills

4. Gene A is found to express the protein amylase, which is used to digest starch. A molecular biologist wants to introduce this gene into bacteria via a plasmid to use it as a marker. The success of the gene transfer and uptake will be determined by growing the bacteria on agar plates containing starch and lactose. Using your knowledge of restriction enzymes, DNA ligases, methylases, plasmids, and selective plating, create a protocol in flow chart form that illustrates the steps involved in producing bacteria that express the amylase gene. Be sure

to include detailed steps, indicating the enzymes that will be used. The restriction enzyme maps for the source fragment and plasmid are depicted in **Figure 2**.

Making Connections

5. Bacteria are ideal organisms to use as "factories" for the manufacture of various genetic products. List the characteristics of bacteria that make them ideal for genetic engineering. Be sure to consider a wide variety of advantages, including scientific and economic.

6. On the Internet and in periodicals, research Canada's present position with respect to the availability of bovine somatotropin to dairy farmers.

GO ➤ www.science.nelson.com

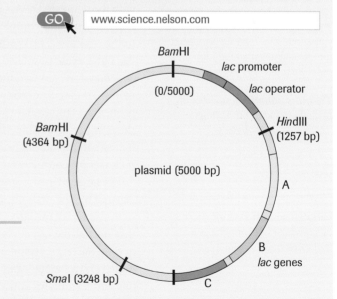

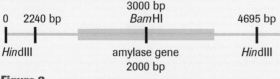

Figure 2
Restriction enzyme maps

Advanced Molecular Biological Techniques

The Polymerase Chain Reaction

Restriction endonucleases, DNA ligase, and plasmids were some of the first tools that biotechnologists used to develop molecular biology protocols. Since the 1970s, numerous other techniques have been developed that have propelled us further into the biotechnological age. One of those techniques is the **polymerase chain reaction (PCR)**.

Before PCR was developed in the late 1980s, it was not possible to make numerous copies of a desired gene fragment unless it was inserted into a plasmid. The problem with this method is that the researcher would have to extract the plasmid DNA from the bacteria, and then excise the fragment in question. PCR is a direct method of making copies of a desired DNA sequence.

The methodology of PCR is closely related to the process of DNA replication that occurs within the nucleus (**Figure 1**). During DNA replication, the two DNA strands are separated using the enzymes DNA gyrase and DNA helicase. In PCR, the strands of DNA are separated using heat. DNA is subjected to a temperature in the 94°C–96°C range. At such a high temperature, the hydrogen bonds are broken between complementary bases, and the strands separate.

Once the strands have been separated, the two single strands can be used as templates to build complementary strands. During DNA replication, DNA polymerase III is responsible for building the complementary strands. DNA polymerase III requires an RNA primer to start adding strands and only works in the 5′-to-3′ direction. In PCR, **DNA primers** replace RNA primers and are used because they are easily synthesized in the laboratory. The DNA primers must be complementary to the target area to be copied; one of the primers is the complement of one end of a DNA strand of the target DNA, and the other primer is the complement of the opposite end of the opposing strand. The primers are complements to the 3′–5′ ends of the template DNA on each opposing strand, resulting in 5′–3′ primers. One of the primers is known as the forward primer while the other is called the reverse primer because they mediate the synthesis of DNA in opposite directions toward each other.

The temperature is brought down to the 50°C–65°C range for the primers to anneal with the template DNA. Once the primers have annealed, *Taq* **polymerase**, a DNA polymerase, can build complementary strands using free nucleotides that have been added to the solution. The synthesis of the DNA strand takes place at a temperature of 72°C. *Taq* polymerase is isolated from ***Thermus aquaticus***, a bacterium that lives in hot springs and, therefore, has enzymes that can withstand high temperatures. Ordinary DNA polymerase III denatures above 37°C, so it is not practical or efficient to use. When the complementary strands have been built, the cycle repeats itself. Each subsequent cycle doubles the number of double-stranded copies of the target area, resulting in an exponential increase in the number of copies of the target DNA. After about 30 cycles, more than 1 billion copies of the targeted area will exist (2^{30}).

The targeted area is not completely isolated in the first few cycles of DNA replication. *Taq* polymerase adds nucleotides until it reaches the end of the DNA, which is not necessarily the end of the target area. After the first cycle, **variable-length strands** are produced that start at the target region on one end (where the primer has annealed) and extend beyond the target region on the other end. In the second cycle, the DNA strands are again heated and separated, and the primers are allowed to anneal. On two of the DNA strands, one end terminates at the target region (from the variable-length strands produced in cycle one), and the primers anneal to the other end of the target area. *Taq* polymerase then

polymerase chain reaction (PCR) amplification of DNA sequence by repeated cycles of strand separation and replication

DNA primers short sequences of DNA nucleotides that are complementary to the opposing 3′-to-5′ ends of the DNA target sequence that is to be replicated

Taq **polymerase** DNA polymerase, extracted from *Thermus aquaticus*, that is able to withstand high temperatures

Thermus aquaticus species of bacteria found in hot springs

variable-length strands mixture of strands of DNA that have been replicated and are of unequal length

starts adding the appropriate nucleotides, commencing from the primer, and ceases when it reaches the end that terminates at the target region (**Figure 1**). These strands are known as **constant-length strands**. The remaining two strands are extended by *Taq* polymerase, as in the first cycle. By the third cycle, the number of copies of the targeted strands begins to increase exponentially.

constant-length strands mixture of strands of DNA that have been replicated and are of equal length

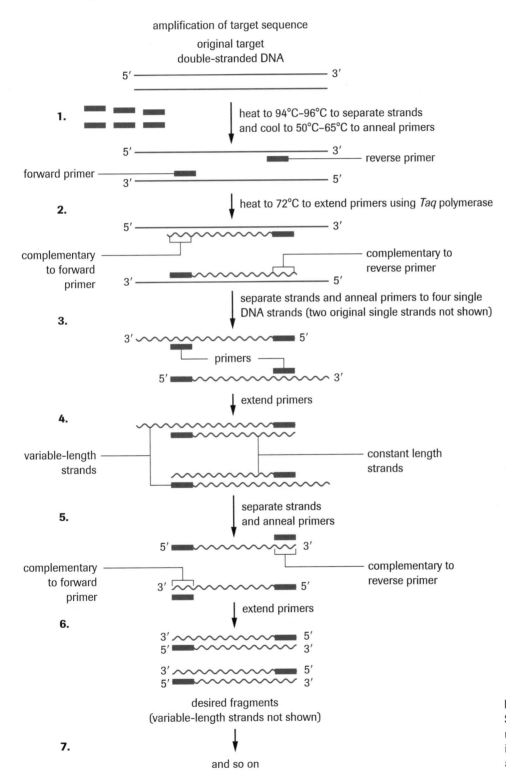

amplification of target sequence

Figure 1
Summary of the polymerase chain reaction. For simplicity, the two original single strands are not shown after step 3.

PCR is useful in forensic criminal investigations, medical diagnosis, and genetic research, and only requires a small amount of DNA to work. In criminal investigations, forensic scientists can find enough DNA in a hair follicle or one cell to use as a starting point for PCR. Therefore, only a small amount of DNA evidence is needed because it can be copied over and over again. PCR can also improve medical diagnoses, such as confirming the presence of the AIDS-causing virus. HIV cannot be detected immediately by looking for antibodies, because it takes time for the body to build antibodies against it. Traditional testing relies on the detection of these antibodies. With PCR, primers can be designed to complement short regions of the DNA of HIV. The DNA can be amplified and then examined for the presence of the HIV genome. Another application of PCR is that researchers can use it to determine, from fossil remains, whether or not two species are closely related.

Kary Mullis first proposed the process of PCR in 1987. It did not take long for the research community to appreciate the widespread applications of the process. Mullis was awarded the Nobel Prize in Chemistry in 1992 for his development of PCR.

> ### ▶ *Practice*

Understanding Concepts

1. Outline the process of PCR. Be sure to describe the characteristics of a PCR cycle.

2. Using the DNA sequence in **Figure 2** as a starting point, draw the successive fragments produced by two cycles of PCR. Explain why the fragments are initially of variable length.

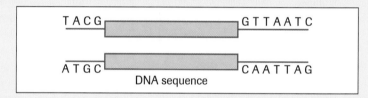

Figure 2

3. Explain why *Taq* polymerase obtained from thermophilic bacteria (bacteria that grow in temperatures above 45°C) must be substituted for DNA polymerase III in the PCR technique.

4. Explain how the following obstacles are overcome in the process of PCR:
   (a) DNA polymerase III is not present.
   (b) *Taq* polymerase only builds complementary strands in the 5′-to-3′ direction.
   (c) DNA helicase is not used to break hydrogen bonds.

5. PCR is based on the natural process of DNA replication that occurs in the cell. In a comparative format, list the steps of DNA replication and PCR. Indicate how the steps of DNA replication are accomplished outside of a cell using PCR.

Restriction Fragment Length Polymorphism

Polymorphism is any difference in DNA sequence that can be detected between individuals. Organisms of the same species carry the same genes but differ in the alleles that they have for each gene. Hence genomes of individuals from the same species are said to be polymorphic unless the individuals are identical twins.

Polymorphisms occur both in coding and noncoding regions. In coding regions, polymorphisms may be used to identify individuals with specific mutations; for example, a person who carries the allele for sickle cell anemia carries a different allele for β globin that gives the red blood cell a sickle shape compared with a person who does not have sickle cell anemia. In noncoding regions, variations may exist in the number of variable number tandem repeats, also known as microsatellites (discussed in more detail in section 5.8 of in Chapter 5). The polymorphic nature of noncoding DNA is exploited in forensic investigations.

polymorphism any difference in DNA sequence, coding or noncoding, that can be detected between individuals

Restriction fragment length polymorphism (RFLP) analysis involves the comparison of different lengths of DNA fragments produced by a restriction enzyme digested and revealed by complementary radioactive probes to determine genetic differences between individuals. The DNA in question is first digested using either one or several different restriction endonucleases. If the digested DNA is then run on a gel, it appears as a long smear because many fragments differing only slightly in size over a wide range are produced. The amount of DNA is usually so large and the bands so numerous that gel electrophoresis is unable to resolve them.

Differences in the pattern of fragments must be detected to distinguish DNA from two different sources. The gel is subjected to a chemical that causes the double-stranded DNA to be denatured into single-stranded DNA. The single-stranded DNA is then transferred to a nylon membrane with the aid of an electric current. This transfer procedure is called **Southern blotting**, named after E.M. Southern, who developed it. The nylon membrane is placed on the gel with a positive electrode behind it. Since DNA is negatively charged, it will be transferred out of the gel and "blotted" onto the nylon membrane, where it will bind. The nylon membrane containing the single-stranded DNA is now immersed in a solution containing radioactive complementary nucleotide probes for specifically chosen regions. These regions may be mutations in an allele that lead to a specific disease, or a variable number tandem repeat found in a noncoding region of the organism's DNA. Depending on where the complementary sequences lie on the nylon membrane, complementary base pairing will occur between the probes and the DNA (a process referred to as **hybridization**). The nylon membrane is then placed against X-ray film. The radioactive probes cause exposure of the X-ray film in the areas where they have hybridized, called an **autoradiogram**. When the film is developed, a pattern is detected. The differences in pattern can be used, for example, to match a suspect's DNA to the DNA found at a crime scene, or to detect a mutation that is believed to cause a genetic disorder.

restriction fragment length polymorphism (RFLP) analysis a technique in which DNA regions are digested using restriction endonuclease(s) and subjected to radioactive complementary DNA probes to compare the differences in DNA fragment lengths between individuals

Southern blotting a procedure that allows the DNA in an electrophoresis gel to be transferred to a nylon membrane while maintaining the position of the DNA band fragments

hybridization complementary base pairing between strands of nucleic acids via hydrogen bonding

autoradiogram gel pattern imprinted on X-ray film by radioactive probes

ACTIVITY 6.3.1

Restriction Fragment Length Polymorphism Analysis (p.317) How does RFLP analysis help identify an individual who left a crime scene? This simulation activity walks you through the process of RFLP analysis and its application in criminal investigations.

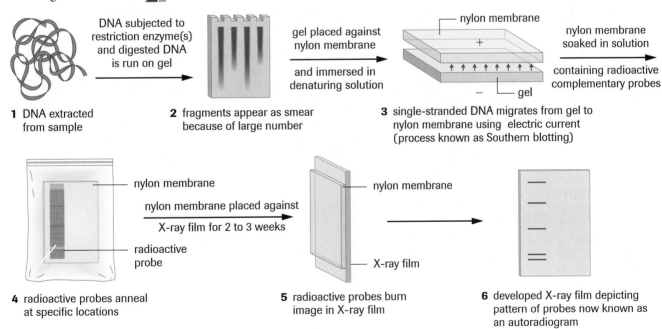

1 DNA extracted from sample

DNA subjected to restriction enzyme(s) and digested DNA is run on gel

2 fragments appear as smear because of large number

gel placed against nylon membrane

and immersed in denaturing solution

nylon membrane

3 single-stranded DNA migrates from gel to nylon membrane using electric current (process known as Southern blotting)

nylon membrane soaked in solution

containing radioactive complementary probes

nylon membrane

nylon membrane placed against X-ray film for 2 to 3 weeks

radioactive probe

4 radioactive probes anneal at specific locations

nylon membrane

X-ray film

5 radioactive probes burn image in X-ray film

6 developed X-ray film depicting pattern of probes now known as an autoradiogram

Figure 3
A summary of restriction fragment length polymorphism analysis. The transfer of the DNA from the gel to the nylon membrane is known as Southern blotting. An example of the type of gel pattern produced can be found in section 6.1, **Figure 7**.

Understanding Concepts

6. The following is a DNA sequence: 5′-ATTAGCATACGGGCAGATTTACCCGATTAATT-3′ Write out the sequence of an eight-base-pair DNA probe that could be used to hybridize it.

7. PCR requires two DNA primers to be used, one that is complementary to the 5′ region of the target area on one strand and the other complementary to the 5′ region of the target area on the opposing strand. Predict what would happen if only one primer were available.

8. DNA from two organisms of the same species is polymorphic. Explain why this is the case. Under what conditions would the DNA from two organisms of the same species not be polymorphic?

9. Outline the process of RFLP analysis.

10. In RFLP analysis, radioactive probes are used that will hybridize to specific sequences within the digested genome. Explain why it is necessary to use radioactive probes as opposed to running the gel and viewing the digested genome directly.

11. Explain why RFLP analysis is not effective when analyzing degraded DNA.

12. Define hybridization. Explain how you can use hybridization to tell whether a certain DNA sequence is present in a DNA sample.

13. In RFLP analysis for a criminal investigation, noncoding regions are used to distinguish between individuals. Explain why noncoding regions are superior to coding regions in this respect.

14. A source of DNA must be available either for PCR or RFLP analysis to produce a DNA fingerprint. Semen samples need not be as large as blood samples to extract comparable amounts of DNA for analysis. Explain why less semen is required than blood for DNA fingerprinting.

Making Connections

15. In the past, phylogenetic trees were developed using techniques such as carbon dating fossils of extinct species, and careful observation of species that are still around. Consideration was also given to the location of the species in question in relation to others around the world. DNA fingerprinting has verified some of the proposed relationships between species, but it also has revealed some errors. Explain why DNA fingerprinting is a more precise technique in developing and identifying phylogenetic relationships. What are some of the limitations of DNA fingerprinting in this respect?

SUMMARY *RFLP and PCR*

Table 1 Comparison of RFLP Analysis and PCR

	RFLP Analysis	**PCR**
State of sample	large (blood: size of quarter, semen: size of dime) and fresh	minute (one cell) and degraded
Size of sample	whole genome	target sequence suffices
Time	three weeks (average)	one day
Basic premise	cleaving DNA using restriction endonuclease(s) followed by subjection to radioactive complementary DNA probes	building complementary strands using principles of DNA replication
Result medium	autoradiogram	gel
Tools	restriction endonucleases, radioactive probes, nylon membrane, X-ray film, gel electrophoresis	DNA polymerase, nucleotides (A,G,C,T), DNA primers, gel electrophoresis
Sensitivity and accuracy	highly sensitive and accurate	sensitive and accurate

DNA Sequencing

To analyze gene structure and its relation to gene expression and protein conformation, biologists must determine the exact sequence of base pairs for that gene. The Human Genome Project was an ambitious project undertaken to decipher the genetic sequence of the 46 human chromosomes. The Human Genome Project used computer technology to read the human DNA sequence. The technology is based firmly on the gene sequencing laboratory techniques developed over the past two decades.

Developed at Cambridge University in Great Britain, the **Sanger dideoxy method** is the most popular and the most easily automated DNA sequencing protocol. In 1977, Frederick Sanger and his colleagues were the first to sequence an entire genome. The genome was of a bacteriophage (viral DNA), which contained 5386 base pairs.

The Sanger dideoxy method utilizes the process of DNA replication. DNA replication requires a single-stranded DNA template, a primer, DNA polymerase, and nucleoside triphosphates. In the Sanger method, the DNA template to be sequenced is treated so that it becomes single stranded. A short, single-stranded, radioactively labelled primer is added to the end of the DNA template. Identical copies of the primed single-stranded DNA are placed in four reaction tubes. For building complementary strands, each tube contains DNA polymerase and a supply of free nucleotides in the form of all four deoxynucleoside triphosphates (dATP, dTTP, dGTP, and dCTP) (**Figure 4**).

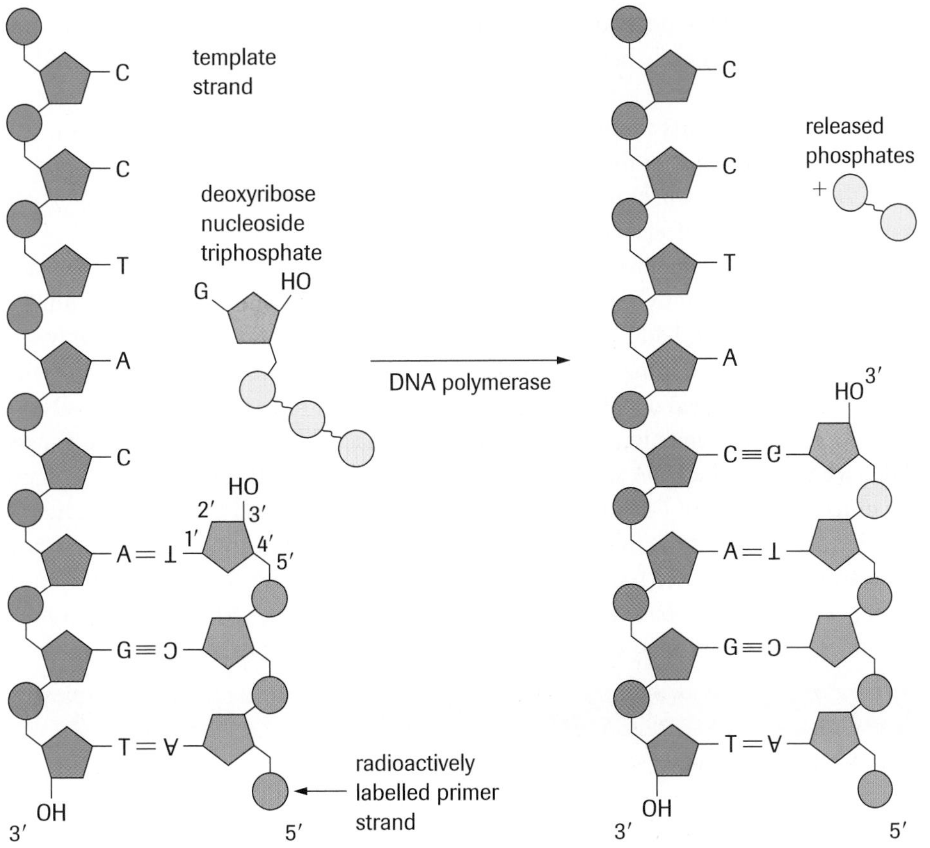

Sanger dideoxy method DNA sequencing technique based on DNA replication that uses dideoxy nucleoside triphosphates

Figure 4
Elongation of complementary DNA strand during Sanger dideoxy sequencing

In addition, each of the four reaction tubes contains a different radioactively labelled **dideoxy analogue** of one of the deoxynucleoside triphosphates (dNTPs) in low concentration. For example, tube 1 may contain all four dNTPs plus dideoxy-adenine (ddATP); tube 2, all four dNTPs plus dideoxy-thymine (ddTTP); tube 3, all four dNTPs plus dideoxy-guanine (ddGTP); and tube 4, all four dNTPs plus dideoxy-cytosine (ddCTP). The dideoxy analogue is a dNTP whose deoxyribose sugar is missing the

dideoxy analogue nucleoside triphosphate whose ribose sugar does not possess a hydroxyl group on the 2′ and the 3′ carbon

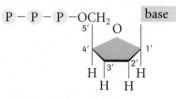

2', 3' dideoxy analog

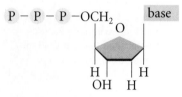

deoxyribonucleoside triphosphate

Figure 5
2', 3' dideoxy analogue versus a 2' deoxy analogue. DNA polymerase requires the presence of the −OH group on the 3' carbon to elongate a DNA strand by creating a phosphodiester bond.

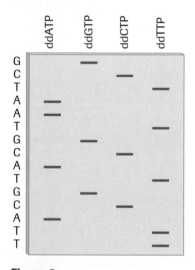

Figure 6
Example of a gel produced using the Sanger dideoxy method of sequencing

−OH group on its 3' carbon. DNA polymerase catalyzes the phosphodiester linkage between the 3'−OH group of the deoxyribose sugar on the growing chain and the phosphate group on the incoming nucleotide (**Figure 5**). Hence, if the 3'−OH group is missing, DNA polymerase cannot add the next complementary base and synthesis stops. Whenever a dideoxy analogue is incorporated, it acts as a chain terminator. Hence, the Sanger method is also called the chain termination technique.

Since only a fraction of the dNTPs are dideoxy analogues, different lengths of complementary DNA will be built before the analogue is incorporated. If the strand to be replicated is of the sequence 3'-AATGCATGCATTAGC-5', and it is part of the mixture that contains the dideoxy analogue adenine, four possible complementary strands may be produced:

5'-TTA
 TTACGTA
 TTACGTACGTA
 TTACGTACGTAA-3'

The incorporation of a dideoxy analogue blocks further growth of the chain. Regular dATPs are in much higher concentration in the solution. If they are incorporated, chain elongation continues, resulting in different lengths of DNA. Because the strands that are built differ in length, they can be separated using polyacrylamide gel electrophoresis, even if they differ by only one nucleotide. After replication is completed, the contents of all four reaction tubes are loaded into four separate lanes of a gel. The fragments differ by only one base pair each; hence, the sequence can be read right off the gel in ascending order (**Figure 6**).

In total, 15 fragments would be generated. The gel is placed against X-ray film. The radioactive decay exposes the X-ray film, and the sequence of the nucleotides is read from the autoradiogram. Using this method, Sanger was able to sequence the 5386 base pairs of a virus that infects bacteria.

The Human Genome Project used a similar sequencing technique, except that each ddNTP was fluorescently tagged. For example, ddGTP may be tagged green, ddATP yellow, ddTTP red, and ddCTP blue. Another modification was that there was only one reaction mixture containing numerous copies of the DNA to be sequenced, along with the fluorescently tagged ddNTPs mixed with dNTPs. The fluorescently tagged ddNTPs incorporated themselves in a manner similar to what occurred in the original Sanger dideoxy method. The DNA sample was then electrophoresed through a gel to separate the different lengths of DNA produced. The computer then read the sequence and the position on the gel of the fluorescence tags and determined the sequence of the DNA. Thousands of automated sequencer machines worked 24 hours a day, 7 days a week to decipher the 3 billion-base-pair human genetic code. Parallel advances in computer technology enhanced the speed at which the project was completed.

> ▶ **Practice**

Understanding Concepts

16. Distinguish between a deoxyribose sugar and a dideoxyribose sugar. Why would a dideoxyribose sugar inhibit DNA chain elongation?

17. Explain why only a small concentration of a dideoxy analogue is needed in each mixture to ensure that sequencing is successful. Predict what would happen if large amounts of the dideoxy analogue were present in the reaction mixture.

18. List the steps that are followed to sequence a fragment of DNA using the Sanger dideoxy method.

19. The Sanger dideoxy method is often referred to as the "chain termination method." Explain why this is an appropriate name.

▶ **Practice** *continued*

Applying Inquiry Skills

20. Record the DNA sequence that resulted in the gel pattern shown in **Figure 7**.

21. The following sequence was determined using the Sanger dideoxy method: 5′-AAT-GCAGCATACC-3′. Draw the gel pattern that would result from this sequence.

22. The gel pattern shown in **Figure 8** was produced after sequencing a certain fragment of DNA. Notice that some bands are missing in the pattern. Provide a reason why this may have happened. What could the molecular biologist do to ensure that this error is not repeated again?

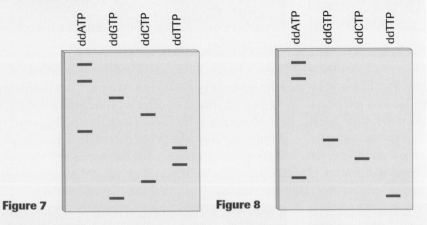

Figure 7   **Figure 8**

Making Connections

23. Another method by which the sequence of DNA can be determined is the Maxam–Gilbert method. Using the Internet and other resources, research this experimental technique. How does the Maxam–Gilbert method differ from the Sanger dideoxy method?

    www.science.nelson.com

---

SUMMARY   *Advanced Molecular Biological Techniques*

Table 2 Summary of PCR and RFLP Analysis

PCR	RFLP
1. DNA sequence to be amplified is identified.	1. Entire genome is subjected to restriction enzyme digestion.
2. Double-stranded DNA is heated to approximately 90°C.	2. Digested DNA is run on an agarose gel, using gel electrophoresis.
3. Temperature is lowered to approximately 60°C. DNA primers that are engineered specifically for target region anneal.	3. Gel is placed against a nylon membrane and subjected to chemical that denatures double-stranded DNA into single-stranded DNA. Current is applied so that single-stranded DNA migrates to nylon membrane and binds to it.
4. Temperature is raised to approximately 70°C. *Taq* polymerase elongates complementary DNA strand.	4. Nylon membrane is placed in solution containing radioactive complementary nucleotide probes engineered for specific regions of the genome.
5. Cycle is repeated 29–30 times (steps 2–4).	5. Radioactive probes hybridize to complementary sequences.
6. Amplified DNA is run on a polyacrylamide gel using gel electrophoresis.	6. Nylon membrane with hybridized radioactive probes is placed against X-ray film.
7. Gel is stained using ethidium bromide and viewed under UV light to visualize the bands of DNA	7. Autoradiogram is developed and band pattern is visualized.

Gene Patent Lawyer

As scientists and biotechnology companies consider the ownership of genetic information and technology, they increasingly turn to gene patent lawyers for guidance. A patent is a contract between an inventor and the government. Gene patent lawyers must have both scientific knowledge and legal expertise to apply for patents that protect such inventions and discoveries as gene sequences, drugs, or new technologies and processes.

Genetic Engineer

Biomedical engineers attack biomedical problems at the cellular level. They design miniature devices that help in the medical treatment of disease by delivering compounds to specific targets in the human body. Biomedical engineers use engineering principles to understand disease processes, design experiments, and invent new techniques and applications in the areas of gene therapy, cloning, and genetic screening.

Bioinformatician

Bioinformatics is a new field that combines chemistry, computer science, and the study of genetics to solve problems in biological and medical research. Bioinformaticians organize large quantities of DNA and protein sequence information, analyze experimental data, and search databases for new connections and facts. They may be involved in the design and evaluation of new drugs, research into anti-cancer or AIDS therapies, and the development of software.

Bioethicist

Bioethicists assist those who work in medicine and science—such as health-care workers, researchers, and companies—to examine and resolve moral issues. They act as consultants, educators, and policy developers. The field of bioethics has traditionally drawn people from all conceivable backgrounds, including law, medicine, philosophy, and religion. Today, educational opportunities are expanding as more universities offer instruction in this interdisciplinary field.

▶ **Practice**

Making Connections

24. Choose a career associated with biotechnology. Either choose one that appears above, or the career of forensic scientist or research scientist (at a university or in private industry).
    (a) Using the Internet, investigate and list the features that appeal to you about this career. Make another list of features that you find less attractive.
    (b) Identify the high school, college, and/or university subjects that are required for this career. Is a postsecondary degree required?
    (c) Using the Internet, survey the employment opportunities in Canada for this career.

 www.science.nelson.com

Medical Applications

The field of medicine has been greatly altered with the advances made in biotechnology over the past 30 years. The study and practice of medicine has two principles: prevention and treatment. The difference today, compared with 40 years ago, is that many medical researchers focus on detecting, treating, and preventing disease at the molecular level.

The earlier a disease is detected, the faster an aggressive treatment may curb its effects. A prime example of a disease treatment that has benefited from the advances of molecular biology is AIDS. The human immunodeficiency virus that causes AIDS can be detected by PCR very soon after infection. The traditional AIDS test detects antibodies in the blood produced by the body against HIV. To be tested by PCR, a person who may have been exposed to HIV need not wait a few months until the antibodies are detectable in the blood. Since PCR relies on the detection of the virus itself as opposed to the body's reaction, it is a direct test.

Genetic screening is the detection of mutations known to be associated with genetic disorders before they manifest themselves in an individual. For example, mutations in the gene for Huntington's chorea can now be detected before the symptoms of the disease surface. Although an effective treatment has yet to be developed, individuals with a family history of the disease can now choose to find out if they will develop it later on in life.

Genetic disorders in the human fetus can also be detected using genetic screening of embryonic cells found in the amniotic fluid during gestation. Such prenatal screens are available for hemophilia, phenylketonuria, cystic fibrosis, and Duchenne's muscular dystrophy. Couples with a family history of genetic disorders who are at risk of passing mutations on to their offspring are offered genetic counselling to better prepare for the birth of a child (**Figure 1**).

Genetic screening and detection techniques are limited only by the lack of knowledge in identifying particular genes or associated gene markers. With the success of the Human Genome Project, it is hoped that the location of more genes will be pinpointed, resulting in better screening and detection of genetic disorders. **Gene therapy**, however, is very much in its infancy; there has been limited success in the treatment of a few diseases. It is still too early to make any far-reaching conclusions. Nevertheless, researchers are optimistic that gene therapy will be a viable medical option in the future.

Currently, research is being conducted in the hopes of developing gene therapy for chronic pain. The body deals with pain in one of two ways, both of which involve transmitter molecules. **Pronociceptive transmitters** induce pain, whereas **antinociceptive transmitters** inhibit the sensation of pain. The effects of these transmitters are eventually relayed to the brain via chemical messengers, resulting in an individual's pain being heightened or dampened. If a therapeutic gene is inserted into a cell that expresses antinociceptive transmitters, more of these molecules will be made, resulting in the minimization of pain. A therapeutic gene can be introduced into adrenal gland cells that normally secrete the antinociceptive transmitters. Several trials using this strategy have been conducted at the Rangueil Hospital Medical School in France. Cells were transplanted to an area just outside the spinal cord in patients with cancer. The patients' pain diminished.

Figure 1
A genetics counsellor helps a couple understand the genetic factors involved in diseases and disorders.

genetic screening process by which an individual's DNA is scanned for genetic mutations

gene therapy the alteration of a genetic sequence in an organism to prevent or treat a genetic disorder

pronociceptive transmitters signal molecules that amplify pain sensation via a neurological pathway

antinociceptive transmitters signal molecules that dampen pain sensation via a neurological pathway

antisense oligonucleotides
complementary DNA or RNA that
anneals to mRNA and inhibits

transgenic organism in which foreign DNA has been artificially incorporated into its genome

Ti plasmid plasmid, found in
Agrobacterium tumefaciens bacteria,
that is able to enter plant cells

Agrobacterium tumefaciens type
of bacteria that infects wounded
plant cells and creates a crown gall

Another strategy to modify gene expression therapeutically, and thereby control pain, is antisense synthetic oligonucleotides therapy. **Antisense oligonucleotides** are short stretches of DNA or RNA that recognize and deactivate complementary mRNA molecules. If antisense RNA is produced that is complementary to the pronociceptive transmitter mRNA and then introduced into the cells, it will hybridize to the mRNA and prevent ribosomes from translating it into protein. In essence, the expression of the pronociceptive gene is blocked. Trials of this strategy using mice have taken place at the University of Iowa and the University of California, San Diego. The results, thus far, have been positive.

Whether molecular biology is used for gene therapy, detection, or prevention, the outcome points in a positive direction. Antibiotics revolutionized medicine half a century ago; molecular biology will define medical advances in this century.

Agricultural Applications

In 1981, Eugene Nestor and Mary Dell Chilton devised a method by which foreign genes could be introduced into plant cells to produce **transgenic** plants. Nestor and Chilton used the **Ti plasmid** (**t**umour-**i**nducing plasmid) that is carried by the soil bacteria **Agrobacterium tumefaciens** as a vector. The bacteria are attracted to chemicals released by a wounded plant and gain access via the wound. The Ti plasmid enters the plant cells through natural transformation, and the result is a bulbous growth called a crown gall (**Figure 2**). The relationship between the plant and the bacteria is not mutually beneficial. The bacteria take nutrients from the plant, yet the plant does not benefit in any way.

The Ti plasmid consists of several regions but only one, called the T region, actually becomes incorporated into the chromosomal DNA of a plant cell. A foreign gene can be inserted in the T region and therefore introduced into a plant cell. Unfortunately, the Ti plasmid can only infect a dicotyledon, such as beans, peas, and potatoes. Gene guns are used to infect a monocotyledon, such as wheat, corn, and rice.

Many new products have been developed using this technology. Crops have been genetically engineered to increase yield, hardiness, uniformity, insect and virus resistance, and herbicide tolerance. One of the first products to be developed and approved was the Flavr Savr tomato. Using antisense technology, the gene that codes for polygalacturonase (the enzyme responsible for ripening and aging of fruits) was minimized to less than 1% expression. The Flavr Savr tomato was one the first transgenic crops to be approved. Other crops are still awaiting approval. An American company has found a way to insert a gene into a cotton plant that results in the production of a polyester polymer inside the cotton fibre. The result is a plant that produces a cotton-polyester blend, reducing the need to manufacture plastic that contributes to environmental pollution.

The effect of transgenic crops on the environment is a huge concern. Scientists are finding it difficult to assess environmental damage because the repercussions may take decades to surface. A current controversy surrounds transgenic corn that has been genetically engineered to produce **Bt toxin**. Bt toxin originates from the bacteria *Bacillus thuringiensis* and is a natural herbicide against the European cornborer. Since corn is pollinated by the wind, the pollen may coat neighbouring noncrop plants. Studies conducted by John Losey at Cornell University in 1999 revealed that only 56% of monarch butterfly larvae survived after eating plants coated with the corn pollen, whereas 100% of the larvae survived when they ate plants coated with nontransgenic corn pollen (**Figure 3**).

Figure 2
A crown gall

DID YOU KNOW ?

**Genetic Research Increases
Agricultural Production**
Worldwide, agricultural production
has drastically increased in the past
50 years. Many factors are responsible for the increase, including
genetic improvement.

Bt toxin poisonous substance
produced by *Bacillus thuringiensis*,
which acts as a natural herbicide

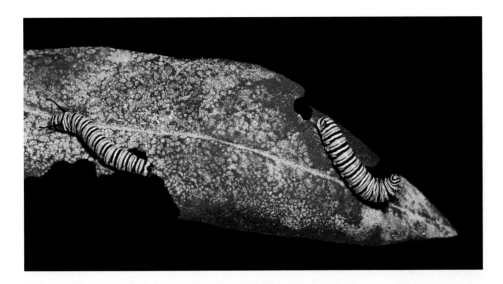

Figure 3
Controversy still surrounds the ecological impact of transgenic corn.
Experiments have been conducted
with monarch butterfly larvae that
eat milkweed dusted with the pollen
from the corn. Additional study is
required to obtain conclusive results.

Conversely, studies have been conducted that reject the effect of Bt toxin on monarch butterfly populations. In September 2001, Mark Sears, an entomologist at the University of Guelph, stated, "The impact of this technology on monarchs is negligible." A key difficulty in ensuring ecological safety of transgenic crops is the time it takes to conduct and evaluate experiments. To get conclusive experimental results, the transgenic plant needs to be grown and studied under different environmental conditions, at different times of the year, in combination with different farming practices.

Concerns with respect to transgenic crops are not limited to ecological ramifications. Economic fallout also plays a large role. For example, BST (bovine somatotropin) has not been approved in Canada. This hormone boosts milk production in cows by up to 10%. However, an increase in the supply of milk would be detrimental to dairy farmers because the price of milk would go down. Many small farming operations would suffer irreparably from the drop in price.

DNA fingerprinting pattern of
bands on a gel, originating from
RFLP analysis or PCR, that is unique
to each individual

Since 1994, approximately 40 genetically modified foods have been approved in Canada. Health Canada and the Canadian Food Inspection Agency share the responsibility for approving and developing food-labelling policies under the *Food and Drug Act*. Mandatory labelling must exist on foods whose compositional or nutritional change is a safety concern. For example, if a gene being expressed in a transgenic crop is an allergy stimulant, that food product would have to be labelled.

Forensics

Genetic engineering has had an enormous impact on the field of forensic science. In 1987, the first court case was heard in Britain in which RFLP analysis was used. The DNA from a suspect was compared with that of the DNA extracted from the semen left behind at the scene of a rape. The suspect was convicted of the crime and sent to prison as a result of the DNA match, along with overwhelming corroborating evidence. In a short time, **DNA fingerprinting** has become one of the most valuable tools investigators can use to determine whether or not the evidence originated from the suspect (**Figure 4**). DNA evidence alone cannot convict a suspect, for it cannot support on its own the claim that the suspect committed the crime. The only conclusion that can be made with very close certainty is that the DNA source came from the suspect in question.

Figure 4
A forensic scientist using DNA
fingerprinting

Either RFLP analysis or PCR can be used in criminal investigations, depending on the state of the DNA sample. RFLP analysis requires large samples that are undegraded, whereas PCR can be performed with minute quantities that are degraded. Both techniques produce a "fingerprint" or pattern of bands on a gel that can be compared with the evidence found at the scene. Both techniques take advantage of the noncoding regions of the human genome. It is in these noncoding regions that most of the variability lies. The noncoding regions differ in the quantity of variable number tandem repeats and, therefore, can be used as a basis for discriminating between individuals. In fact, the probability is very low of matching in six to seven areas of noncoding regions with another person. In some cases, depending on the population and the frequency of the given site, it could be as low as one in 1 billion.

It is important to note that DNA fingerprinting is most often used to illustrate innocence as opposed to guilt. Many cases do not even go to trial because of the lack of DNA evidence, and Ontario has saved large sums in legal costs as a result. Numerous people have been exonerated using this technology. Guy Paul Morin spent approximately 10 years in custody before his conviction of murder was overturned with DNA evidence (**Figure 5**). The murder victim had been raped before being killed. Luckily, the evidence had been kept at the Centre for Forensic Sciences in Toronto. Using PCR, the DNA sample was tested in seven different target regions. Morin did not match up in three of those regions. If he had not matched in only one of the seven regions, that would have been enough evidence to release him, because identical DNA samples should produce identical banding patterns when subjected to the same analysis. What is unfortunate about Morin's case and others like it is that these wrongfully convicted individuals had to wait for the technology to be developed before they could be released.

Figure 5
Guy Paul Morin

Currently, DNA databases are being shared among criminal institutions across the United States and in Canada. In fact, it is now illegal in Canada to refuse to provide a DNA sample to police on arrest if requested. In the past, traditional fingerprints were collected and the future holds the possibility of DNA fingerprints being collected. Many argue that this is an invasion of privacy. The same arguments were made a century ago when police started to collect traditional fingerprints.

The Future

It is impossible to predict the effects that modern biotechnology will have on our lives this century. Society will be faced with moral, ethical, and philosophical dilemmas regarding the use of biotechnology, the products it designs, and the information it provides. Now, more than ever, society needs to be educated about the advantages and disadvantages that biotechnological research and innovation bring with it. The knowledge base that the Human Genome Project has brought forth was unthinkable even 20 years ago. Today, information about the Human Genome Project is almost commonplace. Now we look forward to the information that the Human Proteome Project will reveal, as it provides us with more data on the relationships between genes and proteins.

So many areas of biotechnology are advancing. Microarray technology is accelerating and evolving, and scientists have started to investigate the use of DNA as a computer itself. Laser scissors and tweezers are being used to inactivate parts of chromosomes.

DID YOU KNOW?

Bacterial Cleanup Crews
Biotechnological innovations are helping the environment. *E. coli* strain MV1184 is capable of removing inorganic phosphates from waste water. In 1993, J. Kato and his team used genetic engineering to enhance the genes involved in inorganic phosphate metabolism. The modified bacteria removed three times as much phosphate as the unmodified bacteria.

The genetics of medical disorders and other afflictions, such as chronic pain, are being deciphered. Additional genetic screening tests are being developed, and gene therapy—although in its infancy—holds promise for cures for a number of genetic disorders. Only through education will society be able to assess the ratio of risks and reward that accompanies advances in this exciting yet controversial field. Miescher would be astounded at how far we have advanced since 150 years ago, when he isolated an acidic, phosphorous-rich substance from white blood cells.

▶ *Section 6.4* *Questions*

Understanding Concepts

1. Explain how antisense technology could be used to reduce the expression of a specific gene. Provide examples of the use of this technology.

2. Justify why the pattern on a gel produced by the digestion of an individual's DNA can be considered a "fingerprint."

3. Differentiate between pronociceptive and antinociceptive transmitters. How is gene therapy being used to heighten or dampen their effects in connection with chronic pain?

Making Connections

4. Advances in gene therapy are dependent on the pace of advancement in genetic screening technology. Comment on this statement. Do you agree or disagree? Why or why not? Provide reasons to justify your opinion.

5. The Government of Canada has set regulations that must be met for a new transgenic crop to be approved for distribution in the country. Using the Internet and other resources, research the regulations that have been put into place. Appropriate starting points of research would be Environment Canada, Food and Health Canada, and Agriculture Canada. Do you feel these guidelines are adequate? What modifications would you make to these guidelines if you could? Explain the implications of the guidelines that have been set.

 www.science.nelson.com

6. Using the Internet and other research tools, compile a list of genetic disorders that can currently be screened for.

 www.science.nelson.com

7. The Human Genome Project has provided us with a solid foundation for the elucidation of many genetic disorders and their possible treatments and cures. Using the Internet and print resources, construct a timeline of the major events that have occurred during the Human Genome Project.

 www.science.nelson.com

8. Recombinant DNA has been used to produce human insulin in bacteria. Since millions of people suffer from diabetes, the market for human insulin is enormous and so are the profits. Because of economic pressures, the nature of scientific research has, in part, changed from one of sharing information to one of patents and secretiveness. Companies are unwilling to release any breakthrough for fear that their ideas may be stolen. Scientific research, at least in some fields, is no longer controlled by researchers but by investors and market demand. Should governments play a more active role in regulating private biotechnology enterprise and gene patenting? Support your opinion.

INVESTIGATION 6.1.1

Restriction Enzyme Digestion of Bacteriophage DNA

Inquiry Skills

- ● Questioning
- ○ Planning
- ● Analyzing
- ● Hypothesizing
- ● Conducting
- ● Evaluating
- ● Predicting
- ● Recording
- ● Communicating

In this investigation, bacteriophage lambda DNA will be digested using the restriction endonucleases *Eco*RI, *Hin*dIII, and *Bam*HI. The fragments produced will be separated using gel electrophoresis. Fragment sizes will be calculated from an analysis of the agarose gel. Bacteriophage lambda DNA is obtained from a virus that infects bacterial cells and is 48 514 base pairs in length.

Question

How do the patterns of DNA fragments compare when a piece of DNA is digested using different restriction endonucleases?

Hypothesis

(a) Based on your knowledge of restriction endonucleases and gel electrophoresis, predict the number of bands that will be produced when bacteriophage lambda DNA is digested with *Eco*RI, *Bam*HI, and *Hin*dIII. Use the bacteriophage lambda DNA restriction map (**Figure 1**) to help you make your prediction.

Materials

safety goggles
gloves
70% ethanol solution (or 10% bleach)
four 1.5-mL Eppendorf tubes
waterproof pen for labelling
masking tape
polystyrene cup
freezer
crushed ice
20 μL of 0.5 μg/μL lambda DNA
5 μL 10× restriction buffer

2 μL each of *Bam*HI, *Eco*RI, and *Hin*dIII restriction endonucleases
1.0-20-μL micropipette with tips
microcentrifuge (optional)
37°C water bath (or water heated to, and maintained at, 37°C on a hot plate)
thermometer
1 g agarose
paper boat
electronic balance
500-mL Erlenmeyer flask
250-mL graduated cylinder
microwave or hot plate
flask tongs or oven mitts
gel casting tray and gel electrophoresis box (bought from scientific supply company or made using basic materials, as shown in **Figure 2**)
1L 1× TBE buffer
5 μL loading dye
power supply (45 V)
plastic wrap
25-30 mL 0.025% methylene blue, or enough to cover the gel in the staining tray
light box or overhead projector
acetate sheet

✋ **Wear safety goggles at all times.**

Wear gloves when performing the experiment.

Wipe down all surfaces with 70% ethanol, or 10% bleach, before and after the laboratory exercise.

Wash your hands thoroughly at the end of the laboratory.

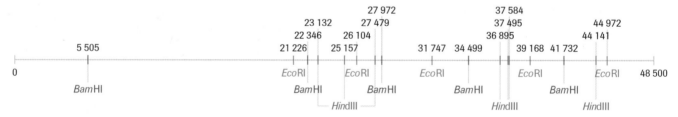

Figure 1
Restriction enzyme map of bacteriophage lambda DNA

INVESTIGATION 6.1.1 *continued*

Procedure

Day 1: Restriction Enzyme Digestion

1. Put on your safety goggles and gloves, and wipe down your bench with a 70% ethanol solution (or 10% bleach).

 Ethanol is highly flammable. Make sure that any flame on your desk or near it is turned off before use.

2. Label four 1.5-mL Eppendorf tubes "*Bam*HI," "*Eco*RI," "*Hin*dIII," and "control." Place the tubes in a polystyrene cup containing crushed ice. **Table 1** outlines the amount of reagents to add to each tube. To keep track of each tube's contents, copy the table into your notebook and check off each reagent as you add it to the tube.

Table 1 Reagents to Add to Tubes

Tube	DNA (μL)	10× buffer (μL)	Water (μL)	*Bam*HI (μL)	*Eco*RI (μL)	*Hin*dIII (μL)
*Bam*HI	4	1	4	1	–	–
*Eco*RI	4	1	4	–	1	–
*Hin*dIII	4	1	4	–	–	1
Control	4	1	4	–	–	–

3. Read down each column, adding the same reagent to all appropriate tubes. Use a fresh tip on the micropipette for each reagent. Add the 4 μL of DNA to each tube first, followed by the 10× reaction buffer, and then the water. *Make sure you add the enzyme last.* Dispense all the contents close to the bottom of the Eppendorf tubes. Ensure that the pipette tip is touching the side of the tubes when dispensing the contents. *Keep everything on ice at all times.*

4. Close the Eppendorf tube tops. Place the tubes in the microcentrifuge, close it, and spin at maximum speed for approximately 3 s. If you do not have access to a microcentrifuge, then just tap the tubes on a soft pad or thick paper towel on the bench, pooling the contents to the bottom.

 When using the microcentrifuge
- **do not open the centrifuge until it stops completely;**
- **if the centrifuge tubes are smaller than the metal holder or holes, use the proper adaptor to accommodate them;**
- **do not unplug the centrifuge by pulling on the cord–pull the plug.**

5. Place the tubes in a 37°C water bath for a minimum of 45 min. Use a thermometer to check the temperature of the water.

6. Once the digestion is complete, place the tubes in the polystyrene cup and put the cup in a freezer until your next class. Make sure you have labelled your cup with your name.

Day 2: Gel Electrophoresis

7. Weigh 0.96 g of agarose powder in a paper boat on an electronic balance and transfer to a 500-mL Erlenmeyer flask.

8. Use a graduated cylinder to add 125 mL of 1× TBE buffer and swirl to mix.

9. Heat the flask on a hot plate or in a microwave until the solution is completely clear. Handle carefully, using tongs or oven mitts.

If the agarose gets too hot it may bubble over. Be sure to observe your Erlenmeyer flask throughout the heating process. If the agarose solution starts to bubble up the neck of the flask, remove it immediately from the heat source using an oven mitt or tongs. Handle all hot glassware with caution.

10. Prepare the gel casting tray. Depending on your gel electrophoresis unit, you may have to tape the gel casting tray. Ensure that the plastic comb is inserted properly (**Figure 2**).

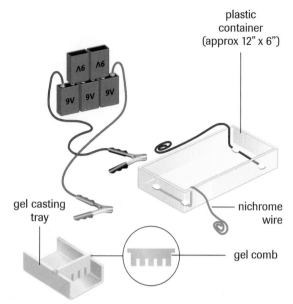

Figure 2
Homemade gel electrophoresis box and casting tray
(a) An inexpensive electrophoresis chamber can be made of a plastic container with two nichrome wire electrodes held in place with silicone caulking. The power supply leads to wire loops with alligator clips (designed for 45 V maximum).
(b) The gel casting tray can be made from a plastic soap dish or playing card container. The ends are cut and taped with masking tape before the gel is cast. The comb should not touch the bottom of the casting tray. It can be cut out of plastic remnants from a soap dish or playing card container.

11. Once the flask with agarose solution is cool enough to handle with bare hands, pour the mixture into the gel casting tray. The comb teeth should be immersed in about 6 mm of agarose. The gel should cover only about one third of the height of the comb teeth. Use a micropipette tip to remove bubbles from the gel as soon as it is poured.

12. Allow the agarose to set for a minimum of 20 min. The gel will become cloudy as it solidifies.

13. Once the gel has set (you may test this by gently touching the lower righthand corner with your finger), flood the gel with 1× TBE running buffer and then pull out the comb gently without ripping any of the wells.

14. Orient the tray containing the gel in the gel electrophoresis box so that the wells made by the comb are at the end with the positive electrode.

15. Add 1× TBE buffer to the gel electrophoresis box until the buffer is approximately 5 mm above the gel. Place the gel electrophoresis box to the side.

16. Add 1 μL of loading dye to each of the Eppendorf tubes. Microfuge for 3 s.

17. Micropipette the full contents of one Eppendorf tube into a well on the gel. Do the same for each tube. Be sure to record the order in which you dispense the tubes. Steady the micropipette over each well using both hands.

18. Close the gel box and connect it to the power supply. If you are using a gel box that you made, set the voltage to 45 V dc and turn it on. Electrophorese for 12 h. Alternatively, if you have a stronger power supply or a store-bought electrophoresis unit, electrophorese at 110 V for 2.5 h.

 When using the power supply
- **be sure the grounding pin in the power supply is not broken;**
- **pull the plug, not the cord, when unplugging the power source;**
- **do not let the wire leads connected to the electric power supply or batteries touch each other.**

19. Unplug the power supply and carefully remove the gel. Wrap the gel in plastic wrap and place it in the refrigerator for a maximum of one day.

Day 3: Staining the Gel

20. Unwrap the gel and place it in the staining tray.

21. Flood the gel with 0.025% methylene blue solution. Let the gel sit in the solution for at least 45 min. Let the gel sit in the solution for at least 20–25 minutes. Pour off the water and replace it with fresh water. Repeat this process 3 more times. Keep an eye on the intensity of the DNA bands. If you destain for too long, you may lose the smaller fragments. If you do not destain for long enough, the whole gel remains blue and the fragments cannot be differentiated.

22. Place the destained gel on a light box or on an overhead projector.

23. Obtain a blank acetate sheet or plastic wrap and place it over the gel. Trace the pattern of bands onto the wrap or sheet. Be sure to draw a line where the bottom of each well starts.

INVESTIGATION 6.1.1 *continued*

Analysis

(b) Carefully measure the distance in millimetres that each band migrated from the well origin. Copy **Table 2** into your notebook and use it to record the distances.

(c) Using the *Hin*dIII digestion as a marker, plot the distance travelled (x-axis) versus the fragment base-pair size (y-axis) on semilogarithmic paper. Please note that the 23 130-base-pair fragment and the 27 491-base-pair fragment do not resolve, but instead travel as one band. Therefore, take an average of their size for graphing purposes.

(d) Using interpolation, determine the fragment size of the bands produced by digestion with *Bam*HI and *Eco*RI.

(e) Enter your calculated base-pair fragment sizes into your table.

(f) Compare the calculated base-pair fragments to the actual base-pair fragments. Use the restriction enzyme map of bacteriophage lambda to determine the size of the actual band fragments for each enzyme. Calculate the percentage error.

(g) What was the purpose of each tube? of the control?

(h) Why do the smaller bands migrate faster than the larger bands?

(i) Some bands that are close in size migrate together. What measures may be taken to resolve bands close in size?

(j) What purpose does the 1× running buffer serve?

(k) Why must the gel be made using 1× TBE buffer?

(l) During electrophoresis, bubbles are produced at the anode and at the cathode. Explain why bubbles appear.

(m) Why must loading dye be added to the samples before they are loaded into the wells of the gel?

(n) Notice on your gel that the larger fragments are stained darker than the smaller fragments. Explain why this is the case.

Evaluation

(o) Suggest possible sources of error in this procedure. Indicate the effects of these sources of error on the results.

(p) Using standard scientific format, prepare a written report.

Table 2 Distance Travelled by Each Band From the Well Origin

*Hin*dIII		*Eco*RI			*Bam*HI		
Actual fragment size	Distance travelled (mm)	Actual fragment size	Distance travelled (mm)	Calculated fragment size	Actual fragment size	Distance travelled (mm)	Calculated fragment size
27 491							
23 130							
9416							
6557							
4361							
2322							
2027							

INVESTIGATION 6.1.2

● Questioning	○ Planning	● Analyzing
● Hypothesizing	● Conducting	● Evaluating
● Predicting	● Recording	● Communicating

Calcium Chloride Transformation of E. coli *with Ampicillin-Resistant Plasmid*

In this investigation, you will transform *E. coli* bacterial cells with a plasmid that contains an ampicillin-resistance gene, using the calcium chloride transformation protocol. Bacteria that have successfully undergone transformation will be identified using selective plating. Ampicillin is an antibiotic that prevents bacteria from building cell walls.

Question

How efficient is the calcium chloride transformation protocol?

Hypothesis

(a) Based on your knowledge of plasmids and the transformation protocol, predict the growth of transformed ampicillin-resistant *E. coli* and non-transformed *E. coli* on ampicillin-containing plates and nonampicillin-containing plates.

Materials

safety goggles
gloves
10% bleach
permanent marker
masking tape
two 15-mL culture tubes
1.5-mL tube rack
10-μL–200-μL
   micropipettes and tips
   (or 1-mL plastic pipettes
   and 0.1-ml disposable
   glass or plastic pipettes)
500 μL of 50-mM calcium
   chloride
sterile water
inoculating loop
Bunsen burner
E. coli (MM294)

0.005 μg/μL pAMP
   (plasmid DNA
   containing ampicillin-
   resistant gene)
crushed ice
polystyrene cup
3 LB plates
2 LB/amp plates
hot water bath
500 μL LB broth
bacterial cell spreader
ethanol in beaker
   (enough to soak the
   end of the spreader)
incubator (37°C)

 Calcium chloride can cause skin irritation, nausea, and vomiting. Wear safety goggles and avoid breathing in the dust.

Procedure

1. Put on your safety goggles and gloves, and wipe down your bench with 10% bleach. The entire experiment must be performed under sterile conditions.

2. Using permanent marker, label 1 sterile 15-mL culture tube "+pAMP" and the other tube "−pAMP." Place the tubes in a tube rack.

3. Using a sterile tip, micropipette 250 μL of 50-mM sterile calcium chloride solution into the −pAMP tube.

4. Using a sterile inoculating loop, transfer 1 or 2 large colonies of *E. coli* MM294 from the starter plate to the culture tubes.

5. Sterilize an inoculating loop by placing it in a nonluminous Bunsen burner flame until it is red-hot.

6. Cool the loop by gently touching it onto the agar plate without touching any of the bacteria.

7. With the loop, scrape up a cell mass, without picking up any agar. Immerse the loop in the calcium chloride solution. Shake the loop back and forth vigorously to dislodge the bacteria. Resterilize the loop before placing it on the bench.

8. Ensure suspension of the bacterial cells in the calcium chloride by micropipetting in and out with a 200-μL micropipette and tip.

9. Using a new disposable pipette or a new pipette tip, repeat this procedure for the +pAMP tube.

10. Add 10 μL of 0.005-μg/μL pAMP directly into the +pAMP tube.

11. Incubate both tubes on ice in a polystyrene cup for 15 min.

12. While waiting, label 2 LB plates, one "−pAMP" and the other "+pAMP."

13. Repeat the labelling procedure in step 12 for 2 LB/amp plates.

14. Once the cells have incubated on ice for 15 min, remove the tubes and place immediately in a 42°C water bath for 90 s.

15. Immediately place the tubes back on ice for 1 min.

INVESTIGATION 6.1.2 *continued*

16. Using a micropipette, transfer 250 μL of LB broth to each culture tube. Gently mix the contents of the tubes by tapping on the sides.

17. Transfer 100 μL of −pAMP cell solution onto an LB plate and 100 μL of −pAMP cell solution onto an LB/amp plate.

18. Using the bacterial cell spreader, spread the bacterial cell suspensions over the whole surface of each plate as described below.

19. Dip the bacterial cell spreader into a 250 mL beaker containing 10 mL of ethanol. Pass the spreader through a nonluminous Bunsen burner flame. Cool the spreader on the edge of the plate away from any of the cell culture suspension.

20. Touch the spreader to the cell suspension and, while moving it back and forth for the full length of the plate, rotate the plate with the other hand, ensuring the cell suspension is spread across the whole plate.

21. Repeat steps 17–20 using the +pAMP solution.

22. Incubate the plates upside down for 12–24 h at 37°C.

23. Wipe the laboratory bench with 10% bleach.

Analysis

(b) Copy **Table 3** into your notebook.

Table 3 Results

	MM294 +pAMP	MM294 −pAMP
LB/amp plate		
LB plate		

(c) Divide each plate into quarters and count the number of colonies on each plate. Record the results in your table. If too many colonies per plate exist, just record "too many to count."

(d) Calculate the transformation efficiency of the calcium chloride protocol by following these steps:
  - Calculate the total mass of pAMP used (concentration × volume = total mass of pAMP).
  - Calculate the fraction of the solution that was spread on the plates (volume spread ÷ total volume in culture tube = fraction spread).
  - Calculate the mass (μg) of pAMP spread on each plate (total mass of pAMP × fraction spread = mass of pAMP spread).
  - Calculate transformation efficiency (number of colonies observed ÷ mass of pAMP spread).

(e) Repeat the transformation efficiency calculation for all plates.

(f) Average transformation efficiency is 10^4 colonies per μg of plasmid DNA (pDNA). Compare your transformation efficiency to the reported average.

(g) Indicate ways that the transformation efficiency might be improved.

(h) Explain the purpose of each of the plates: +LB and −LB; −LB/amp and −LB; +LB/amp and −LB/amp; +LB/amp and +LB.

(i) Do the results support the hypothesis? Explain.

(j) Explain the purpose of placing the bacterial suspensions in a 42°C water bath for 90 s.

(k) Why is it important that the bacterial suspensions are cooled on ice for 15 min in a calcium chloride environment?

(l) Predict the outcome if other salts, such as magnesium chloride or sodium chloride, were used instead of calcium chloride.

Evaluation

(m) Suggest possible sources of error in this procedure. Indicate the effects of these sources of error on the results.

(n) Using standard scientific format, prepare a written report.

Constructing a Plasmid Map

A gene must be inserted into an appropriate location of a plasmid to be expressed (**Figure 3**). For example, if the gene is placed within the promoter region, RNA polymerase will not be able to bind and the gene will never be transcribed. In addition, placing it after an operator allows the new gene can be turned on or off using the chemical that induces or represses the expression of the operon genes. In this activity, you will construct plasmid maps using information gathered from restriction enzyme digestions of plasmids.

The first step in building a plasmid restriction map is to determine which restriction enzymes digest the plasmid and where the digestion sites are located. Molecular biologists incubate plasmid DNA with restriction enzymes one at a time and then in conjunction with each other. The digestions are run on a gel and, once the fragment sizes are determined, the information can be used to build a plasmid map.

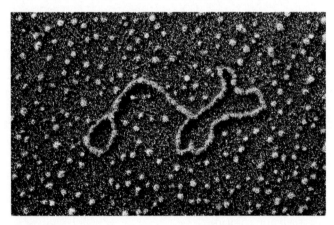

Figure 3
Plasmids are important tools in molecular biology because they can be used as vectors to introduce foreign genes into bacteria.

Materials

calculator   pair of compasses

Procedure

1. With a compass and pencil, draw a circle on a piece of paper to represent plasmid A

2. Using the information from **Table 4**, indicate where the restriction enzyme cuts are located on the plasmid. (Hint: Notice how all the fragments add up to the same total.) Check your answer with your teacher.

Table 4 Results of Restriction Fragment Digestion of Plasmid A

EcoRI (base pairs)	BamHI (base pairs)	EcoRI + BamHI (base pairs)
1800	1200	200
	600	600
		1000

3. Using the information provided in the next 3 digestions (**Tables 5–7**), draw the maps for plasmids B, C, and D.

Table 5 Results of Restriction Fragment Digestion of Plasmid B

EcoRI (base pairs)	BamHI (base pairs)	EcoRI + BamHI (base pairs)
1200	1200	200
600	600	200
		400
		1000

Table 6 Results of Restriction Fragment Digestion of Plasmid C

BamHI (base pairs)	HindIII (base pairs)	BamHI + HindIII (base pairs)
900	500	200
1500	1600	300
3000	3300	400
		700
		1200
		2600

Table 7 Results of Restriction Fragment Digestion of Plasmid D

EcoRI (base pairs)	BamHI (base pairs)	SmaI (base pairs)	EcoRI + BamHI (base pairs)	EcoRI and SmaI (base pairs)	BamHI + SmaI (base pairs)	EcoRI + BamHI + SmaI (base pairs)
800	250	900	200	350	150	150
1500	700	1400	250	450	250	200
	1350		300	450	550	250
			500	1050	600	300
			1050		750	350
						450
						600

Analysis

(a) Why is it necessary to digest a plasmid with individual restriction enzymes first, then follow with a combination of digestions?

(b) How is the number of fragments produced associated with the number of cut-sites available to a restriction enzyme?

(c) Why do all the fragment sizes for each restriction enzyme add up to the same total?

(d) Sketch the pattern that would be found on a gel after the digestions in **Table 6** are completed.

(e) Explain how a molecular biologist could determine whether a cut-site was in the middle of an antibiotic-resistance gene.

(f) Explain how restriction enzyme digestion results would change if a foreign gene were inserted into a plasmid.

🔍 **ACTIVITY 6.3.1**

Restriction Fragment Length Polymorphism Analysis

RFLP analysis generates DNA fingerprints. The DNA in question is digested into various-size fragments using different restriction enzymes. The fragments of the genome are separated using gel electrophoresis. Since the DNA sequence differs from individual to individual, RFLP analysis can be used to identify individuals.

If RFLP analysis is being used in a criminal investigation, the suspect's DNA fingerprint is compared with the DNA fingerprint found as evidence. A match with many restriction enzymes strongly supports the conclusion that the DNA found at the crime scene belongs to the suspect. Other evidence must also be present to support any accusation that the suspect committed the crime. A DNA fingerprint simply ascertains whether or not the DNA at the crime scene belongs to the suspect.

Materials

photocopy of the three suspects' DNA
photocopy of gel templates

Procedure

1. Your teacher will provide you with copies of the three suspects' DNA and the gel templates. Digest each suspect's DNA using restriction enzyme 1. Restriction enzyme 1 cleaves the recognition site 5′-AA/TT-3′.

2. Count the fragment sizes from the digestions for each suspect.

3. Plot the bands that would be produced on the gel template.

4. Repeat the procedure (steps 1–3) for restriction enzyme 2. Restriction enzyme 2 cleaves the recognition site 5′-GAA/TTC-3′.

Analysis

(a) Identify the individual whose DNA matches the DNA found as evidence.

(b) Why is it necessary to use more than one restriction enzyme in a criminal investigation?

(c) Did you digest the DNA of all three individuals with the second restriction enzyme? Why or why not?

Synthesis

(d) In this example, the gel pattern was obtained right away. Reread about the process of RFLP analysis (section 6.3) and describe how the gel pattern is obtained.

(e) Why are noncoding regions of DNA more discriminating than coding regions?

Key Expectations

- Describe the functions of the cell components used in genetic engineering. (6.1)
- Investigate and analyze the cell components involved in protein synthesis, using laboratory equipment safely and appropriately. (6.1)
- Outline contributions of genetic engineers, molecular biologists, and biochemists that have led to further developments in the field of genetics. (6.1, 6.2, 6.3, 6.4)
- Demonstrate an understanding of genetic manipulation, and of its industrial and agricultural applications. (6.2, 6.3, 6.4)
- Describe the main findings that have arisen from the Human Genome Project. (6.3, 6.4)
- Describe, and explain the implications of, the principal elements of the Canadian regulations on biotechnological products. (6.4)

Key Terms

agarose
Agrobacterium tumefaciens
antinociceptive transmitters
antisense oligonucleotides
autoradiogram
blunt ends
Bt toxin
cloned
competent cell
constant-length strands
copy number
dideoxy analogue
DNA fingerprinting
DNA primers
ethidium bromide
fluoresces
gel electrophoresis
gene therapy
genetic screening
host cell
hybridization
methylases
methyl group

multiple-cloning site
plasmids
polyacrylamide
polymerase chain reaction (PCR)
polymorphism
pronociceptive transmitters
recognition site
recombinant DNA
restriction endonucleases
restriction fragment length polymorphism (RFLP) analysis
Sanger dideoxy method
somatropin
Southern blotting
sticky ends
T4 DNA ligase
Taq polymerase
Thermus aquaticus
Ti plasmid
transformation
transgenic
variable-length strands
vectors

▶ **MAKE** a summary

In this chapter, you have been introduced to some of the numerous tools and techniques that molecular biologists, biochemists, and biotechnologists use to investigate and engineer genetic products. In addition, you have been provided with biotechnological examples of medical, agricultural, and forensic applications.

- Starting with the title "Genome" produce a concept map that encompasses the techniques and processes outlined in the chapter.
- Make a list of medical, agricultural, and forensic applications of biotechnology. Choose an example from each of these areas and create a table that addresses the social and economic advantages and disadvantages of the applications.

In your notebook, indicate whether each statement is true or false. Rewrite a false statement to make it true.

1. Restriction endonucleases recognize, bind, and cut specific four to eight nucleotide palindromic sequences of DNA.

2. To produce blunt ends, restriction endonucleases must disrupt both phosphodiester and hydrogen bonds.

3. Methylases add a methyl group to a specific recognition site, preventing the binding of the respective restriction enzyme.

4. DNA ligase is able to join both sticky- and blunt-end fragments together by reconstituting a phosphodiester bond using a condensation reaction.

5. Gel electrophoresis separates DNA fragments according to charge. The larger the fragment, the larger the charge it will carry because of the presence of more phosphate groups.

6. The introduction of a foreign gene into a living cell is known as transformation.

7. Plasmids are single-stranded circular pieces of DNA that are found in eukaryotic cells.

8. The Flavr Savr tomato was developed using antisense technology.

9. Restriction fragment length polymorphism is highly discriminating because it targets polymorphic regions of DNA.

10. The polymerase chain reaction makes a large number of copies of the entire genome.

In your notebook, record the letter of the choice that best completes the statement or answers the question.

11. Agarose gel electrophoresis of DNA involves the movement of DNA in an electric field through the pores of a rectangular slab of agarose. Since DNA has a pronounced negative charge, it naturally moves toward positive electrodes when an electric current is applied to the gel. How do large pieces of DNA move relative to small pieces in such a system?
    (a) slower, because they meet more resistance from the gel matrix
    (b) slower, because their covalent bonds have greater cumulative strength
    (c) faster, because they possess a greater number of negative charges
    (d) faster, because they have a higher ratio of negative-to-positive charges
    (e) faster, because they condense more and can, therefore, move more easily through the gel

12. Which of the following primers, used with PCR, would allow copying of the single-stranded DNA sequence 5′-ATGCCTAGGTC-3′?
    (a) 5′-ATGCC-3′
    (b) 5′-TACGG-3′
    (c) 5′-CTGGA-3′
    (d) 5′-GACCT-3′
    (e) 5′-GGCAT-3′

13. Which of the following tools of recombinant DNA technology is incorrectly paired with its use?
    (a) restriction endonuclease: the production of gene fragments for gene cloning
    (b) DNA ligase: an enzyme that cuts DNA, creating sticky ends
    (c) *Taq* polymerase: copies DNA sequences in the polymerase chain reaction
    (d) electrophoresis: separates DNA fragments according to size
    (e) plasmid: a vector that introduces foreign genes into bacterial cells

14. Which of the following best describes Southern blotting?
    (a) the detection of RNA fragments on membranes by specific radioactive antibodies
    (b) the detection of DNA fragments on membranes by a radioactive DNA probe
    (c) the detection of proteins on membranes using a radioactive DNA probe
    (d) the detection of proteins on membranes using specific radioactive antibodies
    (e) the detection of DNA fragments on membranes by specific radioactive antibodies

15. The restriction enzyme *Bam*HI was purified from which of the following?
    (a) *E. coli*
    (b) lambda bacteriophage
    (c) human cells
    (d) *Bacillus amyloliquefens*
    (e) bovine amniotic cells

16. Which of the following best describes antisense oligonucleotides?
    (a) They are used to promote the expression of a specific gene.
    (b) They are found only in bacterial DNA.
    (c) They hybridize to mRNA and prevent ribosomes from translating the mRNA into protein.
    (d) They can be used to introduce foreign DNA into living cells.
    (e) none of the above

NEL   An interactive version of the quiz is available online.
GO   www.science.nelson.com

Biotechnology **319**

Understanding Concepts

1. Explain why plasmids are good vectors for transformations.

2. Name and describe the four steps involved in the calcium chloride transformation process. Make sure you describe what is happening at the cellular level and why.

3. Describe the appearance of the autoradiogram produced from RFLP analysis radioactive probes as compared with the gel that the autoradiogram was originally produced from. Justify the differences, using your knowledge of the RFLP analysis protocol.

4. Define *restriction enzyme, plasmid, transformation, copy number, antisense oligonucleotides, and recombinant DNA.*

5. Calculate how many copies of an original DNA strand would be synthesized after 20 cycles of PCR.

6. Describe the role that methylases play in a bacterial cell. Describe their role in genetic engineering.

7. Depending on the restriction enzyme used, sticky- or blunt-end fragments may result. Differentiate between the two types of ends, and describe how a restriction fragment with a sticky end and a restriction fragment with a blunt end result. Use diagrams to help you explain your answer.

8. What features of DNA molecules and gels make it possible to separate DNA fragments by length using gel electrophoresis? Why does DNA carry a net negative charge?

9. Why do dideoxy analogues terminate chain elongation? Use a diagram to illustrate your answer.

10. The DNA fragment CGTCATCGATCATGCAGCTC contains a restriction enzyme recognition site. Identify the site.

11. Describe the role that calcium chloride plays in bacterial transformation.

12. Explain why bacterial plasmid vectors cannot be used to transform plant cells. What types of vectors are used instead?

13. Provide an example of a medical, forensic, and agricultural biotechnological application.

14. Explain how the presence of an antibiotic resistance gene marker in a plasmid can be used to determine whether a transformation protocol has been successful.

15. Differentiate between
    (a) T4 ligase and DNA ligase;
    (b) *Taq* polymerase and DNA polymerase.

Applying Inquiry Skills

16. Draw a diagram of a 4627-base-pair plasmid that has the following features:
    (a) a single *Eco*RI site
    (b) a single *Bam*HI site
    (c) two fragments, 1735 base pairs and 2892 base pairs, from *Eco*RI/*Bam*HI digestion
    (d) an antibiotic-resistance gene on the smaller fragment

17. **Figure 1** shows a restriction enzyme map of a plasmid. What would the products of the digestion be if you used
    (a) *Eco*RI and *Bam*HI to cut?
    (b) *Hin*dIII and *Sma*I to cut?
    (c) all four enzymes to cut?

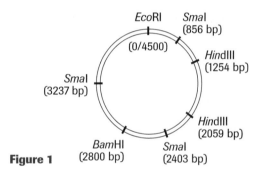

Figure 1

18. From question 17, draw the pattern of bands that would appear as a result of a restriction enzyme digestion with all four enzymes.

19. A piece of linear DNA is 150 000 base pairs in length and it is digested with a restriction enzyme that recognizes a seven-base-pair recognition site. How many fragments do you predict would be produced? Show your calculations.

20. A camper is caught by authorities frying a fish that is believed to be out of season. Luckily, a piece of the fish has yet to be subjected to the barbecue. The authorities want to ensure that this piece of fish is definitely the type of fish in question. DNA fingerprinting is conducted on the fish using RFLP analysis. The gel pattern shown in **Figure 2** is obtained. Determine whether the investigation should continue, given this evidence.

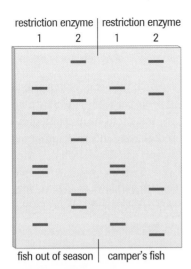

restriction enzyme | restriction enzyme
1   2 | 1   2

Figure 2

fish out of season | camper's fish

21. The DNA fragment 5′-GCAGGCCAGGA-3′ was subjected to the Sanger dideoxy sequencing method. Draw the gel it would produce.

22. The gel pattern shown in **Figure 3** was produced using the Sanger dideoxy sequencing method. Determine the sequence of the fragment that produced this gel.

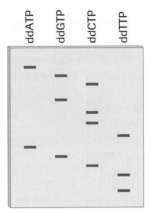

ddATP   ddGTP   ddCTP   ddTTP

Figure 3

23. A genetic engineer wants to isolate gene A and splice it into plasmid B. Write out the steps of the protocol that will be used. How could the genetic engineer experimentally ensure that gene A has been inserted into plasmid B without a bacterial cell undergoing a transformation?

24. You are given an unmapped circular plasmid DNA. After digestion with numerous enzymes, the fragments given in **Table 1** are obtained. Based on this information, draw the plasmid map.

Table 1 Fragments Obtained From Plasmid DNA

Enzyme	Length (base pairs)			
*Eco*RI	2500			
*Hin*dIII	2500			
*Bam*HI	1300	400	800	
*Bam*HI and *Eco*RI	200	400	600	1 300
*Bam*HI and *Hin*dIII	100	800	1 300	300
*Eco*RI and *Hin*dIII	2000	500		

Making Connections

25. In a cloning experiment, a gene for a protein was inserted into a plasmid with a high copy number. Define the term *copy number*. Would a high copy number be beneficial to an experimental transformation protocol? to an industrial process? Explain.

26. In a criminal investigation, PCR is conducted using more than one target sequence. Assume that only one target region is subjected to PCR in the suspect's DNA and the evidence DNA. What arguments could a lawyer make to discredit the process? Why would a match between the suspect's DNA and the evidence DNA after amplification of six or seven sites sufficiently support the conclusion that the evidence DNA came from the suspect? Would this be enough to convict the suspect of the crime? Why or why not?

Extensions

27. *Pseudomonas syringae* is a bacterium found in raindrops and in most ice crystals. Researchers have been able to cleave the frost gene from its genome, thereby preventing the bacteria from forming ice crystals. By spraying the bacteria on tomato plants, scientists have been able to reduce frost damage. The presence of these bacteria can extend growing seasons, thus increasing crop yields, especially in cold climates. However, environmental groups have raised serious concerns about releasing genetically engineered bacteria into the environment. Could these new microbes gain an unfair advantage over the naturally occurring species? What might happen if the genetically engineered microbes mutate? Do you think that genetically engineered microbes should be introduced into the environment? Support your opinion.

28. The gene for growth hormone has been isolated from human chromosomes and cloned in bacteria. The bacteria produce human growth hormone, which can be harvested in large quantities. The hormone is invaluable to people with dwarfism. Before its development, people with dwarfism relied on costly pituitary extracts. Although the prospect of curing dwarfism has been met with approval, some concerns have been raised about the potentially vast supply of growth hormone. Should individuals who do not have dwarfism but who want to grow a few more centimetres have access to the growth hormone biotechnology? Give reasons for your opinion.

PERFORMANCE TASK

Bovine Somatotropin: Canadian Approval or Disapproval

In this unit task, you will explore and research the use of bovine somatotropin in boosting milk production in cows (**Figure 1**). In some parts of the United States, the hormone has been approved for use on dairy farms, but it has been banned in Canada. Your challenge will be to gather, synthesize, and analyze information about the use of bovine somatotropin. Based on your research, you will decide whether the hormone should be approved in Canada. You will present your findings to a Canadian panel of dairy farmers, politicians, and milk consumers to convince them to allow or disallow bovine somatotropin use in Canada. Your final product will include a visual display and a presentation to the class as determined by your teacher.

Figure 1

Table 1 Components and Expectations for Performance Task

Component	Specific expectations and suggestions
Molecular biology	• Research and report on the genetic engineering techniques that have been used to isolate and mass produce bovine somatotropin. • Use a variety of visuals to illustrate the processes, being sure to include all the molecular biology tools and techniques that were used.
Health and safety	• Research the pathway by which bovine somatotropin boosts milk production in cows. • Research any side effects to the health of the cow, or to the consumer of the milk. • Identify any changes in the health care of the cows after they have been placed on bovine somatotropin. • Use a variety of sources and studies to gather your information. • Use diagrams to help illustrate any biochemical pathways affected by bovine somatotropin.
Economics	• Research dairy production in states where the hormone has been approved, before and after the hormone was in use. • Compare the Canadian market to the U.S. market for dairy products. • Explore Canadian export of dairy products to the U.S. and vice versa. • Investigate dairy subsidies to farmers in the U.S. compared with those in Canada. • Research and compare the size of an average dairy farm operation in Canada versus in states where the hormone has been approved. • Evaluate the global significance of the introduction of bovine somatotropin. • Identify the lobby groups and their arguments on both sides of the issue. • Present your findings in graphic format using bar, line, and pie graphs and/or charts. • Support your claims using quantitative data.
Regulations and policies	• Using the Internet, investigate Canadian policy with respect to the approval of biotechnological products. • Justify why bovine somatotropin has been banned in Canada, based on these regulations. • Suggest modifications to the use of bovine somatotropin for it to gain approval in Canada. • Suggest further studies that could be performed to meet the criteria set out by Health Canada.
Social and ethical issues	• Interview a dairy farmer and elicit his/her opinion with respect to a boost in production of milk using bovine somatotropin. • Identify any social or environmental costs associated with the use of bovine somatotropin in Canada.
Current and future research	• Describe any current research initiative associated with the introduction of bovine somatotropin in Canada. • List any questions you would like answered with respect to bovine somatotropin.
Conclusion and synthesis	• Based on your research, make a decision as to whether bovine somatotropin should be introduced into the Canadian dairy farmer market. • Support your opinion with qualitative and quantitative data that you have gathered from your research.

Understanding Concepts

In your notebook write the terms that complete these sentences.

1. Chromosomes consist of DNA and _____.

2. A _____ forms the backbone in a polynucleotide.

3. DNA replication is a _____ process because the molecules produced contain one parent strand and one newly synthesized strand.

4. Each strand acts as a _____ for the construction of a new, complementary strand during DNA replication.

5. DNA is a polymer of _____.

6. Three adjacent nucleotides in mRNA that form the code for a specific amino acid are called a _____.

7. Each tRNA carries a specific _____, which always corresponds to a particular anticodon.

8. Transcription is catalyzed by the enzyme _____.

9. Copying the information in DNA into a molecule of mRNA is called _____.

10. _____ is the process of converting a codon sequence into an amino acid sequence.

11. The _____ is a triplet sequence of nucleotides on the tRNA molecule that base pairs with triplet sequences on the mRNA molecule.

12. _____ are specific enzymes that cut DNA at precise locations known as _____.

13. What type of bond holds the two DNA strands together in the double helix? Why is it important that this bond be a weak bond?

14. Maintaining the correct nucleotide sequence requires that newly formed and existing strands of DNA be checked and mistakes repaired. Explain how these checks and repairs are made. What advantage, if any, would there be to having a mutation perpetuated?

15. Describe the process of semiconservative DNA replication.

16. Briefly describe the events that take place in each of the stages of DNA replication.

17. List three positive results of recombinant DNA technology.

18. Differentiate between transcription and translation.

19. Describe how DNA and RNA differ in their composition, structure, function, and location.

20. Summarize the process of protein synthesis, beginning with the movement of the mRNA to the ribosome.

21. Describe the process of gene regulation, making specific reference to the following terms: *regulator gene, repressor protein, initiator or promoter gene, operator gene or binding site* for the repressor, and *structural gene.*

22. What are some common mutagens? Describe three types of gene mutations.

23. Differentiate between the following terms:
    (a) *exon* and *intron*
    (b) *codon* and *anticodon*
    (c) *missense mutation* and *nonsense mutation*
    (d) *RNA* and *DNA*
    (e) *inducer molecule* and *corepressor*
    (f) *promoter* and *operator*
    (g) *induced mutation* and *spontaneous mutation*
    (h) *SINEs* and *LINEs*
    (i) *mRNA* and *tRNA*
    (j) *restriction endonuclease* and *recognition site*

24. Before an mRNA primary transcript is able to leave the nucleus and enter the cytoplasm, numerous post-transcriptional modifications need to be made. List the modifications that the mRNA primary transcript will undergo in a eukaryotic cell, and explain why each one is necessary.

25. Sickle cell anemia is a genetic disorder. Explain this statement, using the genetic code as a basis for your answer.

26. List seven differences between a prokaryotic genome and a eukaryotic genome.

27. One strand of a DNA molecule contains the nucleotide proportions 15% A, 30% T, 20% G, and 35% C. What proportions of the 4 nucleotides are expected in the double-stranded form of this DNA?

28. Numerous enzymes are involved in DNA replication. Outline the role that the following enzymes play: DNA ligase, DNA gyrase, DNA helicase, DNA polymerase I, and DNA polymerase III.

29. For a gene to be inserted into a bacterium, numerous different biological tools and molecular techniques need to be used. Explain the role that each of the following plays in the genetic engineering process:
    (a) restriction endonucleases
    (b) plasmids
    (c) gel electrophoresis
    (d) DNA ligase

30. Silent mutations do not lead to deleterious effects to an organism. Explain why this is the case.

31. Using diagrams, distinguish between the *lac* operon and the *trp* operon. How are they similar in their mode of action? How are they different?

32. Outline the contributions of the following scientists to molecular biology:
    (a) Hershey and Chase   (d) Rosalind Franklin
    (b) Watson and Crick   (e) Mary Dell Chilton
    (c) Cohen and Boyer

Applying Inquiry Skills

33. Calculate how long a DNA molecule would be if it had 5 million bases on one of its strands. Calculate how many turns would be found in its double-helical structure.

34. The distribution of nucleotides in **Table 1** originated from analysis of single-stranded DNA and double-stranded DNA. Identify which samples are double stranded and which samples are single stranded. Support your answer.

Table 1 Distribution of Nucleotides

	Adenine	Guanine	Thymine	Cytosine
Sample A	15%	45%	10%	30%
Sample B	40%	10%	40%	10%
Sample C	20%	25%	20%	35%

35. Meselson and Stahl's experiment indicated that DNA replicates semiconservatively. Calculate the percentage of DNA double helixes that would not contain any of the original parent strands after three generations of replication. Show all your work.

36. Hypothetically speaking, there are 95 amino acids and only 6 nucleotides available. Calculate how many minimum nucleotides are required to code for all 95 amino acids.

37. The following metabolic pathway exists in a strain of bacteria. The bacteria require product D to survive. The three enzymes required to produce D are controlled by three separate genes (genes 1 to 3).
    A ⇨enzyme 1 B ⇨enzyme 2 C ⇨enzyme 3 D
    (a) If a mutation occurred in the first gene that controls enzyme 1, what effect would this have on

the biochemical pathway? As a researcher, how could you identify that the mutation has occurred in the first gene? For the biochemical pathway to continue to produce D, what would bacteria possessing a mutation in gene 1 have to be supplied with in their nutrient media?

(b) Assume that enzyme 2 and enzyme 3 are both under the control of the same gene. If a mutation in this gene occurred, what would the results be? How would the results differ from the results of a mutation in one of two genes that controlled enzymes 2 and 3 separately? How could you experimentally determine whether there was only one gene controlling enzymes 2 and 3?

(c) Assume that this pathway is under the control of an operon. How would a mutation in the operator affect the pathway? Explain why this would be the case.

(d) Assume that D, the product of the biochemical pathway, is a corepressor. Describe the conditions that must exist for D to be synthesized and for D not to be synthesized.

(e) Draw the operon for this biochemical pathway. Include the operator, the promoter, and all the genes.

(f) Illustrate the following two scenarios using your operon system. Include the repressor protein and RNA polymerase in your illustration.
    (i) the levels of D are high
    (ii) the levels of D are low

38. The DNA fragment shown in **Figure 1** was digested with *Hind*III, *Bam*HI, and *Sma*I.
    (a) What size fragments would be produced if
        (i) it is digested only with *Hind*III?
        (ii) it is digested only with *Bam*HI?
        (iii) it is digested only with *Sma*I?
        (iv) it undergoes a double digestion with *Hind*III and *Bam*HI?
        (v) it undergoes a double digestion with *Bam*HI and *Sma*I?
        (vi) it is digested with all three enzymes?
    (b) Draw the gel electrophoresis pattern that would be obtained from each of the digestions in questions (i)–(vi).

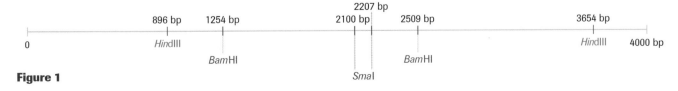

Figure 1

39. An unknown plasmid was digested with *Eco*RI and *Bam*HI. The fragment sizes were determined using gel electrophoresis and the results are shown in **Table 2**.

Table 2

*Eco*RI (base pairs)	*Bam*HI (base pairs)	*Eco*RI + *Bam*HI (base pairs)
1000	850	300
750	900	400
		600
		450

(a) Determine the size of the plasmid.

(b) Draw a restriction enzyme map of this plasmid, indicating the locations of the recognition sites of *Bam*HI and *Eco*RI.

40. The following is a fragment of DNA that codes for part of a protein required in cellular respiration:
3′-TACATAGCATGTATAAGCATAAATGTAGTAC-TAATT-5′

(a) Write the complementary strand for this fragment.

(b) Which of the two strands will RNA polymerase use as the coding strand?

(c) Transcribe the coding strand into mRNA.

(d) Translate the mRNA into amino acids using the genetic code.

(e) Generate a list of anticodons that would be found on tRNAs that would deliver the first two amino acids.

(f) Change one base in the fourth codon that would result in a silent mutation.

(g) If the seventh base is changed from G to A, what effect would this have on the protein synthesized? What class of mutation does this belong to?

(h) If the seventh base is completely eliminated, what effect would this have on the protein synthesized? What class of mutation does this belong to?

(i) Which of the mutations in questions (g) and (h) would be more deleterious to an organism? Why?

41. A bacterium that is not normally tetracycline resistant is transformed using a plasmid that carries the tetracycline-resistance gene. Design an experimental protocol that would allow you to identify colonies that have been conclusively transformed with the plasmid. Include a positive and negative control. Outline the expected results of your protocol if the transformation procedure is successful, and if it is not successful. What additional measures could be taken to ensure that the transformation is successful and that the results are not false positive?

42. **Figure 2** is a sketch of a gel produced when lambda bacteriophage DNA is digested with *Hin*dIII. The fragment sizes for *Hin*dIII are as follows: 25 000, 9416, 6557, 4361, 2322, and 2027. Using these bands as size standards, determine the fragment size of the bands that appear in lanes B and C. You will have to measure the distances that the *Hin*dIII bands have travelled and plot a semilogarithmic graph (distance travelled versus base-pair size).

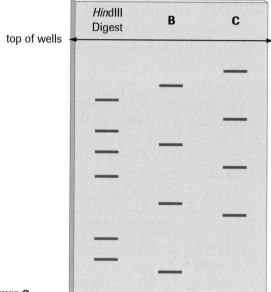

Figure 2

43. Assume that in lane C (**Figure 2**) *Pst*I was used to digest the DNA. Indicate what the banding pattern would look like if the DNA were incubated with M. *Pst*I methylase beforehand. Justify your answer.

Making Connections

44. The luciferase gene produces an enzyme that is responsible for the luminescence of fireflies. With the use of biotechnological techniques, the luciferase gene has been inserted into a tobacco plant. Since the tobacco plant cells are able to express the gene, they glow in the dark. Explain why a tobacco plant can express a firefly gene. Predict what would happen if the luciferase gene were inserted into a bacterium. Support your prediction using your knowledge of eukaryotic and prokaryotic gene expression.

45. To isolate Okazaki fragments, Pauling and Hamm in 1969 observed DNA replication in the presence of a temperature-sensitive DNA ligase. The DNA ligase did not work at temperatures above 40°C, but functioned properly at 25°C. Pauling and Hamm allowed the DNA to replicate at 25°C and then transferred the DNA into a medium that contained radioactive nucleotides, while raising the temperature to 40°C. They repeated the same experiment with DNA ligase that works above 40°C. On the basis of your knowledge of DNA replication, the formation of Okazaki fragments, and the role of DNA ligase, answer the following questions:

(a) What would be a hypothesis for this experiment?

(b) Why was it necessary to repeat the experiment with DNA ligase that functioned above 40°C?

(c) Where (if anywhere) would radioactivity have been found?

(d) Where (if anywhere) would Okazaki fragments have been found?

(e) What conclusions would Pauling and Hamm have been able to make from this experiment?

46. To detect and identify a mutation that causes a genetic disorder, researchers examine families who have a large incidence of this genetic disorder in their family tree. Describe how a mutation that causes a genetic disorder could be pinpointed using the tools of molecular biology that you have read about in this chapter.

47. One of the most common methods of treating cancer is using chemical analogues that mimic the structure of nucleotides. These analogues are incorporated into DNA and prevent DNA polymerase from further building a complementary strand during DNA replication.

(a) Explain why this is an effective strategy against the proliferation of cancer cells.

(b) One of the side effects of chemotherapy is hair loss. Explain why this is the case.

Extensions

48. Telomeres are repetitive units of DNA found at the end of chromosomes. The role of telomeres is to prevent chromosomes from sticking together and to act as a buffer against the shortening of DNA during DNA replication. Studies have shown that telomeres shorten with aging, until the cell eventually dies. In contrast, cancer cells, despite their rapid duplication, generally do not experience a shortening of their telomeres. The enzyme telomerase has been found in large amounts in cancer cells. Telomerase is responsible for extending the telomeres. How could the manipulation of telomerase be used to produce a cure for cancer? Using the Internet and scientific journals, investigate the current research being conducted into this possible treatment. Present your findings in a written report.

 www.science.nelson.com

49. During the past century, humans discovered that antibiotics could be used to combat bacterial infections. The effectiveness of antibiotics against bacterial invasions is slowly diminishing because of the rapid rate that bacteria mutate and develop defence mechanisms. Instead of battling the bacterium itself, some researchers are trying to understand how disease-causing organisms penetrate a cell. By understanding the strategy that the organisms use, researchers hope to build cells that can provide more of a defence against these invading organisms. Hence, instead of developing offensive drugs, researchers are hoping to enhance defence mechanisms. Researchers have found that host cells actively participate in the introduction of pathogens within them. Bacteria and viruses manipulate the host cell so that it allows the pathogen in. Using the Internet, scientific journals, and other print resources, research the methods that invading organisms use to gain access to a host cell. Present the results of your research in a written report.

 www.science.nelson.com

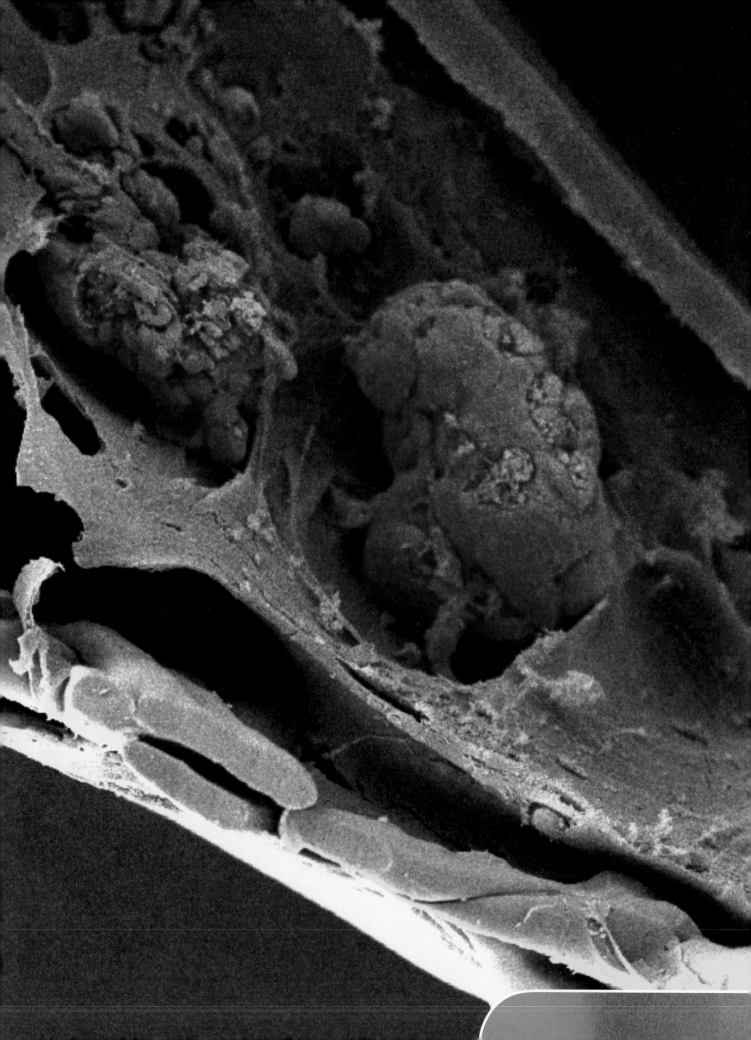

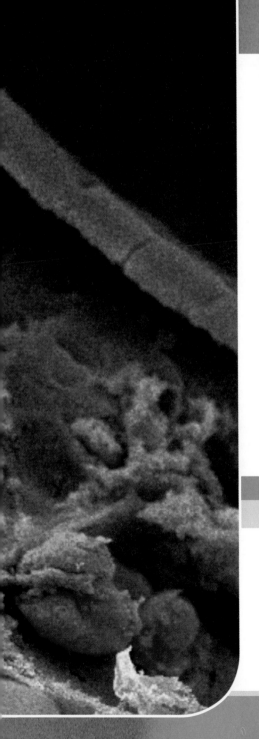

Homeostasis

In the 1987 movie *RoboCop*, a police officer is killed in the line of duty. Scientists use his remains to construct a cyborg—part man, part machine—that is virtually indestructible. Although the film may have been ahead of its time in showing how science can rebuild human body parts, it did reflect developments in a rapidly growing field. Understanding how an organ works and interacts with other body systems provides a basis for engineering new organ systems. Organs do not work independently; rather, they work in coordinated systems that continuously respond and adjust to changing environments.

Researchers are investigating artificial substitutes for many human organs and cells. Artificial cells that attempt to mimic the biological processes of natural cells could one day be used to help build artificial kidneys and livers. Synthetic fabric could temporarily serve as artificial skin for burn victims. A bio-artificial pancreas that is currently being tested in animal models at the University of Alberta could one day provide a cure for diabetes. What characteristics do these substitutes need to function effectively in the body? Could they eventually outperform their biological counterparts? In this unit, you will study homeostasis, the body's attempt to adjust to a fluctuating external environment. You will also learn about the mechanisms that maintain homeostasis in the body and the challenges raised by emerging research in physiology.

▶ Overall Expectations

In this unit, you will be able to

- describe and explain mechanisms involved in the maintenance of homeostasis;
- investigate, through experiments and models, how feedback systems enable the body to maintain homeostasis;
- analyze how environmental factors and technological applications affect homeostasis;
- demonstrate an awareness of socially and personally relevant issues raised by emerging research in physiology.

ARE YOU READY?

Knowledge and Understanding

1. Examine **Figure 1** and explain the function of each of the labelled tissues.

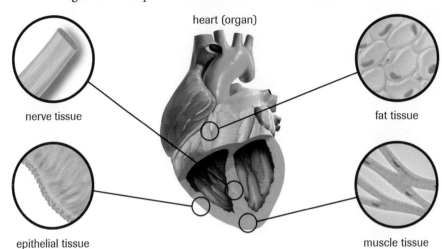

heart (organ)

nerve tissue

fat tissue

epithelial tissue

muscle tissue

Figure 1
Tissues of the human heart

2. Explain what happens to an animal cell placed in
   (a) a hypertonic solution;
   (b) a hypotonic solution.

3. Copy **Table 1** into your notebook and fill in all the blank spaces.

Table 1

Organ system		Excretory system	
organ	ovary		
tissue		epithelial, blood, nerve, fat (adipose), and connective tissues	
cell	egg cell		white blood cell (leukocyte)

4. In **Table 2**, match the organ(s) in the left-hand column with the corresponding regulatory function in the right-hand column. (Note: An organ can have more than one function, and a function can be linked to more than one organ.)

Table 2

Organ(s)	Regulatory function
(a) skin	(i) disease prevention
(b) brain	(ii) thermoregulation (regulation of body temperature)
(c) lymph vessel and lymph node	(iii) maintaining blood sugar
(d) pancreas	(iv) maintaining blood pressure
(e) testes	(v) osmoregulation (regulation of the osmotic balance of body fluids)
(f) heart	(vi) maintaining blood pH
(g) kidney	(vii) growth

▶ **Prerequisites**

Concepts

- levels of cell organization
- factors affecting enzyme activity
- role of hormones in homeostatic control
- male and female reproductive hormones
- antibody-antigen interaction
- effects of prescription and nonprescription drugs on homeostasis

Skills

- design and conduct controlled experiments
- select and use laboratory materials accurately and safely
- organize and display experimental observations and data in suitable tabular and graphical formats
- present informed opinions about personal lifestyle choices related to homeostasis

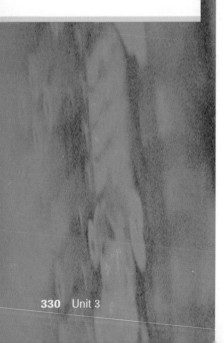

Inquiry and Communication

5. A group of students conducts an experiment to determine how the body responds to stress. An ice cube is placed on the back of a subject's neck while another group member monitors changes in the subject's pulse rate.
   (a) Why is pulse rate used to monitor stress?
   (b) Create a hypothesis for the experiment.
   (c) Identify the dependent and independent variables in the experiment.
   (d) What variables must be controlled to obtain reliable data?
   (e) Design a data table for the experiment.
   (f) Would you expect identical data from different subjects? Explain your answer.
   (g) Through the course of the experiment, when would you take the pulse rate? Give your reasons.
   (h) What practical information might the experiment provide?

Figure 2
Pulse rate is monitored in response to stress.

6. The two photos in **Figure 3** show a white blood cell and *Escherichia coli* bacterial cells.

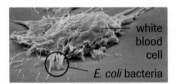 

(a)   **(b)**

Figure 3
A white blood cell and
E. Coli bacteria

   (a) Describe what is happening in **Figure 3**. Draw a conclusion from the two pictures.
   (b) Explain how this process helps to maintain a balanced internal environment.

7. Scientists monitored six types of ions in the cells of *sargassum* (a brown alga often called seaweed) when maintained in brackish water (a mixture of salt and fresh water) and in marine water (seawater). The scientists noted that the concentration of ions did not change. The results are shown in **Table 3**.

Table 3

Type of ion	Cell	Marine environment	Brackish water
calcium	1.7	12	1.7
magnesium	0.005	57	6.5
sulfate	0.01	36	2.8
sodium	90	500	60
potassium	490	12	1.4
chloride	520	520	74

All concentrations are measured in mmol/L.

Assume that the cell membrane is permeable to all of the ions.
   (a) Which ion must be actively transported inside the cell in both brackish water and marine water? Explain your answer.
   (b) Which ion enters the cell by diffusion from marine water, but must be actively transported inside the cell by active transport from brackish water? Explain your answer.
   (c) Explain how a cell could maintain its sodium ion concentration despite living in marine or brackish water environments.

7

Maintaining an Internal Balance

In this chapter, you will be able to

- explain the role of the kidneys in maintaining water and ion balance;

- describe and explain homeostatic processes involved in maintaining water, ionic, thermal, and acid–base equilibriums in response to both a changing environment and medical treatments;

- construct a model that illustrates the essential components of the homeostatic process;

- compile and display information about homeostasis in a variety of formats;

- present informed opinions about problems related to the health industry, health legislation, and personal health;

- describe some Canadian contributions to knowledge and technology in the field of homeostasis.

In August 2000, Canadians Peter Reid and Lori Bowden won the Ironman Canada Triathlon men's and women's titles, completing the gruelling 226-km swim–cycle–run in 8 h, 29 min, and 49 s and 9 h, 17 min, and 23 s, respectively. The husband-and-wife team, dubbed "the world's fittest couple," went on to further victories: Peter once again won the Ironman Canada Triathlon in August 2001, and Lori won her third straight women's crown at the Australian Ironman competition in April 2001.

Imagine completing a 4-km swim and a 180-km bicycle ride only to have a 42-km marathon ahead of you. To meet the high-energy demands of this challenging competition, the body undergoes a series of adjustments to continue operating. One such adjustment is an increase in the rate of cellular respiration. However, the oxidation of glucose generates waste energy, in the form of heat. During severe strenuous exercise, body temperature can increase to more than 39°C, a temperature that under normal circumstances would be associated with a fever. To dissipate heat, sweat is produced.

The evaporation of sweat is a cooling process. However, temperature changes also cause other body systems to require adjustments. The loss of water alters the volume of body fluids, which can cause a drop in blood pressure. The heart and circulatory system respond to changes in blood pressure, while the kidneys conserve water in an attempt to maintain fluid volume.

Water is not the only thing lost with sweating; many salts essential for nerve function and muscle contraction are carried to the skin with the perspiration. The kidneys are also responsible for maintaining the body's electrolyte balance.

With the increase in energy demands, the body must keep blood glucose within normal levels to maintain ATP supplies and sustain exercise. The pancreas uses two hormones, insulin and glucagon, to regulate blood sugar levels. Insulin stimulates the uptake and utilization of glucose by the muscles, and glucagon stimulates glucose release.

The nervous system monitors oxygen levels, increasing the breathing rate. The nervous system also interacts with the circulatory system to divert blood toward the skeletal muscles, brain, and heart and away from organs less essential for exercise, such as the stomach.

💡 REFLECT on your learning

1. What dangers exist if your body is unable to regulate its temperature?
2. How does your body respond to an increase in body temperature?
3. What dangers exist if your body is unable to regulate the fluid balance of your tissues?

Detecting Temperature Changes

Heat and cold receptors, rather than detecting specific temperatures as does a thermostat, are adapted to signal *changes* in environmental temperatures.

Materials: 3 bowls or large beakers, warm water, room-temperature water, cold water

- Fill three bowls or large beakers with water—one with warm, one with room-temperature, and one with cold water.
- Place your right hand in the cold water and your left hand in the warm water (**Figure 1**). Allow your hands to adjust to the temperature and then transfer both hands to the bowl that contains room-temperature water.
  (a) Describe what happens.
  (b) Explain why you might feel a chill when you step out of a warm shower even though room temperature is comfortable.

(c) Explain the following observations: When a frog is placed in a beaker of water above 40°C, the frog will leap out immediately. When the frog is placed in room-temperature water and the temperature is slowly elevated, the frog will remain in the beaker.

Figure 1

The human body works best at a temperature of 37°C, with a 0.1% blood sugar level and a blood pH level of 7.35. However, the external environment does not always provide the ideal conditions for life. Air temperatures in Canada can fluctuate between −40°C and 40°C. Rarely do foods consist of 0.1% glucose and have a pH of 7.35. You also place different demands on your body when you take part in various activities, such as playing racquetball, swimming, or digesting a large meal. Your body systems must adjust to these variations to maintain a stable internal environment.

The term **homeostasis** refers to the body's attempt to adjust to a fluctuating external environment. The word is derived from the Greek words *homoios*, meaning "similar" or "like," and *stasis*, which means "standing still." The term is appropriate because the body maintains a constant balance, or steady state, through a series of monitored adjustments. This system of active balance requires constant monitoring and feedback about body conditions (**Figure 1**). An increase in the heart rate during exercise or the release of glucose from the liver to restore blood sugar levels are a couple of examples of the adjustments made by regulators.

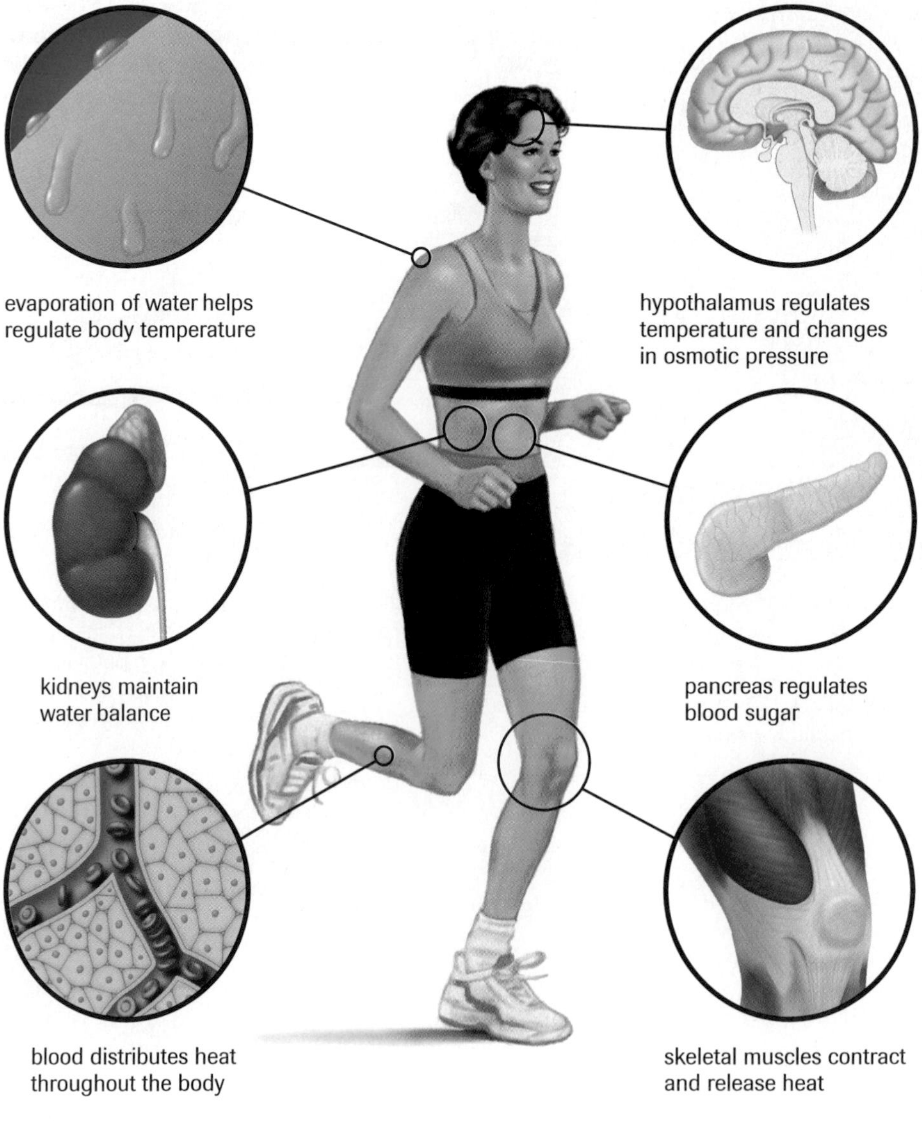

homeostasis process by which a constant internal environment is maintained despite changes in the external environment

evaporation of water helps regulate body temperature

hypothalamus regulates temperature and changes in osmotic pressure

kidneys maintain water balance

pancreas regulates blood sugar

blood distributes heat throughout the body

skeletal muscles contract and release heat

Figure 1
Homeostasis requires the interaction of several regulatory systems. Information about blood sugar, fluid balance, body temperature, oxygen levels, and blood pressure are relayed to a nerve coordinating centre once they move outside the normal limits. From the coordinating centre, regulators bring about the needed adjustments.

All homeostatic control systems have three functional components: a monitor, a co-ordinating centre, and a regulator. Special sensors (monitors) located in the organs of the body signal a coordinating centre once an organ begins to operate outside its normal limits. The coordinating centre relays the information to the appropriate regulator, which helps to restore the normal balance. For example, when carbon dioxide levels increase during exercise, chemical receptors in the brain stem are stimulated. Nerve cells from the brain then carry impulses to muscles that increase the depth and rate of breathing. The increased breathing movements help flush excess carbon dioxide from the body.

A group of chemical receptors in the arteries in the neck can detect low levels of oxygen in the blood. A nerve is excited and sends a message to the brain, which relays the information by way of another nerve to the muscles that control breathing movements. This system ensures that oxygen levels are maintained within a tolerable range. It is for this reason that homeostasis is often referred to as a **dynamic equilibrium**. Although there are fluctuations in blood glucose, body temperature, blood pressure, and blood pH, the homeostatic mechanism ensures that all body systems function within an acceptable range to sustain life (**Figure 2**).

dynamic equilibrium condition that remains stable within fluctuating limits

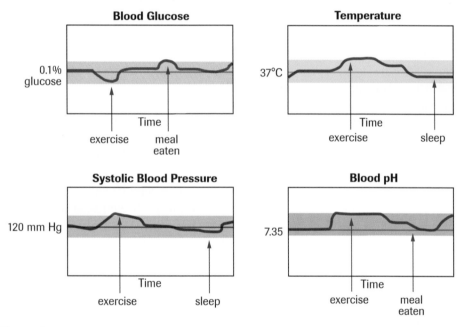

Figure 2
Blood glucose is maintained within a narrow range and movement outside of the range can signal disease. Body temperature can fluctuate by +2°C with exercise and −2°C with sleep. Systolic blood pressure is usually near 120 mm Hg but can move as high as 240 mm Hg in a very fit athlete for a limited time during strenuous exercise. Blood pH operates within a narrow range, and changes of +/−0.2 can lead to death.

Homeostasis and Feedback

Mechanisms that make adjustments to bring the body back within an acceptable range are referred to as **negative feedback** systems. The household thermostat is an example of such a system (**Figure 3**, page 336). In this case, the coordinating centre, called a thermostat, also contains the monitor (a thermometer). When the room temperature falls below a set point, say 20°C, the thermostat switches on the regulator (the furnace). When the thermometer detects a temperature above the set point, the thermostat switches off the furnace. This type of control circuit is called negative feedback because a change in the variable being monitored triggers the control mechanism to counteract any further

negative feedback process by which a mechanism is activated to restore conditions to their original state

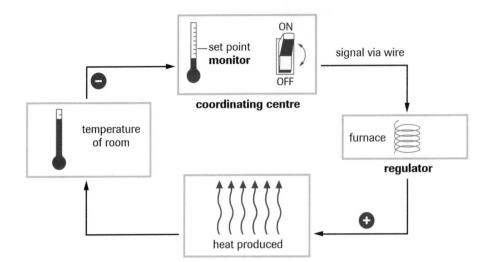

Figure 3
The household thermostat illustrates a negative feedback system. When the variable exceeds the set point, the coordinating centre turns the regulator off. The + indicates "stimulation" or "activation," and the − indicates "inhibition" or "turning off."

positive feedback process by which a small effect is amplified

change in the same direction. Negative feedback mechanisms prevent small changes from becoming too large. Most homeostatic mechanisms in animals operate on this principle of negative feedback.

Positive feedback systems are less common in the body. Whereas negative feedback systems are designed to resist change, positive feedback systems reinforce the change. Positive feedback systems move the controlled variable even further away from a steady state. The value of a positive feedback system is that it allows a discrete physiological event to be accomplished rapidly. Once this event is accomplished, the feedback system stops.

A good example of positive feedback in humans is the birth process. A decrease in progesterone, a hormone associated with pregnancy, is believed to initiate small contractions of the uterus or womb. In turn, the contractions bring about the release of another hormone, oxytocin, which causes much stronger contractions of the uterus. As contractions build, the baby moves toward the opening of the uterus, the cervix. This causes even greater release of oxytocin and stronger contractions until the baby is expelled from the uterus. Once the baby is expelled, the uterine contractions stop, which in turn stops the release of oxytocin. The event is accomplished and the positive feedback system is complete.

SUMMARY *Homeostasis and Control Systems*

- Homeostasis refers to the body's attempt to adjust to a fluctuating external environment.
- All homeostatic control systems have three functional components: a monitor, a coordinating centre, and a regulator.
- Negative feedback mechanisms trigger a response that reverses the changed condition.
- Positive feedback systems move the controlled variable even further away from a steady state.

▶ *Section 7.1* Questions

Understanding Concepts

1. Explain in your own words the concept of homeostasis.
2. Differentiate between positive and negative feedback systems.
3. Explain homeostasis using the example of a thermostat in a home.
4. Explain the negative feedback system shown in **Figure 4**.

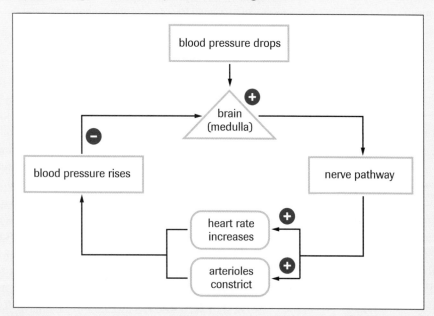

Figure 4

5. During lactation (milk production), the suckling by the baby stimulates the production of oxytocin, which in turn causes contraction of smooth muscle surrounding the milk duct, causing milk to flow. The flow of milk increases the suckling by the baby and more oxytocin is produced.
    (a) Identify the feedback system described above.
    (b) What would end the feedback loop?

Applying Inquiry Skills

6. Draw and label a flow chart of a positive feedback system for the release of oxytocin.
7. Draw and label a negative feedback system for a home heating and cooling system. Include the following in your diagram: furnace, air conditioning unit, and thermostat.

Making Connections

8. Every year some deaths of athletes are linked with extreme exercise in hot, humid conditions.
    (a) Why is exercise at high temperatures dangerous?
    (b) Make suggestions that could help safeguard the health of athletes.
9. Boxers have been known to take laxatives to help them meet the weight restrictions of lower fighting classes.
    (a) How would taking laxatives help them lose weight?
    (b) Why might this practice be dangerous?

thermoregulation maintenance of body temperature within a range that enables cells to function efficiently

hypothalamus region of the vertebrate's brain responsible for coordinating many nerve and hormone functions

Figure 1
A circulatory system designed to conserve heat helps fast-swimming fish, such as tuna, maintain an internal environment that supports metabolic activities at a high rate.

Different species of animals are adapted to different temperature ranges, and each animal has an optimal temperature range. **Thermoregulation** is the maintenance of body temperature within a range that enables cells to function efficiently. To understand the mechanisms of temperature regulation, we first need to consider the exchange of heat between animals and their environment.

Invertebrates, and most fish, amphibians, and reptiles, are referred to as ectotherms. These animals depend on air temperature to regulate metabolic rates; therefore, activity is partially regulated by the environment. To overcome this limitation, some reptiles have developed behavioural adaptations, such as sunning themselves on rocks or retreating to shaded areas, to regulate body temperature. Some very active fish, such as tuna (**Figure 1**), have highly adapted circulatory systems designed to conserve heat. Some of their internal organs are significantly warmer than the surrounding water.

Mammals (including humans) and birds are referred to as endotherms; they are able to maintain a constant body temperature regardless of their surroundings. Endotherms adjust to decreases in environmental temperatures by increasing the rate of cellular respiration to generate heat. In humans, normal body temperature is usually stated as 37°C; however, there is variation within any population. Some humans have their "thermostat," the **hypothalamus**, set a few degrees lower, while others operate at higher temperatures. Recent studies even indicate that body temperatures vary slightly during the day. Temperatures in most individuals fall slightly during the night. It should also be noted that core temperatures and peripheral temperatures of the body tend to vary from each other. Core temperatures, found in the chest cavity, the abdominal cavity, and the central nervous system, remain relatively constant and are usually higher than 37°C. The peripheral temperatures can be as much as 4°C lower on very cold days.

Response to Heat Stress

How does the body protect itself against excessive heating caused by exercise or high environmental temperatures? **Figure 2** shows what it does. When sensors in the brain detect a rise in body temperature, a nerve message is coordinated within the hypothalamus and a signal is sent to the sweat glands to initiate sweating. The evaporation of perspiration from the skin causes cooling. At the same time, a nerve message is sent to the blood vessels in the skin, causing them to dilate. This allows more blood flow to the skin. Because the skin has been cooled by the evaporation of sweat, the blood loses heat to the skin. When blood from the skin returns to the core of the body, it cools the internal organs.

Along with water, valuable salts are also carried to the skin's surface and lost with perspiration. Later in this chapter you will study how the kidneys help regulate the loss of electrolytes.

Response to Cold Stress

In many ways, your response to cold mirrors your response to heat (**Figure 2**). When external temperatures drop, thermoreceptors in the skin send a message to the hypothalamus. Acting as a coordinating centre, the hypothalamus sends messages to the organs and tissues to increase body temperature. Nerves going to the arterioles of the skin cause smooth muscles to contract and the arterioles to constrict, limiting blood flow. This reduces heat loss from the skin and retains heat in the core of the body.

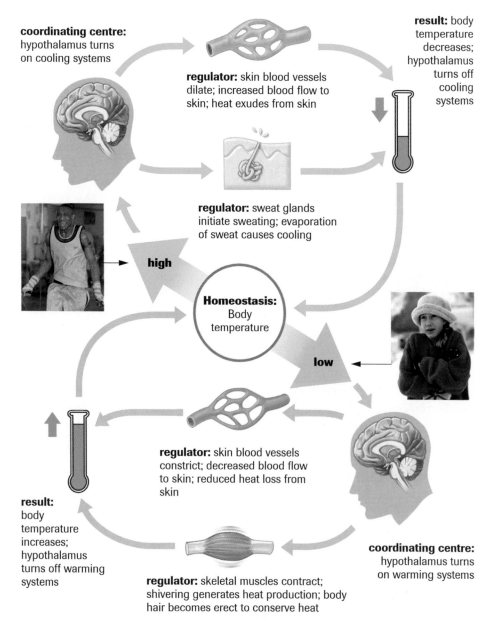

coordinating centre: hypothalamus turns on cooling systems

regulator: skin blood vessels dilate; increased blood flow to skin; heat exudes from skin

result: body temperature decreases; hypothalamus turns off cooling systems

regulator: sweat glands initiate sweating; evaporation of sweat causes cooling

high

Homeostasis: Body temperature

low

regulator: skin blood vessels constrict; decreased blood flow to skin; reduced heat loss from skin

result: body temperature increases; hypothalamus turns off warming systems

regulator: skeletal muscles contract; shivering generates heat production; body hair becomes erect to conserve heat

coordinating centre: hypothalamus turns on warming systems

Figure 2

The evaporation of sweat and dilation of blood vessels provide a negative feedback response by cooling the body. The constriction of blood vessels, shivering, and erection of the body hairs provide a negative feedback response that helps conserve heat.

DID YOU KNOW ?

Self-Healing Behaviour

It has been discovered that sick lizards can intentionally give themselves a fever by resting in hot, sunny places. This behaviour raises their temperature to a slightly higher level, and the resulting fever helps fight infection.

Nerve messages are also carried to the smooth muscle that surrounds the hair follicles in your skin, causing the hair to "stand on end." The small bump made by the contraction of the muscle attached to the hair is often called a "goosebump." The erect hair traps warm, still air next to the surface of your skin and helps reduce heat loss. This response is particularly effective in mammals with a thick coat of body hair.

In addition, the hypothalamus also sends nerve messages that initiate shivering. The shivering response is a rhythmic contraction of skeletal muscle. Cycles of rapid muscle contractions of between 10 and 20 times per minute generate heat production by increasing metabolism.

Prolonged exposure to cold can create a hormonal response that also elevates metabolism. This type of heat production is most often associated with a special adipose tissue called brown fat. Although its role in humans remains controversial, brown fat is especially capable of converting chemical energy into heat. Brown fat is especially important in newborns because they lack the ability to shiver. Babies have a small amount of brown fat in their neck and armpits and near the kidneys that insulates and generates heat.

Mammalian Diving Reflex

What happens to the heart rate when a person falls into cold water? Test the diving reflex with this activity, and see how your body responds.

Materials: pan filled with about 8 cm of cold water, stopwatch or clock with a second hand, towel

• Measure your resting pulse rate in beats per min. Have your partner place two fingers on your wrist and count your heartbeat for 15 s. Multiply this number by four and record your results.

• Now hold your breath and submerge your face in the pan of cold water for 15 s. Before you start, explain what you think will happen to your pulse as a result of this test.

• While your face is in the water, have your partner measure your pulse for 15 s. Multiply this number by four and record your results.

• Switch roles and repeat the procedure.
  (a) What happens to your pulse when your face is immersed in cold water?
  (b) Why do you think this happened?
  (c) How might this response help a person who has fallen into very cold water?

Hypothermia is a condition in which the body core temperature falls below the normal range. A drop in temperature of only a few degrees can lead to a coma and possibly death. However, some people, mainly small children, have survived sustained exposure to cold temperatures. This is often explained by the mammalian diving reflex. When a mammal is submerged in cold water, the heart rate slows and blood is diverted to the brain and other vital organs to conserve heat.

Research in Canada: Freezing Cells

Anthropologists huddle around the figure of a primitive person frozen in ice. As the figure in the ice begins to thaw, one of the scientists comments that the figure must look much like it did nearly 3000 years ago. Suddenly, the figure's arm moves.

The scene just described is that of science fiction—no human has ever returned to life following prolonged freezing. However, the phenomenon of suspended animation can be viewed every spring. Frogs, frozen solid in blocks of ice (**Figure 3**), are capable of continuing their existence once the ice thaws.

As cells or organs freeze, ice crystals form. Acting much like microscopic knives of ice, these ice crystals pierce and slash their way through cell membranes. Many important nutrients and cell organelles leak through the injured membrane. The cells collapse and die. The damage to large organs is especially devastating. Blood vessels rupture, nerves are crushed, and supporting structures are destroyed by ice. Thawing can be even more dangerous. As cells approach melting temperature, ice crystals may melt together, causing cells to fill with water and push against one another. Sturdy cells, such as muscle cells, broaden, while more delicate cells are crushed.

If ice is so destructive, how is it possible for some animals to survive freezing? In 1957, the Norwegian-born American physiologist Per Scholander speculated about a type of antifreeze in fish that he was studying. Scholander found that the temperature of the salt water off the coast of Baffin Island in the Canadian Arctic was often below the freezing point of the blood of the fish. A decade later, other scientists identified a protein in these fish that prevented ice crystals from forming. By interfering with the formation of the ice crystals, the protein was able to prevent cell damage. Although important, the protein is not the only way in which cells can protect themselves from extreme cold.

Figure 3
The frog's cells contain a cellular antifreeze that prevents damage during freezing and thawing.

Dr. Kenneth G. Storey (**Figure 4**) of Carleton University, a world authority on wood frogs, has shown that high levels of glucose, a simple sugar, can act as antifreeze in the blood. Glucose levels of a frog exposed to freezing temperatures can exceed normal levels in humans by as much as 100 times. Scientists have found that the wood frog may lose as much as 60% of its cell water during freezing, reducing the dangers posed by ice crystals.

Figure 4
Dr. Kenneth G. Storey

SUMMARY *Thermoregulation*

Table 1 Summary of Stimulus–Response in Thermoregulation

Stimulus	Physiological response	Adjustment
decreased environmental temperature	• constriction of blood vessels in skin • hairs on body erect • shivering	• heat is conserved • more heat is generated by increased metabolism
increased environmental temperature	• dilation of blood vessels of skin • sweating	• heat is dissipated

▶ *Section 7.2 Questions*

Understanding Concepts

1. How do "goosebumps" help protect against rapid cooling?

2. Why is freezing dangerous for cells? How does cellular antifreeze prevent cell damage?

3. What advantages do endotherms have over ectotherms?

4. What behavioural adjustments affect thermoregulation?

5. Explain why oral and rectal thermometers can give different readings.

6. Heat exhaustion caused by a person's exposure to heat can result in weakness or collapse. It usually involves a decrease in blood pressure. Explain why the homeostatic adjustment to heat can cause a drop in blood pressure.

7. The maximum suggested temperature of the water in a hot tub is about 38°C. A higher temperature can seriously increase the risk of heat stroke. Explain why people will collapse in a hot tub set at 45°C, but can survive temperatures of over 120°C in heated rooms.

Applying Inquiry Skills

8. In **Figure 5a**, beginning at the box labelled increase in body temperature, replace the letters with the following homeostatic feedback mechanisms for temperature control by the body: sweating, shivering, adjustment, evaporation. Do the same in **Figure 5b**, beginning at the box labelled decrease in body temperature.

Making Connections

9. Use the Internet or library to research how rapid cooling of the organs and tissues is used for surgery.

 www.science.nelson.com

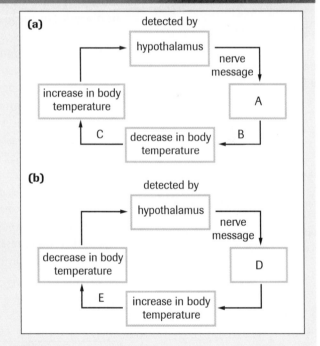

Figure 5

10. Drugs such as ecstasy interfere with the feedback mechanism that helps maintain a constant body temperature. Explain why these drugs are dangerous.

11. Should research on the freezing of animals ever be applied to humans? Justify your answer.

12. Should we try to change how human cells respond to the environment? Defend your answer by completing a risk–benefit analysis.

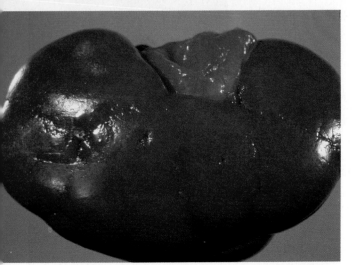

Figure 1
The human kidney is about the size of a fist and weighs approximately 0.5 kg.

deamination removal of an amino group from an organic compound

urea nitrogen waste formed from two molecules of ammonia and one molecule of carbon dioxide

uric acid waste product formed from the breakdown of nucleic acids

The cells of the body obtain energy by converting complex organic compounds into simpler compounds. However, many of these simpler compounds can be harmful. To maintain life processes, the body must eliminate waste products. The lungs eliminate carbon dioxide, one of the products of cellular respiration. The large intestine removes toxic wastes from the digestive system. The liver transforms ingested toxins, such as alcohol and heavy metals, into soluble compounds that can be eliminated by the kidneys (**Figure 1**). The liver also transforms the hazardous products of protein metabolism into metabolites, which are then eliminated by the kidneys. In fact, the kidneys play a crucial role in removing waste, balancing blood pH, and maintaining water balance.

The average Canadian consumes more protein than is required to maintain tissues and promote cell growth. Excess protein is often converted into carbohydrates. Proteins, unlike carbohydrates, contain nitrogen. The nitrogen molecule and two attached hydrogen molecules, which are characteristic of amino acids (the building blocks of protein), must be discarded by the body.

This removal process, referred to as **deamination**, occurs in the liver. The byproduct of deamination is ammonia, a water-soluble gas. However, ammonia is extremely toxic— a buildup of as little as 0.005 mg can kill humans. Fish are able to avoid ammonia buildup by continually releasing it through their gills. Land animals, however, do not have the ability to release small quantities of ammonia throughout the day—wastes must be stored. Once again, the liver is called into action. In the liver, two molecules of ammonia combine with another waste product, carbon dioxide, to form **urea**. Urea is 100 000 times less toxic than ammonia. The blood can dissolve 33 mg of urea per 100 mL of blood. A second waste product, **uric acid**, is formed by the breakdown of nucleic acids. **Table 1** summarizes the roles of the excretory organs in the removal of wastes.

The kidneys help maintain water balance. Although it is possible to survive for weeks without food, humans cannot survive for more than a few days without water. Humans deplete their water reserves faster than their food reserves. The average adult loses about 2 L of water every day through urine, perspiration, and exhaled air. Greater volumes are

Table 1 Removal of Metabolic Wastes

Waste	Origin of waste	Organ of excretion
ammonia	• deamination of amino acids by the liver	kidneys
urea	• deamination of amino acids by the liver • ammonia combined with carbon dioxide	kidneys
uric acid	• product of the breakdown of nucleic acids, such as DNA	kidneys
carbon dioxide	• waste product of cellular respiration	lungs
bile pigments	• breakdown of red blood cell pigment, hemoglobin	liver
lactic acid	• product of anaerobic respiration	liver
solid waste	• by product of digestible and indigestible material	large intestine

lost when physical activity increases. For the body to maintain water balance, humans must consume 2 L of fluids daily. A drop in fluid intake by as little as 1% of your body mass will cause thirst, a decrease of 5% will bring about extreme pain and collapse, while a decrease of 10% will cause death.

> ▶ **TRY THIS** activity

Making a Model of a Filtering Excretory System

You can create a model of a filtering excretory system.

Materials: funnel, aquarium charcoal, 2 small beakers, food colouring, non-absorbent cotton, ring stand

- Place a small piece of non-absorbent cotton in a funnel. Fill the funnel with aquarium charcoal, and put a small beaker beneath the funnel. Fill a second beaker with about 25 mL of water, and add five drops of food colouring.
- Pour the coloured water through the funnel and collect it in the beaker beneath as shown in **Figure 2**.
  (a) Compare the colour of the filtered water with the original coloured water.
  (b) Predict what will happen if the water is filtered once again. Test your prediction.
  (c) How would you improve the filter?

Figure 2

Excretion: From Simple to More Complex Animals

For unicellular organisms, getting wastes out of the cell is just as important as bringing in nutrients. Without a way to get rid of wastes, toxins would build up and the cell would soon die. In these unicellular organisms and in primitive multicellular organisms such as a sponge, where every cell is in direct contact with the external environment, wastes are released directly from the cell. Water currents then carry the wastes away.

A greater challenge for primitive organisms is that of fluid regulation. Unicellular organisms, such as the ameoba and paramecium, are hypertonic to their freshwater surroundings. Without a system of fluid regulation, these cells would draw in water by osmosis, expand, and eventually burst. A **contractile vacuole** expels excess water, preventing many unicellular organisms from swelling (**Figure 3**).

contractile vacuole a structure in unicellular organisms that maintains osmotic equilibrium by pumping fluid out from the cell

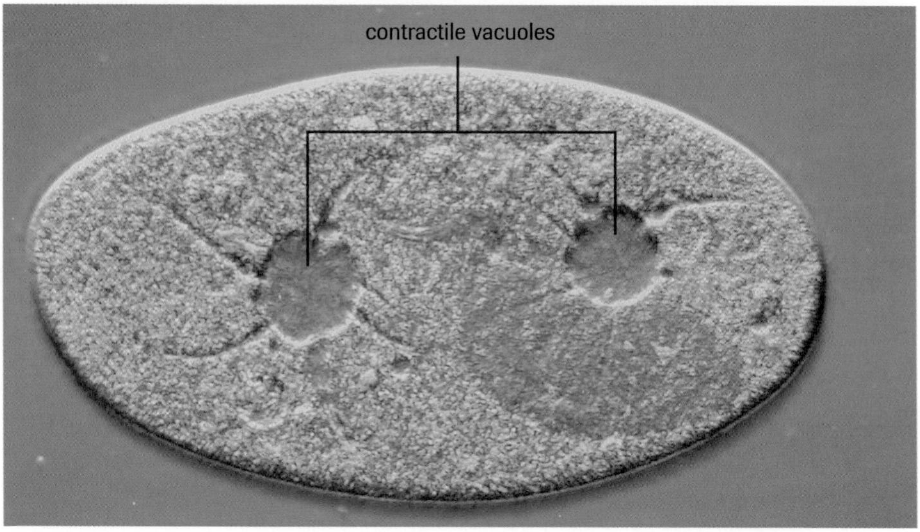

contractile vacuoles

Figure 3
Contractile vacuoles allows a paramecium to maintain a fluid balance. The paramecium is hypertonic to its surroundings.

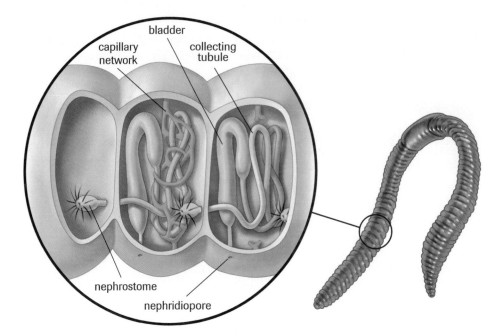

Figure 4
The earthworm uses a series of tubules to remove wastes from the blood and body cavity. Cells lined with cilia surround a funnel-like structure, the nephrostome, and draw fluids from the body cavity into tiny tubules. The wastes are stored as urine and are released through small pores—nephridiopores—along the body wall.

Complex multicellular organisms are faced with the same problem but on a much bigger scale. For organisms that have three distinct cell layers, such as worms (**Figure 4**), insects (**Figure 5**), and mammals, not every cell is in direct contact with the external environment, so wastes must be collected and temporarily stored. A secondary problem arising from greater cell specialization is that not every cell is designed to remove wastes. Wastes must be transported to cells that are capable of excretion. These specialized cells work together in the excretory system to remove wastes from the body or store the wastes until signalled to remove them. The excretory system has another crucial function in most animals: it helps to regulate body water.

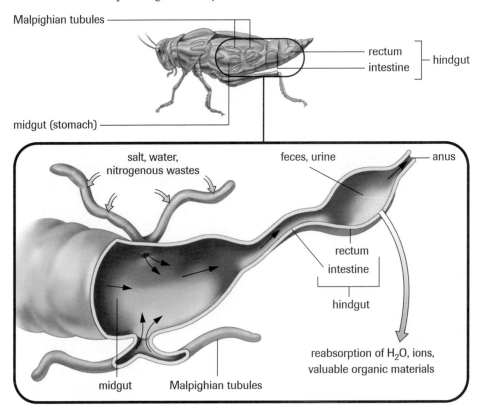

Figure 5
Malpighian tubules that run throughout the body cavity of an insect absorb wastes by diffusion. Wastes are released into the gut and eliminated with solid wastes from the anus.

SUMMARY   *The Importance of Excreting Wastes*

- The kidneys filter wastes from the blood.
- The liver helps to eliminate toxic nitrogen groups from the body by deamination.
- Land animals produce urea.
- Larger, more complex animals generate more wastes; they have specialized cells that allow more efficient waste removal.

▶ *Section 7.3* *Questions*

Understanding Concepts

1. Describe the two main functions of the kidneys.
2. What is deamination and why is it an important process?
3. How does the formation of urea prevent poisoning?
4. Fish are able to excrete ammonia continuously from their gills. Explain why birds and mammals cannot continuously remove this highly toxic waste.
5. Make a table outlining similarities and differences between the excretory systems of earthworms, insects, and mammals.

Applying Inquiry Skills

6. **Table 2** shows the percentage of the body composed of water for various vertebrates.

   Table 2

Animal	% of body composed of water
herring	67
frog	78
chicken	74
kangaroo rat	65

   (a) Which animal would you expect to have the most concentrated urine? Provide a reason for your prediction.
   (b) The kangaroo rat and the chicken feed largely on plant products, while the herring and frog are carnivores. Explain how food sources would affect urine composition.
   (c) Why would the concentration of urea be higher in the kangaroo rat than in the herring?

Renal arteries branch from the aorta and carry blood to the kidneys. With a mass of about 0.5 kg, the fist-shaped kidneys may hold as much as 25% of the body's blood at any given time. Wastes are filtered from the blood by the kidneys and conducted to the urinary bladder by **ureters**. A urinary sphincter muscle located at the base of the bladder acts as a valve, permitting the storage of urine. When approximately 200 mL of urine has been collected, the bladder stretches slightly and nerves send a signal to the brain. When the bladder fills to about 400 mL, more stretch receptors are activated and the message becomes more urgent. If a person continues to ignore the messages, the bladder continues to fill. After about 600 mL of urine has accumulated, voluntary control is lost. The sphincter relaxes, urine enters the **urethra**, and it is voided.

The cross section of the kidney in **Figure 1** reveals three structures. An outer layer of connective tissue, the **cortex**, encircles the kidney. An inner layer, the **medulla**, is found beneath the cortex. A hollow chamber, the **renal pelvis**, joins the kidney with the ureter.

ureters tubes that conduct urine from the kidneys to the bladder

urethra tube that carries urine from the bladder to the exterior of the body

cortex outer layer of the kidney

medulla area inside of the cortex

renal pelvis area where the kidney joins the ureter

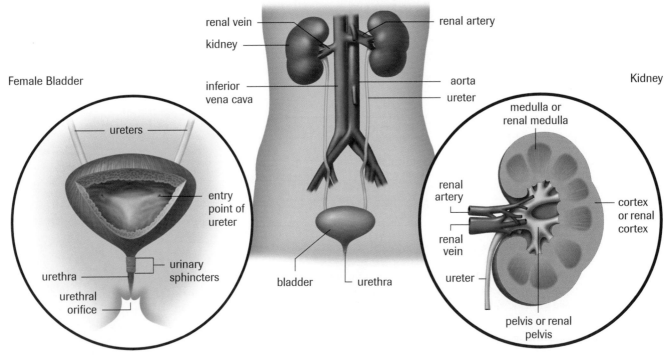

Figure 1
Simplified diagram of the human urinary system

Nephrons

Approximately one million slender tubules, called **nephrons**, are the functional units of the kidneys (**Figure 2**). Small branches from the renal artery, the **afferent arterioles**, supply the nephrons with blood. The afferent arterioles branch into a capillary bed, called the **glomerulus**. Unlike other capillaries, the glomerulus does not transfer blood to a venule. Blood leaves the glomerulus by way of other arterioles, the **efferent arterioles**. Blood is carried from the efferent arterioles to a net of capillaries called **peritubular capillaries** that wrap around the kidney tubule.

nephrons functional units of the kidneys

afferent arterioles small branches that carry blood to the glomerulus

glomerulus high-pressure capillary bed that is the site of filtration

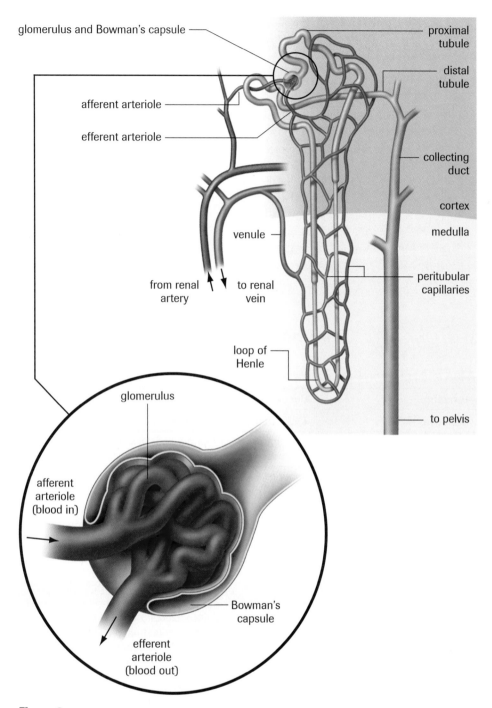

glomerulus and Bowman's capsule

proximal tubule

distal tubule

afferent arteriole

efferent arteriole

collecting duct

cortex

medulla

venule

peritubular capillaries

from renal artery

to renal vein

loop of Henle

to pelvis

glomerulus

afferent arteriole (blood in)

Bowman's capsule

efferent arteriole (blood out)

Figure 2
Diagram of a nephron, showing close-up of the glomerulus and Bowman's capsule

The glomerulus is surrounded by a funnel-like part of the nephron, called the **Bowman's capsule**. The Bowman's capsule, the afferent arteriole, and the efferent arteriole are located in the cortex of the kidney. Fluids to be processed into urine enter the Bowman's capsule from the blood. The capsule tapers to a thin tubule, called the **proximal tubule**. Urine is carried from the proximal tubule to the **loop of Henle**, which descends into the medulla of the kidney. Urine moves through the **distal tubule**, the last segment of the nephron, and into the **collecting ducts**. As the name suggests, the collecting ducts collect urine from many nephrons that, in turn, merge in the pelvis of the kidney.

efferent arterioles small branches that carry blood away from the glomerulus to a capillary net

peritubular capillaries network of small blood vessels that surround the nephron

Bowman's capsule cuplike structure that surrounds the glomerulus

proximal tubule section of the nephron joining the Bowman's capsule with the loop of Henle

loop of Henle carries filtrate from the proximal tubule to the distal tubule

distal tubule conducts urine from the loop of Henle to the collecting duct

collecting duct tube that carries urine from nephrons to the pelvis of a kidney

▶ TRY THIS activity *Observing a Kidney*

Materials: safety goggles, laboratory apron, dissecting gloves, dissection tray, preserved and injected pig kidney, scalpel, hand lens or dissecting microscope

- Obtain a preserved and injected pig kidney and place it in a dissection tray. Use a scalpel to cut the kidney in half with a longitudinal incision.

- Use a hand lens or dissecting microscope to examine the internal structures.
  (a) Describe the appearance of the cortex, medulla, and renal pelvis.
  (b) Describe the appearance of the nephrons.
  (c) Explain the advantage of having approximately 1 million nephrons in a human kidney.

🖐 Wear eye protection and a laboratory apron at all times. Wear plastic gloves when handling the preserved specimen.

Always cut away from yourself and others sitting near you in case the scalpel slips.

SUMMARY *The Urinary System*

Table 1 Summary of Nephron Structure and Description

Structure	Description
afferent arteriole	• carries blood to the glomerulus
glomerulus	• a high-pressure capillary bed enclosed by the Bowman's capsule
efferent arteriole	• carries arteriolar blood away from the glomerulus
peritubular capillary bed	• capillaries that network around the nephron • reabsorbs solute from the nephron into the blood and secretes solute from the blood into the nephron
venule	• carries filtered blood back to the heart

▶ Section 7.4 Questions

Understanding Concepts

1. Why do you think it is beneficial to humans to have two kidneys rather than one. Explain your answer.

2. Explain the function of nephrons.

3. Use the diagram in **Figure 3** to identify the following:

(a) the structure that filters blood
(b) the structure that carries urine from the kidney
(c) the structure that carries blood containing urea into the kidney
(d) the structure that stores urine

4. Athletes now undergo random urine testing for drugs. From your knowledge of excretion, describe the pathway of substances such as drugs through the urinary system, from the time they enter the glomerulus until they are excreted in the urine.

Applying Inquiry Skills

5. An adult under normal conditions will eliminate about 1.5 L of urine daily. Design an experiment that will test how urine output is affected by the consumption of a food containing caffeine (e.g., coffee, tea, chocolate, cola).

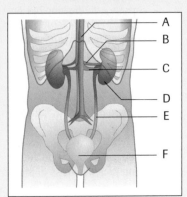

Figure 3

Formation of Urine 7.5

Urine formation depends on three functions. **Filtration** is accomplished by the movement of fluids from the blood into the Bowman's capsule. **Reabsorption** involves the transfer of essential solutes and water from the nephron back into the blood. **Secretion** involves the movement of materials from the blood back into the nephron.

Filtration

Each nephron of the kidney has an independent blood supply. Blood moves through the afferent arteriole into the glomerulus, a high-pressure filter. Normally, pressure in a capillary bed is about 25 mm Hg. The pressure in the glomerulus is about 65 mm Hg. Dissolved solutes pass through the walls of the glomerulus into the Bowman's capsule. Although materials move from areas of high pressure to areas of low pressure, not all materials enter the capsule. Scientists have extracted fluids from the glomerulus and Bowman's capsule using a thin glass tube called a micropipette. **Table 1** compares sample solutes extracted from the glomerulus and Bowman's capsule.

filtration process by which blood or body fluids pass through a selectively permeable membrane

reabsorption transfer of glomerular filtrate from the nephron back into the capillaries

secretion movement of materials, such as ammonia and some drugs, from the blood back into the distal tubule

Table 1 Comparison of Solutes

Solute	Glomerulus	Bowman's capsule
water	yes	yes
sodium chloride	yes	yes
glucose	yes	yes
amino acids	yes	yes
hydrogen ions	yes	yes
plasma proteins	yes	no
erythrocytes (blood cells)	yes	no
platelets	yes	no

Plasma protein, blood cells, and platelets are too large to move through the walls of the glomerulus. Smaller molecules pass through the cell membranes and enter the nephron.

Reabsorption

The importance of reabsorption is emphasized by examining changes in the concentrations of fluids as they move through the kidneys. On average, about 600 mL of fluid flows through the kidneys every minute. Approximately 20% of the fluid, or about 120 mL, is filtered into the nephrons. Imagine what would happen if none of the filtrate were reabsorbed. You would form 120 mL of urine each minute. You would also have to consume at least 1 L of fluids every 10 minutes to maintain homeostasis. Much of your day would be concerned with regulating water balance. Fortunately, only 1 mL of urine is formed for every 120 mL of fluids filtered into the nephron. The remaining 119 mL of fluids and solutes is reabsorbed.

Selective reabsorption occurs by both active and passive transport. Carrier molecules move Na^+ ions across the cell membranes of the cells that line the nephron. Negative ions, such as Cl^- and HCO_3^-, follow the positive Na^+ ions by charge attraction (**Figure 1**, page 350). Numerous mitochondria supply the energy necessary for active transport. However, the energy supply is limited. Reabsorption occurs until the **threshold level** of a substance is reached. Excess NaCl remains in the nephron and is excreted with the urine.

threshold level maximum amount of material that can be moved across the nephron

Figure 1

Overview of the steps in urine formation. The letters in the diagram match the processes in **Table 2**.

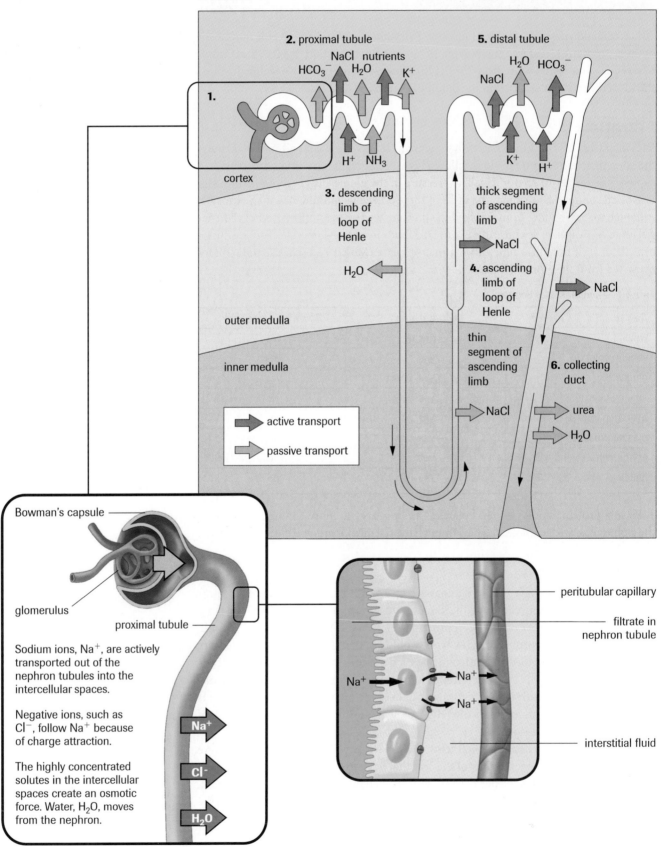

2. proximal tubule

HCO_3^- NaCl nutrients
H_2O K^+

H^+ NH_3

cortex

5. distal tubule

H_2O HCO_3^-
NaCl

K^+ H^+

3. descending limb of loop of Henle

thick segment of ascending limb

NaCl

4. ascending limb of loop of Henle

NaCl

H_2O

outer medulla

thin segment of ascending limb

6. collecting duct

inner medulla

NaCl

urea

H_2O

active transport
passive transport

Bowman's capsule

glomerulus

proximal tubule

Sodium ions, Na^+, are actively transported out of the nephron tubules into the intercellular spaces.

Negative ions, such as Cl^-, follow Na^+ because of charge attraction.

The highly concentrated solutes in the intercellular spaces create an osmotic force. Water, H_2O, moves from the nephron.

Na^+

Cl^-

H_2O

Na^+

Na^+

Na^+

peritubular capillary

filtrate in nephron tubule

interstitial fluid

Table 2 Urine Formation

Site	Description of process	Substances transported
1. glomerulus and Bowman's capsule	• Filtration of water and dissolved solutes occurs as blood is forced through walls of glomerulus into Bowman's capsule by fluid pressure in capillaries.	• sodium ions (Na^+), chloride ions (Cl^-), water (H_2O), hydrogen ions (H^+), glucose, amino acids, vitamins, minerals, urea, uric acid
2. proximal tubule	• Selective reabsorption of nutrients from filtrate back into blood by active and passive transport. • Within proximal tubule, pH is controlled by secretion of hydrogen ions (H^+) and reabsorption of bicarbonate ions (HCO_3^-).	• bicarbonate ions (HCO_3^-), salt ($NaCl$), water (H_2O), potassium ions (K^+), hydrogen ions (H^+), ammonia (NH_3), glucose, amino acids, vitamins, urea
3. descending limb of loop of Henle	• Descending limb of loop of Henle is permeable to water, resulting in loss of water from filtrate by osmosis. • Salt ($NaCl$) becomes concentrated in filtrate as descending limb penetrates inner medulla of kidney.	• water (H_2O)
4. ascending limb of loop of Henle	• Thin segment of ascending limb of loop of Henle is permeable to salt, resulting in diffusion of salt out of ascending limb. • Salt continues to pass from filtrate to interstitial fluid in thick segment of ascending limb.	• salt ($NaCl$)
5. distal tubule	• Selective reabsorption of nutrients from blood into nephron by active transport. Distal tubule helps regulate potassium (K^+) and salt ($NaCl$) concentration of body fluids. • As in proximal tubule, pH is controlled by tubular secretion of hydrogen ions (H^+) and reabsorption of bicarbonate ions (HCO_3^-).	• salt ($NaCl$), potassium ions (K^+), water (H_2O), hydrogen ions (H^+), bicarbonate ions (HCO_3^-), uric acid, ammonia (NH_3)
6. collecting duct	• Urine formation.	• water (H_2O), salt ($NaCl$), urea, uric acid, minerals

Other molecules are actively transported from the proximal tubule. Glucose and amino acids attach to specific carrier molecules, which shuttle them out of the nephron and into the blood. However, the amount of solute that can be reabsorbed is limited. For example, excess glucose will not be shuttled out of the nephron by the carrier molecules. This means that individuals with high blood glucose and those who consume large amounts of simple sugars will excrete some of the excess glucose.

The solutes that are actively transported out of the nephron create an osmotic gradient that draws water from the nephron. A second osmotic force, created by the proteins not filtered into the nephron, also helps reabsorption. The proteins remain in the bloodstream and draw water from the **interstitial fluid** into the blood. As water is reabsorbed from the nephron, the remaining solutes become more concentrated. Molecules such as urea and uric acid will diffuse from the nephron back into the blood, although less is reabsorbed than was originally filtered.

interstitial fluid fluid that surrounds the body cells

Secretion

Secretion is the movement of wastes from the blood into the nephron. Nitrogen-containing wastes, excess H^+ ions, and minerals such as K^+ ions are examples of substances secreted. Even drugs such as penicillin can be secreted. Cells loaded with mitochondria line the distal tubule. Like reabsorption, tubular secretion occurs by active transport, but, unlike reabsorption, molecules are shuttled from the blood into the nephron.

LAB EXERCISE 7.5.1

Comparing Solutes in the Plasma, Nephron, and Urine (p. 363)
In this exercise, you will use experimental data to compare solutes along the nephron.

Formation of Urine

- Urine formation depends on three functions: filtration, reabsorption, and secretion.
- The glomerulus acts as a high-pressure filter.
- Selective reabsorption occurs by both active and passive transport.
- Secretion is the movement of wastes from the blood into the nephron.

▶ **Section 7.5** *Questions*

Understanding Concepts

1. Draw and label the following parts of the excretory system: kidney, renal artery, renal vein, ureter, bladder, and urethra. State the function of each organ.

2. State the function of each part of the nephron: Bowman's capsule, proximal tubule, loop of Henle, distal tubule, and collecting duct.

3. Describe the three main processes that are involved in urine formation.

4. Explain why individuals who consume large amounts of sugars might do the following:
   (a) excrete large amounts of glucose in the urine
   (b) excrete large amounts of urine

5. Use **Figure 2** to answer the following:
   (a) Identify which letters indicate the afferent and efferent arterioles.
   (b) Explain how an increase in blood pressure in area (B) would affect the functioning of the kidney.
   (c) Explain why proteins and blood cells are found in area (B) but not in area (D).
   (d) In which area of the nephron would you expect to find the greatest concentration of glucose?

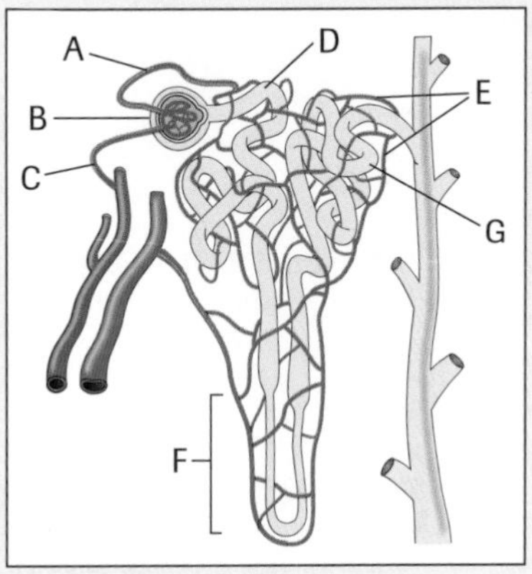

Figure 2

(e) Identify the area(s) in which Na^+ ions are actively transported.
(f) Identify the area of secretion.
(g) In which area(s) of the nephron would you expect to find urea?
(h) In which area of the nephron would you expect to find cells with a great number of mitochondria? Give reasons for your answer.

6. The following is a random list of processes that occur in the formation and excretion of urine once the blood has entered the kidney. Place these subsequent processes in the correct order:
   (a) urine is stored in the bladder
   (b) blood enters the afferent arteriole
   (c) fluids pass from the glomerulus into the Bowman's capsule
   (d) urine is excreted by the urethra
   (e) Na^+ ions, glucose, and amino acids are actively transported from the nephron
   (f) urine passes from the kidneys into the ureters

7. Marine fish, such as herring and cod, live in a hypertonic environment. These fish lose water through their gills by osmosis. To replace the water, the fish drink seawater.
   (a) Explain why these fish must actively transport salt from their bodies.
   (b) Because these fish excrete salt through their gills, kidney function is affected. Explain the effect on the volume of urine excreted and the concentration of solutes in the urine.

Making Connections

8. Explain why the regulation of salt is important for people with renal hypertension.

The body adjusts for increased water intake by increasing urine output. Conversely, it adjusts for increased exercise or decreased water intake by reducing urine output. These adjustments involve the interaction of the body's two communication systems: the nervous system and the endocrine system.

Regulating ADH

A hormone—**antidiuretic hormone (ADH)**—helps regulate the osmotic pressure of body fluids by causing the kidneys to increase water reabsorption. When ADH is released, a more concentrated urine is produced, thereby conserving body water. ADH, produced by specialized nerve cells in the hypothalamus, moves along specialized fibres from the hypothalamus to the pituitary gland, which stores and releases ADH into the blood.

Specialized nerve receptors, called **osmoreceptors**, located in the hypothalamus detect changes in osmotic pressure. When you decrease water intake or increase water loss—by sweating, for example—blood solutes become more concentrated. This increases the blood's osmotic pressure. Consequently, water moves into the bloodstream, causing the cells of the hypothalamus to shrink (**Figure 1**). When this happens, a nerve message is sent to the pituitary, signalling the release of ADH, which is carried by the bloodstream to the kidneys. By reabsorbing more water, the kidneys produce a more concentrated urine, preventing the osmotic pressure of the body fluids from increasing any further.

antidiuretic hormone (ADH) causes the kidneys to increase water reabsorption

osmoreceptors specialized nerve cells in the hypothalamus that detect changes in the osmotic pressure of the blood and surrounding extracellular fluids (ECF)

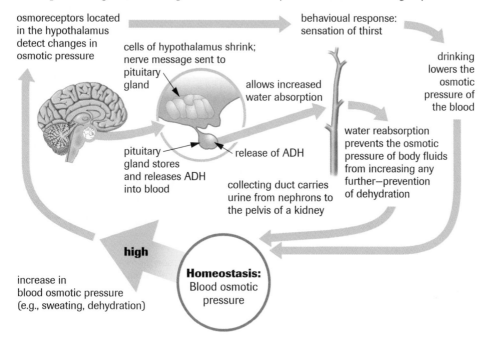

Figure 1
By increasing water reabsorption in the kidneys, ADH helps conserve body water. The osmoreceptors in the hypothalamus initiate the thirst response.

The shrinking of the cells of the hypothalamus also initiates a behavioural response: the sensation of thirst. If more water is taken in, it is absorbed by the blood and the concentration of solutes in the blood decreases. The greater the volume of water consumed, the lower the osmotic pressure of the blood. As the blood becomes more dilute, fluids move from the blood into the hypothalamus. The cells of the hypothalamus swell, and nerve messages to the pituitary stop. Less ADH is released, and less water is reabsorbed from the nephrons.

INVESTIGATION 7.6.1

Do Sports Drinks Really Work? (p. 363)

Are sports drinks any better than water and sugar? How can you determine whether a sports drink is able to restore the electrolytes essential for the operation of nerves and muscles?

aldosterone hormone that increases Na$^+$ reabsorption from the distal tubule and collecting duct

ADH and the Nephron

Approximately 85% of the water filtered into the nephron is reabsorbed in the proximal tubule. Although the proximal tubule is very permeable to water, this permeability does not extend to other segments of the nephron. (Refer to **Figure 1** in section 7.5.) The descending loop of Henle is permeable to water and ions, but the ascending tubule is only permeable to NaCl. Active transport of Na$^+$ ions from the ascending section of the loop concentrates solutes within the medulla of the kidney. Without ADH, the rest of the tubule remains impermeable to water, but continues to actively transport Na$^+$ ions from the tubules. The remaining 15% of the water filtered into the nephron will be lost if no ADH is present.

ADH makes the upper part of the distal tubule and collecting duct permeable to water. When ADH makes the cell membranes permeable, the high concentration of NaCl in the intercellular spaces creates an osmotic pressure that draws water from the upper section of the distal tubule and collecting duct. As water passes from the nephron to the intercellular spaces and the blood, the urine remaining in the nephron becomes more concentrated. It is important to note that the kidneys only control the last 15% of the water found in the nephron. By varying water reabsorption, the kidneys regulate the osmotic concentrations of body fluids.

Kidneys and Blood Pressure

The kidneys play a role in the regulation of blood pressure by adjusting for blood volumes. A hormone called **aldosterone** acts on the nephrons to increase Na$^+$ reabsorption (**Figure 2**). The hormone is produced in the cortex of the adrenal glands which lies above the kidneys. Not surprisingly, as NaCl reabsorption increases, the osmotic gradient increases and more water moves out of the nephron by osmosis.

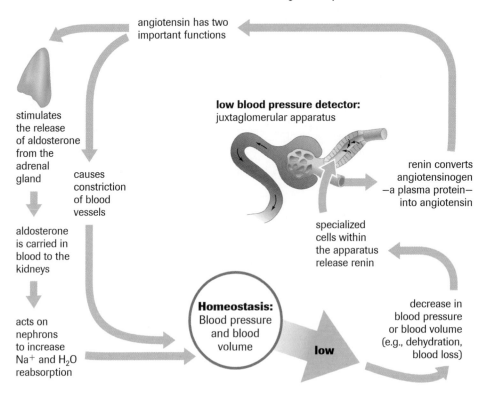

Figure 2
The hormone aldosterone maintains homeostasis by increasing Na$^+$ and water reabsorption.

Conditions that lead to increased fluid loss can decrease blood pressure, reducing the delivery of oxygen and nutrients to tissues. Blood pressure receptors in the juxtaglomerular apparatus (found near the glomerulus) detect low blood pressure. Specialized cells within the structure release renin, an enzyme that converts angiotensinogen, a plasma protein produced by the liver, into angiotensin.

Angiotensin has two important functions. First, the activated enzyme causes constriction of blood vessels. Blood pressure increases when the diameter of blood vessels is reduced. Second, angiotensin stimulates the release of aldosterone from the adrenal gland. Aldosterone is then carried in the blood to the kidneys, where it acts on the cells of the distal tubule and collecting duct to increase Na^+ transport. This causes the fluid level of the body to increase.

pH Balance

In addition to regulating body fluid volumes and maintaining the composition of salts in the blood, the kidneys maintain pH balance. Despite the variety of foods and fluids consumed with varying pH levels, the pH of the body remains relatively constant, between 7.3 and 7.5. In addition, during cellular respiration, cells produce carbon dioxide which forms carbonic acid. Carbonic acid and other excess acids ionize to produce H^+ ions. The buildup of H^+ ions lowers the pH.

An acid–base balance is maintained by buffer systems that absorb excess H^+ ions or ions that act as bases. Excess H^+ ions from metabolic processes are buffered by bicarbonate ions in the blood. Carbonic acid, a weak acid, is produced. In turn, the carbonic acid breaks down to form carbon dioxide and water. The carbon dioxide is then transported to the lungs where much of it is exhaled. The following reaction shows one type of buffer system, called the bicarbonate–carbon dioxide buffer system (**Figure 3**):

$$H_2O + CO_2 \rightleftharpoons \underbrace{\underset{\text{carbonic acid}}{H_2CO_3} \rightleftharpoons \underset{\text{bicarbonate ion}}{HCO_3^-} + H^+}_{\text{carbonic acid/bicarbonate buffer}}$$

Bicarbonate ions, HCO_3^-, eliminate the excess H^+ ions, preventing a change in pH.

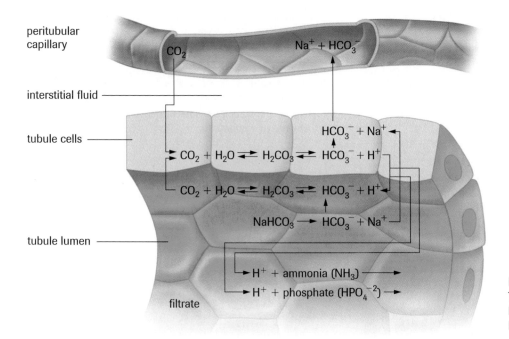

peritubular capillary

interstitial fluid

tubule cells

tubule lumen

CO_2

$Na^+ + HCO_3^-$

$HCO_3^- + Na^+$

$CO_2 + H_2O \rightleftharpoons H_2CO_3 \rightleftharpoons HCO_3^- + H^+$

$CO_2 + H_2O \rightleftharpoons H_2CO_3 \rightleftharpoons HCO_3^- + H^+$

$NaHCO_3 \longrightarrow HCO_3^- + Na^+$

$H^+ + \text{ammonia (NH}_3) \longrightarrow$

$H^+ + \text{phosphate (HPO}_4^{-2}) \longrightarrow$

filtrate

Figure 3
The bicarbonate-carbon dioxide buffer system maintains the pH balance.

The buffer system of the blood removes excess H^+ ions; however, the buffer must be restored if the body is to be protected. The kidneys help restore the buffer by reversing the reaction. As shown in **Figure 3** (page 355), carbon dioxide is actively transported from the peritubular capillaries, which surround the nephron, into the cells that line the nephron. The carbon dioxide combines with water to initiate the reverse reaction, generating HCO_3^- and H^+ ions. The bicarbonate ions diffuse back into the blood, thereby restoring the buffer. The H^+ ions recombine with either phosphate ions or ammonia and are excreted with the filtrate from the nephron.

SUMMARY   *Water Balance*

- Antidiuretic hormone (ADH) helps regulate osmotic pressure of body fluids and fluid volume.
- Aldosterone regulates body fluid volume.
- ADH and aldosterone are regulated by negative feedback.
- Kidneys restore buffers by excreting excess H^+ ions or restoring more HCO_3^- ions.

▶ Section 7.6 Questions

Understanding Concepts

1. What is ADH and where is it produced? Where is ADH stored?

2. Describe the mechanism that regulates the release of ADH.

3. Where is the thirst centre located?

4. Describe the physiological adjustment to increased osmotic pressure in body fluids.

5. Describe the behavioural adjustment in humans to increased osmotic pressure in body fluids.

6. Discuss the mechanism by which aldosterone helps to maintain blood pressure.

7. What role do the kidneys serve in maintaining the pH of body fluids?

8. Using the HCO_3^- buffering system, explain what would happen if the kidneys failed to excrete H^+ ions.

Applying Inquiry Skills

9. Draw a flow chart that shows why the release of ADH is a negative feedback mechanism.

10. A micropipette was used to extract fluids from various structures within the kidney. The data in **Table 1** shows an analysis of substances collected.
    (a) According to the data provided, which substance is not filtered from the blood into the Bowman's capsule? Give reasons for your answer.

Table 1

Substance found in body fluid	Blood plasma from afferent arteriole	Glomerular filtrate from Bowman's capsule	Urine
protein	7.00	0.00	0.00
urea	0.04	0.04	2.00
glucose	0.10	0.10	0.00
sodium ions	0.32	0.32	0.35
chloride ions	0.38	0.38	0.60

Quantities are in g/100 mL.

(b) Which substance provides evidence of secretion? Provide reasons for your response.
(c) Which substance provides evidence of reabsorption? Justify your answer.

Making Connections

11. Using the Internet and other resources, conduct research to explain how urine might be recycled on a space flight.

 GO www.science.nelson.com

Proper functioning of the kidneys is essential for the body to maintain homeostasis. The multifunctional kidneys are affected when other systems break down; conversely, kidney dysfunction affects other systems. Many kidney disorders can be detected by urinalysis (**Figure 1**).

Urinalysis Requisition and Report
Clinical Biochemistry

Pre-Admission
Pre-EOPS } date:
Pre-Surgery

Sample ID#

Collection Date/Time:

Type of Collection: ☐ voided ☐ catheter ☐ mid-stream
Microscopy will be routinely performed if the specimen is fresh and the dipstick screen is positive for blood, protein, nitrite, leukocytes, or glucose.

Requested by Doctor: ☐ STAT Phone:

Clinical Comments: ☐ Workers' Compensation

Dipstick Screen

Glucose	negative	1+	2+	3+	4+	
Bilirubin	negative	1+	2+	3+		
Ketones	neg trace	1+	2+	3+		
Specific Gravity	1.0 _____					
Blood (Heme)	neg trace	1+	2+	3+		
pH	_____					
Protein	neg trace	1+	2+	3+		
Urobilinogen	normal	1+	2+	3+	4+	
Nitrite	negative	positive				
Leukocytes	negative	1+	2+	3+	4+	

Microscopy *(routinely 12 mL centrifuged, sediment resuspended in 0.4 mL supernatant)*

☐ Volume centrifuged only _____ mL   ☐ Heavy sediment – not centrifuged

Casts/low-power field (magnification x 100) [F]-Few [S]-Several [M]-Many [P]-Packed

Granular: hyaline or fine [   ] coarse [   ] heme [   ]
Cellular: erythrocyte [   ] leukocyte [   ] epithelial [   ] bacterial [   ]

Cells/high-power field (magnification x 400)

Leukocytes: < 2   2-5   5-10   10-20   20-50   > 50
Erythrocytes: < 2   2-5   5-10   10-20   20-50   > 50
Epithelial Cells: non-squamous (renal/urothelial) [   ] squamous [   ]
Microorganisms: bacteria [   ] yeast [   ] trichomonads [   ]

Other sediment or comments _____
_____ Tech: _____

Figure 1
Many kidney problems can be diagnosed by analyzing a urine sample.

Diabetes Mellitus

Diabetes mellitus is caused by inadequate secretion of insulin from islet cells in the pancreas. Without insulin, blood sugar levels tend to rise. The cells of the proximal tubule are supplied with enough ATP to reabsorb 0.1% blood sugar, but in diabetes mellitus much higher blood sugar concentrations are found. The excess sugar remains in the nephron. This excess sugar provides an osmotic pressure that opposes the osmotic pressure created by other solutes that have been actively transported out of the nephron. Water remains in the nephron and is lost with the urine. Individuals with diabetes mellitus void large volumes of urine, which explains why they are often thirsty. The water lost with the excreted sugar must be replenished.

Diabetes Insipidus

The destruction of the ADH-producing cells of the hypothalamus or the destruction of the nerve tracts leading from the hypothalamus to the pituitary gland can cause diabetes insipidus. Without ADH to regulate water reabsorption, urine output increases dramatically. In extreme cases, as much as 20 L of dilute urine can be produced each day, creating a strong thirst response. A person with diabetes insipidus must drink large quantities of water to replace what he or she has not been able to reabsorb.

Bright's Disease

Named after Richard Bright, a 19th-century English physician, Bright's disease is also called nephritis. Nephritis is not a single disease but a broad description of many diseases characterized by inflammation of the nephrons. One type of nephritis affects the

Earliest Treatment of Kidney Stones

Operations to remove kidney stones were performed in the time of Hippocrates, the Greek physician considered to be the father of medicine (c. 460–377 B.C.).

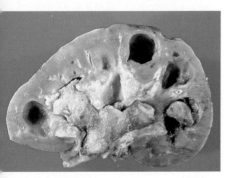

Figure 2

This kidney contained several stones. Most stones consist mainly of calcium oxalate or phosphate, or both.

⚙ INVESTIGATION 7.7.1

Diagnosis of Kidney Disorders (p. 364)

How is urinalysis used to detect various kidney disorders? In this investigation you will test simulated urine for kidney disease.

tiny blood vessels of the glomerulus. It is believed that toxins produced by invading microbes destroy the tiny blood vessels, altering the permeability of the nephron. Proteins and other large molecules are able to pass into the nephron. Because no mechanism is designed to reabsorb protein, the proteins remain in the nephron and create an osmotic pressure that draws water into the nephron. The movement of water into the nephron increases the output of urine.

Kidney Stones

Kidney stones (**Figure 2**) are caused by the precipitation of mineral solutes from the blood. Kidney stones are categorized into two groups: alkaline and acid stones. The sharp-sided stones can lodge in the renal pelvis or move into the narrow ureter. Delicate tissues are torn as the stone moves toward the bladder. The stone can move farther down the excretory passage and lodge in the urethra, causing excruciating pain as it moves.

Frontiers of Technology: Blasting Kidney Stones

The traditional treatment for kidney stones has been surgical removal followed by a period of convalescence. A technique developed by German urologist Dr. Christian Chaussy, called extracorporeal shock-wave lithotripsy (ESWL), has greatly improved prospects for kidney-stone patients with stones less than 2 cm in size.

The nonsurgical technique uses high-energy shock waves to break the kidney stones into small fragments. The shock waves pass through soft tissue and strike the stone. After a few days, tiny granules from the stone can be voided through the excretory system.

Not all stones can be eliminated by shock-wave treatment. The size of the stone, its location in the urinary tract, and the stone composition all determine whether ESWL is an appropriate treatment. In most cases, this technique can be performed on an outpatient basis, and recovery time is greatly reduced from that of surgical removal. ⚙

Dialysis Technology

For people whose kidneys cannot effectively process bodily wastes, a dialysis machine can restore the proper solute balance. Dialysis is defined as the exchange of substances across a semipermeable membrane. Like a kidney that is functioning normally, a dialysis machine operates on the principles of diffusion and blood pressure. However, unlike a kidney, a dialysis machine cannot perform active transport.

There are two types of dialysis: hemodialysis and peritoneal dialysis (**Figure 3**). In hemodialysis, the machine is connected to the patient's circulatory system by a vein. Blood is pumped through a series of dialysis tubes that are submerged in a bath of various solutes. Glucose and a mixture of salts set up concentration gradients. For example, HCO_3^- ions will move from the bath into the blood if it is too acidic. Because the dialysis fluids have no urea, this solute always moves from the blood into the dialysis fluid. Urea will move from the blood into the dialysis fluid until equal concentrations are established. By continually flushing expended dialysis solution and replacing it, urea and other waste solutes are continuously removed. During hemodialysis, the body also receives the hormones the kidneys are unable to produce.

An alternative is peritoneal dialysis, sometimes referred to as continuous ambulatory peritoneal dialysis (CAPD). With this method, 2 L of dialysis fluids is pumped into the abdominal cavity, and the membranes of the cavity selectively filter wastes from the blood. Urea and other wastes diffuse from the plasma into the peritoneum and into the dialysis fluid. Wastes accumulate in the dialysis fluids, which can be drained off and replaced several times a day. As dialysis occurs, the patient may continue with less strenuous activities. Peritoneal dialysis allows for greater independence because patients can perform the procedure on their own at home.

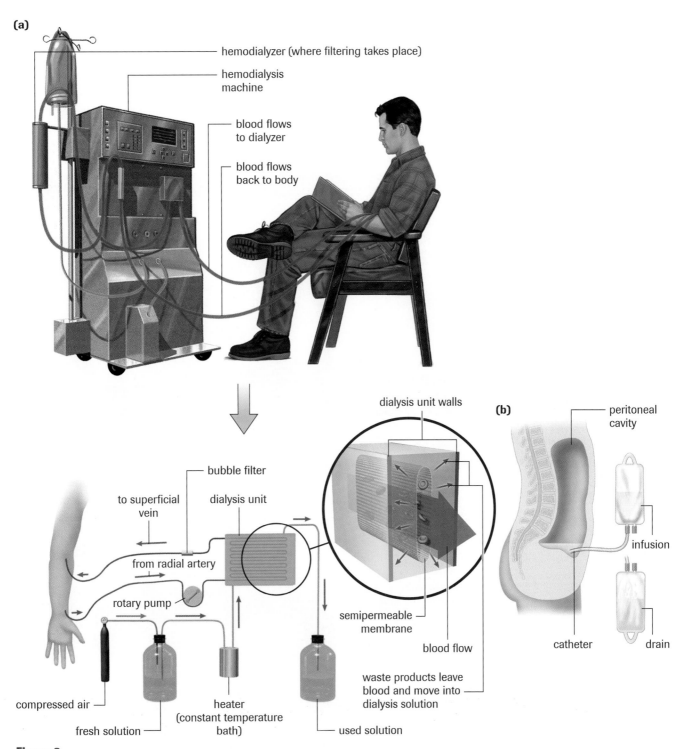

(a)
- hemodialyzer (where filtering takes place)
- hemodialysis machine
- blood flows to dialyzer
- blood flows back to body

dialysis unit walls

(b)
- peritoneal cavity
- infusion
- catheter
- drain

- bubble filter
- to superficial vein
- dialysis unit
- from radial artery
- rotary pump
- semipermeable membrane
- blood flow
- compressed air
- heater (constant temperature bath)
- fresh solution
- used solution
- waste products leave blood and move into dialysis solution

Figure 3

(a) In hemodialysis, a unit called a dialyzer mimics the action of the nephron. For hemodialysis treatments, a person must first have a minor surgical procedure to create an access, a shunt, for the needles and tubing needed to connect the circulatory system to the dialysis machine. Most people need three weekly dialysis sessions of about four hours each.

(b) Peritoneal dialysis is done through the peritoneal membrane, which is the lining of the abdominal cavity. In a minor surgical procedure, a catheter (a thin tube) is first inserted. A solution called the dialysate is then fed into the abdominal cavity through the catheter. The dialysate remains in this cavity for two to six hours. Then the dialysate fluid is drained from the abdomen via the catheter. Once the fluid is drained, new fluid is placed to begin the process anew.

xenotransplants transplants from one species to another; the word xeno means "strange" or "foreign"

transgenic animals animals that have genes from other species incorporated into their DNA

Although dialysis technology can remove toxic wastes from the body and maintain electrolyte balance, it is unable to accomplish other tasks of the kidneys. Dialysis equipment is not able to produce hormones, such as erythropoietin and renin, nor is it able to activate vitamin D.

A new and promising technique involves the transplant of kidney cells from a pig into a dialysis machine. The living cells not only produce renal hormones, but seem to be much better at regulating electrolytes and responding to ingested foods with a wider range of pH.

▶ *EXPLORE* an issue

Debate: Xenotransplants

Decision-Making Skills

○ Define the Issue   ● Analyze the Issue   ● Research
● Defend the Position   ● Identify Alternatives   ● Evaluate

In a year 2000 survey of Canadians

- 94% agreed that organ donation is a positive outcome of a person's death;
- 81% indicated a willingness to donate organs;
- 65% reported having had a discussion about organ donation with loved ones.

In spite of massive education plans, the organ donation rate in Canada is less than 40%. The shortage of organs has spurred scientists to explore new and creative solutions for the many patients awaiting new organs. **Xenotransplants**, animal-to-human transplants, have been attempted for several decades, but scientists have yet to solve the problem of organ rejection. Improvements in immunosuppressive drugs have extended the boundaries of possibility and would relieve the wait for thousands of patients.

A second advancement, the placement of human genes in animals by genetic engineering, has made xenotransplants even more viable. Because **transgenic animals** possess their own genes, plus those of humans, the chances of rejection are reduced. The immune system of the recipient will recognize the human marker on cell membranes as being related to their own tissues.

Although primates were once used as a primary source for xenotransplants, pigs have become the most desired animals (**Figure 4**). The organs of the pig resemble those of humans in both size and structure. In addition, pigs are easier to breed and less expensive. Baboons, the early primate of choice, were found to harbour many viruses that can easily be transferred to humans.

As of 2001, xenotransplants are not allowed in Canada. One of the fears is the introduction of new viruses into humans. Microbes that might be harmless in their natural host could be deadly in a human. Could xenotransplants cause an outbreak of a deadly disease?

Statement
The government should allow xenotransplants in Canada.

- Form groups and research the issue.
- Search for information in newspapers, periodicals, CD-ROMs, and on the Internet.

 www.science.nelson.com

Figure 4
Pigs have become the animal of choice for xenotransplants.

- Discuss the issue with class members and others in preparation for the debate.
- Write a list of points and counterpoints that your group considered.
- Decide whether your group agrees or disagrees with the statement.
- Defend your group's position in a debate.
- What responsibility do governments have to ensure that all groups have a voice in the debate?

Kidney Transplants

According to the Kidney Foundation of Canada, a patient diagnosed with end-stage renal disease (kidney failure) in the 1960s had little chance of surviving. By the 1970s, renal dialysis had changed life expectancy dramatically, but the patient had to spend up to 36 hours each week in treatment. By the 1980s, hemodialysis had reduced treatments to 12 hours a week.

Although dialysis machines are effective, nothing can surpass the workings of a real kidney. Today, kidney transplants are 85% successful and the preferred treatment for many patients. A transplanted kidney produces hormones and responds to the homeostatic adjustment of other body systems. The main disadvantage with any transplant is the immune response of the recipient. The donor kidney is often identified as a foreign invader and the recipient's immune system springs into action in an attempt to destroy it. The immune response will be discussed further in Chapter 10.

A kidney transplant (**Figure 5**) involves placing a new kidney and ureter in the lower abdomen near the groin, where they are surgically attached to the blood vessels and bladder (**Figure 6**). The operation usually takes two to four hours. The old kidney is not usually removed unless it is very large or chronically infected. After surgery, a catheter is inserted into the bladder for several days to drain the urine produced by the new kidney. Sometimes dialysis is required after the transplant until the new kidney can fully function. Immunosuppresive drugs are given after the transplant to help prevent rejection of the new organ.

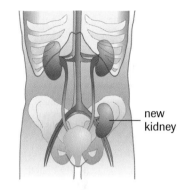

new kidney

Figure 6
Location of new kidney

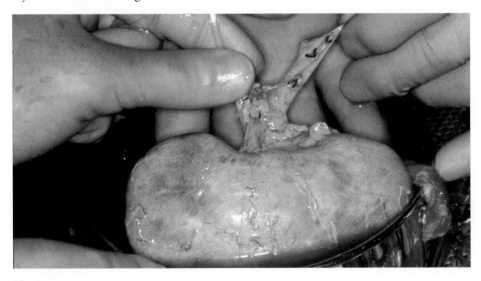

Figure 5
A human kidney being prepared for transplant

SUMMARY *Kidney Disease*

- Proper functioning of the kidneys is essential for homeostasis.
- Many kidney diseases can be detected by urinalysis.
- A number of kidney diseases affect proper kidney function, including diabetes mellitus, diabetes insipidus, Bright's disease, and kidney stones.
- Dialysis and transplants are currently the most common treatments for kidney disease.

Understanding Concepts

1. What are kidney stones?

2. Explain why people with diabetes become dehydrated.

3. Why isn't there a cure for Bright's disease?

4. Sketch a diagram of a kidney dialysis machine and explain how it works.

5. Identify advantages of peritoneal dialysis over hemodialysis.

6. Complete **Table 1**.

Table 1

Kidney disease	Cause of problem	Problem created by disease	Recommended treatment
diabetes mellitus	lack of insulin production	glucose in urine will cause dehydration	
diabetes insipidius			ADH provided by injection
Bright's disease			
kidney stones			

7. What is the most difficult challenge to overcome in achieving successful kidney transplants? Provide a reason.

Applying Inquiry Skills

8. Tests were performed on patients A, B, C, and D. Results from the tests are provided in **Table 2**. The results obtained for patient A are considered normal.

Table 2

Patient	Blood pressure (mm/Hg)	Cardiac output (L/min)	Glucose in urine (g/100 mL)	Urine output (mL/24h)
A	120/70	5.0	0.00	1500
B	130/80	5.5	0.00	1700
C	115/70	4.5	0.06	1950
D	90/55	3.0	0.00	500

(a) Which patient could have a circulatory problem?

(b) Explain how a circulatory problem could affect urine output.

(c) Explain why the urine output of patient C is elevated.

Making Connections

9. Why are some people opposed to xenotransplants?

10. Alcohol is a diuretic, a substance that increases the production of urine. Alcohol also suppresses the production and release of ADH. Should people who are prone to developing kidney stones consume alcohol? Explain.

LAB EXERCISE 7.5.1

Comparing Solutes in the Plasma, Nephron, and Urine

Inquiry Skills

○ Questioning	○ Planning	● Analyzing
○ Hypothesizing	○ Conducting	○ Evaluating
● Predicting	○ Recording	○ Communicating

Micropipettes were used to draw fluids from the Bowman's capsule, the glomerulus, the loop of Henle, and the collecting duct. Solutes in the fluids were measured. The resulting data are displayed in the table below. Some of the data were not taken, as indicated in **Table 1**.

Analysis

(a) Which of the solutes was not filtered into the nephron? Explain your answer.

(b) The test for glucose was not completed for the sample taken from the glomerulus. Predict whether glucose would be found in the glomerulus. Provide reasons for your prediction.

(c) Why do urea and ammonia levels increase after filtration occurs?

(d) Chloride ions, Cl^-, follow actively transported Na^+ ions from the nephron into the blood. Would you not expect the Cl^- concentration to decrease as fluids are extracted along the nephron? What causes the discrepancy?

(e) Is it correct to say that veins carry blood with high concentrations of waste products and arteries carry blood with high concentrations of nutrients? Explain.

(f) Compare the blood found in a renal artery and a renal vein with respect to urea and glucose.

Table 1

Solute	Bowman's capsule	Glomerulus	Loop of Henle	Collecting duct
protein	0	0.8	0	0
urea	0.05	0.05	1.50	2.00
glucose	0.10	no data	0	0
chloride	0.37	no data	no data	0.6
ammonia	0.0001	0.0001	0.0001	0.04
substance X	0	9.15	0	0

Quantities are in g/100 mL.

⚗ INVESTIGATION 7.6.1

Do Sports Drinks Really Work?

Inquiry Skills

● Questioning	● Planning	● Analyzing
● Hypothesizing	● Conducting	● Evaluating
● Predicting	● Recording	● Communicating

Sweating helps to cool the body while exercising. Drinking water during and after exercising helps to restore water balance, but does not, according to many sports drinks advertisers, enable the body to continue operating at peak athletic performance. Sugar and electrolyte levels must be restored. Sugars provide the fuel for cellular respiration. Electrolytes, such as K^+ and Ca^{2+}, are essential for nerve and muscle action.

Nerve and muscle reaction can be measured by monitoring changes in reaction time. In this investigation, you will design ways to test the effects of a sports drink on reaction time.

Question

How do sports drinks affect reaction time?

Hypothesis/Prediction

(a) Before the investigation, predict the effects of the sports drink on reaction time and formulate a hypothesis to explain your prediction.

(b) What criteria did you use to make your prediction?

Experimental Design

(c) Design a controlled experiment to test your hypothesis. Include the following in your design:
- descriptions of the independent, dependent, and controlled variables
- a step-by-step description of the procedure, including the steps for measuring reaction time (one possibility for measuring reaction time is given below)
- a list of safety precautions
- a table to record observations

Materials

(d) List the materials and apparatus needed to carry out the procedure.

Procedure

1. Submit your procedure, safety precautions, data table, and list of materials and apparatus to your teacher for approval. The procedure for measuring reaction time is given below. For the rest of the procedure, use your own approved design.

Measuring Reaction Time

2. Ask your subject to place his or her forearm flat on the surface of a desk. The subject's entire hand should be extended over the edge of the desk.

3. Ask the subject to place his or her index finger and thumb approximately 2 cm apart. Hold a 30-cm ruler vertically between the thumb and forefinger of the subject. The lower end of the ruler should be even with the top of the thumb and forefinger (**Figure 1**).

4. Indicate when ready, and release the ruler within the next 30 s. Measure the distance the ruler falls before

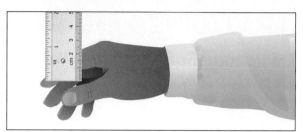

Figure 1

being caught between the subject's thumb and forefinger. Repeat the procedure for the left hand. Record your data in a table similar to **Table 2**.

Table 2

Trial	Distance, right hand (cm)	Distance, left hand (cm)
1		
2		
3		
Average		

Analysis

(e) Explain how the sports drink affected reaction time.

(f) Explain how the data confirmed or disproved your prediction.

Evaluation

(g) Describe any problems you encountered while carrying out the procedure.

(h) Describe how you could improve your current design.

(i) If you were to repeat this experiment, what new factors would you investigate? Write a brief description of the new procedure.

A INVESTIGATION 7.7.1

Diagnosis of Kidney Disorders

Inquiry Skills

○ Questioning	○ Planning	● Analyzing
● Hypothesizing	● Conducting	● Evaluating
● Predicting	● Recording	● Communicating

The identification of proteins and sugars in urine samples can reveal kidney disease. This investigation will involve the use of simulated urine samples to test for indications of disease.

Biuret reagent can be used to identify proteins. It reacts with the peptide bonds joining amino acids together, producing colour changes from blue, indicating no protein, to pink or purple.

Benedict's solution can be used to identify sugars. In this investigation, Benedict's solution will be used to detect glucose in the urine. The colour chart in **Table 3** summarizes the quantitative results obtained when reducing sugars, such as glucose, react with Benedict's solution.

INVESTIGATION 7.7.1 *continued*

Table 3

Colour of Benedict's solution	Approximate % of sugar
blue	negative
light green	0.5%–1.0%
green to yellow	1.0%–1.5%
orange	1.5%–2.0%
red to red-brown	+2.0%

Question

Which of the samples indicates kidney disease?

Hypothesis/Prediction

(a) Before beginning the investigation, predict what type of result will indicate disease.

(b) Formulate a hypothesis to explain your prediction.

Materials

safety goggles
laboratory apron
4 urine samples (simulated), labelled W, X, Y, and Z in dropper bottles
4 small test tubes
wax pencil
distilled water in wash bottle
Benedict's solution (in small dropper bottle)
test-tube clamp
hot water bath
test-tube brush
Biuret reagent (in small dropper bottle)
hydronium pH paper

 Safety goggles and a laboratory apron must be worn for the entire laboratory.

 Benedict's solution is toxic and corrosive. Biuret reagent is toxic. Avoid skin and eye contact. Wash all splashes off your skin and clothing thoroughly. If you get any chemical in your eye, rinse your eye for at least 15 minutes and inform your teacher.

Procedure

1. Label four test tubes W, X, Y, Z. Place 20 drops of urine sample W in test tube W. Repeat the procedure for samples X, Y, and Z in their respective test tubes.

 Handle hot objects and their contents carefully to avoid burns.

2. Add 10 drops of Benedict's solution to each test tube and, using a test-tube clamp, place the test tubes in a hot water bath (approximately 80°C).

3. Observe for 6 min.

(c) Record any colour changes in a table.

(d) Use **Table 3** to identify the values for each sample. Record the values in a table.

4. Wash each of the test tubes and dry them before beginning the protein test.

5. Use your four labelled test tubes. Place 20 drops of each urine sample in their respective test tube. Add 20 drops of Biuret reagent to each of the test tubes, then tap the test tubes with your fingers to mix the contents.

(e) Record your results in a table.

6. Use hydronium paper to determine the pH of each sample. A chart is usually located on the pH paper dispenser.

(f) Record your results in a table.

7. Clean up your work space. Dispose of all chemicals as directed by your teacher

8. Wash your hands thoroughly.

Analysis

(g) Which sample indicates diabetes mellitus? Provide your reasons.

(h) Which sample indicates diabetes insipidus? Give reasons for your response.

(i) Which sample indicates Bright's disease? Provide reasons for your answer.

(j) Which sample indicates a tremendous loss of body water while exercising? Provide your reasons.

Synthesis

(k) What are recommended treatments for diabetes mellitus and diabetes insipidus?

(l) Why is Bright's disease difficult to treat?

Key Expectations

- Describe and explain homeostatic processes involved in maintaining thermal balance in response to changing environments. (7.1, 7.2)

- Construct models and use flow charts to describe feedback control systems. (7.1, 7.2, 7.6)

- Describe and explain homeostatic processes involved in maintaining water, ionic, and acid–base equilibriums in response to changing environments. (7.1, 7.3, 7.4, 7.5, 7.6)

- Describe contributions of Canadian scientists. (7.2)

- Design and carry out an experiment to investigate feedback systems. (7.6)

- Explain the role of hormones and negative feedback actions in the kidney to maintain osmotic balance and body fluid volumes. (7.6)

- Discuss and evaluate opinions, technological advancement, and problems related to the treatment of kidney disease. (7.7)

Key Terms

afferent arterioles

aldosterone

antidiuretic hormone (ADH)

Bowman's capsule

collecting ducts

contractile vacuole

cortex

deamination

distal tubule

dynamic equilibrium

efferent arterioles

filtration

glomerulus

homeostasis

hypothalamus

loop of Henle

medulla

negative feedback

nephrons

osmoreceptors

peritubular capillaries

positive feedback

proximal tubule

reabsorption

renal pelvis

secretion

thermoregulation

threshold level

transgenic animals

urea

ureters

urethra

uric acid

xenotransplants

▸ *MAKE* a summary

In this chapter, you studied how body temperature and body fluids are regulated. To summarize your learning, create two flow charts or diagrams that show how the excretory system and the thermoregulatory system maintain homeostasis. Label the diagrams with as many of the key terms as possible. Check other flow diagrams and use appropriate designs to make your sketch clear.

In your notebook, record the letter of the choice that best completes the statement or answers the question.

1. Homeostasis is best defined as
   (a) a feedback system that prevents a body system from changing;
   (b) a feedback system designed to maintain body systems within an optimal range while responding to environmental or external change;
   (c) a positive feedback control that enables the body to respond to changes in the external environment by changing the internal environment;
   (d) a control system designed to regulate the external environment by making subtle changes to the internal environment;
   (e) a control system that causes body systems to change if the external environment remains constant.

2. A rapid increase in external temperature would be followed by which homeostatic adjustment in humans?
   (a) an increase in blood flow to the arms and an increase in perspiration
   (b) the formation of "goosebumps" and shivering
   (c) decreased blood flow to the arms and shivering
   (d) an increased heart rate and decreased blood flow to the arms
   (e) an increase in urine production and decrease in heart rate

3. Nitrogen wastes from the breakdown of proteins and amino acids are removed from the body by the
   (a) conversion of ammonia to urea in the liver and filtration by the kidney;
   (b) conversion of ammonia to urea in the kidney and filtration by the kidney;
   (c) conversion of urea to ammonia in the liver and secretion by the kidney;
   (d) conversion of urea to ammonia in the kidney and secretion by the kidney;
   (e) conversion of nitrogen to ammonia in the liver and secretion by the kidney.

4. In **Figure 1**, which of the following is the area of the nephron in which ADH has an effect?
   (a) d   (d) h
   (b) f   (e) a
   (c) g

5. In which of the following areas of **Figure 1** would proteins be found?
   (a) b and e   (d) g and h
   (b) b and d   (e) a and f
   (c) c and g

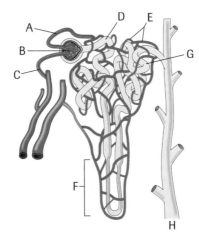

Figure 1
Nephron and associated blood vessels

6. An increase in blood pressure in the glomerulus would cause
   (a) an increase in filtration and increase in urine output;
   (b) a decrease in filtration and increase in urine output;
   (c) an increase in filtration and decrease in urine output;
   (d) a decrease in filtration and decrease in urine output;
   (e) no effect on urine output.

7. Concentrated urine is produced when ADH is
   (a) lacking and the collecting duct is impermeable to water;
   (b) abundant and the loop of Henle is impermeable to water;
   (c) abundant and the collecting duct is permeable to water;
   (d) lacking and the loop of Henle is permeable to water;
   (e) unchanged and the nephron is impermeable to water.

8. After a severe cut to the skin, the production of urine temporarily decreases. This can be explained by the drop in blood pressure, which causes
   (a) the release of ADH, which increases water reabsorption in the nephron;
   (b) the release of ADH, which decreases water reabsorption in the nephron;
   (c) the release of aldosterone, which increases Na^+ reabsorption in the nephron, leading to decreased water reabsorption;
   (d) the release of aldosterone, which increases K^+ reabsorption in the nephron, leading to decreased water reabsorption;
   (e) the release of aldosterone, which increases Na^+ reabsorption in the nephron, leading to increased water reabsorption.

NEL   An interactive version of the quiz is available online.
GO www.science.nelson.com

Maintaining an Internal Balance   **367**

Understanding Concepts

1. In **Figure 1**, labels 1 and 2 represent two hormones that directly affect the permeability of the kidney. What are these two hormones?

2. In **Figure 1**, if hormone 1 increases the permeability of one section of the nephron to water, what action does hormone 2 perform?

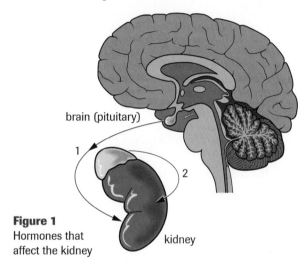

Figure 1
Hormones that affect the kidney

3. A pH analysis of urine reveals that the urine of humans fluctuates between acidic and basic depending on the diet. How does the kidney help to maintain a constant blood pH?

4. In **Figure 2**, as blood moves through blood vessel A, through the dialysis tubing, and into blood vessel B, what happens to the concentration of urea in the blood?

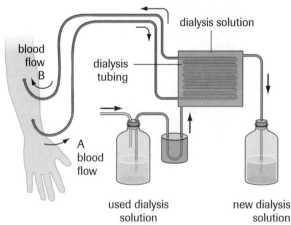

Figure 2

5. For effective dialysis to occur in **Figure 2**, will wastes move by active transport or by diffusion? Which fluid must contain the lower concentration of wastes: the blood or the dialysis solution?

6. Identify which letter(s) in **Figure 3** represent negative feedback. Explain your response.

7. Make a list of problems that occur with heat stroke and identify one way that the body attempts to compensate for each problem.

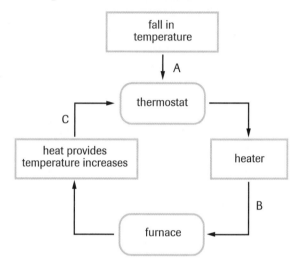

Figure 3
Regulation of household temperatures

8. A drug causes dilation of the afferent arteriole and constriction of the efferent arteriole. Indicate how the drug will affect urine production.

9. Why do the walls of the proximal tubule contain so many mitochondria?

10. How does excessive salt intake affect the release of ADH from the pituitary gland?

11. A drug that inhibits the formation of ATP by the cells of the proximal tubule is introduced into the nephron. How will the drug affect urine formation? Provide a complete physiological explanation.

12. A blood clot lodges in the renal artery and restricts blood flow to the kidney. Explain why this condition leads to high blood pressure.

13. For every 100 mL of salt water consumed, 150 mL of body water is lost. The solute concentration found in seawater is greater than that found in the blood. Provide a physiological explanation to account for the loss of body water. (Hint: Consider the threshold level for salt reabsorption by the cells of the nephron.)

Applying Inquiry Skills

14. Create two separate flow charts showing adjustments to cold temperatures by a frog and a human. Compare each flow chart and describe the differences.

15. Predict how a drop in blood pressure would affect urine output. Give reasons for your prediction.

16. An experiment was performed to determine the effect of a drug, labelled X, on human metabolism. The rate of metabolism can be measured indirectly by monitoring changes in body temperature. Human metabolism is normally regulated by the thyroid gland.

    Procedure
    - Four comparable groups consisting of 50 individuals per group were used in the experiment.
    - At the same time each day, all group members were given a dosage of the drug, except for group 4, which was given a placebo that did not contain the drug.
    - Each group member was monitored for changes in body temperature and in the volume of secretion from the thyroid gland. The observations are recorded in **Table 1**.

Table 1

Group #	Dosage of drug X (10^{-6}g/50 kg of body mass)	Change in body temp. (°C)	Perspiration	Change in thyroid gland output (10^{-6}g/ 50 kg of body mass)
1	1	+0.2	slight increase	−0.9
2	10	+0.9	moderate increase	−9.8
3	100	+1.2	large increase	−97.2
4	placebo	0.0	no change	0.0

   (a) What controls were used for this experiment?
   (b) Identify the dependent and independent variables for this experiment.
   (c) Using the information provided, does the drug increase metabolic rate? Justify your answer.
   (d) What evidence suggests that drug X exerts a negative feedback response?

17. In an experiment, the pituitary gland of an animal is removed. Predict how the removal of the pituitary gland will affect the animal's regulation of water balance.

18. A micropipette is used to collect fluids from three different areas of the kidney. The data in **Table 2** show an analysis of substances in the fluids.

Table 2

Area of the kidney	Protein	Urea	Amino acids	Inorganic salts
glomerulus	8	0.04	0.05	0.72
Bowman's capsule	0	0.04	0.05	0.72
loop of Henle	0	0.04	0.00	0.99
collecting duct	0	2.00	0.00	1.55

Quantities are in g/100 mL of urine.

   (a) Why is protein only found in the glomerulus?
   (b) Explain why no amino acids appear in the filtrate of the loop of Henle and in the collecting duct.
   (c) Which data provide the best evidence for water reabsorption from the nephron?

Making Connections

19. Scientists are interested in storing tissues and cells at low temperatures. Many of these tissues and cells could be used for organ transplant. Outline some benefits and risks of such a practice. Explain why some people may be opposed to this kind of research.

20. Make a chart that identifies the advantages and disadvantages of the following:
    (a) hemodialysis
    (b) peritoneal dialysis
    (c) kidney transplants by living donors, cadaver donors, and xenotransplants

21. In some countries, kidneys are sold for transplant. Do you believe that this practice is acceptable? Explain your answer.

Extension

22. Medical researchers are investigating the possibility of using artificial substitutes for human tissues. For example, fluorocarbon compounds can be used as artificial blood and artificial skin for severe burn victims.
    (a) In what situations would artificial skin or blood be preferred to the natural tissue?
    (b) Make a list of advantages and disadvantages of artificial tissues.
    (c) What other artificial tissues might prove useful?

23. Design the perfect kidney for an animal living in a desert.

chapter

8

Chemical Signals Maintain Homeostasis

In this chapter, you will be able to

- describe the anatomy and physiology of the endocrine and nervous systems, and explain their roles in homeostasis;

- explain the action of hormones in the female and male reproductive systems, including the feedback mechanisms involved;

- describe how disorders of the endocrine system affect homeostasis;

- study the effects of taking chemical substances to enhance performance or improve health;

- describe some Canadian contributions to knowledge and technology in the field of homeostasis;

- present informed opinions about problems related to the health industry, health legislation, and personal health.

Olga Yegorova upset reigning World and Olympic 5000-m champion Gabriela Szabo at the 2001 World Track and Field Championships held in Edmonton. Uncharacteristic of the usually polite Canadian audience, Yegorova was booed as she crossed the finish line. The Russian distance runner was not booed for her performance or demeanor, but rather for what was perceived to be an unfair advantage—taking the banned chemical hormone erythropoietin, or EPO. Following a competition in Paris, Yegorova had tested positive for erythropoietin, but she had been reinstated just before the world championships on a technicality. Although EPO was detected in a urine sample from Yegorova, organizers of the track meet in Paris failed to follow up with a blood test, as required by the International Amateur Athletics Federation. By the time Yegorova was tested again, abnormally high levels of EPO could no longer be identified.

Erythropoietin is a naturally occurring hormone produced by the kidneys. The hormone boosts red blood cell production, increasing the transport of oxygen to the tissues. More oxygen means greater energy for endurance athletes, such as 5000-m runners. Tests in Australia have shown that athletic enhancement gained by using EPO over four weeks would match that of several years of training. However, EPO is dangerous. Increased red blood cell production makes the blood thicker and more difficult to pump. Very high red blood cell counts can increase blood clotting and overwork the heart. According to some doping experts, the deaths of 20 European cyclists between 1988 and 1998 can be linked directly to the use of EPO.

Because the body produces EPO, it is difficult to detect. New tests that analyze blood for abnormally high red blood cell volume and analyze a urine sample for the presence of unusually high levels of EPO are being used to detect the use of banned substances. Unfortunately, the test is not yet foolproof; athletes can still avoid getting caught by stopping EPO treatments a few weeks before the testing.

REFLECT on your learning

1. EPO is one of a few chemicals that have been used to enhance athletic performance illegally. List some other banned chemicals that have been used and describe what types of advantages are achieved.

2. Explain how hormones help the body adjust to stress.

3. Antidiuretic hormone (ADH) and aldosterone are hormones that affect the kidney. Explain why the regulatory systems for osmotic pressure of fluids and for body fluid volumes are controlled by chemicals carried by blood, rather than by nerves.

4. Name reproductive hormones for males and females, and explain the function of each hormone.

TRY THIS activity
Chemical Signals and Sports

Find out more about the use of banned drugs in sports.

GO ▶ www.science.nelson.com

(a) Choose one banned drug and explain the unfair advantage it provides.
(b) What are some of the health risks associated with its use?
(c) Identify some of the technologies used to detect whether an athlete is using the drug.

hormones chemicals released by cells that affect cells in other parts of the body

endocrine hormones chemicals secreted by endocrine glands directly into the blood

The trillions of cells of the body all interact with each other—no cell operates in isolation. The integration of body functions depends on chemical controls. **Hormones** are chemical regulators produced by cells in one part of the body that affect cells in another part of the body. Only a small amount of a hormone is required to alter cell metabolism. Chemicals produced in glands and secreted directly into the blood are referred to as **endocrine hormones** (**Figure 1**). The circulatory system carries these hormones to the various organs of the body.

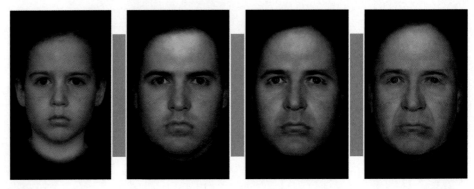

Figure 1
Endocrine hormones are chemical controls involved in the regulation of growth, development, and homeostasis. This sequence of photos is a computer simulation of the aging process based on statistical data.

DID YOU *KNOW*?

Exocrine Glands
Exocrine glands, unlike the endocrine glands, are glands that secrete substances through ducts or tubes onto a body surface or into a cavity. Most of the body's glands are exocrine glands. Digestive, mucous, sebaceous, and sweat glands are included in this category. The pancreas is considered both a digestive gland and an endocrine gland.

growth hormone (GH) hormone, produced by the pituitary gland, that stimulates growth of the body; also known as somatotropin (STH)

insulin hormone produced by the islets of Langerhans in the pancreas; insulin is secreted when blood sugar levels are high

epinephrine hormone, produced in the adrenal medulla, that accelerates heart rate and body reactions during a crisis (the fight-or-flight response); also known as adrenaline

Hormones are classified according to their activation site. They affect many cells throughout the body: **growth hormone (GH)**, or somatotropin (STH), regulates the development of the long bones; **insulin** regulates blood sugar by increasing the permeability of cells to glucose; **epinephrine** (adrenaline) is produced in times of stress. These hormones are called nontarget hormones. Other hormones affect specific cells or target tissues, such as parathyroid hormone, which regulates calcium levels in the body, and gastrin, which stimulates cells of the stomach to produce digestive enzymes.

Chemical Control Systems

Along with the nervous system, the endocrine system provides integration and control of the organs and tissues. The malfunction of one organ affects other organs; however, an animal can continue to function because of compensations made by the two control systems. The nervous system enables the body to adjust quickly to changes in the environment. The endocrine system is designed to maintain control over a longer duration. Such endocrine hormones as growth hormone and the various sex hormones, for example, regulate and sustain development for many years.

The division between the nervous system and endocrine system is most subtle in the hypothalamus. The hypothalamus regulates the pituitary gland through nerve stimulation. However, the endocrine glands, stimulated by the pituitary, secrete chemicals that affect the nerve activity of the hypothalamus (**Figure 2**).

The word *hormone* comes from the Greek *hormon*, meaning "to excite or set into motion." Hormones serve as regulators, speeding up or slowing down certain bodily processes.

The relationship between chemical messengers and the activity of organ systems within the body was established by experiment. In 1889, scientists Joseph von Mering and Oscar Minkowski of the University of Strasbourg in France showed that a chemical messenger produced in the pancreas is responsible for the regulation of blood sugar. After removing the pancreases from a number of dogs, the two scientists noticed that the animals began to lose weight very quickly. Within a few hours, the dogs became fatigued and displayed some of the symptoms that are now associated with diabetes in humans. By chance, the two scientists also noted that ants began gathering in the kennels where the sick dogs were kept. No ants, however, were found in the kennels of healthy dogs. What had caused the ants to gather? The scientists analyzed the urine of the sick dogs and found that it contained glucose, a sugar, while the urine of the healthy dogs did not. The ants were attracted to the sugar. The experiment provided evidence that a chemical messenger, produced by the pancreas, was responsible for the regulation of blood sugar. This chemical is the hormone called insulin.

The experiment by von Mering and Minkowski typifies classical approaches to uncovering the effect of specific hormones. In many cases, a gland or organ was removed and the effects on the organism were monitored. Once the changes in behaviour were noted, chemical extracts from the organ were often injected into the animal, and the animal's activities monitored. By varying dosages of the identified chemical messenger, scientists hoped to determine how it worked. Although effective to a certain degree, classical techniques were limited. No hormonal response works independently. The concentrations of other hormones often increase to help compensate for a disorder.

Some glands produce a number of different hormones. Therefore, the effect cannot be attributed to a single hormone. For example, early experimenters who attempted to uncover the function of the thyroid gland unwittingly removed the parathyroid glands along with the thyroid. It might be expected that these tiny glands are part of the thyroid and are related to its function. However, although the parathyroids are embedded in the tissue of the thyroid, they have a separate function. What the scientists failed to realize was that many of the symptoms they associated with low thyroid secretions are actually created by the parathyroids.

The main problem for early researchers was obtaining and isolating the actual messenger among the other chemicals found within the removed organ. Most hormones are found in very small amounts. Furthermore, the concentration of hormone varies throughout the day. The prediction of site and time was often a matter of mere luck.

In the past few years, technological improvements in chemical analysis techniques and microscopy have vastly increased our knowledge of the endocrine system (**Figure 3**). Radioactive tracers enable scientists to follow messenger chemicals from the organ in which they are produced to the target cells. The radioactive tracers also allow researchers to discern how the chemical messenger is broken down into other compounds and removed as waste. With new chemical analysis equipment, scientists can determine and measure the concentration of even the smallest amounts of a hormone as the body responds to changes in the external and internal environments. In addition, high-power microscopes provide a clearer picture of the structure of cell membranes and allow a better understanding of how chemical messengers attach themselves to target sites.

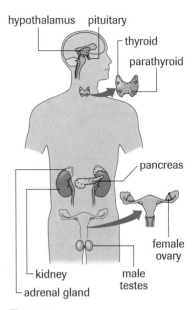

Figure 2

The location and appearance of some important endocrine glands in the human body. Hidden within the thyroid gland are four small glands—the parathyroid glands.

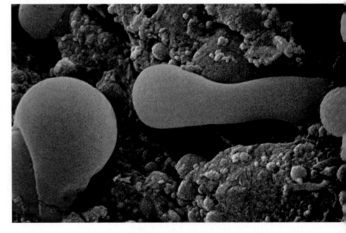

Figure 3

This scanning electron micrograph (SEM) shows thyroid hormones at 8500x magnification. The photo shows the first part of a two-stage hormone production process. Thyroid protein is being secreted into a storage chamber (a follicle), from the cells lining the chamber's walls. The thyroid gland produces hormones that affect every cell in the body, regulating metabolism and contributing to brain and bone development in growing children.

Chemical Signals: Steroid and Protein Hormones

How do hormones signal cells? First, it is important to note that hormones do not affect all cells. Cells may have receptors for one hormone but not another. The number of receptors found on individual cells also may vary. For example, liver cells and muscle cells have many receptor sites for the hormone insulin. Fewer receptor sites are found in less active cells such as bone cells and cartilage cells.

Second, there are two types of hormones, which differ in chemical structure and action. The first group, **steroid hormones**, is made from cholesterol, a lipid compound, and includes male and female sex hormones and **cortisol**. Steroid molecules are composed of complex rings of carbon, hydrogen, and oxygen molecules and are not soluble in water but are soluble in fat. The second group, **protein hormones**, includes insulin and growth hormone. These hormones contain chains of amino acids of varying length and are soluble in water.

Steroid hormones diffuse from the capillaries into the interstitial fluid and then into the target cells, where they combine with receptor molecules located in the cytoplasm. The hormone–receptor complex then moves into the nucleus and attaches to a segment of chromatin that has a complementary shape. The hormone activates a gene that sends a message to the ribosomes in the cytoplasm to begin producing a specific protein (**Figure 4**).

steroid hormones group of hormones, made from cholesterol, that includes male and female sex hormones and cortisol

cortisol hormone that stimulates the conversion of amino acids to glucose by the liver

protein hormones group of hormones, composed of chains of amino acids, that includes insulin and growth hormone

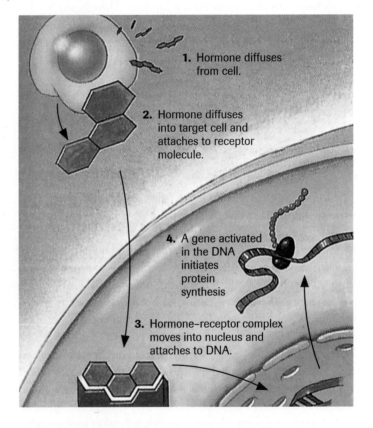

1. Hormone diffuses from cell.

2. Hormone diffuses into target cell and attaches to receptor molecule.

4. A gene activated in the DNA initiates protein synthesis

3. Hormone–receptor complex moves into nucleus and attaches to DNA.

Figure 4
The steroid hormone molecule passes into the cell, combines with the receptor molecule, and then activates a gene in the nucleus. The gene initiates the production of a specific protein.

Protein hormones exhibit a different action. Unlike steroid hormones, which diffuse into the cell, protein hormones combine with receptors on the cell membrane. Specific hormones combine at specific receptor sites. Some of the protein hormones form a hormone–receptor complex that activates the production of an enzyme called adenylyl cyclase. The adenylyl cyclase causes the cell to convert adenosine triphosphate (ATP), the primary source of cell energy, into **cyclic adenosine monophosphate (cyclic AMP)**. The cyclic AMP functions as a messenger, activating enzymes in the cytoplasm to carry out

cyclic adenosine monophosphate (cyclic AMP) secondary chemical messenger that directs the synthesis of protein by ribosomes

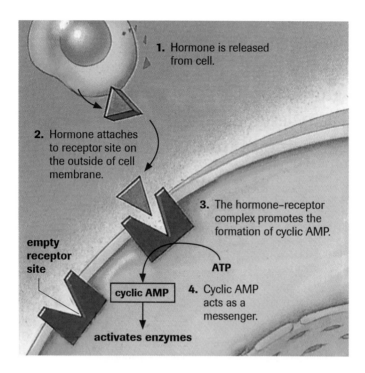

1. Hormone is released from cell.

2. Hormone attaches to receptor site on the outside of cell membrane.

3. The hormone–receptor complex promotes the formation of cyclic AMP.

empty receptor site

ATP

cyclic AMP

4. Cyclic AMP acts as a messenger.

activates enzymes

Figure 5
The protein hormone combines with specific receptor sites and triggers the formation of cyclic AMP from ATP. Cyclic AMP acts as a secondary messenger, activating enzymes within the cell.

their normal functions (**Figure 5**). For example, when thyroid-stimulating hormone (TSH) attaches to the receptor sites in the thyroid gland, cyclic AMP is produced in thyroid cells. Cells of the kidneys and muscles are not affected because they have no receptors for thyroid-stimulating hormone. The cyclic AMP in the thyroid cell activates enzymes, which begin producing **thyroxine**, a hormone that regulates metabolism. Thyroxine will be discussed in greater depth later in the chapter.

thyroxine iodine-containing hormone, produced by the thyroid gland, that increases the rate of body metabolism and regulates growth

The Pituitary Gland: The Master Gland

The **pituitary gland** is often referred to as the "master gland" because it exercises control over other endocrine glands. This small sac-like structure is connected by a stalk to the hypothalamus, the area of the brain associated with homeostasis (**Figure 6**, page 376). The interaction between the nervous system and endocrine system is evident in this hypothalamus–pituitary complex. The pituitary gland produces and stores hormones. The hypothalamus stimulates the release of hormones by the pituitary gland by way of nerves.

The pituitary gland is actually composed of two separate lobes: the posterior lobe and the anterior lobe. The posterior lobe of the pituitary stores and releases hormones such as antidiuretic hormone (ADH) and oxytocin, which have been produced by the hypothalamus. Antidiuretic hormone acts on the kidneys and helps regulate body water. Oxytocin initiates strong uterine contractions during labour. The hormones travel by way of specialized nerve cells from the hypothalamus to the pituitary. The pituitary gland stores the hormones, as shown in **Figure 6(a)**, releasing them into the blood when necessary.

The anterior lobe of the pituitary, unlike the posterior lobe, produces its own hormones. However, like the posterior lobe, the anterior lobe is richly supplied with nerves from the hypothalamus. The hypothalamus regulates the release of hormones from the anterior pituitary, as shown in **Figure 6(b)**. Hormones are secreted from the nerve ends of the cells of the hypothalamus and transported in the blood to the pituitary gland. Most of these hormones activate specific cells in the pituitary, causing the release of pituitary hormones, which are then carried by the blood to target tissues. Two hypothalamus-releasing factors inhibit the secretion of hormones from the anterior lobe of

pituitary gland gland at the base of the brain that, together with the hypothalamus, functions as a control centre, coordinating the endocrine and nervous systems

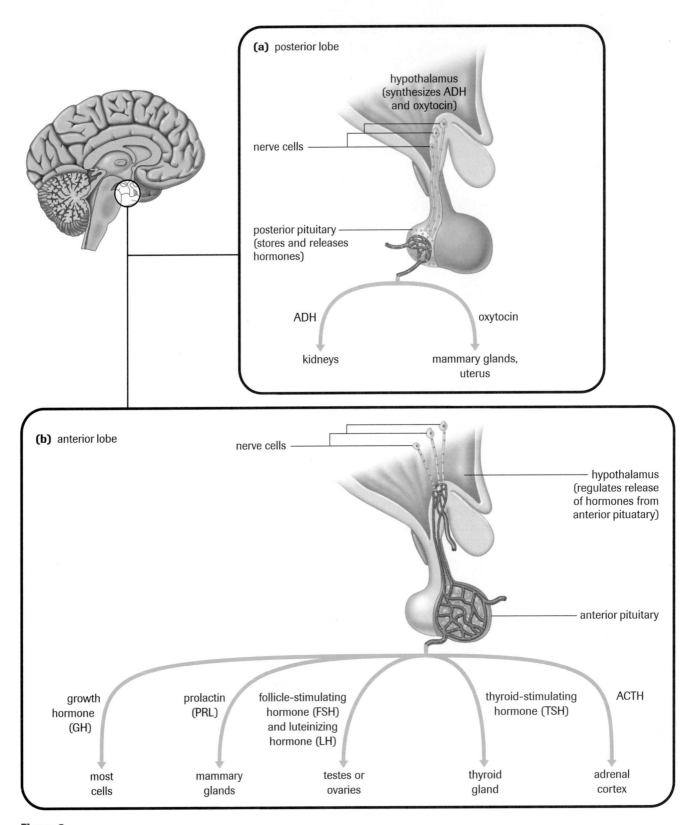

Figure 6

The pituitary gland is composed of two separate lobes: the posterior lobe and the anterior lobe.

(a) The cells of the hypothalamus synthesize antidiuretic hormone (ADH) and oxytocin, which travel from the hypothalamus to the pituitary along specialized nerve cells. The hormones remain in the pituitary gland and are released into the blood when they are needed. ADH acts on the kidneys and helps regulate body water. Oxytocin initiates strong uterine contractions during labour.

(b) Hormones released by nerve cells of the hypothalamus regulate hormones secreted by the anterior pituitary.

the pituitary. The releasing factor, called dopamine, inhibits the secretion of prolactin (PRL), a pituitary hormone that stimulates milk production in women who have given birth. The hormone somatostatin inhibits the secretion of somatotropin, the pituitary hormone associated with growth of the long bones.

Various regulator hormones are stored in the anterior lobe of the pituitary gland. Thyroid-stimulating hormone (TSH), as the name implies, stimulates the thyroid gland to produce its hormone, thyroxine. The anterior pituitary also releases reproductive-stimulating hormones, growth-stimulating hormones, prolactin, and adrenocorticotropic hormone (ACTH), the hormone that stimulates the adrenal cortex.

> ▶ **Practice**

Understanding Concepts

1. What are target tissues or organs?
2. Explain how the nervous system and endocrine system are specialized to maintain homeostasis.
3. Describe the signalling action of steroid hormones and protein hormones.
4. What is cyclic adenosine monophosphate (cyclic AMP)?

SUMMARY   *Importance of the Endocrine System*

Table 1 Pituitary Hormones

Hormone	Target	Primary function
Anterior lobe		
thyroid-stimulating hormone (TSH)	thyroid gland	• stimulates release of thyroxine from thyroid • thyroxine regulates cell metabolism
adrenocorticotropic hormone (ACTH)	adrenal cortex	• stimulates release of hormones involved in stress responses
somatotropin (STH), or growth hormone (GH)	most cells	• promotes growth
follicle-stimulating hormone (FSH)	ovaries, testes	• in females, stimulates follicle development in ovaries • in males, promotes the development of sperm cells in testes
luteinizing hormone (LH)	ovaries, testes	• in females, stimulates ovulation and formation of the corpus luteum • in males, stimulates the production of the sex hormone testosterone
prolactin (PRL)	mammary glands	• stimulates and maintains milk production in lactating females
Posterior lobe		
oxytocin	uterus, mammary glands	• initiates strong contractions • triggers milk release in lactating females
antidiuretic hormone (ADH)	kidneys	• increases water reabsorption by kidneys

The pancreas contains two types of cells: one type produces digestive enzymes; the other type produces hormones. The hormone-producing cells are located in structures called the islets of Langerhans, named after their discoverer, German scientist Paul Langerhans. More than 2000 tiny islets, each containing thousands of cells, are scattered throughout the pancreas. The islets contain beta and alpha cells that are responsible for the production of two hormones: insulin and **glucagon**.

Insulin is produced in the beta cells of the islets of Langerhans and is released when the blood sugar level increases. After a meal, the blood sugar level rises and an appropriate amount of insulin is released (**Figure 1**). The insulin causes cells of the muscles, the liver, and other organs to become permeable to the glucose. In the liver, the glucose is converted into glycogen, the primary storage form for glucose. In this way, insulin enables the blood sugar level to return to normal. As discussed in Chapter 7, insulin helps maintain homeostasis.

glucagon hormone produced by the pancreas; when blood sugar levels are low, glucagon promotes conversion of glycogen to glucose

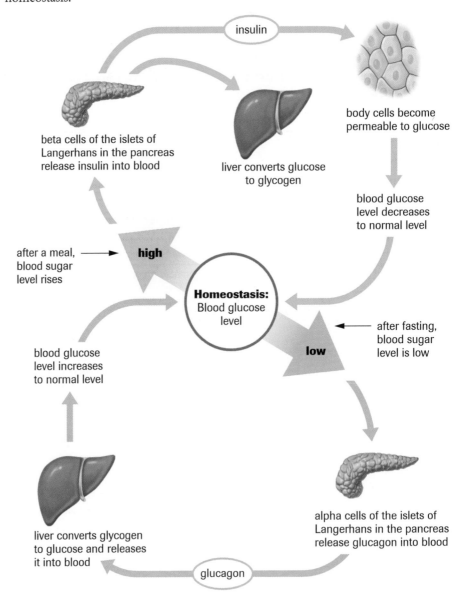

insulin

beta cells of the islets of Langerhans in the pancreas release insulin into blood

liver converts glucose to glycogen

body cells become permeable to glucose

blood glucose level decreases to normal level

after a meal, blood sugar level rises

high

Homeostasis: Blood glucose level

after fasting, blood sugar level is low

low

blood glucose level increases to normal level

liver converts glycogen to glucose and releases it into blood

alpha cells of the islets of Langerhans in the pancreas release glucagon into blood

glucagon

Figure 1
Insulin, released when blood sugar levels are high, increases the permeability of cells to glucose. Glucose is converted into glycogen within the liver, thereby restoring blood sugar levels. Glucagon, released when blood sugar levels are low, promotes the conversion of liver glycogen into glucose, thereby restoring blood sugar levels.

Glucagon and insulin work in a complementary fashion. Insulin causes a decrease in the blood sugar level, and glucagon causes an increase in the blood sugar level. Produced by the alpha cells of the islets of Langerhans, glucagon is released when blood sugar levels are low, such as after periods of fasting. Glucagon promotes the conversion of glycogen to glucose, which is released into the blood. As glycogen is converted to glucose in the liver, the blood sugar level returns to normal.

Diabetes

Diabetes is a chronic disease with no cure that affects more than two million Canadians. It is caused by insufficient production or use of insulin. It is a leading cause of death by disease and, when left untreated, can cause blindness, kidney failure, nerve damage, and nontraumatic limb amputation.

Without adequate levels of insulin, blood sugar levels tend to rise very sharply following meals. This condition is known as hyperglycemia, or high blood sugar (from *hyper*, meaning "too much;" *glyco*, meaning "sugar;" and *emia* referring to a condition of the blood). A variety of symptoms are associated with hyperglycemia. The kidneys are unable to reabsorb all the blood glucose that is filtered through them, so the glucose appears in the urine. Since high concentrations of glucose in the nephrons draw water out of the plasma by osmosis, people with diabetes excrete unusually large volumes of urine and are often thirsty.

People with diabetes often experience low energy levels. Remember that insulin is required for the cells of the body to become permeable to glucose. Despite the abundance of glucose in the blood, little is able to move into the cells of the body. The cells of people with diabetes soon become starved for glucose and must turn to other sources of energy. Fats and proteins can be metabolized for energy, but, unlike carbohydrates, they are not an easily accessible energy source. The switch to these other sources of energy creates a host of problems. Acetone, an intermediate product of excessive fat metabolism, can be produced in an untreated diabetic. In severe cases, the smell of acetone can be detected on the breath of these people.

There are three main types of diabetes mellitus. Type 1 diabetes (formerly known as juvenile-onset diabetes) occurs when the pancreas in unable to produce insulin because of the early degeneration of the beta cells in the islets of Langerhans. It is usually diagnosed in childhood, and people who have it must take insulin to live (**Figure 2**). Approximately 10% of people with diabetes have type 1 diabetes.

Type 2 diabetes (sometimes referred to as adult-onset diabetes) is associated with decreased insulin production or ineffective use of the insulin that the body does produce. Type 2 diabetes is usually diagnosed in adulthood and can be controlled with diet, exercise, and oral drugs known as sulfonamides. It is believed that sulfonamides, which are not effective against type 1 diabetes, stimulate the islets of Langerhans to function in adults. About 90% of people with diabetes have type 2 diabetes.

A third type of diabetes, gestational diabetes, is a temporary condition that occurs in 2% to 4% of pregnancies. It increases the risk of diabetes in both mother and child. 📖

Research in Canada: Banting and Best

Frederick Banting (1891–1941) served in World War I as a doctor. On returning from the war, he became interested in diabetes research. At the time, diabetes was thought to be caused by a deficiency of a hormone located in specialized cells of the pancreas. However, extracting the hormone from the pancreas presented a problem since the pancreas also stores digestion enzymes capable of breaking down the protein hormone.

In 1921, Banting approached John J.R. MacLeod, a professor at the University of Toronto, with his idea for isolating the hormone. MacLeod assigned Banting a makeshift laboratory as well as an assistant, Charles Best, who was a graduate student in biochemistry. Banting

diabetes chronic disease that occurs when the body cannot produce any insulin or enough insulin, or is unable to use properly the insulin it does make

DID YOU KNOW❓

Hypoglycemia
When the blood sugar level falls below normal, a condition known as hypoglycemia occurs. Hypoglycemia can be caused by too much insulin or too little glucagon.

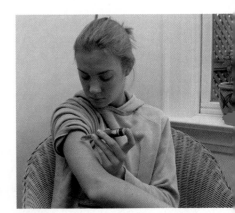

Figure 2
The pen is the newest way of injecting insulin. It is portable, accurate, and easy to use.

📖 INVESTIGATION 8.2.1

Identification of Hyperglycemia (p. 400)
How is urinalysis used to identify diabetes? In this investigation, you will use simulated urine samples to identify diabetes.

Figure 3
Dr. Charles Best (left) and
Dr. Frederick Banting (right)

and Best (**Figure 3**) tied the pancreatic duct of dogs, and waited seven weeks for the pancreas to shrivel. Although the cells producing digestive enzymes deteriorated, cells from the islets of Langerhans remained. The hormone was then extracted from the pancreas. When the hormone was injected into dogs that had had their pancreases removed, symptoms of diabetes ceased. Banting and Best wanted to call the hormone *isletin*, but MacLeod insisted that it be called *insulin*.

In 1923, Banting and MacLeod were awarded the Nobel Prize for physiology and medicine. Banting was furious that Charles Best, his coworker, had not been included, and MacLeod, the professor who had contributed laboratory space, had gotten the glory instead.

Frontiers of Technology: Islet Cell Transplants

Type 1 diabetes is the second leading cause of blindness in Canada. Other side effects of the disease, such as kidney and heart failure, stroke, and peripheral nerve damage, affect more than 50 000 Canadians. Although insulin injections provide some regulation of blood sugar, they do not necessarily prevent many of the serious complications of diabetes, such as blindness and stroke. Insulin therapy requires monitoring of blood sugar level and balancing injections of insulin with carbohydrate intake and exercise.

Transplanted islet cells, however, could replace the body's natural mechanism for monitoring and producing insulin. Unlike insulin therapy, islet cell transplantation holds the potential to reverse the effects of diabetes. One of the main barriers to successful clinical islet transplantation is immune rejection. Current anti-rejection drugs are toxic and harmful to islet function, and immunosuppression leaves the recipients susceptible to invading microbes.

Researchers around the world are searching for solutions. A team of researchers at the University of Alberta has pioneered a treatment—known as the Edmonton Protocol—designed by Dr. James Shapiro, Director and Head of the Clinical Islet Transplant Program (**Figure 4**). The treatment uses a steroid-free combination of three drugs to prevent rejection of the transplanted islets and to prevent diabetes from returning. The success of the treatment depends on new methods of isolating and transplanting pancreatic cells.

Figure 4
Members of the Edmonton Protocol
at the University of Alberta

Unlike other types of transplant surgery, the technique used for islet transplants is noninvasive and presents few risks. Islet cells are extracted from the pancreas of a donor and infused into the recipient's liver by way of a large vein. The surgeon uses ultrasound to see the vein leading into the liver, the skin is frozen, and a syringe is used to put the new cells in place. The patient can usually return home the next day. The liver is used because, when damaged, it is able to regenerate itself by building new blood vessels and cells. New blood vessels and nerve cells connect to the transplanted islets in the liver and eventually produce enough insulin to control blood sugar.

In 1989, Jim Connor became Canada's first patient to receive transplanted islets of Langerhans cells. Dr. Garth Warnock, a surgeon at the University of Alberta Hospital, transplanted millions of isolated cells from the pancreas of a donor. As of late 2001, the research group had performed 15 islet transplants and eight patients were totally insulin-free. The research team is also working on a procedure that would permit the transplant of islet cells into people with type 1 diabetes before the onset of physical complications, such as renal failure. A challenge regarding these transplants, is finding an affordable supply of insulin-producing cells. At present, cadavers provide the only source of cells and the cost of processing islet from donors is formidable.

Adrenal Glands

The adrenal glands are located above each kidney. (The word adrenal comes from the Latin *ad*, meaning "to" or "at," and *renes*, meaning "kidneys.") Each adrenal gland is made up of two glands encased in one shell. The inner gland, the **adrenal medulla**, is surrounded by an outer casing, called the adrenal cortex. The medulla is regulated by the nervous system, while hormones regulate the adrenal cortex.

The adrenal medulla produces two hormones: epinephrine (also known as adrenaline) and **norepinephrine** (noradrenaline). The nervous system and the adrenal medulla are linked by the fact that both produce epinephrine. The hormone-producing cells within the adrenal medulla are stimulated by sympathetic nerves in times of stress.

In a stress situation, epinephrine and norepinephrine are released from the adrenal medulla into the blood. Under their influence, the blood sugar level rises. Glycogen, a carbohydrate storage compound in the liver and muscles, is converted into glucose, a readily usable form of energy. The increased blood sugar level ensures that a greater energy reserve will be available for the tissues of the body. These hormones also increase heart rate, breathing rate, and cell metabolism. Blood vessels dilate, allowing more oxygen and nutrients to reach the tissues. Even the iris of the eye dilates, allowing more light to reach the retina—in a stress situation, the body attempts to get as much visual information as possible.

The **adrenal cortex** produces three different types of hormones: the glucocorticoids, the mineralocorticoids, and small amounts of sex hormones. The **glucocorticoids** are associated with blood glucose levels. One of the most important of the glucocorticoids, cortisol, increases the level of amino acids in the blood in an attempt to help the body recover from stress. The amino acids are converted into glucose by the liver, thereby raising the level of blood sugar. Increased glucose levels provide a greater energy source, which helps cell recovery. Any of the amino acids not converted into glucose are available for protein synthesis. The proteins can be used to repair damaged cells. In addition, fats in adipose tissue are broken down into fatty acids. Thus, a second source of energy is provided, helping conserve glucose in times of fasting. Under the influence of cortisol, blood glucose uptake is inhibited in many tissues, especially in the muscles. The brain is not affected though, since any significant decrease in glucose absorption of the brain would lead to convulsions.

Short-term and long-term stress responses are shown in **Figure 5** (page 382). The brain identifies stressful situations. The hypothalamus sends a releasing hormone to the anterior lobe of the pituitary, stimulating the pituitary to secrete corticotropin, also called adrenocorticotropic hormone (ACTH). The blood carries the ACTH to the target cells in the adrenal cortex. Under the influence of ACTH, the cells of the adrenal cortex secrete mineralocorticoids and glucocorticoids (among them cortisol), which are carried to target cells in the liver and muscles. As cortisol levels rise, cells within the hypothalamus and pituitary decrease the production of regulatory hormones, and, eventually, the levels of cortisol begin to fall. This process is called a long-term stress response. The short-term stress response is regulated by the adrenal medulla, which secretes epinephrine and norepinephrine.

Aldosterone is the most important of the **mineralocorticoids**, the second major group of hormones produced by the adrenal cortex. Secretion of aldosterone increases sodium retention and water reabsorption by the kidneys, thereby helping to maintain body fluid levels.

adrenal medulla found at the core of the adrenal gland, the adrenal medulla produces epinephrine and norepinephrine

norepinephrine also known as noradrenaline, it initiates the fight-or-flight response by increasing heart rate and blood sugar

adrenal cortex outer region of the adrenal gland that produces glucocorticoids and mineralocorticoids

glucocorticoids various hormones, produced by the adrenal cortex, designed to help the body meet the demands of stress

DID YOU *KNOW*❓

Removing the Adrenal Medulla
The adrenal medulla can be surgically removed without any apparent ill effects. If the adrenal cortex were removed, however, death would result.

mineralocorticoids hormones of the adrenal cortex important for regulation of salt–water balance

LAB EXERCISE 8.2.1

The Effects of Hormones on Blood Sugar (p. 401)
What effects do hormones have on blood sugar? In this exercise, you will use experimental data to investigate the effects of various hormones on blood sugar levels.

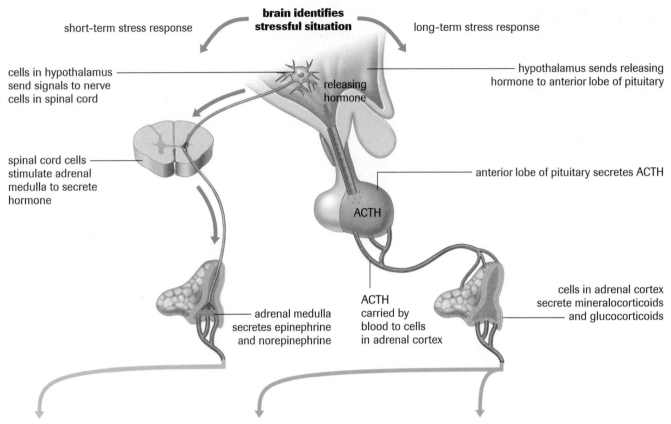

short-term stress response

brain identifies stressful situation

long-term stress response

cells in hypothalamus send signals to nerve cells in spinal cord

releasing hormone

hypothalamus sends releasing hormone to anterior lobe of pituitary

spinal cord cells stimulate adrenal medulla to secrete hormone

anterior lobe of pituitary secretes ACTH

ACTH

adrenal medulla secretes epinephrine and norepinephrine

ACTH carried by blood to cells in adrenal cortex

cells in adrenal cortex secrete mineralocorticoids and glucocorticoids

epinephrine and norepinephrine response
- increase in blood glucose due to glycogen that has been converted into glucose
- increase in heart rate, breathing rate, and cell metabolism
- change in blood flow patterns that direct more blood to heart and muscle cells

mineralocorticoids response
- increase in the amounts of sodium ions and water retained by the kidneys
- increase in blood volume and blood pressure

glucocorticoids response
- increase in blood glucose due to proteins and fats that are broken down and converted into glucose
- suppression of the inflammatory response of the immune system

Figure 5
Stress responses

 **SUMMARY**   *Hormones That Affect Blood Sugar*

Table 1

Hormone	Location of hormone production	Effect
insulin	islets of Langerhans (pancreas)	• increases permeability of cells to glucose; increases glucose uptake • allows for the conversion of glucose to glycogen • brings about a decrease in blood sugar
glucagon	islets of Langerhans (pancreas)	• promotes the conversion of glycogen to glucose • brings about an increase in blood sugar
epinephrine and norepinephrine	adrenal medulla	• promotes the conversion of glycogen to glucose • brings about an increase in blood sugar • brings about an increase in heart rate, and cell metabolism
cortisol (a type of glucocorticoid)	adrenal cortex	• promotes the conversion of amino acids to glucose • promotes the breakdown of fats to fatty acids • decreases glucose uptake by the muscles (not by the brain) • brings about an increase in blood sugar in response to stress

> **Section 8.1 and 8.2** *Questions*

Understanding Concepts

1. List the hormones released from the pancreas and the adrenal glands, and indicate their control mechanisms.

2. How does insulin regulate blood sugar levels?

3. How does glucagon regulate blood sugar levels?

4. Using a flow chart, show a homeostatic adjustment for a person who has consumed a significant amount of carbohydrates in the past hour.

5. What advantage is provided by increasing blood sugar above normal levels in times of stress?

6. How would high levels of adrenocorticotropic hormone (ACTH) affect secretions of cortisol from the adrenal glands? How would high levels of cortisol affect ACTH?

Applying Inquiry Skills

7. A number of laboratory experiments were conducted on laboratory mice. The endocrine system of mice is similar to that of humans. Brief summaries of the procedures are provided in **Table 2**.
   (a) In procedure 1, identify the gland that was removed and explain why the levels of ACTH increased.
   (b) In procedure 2, identify the hormone that was injected and explain why blood sugar levels decreased.
   (c) In procedure 3, identify the hormone that was affected and explain why urine production increased.
   (d) In procedure 4, identify the hormone that was injected and explain why blood glucose levels increased.

Table 2

#	Procedure	Observation
1	gland removed	• urine output increased • Na^+ ion concentration in urine increased • ACTH level increased in blood
2	hormone injected	• blood glucose levels decreased
3	blood flow from the posterior pituitary reduced	• urine production increased
4	hormone injected	• glycogen converted to glucose in the liver • blood glucose increased

Making Connections

8. The North American lifestyle and diet are believed to be major contributors to type 2 diabetes. Many companies know that foods can be made more palatable to consumers by adding fats and sugars. Discuss the practice of adding fats and sugars to food products to increase sales.

9. Cortisone is often prescribed as an anti-inflammatory drug. Why are doctors hesitant to provide this drug over a long duration?

thyroid gland a two-lobed gland at the base of the neck that regulates metabolic processes

parathyroid glands four pea-sized glands in the thyroid gland that produce parathyroid hormone to regulate blood calcium and phosphate levels

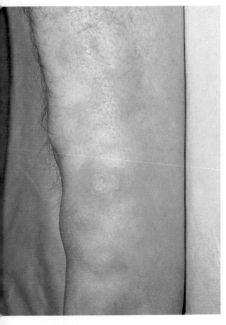

Figure 2
A lack of thyroid hormones can cause conditions such as myxedema, characterized by dry skin, weight gain, and edema (swelling).

In this section, you will explore three different glands that affect metabolism: the **thyroid gland**, which produces the hormones triiodothyronine, thyroxine, and calcitonin; the **parathyroid glands**, which produce parathyroid hormone; and the anterior pituitary gland, which produces growth hormone among many other regulatory hormones. The thyroid gland helps regulate body metabolism or the rate at which glucose is oxidized. The parathyroid glands help regulate calcium levels in the blood and lower phosphate levels. Growth hormone, or somatotropin, one of a multitude of hormones produced by the anterior pituitary gland, influences the growth of long bones and accelerates protein synthesis.

Thyroid Gland

The thyroid gland (**Figure 1**) is located at the base of the neck, immediately in front of the trachea or windpipe. Two important thyroid hormones, thyroxine (T4) and triiodothyronine (T3), regulate body metabolism and the growth and differentiation of tissues. Although both hormones appear to have the same function, approximately 65% of thyroid secretions are thyroxine. Most of this discussion will focus on the principal hormone, thyroxine.

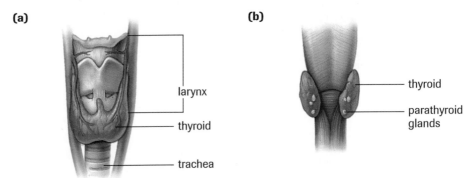

(a)   **(b)**

larynx

thyroid

trachea

thyroid

parathyroid glands

Figure 1
(a) Anterior view of thyroid gland **(b)** posterior view of thyroid gland

Have you ever wondered why some people seem to be able to consume fantastic amounts of food without any weight change, while others appear to gain weight at the mere sight of food? Thyroxine and the regulation of metabolic rate can partly explain this anomaly. Individuals who secrete higher levels of thyroxine oxidize sugars and other nutrients at a faster rate. Approximately 60% of the glucose oxidized in the body is released as heat (which explains why these individuals usually feel warm). The remaining 40% is transferred to ATP, the storage form for cell energy. This added energy reserve is often consumed during activity. Therefore, these individuals tend not to gain weight.

Individuals who have lower levels of thyroxine do not oxidize nutrients as quickly, and therefore tend not to break down sugars as quickly. Excess blood sugar is eventually converted into liver and muscle glycogen. However, once the glycogen stores are filled, excess sugar is converted into fat. It follows that the slower the blood sugar is used, the faster the fat stores are built up. People who secrete low amounts of thyroxine often experience muscle weakness, cold intolerance, and dry skin and hair (**Figure 2**). It is important to note that not all types of weight gain are due to hypothyroidism (low thyroid secretions); in many cases, weight gain reflects a poor diet.

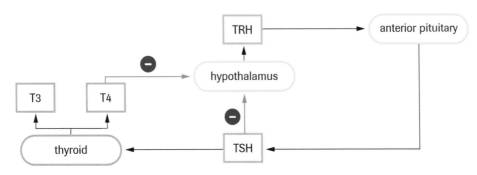

Figure 3
Feedback control loops in the secretion of thyroid hormones

Control of thyroid hormones, like many other hormones, is accomplished by negative feedback (**Figure 3**). Should the metabolic rate decrease, receptors in the hypothalamus are activated. Nerve cells secrete thyroid-releasing hormone (TRH), which stimulates the pituitary to release thyroid-stimulating hormone (TSH). Thyroid-stimulating hormone is carried by the blood to the thyroid gland, which, in turn, releases thyroxine. Thyroxine raises metabolism by stimulating increased sugar utilization by body cells. Higher levels of thyroxine cause the pathway to be "turned off." Thyroxine inhibits the release of TRH from the hypothalamus, thus turning off the production of TSH from the pituitary.

In addition to thyroxine, the thyroid gland contains **calcitonin**. Calcitonin is a hormone that acts on the bone cells to lower the level of calcium found in the blood.

Thyroid Disorders

Iodine is an important component of both thyroid hormones. A normal component of the diet, iodine is actively transported from the blood into the follicle cells of the thyroid. The concentration of iodine in the cells can be 25 times greater than that in the blood. Problems are created when iodine levels begin to fall. When inadequate amounts of iodine are obtained from the diet, the thyroid enlarges, producing a **goiter** (**Figure 4**).

The presence of a goiter emphasizes the importance of a negative feedback control system. Without iodine, thyroid production and secretion of thyroxine drops. This causes more and more TSH to be produced and, consequently, the thyroid is stimulated more and more. Under the relentless influence of TSH, cells of the thyroid continue to develop, and the thyroid enlarges. In regions where the diet lacks iodine, the incidence of goiter remains high. In many countries, iodine is added to table salt to prevent this condition.

Parathyroid Glands

Four small parathyroid glands are hidden within the larger thyroid gland. At one time, surgeons mistakenly removed the unknown glands with sections of a hyperactive thyroid gland to treat goiters. Although the surgery relieved symptoms associated with an overly developed thyroid gland, the patients developed signs of more serious problems. Rapid, uncontrolled muscle twitching, referred to as tetanus, signaled abnormal calcium levels. This condition occurs because the nerves become easily excited.

Usually, nerves or other hormones regulate the endocrine glands. The parathyroid glands are one of the exceptions. These glands maintain homeostasis by responding directly to chemical changes in their immediate surroundings. For example, low calcium levels in the blood stimulate the release of **parathyroid hormone (PTH)** (**Figure 5**, page 386). A rise in PTH levels causes the calcium levels in the blood to increase and phosphate levels to decrease. The hormone does this by acting on three different organs: the kidneys, the intestines, and the bones. PTH causes the kidneys and gut to retain calcium while promoting calcium release from bone (approximately 98% of the body's calcium is held in storage

calcitonin hormone produced by the thyroid gland that lowers calcium levels in the blood

goiter disorder that causes an enlargement of the thyroid gland

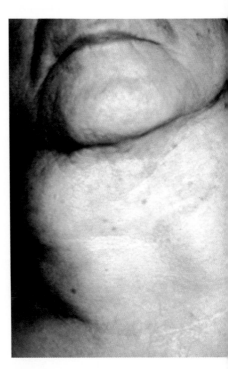

Figure 4
A goiter appears as a swelling in the neck area.

✎ LAB EXERCISE 8.3.1

Hormonal Control of Metamorphosis (p. 402)
What is the effect of thyroid hormones on the rate of metamorphosis in amphibians? In this exercise, you will examine data from several experiments involving amphibians that do and do not metamorphose.

parathyroid hormone (PTH) hormone produced by the parathyroid glands, which will increase calcium levels in the blood and lower the levels of phosphates

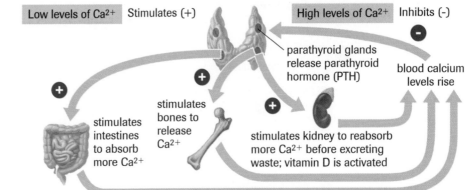

Low levels of Ca²⁺   Stimulates (+)

High levels of Ca²⁺   Inhibits (-)

parathyroid glands release parathyroid hormone (PTH)

blood calcium levels rise

stimulates intestines to absorb more Ca²⁺

stimulates bones to release Ca²⁺

stimulates kidney to reabsorb more Ca²⁺ before excreting waste; vitamin D is activated

Figure 5
Low levels of blood calcium stimulate the release of PTH from the parathyroid glands. PTH causes the kidneys and gut to retain calcium while promoting calcium release from bone. This causes blood calcium levels to rise, which in turn inhibits the release of PTH from the parathyroid glands.

by the skeletal system). The bone cells break down, and calcium is separated from phosphate ions. Then calcium is reabsorbed and returned to the blood while phosphate is excreted in the urine. This helps conserve much of the body's calcium that is dissolved in plasma. Since PTH also enhances the absorption of calcium from undigested foods in the intestine, as PTH levels increase, the absorption of calcium ions will increase.

Once PTH levels have elevated calcium levels, the release of PTH is inhibited. This ensures the blood calcium levels will not increase beyond demand. Abnormally high levels of PTH can cause problems. The prolonged breakdown of bone is dangerous. A strong, rigid skeleton is necessary for support. In addition, high levels of calcium could collect in blood vessels or form stonelike structures in the kidneys. Calcium levels signal both the release and inhibition of PTH.

PTH also helps activate vitamin D. Low levels of vitamin D can cause a disease called rickets (**Figure 6**). With this disease, too little calcium and phosphorous are absorbed from foods and the bones develop improperly.

Growth Hormone

The effects of growth hormone, or somatotropin, are most evident when the body produces too much or too little of it. Low secretion of growth hormone during childhood can result in dwarfism; high secretions during childhood can result in gigantism. Although growth hormone affects most of the cells of the body, the effect is most pronounced on cartilage cells and bone cells. If the production of growth hormone continues after the cartilaginous growth plates have been fused, other bones respond. Once the growth plates have fused, the long bones can no longer increase in length, but bones of the jaw, forehead, fingers, and toes increase in width. The disorder, referred to as acromegaly, causes a broadening of the facial features.

Under the influence of growth hormone, cells of soft tissues and bone begin to grow by increasing the number of cells (hyperplasia) and increasing the size of cells (hypertrophy). Growth hormone increases cell size in muscle cells and connective tissues by promoting protein synthesis while inhibiting protein degradation or breakdown. Proteins in many cells, such as muscle, are in a constant state of breakdown and repair. Amino acid uptake increases, which in turn provides the raw materials for protein synthesis. Growth hormone also stimulates ribosomes to follow the genetic instructions for protein synthesis. This may help explain the link between declines in growth hormone production and the aging process. As a person ages, growth hormone production begins to decline and cellular repair and protein replacement is compromised. As you age, protein is often replaced by fat, causing changes in the body's shape.

Growth hormone stimulates the production of insulin-like growth factors, which are produced by the liver. In response to growth hormone, these insulin-like growth factors

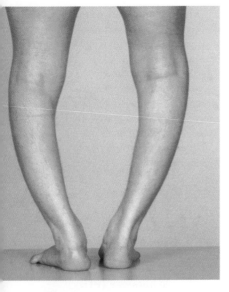

Figure 6
Improper bone formation can result from a diet lacking in fruits and vegetables.

are secreted into the blood, where they stimulate cell division in the growth plates, causing the elongation of the skeleton. Growth hormone promotes the elongation of long bones. In addition, growth hormone increases fatty acid levels in the blood by promoting the breakdown of fats held in adipose tissue. The muscles use the fatty acids instead of glucose as a source of metabolic fuel. By switching fuel sources from glucose to fatty acids, growth hormone causes an increase in blood glucose levels. This is especially important for glucose-dependent tissues, such as the brain. The brain is unable to use fat as an energy source. This metabolic pathway is particularly important in times of prolonged fasting where glucose supplies are limited. Growth hormone increases the utilization of fat stores and promotes protein synthesis, changing body form away from adipose tissue toward an increase in protein and muscle. This may help explain why quick growth spurts are often accompanied by a loss of body fat.

DID YOU KNOW ?

Growth Hormone: Not Just for Growth
Even in adults, growth hormone is the most abundant hormone produced by the anterior pituitary gland. In addition to promoting growth, the hormone helps adjust blood sugar in times of prolonged fasting and enhances the immune system.

SUMMARY *Hormones That Affect Metabolism*

Table 1

Gland	Hormone	Effect on metabolism
thyroid	thyroxine (T4) and triiodothyronine (T3)	• regulates the rate at which glucose is oxidized within body cells
thyroid	calcitonin	• lowers calcium levels in the blood
parathyroid glands	parathyroid hormone (PTH)	• raises calcium levels in the blood
anterior pituitary	growth hormone (GH), or somatotropin	• promotes protein synthesis by increasing the uptake of amino acids by cells • causes a switch in cellular fuels from glucose to fatty acids

▶ *Section 8.3 Questions*

Understanding Concepts

1. How would decreased secretion of growth hormone affect an individual?

2. How would an increased secretion of growth hormone affect an individual after puberty?

3. How does thyroxine affect blood sugar?

4. List the symptoms associated with hypothyroidism and hyperthyroidism.

5. How do the pituitary and hypothalamus interact to regulate thyroxine levels?

6. What is a goiter and why does it create a problem?

7. How does parathyroid hormone (PTH) regulate blood calcium levels?

8. Why would removal of the parathyroid glands lead to tetany?

Applying Inquiry Skills

9. Negative feedback control systems influence hormonal levels. The fact that some individuals have higher metabolic rates than others can be explained by the response of the hypothalamus and pituitary to thyroxine. Some feedback systems turn off quickly. Sensitive feedback systems tend to have comparatively lower levels of thyroxine; less sensitive feedback systems tend to have higher levels of thyroxine. One hypothesis attempts to link different metabolic rates with differences in the number of binding sites in the hypothalamus and pituitary. How might the number of binding sites for molecules along cell membranes affect hormonal levels? How would you go about testing the theory?

Making Connections

10. In July 1990, Dr. Daniel Rudman published a study in the prestigious *New England Journal of Medicine* proposing that injections of growth hormone could slow the aging process. Today, antiaging enthusiasts believe that growth hormone could be an antidote to the effects of decades of aging. Although researchers warn that the drug's long-term effects have not been documented and that the drug may not be suitable for everyone, speculation about the potentials of an antiaging drug abounds in both scientific and nonscientific communities. Comment on the social implications of using a drug to slow aging.

Figure 1
Hans Selye (1907–1982), the Austrian-born Canadian endocrinologist, was an authority on the link between psychological stress, biochemical changes, and disease.

Dr. Hans Selye (**Figure 1**) was one of the first to identify the human response to long-term stress from a noxious stimulus. According to Selye, a general adaptation syndrome results from exposure to prolonged stress brought on by a disruption of the external and internal environment. When an initiator of stress is identified, both the endocrine system and nervous system make adjustments that enable the body to cope with the problem. The nervous system rapidly adjusts to stress by increasing heart rate and diverting blood to the needed muscles. Although somewhat slower in response, hormones from the endocrine system provide a more sustained response to the stimulus (**Figure 5**, on page 382). **Table 1** summarizes some of the hormonal changes in response to stress.

Table 1 Hormonal Changes in Response to Stress

Hormone	Change	Adjustment
epinephrine	increases	• mobilizes carbohydrate and fat energy stores • increases blood glucose and fatty acids • accelerates heart rate and the activity of the respiratory system
cortisol	increases	• mobilizes energy stores by converting proteins to glucose • elevates blood amino acids, blood glucose, and blood fatty acids
glucagon	increases	• converts glycogen to glucose
insulin	decreases	• decreases the breakdown of glycogen in the liver

Stress hormones provide more blood glucose to cope with the elevated energy requirements brought on by stress. Remember that the primary stimulus for insulin secretion is a rise in blood glucose. If insulin release was not inhibited during a stress response, the hyperglycemia caused by stress would lead to an increased secretion of insulin, which would then lower blood glucose. Consequently, the elevated blood glucose would not be sustained to deal with the continued stress.

In addition to hormones that regulate blood sugar during stress, other hormones regulate blood pressure and blood volume. The nervous system activates the renin–angiotensin–aldosterone pathway in response to reduced blood flow to the kidneys. By increasing Na^+ reabsorption, the kidneys help maintain increased fluid volume. This helps sustain adequate blood pressure during stress. In addition, the stressor activates the hypothalamus, which causes an increased release of antidiurectic hormone (ADH). ADH will further increase water reabsorption from the nephron to help maintain body fluids.

During athletic competition, the accelerated cardiovascular activity provides greater oxygen delivery to the tissues for cellular respiration. Increases in blood sugar and fatty acid levels provide more fuel for metabolic processes. In turn, the greater supply of reactants can provide more ATP for activity.

It is more difficult to adjust to emotional or psychological stress because the increased energy supply is not always used. Although increased nerve activity requires greater energy, the ATP provided by homeostatic adjustment often outstrips demand. Prolonged exposure to high blood glucose, high blood pressure, and an elevated metabolic rate

Table 2 Problems Associated with Long-Term Stress

New operating limit	Problem created
higher blood sugar	• alters osmotic balance between blood and extracellular fluids; can lead to increased fluid uptake by the blood and increased blood pressure • increased water loss from nephron
increased blood pressure	• possible rupture of blood vessels due to higher pressure • increased blood clotting
increased heart rate	• can lead to higher blood pressure • possible destruction of heart muscle

often causes a readjustment of control systems to permit the higher operating range. As shown in **Table 2**, operating with elevated blood sugar, blood pressure, and heart rate creates more problems for the body.

Prostaglandins

Local responses to changes in the immediate environment of cells are detected by mediator cells, which produce **prostaglandins**. More than 16 different types of prostaglandins alter cell activity in a manner that counteracts or adjusts for the change. Generally, prostaglandins are secreted in low concentrations by mediator cells, but secretions increase when changes take place. Most of the molecules released during secretion, even in times of change, tend to be absorbed rapidly by surrounding tissues. Few of the prostaglandin molecules are absorbed by capillaries and carried in the blood.

Two different prostaglandins can adjust blood flow in times of stress. Stimulated by the release of epinephrine, the hormones increase blood flow to local tissues. Other prostaglandins respond to stress by triggering the relaxation of smooth muscle in the passages leading to the lung. Prostaglandins are also released during allergic reactions.

prostaglandins hormones that have a pronounced effect in a small localized area

Chemically Enhanced Sports Performance

Strenuous exercise places stress on body systems, which compensate by delivering more fuel and oxygen to the tissues. Long before they were used in sport, ancient people documented how different drugs could mirror hormones produced by the body to affect heart rate, breathing rate, and blood pressure. Caffeine, for example, was found to produce many of the same effects as epinephrine (adrenaline), by increasing heart rate, blood pressure, and alertness.

The quest to gain an advantage began in the 1950s when weight lifters began injecting themselves with **anabolic steroids**. Anabolic steroids are designed to mimic many of the muscle-building traits of the sex hormone testosterone. Although still controversial, some have reported that anabolic steroids can provide athletes with greater lean muscle development and increased strength and, therefore, are advantageous for weightlifting and shorter sprints. However, anabolic steroids do not provide increased agility or skill level, nor do they enhance the ability of the cardiovascular system to deliver oxygen. In fact, they would be detrimental to athletes who need to sustain a high level of aerobic activity over a longer duration, such as marathon runners or cyclists. Although some athletes claim that steroids provide faster recovery from injury, and, therefore, allow more rigorous training, these claims have not been conclusively proven by laboratory studies. Whether advantageous or not, these types of drugs have been banned from competitive sports. During the 1988 Olympics, Canadian sprinter Ben Johnson was disqualified for using Stanozolol, an anabolic steroid.

anabolic steroids substances that are designed to mimic many of the muscle-building traits of the sex hormone testosterone

A number of health risks have been linked to the extended use of large dosages of anabolic steroids (**Figure 2**). Of particular interest to teens is that anabolic steroids prematurely fuse growth plates in the long bones, thereby reducing the height potential of the individual. Psychological effects, such as mood swings and feelings of rage, have also been documented.

Today, athletes have access to a myriad of drugs that do more than just increase strength (**Table 3**). Sharpshooters and archers have used beta blockers to slow the heartbeat, which helps to steady their aim and calm jangled nerves. As we saw in the chapter opener, the endurance athlete can gain an advantage by taking erythropoietin (EPO). Growth hormone decreases fat mass and promotes protein synthesis for muscle development; the enhancement of repair and growth increases strength and permits more vigorous training.

Because the body naturally produces growth hormone and EPO, they are difficult to detect with standard testing methods. More sophisticated methods must be used to detect small chemical differences between natural and artificial growth hormone. (Artificial growth hormone is synthesized by genetically modified bacteria.) Esters of testosterone are another group of muscle-building drugs that are difficult to detect. The esters slow the metabolism of testosterone by the body, keeping it in the body longer. Normally, testosterone would be metabolized in a few hours. The ester and testosterone raise little suspicion when testing is performed because both occur naturally in the body.

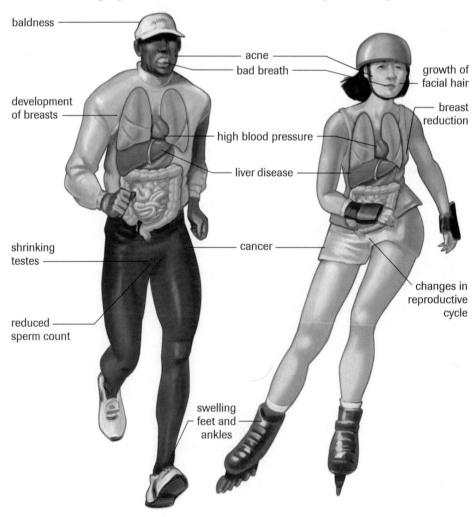

Figure 2
Effects of prolonged anabolic steroid use

Table 3 IOC (International Olympic Committee) Banned Performance-Enhancing Drugs

Drug	Advantage	Side effects
Anabolic steroids		
• Stanozolol, Androstenedoil, Nandrolone	• increases muscle mass and strength	• decreased growth, kidney problems, hair loss, oily skin, acne, shrinking testes, infertility, and cancer
Peptide hormones		
• growth hormone	• decreases fat • improves muscle mass	• diabetes, abnormalities of bones, liver, heart, and kidneys, and liver disease • high blood pressure
• erythropoietin (EPO)	• increases red blood cells that carry greater oxygen	• thickens the blood increasing chances of stroke • heart problems
Beta blockers		
• Atenolol, Bisoprolol, Nandolol	• slows heart rate	• reduces cardiac response time • makes skin more sensitive to sun
Stimulants		
• amphetamine	• increases endurance • relief of fatigue • improves reaction time	• irregular heart beat, nervousness, difficulty sleeping
• caffeine	• increases alertness	• increases blood pressure
• pseudoephedrine	• increases alertness	• narrows blood vessels and increases blood pressure
Masking agents		
• Bromantan	• makes steroid difficult to detect	• unknown
• Probenecid	• stops excretion of steroids for a few hours	• headache, tissue swelling, nausea

▶ *EXPLORE* an issue

Take a Stand: Protecting Athletes

Decision-Making Skills

○ Define the Issue   ● Analyze the Issue   ● Research
● Defend the Position   ● Identify Alternatives   ● Evaluate

A gold medal in most high-profile Olympic events is worth more than $1 million to an athlete in endorsements alone. Add to this the fame of an adoring public and the gratitude of the nation that the athlete represents and it should come as no surprise that many people will do almost anything possible for an added advantage. Careers in most sports are very short and athletes give up many other opportunities by dedicating their early lives to training.

Drug testing began at the Mexico City Olympics in 1968, a year after a British cyclist, who had taken a stimulant, died of heart failure while competing in the Tour de France. Not until 1975 did the International Olympic Committee (IOC) ban the use of anabolic steroids; however, detection methods did not keep pace with masking agents.

The incidents of steroid use in the 1976 Olympics in Montreal were considered the most widespread. A sensitive test for

Figure 3
Some cyclists admitted using recombinant EPO in the 1998 Tour de France, and, as a result, only 96 of the original 189 participants finished the race that year.

▶

steroids was not developed until 1983. Seven years later, the IOC added testosterone and caffeine to the list of banned substances. Today erythropoietin (EPO) and growth hormone are the most prevalent drugs. In 1985, an American biotechnology company produced a recombinant version of EPO to treat anemia associated with renal failure in dialysis patients. Since that time, affordable EPO has been readily available to athletes who are looking for an edge (**Figure 3**). EPO increases the number of red blood cells available to deliver oxygen to the tissues; however, too much of a good thing can be harmful. Increased red blood cells make the blood thicker and more difficult to pump. To date, EPO has been linked with the deaths of cyclists, cross-country skiers, and runners.

Statement

Not enough is being done to prevent the use of banned substances in sports.

- In your group, research the issue. Search for information in newspapers, periodicals, CD-ROMs, and on the Internet.

 www.science.nelson.com

- Prepare a list of points and counterpoints for your group to discuss. You might consider these questions:
  (i) Are some countries complicit in helping athletes hide positive drug tests? Are athletes being sacrificed for national glory?
  (ii) Are organizers of events compromised in their desire to identify users of banned substances by continually pushing for more records?
  (iii) What improvements could be made to help eliminate banned drugs from athletics?

- Develop and reflect on your opinion.

- Write a position paper summarizing your views.

SUMMARY *Adjustments to Stress*

- The endocrine and nervous systems interact to help the body cope with stress.
- More than 16 different types of prostaglandins are secreted in response to changes in the internal environment.
- Anabolic steroids are one of the many types of chemicals used to enhance athletic performance.
- It is difficult to detect banned drugs like growth hormone and erythropoietin.

▶ **Section 8.4 Questions**

Understanding Concepts

1. Both the nervous system and endocrine system respond to stress. Explain the benefits of each system's response.

2. Explain what advantage is gained by elevating blood sugar and blood pressure in times of stress.

3. Why is the secretion of insulin reduced in times of stress?

4. Explain the roles of the adrenal medulla and adrenal cortex in times of stress.

5. What are prostaglandins?

6. What are anabolic steroids? Outline their benefits to an athlete and their dangerous side effects.

7. Explain why a marathon runner would be unlikely to take growth hormone or anabolic steroids.

8. Why would erythropoietin (EPO) give an athlete competing in an endurance event an unfair advantage?

9. Why is it difficult to detect banned drugs like growth hormone and EPO?

Making Connections

10. Some sports writers have identified the increase in home runs in baseball with the use of performance-enhancing drugs. The International Olympic Committee bans many drugs that are not even monitored in professional team sports, such as football, baseball, and hockey. Should the National Hockey League and the National Football League adopt mandatory testing for banned performance-enhancing drugs? Justify your position.

11. Canadian rower Silken Laumann was found to have traces of a banned drug after her quad team placed first at the 1995 Pan American Games in Argentina. Later it was revealed that she'd mistakenly taken a variation of a well-known cough medicine that contains the banned drug. Despite no intention to gain an advantage in the competition, Laumann and her teammates were stripped of their gold medals. Should the IOC ban substances that are found in common over-the counter drugs? Provide supporting arguments for you position.

The human reproductive system involves separate male and female reproductive systems. The male gonads, the testes (singular, testis), produce male sex cells called sperm. The female gonads, the ovaries, produce eggs. The fusion of a male and a female sex cell, in a process called fertilization, produces a zygote. The zygote divides many times to form an embryo, which in turn continues to grow into a fetus.

The Male Reproductive System

Ancient herdsmen discovered that the removal of the testes, known as castration, increased the body mass of their animals, making their meat more tender and savory. The disposition of the castrated males also changed. Steers, which are castrated bulls, tend not to be very aggressive. The castrated animals also lack a sex drive and are sterile.

Figure 1 shows a detailed view of the male reproductive system. The male sex hormones, androsterone and **testosterone**, are produced in the interstitial cells of the testes. As the name suggests, the interstitial cells are found between the seminiferous cells. Although both hormones carry out many functions, testosterone is the more potent and abundant. Testosterone stimulates **spermatogenesis**, the process by which spermatogonia divide and differentiate into mature sperm cells. Testosterone also

testosterone male sex hormone produced by the interstitial cells of the testes

spermatogenesis process by which spermatogonia divide and differentiate into mature sperm cells

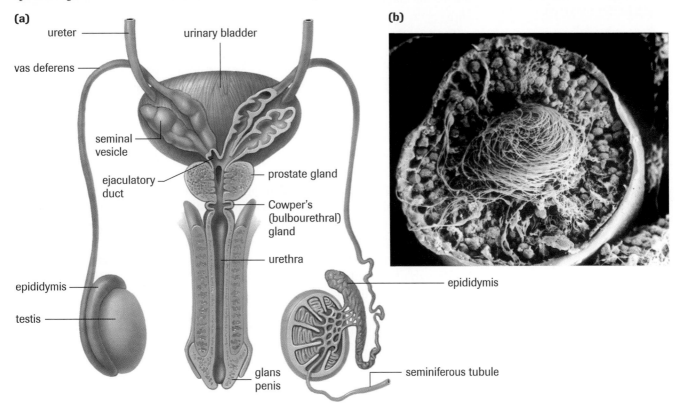

(a)

ureter

urinary bladder

vas deferens

seminal vesicle

ejaculatory duct

prostate gland

Cowper's (bulbourethral) gland

urethra

epididymis

testis

glans penis

seminiferous tubule

(b)

epididymis

Figure 1

(a) View of the male reproductive system

(b) Scanning electron micrograph of a seminiferous tubule. The various structures and functions of the male reproductive system are summarized in **Table 1** (page 394).

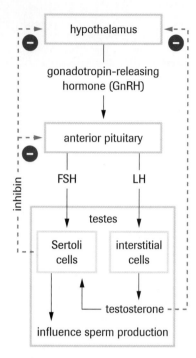

Figure 2
Negative feedback regulatory system for FSH and LH hormones

Table 1 The Male Reproductive System

Structure	Function
testes	• produce sperm cells • produce the hormone testosterone
seminiferous tubules	• produce immature sperm cells
epididymis	• matures and stores sperm cells in coiled tubules
vas deferens	• carries sperm from the epididymis to its junction with the urethra
seminal vesicle	• secretes fructose into the semen, which provides energy for the sperm
prostate gland	• secretes an alkaline buffer into the semen to protect the sperm from the acidic environment of the vagina
Cowper's gland	• secretes mucus-rich fluids into the semen that may protect the sperm from acids in the urethra
urethra	• carries semen during ejaculation • carries urine from the bladder to the exterior of the body
penis	• deposits sperm into the vagina during ejaculation • contains the urethra

gonadotropic hormones
hormones produced by the pituitary gland that regulate the functions of the testes in males and the ovaries in females

follicle-stimulating hormone (FSH) in males, hormone that increases sperm production

luteinizing hormone (LH) in males, hormone that regulates the production of testosterone

gonadotropin-releasing hormone (GnRH) chemical messenger from the hypothalamus that stimulates secretions of FSH and LH from the pituitary

influences the development of secondary male sexual characteristics at puberty, stimulating the maturation of the testes and penis. Testosterone levels are also associated with sex drive.

The male sex hormone also promotes the development of facial and body hair; the growth of the larynx, which causes the lowering of the voice; and the strengthening of muscles. In addition, testosterone increases the secretion of body oils and has been linked to the development of acne in males as they reach puberty. Once males adjust to higher levels of testosterone, skin problems decline. The increased oil production can also create body odour.

The hypothalamus and the pituitary gland in the brain control the production of sperm and male sex hormones in the testes. Negative feedback systems ensure that adequate numbers of sperm cells and constant levels of testosterone are maintained. The pituitary gland produces and stores the **gonadotropic hormones**, which regulate the functions of the testes; the male **follicle-stimulating hormone (FSH)**, which stimulates the production of sperm cells in the seminiferous tubules; and the male **luteinizing hormone (LH)**, which promotes the production of testosterone by the interstitial cells.

At puberty, the hypothalamus secretes the **gonadotropin-releasing hormone (GnRH)**. GnRH activates the pituitary gland to secrete and release FSH and LH. The FSH acts directly on the sperm-producing cells of the seminiferous tubules, while LH stimulates the interstitial cells to produce testosterone. In turn, the testosterone itself increases sperm production. Once high levels of testosterone are detected by the hypothalamus, a negative feedback system is activated (**Figure 2**). Testosterone inhibits LH production by the pituitary by deactivating the hypothalamus. The hypothalamus releases less GnRH, leading to decreased production of LH. Decreased GnRH output, in turn, slows the production and release of LH, which leads to less testosterone production. Testosterone levels thus remain in check. The feedback loop for sperm production is not well understood. It is believed that FSH acts on support cells, known as Sertoli cells, which produce a peptide hormone called inhibin that sends a feedback message to the pituitary, inhibiting production of FSH.

The Female Reproductive System

In many ways, the female reproductive system (**Figure 3**) is more complicated than that of the male. Once sexual maturity is reached, males continue to produce sperm cells at a somewhat constant rate. By contrast, females follow a complicated sexual cycle, in which one egg matures approximately every month (**Figure 4**). Hormonel levels fluctuate through the reproductive years, which end at menopause.

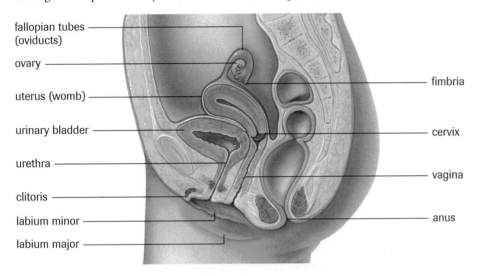

fallopian tubes (oviducts)
ovary
uterus (womb)
urinary bladder
urethra
clitoris
labium minor
labium major
fimbria
cervix
vagina
anus

Figure 3
Female reproductive anatomy, side view. Note that there are two ovaries and two oviducts. **Table 2** summarizes the various structures and functions of the female reproductive system.

Table 2   The Female Reproductive System

Structure	Function
ovaries	• produce the hormones estrogen and progesterone • site of ova (egg cell) development and ovulation
fallopian tubes (oviducts)	• carry the ovum from the ovary to the uterus • usually the site of fertilization
fimbria	• sweep the ovum into the oviduct following ovulation
uterus (womb)	• pear-shaped organ in which the embryo and fetus develop • involved in menstruation
cervix	• separates the vagina from the uterus • holds the fetus in place during pregnancy • dilates during birth to allow the fetus to leave the uterus
vagina	• extends from the cervix to the external environment • provides a passageway for sperm and menstrual flow • functions as the birth canal

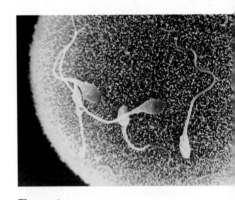

Figure 4
The male sperm cell is dwarfed by the much larger female egg cell. In humans, the egg cell is 100 000 times larger than the sperm cell.

Oogenesis and Ovulation

The ovaries (**Figure 5**, page 396) contain fibrous connective tissue and small groups of cells called **follicles**. The follicles comprise two types of cells: the primary oocyte and the granulosa cells. The oocyte, containing 46 chromosomes, undergoes meiosis and is transformed into a mature oocyte, or ovum. The granulosa cells provide nutrients for the oocyte.

Unlike the testes, which replenish sex cells, the female ovaries undergo continual decline after the onset of puberty. Each of the two ovaries contains about 400 000 follicles at puberty. Many follicles develop during each female reproductive cycle, but usually only a single follicle becomes dominant and reaches maturity. The remaining follicles deteriorate and are reabsorbed within the ovary. Between the ages of about 12 and 50 in a woman's life, approximately 400 eggs will mature. By the time a woman reaches menopause, few follicles remain. It has been suggested that the higher incidence of genetic defects in

follicles structures in the ovary that contain the egg and secrete estrogen

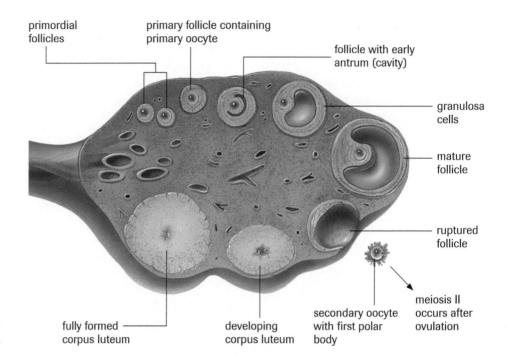

primordial follicles

primary follicle containing primary oocyte

follicle with early antrum (cavity)

granulosa cells

mature follicle

ruptured follicle

meiosis II occurs after ovulation

secondary oocyte with first polar body

developing corpus luteum

fully formed corpus luteum

Figure 5
The process of ovulation. Pituitary hormones regulate the events of follicle development, ovulation, and the formation of the corpus luteum.

ovulation release of the egg from the follicle held within the ovary

corpus luteum a mass of follicle cells that forms within the ovary after ovulation; secretes estrogen and progesterone

flow phase phase of the menstrual cycle marked by shedding of the endometrium

children produced by older women can be linked to the age of the follicles. The longer the follicle lives, the greater the chance of genetic damage. Because female sex hormones are produced within the ovary, menopause marks the end of a female's reproductive life and signals a drop in the production of female hormones.

A hormone produced by the pituitary gland controls follicle development. Nutrient follicle cells surrounding the primary oocyte begin to divide. As the primary oocyte undergoes meiosis I, the majority of cytoplasm and nutrients move to one of the poles and form a secondary oocyte. The secondary oocyte contains 23 chromosomes. The remaining cell, referred to as the polar body, receives little cytoplasm and dies. As the secondary cells surrounding the secondary oocyte develop, a fluid-filled cavity forms. Eventually the dominant follicle pushes outward, ballooning the outer wall of the ovary. Blood vessels along the distended outer wall of the ovary collapse, and the wall weakens. The outer surface of the ovary wall bursts and the secondary oocyte is released. This process is referred to as **ovulation**. Surrounding follicle cells remain within the ovary and are transformed into the **corpus luteum**, which secretes hormones essential for pregnancy; however, if pregnancy does not occur, the corpus luteum degenerates after about 10 days. All that remains is a scar, referred to as the corpus albicans. The secondary oocyte enters the oviduct and begins meiosis II, which is completed after fertilization. Once again, the division of cytoplasm and nutrients is unequal; the cell that retains most of the cytoplasm and nutrients is referred to as the mature oocyte or ovum. As in meiosis I, the polar body deteriorates.

Menstrual Cycle

The human female menstrual cycle, which is repeated throughout a woman's reproductive lifetime, takes an average of 28 days, although variation in this cycle is common. The menstrual cycle can be divided into four distinct phases: flow phase, follicular phase, ovulatory phase, and luteal phase (**Figure 6**). The shedding of the endometrium, or menstruation, marks the **flow phase**. This is the only phase of the female reproductive cycle that can be determined externally. For this reason, the flow phase is used to mark the beginning of the menstrual cycle. Approximately five days are required for the uterus to shed the endometrium.

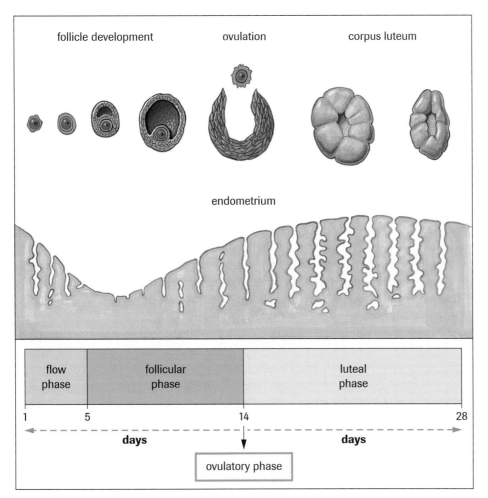

follicle development   ovulation   corpus luteum

endometrium

| flow phase | follicular phase | luteal phase |

1   5   14   28

days   **days**

ovulatory phase

Figure 6
The thickness of the endometrium increases from the beginning of the follicular phase to the end of the luteal phase. The development of blood vessels and glandular tissues helps prepare the uterus for a developing embryo. Should no embryo enter the uterus, menstruation occurs, and the menstrual cycle begins again.

DID YOU KNOW ?

How Do Birth Control Pills Work?
Some birth control pills, also referred to as oral contraceptives, contain high concentrations of progesterone, which inhibits ovulation, thereby preventing conception.

The **follicular phase** is characterized by the development of follicles within the ovary. As follicle cells develop, the hormone **estrogen** is secreted. As follicles continue to develop, estrogen concentration in the blood increases. The follicular phase normally takes place between days 6 and 13 of the female menstrual cycle.

During ovulation, the third phase of the female menstrual cycle, the egg bursts from the ovary and follicular cells differentiate into the corpus luteum. The development of the corpus luteum marks the beginning of the **luteal phase**. Estrogen levels begin to decline when the oocyte leaves the ovary, but are restored somewhat when the corpus luteum forms. The corpus luteum secretes both estrogen and **progesterone**. Progesterone continues to stimulate the endometrium and prepares the uterus for an embryo. It also inhibits further ovulation and prevents uterine contractions. Should progesterone levels fall, uterine contractions would begin. The luteal phase, which occurs between days 15 and 28, prepares the uterus to receive a fertilized egg. Should fertilization of an ovum not occur, the concentrations of estrogen and progesterone will decrease, thereby causing weak uterine contractions. These weak uterine contractions make the endometrium pull away from the uterine wall. The shedding of the endometrium marks the beginning of the next flow phase, and the female menstrual cycle starts all over again. This cycle is summarized in **Table 3** (page 398).

follicular phase phase marked by development of ovarian follicles before ovulation

estrogen female sex hormone that activates the development of female secondary sex characteristics, including development of the breasts and body hair, and increased thickening of the endometrium

luteal phase phase of the menstrual cycle characterized by the formation of the corpus luteum following ovulation

progesterone female sex hormone produced by the ovaries that maintains uterine lining during pregnancy

follicle-stimulating hormone (FSH) in females, a gonadotropin that promotes the development of the follicles in the ovary

luteinizing hormone (LH) in females, a gonadotropin that promotes ovulation and the formation of the corpus luteum

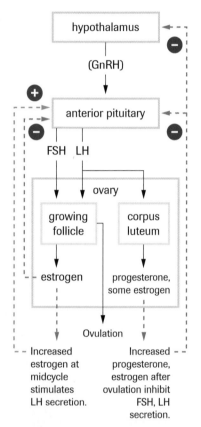

Figure 7
Feedback loop showing the regulation of ovarian hormones

✎ LAB EXERCISE 8.5.1

Hormone Levels During the Menstrual Cycle (p. 404)
How do hormone levels regulate the female menstrual cycle? In this exercise, you will use experimental data to analyze hormone levels during the menstrual cycle.

Table 3 The Female Menstrual Cycle

Phase	Description of events	Hormone produced	Days
flow	• menstruation		1–5
follicular	• follicles develop in ovaries • endometrium is restored	estrogen produced by follicle cells	6–13
ovulation	• oocyte bursts from ovary		14
luteal	• corpus luteum forms and endometrium thickens	estrogen and progesterone	15–28

Hormonal Control of the Female Reproductive System

The hypothalamus–pituitary complex regulates the production of estrogen and progesterone, the hormones of the ovary. Gonadotropins—female **follicle-stimulating hormone (FSH)** and **luteinizing hormone (LH)**—regulate the control of hormones produced by the ovaries: estrogen and progesterone. In turn, ovarian hormones, as part of a complex negative feedback mechanism, regulate the gonadotropins.

The onset of female puberty is signalled by the release of GnRH (gonadotropin-releasing hormone) from the hypothalamus (**Figure 7**). GnRH activates the pituitary gland, which is the storage site of FSH and LH. During the follicular phase of the menstrual cycle, the blood carries FSH secretions to the ovary, where follicle development is stimulated. The follicles within the ovary secrete estrogen, which initiates the development of the endometrium. As estrogen levels rise, a negative feedback message is sent to the pituitary gland to turn off secretions of FSH. The follicular phase of the menstrual cycle has ended. Simultaneously, the rise in estrogen exerts a positive message on the LH-producing cells of the pituitary gland. LH secretion rises and ovulation occurs.

After ovulation, the remaining follicular cells, under the influence of LH, are transformed into a functioning corpus luteum. The luteal phase of the menstrual cycle has begun. Cells of the corpus luteum secrete both estrogen and progesterone. The buildup of estrogen and progesterone will further increase the development of the endometrium. As progesterone and estrogen build up within the body, a second negative feedback mechanism is activated. Progesterone and estrogen work together to inhibit the release of both FSH and LH. Without gonadotropic hormones, the corpus luteum begins to deteriorate, slowing estrogen and progesterone production. The drop in ovarian hormones signals the beginning of menstruation. ✎

Similarities between male and female systems extend beyond the secretion of FSH and LH. Androgens (male sex hormones) and estrogen (female sex hormone) can be produced by either gender. Male characteristics result because the levels of androgens exceed the levels of estrogen. Males are ensured of maintaining low levels of female hormones by excreting them at an accelerated rate. This may explain why the urine of a stallion contains high levels of estrogen.

The importance of balancing androgen and estrogen levels has been demonstrated with roosters. Removal of testes from a rooster makes the flesh more tender. Such animals are called capons. Injections of estrogen will bring about the same effect. In humans, the secretions of androgens will stimulate the development of the male's prostate gland, but injections of estrogen will slow the process. This may explain why cancerous tumours of the prostate can be slowed by injections of estrogen-like compounds.

SUMMARY   *Female Reproductive Hormones*

Table 4   Female Reproductive Hormones

Hormone	Location	Description of function
estrogen	follicle cells (ovary)	inhibits growth of facial hair, initiates secondary female characteristics, and causes thickening of the endometrium
progesterone	corpus luteum (ovary)	inhibits ovulation, inhibits uterine contractions, and stimulates the endometrium
follicle-stimulating hormone (FSH)	pituitary	stimulates the development of the follicle cells in the ovary
luteinizing hormone (LH)	pituitary	stimulates ovulation and the formation and maintenance of the corpus luteum

▶ *Section 8.5 Questions*

Understanding Concepts

1. Outline the functions of testosterone.

2. How do gonadotropic hormones regulate spermatogenesis and testosterone production?

3. Using luteinizing hormone (LH) and testosterone as examples, explain the mechanism of negative feedback.

4. Can a woman who has reached menopause ever become pregnant? Explain your answer.

5. What is menstruation? Why is it important?

6. Describe the process of ovulation. Differentiate between primary oocytes, secondary oocytes, and mature ova.

7. Describe how the corpus luteum forms in the ovary.

8. Describe the events associated with the flow phase, follicular phase, and luteal phase of menstruation.

9. Outline the functions of estrogen and progesterone.

10. How do gonadotropic hormones regulate the function of ovarian hormones?

11. With reference to the female reproductive system, provide an example of a negative feedback control system.

Applying Inquiry Skills

12. Predict how low secretions of gonadotropin-releasing hormone (GnRH) from the hypothalamus would affect the female menstrual cycle.

13. An experiment was performed in which the circulatory systems of two mice (A and B) with compatible blood types were joined (**Figure 8**). The data collected from the experiment is expressed in **Table 5**. (Note that + indicates "found," − indicates "not found"). Explain the data that were collected.

Making Connections

14. To alleviate the symptoms of menopause, many women turn to hormone replacement therapy (HRT). Although

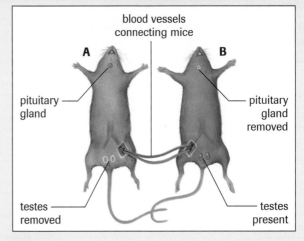

blood vessels connecting mice

A   B

pituitary gland   pituitary gland removed

testes removed   testes present

Figure 8

Table 5

Animal	Testosterone	LH	FSH	Sperm in urethra
A	+	+	+	−
B	+	+	+	+

estrogen—the main hormone in HRT—reduces "hot flashes," prevents bone loss, and may even improve memory, it raises the risk of blood clots and ovarian and breast cancer. Using the Internet and other resources, investigate and write a brief report on what is being done to meet the demand for HRT without subjecting women to its risks. Include some lifestyle changes that might help in the management of menopause.

 www.science.nelson.com

INVESTIGATION 8.2.1

Identification of Hyperglycemia

In this investigation, you will use simulated urine samples to determine how urinalysis is used to identify hyperglycemia and diabetes.

Question

How is urinalysis used to idenfity hyperglycemia diabetes?

Prediction

(a) Before the investigation, predict what colour(s) you will observe in the urine samples using Benedict's test and glucose test tape that will indicate diabetes.

(b) What criteria did you use to make your decision?

Materials

safety goggles	test-tube rack
laboratory apron	400-mL beaker
4 test tubes	beaker tongs
wax pen	hot plate
10-mL graduated cylinder	test-tube clamp
Benedict's solution	distilled water
medicine dropper	forceps
4 samples of simulated urine	glucose test tape

 Benedict's solution is toxic and an irritant. Avoid skin and eye contact. Wash all splashes off your skin and clothing thoroughly. If you get any chemical in your eyes, rinse for at least 15 minutes and inform your teacher.

Procedure

Part I: Benedict's Test

Benedict's solution identifies reducing sugars. Cupric ions in the solution combine with sugars to form cuprous oxides, which produce colour changes (**Table 1**).

Table 1 Benedict's Test Colour Chart

Colour of solution	Glucose concentration
blue	0.0%
light green	0.15%–0.5%
olive green	0.5%–1.0%
yellow-green to yellow	1.0%–1.5%
orange	1.5%–2.0%
red to red-brown	2.0%+

1. Label the four test tubes A, B, C, and D. Use a 10-mL graduated cylinder to measure 5 mL of Benedict's solution into each test tube.

2. With a medicine dropper, add 10 drops of urine from sample A to test tube A. Rinse the medicine dropper and repeat for samples B, C, and D.

(c) Why must the medicine dropper be rinsed?

3. Fill a 400-mL beaker with approximately 300 mL of tap water. Using beaker tongs, position the beaker on a hot plate. The beaker will be used as a hot-water bath. Use the test-tube clamp to place the test tubes in the hot-water bath for 5 min.

4. With the test-tube clamp, remove the samples from the hot-water bath.

(d) Record the final colours of the solutions.

Part II: Glucose Test Tape

The reducing sugar in the urine will react with copper sulfate to reduce cupric ions to cupric oxide. The chemical reaction is indicated by a colour change of the test tape. **Table 2** below provides quantitative results.

Table 2 Glucose Test Tape Colour Chart

Colour of solution	Glucose concentration
blue	0.0%
green	0.25%–0.5%
green to green-brown	0.5%–1.0%
orange	2.0%+

5. Clean the four test tubes and place 10 drops of distilled water into each of them.

6. Add five drops of urine to each of the appropriately labelled test tubes. Place the test tubes in a test-tube rack.

7. Use forceps to dip test tape into each of the test tubes.

(e) Record the final colours of the test tape.

Analysis

(f) Which subjects could have diabetes?

(g) Could there be any other reasons for a positive test?

⚗ INVESTIGATION 8.2.1 *continued*

Evaluation

(h) Describe any difficulties you had in carrying out your investigation.

(i) Explain the advantage of conducting two different tests. Which test was more appropriate? Explain your answer.

(j) Today, people with diabetes test their blood to monitor sugar levels. Explain why blood tests are preferred for people with diabetes. Why weren't blood tests carried out in this investigation?

Synthesis

(k) Why is insulin not taken orally?

(l) Explain why people with diabetes experience the following symptoms: low energy levels, large volumes of urine, the presence of acetone on the breath, and acidosis (blood pH becomes acidic).

(m) Why might the injection of too much insulin be harmful?

(n) Explain how you would help someone who had taken too much insulin.

✎ LAB EXERCISE 8.2.1

Effects of Hormones on Blood Sugar

Blood sugar levels of a person with diabetes mellitus and a person without were monitored over a period of 12 h. Both ate an identical meal and performed 1 h of similar exercise. Use the data in **Figure 1** to answer the questions below.

Observations

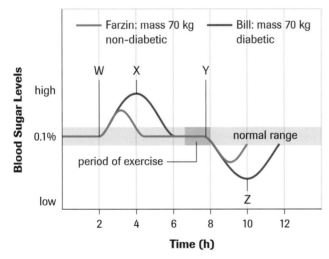

Figure 1
Blood sugar was monitored over 12 h.

Inquiry Skills

○ Questioning	○ Planning	● Analyzing
○ Hypothesizing	○ Conducting	● Evaluating
○ Predicting	○ Recording	● Communicating

Analysis

1. Which hormone injection did Bill receive at the time labelled X? Provide reasons for your answer.

2. What might have happened to Bill's blood sugar level if hormone X had not been injected? Justify your answer.

3. Explain what happened at time W for Bill and Farzin.

4. Explain why blood sugar levels begin to fall after time Y.

5. What hormone might Bill have received at time Z? Explain your answer.

6. Why is it important to note that both Farzin and Bill have the same body mass?

7. What differences in blood sugar levels are illustrated by the data collected from Bill and Farzin?

8. Why do Bill and Farzin respond differently to varying levels of blood sugar?

Hormonal Control of Metamorphosis

Inquiry Skills

● Questioning	○ Planning	● Analyzing
● Hypothesizing	○ Conducting	○ Evaluating
○ Predicting	○ Recording	● Communicating

In amphibians, thyroid hormones, thyroxine (T4) and tri-iodothyronine (T3), regulate the rate at which metamorphosis occurs. Metamorphosis is the change that takes place during the development of a postembryonic form, such as a tadpole, into an adult.

Different species of salamanders show variations in metamorphosis (**Figure 2**). Some salamanders do not metamorphose, retaining the larval characteristics, such as tail and gills, their entire life. They do, however, reach sexual maturity and reproduce—a characteristic of adults.

Figure 2
Metamorphosis in a salamander

The variations of 3 species of salamanders are shown in **Table 3**.

Question

What scientific question could be constructed from the data provided in **Table 3**?

Table 3

Type of salamander	Description of metamorphosis
mud puppy (*Necturus*)	no metamorphosis
Mexican salamander (*Ambystoma mexicanum*)	no metamorphosis
tiger salamander (*Ambystoma tigrinum*)	metamorphosis in most habitats, except glacial lakes

Hypothesis

In an attempt to collect more information about these variations in the metamorphosis of salamanders, an examination was conducted, which revealed that all three species have thyroid glands. There are several hypotheses that could be tested to explain why the mud puppy and Mexican salamander do not metamorphose:

- The thyroid glands do not produce enough T3 or T4 for metamorphosis to occur in these salamanders.
- The environment does not contain enough iodine for the T3 or T4 to be formed.
- The pituitary does not produce enough thyroid-stimulating hormone (TSH).
- The amphibians lack some other substance that is necessary for metamorphosis.

The following experiments were carried out to test these hypotheses.

Procedure I

1. Various concentrations of T3 and T4 were given to a large number of different species of *Necturus*.

Observation

No metamorphosis occurred in any of the salamanders.

(a) Would it have helped at this point to administer either TSH or TRH? Explain.

Procedure II

2. Fifty young frog tadpoles were divided into two equal groups. Group A was given an extract from the pituitary glands of the mud puppy. Group B acted as a control, receiving no extract from the pituitary.

Group A

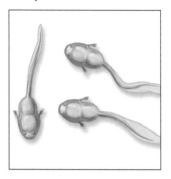   

Group B

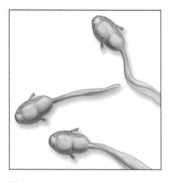 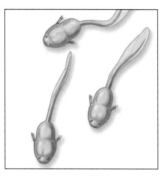 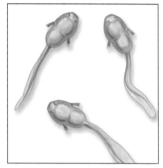 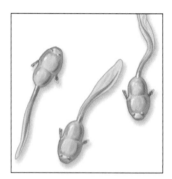

Figure 3
Group A metamorphosed into frogs. The control group B did not change to the adult form.

Observation

Group A metamorphosed into frogs (**Figure 3**), while group B remained tadpoles.

(b) What inference can be drawn from these results?

(c) Which of the above hypotheses might account for the failure of mud puppies (*Necturus*) to metamorphose? Why?

Procedure III

3. A T4 extract was given to the Mexican salamander and tiger salamander.

Observation

In both species metamorphosis occurred.

(d) Using the data provided, which hypothesis would not apply to both the tiger and Mexican salamanders? Explain why.

Procedure IV

4. An iodine compound was added to the water of the tiger and Mexican salamanders.

Observation

Metamorphosis occurred in the tiger salamander but not in the Mexican salamander.

(e) Using the data provided, what inferences can be drawn from the results?

Procedure V

5. The pituitary glands of the tiger and Mexican salamanders were cross-transplanted.

Observation

Metamorphosis did not occur in the tiger salamander but it did occur in the Mexican salamander.

Analysis

(f) Using the data from all 5 experiments, what conclusions can you draw about the tiger and Mexican salamanders?

Hormone Levels During the Menstrual Cycle

Inquiry Skills

○ Questioning	○ Planning	● Analyzing
○ Hypothesizing	○ Conducting	○ Evaluating
○ Predicting	○ Recording	● Communicating

How do hormone levels regulate the female menstrual cycle? Use the following experimental data to analyze hormone levels during the menstrual cycle.

Analysis

1. Gonadotropic hormones regulate ovarian hormones. Study the feedback loop shown in **Figure 4**.

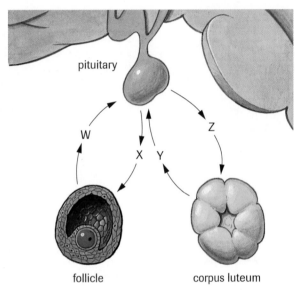

Figure 4
Feedback loop showing the regulation of ovarian hormones by gonadotropic hormones.

(a) Identify as W, X, Y, or Z the two gonadotropic hormones represented in the diagram.

(b) Identify the ovarian hormones shown in the diagram.

(c) Which two hormones exert negative feedback effects?

2. Body temperatures of two women were monitored during their menstrual cycles. One woman ovulated; the other did not. The results are shown in **Table 4**.

(d) Graph the data provided. Plot changes in temperature along the y-axis (vertical axis) and the days of the menstrual cycle along the x-axis (horizontal axis).

(e) Assuming this menstrual cycle represents the average 28-day cycle, label the ovulation day on the graph.

(f) Describe changes in temperature before and during ovulation.

(g) Compare body temperatures with and without a functioning corpus luteum.

Table 4

	Temperature (°C)	
Days	**Ovulation occurs**	**No ovulation**
5	36.4	36.3
10	36.2	35.7
12	36.0	35.8
14	38.4	36.2
16	37.1	36.1
18	36.6	36.0
20	36.8	36.3
22	37.0	36.3
24	37.1	36.4
28	36.6	36.5

3. **Figure 5** shows changes in the thickness of the endometrium throughout the female menstrual cycle.

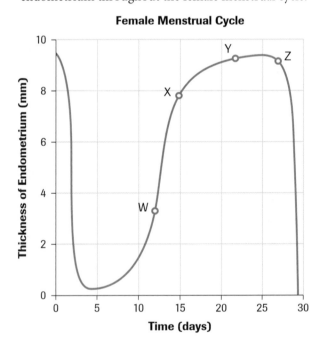

Figure 5

(h) Identify the events that occur at times X and Z.

(i) Identify by letter the time when follicle cells produce estrogen.

(j) Identify by letter the time when the corpus luteum produces estrogen and progesterone.

LAB EXERCISE 8.5.1 *continued*

4. Levels of gonadotropic hormones monitored throughout the female reproductive cycle are shown in **Figure 6**. Levels are recorded in relative units.

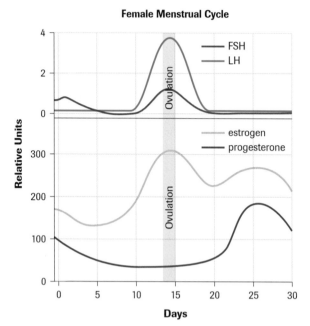

Figure 6

(k) How does LH affect estrogen and progesterone?

Synthesis

(l) **Figure 7** shows estrogen and progesterone levels during three menstrual cycles.

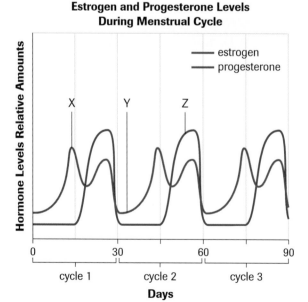

Figure 7

(i) On which day (X, Y, or Z) would ovulation occur? Explain your answer.

(ii) On which day (X, Y, or Z) would you expect to find a functioning corpus luteum? Explain your answer.

(m) Explain why only one corpus luteum may be found in the ovaries of a woman who has given birth to triplets.

(n) Estrogen plays a crucial role in maintaining bone strength and density. This is why women over 50 and women who experience premature menopause are at risk for developing osteoporosis. What can be done to minimize this risk? Investigate both hormone therapies and lifestyle factors.

(o) Cattle are given various steroid hormones to increase meat production. Recently, some scientists have expressed concern that animal growth stimulators might have an effect on humans. Comment on the practice of using hormones in cattle. What potential problems in humans might be associated with such procedures?

Key Expectations

- Describe the anatomy and physiology of the endocrine system and explain the feedback mechanisms involved in maintaining blood sugar. (8.1, 8.2)
- Construct models, use flow charts, and draw graphs to describe homeostatic control systems. (8.2, 8.5)
- Describe the anatomy and physiology of the endocrine system and explain the role of hormones that affect metabolism. (8.1, 8.3)
- Describe contributions of Canadian scientists in diabetes research. (8.2)
- Synthesize data, and discuss issues related to the use of performance-enhancing drugs in sports. (8.4)
- Explain the role of hormones and negative feedback actions that enable the body to adjust to stress. (8.4)
- Explain the action of hormones and feedback mechanisms in the female and male reproductive systems. (8.5)

Key Terms

adrenal cortex

adrenal medulla

anabolic steroids

calcitonin

corpus luteum

cortisol

cyclic adenosine monophosphate (cyclic AMP)

diabetes

endocrine hormones

epinephrine

estrogen

flow phase

follicles

follicle-stimulating hormone (FSH) (male and female)

follicular phase

glucagon

glucocorticoids

goiter

gonadotropic hormones

gonadotropin-releasing hormone (GnRH)

growth hormone (GH)

hormones

insulin

interstitial fluid

luteal phase

luteinizing hormone (LH) (male and female)

mineralocorticoids

norepinephrine

ovulation

parathyroid glands

parathyroid hormone (PTH)

pituitary gland

progesterone

prostaglandins

protein hormones

spermatogenesis

steroid hormones

target tissues

testosterone

thyroid gland

thyroxine

▶ *MAKE* a summary

In this chapter, you studied how chemical signals maintain homeostasis. To summarize your learning, make a sketch of the human endocrine system and show how the system maintains homeostasis in response to stress. Use as many of the key terms as possible.

In your notebook, record the letter of the choice that best answers the question.

1. Which of the following describes a negative feedback reaction?
   (a) Glucagon stimulates the release of glucose from the liver, which increases blood glucose.
   (b) LH stimulates the interstitial cell to produce testosterone, which inhibits the release of LH.
   (c) The hypothalamus releases thyroid-releasing hormone (TRH), which travels to the pituitary gland initiating the release of thyroid-stimulating hormone (TSH), which stimulates the release of thyroxine from the thyroid gland.
   (d) Calcitonin is released from the thyroid gland and blood calcium levels decrease.
   (e) All of the above describe negative feedback reactions.

2. Glucagon is produced in an organ and affects target cells that are in another part of the body. The organ of production and the location of the target cells are, respectively, which of the following?
   (a) adrenal medulla and adrenal cortex
   (b) liver and pancreas
   (c) pituitary and adrenal medulla
   (d) pancreas and liver
   (e) hypothalamus and pituitary

3. Two hormones that adjust body systems for short-term stress and long-term stress, respectively, are which of the following?
   (a) thyroxine and parathyroid hormone (PTH)
   (b) estrogen and growth hormone
   (c) epinephrine and cortisol
   (d) TSH and epinephrine
   (e) testosterone and estrogen

4. In times of stress, under the influence of cortisol, levels of amino acids increase in the blood. Why is this change beneficial as a response to stress?
   (a) The amino acids are converted into proteins, which are used to repair cells damaged by the stress.
   (b) The amino acids are converted to glucose by the liver, raising blood sugar, thereby providing more energy to deal with stress.
   (c) The amino acids are converted into proteins, which provide more energy to deal with stress.
   (d) The amino acids are converted to glycogen by the liver, lowering blood sugar, which stimulates the release of insulin.
   (e) The amino acids are converted to glycogen by the liver, lowering blood sugar, which stimulates the release of TRH.

5. Which of the following would be the result of hyper-secretion of the thyroid gland?
   (a) a tendency not to gain weight, a warm peripheral body temperature, and a high energy level
   (b) a tendency to gain weight, a cold peripheral body temperature, and a high energy level
   (c) a tendency not to gain weight, a cold peripheral body temperature, and a low energy level
   (d) a tendency to gain weight, a warm peripheral body temperature, and a low energy level
   (e) no change in body weight but increased body temperatures and increased urine output

6. Identify a glucocorticoid released by the adrenal cortex.
   (a) epinephrine
   (b) triiodothyronine
   (c) follicle-stimulating hormone (FSH)
   (d) cortisol
   (e) glucagon

7. Which one of the following choices signifies the beginning of menstruation?
   (a) FSH and LH secretions decrease and the corpus luteum deteriorates.
   (b) Estrogen levels rise and progesterone levels begin to decline.
   (c) Progesterone levels increase and positive feedback to the pituitary increases the LH secretions.
   (d) There are no longer any eggs remaining in the ovary.
   (e) Oxytocin and PTH levels decrease because no eggs remain in the ovary.

8. A laboratory animal is accidentally given too much insulin and begins convulsing. What could you do to quickly return the animal to a normal blood sugar?
   (a) provide sugar in a fruit drink
   (b) increase water intake
   (c) inject erythropoietin
   (d) cool the animal as rapidly as possible
   (e) warm the animal as rapidly as possible

NEL   An interactive version of the quiz is available online.
GO www.science.nelson.com

Chemical Signals Maintain Homeostasis **407**

Understanding Concepts

1. List and explain the symptoms experienced by people with diabetes.

2. Use negative feedback diagrams to show how insulin and glucagon regulate blood sugar.

3. With reference to the adrenal glands, explain how the nervous system and endocrine system interact in times of stress.

4. Why do insulin levels decrease during times of stress?

5. A tumour on a gland can increase the gland's secretions. Explain how increases in the following hormones affect blood sugar levels: insulin, epinephrine, and thyroxine.

6. With reference to the importance of negative feedback, provide an example of why low levels of iodine in your diet can cause goiters.

7. Anabolic steroids act in a similar fashion to testosterone by turning off secretions of gonadotropic hormones. What effects do anabolic steroids have on secretions of testosterone? Explain your answer.

8. A physician notes that individuals with a tumour on the pancreas secrete unusually high levels of insulin. Unfortunately, insulin in high concentrations causes blood sugar levels to fall below the normal acceptable range. In an attempt to correct the problem, the physician decides to inject the patient with cortisol. Why would the physician give the patient cortisol? What problems could arise from this treatment?

Applying Inquiry Skills

9. A disorder called testicular feminization syndrome occurs when the receptor molecules to which testosterone binds are defective. Predict the effect of testicular feminization syndrome and explain how normal steroid hormone action is altered.

10. A rare virus destroys cells of the anterior lobe of the pituitary. Predict how the destruction of the pituitary cells would affect blood sugar. Explain.

11. A physician notes that her patient is very active and remains warm on a cold day even when wearing a light coat. Further discussion reveals that although the patient's daily food intake exceeds that of most people, the patient remains thin. Why might the doctor suspect a hormone imbalance? Which hormone might the doctor suspect?

12. Three classical methods have been used to study hormone function:
    (a) The organ that secretes the hormone is removed. The effects are studied.
    (b) Grafts from the removed organ are placed within a gland. The effects are studied.
    (c) Chemicals from the extracted gland are isolated and injected back into the body. The effects are studied.

    Explain how each of these procedures could be used to investigate the effect of insulin on blood sugar.

13. **Figure 1** shows changes in daily testosterone secretion over time in males.

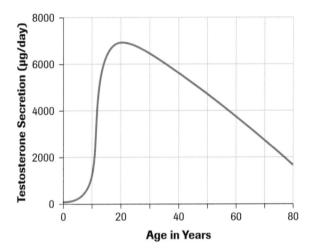

Figure 1

    (a) What evidence links testosterone secretions to the development of skeletal muscle?
    (b) What additional information would you need to support or falsify the inference that you made?

14. Caffeine was one of the first performance-enhancing drugs used by athletes. Some people believe that it can increase endurance. Dramatic increases in caffeine levels and high consumption of caffeine have been linked with sleep disorders, impaired fine motor activities, increased fatty acid levels in the blood, and heart attacks. A study was conducted on a group of elite cyclists and a group of high school students to determine whether caffeine provided any marked advantage. Each group pedalled at 80% maximum capacity for as long as possible. **Table 1** reveals the average amount of time each group was able to pedal after drinking decaffeinated coffee and drinking coffee with caffeine.

Table 1

Group	Average cycling time (min)	
	Decaffeinated coffee (250 mL)	Coffee with caffeine (250 mL, 340 mg caffeine)
Elite athletes	82	123
High school students	41	42

(a) Identify the problem being investigated by the research group.
(b) Identify the control for the experiment.
(c) Identify the independent and dependent variables.
(d) What conclusions can you draw from the data provided?
(e) Why was neither group told which coffee contained caffeine?
(f) The procedure does not indicate how much time passed between each exercise test . Explain why this is an important factor to know.
(g) To learn more about how caffeine helps your body adjust to physical stress, you might want to monitor changes in certain hormones during testing. Identify hormone levels that you might want to monitor in the blood during the testing and explain what information might be gained by monitoring these hormones.

Making Connections

15. Some scientists have speculated that certain young female Olympic gymnasts may have been given growth hormone inhibitors. Why might the gymnasts have been given growth inhibitors? Do you think hormone levels should be altered to regulate growth patterns?

16. The embryo is sensitive to drugs, especially during the first trimester. In the late 1950s, a drug called Thalidomide was introduced in Europe to help prevent morning sickness. Unfortunately, the drug interfered with limb bud formation in developing embryos. Before the drug could be withdrawn from pharmacies, children were born with lifelong disabilities. Although Thalidomide was pulled from Canadian pharmacies in 1962, a great debate centres around drugs that may have less pronounced effects on the fetus. Some evidence suggests that tranquilizers may cause improper limb bud formation. Even some acne drugs have been linked to facial deformities in newborns, and the antibiotic streptomycin

has been associated with hearing problems. Are drugs placed on the market too quickly? Explain your position.

17. On April 26, 1986, a nuclear accident in Chernobyl caused the release of radioactive wastes into the air. The extent of the problem is still unknown, but the effects on children have been the most extreme. One of the most dangerous radioactive materials released was iodine-131, which was absorbed by the thyroid glands of children. Iodine-131 causes inflammation of the thyroid gland and can lead to cancer.
(a) Describe some possible symptoms of children who had their thyroid glands completely or partially destroyed.
(b) Draw a feedback loop showing how thyroxine levels might be affected.
(c) Explain why children were given nonradioactive iodine.
(d) Initially, the government tried to suppress information about the nuclear accident. What should have been done?
(e) Nuclear wastes were carried over the European continent with weather. Should surrounding countries be able to demand financial and medical compensation?

18. Bovine somatotropin (BST) is a growth hormone now produced by gene recombination. BST can increase milk production in cows by as much as 20% by increasing nutrient absorption from the blood stream into the cow's milk. Should BST be used? Why might some individuals be concerned?

Extension

19. Design an experiment to demonstrate the independent functions of the interstitial cells and seminiferous tubules of the testes.

20. Design an experiment to show how ovarian hormones regulate female gonadotropic hormones.

chapter

9

How Nerve Signals Maintain Homeostasis

- describe the anatomy and physiology of the nervous system, and explain its role in homeostasis;

- describe and explain the effects of disorders of the nervous system;

- design and conduct an experiment using invertebrates to study response to external stimuli;

- analyze, compile, and display experimental data that reveal how the nervous system responds to external stimuli;

- evaluate opinions and discuss difficulties in treating neurological diseases;

- describe some Canadian contributions to knowledge and technology in the field of homeostasis;

- present informed opinions about problems related to the health industry, health legislation, and personal health.

In 1998, Michael J. Fox announced that he was leaving the popular television sitcom *Spin City* because of Parkinson's disease. Fox was first diagnosed with early stages of Parkinson's disease in 1991 when he noticed a twitch in a finger. Over the next seven years the disease progressed, making the rigours of a demanding acting schedule very difficult.

Parkinson's disease is a progressive degenerative nerve disorder that affects muscle activity. Brain cells in two areas of the lower brain, the substantia nigra and the locus cerulus, degenerate and die, but no one knows why. Cells from the substantia nigra secrete a transmitter chemical called dopamine, and cells from the locus cerelus secrete the transmitter chemical norepinephrine. Any reduction in these chemicals affects muscle movement. Early symptoms of Parkinson's include muscle tremors, slow body movements, rigidity in one or more joints, and an inability to regain one's balance when body position changes quickly. As the disease progresses, the symptoms become more pronounced and daily activities like washing your face or brushing your teeth become extremely difficult.

First discovered in 1817 by English physician Dr. James Parkinson, the cause of the disease is not really known. In about 15% of all cases, heredity plays a role. A person can inherit one of two genes that produce proteins that destroy cells in the brain. For the remaining 85% of cases, many scientists believe that a dormant gene is triggered by some external influence. Unfortunately, the actual trigger and how the gene is triggered is unknown.

Although the disease is usually associated with people over 50 years of age, Parkinson's does affect younger adults. Former heavyweight boxing champion of the world Muhammad Ali, actor Katharine Hepburn, and country singer Johnny Cash all have Parkinson's.

💡 REFLECT on your learning

1. Do nerves carry electrical current? Explain.

2. Does a nerve that carries information from your eye function any differently from a nerve that sends information to a muscle?

3. A woman touches a hot object and quickly moves her finger away. Does the brain coordinate the movement of the finger away from the hot object?

4. A cougar jumps from behind a bush and startles a man standing nearby. The information is passed to the man's brain. Explain how the nervous system, endocrine system, and urinary system prepare his body for stress.

> ▶ **TRY THIS** activity

Stimulus and Response in Planaria

Planaria are free-living, freshwater flatworms that have a distinct top and bottom, front and back, and head and tail. In this activity, you will observe the response of a planarian to a simple stimulus.

Materials: medicine dropper, planarian, microscope slide, paper towel

- Using a medicine dropper, remove a planarian from a culture and place it on a microscope slide.

- Gently touch the head of the planarian with a piece of paper towel and note its response.
  (a) Explain why the planarian responded as it did.
  (b) What can you infer about the nervous system of the planarian?
  (c) How do you think a planarian would respond to a concentration of salt added to its environment?

Prisoners have often been isolated and placed in dark rooms as a means of punishment. Imagine how you would be affected if you didn't know whether it was day or night, or if you couldn't hear a sound for days.

Even in these extreme conditions, however, your nervous system remains active. Information about your depth of breathing, the physical condition of the breathing muscles, and the amount of water contained in the respiratory tract is continually relayed to the brain for processing and storage. Other nerve cells detect air temperature, light intensity, and odours. Pressure receptors in the skin—known as baroreceptors—inform you of the fit of your clothes and can detect an insect scurrying across your leg. Blinking your eyes or scratching your nose requires coordinated nerve impulses. Memories of happy times and hopes for your future reside in the nervous system.

Both the nervous system and the endocrine system control the actions of the body. Through a series of adjustments, all systems of the body are regulated to maintain the internal environment within safe limits. Responses to change in internal and external environments are made possible by either electrochemical messages relayed to and from the brain, or by a series of chemical messengers, many of which are carried by the blood. The chemical messengers, hormones, are produced by glands and require more time for response than nerves require.

The nervous system is an elaborate communication system that contains more than 100 billion nerve cells in the brain alone. That number exceeds the number of visible stars in the Milky Way galaxy. Although all animals display some type of response to the environment, the development of the nervous system seems to reach its pinnacle in humans. Memory, learning, and language are all functions of the human nervous system.

Vertebrate Nervous Systems

The nervous system has two main divisions: the **central nervous system (CNS)** and the **peripheral nervous system (PNS)** (**Figure 1**). The central nervous system consists

INVESTIGATION 9.1.1

Teaching a Planarian (p. 451)
Do all planaria learn at the same rate? Find out by performing this investigation into the nervous system of a flatworm.

central nervous system (CNS) the body's coordinating centre for mechanical and chemical actions; made up of the brain and spinal cord

peripheral nervous system (PNS) all parts of the nervous system, excluding brain and spinal cord, that relay information between the central nervous system and other parts of the body

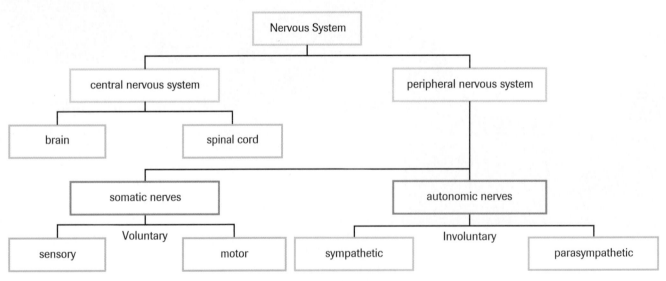

Figure 1
The main divisions of the nervous system

of the nerves of the brain and spinal cord and acts as a coordinating centre for incoming and outgoing information. The peripheral nervous system consists of nerves that carry information between the organs of the body and the central nervous system.

The peripheral nervous system can be further subdivided into somatic and autonomic nerves. The somatic nervous system controls the skeletal muscle, bones, and skin. Sensory somatic nerves relay information about the environment to the central nervous system, while motor somatic nerves initiate an appropriate response. The autonomic nervous system contains special motor nerves that control the internal organs of the body. The two divisions of the autonomic system—the sympathetic nervous system and the parasympathetic nervous system—often operate as "on–off" switches. These two systems will be discussed later in the chapter.

Anatomy of a Nerve Cell

Two different types of cells—glial cells and neurons—are found in the nervous system. **Glial cells**, often called neuroglial cells, are nonconducting cells and are important for the structural support and metabolism of the nerve cells. **Neurons** are the functional units of the nervous system (**Figure 2**). These specialized nerve cells are categorized into three groups: the sensory neurons, interneurons, and motor neurons. **Sensory neurons** (also known as afferent neurons) sense and relay information (or stimuli) from the environment to the central nervous system for processing. For example, special sensory receptors in your eyes, known as photoreceptors, respond to light; some in your nose and tongue, called chemoreceptors, are sensitive to chemicals; others in your skin and hypothalamus, known as thermoreceptors, respond to either warm or cold temperatures. Sensory neurons are located in clusters called **ganglia** (singular, ganglion) located outside of the spinal cord.

glial cells nonconducting cells important for structural support and metabolism of the nerve cells

neurons nerve cells that conduct nerve impulses

sensory neurons neurons that carry impulses from sensory receptors to the central nervous system; also known as afferent neurons

ganglia collections of nerve cell bodies located outside of the central nervous system

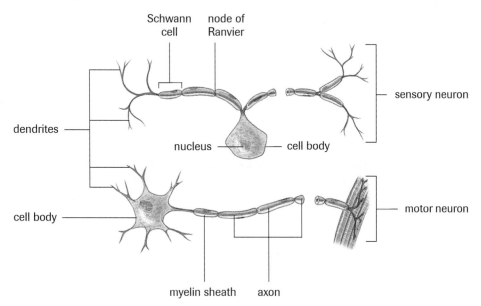

Figure 2
Structure of a motor neuron and a sensory neuron

Interneurons, as the name suggests, link neurons within the body. Found predominantly throughout the brain and spinal cord, the interneurons (also known as association neurons) integrate and interpret the sensory information and connect neurons to outgoing motor neurons. **Motor neurons** (also known as efferent neurons) relay information to the effectors. Muscles, organs, and glands are classified as effectors because they produce responses.

motor neurons neurons that carry impulses from the central nervous system to effectors; also known as efferent neurons

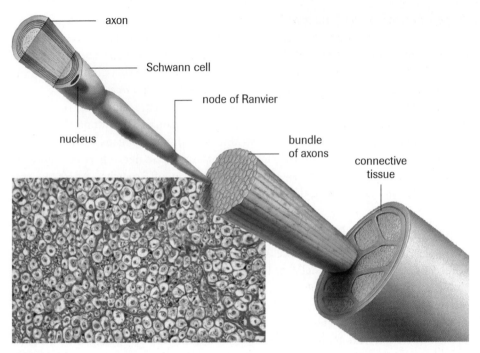

Figure 3
Most nerves are composed of many different neurons.

dendrites projections of cytoplasm that carry impulses toward the cell body

axon extension of cytoplasm that carries nerve impulses away from the cell body

myelin sheath insulated covering over the axon of a nerve cell

Schwann cells special type of glial cell that produces the myelin sheath

nodes of Ranvier regularly occurring gaps between sections of myelin sheath along the axon

neurilemma delicate membrane that surrounds the axon of some nerve cells

All neurons contain dendrites, cell bodies, and axons. The **dendrites** receive information, either from sensory receptors, as in the case of sensory neurons, or from other nerve cells, as in the case of motor neurons. Like all living cells, nerve cells contain a cell body with a nucleus. Dendrites conduct nerve impulses toward the cell body. An extension of cytoplasm, called the **axon**, projects nerve impulses from the cell body. In humans, the axon is extremely thin; more than 100 axons could be placed inside the shaft of a single human hair. The axon carries the nerve impulse toward other neurons or to effectors. A close examination of most nerves shows that they comprise many neurons held together (**Figure 3**). Large composite nerves appear much like a telephone cable that contains many branches of incoming and outgoing lines.

Many axons are covered with a glistening white coat of a fatty protein called the **myelin sheath**, which acts as insulation for the neurons. Axons that have a myelin covering are said to be myelinated. Formed by special glial cells called **Schwann cells**, the myelin sheath insulates by preventing the loss of charged ions from the nerve cell. The areas between the sections of myelin sheath are known as the **nodes of Ranvier**. Nerve impulses jump from one node to another, thereby speeding the movement of nerve impulses. Not surprisingly, nerve impulses move much faster along myelinated nerve fibres than nonmyelinated ones. The speed of the impulse along the nerve fibre is also affected by the diameter of the axon. Generally, the smaller the diameter of the axon, the faster the speed of the nerve impulse.

All nerve fibres found within the peripheral nervous system contain a thin membrane called the **neurilemma**, which surrounds the axon. The neurilemma promotes the regeneration of damaged axons. This explains why feeling gradually returns to your finger following a paper cut—severed neurons can be rejoined. However, not all nerve cells contain a neurilemma and a myelin sheath. Nerves within the brain that contain myelinated fibres and a neurilemma are called white matter, because the myelinated axons are whitish in appearance. Other nerve cells within the brain and spinal cord, referred to as the grey matter, lack a myelin sheath and a neurilemma, and do not regenerate after injury. Damage to the grey matter is usually permanent.

Frontiers of Technology: Reattaching and Regenerating Nerves

For years, scientists have been puzzled about why the central nervous system does not support nerve growth in the same way as the peripheral nervous system. New surgical procedures, the identification of factors that inhibit nerve cell regeneration in the central nervous system, and emerging work with stem cells provide hope for the many people who are paralyzed by spinal cord injury (SCI) (**Figure 4**).

In Norrtalje, Sweden, in the winter of 1993, 25-year-old Thomas Westburg sustained a serious spinal cord injury while snowmobiling. Four nerves were torn from the spinal cord in the area of the neck. The injury left Westburg's left shoulder, arm, and hand completely paralyzed. Surgeons at the Karolinska Hospital in Stockholm reattached two of the nerves. Remarkably, the repair job provided a channel along which new nerves began to grow from cell bodies in Westburg's spinal cord. The slow growth of nerve cells finally connected the spinal cord with muscles that move the arm. In Westburg's case, about 40% of mobility was restored.

Westburg's recovery challenges the conventional scientific understanding that the neurological function of nerves is irretrievably lost when nerves are destroyed. The technique used in Sweden has been shared with surgeons worldwide; however, success has been limited. Grafts from the peripheral nervous system seem to have greater success in the central nervous system. A study on severed optic nerves shows that neurons from the peripheral nervous system grafted into the stalk of the optic nerve regrow approximately 10% of the retinal ganglions. No reconnections were seen when neurons coming directly from the brain to the eye were left alone. This finding suggests that some factors prevent the regeneration of neurons in the central nervous system.

Scientists continue to look for chemical factors that both stimulate and inhibit the growth of new nerve cells. One such factor is myelin-associated glycoprotein (MAG), which is abundant in the myelin sheath of neurons in the central nervous system but is scarce in the myelin of nerve cells of the peripheral nervous system. MAG prevents the growth of axons in tissue cultures. Other growth inhibitors that prevent the regeneration and reconnection of nerve pathways have been isolated from myelinated nerve cells and grey matter. The possibility of "turning off" these growth inhibitors might open the door for nerve-cell regeneration.

Some promising research comes from the use of stem cells. Stem cells are cells that have not yet specialized into tissue cells, such as skin, bone, muscle, or nerve cells. Scientists are experimenting with the possibility of replacing cells that have been damaged by disease or trauma, such as in cases of spinal cord injury or Parkinson's disease.

In October 2000, scientists announced that they had reconnected severed nerves in the spinal cords of rats using spore-like cells from the nervous system of adult rats. Only 3 μm (micrometres) in diameter, these repair cells are so small that some researchers first regarded them as cellular debris. The spore-like cells can be frozen for more than a month and still be retrieved for use. Properly incubated, they grow easily and can withstand a decrease in nutrients and changes of temperature. Placed in the body of a mammal, they are able to survive with limited amounts of oxygen for several days until blood vessels grow into the area. These spore-like cells can only transform into cells associated with nerve conduction.

Figure 4
In Canada, snowmobile accidents account for a high number of spinal cord injuries.

DID YOU *KNOW* ?

Spinal Cord Injury in Canada
According to the Canadian Paraplegic Association (CPA), about 1000 new injuries a year result in some level of permanent paralysis or neurological deficit. Spinal cord injury is predominantly experienced by males in the 15–34 age group.

Figure 5
Spinal cords have been reconnected in rats using spore-like cells from other adult rats.

reflex arc neural circuit through the spinal cord that provides a framework for a reflex action

⚡ **INVESTIGATION 9.1.2**

Reflex Arcs (p. 452)
Reflex arcs provide a framework for reflex actions. Simple physical tests can be performed to test reflexes. In this investigation, you will observe the presence and strength of a number of reflex arcs.

Scientists harvested the spore-like nerve cells from the spines of healthy adult rats (**Figure 5**) and seeded them into the spinal cords of injured rats. Quickly the new cells began to grow in the area of the severed cord. After 10 days, researchers recorded small twitches in the toes of the rats. Within three months, some of the rats could stand on their hind legs.

The use of adult stem cells has also been proposed for this purpose, however, further research is required to determine whether these cells could be used to treat neurological diseases and injuries.

Neural Circuits

If you accidentally touch a hot stove, you probably do not stop to think about how your nervous system tells you that it is hot. The sensation of heat is detected by specialized temperature receptors in your skin, and a nerve impulse is carried to the spinal cord. The sensory neuron passes the impulse on to an interneuron, which, in turn, relays the impulse to a motor neuron. The motor neuron causes the muscles in the hand to contract and the hand to pull away. All this happens in less than a second, before the information even travels to the brain. Very quickly, the sensation of pain becomes noticeable and you may let out a scream.

Reflexes are involuntary and often unconscious. Imagine how badly you could burn yourself if you had to wait for the sensation of pain before removing your hand from the hot stove. The damage would be much worse if you had to go through the process of gauging the intensity of the pain and then contemplating the appropriate action. Even the small amount of time required for nerve impulses to move through the many circuits of the brain and back to the muscle would increase the damage.

The simplest nerve pathway is the **reflex arc**. Most reflexes occur without brain co-ordination. Reflex arcs contain five essential components: the receptor, the sensory neuron, the interneuron in the spinal cord, the motor neuron, and the effector (**Figure 6**). ⚡

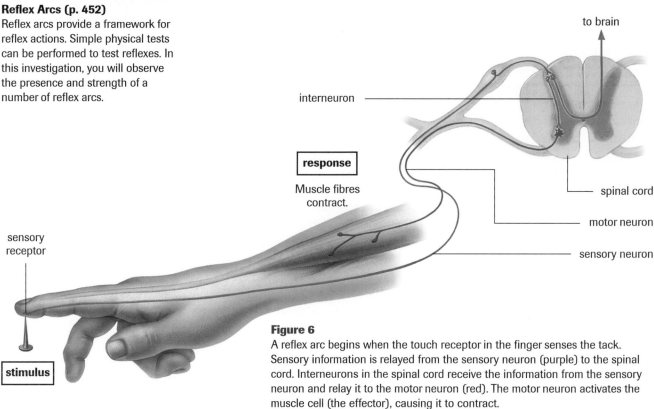

Figure 6
A reflex arc begins when the touch receptor in the finger senses the tack. Sensory information is relayed from the sensory neuron (purple) to the spinal cord. Interneurons in the spinal cord receive the information from the sensory neuron and relay it to the motor neuron (red). The motor neuron activates the muscle cell (the effector), causing it to contract.

 **SUMMARY** *The Importance of the Nervous System*

Table 1

Structure	Function
neuron	• nerve cell that conducts nerve impulses
sensory neuron (afferent neuron)	• carries impulses to the central nervous system
interneuron	• carries impulses within the central nervous system
motor neuron (efferent neuron)	• carries impulses from the central nervous system to effectors
dendrite	• projection of cytoplasm that carries impulses toward the cell body
axon	• extension of cytoplasm that carries nerve impulses away from the cell body
myelin sheath	• insulated covering over the axon of a nerve cell • composed of Schwann cells
nodes of Ranvier	• regularly occurring gaps between sections of myelin sheath along the axon where nerve cells are transmitted
neurilemma	• delicate membrane that surrounds the axon of some nerve cells
reflex arc	• neural circuit that travels through the spinal cord • provides a framework for a reflex action

▶ **Section 9.1** *Questions*

Understanding Concepts

1. Differentiate between the peripheral nervous system (PNS) and central nervous system (CNS).

2. Differentiate between sensory nerves and motor nerves.

3. Briefly describe the function of the following parts of a neuron: dendrites, myelin sheath, Schwann cells, cell body, and axon.

4. Name the essential components of a reflex arc and the function of each.

5. What would happen if neuron I in **Figure 7** was severed?

6. In **Figure 7**, what is the order in which an impulse travels along a reflex arc?

Making Connections

7. Primitive sporelike repair cells have been extracted from adult rats. Discuss some of the benefits of using mature repair cells.

8. The incidence of multiple sclerosis (MS) varies among different regions of Canada. Provide a possible explanation for different distributions of the disease.

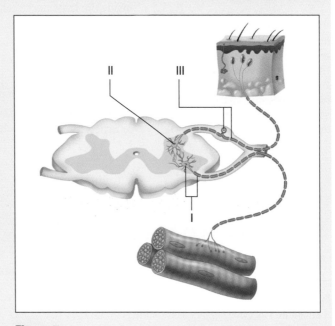

Figure 7
Reflex arc

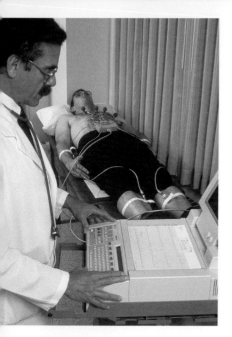

Figure 1
Mapping electrical current has diagnostic value.

In the late 18th century, the Italian scientist Luigi Galvani discovered that the leg muscle of a dead frog could be made to twitch under electrical stimulation. Galvani concluded that the "animal electricity" was produced by the muscle. Although Galvani's conclusion was incorrect, it spawned a flood of research that led to the development of theories about how electrical current is generated in the body. In 1840, Emil DuBois-Reymond of the University of Berlin set about refining instruments that would detect the passage of currents in nerves and muscles. By 1906, the Dutch physiologist Willem Einthoven began recording the transmission of electrical impulses in heart muscle. The electrocardiogram (ECG) has been refined many times since then and is still used today to diagnose heart problems (**Figure 1**). In 1929, German psychiatrist Hans Berger placed electrodes on the skull of a subject and measured electrical changes that accompany brain activity. The electroencephalograph (EEG) is used to measure brain-wave activity.

As research continued, the difference between electrical and neural transmission soon became evident. Current travels along a wire much faster than an impulse travels across a nerve. In addition, the cytoplasmic core of a nerve cell offers great resistance to the movement of electrical current. Unlike electrical currents, which diminish as they move through a wire, nerve impulses remain as strong at the end of a nerve as they were at the beginning. One of the greatest differences between nerve impulses and electricity is that nerves use cellular energy to generate current. By comparison, the electrical wire relies on some external energy source to push electrons along its length.

As early as 1900, German physiologist Julius Bernstein suggested that nerve impulses were an electrochemical message created by the movement of ions through the nerve cell membrane. Evidence supporting Bernstein's theory was provided in 1939 when two researchers at Columbia University, K.S. Cole and H.J. Curtis, placed a tiny electrode inside the large nerve cell of a squid (**Figure 2**). A rapid change in the electrical potential difference—commonly called the potential—across the membrane was detected every time the nerve became excited. The resting membrane normally had a potential somewhere near −70 mV (millivolts); however, when the nerve became excited, the potential on the inside of the membrane registered +40 mV. This reversal of potential is described as an **action potential**. Cole and Curtis noticed that the +40 mV did not last more than a few milliseconds (ms) before the potential on the inside of the nerve cell returned to −70 mV—the **resting potential**.

action potential the voltage difference across a nerve cell membrane when the nerve is excited

resting potential voltage difference across a nerve cell membrane during the resting stage (usually negative)

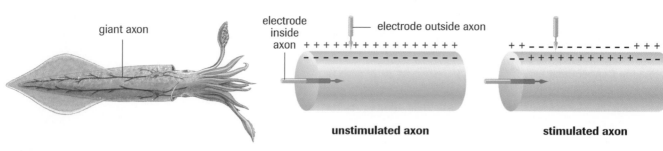

Figure 2
A miniature electrode is placed inside the giant axon of a squid. The inside of the resting membrane is negative with respect to the outside of the membrane. When stimulated, the charges across the nerve membrane temporarily reverse.

How do nerve cell membranes become charged? To find the answer, we must examine the nerve cell on a molecular level. Unlike most cells, neurons have a rich supply of positive and negative ions both inside and outside the cell. Although it might seem surprising, negative ions do little to create a charged membrane. They are mainly large ions that cannot cross the membrane and, therefore, stay inside the cell. The electrochemical event is caused by an unequal concentration of positive ions across the nerve cell membrane. The highly concentrated potassium ions inside the nerve cells have a tendency to diffuse outside the nerve cells. Similarly, the highly concentrated sodium ions outside the nerve cell have a tendency to diffuse into the nerve cell. As potassium diffuses out of the neuron, sodium diffuses into the neuron. Therefore, positively charged ions move both into and out of the cell. However, the diffusion of sodium ions and potassium ions is unequal. The resting membrane is about 50 times more permeable to potassium than it is to sodium. Consequently, more potassium ions diffuse out of the nerve cell than sodium ions diffuse into the nerve cell (**Figure 3**).

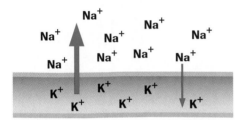

Potassium diffuses out of the nerve cell faster than sodium diffuses into the nerve cell.

Figure 3
The uneven concentrations of Na^+ and K^+ on either side of the neuron membrane maintain an external positive charge.

The more rapid diffusion of potassium ions out of the nerve membrane means that the nerve cell loses a greater number of positive ions than it gains, and the exterior of the membrane becomes positive relative to the interior. Biologists now believe that protein channels, known as ion gates, control the movement of ions across the cell membrane. Excess positive ions accumulate along the outside of the nerve membrane, while excess negative ions accumulate along the inside of the membrane. The resting membrane is said to be charged and is called a **polarized membrane**. The separation of electrical charges by a membrane has the potential to do work, which is expressed in millivolts (mV). A charge of -70 mV indicates the difference between the number of positive charges found on the inside of the nerve membrane relative to the outside. (A charge of -90 mV on the inside of the nerve membrane would indicate even fewer positive ions inside the membrane relative to the outside.)

Upon excitation, the nerve cell membrane becomes more permeable to sodium than potassium (**Figure 4**, page 420). Scientists believe that sodium gates are opened in the nerve membrane, while potassium gates close. The highly concentrated sodium ions rush into the nerve cell by diffusion and charge attraction. The rapid inflow of sodium causes a charge reversal, also referred to as **depolarization**. Once the voltage inside the nerve cell becomes positive, the sodium gates slam closed and the inflow of sodium is halted. A **sodium-potassium pump**, located in the cell membrane, restores the condition of the resting membrane by transporting sodium ions out of the neuron while moving potassium ions inside the neuron, in a ratio of three Na^+ ions out to two K^+ ions in (**Figure 5**, page 420). The energy supply from adenosine triphosphate (ATP) fuels the pump to maintain the polarization of the membrane. The process of restoring the original polarity of the nerve membrane is called **repolarization**.

polarized membrane membrane charged by unequal distribution of positively charged ions inside and outside the nerve cell

depolarization diffusion of sodium ions into the nerve cell resulting in a charge reversal

sodium-potassium pump an active transport mechanism that moves sodium ions out of and potassium ions into a cell against their concentration gradients

repolarization process of restoring the original polarity of the nerve membrane

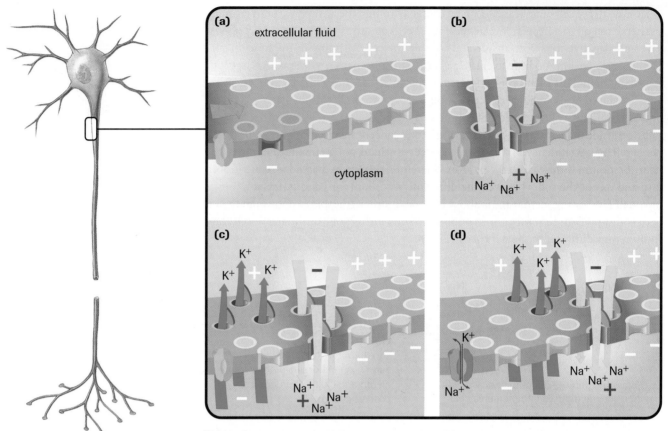

Figure 4

(a) The resting membrane is more permeable to potassium than to sodium. Potassium ions diffuse out of the nerve cell faster than sodium ions diffuse into the nerve cell. The outside of the nerve cell becomes positive relative to the inside.

(b) A strong electrical disturbance, shown by the darker colouration, moves across the cell membrane. The disturbance opens sodium ion gates, and sodium ions rush into the nerve cell. The membrane becomes depolarized.

(c) Depolarization causes the sodium gates to close, while the potassium gates are opened once again. Potassium follows the concentration gradient and moves out of the nerve cell by diffusion. Adjoining areas of the nerve membrane become permeable to sodium ions, and the action potential moves away from the site of origin.

(d) The electrical disturbance moves along the nerve membrane in a wave of depolarization. The membrane is restored, as successive areas once again become more permeable to potassium. The sodium-potassium pump restores and maintains the polarization of the membrane by pumping potassium ions in and sodium ions out of the cell.

Nerves conducting an impulse cannot be activated until the condition of the resting membrane is restored. The period of depolarization must be completed and the nerve must repolarize before the next action potential can be conducted. The time required for the nerve cell to become repolarized is called the **refractory period**. The refractory period usually lasts 1 to 10 ms.

refractory period recovery time required before a neuron can produce another action potential

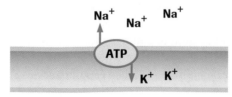

Sodium is pumped out of the nerve cell, while potassium is pumped into the nerve cell.

Figure 5
The potassium and sodium pumps of the nerve cell are highly effective.

Movement of the Action Potential

The movement of sodium ions into the nerve cell causes a depolarization of the membrane and signals an action potential in that area. However, for the impulse to be conducted along the axon, the impulse must move from the zone of depolarization to adjacent regions (**Figure 6**).

It is important to understand the action potential. The action potential (**Figure 7**) is characterized by the opening of the sodium channels in the nerve membrane. Sodium ions rush into the cytoplasm of the nerve cell, diffusing from an area of high concentration (outside the nerve cell) to an area of lower concentration (inside the nerve cell). The influx of the positively charged sodium ions causes a charge reversal, or depolarization, in that area. The positively charged ions that rush into the nerve cell are then attracted to the adjacent negative ions, which are aligned along the inside of the nerve membrane. A similar attraction occurs along the outside of the nerve membrane. The positively charged sodium ions of the resting membrane are attracted to the negative charge that has accumulated along the outside of the membrane in the area of the action potential.

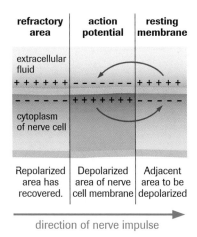

Figure 6
The movement of a nerve impulse

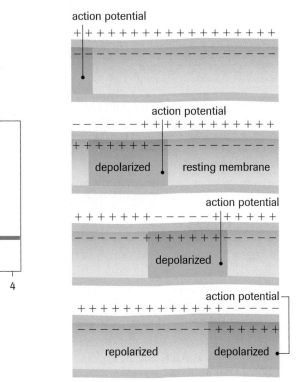

Figure 7
The action potential moves along the nerve cell membrane, creating a wave of depolarization and repolarization.

The flow of positively charged ions from the area of the action potential toward the adjacent regions of the resting membrane causes a depolarization in the adjoining area. The electrical disturbance causes sodium channels to open in the adjoining area of the nerve cell membrane and results in the movement of the action potential. A wave of depolarization moves along the nerve membrane, and then the initiation point of the action potential enters a refractory period as the membrane once again becomes more permeable to potassium ions. Depolarization of the membrane causes the sodium channels to close and the potassium channels to reopen. The wave of depolarization is followed by a wave of repolarization.

Threshold Levels and the All-or-None Response

A great deal of information about nerve cells has been acquired through laboratory experiments. Nerve cells respond to changes in pH, changes in pressure, and to specific chemicals. However, mild electrical shock is most often used in experimentation because it is easily controlled and its intensity can be regulated.

In a classic experiment, a single neuron leading to a muscle is isolated and a mild electrical shock is applied to the neuron. A special recorder measures the strength of muscle contraction. **Figure 8** shows sample data for this experiment. In this example, stimuli of less than 2 mV does not produce any muscle contraction. A potential stimulus must be above a critical value to produce a response. The critical intensity of the stimulus is known as the **threshold level**. Stimuli below threshold levels do not initiate a response. In **Figure 8**, although a threshold level of 2 mV is required to produce a response, threshold levels are different for each neuron.

threshold level minimum level of a stimulus required to produce a response

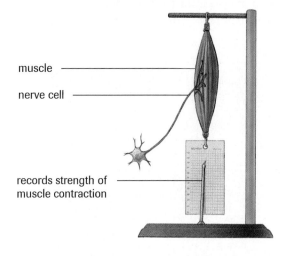

muscle

nerve cell

records strength of muscle contraction

Data	
Strength of stimuli	**Force of contraction**
1 mV	—
2 mV	3 N
3 mV	3 N
10 mV	3 N

Figure 8
The threshold level for this neuron is 2 mV. Different neurons have different threshold levels.

all-or-none response a nerve or muscle fibre responds completely or not at all to a stimulus

A second, but equally important, conclusion can be drawn from the experimental data in **Figure 8**. Increasing the intensity of the stimuli above the critical threshold value will not produce an increased response—the intensity of the nerve impulse and speed of transmission remain the same. In what is referred to as the **all-or-none response**, neurons either fire maximally or not at all.

How do animals detect the intensity of stimuli if nerve fibres either fire completely or not at all? Experience tells you that you are capable of differentiating between a warm object and one that is hot. To explain the apparent anomaly, we must examine the manner in which the brain interprets nerve impulses. Although stimuli above threshold levels produce nerve impulses of identical speed and intensity, variation with respect to frequency does occur. The more intense the stimulus, the greater the frequency of impulses. Therefore, when a warm glass rod is placed on your hand, sensory impulses may be sent to the brain at a slow rate. A hot glass rod placed on the same tissue also causes the nerve to fire, but the frequency of impulses is greatly increased—a difference the brain recognizes.

The different threshold levels of neurons provide a second way for the intensity of stimuli to be detected. Each nerve is composed of many individual nerve cells or neurons. A glass rod at 40°C may cause a single neuron to reach threshold level, but the same glass rod at 50°C will cause two or more neurons to fire (**Figure 9**). The second neuron has a higher threshold level. The greater the number of impulses reaching the brain, the greater the intensity of the response.

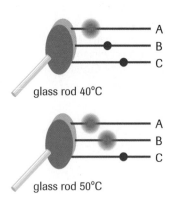

glass rod 40°C

glass rod 50°C

Figure 9
Neuron B has a higher threshold level than neuron A and will not fire until the glass rod is heated above 40°C. The brain interprets both the number of neurons excited and the frequency of impulses.

Synaptic Transmission

Small spaces between neurons, or between neurons and effectors, are known as **synapses**. A single neuron may branch many times at its end plate and join with many different neurons (**Figure 10**). Synapses rarely involve just two neurons. Small vesicles containing chemicals called **neurotransmitters** are located in the end plates of axons. The impulse moves along the axon and releases neurotransmitters from the end plate. The neurotransmitters are released from the **presynaptic neuron** and diffuse across the synaptic cleft, creating a depolarization of the dendrites of the **postsynaptic neuron**. Although the space between neurons is very small—approximately 20 nm (nanometres)—the nerve transmission slows across the synapse. Diffusion is a slow process. Not surprisingly, the greater the number of synapses, the slower the speed of transmission over a specified distance. This may explain why you react so quickly to a stimulus in a reflex arc, which has few synapses, while solving biology problems, which involves many more synapses, requires more time.

synapses regions between neurons, or between neurons and effectors

neurotransmitters chemicals released from vesicles into synapses

presynaptic neuron neuron that carries impulses to the synapse

postsynaptic neuron neuron that carries impulses away from the synapse

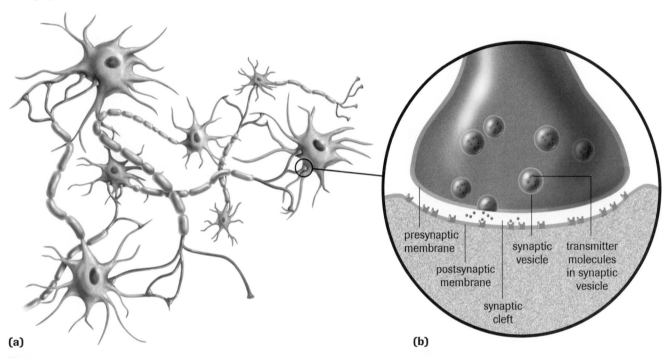

(a) **(b)**

Figure 10
(a) Branching end plates synapse with the cell bodies and dendrites of many different neurons.
(b) Synaptic vesicles in the end plate of the presynaptic neuron release neurotransmitters by exocytosis. The neurotransmitters attach themselves to the postsynaptic membrane, causing it to depolarize. The action potential continues along the postsynaptic neuron.

Acetylcholine is an example of a neurotransmitter found in the end plates of many nerve cells. Acetylcholine can act as an excitatory neurotransmitter on many postsynaptic neurons by opening the sodium ion channels (**Figure 11**, page 424). Once the channels are opened, the sodium ions rush into the postsynaptic neuron, causing depolarization. The reversal of charge causes the action potential. However, acetylcholine also presents a problem. With the sodium channels open, the postsynaptic neuron would remain in a constant state of depolarization. How can the nerve respond to the next impulse if it never recovers? The release of the enzyme **cholinesterase** from the postsynaptic membrane destroys acetylcholine. Once acetylcholine is destroyed, the sodium channels are closed,

acetylcholine neurotransmitter released from vesicles in the end plates of neurons, which makes the postsynaptic membranes more permeable to Na+ ions

cholinesterase enzyme, which breaks down acetylcholine, that is released from postsynaptic membranes in the end plates of neurons shortly after acetylcholine

Myasthenia Gravis

Drugs that temporarily keep the enzyme cholinesterase from working are used to treat myasthenia gravis, a disease of progressive fatigue and muscle weakness caused by the impaired transmission of nerve impulses.

hyperpolarized condition in which the inside of the nerve cell membrane becomes more negative than the resting potential

and the neuron begins its recovery phase. Many insecticides take advantage of the synapse by blocking cholinesterase. The heart of an insect, unlike the human heart, is totally under nerve control. An insecticide causes the insect's heart to respond to the nerve message by contracting but never relaxing.

Not all transmissions across a synapse are excitatory. Although acetylcholine may act as an excitatory neurotransmitter on one postsynaptic membrane, it could act as an inhibitory neurotransmitter on another. It is believed that many inhibitory neurotransmitters make the postsynaptic membrane more permeable to potassium. By opening even more potassium gates, the potassium ions inside the neuron follow the concentration gradient and diffuse out of the neuron. The rush of potassium out of the cell increases the number of positive ions outside the cell relative to the number found inside the cell. Such neurons are said to be **hyperpolarized** because the resting membrane is even more negative. More sodium channels must now be opened to achieve depolarization and an action potential. As the name suggests, these inhibitory neurotransmitters prevent postsynaptic neurons from becoming active.

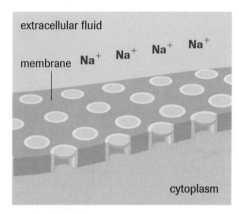 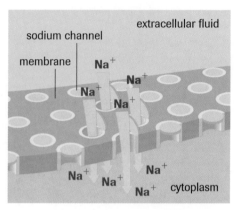

Figure 11
Model of an excitatory synapse. Acetylcholine opens channels for Na$^+$ ions in the postsynaptic membrane.

No acetylcholine is present and sodium channels remain closed.

Acetylcholine is released and sodium channels are opened.

Figure 12 shows a model of a typical neural pathway. Neurotransmitters released from neurons A and B are both excitatory, but neither neuron is capable of causing sufficient depolarization to initiate an action potential in neuron D. However, when both neurons A and B fire at the same time, a sufficient amount of neurotransmitter is released to cause depolarization of the postsynaptic membrane. The production of an action potential in neuron D requires the sum of two excitatory neurons. This principle is referred to as **summation**.

summation effect produced by the accumulation of neurotransmitters from two or more neurons

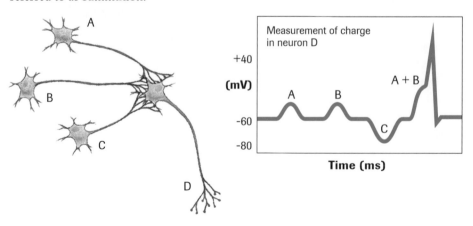

Figure 12
Action potentials must occur simultaneously in A and B to reach the threshold level in D.

The neurotransmitter released from neuron C produces a dramatically different response. Neuron D becomes more negatively charged when neuron C is activated. You may have already concluded that neuron C is inhibitory. But data reveal even more striking information: neurotransmitters other than acetylcholine must be present. A number of neurotransmitters such as serotonin, dopamine, gamma-aminobutyric acid (GABA), and glutamic acid have been identified in the central nervous system. Another common neurotransmitter, norepinephrine (also known as noradrenaline), is found in both the central and peripheral nervous systems. To date, all effects of norepinephrine in the peripheral nervous system appear to be excitatory, while those in the central nervous system can be excitatory or inhibitory.

The interaction of excitatory and inhibitory neurotransmitters is what allows you to throw a ball. As the triceps muscle on the back of your upper arm receives excitatory impulses and contracts, the biceps muscle on the front of your arm receives inhibitory impulses and relaxes. By coordinating excitatory and inhibitory impulses, the two muscles of the arm do not pull against each other.

Inhibitory impulses in your central nervous system are even more important. Sensory information is received by the brain and is prioritized. Much of the less important information is ignored so that you can devote your attention to the more important sensory information. For example, during a biology lecture, your sensory information should be directed at the sounds coming from your teacher, the visual images that appear on the chalkboard, and the sensations produced as you move your pen across the page. Although your temperature receptors may signal a slight chill in the air and the pressure receptors in your skin may provide the reassuring information that you are indeed wearing clothes, the information from these sensory nerves is suppressed. Inhibitory impulses help you prioritize information.

Various disorders have been associated with neurotransmitters. Parkinson's disease, characterized by involuntary muscle contractions and tremors, is caused by inadequate production of dopamine. Alzheimer's disease, associated with the deterioration of memory and mental capacity, has been related to decreased production of acetylcholine.

SUMMARY   *Electrochemical Impulse*

- Nerves conduct electrochemical impulses from the dendrites along the axon to the end plates of the neuron.
- Active transport and diffusion of sodium and potassium ions establish a polarized membrane.
- An action potential is caused by the inflow of sodium ions.
- Nerve cells exhibit an all-or-none response.
- Neurotransmitters allow the nerve message to move across synapses.

Understanding Concepts

1. What evidence suggests that nerve impulses are not electrical but electrochemical events?

2. Why was the squid axon particularly appropriate for nerve research?

3. What is a polarized membrane?

4. What causes the inside of a neuron to become negatively charged?

5. Why does the polarity of a cell membrane reverse during an action potential?

6. What changes take place along a nerve cell membrane as it moves from a resting potential to an action potential to a refractory period?

7. Why do nerve impulses move faster along myelinated nerve fibres?

8. What is the all-or-none response?

9. Explain the functions of acetylcholine and cholinesterase in the transmission of nerve impulses.

10. In **Figure 13**, which area(s) of the graph indicate(s) the opening of Na^+ ion channels and the diffusion of Na^+ ions into the nerve cells? Explain your answer.

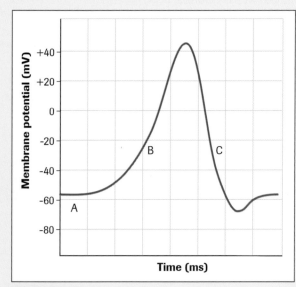

Figure 13
Action potential

11. In **Figure 13**, repolarization occurs in which areas? Explain your answer.

Applying Inquiry Skills

12. Explain summation, using the synapse model in **Figure 14**.

13. Use the synapse model in **Figure 14** to explain why nerve impulses move from neuron A to neuron B, but not from neuron B back to neuron A.

Figure 14
Nerve pathway

Making Connections

14. The action of many psychoactive drugs can be explained in terms of neurotransmitters. Valium, a depressant, interacts with gamma-amino-butyric acid (GABA) transmitter–receptor sites on postsynaptic membranes. The greater the number of receptor sites that are occupied, the more effective the neurotransmitter. LSD and mescaline, both hallucinogenic drugs, are thought to interact with the receptor sites of serotonin.
    (a) Draw a diagram that shows how Valium and hallucinogenic drugs work.
    (b) What dangers exist from taking drugs that interfere with naturally produced neurotransmitter chemicals?

15. The neurotransmitter serotonin is normally involved in temperature regulation, sensory perception, and mood control. A class of compounds known as selective serotonin reuptake inhibitors (SSRIs) has proven highly successful in the treatment of depression, anxiety, and obsessive-compulsive disorder (OCD). How do these therapeutic drugs affect serotonin? Are there any risks involved? Search for information in newspapers, periodicals, CD-ROMs, and on the Internet.

 www.science.nelson.com

16. Research the effects of nerve gas on the human body. Does it make sense to carry a gas mask in case of exposure to nerve gas? Search for information in newspapers, periodicals, CD-ROMs, and on the Internet.

 www.science.nelson.com

The central nervous system consists of the brain and spinal cord. The brain is formed from a concentration of nerve tissue in the anterior portion of animals and acts as the co-ordinating centre of the nervous system. Enclosed within the skull, the brain is surrounded by a tough three-layer protective membrane known as the **meninges**. The outer membrane is called the *dura mater*, the middle layer is the *arachnoid mater*, and the inner layer is the *pia mater*. These three membrane layers form the blood-brain barrier, which determines what chemicals will reach the brain.

Cerebrospinal fluid circulates between the innermost and middle meninges of the brain and through the central canal of the spinal cord. The cerebrospinal fluid acts both as a shock absorber and a transport medium, carrying nutrients to brain cells while relaying wastes from the cells to the blood. Physicians can extract cerebrospinal fluid from the spinal cord to diagnose bacterial or viral infection. The technique, referred to as a lumbar puncture or spinal tap, is used to identify poliomyelitis and meningitis.

meninges protective membranes that surround the brain and spinal cord

cerebrospinal fluid cushioning fluid that circulates between the innermost and middle membranes of the brain and spinal cord; it provides a connection between neural and endocrine systems

The Spinal Cord

The spinal cord carries sensory nerve messages from receptors to the brain and relays motor nerve messages from the brain to muscles, organs, and glands. Emerging from the skull through an opening called the foramen magnum, the spinal cord extends downward through a canal within the backbone (**Figure 1**). A cross section of the spinal cord reveals the two types of nerve tissue introduced earlier in this chapter: white matter and grey matter. Although the central grey matter consists of nonmyelinated interneurons, the surrounding white matter is composed of myelinated nerve fibres from the sensory and motor neurons. The interneurons are organized into nerve tracts that connect the spinal cord with the brain. A dorsal nerve tract brings sensory information into the spinal cord, while a ventral nerve tract carries motor information from the spinal cord to the peripheral muscles, organs, and glands.

DID YOU KNOW?

Meningitis
Meningitis is caused by a bacterial or viral infection of the outer membranes of the brain. Its symptoms include fever, vomiting, an intense headache, and a stiff neck. If left untreated, bacterial meningitis can lead to death.

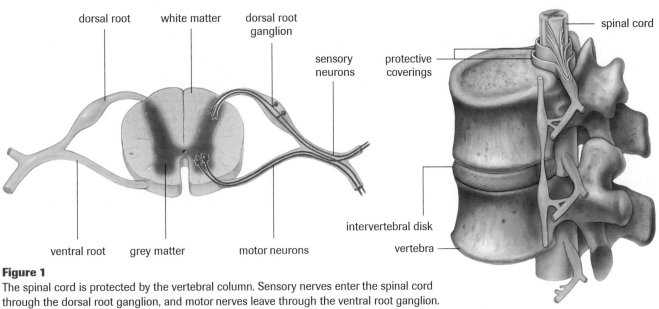

Figure 1
The spinal cord is protected by the vertebral column. Sensory nerves enter the spinal cord through the dorsal root ganglion, and motor nerves leave through the ventral root ganglion.

The Brain

Brain complexity is what distinguishes humans from other animals, although we come up short in many other ways. Humans lack the strength and agility of other mammals of comparable size. Human hearing, vision, and sense of smell are unimpressive when compared with those of many other species. Humans also lag behind in the area of reproduction; in comparison to mice, for example, humans reproduce relatively slowly. What makes *Homo sapiens* unique is intellect and the reasoning functions of the brain. However, despite its apparent uniqueness, the human brain has developmental links with other chordates (**Figure 2**). As in primitive vertebrates, the human brain comprises three distinct regions: the forebrain, the midbrain, and the hindbrain.

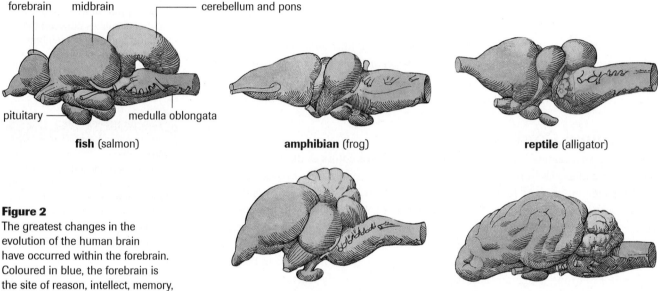

forebrain   midbrain   cerebellum and pons

pituitary   medulla oblongata

fish (salmon)

amphibian (frog)

reptile (alligator)

bird (pigeon)

mammal (dog)

Figure 2
The greatest changes in the evolution of the human brain have occurred within the forebrain. Coloured in blue, the forebrain is the site of reason, intellect, memory, language, and personality.

olfactory lobes areas of the brain that process information about smell

cerebrum largest and most highly developed part of the human brain, which stores sensory information and initiates voluntary motor activities

cerebral cortex outer lining of the cerebral hemispheres

corpus callosum nerve tract that joins the two cerebral hemispheres

The forebrain contains paired **olfactory lobes**, which are centres that receive information about smell. The **cerebrum** is also contained within the forebrain. These two giant hemispheres act as the major coordinating centre from which sensory information and accompanying motor actions originate. Speech, reasoning, memory, and even personality reside within these paired cerebral hemispheres. The surface of the cerebrum is known as the **cerebral cortex**. Composed of grey matter, the cortex has many folds that increase surface area. The deep folds are known as fissures.

Recent research has demonstrated that information stored in one side of the brain is not necessarily present in the other. The right side of the brain has been associated with visual patterns or spatial awareness; the left side of the brain is linked to verbal skills. Your ability to learn may be related to the dominance of one of the hemispheres. A bundle of nerves called the **corpus callosum** allows communication between the two hemispheres. Each hemisphere can be further subdivided into four lobes (**Figure 3**): the frontal lobe, the temporal lobe, the occipital lobe, and the parietal lobe. **Table 1** lists the functions of each of the lobes.

Stimulation of the motor cortex by electrical probes can trigger muscles in various parts of the body. Not surprisingly, the number of nerve tracts leading to the thumb and fingers is greater than the number leading to the arms or legs, since the thumb and fingers are capable of many delicate motor movements. Wrist and arm movements, by contrast, are limited and, therefore, regulated by fewer nerves. **Figure 4** shows parts of the

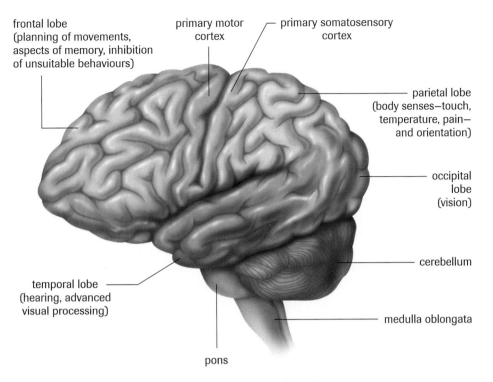

frontal lobe
(planning of movements,
aspects of memory, inhibition
of unsuitable behaviours)

primary motor
cortex

primary somatosensory
cortex

parietal lobe
(body senses–touch,
temperature, pain–
and orientation)

occipital
lobe
(vision)

cerebellum

temporal lobe
(hearing, advanced
visual processing)

medulla oblongata

pons

Figure 3
Primary receiving and integrating centres of the human cerebral cortex. Primary cortical areas receive signals from receptors on the body's periphery. Association areas coordinate and process sensory input from different receptors.

Table 1 The Lobes of the Cerebrum

Lobe	Function
frontal lobe	• Motor areas control movement of voluntary muscles (e.g., walking and speech). • Association areas are linked to intellectual activities and personality.
temporal lobe	• Sensory areas are associated with vision and hearing. • Association areas are linked to memory and interpretation of sensory information.
parietal lobe	• Sensory areas are associated with touch and temperature awareness. • Association areas have been linked to emotions and interpreting speech.
occipital lobe	• Sensory areas are associated with vision. • Association areas interpret visual information.

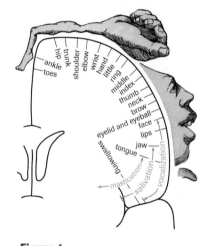

Figure 4
Regions of the body are drawn in proportion to the area of the motor cortex required to control the region.

thalamus area of brain that coordinates and interprets sensory information and directs it to the cerebrum

human body drawn in proportion to the number of motor nerves that control them. Note the size of the tongue and mouth. Human speech depends on subtle changes in the position of the tongue and mouth.

Below the cerebrum is the **thalamus**, and immediately below the thalamus is the hypothalamus. A direct connection between the hypothalamus and the pituitary gland unites the nervous system with the endocrine system.

The midbrain is less developed than the forebrain. Consisting of four spheres of grey matter, the midbrain acts as a relay centre for some eye and ear reflexes. The hindbrain, as the name suggests, is found posterior to the midbrain and joins with the spinal cord. The cerebellum, pons, and medulla oblongata are the major regions of the hindbrain. The **cerebellum**, located immediately beneath the two cerebral hemispheres, is the largest section of the hindbrain. The cerebellum controls limb movements, balance, and muscle tone. Have you ever considered the number of coordinated muscle actions required to pick up

cerebellum part of the hindbrain that controls limb movements, balance, and muscle tone

pons region of the brain that acts as a relay station by sending nerve messages between the cerebellum and the medulla

medulla oblongata region of the hindbrain that joins the spinal cord to the cerebellum; one of the most important sites of autonomic nerve control

a pencil? Your hand must be opened before you touch the pencil; the synchronous movement of thumb and fingers requires coordination of both excitatory and inhibitory nerve impulses.

The **pons**, meaning "bridge," is largely a relay station that passes information between the two regions of the cerebellum and between the cerebellum and the medulla. The posterior region of the hindbrain is the **medulla oblongata**. Nerve tracts from the spinal cord and higher brain centres run through the medulla, which acts as the connection between the peripheral and central nervous systems. The medulla oblongata controls involuntary muscle action. Breathing movements, the diameter of the blood vessels, and heart rate are but a few things regulated by this area of the hindbrain. The medulla oblongata also acts as the coordinating centre for the autonomic nervous system.

How did scientists conclude that certain areas of the brain have specific functions (**Figure 5**)? Patients who had experienced strokes provided them with the evidence. Any factor that reduces blood flow to cells of the brain can cause a stroke. Strokes that occur in the right side of the motor cortex cause paralysis of the left side of the body. A patient's recovery from a stroke provides evidence that other cells can be trained to assume the function of previously damaged cells. With time, speech, hearing, and some limb movements may be partially restored.

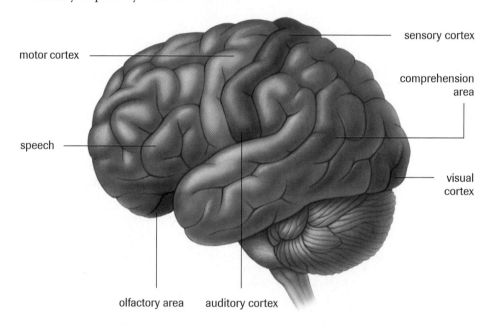

Figure 5
The motor cortex is shown in yellow and the sensory cortex is shown in purple.

Brain Mapping

Now considered the seat of intellect and reason, the human brain is one of the most valued organs of the body. However, the brain was not always thought of in such lofty terms. Around 350 B.C., the Greek philosopher Aristotle believed that the primary function of the brain was to cool the blood. His vision of the brain as a glorified air conditioner persisted for centuries. About 500 years later, the Greek physician Galen speculated that the brain functioned as a hydraulic system, in which fluid spirits were shunted through a series of ducts. In the early 17th century, the French philosopher René Descartes showed Galen's influence by describing the brain in terms of the flow of "animal spirits." However, unlike Galen, Descartes associated the brain with coordinated sensory and motor activity. He believed that sensory information was stored in the brain and that motor function completed the loop. He even speculated that the pineal gland of the brain was the site of the soul.

By 1811, Franz Gall, a German physician who studied in Austria, suggested that sensory nerves terminate in specific sections of the grey matter of the brain. The idea that specific parts of the brain carry out specific functions led some of Gall's followers to suggest that mental ability or personality could be studied by investigating brain development. This train of thought led to the study of a pseudoscience called *phrenology*. Assuming that the brain case could reveal changes in brain development, phrenologists attempted to determine mental capacity by feeling for bumps on the head.

Phrenologists believed that a large brain signified greater intellect. In fact, brain size is more often correlated with body size than with intelligence. Ironically, the fact that Gall himself had a small head seemed to escape the phrenologists. **Table 2** shows the relationship between brain size and body mass for various mammals.

Table 2 Comparison of Brain Size to Body Mass

Animal	Brain size	Ratio of brain size to body mass
chimpanzee	400 g	1:150
gorilla	540 g	1:500
human	1500 g	1:50
elephant	6000 g	1:1000
whale	9000 g	1:10 000

Research in Canada:
Wilder G. Penfield

Wilder G. Penfield (1891–1976), founder of the Montreal Neurological Institute, was the foremost pioneer in brain mapping. Using electrical probes, Penfield (**Figure 6**) located three speech areas within the cerebral cortex. Interestingly, the predominant speech areas reside on the left side of the brain. Penfield's finding dismissed the once-held notion that the two hemispheres were mirror images of each other.

In a classic experiment, Penfield applied an electrode to a particular speech area and then showed the subject a picture of a foot. Although the subject was able to recognize the foot, the word would not come. Once electrical stimulation ceased, the subject could say "foot." Penfield concluded that the thought processes associated with recognizing the foot reside within a particular location separated from the areas of the brain that control the human speech function.

Penfield spent a great deal of his time mapping the cerebral cortex of people with epilepsy. Epilepsy is often associated with injuries to the cerebral cortex; electrical "storms" spread across the damaged tissue, creating anything from a tingly sensation to violent convulsions. Penfield developed a surgical technique that involved removing a section of the skull and probing the brain with electrodes to locate the diseased area. The damaged tissue can be removed, but adjacent functional tissues must not be extracted. Since the brain does not contain any sensory receptors, the surgeon can be guided by the conscious patient—only a local anesthetic is required. During the mapping process, the surgeon stimulates various areas of the brain, and the patient talks to an observer to ensure preservation of functional tissue. Penfield noted that some patients would begin to laugh as they recalled past events. One woman heard songs as clearly as if a record player were in the operating room. Incredibly, the song stopped once the electrode was removed from that area of the brain.

Figure 6

In the 1940s and early 1950s, Dr. Wilder G. Penfield studied brain structure and function in living humans using an elaborate surgical procedure.

Case Study *Phineas Gage*

In September 1848, a thunderous explosion shook the ground near the small town of Cavendish, Vermont. Phineas Gage, the 25-year-old foreman of a railway construction crew, lay on the ground impaled by a tamping iron. Apparently Gage had accidentally set off blasting caps by tamping them with a large iron bar. A closer examination revealed that the metre-long bar had entered his skull immediately below the left eye and exited through the top of the skull (**Figure 7**). Incredible as it may seem, Phineas Gage recovered from the explosion and lived for another 12 years. He showed no signs of physical impairment. His vision, hearing, balance, and speech remained intact. However, he did experience one change: the once quiet and thoughtful Phineas became irresponsible and short-tempered. Spontaneous temper tantrums would send him into a fit of profanity. What could have triggered such changes?

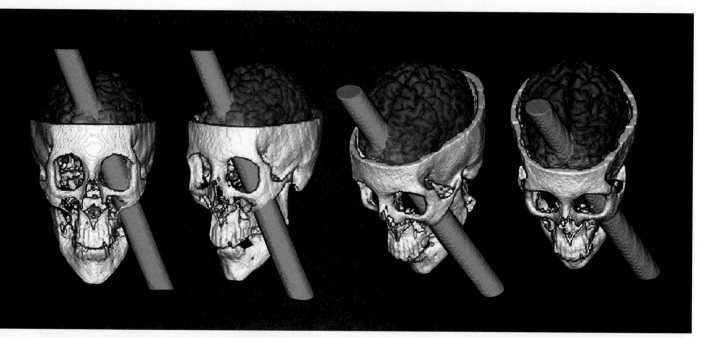

Figure 7
Computer model of the skull of Phineas Gage shown from four angles

▶ Case Study Questions

Understanding Concepts

1. Which lobe of Gage's brain was damaged?

Applying Inquiry Skills

2. Provide a hypothesis to explain why Phineas Gage's personality changed. How would you test your hypothesis?

3. In 1949, Portuguese neurologist Antonio Egas Moniz received the Nobel Prize for his surgical procedure—known as prefrontal leukotomy—in which some of the nerve tract between the thalamus and the frontal lobes is severed. Why might a physician attempt such an operation?

Neuroimaging: Viewing the Living Brain

It was once very difficult to observe brain structure and function in living humans. Noninvasive imaging techniques are now available to researchers studying normal body functions and to physicians diagnosing various disorders, including cancer. These techniques are especially useful in neuroimaging–studying the brain.

One imaging technology widely used in brain research is called positron-emission tomography (PET). PET scans can reveal various physiological and biochemical processes in the body. In preparation for a PET scan, water, glucose, or another molecule (depending on the process to be studied) is labelled with a radioactive isotope and injected into the patient's bloodstream. The radioactive compound goes to the most active areas of the brain, allowing the researcher to map the brain as it is performing particular tasks. The positrons interact with oppositely charged electrons on the body's atoms, and the resulting radiation is detected by a PET camera connected to a computer. The computer then creates a brain map that shows localized brain activity under different conditions. Physicians use the PET scanner to evaluate brain disorders as well as heart problems and certain types of cancer.

Magnetic resonance imaging (MRI) and functional MRI (fMRI) are computer-aided imaging techniques that can produce two- and three-dimensional pictures of the brain. Functional MRI differs from MRI in that it measures brain function rather than structure. MRI takes advantage of the behaviour of hydrogen atoms in water molecules. The nuclei of hydrogen atoms are usually oriented in random directions. MRI uses powerful magnets to align the nuclei, then knocks them out of alignment with a brief pulse of radio waves. Still under the magnets' influence, the hydrogen nuclei spring back into alignment, emitting faint radio signals that are detected by the MRI scanner and then translated into a computer image (**Figure 8**). Soft tissues have a relatively high water content and appear more opaque in the images than dense structures, such as bone, which contain relatively little water. For this reason, MRI is useful for detecting problems in the brain and spinal cord, which are surrounded by bone.

A third imaging technique, computerized tomography (CT), produces images of thin X-ray sections through the body. As the patient is slowly moved through the CT machine, an X-ray source circles around the body, illuminating successive sections from various angles. A computer then produces high-resolution video images of the sections, which can be studied individually or combined into various three-dimensional views. CT is especially useful for detecting ruptured blood vessels in the brain.

The main limiting feature of imaging technology is its cost. Imaging machines cost millions of dollars to buy, maintain, and operate; therefore, access is usually limited to major urban and teaching centres.

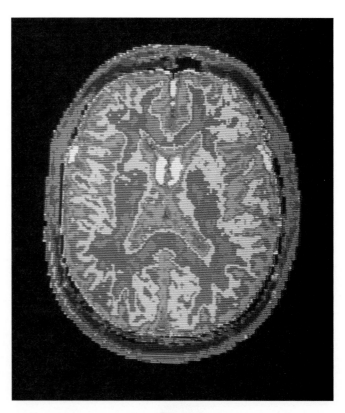

Figure 8
MRI image of a normal human brain

SUMMARY *The Central Nervous System*

Table 3

Structure	Function
meninges	• protective membranes that surround the brain and spinal cord
cerebrospinal fluid	• circulates between the innermost and middle membranes of the brain and spinal cord • acts as a transport medium and shock absorber (cushion)
olfactory lobes	• areas of the brain that detect smell
cerebrum	• the largest and most highly developed part of the human brain • stores sensory information and initiates voluntary motor activities
cerebral cortex	• the outer lining of the cerebral hemispheres
corpus callosum	• a nerve tract that joins the two cerebral hemispheres
cerebellum	• the region of the brain that coordinates muscle movement
pons	• the region of the brain that acts as a relay station by sending nerve messages between the cerebellum and the medulla
medulla oblongata	• the region of the hindbrain that joins the spinal cord to the cerebellum • the site of autonomic nerve control

► Section 9.3 Questions

Understanding Concepts

1. List the four regions of the cerebral cortex and state the function of each.

2. Explain the medical importance of brain-mapping experiments.

3. Name the different areas of the brain labelled on **Figure 9** and indicate the functions of the different areas.

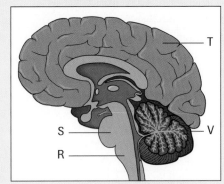

Figure 9
Human brain

4. A physician makes an incision completely through the corpus callosum. How might this affect the patient?

5. Compare the evolutionary development of the human brain to that of a fish's brain. What special advancements are noted in the human brain?

6. The old saying that "an elephant never forgets" seems to have some basis. What area of the brain would you examine to begin researching this question? Explain why.

Applying Inquiry Skills

7. Studies have been conducted to attempt to demonstrate the mental or reasoning superiority of some people based on skull size. Critique these studies.

Making Connections

8. Conduct an information search on strokes, including the causes, risk factors, warning signs, and effects on the various body systems. Include statistics on the incidence of strokes in Canada and on some lifestyle strategies for reducing the risk of stroke. Prepare a poster summarizing your research results in the form of charts, graphs, and tables. Be prepared to share your findings with your class. Search for information in newspapers, periodicals, CD-ROMs, and on the Internet.

 www.science.nelson.com

9. The EEG has been used to legally determine death. Although the heart may continue to beat, the cessation of brain activity signals legal death. Ethical problems arise when some brain activity remains despite massive damage. Artificial resuscitators can assume the responsibilities of the medulla oblongata and regulate breathing movements. Feeding tubes can supply food, and catheters can remove wastes when voluntary muscles can no longer be controlled. The question of whether life should be sustained by artificial means has often been raised. Should a machine like the EEG be used to define the end of life? Explain your answer.

The autonomic nervous system is part of the peripheral nervous system (**Figure 1** in section 9.1). This regulatory system works together with the endocrine system in adjusting the body to changes in the external or internal environment. All autonomic nerves are motor nerves that regulate the organs of the body without conscious control. Motor somatic nerves, by contrast, lead to muscles and are regulated by conscious control.

Rarely do you consciously direct your breathing movements. Blood carbon dioxide and oxygen levels are monitored throughout the body. Once carbon dioxide or oxygen levels exceed or drop below the normal range, autonomic nerves act to restore homeostasis. The autonomic system maintains the internal environment of the body by adapting to the changes and demands of an external environment. During emergencies, your autonomic nervous system diverts blood flow from your digestive organs to the skeletal muscles, increases your heart and breathing rates, and increases your visual field by causing the pupils of your eyes to dilate.

The autonomic system is made up of two distinct, and often opposing, units, the **sympathetic nervous system** and **parasympathetic nervous system** (**Figure 1**, page 436). The sympathetic system prepares the body for stress, while the parasympathetic system restores normal balance. **Table 1** summarizes the effects of the autonomic nervous system. Sympathetic and parasympathetic nerves also differ in anatomy. Sympathetic nerves have a short preganglionic nerve and a longer postganglionic nerve; the parasympathetic nerves have a long preganglionic nerve and a shorter postganglionic nerve. The preganglionic nerves of both systems release acetylcholine, but the postganglionic nerve from the sympathetic system releases norepinephrine. The postganglionic nerves from the parasympathetic system release acetylcholine. The sympathetic nerves come from the thoracic vertebrae (ribs) and lumbar vertebrae (small of the back). The parasympathetic nerves exit directly from the brain or from either the cervical (the neck area) or caudal (tailbone) sections of the spinal cord. Nerves that exit directly from the brain are referred to as cranial nerves. An important cranial nerve is the **vagus nerve** (*vagus* meaning "wandering"). Branches of the vagus nerve innervate the heart, bronchi of the lungs, liver, pancreas, and the digestive tract.

sympathetic nervous system nerve cells of the autonomic nervous system that prepare the body for stress

parasympathetic nervous system nerve cells of the autonomic nervous system that return the body to normal resting levels after adjustments to stress

vagus nerve major cranial nerve that is part of the parasympathetic nervous system

DID YOU KNOW ?

How Polygraphs Work
Lie detectors (also known as polygraphs) monitor changes in the activity of the sympathetic nervous system. One component of a lie detector, the galvanic skin response, checks for small changes in perspiration. In theory, a stressful situation, such as lying, would cause the stimulation of sympathetic nerves, which, in turn, would activate the sweat glands. Increased breathing and pulse rates are also monitored by lie detectors. Because lie detectors cannot always differentiate between lying and other stressful situations, they are not considered 100% accurate.

Table 1 Some Effects of the Autonomic Nervous System

Organ	Sympathetic	Parasympathetic
heart	increases heart rate	decreases heart rate
digestive	decreases peristalsis	increases peristalsis
liver	increases the release of glucose	stores glucose
eyes	dilates pupils	constricts pupils
bladder	relaxes sphincter	contracts sphincter
skin	increases blood flow	decreases blood flow
adrenal gland	causes release of epinephrine	no effect

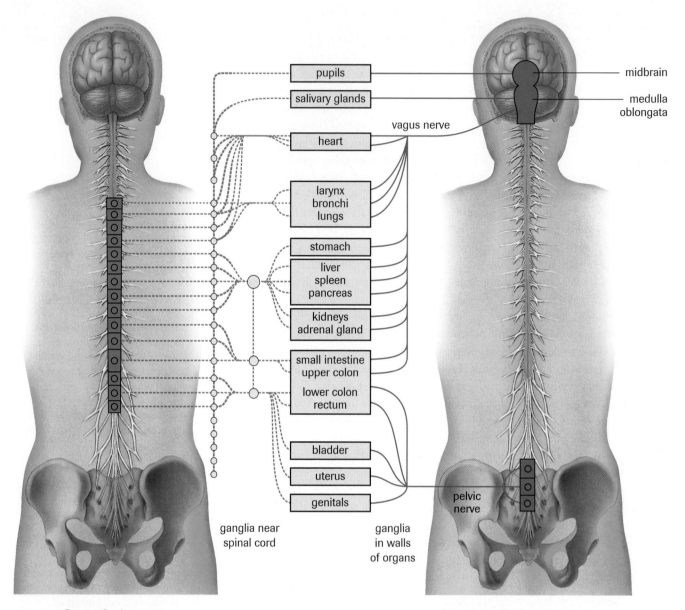

Sympathetic nerves

Parasympathetic nerves

Labels in figure:
pupils
salivary glands
heart — vagus nerve
larynx
bronchi
lungs
stomach
liver
spleen
pancreas
kidneys
adrenal gland
small intestine
upper colon
lower colon
rectum
bladder
uterus
genitals
ganglia near spinal cord
ganglia in walls of organs
midbrain
medulla oblongata
pelvic nerve

Figure 1

The sympathetic nerves are shown in green, and the parasympathetic nerves are shown in purple.

endorphins natural painkillers belonging to a group of chemicals called neuropeptides; contain between 16 and 31 amino acids

enkephalins natural painkillers belonging to a group of chemicals called neuropeptides; contain 5 amino acids and are produced by the splitting of larger endorphin chains

Natural and Artificial Painkillers

Have you ever heard a runner talk about the euphoria he or she feels while running? This feeling is produced by a group of natural painkillers called **endorphins** and **enkephalins**, which are manufactured by the brain. Pain is interpreted by specialized cells in the substantia gelatinosa (SG)—a band of gelatinous grey matter in the dorsal part of the spinal cord. When stimulated, the SG cells produce a neurotransmitter that "informs" the injured organ or tissue of the damage. The greater the amount of pain transmitter attached to the injured organ, the greater the perception of pain. However, when endorphins and enkephalins attach to the receptor sites on the SG cell, the pain transmitter is not produced and pain is reduced.

Opiates (also called sedative narcotics) such as heroin, codeine, and morphine work in much the same way as endorphins (**Figure 2**). Opiates attach to the SG neurons in the central nervous system, preventing the production of the pain transmitters. Heroin and opium not only reduce pain but also create a feeling of tranquility. The intake of opiates causes the production of the body's natural painkillers to decrease. Therefore, to achieve a consistent effect, the user must continue to take the drug. When use of the drug stops, the SG receptors are soon vacant and the pain transmitter is produced in abundance.

Drugs that act as depressants, such as Valium and Librium, enhance the action of inhibitory synapses. The synthesis of inhibitory neurotransmitters like gamma-amino-butyric acid (GABA) often increases under the influence of these drugs. Alcohol differs from other depressants in that it does not act directly on the synapse. Correctly categorized as a depressant, alcohol acts directly on the plasma membrane to increase threshold levels.

Drugs such as ethanol, barbiturates, and benzodiazepines are among the most widely used and abused of this class of drugs. However, little information exists on the effects of long-term use, such as tolerance, dependence, drug-withdrawal seizures, or brain damage. More insight into the molecular activities of such drugs is essential before the effects of long-term exposure can be fully understood.

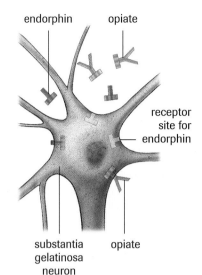

Figure 2
Opiates have a similar structure to endorphins. This allows opiates to occupy SG neuron receptor sites designed for endorphins.

 **SUMMARY**

Homeostasis and the Autonomic Nervous System

- The autonomic nervous system is a motor system.
- The sympathetic nervous system prepares the body for stress; the parasympathetic system returns the body to a resting state.
- The neurotransmitters released from the sympathetic system are acetylcholine and norepinepherine; the parasympathetic system releases only acetylcholine.
- Endorphins and enkephalins are natural painkillers produced by the body.

▶ Section 9.4 Questions

Understanding Concepts

1. Compare the structure and function of autonomic and somatic nerves.

2. State the two divisions of the autonomic nervous system and compare their structures and functions.

3. What are the functions of the vagus nerve?

4. What are endorphins? How do they work?

Applying Inquiry Skills

5. Many prescription drugs affect the autonomic nervous system. **Table 2** describes the action of four different drugs.
   (a) Which drug should not be taken by someone who has high blood pressure? Give reasons for your answer.
   (b) A patient who has taken too much neostigmine is admitted to hospital. What symptoms would be displayed?

Making Connections

6. Draw a diagram that shows how a drug such as Valium or Librium might work.

7. Research has revealed a link between acupuncture and endorphins. It is believed that acupuncture needles stimulate the production of pain-blocking endorphins. Although acupuncture is a time-honoured technique in the East, it is still considered to be on the fringe of modern Western medical practice. Why have Western scientists been so reluctant to accept acupuncture?

Table 2 Drug Actions

Drug	Action
pilocarpine	produces effects similar to the parasympathetic nervous system
resperine	inhibits the activity of the sympathetic nervous system by preventing the synthesis of norepinephrine
ephedrine	stimulates the release of norepinephrine from postganglionic nerves
neostigmine	blocks the action of cholinesterase at neuromuscular junctions

A stimulus is a form of energy. Sensory receptors convert one source of energy into another. For example, taste receptors in your tongue convert chemical energy into a nerve action potential, a form of electrical energy. Light receptors in the eye convert light energy into electrical energy. Balance receptors of the inner ear convert gravitational energy and mechanical energy into electrical energy.

Sensory receptors are highly modified ends of sensory neurons. Often, different sensory receptors and connective tissues are grouped within specialized sensory organs, such as the eye or ear. This grouping of different receptors often amplifies the energy of the stimulus to ensure that the stimulus reaches threshold levels. **Table 1** lists different types of sensory receptors found within the body.

sensory receptors modified ends of sensory neurons that are activated by specific stimuli

Table 1 The Body's Sensory Receptors

Receptor	Stimulus	Information provided
taste	chemical	presence of specific chemicals (identified by taste buds)
smell	chemical	presence of chemicals (detected by olfactory cells)
pressure	mechanical	movement of the skin or changes in the body surface
proprioceptors	mechanical	movement of the limbs
balance (inner ear)	mechanical	body movement
outer ear	sound	sound waves
eye	light	changes in light intensity, movement, and colour
thermoregulators	heat	flow of heat

A network of touch, high-temperature, and low-temperature receptors are found throughout the skin. Brain-mapping experiments indicate that sensations occur in the brain and not the receptor itself. When the neurotransmitter released by the sensory neuron is blocked, the sensation stops. The brain registers and interprets the sensation. When the sensory region of the cerebral cortex is excited by mild electrical shock at the appropriate spot, the sensation returns even in the absence of the stimulus.

Despite an incredible collection of specialized sensory receptors, much of your environment remains undetected. What you detect are stimuli relevant to your survival. For example, consider the stimuli from the electromagnetic spectrum. You experience no sensation from radio waves, or from infrared or ultraviolet wavelengths. Humans can only detect light of wavelengths between 350 nm and 800 nm. Your range of hearing, compared with that of many other species, is also limited.

Most animals can tolerate a wide range of temperatures, but are often harmed by rapid temperature changes. For example, a rapid change in temperature of 4°C will kill some fish. Humans have also died from an unexpected plunge in very cold or very hot water. This principle was introduced at the beginning of Chapter 7 with the description of the "hot frog" experiment. If a frog is placed in a beaker of water above 40°C, the frog will leap out immediately. However, if the frog is placed in room-temperature water, and the temperature is slowly elevated, it will remain in the beaker. The frog's skin receptors have had time to adjust.

DID YOU KNOW ?

Seeing Stars

Occasionally a sensory receptor can be activated by stimuli that it was not designed to detect. Boxers who receive a blow to the eye often see stars. The pressure of the blow stimulates the visual receptors at the back of the eye, and the blow is interpreted as light. Similarly, a blow near the temporal lobe can often be interpreted as a bell ringing.

Sensory adaptation occurs once the receptor becomes accustomed to the stimulus. The neuron ceases to fire even though the stimulus is still present. The adaptation seems to indicate that the new environmental condition is not dangerous. The same principle of adaptation can be applied to touch receptors in the skin. Generally, the receptors are only stimulated when clothes are put on or taken off. Sensory information assuring you that your clothes are still on your body is usually not required.

sensory adaptation occurs once you have adjusted to a change in the environment; sensory receptors become less sensitive when stimulated repeatedly

The Structure of the Eye

The eye comprises three separate layers: the sclera, the choroid layer, and the retina (**Figure 1**). The **sclera** is the outermost layer of the eye. Essentially a protective layer, the white fibrous sclera maintains the eye's shape. The front of the sclera is the clear, bulging **cornea**, which acts as the window to the eye by bending light toward the pupil. Like all tissues, the cornea requires oxygen and nutrients. However, the cornea is not supplied with blood vessels, which would cloud the transparent cornea. Most of the oxygen is absorbed from gases dissolved in tears. Nutrients are supplied by the **aqueous humour**, transparent fluid in a chamber behind the cornea.

sclera outer covering of the eye that supports and protects the eye's inner layers; usually referred to as the white of the eye

cornea transparent part of the sclera that protects the eye and refracts light toward the pupil of the eye

aqueous humour watery liquid that protects the lens of the eye and supplies the cornea with nutrients

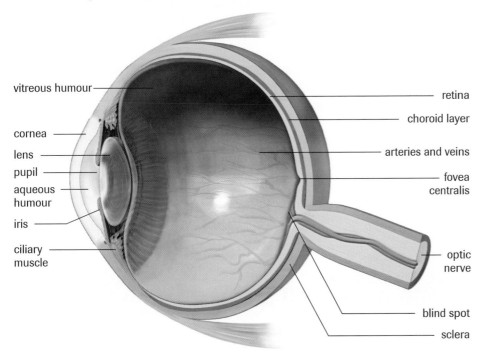

vitreous humour

cornea

lens

pupil

aqueous humour

iris

ciliary muscle

retina

choroid layer

arteries and veins

fovea centralis

optic nerve

blind spot

sclera

Figure 1
Simplified diagram of the human eye

The middle layer of the eye is called the **choroid layer**. Pigmented granules within the layer prevent light that has entered the eye from scattering. Toward the front of the choroid layer is the **iris**. The iris is composed of a thin circular muscle that acts as a diaphragm, controlling the size of the pupil, the opening formed by the iris that allows light into the eye. The lens, which focuses the image on the retina, is found in the area immediately behind the iris. Ciliary muscles, attached to ligaments suspended from the dorsal and ventral ends of the lens, alter the shape of the lens. A large chamber behind the lens, called the vitreous humour, contains a cloudy, jellylike material that maintains the shape of the eyeball and permits light transmission to the retina.

The innermost layer of the eye is the **retina**, which comprises three different layers of cells: light-sensitive cells, bipolar cells, and cells from the optic nerve. The light-sensitive cell layer is positioned next to the choroid layer. There are two different types of light-sensitive

choroid layer middle layer of tissue in the eye that contains blood vessels and dark pigments that absorb light to stop reflection

iris opaque disk of tissue surrounding the pupil that regulates amount of light entering the eye

retina innermost layer of tissue at the back of the eye containing photoreceptors

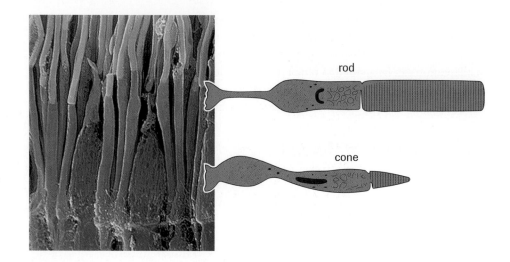

Figure 2
Rods and cones are light-sensitive cells in the retina.

rods photoreceptors that operate in dim light to detect light in black and white

cones photoreceptors that operate in bright light to identify colour

fovea centralis area at centre of retina where cones are most dense and vision is sharpest

cells: the **rods** and the **cones** (**Figure 2**). The rods respond to low-intensity light; the cones, which require high-intensity light, identify colour. Both rods and cones act as the sensory receptors. Once excited, the nerve message is passed from the rods and cones to the bipolar cells, which, in turn, relay the message to the cells of the optic nerve. The optic nerve carries the impulse to the central nervous system (**Figure 3**).

In the centre of the retina is a tiny depression referred to as the **fovea centralis**. The most sensitive area of the eye, it contains cones packed very close together. When you look at an object, most of the light rays fall on the fovea centralis. Rods surround the fovea, which could explain why you may see an object from the periphery of your visual field without identifying its colour. There are no rods or cones in the area in which the optic nerve comes in contact with the retina. Because of this absence of photosensitive cells, this area is appropriately called the blind spot.

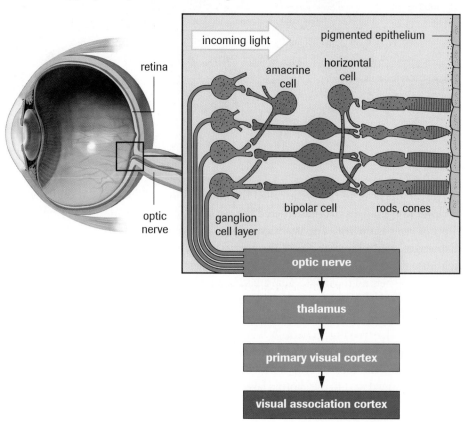

Figure 3
The pathway leading from the retina to the brain

Table 2 Parts of the Eye

Structure	Function
Outer layer	
sclera	• supports and protects delicate photocells
cornea	• refracts light toward the pupil
Middle layer	
aqueous humour	• supplies cornea with nutrients and refracts light
choroid	• contains pigments that prevent scattering of light in the eye by absorbing stray light • also contains blood vessels
iris	• regulates the amount of light entering the eye
vitreous humour	• maintains the shape of the eyeball and permits light transmission to the retina
lens	• focuses the image on the retina
pupil	• the opening in the iris that allows light into the eye
Inner layer	
retina	• contains the photoreceptors
rods	• used for viewing in dim light
cones	• identify colour
fovea centralis	• most light-sensitive area of the retina • contains only cones
blind spot	• where the optic nerve attaches to the retina

Light and Vision

The Greek philosopher Democritus first speculated in the 5th century B.C. about how the eye worked. In his hypothesis, matter was composed of indivisible particles, which he called *atoms*. He reasoned that once the atoms touched the eye, they were carried to the soul and, therefore, could be viewed. Empedocles, a contemporary of Democritus, proposed a different theory of vision. He believed that matter was composed of four essential elements: earth, air, water, and fire. Vision must be linked to fire because it provided light. According to Empedocles, light radiated from the eye and struck objects, making them visible.

Galen, the Roman physician, combined Democritus's theory of the eye and the soul with Empedocles's notion that light was emitted from the eye. Galen believed that the optic nerve conducted visual spirits from the brain. The spiritual link between the soul and vision remains today, for example, in the expression "evil eye."

Today, scientists accept two complementary theories of light first proposed in the 17th century by Sir Isaac Newton and Dutch physicist Christiaan Huygens. Particles of light (called photons) travel in waves of various lengths. Light enters the eye as it is reflected or transmitted from objects. In many ways the eye operates by the same principle as a camera (**Figure 4**). Both camera and eye are equipped with lenses that focus images. The diaphragm of a camera opens and closes to regulate the amount of light entering the camera. The iris of the eye performs an equivalent function. The image of the camera is focused on a chemical emulsion—the film. Similarly, the image of the eye is focused on the retina.

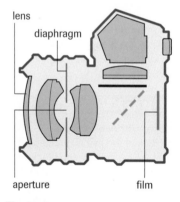

Figure 4
The lens focuses images and the diaphragm controls the amount of light that enters the camera.

Afterimages

Have you ever noticed a trailing blue or green line that stays in your vision after you look into a camera flash? What you see is an afterimage. There are two different types of afterimages: positive and negative. The positive afterimage occurs after you look into a bright light and then close your eyes. The image of the light can still be seen even though your eyes are closed. The more dramatic negative afterimage occurs when the eye is exposed to bright coloured light for an extended period of time.

Focusing the Image

As light enters the eye, it is first bent toward the pupil by the cornea. Light waves normally travel in straight lines and slow down when they enter more dense materials like the cornea. The slowing of light by a denser medium causes bending, which is called refraction. The cornea directs light inward toward the lens, resulting in further bending. Because the biconvex lens is symmetrically thicker in the centre than at its outer edges, light is bent to a focal point. An inverted image is projected on the light-sensitive retina.

Ciliary muscles control the shape of the lens, and suspensory ligaments maintain a constant tension. When close objects are viewed, the ciliary muscle contracts, the tension on the ligaments decreases, and the lens becomes thicker. The thicker lens provides additional bending of light for near vision. For objects that are farther away, relaxation of the ciliary muscles increases the tension of the ligaments on the lens, and the lens becomes thinner. The adjustment of the lens to objects near and far is referred to as **accommodation**. Objects 6 m from the viewer need no accommodation.

The importance of the accommodation reflex becomes more pronounced with age. Layers of transparent protein covering the lens increase throughout your life, making the lens harder. As the lens hardens, it loses its flexibility. By the time you reach age 40, near-point accommodation has diminished and may begin to hinder reading.

A secondary adjustment occurs during the accommodation reflex. When objects are viewed from a distance, the pupil dilates in an attempt to capture as much light as possible. When objects are viewed close up, the pupil constricts in an attempt to bring the image into sharp focus. Test this for yourself by looking at the print on this page with one eye. Move your head toward the book until the print gets very blurry. Now crook your finger until you have a small opening and look through it. Gradually make the opening smaller. The image becomes sharper. Light passes through a small opening and falls on the most sensitive part of the retina, the fovea centralis. The Inuit were aware of this principle when they made eyeglasses for their elders by drilling holes in whalebone. Light passing through the narrow openings resulted in a sharper focus.

Vision Defects

Glaucoma is caused by a buildup of aqueous humour in the anterior chamber of the eye. Although a small amount of the fluid is produced each day, under normal conditions tiny ducts drain any excess. Blockage of these drainage ducts causes the fluid pressure to collapse blood vessels in the retina. Without a constant supply of nutrients and oxygen, neurons soon die and blindness may result.

▶ *TRY THIS* activity

Afterimages

Stare at the cross in **Figure 5** with one eye for 30 s, and then stare at a bright white surface for at least 30 s. The colours will reverse. The afterimage is believed to be caused by fatigue of that particular type of cone in that area of the retina. The horizontal red cones become fatigued, but the complementary green cones continue to fire. The opposite effect occurs for the vertical bar.

Figure 5
The red bar produces a green afterimage; the green bar produces a red afterimage.

---

accommodation adjustments made by the lens and pupil of the eye for near and distant objects

---

DID YOU KNOW ❓

Seeing Ultraviolet Light
The lens of the eye is not clear. A slight yellow colouration blocks out rays from the ultraviolet end of the electromagnetic spectrum. As you age, your lenses become thicker and more yellow, making you less able to see wavelengths from the ultraviolet end of the spectrum. Cataract patients who have had their entire lens removed and replaced with a clear plastic one are able to see into the ultraviolet wavelengths or "black light."

---

glaucoma disorder of the eye caused by buildup of fluid in the chamber anterior to the lens

---

Problems may arise with the lens. Occasionally, the lens becomes opaque and prevents some of the light from passing through. The condition is known as a **cataract**. A traditional solution to the problem has been to remove the lens and to fit the patient with strong eyeglasses.

In most people, the lens and cornea are symmetrical. Incoming light is refracted along identical angles for both the dorsal and ventral surfaces, forming a sharp focal point. In some individuals, however, the lens or cornea is irregularly shaped. This condition is called **astigmatism**.

Two of the more common vision defects are **nearsightedness** (also known as myopia) and **farsightedness** (hyperopia). Nearsightedness occurs when the eyeball is too long. Since the lens cannot flatten enough to project the image on the retina, the distant image is instead brought into focus in front of the retina. Someone who is nearsighted is able to focus on close objects, but has difficulty seeing objects that are distant. Glasses that contain a concave lens can correct nearsightedness (**Figure 7**). Farsightedness is caused by an eyeball that is too short, causing distant images to be brought into focus behind the retina, instead of on it. A farsighted person is able to focus on distant objects, but has trouble seeing objects that are close up. Farsightedness can be corrected by glasses that have a convex lens.

▶ **TRY THIS** *activity*

Testing for Astigmatism

The chart in **Figure 6** will help you determine whether you have astigmatism. Cover one eye and look at the chart. If you have cornea astigmatism, the lines along one plane will appear sharp, but those at right angles will appear fuzzy. Repeat the test with the other eye.

Figure 6
A test for cornea astigmatism

cataract condition that occurs when the lens or cornea becomes opaque, preventing light from passing through

Nearsightedness (myopia)

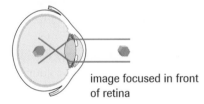

image focused in front of retina

Astigmatism

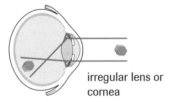

irregular lens or cornea

Farsightedness (hyperopia)

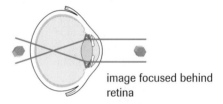

image focused behind retina

Correction for nearsightedness

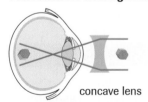

concave lens

Correction for astigmatism

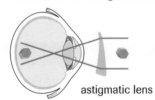

astigmatic lens

Correction for farsightedness

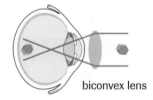

biconvex lens

Figure 7
Visual defects can be improved with corrective lenses.

astigmatism vision defect caused by abnormal curvature of surface of the lens or cornea

nearsightedness condition that occurs when the image is focused in front of the retina

farsightedness condition that occurs when the image is focused behind the retina

Frontiers of Technology: Corneal Surgery

Surgery for treating nearsightedness was first developed in Russia in the mid-1970s by Dr. Svyatoslav Fyodorov. He was inspired by a Russian teenager whose glasses had shattered during a fight, badly cutting his cornea. Remarkably, the eye healed and the boy's myopia seemed cured. An alteration of the cornea had corrected the myopia. Dr. Fyodorov soon developed a procedure called radial keratotomy for correcting myopia. A specially designed scalpel was used to make incisions in the cornea in a spokelike arrangement. The incisions flattened the outer cornea, altering its curvature and enhancing vision.

With the development of laser surgery in the early 1980s, new, less invasive, procedures were developed (**Figure 8**, page 444). Two procedures for nearsightedness are common today, both involving the use of computers and special cold lasers. In photorefractive keratotomy (PRK), a laser removes a microscopic layer of tissue to resculpt the cornea. This procedure changes the way the cornea refracts light and leads to a sharper focus. No incision

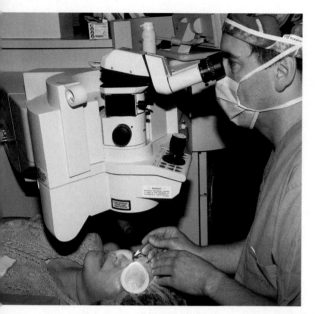

Figure 8
Laser eye surgery procedure

is required. The procedure usually takes less than two minutes, the post-surgical recovery time is less than that for radial keratotomy, and the results are longer lasting. Laser in-situ keratomileusis (LASIK) involves cutting a flap into the centre of the cornea, then flattening the layers underneath with pulses from an excimer laser. The flap is then resealed.

For mild forms of nearsightedness, corneal ring implants can be used to flatten the surface of the cornea without having to make any incisions. The rings are placed in "tunnels" at the edge of the cornea, changing its thickness. Corneal rings can be removed or upgraded at any time, but the recovery period is longer than that for laser procedures.

Another option that is currently under investigation is the use of implantable contact lenses. Two weeks before surgery, the patient undergoes an iridotomy—a small hole is created in the iris—to ensure that no fluid will build up behind the lens once it is in place. During surgery, an incision is made at the edge of the cornea and the corrective lens is inserted behind the cornea but in front of the natural lens of the eye. Implantable contact lenses are for extreme cases only.

ACTIVITY 9.5.1

Human Vision (p. 453)
Is your vision 20/20? Did you know it is possible to see even better than that? In this activity, you will test your eyesight and learn more about human vision.

SUMMARY *Sensory Information: Vision*

- Sensory receptors are highly modified ends of sensory neurons.
- Sensory receptors convert one form of energy into another. For example, the eye converts light energy into an electrochemical impulse.
- Images are displayed on the retina. Rods are photosensitive receptors that detect images in dull light. Cones are photosensitive receptors that distinguish colour in bright light.
- Ciliary muscles change the shape of the lens. A thicker lens permits the greater bending of light for viewing near objects, while a more flattened lens is used to view distant objects.

▶ *Section 9.5 Questions*

Understanding Concepts

1. List the three layers of the eye and describe the function of each layer.

2. Indicate the function of each of the following parts of the eye: vitreous humour, aqueous humour, cornea, pupil, iris, rods, cones, fovea centralis, and blind spot.

3. What are accommodation reflexes?

4. Why do rods not function effectively in bright light?

5. Identify the causes for each of the following eye disorders: glaucoma, cataract, astigmatism, nearsightedness, and farsightedness.

6. Illustrate how corrective lenses provide for normal vision.

Making Connections

7. Laser surgery can provide a cure for myopia (shortsightedness), but skeptics argue that surgery has risks and that shortsightedness can be corrected with glasses. A "halo effect" (circles of light that can distort night vision) may result from laser surgery.

   (a) Research the halo effect. How often does it occur after laser surgery?

   (b) Should people who experience the halo effect be allowed to drive at night? Why or why not?

   (c) Do you believe that surgery should be attempted? Explain your answer.

   (d) Do you think this surgery should be covered by medicare? Justify your answer.

The ear (**Figure 1**) is associated with two separate functions: hearing and equilibrium. Sensory cells for both functions are located in the inner ear. Each tiny hair cell contains between 30 and 150 cilia, which respond to mechanical stimuli. Movement of the cilia causes the nerve cell to generate an impulse. The ear can be divided into three sections for study: the outer ear, the middle ear, and the inner ear. The outer ear comprises the **pinna**, the external ear flap, which collects the sound, and the **auditory canal**, which carries sound to the eardrum. The auditory canal is lined with specialized sweat glands that produce earwax, a substance that traps foreign invading particles and prevents them from entering the ear.

pinna outer part of the ear that acts like a funnel, taking sound from a large area and channelling it into a small canal

auditory canal carries sound waves to the eardrum

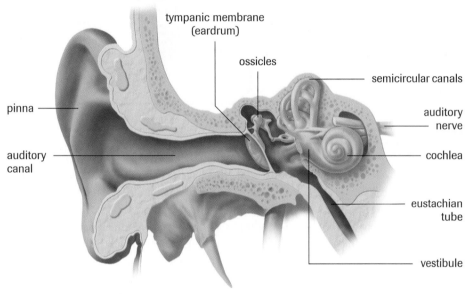

tympanic membrane (eardrum)

ossicles

semicircular canals

auditory nerve

pinna

auditory canal

cochlea

eustachian tube

vestibule

Figure 1
Anatomy of the human ear

The middle ear begins at the eardrum, known as the **tympanic membrane**, and extends toward the oval and round windows. The air-filled chamber of the middle ear contains three small bones, called the ear **ossicles**, which include the malleus (also called the hammer), the incus (the anvil), and the stapes (the stirrup). Sound vibrations that strike the eardrum are first concentrated within the solid malleus, and then transmitted to the incus, and finally to the stapes. The stapes strikes the membrane covering the **oval window** in the inner wall of the middle ear (**Figure 2**, page 446). The amplification of sound is accomplished, in part, by concentrating the sound energy from the large tympanic membrane to a smaller oval window.

The **eustachian tube** extends from the middle ear to the mouth and the chambers of the nose. Approximately 40 mm in length and 3 mm in diameter, the eustachian tube permits the equalization of air pressure. Have you ever noticed how your ears seem to pop when you go up in a plane? The lower pressure on the tympanic membrane can be equalized by reducing air pressure in the eustachian tube. Yawning, swallowing, and chewing gum allow air to leave the middle ear. An ear infection can cause the buildup of fluids in the eustachian tube and create inequalities in air pressure. Discomfort, temporary deafness, and poor balance can result.

The inner ear is made up of three distinct structures: the vestibule and the semicircular canals, which are involved with balance, and the cochlea, which is connected with hearing. The **vestibule**, connected to the middle ear by the oval window, houses two small sacs,

tympanic membrane thin layer of tissue, also known as the eardrum

ossicles tiny bones that amplify and carry sound in the middle ear

oval window oval-shaped hole in the vestibule of the inner ear, covered by a thin layer of tissue

eustachian tube air-filled tube of the middle ear that equalizes pressure between the external and internal ear

vestibule chamber found at the base of the semicircular canals that provides information about static equilibrium

semicircular canals fluid-filled structures within the inner ear that provide information about dynamic equilibrium

cochlea coiled structure of the inner ear that responds to various sound waves and converts them into nerve impulses

the utricle and saccule, which establish head position. There are three **semicircular canals**, arranged at different angles, and the movement of fluid in these canals helps you identify body movement. The **cochlea** (**Figure 2**) is shaped like a spiralling snail's shell and contains two rows of specialized hair cells that run the length of the inner canal. The hair cells identify and respond to sound waves of different frequencies and intensities and converts them into nerve impulses.

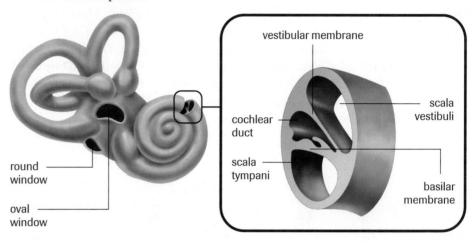

Figure 2
Sound waves are transformed into membrane vibrations in the cochlea.

Hearing and Sound

Sound is a form of energy. Like light, thermal energy, and various forms of chemical energy, sound energy must be converted into an electrical impulse before you can interpret it. The sensitivity of the ear can be illustrated by the fact that you can hear a mosquito outside your window, even though the energy reaching your ear is less than one quadrillionth of a watt. The average light in a house uses 60 W of energy.

The nature of sound became clearer after an experiment by British scientist Robert Boyle in 1660. Boyle wondered whether sound could be produced in a vacuum. In his experiment, Boyle suspended a watch with an alarm inside a bell jar. The alarm was set and the air was pumped out of the jar. Boyle waited for the alarm to ring at the time he had set it, but to his amazement the alarm could not be heard.

What Boyle learned from this experiment was that sound must travel through a medium. Although air is the most common medium, sound also travels through water and solids. Sound travels through water (1480 m/s at 20°C) four times faster than it does through air (370 m/s at 20°C). In fact, dolphins and porpoises rely on sound more than sight, especially in murky water. Solids transmit sound even more rapidly than liquids do; for example, sound travels in steel at a speed of 6096 m/s at 20°C. This explains why movie cowboys were shown listening for the approach of distant trains by putting their ears to a railway track.

Hearing begins when sound waves push against the eardrum, or tympanic membrane. The vibrations of the eardrum are passed on to the three bones of the middle ear: the malleus, the incus, and the stapes. Arranged in a lever system, the three bones are held together by muscles and ligaments. The bones concentrate and amplify the vibrations received from the tympanic membrane. The ossicles can triple the force of vibration from the eardrum; they move a shorter distance but exert greater force by concentrating the energy in a very small area.

Muscles that join the bones of the middle ear act as a safety net protecting the inner ear against excessive noise. Intense sound causes the tiny muscles—the smallest in your body—to contract, restricting the movement of the malleus and reducing the intensity of movement. At the same time, a second muscle contracts, pulling the stapes away from the oval window, thereby protecting the inner ear from powerful vibrations. Occasionally, the safety mechanism doesn't work quickly enough. The sudden blast from a firecracker can send the ossicles into wild vibrations before the protective reflex can be activated.

The oval window receives vibrations from the ossicles. As the oval window is pushed inward, the round window, located immediately below the oval window, moves outward. This triggers waves of fluid within the inner ear. The cochlea receives the fluid waves and converts them into electrical impulses, which you interpret as sound. The hearing apparatus within the cochlea is known as the **organ of Corti** and comprises a single inner row and three outer rows of specialized hair cells (**Figure 3**), anchored to the **basilar membrane**. Covered with a gelatinous coating, the hair cells respond to vibrations of the basilar membrane. Fluid vibrations move the basilar membrane, and the hair cells bend. The movement of the hair cells, in turn, stimulates sensory nerves in the basilar membrane. Auditory information is then sent to the temporal lobe of the cerebrum via the auditory nerves.

organ of Corti primary sound receptor in the cochlea

basilar membrane anchors the receptor hair cells in the organ of Corti

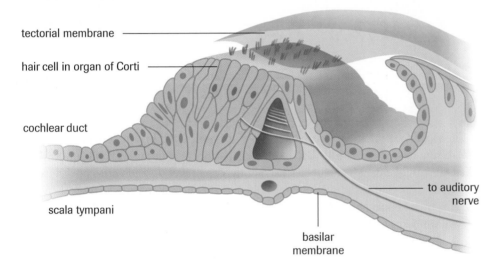

Figure 3
Hair cells within the cochlea are anchored to a basilar membrane. The free ends of the hair cells are embedded in a jellylike substance. Vibrations of the basilar membrane cause the hair cells to bend. Movement of the hair cells stimulates nerves in the basilar membrane.

The inner ear is able to identify both pitch and loudness because of the structure of the cochlea. The cochlea becomes progressively narrower as it coils, and the hairs become shorter. The narrowest area is activated by high-frequency sound waves, which contain enough energy to move the rigid hair receptors. The high-frequency waves are transformed into basilar membrane vibrations, which, in turn, cause the hair cells to move. The receptor hair cells trigger an action potential, which is carried to the area of the brain that registers high-pitched sounds. The high-pitched noise of a police siren dies out quickly in the narrow, rigid part of the cochlea. However, low-frequency waves move further along the cochlea, causing the hair cells in the wider, more elastic area to vibrate. This may explain why you can actually feel the vibrations from a bass guitar. The stimulation of nerve cells in different parts of the cochlea enables you to differentiate sounds of different pitch. Each frequency or pitch terminates in a specific part of the auditory section of the brain.

In addition to responding directly to sound energy, the basilar membrane can respond directly to mechanical stimulation. A jarring blow to the skull sets up vibrations that are passed on toward the cochlea. Aside from the sound created by the blow, the resulting mechanical vibrations of the skull can also be interpreted as sound.

Equilibrium

Balance consists of two components: static equilibrium and dynamic equilibrium. Static equilibrium involves movement along one plane, such as horizontal or vertical. Head position is monitored by two fluid-filled sacs called the saccule and the utricle. Tiny hair-like receptors are found within the saccule and utricle. Cilia from the hair cells are suspended in a gelatinous material that contains small calcium carbonate granules called **otoliths**. When the head is in the normal position, the otoliths do not move; however, when the head is bent forward, gravitational force acts on the otoliths, pulling them downward. The otoliths cause the gelatinous material to shift, and the hair receptors to bend (**Figure 4**). The movement of the hair receptors stimulates the sensory nerve, and information about head position is relayed to the brain for interpretation.

otoliths tiny stones of calcium carbonate embedded in a gelatinous coating within the saccule and utricle

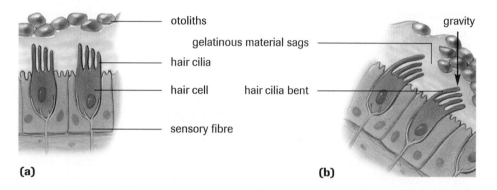

Figure 4
(a) When the head is in the erect position, the cilia from the hair cells remain erect.
(b) Movement of the head causes movement of the hair cells. Any movement of the cilia from the hair cells initiates nerve impulses.

The second aspect of balance, referred to as dynamic equilibrium, provides information during movement. While you are moving, balance is maintained by the three fluid-filled semicircular canals (**Figure 5**). Each of the canals is equipped with a pocket called an ampulla. Rotational stimuli cause the fluid in the semicircular canals to move, bending the cilia attached to hair cells in the ampullae. Once the hair cells bend, they initiate nerve impulses, which are carried to the brain. It is believed that rapid continuous movement of the fluids within the semicircular canals is the cause of motion sickness.

🔬 INVESTIGATION 9.6.1

Hearing and Equilibrium (p. 454)
Have you ever wondered how good your hearing actually is? Do you have trouble with motion sickness? Investigation 9.6.1 will help you learn more about hearing and equilibrium.

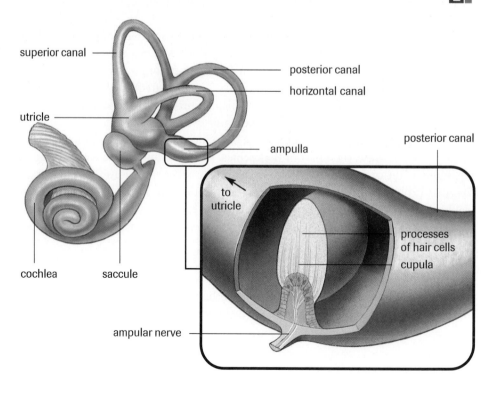

Figure 5
Three semicircular canals provide information about motion. Cilia attached to hair receptor cells in the ampulla respond to the movement of fluid in the semicircular canals.

▶ *EXPLORE an issue*

Debate: Hearing Damage

Canadian scientist Dr. Marek Roland-Mieszkowski has been crusading against loud music in dance clubs for over a decade, suggesting that clubs establish safety levels of no more than 85 dB (decibels). For club patrons (**Figure 6**), listening to loud music for several hours at a time, coupled with having to shout in someone's ear to be heard (often at levels of up to 140 dB), can cause pain, ringing in the ears, and/or impaired hearing for days afterward. Research also shows that hearing loss accumulates, so that more exposure to loud sounds leads to more hearing loss. Exposure to loud sound can also affect heart rate, vision, and reaction time.

Figure 6
Research shows that cumulative exposure to loud sounds, not age, is the major cause of hearing loss.

The most common type of hearing loss is caused by the destruction of the cilia on the hair cells of the cochlea (**Figure 7**). Although the cilia gradually wear away with aging, high-intensity sounds, such as loud music, can literally tear the cilia apart.

The sound level of a normal conversation ranges between 60 dB and 75 dB, heavy street traffic can reach 80 dB, and the crash of thunder usually registers at about 100 dB. Pain begins to be felt at 120 dB. In Canada, no law exists to protect people who frequent entertainment venues from harmful sound exposure. However, many people believe that everyone has a right to decide what sound level they listen to.

Statement
The sound levels at dance clubs should be regulated to protect the club patrons and staff.

- In a group, research the issue. Search for information in newspapers, periodicals, CD-ROMs, and on the Internet.

   www.science.nelson.com

- Prepare a list of points and counterpoints for your group to discuss. You might consider these questions:
  (a) There are laws protecting industrial workers from harmful sound exposure. Should there be similar laws for the entertainment industry?
  (b) Should club owners be held responsible for warning patrons and staff about the dangers of high sound levels?

- Suggest some improvements that would help alleviate the risk of hearing loss from long-term exposure to such high sound levels.

- Develop and reflect on your opinion.

- Prepare to defend your viewpoint in a class debate.

(a)   **(b)**   **(c)**

Figure 7
(a) Hair cells embedded in the basilar membrane
(b) Normal rows of hair cells in the cochlea
(c) Destruction of hair cells following 24-hour exposure to high-intensity music, with noise

SUMMARY

Table 1

Structure	Function
External ear	
pinna	• outer part of the external ear • amplifies sound by funnelling it from a large area into the narrower auditory canal
auditory canal	• carries sound waves to the tympanic membrane
Middle ear	
ossicles	• tiny bones that amplify and carry sound in the middle ear
tympanic membrane	• receives sound waves • also called the eardrum
oval window	• receives sound waves from the ossicles
eustachian tube	• air-filled tube of the middle ear that equalizes pressure between the outer and middle ear
Inner ear	
vestibule	• chamber at the base of the semicircular canals • concerned with static equilibrium
semicircular canals	• fluid-filled structures that provide information concerning dynamic equilibrium
cochlea	• coiled tube within the inner ear that receives sound waves and converts them into nerve impulses

▶ **Section 9.6 Questions**

Understanding Concepts

1. Briefly outline how the external ear, middle ear, and inner ear contribute to hearing.

2. What function do the tympanic membrane, ossicles, and oval window serve in sound transmission?

3. Categorize the following structures of the inner ear according to whether their functions relate to balance or hearing: organ of Corti, cochlea, vestibule, saccule, ampulla, semicircular canals, oval window, and round window.

4. Explain how the ligaments, tendons, and muscles that join the ossicles of the middle ear provide protection against high-intensity sound.

5. Differentiate between static equilibrium and dynamic equilibrium.

6. How do the saccule and utricle provide information about head position?

7. Describe how the semicircular canals provide information about body movement.

Applying Inquiry Skills

8. A scientist replaces ear ossicles with larger, lightweight bones. Would this procedure improve hearing? Support your answer.

Making Connections

9. Should individuals who refuse to wear ear protection while working around noisy machinery be eligible for medical coverage for the cost of hearing aids? What about rock musicians or other individuals who knowingly play a part in the loss of their own hearing? Justify your position.

10. Explore some current techniques for improving hearing. Search for information in newspapers, periodicals, CD-ROMs, and on the Internet.

     www.science.nelson.com

11. Research motion sickness, including its probable causes and some current solutions. You can begin your research on the Internet.

     www.science.nelson.com

INVESTIGATION 9.1.1

Inquiry Skills

● Questioning	● Planning	● Analyzing
● Hypothesizing	● Conducting	● Evaluating
○ Predicting	● Recording	● Communicating

Teaching a Planarian

Flatworms, such as the planarian, have a head region containing sensory cells and groups of longitudinal nerve cords. Normally, a planarian moves toward a habitat with limited light for protection from predators. However, you can teach the planarian to avoid shade by using a weak electrical charge from a 9-V battery. Do all planaria learn at the same rate? Find out by performing this investigation into the nervous system of a flatworm.

✋ **Do not allow the electrodes to touch each other when connected to the battery.**

Question
Do all planaria learn at the same rate?

Hypothesis
(a) Before the investigation, formulate a hypothesis. State what criteria you used to make your hypothesis.

Materials
9-V battery
2 electrodes
planarians
petri dish or glass tubing filled with water

Procedure
Electrical stimulation can be used to teach a planarian. Your procedure must be approved by your teacher before you begin the experiment. Use the experimental design provided in **Figure 1** to set up your procedure and to prepare data tables. Use the analysis questions as a checklist for your procedure.

Analysis
(b) Identify the dependent and independent variables.

(c) What factors must be controlled?

(d) Present your results in table format. What conclusions can you draw?

Evaluation
(e) Critique your experimental design. What improvements would you make if you were to repeat the experiment?

(f) Explain how the duration of the electrical stimuli would affect results.

(g) Explain how the frequency of electrical stimuli during the experiment would affect results.

Synthesis
(h) Do all planarians learn at the same rate?

(i) How would you modify the experiment to demonstrate that planarians can retain what has been learned?

(j) You want to perform a similar experiment to demonstrate that learning is a chemical process. What must you do to show that learning involves both nerves and transmitter chemicals?

(k) How can you teach the planarian to move toward the shade again?

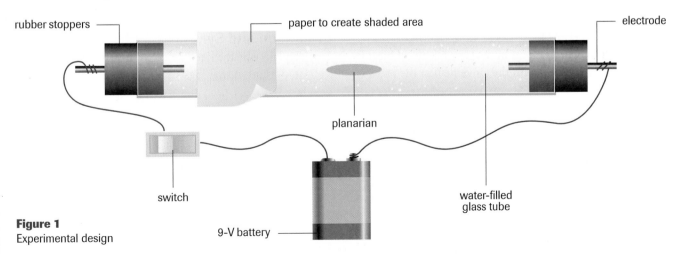

Figure 1
Experimental design

Reflex Arcs

Inquiry Skills

● Questioning	○ Planning	● Analyzing
● Hypothesizing	● Conducting	● Evaluating
● Predicting	● Recording	○ Communicating

Reflex arcs make up the neural circuit that travels through the spinal cord, providing a framework for reflex actions. Simple physical tests are used to check reflexes. In this investigation, you will observe the presence and strength of a number of reflex arcs.

Question

What is the advantage of being able to test different reflexes?

Hypothesis

(a) Using what you know about reflexes, predict what will happen in each part of the procedure.

(b) Formulate a hypothesis to explain your predictions. What criteria did you use to make your prediction?

Materials

rubber reflex hammer
penlight

Procedure

Part I: Knee Jerk

1. Find a partner. You will act as each other's subjects.

2. Have your subject sit on a chair with his or her legs crossed. The subject's upper leg should remain relaxed.

3. Locate the position of the kneecap and find the large tendon below the midline of the kneecap.

4. Using a reflex hammer, gently strike the tendon below the kneecap.

(c) Describe the movement of the leg.

5. Ask the subject to clench a book with both hands, then strike the tendon of the upper leg once again.

(d) Compare the movement of the leg while the subject is clenching the book with the movement in the previous procedure.

Part II: Achilles Reflex

6. Have the subject remove a shoe. Ask your subject to kneel on a chair so that his or her feet hang over the edge of the chair. Push the toes toward the legs of the chair and then lightly tap the Achilles tendon with the reflex hammer.

(e) Describe the movement of the foot.

Part III: Babinski Reflex

7. Now ask the subject to remove a sock. Have the subject sit in a chair, then place the heel of the bare foot on another chair for support. Quickly slide the reflex hammer along the sole of the subject's foot, beginning at the heel and moving toward the toes.

(f) Describe the movement of the toes.

Part IV: Pupillary Reflex

8. Have the subject close one eye for approximately 1 min. Ask him or her to open the closed eye. Compare the size of the pupils.

(g) Which pupil is larger?

9. Ask the subject to close both eyes for one min, then open both eyes. Shine a penlight in 1 of the eyes.

(h) Describe the changes you observe in the pupil.

10. Select a student with light-coloured eyes to be the subject. With at least two observers carefully watching the subject's eyes, gently stroke the fine hairs on the nape of the subject's neck.

(i) Describe any change in the pupils of the subject.

Analysis

(j) From your observations, formulate a hypothesis about the sequence of events that occur in the nervous system in each part of the procedure.

(k) How does the knee-jerk reflex change when the subject is clenching the book? Why do you think this is?

(l) What is the purpose of testing different reflexes?

Evaluation and Synthesis

(m) Explain why the knee-jerk and Achilles reflexes are important in walking.

(n) A person touches a stove, withdraws his or her hand, and then yells. Why does the yelling occur after the hand is withdrawn? Does the person become aware of the pain before the hand is withdrawn?

(o) While examining the victim of a serious car accident, a physician lightly pokes the patient's leg with a needle. The pokes begin near the ankle and gradually progress toward the knee. Why is the physician poking the patient? Why begin near the foot?

ACTIVITY 9.5.1

Human Vision

Do you know if your vision is 20/20 or better? In this activity, you will test your visual acuity and observe some other aspects of human vision.

Materials

Snellen eye chart
ruler
pencil

Procedure

Part I: Visual Acuity

Your ability to distinguish detail or visual acuity can be determined with a Snellen eye chart. Have you ever heard someone say that he or she has 20/20 vision? This means that a person is able to read the "20" line on an eye chart from a distance of 20 ft. (Charts that have been converted to metric measurements show normal vision to be 6/6. This indicates that the line marked "6" can be read from a distance of 6 m.) A person with 20/100 can only read line "100" from 20 ft, which is very poor vision. A person with 20/15 vision has better than 20/20 vision.

1. Stand 20 ft from the eye chart. (Stand 6 m away if a metric chart is available.) Cover your right eye and begin reading letters from each line while a partner observes.

(a) Record the visual acuity of your left eye.

2. Repeat the procedure by covering your left eye.

(b) Record the visual acuity of your right eye.

Part II: Blind Spot

3. Hold this text approximately 50 cm from your left eye and view the black cross in **Figure 2**. Close your right eye and begin moving the text toward your left eye while staring at the cross. Do not allow your eye to leave the cross! At some point you will not see the circle. The blind spot occurs when the image falls on the place where the optic nerve attaches to the retina.

Figure 2

(c) With the help of your partner, measure the distance of the book from your left eye.

4. Repeat the procedure and determine the blind spot for the right eye, keeping your eye on the circle this time.

(d) Measure the distance of the book from your right eye.

Part III: Dominant Eye

5. Face the corner of the room and view the upper corner. Holding a pencil in either your right or left hand, extend that arm, bringing the top of the pencil into the middle of your field of view. Align the pencil with the line where the two walls meet.

6. Close your right eye and note the location of the pencil.

(e) Describe what, if anything, happened to the position of the pencil.

7. Close your left eye and note the location of the pencil.

(f) Describe what, if anything, happened to the position of the pencil.

(g) Which is the dominant eye? (The object appears to move the least in the dominant eye.)

(h) Survey the students in your class to determine whether right-handed people have dominant right eyes and whether left-handed people have dominant left eyes.

Part IV: Stereoscopic Vision

Most of what you see with your right eye is also seen with the left eye. Visual fields overlap. Each eye sees the image from a slightly different angle. The image is created in each retina and stored on the opposing cerebral hemispheres. Information is relayed between the two hemispheres and the two images are superimposed on each other, creating a three-dimensional image. By working together, the eyes provide slightly different angles of view that permit the brain to estimate distance.

8. Stare directly at a distant object as shown in **Figure 3**. Touch the fingertips of your right and left hands, but spread your fingers enough to allow light to pass between them. Extend your arms and slowly bring the fingertips in front of your face, but continue looking at the distant object through your fingers.

(i) Describe what you see.

Figure 3

Analysis

(j) In an attempt to map the position of the blind spot, a student rotates a book 180° and follows the steps described in Part II. How would rotating the text help the student map the position of the blind spot?

(k) In Part III of the laboratory, you discovered that the pencil moved when you viewed it with one of your eyes. Offer an explanation for your observation.

(l) Horses and cows have eyes on the sides of their head, which means their visual fields overlap very little. What advantage would this kind of vision have over human vision?

(m) Like humans, squirrels have eyes on the front of their head. The visual field of the squirrel's right eye overlaps with the visual field of its left eye. Why does a squirrel need this type of vision?

(n) If students place a large piece of cardboard between their eyes while attempting to read two different books at the same time, will they be able to read two books at once? Provide an explanation.

🔬 INVESTIGATION 9.6.1

Hearing and Equilibrium

The inner ear contains the organs of hearing and equilibrium. In this investigation, you will test the effects of environmental factors on both hearing and equilibrium.

Inquiry Skills

● Questioning	● Planning	● Analyzing
● Hypothesizing	● Conducting	○ Evaluating
● Predicting	● Recording	○ Communicating

Questions
What effect will environmental factors have on hearing? on equilibrium?

Hypothesis/Prediction
(a) Using what you know about the structure and function of the human ear, predict what will happen in each part of the procedure.

(b) Formulate a hypothesis and explain your predictions.

Materials
tuning fork
rubber hammer
metre stick
swivel chair

Procedure
Part I: Factors That Affect Hearing

1. Strike a tuning fork with a rubber hammer and listen to the sound. Holding the tuning fork in your left hand, place the *stem* (not the prongs!) of the tuning fork on your forehead. Place the palm of your right hand over your right ear.

(c) From which direction does the sound seem to come?

(d) Describe any changes in the intensity of the sound.

2. Repeat the procedure, but this time hold the tuning fork in your right hand and place your left hand over your left ear.

(e) From which direction does the sound seem to come?

3. Repeat the procedure a third time, but ask your lab partner to cover both of your ears.

(f) Describe any changes in the intensity of the sound.

A | **INVESTIGATION 9.6.1** *continued*

4. Strike the tuning fork with a rubber hammer and hold the tuning fork approximately 1 m from your ear.

5. Ask your lab partner to place a metre stick gently on the bony bump immediately behind your ear. Then ask him or her to place the stem of the tuning fork on the metre stick.

(g) Describe any changes in the intensity of the sound.

Part II: Equilibrium

6. Ask your lab partner to sit in a swivel chair. Have your partner elevate his or her legs and begin slowly rotating the chair in a clockwise direction. After 20 rotations, have the subject stand. (Be prepared to support your partner!)

(h) In which direction did the subject lean?

7. After a 3-min recovery period, repeat the process, but this time rotate the swivel chair in a counterclockwise direction.

(i) In which direction did the subject lean?

8. Ask your lab partner to tilt his or her head to the right, and begin a clockwise rotation of the swivel chair. After 20 rotations, ask the subject to hold his or her head erect and to stand up. (Again, be prepared to catch your lab partner—most people attempt to sit very quickly after they stand.)

(j) Ask the subject to describe the sensation.

Analysis

(k) Provide explanations for the data collected in questions (c), (d), (e), and (f).

(l) Using the data collected, provide evidence to suggest that sound intensity is greater in fluids than in air.

(m) Using the data collected, provide evidence to suggest that the fluid in the semicircular canals continues to move even after rotational stimuli cease.

(n) What causes the falling sensation produced in step 8?

(o) Describe the manner in which the semicircular canals detect changes in motion during a roller-coaster ride.

Key Expectations

- Explain and provide examples of how the nervous system monitors changes in your external environment and makes adjustments in the internal environment to maintain physiological and biochemical systems operating within optimal limits. (9.1, 9.2, 9.3, 9.4, 9.5, 9.6)

- Using invertebrates, design and carry out experiments to study responses to external stimuli. (9.1)

- Analyze, compile, and display experimental data that reveals how the nervous system responds to external stimuli. (9.1, 9.5, 9.6)

- Evaluate opinions and discuss difficulties in treating neurological disease. (9.1, 9.2, 9.3)

- Examine the contribution made by Canadian scientists in health. (9.1, 9.3)

- Present informed opinions about personal health. (9.1, 9.2, 9.6)

Key Terms

accommodation

acetylcholine

action potential

all-or-none response

aqueous humour

astigmatism

auditory canal

autonomic nervous system

axon

basilar membrane

blind spot

cataract

central nervous system (CNS)

cerebellum

cerebral cortex

cerebrospinal fluid

cerebrum

cholinesterase

choroid layer

cochlea

cones

cornea

corpus callosum

dendrites

depolarization

effectors

endorphins

enkephalins

eustachian tube

farsightedness

fovea centralis

glaucoma

glial cells

hyperpolarized

interneurons

iris

medulla oblongata

meninges

motor neurons

myelin sheath

nearsightedness

nerves

neurilemma

neurons

neurotransmitters

nodes of Ranvier

olfactory lobes

organ of Corti

ossicles

otoliths

oval window

parasympathetic nervous system

peripheral nervous system (PNS)

pinna

polarized membrane

pons

postsynaptic neuron

presynaptic neuron

reflex arc

refractory period

repolarization

resting potential

retina

rods

Schwann cells

sclera

semicircular canals

sensory adaptation

sensory neurons

sensory receptors

somatic nervous system

summation

sympathetic nervous system

synapses

thalamus

threshold level

tympanic membrane

vagus nerve

vestibule

▶ **MAKE** a summary

Imagine a cougar or bear suddenly walks in your path. To summarize your learning, create a flow chart or diagram that shows how your nervous system would respond to this stressful situation. Include the coordination of responses from visual and auditory stimuli, and label the diagram with as many of the key terms as possible. Examine a variety of flow diagrams and choose the most appropriate design to make your sketch clear.

1. Match each structure in the following list with the correct function: sensory neuron, action potential, acetylcholine, accommodation, axon, refractory period, occipital lobe of the brain, autonomic nerves.
   (a) part of the neuron; an extension of cytoplasm that carries nerve impulses away from dendrites
   (b) nerve impulses caused by the reversal of charge across a nerve membrane
   (c) motor nerves, not under conscious control, designed to maintain homeostasis
   (d) a neurotransmitter that permits the transmission of an action potential across a synapse
   (e) carries information about the environment to the brain
   (f) a reflex that makes adjustment for near and distant objects
   (g) a sensory area involved with vision
   (h) the time required before another action potential can be produced

In your notebook, record the letter of the choice that best answers the question.

2. What is the primary function of the myelin sheath?
   (a) to supply nutrients to the axon
   (b) to carry wastes from the axon
   (c) to increase the speed at which nerve impulses travel
   (d) to conduct active transport of potassium ions
   (e) to regulate the diffusion of sodium ions across the synapse

3. What makes it possible for an impulse to move from one neuron to an adjacent neuron?
   (a) The axon of one neuron always touches the axon of the adjacent neuron.
   (b) Dendrites of one neuron always touch the axon of the adjacent neuron.
   (c) Neurotransmitters are released from the dendrites of one neuron and diffuse to the axon terminal of the adjacent neuron.
   (d) Neurotransmitters are released from the axon terminal of one neuron and diffuse to the dendrites of the adjacent neuron.
   (e) Cerebrospinal fluids are released from the axon terminal of one neuron and diffuse to the axons of the adjacent neuron.

4. In an experiment, a neuron is stimulated by various electrical charges. The data collected are represented in **Table 1**. Which of the following would be the threshold potential?
   (a) 10 mV   (c) 30 mV   (e) less than 10 mV
   (b) 20 mV   (d) 40 mV

Table 1 Stimulation of a Neuron by Various Electrical Charges

Intensity of the stimulus (mV)	Action of the neuron
10	no action potential
20	action potential
30	action potential
40	action potential

5. In **Table 1**, the scientists noted that in each test that produced an action potential, the membrane charge reached 40 mV during the action potential. Which of the following facts does this demonstrate?
   (a) The threshold potential is 40 mV.
   (b) Anything above threshold produces an all-or-none response.
   (c) Depolarization is dependent on the strength of the stimulus.
   (d) The refractory period is dependent on the strength of the stimulus.
   (e) Repolarization is dependent on the frequency of the stimuli.

6. Which of the following choices describes the structure(s) of the inner ear responsible for the conversion of sound waves into a nerve impulse?
   (a) ossicles   (d) eustachian tube
   (b) semicircular canals   (e) semicircular canals
   (c) cochlea   and eustachian tube

7. Which of the following choices describes the structure(s) of the inner ear primarily responsible for maintaining dynamic equilibrium?
   (a) ossicles   (d) eustachian tube
   (b) semicircular canals (e) ossicles and cochlea
   (c) cochlea

8. If the curvature of the cornea along the horizontal axis of the eye was greater than the curvature along the vertical axis, the result would be which of the following?
   (a) glaucoma   (d) farsightedness
   (b) night blindness   (e) impairment of the
   (c) astigmatism   accommodation reflexes

9. A person suffers a stroke that results in a loss of speech, difficulty in using the right arm, and an inability to solve mathematical equations. Which area of the brain was damaged?
   (a) left cerebellum
   (b) right cerebellum
   (c) left cerebral hemisphere
   (d) right cerebral hemisphere
   (e) medulla oblongata

Understanding Concepts

1. Explain how Luigi Galvani's discovery that muscles twitch when stimulated by electrical current lead to the discovery of the ECG and EEG.

2. Use what you have learned about threshold levels to explain why some individuals can tolerate more pain than others.

3. In **Figure 1**, the neurotransmitter released from neuron X causes the postsynaptic membrane of nerve Y to become more permeable to sodium. However, the neurotransmitter released from nerve W causes the postsynaptic membrane of nerve Y to become less permeable to sodium but more permeable to potassium. Why does the stimulation of neuron X produce an action potential in neuron Y, but the stimulation of neuron X and W together fails to produce an action potential?

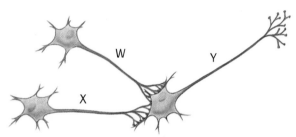

Figure 1
Nerve pathway

4. Botulism (a toxin produced by bacteria that causes food poisoning) and curare (a natural poison) inhibit the action of acetylcholine. What symptoms would you expect to find in someone exposed to botulism or curare? Provide an explanation for the symptoms.

5. A patient complains of losing his sense of balance. A marked decrease in muscle coordination is also mentioned. Which area of the brain might a physician look at for the cause of the symptoms?

6. The retina of a chicken is composed of many cones very close together. What advantages and disadvantages would be associated with this type of eye?

7. Why do people often require reading glasses after they reach the age of 40?

8. When the hearing of a rock musician was tested, the results revealed a general deterioration of hearing as well as total deafness for particular frequencies. Why is the loss of hearing not equal for all frequencies?

Applying Inquiry Skills

9. One theory suggests that painters use less purple and blue as they age because layers of protein build up on the lens in their eyes. As the buildup gradually becomes thicker and more yellow, the shorter ultraviolet wavelengths from the ultraviolet end of the spectrum are filtered. How would you test the theory?

10. A nerve cell that synapses in a muscle is stimulated by electrical current. The strength of the stimulus is increased and the force of muscle contraction is recorded. The results are recorded in **Table 1**.

Table 1

Trial	Strength of stimulus (mV)	Force of contraction of muscle (N)
1	0	none
2	10	none
3	20	4
4	30	not measured

(a) Predict the force of muscle contraction in trial 4. Give your reasons.
(b) Identify the threshold level from the experiment.

11. Three different neurons synapse on a single neuron, as shown in **Figure 2**. The experimental data is recorded in **Table 2**.
(a) From the experimental data, indicate which neuron releases an inhibitory neurotransmitter.
(b) Explain the principle of summation using the experimental data.

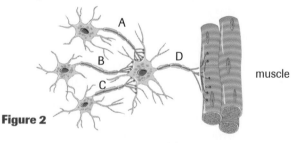

Figure 2

Table 2

Neuron stimulated	Effect on muscle
A	contraction
A and B	no contraction
B	no contraction
A and C	contraction
C	no contraction
B and C	no contraction

12. The following data were collected from an experiment.

Table 3

Age	Near-point accommodation (cm)
10	7.5
20	10.2
30	11.3
40	17.2
50	56.8
60	87.3

(a) What question was being investigated in the experiment?
(b) Write a hypothesis for the experiment.
(c) At what age is near-point accommodation most affected?
(d) Explain what causes the change in near-point accommodation? How does it affect people?

Making Connections

13. During World War I, physicians noted a phenomenon called "phantom pains." Soldiers with amputated limbs complained of pain or itching in the missing limb. Use your knowledge of sensory nerves and the central nervous system to explain this phenomenon.

14. People with Parkinson's disease have low levels of the neurotransmitter dopamine. Researchers have been able to coax rat embryonic stem cells to develop into dopamine neurons. When these neurons were implanted into rats with a rodent version of Parkinson's, the characteristic tremor of the disease disappeared. Conduct library and Internet research to identify the latest research into treating Parkinson's disease.

 www.science.nelson.com

15. For hundreds of years, the people of China have believed that drinking herbal tea can improve one's memory. Researchers have isolated a compound from the tea that inhibits the action of cholinesterase. The compound, called huperzine A, is believed to be the active ingredient.
(a) Why are researchers exploring the use of huperzine A for Alzheimer's patients?
(b) Why do you think that some Western scientists have been reluctant to research medicinal effects of herbal teas? How do you think the research into herbal teas will be received once the action of hyperzine A is known?

Extension

16. Alzheimer's disease is a complex biochemical puzzle of parts that seemed unrelated just a few years ago. Approximately 15% of people who live to 65 years of age will develop some form of dementia; by 85 years of age, it increases to 35%. Beta-amyloid plaques (**Figure 3**) cause the damage in several ways: by interfering with calcium regulation, by causing the formation of free radicals, or by initiating a response by the immune system.

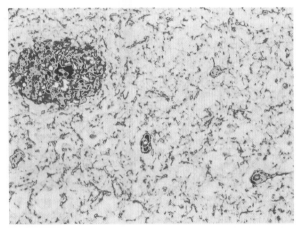

Figure 3
Light microscope view of a human brain showing Alzheimer's disease. The plaque can be identified by the dark patches. The creation of amyloid peptides make up the plaque.

(a) Explore some current advances in Alzheimer's research. Search for information in newspapers, periodicals, CD-ROMs, and on the Internet.

 www.science.nelson.com

(b) How have studies in immunology, biochemistry, and neurology been combined to find answers?
(c) Many times scientists from different disciplines do not share research interests. Using Alzheimer's research as an example, explain how scientific discoveries from unrelated fields can converge to provide answers.
(d) What challenges remain in finding a cure?

10

The Maintenance of Balance by the Immune System

During the 1980s and early 1990s, thousands of people in Britain ate beef infected with bovine spongiform encephalopathy (BSE), often called mad cow disease. Although it causes severe nerve degeneration, little concern was raised because BSE was believed to only affect cows. The extreme neurological damage caused by BSE prevents a normal walking motion. In the later stages, even standing becomes impossible. By the mid-1990s, the first deadly human version of mad cow disease, known as variant Creutzfeldt–Jakob disease (vCJD), surfaced, and an alarm was sounded. For the first time, scientists considered the possibility that the disease could be transferred from cows to people through the food chain.

Great concern arose because Creutzfeldt–Jakob disease—unlike familiar infectious diseases caused by viruses and bacteria—is caused by a neurological invader that does not contain nucleic acids. The disease is caused by a *prion*, an abnormal infectious version of a protein that exists in all healthy cattle, sheep, and humans. Current theory suggests that once the prion enters an organism, it begins to alter the shape of normal proteins, converting them into the disease-causing form that results in sickness and death.

So great is the concern that other countries banned the import of British beef and dairy products. In part, the fear arises from so many unanswered questions. Scientists do not know what causes the protein to alter its shape, nor do they know how many different kinds of potentially harmful prions can infect meat. By 2001, 56 people had died from vCJD; however, little evidence exists on the number of people who have become infected. Even more worrisome is that scientists do not have any clear understanding of the disease's incubation period. We may be seeing only the tip of the iceberg.

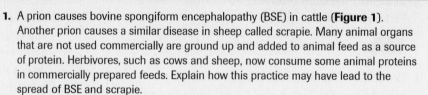

REFLECT on your learning

1. A prion causes bovine spongiform encephalopathy (BSE) in cattle (**Figure 1**). Another prion causes a similar disease in sheep called scrapie. Many animal organs that are not used commercially are ground up and added to animal feed as a source of protein. Herbivores, such as cows and sheep, now consume some animal proteins in commercially prepared feeds. Explain how this practice may have lead to the spread of BSE and scrapie.

2. How do you think the idea that a protein can cause an infectious disease has altered the way we think about disease?

3. Which of the following medical conditions have been linked with bacteria, viruses, or prions?
   - (a) heart attack
   - (b) ulcers
   - (c) AIDS
   - (d) measles
   - (e) diabetes mellitus

4. Allergies are caused by a mistake made by your immune system. Harmless agents, such as proteins in peanut butter, are recognized as harmful invaders and an immune response is mobilized. What other mistakes of the immune system can cause problems?

Figure 1
A normal prion (top) compared to the infectious form (bottom) that causes BSE. The internal structures of the proteins are shown using the ribbon model for ease of comparison.

▶ **TRY THIS** activity   *Examining Brain Tissue*

Figure 2 shows brain tissue of a sheep infected by a prion that causes scrapie, a neurological disease very similar to BSE.

Materials: prepared slide of normal brain tissue, light microscope

- Place a prepared slide of normal brain tissue under your light microscope.
- Examine the slide under low power and then under higher power magnification. Observe the cytoplasm of a healthy neuron of the brain.
  (a) Compare what you see on the slide to the brain tissue shown in **Figure 2**. Describe the differences that you observe.
  (b) Name any cell parts that you recognize.

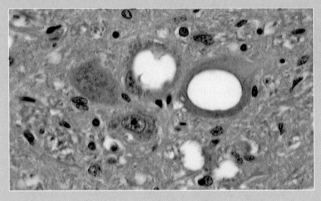

Figure 2

The Maintenance of Balance by the Immune System

To appreciate the importance of the immune system, consider the story of David, the "boy in the plastic bubble" (**Figure 1**). David Vetter was born in Texas on September 21, 1971. He was the second son born to Carol Ann and David Vetter, Jr. The first son, also named David, was born in 1970, but died at six months of severe combined immunodeficiency (SCID) because his body was unable to produce the cells necessary to protect him from disease. The Vetters were told that a gene on the X chromosome caused SCID and that all boys born to a mother with the mutation would have a 50/50 chance of having SCID. However, the Vetters decided to try again. When their second son was born, he was placed in a plastic bubble 10 seconds after birth.

David had to live in a virtually germ-free environment and, therefore, remained in a plastic bubble for 12 years. People who came in contact with him had to take many precautions so that they would not infect him. Doctors hoped that they could cure SCID with a bone marrow transplant—they believed that this would help David grow a new immune system. On October 21, 1983, one month after his 12th birthday, David received a bone marrow transplant from his sister, Katherine. She had a functioning immune system, but the doctors did not know that her marrow contained the Epstein–Barr virus, the virus that causes mononucleosis. Because David had no immune system, the virus grew out of control into a type of cancer called Burkitt's lymphoma. Knowing he was dying, David removed himself from his bubble for the last few weeks of his life. He died on February 22, 1984.

People are born with SCID even today, but the case of David Vetter is by far the most famous. Doctors learned many things from David's case. For instance, they now know that viruses can cause cancer. SCID is now treated within the first three months after birth with a bone marrow transplant that has been screened for **pathogens**. Infants who receive this transplant have a greater than 95% chance of developing an immune system that allows them to have a fairly normal life. Untreated children rarely live to age 2.

The human body must constantly defend itself against the many unwelcome intruders it encounters in the air, in food, and in water. It must also deal with abnormal body cells that sometimes turn into cancer. Three lines of defence have evolved to help resist infection and possible death from fatal illnesses. The first two lines of defence are considered nonspecific immune responses, meaning that they do not distinguish one microbe from another. The third line of defence—the immune system—is a specific immune response that reacts in specialized ways to various invaders. The specific immune response will be discussed in section 10.2.

The First Line of Defence

The body's first line of defence against foreign invaders is largely physical. Like a medieval city that used walls and moats to defend against attack from outsiders, the skin and mucous membranes defend against viral and bacterial invaders. Intact skin provides a protective barrier that cannot normally be penetrated by bacteria or viruses. The skin also has chemical defences in the form of acidic secretions, which keep it at a pH range of 3 to 5, acidic enough to inhibit the growth of microbes. Lysozyme, an antimicrobial enzyme secreted in human tears, saliva, mucous secretions, and perspiration, destroys the cell walls of bacteria, killing them.

In the respiratory passage, invading microbes and foreign debris become trapped in a layer of mucus or filtered by tiny hairlike structures called cilia (**Figure 2**). The cilia move

Figure 1
David, the "boy in the plastic bubble," had severe combined immunodeficiency.

pathogens disease-causing organisms

in waves, sweeping particles toward the entrance where coughing can expel them. Corrosive acids in the stomach and protein-digesting enzymes destroy most of the invading microbes carried into the body with food.

The Second Line of Defence

A second line of defence can be mobilized if the invader takes up residence within the body. **Leukocytes**, or white blood cells, are large opaque blood cells that may engulf invading microbes or produce antibodies. White blood cells have a nucleus, making them easily distinguishable from red blood cells. In fact, the shape and size of the nucleus, along with the granules in the cytoplasm, can be used to identify different classes of leukocytes. One class of leukocytes, called *granulocytes*, contains cytoplasmic granules. Granulocytes are produced in the bone marrow. *Agranulocytes* are white blood cells that do not have a granular cytoplasm. Agranulocytes are also produced in the bone marrow but are modified in the lymph nodes.

The body's nonspecific defence mechanisms rely mainly on the process of **phagocytosis**, the ingestion of invading microbes by certain types of white blood cells. When a foreign particle penetrates the skin through an injury, special leukocytes, known as *monocytes*, migrate from the blood into the tissues, where they develop into **macrophages** (meaning "big eaters"). The macrophages extend long protrusions, called *pseudopods*, that attach to the surface of the invading microbe; the microbe is then engulfed and destroyed by enzymes within the macrophage.

In another phagocytic response, white blood cells called *neutrophils* are attracted to chemical signals given off by cells that have been damaged by microbes. In a process called *chemotaxis*, the neutrophils squeeze out of capillaries and migrate toward the infected tissue. The neutrophils then engulf the microbe and release lysosomal enzymes that digest both the microbe and the leukocyte. The remaining fragments of protein, dead white blood cells, and the digested invader are called *pus*. Tissue damage due to physical injury also initiates a localized **inflammatory response**—a nonspecific immune response resulting in swelling, redness, heat, and pain (**Figure 3**). Pus and accompanying inflammation are sure signs that the second line of defence has been at work.

Figure 2
Special cells in the lining of the trachea produce a mucus that traps microbes before they reach the lungs. The cilia move mucus and trapped microbes upward where they can be expelled.

leukocytes white blood cells that may engulf invading microbes or produce antibodies

phagocytosis process by which a white blood cell engulfs and chemically destroys a microbe

macrophages phagocytic white blood cells found in lymph nodes or in the blood of the bone marrow, spleen, and liver

inflammatory response localized nonspecific response triggered when tissue cells are injured by bacteria or physical injury, characterized by swelling, heat, redness, and pain.

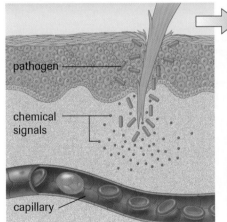

(a) At the first sign of injury, chemical signals are released by the foreign invader. Other chemicals—histamines and prostaglandins—are released by the cells of the body.

(b) Chemical signals cause the capillaries to dilate, which results in increased blood flow, and the permeability of the capillaries increases. Injured cells also release chemicals that attract phagocytic cells and specialized white blood cells.

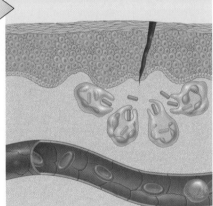

(c) Phagocytes engulf and digest the invaders and cellular debris, which promotes healing of the tissues.

Figure 3
Damage to tissue cells by bacteria or physical injury initiates a localized inflammatory response.

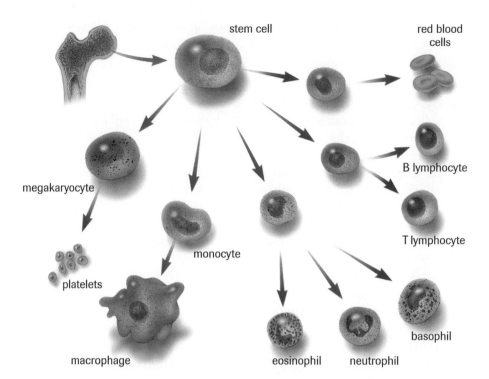

stem cell

red blood cells

megakaryocyte

platelets

monocyte

macrophage

B lymphocyte

T lymphocyte

basophil

eosinophil   neutrophil

Figure 4
Blood cells are formed in the bone marrow and released into the bloodstream. Two classes of white blood cells are shown in this illustration. Granulocytes include eosinophils, basophils, and neutrophils. Agranulocytes include monocytes and lymphocytes.

How the body's defences respond to a localized injury, such as a cut or a puncture, has already been discussed. However, the body can respond with a system-wide defence to more severe damage or infection. Injured cells emit chemicals that stimulate the production of phagocytic white blood cells and increase their release into the bloodstream. **Figure 4** summarizes the development of phagocytic white blood cells, red blood cells, and lymphocytes from the bone marrow. Bone marrow, the source of all blood cells, is found chiefly in the inner, spongy part of the upper leg bone, upper arm bone, breastbone, and shoulder blades. The role of lymphocytes will be discussed in section 10.2.

▶ **TRY THIS** activity   *Observing Phagocytosis*

Protist models can be used to observe phagocytosis, as shown in **Figure 5**.

Materials: prepared slide of amoeba, light microscope, medicine dropper, slide, cover slip, live amoeba culture, live yeast culture

- Obtain a prepared slide of amoeba that shows phagocytosis, and use a light microscope to look at it under high-power magnification. Draw what you see. Label the extension of false feet as "pseudopods" and indicate the food vacuole (if present).
- Using a medicine dropper, make a wet mount from a live amoeba culture. Observe the movement of the amoeba.
- Remove the cover slip from the slide and use a medicine dropper to add a drop from a live yeast culture. Replace the cover slip and observe for phagocytosis.
  (a) Describe the movement of the amoeba.
  (b) Describe the process of phagocytosis.

Figure 5
The macrophage has long, sticky extensions of cytoplasm that draw bacteria toward the macrophage. Once the bacteria come in contact with the macrophage, they are engulfed and destroyed.

A fever is another example of the body's system-wide response to infection. When infectious organisms spread throughout your body, such as when you have a cold or flu, neutrophils and macrophages digest the invaders and release chemicals into your bloodstream. When these chemicals reach your hypothalamus, they reset the body's thermostat to a higher temperature—about 40°C. The conditions in your system during a fever make it difficult for harmful bacteria to survive; thus, the fever helps to prevent the proliferation of the infectious organisms. Reducing your fever by taking aspirin may actually prolong the infection. However, if your body temperature rises above 40°C, it can be unsafe. For example, a fever of 41°C may cause convulsions, especially in young children; human cells cannot survive above 43°C. ▣▮

ACTIVITY 10.1.1

Diagnosing Disease by Examining Blood Cells (p. 492)
In this activity, you will look at prepared slides of blood to identify different types of white blood cells. How are changes in white blood cell counts used as clues to diagnose disease?

SUMMARY *The Body's Lines of Defence*

- Skin and mucous membranes provide physical barriers that prevent most infectious organisms from entering the body.
- Leukocytes (white blood cells), produced in the bone marrow, fight infection in a variety of ways.
- Phagocytosis of invading microbes is one of the main methods used by certain leukocytes to combat infection.
- Tissue damage due to physical injury initiates the inflammatory response, which is a nonspecific immune response resulting in swelling, redness, heat, and pain.

▶ *Section 10.1* **Questions**

Understanding Concepts

1. Using immunodeficiency diseases as an example, explain why an immune system is important.
2. How do lysozymes protect the body against invading microbes?
3. Outline protective mechanisms provided by the respiratory tract and digestive tract.
4. How do monocytes protect against microbes?
5. Explain why the presence of swelling and pus at the site of an injury are signs that the immune system is functioning.

Applying Inquiry Skills

6. The data in **Table 1** were collected from three patients.
   (a) Lead poisoning is characterized by a destruction of bone marrow. Which patient would you suspect has lead poisoning? Give your reasons.
   (b) Predict which patient has a viral infection. Explain your answer.
   (c) Leukemia is a cancer characterized by a proliferation of white blood cells. Which patient would you suspect has leukemia? Give reasons for your response.

Table 1

Data	Normal values	Patient X	Patient Y	Patient Z
red blood cell count	$5 \times 10^6/\mu L$	$2 \times 10^6/\mu L$	$2.5 \times 10^6/\mu L$	$5.1 \times 10^6/\mu L$
white blood cell count	7000/μL	3000/μL	10 000/μL	15 000/μL
body temperature	37°C	37°C	36.5°C	39°C

7. A patient displaying a high fever may be asked by the physician to have blood tests done. One of these tests would likely be a white blood cell count. Explain what an abnormal result might indicate.

Making Connections

8. Compare your childhood with that of David Vetter. How would your childhood have been different if you had an immunodeficiency disease?

Although some macrophages migrate throughout the body, others reside permanently in body tissues, such as the brain, lungs, kidneys, liver, and connective tissues. The fixed macrophages that reside in the spleen, lymph nodes, and other tissues of the lymphatic system trap and filter out microorganisms and foreign invaders that enter the blood. **Figure 1** shows the structures of the lymphatic system.

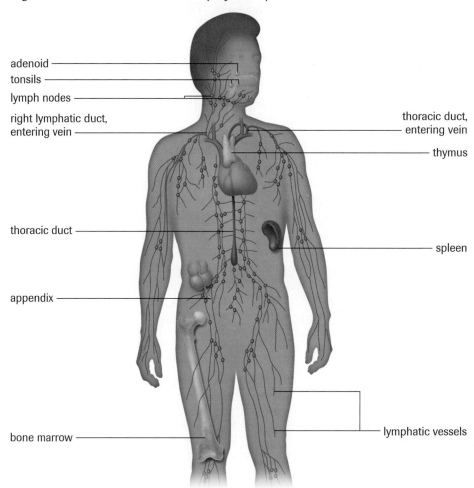

Figure 1
In the lymphatic system, pathogens, foreign cells and material, and debris from the body's tissues are filtered out from the lymph (the fluid found outside capillaries), and the lymph is returned to the circulatory system.

complement proteins plasma proteins that help defend against invading microbes by tagging the microbe for phagocytosis, puncturing cell membranes, or triggering the formation of a mucous coating

The appearance of foreign organisms in the body activates antimicrobial plasma proteins, often called **complement proteins** (**Figure 2**). There are about 20 known types of complement proteins. Under normal conditions, these proteins are present in the circulatory system in an inactive form. Marker proteins from invading microbes activate the complement proteins, which, in turn, serve as messengers. The proteins aggregate to initiate an attack on the cell membranes of fungal or bacterial cells. Some of the activated proteins trigger the formation of a protective coating around the invader, as shown in **Figure 2(a)**. This coating seals the invading cell, immobilizing it. A second group punctures the cell membrane, as seen in **Figure 2(b)**. Water enters the cell through the pore created by the protein, causing the cell to swell and burst. A third group of proteins attaches to the invader, as illustrated in **Figure 2(c)**. The tiny microbes become less soluble and more susceptible to phagocytosis by leukocytes.

Complement Proteins

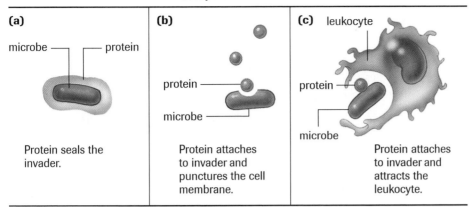

(a)

microbe — — protein

Protein seals the invader.

(b)

protein —

microbe —

Protein attaches to invader and punctures the cell membrane.

(c) leukocyte

protein —

microbe —

Protein attaches to invader and attracts the leukocyte.

Figure 2
Complement proteins aid the immune response.

Another specialized group of white blood cells, called **lymphocytes**, produces **antibodies**. Antibodies are protein molecules that protect the body from invaders. All cells have special markers located on their cell membranes. Normally, the immune system does not react to the body's own markers. However, intruding cells or foreign proteins activate the production of antibodies. The cell membrane of a bacterium and the outer coat of a virus contain many different **antigens**. The antigen (a term derived from *anti*body *gen*erator) may even be a toxin produced by moulds, bacteria, or algae. The toxin presents a danger to the cells of the body because it interferes with normal cell metabolism.

Two different types of lymphocytes are found in the immune system. The first is the **T cell**, which is produced in the bone marrow and is stored in a tiny organ called the thymus gland, from which the T cell receives its name. The T cell's mission is to seek out the intruder and signal the attack. Acting much like a sentry, some T cells identify the invader by its antigen markers (**Figure 3**), which are located on the cell membrane. Once the antigen is identified, another T cell passes this information on to the antibody-producing **B cell**.

B cells multiply and produce chemical weapons: the antibodies. Each B cell produces a single type of antibody, which is displayed along the cell membrane. Eventually, the B cells are released from the bone marrow and enter the circulatory system. Some B cells differentiate into super-antibody-producing cells called *plasma cells*. These plasma cells can produce as many as 2000 antibody molecules every second.

Antigen–Antibody Reactions

Antibodies are Y-shaped proteins engineered to target foreign invaders. Antibodies are specific; this means that an antibody produced against the influenza virus, for example, is not effective against HIV, the virus that causes AIDS. The tails of these Y-shaped proteins are very similar, regardless of the type of antibody. Variations only exist at the outer edge of each arm, the area in which the antibody combines with the antigen (**Figure 4**). Antigen markers found on the influenza virus are different from those found on HIV. Each antibody has a shape that is complementary to its specific antigen. Thus, the combining site of an antibody produced in response to the influenza virus will not complement HIV.

Many different antigen markers are located on the membrane of a virus or bacterium. Although different antibodies can attach to the invader, each antibody attaches only to its complementary marker. The attachment of antibodies to the antigens increases the size of the complex, making the antigen–antibody combination more conspicuous and, therefore, more easily engulfed and destroyed by the wandering macrophages.

lymphocytes specialized white blood cells that produce antibodies

antibodies proteins formed within the blood that inactivate or destroy antigens

antigens substances, usually protein in nature, that stimulate the formation of antibodies

T cell lymphocyte, manufactured in the bone marrow and processed by the thymus gland, that identifies and attacks foreign substances

B cell lymphocyte, made and processed in the bone marrow, that produces antibodies

Antigen Markers

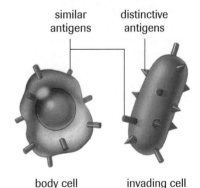

similar antigens   distinctive antigens

body cell   invading cell

Figure 3
Sugar-protein complexes located on the cell membrane act as markers. T cells distinguish the markers on the body's cells from those of invading cells.

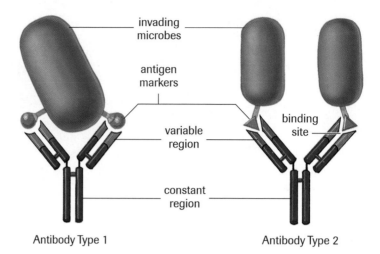

Figure 4
Each type of antibody will only combine with the appropriate antigen.

Antibody Type 1   Antibody Type 2

receptor sites ports along cell membranes into which nutrients and other needed materials fit

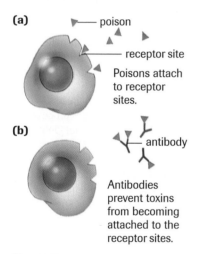

(a)
poison

receptor site

Poisons attach to receptor sites.

(b)
antibody

Antibodies prevent toxins from becoming attached to the receptor sites.

Figure 5
In **(a)**, the poison attaches to the appropriate receptor site and masquerades as a nutrient or hormone. Once attached, the cell engulfs the toxin. In **(b)**, antibodies tie up the poison. The shape of the toxin has been altered so that it can no longer gain access to the receptor site.

How do antibodies prevent poisons or toxins from destroying cells? Specialized **receptor sites** are found on different cells, which may explain why some poisons affect the nervous system while others affect the digestive or circulatory system. The receptor site is designed to accommodate either a hormone or a specific nutrient. Toxins or poisons have a specialized geometry that allows them to become attached to the receptor sites on cell membranes. Unfortunately, the poison has a shape similar to a hormone or nutrient. Once attached, the poison is engulfed by the cell, which assumes that the poison is actually a needed substance. Antibodies interfere with the attachment of the toxins to the cell membranes' receptor sites by binding with toxins, as shown in **Figure 5**.

Viruses also use receptor sites as entry ports. The virus injects its hereditary material into the cell, but most often leaves the outer protein coat in the entry port. Because of this outer coat, different viruses come to rest in distinct locations. For example, the outer coat of the cold virus has a geometry that enables it to attach to lung cells. HIV attaches to the receptor sites of the T cell. Once attached, the T cell engulfs the virus, creating another problem for the immune system. Antibody production requires a blueprint of the invader, but the protein coat of the virus hides inside the very cells assigned as sentries for invading antigens. Does this provide a clue as to why the body experiences difficulty defeating HIV (**Figure 6**)?

Antibodies that attach themselves to the invading viruses alter their shape, thereby preventing access to the entry ports. Misshapen viruses float around the body, unable to find an appropriate entry port. Occasionally, the outer coat of the invader will change shape because of mutations. The mutated microbes may still gain access to the receptor site but are not tied up by the antibody.

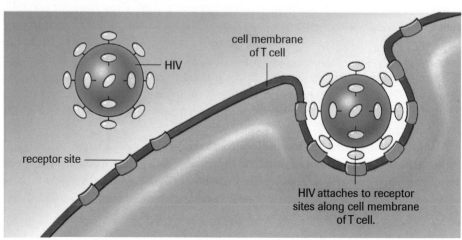

Figure 6
HIV has a shape that provides access to the T cell. However, unlike most other viruses, the T cell engulfs HIV.

How the Body Recognizes Harmful Antigens

Figure 7 illustrates how the body recognizes harmful antigens. The T cells scout the body in search of foreign invaders that pose a threat to your survival. The macrophages attack the invaders and engulf them. As mentioned earlier, the antigen markers are not destroyed with the invader but are pushed toward the cell membrane of the macrophage. Pressing the antigens into its cell membrane, the macrophage couples with the T cells, also referred to as **helper T cells**. The T cells read the antigen's shape and release a chemical messenger called **lymphokine**. The lymphokine causes the B cells to divide into identical cells called clones. Later, a second message is sent from the helper T cells to the B cells, triggering the production of antibodies. Each B cell produces a specific type of antibody. By the time the B cells enter the circulatory system, many antibodies are attached to their cell membranes.

The helper T cells activate an additional defender, the **killer T cells**. As the name suggests, these lymphocytes carry out search-and-destroy missions. Once activated, the killer T cells puncture the cell membrane of the intruder, which may be a fungus, protozoan parasite, or bacterium. Viruses, however, are much more insidious, because they hide within the familiar confines of the host cell. Here, the true value of the killer T cells is demonstrated. Once the viral coat is found attached to the cell's membrane, the T cell attacks the infected cell. By destroying the infected body cell, the killer T cell prevents the virus from reproducing.

helper T cells T cells with receptors that bind to fragments of antigens

lymphokine protein produced by the T cells that acts as a chemical messenger between other T cells and B cells

killer T cells T cells that puncture the cell membranes of cells infected with foreign invaders, killing the cells and the invaders, and which destroy body cells that have become infected by a virus

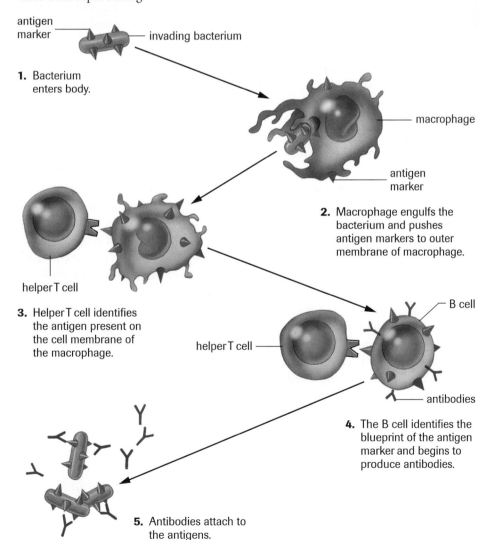

antigen marker

invading bacterium

1. Bacterium enters body.

macrophage

antigen marker

2. Macrophage engulfs the bacterium and pushes antigen markers to outer membrane of macrophage.

helper T cell

3. Helper T cell identifies the antigen present on the cell membrane of the macrophage.

helper T cell

B cell

antibodies

4. The B cell identifies the blueprint of the antigen marker and begins to produce antibodies.

5. Antibodies attach to the antigens.

Figure 7
The body recognizes harmful antigens.

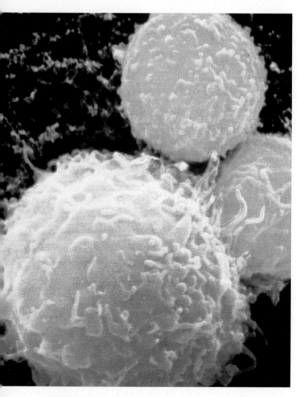

Figure 8
Killer T cells bind with a tumour cell.

suppressor T cell T cell that turns off the immune system

memory B cell cell that retains information about the geometry of an antigen or antigens

⚗ **INVESTIGATION 10.2.1**

Virtual Immunology Laboratory (p. 493)
How does your immune system respond to foreign invaders? How do antibodies know which antigen to attack? Can the formation of antibodies be used to diagnose disease? In this investigation, you will use the ELISA test (**E**nzyme-**L**inked **I**mmuno**s**orbent **A**ssay) to detect either the antigen or antibody associated with a disease-causing agent.

Killer T cells also destroy mutated cells (**Figure 8**). This is an extremely important process because some of the altered cells may be cancerous. Many experts believe that everyone develops cancerous cells, but, in most cases, the T cells eliminate the problem before a tumour forms. Whether or not you develop cancer depends on the success of your killer T cells. Killer T cells may also account for the body's rejection of organ transplants. Antigen markers on the cell membranes of the donor will be different from those of the recipient. Once the foreign markers of the transplanted tissue are recognized, their killer T cells initiate an assault. Immunosuppressant drugs, such as cyclosporin, can slow the killer T cells. Unfortunately, slowing the killer T cells can result in a new set of complications. Individuals who receive these drugs become susceptible to bacterial infections. One of the leading causes of death for an organ transplant patient is pneumonia.

Once the battle against foreign invaders has been won, another T cell, the **suppressor T cell**, signals the immune system to shut down. Communication between the helper T cells and the suppressor T cells ensures that the body maintains adequate numbers of antibodies to contain the invading antigen. Most of the B cells and T cells will die off within a few days after the battle, but a small contingent will remain long after to guard the site. Phagocytes survey the area, cleaning up the debris left from dead and injured cells. Tissues begin the work of repair and replacement.

The Immune System's Memory

The native population of Hawaii was nearly annihilated by measles in the late 18th and early 19th centuries after British explorer James Cook and his sailors unwittingly introduced the disease when they arrived at the Hawaiian Islands. In North America, the native population was decimated by epidemics of smallpox. Because neither group had been exposed to these viruses before, they had no antibodies to fight infection.

At this time, Europeans and Asians, unlike the native populations of Hawaii and North America, had long been exposed to many different types of viruses and were better able to produce antibodies to fight them. Europeans and Asians were among the first to domesticate and live in close proximity to animals, so many diseases had been transmitted via animals to humans. The diseases were easily transmitted among humans because of overcrowding in cities and poor sanitation conditions. The diseases killed those who were susceptible; those who were resistant to disease lived on and conferred their resistance to future generations.

As mentioned earlier, the helper T cells must read a blueprint of the invader before B cells produce antibodies. This blueprint is stored even after the invader is destroyed so that subsequent infections can be destroyed before the microbe gains a foothold. Immunity is based on maintaining an adequate number of antibodies.

It is believed that a **memory B cell** is generated during the infection. Like helper T cells, the memory B cells hold an imprint of the antigen or antigens that characterize the invader. Most T cells and B cells produced to fight the infection die off within a few days; however, the memory B cells remain. The memory B cells identify the enemy and quickly mobilize antibody-producing B cells. Invading pathogens are defeated before they become established. As long as the memory B cell survives, the individual is immune. That is why a person does not usually catch chicken pox more than once. ⚗ ▌

Matching Tissues for Organ Transplant

The main challenge with any tissue or organ transplant is the immune response of the recipient—that is, the immune system's ability to distinguish between "self" and "nonself." The donor organ is often identified as a foreign invader by distinctive protein markers on its cell membrane. The distinctive marker (known as major histocompatability complex, or MHC) is a protein fingerprint unique to each individual. The recipient makes antibodies designed to destroy the foreign invader.

Kidney transplants can be used as an example. Living donor kidneys account for about 15% of all kidney transplants. Because humans are born with two kidneys, the donor is able to give one kidney without significant effects on quality of life. A single kidney can carry out the filtering and osmoregulatory functions of the body. To reduce rejections, attempts are made to match MHC of the tissues of donors and recipients as closely as possible. For living donor transplants, physicians usually look to close relatives because the MHC is genetically controlled. The better the match, the greater the chances of long-term success.

Kidney transplants from recently deceased donors account for the vast majority of transplants. However, the need for organs far surpasses supply (**Figure 9**). Again, as with living donors, close matching is essential. Not every donor kidney is appropriate for a specific recipient. To help reduce the rejection factor, even for close matches, immunosuppressant drugs can be given. However, a drug that minimizes the fight against foreign tissues will also reduce the immune system's ability to fight off invading viruses and bacteria. These drugs place patients at risk of infections.

One of the most promising breakthroughs in organ transplant research comes from a research team working at Ontario's London Health Sciences Centre. The research team has come up with an antibody that targets and disarms immune cells responsible for organ rejection while leaving the remainder of the immune system untouched. Early work suggests that the antibody reprograms the immune system to accept transplanted organs.

Stem Cell Research

The answer for replacing damaged tissues may lie in stem cell research, not transplantation. Stem cells are cells that can differentiate and develop into a variety of different tissues such as epithelial tissue, muscle tissue, or nerve tissue. Intestinal stem cells reline the gut; stem cells of the skin replace cells that are continuously sloughed off; and stem cells give rise to a wide range of blood cells that protect against foreign invaders and identify human cells that have mutated, such as cancer cells. Stem cells are **pluripotent cells** that can give rise to different types of body cells.

In 1998, James Thomson, a researcher at the University of Wisconsin, demonstrated that human stem cells could transform into a variety of cells, such as bone marrow, brain tissue, muscle, skin, pancreas, liver, or practically any human tissue. If it were possible to regulate the differentiation of human stem cells, the cells could replace destroyed islet cells that produce insulin, repair damaged cartilage, or repair cardiac tissue that has been destroyed by heart disease.

Dr. Freda Miller and colleagues at the Montreal Neurological Institute (MNI) have discovered multipotent stem cells in adult skin. These skin cells can be directed to become neurons or even muscle cells. The MNI researchers expect that new findings will confirm the versatility of adult skin stem cells.

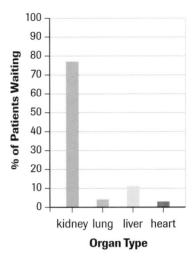

Figure 9
Percentage of patients waiting for a transplant, by organ type. Kidney transplants are the second most common type of transplant in Canada (corneal transplants rank number one) and the most common organ transplant.

DID YOU KNOW ❓

Organ Donation in Canada
Canada's organ donation rate is among the lowest of all the developed countries—more than 3000 Canadians are waiting for an organ transplant. One organ donor can donate numerous organs and tissues including lungs, heart, liver, kidneys, pancreas, bowel, eye tissue, skin, heart valves, bone, tendons, veins, and ligaments. You can indicate your wish to become an organ donor on your health-care card or your driver's license. Discuss this decision with family so your wishes are known.

pluripotent cells cells that are capable of differentiating into a number of different specialized cells, such as neurons or muscle cells

Take a Stand:
The Future of Stem Cell Research

The greatest challenge of organ transplants is to trick the recipient's immune system into accepting the new organ. Finding someone with a close tissue match to donate an organ can be extremely difficult. What if the person who needs a transplant could use his or her own stem cells to repair the damaged organ?

Consider this scenario: Twenty-four people suffering from the same degenerative brain disorder participate in a research project. Scientists inject immature neurons (glial cells) into the patients' brains in the hope that the degenerative effects of the disorder can be reversed by replacing the damaged nerve cells. The glial cells were obtained by culturing the patients' own bone marrow stem cells (stromal cells) with retinoic acid and growth factor, and subjecting them to other conditions that induced them to differentiate into glial cells. After two years, 13 patients show signs of improvement, 7 show no beneficial effects, 2 show accelerated degeneration of brain tissue, 1 experiences an unexplained immune-like reaction, and 1 develops a glioma (a form of brain cancer).

Statement

Governments should redirect some funding from organ transplant research to autologous (i.e., originating from the same individual) stem cell research.

Decision-Making Skills

○ Define the Issue ● Analyze the Issue ● Research
● Defend the Position ● Identify Alternatives ○ Evaluate

- Investigate this rapidly changing field of research. Search for information in newspapers, periodicals, CD-ROMs, and on the Internet.

 www.science.nelson.com

- Prepare a list of points and counterpoints. You might consider these questions:
  (i) Researchers are working on ways to use mature cells as a source for stem cells as well. How will this change the prospects for successful treatment?
  (ii) Companies have applied for patents on specific stem cells and techniques used to culture stem cells and induce differentiation. These companies will own a medical procedure and collect huge royalties. Should this be allowed?
  (iii) Why do many people object to stem cell research?
  (iv) Why are governments regulating this field of research?
- Decide whether you agree or disagree with the statement.
  (a) Prepare an outline.
  (b) Write your position paper.
  (c) Prepare to defend your position in class.

SUMMARY *The Immune Response*

Cell type	Function
lymphocytes	produce antibodies
helper T cells	act as sentries to identify foreign invading substances
B cells	produce antibodies
killer T cells	puncture cell membranes of infected cells, thereby killing the cell
suppressor T cells	turn off the immune system
memory T cells	retain information about the geometry of the antigen

---

▶ **Section 10.2** *Questions*

Understanding Concepts

1. Define and contrast these terms: antigen, antibody; T cell lymphocytes, B cell lymphocytes; macrophages, lymphocytes.
2. Explain how B cell, helper T cell, and killer T cell lymphocytes provide immunity.
3. How do antibodies defeat antigens? Describe four contributions that antibodies make to the immune system.
4. How do memory B cells provide continuing immunity?

Making Connections

5. A research group has begun testing on a potential cure for type 1 diabetes, an inherited disease caused by the destruction of the insulin-producing cells in the pancreas by one's own immune system. An immunosuppressant drug is administered twice daily to a test group of 150 people.
   (a) Why can the immunosuppressant drug prevent diabetes?
   (b) Researchers found that the drug wasn't effective once symptoms for diabetes were expressed in test subjects. What conclusions can you draw about this?
   (c) Explain why researchers are working on a test to identify antibodies that destroy insulin-producing cells.
   (d) List three important research questions that remain to be answered.

Abnormal functioning of the immune system can cause two types of problems: immune-deficiency diseases and inappropriate attacks of the immune system against nonthreatening agents. Immune-deficiency diseases may be caused by a foreign agent, such as the HIV virus, which attacks T cells, or a hereditary condition, such as severe combined immunodeficiency (SCID). The gene mutation that causes SCID results in the inability to produce B cells and T cells. Cancer therapy or prolonged exposure to anti-inflammatory drugs, such as cortisol, can also reduce the effectiveness of the immune system.

Inappropriate or exaggerated immune responses can also create problems. A hypersensitivity to harmless agents, or a response in which the immune system begins to attack normal cells in one's own body, can destroy tissues and organs.

Allergies

Allergies occur when your immune system mistakes harmless cells for harmful invaders. If you are allergic to peanuts, your immune system recognizes one of the proteins in the peanut as dangerous. Although the protein is quite safe, you mobilize the antibody strike force against the peanut. Increased tissue swelling and mucus secretion, and, sometimes, constricted air passages are all part of the immune response. Dust, ragweed, strawberries, and leaf moulds do not pose any direct threat to life, but the immune response itself can sometimes be so severe that it becomes life threatening.

A severe food allergy is an anaphylactic reaction (**Figure 1**), which involves the respiratory and circulatory systems. It often is accompanied by swelling of different body parts, hives, and itching. When you eat a food to which you are allergic, cells that "believe" they are endangered release a chemical messenger, called *bradykinin*, which stimulates the release of another chemical stimulator, *histamine*. Histamine is produced by the circulating white blood cells known as basophils and by mast cells found in connective tissues. Histamine changes the cells of the capillaries, increasing permeability. The enlarged capillary causes the area to redden. Proteins and white blood cells leave the capillary in search of the foreign invader, but, in doing so, they alter the osmotic pressure. The proteins in extracellular fluids create another osmotic force that opposes the osmotic force

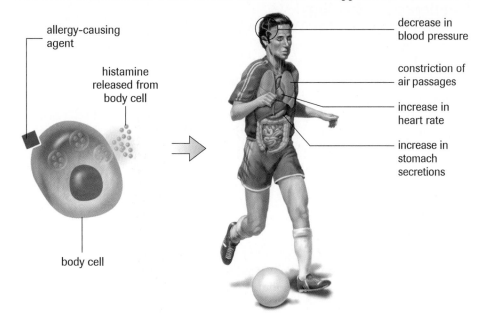

allergy-causing agent

histamine released from body cell

body cell

decrease in blood pressure

constriction of air passages

increase in heart rate

increase in stomach secretions

Figure 1

An anaphylactic reaction to an allergy-causing agent can be life threatening.

in the capillaries. Less water is absorbed into the capillaries, and tissues swell. These reactions can be brought on by drugs, vaccines, and some foods (peanuts, shellfish, eggs, berries, and milk) in individuals who are sensitive to these substances. Anaphylactic shock can occur very quickly. Weakness, sweating, and difficulty breathing are indicators of the condition. Nausea, diarrhea, and a drop in blood pressure may also occur. Medical precautions may range from carrying a kit with adrenaline (epinephrine) to carrying antihistamines. People with severe food allergies should wear a medical alert bracelet or necklace and read all food labels carefully.

Autoimmune Disease

The immune system can make mistakes. As you have already learned, allergies are caused when the immune system perceives harmless substances to be life threatening. The immune system can also go awry and launch an attack on the body's own cells. The renegade lymphocytes treat the body's cells as aliens and make antibodies to attach to their own cell membranes. Many researchers believe that most people have the mutated T cells and B cells that are capable of attacking the body; however, the renegade cells are usually held in check. The suppressor T cells play an important role in recognizing and intercepting the renegade T cells and B cells. One theory suggests that the suppressors secrete a substance that tells the macrophages to engulf the renegade cells.

The failure of the suppressor T cells to control the renegade cells can be seen in rheumatoid arthritis, where an immune response is mounted against the bones and connective tissues surrounding the joints. Rheumatic fever, another autoimmune disorder, results from an exaggerated immune response, which scars the heart muscle. Type 1 diabetes is caused by an immune reaction against the insulin-producing cells of the pancreas, and lupus is caused by the accumulation of antigen–antibody complexes that build up in the walls of blood vessels, joints, kidneys, and skin.

Drugs or serious infections can weaken the suppressor T cells, leaving the body vulnerable to autoimmune diseases. We know that the number of suppressor T cells declines with age, increasing the incidence of rheumatoid arthritis and other autoimmune diseases. Some individuals are born with defective suppressor T cells, and they battle these diseases throughout their lives. Although no single cure exists, artificial immune suppressor drugs have been developed that reduce the intensity of the attack by the renegade cells.

Frontiers of Technology:
Looking for a Cure for Multiple Sclerosis

Multiple sclerosis (MS) is an autoimmune disease in which T cells of the body initiate an attack on the myelin sheath of nerve cells. In the advanced stages of MS, paralysis results from the destruction of the insulation of the nerve cell provided by the myelin sheath.

Promise for a treatment comes from studies on monkeys that have a disease similar to human MS. In January 2001, while studying what triggers abnormal T cell responses, a group of researchers at the National Institute of Allergy and Infectious Disease in Maryland found that exposure of T cells to a small amount of myelin protein initiated an attack; however, exposure to larger amounts of the same protein reduced the immune response. Researchers found when T cells were exposed to large amounts of protein, they self-destructed. The treatment is puzzling because adding more of the antigen prevents the autoimmune response and cures the disease. That would be the equivalent to adding gasoline to a fire.

To test their theory, the researchers injected monkeys with enough myelin protein to promote an immune response. T cells attacked the myelin sheath, producing MS-like symptoms. Next, they divided the monkeys into three groups (**Figure 2**). Each group

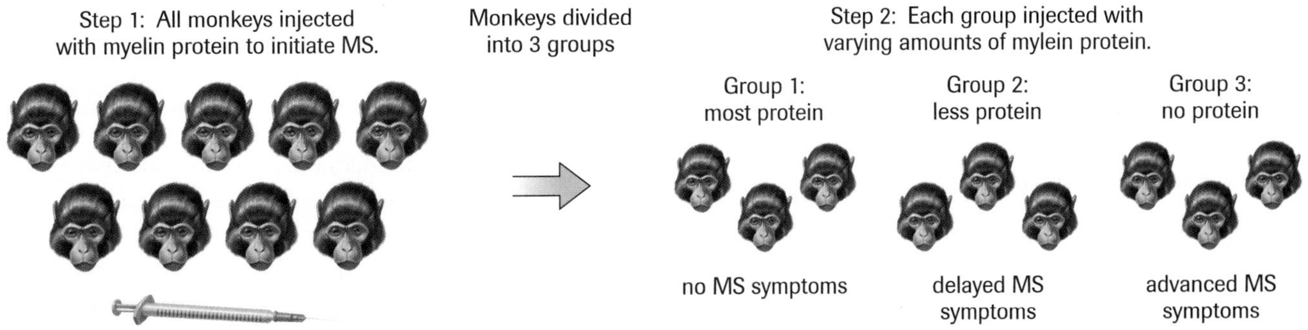

Step 1: All monkeys injected with myelin protein to initiate MS.

Monkeys divided into 3 groups

Step 2: Each group injected with varying amounts of mylein protein.

Group 1: most protein

no MS symptoms

Group 2: less protein

delayed MS symptoms

Group 3: no protein

advanced MS symptoms

Figure 2

MS research revealed that increased amounts of myelin protein reduced the immune response and appeared to cure the disease.

received injections with different amounts of myelin protein. Group 1 received the greater amount of myelin protein; Group 2, less protein; Group 3, no myelin protein.

Researchers noted the following observations:

- Group 1: no symptoms of MS

- Group 2: delayed symptoms

- Group 3: symptoms of MS

Unlike other research, which looks for a cure by suppressing the entire immune system, the targeted T cell experimental therapy does not seem to impair the effectiveness of the immune system in fighting off invading microbes.

SUMMARY *Malfunctions of the Immune System*

- Abnormal functioning of the immune system can cause two types of problems: immune-deficiency diseases and autoimmune diseases.

- Allergies occur when the immune system mistakes harmless cells for harmful invaders.

- Autoimmune diseases occur when lymphocytes treat the body's cells as aliens.

▶ *Section 10.3 Questions*

Understanding Concepts

1. What are allergies?

2. Explain how an allergic reaction to peanuts can be life threatening.

3. Why is adrenaline (epinephrine) administered as a treatment for a severe allergic reaction?

4. What causes autoimmune diseases?

Making Connections

5. What evidence suggests that suppressor T cells may be a significant factor in autoimmune diseases?

6. Give reasons why some people believe potentially deadly allergens, such as peanut butter and peanuts, should be banned from school cafeterias and other public places.

7. Select an autoimmune disease and research the latest medical advancements toward a cure. Search for information in newspapers, periodicals, CD-ROMs, and on the Internet.

 GO www.science.nelson.com

8. Research the role of histamines in an allergic reaction.

E. coli Poisoning in Walkerton, Ontario

One of Canada's worst public health disasters occurred in the Ontario farming community of Walkerton in May 2000, when heavy rains washed bacteria from cattle manure into a shallow town well, causing a deadly outbreak of E. coli poisoning. Before an advisory to boil water was issued, the sewage-contaminated water had killed 6 people and sickened more than 2000. Authorities believe the total costs of Walkerton's E. coli outbreak could potentially reach $100 million, which is nothing compared with the tragic loss of life.

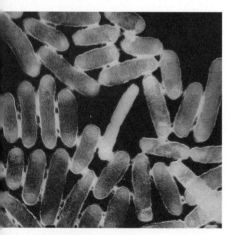

Figure 1
Photomicrograph of *Escherichia coli* 0157:H7, the pathogen that caused the deadly outbreak of E. coli poisoning in Walkerton, Ontario.

Some diseases, such as type 2 diabetes, stroke, and kidney stones are caused by failure in the normal functioning of a body system. Other diseases are associated with a foreign microorganism that invades the body. If the invader causes a disease, the invader is called a pathogen (**Figure 1**). However, not every invading microorganism causes disease. Many microbes inhabit your large intestine, skin, and hair. Microbes in the large intestine help process waste products and provide your body with some needed vitamins. Microbes that live on your skin feed off your body oils and the dead cells that are sloughed off. They protect your skin from harmful microbes that might attempt to colonize.

Communicable or contagious diseases are transmitted from person to person. All are associated with a microorganism that moves from one host to another. The host is the organism in which the microbe lives. Biologists believe that many human diseases originated from domestic animals. A random mutation may have allowed these microbes to jump species.

Rabies, a **virulent virus**, is still a feared disease today, despite a very effective vaccine. The microbe moves from the blood into nerve cells, causing a rapid deterioration of the brain. The image of a mad dog biting you is enough to send chills down your spine. However, rabies should not be thought of as a highly contagious disease. It is not transmitted by direct contact, nor can you get it by washing in water from which a rabies-infected animal drank. The microbe that causes rabies must be introduced directly into your blood stream.

AIDS, or acquired immune deficiency syndrome, is another disease that is not highly contagious. The HIV virus, responsible for AIDS, cannot be transferred by shaking hands with someone who is infected, nor is it airborne or transferred in water. Although some pathways in the transfer of this disease are still unknown, most experts suggest that the HIV microbe cannot pass from one human to another by casual contact. AIDS can, however, be transferred by a blood transfusion or through direct sexual contact.

Cholera is an example of a disease that is both virulent and highly contagious. It is still responsible for many deaths, especially in Asia and Africa. This highly infectious bacterium is released with the wastes from the bowel and can be transmitted to a new host through water or food. Influenza—or as it is more commonly known, the flu—may be an all too familiar contagious disease to you. Its symptoms are stomach upset, muscle ache, and high fever. This highly infectious virus was responsible for 21 million deaths during the great flu epidemic of 1918–19. The common cold is another contagious microbe that can make life unpleasant. Fortunately, the common cold is not a virulent pathogen. Most people recover from the virus.

Bubonic Plague: The Black Death

Historically, nothing has equaled the destruction and fear that the bubonic plague brought to Europe in the 14th century, and again in the 17th century. Better known as the Black Death, bubonic plague most likely originated in China, where there was an outbreak of the disease in 1330. Caused by a bacterium known as *Yersinia pestis*, the plague was carried by rats and transmitted to humans by flea bites. The symptoms of the Black Death began with headache, nausea, aching joints, vomiting, and a general malaise. A fever then set in, along with painful swelling of the lymph nodes in the neck, armpits, and groin. (This swelling is referred to as buboes, from which the name *bubonic plague* is derived.) Red spots then turned up on the skin, which later turned black.

virulent virus a harmful microbe that is effective at overcoming the defence mechanisms of the host

The plague made its way from China to Europe in October of 1347 on merchant ships. The disease spread along trade routes through France and on to other countries. By the following summer, the disease had reached England, where it got the name the Black Death. Bubonic plague killed so many Europeans that it was credited with ending feudalism, because the cheap labour force was seriously depleted. In 1348, more than half of the urban population of Europe died from it. The epidemic of 1665 turned the city of London into a morgue (**Figure 2**). Some historians estimate that 60% of the population died of the plague.

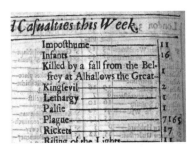

Figure 2
Extract from a weekly death register, London, 1665

How Do You Catch a Disease?

Pathogens can be transmitted by various means (**Table 1**). **Figure 3** shows one of the most common ways in which infections are transmitted. A cough or sneeze discharges millions of tiny water droplets into the air, each of which houses thousands of tiny microbes. This is called a droplet infection. The tiny droplet can remain suspended in the air for a long period of time, especially if it is very small. Unfortunately, the droplet can be inhaled by another host. Another common port of entry is through the mouth, when the droplet lands on food, water, or an eating utensil.

Table 1 Transmission of Some Common Pathogens

Disease	Pathogen	Transmission route
AIDS	HIV virus	body fluids
typhoid fever	bacteria	food or water
influenza	virus	droplet infection
common cold	virus	droplet infection
bubonic plague	bacteria	insect vector
athlete's foot	fungus	direct contact
ringworm	fungus	direct contact

Figure 3
Some infections can be passed by water droplets in a cough or a sneeze.

Some infections enter your body through water. These are known as waterborne infections. A number of these infections are associated with the unsanitary disposal of human wastes, which is why water treatment is essential, particularly in communities that depend on underground springs for drinking water. If the water becomes fouled by sewage, the disease can be transmitted to everyone who uses the groundwater. The disease may spread, often as far as a kilometre, from the source of infection. It is for this reason that many lakes in Ontario forbid cottage owners to use outhouses.

Direct contact infections are slower in spreading disease. A person serving or preparing food may pass on the infection. An animal bite, or physical contact with the infected person's clothing or utensils, can also cause the infection. Most contact infections only enter the body at specific sites. Gonorrhea, a serious venereal disease, is passed from one carrier to another during sexual intercourse. The microbe must enter the body through a mucous membrane. Most commonly, the microbe enters the body through the genitals; however, the microbe can enter through other mucous membranes, such as the mouth. As we learned earlier in this section, the rabies virus must pass directly into the blood before it can infect your body. Neither rabies nor HIV, the virus that causes AIDS, can gain access to your body by resting on the skin. That is why casual contact with either infected persons or their clothes cannot infect you with the disease.

Another way that diseases are spread is by a **vector** (**Figure 4**). A vector is a disease-carrying animal, most often an insect. One type of vector is the housefly. The fly is attracted to animal wastes from which it may pick up microbes. It then may transfer these microbes to foods. Lice, ticks, fleas, and mosquitoes also act as vectors for pathogens.

Figure 4
Flies, lice, ticks, fleas, and mosquitoes all act as vectors for pathogens.

vector organism—most often an insect—that carries and transmits diseases to other organisms

▶ TRY THIS activity *Following an Infection*

In this activity, you will simulate the spread of an infection. Each member of your class will be provided with a numbered plastic cup filled with a mystery fluid. One of these cups will contain an "infection."

 Safety goggles and a lab apron must be worn for the entire laboratory.

Materials: index card, numbered plastic cup of mystery fluid

- Write your name and cup number on your index card.
- Share your mystery fluid with a classmate. Pour all of your fluid into your partner's cup. Then your partner pours half of the combined liquids back into your cup. Each partner then records the other's cup number on his or her index card.
- Repeat the previous step until each student has shared fluids with exactly three other students.
- Once all exchanges have occurred, your teacher will add a drop of phenolphthalein indicator to each of the cups. A pink colour indicates an infection.
  (a) Can you identify the origin of the infection?
  (b) If so, then identify the source. If not, why not?

Case Study Disease Transmission

Even before Louis Pasteur proposed the germ theory, a Hungarian physician, Ignaz Semmelweis (1818–65), made an important connection between sanitation and disease. Semmelweis was distressed by the high death rate from childbed fever among women who gave birth to babies in hospitals in Vienna, while the death rate among women who gave birth at home was considerably lower. Suspicions were further peaked when it was discovered that a physician who had cut himself while dissecting a cadaver developed childbed fever, the same fever linked to the death of mothers who had given birth in hospitals.

Semmelweis insisted that the doctors wash their hands with a solution of chlorinated lime before attending to women giving birth. Within a year, the death rate in maternity wards fell from 12% to 1.5%. However, many Austrian doctors were humiliated by the inference that they had been unwilling accomplices in the deaths of women in labour. Semmelweis was driven out of Vienna and returned to Budapest, where he again reduced the incidence of childbed fever. His recommendations were accepted in Hungary and the government instructed all district authorities to adopt his measures. Once again, Semmelweis encountered opposition from many prominent doctors. In 1865—the same year that English doctor Joseph Lister independently demonstrated how germs could be controlled by chemical agents—Semmelweis suffered a mental breakdown and was admitted to a mental institution where he soon died.

One of the most famous cases of disease transmission involves a cook who worked for eight families in the state of New York in the early 1900s. Known as "Typhoid Mary," this middle-aged woman showed no symptoms of typhoid but spread it to those who ate the food she prepared. Over a 10-year period, Mary Mallon was responsible for seven epidemics that infected more than 50 people with typhoid fever. One estimate indicates that she may have been indirectly responsible for as many as 200 cases. Authorities first tracked her down in 1907 and isolated her in a cottage on an island. In 1910, she was released after agreeing not to take a food-handling job again, but she did, causing more typhoid outbreaks. The New York City health authorities found her again in 1915 working as a "Mrs. Brown" in a maternity hospital in New York, where she had infected another 25 people (of whom two died). She was sent back to the island and remained there for the remaining 23 years of her life.

The carrier is the key link in the transmission of typhoid. The convalescent patient continues to release both solid and fluid wastes containing the harmful microbe. The pathogenic microbe is transferred to food and finally to a new host. The bacteria then incubates in the new host, usually for 14 days, and then symptoms begin to appear. High fever is followed by chills. In most cases, typhoid fever causes mucus to build up in the respiratory tract. The patient experiences difficulty breathing and, in many cases, the disease produces rosy spots on the abdomen. The greatest damage that the microbe does is to the digestive tract; small holes or perforations in the small intestine often begin to bleed.

▶ **Case Study** *Questions*

Understanding Concepts

1. What might account for a higher death rate in hospitals? Where would you begin to look for the source of the problem?

2. In the Semmelweiss case study, what evidence suggests that the doctors themselves were transmitting childbed fever?

3. What was the source of the disease?

4. Speculate as to why Mary Mallon did not show any symptoms of typhoid.

5. Why are perforations to the intestine dangerous?

6. In the case of typhoid fever, how might the cycle of disease transmission be broken?

Discovery of Pathogens

When someone comes down with a fever and chills, the first thing that person suspects is the flu. The comment you might hear is "Yes, I believe a new bug is making the rounds." The person who made the comment may not know how a virus differs from bacteria but does have a general idea of what causes disease. Remarkable as it might seem, that is a great deal more than people once knew. Diseases were thought to be caused by foul air, evil spirits, or bad blood.

A direct link between disease and microbes was not established until French chemist Louis Pasteur (**Figure 5**) began experimenting with yeast in 1854. Although many people suspected that invisible organisms cause disease, no one had proven the role of microbes. The earliest description of bacteria came from Dutch physicist Anton van Leeuwenhoek, the inventor of the microscope. Although van Leeuwenhoek described bacteria in 1683, he did not know what they were.

Louis Pasteur was interested in discovering why wine often went bad during the fermentation process. While studying wine under a microscope, Pasteur identified tiny round spheres suspended in the wine. The spheres, he discovered, were living yeast cells. As the population of yeast cells grew, the amount of alcohol in the wine increased. How did the presence of yeast affect the grape juice? Pasteur concluded that yeast cells produce alcohol.

The spoiled wine did not contain alcohol but another chemical, called *lactic acid*. When Pasteur looked at the spoiled wine under a microscope he found rod-shaped bacteria. He concluded that the rod-shaped bacteria were causing the wine to go bad. The microbe was the villain!

It wasn't long before Pasteur and other scientists began looking for the villainous microbe as the cause for human diseases. Pasteur's discovery led to what has come to be

Figure 5
French chemist and bacteriologist Louis Pasteur (1822–95) proved that the fermentation process was caused by a specific microbe. Best known for introducing the process of pasteurization, Pasteur also isolated the bacterium that causes anthrax and pioneered the vaccine against rabies.

known as the germ theory of disease. However, many scientists around the world did not accept the germ theory. Pasteur was seen as an opportunist by many of his opponents who saw him as a chemist meddling in the field of medicine. They believed that a chemist should leave the study of health and disease to biologists and doctors. Pasteur was treading on their ground. Even after his theories were supported by an abundance of experimental evidence, some of his opponents continued to disagree.

Bacteria

Bacteria are single-celled organisms that can damage the tissues of the body. Some bacteria, such as the microbe that causes typhoid fever, attack the cells of the stomach, while others, like tuberculosis, destroy the tissues of the lung. Some bacteria cause damage by producing poisons or toxins. The toxins are metabolic wastes, harmless to the microbe but poisonous to the human host. The toxins are released from the microbe and dissolved in body fluids, then are carried by the blood to vulnerable tissues often a great distance from the site of the primary infection. Diphtheria, a bacterial disease, is caused by a bacterium that lodges in the throat and then produces toxins that affect tissues throughout the body.

Two forms of food poisoning are also caused by toxins. The more serious and often fatal form of food poisoning is botulism, a disease caused by a toxin produced by bacteria that grow in environments that lack oxygen. The bacteria that cause botulism can survive in certain improperly processed canned foods or home preserves. Another type of food poisoning, salmonella, is much more common, but less likely to be fatal.

Viruses

Viruses are much smaller than bacteria (**Figure 6**). These tiny pathogens were not seen by scientists until 1927, although they speculated about their presence. Unlike the much larger bacteria, a virus is not a cell. It does not feed and does not excrete wastes. A virus has none of the structures commonly associated with living cells. Biologists do not even describe the virus as a living thing, which might make you ask how it causes so much harm. In fact, isolated from living cells, viruses cannot even reproduce.

The virus becomes active once it enters a living cell. Once inside, it takes over the machinery of the cell. The virus uses the cell's raw materials and energy to make new virus particles. The virus is able to reproduce inside another cell. The host cell becomes a slave to the virus. Many viruses have their host cell produce virus particles until the host cell dies.

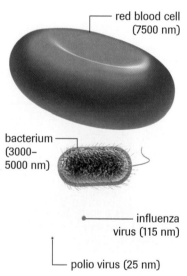

Figure 6
Comparative sizes of disease-causing pathogens

red blood cell (7500 nm)

bacterium (3000–5000 nm)

influenza virus (115 nm)

polio virus (25 nm)

HIV and AIDS

Acquired immune deficiency syndrome, or AIDS, describes several disorders associated with the human immunodeficiency virus, or HIV. Two different types of HIV have been identified: HIV-1 was discovered in 1981, HIV-2, in 1985. As of early 2002, no cure for AIDS existed. However, advances in microbiology, genetics, and molecular biology, along with improvements in the microscope, are causes for cautious optimism. For the time being, an educated public may prove to be the best defence against the virus.

HIV must be directly transmitted. Unlike chicken pox and flu viruses, which can be transmitted through the air, HIV must enter the bloodstream. HIV has been found in human body fluids. It is spread primarily through sexual contact and by the introduction of blood or blood components into the bloodstream through the sharing of needles or syringes for injection drug use. HIV can also be transmitted to infants during pregnancy or at the time of birth. In rare cases, HIV has been transmitted through the breast milk of an infected mother.

Although HIV is very tiny, the damage it causes can be devastating. What makes HIV more insidious than other viruses is that it attacks the immune system directly. The target of HIV are the helper T cells (sometimes referred to as T4 lymphocytes), the cells that act as guards against invading pathogens. Thus, HIV destroys the body's own defences, rendering it incapable of defeating other pathogens.

Prions

In the 1970s, while studying diseases that cause brain tissue to appear spongy, Dr. Stanley Prusiner of the Department of Neurology at the University of California, San Francisco, was unable to find any evidence of foreign nucleic acids or virus particles. By 1982, he had determined that the infectious agent was not a microbe but a protein, which he named a *prion* (derived from "proteinaceous infectious particle"). The news challenged Pasteur's theory that only microbes cause disease. When Prusiner injected prions of different abnormal configurations into animals, he found that they began to develop abnormal proteins that resembled the injected prions (**Figure 7**).

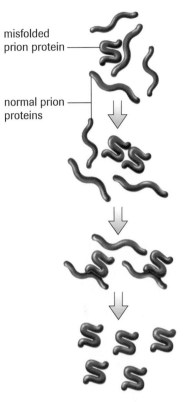

misfolded prion protein —

normal prion — proteins

Figure 7
Prions are abnormally folded proteins (shown in blue) that can influence normal proteins (shown in grey) to assume the abnormal shape.

SUMMARY   *Pathogens and Disease*

- Not all microbes cause disease, and not all diseases are caused by microbes.
- Contagious diseases are transmitted from host to host. Disease can be transmitted by
  – droplet infection
  – water
  – direct contact
  – vectors
- Virulent diseases are those that are difficult for the immune system to defeat.

▶ *Section 10.4 Questions*

Understanding Concepts

1. Which of the following diseases are communicable?
   (a) typhoid fever   (d) AIDS
   (b) heart disease   (e) influenza
   (c) cancer

2. Distinguish between pathogenic and nonpathogenic microorganisms.

3. Most people would be flattered if you referred to them as the perfect host. Explain how your body acts as the perfect host for microbes.

4. Compare AIDS with the common cold. Which is more virulent? Which is more contagious? Explain your answers.

5. What animal served as the vector for the bubonic plague? Explain this animal's role in the disease.

6. Why was Typhoid Mary so dangerous?

Making Connections

7. In 1991, David Acer, a Florida dentist with AIDS, was thought to have infected five of his patients with HIV. It is not known whether all were infected by the dentist himself, or whether he failed to use new gloves and sterile instruments after treating other patients who were HIV positive. This widely publicized case prompted demands for compulsory HIV testing for people in the health-care professions. Some people were concerned that, during a medical procedure, an HIV-positive health-care worker might be accidentally cut and bleed into a patient, thereby transmitting the virus.
   (a) What other careers present dangers for HIV infection? Explain where the danger lies.
   (b) Choose one of the occupations and indicate what measures would help reduce HIV and AIDS.
   (c) Although there is no requirement at the moment that medical workers be tested, many health-care practitioners are concerned about possible mandatory testing. Why might they be concerned?
   (d) Should AIDS testing be mandatory for all people? Justify your position.
   (e) Is HIV transmittable by blood-sucking insects, as is malaria? Explain.

active immunity occurs when the body produces antibodies against an invading antigen

passive immunity occurs when antibodies are introduced into the body directly; provides only temporary protection from the agent, unlike active immunity

Induced immunity is brought about by intervention from either within the individual's own body or from an outside source. Immunity may be acquired naturally when antibodies are produced in response to a specific infectious agent. Since the body takes an active role in producing antibodies that identify and destroy the invading antigen, this type of immunity is referred to as **active immunity**. Contact with one strain of influenza (flu) assures continued protection as long as the antibodies remain in the body's system.

However, immunity to one strain of flu does not mean immunity to every type. Because each strain of flu virus has a characteristic shape, antibodies having a geometry that allows them to destroy one strain may prove ineffective against another.

Passive immunity occurs when antibodies are introduced into the body directly. The transfer of antibodies from mother to baby through the placenta provides temporary protection for the baby. The baby maintains a degree of resistance for about six months. Another form of passive immunity can occur when antibodies from the plasma of an animal that has been exposed to a disease are transferred into the blood of another animal. This temporary measure will provide a few weeks of protection. A tetanus shot given after suffering a puncture wound from stepping on a rusty nail, for example, contains antibodies produced by a horse. You will then be protected from the toxin for a short time. This may explain why doctors routinely ask you when you had your last tetanus shot.

The first disease to be controlled by a chemical agent was syphilis, a crippling venereal disease. Mercury was used to treat it as early as 1495; however, the treatment was often more devastating than the disease. Mercury proved to be toxic to both the microbe causing the disease and the patient.

Vaccination

vaccines antigen-containing substances that can be swallowed or injected to provide continued immunity to specific diseases

It is difficult for us to understand what it must have been like before **vaccines**. Large families were the only hedge against epidemics of the bubonic plague and smallpox. Microbes killed both the rich and poor without discrimination. Mary II, Queen of England, was a victim of smallpox. More than 60 million Europeans died from the smallpox virus during the 18th century.

Vaccines are not a Western invention. The ancient Chinese scraped dried bits of skin from smallpox victims and then blew the powder into the nose of healthy people. Records show that Greeks and Turks also experimented with innoculations long before western Europeans thought of vaccinations. They pricked the sores of smallpox patients with a needle, and then poked it into the skin of healthy people. The people who received the needle usually demonstrated some resistance to the disease, but not all were so fortunate; the introduction of live active smallpox was occasionally fatal. Also, scratching the skin with unsterilized needles often created as many problems as it solved. A partial immunity from one microbe often meant the introduction of another.

Vaccines are one of the most effective methods of inducing active immunity. Vaccination involves introducing a weakened or dead microbe, rather than preventing it from entering (**Figure 1**). (In some cases, even a small portion of protein from the microbe is used to create the vaccine.) A microbe that is full strength might reproduce in the body faster than the immune system can produce antibodies against it. Such a microbe would cause the disease before the immune system could provide immunity. Today, genetic engineering technologies, such as recombinant DNA, have been used to provide active immunity.

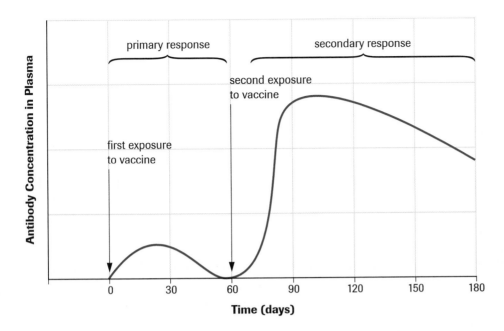

Figure 1
The first exposure to a vaccine stimulates the body to produce some antibodies against the antigen. The second stimulates greater antibody production.

Proteins developed in the laboratory have a similar geometry to the harmful antigen and stimulate the immune response without presenting much of a risk. Hepatitis B vaccines have been produced using this new approach.

No antibody lives in the body forever. This is why some vaccines require a booster. The booster provides a second exposure to the antigen. Once the antigen is reintroduced, antibody production accelerates again to defeat the harmful agent (**Figure 1**).

Smallpox and the First Vaccine

In England, a country doctor named Edward Jenner developed the first vaccine. His technique of injecting a less harmful virus into the body to stimulate the immune system was truly remarkable. The smallpox vaccine was developed nearly 150 years before the virus could be seen with an electron microscope. The action of white blood cells was not understood until nearly 100 years after the administration of the smallpox vaccine.

Despite a lack of background information, Edward Jenner developed a successful vaccine in 1796. By noticing that rural townsfolk seemed less susceptible to the deadly smallpox, Jenner began to formulate a hypothesis. He believed that the immunity of the country people must be related to environment. Jenner noted that dairymaids had a particularly low incidence of smallpox, but a high incidence of a much less harmful disease called cowpox (**Figure 2**). Cowpox produces many of the same symptoms as smallpox.

Jenner reasoned that cowpox must have provided some immunity to the more virulent smallpox virus. To test his theory, Jenner injected the pus from a festering wound on the arm of a dairymaid, Sarah Nelms, into a young boy named James Phipps. Not suprisingly, James developed cowpox. However, he recovered from the mild infection very quickly. Two months later, Jenner inoculated James once again, but this time with the more virulent smallpox. When James failed to develop smallpox, Jenner declared that a successful vaccine had been developed.

When the helper T cells identify cowpox, they signal the B cells to produce antibodies against them. The cowpox virus has a shape that is very similar to that of smallpox. Fortunately, the antibodies for cowpox also prove successful against smallpox. If Jenner had injected smallpox first, James might have died. Smallpox reproduces much faster than cowpox. Fortunately for James, the antibodies against cowpox lay in wait for the more dangerous smallpox.

Figure 2
The cowpox virus has a similar shape to that of smallpox.

Pasteur and the Rabies Vaccine

Rabies, although never the mass killer of smallpox, was greatly feared for its devastating effects. The tiny virus migrates from the blood into the nervous system, where it destroys cells, causing convulsions and great suffering.

The technique used for smallpox cannot be applied to all microbes. Most virulent microorganisms do not have a harmless twin like cowpox. Injecting a full-strength virus means developing the disease. Louis Pasteur was able to grow the virus in tissue cultures. Using trial-and-error testing, he found ways of weakening the virus.

Because the rabies virus remains dormant in the body for 14 days, a weakened virus stimulates the production of antibodies that lie in wait for the virulent rabies. Once the dormant virus emerges, the antibodies destroy it before it gains a foothold in the body.

In 1885, Pasteur administered the vaccine to Joseph Meisner, a nine-year-old boy who had been bitten by a dog carrying rabies. Joseph lived and eventually became the gatekeeper at the Pasteur Institute in Paris. His devotion to the chemist was so great that he committed suicide to avoid opening Pasteur's crypt for the invading Nazi army in 1940.

Jonas Salk and the Polio Vaccine

Poliomyelitis (or polio) was responsible for many deaths before the Salk vaccine of 1955. Many died, but far more were left crippled for life by this devastating disease.

Microbiologist Jonas Salk (**Figure 3**), the son of a Russian immigrant garment worker, became an overnight celebrity with the development of a polio vaccine. By killing the polio virus in a bath of formaldehyde, Salk was able to inject the polio into test animals without causing the disease. The viral coats stimulated the production of antibodies. Subsequent invasions of any virulent polio came up against these antibodies. Salk later founded the Jonas Salk Institute for Biological Studies in California, where he continued his research into the causes, prevention, and cure of diseases such as cancer and AIDS. Dr. Salk never patented his polio vaccine but distributed the formula freely, so the whole world could benefit from his discovery.

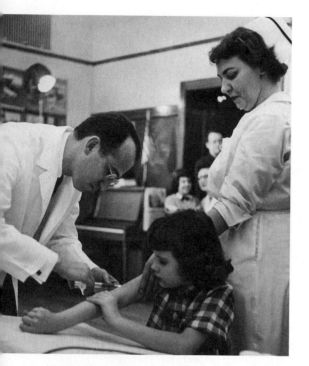

Figure 3
Dr. Jonas Salk (1914–95)

Chemical Controls

Dr. Paul Ehrlich (1854–1915), a German bacteriologist, is considered one of the founders of modern immunology and a pioneer in the study of chemotherapy. His work on immunity earned him a joint Nobel Prize (with Ilya Ilyich Mechnikov) in Physiology or Medicine in 1908. One of Ehrlich's many and varied areas of research centred on finding chemical substances that would go straight to the pathogenic organisms at which they were aimed and leave host cells undamaged. He believed that these "magic bullets" existed because there are stains that selectively tint certain organelles or cells and leave others uncoloured.

When the organism that causes syphilis was discovered, Ehrlich chose to look for a "magic bullet" for this disease because it was a particularly serious problem in the 19th century. Ehrlich tested hundreds of chemicals before selecting salvarsan in 1910. This arsenic compound selectively blocks chemical reactions of the microbe without interfering to any great extent with those of the patient. The idea of using a chemical to treat a patient was revolutionary and, therefore, not well accepted by some members of the scientific community. The great success that people such as Edward Jenner and Louis Pasteur had had with vaccines fuelled the opposition to Ehrlich's new approach. Vaccines stimulate the immune system, which then destroys the microbe. Chemical therapy, by contrast, attacks the organism directly.

Chemical therapy was largely ignored until 1935. A drug called sulfanilamide, one of the first antibiotics, was found to selectively block a chemical reaction essential to bacteria. **Figure 4** shows how the drug interferes with the enzyme catalyzed reaction of the microbe. The drug attaches itself to the enzyme, thereby preventing it from combining with its substrate.

Antibiotics

Antibiotics are a special kind of chemical agent, usually obtained from living organisms. It has been long known that living things compete with each other for food and space. Some microbes gain an advantage by producing substances toxic to their competitors.

Long before European and North American physicians hailed penicillin as a wonder drug, the Chinese used mouldy soybean curds to treat boils. For many years, people have buried diseased organisms. In 1924, it was discovered that a soil organism called actinomycetes produced a bacteria-killing substance called *actinomycetin*.

In 1928, Alexander Fleming (**Figure 5**) observed that agar plates inoculated with bacteria had become contaminated with mould. The mould gradually grew over the bacteria on the plate. A bacteria-destroying secretion, now called penicillin, was destined to become the miracle drug of the 1940s. Fleming was jointly awarded the 1945 Nobel Prize in Medicine or Physiology "for the discovery of penicillin and its curative effect in various infectious diseases."

Penicillin interferes with bacterial cell walls. The weakened walls are incapable of withstanding the pressure created by their own cytoplasm. Like a flawed dike, the cell walls of the bacteria eventually burst and the cells die.

Since the 1940s, many new antibiotics have been uncovered. Ideally, the antibiotic should kill the invading bacteria without interfering with a person's normal body functions. However, individuals with very sensitive helper T cells identify the antibiotic as a harmful antigen and the immune response is activated. Allergies to an antibiotic can often be even more life threatening than the invading bacteria. Everything from stomach upset to puffiness around the eyes may be an indication of an allergic reaction to an antibiotic. You should contact a physician if you experience an allergic reaction to an antibiotic.

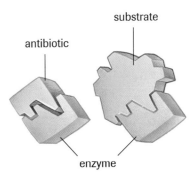

Figure 4
When the antibiotic combines with the enzyme, it prevents the enzyme from binding to its substrate molecules. Enzyme-catalyzed reactions are essential for the microbe's survival.

Figure 5
In 1928, British bacteriologist Sir Alexander Fleming (1881–1955) accidentally discovered the effect of penicillin on bacteria.

Antibiotic-Resistant Bacteria

Bacteria are capable of multiplying so rapidly that favourable mutations can change the characteristics of a bacteria population in a relatively short time. For example, the bacterium *Escherichia coli (E. coli)* contains about 5000 genes and it can be expected that a mutation will occur by chance in every 1 million copies. Considering that populations can double every 20 minutes under optimal conditions, that results in 1 mutant gene in every 200 bacteria. A typical spoonful of soil, containing over a billion bacteria, would contain over 5 million mutant bacteria.

Unfortunately, some of the random mutations have allowed microbes to become resistant to penicillin and a wide range of other antibiotics. Some strains of *Staphylococcus aureus* have created huge problems in hospitals because of their ability to withstand a large number of antibiotics. Particularly resistant strains of tuberculosis have also emerged among homeless people in crowded cities.

donor cell   **recipient cell**

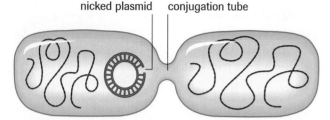

nicked plasmid   conjugation tube

(a) A conjugation tube has already formed between a donor and a recipient cell.

donor cell   **recipient cell**

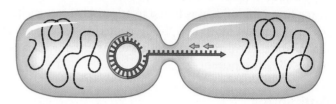

(c) In the recipient cell, replication starts on the transferred DNA.

(b) DNA replication starts on the nicked plasmid. The displaced DNA strand moves through the tube and enters the recipient cell.

(d) The cells separate from each other; the plasmids circularize. When the process is complete, both cells contain a copy of the plasmid.

Figure 6
The process of conjugation

The insertion of genetic material into a plasmid of bacteria has allowed antibiotic-resistant genes to spread rapidly within a population. When plasmids containing genetic mutations that enable drug resistance are transferred during conjugation, new microbes acquire the genes (**Figure 6**).

The exposure of large populations of relatively harmless bacteria to low dosages of antibiotics is believed to have caused much of the problem. Low dosages of antibiotics have been added to cattle feed to prevent mastitis, a bacterial udder disease that affects milk quality and production. The antibiotic kills the vast majority of bacteria that could cause problems. Unfortunately, those bacteria that survive have mutations that enable them to deal with the antibiotic in low concentrations. Now because competing bacteria have been killed, the growth and reproduction of the drug-resistant bacteria increases—they have more food and space. To kill those that remain, a higher concentration of antibiotic is required or a new drug that attacks a different metabolic site of the bacteria must be employed.

A secondary problem arises in this situation. Milk, laced with antibiotics, is consumed by humans. Because humans naturally harbour many microbes, the same process of natural selection is repeated. Microbes containing mutated genes able to withstand the antibiotic begin changing the genetic makeup of the population.

One of the main factors in the development of drug-resistant bacteria is the overuse of antibiotics by humans to prevent disease. Antibiotics can eliminate harmful bacteria, but do nothing to eliminate the noncellular virus. Sore throats, for example, can be caused by either a viral or bacterial infection. To identify the bacteria, a culture must be grown on a petri dish, but the procedure takes at least 36 hours. During that time, the infection might spread even further. To safeguard against further complications, antibiotics are often prescribed; however, a myriad of other naturally occurring bacteria are exposed to the drug. The few that remain carry a genome that provides resistance to the drug. Should the actual source of the sore throat be viral, the antibiotic has no beneficial effect.

Guided Missiles: Monoclonal Antibodies

When a harmful invader enters the body, the immune system releases antibodies. However, of the great number of antibodies released into the system, only a few may have the correct shape to tie up the invader. Your immune system releases many antibodies on the assumption that other invaders may be present.

A new technology is able to make a large number of antibodies in pure form by taking advantage of cloning, a technique that permits the exact duplication of identical cells. Developed in 1975 by George Kohler and Cesar Milstein of Cambridge University, England, **monoclonal antibodies** have become a promising weapon in the fight against cancer and other diseases. In this procedure, an antibody-producing cell from a mouse is fused with a cancer cell. The new cell, called a hybridoma, inherits the characteristics of the two individual cells—producing antibodies like a lymphocyte, while reproducing at the rate of a cancer cell (**Figure 7**). Each clone produces a specific antibody.

monoclonal antibodies antibodies secreted by hybridomes; new cells created by the fusion (or cloning) of two cells

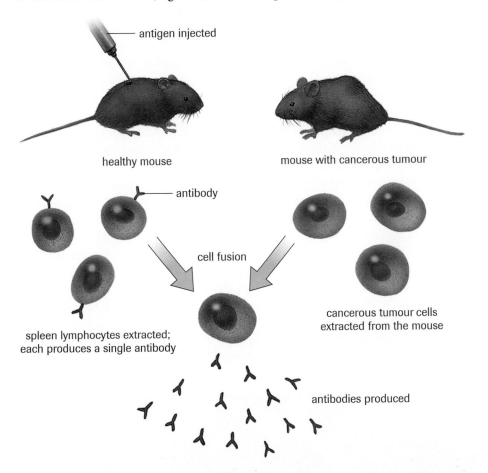

antigen injected

healthy mouse

mouse with cancerous tumour

antibody

cell fusion

spleen lymphocytes extracted; each produces a single antibody

cancerous tumour cells extracted from the mouse

antibodies produced

Figure 7
The production of hybridomas

The possibilities of monoclonal antibodies are intriguing. Antibodies could be grown in tissue cultures, collected, and given to patients who have a particular disease. For example, one culture could produce antibodies against hepatitis, while another might provide antibodies for specific cancer cells. Cancer cells have marker antigens on their cell membranes that distinguish them from normal cells. If the monoclonal antibody could identify the cancer antigen, it could carry a poison to the cancer cell. The potential of using monoclonal antibodies to carry radioisotopes, drugs, or killer toxins to cancerous tumours has sparked a great deal of research. Since cancer cells and normal cells are very similar, many of the undesirable side effects of chemotherapy involve the interference of normal cell action. The possibility of having a cancer drug that can be delivered to cancer cells without destroying normal cells now looks promising.

Biological Warfare

The recorded history of biological warfare reaches back to about 600 B.C. Solon, an Athenian legislator, poisoned the water supply of the city of Kirrha. Roots from the *Helleborus* plant, which contain a particularly potent toxin, were placed in the water. Another example of biological warfare comes from the 14th century, when the Tartar army hurled the dead bodies of plague victims over the walls of the city of Kaffa, hoping that the disease-ridden corpses would ignite an epidemic.

During World War II, British, American, and Canadian armies were involved with the development of biological weapons. Porton Down in England, Suffield in Alberta, and Camp Detrick in Maryland were designated as research and testing stations. The deadly anthrax bacterium, which affects both cattle and people, was the preferred microbe (**Figure 8**). Anthrax meets all the criteria of a biological weapon: the deadly spores of the rod-shaped bacterium live for long periods of time, are highly contagious, and are resistant to many environmental factors. American and British armies drew plans to make thousands of anthrax bombs, which would disperse the microbes on impact. Fortunately, World War II ended before the plan was implemented.

The Japanese army was also engaged in the development of biological weapons. Pingfan, a small village near Harbin, China, housed more than 3000 researchers, technicians, and soldiers dedicated to exploiting the disease-causing properties of typhoid fever, anthrax, cholera, and the plague.

Almost any disease-causing agent can be exploited for biological weaponry. The microbe, or toxin produced by the microbe, need not even be harmful to humans. The destruction of livestock, cereal grains, or bacteria found in the soil could create food shortages and cause economic ruin. Fortunately, few organisms are well suited for mass destruction. For example, HIV, the virus that causes AIDS, is not highly contagious. HIV cannot be transmitted through the air or by casual contact. Therefore, releasing the virus into a city's drinking water will not create an epidemic. HIV must be transmitted in body fluids such as blood. On the other hand, the bacterium *Clostridium botulinum* produces one of the most powerful poisons known to humans. It has been estimated that 1 kg of the toxin placed in a typical water reservoir could kill 50 000 people. Sixty percent of the population would die in less than 24 hours. However, this microbe would have little effect if released into the air. *Clostridium* must be cultured in oxygen-free environments, where it is capable of producing the harmful toxin. It seems, then, that even the most dangerous of microbes have some natural controls that have evolved over time. No disease-causing agents can survive if their hosts are eliminated.

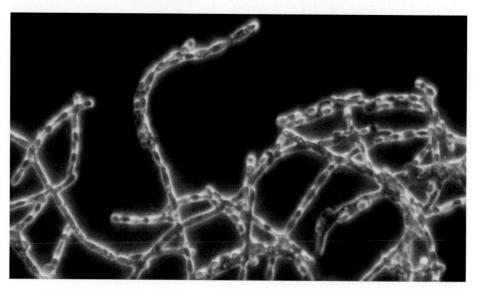

Figure 8
An electron micrograph of *Bacillus anthracis*, the bacteria that causes anthrax

The act of combining genes for weaponry could pose a serious threat. Merging genes that permit rapid reproduction, for example, with those that demonstrate a resistance to environmental factors could create what is known as a "superbug." Bacteria that carry drug-resistant genes in their plasmids already pose problems. The genetic information can be duplicated and passed between microbes during sexual reproduction. Consider the possibility of a microbe, which is resistant to a battery of antibiotics, being introduced to an enemy population. The attacking army might be able to construct a secret drug that would be capable of protecting their allies from the microbe. They could then remove the enemy selectively. By the time the enemy found an antibiotic for the disease, a significant number would have been killed and their resistance weakened. The hybrid microbe might be ideal for biological warfare in another way. It is quite possible that disease-causing microbes could escape detection by both the body's immune system and physicians. For example, *E. coli* bacteria are a normal fauna of the human gut. The body will not mobilize an immune response against this organism. By slipping disease-causing genes into the *E. coli*, the source of the infection would be masked. Most physicians would have little difficulty identifying an anthrax infection, but associating the symptoms of a disease with the *E. coli* microbe would be much more difficult. What might be considered yet another advantage of biological warfare over conventional warfare is that humans would be harmed without the accompanying destruction of property.

Advanced Drug Delivery Systems

An exciting and expanding field in the treatment and control of disease is that of advanced drug delivery systems. Drugs, such as insulin, or hormones, such as human growth hormone, are placed in tiny, synthetic polymer "capsules" (**Figure 9**). The capsules are designed to release their enclosed molecules at specific target tissues at optimum rates over long periods of time. The "patches" for nicotine, nitroglycerine, and, more recently, the female contraceptive are examples of a type of advanced drug delivery system that releases drugs or hormones through the skin.

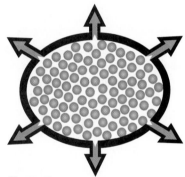

Figure 9
A type of delivery system through which a drug or hormone is released from a custom-made, nontoxic polymer capsule is changing the way that these substances are administered.

One of the problems with traditional drug or hormone therapy is in maintaining an optimum level of the substance in the body over long periods of time. Often a high dose of a drug is needed to ensure that the proper level reaches the intended tissues; adverse and sometimes deadly side effects are not uncommon. Another problem is that it is sometimes difficult to deliver drugs to certain tissues. Deep lung and nasal sinus tissues are particularly difficult to reach. As a consequence, the patient needs to take high doses of the drug. A particularly difficult problem arises in hormone therapy. Hormones are proteins, sometimes very large. Taking hormones orally exposes them to digestive enzymes that will degrade them to simpler molecules. In addition, large protein molecules do not pass readily from the digestive tract into the bloodstream. Bioengineers are designing polymer systems that will readily pass into the bloodstream and deliver their encapsulated hormone or drug only to the specific target tissue at a predetermined rate over an extended time period. In some cases, by implanting or injecting the capsule at the intended site—for example, the brain—the drug will be released only in that tissue and at the proper dose.

Scientists have designed polymers with specific antibodies incorporated into their surface that will respond to a particular foreign molecule that may be present in the body at certain times. The resulting antibody–antigen interaction triggers the release of a drug held within the polymer capsule that counteracts the foreign molecule.

Hormone levels are very finely regulated in the body. Insulin levels, for example, increase automatically when blood sugar levels rise. Although diabetics undergoing traditional insulin therapy are able to manage their blood sugar concentrations, their insulin levels are not homeostatically regulated. There are a host of problems, such as dosage control, diet management, delivery methods, and risk of infections, associated with insulin therapy that diabetics face throughout their lives. Bioengineers are working on polymer systems that respond to changing blood glucose levels, thereby releasing their insulin only when needed.

Annual sales of advanced drug delivery systems in North America are near US$20 billion. The future of such systems is promising and is one in which many lives will be saved because of safer, more efficient delivery of drugs and hormones.

SUMMARY *Active and Passive Immunity*

- Active immunity occurs when the body produces antibodies against an invading antigen.
- Passive immunity occurs when antibodies are introduced into the body directly. Unlike active immunity, passive immunity provides only temporary protection from the agent.
- Vaccines are antigen-containing substances that can be swallowed or injected to provide continued immunity to specific diseases.
- Antibiotics are a special kind of chemical agent, usually obtained from living organisms.

▶ Section 10.5 Questions

Understanding Concepts

1. Why doesn't exposure to the influenza virus provide immunity to other viruses?

2. How is it possible to catch the influenza virus more than once?

3. Why can't antibiotics be used to cure AIDS?

4. Explain in your own words how vaccines work.

5. Differentiate between active and passive immunity.

6. Explain the difference between using antibiotics and using vaccines to control disease.

7. What are monoclonal antibodies?

Making Connections

8. Smallpox vaccinations once involved scratching the skin and then swabbing it with an antigen.
   (a) Refer to **Figure 10** and state a conclusion about using only one vaccine.
   (b) What advantage would be gained from a booster (second) vaccine?

9. Chemotherapy works by destroying cancer cells. An unfortunate side effect is that normal, healthy cells are destroyed as well. Research some common side effects of chemotherapy drugs, and explain how these drugs affect homeostasis in the various body systems (blood cell development, the nervous system, the endocrine system,

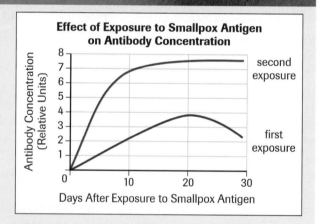

Figure 10

and the excretory system). Search for information in newspapers, periodicals, CD-ROMs, and on the Internet.

 www.science.nelson.com

10. Conduct research into the potential for development of biological weapons. Search for information in newspapers, periodicals, CD-ROMs, and on the Internet.

 www.science.nelson.com

▸ *CAREERS in Biology*

Sports Nutritionist

Sports nutritionists assist athletes or anyone who wants to improve his or her physical performance. These health-care professionals work in gyms, university health centres, private clinics, hospitals, and in business or industry. Sports nutritionists must first obtain a four-year Bachelor of Science degree in foods and nutrition and complete an accredited internship through the Dietitians of Canada before joining the College of Dietitians of Ontario.

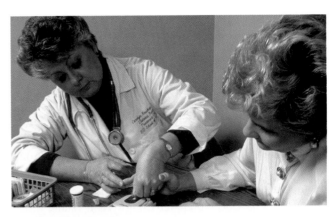

Endocrinologist

An endocrinologist is a medical specialist with a doctorate in endocrinology. To treat animals, you must first obtain a Doctor of Veterinary Medicine degree instead of a medical degree. These specialists study all the glands in the body to diagnose and treat endocrine disorders. Endocrinologists work in hospitals, government laboratories, veterinary clinics, and research centres in both public and private sectors.

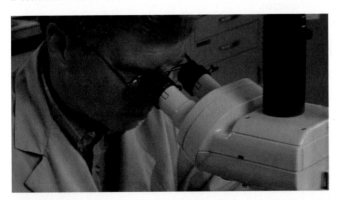

Immunologist

An immunologist studies the human immune system and the diseases that affect it. A strong high-school background in biology, chemistry, and math is necessary to enter a four-year Bachelor of Science program with a concentration in immunology. Many immunologists work in the biotechnology and pharmaceutical fields, or for hospitals and government research centres.

Dialysis Technologist

To become a dialysis technologist requires either a Bachelor of Science degree or a technologist's diploma from a community college, either in computer science, electrical engineering, or electronics engineering technology. You must then complete an intensive three-month training program at a hospital. Dialysis technologists work in hospitals or in dialysis clinics.

▸ *Practice*

Making Connections

1. Identify several other careers that require knowledge about homeostasis. From your list, select a career that interests you. What university program leads to that career? Which universities offer degree programs in this area? Which high-school subjects are required to enter the program? Does the career require further training beyond university?

2. Survey newspapers or conduct an Internet search to investigate and describe the responsibilities and duties involved in this career. What appeals to you about the career? What do you find less attractive?

 www.science.nelson.com

🔍 ACTIVITY 10.1.1

Diagnosing Disease by Examining Blood Cells

White blood cell counts can be used as clues in the diagnosis of disease. In this activity, you will examine prepared slides to identify different types of white blood cells and to determine how changes in blood cell counts are used to diagnose disease.

The slides have been prepared using Wright's stain, which allows you to clearly view cells and many types of microorganisms. Wright's and similar stains for blood and bone marrow smears are mixtures of basic and acidic dyes. According to the number of acid and basic groups present, cell components absorb the dyes from the mixture in various proportions.

Materials

prepared slide of human blood
light microscope
lens paper

Procedure

1. Before beginning the investigation, clean all microscope lenses with lens paper and rotate the nosepiece to the low-power objective. Place the slide of blood on the stage, and focus under low power. Locate an area in which individual blood cells can be seen.

2. Rotate the revolving nosepiece to the medium-power objective, and focus. Red blood cells greatly outnumber white blood cells.

(a) Draw a single human red blood cell.

(b) Estimate the size of the human red blood cell. Show your calculation.

3. Scan the field of view for different white blood cells. Using the classification of leukocytes provided in **Table 1**, classify the leukocytes and record your results in a table similar to **Table 1**.

4. Repeat the procedure by scanning 10 different visual fields. Record the data in your table.

Analysis

(c) Explain why few blood tests provide a diagnosis of disease.

Synthesis

Blood tests are used to help diagnose diseases. **Table 2** shows a few representative diseases. Use **Table 2** to answer the following questions:

(d) Why would a physician not diagnose leukemia based on a single blood test?

Table 1 Classification of Leukocytes

Type	Description	Normal % of total	Observed number	Observed % of total
Granulocyte	granular cytoplasm			
neutrophil	3-lobed nucleus, 10 μm (Wright's stain: purple nucleus, pink granules)	65%		
eosinophil	2-lobed nucleus, 13 μm (Wright's stain: blue nucleus, red granules)	2%–4%		
basophil	2-lobed nucleus, 14 μm (Wright's stain: blue-black nucleus, blue-black granules)	0.5%		
Agranulocyte	nongranular cytoplasm			
monocyte	U-shaped nucleus, 15 μm (Wright's stain: light bluish-purple nucleus, no granules)	4%–7%		
lymphocyte (small)	large nucleus, 7 μm (Wright's stain: dark bluish-purple nucleus, no granules)	2%–3%		
lymphocyte (large)	large nucleus, 10 μm (Wright's stain: dark bluish-purple nucleus, no granules)	20%–25%		

ACTIVITY 10.1.1 *continued*

(e) What information might a blood test provide a physician about a patient being treated for the lung disease tuberculosis? Why would blood tests be taken even after the disease has been diagnosed?

(f) Leukemia can be caused by the uncontrolled division of cells from two different sites: the bone marrow or lymph nodes. Indicate how blood tests could be used to determine which site harbours the cancerous tumour.

(g) Do blood donors need to have their blood counts taken? Why or why not?

Table 2

Leukocyte change	Associated conditions
increased eosinophils	allergic condition, cholera, scarlet fever, granulocyte leukemia
increased neutrophils	toxic chemical, newborn acidosis, hemorrhage, rheumatic fever, severe burns, acidosis
decreased neutrophils	pernicious anemia, protozoan infection, malnutrition, aplastic anemia
increased monocytes	tuberculosis (active), monocyte leukemia, protozoan infection, mononucleosis
increased lymphocytes	tuberculosis (healing), lymphocyte leukemia, mumps

INVESTIGATION 10.2.1

Virtual Immunology Laboratory

Inquiry Skills

- ● Questioning ○ Planning ● Analyzing
- ● Hypothesizing ● Conducting ● Evaluating
- ● Predicting ● Recording ● Communicating

How does your immune system respond to foreign invaders? How do antibodies know which antigen to attack? Can the formation of antibodies be used to diagnose disease? In this investigation, you will use the ELISA test (**E**nzyme-**L**inked **I**mmuno**s**orbent **A**ssay) to detect either the antigen or antibody associated with a disease-causing agent.

Begin the simulation.

 www.science.nelson.com

Locate and click on the Virtual Immunology Laboratory hyperlink. Select " Find" and read more about the diagnostic test called ELISA. (Note: To view the graphics from the simulation, the Shockwave plug-in is required.)

Question

Construct a question that will be investigated.

Procedure

1. Select ENTER and begin reading about the protocol.

2. Select "Figure 1" to view the procedure. Print the schematic to use throughout the investigation and while answering the Analysis questions.

3. Use the BACK key to return to the protocol.

4. At the top of the screen under "HHMV1 Virtual Laboratory," scroll down and select BACKGROUND. Read the entire section.

5. Use the BACK key and return to the protocol.

6. Select CLICK HERE TO START under the graphic on the right. Follow the instructions under each graphic.

7. As you go through all 11 steps, there will be a description of the lab protocol in the lab book on the left of the screen. To get further information, select WHY?

8. At the end of the experiment, calculate the titre of all three patients and make your diagnosis.

Analysis

(a) What diseases are you attempting to diagnose?

(b) Are you using antibodies or antigens in this ELISA?

(c) List the components of the following:
    (i) a positive control   (ii) a negative control

Evaluation

(d) Identify any errors you made in conducting the protocol and describe how they may have affected your results.

Key Expectations

- Explain and provide examples of how the immune system monitors changes in your external environment and makes adjustments in the internal environment to maintain homeostasis. (10.1, 10.2, 10.3)
- Present informed opinions about personal health. (10.1, 10.2, 10.5)
- Examine the contribution made by Canadian scientists in the field of health. (10.2)
- Evaluate opinions and discuss difficulties in treating disease. (10.3, 10.4, 10.5)
- Describe immunological responses to viral and bacterial infections. (10.4, 10.5)

Key Terms

active immunity

antibodies

antigens

B cell

complement proteins

helper T cells

inflammatory response

killer T cells

leukocytes

lymphocytes

lymphokine

macrophages

memory B cell

monoclonal antibodies

passive immunity

pathogens

phagocytosis

pluripotent cells

receptor sites

suppressor T cell

T cell

vaccines

vector

virulent virus

▸ *MAKE* a summary

Imagine a microbe entering your blood. To summarize your learning, create a flow chart or diagram that shows how the immune system would respond to this potentially dangerous situation. Label the diagram with as many of the key terms as possible. Refer to other flow diagrams for ideas, and use appropriate designs to make your sketch clear.

In your notebook, record the letter of the choice that best answers each question.

1. Which of the following characteristics of an invader can be recognized by antibodies of the immune system?
   (a) nuclear material
   (b) cell membrane shape
   (c) cell size
   (d) the electrical charge of the cell
   (e) poisons

2. What is an antibody?
   (a) a white blood cell that engulfs an invading microbe
   (b) a protein that reacts with an antigen
   (c) a white blood cell that identifies a foreign invader
   (d) a protein that is attached to the cell membrane of a microbe
   (e) a protein that is always attached to the cell membrane of an antibody-producing cell

3. Which of the following is the cellular component responsible for antibody production?
   (a) B cell   (d) platelet
   (b) T cell   (e) mast cell
   (c) macrophage

4. Which part of your immune system is charged with recognition of invading pathogens?
   (a) antibodies   (d) phagocyte
   (b) B cell lymphocyte   (e) all of the above
   (c) T cell lymphocyte

5. Which of the following is sometimes a glycoprotein complex found on the surface of cell membranes that is not found in one's body?
   (a) a toxin   (d) a virus
   (b) an antibody   (e) a prion
   (c) an antigen

6. When do B cells turn into super antibody-producing cells?
   (a) when macrophages engulf foreign invading materials
   (b) when T cells identify the invader and release lymphokine
   (c) when B cells identify the invader and release histamine
   (d) when plasma proteins identify the invader and release lymphokine
   (e) when platelets identify the invader and release histamine

7. What is an important defence against invading microbes that can be found in tears?
   (a) Macrophages from the tears engulf foreign invaders by phagocytosis.
   (b) Lysozyme, special enzymes secreted in tears and sweat, destroy bacteria by dissolving their cell walls.
   (c) Mucus produced by tears adheres to microbes and targets them for antibody production.
   (d) Cilia, found along the eyelashes, sweeps microbes from the eye.
   (e) Water washes away the microbes, preventing them from entering the tear duct.

8. Which of the following does not describe the action of complement proteins produced by the immune system?
   (a) Some of the proteins act as markers by causing antigens to aggregate.
   (b) Some of the proteins attach themselves to the cell membrane of the antigen, immobilizing the antigen.
   (c) Some of the proteins engulf the foreign invader by phagocytosis.
   (d) Some of the proteins form a protective coating around the microbe.
   (e) Some of the proteins dissolve or puncture the cell membrane of bacterial invaders.

9. Why is the transplantation of donor organs affected by the immune response?
   (a) Transplanted organs contain antigens that T cells recognize as harmful.
   (b) Transplanted organs contain antibodies that recognize cells in your body as being harmful.
   (c) Transplanted organs contain antigens that stimulate the production of antigens in your own body.
   (d) Transplanted organs contain antibodies that stimulate the production of antigens in your own body.
   (e) Transplanted organs contain antibodies that B cells recognize as harmful.

10. How can transplant rejection be reduced?
   (a) Keep the area clean to prevent microbe infections.
   (b) Transplant organs with a close tissue match.
   (c) Inject antibodies into the recipient's body for the first few weeks after the operation.
   (d) Choose transplanted organs from an entirely different species so that they cannot be recognized by the immune system.
   (e) Either (c) or (d).

Understanding Concepts

1. (a) Make a drawing of a Y-shaped antibody, showing how it attaches to specific antigens.
   (b) Label the receptor sites on the cell membrane.
   (c) Use the diagram to explain why an antibody produced in response to the mumps virus would have no effect against influenza.
   (d) Use the diagram to explain how antibodies target antigens for phagocytosis.

2. Explain why when the second time an organism invades the body, the patient is not likely to get seriously ill.

3. Differentiate between T cell lymphocytes and B cell lymphocytes.

4. How do viruses use the receptor sites to gain access into the cell?

5. Explain why T cells have difficulty identifying antigens from HIV.

6. Draw a diagram showing how antibodies attach themselves to a virus, preventing them from gaining access into a cell.

7. Describe the function of lymphokine.

8. Explain in your own words the function of each of the following:
   (a) killer T cells   (c) suppressor T cells
   (b) helper T cells   (d) memory T cells

9. Explain why the Hawaiian population was so severely affected by measles and the Aboriginal population of North America was so much more susceptible to smallpox than Europeans were?

10. What are pluripotent cells?

11. What are autoimmune diseases?

12. Why does the likelihood of autoimmune disease increase with age?

13. Explain how a food allergy can threaten life.

14. Copy **Table 1** into your notebook and complete it by placing a checkmark (✔) in the correct boxes.

Table 1

Disease	Highly virulent	Highly contagious
rabies		
HIV		
influenza		
common cold		
cholera		
chicken pox		

15. Explain how the discovery of prions has changed Pasteur's germ theory for explaining disease.

16. Explain why viral infections cannot be treated by antibiotics.

17. How does proper sewage treatment help reduce the transmission of disease?

18. Why is it difficult to develop a vaccine against HIV?

19. Differentiate between active and passive immunity.

20. Describe how Edward Jenner developed the first vaccine.

21. How did the development of a vaccine for rabies differ from the approach used by Edward Jenner?

22. What are monoclonal antibodies?

23. Transmissible spongiform encephalopathies (TSEs) are brain diseases caused by prions. Mad cow disease in cattle, scrapie in sheep, and Creutzfeldt–Jakob disease in humans are all known to be prion diseases. In 1997, Dr. Stanley Prusiner was awarded the Nobel Prize in Physiology or Medicine for his work on prions. Prusiner showed that TSE can be contracted by injecting infected brain tissue into the healthy brain of a recipient animal.
    (a) In the 1960s, British researchers T. Alper and J. Griffith noted that TSE preparations remained infectious even after exposure to radiation. Why does radiation destroy viral infections but have little effect on TSE?
    (b) How does this research challenge the tenant of molecular biology that only nucleic acids provide directions for cells?
    (c) In the late 1990s, Professor Charles Weissmann of the Imperial College School of Medicine in London showed that mice genetically engineered to lack Prusiner's prion would not develop TSE. Why don't these mice develop TSE when fed infected brains?
    (d) In a subsequent experiment, Weissmann grafted brain tissue infected with TSE into a healthy brain of a mouse genetically engineered not to produce the Prusiner protein. The grafted brain displayed symptoms of TSE, while the rest of the brain remained healthy. Explain these observations.

Applying Inquiry Skills

24. Twenty-five rabbits of the same sex, age, and breed were divided into four groups. Each group was given an injection of sheep red blood cells every two weeks

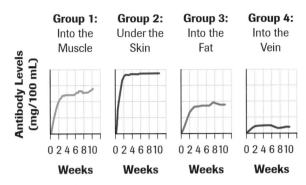

Figure 1
Antibody levels after vaccination with sheep blood

for a total of three injections. The injections were delivered into different areas of the body. During a six-week trial period, antibody formation was monitored in each of the rabbits. The results of the tests are displayed in **Figure 1**.

(a) Suggest how a control group could be established for the experiment.

(b) What conclusions can you draw from the graphic data?

(c) Provide a theory that explains why it is more effective to give injections into a vein than into other areas of the body.

25. Mumps and measles vaccines are prepared in chick embryos. The weakened forms of the virus are grown in living tissues.

(a) Why aren't the viruses grown in agar plates?

(b) Why do physicians often have epinephrine (adrenaline) on hand following the vaccination? (It is not usually necessary for the epinephrine to be administered.)

Making Connections

26. Before the work of Dr. Barry J. Marshall and Dr. Robert Warren of Perth, Australia, most doctors believed that stress was the cause of ulcers. In 1983, Marshall and Warren reported that *Helicobacter pylori*, a bacterium living in the stomach, is the most common cause of ulcers. Today, researchers around the world are turning to another bacteria, *Chlamydia pneumonia*, as the main culprit in triggering coronary heart disease.

(a) If coronary heart disease is caused by a bacterium, how might this affect the search for treatment?

(b) Explain why physicians attempting to diagnose coronary heart disease may be monitoring antibodies.

(c) Use the information about chlamydia provided below to explain why scientists are concerned about disease transmission.
   • Chlamydia is spread by coughs and sneezes.
   • By the time you are 20, there is a 1 in 50 chance that you are carrying the infection.
   • Incidence of disease increases with age.

(d) The DNA from a strain of *Chlamydia pneumonia*, called TWAR, has been found in the fatty deposits that block coronary arteries. How does this support the theory that some forms of heart disease can be caught?

27. Viruses can cross species, and influenza, which originated from domestic birds, is a result of this process. A genetic mutation changed the shape of the virus, allowing influenza to gain access to the receptor sites on human tissues. Today, some of the most deadly viruses, known as filoviruses, are believed to have appeared because they crossed over from primates to humans.

(a) Draw a diagram showing how a virus can mutate and become transgenic (move from one species to another).

(b) One filovirus, called *Ebola*, has an extraordinary killing rate in excess of 90% fatality. It is believed that many filoviruses come from other primates. Explain how researchers who use primates may expose others to dangers. Should researchers be allowed to use primates?

(c) Explain how the clearing of frontier jungle areas may contribute to the spread of ebola and other filoviruses.

27. How has air travel affected the transmission of disease?

Extension

28. How is normal rapid cell division different than cancer?

Determining the Effects of Caffeine on Homeostasis

Caffeine is considered the world's most popular drug. Some surveys indicate that over 50% of North Americans begin their day with a cup of coffee, a well-known source of caffeine. Caffeine is also found in tea, soft drinks, chocolate, and some headache and cold remedies.

Many people report that caffeine prevents them from sleeping and causes feelings of anxiety. Clinically, caffeine has been shown to cause an increase in heart rate, an elevation in blood pressure, and an increase in urine output. In high concentrations, caffeine can raise an individual's blood sugar level. Although the research is inconclusive, many physicians discourage people with stomach disorders, such as ulcers, from consuming large quantities of coffee and tea. Current evidence suggests that a daily caffeine intake of up to 450 mg (roughly equivalent to the caffeine in four cups of brewed coffee) will pose no threat to a healthy adult. However, some individuals are sensitive to much lower quantities.

This task is intended to investigate the effects of caffeine on different human body systems. The task should demonstrate how the homeostatic feedback adjustment operates. For example, you might study the effects of caffeine on heart rate or blood pressure. However, because large quantities of caffeine can be harmful to some people, the experiment will not use human subjects. Instead, you will select an invertebrate or a protist (such as one of those shown in **Figure 1**) for testing the effect of caffeine on either a type of cell, tissue, or organ operating within a physiological system. The invertebrate or protist will serve as a model to provide information that may be applicable to human physiological systems; from it you will infer how the human system is affected by and adjusts to caffeine.

earth worm

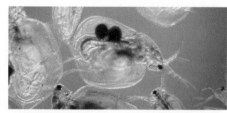
daphnia

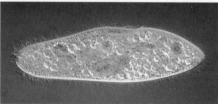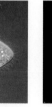
paramecium

planarian

Figure 1
A protozoan (or protist) and some invertebrates

Questions
(a) Create two questions that could form the basis of investigation into the effects of caffeine on different physiological systems. Select the organism that you think will be the most appropriate model for the investigation.

Hypotheses/Predictions
(b) For each of your questions, make a prediction about how caffeine will affect the physiological system. Formulate and justify a hypothesis to explain each prediction.

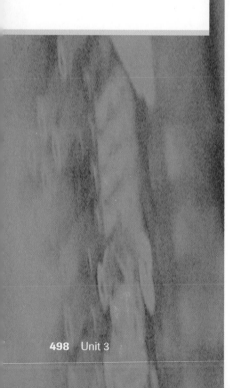

Experimental Design

(c) Select one of the questions from (a) and describe an experimental design.
- (i) Identify the dependent and independent variables.
- (ii) Indicate how the variables will be observed or measured, or both.
- (iii) List the variables that must be controlled.
- (iv) Create a data table.

(d) Explain which human system is being modelled, and why you think your selected organism is the most appropriate.

Materials

(e) Provide a list of required materials and equipment.

Procedure

(f) Outline a detailed step-by-step procedure for carrying out the experiment. Provide instructions on the use of all equipment, and include safety precautions where appropriate. Have your teacher approve your procedure, and then carry it out.

Observations

Include baseline measurements and other appropriate data when you record your observations. Use the correct SI units for all measurements.

(g) Display your data in a table, and graph the data in an appropriate format.

Analysis

(h) Describe the effect of caffeine on the cell, tissue, organ, or body system. How is homeostasis maintained?

(i) Compare your results with available standards, either determined by your group or from another source, such as the Internet. Describe the comparison.

 www.science.nelson.com

(j) Identify and describe any sources of error in your measurements. Indicate the uncertainty in your measurements as a percentage, if necessary.

Evaluation

(k) Did you experience difficulties in obtaining data? If so, suggest ways to improve the design and/or procedure.

(l) Evaluate the standards used for comparison. What are the limitations of using these standards?

(m) What assumptions must be made in using the invertebrate as a model to draw conclusions about human body systems? Is the model appropriate? Explain.

Synthesis

(n) Caffeine was one of the first drugs taken by athletes to improve performance. Outline some of the dangers in taking caffeine prior to exercise.

(o) Some individuals consume more than 1500 mg of caffeine daily, either in coffee or soft drinks. List and briefly describe some of the problems associated with a high daily consumption of caffeine.

(p) If you were a physician, what categories of people would you discourage from consuming caffeine? Give reasons for your recommendations.

Understanding Concepts

1. In a table, summarize the three processes of urine formation. In your summary, include the name and a brief explanation of each process, including the organ(s) in which it occurs and the substances involved.

2. Explain three functions performed by the kidneys.

3. Draw a diagram to illustrate the path of a molecule of urea, beginning in the renal artery and ending when it passes out of the body. Include the structures of the nephron and the organs of the urinary system.

4. Draw a flow chart to illustrate how the kidney regulates blood pressure.

5. Draw a flow chart to illustrate how the hypothalamus and ADH regulate the water content of blood.

6. List four characteristics of the endocrine system.

7. Due to chemical differences, hormones can be divided into two categories. These differences affect the function of the hormones when they meet a target cell.
   (a) List the two different categories of hormones.
   (b) Describe the chemical differences between the two types of hormones, and explain how each type functions when it meets its target cell.
   (c) In point form, describe how the two types of hormones activate their respective target cells.

8. In the homeostatic process of negative feedback, the secretion of most hormones is regulated by other hormones. Using thyroxine as an example, draw a flow chart to show this process of regulation by negative feedback.

9. The length of the follicular phase, ending with ovulation, may vary considerably, making the prediction of a woman's most fertile time quite difficult. Describe the hormonal controls of the follicular phase and the factors that may cause its length to vary.

10. During pregnancy, several hormonal changes occur.
    (a) Explain why it is necessary that estrogen and progesterone inhibit the secretion of FSH and LH.
    (b) Explain why the level of human chorionic gonadotropin (HCG) decreases during the second trimester, starting at about week 13 or 14 of pregnancy.

11. Explain why depolarization, which results in a nerve impulse, is an example of positive feedback.

12. Distinguish between excitatory and inhibitory synapses.

13. Explain why two stimuli, applied 0.0001 s apart, would produce only one nerve impulse along the fibre of a specific neuron.

14. Cerebral palsy is a group of disorders that affects body movement and muscle coordination. It is caused by damage to, or malformation of, the brain during development in the womb or in the first few years of life. The effects of cerebral palsy vary widely, from slight awkwardness of movement or hand control to eating difficulties, poor bladder and bowel control, and breathing problems. Which area(s) of the brain are most likely affected? Explain your answer.

15. In what ways are all vertebrate brains similar in their basic structures?

16. Describe how the eye adjusts its focus when changing from looking at an object far away to looking at an object close by. Include the following terms in your description: *point of focus, retina, ciliary muscle, suspensory ligaments,* and *lens.*

17. Explain how sounds are heard. In your explanation, describe the structures through which the sound waves and nerve impulses travel, beginning with the outer ear and ending with the cerebrum. State the function of each structure.

18. The immune system can respond to millions of different antigens. Explain how the B cells and the four different T cells respond to foreign antigens.

19. Explain how the immune system causes autoimmune disease.

20. Describe how vaccinations provide protection against disease-producing microbes.

21. Allergies are the most common type of immune disorder. Describe an allergic reaction and explain why it may be harmful.

22. **Figure 1** shows antigen–antibody complexes.
    (a) Use the following list to identify the labelled structures: *microbe, antigen on microbe, binding site on antibody, constant area of antibody,* and *variable region of antibody.*
    (b) How are antibodies able to recognize different antigens?
    (c) How do antibodies help prevent the antigen from causing disease?

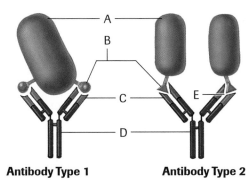

Antibody Type 1   Antibody Type 2

Figure 1

Applying Inquiry Skills

23. The following data was collected from a subject admitted to hospital:

Gender:	female
Age:	34
Blood type:	O
WBC count:	$2.5 \times 10^4/\mu l$ (normal = $7.0 \times 10^3/\mu l$)
RBC count:	$4.5 \times 10^6/\mu l$ (normal = $5.0 \times 10^6/\mu l$)
Body temperature:	38.4°C (normal = 37.0°C)
Pulse rate:	72 beats/min (normal = 75 beats/min)

(a) Use the information provided to suggest a possible diagnosis.

(b) What further test could be conducted to confirm your diagnosis?

24. The data in **Table 1** on the relationship between ambient air temperature and internal body temperature was obtained from two different animals: a snake and a cougar. Unfortunately, the researcher did not take proper care in labelling the data.

Table 1

Ambient air temperature (°C)	Internal body temperature of organism X (°C)	Internal body temperature of organism Y (°C)
2	3	36
10	10	38
15	15	38
20	20	38
25	25	38
30	30	39

(a) Using the data provided, identify the correct animal, and classify each as either an endotherm or an ectotherm.

(b) Explain why there is little fluctuation in the body temperature of organism Y.

(c) Predict how the data might be different for a field mouse. Give your reasons.

25. Answer the following questions using **Table 2**, which compares the plasma components of several patients.

Table 2

	Urea	Uric acid	Glucose	Amino acids	Proteins
normal readings	0.03	0.004	0.10	0.05	8.00
patient 1	0.03	0.004	0.50	0.05	8.00
patient 2	0.03	0.005	1.70	0.05	8.00
patient 3	0.03	0.050	0.10	0.09	8.00
patient 4	0.07	0.004	0.10	0.06	4.00

All quantities are in g/100 mL of plasma.

(a) According to the data provided, which patient(s) might have diabetes mellitus? Give your reasons.

(b) Which patient(s) may have experienced a failure in the glomerulus? Give your reasons.

(c) Which patient(s) may have eaten a protein-rich meal prior to giving the blood sample? Provide reasons for your answer.

26. A small glass tube called a micropipette can be used to collect fluids from various parts of the excretory system. Fluid components were collected from three different areas of a subject and analyzed for various components. The compositions of all components are recorded in **Table 3**.

Table 3

Component	Plasma	Filtrate	Urine
urea	0.030	0.030	2.00
uric acid	0.004	0.004	0.05
glucose	0.100	0.100	0.00
amino acids	0.050	0.050	0.00
salts	0.720	0.720	1.50
proteins	8.000	0.000	0.00

All quantities are in g/100 mL.

(a) Explain changes in urea and uric acid concentrations between the filtrate and urine.

(b) Explain why glucose and amino acids are found in the plasma and filtrate but not in the urine.

(c) Why do salts become more concentrated in the urine?

(d) Why are proteins not found in the filtrate or urine?

27. **Figure 2** shows the relationship between hormone levels and body temperature during the menstrual cycle.

Hormone Concentration and Body Temperature During a Typical Menstrual Cycle

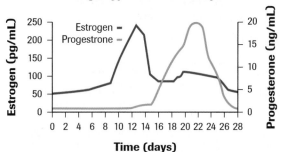

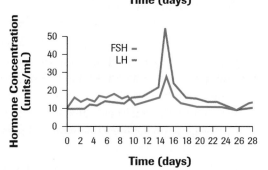

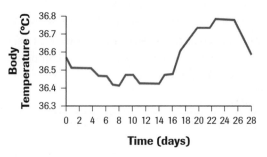

Figure 2

(a) According to the data, at what point in the menstrual cycle does temperature increase?

(b) Which hormone seems to have the greatest effect on temperature? Why?

(c) Describe an experiment that could help prove the relationship between the hormone you selected and an increase in body temperature.

28. A laboratory experiment was conducted to determine the effect of thyroxine on metabolic rate. Four groups of adult male rats were used. All groups were maintained in similar environments, designed to provide maximum physical activity. Each group was provided with adequate supplies of water and one of the following diets:

Diet A: food containing all essential nutrients

Diet B: food containing all essential nutrients and an extract of thyroxine

Diet C: food containing all essential nutrients and a chemical that counteracts the effect of thyroxine

Diet D: food containing all essential nutrients, except iodine

The results of the experiment appear in **Table 4**.

Table 4

Group	Average initial mass (g)	Average mass after 2 weeks of treatment (g)	Final average oxygen consumption (mL/kg/min)
I (diet A)	310	312	4.0
II (diet ?)	320	309	10.1
III (diet ?)	318	340	2.7
IV (diet ?)	315	400	2.0

(a) Formulate a hypothesis for this experiment.

(b) Identify the dependent and independent variables.

(c) Which group was most likely used as a control? Explain your response.

(d) Diet B was most likely fed to which group(s)? Explain your answer.

(e) Diet D was most likely fed to which group(s)? Explain your answer.

29. Serotonin is a naturally occurring neurotransmitter that has a role in determining mood and emotions. A shortage of serotonin has been linked to phobias, schizophrenia, aggressive behaviour, depression, uncontrolled appetite, and migraine headaches. Several types of drugs, shown in **Table 5**, affect serotonin levels.

(a) Why is Prozac (fluoxetine hydrochloride) prescribed for people with depression?

(b) Which drug should not be taken by someone experiencing clinical depression?

Table 5

Drug	Effect
Prozac, Paxil, Zoloft	cause serotonin to remain longer in the brain
clozapine	prevents serotonin from binding to the postsynaptic membranes
hallucinogens (LSD, ecstasy)	react directly with the serotonin receptors to produce the same effect as serotonin

(c) Draw a diagram showing the effect of LSD or ecstasy on serotonin.

(d) Explain how taking hallucinogens over time could reduce serotonin levels.

30. After being in a brightly lit environment for one hour or more, the human eye takes at least 20 minutes to become adapted to a dimly lit room. Explain why.

31. Assuming that immunity memory is intact, how would you explain people getting colds or flu year after year?

32. Explain specifically how the human immunodeficiency virus (HIV) infects the immune system.

Making Connections

33. High temperatures can seriously increase the risk of heat stroke. The maximum suggested temperature of the water in a hot tub is about 38°C.
    (a) Indicate what can happen to a person who sits in a hot tub for an extended period.
    (b) Explain why public hot tubs can present a threat to health. What safety features should be in place?
    (c) Eventually, all thermostats fail. Should hot tubs be equipped with additional safety equipment? Give reasons for your response.

34. Describe two disorders of the urinary system. Explain how these disorders affect lifestyle.

35. Athletes are now undergoing random urine testing for drugs following competitions.
    (a) Use your knowledge of excretion to describe the pathway of substances, such as drugs, through the urinary system, from the time they enter the glomerulus until they are excreted in the urine.
    (b) How would it be possible for athletes to take banned drugs and still test negative?
    (c) Make suggestions that would prevent athletes from being able to undermine drug tests.

36. Prior to Banting and Best, patients with type 1 diabetes mellitus usually died within months of the onset of the disease. Today, patients are treated with insulin injections; however, many of these patients do not escape insulin-related disorders brought on by the fluctuations in blood glucose levels. Changes in blood glucose may damage blood vessels, limiting circulation, which in turn can lead to blindness, kidney failure, and the destruction of muscle and nerve tissue in the hands and feet.
    (a) Explain why the transplant of islet cells from a donor's pancreas into a diabetic patient is a promising option.
    (b) What are two technological challenges presented by islet transplants?
    (c) Describe one societal issue that researchers must face as islet transplants become more common.

37. Use what you know about the transmission of nerve impulses to formulate a hypothesis about how local and general anaesthetics work.

Extension

38. Imagine that you have just been offered your dream job. Your prospective employer requires a urine drug test at the time of hiring and at intervals during the term of employment. Any employee who fails a drug test can be dismissed from his or her job. What are some arguments for and against drug testing on the job?

39. In June, 2000, a team of researchers at the Massachusetts Institute of Technology and Bell Labs reported that they had created an electronic circuit based on the cerebral cortex, the brain's centre of intelligence. Lauded as the first artificial circuit to mimic brain activity, it was modelled on the much larger network of natural neurons and feedback loops in the human brain. Use several sources to research artificial intelligence or neuromorphic engineering, and find out how this circuitry works. Search for information in newspapers, periodicals, CD-ROMs, and on the Internet.

 www.science.nelson.com

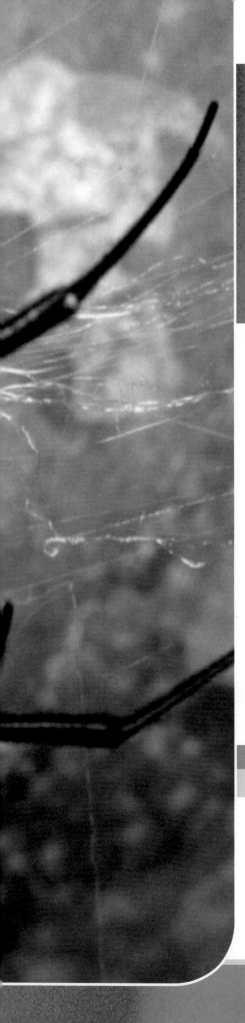

Evolution

The natural world holds many puzzles and surprises. For example, the black widow spider comes by its name quite honestly, as females often consume their mates during or immediately after mating. How can such unusual behaviour be explained? Mysteries in nature give rise to a multitude of questions. To answer these questions, biologists rely on an understanding and application of the theory of evolution.

A species of black widow, the Australian redback (*Latrodectus hasselti*), exhibits voluntary cannibalism. Dr. Maydianne Andrade, a biology professor at the University of Toronto, has used evolutionary biology to explain this odd behaviour. During courtship, some males intentionally sacrifice their body as food to the female. Prior to Dr. Andrade's research, many biologists believed that, by sacrificing themselves, the male spiders provided important nutrients to the females that would benefit their offspring. Instead, the reason appears to involve sperm competition. Males offer their abdomens as food in order to extend their mating time. Females normally mate with several males, store their sperm, and use it to fertilize up to 3000 eggs during the two years of their life. Males, by contrast, are unlikely to live long enough to mate with more than one female. Dr. Andrade found that males eaten during mating were able to transfer more sperm to the female, thereby increasing their chances of fertilizing more eggs. The result? Males that are eaten have more offspring than those that are not.

▶ Overall Expectations

In this unit, you will be able to

- evaluate the scientific evidence that supports the theory of evolution;
- analyze evolutionary mechanisms, and the processes and products of evolution;
- analyze how the science of evolution can be related to current areas of biological study, and how technological developments have extended or modified knowledge in the field of evolutionary biology.

▸ **Prerequisites**

Concepts

- roles of meiosis and mitosis
- DNA, genes, alleles
- relationship between genotypes, phenotypes and the environment
- types of mutations
- role of genetic recombination including crossing over
- Punnett squares and inheritance
- adaptations to the environment
- comparison of eukaryotes and prokaryotes

Skills

- use Punnett squares to predict the outcome of genetic crosses.
- apply an understanding of inheritance to selective breeding
- relate an understanding of biological diversity to genetic diversity

Knowledge and Understanding

1. In your notebook, indicate whether the statement is true or false. Rewrite a false statement to make it true.
   (a) Mutations are changes in DNA that are harmful to the cell.
   (b) Sexual reproduction is disadvantageous because the offspring show little or no genetic diversity.
   (c) Through the process of genetic recombination, meiosis produces diploid cells and increases the potential diversity of offspring.
   (d) In species that reproduce asexually, offspring are always genetically identical to their parent.
   (e) Over time, dominant alleles will tend to become more common in a population, while recessive alleles will become more rare.
   (f) An organisms genotype refers to its genetic makeup, which is unaffected by the environment.
   (g) An organism's phenotype refers to traits that are expressed in the organism and affected by both its genotype and the environment.
   (h) A species is a population or populations of organisms that are able to inter-breed under natural conditions and produce fertile offspring.
   (i) Harmful or lethal mutations have little or no effect on the health of large multicellular organisms.
   (j) Virtually all large populations exhibit genetic variation among individuals.
   (k) In eukaryotic organisms, DNA is found not only in the nucleus but also in mitochondria and chloroplasts.

2. Study the process of genetic recombination illustrated in **Figure 1**. For each of the four steps, explain what event is occurring, as well as how the event is significant to reproduction and genetic variation.

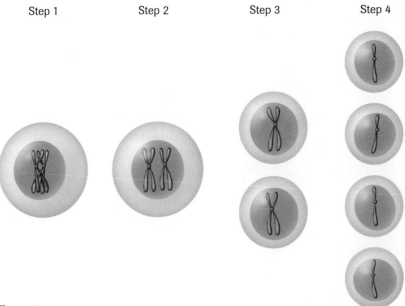

Step 1   Step 2   Step 3   Step 4

Figure 1
Genetic recombination, shown in four steps

3. Identify the likely effect of each of the following types of mutation on a cell. Explain each answer.

(a) a silent point mutation

(d) an inversion

(b) a point mutation causing a frame shift

(e) a translocation

(c) a point mutation in an intron region

4. Each of the organisms in **Figure 2** exhibits special adaptations. For each species, describe two obvious adaptations and state how they enhance the biological success of the organism.

(e)

(a)   (b)   (c)   (d)

(f)

Figure 2
(a) morning glory; **(b)** kangaroo; **(c)** sea nettle; **(d)** bull elk; **(e)** luna moth; **(f)** blue-footed booby

Inquiry and Communication

5. Examine the Punnett square in **Figure 3**, which represents a cross between a male that is homozygous for two dominant alleles, *AA* and *HH*, and a female that is homozygous for two recessive alleles, *aa* and *hh*.
   (a) Has the variety and ratio of genotypes changed in the F_1 generation?
   (b) In the parent generation, state the ratio of the specific alleles for each gene (i.e., *A:a* and *H:h*).
   (c) Taking all F_1 individuals into consideration, do the allele ratios change in the F_1 generation? Explain your reasoning.
   (d) Draw a Punnett square and determine the allele ratios for the F_2 generation.

gametes ➤	*ah*	*ah*
AH	AaHh	AaHh
AH	AaHh	AaHh

Figure 3

Making Connections

6. For each of the following, list two examples, one that is not genetically inherited and one that might have been genetically inherited:
   (a) physical characteristics
   (b) diseases and medical conditions
   (c) behaviours, and likes and dislikes

7. Drawing from your understanding of genetics, explain the biological mechanisms and processes that dog breeders have used to produce such a diverse variety of domesticated dogs (**Figure 4**) from an original population of wild dogs that showed little outward variation.

8. In January 2002, the North American Commission for Environmental Cooperation released a report that declared that North America is facing a "widespread crisis" because of its shrinking biodiversity.
   (a) How is the loss of biodiversity related to the loss of genetic diversity?
   (b) Why do so many scientists, government agencies, and members of the public consider it an important issue?

Figure 4

chapter

11

Origins of Evolutionary Science

In this chapter, you will be able to

- describe, in both a historical and a cultural context, how various scientists' contributions have altered our understanding of evolution;

- formulate your own hypothesis and judge different hypotheses that have influenced the development of the theory of evolution;

- explain the mechanism of natural selection and how species adapt to their environment;

- evaluate the impact technology has had on our understanding of Earth's history.

Scientists have identified more than a million different species, past and present, on Earth, and millions more await study. Despite the extraordinary diversity, all life forms are fundamentally similar at the cellular and molecular level. How have cells transformed into the current diversity of life? How are fossilized life forms related to living organisms? Solving such puzzles is one of science's greatest achievements.

Evolution is the compelling theory that explains the origin of species and the history of life on Earth. Geneticist Theodosius Dobzhansky expressed the significance of this theory when he said, "Nothing in biology makes sense except in the light of evolution." To appreciate the significance of evolution, it is important to study the evidence that supports the theory as well as the mechanisms by which it operates. Evolutionary biology is a modern science that, as well as providing answers to questions about the past, has direct applications in the health sciences, agriculture, industry, and conservation.

💡 REFLECT on your learning ▼

1. Domesticated animals, such as dogs (**Figure 1**), cats, horses, and cattle are descended from wild animals. Explain how so many different domesticated breeds now exist. How might breeders have produced animals with traits that their wild ancestors did not have?

Figure 1

2. What kinds of information might scientists obtain from fossils, such as the one on the next page? How can the age of a fossil be determined?

3. Offer some explanations for penguins, ostriches, and eagles sharing so many similar features, while fish, antelope, and bats do not.

Puzzling Over Evidence

Biologists investigate the natural world by collecting evidence through methodical observation and experimentation. With accumulating evidence, they continually form and test hypotheses, moving toward the development of a theory. In this activity, you model this process by arranging evidence, making observations, drawing inferences, and developing hypotheses to solve a puzzle.

Materials: large envelope (holding at least 40 jigsaw pieces), sheet of cardboard

- Remove 10 puzzle pieces at random from your group's envelope. (Unused evidence must remain in the envelope.)

- Arrange the pieces on the cardboard. Within your group, discuss what evidence you observe and make a prediction about the completed puzzle. Record this prediction in your notebook as your group's first hypothesis, along with any other hypotheses you develop.

- Show your first set of hypotheses to your teacher. With your teacher's approval, remove 15 more puzzle pieces from the evidence envelope.

- With this new evidence, add to and modify your original arrangement of puzzle pieces. Make and record your second set of hypotheses. What specific pieces of evidence do you expect to find in the envelope that would confirm your hypotheses?

- On your teacher's instruction, visit other groups and carefully observe the evidence that they hold. Record information that you think is useful for your group.

- Return to your puzzle and discuss your findings. Choose a representative to present your group's best hypothesis about the completed puzzle to the class.

- How did "gaps" in your evidence influence the confidence you had in your hypotheses?

evolution the process in which significant changes in the inheritable traits (i.e., the genetic makeup) of a species occur over time

immutable unchanged and unchanging, believed (before evolutionary theory became accepted) to be characteristic of life forms

For centuries, scientists have been gathering and piecing together evidence to solve the puzzle of the origin and history of life on Earth. They have made careful observations, formulated and tested hypotheses, analyzed data from diverse sources, and drawn inferences to develop the theory that is now generally accepted as the solution to the puzzle of **evolution**. Acceptance of theories of evolution has never been universal. Many people hold cultural and religious beliefs about the origin of life that may not be in harmony with scientifically accepted reasoning and conclusions.

In 1650, for example, a highly regarded Irish scholar and theologian published his calculations for the age of Earth based on astronomy, history, and biblical sources. Archbishop James Ussher of Armagh (**Figure 1**) declared that Earth was created on Sunday, October 23, 4004 B.C., and he went on to calculate the dates of other significant biblical events. In 1701, his chronology was printed as marginal notes in an authorized version of the Bible. About 150 years later, most Europeans mistakenly thought his chronology was part of the original scripture. Today, most people understand that the world is a dynamic environment in which change is both natural and unavoidable, but 500 years ago most people thought that their natural surroundings changed very little and life forms were thought to be **immutable**. Before the technology was available to show otherwise, Earth was thought to be of recent origin.

Figure 1
Archbishop James Ussher of Armagh (1581–1656)

Fossil evidence provided important scientific insights into the past, as a record of both the great diversification of species and the extinction of many others. As well as the fossil record, the geographic distribution of living species began to give scientists valuable clues to patterns of evolution. By the 19th century, the scientific community had accumulated sufficient evidence for general agreement that Earth was very old and that life forms on Earth had undergone and continued to undergo changes. However, not until 1859 did scientists formulate a viable explanation for the mechanism of evolution.

Like all scientific theories, the validity and value of evolutionary biology are based on rigorous and continual analysis and interpretation of accumulating evidence. Today there is a broad consensus among scientists on the facts of evolution—that the history of life on Earth has been one of continual change over billions of years. Although modern knowledge of molecular biology and genetics offers additional evidence and support for evolution, as with other fields, many questions remain open for study.

> **Section 11.1 Question**

Making Connections

1. North American Aboriginal cultures, like societies elsewhere in the world, have stories about their people's origin that have been repeated orally from generation to generation to this day. Find out about and compare some creation stories from Aboriginal cultures in North America, from ancient Sumer and Babylon, from China, Samoa, Persia, and Japan or from other cultures of your choice. Consult print material or the Web for textual and visual sources, or spiritual leaders as oral sources. Discuss possible reasons for different societies having developed different beliefs about the origins of life.

 www.science.nelson.com

Strong evidence for a changing Earth began with a careful examination of **fossils**. Near the end of the 15th century, Leonardo da Vinci pondered the numerous seashell remains he found high in the Tuscany mountains, hundreds of kilometres from any sea. He became convinced that these very old shell deposits had been formed in an ancient ocean and concluded that Earth's surface had changed dramatically over time.

The Fossil Record

The fossils da Vinci had been examining were traces of organisms from the past. By the late 17th century, geologists had begun to map locations where exposed layers of rock contained distinctive and remarkable fossils that were considered to be evidence of prehistoric life. In 1669, Nicholas Steno's detailed and impressive analyses of fossils clearly demonstrated that they were mineralized remains of living organisms. These ideas were supported by such respected scientists as Robert Hooke, who was among the first scientists to suggest that the surface of Earth had changed over time.

The most common and easily recognized fossils are such hard body parts as shells, bones, and teeth (**Figure 1**). Fossils also include impressions of burrows, footprints, and even chemical remains. Fossils are commonly formed when the bodies of organisms become trapped in sediments, which become compressed into strata, or layers, and eventually harden into sedimentary rock. Many fossils have been unearthed by digging, quarrying, and such natural causes as erosion and Earth movements or slides (**Figure 2**, page 512). An organism may simply leave an impression in hardened material, or, if the rate of decomposition is very slow, the organism's cells may be replaced by minerals, resulting in a **permineralized fossil**. On rare occasions, when conditions

fossils any preserved remains or traces of an organism or its activity; many fossils are of such hardened body parts as bone

permineralized fossil a fossil formed when dissolved minerals precipitate from a solution in the space occupied by the organism's remains

Fossils Offer Evidence of Environmental Change
Just as da Vinci found seashell fossils in mountain rock formations, palaeontologists have found 40- to 50–million-year-old whale, clam, snail, and other marine species remains in the deserts of Egypt—clear indications that the environment was considerably different when they were living. Fossilized pollen grains can provide excellent information about the climate that existed when the pollen was deposited.

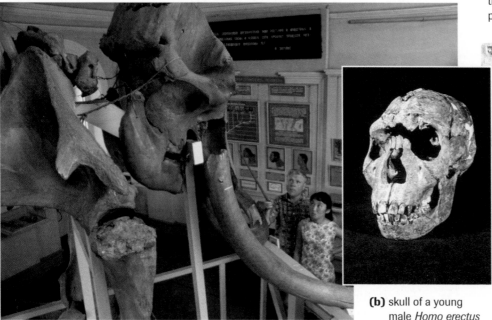
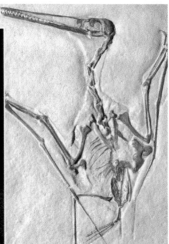

(b) skull of a young male *Homo erectus*

(c) *Pterodactylus kochi* from the Jurassic period

(a) reconstructed mammoth skeleton

Figure 1
Fossils may be organisms that have been **(a)** preserved intact, **(b)** hard body parts, or **(c)** impressions.

(a) dead organism

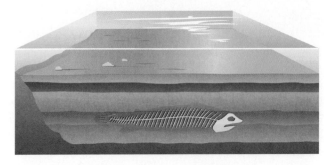

(b) organism is buried and compressed under many layers of sediment

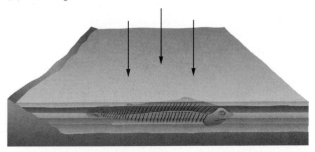

(c) under high pressure deposits harden to form sedimentary rock and the fossil remains become mineralized

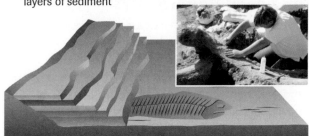

(d) erosion or excavation of sedimentary rock exposes fossil remains on the surface

Figure 2
Long after fossils form in sedimentary rock, mechanisms such as changing sea level, faults, erosion, and human excavation may bring these fossils to the surface (see inset photo).

fossilization the process by which traces of past organisms become part of sedimentary rock layers or, more rarely, hardened tar pits, volcanic ash, peat bogs, or amber

microfossils microscopic remains of tiny organisms or structures that have hard and resistant outer coverings

prevent most decomposition, organisms may be preserved nearly intact; such fossils have been found in tar pits, volcanic ash, peat bogs, permanently frozen ground, and amber (**Figure 3**), or hardened tree sap.

The ideal conditions for **fossilization** are rare. After an organism dies, its soft parts usually are consumed or decompose quickly. Consequently, organisms that have hard shells or bones and that live in or near aquatic environments are much more likely to become fossilized than soft-bodied and land-dwelling species. Although fossils of large land animals such as dinosaurs and sabre-toothed cats are well known, they are rare. Much more common are fossils of hard-bodied marine organisms such as clams and snails. The most abundant are **microfossils**, microscopic remains, such as those of pollen and foraminifera. Regardless of the size of a species, fossils offer unique opportunities to observe evidence of past life directly.

Figure 3
Insect fossilized in amber

▶ *TRY THIS* activity   ***Visit the Tyrrell on the Web***

The Royal Tyrrell Museum in Alberta is one of the world's foremost fossil research centres. At the museum's web site you can

• take the virtual tour of the museum to see what fossils are on display;

• check the "What's Hot" feature and write a short report on the latest fossil discoveries;

• submit a question to "Ask a Palaeontologist;"

• find out the education and training needed to be a palaeontologist.

    www.science.nelson.com

The Study of Fossils

The systematic study of fossils—now referred to as **palaeontology**—began in the 18th century, most notably by the respected anatomist, Baron Georges Cuvier. His extensive investigations of fossils revealed that many fossils were of extinct species and that different sedimentary strata contained distinctive fossilized species. His evidence for extinction challenged the widely held view of his time that all fossils were of living species, some of which had yet to be discovered. Today's fossil record comprises more than 250 000 identified species, a number thought to be only a tiny fraction of all species that have lived on Earth. Less than 1% of species in the fossil record are living today.

Cuvier observed that while fossils of simple organisms could be found at all depths, more complex forms were found in the shallower deposits. In addition, fossils contained in shallower deposits were more likely to resemble living species. Each layer seemed to contain many distinct species not found in layers above or below it. He was puzzled by such findings, as he did not believe that species changed over time. Like others of the time, Cuvier believed that all life on Earth had been created. However, rather than believing in a single creation, he suggested that multiple events had occurred at different times. According to his theory of **catastrophism**, global catastrophes, such as floods had caused widespread extinctions. These extinct life forms would then be replaced by newly created species. Although this explanation could account for the fossils of new species being found above fossils of extinct species in the same location, it could not account for the progressive complexity of species.

Cuvier determined the **relative age** of fossil deposits by assuming a chronology for rock strata and a corresponding sequence for the location of fossils in the layers. Based on his observations of sediment deposition, he reasoned that deeper deposits were older and, therefore, contained more ancient fossils, while shallower deposits were more recently formed and contained younger fossils. However, Cuvier and others of his time had no precise method for calculating an **absolute age** for the rock or the fossils embedded in them. Not until nearly a century later did scientists have the means to determine such ages.

palaeontology the scientific study of fossil remains

catastrophism Cuvier's theory that numerous global catastrophes in the past had repeatedly caused the extinction of species that were then replaced by newly created forms

relative age an estimate of the age of a rock or fossil specimen in relation to another specimen

absolute age an estimate of the actual age of a rock or fossil specimen

ACTIVITY 11.2.1

Applying Fossil Evidence (p. 534)
How can you apply fossil evidence to test hypotheses about the history of life on Earth? In this activity, you can test your own hypotheses about fossil evidence much as Cuvier did.

radioactive decay the release of subatomic particles from the nucleus of an atom, which results in the change of a radioactive parent isotope into a daughter isotope; when the number of protons is altered, a different element is formed

The Age of Earth

Physicist Lord William Thomson Kelvin was the first to try to determine the age of Earth through rigorous mathematical and scientific calculations. In 1866, based on the assumption that Earth was gradually cooling down, he assigned it an absolute age of 400 million years; he later revised his estimate to 15 million to 20 million years. As Kelvin's abilities were very highly regarded, few thought to question his estimates. The discovery, by Pierre Curie in 1903, that **radioactive decay** produces heat suggested that radioactive decay within Earth was a major heat source, so Kelvin's model of a once-molten Earth simply cooling off was no longer considered valid. The study of radioactivity also provided geologists with the means to estimate the absolute age of Earth with much greater precision.

Radiometric and other techniques have been used to date meteorites samples (**Figure 4**). Virtually every meteorite that has struck Earth has yielded an age of 4.6 billion years. Moon rocks collected during Apollo missions have all been dated at about 4.53 billion years in age. The oldest rock on Earth—from the Canadian Shield north of Yellowknife—has been dated at about 3.9 billion years. Scientists now believe that this is the age at which the Earth cooled enough that the oldest solidified rock did not undergo further remelting. It is therefore thought that Earth may be about 4.6 billion years old.

Figure 4
Meteorites provide evidence about the age of other bodies in the solar system and universe.

Radiometric Dating

Radioisotopes are atoms that undergo radioactive decay, and radioactive decay rates can be measured very accurately. The decay of radioactive materials changes a **parent isotope** into a **daughter isotope** of the same element or of a different element. For example, radioactive potassium 40 (^{40}K), can undergo two forms of decay: it can change into either argon 40 (^{40}Ar), or calcium 40 (^{40}Ca). Each radioisotope decays at its own constant rate, measured in a unit called a **half-life**. A half-life is the time it takes 50% of a sample of a parent isotope to decay into a daughter isotope (**Figure 5**). Physicists have determined the half-lives of many radioactive materials (**Table 1**), which are considered accurate to within a few percentages. The rate of half-life decay for different isotopes varies considerably

radioisotopes atoms with an unstable nuclear arrangement that undergo radioactive decay

parent isotope changes into a daughter isotope as radioactive decay occurs

daughter isotope what a parent isotope changes into during radioactive decay; may be stable or may be radioactive and capable of further decay

half-life the time required for half a radioactive material to undergo decay; for any given isotope the half-life is constant

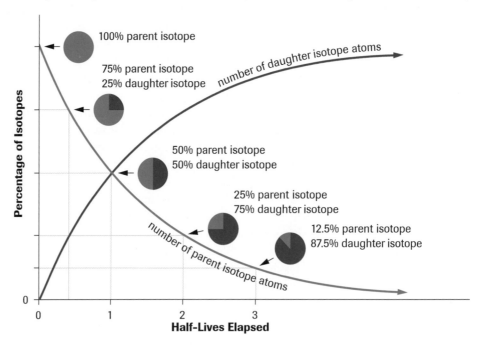

Figure 5
During each half-life of radioactive decay, 50% of a parent isotope decays into a daughter isotope. The result is a quantitative, predictable relationship between the ratio of parent-to-daughter isotopes and elapsed time.

Table 1 Radioisotopes Used in Radiometric Dating

Parent isotope	Daughter isotopes	Half-life (years)	Effective dating range (years)
^{14}C (carbon 14)	^{14}N (nitrogen 14)	5730	100 to 100 000
^{235}U (uranium 235)	^{207}Pb (lead 207)	713 million	10 million to 4.6 billion
^{40}K (potassium 40)	^{40}Ar (argon 40) and ^{40}Ca (calcium 40)	1.3 billion	100 000 to 4.6 billion

but is unaffected by temperature, moisture, or other environmental conditions. As the half-life for a given isotope is constant over time, it can be used as a naturally occurring radiometric clock. Since modern palaeontologists are able to measure the age of rock through **radiometric dating**, they can estimate much more accurate ages for fossils found between layers of rock that contain radioactive materials. Because carbon is found in all living things, carbon-14 dating can be used to determine the age of organic materials directly. The relatively short half-life of carbon 14—at 5730 years—makes it unreliable for testing objects more than 100 000 years old; it is, however, ideally suited for testing such archaeological finds as human remains.

radiometric dating calculation of the age of rock—and of embedded fossils or other objects—through the measurement of the decay of radioisotopes in the rock

Potassium–Argon Dating

When a volcano erupts, gases in the lava escape. As a result, when the lava cools to form igneous rock, it may contain radioactive potassium, ^{40}K, but it will not contain argon gas, ^{40}Ar. As radioactive decay occurs, some argon daughter isotopes will accumulate within the solid rock. Thus, the ratio of the amount of ^{40}K to the amount of ^{40}Ar present when samples are taken can be used to determine how long ago the rock in the sample formed.

Example

A sample of igneous rock contains small amounts of radioactive potassium and argon. Using the ratio of ^{40}Ar to ^{40}K, it is determined that only 25% of the original parent potassium isotope remains. How old is this rock sample?

Solution

Figure 5 shows that it requires two half-lives for the ratio of parent-to-daughter isotopes to change from 100%:0% to 25%:75%. Given a half-life for ^{40}K of 1.3 billion years (**Table 1**), this igneous rock formed 2.6 billion years ago.

▶ Practice

Understanding Concepts

1. A sample of igneous rock is found to contain the radioactive parent and daughter isotopes uranium, ^{235}U, and lead, ^{207}Pb, in the ratio of 12.5%:87.5%. Assuming that no ^{207}Pb was present when the rock first formed, estimate the age of this sample.

2. A fossil skull of *Homo neanderthalensis*, is discovered in northern Europe and is tested using carbon-14 dating. Palaeontologists are curious about whether the Neanderthal was living at the same time as members of *H. sapiens*, thought to have been living in the same area of northern Europe 45 000 years ago. Measurements suggest that, of the original amount of carbon-14 isotope present in the skull when the Neanderthal died, only 1.56% remains in the fossil fragment.
   (a) How old is the fossil?
   (b) Could this Neanderthal have been a contemporary of *H. sapiens* in this area?

Answers

1. 2.14 billion years old
2. (a) 34 380 years old

▶ **EXPLORE** an issue

Take a Stand: The Economics of Fossil Study

Fossil-hunting expeditions are costly and the rarest fossils are very valuable. In October 1997, Sotheby's auctioned an almost perfect 65-million-year-old *Tyrannosaurus rex* fossil for U.S.$8.4 million (**Figure 6**). Sotheby's described the fossil as a world treasure. The buyer, the Field Museum of Natural History in Chicago, received donations from Ronald McDonald House, Walt Disney, and others to buy the fossil. The sale ended a legal, scientific, and ethical controversy revolving around the rightful ownership of the fossil, which was found by a commercial fossil hunter.

Statement: Fossils should be donated to research institutions because their investigations benefit everyone.

- In a group, research issues for this statement using print and electronic resources.
- Develop a list of points and counterpoints on these issues.

Decision-Making Skills

- ● Define the Issue
- ○ Defend the Position
- ● Analyze the Issue
- ● Identify Alternatives
- ● Research
- ● Evaluate

- Write recommendations on the economics of fossil research.

 www.science.nelson.com

Figure 6
Museums must bear the costs of equipment, research staff, facilities, and the procurement of fossils through purchases or field expeditions.

SUMMARY   *The Fossil Record*

- Early scientific thinking about the origins of life on Earth focused on the study of fossils embedded in rock. Fossils were readily identified as mineralized remains or traces of organisms that are now extinct. Fossils contained within older (usually deeper) rock formations were observed to be simpler in structure and less similar to modern species than those found in younger (usually shallower) rock deposits.

- By the 19th century, scientists began to gather evidence that Earth could be extremely old. As fossils were part of rock that might be of great antiquity, the possibility was raised that life on Earth might also have an ancient past.

- The ability to measure relative amounts of radioactive parent and daughter isotopes in rock provided an accurate and reliable method of determining the age of both rock and fossil remains.

▶ Section 11.2 Questions

Understanding Concepts

1. Outline the processes of fossil formation in rock.

2. In what materials might fossil remains be found besides sediments? Explain why such finds are rare.

3. On what evidence did Cuvier base his theory of cata-strophism? Cuvier compared fossil remains with species living at his time. Explain his reasoning.

4. What property of radioactive material invalidated Kelvin's assumptions? How?

5. How are scientists able to assume that igneous rock contains no ^{40}Ar when it first forms?

6. How old is Earth now thought to be, and how was this estimate derived? How much greater is this age than the calculations of Lord Kelvin and Archbishop Ussher?

7. The Moon and most other solid bodies in the solar system, such as meteors, do not exhibit volcanic activity. How does this factor affect reasoning about the age of Earth?

Applying Inquiry Skills

8. Outline the steps you might use to determine the age of a fossil embedded between two layers of igneous rock.

9. Explain how carbon-14 might be used to date relatively recent organic remains or archaeological finds.

10. Before the discovery of radioactivity, what doubts might scientists of Kelvin's day have raised about his estimates? Research Kelvin's investigations into evidence that Earth was cooling and prepare an assessment of his reasoning from his experiments. How might his approach have given credibility to his theories?

11. Examine **Figure 7**, which shows one of the world's most famous fossil species. The first *Archaeopteryx* was discovered in 1861 in a quarry in Solnhofen, Germany. What features are visible on this fossil that might have interested palaeontologists studying the evolution of reptiles and birds?

Figure 7
A 140-million-year-old fossil of *Archaeopteryx*

Making Connections

12. Cuvier and da Vinci both approached their interpretations about Earth's past from acute observations. Find out more about their scientific studies that supported their interpretations. Compare their theories and beliefs, including ways in which they were influenced by nonscientific thinking.

13. Fossils have been unearthed by many people who are not scientists. Find articles in print and online about such finds and compare them with scientific fieldwork. Prepare a class presentation of your comparison. Be sure to take into account societal and technological differences.

 www.science.nelson.com

14. Microfossils of the foraminifera, *Neogloboquadrina pachyderma* are excellent recorders of relative ocean temperatures. As they grow, these marine microorganisms build an outer shell that generally coils to the right under warm water conditions and to the left under cooler conditions.
    (a) How might such differences in coiling be of value to scientists studying past climate changes or to oil companies in search of new fuel deposits?
    (b) Search electronic sources for further examples of applications of fossil study to significant present-day issues.

 www.science.nelson.com

Eighteenth-century scientists were starting to understand pieces of the evolution puzzle by drawing on early physical evidence from the fossil record. Geology offered the foundation for new and tentative, but scientific, hypotheses of the age of Earth and the origins of life. In 1795, Scottish geologist James Hutton proposed a theory he called **actualism** to contrast with Cuvier's catastrophism. Hutton explained the geological formation of landforms as the result of slow processes, such as erosion, that were ongoing and observable in his day (**Figure 1**). Building on Hutton's ideas, Sir Charles Lyell revolutionized geology with his principles of **uniformitarianism**. In his now-famous *Principles of Geology*, published in 1830, Lyell made the following arguments:

- Earth has been changed by the same processes in the past as can be observed occurring in the present.

- Geological change is slow and gradual rather than sudden and catastrophic.

- Natural laws and processes are constant and eternal, and they operated with the same kind of intensity in the past as they do in the present.

Lyell had based his theories on extensive examination of fossil deposits and such processes as erosion and sedimentation. He championed the need for direct and systematic observation of nature. While travelling through North America in 1841, he estimated the age of Niagara Falls at 35 000 years. In a later trip, he discovered important fossils at Joggins, Nova Scotia.

The work of Hutton and Lyell led to two significant conclusions: that Earth must be unimaginably ancient and that dramatic change could result over such extremes of time through slow, seemingly slight processes. These conclusions provided the foundation on which other scientists could build theories about the history of life forms on Earth.

During the second half of the 18th century, scholarly interest in observable changes to species intensified. A leading naturalist, Georges Buffon, proposed that species could change over time and that these changes could lead to new organisms. Carl Linnaeus, the founder of biological nomenclature, and Erasmus Darwin, a well-respected physician and poet who was the grandfather of Charles Darwin, both proposed views similar to those of Buffon. Linnaeus proposed that a relatively few species had formed many new species through hybridization and interbreeding. Erasmus Darwin wrote the first detailed treatise on evolution, in which he asserted strong evidence for the idea that all life had developed from a single source. Along with others, he believed that humans may be closely related to primates. But scientists of this time could offer no mechanisms to explain how evolution might occur.

Adaptation and Inheritance

A student of Buffon made significant contributions to the 19th-century debate on evolution. Jean Baptiste Pierre Antoine de Monet, Chevalier de Lamarck was the first prominent biologist to recognize the key role played by the environment in evolution. Lamarck reasoned that for species to survive over long periods of time, they must be able to adapt to changing environmental conditions. Unlike Buffon and his contemporaries, Lamarck did not believe that a single species could give rise to additional species; instead, he argued that each species gradually became more complex and that new very simple species were continually being created by **spontaneous generation**. Lamarck believed in the evolutionary change and improvement of individual species.

actualism the theory that the same geological processes occurring in the present also occurred in the past

uniformitarianism the theory that Earth's surface has always changed and continues to change through similar, uniform, and very gradual processes

spontaneous generation the idea that living organisms arise from non-living matter

DID YOU *KNOW* ?

Lyell in Nova Scotia
Nova Scotia became a famous place for fossils when, in 1851, Sir Charles Lyell and the principal of McGill University, Sir William Dawson, visited the Joggins cliffs. There, they found many fossils, including the remains of large plant trunks, some tiny bones, and shells. The Joggins fossils are 280 million years old and are among the earliest of all reptiles, land snails, and millipedes.

Figure 1
Close observation and study of changes to such landforms as the cliffs of Dover in England sparked the theoretical developments credited to James Hutton, Sir Charles Lyell, and others.

acquired traits those changes in an individual resulting from interaction with the environment

Building on others' ideas, he postulated the inheritance of **acquired traits**. By this hypothesis, changes acquired by an organism as a result of adaptation to environmental conditions during that organism's lifetime could be inherited by future generations. One generation of giraffes, for example, might have had to strive to obtain food higher in trees, so that, over their lifetime, continual stretching might have led to a slight elongation of the neck. For Lamarck, this slightly longer neck became an acquired trait to be inherited by offspring. Since then, it has been established that such acquired traits cannot be inherited. Ridiculed by contemporaries—including the influential Cuvier who rejected evolution outright—Lamarck deserves credit for his recognition of the role of the environment in driving evolutionary change. His attempt to formulate a mechanism for evolution smoothed the way for succeeding theories.

SUMMARY   *Early Ideas About Evolution*

Table 1

Scientist	Contribution to development of theory of evolution
Sir Charles Lyell (1797–1875)	suggested Earth had undergone and continues to undergo slow, steady, and very gradual changes
Comte Georges Louis Leclerc de Buffon (1707–1788)	suggested that similar organisms may have a common ancestor
Erasmus Darwin (1731–1802)	proposed that all life may have a single source
Jean Baptiste Pierre Antoine de Monet, Chevalier de Lamarck (1744–1829)	was the first scientist to recognize that the environment plays a key role in the evolution of species. He further postulated the theory of inheritance of acquired traits

▶ Section 11.3 Questions

Understanding Concepts

1. How did evidence about the age of Earth support thinking about the origin of life forms?

2. Give an example of the kind of evolutionary change that Buffon thought might occur.

3. Describe the theory of the inheritance of acquired traits, including an example.

4. Lamarck postulated the adaptation of species to a changing environment, spontaneous generation, and the inheritance of acquired traits. How did his thinking contrast with that of his contemporary Cuvier?

5. In your notebook, make a list of inherited and acquired traits.
   (a) Write examples of your own traits (physical or mental).
   (b) Which, if any, of your inherited traits could be altered during your life? How?
   (c) Which, if any, of your acquired traits can be passed on to your offspring? How?

Applying Inquiry Skills

6. Lyell estimated that Niagara Falls was about 35 000 years old. Find out the current scientific estimate. How close was Lyell?

7. How might you investigate scientific evidence to support your response to question 5(c)?

Making Connections

8. Much scientific investigation and debate preceded Lyell's principles in support of an ancient history for Earth. Explain why the conceptual leap in support of a lengthy history for organisms—and especially for humanity—sparked even greater investigation and debate in 19th-century Europe.

9. Some medical conditions, such as viral infections, are acquired while others such as Huntington's disease are inherited. An acquired disease results from environmental factors, which can be monitored and, possibly, altered or eliminated. An inherited disease results in fundamental physiological changes within the body that require very different medical treatment. Research one of the following diseases using electronic and print sources. Determine whether it is acquired, inherited, or a combination.
   (a) breast cancer
   (b) diabetes
   (c) cystic fibrosis
   (d) multiple sclerosis
   (e) Duchenne's muscular dystrophy

 www.science.nelson.com

On August 24, 1831, Charles Darwin (**Figure 1**) received a letter from his former botany professor, the Reverend John Henslow, who informed the 22-year-old he had been recommended as a naturalist on a voyage to South America. Henslow said that, despite Darwin's lack of formal training, he considered Darwin to be "the best qualified person I know of … for collecting, observing, and noting anything worthy to be noted in natural history." As a boy, Darwin had preferred collecting insects, hunting, and fishing to attending school lessons. Although his father—a wealthy physician like his father before him—had encouraged Darwin to study medicine, the young man's time at medical school was short-lived; he was so nauseated by the operating room, where anaesthetics were not yet in use, that he quickly gave up the idea of becoming a physician. He had just graduated from his studies for an alternate career in the clergy, hoping that the quiet life would afford him ample time to spend outdoors, when he was recommended as naturalist for the trip to South America. In December 1831, he set sail on the *HMS Beagle* (**Figure 2**) on what became one the most influential voyages in human history.

Figure 1
Charles Darwin

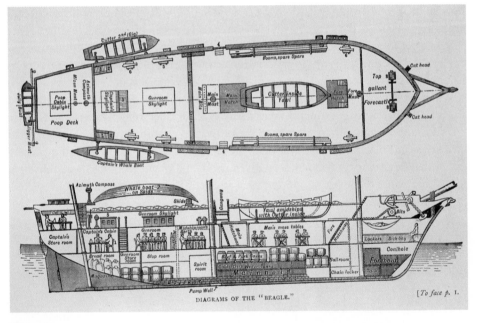

Figure 2
HMS Beagle was a 10-gun brig, 27.5 m long, that offered little room or comfort for the 74 people onboard.

Initially scheduled to take two years, the voyage lasted almost five (**Figure 3**, page 520). Captain Robert FitzRoy had a primary mission to survey and map the coastal waters of South America for the British navy. FitzRoy accepted Darwin onboard in the hope that he would find evidence to support biblical creation. Darwin's role was to observe, record, and collect specimens of rocks and minerals, plants and animals. His supplies included jars of spirits to preserve specimens, a microscope, binoculars, a compass, many notebooks, pistols, a rifle, and a gift from Henslow—the first volume of Lyell's *Principles of Geology*. Darwin collected thousands of specimens, which he shipped to England at

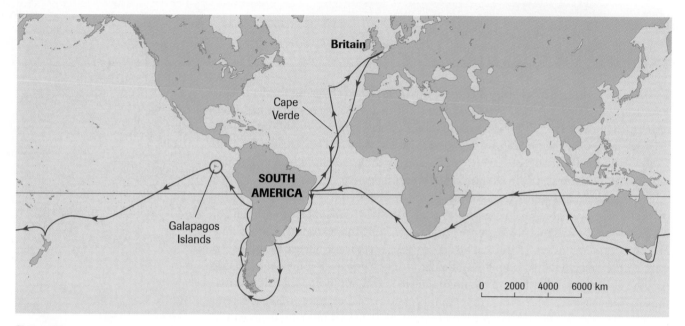

Figure 3
Route of the five-year voyage of the *Beagle*

intervals. In an 1832 letter to Henslow, he wrote: "I am afraid you will groan, or rather the floor of the Lecture room will, when the casks arrive." His detailed observations covered more than 2000 journal pages, in which he speculated about his unusual findings.

While the ship's crew surveyed the coast of South America, Darwin was put ashore to explore the continent. His vivid descriptions convey his immense sense of wonder as he discovered ecosystems utterly unfamiliar to him. While in Patagonia, he unearthed fossils of giant *Glyptodon* (**Figure 4b**) and *Megatherium* (**Figure 5a**). Later, he recognized the resemblance of these extinct animals to the much smaller modern armadillo (**Figure 4a**) and sloth (**Figure 5b**), respectively. These two sets of animals, modern and extinct, share the same patterns of distribution in South America. After a severe earthquake, he noted that one section of shoreline, with mussels still intact, was lifted 3 m above sea level. This evidence, he felt, supported Lyell's assertions that such events could eventually lead to the formation of entire mountain chains.

(a) armadillo　　　　**(b)** *Glyptodon*

Figure 4
A present-day armadillo **(a)**, which Darwin recognized as related to the extinct *Glyptodon* **(b)**.

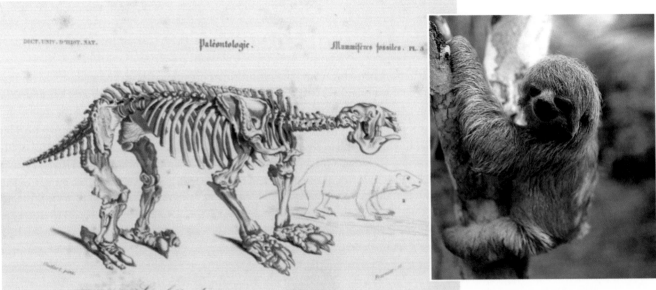

(a) *Megatherium*

(b) sloth

Figure 5
Fossil of the extinct *Megatherium*, which Darwin unearthed in South America and recognized as related to the modern sloth, shown above.

Sailing west from South America, the *Beagle* reached the volcanic Galapagos Islands (**Figure 6**), where Darwin noted that, although the flora and fauna closely resembled species in South America, those of the Galapagos had unique traits. While there, Darwin failed to realize that each island supported unique species of plants and animals. He collected no tortoises, for instance, and gathered only a few specimens of finch, which he failed to label by island. He did collect mockingbird specimens, which he noted showed distinct variation from island to island. It would only be much later, after expert analysis, that he would realize the significance of these organisms and write: "Several of the islands possess their own species of the tortoise, mocking-thrush, finches, and numerous plants, these species … obviously filling the same place in the natural economy." Similarly, when visiting the Cape Verde Islands, Darwin noticed that species there closely resembled those on the closest landmass, Africa.

Darwin had left England believing in the immutable nature of species. He returned in 1836 filled with enthusiasm and questions about what he had seen and experienced. In the years to follow, his analyses of the evidence he had collected on the voyage, as well as those of his colleagues, led Darwin to conclusions that have placed him among the intellectual giants of science.

DID YOU *KNOW* ❓

Darwin's Writings
Late in the voyage, Darwin lavished attention on coral reefs. In 1842, he published his theories—still accepted today—on their formation in his first scientific book, *On the Nature of Coral Reefs.* Darwin wrote his own accounts of his travels in *The Voyage of The Beagle*, which can be read in print or online.

GO   www.science.nelson.com

Figure 6
Darwin, on the Galapagos Islands: "Hence, both in space and time, we seem to be brought somewhat near to that great fact—mystery of mysteries—the first appearance of new beings on this Earth."

biogeography the careful observation and analysis of the geographic distribution of organisms

The earliest theories regarding the evolution of life met with stiff opposition. Many scientists and most members of the public were not readily swayed by evidence as unfamiliar to them as fossilized remains or descriptions of geologic processes. However, a new and powerful line of evidence would soon be provided by Darwin, who began to wonder why continents separated by large distances would have entirely different species occupying similar ecological niches. Why, for example, would animals living in the grasslands of South America or Australia be more similar to animals living in a nearby tropical rain forest than to grassland species on other continents? His careful observation and analysis of the geographic distribution of organisms provided a foundation for the later scientific study of **biogeography**.

grasping, probing bill
eats insects

large crushing bill
eats seeds

Chisels through tree bark to find insects. Uses a tool (a cactus spine or small twig) to probe insects.

parrotlike bill
eats fruit

Figure 1
Ornithologist John Gould identified at least 13 distinct species of finches from the specimens Darwin and others collected on the Galapagos Islands.

> **▶ TRY THIS** activity   *Darwin's Seeds at Sea*

Darwin was interested in how plant species came to populate remote islands. He suspected that many plant seeds had floated long distances in the open ocean, landing eventually on islands distant from their origins. If his hypothesis were correct, the seeds must have been able to endure long periods of exposure to salt water. In this activity, you can replicate one of Darwin's experiments to test this hypothesis.

 Some seeds are potentially toxic. Wear gloves when handling seeds. Do not ingest seeds.

- Working in pairs, design and conduct a controlled experiment to test Darwin's hypothesis. Consider such factors as seed type, other materials, temperature, and number of days seeds will be immersed in salt water.
- After your teacher approves your design, carry out your experiment.
- Analyze your results and share them with your class. Do the class results support Darwin's original hypothesis?
- Read the published results of Darwin's own seed trials ("Does Sea-Water Kill Seeds?") and compare your findings with his. What other questions did Darwin raise in his report?

Back in England, Darwin sent specimens collected on his voyage to experts for examination. His fossils were sent to Richard Owen, Britain's most eminent palaeontologist, who confirmed Darwin's suspicions that *Glyptodon* and *Megatherium* fossils were large versions of the armadillo and sloth that still inhabited the continent (**Figures 4** and **5** in section 11.4 on pages 520–521). Darwin recognized that there were many similar instances where unusual fossils bore a close resemblance to species currently living in the same region. He thought the best explanation for this fact was that the fossils represented ancestral forms of the living organisms.

Darwin's bird collections had been sent to ornithologist John Gould, who informed him that 25 of his 26 Galapagos birds were different species, including three species of mockingbird and a large number of finches (**Figure 1**). Unfortunately, as Darwin had believed their differences were just variations within a very few species, he did not separate them in his specimen collection and did not label them by island in his notes. Even so, for Darwin, Gould's evidence suggested that perhaps a single ancestral species transported from a nearby land might give rise to a number of similar but distinct new species, especially when isolated on separate islands. Darwin had noted that remote islands were inhabited by birds and other flying organisms, and such land animals as lizards and turtles that could survive long periods at sea with no food or fresh water. Mammals unable to survive such travel were entirely absent from these remote islands—with the exception of rats and domesticated animals from ships. These islands certainly appeared to have been initially populated by species that originally travelled from the closest landmass.

In July 1837, Darwin began his first notebook on what he called the "transmutation of species" with his observations about Galapagos species and South American fossils. In a letter to his friend J.D. Hooker in 1844, he commented: "I was so struck ... that I determined to collect blindly every sort of fact, which could bear any way on what are species. I have read heaps of agricultural and horticultural books, and have never ceased collecting facts." Darwin committed his life to this investigation.

Homologous and Analogous Features

Darwin relied on specialists to discern the subtle differences between closely related species. But he and others were very interested in what could be learned by comparing species that were very different in outward appearance. Like Cuvier and Owen, Darwin was struck by the numerous instances in which body parts of organisms with entirely different functions were similar in structure (**Figure 2**). These **homologous features** were

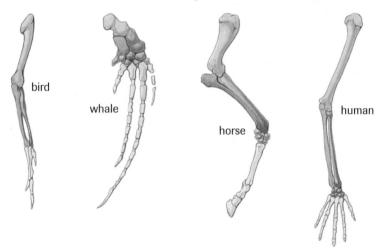

a puzzle to Darwin, who wrote: "What can be more curious than that the hand of a man, formed for grasping, that of a mole for digging, the leg of the horse, the paddle of the porpoise, and the wing of the bat, should all be constructed on the same pattern, and should include the same bones, in the same relative positions?" By contrast, **analogous features** shared by many organisms and serving a common function—such as the wings of birds and butterflies, the eyes of lobsters and fish—were observed to be different in internal anatomy. Darwin came to the conclusion that organisms with homologous features likely shared a more recent common ancestor, while those with analogous features did not.

Homologous features can also appear during embryonic development (**Figure 3**, page 524). Such features serve no function as the organisms grow. In the early weeks of development, human embryos possess a tail similar to that in chicken and fish embryos, while the embryos of anteaters reveal the presence of rudimentary teeth, which are lost before birth. In certain terrestrial frogs, all early development occurs within the eggs so that fully formed frogs hatch ready for life on land; within the egg, however, the larvae pass through a tadpole stage that includes a tail, gills, and other features clearly adapted for life in the water.

DID YOU KNOW ?

Odd Places for Fossils
While in South America, Darwin found fossilized remains of tropical forests at an elevation of over 3000 m in the mountains. He was even more surprised to discover fossilized seabed specimens 300 m higher up the same sedimentary deposit. For Darwin, this was evidence that the Earth's environment had undergone continuous and dramatic change.

homologous features structures that share a common origin but may serve different functions in modern species (e.g., dolphin flippers and human hands)

Figure 2
The forelimbs of these organisms are each adapted to carry out very different functions, yet they all possess very similar bone structure.

analogous features structures similar in function but not in origin or anatomical structure (e.g., wings of birds and bees)

ACTIVITY 11.5.1

Looking for Homologies (p. 535)
How can homologous and analogous features be recognized? In this activity, you compare five organisms based on outward appearances and habits, and on internal anatomical and physiological features.

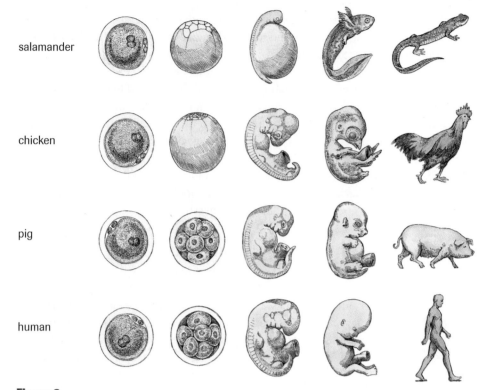

salamander

chicken

pig

human

Figure 3
Homologous features such as tailbones are clearly evident in the embryos of these organisms.

Vestigial Features and Anatomical Oddities

vestigial features rudimentary and nonfunctioning structures that are homologous to fully functioning structures in closely related species

Darwin and his contemporaries also studied numerous **vestigial features**, structures that serve no useful function in a living organism, such as digits in dogs, pigs, and horses (**Figure 4**). Some beetles have fully developed membranous wings underneath fused covers, so that it is impossible for them to open these wings. Humans have muscles for moving ears, much as dogs and other mammals do. With practice, some people can make slight use of these muscles, but they do not routinely flip their ears back when they hear a noise behind them. Some snakes and large whales have vestigial hips homologous to the functioning hips in animals with hind limbs (**Figure 5**).

dog

pig

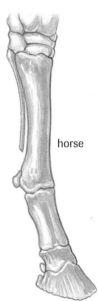

horse

Figure 4
Dogs have a vestigial toe. Although the bones remain, this digit serves no present purpose. Pigs' feet have two well-developed digits; the others are vestigial. The horse has only one enlarged digit; the others are vestigial or have been lost entirely.

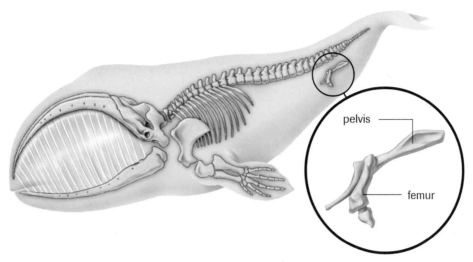

Figure 5
Hipbones are used to transfer body weight to the hind legs. The existence of these bones in modern whales is compelling evidence that these mammals evolved from ancestors that possessed hind legs.

Many anatomical oddities observed by Darwin and his colleagues also suggested an evolutionary past. Upland geese and frigate birds, for example, have fully webbed feet but never enter water. A most striking example is the fourth vagus nerve. In fish, this nerve passes directly from the brain to the gill located nearest the heart, but in all mammals, it serves the larynx but follows the same pathway. In the giraffe, the nerve exits from the brain, bypasses the nearby larynx, travels the full length of the neck down to the heart, turns around and travels all the way back for an added distance of about 4 m.

In addition to morphological features, modern researchers have found very large numbers of vestigial genes in the DNA of living organisms; these genes do not function, yet they bear a striking resemblance to functioning genes. Such molecular evidence was not available to scientists of Darwin's day.

Artificial Selection

Darwin became increasingly sure that some mechanism of inheritance must be the key to evolutionary processes. As well as studying the structures, functions, and behaviour of species, he looked for evidence of changes in populations of organisms that were reproducing. Darwin became keenly interested in artificial selection, breeding his own pigeons, conducting thousands of experiments in his gardens and greenhouse, and bombarding professional breeders with questions.

Humans have been improving domesticated plant and animal species for thousands of years. By selecting offspring with desirable traits as breeding stock for succeeding generations, people have produced an extraordinary diversity of breeds of farm animals, cats, and dogs. Plant breeders have produced a diversity of flowering plants, crops, and improvements to crops. For example, in 1896, researchers began artificial selection experiments that increased the oil content of corn kernels from 4% to 6% to more than 14% in only 60 generations. The researchers selected seeds for each new generation from plants estimated to have the highest average oil content per seed. The process of artificial selection resulted in corn seeds with oil far above and outside the range of variability within the original plants (the original minimum and maximum oil content is indicated in red in **Figure 6**, page 526). Later generations of species are unlike their original sources.

Darwin recognized that all species possess inherited variations that can be selected to change the species in desirable ways. Darwin reasoned that if people could alter the appearance and behaviour of species through artificial selection, then the environment could have a similar selective effect on wild species. He felt that if Lyell were correct about the age of Earth and if fossil evidence supported a comparable age for organic

DID YOU KNOW ?

Examples of Vestigial Features
Many living organisms, including humans, have vestigal features. Here are some examples:
- pelvic bones and rudimentary leg bones in some pythons and boas
- dew claws in dogs and toes elevated off the ground in pigs, cattle, and deer
- the appendix in humans
- human "goosebumps"
- rudimentary wings in many flightless insects such as earwigs
- rudimentary eyes (or empty eye sockets) in blind cave-dwelling fish and amphibians

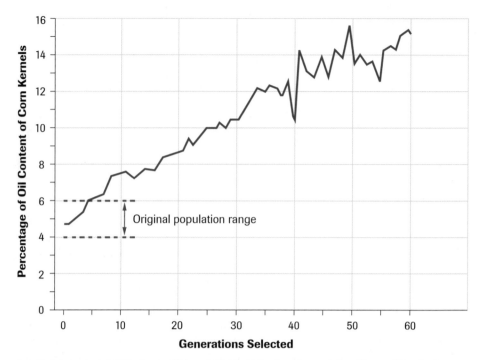

From *Evolutionary Analysis* 2/E by Freeman/Herron, © 2001, p. 69. Reprinted by permission of Pearson Education, Inc., Upper Saddle River, N.J.

Figure 6
Artificial selection in corn plants produced a dramatic increase in corn oil content in only 60 generations.

life forms, there could be time for even small changes in species to accumulate over many thousands of generations.

The Struggle for Survival

Darwin was accumulating a large amount of physical evidence from his own and colleagues' observations and experiments in support of the evolution of life forms. He had become convinced from the evidence of biogeography and comparative anatomy that life had evolved. He believed that artificial selection offered a model for how evolution might operate in nature, but he did not know how nature chose individuals with particular desirable variations for reproduction—as breeders do with plants and dogs. He found the answer in the mathematical treatise written in 1798 by Thomas Malthus, *Essay on the Principle of Population*.

Malthus postulated that in nature both plants and animals produce far more offspring than are able to survive. The final piece of Darwin's puzzle slipped into place, as he stated in his autobiography: "In October 1838 ... I happened to read for amusement Malthus on population, and ... it at once struck me that under [these] circumstances favourable variations would tend to be preserved, and unfavourable ones to be destroyed. The results of this would be the formation of a new species. Here, then I had at last got a theory by which to work." Darwin realized that if far more offspring are born than can survive and reproduce, there must be intense competition among individuals of the same species to survive. He also realized that most never survive to reproduce. Darwin applied Malthus's principles to the theory he was formulating on the origin and evolution of species.

▶ *TRY THIS* activity   *The Odds of Survival*

On reading the Malthus essay, Darwin immediately realized the significance for individual members of a population. The number of individuals being born in any population far exceeds any increase in the overall size of the population. In ideal conditions, every individual would live to reproduce and die of old age. In natural conditions, population size does not usually change significantly over time.

- Examine **Tables 1** and **2**, in which two scenarios are presented for four populations.
  (a) What might cause populations of species to change in size as they do in **Table 1**?

Table 1 Changes in Population Size in Natural Conditions

Initial population	Final population*
100 elephants	750 years later: population has increased to 130
500 starfish	16 years later: population has declined to 405
1000 house flies	5 months later: population reduced to 890
1 million staphylococcus bacteria cells	48 hours later: population has risen to 1.5 million

(b) The second scenario calculates the population sizes that would result if all offspring survived, reproduced successfully during their life, and died of old age. What factors prevent this from happening?

(c) Considering the results in **Table 2**, is it reasonable to suggest that few, if any, species ever experience ideal conditions? Explain your answer.

(d) Darwin concluded that, in all species, many individuals do not survive to reproduce. How did he directly link this struggle for survival with the inherited variations of individuals?

Table 2 Changes in Population Size in Ideal Conditions

Initial population	Final population*
2 elephants	750 years later: 19 million elephants
2 starfish	16 years later: 10^{79} starfish**
2 house flies	5 months later: 190×10^{18} houseflies
1 staphylococcus bacteria cell	48 hours later: cells cover Earth's surface to a depth of 2 m

* From *Evolutionary Analysis*, 2/E. by Freeman/Herron, © 2001, p. 55. Reprinted by permission of Pearson Education, Inc., Upper Saddle River N.J.

** A female starfish can produce more than 2 billion eggs a year.

SUMMARY   *Accumulating Evidence for an Evolutionary Mechanism*

- Darwin recorded detailed observations and collected large numbers of specimens during his voyage as a naturalist on the *Beagle*. He ensured that his specimens were sent to highly respected experts for analysis and, on his return, conducted investigations of his own.

- Darwin suspected that many species bearing a striking similarity to other species, both extinct and living, in the same region were related to one another. He also drew inferences about evolutionary ancestors from homologous, analogous, and vestigial features.

- Artificial selection was used to alter the appearance and behaviour of domesticated plants and animals. Darwin postulated that natural processes could act as agents of natural selection in much the same way.

- Darwin concluded that, as species produce many more offspring than can survive, there must be intense competition among individuals in each species and that, in this struggle to survive, the most favourable traits are inherited by succeeding generations.

Understanding Concepts

1. Less than a year after returning from his famous voyage, Darwin was beginning to believe that species could change. What steps did he take to gather evidence to support this possibility?

2. What did Darwin observe to convince him that Lyell's ideas about Earth were accurate?

3. Which investigations by Darwin can be considered as preliminary efforts in biogeography?

4. Explain how Darwin's reasoning was affected by his identification and analysis of homologous, analogous, and vestigial features. Provide examples with your explanation.

5. On the Galapagos Islands, Darwin witnessed the flightless cormorant. What might its vestigial wings have suggested about this species in the past?

6. For each of the following vestigial features, write your hypothesis for its ancient function, and indicate a change that may have led to the loss of this function. Show your answers in a graphic organizer.
   (a) pelvic (hip) bones in some whales
   (b) elevated toe on hind leg of deer
   (c) human "goosebumps" and muscles to make human ears move
   (d) rudimentary wings in many flightless insects such as earwigs
   (e) webbed feet of frigate and upland goose (birds that never enter the water)

7. How can one infer from the history of artificial selection the possibility that species can evolve over time? Support your reasoning with examples.

8. Describe the impact that the principles in Malthus's essay had on Darwin's thinking.

9. Describe the puzzle created by the presence of hipbones for walking on land in the anatomy of living whales.

10. Would you consider human body hair to be a vestigial feature? Support your answer.

Applying Inquiry Skills

11. The hoary bat (*Lasiurus cinerus cinerus*) is the only mammal native to the islands of Hawaii. Account for this observation.

12. In summer months, female aphids reproduce by parthenogenesis, giving birth to females from unfertilized eggs. These females mature within 24 h and begin reproducing in the same way. Assume that each female aphid produces two females the day after she herself is born, and then she dies.
   (a) If there is a single female aphid on June 1, how many aphids would there be under ideal conditions by August 31?
   (b) What prevents such high population numbers from actually occurring?

   (c) When lacebugs and ladybird beetles (ladybugs), the principal predator of aphids, are absent, aphid populations increase dramatically. Explain why.

13. Based on what you have learned in this section, predict the main features of Darwin's theory of the evolution of species.

Making Connections

14. Discuss examples from animal breeding that suggest some behavioural traits can be inherited.

15. Study the map showing the numbers of mammal species on the islands of Newfoundland, Cape Breton, and Prince Edward Island (**Figure 7**).

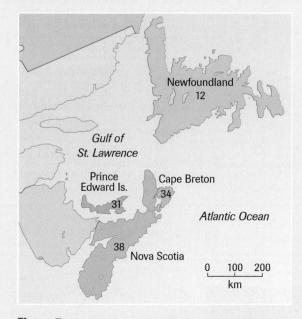

Figure 7
Numbers of mammal species found in Newfoundland, Prince Edward Island, Nova Scotia, and Cape Breton.

   (a) What physical feature of Newfoundland might account for its having far fewer species of mammals than the other islands have?
   (b) Why might most of the mammals found in Newfoundland be relatively large, in general, while those found on the other islands include a wider range of sizes?

16. Brainstorm ways that modern travel has affected the ability of species to reach previously remote or isolated islands.

17. List and discuss several species that humans have released, either intentionally or unintentionally, into new locations in the wild. What implications might such releases have on the process of evolution?

A little more than seven years after his voyage on the *Beagle*, Darwin put together a workable theory on the origin of species by natural selection. In a letter to Hooker he said, "At last gleams of light have come, and I am almost convinced (quite contrary to the opinion I started with) that species are not (it is like confessing a murder) immutable.... I think I have found out ... the simple way by which species become exquisitely adapted to various ends." In 1844, Darwin gave his wife the 213-page manuscript in which he had outlined his theory, so that it might be published if he were to die suddenly.

Aware that his ideas would be controversial, Darwin continued to gather evidence and look for flaws in his reasoning rather than make the theory public. By June 1858, he had written a quarter of a million words toward a major treatise when he received a letter from Alfred Russell Wallace, a brilliant naturalist working in Malaysia. Wallace had independently arrived at the same conclusions as Darwin and described his own theory in the letter.

It was an odd moment in history. Wallace, with much less supporting evidence, was anxious to present the new theory he had written in just two days, while Darwin, having spent 20 years amassing evidence in support of the same theory, had not been willing to publish. Learning of the situation, Lyell and Hooker urged Darwin to submit a paper along with Wallace's at the next meeting of the Linnaean Society. Their theory was presented jointly on July 1, 1858. A year and a half later, Darwin published a condensed version of his treatise, entitled *On the Origin of Species by Means of Natural Selection or The Preservation of Favoured Races in the Struggle for Life*. It sold out on the first day. The response in society was as dramatic as predicted (**Figure 1**).

MR. BERGH TO THE RESCUE.

THE DEFRAUDED GORILLA. "That *Man* wants to claim my Pedigree. He says he is one of my Descendants."

Mr. BERGH. "Now, Mr. DARWIN, how could you insult him so?"

Figure 1
Darwin was often vilified by the press.

▶ *EXPLORE* an issue

Role-Play: Science in Society

Decision-Making Skills

○ Define the Issue ● Analyze the Issue ● Research
○ Defend the Position ○ Identify Alternatives ○ Evaluate

The scientists in this chapter were influenced in their thinking by their social upbringing and by public opinion, as well as by their peers in the scientific community. Imagine a gathering at the home of Charles and Emma Darwin of the following individuals (some of whom would have been long dead by Darwin's time): Leonardo da Vinci, Georges Cuvier, Charles Lyell, Georges Buffon, Jean Lamarck, William Thomson Kelvin, Erasmus Darwin, Captain Robert FitzRoy, Thomas Huxley, Samuel Wilberforce, Frederick Temple, Alfred Russell Wallace, and Emma Darwin.

Question: Are science and scientists subject to societal influences?

- Research the life of one of these individuals. You will role-play this person at the gathering described above. In the role-play, include a summary of his or her ideas, the circumstances that most influenced his or her life, and the ways in which he or she influenced others' thinking, particularly Darwin's. As you

research print and electronic sources, consider social beliefs and customs and their influence on scientific thinking.

 www.science.nelson.com

- In a group, briefly outline the life story of the individual you will be role-playing. Describe this person's beliefs about the origin of life, Earth's history, and the idea of evolution—indicating whether this individual's beliefs changed over time. Was this person's thinking similar to or different from others' in the scientific community? How did his or her views compare with public opinion? In what specific way(s) did this person or his or her ideas influence Darwin in his development of the theory of evolution? Use this information to explain your answer to the question and for your role-play.

Natural Selection

Darwin summarized natural selection, the mechanism for his theory of evolution, in these words: "…can we doubt (remembering that many more individuals are born than can possibly survive) that individuals having any advantage, however slight, over others, would have the best chance of surviving and procreating their kind? On the other hand, we may feel sure that any variation in the least degree injurious would be rigidly destroyed. This preservation of favourable variations and the rejection of injurious variations I call Natural Selection." Darwin's theory of natural selection, based on some basic observations and inferences, is outlined in **Table 1**.

Table 1 The Theory of Evolution by Natural Selection

Observation 1	Individuals within a species vary in many ways.
Observation 2	Some of this variability can be inherited.
Observation 3	Every generation produces far more offspring than can survive and pass on their variations.
Observation 4	Populations of species tend to remain stable in size.
Inference 1	Members of the same species compete with each other for survival.
Inference 2	Individuals with more favourable variations are more likely to survive and pass them on. Survival is not random.
Inference 3	As these individuals contribute proportionately more offspring to succeeding generations, the favourable variations will become more common. (This is natural selection.)

Given the extensive examples from Darwin's observations of natural populations, as well as his experimentation and the experience of plant and animal breeders, Darwin's theory seemed very difficult to refute. Clearly species did exhibit inherited variations, which were obvious even to casual observers. The fact that all species have a high rate of reproduction was also self-evident. Although the idea of evolution was not new, there had been no accepted explanation for how evolution occurred until Darwin and Wallace introduced the mechanism of natural selection. Darwin offered detailed evidence and examples in support of his theory and, although his ideas were hotly debated in public after the release of his book, many scientists were swayed by his compelling arguments. He had explained his theory so simply and clearly that his friend Thomas Huxley, after reading it, exclaimed, "How extremely stupid not to have thought of it myself!"

> ## ▶ Practice
>
> **Understanding Concepts**
>
> 1. Propose an evolutionary scenario in which a species of ancient chameleons could evolve into a species with an unusually long tongue **(Figure 2)**. Keep in mind the two key elements required for natural selection: inherited variation and an environment that favours certain traits over others.

Figure 2

Unanswered Questions

Even with substantive scientific evidence in support, the theory of evolution by natural selection had many opponents. Some genuine difficulties still needed to be overcome. For natural selection to result in the evolution of entirely new species with new adaptations and structures, a great length of time was required. Not everyone shared the view that Earth was many millions of years old. And despite the unearthing of many deep fossil beds containing extinct life forms that were suggestive of an ancient past, the limited fossil record in Darwin's and Lyell's day held many gaps. Further, as opponents pointed out, the fossil record showed no transitional forms from ancient to more modern species. Not until radio-metric dating technology was available could scientists provide conclusive evidence that life on Earth began at least 3.5 billion years ago—ample time for Darwinian natural selection to work. Fossil records now are much more extensive and many of the gaps have been filled, in particular for intermediate species of verte-brates. In recent years, palaeontologists have unearthed 50-million-year-old fossils of whale ancestors with fully functioning large hind limbs and 38-million-year-old ancestors that retain tiny nonfunctional hind limbs (**Figure 3**).

The role of environmental change, too, presented scientists with opportunities for investigation of variations in species. In England in the early 1800s, light-coloured lichen covered many surfaces, providing camouflage for light-coloured forms of the peppered moth (*Biston betularia*). But the growing Industrial Revolution changed the English countryside in many ways; the production of pollutants blanketed some areas in soot, which offered new opportunities for some species to camouflage themselves from pred-ators. In 1848, near Manchester, the first appearance of a melanic (almost black) form of peppered moth was recorded. By the late 1800s, much of the lichen was dead and the trees, rocks, and other hard surfaces were covered in dark soot. By the 1920s, the moth population was almost entirely of melanic individuals. Scientists suspected that the melanic moths had gained a selective advantage. 🔬▮

A more daunting challenge to the theory proposed by Darwin and Wallace was the puzzle of variation. Many scientists were persuaded by the evidence that evolution was occurring but doubted the mechanism that these two scientists proposed. Although they admitted that species demonstrated subtle variations, many questioned how apparently insignificant variations could ever lead to the production of an entirely new structure, such as an eye or a wing, where none had existed before. At the time, it was widely believed that inherited traits from each parent were somehow blended in offspring and, as a result, it was thought that unusual or rare variations would become diluted over time rather than becoming prominent. Even Darwin accepted these assumptions and openly admitted that this was the greatest weakness in his theory. Darwin and his sup-porters could offer no explanation for the source of new variations.

As it turned out, a contemporary scientist had the beginnings of the answer. Just six years after Darwin's publication of *On the Origin of Species*, Gregor Mendel presented and, in 1866, published his investigations of inheritance in garden peas. Unfortunately, his work was largely ignored until 1900. Although Mendel and others provided a foundation for an understanding of inheritance, discoveries in modern molecular genetics have revealed that mutation and recombination provide the source for new inheritable variations. As you will learn in Chapter 12, they also offer the richest and most compelling evidence for biological evolution of life on Earth.

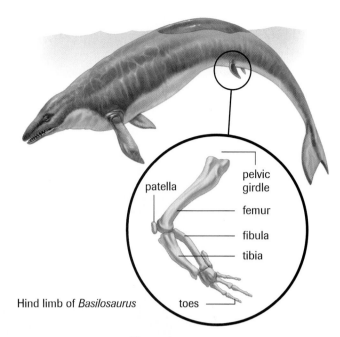

Hind limb of *Basilosaurus*

patella
pelvic girdle
femur
fibula
tibia
toes

Figure 3
Basilosaurus, an ancient relative of the modern whale, had rudimentary hind limbs.

🔬 ▌ **INVESTIGATION 11.6.1**

Industrial Melanism and the Peppered Moth (p. 536)
How might camouflage offer species a selective advantage? In this inves-tigation, you can conduct simula-tions to model and investigate the effects of camouflage on predator success.

Figure 4 presents a simplified model of a hypothetical evolutionary scenario for the evolution of larger claws in crabs over time. The questions below model Darwin's thinking.

- Examine **Figure 4**. Assume that any observed variations can be inherited.
  (a) Describe the original crab population as shown in **Figure 4(a)**. What trait shows variation?
  (b) Crabs produce many offspring. Considering the population size from **Figures 4(a)–(d)**, what must be happening to many of the offspring?

(c) Are the crabs competing? Explain.
(d) From your observations of **Figures 4(a)** and **(b)**, how does claw size influence competition?
(e) What overall change is happening to the population, as depicted in **Figure 4(c)**? Is your answer an inference or an observation? Explain.
(f) What do you notice about the claw sizes in **Figure 4(d)**? This was a problem for Darwin's theory, as he had no satisfactory explanation for the appearance of new traits. Where do you think this new variation came from?

(a)

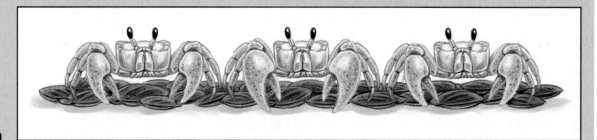

(b)

(c)

(d)

Figure 4
The evolution of large claws in crabs

SUMMARY | *Natural Selection as the Mechanism for Evolution*

- Darwin and Wallace independently developed the theory that the mechanism causing evolution was natural selection.

- According to their theory, all species exhibit inheritable variations that are selected through the struggle by individuals for survival in competition within their population. Individuals with more favourable traits produce more offspring that survive than others and pass along those favourable traits. Over many generations, this process results in a change in the inheritable traits of the population.

- When natural selection was first proposed, scientists had no way to date rocks or fossils with precision, little fossil evidence for species in transition from ancient to modern, and no understanding of the genetic basis of inheritance and variation.

▶ **Section 11.6 Questions**

Understanding Concepts

1. Assess Darwin's contribution to the development of scientific ideas about the origin of life. What aspects of the controversy did his theory and evidence not satisfy?

2. It is evident by his comments that Thomas Huxley did not find Darwin's new theory difficult to understand. Explain how the theory, although of great significance, is a rather simple one.

3. Describe the concept of "blended inheritance," which Darwin himself believed to be valid. Provide examples that appear to demonstrate the blending of inherited features.

4. According to Darwinian evolution, there must be variation and selection. In the evolution of large claws in lobsters,
   (a) what trait(s) might have been variable?
   (b) what factors might have resulted in members of the population being selected?

5. Speculate about why predatory cats such as the lion and the leopard have not evolved to be as fast as the cheetah.

6. The elephant has evolved to be a great size, while the mouse has evolved to be relatively small. Explain how natural selection might favour a different size in each mammal species.

7. Study **Figure 5**. The peppered moth caterpillar avoids detection by extending its body out from a branch and remaining still. Explain how mimicry also contributes to natural selection.

Figure 5
The peppered moth caterpillar

Applying Inquiry Skills

8. Develop reasoning to explain how an ancestor with relatively short legs and neck might have evolved into the present-day great blue heron (**Figure 6**).

Figure 6
Great blue heron

9. Cheetahs have so little genetic variation from individual to individual that even skin grafts among unrelated individuals from the South African subspecies are not rejected. Based on the theory developed by Darwin and Wallace, how might this influence the evolution of this species?

Making Connections

10. Find out through print and electronic sources the many contributions to science of Alfred Russell Wallace. Describe some of the personal, social, and financial challenges he faced in conducting his investigations.

    GO www.science.nelson.com

11. In an 1844 letter to his friend Hooker, Darwin noted that suggesting species are not immutable was like "confessing a murder." Why would Darwin have felt this way? Relate your response to public reaction to *On the Origin of Species*.

🔍 ACTIVITY 11.2.1

Applying Fossil Evidence

Theories about this planet's past were developed as a result of close observation of records left in stone. It is useful to know that palaeontologists and geologists use the following principles to obtain information from such records:

- Sedimentary rock is laid down horizontally as a result of the force of gravity acting on sediments.
- Younger rock is deposited above older rock.
- Rock will not contain fossils for every species living when sediments were deposited.

In this activity, you can test your own hypotheses about when and in what order events in Earth's past may have occurred.

Materials
major event card (1 per group)
student worksheet
set of 5 fossil bed cards (1 set per group)

Procedure/Analysis
1. Examine the 10 major events on the card (**Figure 1**) distributed by your teacher. With your partner or group, discuss the order in which these events might have occurred, from most recent to earliest.

Figure 1 Sample major events card for Activity 11.2.1

Group A Major Events
early apes
pterodactyl
earliest photosynthetic organisms
earliest dinosaur
earliest ancient sharks
sabre-toothed cat
jellyfish
earliest flowering plants
earliest mammal
earliest stone tools

2. On your worksheet (**Figure 2**), write the major events to match your proposed order, starting with the most recent at the top, in the first column, "Proposed event sequence."

Figure 2 Sample worksheet for Activity 11.2.1. The abbreviation for years ago is ya.

Initial Hypotheses		Fossil Evidence	
Proposed event sequence	*Proposed absolute age (ya)*	*Relative age*	*New absolute age estimates (ya)*

3. How many years ago do you think each major event might have taken place? Record your estimates in the second column, "Proposed absolute age (ya)."
4. Circle the major events in the first column of your worksheet about whose place in the sequence you are most confident.
5. Circle those absolute time estimates in the second column of your worksheet that you think are the most accurate.
6. Examine your set of five fossil bed cards. Determine the relative age of fossils in your deposits by arranging the cards in a sequence from most recent to earliest.
7. On your worksheet, record your sequence in the third column, "Relative age." Place the most recent at the top of the column.
8. Compare your hypotheses in column 1 with your empirical data in column 3.
9. Write your estimates for absolute time in the fourth column, "New absolute age estimates (ya)."
10. Where necessary, revise the estimates you recorded in the second column.

Synthesis
(a) How accurate were your initial hypotheses? If you needed to make changes, which surprised you the most? Why?
(b) How confident do you feel about your revised relative ages, based on fossil bed data? How confident do you feel about your revised absolute time estimates? Explain.
(c) Research your list of "major events" and record the latest radiometric date estimates of their absolute ages.

ACTIVITY 11.5.1

Looking for Homologies

In this activity, you compare a group of 6 organisms based on outward appearances and habits, and internal, anatomical, and physiological features.

Procedure

1. Consider and compare traits for the following six organisms: the little brown bat, the orca, the koala, the tiger shark, the emperor penguin, and the black-capped chickadee.

2. Use **Table 1** as a model to create a table for the general appearance and habits of these six selected species. Include an additional column to be filled in later. Traits and habits to be listed in rows in this table include:
   - possessing wings
   - feeding on insects (among other foods)
   - roosting in tress (often)
   - living on land
   - eating mostly plant material
   - living mostly in water
   - possessing structures for swimming
   - feeding on fish (often or usually)
   - having smooth skin and lacking body hair
   - lacking earlobes
   - having relatively large size

Table 1 General Appearance and Habits of Selected Species

Features	Bat	Orca	
possessing wings			
feeding on insects			

3. Create a second table for the anatomical and physiological features of these six selected species. Include an additional column to be filled in later. Features to list in rows in this table include:

- having gills to obtain oxygen
- having lungs to obtain oxygen
- having skeleton comprised of cartilage
- having skeleton comprised of bone
- being warm-blooded (homeothermic)
- having feathers arising from the epidermis
- having a four-chambered heart
- producing hard-shelled eggs
- having true bony teeth
- giving birth without laying eggs
- producing milk in mammary glands
- having young that develop internally through placenta

4. Complete the tables by placing an X in the column for each organism that possesses the feature described in each row. If necessary, consult classmates or research materials.

Analysis

(a) Which species share the greatest number of features listed in the first table?

(b) Would you suspect these features to be homologous or analogous? Explain.

(c) Which species share the greatest number of features listed in the second table?

(d) Would you suspect these features to be homologous or analogous? Explain.

Synthesis

(e) Your additional column in both tables will be for "Humans." Put an X in the appropriate rows.

(f) With which species do humans share the greatest number of analogous features?

(g) With which species do humans share the greatest number of homologous features?

(h) Write a brief assessment to relate your findings to the theory of evolution of species.

Industrial Melanism and the Peppered Moth

Inquiry Skills

○ Questioning	○ Planning	● Analyzing
○ Hypothesizing	● Conducting	● Evaluating
● Predicting	● Recording	● Communicating

Biologist H.B.D. Kettlewell formed and tested his hypothesis that the observed evolutionary change in the colouration of the peppered moth throughout the 1800s was a result of changes in the environment (**Figure 1**). In this investigation, you will test Kettlewell's hypothesis by conducting simulations involving three colours of tree bark to model the effects of camouflage on predator success. You will then analyze your data graphically.

Figure 1
A lighter-coloured peppered moth and a melanic peppered moth

Questions

How might the appearance of individuals influence their likelihood of being seen and eaten by predators? What influence might natural selection have on survival from predators?

Prediction

(a) Make a prediction about the evolution of the moth population in each of three different-coloured habitats.

Materials

light-coloured "tree bark" (pre-Industrial Revolution habitat)
intermediate-coloured "tree bark" (early Industrial Revolution habitat)
dark-coloured "tree bark" (post-Industrial Revolution habitat)
30 light-coloured paper "moths"
30 melanic paper "moths"
timer
graph paper

 Simulations are most effective with low light levels in the room. Remain at your lab station when the light is dim.

Procedure

1. Prepare a table to record your data for each of the three simulations. (Each simulation consists of five trials.) Each table should show the trial number, the number of melanic moths, the number of lighter-coloured moths, the percentage of melanic moths, and the percentage of lighter-coloured moths. Give each table a title.

2. In groups of three, each student should assume the role of Predator, Timer, or Assistant.

3. For each simulation, the student in the role of Predator is seated at arm's length from the simulated bark habitat. This student looks away while the student in the role of Assistant places 15 lighter-coloured moths and 15 melanic moths at random onto one of the simulated tree barks.

4. The student in the role of Timer controls each of the five trials by directing the Predator to "hunt" or "stop" at intervals of 4 s. At the "hunt" direction, the Predator looks at the simulated habitat and removes as many moths as possible one at a time in 4 s. At the "stop" direction, the Predator looks away immediately.

5. Moths removed are presumed to have been consumed. After each trial, the Timer records the number of each coloured moth that remains on the simulated tree bark.

6. The Assistant then replaces each consumed light-coloured moth with a melanic moth, and each consumed melanic moth with a lighter-coloured moth. These represent the new population (always of 30 moths) that has been produced.

7. The Timer records the number of each coloured moth and ensures that there are 30 moths on the simulated tree bark before every trial starts.

8. Steps 2 to 7 are repeated five times for each different-coloured tree bark.

9. Students can switch roles for each simulation, so that each group member has the opportunity to simulate predation for one habitat.

INVESTIGATION 11.6.1 *continued*

Analysis

(b) Calculate the percentage of each of the two moth forms for each trial. Record these in your data tables.

(c) Plot a graph of percentage values versus trial number (generation) for each of the three simulations. Show a different curve for each type of moth on each graph.

(d) Which moths were most successful at avoiding predation in the pre-Industrial Revolution bark habitat? in the post-Industrial Revolution bark habitat?

(e) Based on your observations, do you think that natural selection occurred in these moth populations? Explain with reference to your data.

Evaluation

(f) What was the purpose of using an intermediate-coloured background in this investigation?

(g) Identify and explain any source of potential error in your investigation.

(h) Suggest improvements in the model or propose an alternative model for examining the moth–predator interaction to test this hypothesis.

Synthesis

(i) Some poison-arrow frogs and poison-dart frogs avoid predation through very striking colouration. Suggest how it would not be advantageous for such poisonous frogs to remain green and blend into their surroundings.

(j) Lamarck was the first prominent scientist to argue that species must adapt to changing environments. Discuss the peppered moth example in relation to Lamarck's position.

(k) Predict what might have happened to the peppered moth if the original population had shown little variation and evolution did not take place.

(l) The hypothesis regarding the peppered moth has become very well known. The change in colouration of this moth population in Manchester is referred to as *industrial melanism*. Explain this term.

(m) Biologist Laurence Cook has reported that, since 1975, the prevalence of melanic moths around Manchester has decreased significantly and lighter-coloured moths are making a dramatic comeback. Tree bark is becoming lighter again; however, during the study, lighter-coloured lichens that supposedly camouflage the moths did not seem to be more common and the moths were seldom seen on trees. Assess these new findings in the ongoing evolution of the peppered moth. How do you think they affect Kettlewell's hypothesis?

Key Expectations

- Formulate and weigh hypotheses that reflect the various perspectives that have influenced the development of the theory of evolution. (11.1, 11.2, 11.3, 11.4, 11.5, 11.6)

- Describe, and put in historical and cultural context, some scientists' contributions that have changed evolutionary concepts. (11.1, 11.2, 11.3, 11.4, 11.6)

- Outline evidence and arguments pertaining to the origin and evolution of living organisms on Earth. (11.1, 11.2, 11.3, 11.5, 11.6)

- Describe how the use of radiometric dating technology has extended or modified the scientific understanding of evolution. (11.2)

- Evaluate scientific evidence that supports the theory of evolution. (11.2, 11.3, 11.4, 11.5, 11.6)

- Analyze how science and technological development have extended or modified our understanding of evolution. (11.2, 11.6)

- Explain and analyze the process of evolution by natural selection. (11.5, 11.6)

- Analyze evolutionary mechanisms and their effects on populations. (11.6)

- Explain, using examples, the process of adaptation of individual organisms to their environment. (11.6)

Key Terms

absolute age	microfossils
acquired traits	palaeontology
actualism	parent isotope
analogous features	permineralized fossils
biogeography	radioactive decay
catastrophism	radioisotopes
daughter isotope	radiometric dating
evolution	relative age
fossilization	spontaneous generation
fossils	
half-life	uniformitarianism
homologous features	vestigial features
immutable	

▶ *MAKE* a summary

To accumulate evidence to support his reasoning, Darwin made many observations of nature, conducted experiments, and consulted experts. Create a flow chart to illustrate how Darwin's thinking changed as he analyzed this evidence. Write Darwin's main ideas in boxes drawn down the middle of a page, beginning at the top with his beliefs before he went to South America. Then note each significant change in Darwin's thinking until his theory for the mechanism of evolution is in place. Add side boxes with arrows to indicate some major evidence that triggered each of the significant changes in Darwin's thinking.

In your notebook, record the letter of the choice that best completes the statement or answers the question.

1. Which of the following are analogous structures?
   (a) an eagle's beak and a lobster's claw
   (b) an elephant's ear and a rabbit's ear
   (c) a porcupine's quills and a sheep's wool
   (d) a human leg and a kangaroo leg
   (e) a salamader's tail and monkey's tail

2. According to Darwin's theory of natural selection, a struggle for survival is a result of which one of the following?
   (a) the environment constantly changing
   (b) many species competing with one another
   (c) the large number of offspring born in each generation
   (d) the variable traits the individuals have
   (e) the occurrence of mutations

3. Fossils cannot be formed in which one of the following?
   (a) sedimentary rock   (d) amber
   (b) peat deposits   (e) tar pits
   (c) igneous rock

4. The perfectly formed wings of insects that are reduced in size and never used for flying can be described as which one of the following?
   (a) poorly adapted   (d) fossilized
   (b) vestigial   (e) a new trait
   (c) evidence of an acquired trait

5. Which statement is false?
   (a) Fossils that most closely resemble modern species are found in the upper strata of sedimentary deposits.
   (b) Microfossils are very common.
   (c) Organisms sometimes leave behind fossils of footprints and burrows.
   (d) The most common fossils are of large land animals, such as dinosaurs.
   (e) The oldest fossils are of simple organisms.

6. Which is not a factor in the hypothesis for industrial melanism in the peppered moth?
   (a) On polluted surfaces, it was easier for predators to find lighter-coloured moths.
   (b) Melanic moths were more likely to survive to produce offspring than lighter-coloured moths.
   (c) The allele for melanic colouration was beneficial.

   (d) The environment around Manchester changed during the Industrial Revolution.
   (e) As lighter-coloured moths came into contact with soot, they became darker in colour.

7. Radiometric dating provided scientists with which one of the following?
   (a) a method for determining the exact age of sedimentary deposits
   (b) a method for determining the age of a fossil by measuring its radioactivity
   (c) a method for accurately dating igneous rock and for estimating the age of Earth
   (d) a method of dating that is not accurate for objects only a few thousand years old
   (e) the primary method for determining relative ages

8. Which of the following is not an acquired trait?
   (a) the ability to read   (d) sense of smell
   (b) a suntan   (e) tuberculosis
   (c) a tattoo

9. During his five-year voyage aboard *The Beagle*, which one of the following did Darwin not do?
   (a) develop his theory of natural selection
   (b) witness examples of the geological change in the Earth's crust
   (c) investigate the formation of coral reefs
   (d) collect large numbers of specimens
   (e) examine fossil evidence

10. If a sample of radioactive material with a half-life of 600 000 years is known to be 3 million years old, what fraction of the original parent isotope remains in the sample?
    (a) one fifth   (d) one thirty-second
    (b) one tenth   (e) none
    (c) one sixteenth

11. If you explore a remote oceanic island, what should you expect to observe about the local species?
    (a) that they closely resemble the species living on very similar islands
    (b) that they show little variation
    (c) that they closely resemble the species living on the nearest large landmass
    (d) that they closely resemble the species living in very similar habitats elsewhere in the world
    (e) that they closely resemble no other species anywhere

Understanding Concepts

1. What is meant by *immutable* species?

2. Describe the effect of fossil discoveries on European thinking about the origin and history of life on Earth.

3. Trace the development of reasoning and methods for determining the age of fossil remains.

4. Explain how the following discoveries about fossils offer compelling evidence for evolution:
   (a) the relationship between the age of fossils and the kinds and complexity of fossils
   (b) the relationship between the geographical location of both fossils and living species

5. Discuss some limitations of relying on fossil records to ascertain the history of life on Earth.

6. Based on radiometric analysis, what is the approximate age of each of the following?
   (a) Earth
   (b) the earliest bacterial cells
   (c) the first land plants
   (d) the first mammals

7. The igneous layer above a stratum bearing fossils contains 50% of the original parent uranium isotope. Use the information in **Figure 5** and **Table 1** (section 11.2 page 514) to determine the minimum age of the fossils in this deposit.

8. Outline the contributions of Lyell, Buffon, and Lamarck to an understanding of life on Earth.

9. Describe the theory of the inheritance of acquired traits, often referred to as Lamarckian evolution, and explain how this theory is flawed.

10. Give three examples each of homologous features, vestigial features, and analogous features.

11. Which of the listed features do not indicate an evolutionary relationship? Explain your answer.
    (a) the shell of a clam and the shell of a nut
    (b) the heart of a fish and the heart of a bat
    (c) the eyes of a dragonfly and the eyes of a hawk
    (d) the legs of a spider and the legs of a starfish
    (e) the stem of a cactus and the trunk of a pine
    (f) the seed from a dandelion and a coconut

12. Many plants use animals to aid in the dispersal of their seeds embedded in fruit. What traits of fruit have proven advantageous for the dispersal or survival (or both) of seeds?

13. Use several examples of domesticated animals or plants (or both) to illustrate the great range of traits that can result from artificial selection.

14. What principles from Malthus's essay on populations did Darwin apply to his reasoning about evolution of species?

15. Match the following scientists with the key accomplishments and ideas listed below: Charles Darwin, Gregor Mendel, Jean Lamarck, Lord Kelvin, Pierre Curie, Charles Lyell, Alfred Russell Wallace, Georges Buffon, Thomas Malthus, Georges Cuvier.
    (a) made significant discoveries in the field of radioactivity
    (b) was the first to use scientific methods to estimate the age of Earth
    (c) proposed the theory of uniformitarianism
    (d) was the first to prove that many fossil remains are of extinct species
    (e) was the first naturalist to suggest that species change over time
    (f) proposed that species adapt to their environment by the inheritance of acquired traits
    (g) used mathematical models to demonstrate the tremendous reproductive potential of species
    (h) developed a well-documented theory on the mechanism of evolution
    (i) formed the basic understanding of the inheritance of traits
    (j) was a brilliant naturalist who proposed a theory similar to Darwin's

16. Compare and contrast an explanation by Lamarck and by Darwin for each of the following:
    (a) how the giraffe evolved a long neck (**Figure 1**)
    (b) how the cheetah became an extremely fast runner

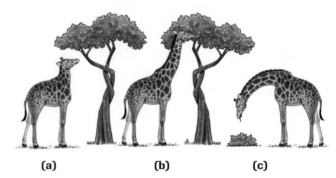

(a)   **(b)**   **(c)**

Figure 1
This giraffe represents an entire population over a long period of time.

17. Briefly describe how advances in science and technology have modified our understanding of the age of Earth. What role did the discovery of radioactivity play in palaeontology?

Applying Inquiry Skills

18. Develop a hypothesis to test Lamarck's theory of inheritable acquired traits. Suggest an investigation design by which your hypothesis might be tested.

19. The change in the peppered moth is thought to have been due to the production of pollutants during the Industrial Revolution. Consider another example of observable change in species that has resulted from human intervention in the environment. How might a hypothesis or prediction about this change be investigated?

20. Based on what was covered in this chapter, complete a time line for the development of scientific thinking about evolution.

Making Connections

21. Imagine you are living in Upper Canada in the 19th century. In your lifetime, would you be able to observe dramatic changes in landforms and geology (e.g., ice ages, mountain formation, or the creation of new landmasses)? Given the lack of scientific information available to 19th-century scholars and society, do you think it was unreasonable for people to think that the world was only a few thousand years old? Explain.

22. Scientists rarely work alone. Sir Isaac Newton, perhaps the best-known scientist in history, credited the work of Galileo and others for his own successes: "If I have seen farther, it is by standing on the shoulders of giants." Describe the contributions of six scientists on whose work Darwin relied for his achievement.

23. Immature fruits are often green in colour and turn bright colours only as they ripen. What selective advantage might this trait have? Can you think of a reason why it is unwise to eat fruits before they ripen? Explain your thinking.

24. Today, many scientific expeditions are funded by such organizations as the Royal Ontario Museum, Tyrrell Museum, Royal National Geographic Society, Geological Survey of Canada and, in the United States, the Smithsonian Institute and National Geographic Society. Find information about major

projects one such organization is funding now and report your findings to your class.

 www.science.nelson.com

Extensions

25. Research carbon-14 (^{14}C) dating techniques. Report your findings to the class, including how the percentage of ^{14}C begins to change after an organism dies.

26. In the evolution of limbs, many very efficient and fast-running species have evolved large prominent central toes, with vestigial outer toes elevated off the ground. Research to determine whether, over time, the outer toes moved up the leg leaving the middle toes where they were or the middle toes elongated, leaving the outer toes behind.

27. It is becoming common for private collectors to compete with scientists for important fossil finds. Southern Alberta is a known source for ammonite fossils (**Figure 2**). As the surface is covered in ammolite, the world's rarest gemstone, large specimens can sell for over $100 000. High costs can make it very difficult for scientists to obtain access to important specimens. Develop a list of recommendations to ensure scientists continue to get reasonable access to specimens for research.

Figure 2
Ammolite gemstone found on an ammonite.

28. Darwin's writings are rich in detail and explanation. Read some of his work in print or online and prepare a report on your assessment of the evidence and arguments he gives in support of his theory.

 www.science.nelson.com

12

Mechanisms of Evolution

In *On the Origin of Species*, Charles Darwin provided compelling arguments and evidence for evolution and natural selection. As scientific knowledge expands—particularly in the field of genetics—even more convincing evidence is being offered for the mechanisms of natural selection. About 70 years after Darwin's work, scientists finally understood how parents' traits are inherited by offspring. Now, geneticists can distinguish the genetic code of one individual from among those of a whole population. This powerful science is also being used to measure the genetic differences between entire species. Although some aspects of evolution continue to be debated by biologists, much of the puzzle has been solved.

💡 ***REFLECT*** on your learning ▼

1. Examine the unusual adaptations of the organisms in **Figure 1**. What function(s) are served by these adaptations? Do they improve each individual's chances of survival? Can you explain how each adaptation might have evolved?

(a) great grey owl   **(b)** nudibranchs

Figure 1

2. In terms of evolution, these organisms are successful if they pass on their genetic code to future generations. Explain the significance of this success.

3. Is it possible for individuals to pass on their genetic code to future generations without reproducing themselves? If so, how?

4. What mechanisms might result in the formation of new species?

> ▶ **TRY THIS** activity **Distinguishing Traits**

Study the photo of the group of people on this page. These individuals exhibit variations, but they also share such inherited physical features as limbs, internal organs, and paired eyes and ears. Yet pigeons, alligators, horses, and toads also possess these features.

(a) List inheritable features by which you can distinguish human beings from all other species.

(b) List inheritable features by which you can distinguish human individuals from one another.

(c) On each list, circle the two or three most significant distinguishing traits.

Do these traits represent unique features found in no other individual or species, or do they represent variation of shared features?

In the 1930s, biologists began integrating the growing body of knowledge in the field of genetics with Darwin's theory of evolution. They established a new understanding of natural selection, known as the modern synthesis, which answered some puzzling questions concerning inheritance and variation.

Geneticists study changes in the inheritable traits of organisms. Such traits that distinguish individuals from one another represent genetic diversity, which varies from species to species (**Figure 1**). The biological code for inheritable traits is in the form of DNA, and the units of DNA are **genes**. Genes are located at specific **loci**, or positions, on DNA molecules of chromosomes. Most eukaryotic organisms are diploid, having two sets of chromosomes, which they inherit, one set from each parent. With the exception of sex chromosomes (where present), the chromosomes appear as homologous pairs, with each pair possessing copies of the same set of genes.

genes portions of the DNA molecule within a chromosome coding for particular polypeptide products

loci the locations along a DNA molecule of a particular gene (singular: locus)

Figure 1
The genetic diversity of many populations, such as this one of long-nosed bats, may not be readily apparent to human observers.

alleles particular forms of a gene; many genes have two or more alleles

homozygous descriptive of a gene for which the paired alleles are identical

heterozygous descriptive of a gene for which the paired alleles differ

genome the complete set of chromosomes of an organism, containing all its genes and associated DNA

genotype the set of all alleles possessed by an individual organism (also, a specific set of alleles at one or more loci)

phenotypes observable traits of an organism that result from interaction between genes and the environment

These genes may come in two or more different forms, called **alleles**. For example, an inheritable trait may come from paired genes with alleles represented by the symbols *BB* or *bb*, if the two sets of chromosomes have copies of the same allele, or by *Bb*, if the two alleles are different. An individual with identical alleles for the same gene (i.e., *BB* or *bb*) is said to be **homozygous** for that trait. An individual that possesses two different alleles for the same gene (i.e., *Bb*) is said to be **heterozygous** for that trait.

All individuals of the same species possess a common **genome** with the sole exception of sex chromosomes, when present. However, each will have a different **genotype**, the combination of alleles at specific loci. Differences in genotypes and environmental influences account for differences among the **phenotypes** of individuals. These different phenotypes are then acted on by natural selection. An understanding of the genetic origin and diversity of phenotypes provides important insights into the mechanisms of evolution.

With techniques such as DNA sequencing and gel electrophoresis, geneticists have begun to analyze and compare the genetic code of individuals, populations, and entire species. One finding is that the amount of DNA present in different species varies dramatically, as shown by the examples in **Table 1** on the next page. Organisms with larger genomes have the potential for greater genetic diversity and present more targets for mutation.

Table 1 Total Amount of DNA in the Genomes of Selected Species

Species	Common name	DNA kilobases	Estimated number of genes
Mycoplasma genetalium	bacterium	580	470
Saccharomyces cerevisiae	yeast	1200	6500
Drosophila melanogaster	fruit fly	180 000	13 000
Xenopus laevis	toad	3 100 000	unknown
Macaca nigra	macaque	3 399 900	unknown
Homo sapiens	human	3 400 000	42 000
Necturus maculosus	mud puppy	81 300 000	unknown
Amphiuma means	newt	84 000 000	unknown
Trillium, species	trillium	100 000 000	unknown
Amoeba dubia	amoeba	670 000 000	unknown

The size of genomes, however, does not provide an accurate comparison of a species genetic diversity. Geneticists report that many genomes of eukaryotic organisms, for instance, comprise DNA that is not transcribed. Some noncoding sequences in the genome of humans, as well as other organisms, may be repeated as many as 500 000 times. Some species, such as maize or wheat, (known as **polyploids**) have more than two copies of each chromosome, resulting in multiple, often identical, copies of the same genes.

Regardless of the total quantity of DNA present, most species, other than some microorganisms, have large numbers of different genes—usually numbering in the thousands. Species that possess a greater number of genes have the potential for increased genetic diversity. Similarly, the greater the number of different alleles for these genes, the greater the extent of genetic variation there will be within a species and from individual to individual. Within a species, the inheritable variation from individual to individual is a result of the different combinations of alleles they possess. Although many plant species are heterozygous at more than 20% of their gene loci, for example, individual humans are heterozygous at an estimated 7% of their loci (i.e., about 3000 separate genes).

Genetic diversity within a **population** increases enormously through sexual reproduction, as the various alleles from two parents recombine in each offspring. For example, an individual of a species with 10 000 genes that is heterozygous at only 10% of these loci could produce a staggering 2^{1000} different genetic recombinations in their gametes. This number is larger than the number of atoms in the universe and does not take into consideration potential variations resulting from crossover and mutation. It should not be surprising, therefore, that no two offspring—except for identical twins or clones—will ever have the same genotype.

polyploids individuals with three or more complete sets of chromosomes

population all members of the same species living in the same region

SUMMARY   *Genetic Variation*

- The quantity of DNA and the number of genes are highly variable among species.
- Variation within a species is a result of the variety and combinations of alleles possessed by individuals.
- Sexual reproduction results in the random recombination of often thousands of different alleles and results in a high degree of genetic diversity within most populations.

Understanding Concepts

1. Describe how the genetic diversity of a population of loons is influenced by recombination and heterozygosity.

2. Study **Table 1** on the previous page.
   (a) Account briefly for the range of values of DNA bases.
   (b) Why might there be estimated numbers of genes for only a few species?
   (c) Find species that are in the same taxonomic class, such as the amphibians or the primates. Compare data presented for them, and contrast them with data presented for species in different classes.

3. Why is a comparison of the size of two genomes not entirely valid as a way to judge their genetic diversity?

4. What contribution do you think that a knowledge of genetics would have made to Darwin's understanding of the theory of evolution?

Applying Inquiry Skills

5. Briefly explain, using the Punnett square in **Figure 2**, how sexual reproduction increases the potential for genetic variation.

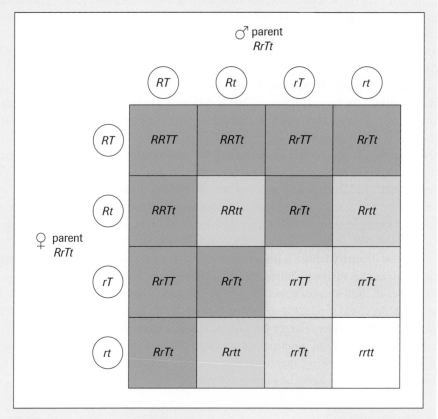

Figure 2
A Punnett square showing the offspring of parents with identical genotypes and phenotypes. The possible phenotypes are represented in different-coloured cells.

Population geneticists have developed a method to quantify a **gene pool**—the genetic information of an entire population—by measuring each **allele frequency**. Thus, evolutionary changes in populations can be measured in part by looking for changes in allele frequencies. Consider a population of moths for which there are two alleles, *A* and *a*, where *A* represents the allele for dark brown wings, which is dominant, and *a* represents the allele for light brown wings, which is recessive. A population of 500 comprises 320 moths with *AA* homozygous dark wings, 160 moths with *Aa* heterozygous dark wings, and 20 moths with *aa* homozygous light brown wings. If each individual proportionately contributes 2 alleles to the gene pool, there would be 640 *A* (from *AA* genotype), 160 *A* + 160 *a* (from *Aa* genotype), and 40 *a* (from *aa* genotype). The allele frequency for *A* is 800 ÷ 1000 = 0.80, or 80%, and that of *a* is 200 ÷ 1000 = 0.20, or 20% (**Figure 1**). Where only a single allele exists for a particular gene, that allele's frequency is 100% and it is described as **fixed**. In terms of evolution, would the dominant form of moth wing become more and more common over time? Do allele frequencies remain constant or change over time?

gene pool total of all alleles within a population

allele frequency the proportion of gene copies in a population of a given allele

fixed manner of describing the frequency of an allele within a population when only a single allele is present for a particular gene (i.e., the allele's frequency is 100%)

Genetic structure of parent population

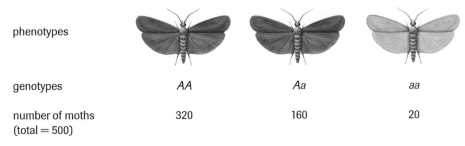

phenotypes			
genotypes	*AA*	*Aa*	*aa*
number of moths (total = 500)	320	160	20

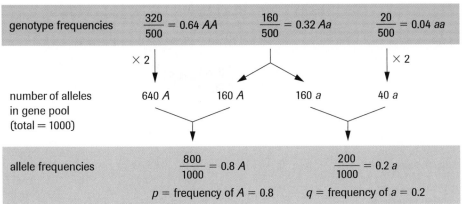

genotype frequencies: $\frac{320}{500} = 0.64\ AA$ $\frac{160}{500} = 0.32\ Aa$ $\frac{20}{500} = 0.04\ aa$

×2   ×2

number of alleles in gene pool (total = 1000): 640 *A*   160 *A*   160 *a*   40 *a*

allele frequencies: $\frac{800}{1000} = 0.8\ A$ $\frac{200}{1000} = 0.2\ a$

p = frequency of *A* = 0.8 q = frequency of *a* = 0.2

Figure 1
The allele frequencies of a population of moths

These questions interested Reginald Punnett. Over a meal in 1908, Punnett posed them to Godfrey Hardy, an eminent mathematician, who, without hesitation, wrote a solution on a napkin. Soon after, working independently, German physician Wilhelm Weinberg formulated the same solution. Now referred to as the Hardy–Weinberg principle, this mathematical relationship, outlined below, shows that allele frequencies will not change from generation to generation, as long as the following conditions are met:

- the population is very large
- mating opportunities are equal
- no mutations occur
- no migration occurs
- no natural selection occurs—all individuals have an equal chance of reproductive success

For a gene with only two different alleles (*A* and *a*), the Hardy–Weinberg principle can be expressed using the equation below:

If p = frequency of allele *A* and q = frequency of allele *a*, then

$$p + q = 1$$
$$(p + q)^2 = 1^2$$
$$p^2 + 2pq + q^2 = 1$$

p^2 = frequency of *AA* genotype; $2pq$ = frequency of *Aa* genotype; and q^2 = frequency of *aa* genotype

Genetic structure of second generation

Recombination of alleles from first generation (parents)

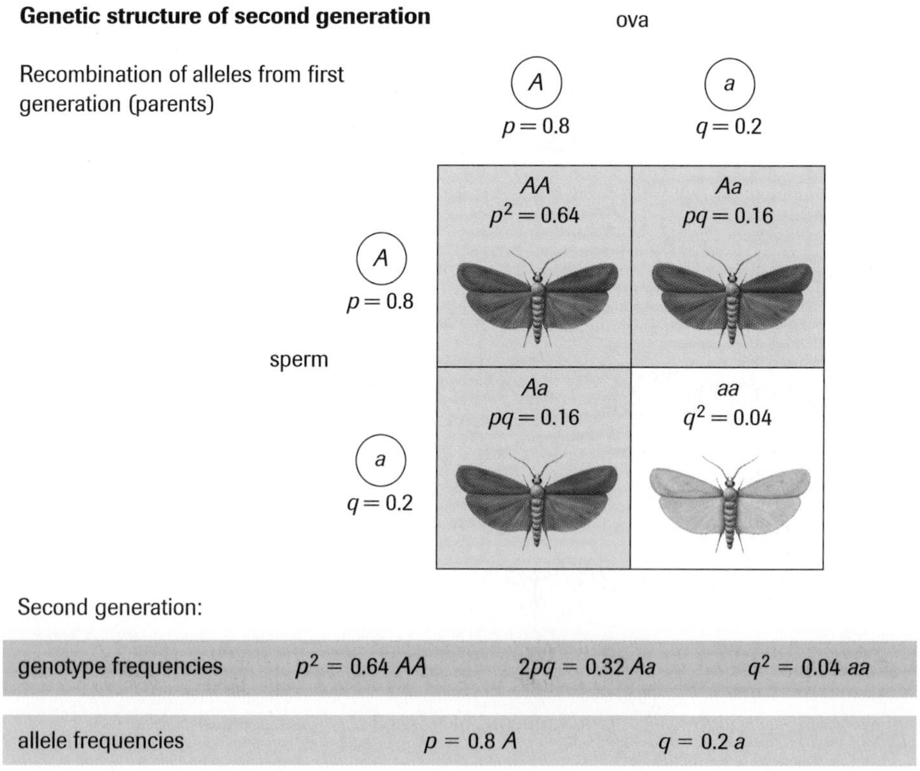

Figure 2

A typical Punnett square shows a cross between two individuals. This one depicts the allele frequencies of offspring within the moth population in **Figure 1**.

Second generation:

genotype frequencies	$p^2 = 0.64$ *AA*	$2pq = 0.32$ *Aa*	$q^2 = 0.04$ *aa*
allele frequencies		$p = 0.8$ *A*	$q = 0.2$ *a*

Applying the Hardy–Weinberg Principle

For the moth population in **Figure 1** on the previous page, the allele frequency of the *A* allele is 0.80, or 80%, and the frequency of the *a* allele is 0.20, or 20%. If mating is random, when the population reproduces, 80% of all gametes will bear the *A* allele, while the remaining 20% of gametes will bear the *a* allele. The genetic recombination that occurs in the next generation is shown in **Figure 2** on the previous page. Substituting these values into the Hardy–Weinberg equation, we get the following:

$$0.80 + 0.20 = 1$$
$$(0.80)^2 + 2(0.80)(0.20) + (0.20)^2 = 1$$
$$0.64 + 0.32 + 0.04 = 1$$

Therefore, for 50 offspring (100 alleles), the frequency of the *AA* genotype is 0.64, or 64%, the frequency of the *Aa* genotype is 0.32, or 32%, and the frequency of the *aa* genotype is 0.04, or 4%. The **genotype frequency** values of offspring generations are the same as those for the parent generation. If random mating continues to occur, allele frequencies are likely to remain constant from generation to generation.

genotype frequency for a particular pair of homologous genes, the proportion of those of a particular genotype (e.g., frequency of *AA*, *Aa*, and *aa*)

Using the Hardy–Weinberg Principle

SAMPLE problem ◄

Apply the Hardy–Weinberg principle equation to solve the following problems in population genetics, assuming that all five of the Hardy–Weinberg conditions are met:

A population has only two alleles, *R* and *r*, for a particular gene. The allele frequency of *R* is 20%. What are the frequencies of *RR*, *Rr*, and *rr* in the population?

Solution

If *p* represents the frequency of allele *R*, *q* the frequency of allele *r*, and $p = 0.20$, then $q = 0.80$. Using the equation for the Hardy–Weinberg principle, we get the following:

$$(0.20)^2 + 2(0.20)(0.80) + (0.80)^2 = 1$$
$$0.04 + 0.32 + 0.64 = 1$$

frequency of *RR* genotype = 0.04, or 4%

frequency of *Rr* genotype = 0.32, or 32%

frequency of *rr* genotype = 0.64, or 64%

▶ Practice

Understanding Concepts

For all questions, assume that the conditions for the Hardy–Weinberg principle are being met.

1. A large population consists of 400 individuals, of which 289 are homozygous dominant, 102 are heterozygous, and 9 are homozygous recessive. Determine the allele frequencies of *M* and *m*.

2. The gene pool of a large population of fruit flies contains only two eye-colour alleles: the dominant red allele, *W*, and the recessive white allele, *w*. Only 1% of the population has white eyes. Determine the allele and genotype frequencies of this population.

3. Manx cats have no tails (or have very short tails) and have large hind legs. The no-tail trait results from a heterozygous genotype, *Tt*. Interestingly, *TT* genotypes are normal cats, while the *tt* genotype is lethal and cat embryos that possess it do not survive. In a large population of 1000 cats, only 1% are Manx and 99% are normal.
   (a) What are the allele frequencies in this population?
   (b) Determine the expected frequency of each genotype in the next generation.
   (c) Determine the allele frequencies of the population of cats from (b).
   (d) What influence do homozygous recessive genotypes have on allele frequencies in this generation?
   (e) Predict the long-term result of a lethal homozygous recessive trait in a wild population.

Answers

1. $M = 0.85$, $m = 0.15$
2. $W = 0.9$, $w = 0.1$, $WW = 0.81$, $Ww = 0.18$, $ww = 0.01$
3. (a) $T = 0.995$, $t = 0.005$
   (b) $TT = 0.99$, $Tt = 0.01$
   (c) $T = 0.995$, $t = 0.005$

The Hardy–Weinberg principle demonstrates that, under a set of specific conditions, a given gene pool remains unchanged from generation to generation. The underlying conditions are critically important. By providing the set of conditions under which genetic change would not occur, the Hardy–Weinberg principle helps identify key factors that can cause **evolution**, a change to the gene pool of a population or a species. The following are the key factors:

evolution defined in genetic terms as any change in gene (and allele) frequencies within a population or species

- When a population is small, chance fluctuations can cause changes in allele frequencies.
- When mating opportunities are nonrandom, individuals that are preferred as mates will pass on their alleles in greater numbers than less preferred mates.
- When genetic mutations occur, new alleles may be created or one allele may be changed into another, thereby changing the frequencies of both new and original alleles.
- When individuals migrate, they remove alleles from one population and add them to another.
- When natural selection occurs, individuals with certain alleles have greater reproductive success than others do, thereby increasing the relative frequency of their alleles in the next generation.

Real populations can be affected by any of these situations, resulting in changes to allele frequencies.

Genetic Drift

When populations are small, chance can play a significant role in altering allele frequencies. For example, assume only 1 in 50 cricket frogs carries a particular allele, C_1 (**Figure 1**). In a large population of 10 000 individuals, you would expect 200 to carry the allele. If severe weather conditions led to the random deaths of half the population, you would expect about 100 of the 5000 survivors to be carrying the C_1 allele; therefore, the allele frequency would not be expected to change. However, if the initial frog population were endangered, with only 100 individuals, you would expect only two to possess the C_1 allele. If half the members of this population died, there would be a good chance that either both the C_1 carriers would die—thereby eliminating the C_1 allele entirely— or both would survive, thereby instantly doubling the allele frequency of C_1. This pattern, while an extreme example, demonstrates **genetic drift**, a change in the genetic makeup of a population resulting from chance.

genetic drift changes to allele frequency as a result of chance; such changes are much more pronounced in small populations

Figure 1
The remaining populations of the endangered Blanchard's cricket frog, *Acris crepitans blanchardi*, once found on Pelee Island in Lake Erie, are very vulnerable to the effects of genetic drift.

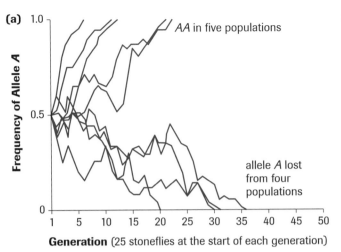

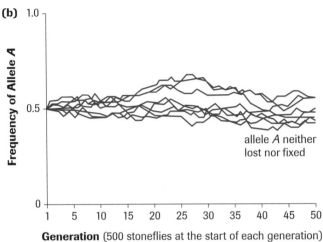

Figure 2
In small populations **(a)**, genetic drift can result in dramatic changes in allele frequencies, while in larger populations **(b)**, genetic drift is not usually significant.

Figure 2(a) illustrates genetic drift in a population of 25 stoneflies. The frequency of allele *A* fluctuates wildly from generation to generation. In five trials, the *A* allele became fixed at 100% in 22 generations or fewer, while in the other four trials, the *A* allele was lost entirely, being reduced to 0 in 36 generations or fewer. In a larger population of 500 stoneflies, as shown in **Figure 2(b)**, the allele frequency remained relatively stable even after 50 generations had passed; there was no trend toward fixing of the allele. Significantly, in small populations, genetic drift can lead to fixation of alleles, thereby increasing the incidence of homozygous individuals within a population and reducing its genetic diversity.

When a severe event results in a drastic reduction in numbers, a population may experience a **bottleneck effect** (**Figure 3**). When this form of genetic drift occurs, a very small sample of alleles survives to establish a new population. Their relative frequencies may differ from those of the original population and additional genetic drift may result in further deviations in the gene pool. This is known to have occurred with the northern elephant seal (**Figure 4**).

bottleneck effect a dramatic, often temporary, reduction in population size usually resulting in significant genetic drift

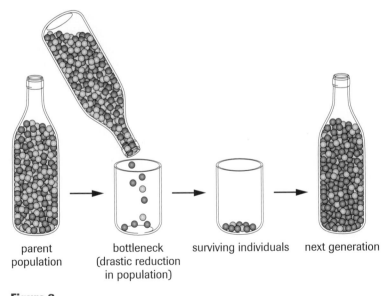

parent population   bottleneck (drastic reduction in population)   surviving individuals   next generation

Figure 3
A dramatic, sometimes temporary, reduction in the size of a population can result in a bottleneck effect.

Figure 4
The northern elephant seal population was reduced by overhunting to 20 individuals in the 1890s. Although the population had rebounded to over 30 000 individuals by 1974, genetic testing of 24 loci exhibited total homozygosity.

Demonstrating Chance

See for yourself how random chance works in small populations.

- Model the allele frequency in a new population by tossing a six-sided die 30 times. Record your results.
- Examine the results of your classmates.
  - (a) How often did you roll a three? Did your response differ from the one sixth you would expect by chance?
  - (b) How many times did you roll a three in your first six tosses? What ratio did this produce?
  - (c) Relate the variations in the frequencies of the number three to variations in allele frequencies that occur when small founder populations form.

founder effect genetic drift that results when a small number of individuals separate from their original population and find a new population

self-pollinating plants plants that habitually fertilize themselves and produce viable offspring

gene flow the movement of alleles from one population to another through the movement of individuals or gametes

neutral mutation has no immediate effect on an individual's fitness; most neutral mutations are silent or occur in noncoding DNA

When a few individuals from a large population leave to establish a new population, the resulting genetic drift is a **founder effect**. The allele frequencies of the new population will not be the same as those of the original population and may deviate further as the new population expands. Founder effects seem to be common in nature, such as when a few seeds carried by a bird or by winds to a distant volcanic island may germinate and rapidly establish a large population. With **self-pollinating plants**, an entire population can be established from a single fertile seed. Founder effects can also be seen in human populations. Members of the Amish community in Pennsylvania are all descendants of about 30 people who emigrated from Switzerland in 1720. One of the founders had a rare recessive allele that causes unusually short limbs. The frequency of this allele in the current Amish population is about 7%, compared to a frequency of 0.1% in most populations.

Founder effects have been documented in the wild. In 1982, Peter Grant and Rosemary Grant from Queen's University witnessed the establishment of a new population of large ground finches (*Geospiza magnirostris*) in the Galapagos Islands. The Grants had been studying Darwin's finches on one island, Daphne Major, and had observed juvenile large ground finches visiting the island every year for 10 years. In 1982, however, three males and two females remained on the island to breed. In early 1983, they produced 17 young birds, which became the founders of a new population. The population that they established has remained ever since. Careful measurements of inheritable traits by the Grants indicated that the founding population has a different genetic composition from that of the original large population of *Geospiza magnirostris* from which the founders came.

Gene Flow

When organisms migrate, leaving one population and joining another, they alter the allele frequencies of both. Such **gene flow** occurs frequently in most wild populations. For example, prairie dogs live in dense colonies consisting of a few dozen members. For much of the year they prevent other prairie dogs from joining their colony. In late summer, however, mature male pups are permitted to enter new colonies, thereby affecting both gene pools. Gene flow can also occur when individuals of adjacent populations mate without moving permanently. In these ways, genetic information is shared between populations. Unlike genetic drift, gene flow tends to reduce differences between populations.

Mutation

Mutations are the only source of additional genetic material and new alleles. Mutations may arise as a result of unrepaired changes in DNA sequences or chromosome breakage and rejoining. Although most mutations occur in somatic (body) cells, these mutations cannot be inherited and, therefore, do not play a role in evolution. However, any mutation that occurs in a gamete has the potential to be passed on to later generations, thereby entering the gene pool. These mutant alleles and any new phenotypes they produce become the source of new raw material for natural selection. What effects can mutations have and how frequently do they occur?

Inheritable mutations can be neutral, harmful, or beneficial. Because mutations are random changes to the genetic code, they are much more likely to be neutral or harmful than they are to be beneficial. A **neutral mutation** is one that has no immediate effect on an individual's **fitness**, or reproductive success. A **harmful mutation** reduces an individual's fitness and usually occurs when a cell loses the ability to produce a properly functioning protein or when major chromosomal changes adversely affect meiosis and mitosis. A **beneficial mutation**, which occurs when a cell gains the ability to produce a new or improved protein, gives an individual a selective advantage: increased reproductive success.

Types of Mutations

Different types of mutations vary in their ability to affect the phenotypes of individuals and their impact on the evolution of populations. Point mutations are changes in single base-pairs along the DNA molecule. When a point mutation occurs in a eukaryotic organism's genome where DNA is noncoding, it will be neutral. When a point mutation causes an amino acid substitution in a coding region of the DNA, a new gene product—and, therefore, a new phenotype—is produced. The change might be deleterious (or lethal), hindering the proper functioning of the final protein product. In other cases, the change may have no significant effect on the protein's function. Rarely, a point mutation could result in a protein with an improved or new function that benefits the individual.

Small insertions and deletions that occur within functioning genes almost always produce a nonfunctioning gene; such mutations are usually harmful. Because they are rarely beneficial, these mutations do not play a major role in evolution. While large-scale inversions are often neutral mutations, they are a useful tool for evolutionary biologists as their presence can be inferred by examining chromosome banding patterns (**Figure 5**).

Gene duplication occurs when unequal crossing over during meiosis results in an additional copy of one or more genes being inserted into a chromosome. This kind of mutation is important because it is a source of new genes. At first, these duplicated genes are just extra copies and add redundancy to the genome, providing no advantage to the individual. However, the new DNA is then free to mutate and, potentially, gain a new function. Such duplication events can ultimately produce entire gene "families" with very similar structures but altered functions. For example, many species have a family of genes called *histones*. These gene copies, numbering in the hundreds, are all very similar in structure and located very close together on the same chromosome. The small differences in their DNA sequences are thought to result from point mutations that take place after duplication events. In addition, nonfunctioning **pseudogenes**—genes that are duplicated and later lose their ability to be transcribed—provide very strong evidence for evolution.

How common are mutations? The best experimental evidence suggests that point mutation rates range from about 1 in 10 000 cell divisions in species with a very small genome (e.g., bacteria) to one or more in each gamete in species with a large genome. Because they rarely result in obvious changes to an individual organism's phenotype, they are not readily observable. Many harmful mutations result in the death of the gamete or individual before birth. Despite these difficulties, researchers sequencing an entire genome are finding evidence of frequent gene duplication. In humans, a gene coding for one enzyme—glyceraldehyde-3-phosphate dehydrogenase—used in glycolysis occurs in a single functioning copy and 20 nonfunctioning pseudocopies. The 300-base-pair sequence called *Alu*, which appears to have originated as a copy of a gene coding for ribosomal RNA, serves no function but is present in 500 000 copies and constitutes about 5% of the entire genome.

fitness general term referring to lifetime reproductive success of an individual

harmful mutation an inheritable change in a cell's DNA that impairs the proper operation of a gene product or regulatory function or adversely affects mitosis or meiosis

beneficial mutation an inheritable change in a cell's DNA that results in an additional or enhanced gene product or regulatory function

gene duplication a mutation leading to the production of an extra copy of a gene locus, usually resulting from unequal crossing over

pseudogenes DNA sequences that are homologous to functioning genes but are not transcribed

(a) **(b)**

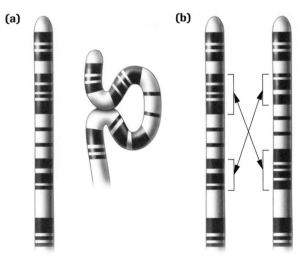

Figure 5
A chromosome before **(a)** and after **(b)** a mutation. Large inversions can result in the reversal of banding patterns along chromosomes.

INVESTIGATION 12.3.1

Agents of Change (p. 577)
Population size, genetic drift, and natural selection all affect allele frequencies. How can their influence be predicted?

Polyploidy, a mutation that results in three or more sets of chromosomes, occurs when unreduced (diploid) gametes join to form a cell containing one or more entire extra sets of chromosomes. Instead of being diploid ($2n$), some organisms are tetraploid ($4n$) or even octoploid ($8n$) or more. The fern, *Ophioglossum reticulatum*, contains an astonishing 1260 chromosome pairs ($630n$). Polyploidy, the most dramatic form of mutation, provides an organism with an immediate doubling of genetic material. This type of mutation has played a major role in the evolution of plants; the majority of ferns and almost half of all flowering plants are polyploids. Animals are rarely polyploids.

▶ *SAMPLE* problem

E. coli *Mutation Rate*

The population of the microorganism **E. coli** *living in your large intestine could be conservatively estimated at 10 billion bacteria. Experimental evidence suggests that these bacteria undergo cell division at a rate of once every hour and experience a mutation rate of 1 per 5000 divisions. How many mutations can this population of bacteria expect to experience in one year?*

Solution

$$\frac{10^{10} \times 24 \text{ divisions}}{d} \times \frac{365 \text{ d}}{y} \times \frac{1 \text{ mutation}}{5000 \text{ divisions}} = 1.7 \times 10^{10} \text{ (17 billion) mutations}$$

Note: It is assumed that the population does not increase in size and that the number of surviving bacteria remains constant. Therefore, the number of cell divisions in each generation remains the same.

▶ Practice

1. A population of 10 million free-tailed bats lives in a large cave. Assume that each year 5 million baby bats are born but the population remains the same. If the mutation rate averages 0.4 mutations per gamete, how many mutations would likely occur in 200 years?

Answer

1. 800 million

DID YOU *KNOW* ❓

Miracle Grain
Triticale is a grain of hybrid origin produced by crossing rye and wheat. The hybrid lacked homologous pairs of chromosomes and was infertile. Researchers treated the hybrid with the drug colchicine, which prevented the first meiotic cell division and resulted in the production of diploid ($2n$) gametes. When these gametes fertilized each other, they produced synthetic tetraploids ($4n$). These tetraploids contain a diploid set of homologous chromosomes for both rye and wheat within the same individuals. The hybrids are fertile and yield a high-quality grain.

SUMMARY   *Random Change*

- Evolution occurs when the allele frequencies of a population change over time.

- Genetic drift and gene flow produce changes in allele frequencies and affect genetic diversity.

- The source of all new genetic information is mutation. Gene duplications are the main source of new genetic material. As extra copies, they are free to mutate without the likelihood of causing harm.

- Although rare in individual cells, mutations are numerous in large populations over many generations.

▶ *Section 12.2–12.3 Questions*

Understanding Concepts

1. Use the term *allele frequency* to explain how biologists define and quantify evolution within a population.

2. Relate two ways in which alleles can become fixed in a population.

3. Define genetic drift and genetic flow, offering two examples to illustrate each definition.

4. Suggest three types of organisms that might produce founder populations. Explain the process that results in this effect.

5. Explain why harmful mutations play virtually no role in the evolution of populations.

6. When a mutation causes gene duplication, it often has little or no immediate effect. How do such mutations play a role in evolution over longer periods of time?

7. For each of the following situations, explain whether the Hardy–Weinberg equilibrium would be maintained generation after generation:
   (a) a population of African violets maintained by a plant breeder
   (b) the population of the black fly, *Simulium venustum*, in northern Ontario
   (c) a racoon population living in the Humber Valley in west Toronto
   (d) a newly discovered bird population on a remote island off the coast of British Columbia

8. How do pseudogenes offer compelling evidence in support of evolution?

9. Consider the Practice Question about mutation in bats on the previous page. If beneficial mutations are very rare, accounting for only one in every 3 million, how many beneficial mutations are likely to occur in the bat population for the same 200-year period?

10. The world's population of cheetahs is almost identical genetically (**Figure 6**). Male cheetahs are known to have low sperm counts and the species in general has a low resistance to many infectious diseases. All cheetahs are thought to be homozygous at over 99.9% of their gene loci.

Explain how a severe genetic bottleneck effect in the past could account for these observations.

Applying Inquiry Skills

11. During the fall migration, several Canada geese stop at a river near a good food source and then nest there the following spring. Because of the abundance of food, this population of geese stops migrating. What effects, both immediate and long-term, might this situation have on the gene pools of the original and founder populations?

12. If variation in species were solely a result of genetic recombination during sexual reproduction, how would that limit the evolution of species?

13. It is thought that a billion prairie dogs once populated an area of more than 100 million ha. Their current territory has been reduced and fragmented to less than 1% of this original space. Predict the impact of these changes in habitat on the prairie dog gene pool, as well as on the evolution and survival of the species.

Making Connections

14. Find and describe an example that does not appear in this text in which the founder effect has altered the allele frequency of a human population.

15. Why might evolutionary biologists be more concerned with the study of population genetics than the study of the simple inheritance of alleles by offspring from their parents?

16. Wildlife biologists in British Columbia estimate that fewer than 100 Vancouver Island marmots, *Marmota vancouverensis*, were alive in 2001.
   (a) Research this endangered species using print and electronic sources to determine the cause(s) of the severe bottleneck effect in their population.
   (b) What efforts, if any, are being made to maintain the genetic diversity of this species?

 www.science.nelson.com

Figure 6
All cheetahs today are virtually identical genetically.

What is the source of variation? How are subtle differences passed from generation to generation? These questions that puzzled Darwin have been answered by the scientific understanding of genetics and mutations. Mutations provide a continuous supply of new genetic variations, which may be inherited and expressed as different phenotypes. Natural selection leads to a variety of outcomes when this genetic variation occurs within competitive populations living under diverse environmental conditions. Sickle-cell anemia, a potentially serious blood disorder, is a useful example of how mutation, genetic variation, and the environment result in different patterns of natural selection.

The allele for sickle-cell anemia differs from the normal hemoglobin gene by having a single base-pair mutation. Individuals homozygous for the sickle-cell allele are severely afflicted with this disorder. Heterozygous individuals are only mildly affected by sickle-cell anemia; however, they are much more resistant to malaria than are people with normal hemoglobin. In regions where malaria is uncommon, individuals with the sickle-cell allele are at a disadvantage and their phenotypes are less likely to contribute alleles to the gene pool. But in regions where malaria is common (**Figure 1**), heterozygous individuals are strongly favoured; they are much more likely to survive and pass on their genes to the next generation. The environment selects the best-adapted phenotype and, in so doing, favours a particular set of alleles.

The sickle-cell allele is only common where it provides an overall advantage to the individual. In populations where it has an overall harmful effect, it does not persist. This pattern establishes an important relationship between mutations and evolution:

- Harmful mutations occur frequently but they are selected against and, therefore, these mutant alleles remain extremely rare.

- Beneficial mutations are rare but they are selected for and, therefore, these mutant alleles accumulate over time.

Although genes provide the source of variation, natural selection acts on individuals and their phenotypes. As a result, particular alleles are most successful and passed on when they enhance the phenotype of the individual and, thereby, contribute to their reproductive success. Selective forces can favour particular variations in the phenotype of individuals in a number of ways.

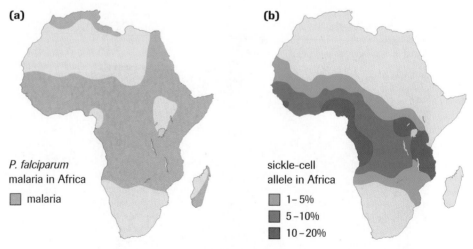

(a)

P. falciparum malaria in Africa
- malaria

(b)

sickle-cell allele in Africa
- 1–5%
- 5–10%
- 10–20%

Adapted from *Biology* 5/E, fig. 20.4 p. 407 Peter H. Raven and George B. Johnson, the McGraw-Hill Companies, Inc.

Figure 1
Of the 120 million new cases of malaria each year, about 1 million are fatal. The prevalence of malaria in Africa **(a)** closely matches the distribution of the sickle-cell allele **(b)**.

Types of Selection

Many factors influence how selection can operate on individual phenotypes in a population. Some populations live for long periods of time under stable conditions, while others live in a constantly changing environment. At times, a variable or diverse environment may favour multiple phenotypes. Within populations themselves, competition for mates adds further selective pressures that can and do influence the evolutionary pathway followed by populations. These different sets of conditions result in a number of general selection types.

Stabilizing Selection

In your lifetime, you are unlikely to see a dramatic change in the appearance of any species as a result of natural selection. Indeed, most species show little change over periods lasting thousands of years. Although these observations suggest the absence of evolutionary processes, theoretical models predict that once species become well adapted to their environment, selection pressures will tend to prevent them from changing. **Stabilizing selection** occurs when the most common phenotypes within a population are most favoured by the environment. For example, human birth weights are variable and part of this variability is inheritable. According to the theory of natural selection, babies born at weights that offer the best chance of surviving birth should be more numerous, and research shows that far more human babies are born weighing just over 3 kg than any other weight. Babies with significantly lower weights are often developmentally premature and less likely to survive, while heavier babies often experience birth-related complications that threaten the life of both baby and mother. The evidence suggests that natural selection eliminates extreme variations of a particular trait, resulting in a population in which most human births are near the ideal weight.

Stabilizing selection is by far the most common form of selection. Once a species becomes adapted to its environment, selective pressures maintain their evolved features and traits. For example, the hummingbird draws nectar from flowers with a long bill and tongue (**Figure 2**). Most birds of this species will succeed if their bill and tongue length are well adapted for the size of flowers that they feed on in their local environment. A longer bill requires more nutrients and energy to grow and carry around, while a shorter bill may reduce a bird's ability to reach food. The ideal bill length also increases the success of flower pollination, thereby enhancing the reproductive success of the plant species as well as that of the hummingbird. The environment will select against mutations that occur to produce birds with a bill length that differs from the best-adapted length. The graphs in **Figure 3** show the effects of stabilizing selection based on bill length over successive generations of hummingbirds.

stabilizing selection selection against individuals exhibiting variations in a trait that deviate from the current population average

Figure 2

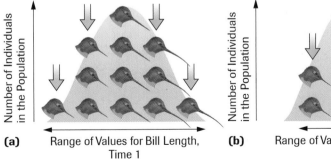

(a) Range of Values for Bill Length, Time 1

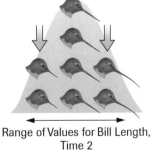

(b) Range of Values for Bill Length, Time 2

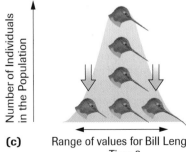

(c) Range of values for Bill Length, Time 3

(In all three graphs the vertical axis reads: Number of Individuals in the Population)

Figure 3
Stabilizing selection in a population of hummingbirds. The bell-shaped curve shows the frequency and variation in hummingbird bill lengths. The arrows indicate the less successful forms.

Directional Selection

Directional selection occurs when the environment favours individuals with more extreme variations of a trait. When an organism migrates to a new environment, or when aspects of its habitat change, it will encounter new forces of natural selection. If the hummingbird population moves to a new habitat with longer flowers, individuals with bills that were best adapted to medium-length flowers will no longer be ideal (**Figure 4**). Birds that inherited longer bills will then be more successful than those with medium-length and short bills; longer-billed birds will obtain more food and contribute more offspring to later generations (**Figure 5**). Directional selection may result in an observable change in a population.

Figure 4

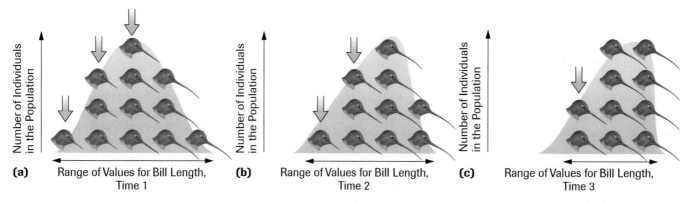

(a) Range of Values for Bill Length, Time 1

(b) Range of Values for Bill Length, Time 2

(c) Range of Values for Bill Length, Time 3

Figure 5
Directional selection shown in a population of hummingbirds. In a new environment with longer flowers, which hummingbirds have an advantage?

directional selection selection that favours an increase or decrease in the value of a trait from the current population average

In species with large populations and short generation times—such as salmon—many offspring are produced. In such species, the amount of genetic variation from both recombination and mutation is increased. Gill-net fishing in the Bella Coola River and the Upper Johnston Strait from 1950 to 1974 had a dramatic effect on the pink salmon population: the average weight of the salmon decreased by about one third (**Figure 6**).

Figure 6
Gill-net fishing substantially decreased the size of adult salmon on a British Columbia river over 25 years. Larger fish were less likely to escape gill nets and, therefore, less likely to contribute to later generations. Distinctly different salmon populations breed in alternating years in each location. At the time of this study imperial measurements were in use.

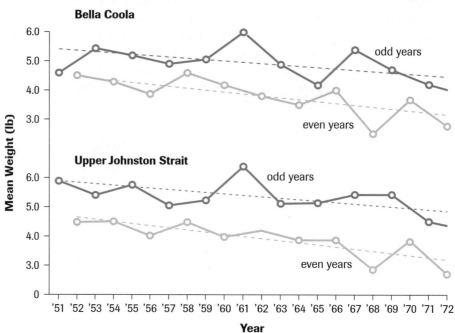

In these years, about 75% of all adult salmon were being caught and, as a result, smaller adults were the most likely to escape the nets and contribute offspring to the next generation. In this way, human activity resulted in directional selection.

Disruptive Selection

Disruptive selection favours individuals with variations at opposite extremes of a trait over individuals with intermediate variations. Sometimes, environmental conditions may favour more than one phenotype. For example, two species of plants with different-sized flowers may be available as a food source for the hummingbird population (**Figure 7**). Each species is a good source of nectar, but neither is well suited to a hummingbird with an average-length bill. Birds with longer and shorter bills will be more successful and will contribute more offspring to later generations (**Figure 8**).

Figure 7

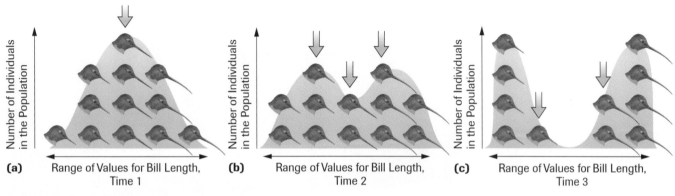

(a) Range of Values for Bill Length, Time 1

(b) Range of Values for Bill Length, Time 2

(c) Range of Values for Bill Length, Time 3

Figure 8
Disruptive selection in a hummingbird population. In a habitat with both shorter and longer flowers, which birds have the advantage?

A living example of disruptive selection in a bird population is provided by the black-bellied seedcracker finch (*Pyrenestes ostrinus*). These African finches depend on the seeds of two different types of sedge, one that produces a soft seed and the other a much harder seed. Finches with small bills, shown in **Figure 9(a)** are efficient at feeding on soft seeds, while birds with larger bills, shown in **Figure 9(b)** are able to crack the hard seeds. A study of 2700 finches produced the findings depicted in the graph shown in **Figure 9**, which appears to be observable evidence for disruptive selection. Disruptive selection is a significant evolutionary mechanism for the formation of distinctive forms within a population. Distinctive groups may eventually become isolated breeding populations with separate gene pools.

disruptive selection selection that favours two or more variations or forms of a trait that differ from the current population average

(a) narrow bill

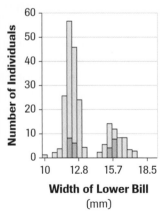

Width of Lower Bill
(mm)

(b) wide bill

From *Biology: The Unity and Diversity of Life* w/CD ROM, 9/E, by C. Starr and Taggart © 2001. Reprinted with permission of Wadsworth, an imprint of the Wadsworth Group, a division of Thomson Learning. Fax 800-730-2215.

Figure 9
The environment has strongly favoured the evolution of two different bill sizes among black-bellied seedcracker finches. They display variation in bill sizes, but during the dry season, survival is limited to two narrow size ranges (orange bars on graph). This is an observable product of disruptive natural selection.

Sexual Selection

Individuals that mate and reproduce frequently make a substantial contribution to the gene pool of later generations. **Sexual selection** favours the selection of any trait that influences the mating success of the individual. The traits favoured in sexual selection include **sexual dimorphism** (i.e., striking differences in the physical appearance of males and females) and behavioural differences between the sexes. The most common forms of sexual selection result from female mate choice and from male-versus-male competition. In some species, females choose mates based on physical traits, such as bright colouration, or behavioural traits, such as courtship displays and song. In other species, males are equipped with physical features that assist them in establishing control of and defending their territory against other males (**Figure 10**). This territory provides an area to which they can attract, and sometimes forcibly detain, the females with which they mate. Such traits are not produced by selective pressures from environmental conditions; if they were, both sexes would be expected to possess them.

Figure 10
Sexual dimorphism may take the form of a physical advantage, such as a much larger size in males, or an enlarged limb, as in fiddler crabs.

Figure 11

Many species have evolved features that are a compromise between different selective pressures. Sexual selection has produced traits that are beneficial for mating but are otherwise detrimental. Avoiding predators is not made easier, for instance, by brilliant plumage or a distinctive song (**Figure 12**). A bizarre extreme—sometimes called runaway selection—can be illustrated by the stalk-eyed fly (**Figure 13**). Although males and females both have eye stalks, females have preferentially mated with males with the longest eye stalks to the point where the feature has become very exaggerated.

Sexual diversity is not limited to animal populations. Most plants do not select mates but they do need to attract or use various agents—such as insects, birds, and bats—to assist in pollination. Flowers and scents are the most obvious sexual features that have evolved through natural selection. By maximizing their chances of being pollinated, plants have a greater likelihood of contributing more alleles to the next generation's gene pool.

Chance and selection play important roles in evolution. Scientists accept that anatomical, morphological, and behavioural traits of organisms evolve chiefly through selection mechanisms. These traits directly influence the daily survival and reproductive success of the individual; at the molecular level, the roles of random chance and selection are less clear. Some biologists argue that most of the differences in genomes result from neutral mutations, while others hold that most differences are responses to subtle effects from selective pressures. Although the debate over the details of molecular diversity continues, the basis for evolution is agreed: the ultimate fate of all genetic combinations rests in their ability to produce individuals best adapted to survive and reproduce in their habitat. Natural selection is the mechanism that drives the evolution of species.

Figure 12
The mating game is risky for male túngara frogs. When calling for a mate in the dark, they run the risk of giving away their location to the deadly frog-eating bat.

Figure 13
This remarkable male fly continues to contribute alleles for eyes on excessively long stalks because they attract females of the species. What might change this situation?

Case Study *The Evolution of Antibiotic Resistance in Bacteria* ▼

An article in the *Canadian Medical Association Journal* (July 2001) reported an alarming six-fold increase in the rate of antibiotic resistance in Canada between 1995 and 1999. In addition to added health risks, fighting antibiotic resistance can be expensive. Sunnybrook and Women's College hospitals in Toronto reported spending $525 000 in two years on fighting resistant bacteria and, across Canada, the cost is estimated at $50 to $60 million a year.

In 1996, doctors took samples of bacteria from a patient suffering from tuberculosis, a lung infection caused by the bacterium *Mycobacterium tuberculosis*. Cultures of the bacteria found it to be sensitive to a variety of antibiotics, including rifampin. The patient was treated with rifampin and initially responded so well that the lung infection seemed to be over. Soon after, however, the patient had a relapse and died. An autopsy revealed that bacteria had invaded the lungs again in large numbers. Cultures of these bacteria were found to be sensitive to many antibiotics, but resistant to rifampin. DNA sequencing revealed that a certain bacteria's gene had a single base-pair mutation that was known to confer resistance to rifampin. Doctors compared the new bacteria culture with the original culture and found that the sequences were identical except for this single mutation. Researchers then examined more than 100 strains of bacteria from other tuberculosis patients living in the same city at the same time. None of these bacteria had the same genetic code as the rifampin-resistant bacteria obtained in the autopsy. When doctors had begun administering rifampin, the bacteria in the patient had been subjected to a new environmental selective agent, one that gave the mutant strain a major adaptive advantage.

The pattern in this story is not uncommon, but evolution offers some hope as well as alarm. Many traits that provide antibiotic resistance are harmful to the bacteria. For example, a strain of *E. coli* bacteria possesses a plasmid with a gene that enables it to produce an enzyme called β-lactamase. This enzyme gives the bacterium resistance to the antibiotic ampicillin. However, there is a cost for this resistance: to maintain its antibiotic resistance, the bacterium must devote cellular resources to producing the enzyme and to making copies of the plasmid before cell division. In another example, the bacterium *Mycobacterium tuberculosis* normally produces catalase, a beneficial but nonessential enzyme. This enzyme, however, activates the antibiotic isoniazid, which destroys the bacterium. Bacteria that have a defective catalase gene are, therefore, resistant to isoniazid—as it cannot be activated in the absence of catalase—but they lack the benefits normally provided by the enzyme. As a result of these costs of resistance, when an antibiotic is not present, natural selection often favours those bacteria that do not carry antibiotic-resistant alleles.

▶ Case Study Questions

Understanding Concepts

1. Did the rifampin-resistant bacteria found in the autopsy evolve within the patient's lungs or did they result from a brand new infection? Explain the evidence.

2. Most antibiotics are derived from microorganisms that do not occur naturally in the human body. Most infectious bacteria showed no resistance to these antibiotics when they were first used in the 1940s, because pathogens (disease-causing organisms) did not already have antibiotic resistance to them then. Why?

3. Bacteria that are sensitive to antibiotics usually out-compete resistant strains in the absence of antibiotics. Account for this observation.

Applying Inquiry Skills

4. Tuberculosis patients are now routinely given two different antibiotics at the same time. Why might this approach be more effective than administering a different antibiotic only after bacteria develop resistance to the first?

Making Connections

5. Suggest some strategies that could help reduce the incidence of antibiotic resistance in your own home, your school, and in society at large.

▶ EXPLORE an issue

Decision-Making Skills

○ Define the Issue	● Analyze the Issue	● Research
● Defend the Position	● Identify Alternatives	○ Evaluate

Debate: Treatment of Drug-Resistant Tuberculosis

In October 1999, the Harvard Medical School reported that a deadly strain of tuberculosis (TB) was spiralling out of control. This was six years after TB had been declared a global health crisis. TB has been around for a long time; evidence of the bacterium that causes the lung infection *Mycobacterium tuberculosis* has been found in 4000-year-old Egyptian mummies. It used to have a 50% mortality rate. Presently, it is unlikely to develop in people who have a healthy immune system capable of keeping the bacterium in check, but TB is common among people with poor nutrition, a weakened immune system, and inconsistent access to shelter and medical care. As much as one third of the global human population is thought to be infected by TB. Until the late 1990s, modern drug treatments—usually with the use of the antibiotics isoniazid and

rifampin—had produced a 98% cure rate. The 1999 report was of a worldwide outbreak of a strain of TB that was resistant to this antibiotic treatment (**Figure 14**).

Modern medical treatment of TB requires antibiotic use from a few months to more than a year. Because these long exposures to antibiotic microorganisms increase the opportunity for TB bacteria to develop resistance, doctors have been giving patients at least two antibiotics from the outset. For people most at risk from TB, this lengthy treatment can be difficult to maintain and a challenge for medical officials to monitor, especially in regions with few clinics and trained staff. In their efforts to prevent the spread of the lethal **multidrug-resistant tuberculosis**, some authorities in North America are resorting to legally enforced quarantine in patients' homes.

EXPLORE an issue *continued*

Statement

Individuals being treated for multidrug-resistant TB should be placed under medical house arrest to ensure they complete their entire treatment under medical supervision.

- In a group, share responsibility for researching this issue, using written and electronic resources.

  GO www.science.nelson.com

- You may want to consider the current status of the TB outbreak. Some TB strains are resistant to four or more of the

most effective antibiotics, which increases the cost for treating such patients at a time when medical expenses are escalating. Find out how house-arrest programs in North America operate and how effective they have been in the past. Can individual rights be balanced against society's need to prevent the spread of antibiotic-resistant diseases? Consider other alternatives to enforced confinement that could guarantee public safety.

- Prepare arguments and counterarguments to debate your position on this issue.

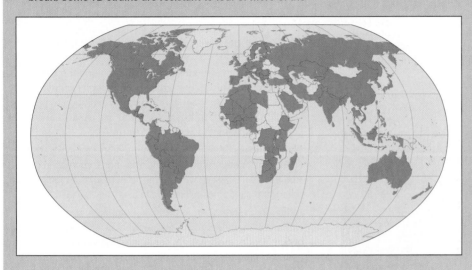

Figure 14
The purple areas on the map indicate countries reporting cases of multidrug-resistant tuberculosis as of 1999

SUMMARY *Patterns of Selection*

- Neutral mutations—the most common type in eukaryotic organisms—usually occur in noncoding regions of genetic material but can provide additional genetic material.

- Harmful mutations occur frequently but the environment selects against them and, therefore, alleles that result from them remain rare.

- Beneficial mutations are rare but the environment selects them and, therefore, alleles resulting from them accumulate over time.

- The frequency of mutations in a large population is substantial.

- Stabilizing selection tends to maintain allele frequencies in stable environments over long periods of time.

- Directional selection occurs when an extreme of an inheritable trait is favoured, usually when a population experiences a change in habitat or environmental conditions.

- Disruptive selection favours two or more distinct forms, often as a result of a change in habitat or environmental conditions.

- Sexual selection favours inherited traits that enhance mating success but may reduce an individual's chances of survival.

Understanding Concepts

1. Identify and explain the type of selection that accounts for each of the following:
   (a) the hollow bones of birds
   (b) the light emitted by fireflies
   (c) the smell of a skunk
   (d) the body diameter of snakes

2. Suggest how large antlers or bright colouration could be a disadvantage for males of some species.

3. Identify and explain whether stabilizing, directional, or disruptive selection is the most likely cause of each of the following:
   (a) the evolution of bill size in black-bellied seedcrackers
   (b) the consistent colour of blue jays for the past 100 years
   (c) the evolution of the elephant's trunk
   (d) the evolution of human brain size

4. Study **Figure 15**. Describe the sexual dimorphism shown and its evolutionary role.

Figure 15
Male (with inflated throat sac) and female frigate birds

5. Respond to this statement: Harmful mutations are more common than beneficial ones, and as the many harmful effects of mutations accumulate, species become weaker and less able to adapt to changes in their environment.

Applying Inquiry Skills

6. Consider the example of natural selection and evolution in the African seedcracking finches. What specific technologies do you think might have been used by the researchers to gather, analyze, and interpret appropriate data? Be as detailed as possible.

Making Connections

7. Many insect species have evolved resistance very rapidly to a range of pesticides. Like other species, insects exhibit variation in physical, chemical, and behavioural traits.
   (a) Describe how an insect species would evolve resistance to a pesticide newly introduced into its environment.
   (b) Identify and describe the type of selection the insects exhibit.
   (c) How might high rates of reproduction and the short duration of insect generations affect their evolution?
   (d) How might an understanding of the evolution of pesticide resistance influence how you use pesticides or alternative methods of insect control?

8. Many Africans who are carriers of the allele for sickle-cell anemia have emigrated from malaria-stricken areas in Africa to North America. Has this influenced the biological role of the sickle-cell allele? Explain.

9. Although, in theory, an individual could mate at random with other members of a large population, this seldom occurs. Under most natural conditions, individuals tend to mate with nearby members of the same species, especially if they are not very mobile. Alternatively, individuals choose mates that share similar traits; for example, toads (and often humans) tend to pair according to size.
   (a) How might inbreeding (the mating of closely related individuals) lead to an increase in recessive phenotypes? Relate your answer to either a population of cheetahs in the wild or a population of golden retrievers in a breeding kennel.
   (b) Does nonrandom mating result in changes to population phenotype frequencies, genotype frequencies, or allele frequencies?

The work of natural selection that is directly observable—such as the change in the coloration of moths or the production of multidrug-resistant bacteria—is readily understandable. But what about the evolutionary explanations for more extraordinary changes in species, such as the development of the complex relationship between colourful flower blossoms and their pollination by insects, or the evolution of organisms with complex eyes from organisms with no eyes? It has been much more difficult to gain acceptance for the role of natural selection in these types of changes.

The Blind Watchmaker?

In 1802, theologian William Paley questioned how evolution could possibly result in the formation of a structure as complex and seemingly well designed as an eagle's eye. He used the analogy of a pocket watch, arguing that if one examines the complex set of gears and mechanisms within a watch, one concludes that there must have been a watchmaker. Similarly, he argued, an examination of the complex structures of an eye—much more intricate than a pocket watch—must lead to the conclusion that there had to be a Creator. Paley assumed, mistakenly, that the theory of evolution was based on pure chance. He could not believe that random chance could produce the eye of an eagle any more than randomly throwing together watch parts could produce a functioning watch. But while mutations are the result of chance, natural selection is not. Natural selection favours certain individuals over others, a process that is the opposite of chance. This nonrandom process can account for the eagle's eye that so puzzled Paley.

The Evolution of Complex Features

Consider how an organism with a complex eye may have evolved from an ancestor with no eyes at all. Assume that the early ancestor was a wormlike animal. How might mutation and selective pressures have initiated the evolution of an eye in a previously sightless organism?

Two scenarios can be put forward. One is the sudden production by chance of a fully functional eye. A worm born with a complex, fully functioning eye might have improved chances of survival and passed on this benefit to later generations. But the making of a whole eye by pure chance, genetically speaking, would require many thousands of beneficial mutations to occur at once in one pair of worm gametes that bear no previous genetic code for an eye. An individual gamete is very unlikely to possess even a single beneficial mutation, so the likelihood of one having thousands is very remote. Therefore, this scenario could not account for the existence of complex eyes.

A second scenario is the gradual production of an eye in a series of accumulated beneficial mutations. In this case, it would have to be demonstrated that each stage in the evolution of a complex eye must have benefited the organism.

The second scenario is a plausible process that draws on a sound understanding of genetics. Suppose that the wormlike ancestor experienced a mutation that resulted in the development of light-sensitive skin cells, as shown in **Figure 1(a)** on the next page. In such cells, similar to the light-sensitive skin in present-day earthworms, light would have triggered a chemical change. Perhaps the altered chemical could not participate in some chemical pathway and the organism's behaviour was affected in some way; for instance, the primitive worm may simply have stopped moving. The current understanding of biochemistry suggests that this is a plausible hypothesis, as organisms exhibit responses

Figure 1

(a) A small mutation could produce a patch of cells with chemical sensitivity to light.

(b) Any mutation resulting in the formation of a pit would provide directional information.

(c) Visual ability would be enhanced if the pit deepened and an image would be formed if the opening narrowed.

(d) Accumulated epidermal excretions may have been the source of material that formed the first crude lens.

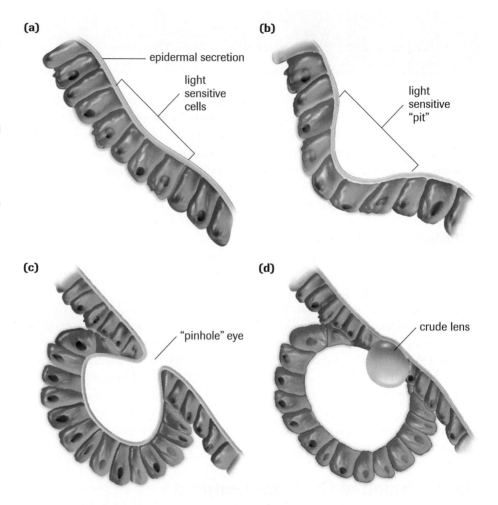

(a) epidermal secretion / light sensitive cells

(b) light sensitive "pit"

(c) "pinhole" eye

(d) crude lens

to minute changes in internal chemistry. In addition, small changes in molecular structure can dramatically affect their chemical sensitivity to light. This initial mutation would have had no long-lasting consequences in evolutionary terms, however, unless it provided some benefit to the wormlike organism. Not moving in the light may have reduced the organism's chances of attracting the attention of predators, or it may have helped prevent its skin from drying.

Although such a mutation might have altered only a single chemical step, its evolutionary impact would have been monumental. The resulting organisms would have been strongly favoured, out-competing organisms that were not light sensitive. After a number of generations, the light-sensitive individuals would compose the entire population.

Mutations would have continued to occur in the population of successful wormlike organisms. Most would have been neutral or harmful and would have gone unnoticed. Suppose that, a million years after the occurrence of the initial mutation, a new mutation occurs that, as illustrated in **Figure 1(b)**, results in a wormlike organism with a pitted skin surface. The surface structure of many organisms varies considerably, so this is a reasonable hypothesis. For this mutation to be selected, it must benefit the organism. Having the light-sensitive cells in an indentation or pit could have provided the major benefit of allowing a worm to distinguish the direction of a light source. Being able to move away from light is a clear advantage over simply stopping in response to it. Over a number of generations, in which natural selection favours this trait, the population would have been replaced by worms with light-sensitive cells positioned in a pit.

As time passed, other mutations may have influenced the skin's surface. As pictured in **Figure 1(c)**, any mutation that caused the indentation or pit to become exaggerated to form a deeper pit would have enhanced directional sensitivity. In combination with natural selection, mutations could also have led to the evolution of better optical traits. A mutation that resulted in a narrowing of the entrance to the pit could have been beneficial, resulting in an eye capable of forming a crude image. This increased visual acuity would have provided a major advantage to the organism. Many organisms produce and deposit transparent protective coatings on their epidermal surfaces. Having a deep pit and narrow opening may also have led to the accumulation and hardening of such material within the pit. As **Figure 1(d)** illustrates, this could have formed a crude lens, as well as a transparent coating over the opening of the pit.

Evolution of a complex structure is a cumulative process. Rare beneficial mutations may be separated by vast amounts of time but, when they do occur, natural selection favours them, and the **adaptations** they produce—no matter how small—accumulate one by one. This slow process of **cumulative selection** could have produced complex structures, such as eyes, and continues to produce them.

adaptation any trait that increases an individual's ability to survive or reproduce compared to organisms that do not have the trait

cumulative selection the accumulation of many small evolutionary changes over long periods of time and many generations, resulting in a significant new adaptation relative to the ancestral species

▶ **TRY THIS** activity　　*Darwinian Challenge*

- In a group, choose one of the challenges depicted in **Figures 2, 3,** and **4**.
- Brainstorm an evolutionary scenario that might account for the feature in your challenge. (If you have difficulty identifying advantages to the feature, ask your teacher for a hint.)
- Present your solution to your challenge to the class. Assess different solutions proposed by other groups that selected the same challenge.

Figure 2
The skin of this poison-arrow frog produces substances that are very toxic. How might the bright skin coloration of such frogs have evolved?

Figure 3
Eyes positioned on the sides of the head provide a wide field of vision, more helpful for detecting predators. This primate has the trait of forward-positioned eyes, shared by humans, which results in a very large blind spot. How might forward-positioned eyes have evolved?

Figure 4
The flamingo feeds on tiny aquatic organisms with its bill held in an "upside-down" position. How might this bill shape and feeding pattern have evolved?

Evolution of Insect Pollination

Just as cumulative selection can gradually produce a complex structure in a population, this mechanism can also promote evolutionary relationships between two different species. The most complex plants on Earth are flowering plants, which number more than 250 000 species. Their most significant evolutionary advances are their flowers, many of which are adapted for pollination by insects. Flowering plants likely evolved from simpler, nonflowering seed plants.

(a)

(b)

Figure 5

(a) Wind-pollinated grasses have inconspicuous flowers with large amounts of pollen that is not sticky.

(b) By contrast, insect-pollinated plants have small amounts of very sticky pollen but have large and colourful blossoms that are highly attractive to pollinating insects.

The first seed plants, like present-day gymnosperms, probably reproduced solely by wind dispersal of pollen. They would have produced large quantities of pollen because the chances of wind-borne pollen reaching an egg would have been slim, especially for plants that were widely separated or in a habitat little affected by air movement. The pollen was likely nonsticky, as would be the case with the grasses in **Figure 5(a)**, so that winds could easily pick it up and carry it.

Imagine the effects of a mutation in a plant that resulted in the production of pollen with a slightly sticky surface. This adaptation would have meant a reduction in pollen that could be dispersed by wind, but would have had the advantage of adhering to the legs and backs of insects visiting the plants in search of pollen as food. As the insects moved to another plant, they might have carried some pollen with them, depositing it directly onto a female flower part. These plants might have experienced much greater reproductive success. Of course, the insects would have benefited as well by increasing the numbers of the very plants they preferred to feed on. Given the benefits, this unintended cooperation in pollination could have been strongly selected. The evolution of sticky pollen was likely followed by the evolution of additional features to attract insects, such as colourful leaves or petals, fragrance, and nectar (**Figure 5(b)**). Insect-pollinated plants would have benefited from any mutations that made their flowers attractive and easier to find. For example, large, bright petals and a strong scent would have been favoured over small, dull flowers with no fragrance. Some adaptations are very highly specialized, such as the orchid blossom that resembles the body of a queen bee to attract the male agent of pollination (**Figure 6**). Any mutation that resulted in a change to exaggerate the appearance of flowers would have had the potential to increase pollination success by attracting more insects. Plants have evolved similar relationships with a wide variety of animal pollinators (**Figure 7**).

Figure 6

A male bee grasps this orchid's flower, which resembles a female bee's abdomen, and tries to copulate. When he flies away, he carries pollen that has adhered to his body to the next orchid he tries to mate with.

Figure 7

Birds can also be agents of pollination.

▶ **TRY THIS** activity   *Cooperation*

Demonstrate the evolutionary benefits of cooperation. For this activity, assess and improve your strategies for cooperation, just as natural selection favours beneficial advantages.

- Find a partner to form a "population" of two. Prepare to cooperate or compete with the other member of your population for 20 encounters. Sit opposite one another, folding your hands loosely in front of you. The purpose of this activity is to use a behaviour pattern that will maximize YOUR score.

- The activity is similar to the game "rock, paper, scissors." Each round you and your partner display your hands in one of two positions: "cooperate" with hands spread open, or "compete" with hands as fists.

- Keep a tally by recording the outcome of each encounter. If you both cooperate, award each of you three points. If you

both compete, award each of you two points. If one cooperates and one competes, award the cooperator one point and the competitor four points.

- Attempt to devise a strategy that will maximize your score. Test your strategy by playing other class members.

- As a class, discuss how the point system in this activity represents the benefits and costs of cooperating and competing. Relate the strategies used in the activity to the potential for evolutionary success of cooperative behaviour in the wild. (For instance, does it surprise you that birds are often seen preening one another in a cooperative manner?)

Altruism

Natural selection can result in mutual benefits for two different species, as in the case of insects and flowering plants. There are also examples, however, in which one organism benefits much more from the behaviour of another individual. With such **altruism**, the recipient is better off than the helper. Observation of, and evidence for, what appears to be altruistic behaviour prompted Darwin to comment that this "special difficulty, which at first appeared to me insuperable, [was] actually fatal to my whole theory." Science took more than a century to address his concern, as the underlying basis for this behaviour is genetic.

The most striking examples of altruism are among the social insects; about 9000 species of ants, 2000 species of termites, and hundreds of species of bees and wasps demonstrate altruistic behaviour. Living in colonies that can number in the millions, they often consist of groups of nonreproducing sisters that cooperate by raising offspring bred by a queen. Termites of the order *Isoptera* and many species of the order *Hymenoptera* have a haplodiploidy genetic system: the males are haploid and develop from unfertilized eggs, while the females are diploid and develop from fertilized eggs. This reproductive arrangement is very unusual. Female *Hymenoptera* with this genetic system— more closely related to their sisters than to their own offspring—share 50% of their genes with their mother, father, and offspring and 75% of their genes with their sisters (i.e., those with the same father). They help their mother, the colony queen, produce more of their sisters rather than their own offspring.

Female *Polistes meticus* wasps sometimes nest on their own and sometimes in sister pairs. When they nest in pairs, one female lays the eggs and the other gathers food. Studies show that sister-pair nests are three times as successful as single-female nests. Females who act as reproductive helpers and lay no eggs themselves may realize a contribution of 30% more of their alleles to the next generation than when they nest alone. This process is referred to as **kin selection**. Female wasps receiving cooperation increase their contribution of alleles by 77%. The helper only *appears* to be acting in an altruistic way, as she is improving her chances of passing on copies of alleles identical to her own to the next generation.

Altruism can be seen beyond these insect species. A human parent who sacrifices her life to save that of her child reduces her own fitness as an individual, but increases the fitness for survival of her own genes, half of which reside in her child.

DID YOU *KNOW* ❓

Selfish Genes?
Richard Dawkins described beneficial alleles that increase their own chances of success by improving the individual's chances of survival. He called them "selfish genes."

altruism behaviour that decreases the fitness of an individual that is assisting or cooperating with a recipient individual whose fitness is increased; such behaviour can evolve only when the assisting or cooperating individual increases its fitness indirectly (i.e., through the fitness of its kin)

kin selection the natural selection of a behaviour or trait of one individual that enhances the success of closely related individuals, thereby increasing the first individual's fitness indirectly

- Chance alone cannot account for the evolution of complex features. Natural selection favours particular phenotypes, which is the opposite of chance.

- Complex features arise through the accumulation of many small changes over long periods of time.

- These small changes, the result of mutations, provide an adaptive advantage to the individual and are favoured by natural selection.

- Many behaviours that reduce the fitness of a helpful individual and enhance the fitness of the recipient individual can be explained by evolutionary models.

- Kin selection is widespread among social insects.

▶ *Section 12.5 Questions*

Understanding Concepts

1. Define and give an example of kin selection.

2. Describe two adaptations early in the evolution of a complex eye and the possible advantages they may have offered to an organism's eyeless ancestor.

3. Compare and contrast random chance and cumulative selection in evolution.

4. Can mutations by themselves lead to the development of new inheritable features in a species? Relate your response to the role of mutations in cumulative selection.

5. Draw a sequence of sketches to show how cumulative selection might have resulted in the development of insect-pollinated flowering plants (**Figure 8**) from ancient wind-pollinated ancestors.

Figure 8
An insect-pollinated plant

Making Connections

6. Hermaphroditic molluscs try to impregnate one another with their outstretched penis while trying to escape being fertilized themselves. What possible advantage might such molluscs have in avoiding the production of eggs while fertilizing other individuals instead?

7. What environmental conditions and/or mutations might have led to the evolution of flowers that attract birds or bats, rather than insects, as agents in pollination (**Figure 9**)?

Figure 9
This very fragrant night-blooming cactus attracts bats.

8. Orb-web spiders have evolved the ability to spin complex netlike webs. Their evolutionary ancestors must have spun less complex webs and gradually gained the ability to spin more complex webs.
   (a) Find out what advantage highly sophisticated webs provide to orb-web spiders.
   (b) Explain how less complex webs would have been advantageous to spider ancestors.

9. Suppose a new species is found with a very unusual feature. If no biologist can suggest an evolutionary advantage for such a structure, is this evidence against evolution? Explain.

Natural selection, which favours beneficial variations, is able to account for the adaptations exhibited by individuals within populations. These same selective mechanisms are able to account for **speciation**—the formation of entirely new species—and, from there, the evolution of new groups of living organisms. Evolutionary changes that occur at the species level are referred to as **microevolution**.

speciation the evolutionary formation of new species

microevolution changes in gene (allele) frequencies and phenotypic traits within populations and species; microevolution can result in the formation of new species

What Is a Species?

A biological **species**, according to a widely accepted definition, includes the members of groups or populations that interbreed or have the ability to interbreed with each other under natural conditions. Thus, members of different species are reproductively isolated from one another. Populations of different species will have no gene flow between them. Although this definition is sound theoretically, it is not always possible to apply. Evidence of interbreeding and gene flow between populations can be difficult or impossible to examine; little data exist, for example, for most populations of wild animals. For these reasons, biologists need other methods to determine the species status of individuals and populations.

species members of interbreeding groups or populations that are reproductively isolated from other groups and evolve independently

Differentiating species based on physical appearance, or morphology, works well in many cases. Palaeobiologists have relied on morphological features from the diverse record of fossilized organisms to distinguish thousands of different species. For living populations that are morphologically indistinguishable (**Figure 1**), behavioural or other biological methods of identification are needed.

Figure 1
Two species of grey tree frogs, *Hyla versicolor* and *Hyla chrysoscelis*, cannot be distinguished by their appearance but can be differentiated by their mating calls. The trill rate of the *H. versicolor* ranges from 17 to 35 notes per second compared to the trill rate of the *H. chrysoscelis* at 34 to 69 notes per second.

Species can be differentiated on a biological basis by identifying their **reproductive isolating mechanisms**; for species that reproduce sexually, genetic isolation results from reproductive isolation. Although not applicable to such asexual organisms as prokaryotic organisms, some fungi and plants, and even some vertebrates, this method of differentiation is practical for most other species. Mechanisms that isolate populations through reproduction offer a significant key to speciation.

reproductive isolating mechanisms any behavioural, structural, or biochemical traits that prevent individuals of different species from reproducing successfully together

Reproductive Isolating Mechanisms

A reproductive isolating mechanism is any behavioural, structural, or biochemical trait that prevents individuals of different species from reproducing successfully together. Some result from traits belonging to the species themselves, while others result from environmental factors.

There exists a wide range of **prezygotic mechanisms** that prevent fertilization, thereby maintaining species isolation (**Table 1**). Three such mechanisms prevent mating. Species that occupy separate habitats or separate niches of the same habitat do not encounter one another to reproduce, so they are said to be in *ecological isolation* (**Figure 2**). Otherwise compatible species that inhabit an overlapping range can remain in *temporal isolation* because their reproductive cycles for flowering or mating occur at different times. A distinct mating ritual by one species may prevent members of another species from recognizing or selecting a mate (**Figure 3**); this is called *behavioural isolation*.

Figure 2
The hoary marmot lives at high elevations in alpine meadows in the Canadian Rocky Mountains.

prezygotic mechanisms reproductive isolating mechanisms that prevent interspecies mating and fertilization (e.g., ecological isolation, temporal isolation, and behavioural isolation)

Table 1   Prezygotic Reproductive Isolating Mechanisms

Mechanism	Example
Prevention of mating	
ecological isolation	Ground squirrel species occupy different habitats. Woodchucks (ground hogs), for example, live in fields at lower elevations while marmots (**Figure 2**) live in alpine meadows.
temporal isolation	Similar plant species may bloom at different times of the day (e.g., day- and night-blooming cacti) or in different seasons (e.g., spring- and summer-blooming irises).
behavioural isolation	Each species may use different signals for attracting a mate. The mating behaviour of male jumping spiders is an elaborate dance in which they shake their legs and wave their palps (**Figure 3**). Females of different species do not respond to the dance.
Prevention of fertilization	
mechanical isolation	Pollen sacs in a lady's slipper orchid become attached to an insect, but they are not removed by any other kind of flower (**Figure 4**).
gametic isolation	Giant clams release sperm and eggs into open water; gametes recognize one another by molecular markers.

Figure 3
The jumping spider from Arizona is an example of a species that experiences behavioural reproductive isolation.

Figure 4
This pink lady's slipper orchid exhibits a mechanical reproductive isolating mechanism.

Two prezygotic mechanisms restrict fertilization. *Mechanical isolation*, or structural differences in reproductive organs, can prevent copulation. For instance, as a result of the complex shape of the penis in certain arthropod species, many closely related species cannot interbreed with them. Similarly, the shape of floral features in many plants can affect the transfer of pollen; orchids exhibit a variety of complex features of this type (**Figure 4**). *Gametic isolation* may prevent fertilization at a molecular level. In coral reefs, many species with external fertilization may release their gametes simultaneously, so trillions of sperm and eggs may enter the shallow water at one time. Sperm and eggs of the same species recognize each other by molecular markers. In other cases, the male gamete cannot survive inside the female, as often occurs with internal fertilization in animals and with germinating pollen tubes in plants.

Some reproduction between species does occur to produce **hybrids**, but **postzygotic mechanisms** prevent the hybrid zygote from developing into a healthy and fertile adult (**Table 2**). When fertilization occurs successfully between closely related species, some incompatibility (usually chromosomal) results in zygotic mortality; the zygote or embryo fails to develop properly. In rare cases, an embryo does develop, but the resulting hybrid will be inviable, experiencing severely reduced fitness and, often, an early death. Hybrid offspring that become strong and fit adults are likely to be infertile, failing to undergo successful meiotic division (**Figure 5**). Therefore, postzygotic mechanisms ensure reproductive isolation of a gene pool by preventing allele exchange between the parental species.

hybrids the offspring of genetically dissimilar parents or breeding stock

postzygotic mechanisms reproductive isolating mechanisms that prevent maturation and reproduction in offspring from interspecies reproduction

Table 2 Postzygotic Isolating Mechanisms that Prevent Hybrids from Reproducing

Mechanism	Result
zygotic mortality	No fertilized zygotes or embryos develop to maturity.
hybrid inviability	Hybrid offspring are unlikely to live long.
hybrid infertility	Offspring of genetically dissimilar parents are likely to be strong but sterile. An example is the mule (**Figure 5**).

Figure 5
Mules are bred from a female horse and male donkey.

Modes of Speciation

allopatric speciation the evolution of populations into separate species as a result of geographic isolation

sympatric speciation the evolution of populations within the same geographic area into separate species

Any series of events that results in the reproductive isolation of two populations may also lead to the formation of new species. Within isolated gene pools, any mutations and subsequent selection processes that occur in one population can no longer be shared with others. Significant subsequent evolutionary changes that occur in either population will result in the formation of two separate species. Such events are most commonly a result of geographical isolation.

When two populations become geographically separated from one another—especially by a physical barrier—they experience **allopatric speciation** (**Figure 6**). A large river or canyon may act as a barrier to small rodents or snakes. Birds that are not strong flyers will not cross mountain ranges or large bodies of water. Aquatic species can experience physical separation, perhaps as a result of human constructions, such as dams, canals, or major highways. Founder populations may demonstrate speciation as a result of being separated from their original populations. When strong winds carry insects, birds, or seeds across expanses of water, they immediately become isolated. Once separation has occurred as a result of any of these causes, it is a matter of time before the isolated population becomes extinct or evolves into a different species.

There is evidence to suggest that some populations split into separate gene pools and continue to share a similar geographic location. This process, called **sympatric speciation**, can be illustrated by the two species of grey tree frog, *H. versicolor* and *H. chrysoscelis* (**Figure 1**, page 571), whose territories overlap (**Figure 7**). The two species share a very similar genetic code, except that *H. versicolor* is tetraploid, while *H. chrysoscelis* is diploid. Very likely, the original population of *H. chrysoscelis* split into two species when a mutation produced the tetraploid *H. versicolor*. The extra set of chromosomes, being identical,

(a)

(b)

Figure 6
The Atlantic blue-headed wrasse **(a)** and the Pacific Cortez rainbow wrasse **(b)** are probably descendants of a single species that underwent allopatric speciation after the formation of the Isthmus of Panama separated the Atlantic and Pacific oceans.

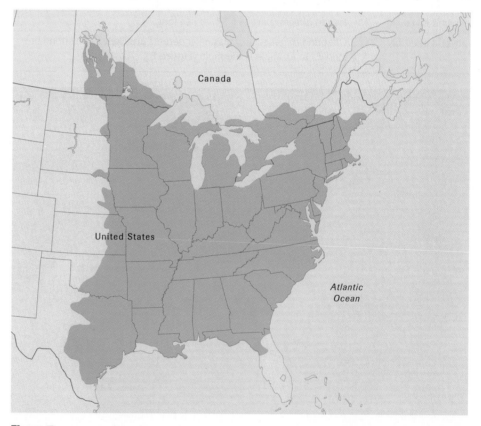

Figure 7
Two species of grey tree frogs (*H. versicolor* and *H. chrysoscelis*) share the same large-scale geographic distribution.

Figure 8
Evidence strongly suggests that a number of species of sticklebacks evolved by sympatric speciation.

produced a similar phenotype that was well adapted to the same environmental conditions as the original one. Their new genetic differences resulted in immediate reproductive isolation of the two populations into two new species but did not result in ecological isolation.

On occasion, individuals or groups can become reproductively isolated while living within a large population. A well-documented example of sympatric speciation exists in some lakes in British Columbia. Researchers have investigated two species of sticklebacks that occupy different habitats of the same lake (**Figure 8**).

The modern science of genetics has added greatly to our understanding and quantification of evolutionary mechanisms and processes, as it provides answers to the most puzzling questions of Darwin's time. A variety of different types of mutations can and do produce new genetic sequences, the ultimate source of all new inheritable variations within species. Many environmental and competitive factors direct the course of natural selection resulting in the evolution of modified or new adaptations, as well as the formation of entirely new species through reproductive isolation. Evolutionary biology is being applied to enhance the understanding of the history and origins of life on Earth, including that of humans. 🔬⬛

🔬 INVESTIGATION 12.6.1

Evolution in Motion (p. 578)
How can seemingly opposing processes of random mutation and nonrandom natural selection work at the same time? In this investigation, you will produce a quantitative model of speciation.

 **SUMMARY** *Speciation*

- A species can be defined as all members of groups or populations that interbreed or have the ability to interbreed with each other under natural conditions.

- The formation of new species is a result of the evolution of one or more reproductive isolating mechanisms. Such mechanisms prevent or restrict mating, fertilization, or successful development of zygotes.

- Geographic barriers are an important cause of the reproductive isolation of two populations by allowing accumulation of differences in mating systems, etc.

- Once two populations become reproductively isolated, they no longer share mutations and subsequent selection processes.

- Significant evolutionary changes that occur in either population will result in differences between the two gene pools, creating separate species.

Understanding Concepts

1. Give an example for each of the following reproductive isolating mechanisms:
   (a) temporal isolation
   (b) behavioural isolation
   (c) ecological isolation
   (d) gametic isolation

2. What factors might prevent palaeontologists from differentiating species of fossilized organisms?

3. Describe reproductive isolating mechanisms operating in each situation below:
   (a) Clouds of gametes are expelled by organisms in a coral reef for fertilization that involves fusing in open water. Sperm and eggs of individual species rely on molecular markers to recognize one another so that they do not fuse with gametes of another species.
   (b) Sticklebacks spend most of their time feeding close to the shoreline of a lake that is also inhabited by sticklebacks that feed mostly in deep water near the lakebed.
   (c) Songbirds of diverse species in captivity inhabit the same enclosure.
   (d) During spring nights, the distinctive calls of male wood frogs and spring peepers can be heard coming from the same pond.
   (e) A bee visits the flowers of a lady's slipper orchid (*Cypripedium acaule*), a trillium (*Trillium grandiflora*), and then another *Cypripedium acaule*.
   (f) In some bird species, such as the flamingo, the male possesses a penis, while in others, such as the heron, the male does not. A male heron cannot mate successfully with a female flamingo.
   (g) A ewe impregnated by a male goat suffers a miscarriage.
   (h) An experiment fails when a fish breeder tries to fertilize trout eggs with salmon milt. A few eggs hatch, but the fry lack vigour and many are deformed.
   (i) The periodic cicada (*Magicicada septendecim*) lives underground as a nymph for 17 years, then emerges above ground to mature and reproduce within a few weeks before dying. It shares its habitat with *Magicicada tredecim*, which reproduces every 13 years. The two species differ in size, colour, and song.
   (j) Lions and tigers will breed in captivity but produce sterile offspring.

4. A male liger—the hybrid offspring of a male lion and female tiger—was kept in captivity with a female tigon—the hybrid offspring of a female lion and male tiger—for their entire lives, but they failed to produce offspring. Provide some possible explanations for this lack of reproductive success.

5. Genetic studies of small populations of the snail, *Helix aspersa*, living in backyard gardens in the town of Byran, Texas, show dramatically different allele frequencies. These populations are separated by alleys and city streets. Suggest some of the mechanisms that might be contributing to the observed differences in their gene pools.

Applying Inquiry Skills

6. Would you expect to find a greater number of unique species on islands that are remote, such as Iceland, or on islands, such as Prince Edward Island, that are close to a large landmass? Explain your reasoning.

7. In some populations of wild deer, selection by humans favours the survival of smaller males, as large bucks are preferentially hunted.
   (a) What would you expect might happen to the genetic makeup of this population over time?
   (b) In these populations, will mutations for small size occur more often than for large size? Explain.

Making Connections

8. Construction of a canal through Panama has had effects on marine and terrestrial species. Consult an atlas to see the extent of the canal.
   (a) Comment on the canal's likely effects on the evolution and speciation of Atlantic and Pacific marine organisms in the vicinity of the Panama Canal.
   (b) How might the construction of the Panama Canal have influenced the evolution of terrestrial species?

9. Plant breeders routinely create hybrids. Research and report on hybrid production, especially from the wild stock of flowering plants. Include your assessment of the significance of proliferating hybrids in urban landscapes.

10. Suggest human activities that have directly created or eliminated reproductive isolating mechanisms. Offer some consequences of these actions.

⚗ INVESTIGATION 12.3.1

Agents of Change

- ● Questioning
- ● Hypothesizing
- ● Predicting
- ● Planning
- ● Conducting
- ● Recording
- ● Analyzing
- ○ Evaluating
- ● Communicating

In this investigation, you will design and conduct experiments to examine the influence of population size, genetic drift, and natural selection on the rate of evolution by measuring changes in allele frequencies.

Question

How do genetic drift and natural selection influence the allele frequency within a population?

Hypotheses

(a) Work with a partner to develop hypotheses to explain your predictions about how genetic drift and natural selection influence the allele frequency within a population.

Materials

80 or more beads in colour A (to represent allele R)
80 or more beads in colour B (to represent allele r)
large opaque container (to represent a gene pool)

 Immediately pick up any beads that drop on the floor, where they might cause someone to slip.

Experimental Design

(b) Work with your partner to conduct experiments for Parts I, II, and III by using or modifying the sample procedures below. Record a complete set of steps and prepare data collection tables. Submit your modified experimental design to your teacher for appproval before conducting each of your three experiments.

Sample Procedure

1. Place 40 beads of each colour (80 beads in total) in the large opaque container.

2. Thoroughly mix the "alleles" (beads) in the "gene pool" (container).

3. At random, reach into the gene pool and pick out 20 pairs of alleles to represent offspring genotypes that contribute to the next generation.

4. Determine and record the number of each genotype (e.g., 5 RR, 7 Rr, 8 rr).

5. Record the F_1 allele frequencies as decimal values. For example, divide 17 R and 23 r by 40 to get the frequencies of 0.425 R and 0.575 r, respectively.

6. Place the next generation of 80 beads in the "gene pool" container in the same proportions of allele frequencies as the "offspring" (e.g., 0.425 × 80 = 34 R, 0.575 × 80 = 46 r).

7. Repeat steps 3–6 for four additional generations.

8. Plot a graph of allele frequency versus generation to present your data. Use two different colours on the same set of axes to represent the R and r alleles.

Part I: Random Mating, No Selection

9. Run at least two trials in which you use large populations and meet the conditions of the Hardy–Weinberg principle. These trials act as your control in Part II and Part III.

Part II: Genetic Drift

10. Run at least two trials in which you examine the influence of population size on the degree and rate of genetic drift. Choose two or more starting populations of different sizes. As an option, you may also wish to model a founder effect.

Part III: Natural Selection

11. Run at least two trials in which natural selection occurs. You might model a favoured homozygous genotype in which, for example, RR offspring might be twice as successful as other genotypes. If so, you would need to allow for the increased ratio of offspring contributing to the next generation, while maintaining a stable, large population. As an option, you could investigate selection against a homozygous lethal allele by assuming that each time a specific homozygous allele pair is selected, it dies, and you have to keep adding pairs until you have 20 offspring. Another option is to investigate a selective advantage for a dominant phenotype.

Analysis

(c) Prepare and submit completed laboratory reports for your three investigations, including the presentation of your results.

(d) How did population size influence the degree and rate of evolutionary change? Did any alleles become fixed in a population? In what size populations might you expect it to be relatively common for alleles to become fixed? Why?

(e) What conditions occur in nature that result in small populations?

(f) How did natural selection influence the degree and rate of evolutionary change? Did any alleles become fixed in the population?

(g) Were your results unusual compared with similar conditions in other groups?

(h) For each of your experiments in which evolution did occur, which of the five conditions of the Hardy–Weinberg principle was not met?

Synthesis

(i) Assume you introduced a single new mutant allele to your population. Explain what you expect would happen under each of the following conditions:

(i) The mutant is harmful and the population size is large.

(ii) The mutant is harmful and the population size is small.

(iii) The mutant is beneficial and the population size is large.

(iv) The mutant is beneficial and the population size is small.

(v) A beneficial mutant is introduced and the population is observed four generations later.

(vi) A beneficial mutant is introduced and the population is observed 400 generations later.

INVESTIGATION 12.6.1

Evolution in Motion

Inquiry Skills

○ Questioning	● Planning	● Analyzing
● Hypothesizing	● Conducting	○ Evaluating
● Predicting	● Recording	● Communicating

In this investigation, you will use a computer simulation to explore the influences of random mutation and natural selection on the evolutionary change in a species. You will test the effect of modifying the environment on the potential for new species to form.

Question

How can random mutation and nonrandom natural selection work at the same time to cause evolutionary change in a species?

Predictions

As you follow the steps in the procedure, you will be directed to make predictions about what you observe.

Procedure

1. Log on to your computer and follow instructions from your teacher to access the "Evolution in Motion" simulation.

 www.science.nelson.com

Part I: Cumulative Selection

2. From the main menu, choose "Rover Biology." Read the introduction and become familiar with the the assorted rover species, their gene values, and their environment.

3. Return to the main menu and choose "Cumulative Selection."

4. Record in your notebook the initial population size, gene values, and environmental conditions.

(a) Make a prediction about what changes, if any, you expect to occur in gene values over time.

5. Watch the program for two or three generations.

6. Describe in your notebook the general motion of the rovers, and identify those species present.

7. Continue the simulation. While the program is running, carefully monitor the gene pool values over the next 5 to 10 generations.

8. Record in your notebook any noticeable changes or trends in the population gene values.

(b) Is there any evidence of evolution?

⚗ INVESTIGATION 12.6.1 *continued*

9. Once directional selection has ended and the gene values are relatively stable, stop the motion by adjusting the rover speed to zero. You may wish to first use the "turbo" mode to speed up the simulation and view the gene values over a number of generations.

10. Describe in your notebook the evolution that has occurred.

(c) What gene values seem to be ideally suited to this environment?

(d) What name is given to this species of rover?

(e) Clearly explain in your notebook the roles that mutation and selection played in the evolution of the rovers. Why do you think evolutionary biologists like Richard Dawkins refer to this process as cumulative selection?

11. Run the program in turbo mode until at least 50 generations have elapsed. Then select "Graph" and view the evolutionary history of the rover species during your cumulative selection simulation.

(f) Is there any evidence that stabilizing selection pressures are at work? Explain.

(g) Was your initial prediction correct? If not, what happened that you did not expect?

Part II: Speciation

12. From the main menu, select "Speciation."

(h) Describe in your notebook the changes that have been made to the environment. What species of rover(s) might the new environment of "Darwin Island" favour? Explain.

(i) Make a prediction about how you think rovers will evolve in the new environment of "Darwin Island."

(j) Make a prediction about how you think rovers will evolve using both the Darwin and Wallace Islands simulation.

13. Run the simulation with Darwin Island. Record your results, including population gene values and the type(s) of species that evolve in both the "open" environment and the richer island environment.

(k) Identify and explain which type of selection was in operation: stabilizing, directional, or disruptive.

(l) Is this an example of allopatric or sympatric speciation?

(m) Does evolution ever come to an end in these simulations? in nature?

(n) Does natural selection or mutation ever come to an end in these simulations? Explain.

14. Run the simulation with both Darwin and Wallace Islands. Record your results.

(o) Was your Darwin Island prediction correct? If not, what happened that you did not expect?

(p) Was your Darwin and Wallace Islands prediction correct? If not, what happened that you did not expect?

Synthesis

(q) Write a conclusion that describes how mutation, natural selection, and environmental diversity each contribute to the origin and evolution of species.

Part III: The Evolution of Antibiotic Resistance (optional)

15. Return to the Main menu and select the antibiotic resistance simulation. Read the instructions and begin the simulation.

16. Without adding any antibiotic, allow the simulation to run for a number of generations (you may wish to use turbo mode) while monitoring the magnitude of the antibiotic resistance gene.

(r) Record the range of the gene value. Why is the gene not eliminated completely from the gene pool? If it were eliminated could it recur? Explain.

17. Begin adding antibiotics to the environment. Attempt to destroy all the rovers.

(s) Were you able to eliminate all rovers? Explain how this occurs.

18. Repeat step 17 two or three times.

(t) Can you always eliminate the rovers?

19. Attempt to evolve a species of rovers that are highly resistant to antibiotics.

(u) Were you able to modify the environment to evolve such species? Explain your method.

(v) How does this method model what occurs in the real world?

20. Run the program at high speed or on turbo without adding any more antibiotic to the environment. Monitor the antibiotic resistance gene value.

(w) Account for your observations.

Key Expectations

- Evaluate scientific evidence that supports the theory of evolution. (12.1, 12.2, 12.3, 12.4, 12.5, 12.6)

- Describe and put in historical context the contributions of Hardy and Weinberg to the understanding of evolutionary mechanisms. (12.2)

- Solve problems related to the Hardy–Weinberg equation. (12.2)

- Analyze evolutionary mechanisms and their effects on populations, including mutation, genetic drift, natural selection, and sexual selection. (12.1, 12.2, 12.3, 12.4, 12.5, 12.6)

- Define the concepts of species and speciation, and explain the mechanisms of speciation. (12.6)

- Explain, using examples, the process of adaptation of individual organisms to their environment, including the evolution of antibiotic resistance in microorganisms. (12.3, 12.4, 12.5, 12.6)

- Outline evidence and arguments pertaining to the origin and diversity of living organisms on Earth. (12.1, 12.2, 12.3, 12.4, 12.5, 12.6)

- Describe and relate advances in molecular genetics and biotechnology to the scientific understanding of evolution. (12.1, 12.2, 12.5, 12.6)

- Conduct investigations and collect data on allele frequencies in a population and relate these to evolutionary processes. (12.3, 12.6)

- Communicate the procedures and results of investigations and research for specific purposes, using data tables and laboratory reports. (12.3, 12.6)

- Locate, select, analyze, and integrate information on topics under study, working independently and as part of a team, and using library and electronic research tools, including Internet sites. (12.4)

- Compile, organize, and interpret data, using appropriate formats and treatments, including tables, flow charts, graphs, and diagrams. (12.3, 12.4)

Key Terms

adaptation	bottleneck effect
alleles	cumulative selection
allele frequency	directional selection
allopatric speciation	disruptive selection
altruism	evolution
beneficial mutation	fitness

fixed	neutral mutation
founder effect	phenotypes
gene duplication	point mutations
gene flow	polyploids
gene pool	population
genes	postzygotic mechanisms
genetic drift	prezygotic mechanisms
genome	pseudogenes
genotype	reproductive isolating mechanism
genotype frequency	
harmful mutation	self-pollinating plants
heterozygous	sexual dimorphism
homozygous	sexual selection
hybrids	speciation
kin selection	species
loci	stabilizing selection
microevolution	sympatric speciation
multidrug-resistant tuberculosis	

Key Equation
Hardy–Weinberg Equation

If p = frequency of allele A and q = frequency of allele a, then

$$p + q = 1$$
$$(p + q)^2 = 1^2$$
$$p^2 + 2pq + q^2 = 1$$

where p^2 = frequency of AA genotype; $2pq$ = frequency of Aa genotype; and q^2 = frequency of aa genotype

▶ **MAKE** a summary

Some people think that the theory of evolution implies that all life on Earth has evolved as a result of random chance acting alone. Write a detailed article to address this fundamental misconception by explaining these key aspects of evolutionary biology:

- the role of random mutation
- the potential effect a mutation can have on an individual
- the role that selection plays in the inheritance of a mutation
- the impact of cumulative selection on a population over a long period of time

In your notebook, record the letter of the choice that best answers the question.

1. Which of the following is true?
   (a) Individuals evolve but species do not.
   (b) Populations evolve but species do not.
   (c) Populations evolve but individuals do not.
   (d) Individuals and species evolve.
   (e) Individuals and populations evolve.

2. Which of the following would not be considered a reproductive isolating mechanism?
   (a) differences in the timing of mating season
   (b) differences in mating behaviours
   (c) differences in the gamete recognition
   (d) differences in outward appearance
   (e) differences in age or sexual maturity

3. In a population of snakes, 216 are *RR* with yellow stripes, 288 are *Rr* with yellow stripes, and 96 are *rr* with red stripes. What are the population allele frequencies?
   (a) *R* = 0.4, *r* = 0.6   (d) *R* = 0.84, *r* = 0.16
   (b) *R* = 0.6, *r* = 0.4   (e) *R* = 0.16, *r* = 0.84
   (c) *R* = 0.72, *r* = 0.48

4. Does the population of snakes described in question 3 appear to be in Hardy–Weinberg equilibrium?
   (a) Yes. The allele frequency will not change in the next generation.
   (b) No. The allele frequency will change in the next generation.
   (c) Yes. The allele frequency is changing over time.
   (d) No. At equilibrium, the frequency of *R* will equal the frequency of *r*.
   (e) There is not enough information to tell.

5. Which of the following does not contribute to the rapid evolution of bacteria?
   (a) Bacteria reproduce very rapidly.
   (b) Bacteria can be subjected to changes in their environment.
   (c) Many harmful mutations occur over time.
   (d) Bacteria populations are extremely large.
   (e) Bacteria have smaller quantities of DNA than eukaryotic organisms.

6. The founder effect occurs when which of the following takes place?
   (a) Small populations become isolated.
   (b) A large population is able to move into a new environment.
   (c) There is gene flow between two previously separated populations.
   (d) A geographic barrier is removed.

(e) A small sample of a population survives a drastic reduction in numbers and establishes a new population.

7. The following is the type of mutation least likely to influence the evolution of a species:
   (a) a neutral mutation in which a number of genes are duplicated
   (b) a beneficial mutation that occurs extremely rarely
   (c) a mutation that is harmful but not lethal
   (d) a mutation leading to the formation of a polyploid individual
   (e) a mutation that is lethal

8. The following condition describes when a new species is most likely to form:
   (a) An animal's alleles become fixed at 100%.
   (b) A plant becomes a polyploid.
   (c) There is a physical environmental barrier.
   (d) A species experiences runaway selection.
   (e) Gene flow occurs between adjacent populations.

9. The mutation that produced the sickle-cell anemia allele is best described as which one of the following?
   (a) a beneficial mutation
   (b) a harmful mutation
   (c) a neutral mutation
   (d) only neutral in a nonmalarial region
   (e) beneficial or harmful depending on the environment

10. Which of the following is false?
    (a) Directional selection is responsible for increased running speed in the cheetah.
    (b) Disruptive selection resulted in the evolution of Darwin's finches.
    (c) Stabilizing selection can account for the little change in crocodiles over millions of years.
    (d) Sexual selection is primarily responsible for bright coloration in poison-arrow frogs.
    (e) Directional selection accounts for the change in body size of salmon populations under heavy fishing pressure.

11. Which of the following is true in small populations?
    (a) Genetic drift is rare and has little influence over evolutionary change.
    (b) Bottlenecks result in an increase in genetic diversity in the population.
    (c) Chance leads to increased homozygosity.
    (d) Alleles have little chance of becoming fixed in the population.
    (e) Chance leads to increased heterozygosity.

Understanding Concepts

1. Describe how gene duplication and recombination during meiosis contribute to the overall genetic diversity of populations.

2. Identify and describe the evolutionary role of the selection mechanism in **Figure 1**.

Figure 1
Sage grouse

3. Mutations are very rare events.
   (a) Explain whether you agree or disagree with the above statement from the perspective of an individual organism.
   (b) How would your answer differ if you were referring to mutation events in entire populations over long periods of time?
   (c) How would your answer differ if you were comparing elephant and bacteria populations?

4. List the conditions that must be met to satisfy the Hardy–Weinberg equilibrium in a population. Clearly explain how the failure to meet anyone of the conditions leads to evolutionary change.

5. For black-bellied seedcrackers (**Figure 9**, page 559), does the environment continue to favour two different bill sizes?
   (a) Explain whether birds born with the largest or smallest bills were the most successful.
   (b) What might happen to the bird populations if the plants producing hard seeds became extinct or evolved softer seed coverings?

6. Explain why the evolutionary cost of failing to mate is extreme.

7. If beneficial mutations are much more rare than harmful ones, how can they have such an important role in evolution?

8. What role does stabilizing selection have on a population when the environment is not changing? Explain

in terms of the selective pressure on mutations that might occur.

9. Relate, with an example, how random chance can have a greater effect on small populations than on larger populations.

10. Describe some of the many reproductive isolating mechanisms that might prevent Canada geese from successfully reproducing with penguins on the Galapagos Islands.

11. Provide three examples for each of the following:
    (a) directional selection   (c) disruptive selection
    (b) stabilizing selection   (d) sexual selection

12. Copy **Figure 2** into your notebook, adding relevant labels. Write a description of the effect this diagram illustrates.

Figure 2

13. List and explain some of the factors that have lead to the rapid evolution of antibiotic resistance among many bacteria.

14. Would you expect bacteria occurring in wildlife, or in domesticated animals, to show signs of antibiotic resistance? Explain your answer.

15. A small population of pygmy mammoth measuring only 2 m in height once lived on a small island off the coast of California. Biologists believe this is an example of a population that descended from a few large mammoth that reached the island more than 50 000 years ago. Explain how the small founding population, remote location, and natural selection on this island might have each contributed to the formation of this unusual species.

16. How would you expect the rate of mutations in the population of black bear living in Ontario to compare with that of the bacteria inhabiting the gut of a single black bear? Why?

17. Describe a scenario through which disruptive natural selection might lead to the formation of a new species.

18. Before the large-scale movement of people around the world, including the slave trade, the sickle-cell allele was extremely rare except in regions where malaria occurs. Based on this distribution, is it accurate to describe the allele as harmful? Explain.

Applying Inquiry Skills

19. In a population of 40 000 bats, you have identified two distinct phenotypes that result from two alleles at a single gene locus. One allele (*C*) produces dark brown hair and the other (*c*) produces cinnamon-coloured hair. If only 16 bats are cinnamon-coloured, estimate the allele frequencies in the population. Assume the population is in Hardy–Weinberg equilibrium.

20. Suppose that 1 in 400 people in a large population have a recessive disorder. Apply the Hardy–Weinberg principle to estimate the proportion of individuals who are carriers of (i.e., heterozygous for) this disorder.

21. Predict how the genetic diversity of a population of lake trout from a small lake in northern Ontario would compare with the genetic diversity of a population of lake trout in Lake Superior. How might you investigate your prediction?

Making Connections

22. Some behaviours are instinctive, meaning inherited, while others are learned.
    (a) List four examples of instinctive behaviour and four examples of learned behaviour.
    (b) Instinctive behaviours have evolved while learned behaviours are acquired. Looking at your own examples, suggest how the instinctive behaviours might have been advantageous over long periods of time.
    (c) Which behaviours do you think are more essential to the human species, instinctive or learned? Support your answer with examples and comparisons.

23. In a group, suggest a possible scenario for the evolution of one of the following:
    (a) long manes in male lions
    (b) human gullibility
    (c) nest-building behaviour in birds
    (d) echolocation in bats
    (e) deciduous trees

24. In what way do cosmetics and clothing companies take advantage of evolutionary mechanisms to market and sell their products?

25. In many zoos, artificial insemination of female tigers is becoming common practice. Semen may be collected from male tigers in various zoos around the world, frozen in liquid nitrogen, and shipped to zoos where it is used to inseminate female tigers in estrus.
    (a) Why do you think this is being done?
    (b) How might this affect the gene pool of tiger populations?
    (c) Do you think these efforts are enough to prevent a genetic bottleneck from occurring? Explain.
    (d) What conditions do you think are necessary to ensure the genetic diversity of zoo populations?

26. The majority of the Afrikaner population in South Africa is descended from a single shipload of Dutch immigrants in 1652. Compared to the Dutch population, these descendants have a much higher incidence of such rare genes as the ones that cause Huntington's disease and the enzyme defect *variegate porphyria*. What is the most likely explanation for this observation?

Extensions

27. Research Richard Dawkins's theory about "selfish genes." Prepare arguments and counterarguments for a class debate on his theory.

     www.science.nelson.com

28. Some prions that cause diseases—such as bovine spongiform encephalopathy (mad cow disease) and the inherited form of human spongiform encephalopathy (Creutzfeldt-Jakob disease)—are known to result from point mutations. After researching the genetics of these diseases, explain how they are transmitted and suggest why they are not eliminated by natural selection.

29. The strange eyes of stalk-eyed flies (**Figure 13**, page 561) have a complicated basis for their sexual selection. Using print or electronic sources, find out what this basis is.

     www.science.nelson.com

30. Try one of the following online activities from the public broadcasting service series on evolution. Summarize your results and findings in a brief report.

     www.science.nelson.com

    (a) "Sex and the Single Guppy"
    (b) "The Mating Game"
    (c) "The Microbe Clock"
    (d) "The Advantage of Sex"

chapter

13

The Evolutionary History of Life

▶ In this chapter, you will be able to

- analyze evolutionary mechanisms, and the processes and products of evolution from the earliest origins of life on Earth;

- describe some scientists' contributions that changed evolutionary concepts;

- outline evidence about the origin, development, and diversity of living organisms on Earth;

- describe, construct, and analyze phylogenetic trees;

- evaluate current evidence that supports the gradualism theory and punctuated equilibrium theory on the rate of evolution;

- explain, using examples, the process of adaptation of organisms, including humans and their recent ancestors, to their environment;

- identify questions that arise from concepts of evolution and diversity, including human health;

- formulate and weigh hypotheses that have influenced the development of the theory of evolution;

- analyze how the science of evolution is related to current areas of biological study, and how technological development has extended or modified knowledge in the field of evolution.

The universe is estimated to be 12 billion to 15 billion years old, and the Earth about 4.6 billion years old. The biochemical processes that resulted in the first life forms are thought to have begun shortly after the Earth cooled, and single-celled life probably began only a few hundred million years later. For much of Earth's history, life was single celled. Then, just over half a billion years ago, complex life forms came into existence and diversified. Only in the last brief fraction of Earth's past did a species evolve that was capable of investigating and compiling this history. Today, scientists understand the general patterns and the broad outline of the evolutionary history of life, although they still have much to learn. The initial complex chemical arrangements, as well as the patterns and processes throughout the appearance of life forms, continue to be the subjects of scientific discussion and investigation.

💡 REFLECT on your learning

1. All organic compounds are composed of atoms. List the four most abundant chemical elements found in all living things. Which of the four is the most chemically reactive? How do the masses and densities of these four elements compare with most elements in the periodic table?

2. Other common elements in living things include sodium, phosphorus, sulfur, chlorine, and calcium. Do these elements have a relatively high or low density?

3. Assuming Earth was once in a molten liquid state, would you expect compounds containing the elements in questions 1 and 2 to be found floating near Earth's surface or deep within Earth's core? Explain.

4. The cell is considered the fundamental unit of living organisms. Describe and compare a simple prokaryotic cell and a complex eukaryotic cell with reference to the following:
   (a) presence or absence of an outer cell membrane
   (b) presence or absence of internal membranes and organelles
   (c) DNA structure and location, including that of mitochondria and chloroplasts

5. Certain bacteria appear very similar to fossils of ancient life dating to more than 3.5 billion years ago. Considering your response to question 4, explain whether this is surprising.

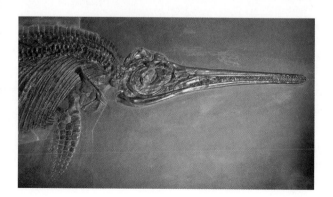

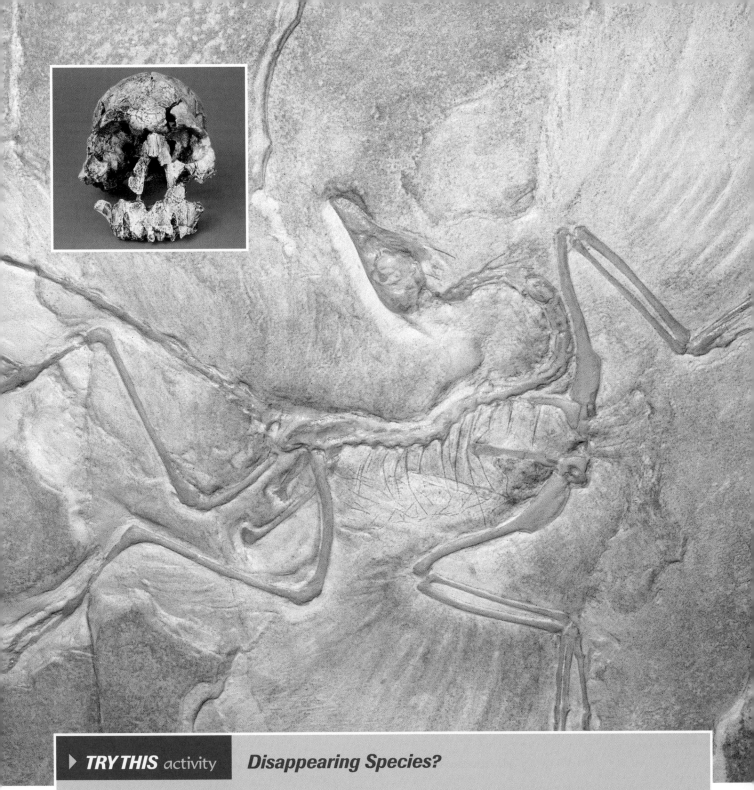

▶ TRY THIS activity *Disappearing Species?*

The fossils shown on these pages represent species that lived on Earth millions of years ago: The photo on the left depicts an ichthyosaurus. The fossil was found in Southern Germany and dates from the Jurassic period, 160 million years ago. The large photo on this page shows a fossil that is sometimes called the link between reptiles and birds. *Archaeopteryx* lived during the Upper Jurassic 152 million years ago. The skull shown in the inset belonged to a *Homo habilis* who lived 2 million years ago. None of these species are living now.

With one or two other students, brainstorm and suggest answers to these questions:

(a) Consider the key abiotic and biotic factors that affect the survival of all species. Would these factors have influenced the species shown on this page in the same way or in different ways? What environmental factor(s) might have resulted in the disappearance of these species?

(b) Could evolution be responsible for the disappearance of some of these species? How?

(c) What effects do you think the sudden extinction of one or many species might have on the evolution of other species?

Earth, and life on it, originated billions of years ago. Scientists have pieced together a scientific description of the initial conditions and events that may have resulted in the origin of life. Much ongoing research focuses on testing hypotheses about the origins and evolution of the earliest life forms.

Primordial Earth

Earth, when it formed some 4.6 billion years ago, was extremely hot. Heat generated by asteroid impacts, internal compression, and radioactivity melted most of the rocky material. Dense materials, composed of such heavy elements as iron and nickel, formed Earth's inner core, while less dense materials formed a thick mantle. The least dense rock, composed mostly of lighter elements, floated on the surface and cooled to form a crust (**Figure 1**).

Hot gases formed Earth's primitive atmosphere. When, after some 500 million to 800 million years, the asteroid bombardment slowed and surface temperatures cooled below 100°C, vast quantities of water vapour condensed. Hundreds of years of torrential rains pooled in surface depressions to form ocean basins. The atmosphere of primordial Earth would have contained large amounts of nitrogen gas, carbon dioxide, carbon monoxide, and water vapour. Other hydrogen compounds—such as hydrogen sulfide, ammonia, and methane—would have been present. It is probable, though not certain, that this early atmosphere also contained hydrogen gas. Oxygen gas is highly reactive and, with the high temperatures present then, would have combined with many other elements to form oxides; for this reason, the atmosphere would have contained little, if any, free oxygen gas. The surface of Earth would have been exposed to many intense sources of energy: radioactivity, intense ultraviolet light, visible light, and cosmic radiation from a young Sun; heat from volcanic activity; and electrical energy from violent lightning storms.

Figure 1
Thermal mud pools, such as those in Yellowstone National Park, are suggestive of conditions on the surface of primordial Earth.

Organic Molecules

In the mid-1930s, the Russian biochemist Alexander Oparin and British biologist J.B.S. Haldane independently proposed the theory of **primary abiogenesis**—that the first living things on Earth arose from nonliving material. They reasoned that the first complex chemicals of life must have formed spontaneously on a primordial Earth and, at some point, arranged themselves into cell-like structures with a membrane separating them from the outside environment.

primary abiogenesis theory that the first living things on Earth arose from nonliving material

Although extremely harsh, the early conditions on Earth were ideal for triggering chemical reactions and the formation of complex organic compounds. What molecules might have formed from the reactions of gases in the primordial atmosphere? In 1953, the Nobel Prize–winning astronomer Harold Urey and his student, Stanley Miller, investigated possible reactions. Their apparatus modelled the water cycle by using a condenser to produce precipitation and a heater to cause evaporation. Since Urey and Miller suspected that the early atmosphere would have contained water vapour, ammonia, and methane and hydrogen gases, they combined these gases and exposed them to electrical sparks, thereby modelling early conditions on Earth (**Figure 2**). After one week, 15% of the original carbon in the methane had been converted to a variety of compounds, including aldehydes, carboxylic acids, urea, and—most interestingly—two amino acids: glycine and alanine.

More recent evidence suggests that the specific combination of gases chosen by Urey and Miller was not likely to have existed in the primordial atmosphere. In response, many other scientists have continued this investigation with experiments that use the combination of gases now thought to have been present. These experiments have produced an even greater variety of simple organic compounds, including essential sugars, all 20 amino acids, many vitamins, and all four nitrogenous bases found in RNA and DNA. The most abundant nitrogenous base, adenine, was the easiest to produce under laboratory conditions. These results suggest that many of the building blocks of life likely formed spontaneously in Earth's primordial environment.

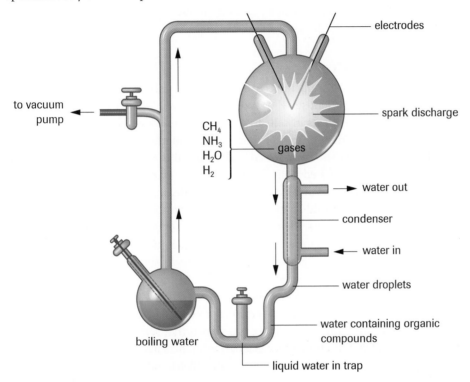

Figure 2
The Miller and Urey experimental apparatus

Chemical Evolution

For the first molecules to have produced living cells, they had to have been able to form more complex chemical and physical arrangements. Polymerization of early monomers may have occurred in numerous ways. Monomers may have become concentrated on hot surfaces as water evaporated, and the increased concentrations and heat energy may have triggered polymerization reactions. Under similar conditions, in 1977, Sidney Fox at the University of Miami was able to trigger the spontaneous production of **thermal proteinoids** consisting of chains of more than 200 amino acids. Other scientists have discovered that such materials as clay particles and iron pyrite form electrostatically charged surfaces that are also capable of binding monomers and catalyzing polymerization reactions. These findings suggest mechanisms for the formation of the first polymers. Could any polymers have then influenced their own formation?

thermal proteinoids a polypeptide that forms spontaneously when amino acids polymerize on hot surfaces; proteinoids are able to self-assemble into small cell-like structures in cool water

ribozymes an RNA molecule able to catalyze a chemical reaction

The most fundamental characteristic of living things is organized self-replication. To self-replicate, molecules must demonstrate catalytic activity, that is, the ability to influence a chemical reaction. But can a molecule act as a catalyst for its own formation? In the 1980s, Thomas Cech, working at the University of Colorado, discovered RNA molecules, called **ribozymes**, that act as catalysts in living cells. In other experiments, simple systems of RNA molecules have been created that are able to replicate themselves. In 1991, while working at the Massachusetts Institute of Technology, chemist Julius Rebek, Jr., created synthetic nucleotidelike molecules that could replicate themselves—and make mistakes, which resulted in nonliving molecular systems that mutated and underwent a form of natural selection in a test tube. As demonstrated by such experiments, Earth's first self-replicating and evolving systems may have been RNA molecules. RNA is also likely to have been the first hereditary molecule. Its catalytic activity and the roles of tRNA, mRNA, and rRNA suggest that it is likely to have played a direct role in the synthesis of proteins. Current scientific thinking about DNA is that it evolved later, perhaps by the reverse transcription of RNA.

Formation of Protocells

The evolution of self-replicating molecular systems and cell-like structures is a vital area of investigation among scientists who study the origin of life. All living things are composed of cells. For chemicals in cells to remain concentrated enough for metabolic processes to occur, they must be separated from the surrounding dilute environment. How might the first membranes have formed and arranged themselves into cell-like packages with an interior separated from the surrounding environment?

liposomes spherical arrangements of lipid molecules that form spontaneously in water

Lipid membranes can and do form spontaneously. Because of their hydrophobic tails, fatty acids and phospholipids naturally arrange themselves into spherical double-layered **liposomes**, or clusters. These can increase in size by the addition of more lipid and, with gentle shaking, can form buds and divide. Their membranes also act as a semipermeable boundaries, so that any large molecules initially trapped within them, or produced by internal chemical activity, are unable to escape, thereby increasing in concentration. Although they are not alive, they can respond to environmental changes or reproduce in a controlled way, which means protocells do share many traits of living cells. Additional experimental evidence has shown that semipermeable liquid-filled spheres can also form from proteinlike chains. Researchers have discovered that if amino acids are heated and placed in hot water, they form proteinoid spheres, which are capable of picking up lipid molecules from their surroundings, as shown in **Figure 3**. These protocells are also able to store energy in the form of an electrical potential across their membrane, a trait found in all living cells. Although these findings are the subject of debate and many unanswered questions remain, there is evidence that chemical evolution could have given rise to molecular systems and cellular structures that are characteristic of life. 🔘

🔘 **ACTIVITY 13.1.1**

Observing Liposome Formation (p. 629)
How do nonliving collections of phospholipids behave in water? In this activity, you can model the formation of cell-like structures as the process may have occurred on primordial Earth.

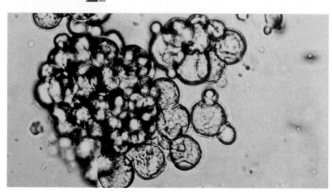

Figure 3
Liquid-filled spheres can form spontaneously by various protein mixtures in water.

Prokayotic Organisms: The First True Cells

The oldest known fossils of cells on Earth—accurately dated to 3.465 billion years ago—were found in western Australia in layered formations called **stromatolites**. These microscopic fossils resemble present-day anaerobic cyanobacteria (**Figures 4 and 5**). Even the world's oldest-known sedimentary rock formations located in Greenland—dating to 3.8 billion years ago—show chemical traces of microbial life and activity.

stromatolites shaped rock formations that result from the fossilization of mats of ancient prokaryotic cells and sediment

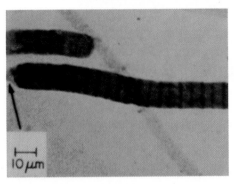

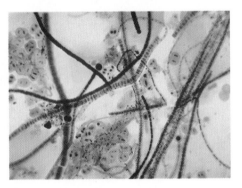

(a)   **(b)**

Figure 4
(a) Microfossils dating to 3.5 billion years ago were discovered by J. William Schopf of UCLA. They closely resemble present-day cyanobacteria **(b)**.

Figure 5
The actions of cyanobacteria began forming these stromatolites in western Australia 2000 years ago. They are almost identical to those containing the fossils discovered by Schopf (**Figure 4**).

Although the oldest fossil bacteria resemble photosynthetic cyanobacteria, which use oxygen, the very first prokaryotic cells would certainly have been anaerobic, as the atmosphere would then have contained little or no free oxygen. These first prokaryotic organisms would likely have relied on abiotic sources of organic compounds. They would have been **chemoautotrophic**, obtaining their energy and raw materials from the metabolism of such chemicals in their environment as hydrogen sulfide, released at high temperatures and in large quantities from ocean-floor vents. These organisms would have adapted to living under harsh conditions of extreme heat and pressure and may have resembled present-day thermophilic archaebacteria. As the first cells reproduced and became abundant, these chemicals would have gradually become depleted. Any cell that was able to use simple inorganic molecules and an alternative energy source would have had an advantage. Fossil evidence suggests that, by 3 billion years ago, photosynthetic autotrophs were doing just that.

chemoautotrophic describes an organism capable of synthesizing its own organic molecules with carbon dioxide as a carbon source and oxidizing an inorganic substance as an energy source

Although the first photosynthetic organisms may have also used hydrogen sulfide as a source of hydrogen, those that used water would have had a virtually unlimited supply. As they removed hydrogen from water, they would have released free oxygen gas into the atmosphere—a process that would have had a dramatic effect. The accumulation of oxygen gas, which is very reactive, would have been toxic to many of the anaerobic organisms on Earth. While these photosynthetic cells prospered, others would have had to adapt to the steadily increasing levels of atmospheric oxygen or perish. Some of the oxygen gas reaching the upper atmosphere would have reacted to form a layer of ozone gas, having the potential to dramatically reduce the amount of damaging ultraviolet radiation reaching Earth. At the same time, the very success of the photosynthetic cells would have favoured the evolution of many **heterotrophic** organisms.

These early life forms and evolutionary stages produced the necessary conditions to support the dramatic success of life on Earth powered and supplied by energy from the sun and the chemical products of photosynthesis.

heterotrophic describes an organism unable to make its own primary supply of organic compounds (i.e., feeds on autotrophs, other heterotrophs, or organic material)

The Panspermia Theory

Proponents of the panspermia theory suggest that life may have arisen elsewhere in this solar system and travelled to Earth on a meteorite or comet. How much evidence is there to support this theory? A meteor from Mars made headlines in 1996 when an examination with a scanning electron microscope of samples revealed objects resembling bacteria fossils (**Figure 6**). Many scientists, however, suspect these structures are inorganic in origin. What conditions on Mars might have permitted or fostered abiogenesis?

 www.science.nelson.com

Figure 6

SUMMARY   *Earliest Evolutionary Processes*

- Earth formed about 4.6 billion years ago. By about 4 billion years ago, less dense compounds had cooled to form a solid crust, water vapour had condensed, and ocean basins had filled.

- Early anaerobic conditions on Earth likely resulted in the formation and polymerization of many small organic molecules.

- Some RNA molecules act as catalysts for various reactions, including their own replication. As such, they are likely candidates for the first hereditary molecular systems.

- Both lipid and protein compounds likely formed liquid-filled semipermeable vessels spontaneously. These vessels have some of the same properties as cells.

- The first cells, which evolved at least 3.8 billion years ago, resembled modern prokaryotic cells. After photosynthetic prokaryotic cells evolved, at least 3 billion years ago, oxygen gas began to accumulate in Earth's atmosphere.

► Section 13.1 Questions

Understanding Concepts

1. Review the theory of natural selection as described by Darwin (Chapter 11, section 11.6). Given that chemical evolution occurred before life existed on primordial Earth, explain when you consider the process of natural selection to have begun.

2. Many scientists study chemical evolution. Suggest ways in which selective forces might have acted on chemicals before living cells existed on Earth.

3. Compare and contrast thermal proteinoids and liposomes.

4. In what way might the lack of oxygen in the early atmosphere have influenced the formation of both complex organic molecules and the first living cells?

Applying Inquiry Skills

5. Suggest reasons that Urey and Miller selected the combination of gases they did for their experiment in which they

attempted to model chemical reactions in the atmosphere of ancient Earth. Even though other scientists have cast doubts on this particular combination of gases, why do we consider the findings of the Urey–Miller experiment to be relevant?

6. (a) In Activity 13.1.1, why was lecithin used to model the formation of protocells?

   (b) Did your observations permit you to determine whether the vesicles have double-layered membranes?

Making Connections

7. Some scientists have different perspectives on the earliest evolutionary history on Earth. In print and electronic sources, find out more about their research, evidence, reasoning, and differing interpretations of experimental results.

 www.science.nelson.com

The evolutionary history of life on Earth has involved the increasing complexity and diversity of living species. New species have formed, while many have changed very little and others have disappeared. In some cases the evolutionary changes have been on a large scale, giving rise to higher taxa such as new genera, orders, and even kingdoms. These continuous patterns of **macroevolution** have been punctuated by episodes of mass extinction and subsequent very rapid diversification. Considerable scientific investigation has been conducted to find causes for these patterns and processes.

macroevolution large scale evolutionary change significant enough to warrant the classification of groups, or lineages into distinct genera or even higher-level taxa

Endosymbiosis in Eukaryotic Cells

Not much fossil evidence of the early evolution of single-celled organisms exists. Comparisons of present-day prokaryotic and eukaryotic DNA, however, suggest that the earliest prokaryotic cells probably gave rise to **eubacteria** and **archaebacteria**. It is likely that photosynthesis and aerobic respiration first evolved among eubacteria. Present-day archaebacteria are adapted to survive in extreme environments not unlike those that may have been widespread on ancient Earth. Archaebacteria may have then given rise to eukaryotic cells. Although present-day eukaryotic organisms still share many genetic traits with modern archaebacteria, the eukaryote **lineage** and archaebacteria lineage are thought to have separated about 3.4 billion years ago. While eventually evolving into eukaryotes, this lineage still consisted of prokaryotic organisms for another billion years. These proposed lineages are shown in **Figure 1**.

eubacteria prokaryotic organisms that are distinguished from archaebacteria and eukaryotic cells by differences in genomic, chemical, and structural traits

archaebacteria prokaryotic organisms with greater genetic similarity to eukaryotic cells than eubacteria (includes methanogens, extreme thermophiles, and extreme halophiles)

lineage the descendants of a common ancestor

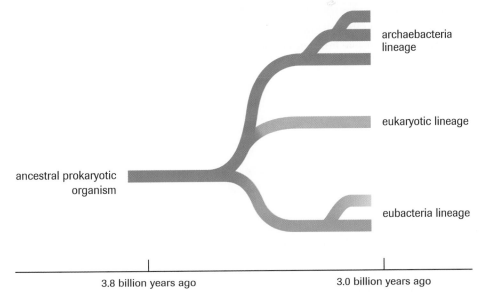

archaebacteria lineage

eukaryotic lineage

ancestral prokaryotic organism

eubacteria lineage

3.8 billion years ago   3.0 billion years ago

Figure 1
This diagram depicts the most widely accepted hypothesis about the origins of the first major lineages.

The appearance of eukaryotic cells marks a key event in the evolutionary history of life. In rock older than 1.5 billion years, most fossils are of microorganisms that appear to be very similar and small in size (**Figure 2**, page 592). Recent fossil discoveries from the Empire Mine in Michigan appear to be those of early eukaryotic algae, dating between 1.85 and 2.1 billion years old. Although eukaryotic cells likely evolved more than 2 billion years ago, rock dated at about 1.4 billion years old offers the earliest clear evidence of much larger cells that appear to have membrane-bound internal structures and elaborate shapes.

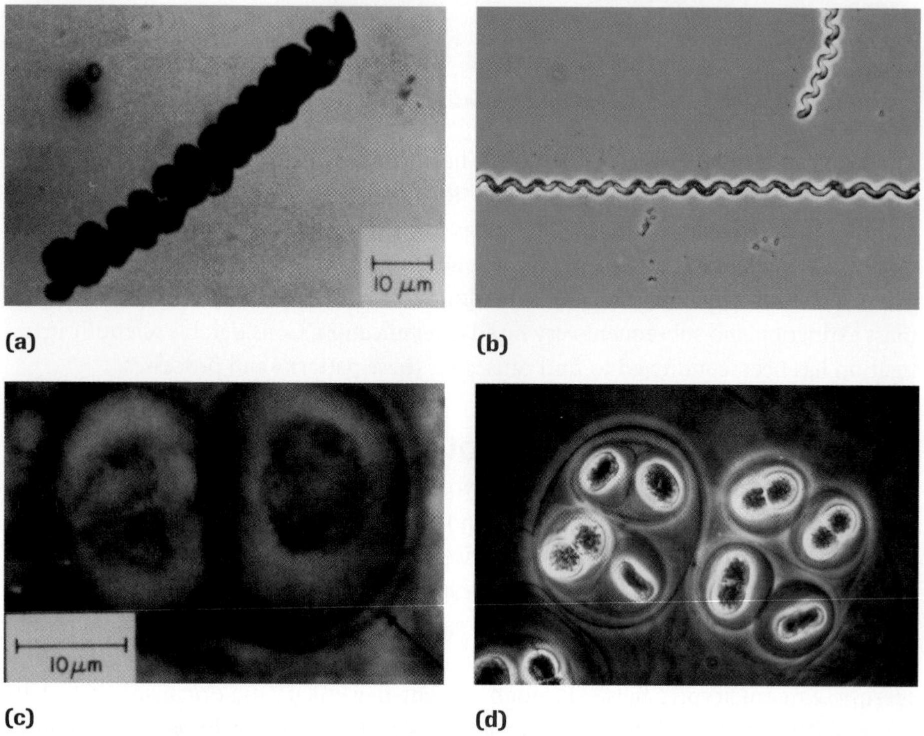

Figure 2
Fossil cyanobacteria and their living relatives:
(a) *Heliconema* (1.5 billion years old) **(b)** *Spirulina* (living)
(c) *Gloeodiniopsis* (950 million years old) **(d)** *Gloeocapsa* (living)

The distinguishing feature of eukaryotic cells is the presence of membrane-bound organelles, such as the nucleus and vacuoles. A nuclear membrane and the endoplasmic reticulum may have evolved from infolding of the outer cell membrane (**Figure 3**). Initially, such folding may have been an adaptation that permitted more efficient exchange of materials between the cell and its surroundings by increasing surface area, and it may also have provided more intimate chemical communication between the genetic material and the environment.

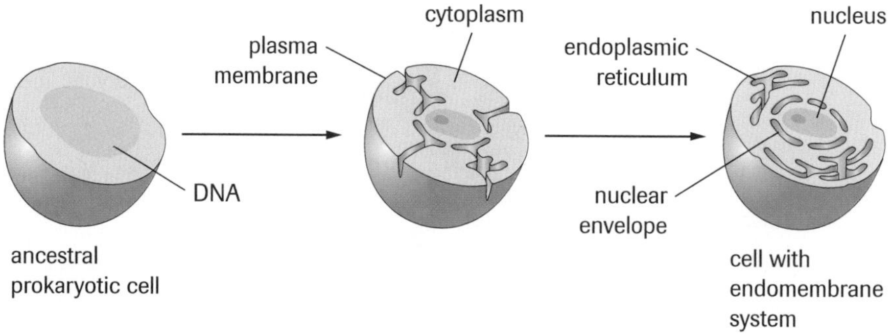

Figure 3
The probable origin of endoplasmic reticulum and nucleus: infolding of the cell membrane of a prokaryotic ancestor could have resulted in the formation of some internal membrane-bound organelles.

endosymbiosis relationship in which a single-celled organism lives within the cell(s) of another organism; recent findings suggest this may be very common

Researchers postulate that a process of **endosymbiosis** may have given rise to mitochondria and chloroplasts, two unusual organelles. According to this now widely accepted theory, early eukaryotic cells engulfed aerobic bacteria in a process similar to phagocytosis in amoeba (**Figure 4** on the next page). Having been surrounded by a plasma membrane, the bacteria were not digested but, instead, entered into a symbiotic relationship with the host cell. The bacteria would have continued to perform aerobic respiration, providing excess ATP to the host eukaryotic cell, which would have continued to seek out and acquire energy-rich molecules from its surroundings. Endosymbiotic bacteria, benefiting from this chemical-rich environment, would have begun to reproduce independently within this larger cell. Subsequently, photosynthetic bacteria—such as cyanobacteria—

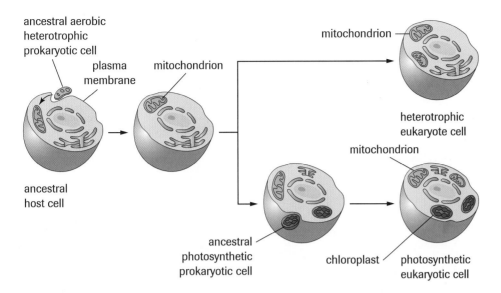

Figure 4
Endosymbiosis. Strong evidence suggests that mitochondria and choloroplasts originated when aerobic and photosynthetic prokaryotic cells began living as symbiotic organisms within ancestral eukaryotic cells.

may have become endosymbiotic in a similar way within aerobic eukaryotic cells. Such a relationship would have benefited the bacteria by providing a richer supply of carbon dioxide for photosynthesis, and the eukaryotic cells by providing excess glucose or other energy-rich products of photosynthesis.

The theory of endosymbiosis is supported by examinations of the organelles themselves. Mitochondria and chloroplasts have features that are different from those of other organelles. They are typically surrounded by two membranes. Although the outer membrane is similar to all other eukaryotic cellular membranes, the chemistry of the internal membrane resembles that of eubacteria plasma membranes. These organelles also have their own DNA, which appears to be remnants of circular eubacterial chromosomes, and contains genetic coding sequences for various proteins and RNA which resemble bacterial genes more than eukaryotic genes. Mitochondria and chloroplasts replicate their own DNA and undergo division independently of their host cell's division. They have, however, lost many vital genes and are no longer able to live independently of the host cell.

The evolution of both aerobic heterotrophic and aerobic photosynthetic eukaryotic cells likely occurred through endosymbiosis. Heterotrophic eukaryotic cells could have evolved into various protists and, later, into fungi and animals, while photosynthetic eukaryotic cells could have been the ancestors of photosynthetic protists and, eventually, plants. It is probable that chloroplasts originated by endosymbiosis in more than one lineage of eukaryotic organisms. One way to represent these hypothetical evolutionary steps is shown in **Figure 5** (page 594).

Endosymbiosis has been discovered to occur in many modern organisms. Some ciliates and marine slugs are known to ingest algae and store their chloroplasts, which continue to perform photosynthesis for a few weeks. Coral organisms house living photosynthetic protists within their tissues, and many insects are now known to host prokaryotic cells within their cells. One protozoan, *Pelomyxa*, relies on three different endosymbiotic bacteria species for respiration. Researchers have even documented the engulfing of one eukaryotic cell by another; for example, the cryptomonad *Guillardia theta*, a eukaryotic alga, contains chloroplasts that are surrounded by four membranes rather than the usual two. Between the outer and inner pairs are remnants of the first host cell, including a small but functioning nucleus complete with eukaryotic DNA. In this case, photosynthetic eubacteria became endosymbiotic within eukaryotic cells, which later also became endosymbiotic (**Figure 6**, page 594).

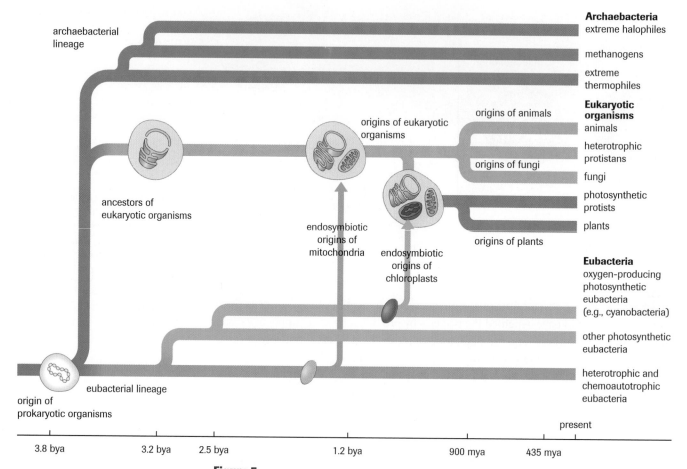

Figure 5

An evolutionary tree depicting a widely accepted hypothesis for major lineages of life forms. Note that the scale is nonlinear.

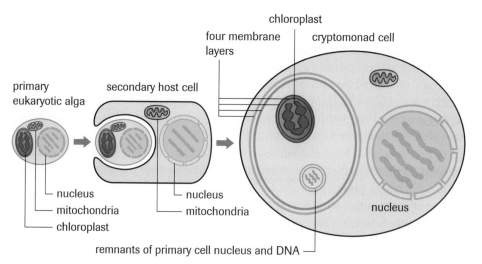

Figure 6

Present-day cryptomonads show evidence of secondary endosymbiosis. The innermost membrane of the chloroplast is chemically similar to that of prokaryotic cells. The outer three layers all resemble eukaryotic membranes. A small nucleus containing 531 genes is sandwiched between the two inner membranes. Analysis of the genes suggests the nucleus originated in the previous primary host cell.

Multicellular Organisms and the Cambrian Explosion

For the first 3 billion years of life on Earth, all organisms were unicellular. Eubacteria gave rise to aerobic and photosynthetic lineages, while archaebacteria evolved into three main groups: **methanogens**, **extreme halophiles**, and **extreme thermophiles**. Once eukaryotic organisms evolved complex structures and processes, including mitosis and sexual reproduction, they would have had the benefit of much more extensive genetic recombination than would have been possible among prokaryotic cells. Photosynthesis continued to increase the oxygen concentration in the atmosphere to the benefit of aerobic organisms. Multicellular organisms, including plants, fungi, and animals, are thought to have evolved less than 750 million years ago.

The oldest fossils of multicellular animals date from about 640 million years ago. However, during a 40-million-year period beginning about 565 million years ago, a massive increase in animal diversity occurred, referred to as the **Cambrian explosion**. Fossil evidence dating from this period shows the appearance of early arthropods, such as trilobites, as well as echinoderms and molluscs; primitive chordates—which were precursors to the vertebrates—also appeared. Animals representing all present-day major phyla, as well as many that are now extinct, first appeared during this period, a time span that represents less than 1% of Earth's history.

methanogens species of archaebacteria that produce methane as a waste product

extreme halophiles species of archaebacteria that live in such extremely saline environments as landlocked saline lakes

extreme thermophiles species of archaebacteria that live in very hot aquatic environments, such as hot springs and hydrothermal vents on the ocean floor

Cambrian explosion a period, beginning about 565 million years ago and lasting about 40 million years, during which the animal kingdom underwent rapid speciation and diversification; the origin of almost all major groups of animals can be traced to this period

DID YOU *KNOW* ?

B.C.'s Burgess Shales
Fossil beds in the Burgess Shale in British Columbia likely formed through a series of mudslides, which buried living organisms in moving sediment. The resulting fossils of early Cambrian animals are superbly preserved (**Figure 7**). Many of the animals had unusual body forms not found in present-day animals. Discovered in 1909 by Charles D. Walcott of the Smithsonian Institute, it was immediately declared the most important geological find in the world. The Burgess Shale is now a World Heritage Site.

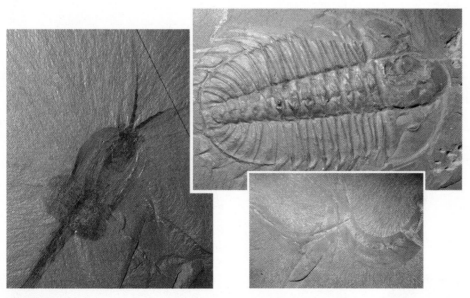

Figure 7
Creatures from Burgess Shale in Yoho National Park, B.C.

Diversification and Mass Extinctions

Figure 8 (page 596) provides a geological time scale and summarizes some of the most significant events in the evolutionary history of Earth since the Cambrian explosion. Geologists have established a geological time scale divided into five eras, each of which is further subdivided into periods and, in some cases, epochs. These time intervals are based on their distinctive fossil records, and dramatic changes in the fossil records mark the boundaries between these intervals. The eras of the Paleozoic (ancient life), Mesozoic (middle life), and Cenozoic (recent life) are remarkable for rapid diversification of life forms, as well as widespread extinctions. The Paleozoic era, for instance, begins with the Cambrian explosion and ends with the Permian extinction, believed to be the most massive extinction in Earth's history.

Moving Continents	Era	Period	Epoch	Age millions of years ago (mya)
10 mya	Cenozoic	Quaternary	Recent	0.01
			Pleistocene	1.8
		Tertiary	Pliocene	5
			Miocene	24
			Oligocene	37
			Eocene	58
			Paleocene	65
65 mya	Mesozoic	Cretaceous	Late	100
			Early	144
		Jurassic		
		Triassic		208
240 mya	Paleozoic	Permian		245
				286
		Carboniferous		
				360
370 mya		Devonian		408
		Silurian		438
		Ordovician		505
420 mya		Cambrian		570
	Proterozoic	Oxygen(O$_2$) abundant		2000
				2500
540 mya	Archean	Oldest fossils known		3500
		Oldest dated rocks		3800
		Approximate origin of the earth		4600

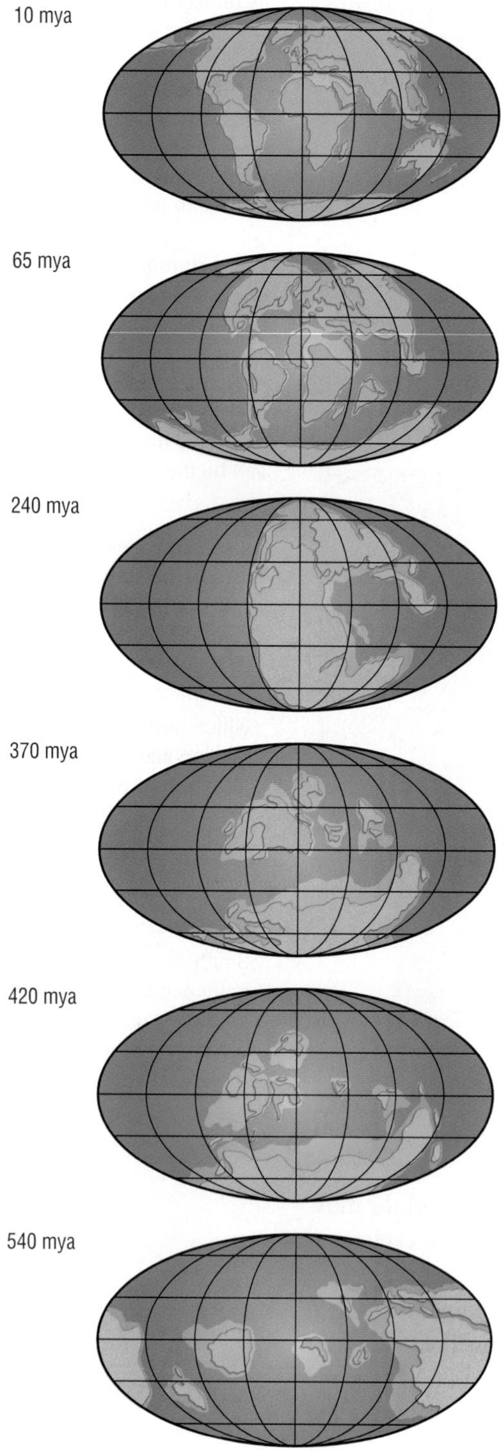

Figure 8
Major events of life on Earth. Mass extinction events have occurred along with a general trend of increasing biological diversity.

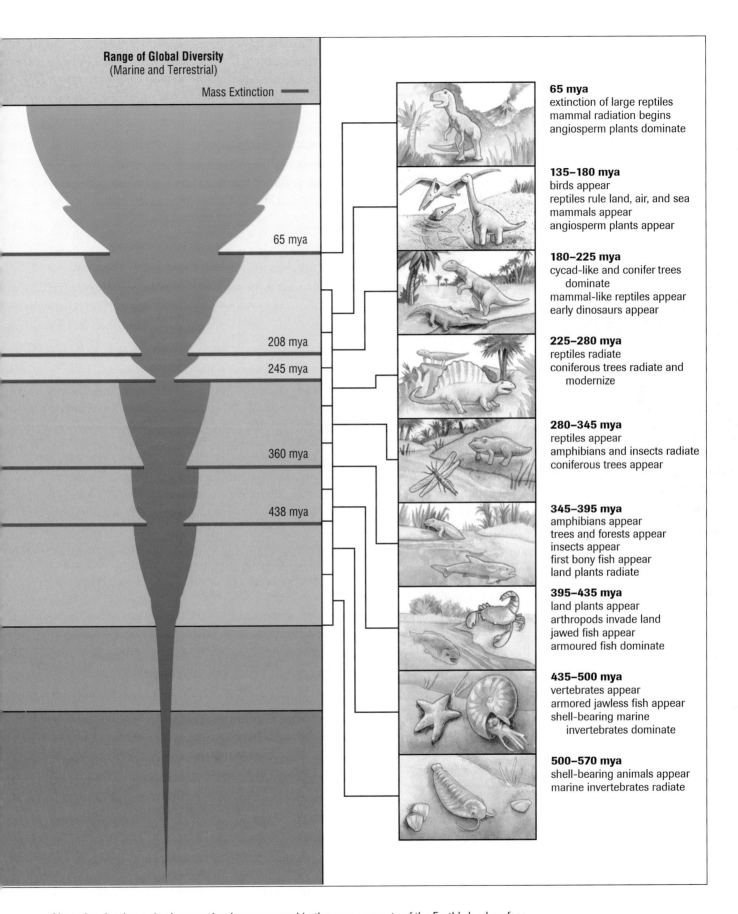

Range of Global Diversity
(Marine and Terrestrial)

Mass Extinction ——

65 mya

208 mya

245 mya

360 mya

438 mya

65 mya
extinction of large reptiles
mammal radiation begins
angiosperm plants dominate

135–180 mya
birds appear
reptiles rule land, air, and sea
mammals appear
angiosperm plants appear

180–225 mya
cycad-like and conifer trees
    dominate
mammal-like reptiles appear
early dinosaurs appear

225–280 mya
reptiles radiate
coniferous trees radiate and
    modernize

280–345 mya
reptiles appear
amphibians and insects radiate
coniferous trees appear

345–395 mya
amphibians appear
trees and forests appear
insects appear
first bony fish appear
land plants radiate

395–435 mya
land plants appear
arthropods invade land
jawed fish appear
armoured fish dominate

435–500 mya
vertebrates appear
armored jawless fish appear
shell-bearing marine
    invertebrates dominate

500–570 mya
shell-bearing animals appear
marine invertebrates radiate

Note also the dramatic changes that have occurred in the arrangements of the Earth's land surface.

Fossil evidence of diversification of marine invertebrates early in the Paleozoic era is very extensive. The first vertebrates are thought to have evolved later, followed by bony fish and amphibians. By the mid-Paleozoic era, plants had invaded land surfaces and the first reptiles and insects had evolved. Around 245 million years ago, a series of cataclysmic events eradicated more than 90% of known marine species, as indicated by their disappearance from the fossil record after this period. Although uncertainty remains about causes of this Permian extinction, many scientists suspect that tectonic movements were a primary cause. The formation of the supercontinent Pangea, which occurred during the Permian period, would have produced major changes in terrestrial and coastal environments as well as in global climate (**Figure 8**, page 596). Ongoing research by Kunio Kaiho of Tohoku University, Japan, and his colleagues has uncovered evidence in southern China of a 60-km-wide asteroid that may have collided with Earth in this period. These researchers believe that the impact may have vaporized enough sulfur to consume a third of the atmospheric oxygen and generate enough acid rain to make the ocean surface water as acidic as lemon juice. If such a catastrophic impact did occur, it would have been the primary cause of the biggest extinction event in history.

Despite the harsh conditions responsible for mass extinctions, life on earth continued. The Mesozoic era is well known for dinosaurs, a diverse group of often very large animals that dominated earth from about the mid Triassic to the late Cretaceous period. Oceans were home to many bony fish, hard-shelled molluscs, and crabs. On land, at first dominated by gymnosperms, early mammals evolved alongside dinosaurs and insects. Placental mammals, birds, and flowering plants also evolved within the Mesozoic era. After this time, the remaining dinosaurs and many other species suddenly disappear from the fossil record. Considerable evidence supports the hypothesis that an asteroid impact caused this best-known mass extinction. The Chicxulub Crater, almost 10 km deep and 200 km in diameter at the edge of the Yucatan peninsula, is thought to be the impact zone for such an asteroid (**Figure 9**). Some theorize that the asteroid would have been moving at about 160 000 km/h and would have blasted 200 000 km^3 of vaporized debris and dust into Earth's atmosphere. The debris and energy released in the resulting fireball—equivalent to 100 million nuclear bombs—would have killed most of the plants and animals in the continental Americas within minutes. Tidal waves 120 m high would have devastated coastlines around the world and atmospheric debris would have blocked out much of the sunlight for months. Among the strong evidence for the impact hypothesis is the presence of unusually high concentrations of iridium in sedimentary rock dated at 65 million years old, the boundary between the Mesozoic and Cenozoic eras. Rock samples from 95 locations worldwide show these same elevated levels. Iridium, a rare metal in the Earth's crust, is abundant in meteorite samples. These findings suggest that a large asteroid may have been the source of a great quantity of iridium-bearing dust, deposited on a global scale.

Although the mass extinctions that ended the Permian and the Mesozoic eras are dramatic in scope, it is important to keep in mind that most species extinctions result from ongoing evolutionary forces of competition and environmental change. Amazingly, even the five major mass extinction events since the Cambrian explosion account for about only 4% of all extinctions that took place during this time. Scientists have also noted that periods of widespread extinction are followed by periods of very rapid diversification. In the present Cenozoic era, life forms have attained the greatest diversity in Earth's history. Flowering plants have out-competed gymnosperms in many habitats and now number more than 250 000 species. Millions of species of insects now dominate the animal kingdom. Are natural extinctions as much a part of evolution as diversification? It is probable that, had the dinosaurs not become extinct, the ancestors of humans may not have met with later successes—which means that humans might have never existed.

Figure 9
A computer-generated image of the 200-km-wide Chicxulub Crater located in the ocean floor at the edge of the Yucatan peninsula. It is thought to be the impact site of an asteroid that caused a mass extinction 65 million years ago, ending the Cretaceous period and the domination of the dinosaurs.

Take a Stand: The Human Meteorite?

Decision-Making Skills

● Define the Issue   ● Analyze the Issue   ● Research
● Defend the Position   ○ Identify Alternatives   ○ Evaluate

Evaluate

Many people are concerned about species now at risk for extinction. Since the 17th century, scientists have documented the extinction of more than 1000 plants and animals. The current list of endangered species worldwide is greater than 25 000. The dodo, great auk, passenger pigeon, Stellar's sea cow, and Banff long-nose dace are examples of species that have become extinct in recent time as a result of human activity. The causes of such human-driven extinctions are numerous, including habitat destruction, introduction of exotic species, overhunting, and commercial harvesting.

Statement

Because extinction is a natural process of evolution, and because the extinction of one species can benefit others, people should not be concerned about the loss of species even as a result of human activity.

- In your group, define the issue within the scope of human-caused extinctions.

- Research the issue, searching for information in print and electronic resources.

 GO www.science.nelson.com

- You might consider how the current pace of human-caused extinction compares with extinction rates in nature. Find out what tropical and conservation biologists, such as E.O. Wilson, think about this issue. What impact might the preservation of the genomes of endangered species have? What are possible effects of extinctions on ecosystems and on human health and welfare? What species are likely to benefit? How long might it take new species to fill the ecological gap left by species that become extinct?

- Write a position paper outlining your stand and be prepared to present your ideas to the class.

The Rate of Evolution

Biologists are keenly interested in the pace at which evolution may be occurring. Until recently, most supported the idea that changes to species were slow and steadily paced over time. The **theory of gradualism** contends that when new species first evolve, they appear very similar to the originator species and only gradually become more distinctive, as natural selection and genetic drift act independently on both species. One would expect to find, according to this theory, as a result of slow incremental changes, numerous fossil species representing **transitional forms** (also called intermediate forms). Many very distinct species, however, seem to appear suddenly in the fossil record with little evidence of gradual transitions from one species to another. Their sudden appearance is often followed by little change over very long periods of time. The most accepted explanation for these deviations from a gradualism model was that the fossil record is incomplete, and intermediate forms may not have been preserved.

Niles Eldredge of the American Museum of Natural History and Stephen Jay Gould of Harvard University rejected this explanation and, in 1972, proposed an alternative theory called the **theory of punctuated equilibrium**. It consists of three main assertions:

- Species evolve very rapidly in evolutionary time.
- Speciation usually occurs in small isolated populations and thus intermediate fossils are very rare.
- After the initial burst of evolution, species do not change significantly over long periods of time.

These contrasting theories about the rate of evolution are represented in **Figure 10** (page 600). To some extent, the differences between them are a matter of perspective. To many population biologists, the word *rapid* in relation to species evolution suggests changes that can be measured in a few generations or, perhaps, decades. To paleontologists, *rapid* might represent the appearance of a new species in the fossil record within a thousand generations or 100 000 years. In fact, both theories are needed to understand the fossil record while remaining compatible with many other forms of evidence. Consider, for instance, how both theories apply to the evolution of species before and after a major extinction event.

theory of gradualism a theory that attributes large evolutionary changes in species to the accumulation of many small and ongoing changes and processes

transitional forms a fossil or species intermediate in form between two other species in a direct line of descent

theory of punctuated equilibrium a theory that attributes large evolutionary changes to relatively rapid spurts of change followed by long periods of little or no change

- Before the event, an environment might be host to many well-adapted species that have evolved to occupy specific ecological niches. They are largely exposed to the pressures of stabilizing selection and evolutionary change would be very slow.

- An environmental crisis results in the extinction of most species, leaving many niches empty.

- Surviving species have many new opportunities and experience strong disruptive selection. These survivors can evolve rapidly into many new species, filling these empty niches.

- Once the new species become well adapted to their new niches in a relatively stable environment, they again experience stabilizing selection pressures. Thereafter, they show little, or more gradual, change until another crisis opens opportunities for diversification.

It is now widely accepted that both gradual and rapid evolutionary processes are at work. Although many species have evolved rapidly at times, the fossil record of some organisms show very gradual change over extended periods of time.

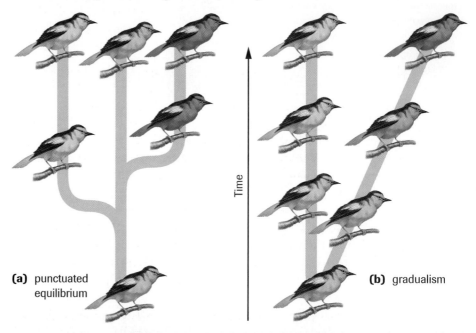

(a) punctuated equilibrium

(b) gradualism

Time

Figure 10
Two theories of the rate of evolution

▶ **TRY THIS** activity

Gradual? Or Rapid and Punctuated?

- Analyze and outline the evidence in sections 13.1 and 13.2 that supports and refutes the roles that both gradualism and punctuated equilibrium may have played in each of these instances:
  (i) the earliest evolution of single-celled organisms (from 3.5 to 1.0 billion years ago)
  (ii) the Cambrian explosion
  (iii) the evolution of photosynthetic eukaryotic cells
  (iv) the general evolutionary pattern from 350 to 250 million years ago (**Figure 8**, page 596)
  (v) the general evolutionary pattern from 55 to 38 million years ago (**Figure 8**)
  (vi) the early evolution of terrestrial plants and animals
  (vii) changes in species diversity before, during, and after a period of glaciation

- Be prepared to defend in a class discussion your conclusions about the roles gradualism and punctuated equilibrium have played in the evolution of life on earth.

Earliest Evolutionary Patterns of Life

- Evidence suggests that, under the conditions present on Earth more than 4 billion years ago, organic molecules formed spontaneously, some able to form cell-like structures, while others had enzymatic properties.

- Endosymbiosis is likely responsible for the evolution of both aerobic and photosynthetic eukaryotic cells.

- All life was unicellular for at least 2 billion years until about 600 million years ago, when life diversified dramatically.

- The history of multicellular life is a record of the ongoing evolution of millions of new species and the extinction of many others.

- Both gradualism and punctuated equilibrium account for the patterns seen in the evolution of life on Earth.

▶ *Section 13.2 Questions*

Understanding Concepts

1. Draw to scale a time line to show key evolutionary processes and events as presented in this chapter, beginning with biochemical reactions 4 billion years ago and ending with the events of 65 million years ago.

2. Below are listed two time periods during which significant evolutionary events occurred. Describe at least two different kinds of evidence scientists used to develop hypotheses about these evolutionary events:
   (a) between 3 and 1 billion years ago
   (b) between 650 and 50 million years ago

Applying Inquiry Skills

3. The Barringer Meteorite Crater, also called the Meteor Crater, in Arizona (**Figure 11**) is thought to have been created about 25 000 years ago by an asteroid about 45 m in diameter. Brainstorm some local and global evolutionary impacts that might have resulted from this asteroid's collision with Earth. Note scientific evidence that might support your reasoning.

4. The ability of ribozymes to recognize and cut specific RNA molecules makes them exciting candidates for human therapy. For example, one target for ribozymes might be the mRNA that encodes vascular endothelial growth factor (VEGF). VEGF stimulates the production of blood vessels necessary for the rapid growth of cancer tumours. A ribozyme that destroys this mRNA might prove valuable in the treatment of many cancers. Find out what other exciting research is underway regarding potential applications of ribozymes. Report your findings to the class.

 www.science.nelson.com

5. Draw a fully labelled set of diagrams to illustrate and describe the evolution of an aerobic eukaryote by the process of endosymbiosis. Draw and clearly label all membrane and chromosomes.

Figure 11

Large-scale evolutionary patterns help to outline the probable evolutionary history of life on Earth. Patterns that occur on a more local scale can demonstrate how processes of change among species may have contributed to and been influenced by these large-scale events.

Divergent and Convergent Pathways

Once a new species forms, its evolutionary pathway may diverge from that of the original species. Disruptive selection may continue long after speciation has occurred, resulting in a pattern of **divergent evolution**. Species with significantly different morphological and behavioural traits may arise, as shown in the various modifications of vertebrate limbs and the activities that would accompany them (Chapter 11, **Figure 2** in section 11.5). Natural selection can also operate to produce striking similarities among distantly related species (**Figure 1**). An excellent example of this pattern of **convergent evolution** can be seen among mammals. For about 50 million years, marsupial mammals in Australia have evolved in isolation from placental mammals throughout the rest of the world, yet natural selection has favoured the evolution of species with similar traits among mammals of both groups (**Figure 2**). Convergent evolution is not restricted to organisms that are geographically isolated. Sharks and dolphins, for example, share wide overlapping geographic distributions and have both evolved very similar streamlined bodies well suited for their high-speed carnivorous behaviour. Traits that are similar in appearance but that have different evolutionary origins are referred to as **homoplasies** (also known as analogous features).

divergent evolution occurs when two or more species evolve increasingly different traits, resulting from differing selective pressures or genetic drift

convergent evolution occurs when two or more species become increasingly similar in phenotype in response to similar selective pressures

homoplasies similar traits found in two or more different species, resulting from convergent evolution or from reversals, not from common descent; also called analogous features

(a)

(b)

Figure 1
Although the shark **(a)** and the dolphin **(b)** are similar morphologically, their genetic history differs. The shark is a fish while the dolphin is a mammal.

adaptive radiation process in which divergent evolution occurs in rapid succession, or simultaneously, among a number of groups to produce three or more species or higher taxa

Sometimes divergent evolution occurs in rapid succession, or simultaneously, among a number of populations. This process, known as **adaptive radiation**, results in one species giving rise to three or more species. The best-documented examples are found on remote archipelagos where the first organisms to arrive have a choice of resources and few or no competitors. Consider, for example, the evolutionary path thought to have been followed by the ground finches that migrated from South America to the Galapagos Islands millions of years ago. While living in South America, these finches would have become ideally adapted for eating medium-size seeds. Finches born with an unusually small bill might have tried to feed on smaller seeds but would have faced fierce competition from other birds that fed on small seeds. Similarly, larger-billed ground finches would have had to compete with large-seed feeders. For the few finches that first reached the Galapagos Islands, the competition would have been eliminated. Assuming that the

Niche	Placental mammals	Australian marsupials

Figure 2
The marsupials of Australia and the placental mammals in other parts of the world have undergone convergent evolution resulting in species that appear similar occupying similar ecological niches.

Galapagos at the time bore various plants bearing different-size seeds, the founding finch population would have been very successful. Individuals born with different-size bills would have been able to find a rich supply of food. But seed size would not have been the only environmental variable directing the evolution of the finches. Although the founding population was composed of ground feeders, the Galapagos provided empty niches for birds to feed from trees, on insects, from cactus, and from other specialized sources. In fact, these niches were eventually filled by 13 species descended from the founding ground-finch population.

The Hawaiian islands are the location of many excellent examples of adaptive radiation. For example, numerous species of honeycreepers, a group of birds with a wide array of bill shapes and sizes, are also thought to have evolved from a single ancestral species (**Figure 3**). Also, 30 species of silverswords, a group of herbaceous plants, are thought to have evolved from a single North American ancestor. The most dramatic example, however, is that of the 800 species of fruit flies of the genera *Drosophila* and *Scaptomyza*. They are so similar that scientists believe they may all have evolved from a common ancestor. Fruit flies are likely to continue to evolve rapidly within the Hawaiian archipelago. As each new volcanic island forms, it can be invaded by organisms from older adjacent islands. The new founding populations will evolve to form additional new species.

The term *adaptive radiation* is also applied to the evolution of entire groups of species. For example, the mammals, which all share a single common ancestor, have undergone adaptive radiation, filling many different feeding niches represented by such major taxa as rodents, carnivores, whales, bats, primates, and ungulates.

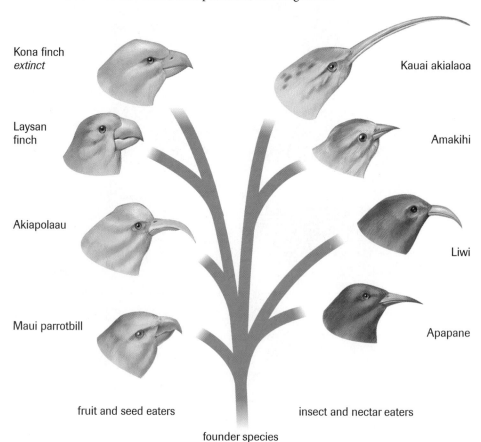

Figure 3
The Hawaiian honeycreepers provide an excellent example of adaptive radiation. Selective pressures on a single founding species have produced many species with bills of different shapes and sizes.

Coevolution

When two species are completely dependent on one another for survival, their evolutionary pathways become linked. This fascinating pattern, called **coevolution**, can be found, for example, in figs that are dependent on a specific wasp for pollination. Without the wasp, the fig cannot reproduce; in turn, the fig wasp can only reproduce within specially modified sterile fig flowers. Highly specialized flower structures have coevolved in the fig alongside unusual behaviour among the wasp pollinators (**Figure 4**). In a similar way, leaf-cutter ants have coevolved with a fungus that is their sole food source. The ants harvest leaves, which they do not eat but bring to underground chambers, where the leaves nourish the growth of the fungi. The ants cannot survive in the absence of their symbiotic fungi; in turn, the fungi are found nowhere else on Earth and, therefore, have become dependent on cultivation by the ants for their survival. Coevolution is widespread among flowering plants and their pollinators, and among parasites and their hosts.

coevolution process, sometimes referred to as reciprocal adaptation, in which one species evolves in response to the evolution of another species

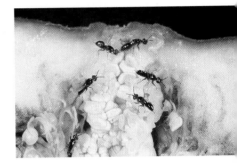

Figure 4
Each of the roughly 900 species of fig is pollinated by its own unique fig wasp species.

SUMMARY *Local Patterns of Evolution*

- Depending on the selective forces at work, the evolution of two different species can result in either a divergence or a convergence of traits.

- Adaptive radiation, which occurs when the environment favours a variety of divergent traits, is prevalent in diverse environments when competitive pressures from other species are reduced or absent.

- Species coevolve when they are dependent on one another for survival.

▶ *Section 13.3* Questions

Understanding Concepts

1. Ongoing disruptive selection results in what evolutionary pattern(s)?

2. Compare and contrast divergent and convergent evolution. Provide an example to support each point you make.

3. Provide an explanation for the vulnerability of populations indigenous to a remote island.

4. Leaf-cutter ants obtain food through the leaves they carry to their subterrestrial habitat, but do not eat the leaves. How does the pattern of coevolution offer an explanation?

Applying Inquiry Skills

5. Many species of fish and aquatic birds exhibit counter shading—their upper dorsal surfaces are dark, while their lower ventral surfaces are much lighter.
   (a) What pattern of selective pressure is most likely at work?
   (b) What environmental factors might be causing this selection?

6. Before the arrival of humans, almost every remote island on the planet had one or more species of flightless bird. How does evolution account for this?

7. Evolution predicts that each flightless bird species evolved on a single island. Therefore, no two islands should be home to the same species of flightless bird. How could biologists test this prediction? Research this topic to find out whether this prediction has been tested.

8. In 1862, after examining the star orchid of Madagascar Island, Darwin predicted that a pollinating hawkmoth with a tongue of just less than 30 cm in length would be discovered. Forty years later, a pollinating hawkmoth with a 25-cm-long tongue was found. What specific understanding of evolution would Darwin have used to make his prediction?

Making Connections

9. Brazil nuts are harvested for human consumption and for the production of Brazil nut oil, a very valuable oil that is often used in soaps and shampoos. The trees that produce Brazil nuts are indigenous to tropical rain forests of South America. However, when these trees were planted in huge monoculture plantations, they failed to become pollinated or produce many nuts. How might evolutionary history account for this observation?

10. The use of an insecticide or a fungicide could potentially harm an entire ecosystem, rather than just target specific organisms. Discuss this statement, using the leaf-cutter ants as a model.

11. Many populations of indigenous peoples, following their initial contact with nonindigenous peoples, have suffered devastating losses as a result of previously unknown diseases. How might evolutionary biology account for their low resistance to these diseases?

For centuries, scientists have used a variety of methods to classify organisms within taxonomic systems. In classical Linnaean taxonomy, organisms are grouped according to their degree of morphological, or structural, similarity. This approach was improved in the 1950s by using a quantitative system and by including many physiological and biochemical features in comparisons, rather than relying exclusively on traditionally accepted traits. In contrast, Darwinian classification systems group organisms based on their **phylogeny**, or evolutionary history. In these systems, closely related organisms are classified together with all organisms that share a common ancestor. These related organisms are placed in the same **monophyletic group**, called a **clade**. Surprisingly, classical taxonomy does not necessarily reflect the degree of relatedness between members of a group. For example, turtles and crocodile, which share many readily observable traits, are both traditionally placed together in the reptiles class, as shown in **Figure 1(a)**, while birds are not. Birds and crocodiles, however, are more closely related to each other—they share a more recent common ancestor—than crocodiles are to turtles. In a phylogenetic classification system, shown in **Figure 1(b)**, birds and crocodiles could be placed in a monophyletic group that does not include turtles. Many taxonomists believe that biological classifications should be based solely on phylogenetic relationships.

phylogeny the theoretical evolutionary history of a species or group

monophyletic group (clade) all the descendants of a single common ancestor

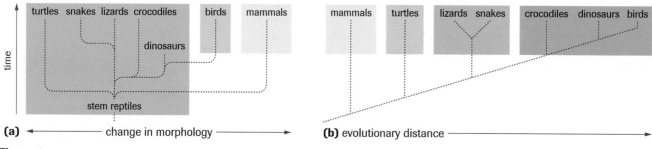

(a) ← change in morphology →   **(b)** evolutionary distance →

Figure 1
Classification systems vary. Although the traditional Linnaean system places turtles and crocodiles within the reptiles class **(a)**, a phylogenetic classification system would place birds and crocodiles together in the same monophyletic group **(b)**.

phylogenetic tree (cladogram) diagram depicting the proposed evolutionary relationships, or common ancestry by descent, of groups of species or populations

cladistics a phylogenetic system of classification used to infer and construct cladograms based on shared derived traits

Phylogeny and Cladistics

From the assumption that all species have evolved from a common ancestor, it follows that all members of a group must have evolved through a series of events in which ancestral species gave rise to new species, each giving rise to a new lineage. The phylogeny can be illustrated in the form of a **phylogenetic tree** (or **cladogram**), on which the branches represent the theoretical sequence of events. The phylogenetic tree in **Figure 2** shows the probable evolutionary relationships of some major groups of vertebrates. Note that A represents the point in time when a species ancestral to the displayed groups split, resulting in two species. One species gave rise to the lineage that includes the ray-finned fish, while the other is an ancestor to all other groups. Similarly, C indicates a time at which an ancestral species split into two main lineages, one giving rise to the mammals. The inference can be made from this phylogenetic tree that birds and crocodiles share a more recent common ancestor with each other at E than crocodiles do with turtles at D. Therefore, crocodiles, lizards, snakes, and turtles could not be grouped into a clade that did not include birds. One can also infer from this diagram that mammals are as closely related to turtles as they are to birds. All three groups share the same most recent ancestor at C.

The reconstruction of the phylogenetic relationships of evolutionary histories is based on a careful evaluation of a wide range of evidence, including the fossil record, comparative anatomy, biochemistry, and genetics. Through a system of classification called **cladistics**, scientists construct phylogenetic trees to group organisms based on an analysis

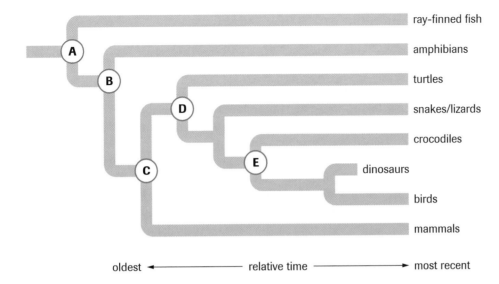

Figure 2
This phylogenetic tree depicts the evolutionary relationships among major groups of vertebrates.

of shared derived (apomorphic) traits, or synapomorphies. **Synapomorphies** are features that are shared only by members of the group and that have been inherited from a single common ancestor in which the features first evolved. Species or groups that share recently derived traits are more closely related than those that do not. For example, most vertebrates have paired appendages: fish have fins, while amphibians, reptiles, birds, and mammals are tetrapods, having four limbs. Fins are considered to be of more ancient origin, while limbs are thought to be a more recently derived condition, as confirmed by such evidence as comparative anatomy and fossil records. Therefore, fish are not thought to have evolved from four-limbed terrestrial vertebrates; terrestrial vertebrates are thought to have evolved from organisms with fins. In **Figure 2**, the ancestor to all vertebrates, at position A, had fins, while the ancestor of all nonfish vertebrates, at position B, had evolved four limbs.

Ancestral features cannot be used to determine the relationship of species or groups; for example, the vertebrate tail is considered the ancient condition, while the lack of a tail is considered to be a derived condition. The fact that monkeys and salamanders both possess tails does not indicate that monkeys are more closely related to salamanders than they are to chimpanzees, which do not have tails.

Cladograms can provide information about the relative sequence in which species split. In **Figure 2**, the cladogram shows that all the named groups are alive today with the exception of the dinosaurs, whose branch ends before reaching most recent times. One can infer that ancestral species A is more ancient than B, which in turn is more ancient than C, and so on. Although a branch to the mammal lineage occurs at position C, this does not imply that mammals existed at that time, only that a common ancestor to the mammals had branched from the previous lineage.

A cladogram can be a valuable tool for posing and testing hypotheses concerning evolutionary events. For example, birds and mammals are endothermic and have relatively large brains, yet, as shown in **Figure 2**, mammals are no more closely related to birds than they are to turtles or snakes. Consider the hypothesis that species B was an ancient large-brained endotherm. If this were the case, these traits must have been lost independently in the organisms that gave rise to turtles, lizards and snakes, crocodiles, and most dinosaurs. Even without examining the fossil evidence, it would seem highly unlikely that having a large brain and being endothermic are ancient conditions. A more likely hypothesis is that having relatively large brains (in relation to other non-mammalian vertebrates) and being endotherms are analogous features, or homoplasies, having evolved independently and relatively recently in both mammals and birds.

synapomorpies shared traits that evolved only once and have been inherited by two or more species

Applying Cladistics

ingroup members of a clade having one or more synapomorphies

outgroup the first group to have diverged from the other members of a clade being considered in a phylogenetic analysis

Given appropriate data for a number of species or groups, it is possible to construct and interpret a cladogram based on an analysis of derived characters (i.e., cladistic traits). For a cladistic assessment, the species or groups chosen for study are called the **ingroup** and a similar but more distantly related group or species is called the **outgroup**. Each member of the ingroup is compared to the outgroup. Because the outgroup is distantly related, any trait shared by two or more members of the ingroup that is not found in the outgroup is assumed to be a shared derived trait.

For a cladistic study in the evolution of terrestrial plants, for example, eight plants were selected for the ingroup: orchid, pine, fern, moss, wheat, maple, liverwort, and spruce; while kelp, a marine plant, was chosen as the outgroup. **Table 1** documents the prevalence of seven derived characters in the ingroup.

Table 1 Incidence of Selected Derived Traits among Terrestrial Plants

Plants	Derived traits						
	vascular tissue	flowers	seeds	single cotyledon	multi-cellular embryo	needle leaves	stomata
Ingroup							
orchid	+	+	+	+	+	−	+
pine	+	−	+	−	+	+	+
fern	+	−	−	−	+	−	+
moss	−	−	−	−	+	−	+
wheat	+	+	+	+	+	−	+
maple	+	+	+	−	+	−	+
liverwort	−	−	−	−	+	−	−
spruce	+	−	+	−	+	+	+
Outgroup							
kelp	−	−	−	−	−	−	−

− indicates absence of the trait   + indicates presence of the trait

To construct a cladogram for this study, a large V is drawn, in which the base, or node represents the most recent ancestor common to all plants. The outgroup, kelp, with no derived traits, is placed at the end of the left branch of the V, as illustrated in **Figure 3(a)**. From **Table 1**, it is apparent that all plants in the ingroup share a multicellular embryo as a derived trait and that all but one, the liverwort, have stomata. Therefore, we can conclude that the liverwort lineage split from the remaining members of the ingroup at B, after the evolution of a multicellular embryo and before the evolution of stomata, as illustrated in **Figure 3(b)**. Note that the evolution of derived characters occurs between nodes, while speciation events, which give rise to new lineages, are represented by nodes. In **Figure 3(c)**, additional nodes and branches are added to the cladogram for the moss and the fern to correspond with the data in **Table 1**. After the rise of the fern lineage, the remaining five plants form two groups, those with needle leaves and those with flowers. Therefore, the next branch must split off one of these groups and then undergo further splitting. The completed cladogram is shown in **Figure 3(d)**. Note that without additional information, it cannot be determined whether the pine and the spruce share a more recent ancestor than the wheat, orchid, and maple or the wheat and the orchid.

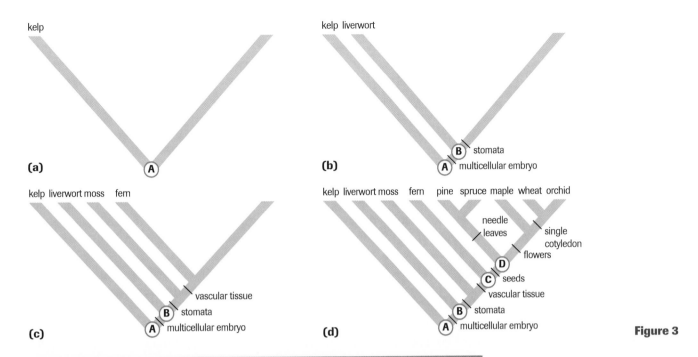

Figure 3

▶ *Practice*

1. In **Figure 3(d)** on page 608, what is happening at C and D?

2. Refer to the nodes on **Figure 3(d)** to describe the possible position at which seed-bearing plants evolved.

3. Assume that this cladogram represents the true evolutionary history of these groups. Is it possible for maple to be more closely related to spruce than to pine? Give reasons for your response.

4. Why is it acceptable to place wheat and orchid on separate branches when they share all the derived characters listed in **Table 1** (page 608)?

5. Are pine and fern more closely related than spruce and orchid? Explain your answer.

6. To what other plant(s) is the maple most closely related? How do you know?

7. Copy the cladogram in **Figure 3(d)** into your notebook, but replace the single fern by two very closely related species of fern.

▶ **TRY THIS** activity   *Construct a Cladogram*

Table 2   Selected Derived Traits for Ingroup and Outgroup

Group	Derived Traits				
	hair	**lungs**	**bony shell**	**grasping hands**	**jaws**
lamprey	−	−	−	−	−
turtle	−	+	+	−	+
gorilla	+	+	−	+	+
lungfish	−	+	−	−	+
pike	−	−	−	−	+
wolf	+	+	−	−	+
human	+	+	−	+	+

− indicates absence of the trait   + indicates the presence of the trait

- Construct a cladogram from the derived traits in **Table 2**. The lamprey is the outgroup.
  (a) Which two species are most closely related?
  (b) Are pike more closely related to lungfish or to gorillas?
  (c) Which of the derived traits came into existence before the bony shell?
  (d) List all the derived characters in the table that would have been present in the ancestor to gorillas, wolves, and humans.

The Evolutionary History of Life   **609**

Applying Genetic Evidence

Modern genetics and biochemistry provide explanations for the mechanism of inheritance and the formation of new variations; they also offer evidence for reconstructing phylogenetic relationships. Although scientists in Darwin's time could compare species based only on their morphology and behaviour, modern biologists can compare their chemical and genetic makeup. In a study of selected organisms, a protein common to them can be analyzed and the resulting amino acid sequences compared. For example, **Figure 4** shows an alignment of the amino acid sequences of the protein cyctochrome *c*, a protein of the electron transport chain, in six organisms. Two patterns in these data are apparent: these organisms share very similar amino acid sequences, and organisms thought to be more closely related to one another have fewer differences. Cytochrome *c* in humans and in rhesus monkeys differs by only a single amino acid, while the same protein in ducks and turtles has seven differences in its sequence, and in humans and yeast, 56 differences in its sequence. A single base-pair substitution in the DNA of an ancestor would account for the observed difference between humans and rhesus monkeys. It is important to note that portions of the amino acid sequence have been preserved among all species. This suggests that mutations that have occurred in these regions have been harmful. These data support the theory that, over long periods of time, mutations in DNA coding for protein can and do occur, and that natural selection has mediated changes to the genetic makeup of species.

Table 2 Amino Acid Abbreviations

A	Alanine	I	Isoleucine	R	Arginine
C	Cysteine	K	Lysine	S	Serine
D	Aspartic Acid	L	Leucine	T	Threonine
E	Glutamic Acid	M	Methionine	V	Valine
F	Phenylalanine	N	Asparagine	W	Tryptophan
G	Glucine	P	Proline	Y	Tyrosine
H	Histidine	Q	Glutamine		

```
        1         10        20        30        40        50        60        70
        .         .         .         .         .         .         .         .
human   ........GDVEKGKKIFIMKCSQCHTVEKGGKHKTGPNLHGLFGRKTGQAPGYSYTAANKNKGIIWGEDDTLME
rhesus  ........GDVEKGKKIFIMKCSQCHTVEKGGKHKTGPNLHGLFGRKTGQAPGYSYTAANKNKGITWGEDDTLME
duck    ........GDVEKGKKIFVQKCSQCHTVEKGGKHKTGPNLHGLFGRKTGQAEGFSYTDANKNKGITWGEDDTLME
turtle  ........GDVEKGKKIFVQKCAQCHTVEKGGKHKTGPNLHGLIGRKTGQAEGFSYTEANKNKGITWGEEETLME
wheat   ASFSEAPPGNPDAGAKIFKTKCAQCHTVDAGAGHKQGPNLHGLFGRQSGTTAGYSYSAANKNKAVEWEENNTLYD
yeast   ...TEFKAGSAKKGATLFKTRCELCHTVEKGGPHKVGPNLHGLFGRHSGQAQGYSYTDANIKKNVLWDENNNMSE

        80        90        100       110
        .         .         .         .
human   YLENPKKYIPGTKMIFVGIKKKEERADLIAYLKKATNE
rhesus  YLENPKKYIPGTKMIFVGIKKKEERADLIAYLKKATNE
duck    YLENPKKYIPGTKMIFAGIKKKSERADLIAYLKDATAK
turtle  YLENPKKYIPGTKMIFAGIKKKAERADLIAYLKDATSK
wheat   YLLNPKKYIPGTKMVFPGLKKPQDRADLIAYLKKATSS
yeast   YLTNPKKYIPGTKMAFGGLKKEKDRNDLITYLKKACE.
```

Figure 4

The amino acid sequences of the cytochrome *c* protein for selected organisms. Single letters are used to represent each of the amino acids in the sequence. Red letters indicate differences from the human sequence while blue letters indicate amino acid residues common to all species. The amino acid abbreviations are shown in **Table 2** (above).

Two species that share a recent common ancestor will have had less time and opportunity for their genomes to diverge; as a result, the codes will exhibit fewer differences. **Figure 5** illustrates this pattern with a comparison of the amino acid sequences of a hemoglobin polypeptide in humans with those of five other species. Although amino acids provide a sound basis for comparison, modern DNA sequencing technologies permit biologists to compare detailed DNA sequences of organisms. As evolution is the change over time in the genetic makeup of a population, such technologies may offer the ultimate tool for quantifying evolutionary relationships among organisms.

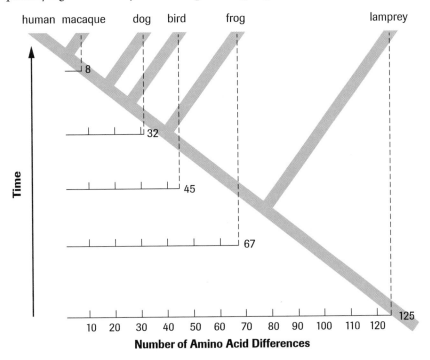

Figure 5
Differences in amino acid sequences in a hemoglobin polypeptide reflect the degree of evolutionary relatedness among different species. More closely related species exhibit greater similarity in their sequences.

Organisms that are most closely related share very similar DNA sequences, while those more distantly related show greater genetic differentiation. Consider the aligned DNA base-pair sequences for selected mammals in **Figure 6** for evidence of the phylogeny of these mammals. Cows and deer share nine synapomorphies; pigs and peccaries share eight; and whales and hippopotamuses share seven—and have two sequence changes in common with cows and deer. The close kinship suggested by the data from these three pairings can be presented in a cladogram (**Figure 7**, page 612).

	143	162 166	182
cow	AGTCCCCAAAGTGAAGGAGA	CTATGGTTCCTAAGCACAAG	GAAATGCCCTTCCCTAAATA
deer	AGTCTCCGAAGTGXAGGAGA	CTATGGTTCCTAAGCACGAA	GAAATGCCCTTCCCTAAATA
whale	AGTCCCCAXAGCTAAGGAGA	CTATCCTTCCTAAGCATAAA	GAAATGCGCTTCCCTAAATC
hippopotamus	AGTCCCCAAAGCAAAGGAGA	CTATCCTTCCTAAGCATAAA	GAAATGCCCTTCTCTAAATC
pig	AGATTCCAAAGCTAAGGAGA	CCATTGTTCCCAAGCGTAAA	GGAATGCCCTTCCCTAAATC
peccary	AGACCCCAAACCTAAGGAGA	CCGTTGTTCACAAGCGTAAA	GGAATGTCCTCCCCTAAATC
outgroup	AGTCCTCCAAACTAAGGAGA	CCATCTTTCCTAAGCTCAAA	GTTATGCCCTCCCTTAAATC

Figure 6
These sequence data of selected mammals includes nucleotides 141 to 200 for the beta-casein gene (encoding a milk protein). Coloured bases indicate derived genetic changes shared by more than one species. A single unresolved base position is marked with an X. Only whales and hippopotamuses share the sequence change at position 166. The outgroup is a distantly related mammal.

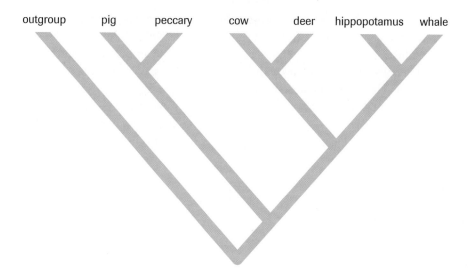

Figure 7
A cladogram based on the DNA sequence data for gene encoding the beta-casein milk protein (**Figure 6**)

Homoplasy can make the analysis of DNA sequence data problematic. Long after speciation, two new species continue to experience mutation and selection, which may cause their sequences to converge. For example, in **Figure 7**, deer and pigs both have a T at position 145, but the remainder of the data suggest that deer and pigs are not very closely related. Therefore, rather than being inherited from a common ancestor, these changes from the original C are thought to have occurred independently after their lineages diverged. Similarly, DNA changes may undergo reversals: pigs and peccaries most likely inherited the G at position 182 from a common ancestor. If, in the future, this G is replaced by an A in peccaries, then this piece of evidence that shows their relationship to pigs will be lost.

Interspersed Elements: Ideal Phylogenetic Traits?

A newly discovered and potentially ideal genetic trait is now being used in the study of phylogeny. These traits are sequences of DNA that occasionally become inserted into a species' genome; they are known as SINEs (**s**hort **in**terspersed **e**lements) and LINEs (**l**ong **in**terspersed **e**lements). Suppose a retrovirus infects an organism and copies of DNA corresponding to the virus' genome are randomly inserted into the host cells' own DNA. If these sequences are inserted into germ cells (precursors of sperm or eggs) in noncoding regions of DNA, they will become a permanent and inheritable part of the junk DNA. As such events are relatively rare, if two species have the same SINE or LINE located at precisely the same position in their DNA, then it can be assumed that the insertion occurred only once in a common ancestor. For a number of reasons, SINEs and LINEs make ideal traits for tracing evolutionary pathways. They are easy to find and identify, even if they undergo small mutational changes, because they are insertions of relatively large and recognizable segments of DNA, often hundreds of base pairs in length. The possibility of homoplasy is extremely remote, as the chances of a SINE or LINE being inserted into exactly the same location in two different species is highly unlikely. Studies of retroviruses have found that the insertion of their DNA elements appears to occur at random. As well, many viruses are host-specific and unlikely to infect other species. Thus, dolphins in the Amazon River are not likely to be infected by a virus that infects white-tailed deer in Ontario. ✎▮

LAB EXERCISE 13.4.1

Looking for SINEs of Evolution (p. 629)
How can SINEs be identified and their presence or absence applied to phylogeny? In this lab exercise, you will use the presence or absence of SINEs and LINEs to construct cladograms and apply the resulting analysis to related issues in evolutionary biology.

Making Phylogenetic Predictions

Molecular and genetic evidence suggests that whales and hippopotamuses are more closely related than either is to deer or cows (**Figure 7**). All four mammals share a more recent common ancestor than they do with pigs. From these findings, a number of testable predictions can be made regarding other existing evidence linking whales to land mammals and, in particular, to hippopotamuses. In fact, this evidence is extensive. Whales are warm-blooded placental mammals that breath air and nurse their young with milk from mammary glands. Interestingly, whales and hippopotamuses are the only mammals that nurse their young underwater (**Figure 8**). Some large whales have vestigial hipbones, suggesting that their ancestors walked on four legs. Fossils of seven prehistoric whale species provide an interesting record of the transition from land to water. Of these, *Pakicetus* (50 million years ago)—the "terrestrial whale"—was a land animal, while *Ambulocetus* (48 million years ago)—the "walking whale"—likely spent much of its time in the water. Both had well-developed limbs and hipbones. *Rodhocetus* (46 million years ago), a truly marine whale, had stubby legs and hips barely attached to the spinal cord, while *Basilosaurus* (40 million years ago) still possessed small rudimentary hind limbs but otherwise closely resembled modern whales (Chapter 11, **Figure 3**, page 531). Scientists of Darwin's time would likely be amazed by the genetic tools now available and being used to unravel the originating events and evolutionary history of life. Scientists are now applying this arsenal of modern tools to tackle the most fascinating and sensitive question in all of evolutionary biology: the evolution of humans.

Highly specific predictions can be made concerning the genomes of other species. If the ancestor that gave rise to cows, deer, whales, and hippopotamuses already had a number of insertions, then these SINEs and LINEs should be found in all close relatives of these four species. For example, a LINE called *aaa228* is known to occur at the identical gene locus in goats and blue whales. Because the possibility of homoplasy is extremely remote, it can be concluded that an ancestor that already had this LINE gave rise to two separate species, one that eventually evolved into whales (and hippopotamuses) and a second that gave rise to the group that includes the goats. Therefore, if the assumption that these mammals share a common ancestor is correct, then all whale species should have the *aaa228* sequence in exactly the same location within their genomes. In addition, all close relatives of the goat—including deer, cows, giraffes, sheep, and antelope families—should also have the *aaa228* sequence in the identical locus in their genomes. To date, many of these predictions have been tested and have always proven correct. Minke whales, Baird's beaked whales, Dall's porpoises, short-finned pilot whales, and bottlenose dolphins all have the predicted LINE in the identical locus in their genomes—as do sheep, reticulated giraffes, axis deer, and the lesser Malayan chevrotain.

Figure 8
A killer whale nursing under water

Cladistics

- Phylogeny, the evolutionary history of life, can be analyzed through cladistics, a method of assessing evolutionary relationships by comparing shared derived traits.

- The sharing of derived traits by a unique set of organisms suggests they belong to a distinct monophyletic group, one having a common ancestor, while those traits shared by organisms outside the group provide no information regarding evolutionary relationships.

- Comparisons of amino acid sequences and DNA sequences can provide detailed phylogenetic relationships by revealing the specific changes in the genetic makeup of species and populations. However, more distantly related species may exhibit a degree of homoplasy that can obscure these data.

- SINEs and LINEs provide excellent inheritable markers for tracing the evolution of species' lineages, allowing for testing of phylogenetic predictions.

▶ Section 13.4 Questions

Understanding Concepts

1. What is the key purpose of classifying organisms within a phylogenetic context?

2. Identify the main factors on which cladistic taxonomy is based, and describe the significance of each.

3. Explain how homoplasy of SINEs and LINEs is extremely rare. How does this factor help in the analysis of the evolutionary history of organisms?

4. Examine **Figure 1** (page 606). Explain which of the classification systems is based on a Linnaean approach and which is based on a cladistic approach.

Applying Inquiry Skills

5. Without doing a lot of research, you can be fairly confident of the following:
   - Wolves and foxes have a relatively recent common ancestor.
   - Wolves are more closely related to jaguars that they are to moose.
   - Moose are more closely related to caribou than they are to wolves or jaguars.

   Using these data, draw an evolutionary tree for these five species.

6. Examine the cyctochrome *c* sequence for only humans and rhesus monkeys (**Figure 4**, page 610). They differ by a single amino acid residue.
   (a) State the position number and the specific amino acid residue for each species.
   (b) For each species, list all possible DNA codons that could specify the amino acid from (a).
   (c) Compare the list of possible human and rhesus monkey codons. What single point mutations could have changed a human codon version into a rhesus monkey codon version, or vice versa?

Making Connections

7. Many plants exhibit very similar morphological features. If tropical rain-forest biologists are trying to measure the biodiversity of a region accurately, what classification system and what technological tools might they use in their research? Explain.

8. In trying to understand human biology better, scientists often use other animal models for their research, such as rats, fruit flies, and even yeast! How would researchers apply their understanding of classification systems to determine the value of using a particular organism for study?

9. Why must museum curators and palaeontologists foster, and rely on, their expertise in classical taxonomy?

Human evolution began about 60 million years ago with the earliest primates: mammals with long snouts, sharp teeth, and large eyes. They probably lived in trees, feeding mostly on insects. Gradually, these ancestral mammals evolved at least three notable traits:

- More-flattened molars that were better suited to a plant diet.

- Grasping hands and feet with opposable first digits that were obvious advantages in an arboreal habitat, enhancing movement and agility and, accompanied by a greater range of shoulder movement, providing the potential to brachiate (i.e., swing hand over hand along a branch or vine).

- Forward-directed eyes that provided binocular vision and depth perception, critical for making accurate decisions when moving quickly among trees.

Primate Phylogeny

During the next 20 million to 30 million years, these mammals evolved into prosimian and anthropoid organisms. The prosimian lineage gave rise to the present-day lemurs, lorises, pottos, and tarsiers, characterized by very large eyes, nocturnal activities, and, often, a long tail for balancing (**Figure 1**). The **anthropoid** lineage has given rise to the present-day monkeys, apes, and humans (**Figure 2**). Almost all anthropoid organisms are diurnal and most feed mainly on fruits and leaves. They have evolved an enlarged brain for processing the information they receive from their enhanced senses, such as colour vision and sensitive touch in their digits. Most live in complex social groups, and have a prolonged period of caring for their young.

anthropoid the higher primates, including all extinct and living monkeys, apes, and humans

Figure 1
A ring-tailed lemur, indigenous to the island of Madagascar

(a)

(b)

Figure 2
(a) Gorillas (*Gorilla gorilla*), **(b)** common chimpanzees (*Pan troglodytes*), pygmy chimpanzees or bonobos (*Pan paniscus*), and orangutans (*Pongo pygmaneus*) are commonly referred to as the great apes.

About 25 to 30 million years ago, the anthropoid lineage split a number of times to give rise to the Old World and New World monkeys and the hominoids (**Figure 3**). Present-day Old World monkeys include both arboreal and ground-dwelling species. They have nostrils that are close together and directed forward and have tails that are not prehensile. Early ancestors of New World monkeys migrated to what is now South America, where they evolved in isolation. Present-day species are readily distinguishable from other primates, as they are exclusively arboreal and have widely separated flared nostrils and prehensile tails. Present-day **hominoids**—that is, gibbons, orangutans, gorillas, two species of chimpanzee, and humans—all lack tails. Although gibbons and orangutans are arboreal, the others are mostly ground dwelling. Hominoid primates have longer front limbs than hind limbs and larger brains relative to body size than do monkeys.

hominoids the apes, humans, and extinct members of their clade

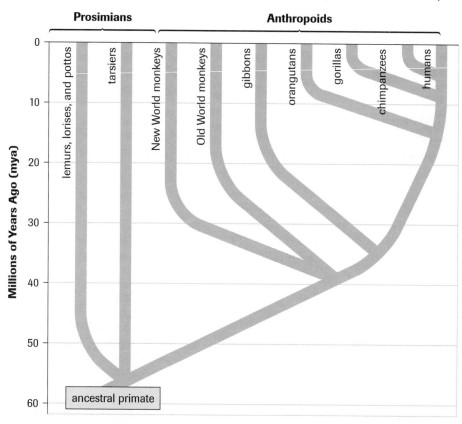

Figure 3
This cladogram depicts the most current thinking on the phylogeny of extant, or living, primates. New World monkeys are indigenous only to the Americas. Of the great apes, only orangutans are not indigenous to Africa.

✍ LAB EXERCISE 13.5.1

Comparing Hominoid Chromosomes (p. 631)
How do the chromosomes of humans compare with those of chimpanzees and other great apes? In this lab exercise, you will examine some genetic data from the 1980 and 1982 published research results of Jorges Yunis and his team, and you will compare the chromosomes.

Humans share a very wide range of physical and behavioural traits with other hominoids and exhibit dramatic genetic similarity to them. The key distinguishing features of humans include bipedal motion (and such related anatomical features as a wide pelvis and curved vertebral column); a greatly enlarged brain; the use of complex language; and the construction and use of complex tools. By comparison, the great apes have a narrow pelvis, smaller brain, and use only simple tools. Although they demonstrate complex social interactions, the great apes do not use a structured language. Until recently, there was fierce scientific debate over the evolutionary relationship of hominoids. A range of evidence now indicates that the earliest ancestor of gibbon species appeared more than 10 million years ago, followed by the evolution of the orangutan lineage. Possible phylogenies of humans and the great apes are presented in **Figure 4**. ✍

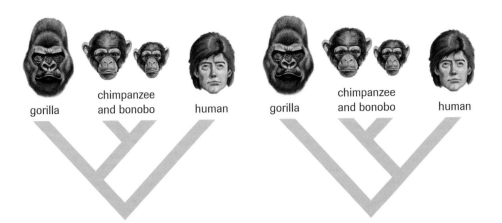

gorilla | chimpanzee and bonobo | human | gorilla | chimpanzee and bonobo | human

Figure 4
These cladograms represent the most common hypotheses of the phylogenetic relationship between the great apes and humans. Extensive molecular data now strongly support the view that humans and chimpanzees are more closely related to each other than either are to gorillas.

Emergence of Humans

In Darwin's day, no fossils had been discovered of early humans, prehumans, or great apes, except for a few fragments of Neanderthal skulls. However, the careful comparison of similarities between living great apes and humans had led Darwin to suggest that "man still bears in his bodily frame the indelible stamp of his lowly origin." Darwin also correctly predicted that fossils of early human ancestors would be found in Africa.

The current fossil record shows that, rather than a step-by-step lineage from ancient ape to human, the **hominid** clade has a rich history of many branching lineages and related species. The hominid clade includes all descendants of the most recent common ancestor of humans and apes. Hundreds of specimens indicate that at least six different species of australopithecines lived in Africa between 4.2 and 1.0 million years ago (**Figure 5**). An even more ancient species, *Ardipithecus ramidus*, recently dated at 5.2 to 5.8 million years, represents one of the oldest known bipeds. The fossil remains of australopithecines, along with the remarkable trace fossils in Laetoli, Tanzania—a set of 69 footprints dated to 3.7 million years ago—show that human ancestors evolved the ability to walk upright long before they had large brains. Very recent evidence from fossils of *A. afarensis* and a later species (*A. africanus*) suggests that both may have been knuckle walkers, a trait previously thought to be unique to chimpanzees and gorillas. The selective advantage gained by bipedal motion is uncertain—possibilities include greater speed and efficiency, enhanced ability to gather and carry food, as well as to see over tall grass. About 3 million years ago, it is thought that an australopithecine ancestor gave rise to the subsequent ancestors of two lineages. Although the precise relationships among them remain unclear and controversial, one branch probably gave rise to a number of robust species with heavy jaws and relatively small brains, while the other ultimately gave rise to the first members of the genus *Homo*.

"Lucy"
The most dramatic early fossil found to date is "Lucy," a skeleton of *Australopithecus afarensis*, dated to 3.2 million years ago. Bipedal and standing about 1 m tall, this species had many apelike features, such as large canine teeth, long arms, finger and toe bones well suited to climbing, and a relatively small brain (about 400 cm^3). "Lucy" and other early human ancestral fossils can be viewed online.

 www.science.nelson.com

hominid humans and other extinct members of their lineage arising from the most recent common ancestor that humans share with the apes

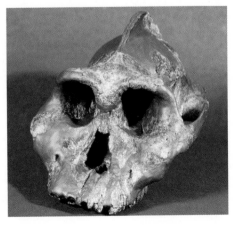

Figure 5
Australopithecine cranial fossils, c. 4.4 million to 1.4 million years ago. The genus name *Australopithecus* is derived from *australo*, meaning southern, and *pithecus*, meaning ape. The name was first applied to fossils found in southern Africa.

As more fossils have been found, palaeontologists have been able to propose a hypothetical phylogenetic tree for humans (**Figure 6**). The oldest fossils of *Homo*—all found in Africa—date from 2.4 to 1.6 million years ago. They are generally classified as *Homo habilis*, although they may represent two different species (i.e., *H. habilis* and *H. rudolfensis*). The cranial capacity of *H. habilis* skulls found to date suggests that they had a brain size of about 600 cm³ to 750 cm³, proportionately smaller jaws and teeth, and longer legs than australopithecine ancestors. Palaeontologists believe that *H. habilis* was the first human ancestor to use stone tools routinely (**Figure 7**).

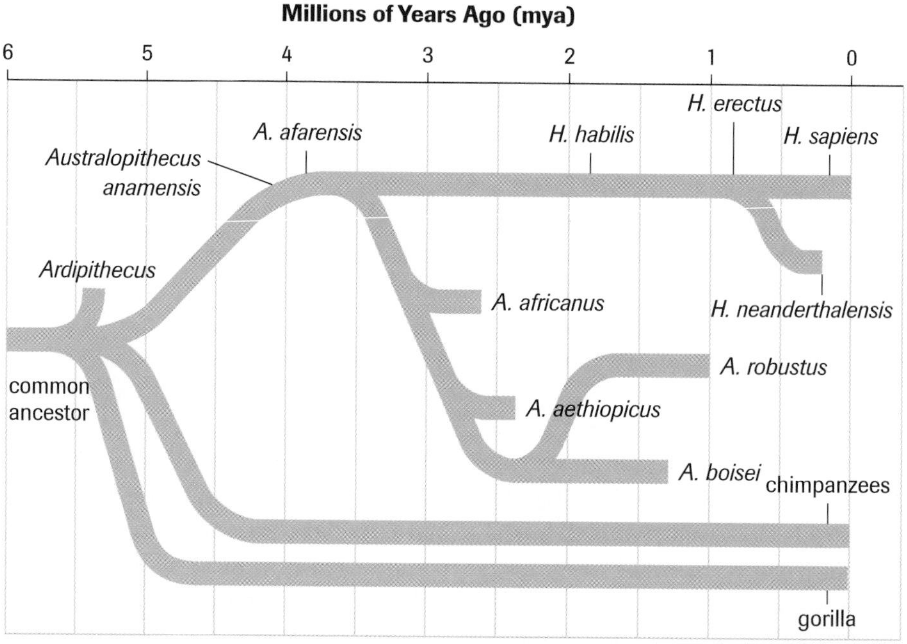

Millions of Years Ago (mya)

Figure 6
This phylogenetic tree represented a widely accepted, though simplified and still hypothetical, version of the hominoid family tree. Many scientists now believe this tree needs revision and that there may have been as many as seven different, closely related *Homo* species. (Extinct and extant species of apes are not included in this diagram.)

Figure 7
Tools used by early human ancestors found at the Olduvai Gorge in Africa:
(a) a cleaver
(b) a crude chopper

(a)　　　　　　**(b)**

Although the number and precise kinship of early *Homo* species is uncertain, it is widely accepted that *H. habilis*, or a closely related species, gave rise about 1.6 million years ago to *H. erectus* (or *H. ergaster*), a species that used a variety of stone tools and fire. This species shows a strong trend toward current human features: brain size averaging about 1000 cm³, a rounded head, and smaller teeth. The fossil record to date suggests that, close to 2 million years ago, *H. erectus* gradually spread out from Africa into Europe and Asia as far as China and Java. Within the last 600 000 years, *H. erectus* may have evolved into two or three species of early humans—*H. heidelbergensis*, *H. neanderthalensis*, and *H. sapiens*—although distinctions among and classification of these groups remain controversial. (The progression in the brain size and skull characteristics of the genus *Homo* can be seen in **Figure 8**.) What is known for certain is that, by 130 000 years ago, *H. sapiens* first appeared in Africa when *H. neanderthalensis* was already living in parts of Europe and eastern Asia. Heavily built individuals with large brains, Neanderthals were skilled at fashioning tools, they performed burial ceremonies, and, as suggested by fossil evidence, they may have been capable of complex speech.

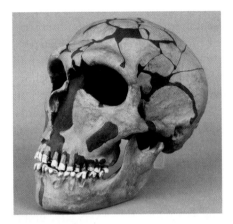

(a) Neanderthal

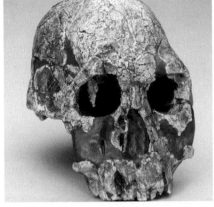

(b) *Homo habilis*

(c) *Homo erectus*

Figure 8
Some fossilized remains of early humans, some as old as 2.4 million years

Competing Theories

Two competing theories attempt to describe the most recent chapter of human evolution. The multiregional hypothesis proposes that anatomically modern humans evolved in parallel in a number of places including Africa, Europe, Asia, and possibly Australia. Much of human regional genetic diversity would have arisen in response to different regional selective pressures over about 1 million years. Continuous gene flow between the populations would account for them not becoming distinct species. The monogenesis (African replacement) hypothesis proposes that *H. sapiens* evolved only in Africa and then migrated to other continents, displacing the Neanderthal and other descendants of the earlier *H. erectus* populations. According to this theory, regional genetic differences in humans evolved within the past 80 000 to 100 000 years, since *H. sapiens* left Africa (**Figure 9**). The most current molecular evidence, including recent findings from the Human Genome Project, strongly favours the monogenesis hypothesis.

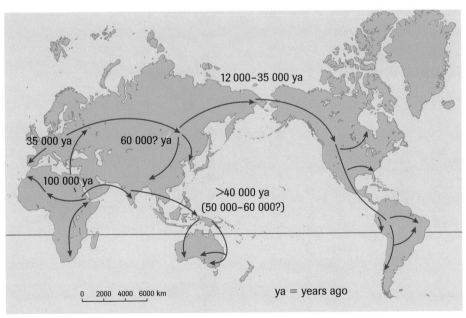

12 000–35 000 ya

35 000 ya   60 000? ya

100 000 ya

>40 000 ya
(50 000–60 000?)

0   2000 4000 6000 km   ya = years ago

Figure 9
The arrows and dates shown for the monogenesis theory are very tentative and likely to be revised as more evidence accumulates.

One would expect less overall genetic diversity among descendants of small founder populations that had spread around the globe relatively recently, compared with that of long-established populations. Indeed, studies of genetic diversity among regional populations indicates that greater genetic diversity exists within sub-Saharan African populations than in other parts of the world. A study of 1600 individuals from 42 populations involved the examination of two highly variable genes on chromosome 12. Of the 24 different versions discovered, 21 were present in African populations, three were found in European populations, and only two were present in Asian and Australian populations. This finding suggests that humans have existed in Africa far longer than in other parts of the world (**Figure 10**). In addition, the recent examination of mitochondrial DNA and studies of the Y chromosome also reveal greater variability among African populations. Other studies suggest that a population of archaic humans in Asia may have contributed to the modern-day human genome. If so, some interbreeding of the African *H. sapiens* with other populations may have taken place.

The genome that humans now possess is the product of a rich evolutionary history. Over a few million years, the forces of natural selection favoured those physical and behavioural traits best suited to a hunter-gatherer way of life. The evolutionary product—*Homo sapiens*—has evolved bipedalism, a powerful and precise grip, a very large brain, and a greatly extended period of dependent childhood. Yet, during the past 40 000 years, most of the successes of *H. sapiens* have resulted from cultural and technological advances rather than biological evolution. Rather than adapting physically to the environment—as other organisms do—humans have used tools to alter the environment and adapt it to their advantage. Does this mean that evolution is no longer important to humans? What aspects of human evolutionary history are relevant today?

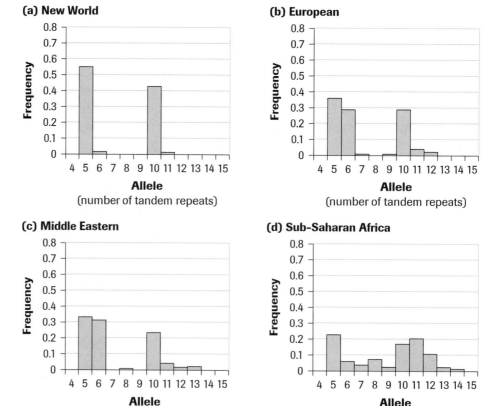

Figure 10

In **(a)** populations of the Americas, gene diversity is limited to four alleles at this locus. By comparison, **(b)** European, **(c)** Middle Eastern, and **(d)** sub-Saharan populations exhibit increasing levels of diversity.

From *Evolutionary Analysis* 2/E by Freeman/Herron, © 2001. Reprinted by permission of Pearson Education, Inc., Upper Saddle River, N.J.

▶ **TRY THIS** activity

Were Neanderthals Homo sapiens?

For many years, palaeontologists disagreed over the classification of Neanderthal. Although they were shorter and heavier than modern humans, their fossil remains are very similar to other recent hominids that have been classified as archaic *H. sapiens*. Many palaeontologists suggested Neanderthal, be classified as a subspecies of humans (i.e., as *H. sapiens neanderthalensis*), while others concluded they should be placed in a separate species (i.e., as *H. neanderthalensis*). Fossil evidence does not provide much information concerning mating patterns and other reproductive isolating mechanisms between very similar populations. However, in the late 1990s, several different samples of mitochondrial DNA were actually extracted from well-preserved Neanderthal fossils, amplified, using polymerase chain reaction, and sequenced.

- Compare an actual Neanderthal mtDNA sequence to that of modern humans.

 www.science.nelson.com

(a) How does the mitochondrial DNA of humans compare with chimpanzees?
(b) How many base-pair differences are typically found between the mtDNA of any two distantly related people?
(c) How many base pairs could be sequenced in the Neanderthal mtDNA?
(d) How many differences were detected between the Neanderthal sequence and that of a modern human?
(e) Does this DNA evidence suggest that Neanderthal are archaic members of the modern human species or members of a different species? Explain.

SUMMARY ## Human Evolution

- Human evolution has followed the same pattern as that of other life forms on Earth.
- The fossil record for humanity is a branching tree of different species.
- The hominid clade includes a diverse group of bipedal species, each adapted to different environments and niches.
- Only a single member of this recent clade remains, of which you are a member: *Homo sapiens*.
- Of the two theories for the existence of *H. sapiens* around the world, the one most supported by genetic evidence proposes that the species evolved in Africa and migrated to other continents, displacing the hominids who were there before them.

▶ **Section 13.5** *Questions*

Understanding Concepts

1. State the key adaptive traits that distinguish each of the following from their predecessors:
   (a) early primates
   (b) anthropoids
   (c) hominoids
   (d) *Australopithecus* and *Ardipithecus*
   (e) *Homo habilis*
   (f) *Homo erectus*
   (g) *Homo sapiens*

2. Create a graphic organizer for the traits you listed in question 1 that would apply to these adaptations:
   (a) an arboreal lifestyle
   (b) diurnal activity
   (c) a ground-level habitat
   (d) communication and tool use

3. What are considered to be the most significant adaptations that distinguish humans from chimpanzees?

4. What evidence suggests that human ancestors evolved bipedal locomotion before large brain size?

5. Examine the human phylogenetic tree (**Figure 6**, page 618).

(a) According to the evolutionary hypothesis presented in this diagram, which species could be described as transitional (or a missing link) between present-day humans and the most recent common ancestor of humans and present-day apes?

(b) Which species became extinct, leaving no living descendants?

(c) List the species that were contemporaries of *Homo habilis*.

6. Sketch a copy of **Figure 6** in your notebook. Mark locations that correspond to

(a) the most recent common ancestor that present-day humans share with *Australopithecus boisei*;

(b) the most recent common ancestor of both humans and chimpanzees;

(c) a possible position at which a species began to use relatively complex tools;

(d) a possible position at which the fusion mutation that resulted in the creation of human chromosome 2 might have occurred.

Applying Inquiry Skills

7. If the multiregional hypothesis regarding human origins were correct,

(a) what fossil evidence might one expect to find outside Africa?

(b) why would geneticists expect to see greater genetic diversity in non-African populations than they have so far?

8. Examine **Figure 11**, a set of aligned DNA sequences. What does this data suggest about the relationships between humans and great apes? Is there evidence of homoplasies in this data? Explain.

9. Review your understanding of carbon-14 dating. What chemical and physical characteristics of carbon make it well suited to radiometric dating of recent human fossils? What makes it of little value in dating fossils of the earliest hominids?

10. Although evidence shows a very close evolutionary relationship between humans and chimpanzees, humans do exhibit marked differences in mental and physical attributes. How might bipedalism have influenced the success of humans as toolmakers? How might this same trait be related to the long period of infant dependency unique to humans?

11. As the rapid pace of fossil discoveries continues, new members are being added to the human family tree. In 2001, palaeontologists reported the discovery of what they believe to be a new species of early hominid. The species, named *Kenyanthropus platyops*, is 3.5 million years old. In addition, evidence suggests that *Homo rudolfensis* may belong in the same genus as *Kenyanthropus*. Work the Web and construct a revised version of the hominid family tree (**Figure 6**, page 618) based on the latest reported fossil evidence.

 www.science.nelson.com

Making Connections

12. (a) Briefly outline the contribution of palaeontology to the present understanding of human origins.

(b) How has this evidence been further supported by advances in genetics? Provide specific examples.

13. *Homo sapiens* is a young species, having existed for only a few hundred thousand years. Most of the recent success of humans can be attributed to cultural and technological advances.

(a) Do you think humans will continue to evolve as a species? Support your ideas.

(b) Brainstorm some of the selective pressures that you think humans may be experiencing now and may experience in the future.

(c) Hypothesize about potential human adaptations that could result.

```
AAGCTTCACCGGCGCAGTCATTCTCATAATCGCCC    human
. . . . . . . . . . . . . . . . A . T . . C . . . . . . . . . . . . .    chimpanzee
. . . . . . . . . . . . . . . TG . . . . T . . . . . T . . . .    gorilla
. . . . . . . . . . . . . . AC . . CC . . . . . G . . T . . . .    orangutan
. . . . . . TT . . . . . . . . AC . . . C . . T . . G . . . . . T .    Old World monkey
. . . T . . . . TT . . A . . CAC . . C . . . T . . . . . T . . . .    New World monkey
```

Figure 11
Note that the dots represent the amino acids shared by humans and each type of great ape.

Evolutionary Biology and Medicine   13.6

Most unique human physiological, anatomical, and behavioural traits evolved during the past few million years. In the last several thousand years, human lifestyles have changed dramatically. What possible impact might such changes have had—and continue to have—on the adaptive value of these highly evolved traits? Biologists who research human health and medicine are finding significant applications from evolutionary biology. The treatment of anemia, for instance, has been improved through a consideration of evolutionary factors.

Anemia is a common sign of infectious disease, as it results from a reduction in hemoglobin-carrying red blood cells. Physicians have routinely administered an iron supplement to correct anemic condition in their patients; if anemia benefits the bacteria by weakening the patient's defences, then an iron supplement should help. There have been cases in which this treatment was not effective. Patients who are also experiencing severe protein malnutrition—lacking transferrin, for example—often acquire fatal infections if they are given iron supplements. Researchers in Africa found that a very high proportion of Masai, who routinely drink large quantities of milk, developed an amoebic infection after being given an iron supplement (**Figure 1**). In addition, Zulu men who consume beverages made in iron pots often get serious liver infections. What is the link?

These circumstances suggest that hosts for pathogens that apparently need iron have evolved mechanisms to keep them from getting it. These mechanisms involve proteins that aggressively bind iron, so that it is unavailable to potential pathogens. The Zulu were unknowingly providing a rich supply of iron to potential pathogens. When the milk protein lactoferrin binds iron, it prevents bacteria from getting it and, thereby, infecting mammal infants—and Masai adults. If an iron supplement is ingested, it provides more iron than milk can bind and this excess becomes available to iron-hungry pathogens. Similarly, transferrin is a blood protein that binds free iron and releases it only to cells bearing special molecular markers. Malnourished individuals who lack adequate supplies of transferrin allow pathogens access to free iron in the blood. More detailed knowledge of patients' protein composition can help physicians decide whether to treat an infection that produces anemia with an iron supplement.

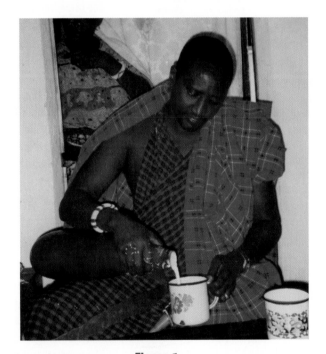

Figure 1
Masai adults consume a diet high in milk protein. Lactoferrin in the milk binds iron, which may help to lower the incidence of intestinal infections among these people.

Case Study   Desert Iguanas and Fever

Matthew Kluger, Linda Vaughn, and other researchers wanted to test the hypothesis that fever is an evolved defence mechanism. In 1974, they began a study of desert iguanas (**Figure 2**, page 624). Like all lizards, these iguanas are ectotherms—their body temperature is determined by the ambient conditions rather than by metabolic activity. They are, however, able to moderate their body temperature through behaviour. They seek out warmer locations to increase their body temperature and cooler locations to lower it. The researchers wanted to determine whether desert iguanas would intentionally raise their body temperature in response to an infection and, if so, whether this influenced their ability to combat the disease.

Figure 2
A desert iguana, *Dipsosaurus dorsalis*

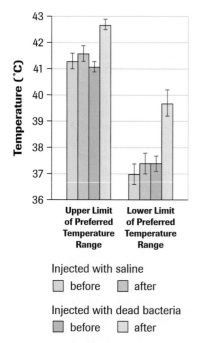

Injected with saline
☐ before ☐ after

Injected with dead bacteria
☐ before ☐ after

Figure 3
This graph shows that none of the iguanas in this experiment chose temperatures lower than about 37°C at any time, but after the infection, the test group never chose temperatures lower than about 39°C.

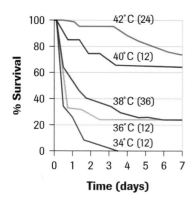

Reprinted with permission from Kluger, Ringler and Aver 1975. M.J. Kluger, P.K. Ringler and M.R. Aver. Fever and Survival. *Science* 188: 166-168. Copyright 1975, American Association for the Advancement of Science.

Figure 4
The percentage of surviving infected individuals maintained at different ambient temperatures over time

As a sign of disease, fever can be uncomfortable and potentially harmful. Does a fever help or hurt a host body's fight against the disease? Evolutionary biology offers two possible interpretations. One is that the invading virus or bacteria releases chemicals that cause the host to increase its body temperature and that this elevated temperature is a benefit to the pathogen, enabling it to grow and/or reproduce with greater success. In this case, the fever provides a selective advantage to the pathogen. Another interpretation is that the host intentionally elevates the body temperature to help combat the infection; in this case, the high temperature may reduce the growth rate of the pathogen or increase the effectiveness of the host's immune system—in either case, providing a selective advantage to the host.

In the first experiments, the researchers recorded the range of temperatures chosen by the 19 iguanas in the study. They then injected 10 iguanas with a solution containing dead bacteria (*Aeromonas hydrophilia*, a normally virulent organism) and, as a control, they injected nine iguanas with saline solution. They then monitored the temperature choices of the two different groups of iguanas to see whether there were differences in the preferred range of temperatures (**Figure 3**).

Next, the biologists wanted to determine whether a behavioural fever could benefit the pathogen, the host iguana, or neither. In 1975, a larger study was conducted, in which 96 desert iguanas were intentionally infected with live bacteria. The iguanas were placed in incubators set at five different fixed temperatures. This experimental design prevented them from using a behavioural response to influence their body temperature. The survival rate of iguanas at each temperature was recorded (**Figure 4**).

In a follow-up project, researchers administered the aspirinlike drug sodium salicylate to some infected iguanas. Sodium salicylate reduces behavioural fever in desert iguanas; when medicated, the iguanas are less likely to seek out warmer locations. All 12 of the infected control group that were not given the drug displayed behavioural fever, and 11 survived. Of the 12 infected iguanas given sodium salicylate, the seven that did not display behavioural fever all died, while the remaining five did display behavioural fever and survived. Since the 1970s, several other studies—all on ectotherms—have shown that fever increases survival. It is much more difficult to study the effects of fever on the survival of endotherms, as their body temperatures cannot be controlled simply by placing them in an incubator. Drugs such as aspirin, ibuprofen, or acetaminophen, which reduce fever in endotherms, can also have other unrelated effects on the immune system and physiological processes.

Case Study *Questions*

Applying Inquiry Skills

1. Examine the upper temperature limits in **Figure 3**. What was the temperature difference for the control group (i.e., the iguanas injected with saline)? for the treatment group (i.e., the iguanas injected with dead bacteria)?

2. Examine the lower temperature limits in **Figure 3**. What was the temperature difference for each group?

3. From **Figure 4**, what temperature offered the best chance of survival to the iguanas?

4. Explain whether the injection of dead bacteria caused an increase in behavioural fever in the desert iguanas.

5. The Kluger–Vaughn team wanted to test the hypothesis that fever is an evolved defence mechanism. Do the results of these experiments support this hypothesis? Explain.

6. Based on these findings, if your pet iguana shows signs of illness, would you rearrange the cage to allow the iguana to move to a higher temperature and create a fever, or give the iguana a drug similar to aspirin to prevent a fever?

7. When people become ill and run a fever, drugs such as aspirin, ibuprofen, and acetaminophen are often taken to reduce the fever and provide comfort to the patient. These drugs do not combat the virus that is causing the cold or influenza. Based on the information provided in this case study, can you assume that taking such a drug in this situation is beneficial?

8. (a) If fever in humans has evolved to benefit the host rather than the pathogen, how might the recovery of patients with the common cold who take aspirin compare with those who do not take aspirin?

   (b) If fever were an adaptive advantage to the pathogen rather than to the host, how would your expectations change?

DID YOU *KNOW* ❓

Fighting Disease with Disease
In the early 1900s, Julius Wagner-Jauregg noted that syphilis was rare in areas where malaria was common. At the time, only 1% of syphilis patients survived—while adults suffering from malaria often lived long lives. Wagner-Jauregg intentionally infected thousands of syphilis patients with the malarial parasite, which produced very high fever. The treatment was effective for about 30% of the patients, who recovered from the syphilis. For his work, he was awarded the Nobel Prize in Medicine in 1927. Considering the ethical implications of his procedure, do you think his work would get the same kind of recognition today? Be sure to take into consideration the prognosis of his patients.

Evolutionary Approaches to Medicine

This Darwinian approach to medicine is not limited to the study of infectious diseases. The idea that many medical conditions have a genetic basis may seem puzzling. One would expect harmful alleles that cause disease or reduce fitness to have become greatly reduced in frequency in a population. Yet, such relatively common conditions as Tay-Sachs disease, myopia (near-sightedness), and breast cancer each have a genetic component. Why have selective pressures not eliminated them?

Theories include possible benefits to a host organism from the alleles that cause the disease or condition. As discussed in Chapter 12, the allele that causes sickle cell anemia also confers some resistance to malaria. The allele responsible for Tay-Sachs disease is found in 3% to 11% of Ashkenazi Jews. The disease is fatal for individuals that are homozygous for the allele; heterozygous individuals not only survive, but have greater resistance to tuberculosis. This allele may be providing an adaptive advantage to heterozygous individuals. This idea is further supported by the fact that, historically, tuberculosis was a major disease agent in this population.

A different evolutionary relationship may be responsible for the high incidence of myopia and breast cancer worldwide. In many populations, 25% or more of the population suffers from myopia. In the 1960s, Francis Young and others examined the incidence of myopia among members of a small indigenous population in Alaska. The researchers found that 42% of individuals aged 6 to 35 suffered from myopia, while only 5% of individuals aged 36–88 suffered from the disorder. Some new environmental factor was causing the increase in myopia among young people in this population. One

pandemic an epidemic that spreads over a very wide geographical area, in some cases sweeping the entire globe

possibility was that their changing lifestyle was exposing younger people to many more hours of indoor close-up viewing activities. Many other research programs have confirmed that the likelihood of developing myopia is increased by a combination of genetic susceptibility and a lifestyle that includes a lot of close-up visual work when the eyeball is actively growing. Although the story is not complete, it would appear that the alleles that cause myopia were not eliminated by natural selection because they did not cause visual impairment in previous generations. It remains unclear what adaptive advantage, if any, is provided by the allele that predisposes individuals to myopia.

Breast cancer, a very serious disease affecting about one in eight North American women, is the focus of intense research. It too may be influenced by changing lifestyles. A growing body of evidence suggests that continual menstrual cycling increases a woman's risk of developing breast cancer. In 1999, Beverly Strassman published a study of menstrual cycling among the Dogon women of Mali who continue to live a traditional lifestyle and use no contraceptives. Strassman found that Dogon women between the ages of 20 and 35 had very few menstrual cycles, as they were either pregnant or experiencing the suppression of menstruation during breast feeding. She calculated that Dogon women had fewer than one-third the number of menstrual cycles experienced by most North American women who live a modern lifestyle. Other studies of women who have menstrual patterns similar to those of Dogon women showed an incidence of breast cancer about one twelfth that of North American women. These findings suggest that the menstrual patterns of modern women may be connected to the high incidence of breast cancer in North American women. This connection may offer insights into new approaches to the prevention of this disease.

As well as considering human evolution, researchers are studying the evolution of pathogens. (See Chapter 12, where antibiotic resistance in bacteria is discussed.) As humans evolve mechanisms to prevent disease and to adapt to the environment, pathogens also evolve to get around these defences. Widespread health risks are posed yearly by such diseases as influenza, informally called the flu. Constantly evolving, the influenza viruses mutate and produce new strains that infect people. The human immune system recognizes and destroys new strains that are similar to previous strains. The 1968 strain contained a new gene previously known to occur only in the bird influenza virus. Recent genetic analysis suggests that, in very rare cases, bird and human strains of the virus simultaneously infect pigs and swap genes. When the mutated strain reinfects people, it is much more difficult for the immune system to recognize and, therefore, is far more harmful. A particularly virulent strain may become **pandemic**, affecting an entire country or much of the world. Current applications of evolutionary biology in medicine include research into such pathogens, as well as the role of evolution in aging, allergies, autoimmune diseases, the repair of injury, mental health, and human behaviour.

Evolution: The Unifying Theory

Scientific research into evolutionary history will continue to add knowledge and understanding to the beliefs that people hold about the origins of life on Earth. With its recent genetic underpinnings, evolution has become the unifying theory in the biological sciences. It provides the foundation for the vast biological diversity evident today. Evolutionary biology offers significant insights into the well being of this dynamic environment and human health. The human brain, which evolved over a hundred thousand years in what must be the most extraordinary development in human evolutionary history, is unravelling these mysteries.

Plant Breeder

Plant breeders develop new plant varieties for agricultural, horticultural, and forestry industries and government agencies. Their primary tool is artificial selection—evolution directed by humans. Recently, they have applied major advances in genetics, biotechnology, and evolutionary biology in their work. Individuals working in this field include highly skilled lab technicians and research scientists. A university degree in plant biology and genetics is advisable. A research scientist requires a four-year university degree, a master's degree, and a doctorate degree.

Conservation Biologist

Conservation biologists study human impacts on the environment. An understanding of evolutionary principles permits conservation biologists to identify species, monitor their genetic diversity, and create programs to prevent the loss of habitats and species. Conservation biologists need a university degree with emphases in environmental science, ecology, and population genetics, or a specialist diploma and training in the field. Conservation biologists work in universities, governments, and such nonprofit organizations as the World Wildlife Fund.

Natural-Product Research Chemist

Research chemists may apply evolutionary approaches to find potentially valuable compounds in the DNA of millions of living species. Thousands of natural chemicals, such as rubber and penicillin, have proven extremely valuable in industry and medicine. Researchers who needed an enzyme that worked at very high temperatures without denaturing found it in thermophilic bacteria that live in hot springs. Most research chemists work in private industry for chemical and pharmaceutical companies. Research chemists interested in natural products have a university degree with a major in biochemistry and benefit from a background in evolutionary biology.

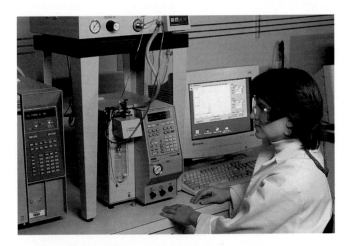

► Practice

Making Connections

1. Suggest several additional careers in which an understanding of evolutionary biology would be beneficial.

2. Select a career that interests you from this page or the list you made in question 1.
   (a) What university program(s) can provide the necessary background for this career?
   (b) What Canadian universities offer degree programs in this area?
   (c) In addition to biology, what other high-school subjects are needed to enter these university programs?

3. What appeals to you about the career you selected in question 2? How might your work in this career benefit society?

4. Conduct an online search to identify employment opportunities in your selected career. Describe a typical day in the career you chose. Explain what the duties are and describe some of the working conditions. Do you work mainly indoors or outdoors? Do you work mainly with people? What materials do you handle?

 www.science.nelson.com

Understanding Concepts

1. In what way can anemia sometimes be of benefit to patients suffering from an infectious disease?

2. If a patient is found to have a genetic condition resulting in reduced levels of transferrin in the blood, how might this influence a doctor's decision regarding the use of iron supplements for this patient?

3. Lactoferrin is found not only in milk, but in tears, saliva, and especially at the site of wounds. Are these findings surprising? Explain.

4. What is behavioural fever and how does it benefit desert iguanas?

5. Name two alleles that have been favoured by natural selection, yet cause disease. State the adaptive advantages they provide.

6. Describe the changes in human behaviour and living conditions that may be partially responsible for elevated rates of breast cancer in modern North American women.

Applying Inquiry Skills

7. A tuberculosis patient was found to be anemic. A physician decided to correct the condition by giving him an iron supplement in hopes of increasing his resistance to the infection. The patient's infection got worse. Provide an explanation for this result in the context of evolutionary biology.

8. Design an experiment that you think could be used to test the Kluger-Vaughn team's hypothesis on humans, rather than on desert iguanas. Include a description of your control group.

9. As you are well aware, humans often cough when they are sick.
   (a) What possible benefit might coughing provide to pathogens living in your lungs or air passages?
   (b) What possible benefit might coughing provide to you as a defence mechanism?
   (c) When patients are suffering from a possible lung infection, would it be wise for a physician to administer a cough suppressant routinely? Explain.
   (d) Outline the procedures you would use to determine whether or not suppression of coughing is of benefit to patients suffering from a lung infection.

Making Connections

10. Recent studies of children and juveniles of other species show that when their eyeballs are actively growing, frequent exposure to blurred images can permanently alter the shape of their eyeball. The blurring results when objects in the field of view are at various distances and is accentuated in low light conditions. Suggest a possible design for children's books and suggest reading habits that might reduce the degree of blurring. What inherited condition might these suggestions help reduce?

11. If fever is of benefit to certain organisms in the fight against infection, suggest a possible reason why they do not maintain an elevated body temperature at all times.

12. Pregnant women often experience morning sickness and food aversions, and may have a particularly strong dislike for spicy foods. In addition, these "symptoms" are most prevalent during the period of pregnancy between one and three months when the developing fetus is most vulnerable to toxins. Given that spices are plant products associated with seeds and flowers, suggest an evolutionary benefit to this malady.

13. Although the Kluger–Vaughn investigations of adaptive fever in the desert iguanas were technically and scientifically very well designed, many people might object to the way in which the animals in this experimental research were used. With two or three other students, discuss the potential benefits of, and ethical concerns about, such research.

14. As a conservation biologist, you monitor and report on the health of salmon stocks. You know that the egg white protein, conalbumin, binds iron. What might this information suggest about the relationship between the nutritional health of adult salmon and successful hatching of their eggs? Explain your reasoning fully.

15. Doctors must decide whether to prescribe an antidiarrhetic to patients with an intestinal bacterial infection. They are unsure whether diarrhea benefits the bacteria by increasing its ability to spread from host to host, or benefits the patient by flushing harmful bacteria out of the intestinal tract. Outline an experimental design that could help the physicians with their decision.

ACTIVITY 13.1.1

Observing Liposome Formation

In this activity, you can model the formation of cell-like structures as the process may have occurred on primordial Earth.

Materials

slides	dropper bottles
lecithin	cover slips
water	microscope
food colouring	paper towels

✋ **Wear safety goggles and aprons.**

Procedure

1. On a clean microscope slide, place one small drop of lecithin and one small drop of water coloured with food colouring a few centimetres apart from each other. Use separate dropper bottles for each.

2. Lower a single cover slip onto the two drops. The cover slip should cause the lecithin and water mixture to contact each other. (If the drops remain separated, add a drop of water at the edge of the cover slip.)

3. Using a microscope, observe the boundary between the lecithin and water mixture. Select the objective magnification that provides the best view. Make a drawing of what you observe, noting the magnification you used.

4. Observe again after 15 min. Record any changes.

5. After 30 min, remove the coloured water mixture and replace it with pure water. Do this by placing a drop of pure water along the side of the cover slip nearest the lecithin. Then place a paper towel at the edge of the cover slip next to the coloured water mixture. Repeat this procedure until the water is relatively clear and can be seen surrounding coloured vesicles, which will have formed.

6. Observe the coloured vesicles. Make drawings of the liposome vesicles that have formed.

Analysis

(a) Lecithin is a mixture of phospholipid molecules. What evidence is there that these molecules are not soluble in water?

(b) Is there any evidence that some or all of the liposome vesicles contain water? Explain.

(c) Were all the liposome vesicles identical in size?

(d) How did the liposomes compare with living cells? What features did they have in common and how did they differ?

Synthesis

(e) Briefly outline how your results model the possible formation of protocells on ancient Earth.

LAB EXERCISE 13.4.1

Looking for SINEs of Evolution

Inquiry Skills

● Questioning	○ Planning	● Analyzing
● Hypothesizing	○ Conducting	○ Evaluating
○ Predicting	○ Recording	● Communicating

In this lab exercise, you can use the presence or absence of SINEs and LINEs to construct cladograms and apply the resulting analysis to related issues in evolutionary biology.

Most of the DNA in a eukaryotic genome is junk, so called because these DNA sequences are never used to regulate gene expression, or code, for a product, such as a polypeptide or ribosomal RNA. In either case, evidence suggests junk DNA has little or no effect on the fitness of the individual. Although it is not used, it is inherited; therefore, changes to these DNA sequences, such as insertions, are passed to succeeding gen

erations. SINEs are a form of insertion often resulting from the reverse transcription of viral RNA into DNA.

Suppose you find a pattern in the junk DNA of two different species, and do not find that pattern in other species. Evolution can explain the situation by saying that the two species recently had a common ancestor, and both species inherited this pattern from their ancestor. The resulting family tree is shown in **Figure 1** (page 630).

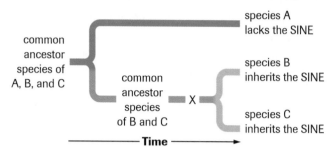

Figure 1

X indicates the time when the SINE became inserted into the genome. Since the SINE insertion occurs only once, at time X, the size and precise location of the SINE will be identical in species B and C.

Hypotheses

(a) Formulate three different hypotheses regarding the evolutionary relationships of cows, hippopotamuses, pigs, deer, camels, and whales. Predict which animals you think are more closely related.

Procedure

Part I: Looking for a SINE

1. Examine the hypothetical DNA code from four different species (**Figure 2**). These species have large sections of DNA that appear to be homologous. These homologous sequences have been aligned vertically so that similarities and differences can be easily seen and colours are used to highlight those nucleotides that are not matches (**Figure 3**).

```
Species  W     AGATAGCGCGTAAAAAG
Species  X     AAATAGCGCGTAAATAG
Species  Y     AAATAGTTAAAGTTACGCATAAATAC
Species  Z     AGATAGCGCGTAAATGG
```

Figure 2

Sequenced DNA fragments from four distantly related species

```
Species  W-    AGATAG.........CGCGTAAAAAG
Species  X-    AAATAG.........CGCGTAAATAG
Species  Y-    AAATAGTTAAAGTTACGCATAAATAC
Species  Z-    AGATAG.........CGCGTAAATGG
```

Figure 3

DNA sequences from **Figure 2** aligned for comparison. Note that spaces appear in the sequences only to facilitate comparisons.

```
Species P  ...AAATTGCTTCGTATTTTCGAATTGCCCCGCTAAAGCGCTTTAGC.......
Species Q  ...AACTTGCTTCGTATTAAGCTGTTGCGTAAAGTTAGTACGAATTGCCCCGGTGAAGCGCTTTAGC......
Species R  ...AATTTGCTTCGTTTTTTCGAATTGCCCCGCTAAAGCGCTTTAGC.......
Species S  ...AACTTGCTACGTATTAAGCCGTTGCGTAAAGTTAGGACGAATCGCCACGGTGACGCGCTTGAGC......
```

These species have large sections of DNA that appear to be homologous. The single nucleotide differences have most likely resulted from point mutations, while the nine-nucleotide segment in species Y is probably the result of an insertion. (Note that this is much more likely than the alternative possibility—that each of the other species experienced an identical deletion event in its past.) This type of pattern often results from a SINE or LINE insertion.

2. Copy the DNA sequences in **Figure 4** into your notebook. Align the homologous sections vertically.

3. Use a highlighter to colour all positions that have the same nucleotide in all four species.

4. Use a different colour to highlight the SINE insertion.

Analysis

(b) Are there any nucleotide differences in the SINE sequences? How might these differences have occurred?

(c) Are mutations that occur within the SINE likely to be harmful, beneficial, or neutral? Explain.

(d) Based on the data alone, draw a cladogram showing the phylogenetic relationship of these species.

Part II: Evolution Displayed by SINEs and LINEs

5. Study the data in **Table 1**. DNA sequencing was used to document the presence or absence of interspersed elements A through I in five mammals. Camels are included as the outgroup.

Table 1 Molecular Evidence for the Evolution of Whales*

Group	SINE or LINE								
	A	**B**	**C**	**D**	**E**	**F**	**G**	**H**	**I**
cow	+	+	−	−	−	−	+	−	+
pig	−	−	−	+	+	−	−	−	+
whale	−	+	+	−	−	−	+	+	+
deer	+	+	−	−	−	−	+	−	+
hippopotamus	−	+	+	−	−	+	+	+	+
camel	−	−	−	−	−	−	−	−	−

+ indicates presence of element   − indicates absence of element
* Data modified from Nikaido 1999

Figure 4

Homologous DNA sequences from four species

LAB EXERCISE 13.4.1 *continued*

6. Use the data to draw a cladogram, and indicate the relative positions on the branches at which each insertion most likely occurred.

Analysis

(e) Were your hypotheses correct? If not, on what type of evidence did you base your hypotheses? Explain how reliable you think that evidence is compared to the data presented here. Revise your hypotheses, if necessary.

(f) Are whales more closely related to cows or hippopotamuses? Explain how you know.

(g) Which insertion happened first: A or B? Explain your reasoning.

Synthesis

(h) Explain whether pigs and camels are more closely related than hippopotamuses and camels.

(i) What must be true about the genomes of all whale species (i.e., which SINEs must they all contain)? Explain.

(j) The genome of a chevrotain—a pigmy deer of African origin—is investigated and found to contain only insertion B, G, and I. Other anatomical evidence suggests that it is closely related to sheep and deer. Based on this information, add a labelled line to your cladogram to represent the chevrotain.

(k) A researcher interested in the evolution of whales wants to know whether orcas are more closely related to white-sided dolphins or to pilot whales. Describe a way to answer this question.

(l) A pygmy hippopotamus is found to contain two new SINEs. The first, M, is discovered to be shared only with the common hippopotamus. The second, N, is not found in the common hippopotamus. Is it possible for any other organism in your cladogram to have this same N insertion? Explain.

LAB EXERCISE 13.5.1

Comparing Hominoid Chromosomes

In this lab exercise, you examine some genetic data from the 1980 and 1982 published research results of Jorges Yunis and his team and perform comparisons of chromosomes. The data appear in the form of karyograms, which are detailed diagrams showing the relative positions and intensities of banding patterns along a chromosome.

Hypotheses

(a) How do the chromosomes of humans compare with those of chimpanzees and other great apes? What mutation(s) might account for some chromosomal differences between present-day humans and living apes? Write some general statements regarding the similarities and differences in chromosome number, structure, and banding patterns that you expect to find.

Materials

complete set of human karyograms
complete set of chimpanzee karyograms
karyograms of the other great apes (optional)
scissors
glue stick

Procedure

Part I: Chromosome Features

1. Examine **Figure 5**. Note the light and dark staining bands along the length of the chromosome. The light bands are thought to contain the coding regions of DNA. These patterns are the same for all members of the same species; therefore, chromosome 1 in all humans exhibits this identical pattern. (The sole exception would be a rare individual who inherited a major chromosomal mutation.) Record the number of bands on human chromosome 1.

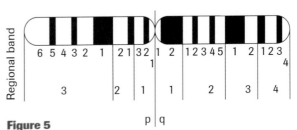

Figure 5
Human chromosome 1

2. Locate 1p3.4 on **Figure 5** (page 631). The place where a chromosome narrows is called the *centromere*. The sections, or arms, on either side of the centromere are referred to as p (the shorter arm) and q (the longer arm). Each arm is divided into regions numbered sequentially from the centromere toward the arms. The light and dark bands are numbered sequentially within a region. Using this system, any band can be identified by the chromosome (1), arm (p), region (3), and band (4).

3. Locate bands 1p3.3 and 1q1.2 on **Figure 5**. Compare the position, colour, and width of each band.

4. Examine the banding patterns in **Figure 6**. Record similarities and differences.

Analysis

(b) Refer to Chapter 12, section 12.3, to review the types of large-scale chromosomal mutations and rearrangements that can occur. Would it be possible for a single chromosomal mutation to make the banding patterns you noted in **Figure 6** match perfectly? If so, describe the mutation.

Part II: Matching Karyograms

5. Obtain from your teacher a set of human chromosome karyograms and a set of chimpanzee karyograms. Note that, as each pair of homologous chromosomes has an identical banding pattern, it is only necessary to compare the 22 autosomal and two sex chromosomes. Record the difference in the number of chromosomes.

6. Cut out the chimpanzee karyograms. Try to match each chimpanzee chromosome with a human chromosome, using size and banding patterns. When you are satisfied with your alignment, glue the chimpanzee chromosomes next to their matching human chromosomes.

7. Chimpanzees have one extra chromosome.

8. Carefully examine chromosome 5 in each of your aligned karyograms. A segment of chromosome that has bands in the reverse order is said to be inverted. (Inversions result from chromosomal mutations.) Locate the section on each chromosome 5 that is reversed and highlight it.

9. Examine the other pairs. Locate and highlight the eight additional cases of chromosome inversion.

Analysis

(c) In general, how would you describe the degree of similarity between the human and chimpanzee chromosomes? Compare your findings with your original hypotheses.

(d) Does the different number of chromosomes in each of these two species represent a significant difference? Explain.

(e) Which two chromosomes are virtually identical?

(f) What obvious difference can you find with chromosome 7? Circle this difference. Does this affect the overall quantity of DNA in the coding region?

(g) How many inversions did you locate?

Part III: Revealing Chromosome 2

10. Chimpanzees, gorillas, and orangutans have 23 autosomal chromosomes, while humans have 22. Potential explanations for this difference include a fusion of two separate chromosomes that occurred in the human line or one or more fissions of a large chromosome that occurred among the ape lines. Does your understanding of cell division provide any theoretical basis for the rejection of the fission hypothesis? Consider the role of the centromere. (Hints: If a chromosome splits in two before or during meiotic division, what

(a)

(b)

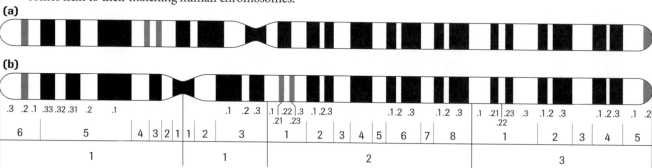

Figure 6
(a) Human and **(b)** chimpanzee karyograms of chromosome 4

serious problem would be faced by the cell in manipulating the new pieces? If a gamete did receive a copy of both pieces of the original chromosome, what problem would persist during the countless mitotic divisions that follow syngamy? Syngamy is the fusion of gametes to form a new cell.)

11. Duplication events occasionally result in the formation of a chromosome with two centromeres. How would the presence of such a chromosome influence your response to question 10?

12. Humans and great apes have, at the ends of their chromosomes, distinctive base sequences called telomeres. Human and chimpanzee telomeres are known to consist of head-to-tail repeats of the bases 5′-TTAGGG running toward the ends of the chromosome. If the fusion hypothesis is correct, where would you expect to find the original telomere sequences? On your karyogram of human chromosome 2, locate and label the predicted position of these imbedded regions.

13. Write out a series of 24 base pairs as they might occur precisely at the point of fusion. (Hint: Draw 12 base pairs as they might exist at the ends of two separate chromosomes, then consider how they would appear when joined.)

14. In 1991, researchers announced the discovery of just such a telomere–telomere union in chromosome 2. Review your hypotheses and findings in view of this information, and revise as necessary.

15. Centromeres are also associated with repetitive sequences. In human chromosome 2, these sequences occur in two locations: one at the centromere, and one at the precise location that would be expected from a fusion event between the two smaller chromosomes. Locate and label the position of this centromere remnant on your karyogram.

Optional: Make a hypothesis for a comparison of the human karyogram with those of gorillas and orangutans. Follow the procedure for Part II and then compare your findings with your hypothesis.

Synthesis

(h) Overall, the DNA sequences of gorillas, chimpanzees, and humans differ by only about 1%. Even in pseudogene regions, where mutations accumulate more rapidly, fewer than 2% of the nucleotide positions differ. Does this evidence support or conflict with what you observed in the banding patterns of human and chimpanzee chromosomes?

(i) A growing body of genetic evidence strongly suggests that chimpanzee species are more closely related to humans than to gorillas and orangutans. Draw a phylogenetic tree to model this relationship.

(j) Assume that each of the nine inversions resulted from a separate mutation. If humans and chimpanzees share a common ancestor from 7 million years ago, what is the average number of years between inversion mutations per species?

(k) Evidence from these and other comparisons indicates that, at some point in the last 7 million years, a human ancestor was born with a fused chromosome 2. Of course such an individual could not pass on this new trait without reproductive success. How could such an individual pass on its fused chromosome, when all potential mates had a different number of chromosomes?

 www.science.nelson.com

Key Expectations

- Select appropriate instruments and use them effectively in collecting and observing data. (13.1)

- Select and use appropriate numeric, symbolic, graphic, and linguistic modes of representation to communicate scientific ideas, plans, and experimental results. (13.1, 13.2)

- Evaluate the scientific evidence that supports the theory of evolution. (13.1, 13.2, 13.3, 13.4, 13.5, 13.6)

- Analyze how the science of evolution can be related to current areas of biological study, and how technological development has extended or modified knowledge in the field of evolution. (13.1, 13.2, 13.3, 13.4, 13.5, 13.6)

- Outline evidence and arguments pertaining to the origin, development, and diversity of living organisms on Earth. (13.1, 13.2, 13.3, 13.5)

- Formulate and weigh hypotheses reflecting the various perspectives that have influenced the theory of evolution. (13.1, 13.2, 13.5)

- Evaluate current evidence concerning the theories of gradualism and punctuated equilibrium. (13.2)

- Locate, select, analyze, and integrate information on topics under study, working independently and as part of a team, and using library and electronic research tools, including Internet sites. (13.2)

- Identify questions that arise from concepts of evolution and diversity, including health-care implications for humans. (13.2, 13.6)

- Explain, using examples, the process of adaptation of individual organisms, including humans and their recent ancestors, to their environment. (13.3, 13.5, 13.6)

- Describe and analyze examples of technology that have extended or modified the scientific understanding of evolution. (13.4, 13.5)

- Relate present-day research and theories on the mechanisms of evolution to current ideas in molecular genetics. (13.4, 13.5, 13.6)

- Analyze evolutionary mechanisms that have occurred during the course of human evolution. (13.5, 13.6)

- Identify and describe science- and technology-based careers related to the subject area under study. (13.6)

Key Terms

adaptive radiation

anthropoid

archaebacteria

Cambrian explosion

chemoautotrophic

cladistics

cladogram

coevolution

convergent evolution

divergent evolution

endosymbiosis

eubacteria

extreme halophiles

extreme thermophiles

heterotrophic

hominid

hominoids

homoplasies

ingroup

lineage

liposomes

methanogens

monogenesis (African replacement) hypothesis

monophyletic group (clade)

multiregional hypothesis

outgroup

pandemic

phylogenetic tree (cladogram)

phylogeny

primary abiogenesis

ribozymes

stromatolites

synapomorphies

theory of gradualism

theory of punctuated equilibrium

thermal proteinoids

transitional forms

► **MAKE** *a summary*

Evolving from Puddles to People

- Construct a chronological representation of 15 of the most significant events in the evolution of life from primordial Earth 4.5 billion years ago to the rise of *Homo sapiens*. Present your timeline in a creative manner, for instance, as a calendar, clock, computerized slide show, televised docudrama, video storyboard, or a series of theatrical sketches. Be sure to represent mutation and natural selection throughout as agents of change. For each event, incorporate a justification for its significance to human evolution.

In your notebook, indicate whether the sentence or statement is true or false. Rewrite a false statement to make it true.

1. The Cambrian explosion, about 240 million years ago, was one the greatest extinction events of all time. Most species went extinct at this time.

2. The closest common ancestor to humans is the chimpanzee.

3. Evolutionary biologists disagree about the pace of evolution.

4. No examples of endosymbiosis can be found among species alive today.

5. A cell alive 2 billion years ago did not resemble single-celled organisms alive today.

6. Evolutionary biology is used to study the past history of life on Earth and, therefore, cannot be used to study species alive today.

7. Earth's early atmosphere contained lots of oxygen because there were no aerobes using oxygen for respiration.

8. Eukaryotic cells, archaebacteria, and eubacteria were probably the only life forms on Earth for more than 2 billion years.

9. Transitional fossils in the reptile-to-mammal lineage are relatively common in the fossil record.

10. The earliest known fossilized living cells resembled simple protozoans that were free swimming.

11. In the past 3 billion years there have been only two major mass extinctions but these have caused almost all of the extinctions in the history of life on Earth.

12. The earliest known organisms were probably phototrophs, able to make their own food using sunlight, water, and carbon dioxide.

13. Human chromosome 2 could not have arisen by a fusion mutation, because an individual with it would not have been able to reproduce.

14. Large brains evolved after bipedalism and before the opposable first digit.

15. Some organelles may have originally formed by the infolding of the outer cell membrane.

16. Match the following terms with the phrases below: primary abiogenesis, adaptive radiation, clade, coevolution, convergent evolution, divergent evolution, gradualism, hominid, punctuated equilibrium, synapomorphy.
    (a) a group with a single common ancestor
    (b) theory that evolution is very slow at times
    (c) a shared derived trait
    (d) theory that life arose from inorganic molecules
    (e) very common evolutionary pattern in archipelagos
    (f) species usually dependent on each other for survival
    (g) species exhibit different genetic codes but similar outward appearance and behaviour
    (h) theory that the pace of evolution does not change significantly
    (i) species may exhibit similar genetic codes but have different outward appearances and behaviour
    (j) includes *Australopithecus afarensis*

Examine the phylogenetic tree in Figure 1 and answer the following questions about it:

17. Which grouping of circles represent clades?

18. Which species is/are most closely related to the following species?
    (a) V   (b) T   (c) Q

19. If a new character (or trait) evolved at position 4, which of the species in the tree would definitely not have the new trait?

20. List all the numbered positions at which a derived (apomorphic) character may have evolved.

21. Should you expect the ancestral species at position 5 to resemble S or T more closely?

22. Along which path could a derived character evolve that would eventually be inherited by S?

23. Which species would you expect to be most ancient in appearance?

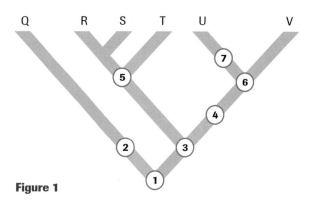

Figure 1

NEL   An interactive version of the quiz is available online.   The Evolutionary History of Life **635**

GO www.science.nelson.com

Understanding Concepts

1. Describe the general chemical and physical composition of Earth shortly after the initial cooling and the formation of a solid outer crust.

2. Describe evidence that supports the theory that simple molecules in the primordial atmosphere were capable of reacting to form complex organic molecules.

3. In what way do the evolutionary processes and products surrounding the Cambrian explosion favour Eldredge and Gould's theory of punctuated equilibrium?

4. Explain whether you would expect gradualism or punctuated equilibrium to best describe the expected pace of evolution following these events:
   (a) A remote island is devastated by a volcanic eruption.
   (b) A national park is created to preserve a remote region of the Amazon rain forest and maintain very high biodiversity.
   (c) An aggressive carnivore is introduced to a remote but biologically diverse island.
   (d) Ice receded in northern Ontario with the end of the last period of glaciation.

5. You want to examine and compare the general habits of monkeys from South America and from Africa. What evidence from Australian mammals would be important to consider in developing your hypotheses?

6. Plant (a) in **Figure 2** is a true cactus, while plant (b) is a distantly related euphorbia. The spines on the cactus are highly modified leaves, while the spines on the euphorbia are not.
   (a) What pattern of evolution do these plants exhibit?
   (b) Describe the role you think the environment would have played in this evolution.
   (c) In general, if you were to compare the desert plants in Africa with the desert plants in Australia, would you expect them to be very similar in their genetic makeup, in their physical appearance, or in both? Explain.

Figure 2

7. Hypothesize an explanation for the extraordinarily rapid and successful adaptive radiation of *Drosophila* (fruit flies) in the Hawaiian island archipelago.

8. Which of the shared traits in the following scenarios would be useful for separating the species into two clades?
   (a) Orchids and roses have petal-bearing flowers. Pine trees do not.
   (b) Ducks and snakes have tails. Humans do not.
   (c) Turtles and pike have mineralized skeletons. Snails have a mineral shell.

9. Homoplasy is fairly common for individual nucleotide bases in distantly related species but is extremely unlikely in the case of SINEs and LINEs. Give several reasons that support this assertion.

Applying Inquiry Skills

10. **Figure 3** depicts a series of early Eocene mammals as they are thought to have appeared, based on fossil evidence. Assume these species represent a set of direct descendants.
    (a) Does this evidence support the theory of gradualism or the theory of punctuated evolution? Explain.
    (b) In what age of sedimentary rock would you expect to find relatives with moderate-size horns?
    (c) How would your answer to (a) change if a fossil resembling the uppermost specimen were found in rock dated at 45 million years old?

Figure 3

| 35 | 40 | 45 | 50 |

Millions of Years Ago (mya)

11. Sketch a simple phylogenetic tree that shows the probable relationships between the earliest prokaryotic cells, eukaryotic cells, eubacteria, and archaebacteria.

12. **Table 1** shows the number of amino acid differences between the composition of human hemoglobin and that of the hemoglobin of six other organisms.

    Hemoglobin contains two alpha and two beta polypeptide chains. These data refer to the beta polypeptide chain, which consists of 146 amino acids.

Table 1 Number of Amino Acid Differences from Human Sequence for Selected Organisms

Organism	Amino acid differences
human	0
gorilla	1
mouse	27
chicken	45
frog (Cordata)	67
lamprey (Cordata)	125
sea slug (Mollusca)	127

(a) What type of evidence do these data provide for evolution?

(b) Based on these data, what generalizations can be made about the degree of common ancestry between humans and gorillas, and between humans and mice?

(c) This data compares other species to humans. What evidence, if any, can it provide about the relationship of these species to one another?

(d) Hypothesize the number of amino acid differences there would be between humans and chimpanzees, orangutans, and eagles.

(e) What information did you use to develop your hypotheses in (d)?

(f) If mutations result in homoplasy, explain how homoplasy would influence your interpretation of the above data. Would you expect the actual number of sequence changes to have been greater or fewer than the values in the table? Explain.

13. Graph the data in **Table 2**. Describe the trend on your graph. Which species had a brain that was large in comparison to the cranial capacity of their contemporary hominids? Which species had a small brain relative to that of their contemporaries?

Table 2 Age and Cranial Capacity of Selected Hominid Fossils

Species	Age (million years ago)	Cranial volume (cm^3)
A. afarensis	3.8	400
A. africanus	2.7	450
H. ergaster	1.8	700
H. erectus	1.0	1000
H. sapiens	0.1	1500
A. robustus	1.9	550
H. rudolfensis	2.5	650
H. habilis	1.9	580
A. boisei	2.3	550
H. neanderthalensis	0.3	1700
H. heidelbergensis	0.6	1200

Making Connections

14. Often, when individuals have a severe malarial attack, they become anemic.
    (a) How could this condition benefit the patient?
    (b) How could this relationship have evolved?

15. Humans and chimpanzee species are thought to have shared a common ancestor about 5 million years ago. In comparisons of human and chimpanzee chromosomes, the base-pair sequences differ by about 1%. Write a paragraph to express your perspective on the rate of human and chimpanzee evolution.

16. In 1996, a fossil was found of an early North American human dating about 9500 years ago. State-of-the-art computer modelling was used to reconstruct the approximate appearance of this "Kennewick Man" based on the shape and surface features of the skull. Researchers tentatively concluded that he most resembles the Ainu people, who are indigenous to northern Japan. Describe the steps you would follow to make use of mitochondrial DNA analysis to further support or refute this hypothesis.

PERFORMANCE TASK

The Evolution of Lactose Intolerance

▶ Criteria

Process

- Develop and submit hypotheses.
- Gather evidence for your hypotheses from a variety of sources-print, electronic, and multimedia and direct observations and experiments.
- Organize, analyze, and interpret your evidence.

Product

- Prepare a suitable report in the form of a journal article.
- Present your evidence effectively using tables, maps, and diagrams.
- Justify the acceptance or rejection of each of your hypotheses and conclusions.
- Include properly documented references.
- Submit your report to a team member for peer review.
- Complete the peer review of a team member's report.
- Revise your report in response to the comments of your reviewer.
- Submit the following to your teacher for assessment: your original report, the completed peer-review assessment form, and your revised final report.

Lactose is a disaccharide found in mammalian milk. The digestive enzyme lactase (β-galactosidase), produced by cells on the villi lining the small intestine, hydrolyzes the disaccharide lactose into glucose and galactose. In most mammals, the ability to digest lactose is lost in adults. In humans, lactase production usually begins to decrease after about the age of two years. This inability to produce high amounts of lactase results in the common condition known as lactose intolerance. In January 2002, Dr. Leena Peltonen of the University of California and Nabil Sabri Enattah of the University of Helsinki, Finland, reported the discovery of a gene responsible for lactose intolerance on chromosome 2. This discovery has shed new light on the evolution, potential diagnosis, and treatment of this condition.

People with lactose intolerance may suffer from nausea, cramps, gas, and diarrhea after consuming foods—dairy products, in particular—that contain lactose. The severity of symptoms varies from individual to individual and depends, in part, on the amount of lactose ingested.

For this task, you and two other students will investigate lactose intolerance and tolerance. Review the following three components of this task and decide as a team which of you will complete which one of the three parts (I, II, or III). Each member is responsible for completing and submitting his or her assigned task, along with a report that includes responses to the Analysis questions. Each team member should be familiar with all the other members' reports. Present your survey and other plans for your procedures to your teacher for approval before carrying them out. Following completion of the procedure and analysis, you will be asked to review one of your team member's reports, and you will submit your report to a team member for peer review.

Questions

As a team, find answers to the following questions:

(a) What is the incidence of lactose intolerance in human populations?

(b) What gene(s) are responsible for the regulation and production of the lactase enzyme?

(c) What role has natural selection and mutation played in the evolution and prevalence of lactose tolerance and intolerance?

(d) What technology and methods are currently used to diagnose, avoid, and or manage lactose intolerance?

(e) How might new advances in technology and in knowledge about lactose intolerance be used to improve its diagnosis and treatment?

Part I: The Genetics and Prevalence of Lactose Tolerance and Intolerance

1. In this investigation, your goal is to obtain a crude lower estimate of the incidence of lactose intolerance in your school or community. Conduct a survey of at least 40 students and/or adults in your school or community.

2. After completing your survey, generate some hypotheses regarding the incidence of lactose intolerance in the total population of surveyed participants.

3. Use print and electronic sources to determine what gene or genes are responsible for the production of lactase, and find out the occurrence of lactose tolerance and intolerance in human populations of the world.

 www.science.nelson.com

Analysis

(f) Does lactose intolerance result from a mutation in the gene that codes for the lactase enzyme? Is this surprising? Explain.

(g) Apply the Hardy-Weinberg equation to estimate the incidence of "carriers" in the population you surveyed and in the populations you researched.

Part II: Natural Selection and Lactose Intolerance

4. Most adult mammals do not produce the enzyme lactase. Apply your understanding of evolution and natural selection to develop and test hypotheses about this trait. Research print and electronic sources to find supporting evidence for your hypotheses and conclusions.

Analysis

(h) What might be the selective advantages of not producing lactase as an adult?

(i) Other than humans, would adult mammals ever suffer from the symptoms of lactose intolerance? Explain.

(j) Would you expect lactose intolerance to be an ancestral or a derived condition? Explain.

(k) What change in the environment and/or human behaviour would create a selective advantage for those individuals who continued to produce lactase as adults?

(l) When and where did humans begin to domesticate cattle, camels, and goats? When and where did humans begin using milking as a source of animal protein? How is this information significant?

(m) Based on this data and evidence collected in Part I, does natural selection appear to have played a role in the evolution of lactose tolerance in humans?

Part III: Options and Opportunities in the Control and Treatment of Lactose Intolerance

5. Human societies have been consuming milk and milk products from large domesticated mammals for thousands of years. When this practice began, humans would not have had the gene for lactose tolerance. Research in print and electronic sources the technologies that have and could be applied to treat lactose intolerance.

 www.science.nelson.com

Analysis

(n) What products have humans used as alternatives to milk to avoid or reduce the incidence of lactose intolerance? Explain how the lactose is removed or reduced in these products.

(o) What is Lactaid? Design and conduct a simple experiment to demonstrate how this product functions. Explain how people with lactose intolerance could use the product.

(p) How might recent discoveries in genetics be used to develop new diagnostic tools for detecting lactose intolerance?

(q) Investigate and report on past research and potential new developments in the areas of lactose intolerance. Could lactose-free cows be genetically engineered? Could lactose-intolerant individuals undergo gene therapy?

Understanding Concepts

1. Identify the major contribution to evolutionary biology of each scientist listed below and briefly describe the evidence on which the contribution was based:
   (a) Lyell   (c) Cuvier
   (b) Lamarck   (d) Eldredge and Gould

2. (a) List the conditions required for the formation of a fossil, as well as the key stages in fossil formation.
   (b) How can direct observation of a fossil bed be used to determine the relative age of fossils? What assumption(s) have to be made when using this technique?

3. Explain why extinct species are not equally represented in the fossil record. What might improve an organism's chances of leaving fossil evidence?

4. Fossils can be described as direct evidence of evolution. Explain this statement.

5. How did the discovery and scientific understanding of radioactivity and radioactive decay improve the ability to estimate the age of Earth and the absolute ages of fossils?

6. Jean Baptiste Lamarck believed that living species evolved over time. Compare and contrast his theory of evolution with that of Charles Darwin.

7. In September 1835, Charles Darwin reached the Galapagos Islands. What did he observe and discover in this archipelago that provided important evidence for his theory of evolution?

8. Brainstorm examples of features that are homologous and/or analogous to
   (a) a bird's wing;
   (b) an elephant's trunk;
   (c) cactus spines;
   (d) cat whiskers;
   (e) a snail's shell.

9. Vestigial features and anatomical oddities can provide compelling evidence of an evolutionary past. List four such features and provide an explanation for their occurrence.

10. Darwin recognized that natural selection by the environment could produce change in a way similar to the artificial selection used by plant and animal breeders. Compare and contrast these two processes, using a specific example for each. Your analysis should answer the following questions:

    (a) What is the source of new variation in each process?
    (b) Does each process involve selection for certain features?
    (c) Does each process involve selection against certain features?
    (d) What types of selection are most at work (disruptive, directional, and/or stabilizing)?
    (e) What amount of time is required for noticeable differences in a phenotype?

11. Clear examples of the competition between different species for resources can be seen in plant succession, as well as in relationships between predators and their prey, and among parasites, diseases, and hosts. Darwin recognized that evolution was driven by competition and selective forces between members of the same species. How did Thomas Malthus' essay on populations provide Darwin with the mathematical basis for this idea?

12. Before the modern synthesis with genetics, what key elements of evolution were unanswered by Darwin's and Wallace's theory of evolution by natural selection?

13. Explain how the genetic variability between and within species is affected by
    (a) genome size;
    (b) number of genes;
    (c) homozygosity of loci;
    (d) genetic recombination during meiosis;
    (e) mutations;
    (f) population size.

14. Define *evolution* in terms of gene frequencies.

15. What conditions must be met to maintain Hardy-Weinberg equilibrium?

16. Genetic changes resulting from mutation can be harmful, beneficial, or neutral. Explain how harmful mutations have little or no influence on the evolution of species, while the rarer, beneficial mutations can have a dramatic long-term impact.

17. Identify the type of selection (disruptive, stabilizing, directional, sexual, runaway, or cumulative) operating in these examples:
    (a) diversity among honey creepers
    (b) a peacock's tail
    (c) the central nervous system
    (d) egg size in robins
    (e) body size in the blue whale
    (f) the courtship dance of the blue-footed booby

18. Describe one of more sexual selection traits in
    (a) moose;
    (c) chorus frogs;
    (b) mallard ducks;
    (d) lions.

19. With regards to the evolution of altruistic behaviour, what genetic relationship typically exists between the helping individual and the individual being helped?

20. Explain, using an example for each, how reproductive isolation of two different populations can occur as a result of
    (a) mechanical isolation;
    (b) ecological isolation;
    (c) behavioural isolation;
    (d) gametic isolation;
    (e) temporal isolation.

21. Compare and contrast allopatric speciation when two populations are isolated from each other on separate large and ecologically *distinct* islands, and on separate large and ecologically *similar* islands.

22. What organic materials other than living things may be capable of limited self-replication? What is the biological role of these kinds of material in living cells?

23. What key pieces of evidence strongly support the idea that mitochondria and chloroplasts evolved through endosymbiosis?

24. (a) What evidence exists that Earth's history has included a number of mass extinction events? What evidence offers clues to the cause(s) of these extinctions?
    (b) Mass extinction events are followed almost immediately by periods of very rapid speciation. Explain this observation.

25. Rearrange these events from most ancient to most recent: first eukaryotic organism, first land plant, first dinosaur, Cambrian explosion, largest mass extinction event, earliest life evolves, first flowering plant.

26. Explain how each species in these pairs benefits from coevolution:
    (a) fig trees and wasps
    (b) leaf-cutter ants and fungi

27. Two new bird species discovered in the Borneo rain forest are very similar in appearance. Describe two different evolutionary histories that could account for their observed similarity.

28. Identify key traits that were strongly selected for as humans evolved from earliest primate ancestors to modern humans. For each trait, suggest a selective advantage that it offered.

29. Examine the phylogenetic tree in **Figure 1**.

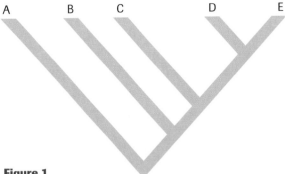

Figure 1

(a) Which two species are most closely related? Explain.
(b) Species B and D share a SINE which is not found in species A. What other species would you expect to also share this trait? Explain.
(c) List all species to which species C is most closely related.

Applying Inquiry Skills

30. A layer of sedimentary rock containing numerous fossils is found between two layers of igneous rock. The layer below the fossil bed contains ^{40}K and ^{40}Ar and daughter isotopes in a ratio of 23%:77%, while the layer above contains ^{40}K and ^{40}Ar and daughter isotopes in a ratio of 25%:75%. Estimate the absolute age of the fossils in this deposit.

31. Explain how each of the following provides evidence for evolution:
    (a) sea shells discovered at high elevations
    (b) the fossil record of whales
    (c) chromosome banding patterns of primates
    (d) homologous anatomy of mammalian limbs
    (e) nonfunctioning rudimentary eyes of cave-dwelling fish
    (f) antibiotic resistance in bacteria

32. Outline basic steps used by breeders to alter a trait such as the nutritional content of the edible component of a plant.

33. Darwin was struck by the fact that the Galapagos and Cape Verde islands were very similar in climate and topography but completely different in flora and fauna. Explain his observations.

34. Islands are recognized for their unique and diverse biological communities as well as for their frequent and rapid species extinctions. What features of islands and the species on them have produced these seemingly contradictory conditions?

35. Study **Table 1** and answer the questions that follow.

Table 1 Genotype Frequencies for Three Selected Populations

Population	AA	Aa	aa
X	25	50	25
Y	40	20	40
Z	2	16	32

(a) For each population, determine the expected genotype frequencies of the next generation.

(b) Which population is not in Hardy-Weinberg equilibrium? Explain whether this population is being influenced by disruptive selection.

36. In parts of Africa, the incidence of sickle cell anemia is greater than 1 in 64. Use the Hardy-Weinberg equation to estimate the genotype frequencies in this population.

37. With reference to the Hardy–Weinberg equilibrium, explain how each of the following situations would result in evolutionary change:

(a) In 1996, only 80 adult piping plovers, an endangered bird species, nested in the entire province of Nova Scotia.

(b) Each year, new strains of the influenza virus cause human disease.

(c) In 1976 and 1977, a severe drought and change in food availability favoured the survival of ground finches (*Geospiza fortis*) with larger bill sizes over the survival of those with smaller bills.

38. (a) Female praying mantises may consume males during or immediately after mating. What selective advantage does this behaviour have to females? How could it benefit males?

(b) Male elephant seals often compete violently for the chance to fertilize females and their injuries can be serious, sometimes fatal. Explain how such behaviour evolves, even though it lowers the average life expectancy of the males and reduces their chances of survival?

39. In many species that reproduce sexually, including humans, sexual selection can be influenced by symmetry. Design an investigation in which you could determine whether humans find symmetrical faces more attractive than asymmetrical faces.

(a) Would you expect human males or females to be influenced more strongly by the symmetrical appearance of prospective mates? Explain your reasoning.

(b) If your teacher approves, conduct your experiment and report the results.

40. In courtship, male scorpion flies (*Panorpa latipennis*) offer females gifts of dead prey. Females feed on these gifts during mating and copulate longer with those males that offer the largest gifts. How might the size of the gift indicate to females that the male is a better mate?

41. In most species that reproduce sexually, selection involves a female choosing a male and male-male competition. Account for the existence of this pattern as opposed to a male choosing a female and female-female competition.

42. What correlation would you expect to find between the genetic diversity of human populations and length of time they have been in a particular geographical location? How does the concept of the founder effect account for this correlation?

Making Connections

43. Most crop plants and livestock breeds have been altered dramatically from their wild origins through artificial selection and breeding. They have been bred to exhibit different nutritional content, behaviours, and size and have been adapted to different climates, pests, and diseases. With two or three other students, consult print and electronic sources to determine the original geographic location and, if possible, the original physical characteristics of corn, wheat, rice, cattle, peanuts, chickens, pigs, rubber, cotton, and sheep.

 www.science.nelson.com

44. Alfred Russell Wallace is not a household name, although, at the same time as Charles Darwin, he presented a virtually identical theory of evolution by natural selection. What scientific and social factors might account for the differences in recognition?

45. Mine wastes often contain high concentrations of toxic metals, such as copper and lead, in their tailings. Few plants can grow on them, although some hardy plants may colonize the edges. Over many generations, these plants may develop increasing resistance to the toxic metals, while their ability to grow on uncontaminated soil decreases.

    (a) Explain what selective processes are occurring in this situation. Can these processes lead to speciation? If so, which type of speciation is occurring?

    (b) How might this evolutionary process be used to rehabilitate certain environments?

    (c) Could the information gained from this situation be applied to artificial selection and plant breeding methods?

46. As a result of human activity, extensive forests are becoming fragmented into small forest islands. How might the increasing isolation of populations in these forests influence their success and evolution? How might the effects differ for a large mammal species, such as the lynx, compared to an insect species, such as a beetle?

47. In June 2000, Dr. Jan M. Hollis and coworkers of the NASA Goddard Space Flight Center analyzed data from telescopic images and reported the first evidence of sugar molecules in space. In December 2001, Dr. George Cooper and his team at this center reported the recovery of sugar molecules from a Murchison meteorite. How might such discoveries influence theories regarding the early evolution of life on Earth? How might they affect future space exploration?

48. Like bacteria, viruses can reproduce, mutate, and evolve rapidly in response to a changing environment. The HIV virus, which causes AIDS, is no exception. Many strains have evolved resistance to the best available drugs. One strategy that has proven effective in many cases is to stop drug treatment temporarily. Although this strategy has no immediate benefit,

when patients are put back on drug therapy, there is little sign of drug resistance and patients show marked improvement. Suggest an explanation for these observations.

49. Insect resistance to pesticides is estimated to cost tens of millions of dollars per year in Canada. The Colorado potato beetle, for example, developed resistance to five different pesticides over a period of only 15 years. How might such evolutionary consequences affect and concern consumers, ecologists, pesticide companies, organic farmers, and plant breeders?

50. A patented technology that a new biotechnology company calls *morphogenics* can accelerate the mutation rate and evolution of organisms, including cell cultures, microbes, plants, and mammals. The resulting organisms are then screened to find the offspring that exhibit viable traits. How might this technology be applied in medicine and/or agricultural breeding programs?

Extension

51. Evolutionary biologists have hypothesized that many epidemics—widespread diseases that usually kill their hosts—such as smallpox or plague, could only have evolved in large human populations. Further, they hypothesize that these diseases originated in mammals that were domesticated. Consider these hypotheses in relation to contact between European explorers and Indigenous peoples, such as the Arawak, Aztec, Maya, Inca, Aboriginal peoples in North America, Aborigines in Australia, and Maori in New Zealand. Consult print and electronic sources to determine whether the exchange of diseases between Europeans and any two of the Indigenous peoples listed above supports one or both of these hypotheses. Report your findings to your class.

 www.science.nelson.com

Population Dynamics

Easter Island, Rapa Nui, in the Pacific Ocean 3000 km off the coast of Chile was first inhabited by about 400 Polynesians around the 5th century A.D. They harvested fish and seabirds, cleared forests, and grew crops inside volcanic craters. They are mostly known now for hundreds of human-faced statues, called *moai* (mow-a), that they created—each *moai* being up to 10 m in height with a mass of up to 85 t. With plentiful food supplies and war unknown, the population on remote Easter Island expanded over several centuries to approximately 15 000. By then, the people could no longer harvest and grow enough food to feed everyone. The Polynesian rat, a foreign species introduced by the settlers, fed on palm seeds and had no natural predators to limit their population growth. Stripped by people and rats, the forests became extinct. There is evidence that the people fought over their dwindling resources, toppled their statues, and fled to caves. By 1722, both the human population and the natural environment had been devastated. The history of Easter Island can serve as a lesson on the importance of sustainable use of resources by human populations.

▶ Overall Expectations

In this unit, you will be able to

- analyze the components of population growth, and explain factors that affect the growth of various populations of species;

- investigate, analyze, and evaluate populations, their interrelationships within ecosystems, and their effect on the sustainability of life on this planet; and

- evaluate the carrying capacity of Earth, and relate the carrying capacity to the growth of populations, their consumption of natural resources, and advances in technology.

ARE YOU READY?

Concepts

- food chains and food webs
- process of bioaccumulation
- resource limits of an ecosystem
- abiotic and biotic interactions
- trophic levels

Skills

- analyze a population case study
- select and use appropriate vocabulary to communicate scientific ideas
- investigate factors that affect ecological systems and the consequences of changes in these factors
- use graphs to compile, organize, and interpret data

Introducing Zebra Mussels

Zebra mussels were unknown in the Great Lakes until the mid-1980s, when they were likely introduced through bilge waters discharged from ships from the Caspian Sea, where the mussels are indigenous. The rapid expansion of the zebra mussel population throughout the Great Lakes ecosystem has had a significant impact on other species. Although zebra mussels are eaten by a few waterfowl, such as the common merganser and goldeneye, and fish such as goby and drum, they are only a tiny component of their diets. Zebra mussels seem to make the water more clear but that change happens because the zebra mussel remove the microorganisms which make the water cloudy. Zebra mussels also remove some toxic chemicals from the water but this action does not make the water more clear. These contaminants are not destroyed; they bioaccumulate in the zebra mussels and remain in the ecosystem. However, as an exotic or foreign species, zebra mussels have triumphed over other native organisms for food and space. They reproduce at an extremely fast rate: one female mussel can produce 30 000 to 40 000 eggs a year. The adult zebra mussels cling together in clusters, clogging pipelines and other underwater structures. Suggested solutions to some of the challenges posed by the growth of the zebra mussel population include using ultrasonic vibrations, adding chlorine or other chemicals to the water to kill the mussels, using chemicals to alter their development, depriving them of oxygen, using dessication, and manual removal.

Knowledge and Understanding

1. Define these terms:
   (a) exotic species
   (b) carrying capacity
   (c) food chain
   (d) food web
   (e) biotic factors
   (f) abiotic factors

2. The carrying capacity of Lake Ontario for the zebra mussel is still unknown. What biotic and abiotic factors might affect the growth—in positive and negative ways—of the zebra mussel population?

3. Zebra mussels are filter feeders that consume microscopic organisms, such as bacteria and algae. If the algae get 70 kJ of energy from the sun, how much of that energy is transferred to the zebra mussel?

4. Is the zebra mussel a producer or a consumer? What trophic level does it occupy?

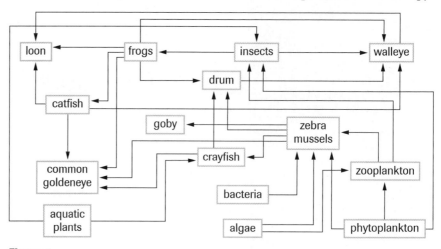

Figure 1
Note the relationship between zebra mussels and the other organisms in this partial food web of the Great Lakes.

Inquiry and Communication

5. Zebra mussels filter some toxic chemicals out of the water. These chemicals can poison the predators that consume them. Suppose that the population of catfish that occupies the same ecosystem as the zebra mussel becomes poisoned by such pollutants and many catfish die. Hypothesize impacts on the food web for this ecosystem (**Figure 1**). Explain the reasoning behind your hypotheses.

6. Zebra mussels often grow in high concentrations on water intake pipes. Assume that **Table 1** represents the number of mussels per square metre of pipe surface sampled over a 10-year period.

Table 1 Zebra Mussel Population in a Small Water Body, 1991–2000

Year	1991	1992	1993	1994	1995	1996	1997	1998	1999	2000
Population (per m²)	400	520	676	879	1142	1485	1930	2509	3262	4241

(a) Draw a population curve for the zebra mussel population from 1991 to 2000. Label your axes, and give your graph an appropriate title.
(b) Describe the growth illustrated by your graph.
(c) Calculate the population growth rate of zebra mussels from 1996 to 2000.
(d) If no measures are taken to control the zebra mussel population in this location, hypothesize what the population would be in 2010. Add these data to your graph.
(e) What factors are likely to limit such growth?

Making Connections

7. Zebra mussels clog pipelines and other underwater structures (**Figure 2**). What social and economic impacts could this problem cause?

Figure 2

Population Ecology

In this chapter, you will be able to

- describe characteristics of a population, including density, distribution and dispersion, growth, carrying capacity, and minimum viable size;

- compare and explain, in terms of carrying capacity, reproductive success and predation, and the fluctuation of populations of plants, animals, and microorganisms;

- use conceptual and mathematical models to determine exponential, sigmoidal, and sinusoidal growth of different populations within ecosystems;

- explain the concepts of competition, predation, defence mechanisms, and symbiotic and parasitic relationships among different organisms;

- using the ecological hierarchy for living things, evaluate how a change in one population can affect the entire hierarchy, both physically and economically;

- determine experimentally the characteristics of population growth of predator and prey populations.

Populations of organisms are dynamic. Some populations, such as those of the African black rhinoceros and the Vancouver Island marmot, are in serious decline and threatened with imminent extinction unless drastic action is taken. Other populations, such as those of the ring-billed gull in Ontario and the cane toads in Australia, are experiencing unprecedented growth. While the number of chimpanzees, our closest living relative, has declined from about 2 million in 1900 to less than 150 000 at present, our own population has increased by more than 4 billion in the same time frame. Can the extinction of an entire species be avoided? What are the consequences of rapid population growth? To answer these questions, biologists must study populations carefully and observe and monitor changing environmental conditions. Changes in population numbers and in the patterns of distribution of individuals can have direct effects on the local ecosystem and may affect the well-being of other species within the ecological community.

Population ecologists use specialized methods to monitor, quantify, and model changes in populations. They also study the interrelationships between different species. In this way, they gather data necessary to predict future trends in the growth of populations. This information can be used to assess the health of individual species and entire ecosystems, to develop policies and plans of action to save species from extinction, and to address the impacts of rapidly growing populations.

REFLECT on your learning

Study the photo on the opposite page and reflect on the following:

1. What possible relationships might exist among these animals?

2. List and explain factors in this environment that might be responsible for the organisms present there.

3. What conditions in ecosystems and which activities by human and other animals might affect the number of individuals within each population?

TRY THIS activity   *Interaction*

- Study **Figure 1** and answer the following questions
  - (a) What do you think happened in this picture? Be prepared to explain your reasoning.
  - (b) What relationships exist between the organisms?
- With two or three other students, share your explanations about what happened in the picture.
  - (c) Were your explanations similar? different? How?
  - (d) Explain whether you think there is one "right" explanation.

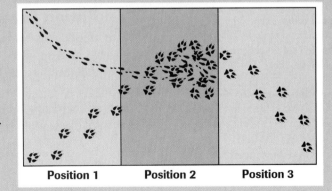

| Position 1 | Position 2 | Position 3 |

Figure 1

Canadian wildlife biologists have expressed concern over the increase in the Greater Snow Goose population in the eastern Canadian Arctic from 50 000 in the late 1960s to about 700 000 in the spring of 1997. It is predicted that the population, which inhabits the Canadian Arctic in summer to breed, will have reached 1 million by 2002 (**Figure 1**). The presence of increasing numbers of these snow geese has affected other species within the **habitat**. Other populations of the same **species** occupy habitats elsewhere in the Canadian Arctic during the summer. Overgrazing and feeding on plants by a closely related species, the Lesser Snow Goose, has caused widespread damage to vegetation of Arctic coastal sites, resulting in a decline in the abundance of other bird and wildlife species that also depend on these coastal habitats for resources. In the fall, both species of snow geese migrate south, stopping to feed on agricultural crops in central and eastern Canada and the United States, so many farmers regard the geese as pests. Members of the Arctic Goose Habitat Working Group, a consortium of Canadian and American wildlife biologists, have recommended that, to restrict damage to Arctic ecosystems, the total population of each of these species be reduced. How do biologists count huge populations of birds that migrate each fall, produce young each spring, and die at different times? How can they determine what population size might be ideal for a particular habitat and how can they tell when a population reaches this ideal size?

habitat the place where an organism or species normally lives

species organisms that resemble one another in appearance, behaviour, chemistry, and genetic makeup, and that interbreed, or have the ability to interbreed, with each other under natural conditions to produce fertile offspring

Figure 1
Greater Snow Geese are endangering their own survival by exceeding the carrying capacity of the natural marshes along the St. Lawrence River.

Population Size and Density

To study populations, scientists measure such characteristics as **population size**, or the estimated total number of organisms, as well as the density and dispersion of organisms within their habitat. The **population density** (D) of any population is calculated by dividing the total numbers counted (N) by the space (S) occupied by the population. For example, the population density of 480 moose living in a 600 hectare (ha) region of Algonquin Park would be calculated as follows:

population size the number of individuals of a specific species occupying a given area/volume at a given time

population density the number of individuals of the same species that occur per unit area or volume

$$D = \frac{N}{S}$$

$$D = \frac{480 \text{ moose}}{600 \text{ ha}}$$

$$= 0.8 \text{ moose/ha}$$

KEY EQUATIONS

$$D = \frac{N}{S}$$

Populations vary widely among different species occupying different habitats. As shown in **Table 1**, small organisms usually have higher population densities than larger organisms. These widely ranging densities pose different challenges to biologists attempting to gather data on a particular species.

Table 1 Examples of Population Densities

Population	Density
jack pine	380/ha
field mice	250/ha
moose	0.8/ha
soil arthropods	500 000/m^2
phytoplankton (in a pond)	4 000 000/m^3

Population density can be deceiving because of unused or unusable space within a habitat. Ecologists, therefore, distinguish between crude density and ecological density. **Crude density** is the number of individuals of the same species per total unit area or volume, and **ecological density** is the number of individuals of the same species per unit area or volume actually used by the individuals. For example, the moose in Algonquin Park do not utilize open lake water. Therefore, if the 600 ha of total habitat included 70 ha of open lake water, the ecological density of the moose population would be 480 moose/(600 − 70) ha, or 0.9 moose/ha. The crude density, however, is 0.8 moose/ha, the same as the population density value we calculated earlier.

crude density population density measured in terms of number of organisms of the same species within the total area of the entire habitat

ecological density population density measured in terms of the number of individuals of the same species per unit area or volume actually used by the individuals

▶ *Practice*

Understanding Concepts

1. Calculate the crude density of a population of painted turtles (**Figure 2**) if 34 turtles were counted in a 200-ha park.

Answer

1. 0.17 turtles/ha

Figure 2

2. Speculate about areas within the park that might not be used by the painted turtles.

3. Suggest a possible proportion (%) of the park that is not used by the turtles, and calculate the ecological density.

population dispersion the general pattern in which individuals are distributed through a specified area

clumped dispersion the pattern in which individuals in a population are more concentrated in certain parts of a habitat

uniform dispersion the pattern in which individuals are equally spaced throughout a habitat

random dispersion the pattern in which individuals are spread throughout a habitat in an unpredictable and patternless manner

Environmental conditions and suitable niches will differ throughout a population's geographic range. For this reason, the **population dispersion** of groups of organisms within a population will vary throughout the range. Biologists have identified three main dispersion patterns among wild populations: clumped, uniform, and random. Most populations exhibit patchy or **clumped dispersion**, in which organisms are densely grouped in areas of the habitat with favourable conditions for survival. Cattails are an example of an organism that exhibits clumped dispersion. They are usually restricted to growing along the edges of ponds and lakes, or in other wet soils (**Figure 3**). Clumped dispersion may also be the result of social behaviour, such as fish swimming in large schools to gain protection from predators, as shown in the photo in **Figure 4(a)**. In contrast to clumped dispersion, organisms may exhibit **uniform dispersion** in which individuals are evenly distributed throughout the habitat. This pattern may result from competition between individuals that set up territories for feeding, breeding, or nesting. When King penguins nest on South Georgia Island in the South Atlantic Ocean, they often exhibit a nearly uniform dispersion pattern, as shown in **Figure 4(b)**. Although wild species rarely exhibit uniform dispersion, farmers' fields, orchards, and tree plantations are often uniformly dispersed. Individuals exhibit **random dispersion** when they are minimally influenced by interactions with other individuals and when habitat conditions are also virtually uniform. As shown in **Figure 4(c)**, some species of trees in tropical rain forests exhibit random dispersion, although this pattern is also rare in nature.

Figure 3
Cattails thrive in wet nutrient-rich environments.

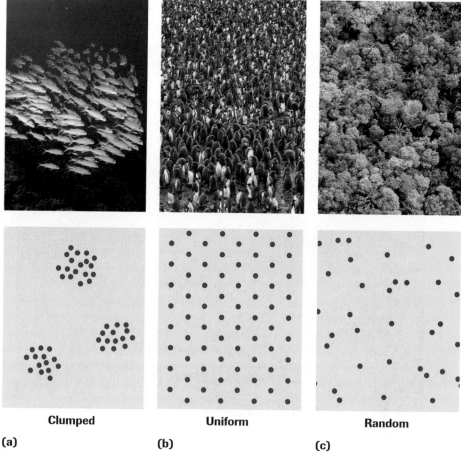

Clumped **(a)**   **Uniform** **(b)**   **Random** **(c)**

Figure 4
Populations generally exhibit one of three patterns of dispersion: **(a)** Yellow goatfish are often found clumped in schools. **(b)** Nesting King penguins exhibit a uniform pattern. **(c)** In tropical rain forests, trees of the same species can be randomly dispersed.

Measuring Population Characteristics

In very rare instances, biologists can make an exact count of the total number of individuals in a population. If, for example, organisms are isolated in a laboratory population, or if organisms are large and found in a relatively small area, counting all individuals might be possible. However, populations are dynamic and their numbers and geographic locations change over time, generally making a precise count impractical. Instead, biologists count a sample of the population at a particular time, then estimate a total size. In some cases, an estimate or index of population size can be made using indirect indicators, such as the number of fecal droppings, the number of tracks, or the number of nests or burrows (**Figure 5**). Biologists use a variety of sampling techniques to estimate the size and density of wild populations.

Accurate estimates of population size and density can be extremely valuable. In the forestry industry, staff must be able to determine the population density and size of valuable tree species. This information is essential in determining allowable harvest rates while still maintaining healthy and viable populations.

A common sampling technique for stationary or small organisms, such as plants and insects, is to isolate a defined area and count samples of one or more populations within that area. A large area, such as a tract of grassland or forest, can be sampled using small, selected plots, in which a sampling frame, or **quadrat**, is placed (**Figure 6**). The number of individuals of one or more species can be counted within each quadrat. Population size and density can be estimated through calculations based on counts within a representative area or all of the selected areas. Average estimates of population size and density for the entire area can then be extrapolated based on these calculations. Quadrat sampling is most effective for stationary species, such as the populations of different tree species in a forest.

Figure 5
Counting the number of nests made by osprey can be a useful indicator when estimating population size.

quadrat a sampling frame used for estimating population size; frames can be real or virtual

Figure 6
Quadrats can be used to estimate the population of different species in a large tract of grassland

Estimating Population Size and Density

Example

A student wants to estimate the population size and density of ragweed plants in a large field measuring 100 m × 100 m. She randomly places three 2.0 m × 2.0 m quadrats in the field. Estimate the population density and size if she finds 18, 11, and 24 ragweed plants in the three quadrats.

Solution

An estimate of the ragweed plant's population density may be obtained by calculating the average density of the sampled quadrats.

$$\text{average sample density} = \frac{\text{total number of individuals}}{\text{total sample area}}$$

$$= \frac{18 + 11 + 24}{(2.0 \text{ m})^2 + (2.0 \text{ m})^2 + (2.0 \text{ m})^2}$$

$$= 4.4 \text{ ragweed plants/m}^2$$

KEY EQUATIONS

$$\text{average sample density} = \frac{\text{total number of individuals}}{\text{total sample area}}$$

Assuming that the sample area is representative of the total study area, the estimated population density of ragweed is 4.4 plants/m².

An estimate of the population size may be obtained by multiplying the estimated population density by the total study area.

$$\text{estimated population size} = (\text{estimated population density})(\text{total size of study area})$$

$$= 4.4 \text{ plants/m}^2 \times 10\ 000 \text{ m}^2$$

$$= 44\ 000 \text{ plants}$$

▶ Practice

Answers

4. estimated population density
   = 5.4 slugs/m²

   estimated population size
   = 540 slugs

Understanding Concepts

4. To estimate the size of the slug population on a golf course, biology students randomly selected five 1.0-m² quadrats in a 10 m × 10 m site. The numbers of slugs in each quadrat were 4, 8, 9, 5, and 1. Estimate the population density and size of slugs in this study site.

---

mark–recapture method
sampling technique for estimating population size and density by comparing the proportion of marked and unmarked animals captured in a given area; sometimes called capture–recapture

Freshwater biologists face the special challenge of not being able to readily observe populations of aquatic organisms, such as valuable fish species. In addition, such populations often demonstrate a clumped dispersion. A common sampling technique for estimating the size and density of mobile wildlife populations, such as fish, is the **mark–recapture method**, in which a sample of animals is captured, marked in some way and then released. By using the mark–recapture method, often with the assistance of commercial and sport fishers, fisheries scientists are able to make good estimates of population sizes and densities on which to base management programs, such as restocking programs.

The mark–recapture method is also used on large animals or those that pose potential danger to researchers, such as polar bears. The animals may be tranquillized, captured, and then marked with tags or bands with sequential numbers on them, or with coloured dye. They are then released. Technicians may also use such opportunities to weigh, measure, and examine tranquillized individuals. After a period of time that allows marked animals to mix randomly with unmarked animals in the population, researchers capture a second sample. In this sample, the proportion of marked animals to unmarked animals provides a basis for estimating the size of the entire population.

The techniques for capturing and marking individual organisms must be carefully planned so that the chances of each individual being caught are equal. Marking techniques must not harm the organism or restrict its normal activities and must remain clearly visible for the duration of the study. Marks must also not alter the chances of being recaptured. The accuracy of mark–recapture as a sampling method for estimating population numbers and/or density depends on certain assumptions:

- Every organism in a population has an equal opportunity of being captured.
- During the time period between the initial marking and the subsequent recapture, the proportion of marked to unmarked animals remains constant.
- The population size does not increase or decrease during the sampling study.

Under ideal conditions, marked and unmarked individuals of the population are captured at random and, during the period of the sampling, no new individuals enter the population, no marked animal dies, and none of the marked animals leave the population.

Mark–Recapture Sampling

Assume a lake is home to a population of 3000 pickerel. If a fisheries biologist captured and marked 300 pickerel and released them back into the lake, the proportion of marked fish in the population would be 0.10 or 10%. Therefore, if the marked fish mix randomly in the population we would expect that if a second sample were taken 0.10 or 10% of the sample would be marked "recaptures." The proportion of marked fish in the entire population is expected to equal the proportion of marked recaptures in a sample.

$$\frac{\text{total \# marked } (M)}{\text{total population } (N)} = \frac{\text{\# of recaptures } (m)}{\text{size of second sample } (n)}$$

KEY EQUATIONS

$$\frac{M}{N} = \frac{m}{n} \text{ or } N = \frac{Mn}{m}$$

Estimating Population Size: Mark–Recapture Sampling

SAMPLE problem ◀

Consider a fish population of unknown size from which 26 individuals are randomly captured, marked, and released, as shown in **Figure 7(a)–(c)**, page 656. Assume that the released individuals move randomly through the population, as illustrated in **Figure 7(d)**. If a second sample of 21 individuals is captured sometime later in which three individuals are found to be marked, as shown in **Figure 7(e)**, we may use these values to estimate the population size.

We assume that the ratio of marked individuals to the size of the total population equals the ratio of marked individuals recaptured in the second sample to the size of the second sample.

If we let N represent the population size, then

number of marked individuals in the population ⟶ $\dfrac{26}{N} = \dfrac{3}{21}$ ⟵ number of marked individuals in second sample (recaptures)

population size ⟶   ⟵ size of second sample

$$N = \frac{26 \times 21}{3}$$

$$N = 182$$

Therefore, the estimated population size is 182.

▶

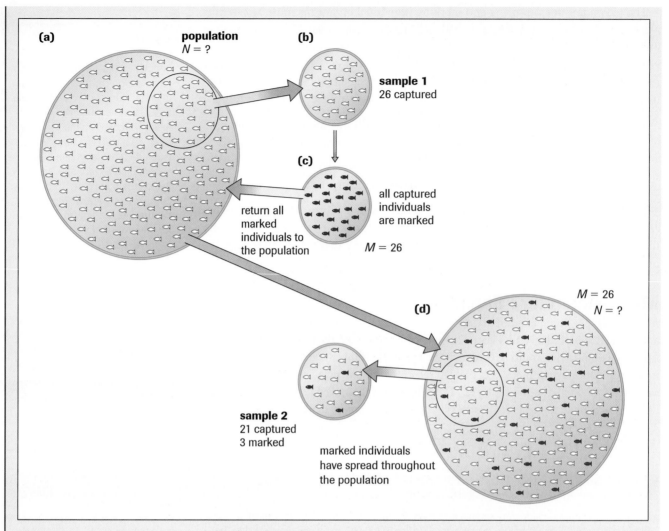

Figure 7
Mark–recapture sampling of a
fish population

Example

Using the mark–recapture method, wildlife researchers surveyed an area of wetlands where 80 wood ducks were captured in traps, marked with permanent metal bands, and then released. Two weeks later, 110 wood ducks were captured. Of the wood ducks recaptured, 12 were marked. Estimate the total size of the wood duck population in the survey area.

Solution

$$N = \frac{Mn}{m}$$

$M = 80$

$m = 12$

$n = 110$

$$N = \frac{(80)(110)}{12}$$

$N = 733.33$

The estimated size of the wood duck population is 733.

Understanding Concepts

5. In a river in British Columbia, 430 sockeye salmon were captured and marked on the fin with a uniquely numbered T-bar anchor tag. Two weeks later, a total of 154 sockeye salmon were recaptured and 15 bore the tags on their fin.
   (a) Estimate the sockeye salmon population in this river during this study.
   (b) Identify conditions that must be met in this study to obtain a valid estimate.

Answer

5. (a) N = 4415

▶ **TRY THIS** activity   *Mark–Recapture Analysis*

You and the students in your class will represent a "captured" sample of students from your school's grade 12 population. Imagine that all of you have been marked in some way and released back into the population at large (at the end of class). Now assume that one of your other grade 12 classes represents a second sample from the same grade 12 population. Any students from your biology class who are in this second class will be considered "recaptures."

• Record the number of students in your biology class and the number of "recaptured" students in your second class.
   (a) Use this data and the appropriate formula to estimate the total population of grade 12 students in your school.
   (b) Compare your estimate to the actual population of grade 12 students and account for any difference in the values.
   (c) Should you count yourself in the first and/or second samples? Explain. ▨

⚗ INVESTIGATION 14.1.1

Estimating Population Size (p. 689)

What methods can be used to estimate the size of a population? In this investigation, you will apply quadrat and mark–recapture sampling methods to test your predictions of the estimated size of a population.

Technological Tracking of Wild Populations

When capturing organisms for estimating population size, researchers may also attach radio collars, satellite-linked devices, or other technological equipment to selected individuals for tracking their migration and/or behaviour patterns (**Figure 8**). It is essential that radio- and satellite-linked devices be firmly attached but that they not harm animals or restrict their activities. Using these technologies, animals can be tracked across their geographic range and the patterns can be mapped in geographic information systems (GIS). It becomes possible to locate dens of hibernating animals with great accuracy. Information about population dispersion patterns and migration activities can be greatly enhanced by tracking animals through such technologies.

DID YOU KNOW ?

Tiny Monitors

Satellite-linked devices weighing less than 0.5 g and measuring about half the size of a person's smallest fingernail can be attached to a variety of species, such as lizards or salamanders, to study their movement and/or behaviour patterns.

Figure 8
Radio collars on a fox and a moose calf

It is difficult to study populations and migration patterns of open-ocean predators, such as sharks, blue marlin, and tuna, using tagging and recapturing methods. In one study, fewer than 1% of over 20 000 tagged blue marlin were recaptured. Researchers now use specially designed tags that send data via satellites to monitor migration habits of these open-ocean predators. The tags are torpedo-shaped microcomputers about 13 cm in length. They are attached into the dorsal area of the fish with a harpoon designed for this purpose. At predetermined intervals, the tag records light levels and duration, temperature, and pressure, which allow scientists to pinpoint the precise location of the fish. Held in place by a short fishing line that dissolves on a signal programmed into the tag, the tag can then float to the ocean surface with the antenna upward, where it transmits data to satellites. From this data, biologists are better able to study migration patterns and populations. For example, recent data from satellite-transmitting tags indicate that great white sharks and blue marlin migrate across the Pacific Ocean and back, whereas scientists previously believed them to move only in coastal waters.

Ethics of Studying Wild Populations

As with any type of animal research, concerns have been raised over the ethics of studying wild populations, especially through mark-recapture sampling and tracking techniques. What are the effects on animals being pursued, captured, and marked? What are long-term effects of repeated tranquillizing of large animals? Are marking techniques humane? Can collars and tags harm animals? Some scientists have expressed concern that the handling of animals during such experiments can affect the animals' behaviour, such that they act differently after their release. Others insist that traps should be designed to capture animals in the least harmful and damaging fashion.

The Canadian Council on Animal Care (CCAC) is developing a set of ethical guidelines for wildlife research based on the principles of the three *R*s: reduction, refinement, and replacement. The CCAC encourages researchers to *reduce* their use of animals as much as possible in their studies; the CCAC supports the use of techniques *refined* to minimize pain and distress; and the CCAC advises biologists to *replace* the trapping of animals wherever possible with computer models to estimate population size. All wildlife research must receive approval from CCAC committees before a project can be started.

Measuring or estimating population size and density are valuable tools used in the development of strategies for reducing effects of environmental change, whether natural or human influenced. Data for changes to population size and density alone do not provide a full understanding of population characteristics and interrelationships. Populations do not remain static. The variety of factors that cause them to continually change is also an important area of study.

SUMMARY   *Scientific Measurement of Populations*

- Biologists use different measurements, such as population size and population density, to describe and monitor populations, and to contribute to the scientific management and conservation of species.

- Populations in a given geographical range exhibit one of three distinct dispersion patterns: clumped, uniform, or random.

- Quadrat analysis and the mark–recapture method are two sampling techniques that biologists use to estimate population size and density.

Understanding Concepts

1. The Arctic Goose Habitat Working Group recommended that the eastern arctic Greater Snow Goose population be held between 800 000 and 1 million birds by 2002. This reduced population would still be 15 to 20 times greater than the population in the late 1960s.
   (a) What are some consequences of the population remaining so large?
   (b) Discuss some ways in which reductions of geese populations may be achieved.

2. In a group, brainstorm and discuss challenges that biologists encounter in estimating population characteristics for wild populations of
   (a) whales that migrate along the western coasts of North and South America;
   (b) algae that live in water bodies receiving excess fertilizers in runoff from cropland;
   (c) caribou that inhabit an Arctic tundra environment; and
   (d) amphibians that live in marshes.

3. A freshwater biologist studying a population of yellow perch in a large, shallow lake found the fish scattered in reed beds over 25% of a 900-ha area. She caught, tagged, and released 195 perch. Three days later, she resampled the lake and caught 210 fish of which 10 were recaptures.
   (a) What is the dispersion pattern of the yellow perch population in this sanctuary?
   (b) Estimate the population size of the yellow perch in this wildlife sanctuary, and calculate the population density.

4. Prairie dogs live in grassland habitats where they form large burrowing colonies. Describe the method you would use to estimate the population size and give reasons for your choice.

Applied Inquiry Skills

5. For a grade 12 biology project, students decided to investigate the effects of pollution on a population of grasshoppers. The students investigated a population of grasshoppers in the 2.0-ha school yard and in a 1.0-ha field near a local oil refinery. Estimates of population size were used as indicators of survival. Data were collected using the mark–recapture method in which a tiny drop of coloured nail polish was used to mark captured grasshoppers.
   (a) From the data in **Tables 1** and **2** calculate the population size and density for each area.
   (b) What conclusions can be drawn from the data?

Table 1 Mark–Recapture of Grasshoppers in School Yard

Number of grasshoppers marked	280
Number of grasshoppers captured in a second sample	130
Number of marked grasshoppers recaptured	8

Table 2 Mark–Recapture of Grasshoppers Near Oil Refinery

Number of grasshoppers marked	150
Number of grasshoppers captured in a second sample	70
Number of marked grasshoppers recaptured	26

6. There are large wildlife populations in the Masai Mara Ecosystems of Kenya in North Africa. Some populations, like that of the elephant, are relatively stable while others, such as the giraffe, are experiencing a serious decline.
   (a) Calculate the population density of elephants in the Masia Mara where 1800 elephants occupy a region measuring approximately 40 km by 60 km.
   (b) What area was occupied in 1977 by the population of 5306 giraffes then having a density of 1.4 animals per km^2?
   (c) Determine the population density after the number of giraffes in the same area declined to just 1050 individuals by 1998?

Making Connections

7. A team of wildlife biologists plans to research the dispersion, behaviour, reproduction, and migration patterns of the rare and endangered North Atlantic right whale population. Serious threats to this population require that its health and status be studied. A detailed project proposal for this research must be approved by the Canadian Council on Animal Care. Draft this proposal, which should include
   (a) the rationale for the research;
   (b) a description of the techniques and technology to be used for the project;
   (c) a discussion of potential long- and short-term effects from using these techniques and/or technology;
   (d) an outline of how the proposed research is consistent with the CCAC three *R*s philosophy.

An ecosystem has finite biotic and abiotic resources at any given time. Biotic resources, such as prey, vary in availability. Some abiotic resources, such as space and light, vary little, while others, such as temperature and water availability, vary greatly. There is, therefore, a limit to the number of individuals that an environment can support at any given time. The **carrying capacity** of an ecosystem is the maximum number of organisms that can be sustained by available resources over a limited period of time. Carrying capacity is dynamic, since environmental conditions are always changing. Two populations of the same species of fish, for example, might occupy quite different ecosystems with different carrying capacities, due to biotic and abiotic variations in the environment. A large, nutrient-poor, oligotrophic lake (**Figure 1(a)**) would have a smaller carrying capacity per unit area than a much smaller, nutrient-rich eutrophic environment (**Figure 1(b)**). Therefore, the population density of fish in the oligotrophic lake would be much lower than that of the fish in the eutrophic lake. The population size in the oligotrophic lake might be limited by available food while the population size in the eutrophic lake might be limited by available space. When populations increase in size, the amount of resources available per individual decreases. When populations change in density, their new density may exceed the available supply of resources. A variety of factors affect populations and influence how rapidly populations can grow before they meet or exceed the carrying capacity of their environment.

carrying capacity the maximum number of organisms that can be sustained by available resources over a given period of time

(a)

(b)

Figure 1
Carrying capacity is determined by the environment in which a population lives.

Factors That Affect Population Growth

Populations are always changing. Depending on the species and on environmental conditions, populations experience natural hourly, daily, seasonal, and annual fluctuations in numbers. Size can change when individual organisms are added to the population through births or removed through deaths. In addition, population size may increase when individuals immigrate and decrease when individuals emigrate. Changes in population characteristics are known as **population dynamics**. The main natural determinants, measured per unit of time, are natality (the number of births), mortality (the number of deaths), immigration (the number of individuals that move into an existing population), and emigration (the number of individuals that move away from an existing population). These determinants may vary from species to species. The females of some species have a high **fecundity**, having the potential to produce very large numbers of offspring in their lifetimes—many species of star fish for example can lay over

population dynamics changes in population characteristics determined by natality, mortality, immigration, and emigration

fecundity the potential for a species to produce offspring in one lifetime

1 million eggs per year. In contrast, a female hippopotamus may have the potential to give birth to just 20 young in an entire lifetime of 45 years. While longer lived species have the potential to complete more reproductive cycles—an adaptation that increases their fecundity—many such species produce fewer offspring per reproductive cycle.

Biologists recognize three general patterns in the survivorship of species (**Figure 2**). Species exhibiting a type I survivorship pattern have very low mortality rates until they are beyond their reproductive years and have a correspondingly long life expectancy. Such species are typically slow to reach sexual maturity and produce relatively small numbers of offspring. At the other extreme, species with a type III survivorship curve have very high mortality rates when they are young while those individuals that do reach sexual maturity have a greatly reduced mortality rate. They often produce large numbers of offspring. The result is a very low average life expectancy for the species. The green sea turtle is an excellent example of a species with many type III characteristics. Females each lay hundreds of eggs but less than 1 % reach adulthood. Those that become adults have few predators and live long lives. Type II species are intermediate between these forms and tend to show a uniform risk of mortality throughout their life.

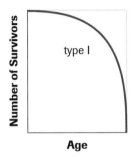
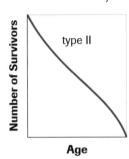
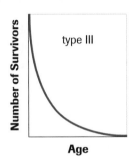

Figure 2
Three generalized types of survivorship curves. Type I populations show high survivorship until fairly late in life and then show high mortality. Type II populations show a fairly constant death rate. Type III populations show low survivorship early in life.

Under natural conditions, the number of offspring that are actually produced in an individual's lifetime, their fertility is often significantly less than their fecundity. Food availability, mating success and disease are just a few of the many factors that act to limit reproductive potential. In some but not all populations, immigration and emigration can also be significant factors. Species that are very mobile are able to quickly move into new environments. While mature plants are not mobile, their offspring often are; small wind born seeds allow for mass movement of new individuals into an area—such as after a forest fire or other major disturbance. Populations of other species that have little mobility or live in very isolated environments are unaffected by immigration or emigration.

In addition to natural factors, human actions also affect birth, death, immigration, and emigration rates. But regardless of the specific forces responsible, population ecologists must be able to quantify these rates of change in order to monitor and evaluate population dynamics. Mathematical models provide the underlying foundation for this science.

Calculating Changes in Population Size

Births, deaths, immigration, and emigration can be used to determine the growth rate of a population in a given period of time. The population growth of any given population is calculated mathematically using the following formula:

$$\text{population change} = \frac{[(\text{births} + \text{immigration}) - (\text{deaths} + \text{emigration})]}{\text{initial population size } (n)} \times 100$$

KEY EQUATIONS

$$\text{population change} = \frac{[(b + i) - (d + e)]}{n} \times 100$$

The solution is expressed as a percentage. This formula is useful for biologists to determine the changes in population size over time. If the number of individuals that were born and migrated into the population is higher than the number of individuals that died and emigrated, the population will have positive growth, increasing in size. Conversely, if the number of deaths and emigration exceeds the number of births and immigration, the population will experience negative growth, decreasing in size. If the number of births and immigration equals the number of deaths and emigration, the population is said to have zero growth.

The type of growth exhibited by a population also depends on whether the population is open or closed. An **open population** refers to a population that is influenced by the factors of natality, mortality, and migration. Most wild populations are open, since they have the ability to immigrate and emigrate between populations that exist in different locations. A **closed population** refers to one in which only natality and mortality determine population growth, since immigration and emigration do not occur. Closed animal populations are rare. Land-based populations that exist on secluded islands, such as the Peary caribou herd that inhabit an Arctic Ocean island, can be thought of as closed because they have no easy means to travel to other populations. (The animals are able to move between islands in winter.)

Some populations under study by scientists become closed for the purposes of their research. Biologists engaged in mark–recapture sampling leave only a short period of time between captures, so that the population being studied is effectively closed. To get an accurate estimate of population size, it is important that no new individuals enter or leave the population during the study. Bacteria and other microscopic organisms being monitored in laboratories have no ability to emigrate, nor can new bacteria immigrate into the culture. Thus, these populations are closed.

Biotic Potential

A basic characteristic of living things is their ability to reproduce. Consider a single *Escherichia coli* (*E. coli*) bacterium on a hamburger patty. Under ideal conditions, *E. coli* can reproduce by binary fission every 12 min. After 12 min there would be two bacterial cells, and after 24 min there would be four cells. If this doubling continued unchecked for the next 24 h, there would be enough *E. coli* cells to cover the entire surface of Earth to a depth of more than 1 m! For any organism, the maximum reproductive rate that could be achieved under ideal conditions is called the **biotic potential**, or the intrinsic rate of natural increase, and is represented by r.

Population Growth Models

A simple model of population growth can be illustrated by graphing the change in population size over time. **Figure 3** shows how this is done for a hypothetical example of White-tailed deer, using the data from **Table 1**.

When birth rates and death rates per individual remain constant, populations grow at a fixed rate in a fixed time interval. Such a fixed rate of growth can be expressed as a ratio or percentage per unit time such as 1.05 (or a growth rate of 5%) per year.

In human populations, growth is continuous because deaths and births occur at all times of the year. For many species, however, while deaths can occur throughout the year, births are restricted to a particular time of the year called the breeding season. In such species, the population typically grows rapidly during the breeding season and then declines throughout the remainder of the year until the next breeding season begins. These populations exhibit **geometric growth**. Their growth rate is a constant and can be determined by comparing the population size in one year to the population size at the same time the previous year. This ratio is the geometric growth rate and is symbolized by the Greek letter lambda (λ).

open population a population in which change in number and density is determined by births, deaths, immigration, and emigration

closed population a population in which change in size and density is determined by natality (birth rate) and mortality (death rate) alone

biotic potential the maximum rate a population can increase under ideal conditions

geometric growth a pattern of population growth where organisms reproduce at fixed intervals at a constant rate

KEY EQUATIONS

$$N(t + 1) = N(t)\lambda$$

$$\lambda = \frac{N(t + 1)}{N(t)}$$

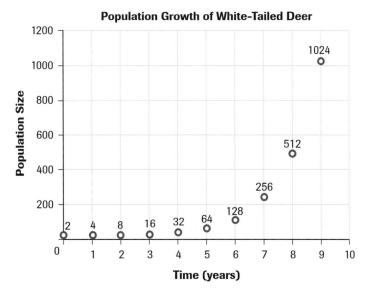

Table 1 Growth of a White-Tailed Deer Population

Time (years)	Population size
0	2
1	4
2	8
3	16
4	32
.	.
.	.
.	.
8	512
9	1024
100	$> 10^{30}$

Figure 3
A simple model of geometric population growth of White-tailed deer, where the population increases at a constant rate (it doubles every 12 months) in an unlimited environment.

Where λ is the fixed growth rate and N is the population size in year $(t + 1)$ and year (t) respectively. By rearranging we find population size at any time (t) as follows:

$N(1) = N(0)\,\lambda$

$N(2) = N(0)\,\lambda \times \lambda$

$N(3) = N(0)\,\lambda \times \lambda \times \lambda$

or

$N(t) = N(0)\lambda^{t}$

In the above equations $N(0)$ represents the initial population size and $N(1)$, $N(2)$, $N(3)$ and $N(t)$ are the population sizes at the end of 1, 2, 3 and t years respectively.

Calculating Geometric Growth

SAMPLE problem ◀

Example
Each May, harp seals give birth on pack ice off the coast of Newfoundland. In a hypothetical scenario, an initial population of 2000 seals gives birth to 950 pups, and during the next 12 months, 150 seals die.

(a) Assuming that the population is growing geometrically, what will the harp seal population be in two years?

(b) Assuming the same geometric growth rate, calculate the population size after eight years.

Solution

(a) In year 1, the population change = 950 seals (births) − 150 seals (deaths)

$$= 800 \text{ seals}$$

Initial population $N(0) = 2000$ seals,

Population at end of year 1, $N(1) = 2000 + 950 - 150$

Geometric growth rate $(\lambda) = \dfrac{N(t+1)}{N(t)} = \dfrac{2800}{2000} = 1.4$

▶

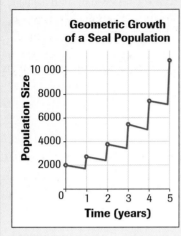

Geometric Growth of a Seal Population

Figure 4

Since the growth rate remains constant in geometric growth, the population size at the end of year 2 is as follows:

$$N(t + 1) = N(t)\lambda$$
$$N(2) = 2800 \times 1.4$$
$$= 3920$$

Therefore, after 2 years of geometric growth, the seal population would reach 3920. **Figure 4** graphically illustrates this population growth.

(b) $N(0) = 2000$, $\lambda = 1.4$, $t = 8$ years

$$N(8) = N(0) \lambda^8$$
$$= 2000 \times (1.4)^8$$
$$= 2000 \times 14.76$$
$$= 29\ 520$$

With the same growth rate the seal population will have reached 29 520 at the end of 8 years.

> ▶ **Practice**

Understanding Concepts

Answers

1. (a) $\lambda = 1.06$
   (b) $N(2) = 56\ 180$,
       $N(10) = 89\ 540$

1. A nesting colony of gannets on Isle Bonaventure exhibits geometric growth. During the year a initial population of 50 000 birds had 32 000 births and 29 000 deaths.
   (a) Calculate the geometric growth rate (λ).
   (b) Estimate the population sizes after 2 and 10 years.

exponential growth a pattern of population growth where organisms reproduce continuously at a constant rate

A wide variety of species are able to reproduce on a continuous rather than intermittent basis. **Exponential growth** describes populations growing continuously at a fixed rate in a fixed time interval. For such populations the fixed growth rate for a fixed time interval (λ) can be determined using the same formula as for populations growing geometrically. However, unlike geometric growth, the chosen time interval is not restricted to that of a particular reproductive cycle. For this reason ecologists are able to determine the instantaneous growth rate of the population expressed in terms of the intrinsic (per capita) growth rate (r). The intrinsic growth rate is the difference between the per capita birth rate, b, and the per capita death rates, d, where $r = (b - d)$. The population growth rate is given by the expression.

$$\frac{dN}{dt} = rN$$

Where dN/dt is the instantaneous growth rate of the population, r is the growth rate per capita and N is the population size.

This is a more mathematically complex model than those we have examined previously and is derived using calculus. While we will not examine this mathematical model in detail, it can be shown that, for any population growing exponentially, the time needed for the population to double in size (the doubling time) is a constant. A useful approximation of such a population's doubling time (t_d) is given by the formula:

$$t_d = \frac{0.69}{r}$$

For example, if a population has a per capita growth rate of 0.020 per year (a 2% growth rate), the approximate time needed for the population to double would be 0.69/0.020 or 34.5 years.

Calculating Exponential Growth

Example
A population of 2500 yeast cells in a culture tube is growing exponentially. If the intrinsic growth rate *r* is 0.030 per hour, calculate:
(a) the initial instantaneous growth rate of the population
(b) the time it will take for the population to double in size
(c) calculate the size of the population after each of four doubling times.

Solution
(a) $r = 0.030$ per hour and $N = 2500$

$$\frac{dN}{dt} = rN$$

$$= 0.030 \times 2500$$

$$= 75 \text{ per hour}$$

When the population size is 2500 the instantaneous population growth rate is 75 per hour.

(b) $r = 0.030$

$$t_d = \frac{0.69}{r}$$

$$= \frac{0.69}{0.030}$$

$$= 23 \text{ hours}$$

The yeast population will double in size every 23 hours.

(c) $t_d = 23$ hours, initial population size is 2500

Table 2

Doubling times	Time (hours)	Population size
0	0	2500
1	23	5000
2	46	10 000
3	69	20 000
4	92	40 000

▶ Practice

Understanding Concepts

2. After the rainy season begins in the tropics, a small population of mosquitoes exhibits exponential growth. The initial population size is 980 and their intrinsic growth rate is 0.345 per day. Calculate:
   (a) the populations initial instantaneous growth rate.
   (b) the doubling time for the population.
   (c) How many doubling times will have to pass in order for the population to exceed 2 000 000? How many days is this?

Answers

2. (a) 338 per day
   (b) 2 days
   (c) 11 doubling times, 22 days

(a)

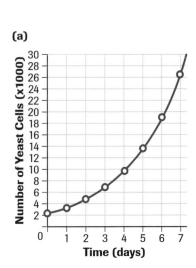

(b)

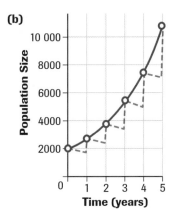

Figure 6
(a) Exponential growth curve
(b) Geometric growth curve

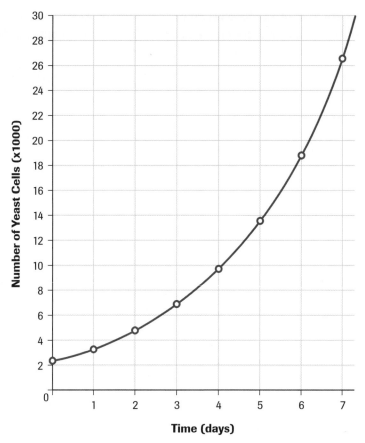

Figure 5
Exponential growth curve

Populations that grow exponentially increase in numbers rapidly, resulting in a J-shaped growth curve (**Figure 5**).

Note that the exponential growth curve is smooth as a result of continuous reproduction in contrast to the geometric growth curve, shown in **Figure 6(b)**, that fluctuates as a result of seasonal or intermittent reproductive cycles. However, the overall trend is the same in both cases, as **Figure 6(a)** and **(b)** illustrate.

Biologists studying populations are often more interested in their long-term growth patterns than their short-term seasonal fluctuations. For this reason, as **Figure 6(b)** shows, most populations' growth graphs are drawn as smooth curves indicating the changes in average population size over time.

Modelling Logistic Growth

Geometric and exponential models of population growth assume that a population will continue to grow at the same rate indefinitely. This implies that the population has continuous access to an unlimited supply of resources. Under these conditions, the intrinsic growth rate, r, is a maximum and is often symbolized as r_{max}. Of course, an unlimited resource supply is never the case in the real world. However, when a population is first starting out, resources are often plentiful and the population grows at a rapid rate.

As the population continues to grow, food, water, light, and space in the ecosystem can become factors that limit population growth as the resources are consumed and the population nears the ecosystem's carrying capacity. Under these conditions, the growth rate drops below r_{max}. As a result, reproduction slows down and the number of deaths closely resembles the number of births, and stable equilibrium is achieved where there is no net increase in population numbers (i.e., births equal deaths). This is the carrying capacity of the environment, and the population number at the carrying capacity is represented by K. **Logistic growth** represents the effect of carrying capacity on the growth of a population. Logistic growth is the most common growth pattern seen in nature.

This pattern in population change of a species over time can be represented with a mathematical model called the logistic growth equation, where the instantaneous growth rate of the population is calculated as follows:

logistic growth a model of population growth describing growth that levels off as the size of the population approaches its carrying capacity

$$\frac{dN}{dt} = r_{max}N\left[\frac{(K - N)}{K}\right]$$

$\frac{dN}{dt}$ is the population growth rate at a given time.

r_{max} is the maximum intrinsic growth rate.

N is the population size at a given time.

K is the carrying capacity of the environment.

Notice that if the population size (N) is close to the carrying capacity (K), then the mathematical expression $(K - N)/K$ approaches 0, and the population growth rate, $r_{max}N[(K - N)/K]$, also nears 0, which means growth virtually ceases. Thus, the equation takes into account declining resource availability as the population grows. The equation predicts the realized population growth rate from the maximum possible growth rate for a given population multiplied by the proportion of resources not yet used.

KEY EQUATIONS

$$\frac{dN}{dt} = r_{max}N\left[\frac{(K - N)}{K}\right]$$

Calculating Logistic Growth

SAMPLE problem ◄

Example

A population is growing continuously. The carrying capacity of the environment is 1000 individuals and its maximum growth rate, rmax, is 0.50.

(a) Determine the population growth rates based on a population size of 20, 200, 500, 900, 990 and 1000.

(b) Describe the relationship between population size and growth rate.

Solution

(a) $\frac{dN}{dt} = r_{max} N \times \left[\frac{(K - N)}{K}\right]$. See **Table 3** for calculated growth rates.

Table 3

r_{max}	Population size N	$\left[\frac{(K - N)}{K}\right]$	Population growth rate
0.50	20	$\frac{980}{1000}$	9.8
0.50	200	$\frac{800}{1000}$	80
0.50	500	$\frac{500}{1000}$	125
0.50	900	$\frac{100}{1000}$	45
0.50	990	$\frac{10}{1000}$	4.95
0.50	1000	0	0

▶

(b) When the population is small the population has a slow rate of growth. It increases as the population increases and then, as it approaches the carrying capacity, the growth rate declines.

> ▶ *Practice*

Understanding Concepts

3. Repeat the sample problem using an r_{max} value of 1.00.

4. Does the maximum rate of growth influence the relationship between population size and the environment's carrying capacity?

Logistic Growth Curve

lag phase the initial stage in which population growth rates are slow as a result of a small population size; characteristic of geometric, exponential, and logistic population growth

log phase the stage in which population growth rates are very rapid; characteristic of geometric, exponential, and logistic growth

environmental resistance any factor which limits a population's ability to realize its biotic potential when it nears or exceeds the environment's carrying capacity

stationary phase the phase in which population growth rates decrease as the population size reaches the carrying capacity and stabilizes; the defining characteristic of logistic population growth

dynamic equilibrium the condition of a population in which the birth rate equals the death rate and there is no net change in population size

The curve formed by a logistic growth pattern on a graph resembles the letter S. As a result, it is referred to as an S-shaped or sigmoidal curve, which has three distinct phases (**Figure 7**). The first, called the **lag phase**, occurs when the population is small and is increasing slowly. The second phase, called the **log phase**, occurs when the population undergoes very rapid growth. As available resources become limited, the population experiences **environmental resistance** and cannot continue rapid growth; therefore, reproduction slows, and the number of deaths increases. This is the **stationary phase**, which occurs at or close to the carrying capacity of the environment. At the stationary phase, a population is said to be in **dynamic equilibrium** because the number of births equals the number of deaths, resulting in no net increase in population size.

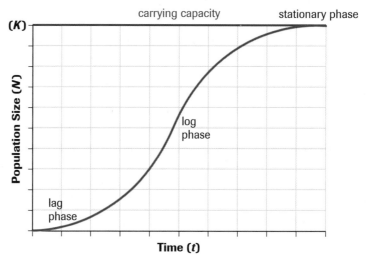

Figure 7
The logistic growth model results in a sigmoidal curve.

Logistic growth can be seen in a population of fur seals on St. Paul Island, Alaska. In 1911, fur seal hunting was banned on St. Paul Island, since the population had become extremely low. Because their numbers were so severely depressed, the seals had many unused resources to support the recovering population. The population began to grow rapidly until it stabilized around its carrying capacity, as shown on the graph (**Figure 8**). The logistic growth model works for some populations growing under suitable conditions but fits few natural populations perfectly.

Scientists studying wild populations draw on mathematical models, as it is impractical to conduct frequent field surveys of a species that interests them. Models can help

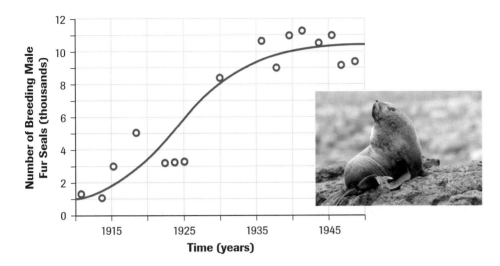

Figure 8
A population graph from 1910 to 1950 for fur seals on St. Paul Island, Alaska

researchers to predict patterns in population change based on data collected in the field. They can also show predictions for a variety of changes in conditions. The visual formats of the models are useful for presenting potential changes in the population of a species. Biologists can predict possible increases or decreases in the number of organisms of a species at risk. They base their predictions on improvements to or further degradation of a population's environment. No population exists by itself, however. In nature, there are interactions between individuals of different species, as well as among members of a single population.

SUMMARY Measuring and Modelling Population Change

- Mathematical and graphed models are used to predict trends in population growth.
- Exponential and geometric growth demonstrate growth without limits.
- Logistic growth, limited by carrying capacity, is most commonly seen in wild populations.

▶ *Section 14.2* Questions

1. Researchers studied a population of 34 peregrine falcons for one year to analyze the effect of pesticides on population growth. In the first three months, 57 eggs were laid. Owing to thin shells suspected to have resulted from pesticide damage, 28 eggs broke. Of the remainder, 20 hatched successfully and lived (**Figure 9**). However, nine baby falcons died from severe birth defects. During the next six months, 11 birds died as a result of direct pesticide exposure, and eight were captured and taken to a conservation area. During the last three months, 4 birds migrated into the area. Calculate the population growth of peregrine falcons in this study.

2. The growth rate for a population of 90 field mice in six months was 429%. If the number of births was 342, the number of deaths was 43, and there was no emigration, calculate the number of mice that migrated into the field.

Figure 9

3. In April, a population of frogs in a local ravine had a population of 42, which was increased in May by 263 tadpoles. Pesticide runoff killed 26 tadpoles and predators comsumed eight frogs. In July, construction workers began clearing the ravine and ran over 12 frogs. Calculate the population growth rate of the frogs in the ravine from April to August. ▶

4. In an attempt to save a rare endangered species, a number of zoos agree to begin a breeding program. A single pair of endangered species is expected to successfully rear an average of nine offspring per year, all of which survive to maturity (i.e., 4.5 per parent per generation).
   (a) What type of growth is seen in this zoo breeding population?
   (b) Use the appropriate equation to calculate the number of individuals in the population after eight years.

5. In many rural areas, stray cats are a problem as they may return to being wild (also known as feral) (**Figure 10**). Feral cats that have not been spayed or neutered can reproduce, which may result in a population of feral cats. One pair of cats can produce 12 kittens in one year. If half these kittens are female, this increased population could potentially produce 84 kittens in the second year. In five years, the population could reach almost 33 000 feral cats.
   (a) What kind of growth is occurring?
   (b) What conditions or factors would have to be in place for the population to achieve its biotic potential?
   (c) Suggest various types of environmental resistance that might restrict the feral cats from reaching their biotic potential.

Figure 10

6. Although zebra mussel populations are growing exponentially in many parts of the Great Lakes, their numbers are decreasing naturally in certain locations. Suggest possible reasons for such a decline.

Applying Inquiry Skills

7. A biologist determines the growth rate of a population of 198 frogs in a marsh in Haliburton to evaluate the quality of the environment. The researcher finds that, in one year, 34 were born, 86 died, 12 migrated into the marsh, and there was no emigration.
   (a) Determine the growth rate of the population.
   (b) Frogs are considered an indicator species for the quality of the environment. Do research to determine whether the growth rate of this population in one year is a cause for concern. If so, what might the concerns be?
   (c) Do you think that tracking the population growth rate of one population of frogs over one year in this marsh is adequate to make a conclusion about the environment? Explain your reasoning.

8. Snakes are released in an effort to repopulate a marsh in which the species had become extirpated in the past. Each original pair produces an average of six offspring that survive to maturity, although many other young snakes are eaten by predators.
   (a) If the population increase of six offspring per pair per generation continues, how many snakes will there be in the fifth generation?
   (b) On a graph, plot the number of snakes against generation, and explain what kind of growth has occurred.
   (c) How would the growth curve differ if there were only three, rather than six, offspring that survived to maturity? Explain the reasoning that supports your answer.

Making Connections

9. To try to limit effects from the exponential growth of populations of zebra mussels, scientists are proposing such strategies as using ultrasonic vibrations, adding chlorine or other chemicals to the water to kill the mussels, and manual removal.
   (a) Research three potential action plans to deal with effects from zebra mussel growth.
   (b) Summarize your findings in a PMI chart. You should include social, scientific and/or technological, and environmental criteria for each plan.

Density-Dependent Factors

In 1993, zebra mussel populations in the lower Illinois River, which had exploded to a density of nearly 100 000 per m², were causing significant harm to aquatic ecosystems. The rapidly increasing zebra mussel population led to severe depletions in the amount of dissolved oxygen available to the entire ecosystem, and increased competition for food resources. The resulting conditions were stressful for other species, but also affected the survival of the zebra mussels. Scientists observed a dramatic decline in these populations. Researchers now believe that the increased density of the zebra mussel population led to increased competition among members of the population (**Figure 1**).

With an increase in population size—for example, following the yearly breeding season for a species—population density increases. To avoid the adverse conditions that result from a high density of organisms, many individuals may disperse. After the breeding season, the young of many species may find it difficult to obtain food. As a result, the young will leave the immediate area where they were born. A **density-dependent factor** that limits population growth is one that intensifies as the population increases in size. In the 19th century, Charles Darwin, relying on the work of Thomas Malthus, recognized that the struggle for available resources within a growing population would inherently limit population size. The struggle for survival involves such factors as competition, predation, disease, and other biological effects.

When the individuals of a population of the same species rely on the same resources for survival, **intraspecific competition** occurs. As the population density increases, there is more competition among individuals for resources and so the growth rate slows. Consider a population of deer in a forested area. The forested area has a carrying capacity of 85 deer. When this population exceeds this carrying capacity, there is intense competition among the deer for remaining resources. Stronger deer that are able to obtain food will survive, while weaker deer will starve or risk death by moving from the area to seek another habitat with adequate resources. This intraspecific competition can have a pronounced effect on the reproductive success of individuals (**Figure 2**). As competition for food increases, the amount of food per individual often decreases. This decrease in

Figure 1
Competition for resources among zebra mussels will eventually limit growth in these populations.

density-dependent factor a factor that influences population regulation, having a greater impact as population density increases or decreases

intraspecific competition an ecological interaction in which individuals of the same species or population compete for resources in their habitat

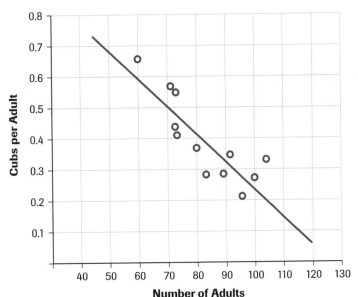

Figure 2
This graph illustrates the relationship between fecundity and population density in a population of grizzly bears. As population density increases, the number of cubs per adult decreases.

predation an ecological interaction in which a predator (a member of one species) catches, kills, and consumes prey (usually a member of another species)

Figure 3
Caribou populations are threatened by loss of habitat and the resulting increase in wolf predation

nutrition results in a decrease in an individual's growth and reproductive success. Harp seals, for example, reach sexual maturity when they have grown to 87% of their mature body weight. When the population density increases, the seals do not get as much to eat and do not reach the 87% body weight as fast as they would if the population density were lower. As a result, they reach sexual maturity at a slower rate, which decreases the potential number of offspring they might have.

Another major density-dependent factor that limits population growth is **predation**, the consumption of prey by carnivores. Some predators prefer one type of prey over another if that prey has a larger population and is easier to catch. This situation can result in density-dependent regulation of the preferred species of prey. A population of sharks searching for prey can hunt several species of fish for food. However, faster prey and those that are more apt to hide from the sharks may escape the shark predators and survive. Fish that are slower, and those that do not have places to hide, may be easier prey for the sharks and be their food of choice. The shark population, therefore, will regulate fish populations that are their prey.

In 1983, a research team led a study to determine controls that regulate moose populations in Ontario. The study reported that wolves, a predator of moose, were killing more moose than the number of calves born and were limiting the size of the moose population. This study corroborated previous studies that demonstrated a corresponding increase in moose populations when wolf populations decrease.

Studies performed in Alberta have also identified wolves as a predator of caribou (**Figure 3**), which explains why caribou populations have become endangered. Researchers have identified wolves as a major cause of death in undisturbed habitats inhabited by caribou. Although caribou migrate—potentially decreasing chances for predation— migration routes and destinations have become disturbed by human activities. Thus, as caribou are compelled to remain within a geographic range shared with wolves, the opportunities for predation increase and the number of caribou decreases.

Disease can also be a significant density-dependent factor that limits population size. In dense or overcrowded populations, pathogens are able to pass from host to host with greater ease because there are more hosts available in close proximity to one another. The population declines in size as a result of increased mortality. The overcrowding of farm animals can lead to the spread of disease, such as foot-and-mouth disease in cattle and encephalitis in poultry (**Figure 4**). Researchers in Britain tracked the spread of a common poultry pathogen, *Mycoplasma gallisepticum*, through the house finch, *Carpodacus mexicanus*. They were able to demonstrate a relationship between population size and the incidence of the disease, concluding that the spread of this disease was density dependent.

Figure 4
A potential risk of increased population density is that it can allow disease to spread more rapidly, resulting in a reduction in population size.

Some density-dependent factors reduce population growth rates at low rather than high densities. Scientist Warder Allee first described the phenomenon. Allee discovered that all plant and animal species suffer a decrease of their per capita rate of increase as their populations reach small sizes or low densities. This phenomenon, named after him, is called the **Allee effect**. For example, if a population is too small, it may become difficult for individuals to find mates. Some species exhibit social interactions that require higher population densities. The Allee effect has implications for threatened populations with low reproductive success. If species have low reproductive rates, then they need increased numbers to perpetuate the population.

Allee effect density-dependent phenomenon that occurs when a population cannot survive or fails to reproduce enough to offset mortality once the population density is too low; such populations usually do not survive

minimum viable population size the smallest number of individuals needed for a population to continue for a given period of time

Figure 5
The passenger pigeon became extinct in part as a result of the Allee effect.

The Allee effect is thought to have played a large role in the extinction of the passenger pigeon (**Figure 5**). Three centuries ago, passenger pigeons were the most common bird in North America, with a population estimated at between 3 billion and 5 billion. The birds were hunted for their soft feathers and inexpensive meat. As a result of uncontrolled hunting and destruction of the birds' habitats, the passenger pigeon drastically declined in numbers. The population could not recover since the pigeons laid only one egg per nest and would breed only in large colonies. The last passenger pigeon reportedly died in the Cincinnati Zoo in 1914.

A small population size can also result in inbreeding and the loss of genetic variation, which can threaten a population's continued survival. The **minimum viable population size** is the smallest number of individuals that ensures the population can persist for a determined interval of time. The minimum viable population size consists of enough individuals so that the population can cope with variations in natality and mortality, as well as environmental change or disasters. The minimum viable population size varies among species. Scientists use it as a model to estimate the size at which a population would be considered at risk. In 1941, biologists were concerned that the whooping crane would become extinct, as the wild population worldwide had decreased to 21 birds, plus two birds kept in captivity (**Figure 6**). Hunting of the birds for meat and eggs, as well as disturbance of their wetland habitats in Wood Buffalo National Park in Canada and along the Texas coast in the United States, had reduced the number of whooping cranes to well below the minimum viable population size predicted by biologists. An ambitious breeding program, along with legal protection of the whooping crane and its winter and summer habitats, have restored the population to 300 birds, although the species is still considered endangered.

Figure 6
Minimum viable population size is only a prediction. The whooping crane did not become extinct, even when there were only 23 birds left.

Density-Independent Factors

Populations may also experience changes in size that are not related to population density. These **density-independent factors** can limit population growth through human intervention, or other extreme weather changes in environmental conditions. For instance, some organisms may not breed in extreme temperatures. This phenomenon occurs with certain species of thrips, a small insect considered a common plant pest. These insects feed on so many different plant species that food supply is rarely a limiting factor of population growth. Cooler temperatures, however, reduce the reproductive success of these species. With a reduced birth rate, the population size declines. With the return of warmer temperatures, reproductive success improves and populations expand once again.

Insecticide application is an example of a density-independent factor in the form of human intervention. Biomagnification as a result of insecticide application can lead to decreases in populations up the entire food chain, starting with the insects the chemicals are intended to harm. Swainson's hawks are raptors that migrate between the grasslands of the Canadian prairies and South America (**Figure 7**). Wildlife biologists noted a drastic decrease in the population of these birds throughout North America, but had not found conditions that might be contributing to the population decline. Through satellite tracking, wildlife biologists determined that the hawks migrate to Argentina and roost in flocks of more than 7000 birds. It was estimated that, during the day, up to half of these hawks hunted for insects, mostly grasshoppers, in farmers' fields. The crops in these fields were regularly sprayed with highly toxic pesticides to control the grasshopper population and preserve the crops for human food. These particular pesticides are banned in North America.

In 1996, 12 Swainson's hawks were captured in Alberta and tagged with satellite transmitters before they migrated to Argentina. Biologists flew to the migration destination to observe the tagged birds and counted more than 5000 dead hawks. These hawks were killed either from direct pesticide exposure or as a result of biomagnification in the food chain. Currently, joint efforts are being made between government agencies and nongovernmental environmental organizations in North and South America to find solutions to the pesticide problem that is destroying such large numbers of the Swainson's hawk.

Limiting factors prevent populations from achieving their biotic potential. These limiting factors may be shortages of such environmental resources as light, space, water, or nutrients, and will determine the carrying capacity of the populations. Of all the resources that individuals or populations require for growth, the resource in shortest supply is called the **limiting factor**, and it determines how much the individual or population can grow. For example, a plant requires nitrogen, carbon dioxide, and sunlight for it to be able to grow and reproduce (**Figure 8**). If it uses up all available nitrogen, it can no longer grow, even if there is still an abundance of sunlight and carbon dioxide. In this case, nitrogen is the limiting factor.

density-independent factors factors that influence population regulation regardless of population density

Figure 7
Swainson's hawks are decreasing in numbers as a result of pesticide use.

limiting factor any essential resource that is in short supply or unavailable

Figure 8
Plants need many different resources for growth and survival. The resource in shortest supply is considered the limiting factor to growth. The orange hawkweed plants in **(a)** are flourishing while the spatial spread of the plants in **(b)** is limited.

(a)

(b)

Increases and decreases in the size of populations can be minor with no serious risks to the survival of the species. More drastic population growth or decline can result in adverse changes to the habitat or even to the population itself. Too many births or too much immigration may result in the population overshooting its carrying capacity. When a population surpasses the carrying capacity of its environment, the number of deaths starts to increase and the number of births starts to decline, resulting in a subsequent reduction in population size. Very high rates of mortality in a population may threaten a species with extinction. Population biologists monitor natural fluctuations in the size and density of populations. They do this, in part, to understand natural patterns better and, in part, to try to predict the organisms' ability to withstand such impacts as natural disasters or destructive human activities.

SUMMARY | *Population Change*

- Populations are dynamic.
- Carrying capacity determines the number of individuals an area can accommodate.
- The growth rate of a population is determined by natality, mortality, immigration, and emigration.
- Density-dependent and density-independent factors can limit the growth of populations.

▶ *Section 14.3 Questions*

Understanding Concepts

1. Describe how natality, mortality, immigration, and emigration affect a population.

2. Explain the difference between density-dependent and density-independent factors.

3. Classify the following scenarios as density dependent or density independent:
   (a) A forest fire destroys a great deal of habitat in Algonquin Park.
   (b) Many aquatic organisms die as a result of changes in adverse weather conditions during the breeding season.
   (c) A young aggressive hawk invades the geographic range of established hawks, driving weaker birds from the geographic range.

4. Identify one density-dependent and one density-independent limiting factor that were not discussed in this section. Explain how they might affect the growth of a population.

Applying Inquiry Skills

5. Study the graph in **Figure 9** regarding a woodland bird population of the great tit, which is a European bird similar to the chickadee. The graph illustrates population density versus clutch size (the number of eggs to be hatched at one time).

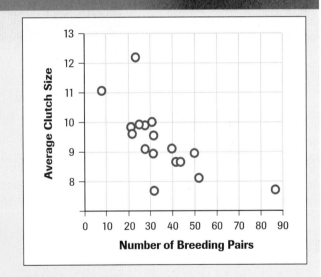

Figure 9

(a) Is this a case of density-dependent or density-independent regulation?
(b) Draw a corresponding graph to illustrate population density versus food supply. Explain the reasoning behind the shape of your graph.

community all populations in a given ecosystem at a given time

Populations do not live in isolation. Within a given ecosystem, populations of different species interact in what is called a **community** (**Figure 1**). Like most scientists, biologists look for patterns within and among organisms. Biologists have identified several levels of organization in which cells combine to form tissues, tissues combine to form organs, organs combine to form organisms, organisms combine to form populations, and populations combine to form communities. Some organisms within communities cannot exist independently of one another and work together for survival. For example, some species of flowering plants rely on insects to pollinate them so they can reproduce. In nature, there are systems working within other systems.

(a) **(b)**

Figure 1
Interactions between populations are common within communities, such as cabbage butterflies and cabbages **(a)**, oxpecker birds and impala **(b)**.

Ecological Niches

On the African savannah, a variety of interactions occur among organisms (**Figure 2**). While the lions, vultures, hyenas, and zebras all occupy the same habitat, each member of this community uses different mechanisms to survive. Animals can either adapt constantly to changes in their community or survive by occupying an **ecological niche**. Ecologist Eugene Odum describes an organism's habitat as its "address" and its ecological niche as its "occupation." The African lion's ecological niche, for example, includes what it eats, what

ecological niche an organism's biological characteristics, including use of and interaction with abiotic and biotic resources in its environment

Figure 2
An African grassland community

eats it, the way it reproduces, the temperature range it tolerates, its habitat, behavioural responses, and any other factors that describe its pattern of living. Ideally, if resources were abundant and no competition with other species existed, the African lion would come close to occupying its **fundamental niche**. A fundamental niche comprises the biological characteristics of an organism and the set of resources that individuals in the populations are theoretically capable of using under ideal conditions. In reality, the lion faces **interspecific competition** with vultures and hyenas for similar resources and occupies only a portion of its fundamental niche, what ecologists call its **realized niche**. As an analogy, you might be very capable of becoming captain of a sports team (your fundamental niche), but competition from others may mean that you actually become assistant captain (your realized niche).

Interactions between individuals of the same species (intraspecific) and among individuals of different species (interspecific) in a community have important influences on population dynamics of individual species. Although species interact in diverse ways, interactions between two species and their effects on population density can be classified into five categories (**Table 1**). **Symbiosis** includes a variety of interactions in which two species live together in close, usually physical, association. Parasitism, mutualism, and commensalism are three general types of symbiotic interactions.

fundamental niche the biological characteristics of the organism and the set of resources individuals in the population are theoretically capable of using under ideal conditions

interspecific competition interactions between individuals of different species for an essential common resource that is in limited supply

realized niche the biological characteristics of the organism and the resources individuals in a population actually use under the prevailing environmental conditions

symbiosis various interactions in which two species maintain a close, usually physical, association; includes parasitism, mutualism, and commensalism

Table 1 Classification of Interactions Between Two Species

Interaction		Effect on populations
competition		Interaction may be detrimental to one or both species.
predation		Interaction is beneficial to one species and usually lethal to the other.
symbiosis	• parasitism	Interaction is beneficial to one species, and harmful but not usually fatal to the other.
	• mutualism	Interaction is beneficial to both species.
	• commensalism	Interaction is beneficial to one species and the other species is unaffected.

Interspecific Competition

Interspecific competition occurs between individuals of different populations and, like intraspecific competition, it serves to restrict population growth. Interspecific competition can occur in two ways. Actual fighting over resources is called **interference competition**, while the consumption or use of shared resources is referred to as **exploitative competition**. An example of interference competition is the fighting that sometimes occurs between tree swallows and bluebirds over birdhouses. An example of exploitative competition occurs when both arctic foxes and snowy owls prey on the same population of arctic hares.

The strongest competition occurs between populations of species that experience niche overlap. The more niches that overlap, the greater the competition between species, as demonstrated by Russian ecologist G.F. Gause. In a 1934 experiment, summarized in his book entitled *The Struggle for Existence*, Gause tested the logistic population growth theory—that two species with similar requirements could not coexist in the same community. He predicted that one species would consume most of the resources, reproduce efficiently, and drive the other species to extinction. Gause's experiments led to the conclusion that if resources are limited, no two species can remain in competition for exactly the same niche indefinitely. This became known as Gause's principle, or the principle of competitive exclusion.

interference competition interspecific competition that involves aggression between individuals of different species who fight over the same resource(s)

exploitative competition interspecific competition that involves consumption of shared resources by individuals of different species, where consumption by one species may limit resource availability to other species

Ecologists, such as Gause, have described the struggle for survival in terms of the competitive interactions between different species in nature. Gause studied the effects of competition on two closely related species, *Paramecium aurelia* and *Paramecium caudatum*. He grew each species in a separate culture tube with bacteria and yeast provided daily as food. Then, the two paramecium cultures were grown together in the same tube. Population size data were collected from the resulting interactions and plotted on graphs. When they were grown separately, stable populations were established as indicated by the logistic growth curves, shown in **Figure 3(a)**. When grown together, one species outcompeted the other for food. As a result of sharing realized niches, one species drove the other to extinction, as **Figure 3(b)** illustrates.

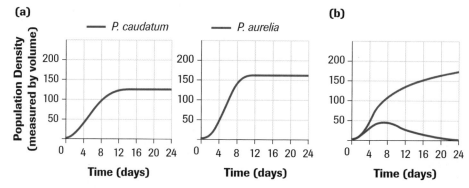

Figure 3
Gause's results for **(a)** *P. caudatum* and *P. aurelia* cultured individually, and **(b)** *P. caudatum* and *P. aurelia* cultured together

Gause's experiment has been replicated many times to support his theory on competition. In a separate experiment, Gause challenged *P. caudatum*, the defeated species of his earlier experiment, with competition from a third species, *Paramecium bursaria*. *P. bursaria* was cultured independently and then added to a new culture of *P. caudatum*. The results of these additional experiments are shown in the graphs in **Figure 4(a)** and **(b)**.

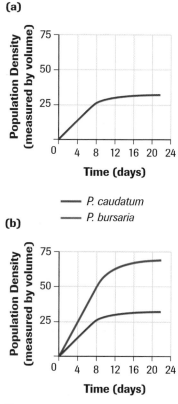

Figure 4
Gause's results for **(a)** *P. bursaria* cultured individually, and **(b)** *P. bursaria* and *P. caudatum* cultured together

> ▶ **Case Study** *Questions*

Understanding Concepts

1. Compare the population densities of the paramecium populations in both experiments. In each of the paired experiments, do species grown in combination reach the same density as species grown independently?

2. What are the limits in population density achieved by each of the species when grown independently and when combined with another species?

3. The ability of *P. aurelia* to grow faster in the presence of *P. caudatum* is an example of exploitative competition. Explain what this statement means.

4. The second experiment, shown in **Figure 4(b)**, did not produce similar results. Propose an explanation for why both species of paramecium survived.

5. Research the food requirements for each of these paramecium species. Are there any differences among these species that might enable one species to acquire food more effectively?

The results of interspecific competition take on several forms: The population size of the weaker competitor could decline. One species could change its behaviour so that it is able to survive using different resources. Individuals in one population could migrate to another habitat where resources are more plentiful. In any of these cases, competition would decline.

One way that individuals of species occupying the same niche can avoid or reduce competition for similar resources is by **resource partitioning**. To minimize competition for food, several species of *Anolis* lizard, partition their tree habitats by occupying different perching sites (**Figure 5**). Some of the lizards use the canopy, while others occupy twigs on the periphery of the forest, the base of trunks, or even the grassy areas near trees. Resource partitioning enables *Anolis* lizards to reduce interspecific competition so that they can coexist in the same geographic range.

resource partitioning avoidance of, or reduction in, competition for similar resources by individuals of different species occupying different nonoverlapping ecological niches

Figure 5
Resource partitioning among several species of *Anolis* lizards of the Caribbean

Resource partitioning is also observed among three species of annual plants—the foxtail, mallow, and smartweed—that grow in abandoned agricultural fields. All three species require water, sunlight, and dissolved minerals, but they have evolved specific adaptations that enable them to acquire these nutrients and reduce competition. The foxtail plant has a shallow fibrous root system, **Figure 6(a)**, that quickly absorbs water from the surface soil, allowing these plants to grow where the moisture in the soil varies on a daily basis. In contrast, mallow plants have a deep taproot system, **Figure 6(b)**, which allows them to grow more deeply into soil and obtain moisture later in the growing season. The taproot system of the smartweed plant, illustrated in **Figure 6(c)**, branches in the topsoil and in the soil below the roots of the other species, so that it exploits water where it is available.

Interspecific competition is a driving force for populations of species to evolve adaptations that enable them to use resources for continued survival. Dr. Dolph Schluter, a zoologist with the University of British Columbia, and his research colleagues have studied the three-spined sticklebacks. Dr. Schluter and his colleagues have found two different species of sticklebacks in each of five different freshwater lakes. In the shallow waters close to shore, an oversize bottom-feeding species with a large mouth feeds on large prey. A smaller stickleback species with a small mouth feeds on plankton in open surface waters farther from shore (**Figure 8** in Chapter 12, page 575). Evidence from DNA analysis shows that, in each lake, a single population of sticklebacks faced competition for limited resources that favoured fish with extremes of body and mouth sizes, resulting in two populations that exploit different foods in different parts of each lake. Increased disruptive selection pressures in this population of sticklebacks has led

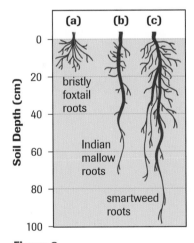

Figure 6
An example of resource partitioning in three species of annual plants—foxtail **(a)**, mallow **(b)**, and smartweed **(c)**—that differ in the way they acquire water and mineral ions from the same habitat.

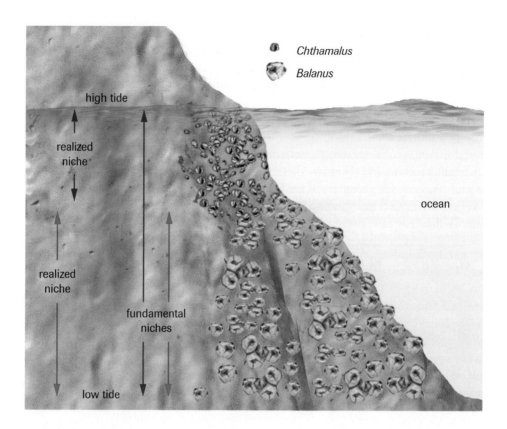

high tide

Chthamalus

Balanus

realized niche

realized niche

fundamental niches

ocean

low tide

Figure 7

Connell's field experiments with barnacle species provides evidence for competition in nature.

to character displacement, which has lessened the competition between the two. Many scientists believe that this evolutionary process has resulted in much of the biodiversity now in existence on Earth.

Although resource partitioning and character displacement provide indirect evidence for competition, field experiments demonstrate more clearly the effect of competition on the density and dispersion of populations. In a classic experiment, Joseph H. Connell, Research Professor at the University of California Santa Barbara, manipulated two barnacle species, *Balanus balanoides* and *Chthamalus stellatus*, that grow on rocks along the Scottish coast. The barnacles have a stratified dispersal: *Balanus* concentrate on rocks on the lower portion of the shore and *Chthamalus* on the rocks on the upper section of the shore (**Figure 7**). On the lower rocks, *Balanus* outcompetes *Chthamalus* by crowding them off the rocks, undercutting them, and even replacing them where they have begun to grow. When Connell removed *Balanus* from the rocks on the lower shore, he observed the *Chthamalus* population spread into that area. *Chthamalus* could survive on rocks lower down the shore in the absence of *Balanus*. This suggests that, under normal conditions, *Balanus* outcompetes *Chthamalus* for space in the lower region of the rock.

Additionally, the larvae of both species of barnacles are capable of growing in either the higher or lower portions of the shore in their ecological niche, but Connell discovered that adult *Balanus* could not survive on rocks in the upper shore. The barnacles readily dry out when this region is exposed to air for a long time during low tides. Connell concluded that the fundamental and realized niches for *Balanus* are similar, whereas the realized niche for *Chthamalus* is only a fraction of its fundamental niche.

Interspecific interactions between species can have different effects on the population density of the species involved. In competition, the population density of both species is affected. For other types of interspecific interactions, other changes to population density may occur.

Predation

Predation is an example of an interspecific interaction in which the population density of one species—the predator—increases while the population density of the other species—the prey—declines. Predator–prey relationships can have significant effects on the size of both predator and prey populations. When the prey population increases, there is more food for predators and this abundance can result in an increase in the size of the predator population. As the predator population increases, however, the prey population decreases. The reduction of prey then results in a decline in the predator population, unless it has access to another food source. There are time lags between each of these responses, as the predator population responds to changes in prey abundance.

Some predator–prey relationships coexist at steady levels and display a cyclical pattern. The two species tend to cycle slightly out of synchronization, with the predator patterns lagging behind the prey patterns (**Figure 8**). In this model of a predator–prey cycle, adjustments to population size can be seen during the time intervals from A to E. This graph is referred to as a sinusoidal curve. At time A, when the prey population density is low, the predators have little food and their population declines. A reduction in the predator population allows the prey population to recover and increase. The predator population does not increase again until they begin to reproduce (at time B). Prey and predator populations grow until the increase in the predator population causes the prey population to decline (from time C to time E). As the predator population increases, more of the prey population is devoured. The resulting low density in the prey population leads to starvation among predators, slowing its population growth rate (at time D).

In nature, many factors can influence this model of the sinusoidal predator–prey cycle. In 1831, the manager of the Hudson's Bay Company in northern Ontario reported that there was a scarcity of rabbits and the local Ojibwa population was starving as a result. These were not rabbits, in fact, but snowshoe hares, which experience a population decline at 10-year intervals. Wildlife biologists began analyzing these 10-year cycles by plotting the fur-trading records of the Hudson's Bay Company in the early 1900s. The most well-known quantitative analysis was derived from records of the Canadian lynx, which is a significant predator of snowshoe hares. Adjustments to its population mirrors, with a slight time lag, changes in the snowshoe hare population (**Figure 9**).

To try to understand the sources of such cycling, Charles Krebs, a professor of Zoology at the University of British Columbia, led a 10-year study of lynx and snowshoe hares. To identify the sources of this cycling, the research team tracked hare population densities in a 1-km² control plot and three experimental plots, each with different characteristics. The experimental plots had additional food or were fertilized to increase plant growth. All plots were surrounded by electrified fences that prevented hares from leaving and prevented lynx and other mammalian predators from entering or leaving. Selected results are summarized in **Table 2**. These manipulations of conditions in the field did not prevent declines in the hare population within the cycle, but simply delayed them. Additional predators, such as owls and other raptors, as well as changes in food supply, also affected the cycle. Other variables—such as changes in weather, presence of alternative prey during the low points of the hare cycle, and changes in the extent of trapping by humans—could also potentially influence this cycle in real-world conditions.

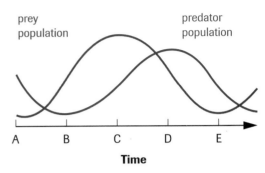

Figure 8
A model of the predator–prey cycle. Because of the oscillations in the populations, the line on this graph is referred to as a sinusoidal curve.

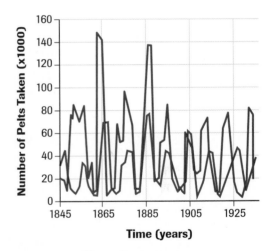

Figure 9
A 10-year cycling pattern in the population of Canadian lynx (blue line) and the population of snowshoe hare (red line)

🖉 LAB EXERCISE 14.4.1

Investigating Predator–Prey Cycling (p. 691)
What effect on a white-tailed deer population might result from the introduction of wolves into their habitat? In this investigation, you can test your hypothesis of this effect on predator–prey cycles.

Table 2 Selected Results From Kreb's Lynx–Hare Cycling Study

	Plot characteristics	Result
Plot 1	predator-free	hare density doubled
Plot 2	extra food	hare density tripled
Plot 3	predator reduction and extra food	hare density increased elevenfold (1100%)

Defence Mechanisms

Predator–prey interactions have resulted in the diverse defence mechanisms that plant and animal species have evolved through repeated encounters with predators over time. Plants use both morphological defences—such as thorns, hooks, spines, and needles—and chemical defences against herbivores. The mustard family of plants, for example, contains oils that give off a pungent odour and make them distasteful and toxic to some insects. Some plants, such as balsam fir, produce chemicals that mimic an insect growth hormone. When linden bugs feed on balsam fir, they remain in their juvenile stage and eventually die.

The defensive systems in plants act as selective agents initiating the evolution of counter-adaptations in herbivore populations. Cabbage butterfly caterpillars, for instance, have evolved mechanisms that enable them to break down the mustard oils that protect cabbage plants against most herbivores. The green coloration of the cabbage butterfly caterpillar camouflages it on the leaves of cabbage and other plants and acts as protection from its predators. Some insects use chemicals produced by their food as a defence against their own potential predators. For instance, the monarch butterfly uses potent plant toxins to make itself distasteful to its predators (**Figure 10**). Caterpillars of the monarch butterfly obtain these toxins by feeding on plants of the milkweed family. After consumption, the chemicals are stored in fatty tissues of the caterpillar, making both it and the adult butterfly unpalatable. Blue jays have been known to regurgitate a monarch butterfly after swallowing it.

Figure 10
The monarch butterfly and its predator (the blue jay) provide an example of defence mechanisms by prey against predators.

Changes brought about by coevolution between plants and insects can, in some cases, also affect competition. Passionflower vines, for example, produce toxic chemicals that protect it from predators, but the *Heliconius* butterfly is able to feed on the vines because of digestive enzymes it possesses that break down these toxic chemicals. The females of *Heliconius* lay bright yellow eggs on passionflower leaves as a signal to other females to avoid depositing their eggs there, thereby reducing intraspecific competition. Yellow nectaries that resemble eggs grow on some species of passionflowers, discouraging female *Heliconius* from laying their eggs there. These nectaries also attract ants and other insects that prey on butterfly eggs, thereby reducing predation on passionflower plants.

Animals sometimes employ passive defence mechanisms, such as hiding, or active defences, such as fleeing from their predators. Active defences are more costly to the prey in terms of energy use than are passive defences. Other effective behavioural defences include alarm calls that may signal members of the prey species to mob the predator (**Figure 11**). A number of species use camouflage, also called cryptic colouration, that enables prey to blend with their surroundings, as a passive defence mechanism to hide from predators (**Figure 12**). Some animals give a visual warning to predators of chemical defence mechanisms, such as poisons (**Figure 13**). The saliva of the small blue-ringed octopus (*Haplochlaena maculosa*) contains a highly venomous neurotoxin that is lethal to many animals, including humans, when bitten. The rings contract and expand as a warning signal to potential predators.

Figure 11
A crow circles a great horned owl, a predator that may kill and eat the crows' eggs and nestlings.

Figure 13
The iridescent blue rings of the world's only lethal octopus, *Haplochlaena maculosa*, warn potential predators.

Both predators and prey species can also protect themselves through mimicry. The mimic may gain an advantage by resembling a distasteful species. Two types of mimicry, Batesian and Mullerian, are named after the scientists who initially observed and described them. In Batesian mimicry, a palatable or harmless species mimics an unpalatable or harmful organism, a phenomenon most often observed in moths and butterflies. Batesian mimicry is named after the naturalist Henry Bates who, in his journeys to the Amazon region of South America in 1857, identified several palatable insects that resembled distasteful species. He reasoned that predators would be fooled by such mimics who, as a result, would avoid predation. A typical example is the Viceroy butterfly's mimicry of the poisonous monarch butterfly (**Figure 14**, page 684). Mullerian mimicry was named after Fritz Muller, a German biologist, who, in 1878, described several unrelated but protected

(a)

(b)

Figure 12
In two examples of defense mechanisms, the inchworm caterpillar closely resembles a twig **(a)**, and the canyon tree frog uses coloration to make itself disappear into a granite background **(b)**.

(a)   **(b)**

Figure 14
As an example of Batesian mimicry, the unpalatable Monarch butterfly **(a)** appears almost identical to its edible mimic, the Viceroy butterfly **(b)**.

animal species that resembled one another and are all poisonous or dangerous. For example, the butterflies *Heliconius sapho* and *H. cydno* are pairs of Mullerian mimics, which have evolved similar coloration to minimize predation. In this case, the pooling of numbers caused predators to learn more quickly to avoid such species. Sometimes predator populations evolve an innate avoidance of prey species.

Symbiosis

Symbiosis, meaning "living together," refers to a relationship in which individuals of two different species live in close, usually physical, contact. At least one of the two species benefits from the association. One type of symbiotic relationship is **mutualism**, which occurs when both species in the relationship benefit and neither is harmed. Most biologists also include as symbiotic relationships **commensalism**, which occurs when one organism benefits and the other neither benefits nor is harmed, and **parasitism**, which occurs when one organism benefits at the expense of another organism's well-being.

Mutualism

There are many common examples of mutualistic relationships in which both organisms benefit. Bacteria live in the guts of herbivores, such as cows, deer, and sheep. These animals do not produce the enzymes required to digest plant products such as cellulose and lignin. The bacteria secrete enzymes to break down these products into useable nutrients for the animals. In return, the bacteria are provided with nutrition themselves. This type of relationship is referred to as **obligatory mutualism**, because neither organism could grow or reproduce without the other. Beneficial bacteria also live in the large intestines of humans, producing nutrients such as vitamins B and K, which our cells can use. Oxpecker birds feed on the backs of grazing animals, as shown in **Figure 1(b)**, page 676. They pick off parasites on the skin of the animals. The birds are provided with a source of food, and the animals are protected from the parasites. Pollination is another example of mutualism. Many insects, bats, and birds ingest the nectar and pollen of flowers. The pollen attaches to their bodies, and as they fly away, the pollen is spread so the plants can reproduce. Another common example of mutualism in nature is provided by lichens, which display a symbiotic relationship between fungi and algae or cyanobacteria. Photosynthesis in the algae or cyanobacteria benefits the fungi, and the fungi provide the algae or cyanobacteria with a moist habitat and improved access to nutrients that they require.

mutualism a symbiotic relationship in which both organisms benefit; as neither is harmed, it is categorized as a +/+ relationship

commensalism a symbiotic relationship in which one organism benefits and the other organism is unaffected; it is categorized as a +/0 relationship

parasitism a symbiotic relationship in which one organism (the parasite) benefits at the expense of another organism (the host), which is often harmed but usually not killed; it is categorized as a +/− relationship

obligatory mutualism a symbiotic relationship in which neither species involved could survive without the other

Commensalism

Commensalistic relationships are difficult to classify. Some biologists argue that they do not exist at all. In the absence of clear observations, it is very difficult to ascertain whether an individual of the unaffected species is, in fact, being harmed or benefited. Caribou and arctic foxes interact in a way that has been classified as commensalistic (**Figure 15**). The foxes follow the caribou herds when they forage for food in their wintering grounds. The caribou have shovel-like feet that can remove snow from lichens on the ground, which is the caribou's primary food source. By removing the snow, caribou expose many small mammals, which are eaten by the foxes. Thus, the foxes benefit from this interspecific interaction and the caribou neither benefit nor are harmed by it. In a similar way, tropical "ant birds" follow army ant colonies through the rain forest, feeding not on the ants but on the other insects and small animals that are disturbed by and flee from the ants.

Other possible examples of commensalism occur in interactions between aquatic organisms. Small fish called remora have suction disks on the back of their heads with which they attach themselves to sharks (**Figure 16**). When the sharks travel, the remora do not expend much energy by swimming and are able to feed on small pieces of the sharks' prey. Adult barnacles do not have the ability for locomotion, which restricts their opportunities to seek new food sources. By attaching themselves to actively moving marine organisms, such as whales, they are able to find new food sources and are afforded protection from predators. The remora and barnacles benefit from these relationships and have very little effect on the organisms to which they are attached.

Parasitism

Parasites live and feed on the most nutritious environments on Earth—the bodies of other living organisms—and cannot complete their life cycle in the absence of their hosts. By living on or in a host organism, parasites have access to a continuous supply of nutrients. For this reason it should not be surprising that parasitism is extremely common. Biologists estimate that as many as one in four animal species may be parasites. Virtually all species of plants and animals are hosts to one or more species of parasite. While the best-known and perhaps most important parasites are responsible for serious human diseases—such as malaria, schistosomiasis, and African sleeping sickness—the vast majority of parasites cause little or no significant harm to their host. This makes sense since it would mean harming the environment on which their own health relies.

Parasites can be **microparasites**, microscopic in size with a rapid reproduction rate. Examples are the blood protozoans: *Plasmodium*, which causes the disease malaria, and *Trypanosoma*, which causes sleeping sickness. Parasites can also be **macroparasites**, larger organisms, such as tapeworms, fleas, and lice. **Endoparasites** are organisms that live inside the body of their hosts. Adult tapeworms live inside the digestive systems of different organisms, using their digested nutrients as food. Many endoparasites live and feed within host tissues. Assorted species of flukes are commonly found living in the liver and lung tissues of their hosts, while many single-celled protozoans live within the bloodstream. **Ectoparasites** are organisms that live and feed on the outside surface of their host. The best-known examples include lice, ticks, and fleas. The beaver, *Castor canadensis*, is host to an interesting ectoparasitic beetle that lives on the beaver's skin surface and is kept dry and protected by the beaver's thick, oily, waterproof coat.

Figure 15
Caribou unknowingly help arctic foxes by exposing snow-covered habitats of fox prey.

Figure 16
Remora get a free ride and access to new food sources by attaching themselves to sharks.

microparasites parasites, such as plasmodia or trypanosomes, that are too small to see with the naked eye

macroparasites larger parasites, such as tapeworms, fleas, and ticks, that are readily visible

endoparasites parasites that live and feed within the host's body

ectoparasites parasites that live and feed on the outside surface of the host, such as lice, ticks, and parasitic mites

TRY THIS activity *Finding a Host*

All parasites live in hosts that will eventually die. For this reason it is essential that parasites are able to find and invade new host individuals on a regular basis. This is particularly challenging for internal parasites and results in very complex life cycles for these species. Choose an endoparasite and research the ways it is able to get from one host to another. Common endoparasites include flukes, tapeworms, *plasmodium*, and *trypanosome* species.

- Draw the life cycle of your chosen parasite, indicating how it gets from one host to another.

social parasites parasites that complete their life cycle by manipulating the social behaviour of their hosts

Some species are classified as **social parasites**. These organisms manipulate the social behaviour of another species so that they can complete their life cycle. North American cowbirds are social parasites that lay an egg in the nests of other smaller birds and, therefore, do not have to expend energy building their own nest or feeding their own young. The cowbird egg usually hatches earlier and the larger newborn cowbird monopolizes the food resources. The other newborn birds are usually killed, resulting in a very high survival rate for the young cowbird.

Symbiotic relationships describe how many populations interact together within communities for growth and survival of all participants. In contrast, some relationships can disrupt the equilibrium of entire communities and ecosystems.

Disruption of Community Equilibrium

Stability in biological communities exists when the resources necessary for survival are sustained, populations do not exceed their environment's carrying capacity, and interspecific interactions contribute to biodiversity. Interspecific interactions help to maintain the necessary equilibrium within the complex and dynamic natural systems that sustain communities. A variety of disturbances can affect this equilibrium in drastic ways. A natural disaster can disturb most populations within a community and can break down the intricate interactions among its organisms. The introduction of exotic, also called nonindigenous, species into a community can disrupt ecosystems and displace indigenous species to such an extent that they pose a serious threat to the preservation of biodiversity. Such exotic species can have devastating biological and economic effects on the habitats they invade. These nonindigenous species, often with few predators, may reduce or eliminate indigenous species by outcompeting them for food and habitat, or by preying on them.

A classic example of the ecological disturbance that can follow the introduction of an exotic species is the case of the European rabbit in Australia. In 1859, an English immigrant to Australia set free two dozen rabbits. Rabbits breed so rapidly that, within a few decades, tens of millions of rabbits had devastated grasslands, depriving indigenous species of food, water, and shelter. Efforts to control the rabbit population began with the use of poisons that killed some rabbits but also killed many indigenous species; worse, the rabbit population rebounded while the indigenous populations did not. The release of European red foxes to prey on the rabbits also failed, as the foxes targeted small mammals, reptiles, and birds instead of the rabbits. Finally, the introduction of a virus that selectively killed the rabbits managed to bring the invader population under control. More recent examples of the harmful effects of introducing an exotic species into an ecological community are shown in **Table 3**.

Table 3 Selected Invading Species

Purple loosestrife (*Lythrum salicaria*)	A tall perennial herb that grows in wet or moist habitats, it is commonly found along roadside ditches, swamps, marshes, and in open meadows. Its widespread presence has altered aquatic ecosystems in North America and interfered with interactions among many native aquatic species.
African killer bees (**Figure 17**)	Imported intentionally to Brazil from Africa by beekeepers for their high honey production, these aggressive bees sting at the slightest provocation. They attack in larger numbers and much faster than the common honeybee. In some instances, humans have died from these bee stings. Aside from public safety, African killer bees have a significant economic impact on commercial beekeepers and food production.
West Nile virus	By 2001, more than 150 dead birds had tested positive for the West Nile virus. The virus, detected in wildlife populations throughout North America, was first identified in the West Nile region of Uganda in 1937. It can be transmitted to humans by three species of mosquitoes: *Culex pipiens* (the common household mosquito), *Aedes vexans* (an indiscriminate feeder) and *A. japonicus*. It is responsible for serious wildlife population losses in many parts of the world. The virus is believed to have been accidentally introduced to North America in an exotic frog species. It is notable that a species of bird and mosquito, both involved in the transmission of this virus, were also introduced as exotic species.

Figure 17
Some African killer bees escaped from Brazilian beekeeping operations and have spread accidentally into North America

▶ *EXPLORE* an issue

Profile of an Exotic Invader

Environment Canada has issued an alert on the invasion of foreign species. You have been hired to research information regarding one of these foreign species and its potential effects on Canada.

- Research and select a foreign species with potential negative impact.
- Investigate the effects of this species on ecosystems throughout Canada.
  (a) Identify specific effects it has on interspecific interactions (such as competition and predation).

Decision-Making Skills

- ● Define the Issue   ● Analyze the Issue   ● Research
- ○ Defend the Position   ● Identify Alternatives   ○ Evaluate

(b) Describe the current and potential economic and health effects resulting from the introduction of this species.
(c) Outline strategies and/or technologies proposed as potential solutions to deal with the invasion.
(d) Prepare a one-page press release to be delivered to the Canadian public informing them about vital information regarding the foreign species.

The complex interactions among interdependent species of a community are subtle yet essential for sustaining biodiversity. Often, the intricacies of these complex interactions are revealed only when conditions in ecosystems deteriorate. Disruption of such interactions can pose serious threats to the ecosystem and the species that inhabit it. Humans are by no means exempt from interventions into biological interactions. As human populations expand, their activities affect ecological communities and, increasingly, their own.

SUMMARY   *Population Interactions Within Communities*

- Many different kinds of interactions occur among and between species that affect population growth.
- The population dynamics of predator–prey interactions are affected by a wide range of factors. Both predators and prey have evolved diverse adaptations that enhance survival.

- Symbiotic interactions may result in mutually beneficial relationships between species, while commensalistic interactions may result in one species indirectly benefiting another. Parasitism is an interaction in which one species feeds and lives in or on a host organism to the detriment of the host.

- Invasions by exotic species can disrupt the stability, or dynamic equilibrium, within an ecological community.

▶ Section 14.4 *Questions*

Understanding Concepts

1. For each of the following examples, identify what type of competition is occurring and provide reasons that support your answer:
   (a) Argentine ants can displace indigenous ants from a community by rapidly depleting resources.
   (b) Plants release toxins that kill or inhibit the growth of other plants, thereby preventing them from growing in close proximity where they may compete for space, light, water, and food.
   (c) In the Kibale Forest in Uganda, mangabey monkeys, a large species, drive away the smaller blue monkeys.
   (d) Hawks and owls rely on similar prey, but hawks feed during full daylight while owls hunt and feed from dusk to dawn.

2. Using the example of the *Balanus balanoides* and the *Chthamalus stellatus* barnacles, explain the difference between a fundamental and a realized niche.

3. A study was conducted of mussels and the starfish *Pisaster* in the intertidal area along the shore. Results showed that greater diversity of marine invertebrates was found in the area where *Pisaster* and mussels were present together compared to where mussels were found alone. Explain this observation, using principles of competition.

4. Two species of ground finch, *Geospiza fuliginosa* and *Geospiza fortis*, living in the Galapagos islands display interesting variations in bill size. On the island of Santa Maria, inhabited by both species, the average bill depth of *G. fuliginosa* is 8.1 mm while that of *G. fortis* is 12.7 mm. When inhabiting separate islands, however, their average bill depths were 9.8 mm and 10.1 mm respectively. Explain how the concept of character displacement can account for these observations.

5. Identify the type of defence mechanism in each of the following examples:
   (a) Tiger moths have a highly detailed wing pattern that makes them virtually undetectable against tree bark.
   (b) When attacked by ants, ladybugs secrete a sticky fluid that entangles ant antennae long enough to allow the ladybug to escape.

6. Explain how predation differs from parasitism.

7. Termites eat wood but cannot digest it. They have unicellular, heterotrophic organisms called zoomastigotes living inside their digestive tract to do this for them. Identify the type of interspecific interaction between the termites and the zoomastigotes.

Applying Inquiry Skills

8. In the Great Smoky and Balsam Mountains, ecologists are studying two species of salamander. *Plethodon glutinosus*, usually lives at lower elevations than its relative, *P. jordani*, shown in **Figure 18**, although the researchers have found some areas inhabited by both species. As part of the study, the scientists established different test plots from which one of the species was removed and control plots in which the populations remained untouched. After five years, no changes were observed in the control plots, but in the test plots, salamander populations were increasing in size. For instance, if one of the test plots was cleared of *P. jordani*, it would have a greater population density of *P. glutinosus* and vice versa. What inferences or conclusions might be drawn from this investigation?

Figure 18
Salamander *Plethodon jordani*

9. In a laboratory, researchers placed a paramecium species in a test tube with its predator protozoan. After a given time, predator–prey cycles became shortened and the system collapsed. Researchers repeated the experiment with new paramecia being added to the test tube every few days. What do you expect the researchers observed about the predator–prey cycle during this second experiment? Provide reasoning for your answer.

Making Connections

10. Parasites are often used as biological controls, replacing chemical pesticides to control agricultural pests. For example, in Leamington, Ontario, 40% of commercial greenhouse tomato growers use a parasitic wasp, *Encarsia formosa*, to control the whitefly, a pest that damages tomato crops. Research some impacts of using parasites as biological controls. Summarize your research on the societal, economic, ecological, and environmental impacts in a PMI chart.

 www.science.nelson.com

△ INVESTIGATION 14.1.1

Estimating Population Size

Inquiry Skills

● Questioning	● Planning	○ Analyzing
○ Hypothesizing	● Conducting	● Evaluating
● Predicting	● Recording	● Communicating

In this investigation, you will apply quadrat and mark–recapture sampling methods to estimate the size of a weed population.

Ecologists need to count organisms for different reasons. They may want to count cells on a microscope slide to test whether food-processing plants are meeting health standards, or count trees in a forest to assess the health of an ecosystem, or count salamanders in a stream to assess the impact of pollutants. For many kinds of organisms, it is virtually impossible to physically count each member of a population.

Materials

weed (plant) identification guide
tape measure
wooden mallet
stakes (18 per group)
string
blunt-tip scissors
piece of white chalk
foamed polystyrene "peanuts" in a plastic bag
black marker

Question

How can you test predictions of the estimated size of a population?

 Be sure not to touch the weeds, especially if you have allergies to weeds. Take care not to trip over the string around your selected study site.

Procedure

Part I: Quadrat Study

1. With two or three other students, choose a study site around your school and select a local weed species within this site. Do not touch any weeds. Identify the species using field guidebooks or other references.

(a) Predict the estimated size of the weed population. Make a note of your reasoning.

2. In your schoolyard, use a tape measure to mark off a 10-m × 10-m square and use a mallet to drive a stake into the ground at each corner.

3. Loop string around each of the four stakes to mark the boundaries of the site. Be sure the string is tightly

attached. Exercise care not to trip over the string during this procedure.

(b) Describe the location of your site, including such abiotic factors as light exposure and soil conditions. Record these observations in an accurately labelled data table.

4. To avoid bias in site selection, randomly select an area within the site to represent the first quadrat by closing your eyes and tossing a piece of white chalk into the square. Make the point where the chalk landed the centre, and use the tape measure to mark off a 1-m × 1-m square. Use the stakes and strings to complete the quadrat.

(c) Record the number of individuals of your chosen plant species within the quadrat in an accurately labelled data table.

5. Repeat step 4 and (c) twice.

(d) Calculate the average population density per square metre using the data obtained from all three quadrats.

(e) Multiply the population density by 100 to estimate the total number of organisms in the larger site. Record this information in your data table.

Analysis

(f) In the quadrat sampling method, why is it necessary to close your eyes when selecting a quadrat site?

(g) Compare the average population size to the size counts obtained for each individual quadrat.

(h) Why is it important to use more than one sample?

Evaluation

(i) Compare your group's results with those of other groups. How close were your predictions? Account for any differences observed in the results.

(j) Explain why sampling an animal population using the quadrat method would be more challenging than sampling a plant population using this method.

Part II: Mark–Recapture Method

6. Obtain a plastic bag of foamed polystyrene "peanuts" ▸ from your teacher.

(k) Predict the number of peanuts in the bag and record your prediction.

7. Mark each peanut with black ink and return them all to the bag. Shake and stir the contents of the bag to mix the peanuts thoroughly.

8. Capture 40–50 peanuts from the bag.

(l) Record in your data table the number you have captured.

9. Recapture a sample of similar size without looking into the bag to see whether you're recapturing peanuts with or without marks.

(m) From your recaptured sample, record in your data table the total number of recaptured peanuts and the number of those that are marked.

10. When you have finished recording, return all the peanuts to the bag. Shake the bag to mix the peanuts.

11. Obtain data for two additional recaptures by repeating steps 9 and 10. Be sure to repeat the records required by (m).

Analysis

(n) From each of your three sets of data from your table, one set from each recapture, calculate a population estimate (N) for the peanuts in the bag by using the following formula:

$$N = \frac{\text{total number marked} \times \text{total recaptured in each sample}}{\text{total number of recaptured which were marked}}$$

Calculate an average population estimate from your three answers.

Evaluation

(o) Compare your three values for the peanut population in your bag: your original prediction (k); your average estimate (n); and the actual population size as revealed by your teacher.

(p) List three factors that could have reduced the accuracy of your total estimate.

(q) Change one of the factors you listed in (p) and perform two more counts using the mark–recapture method. Analyze the new data and discuss how this factor influenced your population estimate.

(r) If migration occurred in this population, how would this influence the reliability of your estimate?

Synthesis

(s) Weed growth in many areas of the province are perceived to be a significant problem. The Ministry of the Environment has asked that some school yards be surveyed for local weed species in an effort to assess the problem. Your school has been selected to share its quantitative data from the quadrat sampling study. Write a report to the Ministry of the Environment outlining your school's weed situation. Include the following in your report:
   (i)   a summary of the quantitative data obtained in Part I of this investigation
   (ii) an interpretation of this data in relation to the local weed species problem
   (iii) an assessment of the sampling technique in terms of its reliability and accuracy, and any improvements you believe would increase the accuracy of your data.

LAB EXERCISE 14.4.1

Investigating Predator–Prey Cycling

Inquiry Skills

● Questioning	○ Planning	● Analyzing
● Hypothesizing	○ Conducting	○ Evaluating
○ Predicting	○ Recording	● Communicating

The large white-tailed deer population in a forest reserve in Ontario has caused concern about overgrazing that might lead to the extinction of plant and animal species found there. To manage this excessive deer population, forest personnel decided to introduce its natural predator, the wolf (**Figure 1**). In the year 1990, 2000 deer lived within the reserve, and 10 wolves were flown into this reserve. Population densities of white-tailed deer and wolves were monitored for a 10-year period.

Figure 1

Table 1 Changes in White-Tailed Deer and Wolf Populations

Year	Whitetailed deer	Wolves
1990	2000	10
1991	2300	12
1992	2500	16
1993	2360	22
1994	2244	28
1995	2094	24
1996	1968	21
1997	1916	18
1998	1952	19
1999	1972	19

Question

What effect does the introduction of a natural predator, the wolf, into a habitat have on the white-tailed deer population?

Hypothesis

(a) Develop a hypothesis about the effect on the white-tailed deer population as a result of the introduction of wolves into their habitat.

Procedure

1. Plot the changes in the white-tailed deer and wolf population using the data in **Table 1**, including both sets of data on one graph and using an appropriate labelling method.

Analysis

(b) Is wolf predation a limiting factor in this forest reserve? Explain your reasoning.

(c) What other factors might limit the deer population?

(d) Explain how the number of wolves in the reserve is influenced by the size of the deer population.

Synthesis

(e) The Atlantic cod population was an extremely abundant stock of primary economic importance to fishing communities throughout the Atlantic provinces. The Department of Fisheries and Oceans has stated that the collapse in Atlantic cod stocks can be attributed to overfishing. Others claim that the use of equipment that disturbs fish spawning sites on the ocean floor is primarily responsible, and still others argue that the harp seal, a predator of Atlantic cod, is responsible for this mass reduction in the cod population. One suggestion to help cod stocks recover is for large numbers of harp seals to be killed. Suggest some ways that marine biologists might study changes to the Atlantic cod population to determine whether the reduction of the harp seal population would be an effective solution.

Key Expectations

- Express the result of any calculation involving experimental data to the appropriate number of decimal places or significant figures. (14.1, 14.2)
- Describe characteristics of a population, such as growth, density, distribution and dispersion, carrying capacity, and minimum viable size. (14.1, 14.2, 14.3)
- Compare and explain the fluctuation of a population of a species of plant, wild animal, and microorganism, with an emphasis on such factors as carrying capacity, reproductive success, and predation. (14.1, 14.2, 14.3, 14.4)
- Use conceptual and mathematical models to determine the growth of populations of various species in an ecosystem. (14.1, 14.2, 14.3, 14.4)
- Using the ecological hierarchy for living things, evaluate how a change in one population can affect the entire hierarchy, both physically and economically. (14.1, 14.3, 14.4)
- Explain the concepts of interaction among different species of animals and plants. (14.3, 14.4)
- Determine experimentally the characteristics of population growth of two populations. (14.4)

Key Terms

Allee effect	environmental resistance
biotic potential	exploitative competition
carrying capacity	exponential growth
character displacement	fecundity
closed population	fundamental niche
clumped dispersion	geometric growth
commensalism	habitat
community	interference competition
crude density	interspecific competition
density-dependent factors	intraspecific competition
density-independent factor	lag phase
	limiting factor
dynamic equilibrium	logistic growth
ecological density	log phase
ecological niche	macroparasites
ectoparasites	mark–recapture method
endoparasites	microparasites

minimum viable population size	predation
mutualism	quadrats
obligatory mutualism	random dispersion
open population	realized niche
parasitism	resource partitioning
population density	social parasites
population dispersion	species
population dynamics	stationary phase
population size	symbiosis
	uniform dispersion

Key Symbols and Equations

- population density: $D = \dfrac{N}{S}$

- average sample density $= \dfrac{\text{total number of individuals}}{\text{total sample area}}$

- estimating population size using mark–recapture sampling:

  $N = \dfrac{Mn}{m}$

- population change $= \dfrac{[(b + i) - (d + e)]}{n} \times 100$

- geometric growth: $N(t + 1) = N(t)\,\lambda$

- exponential growth: $\dfrac{dN}{dt} = rN$

- logistics growth: $\dfrac{dN}{dt} = r_{max}N[\dfrac{(K - N)}{K}]$

▸ *MAKE* a summary

Select a species of microorganism, plant, or animal and consider its role and functions as an individual, as a part of a population, and as part of an ecological community. Describe what life would be like for an individual of this species in terms of the following:

- current population status
- interspecific interactions within the community
- potential changes to population status and size as a result of interspecific interactions
- factors (density dependent and density independent) that might affect species
- any other population dynamics you have learned in this chapter

In your notebook, indicate whether each statement is true or false. Rewrite a false statement to make it true.

1. Monitoring population size and population density is important in preventing the loss of species.

2. Population density provides meaningful information about changes in the population status of a species.

3. In nature, species often exhibit random dispersal.

4. An organism facing interspecific competition always occupies its fundamental niche.

5. Species occupying the same niche can reduce competition by resource partitioning.

6. Population dynamics of predator–prey interactions show that, in the presence of a predator, prey species often become extinct.

7. Prey species use multiple defence mechanisms to provide some protection against their predators.

In your notebook, record the letter of the choice that best completes the statement or answers the question.

8. Which of the following is most true about interspecific competition?
   (a) It involves interaction between individuals of the same species.
   (b) It is the greatest between organisms that occupy different ecological niches.
   (c) It can be easily demonstrated to occur in natural environments.
   (d) It is the greatest between organisms that share similar ecological niches.
   (e) It increases with resource partitioning.

9. Which of the following is true in relation to predation:
   (a) The predator is negatively affected and the prey is positively affected.
   (b) The predator population always increases and the prey population always decreases in size.
   (c) The predator–prey population cycle exhibits a sinusoidal pattern on a graph.
   (d) The prey can never be protected against its predator.
   (e) The snowshoe hare and lynx cycling study demonstrated that predator–prey cycling is not influenced by any other factors.

10. What is the term for a plant or animal species introduced into a community where it may disrupt balance in an ecosystem and threaten biodiversity?
    (a) parasite   (d) indigenous species
    (b) predator   (e) nonindigenous species
    (c) prey

11. Which of the following is an example of a closed population?
    (a) a group of frogs living in a marsh in Haliburton
    (b) a herd of dairy cows living on an Ontario farm
    (c) a pride of lions living in the African savannah
    (d) a school of fish in Lake Ontario
    (e) a flock of geese in a large field

12. Which of the following is a density-independent factor that limits population growth?
    (a) a tornado destroying a stand of pine trees
    (b) two species competing for the same food source
    (c) a disease passing through a population of pigs on a farm
    (d) intraspecific competition
    (e) snowy owls preying on arctic hares

13. What is the term for the number of offspring a population could produce if no limits were placed on it?
    (a) natality
    (b) the intrinsic growth rate
    (c) biotic potential
    (d) a density-dependent limiting factor
    (e) a density-independent limiting factor

14. When does stable population size occur?
    (a) when natality and immigration are greater than mortality and emigration
    (b) when mortality and emigration are greater than natality and immigration
    (c) when there is no emigration or immigration
    (d) when there are no births or deaths
    (e) when natality and immigration are equal to mortality and emigration

15. Birds find shelter and build nests in trees without harming the trees. What is the term for the symbiotic relationship between the birds and the trees?
    (a) mutualism
    (b) parasitism
    (c) commensalism
    (d) predation
    (e) There is no relationship.

Understanding Concepts

1. Using two examples, explain why it is important for scientists to track the population status of Canadian species.

2. Calculate the density of a population of southern flying squirrels, if 940 squirrels were counted in an 68-ha area.

3. Describe three patterns of population dispersion, with an example of a population that exhibits each pattern.

4. Compare and contrast the mark–recapture and quadrat methods for estimating population size. Include sampling issues and, for the mark–recapture method, ethical issues, and underlying assumptions.

5. Explain, using an example, the difference between ecological density and crude density.

6. Describe the concept of carrying capacity and explain its role in population dynamics.

7. Identify and describe ways in which the decline of resources in an ecosystem can affect the growth rate of a population in that ecosystem.

8. What is the Allee effect? Give an example of a population that is affected by this density-dependent limiting factor.

9. For each of the photographs in **Figure 1**, identify the defence mechanism used by each species.

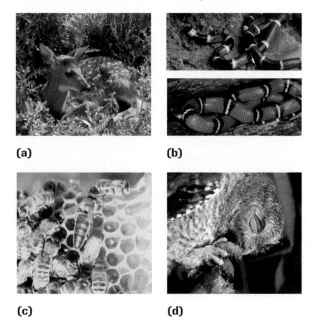

(a)　　　　　　　　　**(b)**

(c)　　　　　　　　　**(d)**

Figure 1
(a) white-tailed deer; **(b)** venomous coral snake (top) and harmless milk snake (bottom); **(c)** African killer bees; **(d)** screech owl feeding on hornworm

10. Explain why it would be a mistake to eliminate a major predator from a community.

11. Identify similarities and differences between predators and parasites.

12. During the past 15 years, white-tailed deer populations in Canada have significantly affected woody plants in forest ecosystems. To estimate white-tailed deer populations in four tracts of the Carolinian forest in southern Ontario, 175 white-tailed deer were captured and tagged with bands.
   (a) After three weeks, 135 deer were recaptured and 21 had bands. Calculate the population density.
   (b) Identify potential defence mechanisms used by the woody plants on which the deer feed.

13. Large groups of sea urchins form spine canopies to protect and shelter their larvae. This defence mechanism requires cooperation among the sea urchins. Explain how the Allee effect is an issue for a population that exhibits this kind of behaviour.

Applying Inquiry Skills

14. Scientists conducted a study into the competition between two species of rodents—the woodland jumping mouse, *Napaeozapus insignis*, and the meadow jumping mouse, *Zapus hudsonius*. The meadow jumping mouse is known to be able to exist in both field and forest habitats. Both species of mice are seed feeders. The experimental design included the selection of three approximately 100-ha plots with similar plant cover. Plot 1 supported a population of *N. insignis*, Plot 2 supported a population of both *N. insignis* and *Z. hudsonius* while Plot 3 supported a population of only *Z. hudsonius* (**Table 1**). The populations of mice were monitored using mark and recapture techniques over a period of four years (**Table 2**).
   (a) Calculate the average population density for each population in each plot over the four year study.
   (b) Based on these results, what conclusions can you infer regarding the interactions between these two rodent species?
   (c) Identify the type of interaction occurring in Plot 3
   (d) Some biologists might argue that the evidence from this study is inconclusive due to the assumptions being made by the researchers. Identify and comment on the acceptability of three such assumptions?
   (e) What improvements could be made to the experimental design used in this study?

Table 1 Experimental Plot Characteristics

Plot 1	Plot 2	Plot 3
Mixed moist woodland	Mixed moist woodland	Mixed moist woodland
100 ha	92 ha	104 ha
Species list:	Species list:	Species list:
N. insignis	*N. insignis, Z.hudsonius*	*Z.hudsonius*

Table 2 Analysis of Experimental Data

	Plot 1				Plot 2				Plot 3			
Year	1	2	3	4	1	2	3	4	1	2	3	4
Napaeo-zapus insignis	632	788	840	671					610	559	663	601
Zapus hudsonius					345	461	509	328	102	188	173	80

15. Study the graphs of two different populations (**Figure 2**): a population of bacteria growing in a laboratory and a population of owls living in a forest. Which graph represents which population? Explain your reasoning.

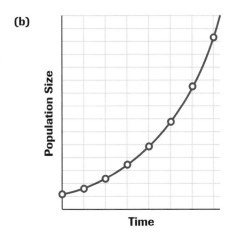

Figure 2

16. Six ground finches began nesting on an island in 1990. Biologists monitored numbers in this population for years, compiling their data as shown in **Table 3**.
    (a) Graph the changes in this population over the 9-year period. Label the various population growth phases on your graph.
    (b) Estimate the carrying capacity of the island. Label this value on your graph.

Table 3 Data on Ground Finch Population

	1990	1991	1992
Finch population	18	35	58
	1993	**1994**	**1995**
Finch population	170	280	477
	1996	**1997**	**1998**
Finch population	359	296	283

Making Connections

17. Research the sea lamprey, a nonindigenous species, which has had a great impact on the fish communities of the Great Lakes.
    (a) Describe the fundamental and realized niche of the sea lamprey.
    (b) Find out how sea lampreys may have entered the Great Lake ecosystem.
    (c) Identify the interspecific interactions of the sea lamprey and its effects on the Great Lakes.
    (d) Describe some economic setbacks faced by Canadian fisheries as a result of sea lamprey invasion and outline any control efforts.

Extension

18. Beginning in 1937, researchers Fred and Norah Urquhart of the University of Toronto began efforts to discover the wintering location of the migratory monarch butterfly. It was not until 1975 that the wintering location was finally discovered.

    Use the Internet to explore the science of marking and releasing organisms. How did the Urquharts tag butterflies? How did their persistent efforts play a role in the discovery of the monarch migration route? Gather satellite data on species tagged with radio transmitters and plot their movements on a map. What information can be obtained from this data regarding population densities, distribution and dispersion patterns, and behaviours?

 www.science.nelson.com

15

Human Populations

▶ **In this chapter, you will be able to**

- explain demographic changes that have affected populations over the past 10 000 years;

- explain problems related to the rapid growth of human populations and the effects of that growth on future generations;

- outline the advances in medical care and technology that have contributed to an increase in life expectancy, and relate these developments to demographic issues;

- analyze the carrying capacity of Earth in terms of the growth of populations and their consumption of resources;

- use examples of the energy pyramid to explain production, distribution, and use of food resources;

- analyze Canadian investments in human resources and agricultural technology in a developing country;

- describe examples of stable food-production technologies that nourish a dense and expanding population;

- describe general principles and strategies for achieving sustainable development.

Humans are an intelligent and successful species. Our use of tools has permitted us to live in conditions and environments well outside our biological range of tolerances. We inhabit almost every terrestrial ecosystem on Earth—from equatorial tropical rain forests and savannahs, and temperate forests and grasslands, to the extreme conditions of deserts and the high Arctic. Even so, human populations exhibit many of the characteristics and limitations that other organisms do. Like all populations, humans live in a finite environment with a limited carrying capacity, and our successes and failures are closely linked to the biotic and abiotic features of the environments in which we live. Today, the rapidly growing human population and its consumption of global biological and mineral resources, such as water, air, and soil for food production, is placing significant stresses on these environments, on which people and other organisms depend for survival. Scientific study of global environmental patterns over the past century has raised serious concerns about the well-being of life support systems worldwide. Scientific and technological advances seek to improve medical and health care, human access to food and clean water, and strategies for sustaining Earth's carrying capacity in order to support the well-being and future of human populations.

💡 REFLECT on your learning

Study the images on the opposite page and reflect on the following questions:

1. What are some factors that enable people to live in different environments?

2. What are the key resources needed to support any human population? Where and how are these resources obtained?

3. Compare and contrast the ways in which urban populations differ from rural populations in how they obtain resources.

4. What are potential ecological effects on the environment of the activities of large human populations?

> ▶ **TRY THIS** activity

Effects and Consequences

With two or three other students, brainstorm some possible effects of the human population on each of the following:

- forests
- wildlife
- biodiversity
- harvesting of seafood
- grain and livestock production
- water supply and water quality
- the atmosphere

(a) Design a concept map to illustrate these effects.

(b) Discuss your concept map with the class.

The United Nations now estimates the world human population to be in excess of 6 billion people. This value is in sharp contrast to the few thousand people that are thought to have lived 100 000 years ago. How did humans become such a successful species, able to inhabit almost every conceivable ecosystem and outnumber every other large terrestrial vertebrate? In what manner are these 6 billion people distributed over the surface of Earth? How do we obtain the necessary resources to feed, clothe, and shelter ourselves? What impact do we have on the natural environment? **Figure 1** illustrates the dramatically uneven distribution of the human population by geographic region.

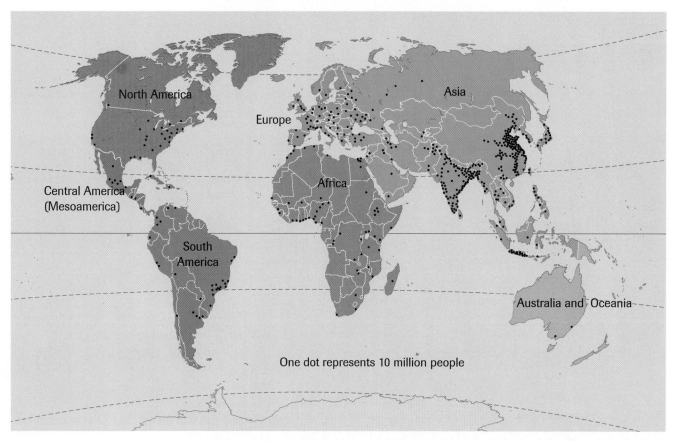

Figure 1

The distribution of people on Earth. Each dot represents 10 million people. Dots are placed at or near the greatest population centres in each region.

Wheat, rice, and corn (maize) provide more than 60% of the total food energy consumed by the world's population. In many parts of the world, these crops are cultivated intensively, as illustrated in **Figure 2(a)**. The primary regions of wheat and rice production are shown in **Figure 2(b)**.

(a)

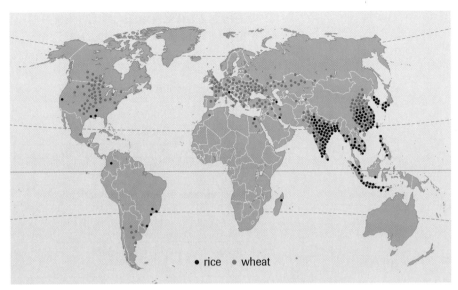

• rice   • wheat

(b)

A Brief History of Human Population Growth

More than 1 million years ago, long before humans evolved, our early ancestors began to walk upright, had large brains, and could fashion and use tools, including fire. By about 100 000 years ago, anatomically modern humans were living as successful hunter-gatherers and were rapidly spreading across the globe. Their extraordinary ability to learn and communicate, make tools, and control fire enabled them to invade and live in environments unsuitable to their earlier ancestors. Their great success as hunters enabled them to feed on herds of large wild animals, reducing their dependence on wild plants that had once supplied the bulk of their natural diets (**Figure 3**). By about 12 000 years ago, humans had reached and inhabited all the world's major landmasses, with the exception of Antarctica. The entire world's human population probably numbered at little more than 5 million.

Groups of hunter-gatherers could not increase quickly in number. A nomadic, wandering lifestyle with no permanent residence placed a variety of biological and social restrictions on average family size. Although birth rates were low, death rates were often very high, especially among the very young, the old, and the ill. As a result, the populations dependent on hunting, while often very successful, tended to remain small.

Figure 2

(a) Terraced hillsides used for rice paddies. Rice is the main food source for more than half the people in the world.

(b) The two most important food crops, wheat (in red) and rice (in black), are produced in very large quantities in different regions of the world. Each dot represents an annual production of 2 million tonnes of grain.

Figure 3
Skills, such as tool-making, enabled hunter-gatherers to move into environments unsuitable for their earlier ancestors.

Seeds and Breeds of Change

One of the most profound changes in human history came with the intentional planting of seeds and the beginnings of agriculture. In different places and at different times, humans discovered that certain plants and animals were ideally suited for cultivation and domestication (**Table 1**).

Table 1 Earliest Known Times and Locations of Domestication of Important Plant and Animal Species

Area	Domesticated		Earliest attested date
	Plants	*Animals*	of animal domestication
Independent origins of domestication			
Southwest Asia	wheat, pea, olive	sheep, goat, cow	8500 B.C. 6000 B.C.
China	rice, millet	pig, silkworm water buffalo	by 7500 B.C. 4000 B.C.
Ukraine/central Asia	none	horse	4000 B.C.
Mesoamerica	corn, beans, squash	turkey	by 3500 B.C.
Andes and Amazonia	potato, cassava	llama	by 3500 B.C.
eastern United States	sunflower	none	2500 B.C.
Sahel	sorghum, African rice	guinea fowl	by 5000 B.C.
tropical West Africa	African yams, oil palm	none	by 300 B.C.
New Guinea	sugar cane, banana	none	7000 B.C.

In Asia, native species of wheat and oats were first cultivated about 10 500 years ago. Within 1000 years, native species of rice were being domesticated in China, and 5500 years ago, corn was successfully cultivated in Mesoamerica. The most important examples of the domestication of animals include the following: sheep, goats, and pigs about 10 000 years ago in Asia; cattle 8000 years ago in Southwest Asia, India, and possibly North Africa; and the horse 6000 years ago in Eastern Europe.

The dramatic impact on human history of the adoption of agricultural practices cannot be overstated. Human populations dependent on the cultivation of plants were stationary. As a consequence, having large families was now possible. In addition, growing crops and raising livestock could support much higher population densities than hunting and gathering. Only a relatively small number of individuals were needed to plant, tend, and harvest the bulk of the food needed to feed a large population. This altered the pattern of life for most. Many people, freed from the need to obtain food for themselves and their families, became specialized artisans, scholars, and merchants. Free time and a stationary dwelling allowed for major advances in technology. The combined effect of large food surpluses, stationary living, large populations, and specialization of the workforce led to the establishment of complex political organizations and the creation of large states.

The adoption of agricultural practices was neither uniform nor predictable. Indeed, the fate of much of human history has been a direct consequence of geographic accident rather than human planning. The geographic distribution of suitable wild plants and animals largely determined the regions in which agriculture would arise and where human societies would follow. The vast majority of wild plant and animal species are unfit for domestication (**Table 2**) and those that are fit are not uniformly distributed around the globe.

Table 2 Mammalian Candidates[1] for Domestication

	Continent			
	Eurasia	*Sub-Saharan Africa*	*The Americas*	*Australia*
Candidates	72	51	24	1
Domesticated species	13 (pig, goat, sheep, horse, cow, water buffalo, camels (2), donkey, reindeer, yak, guar, banteng)	0	1 (llama)	0
Percentage of candidates domesticated	18%	0%	4%	0%

[1] A candidate is defined as a terrestrial herbivore or omnivore with an average mass of 50 kg or more.

Almost half of the 148 large mammalian herbivore and omnivore species of Earth are native to Europe and Asia. Of these 72 species, 13 have been successfully domesticated. Of the remaining 76 species distributed among sub-Saharan Africa, the Americas, and Australia, only one, the llama, has been domesticated. In sub-Saharan Africa, of 51 candidate species, not a single mammal has been successfully domesticated. The vast majority of large mammals are simply unsuitable for domestication. Although many individual large mammals can been "tamed," they are often difficult or impossible to breed in captivity, do not exhibit social behaviour suitable to "herding," or are simply too dangerous. Grizzly bears for example, are mostly vegetarian, able to consume a wide variety of foods, and grow quickly. However, it is very unlikely that they will ever be reared in large numbers for food!

An uneven distribution is also the case for large-seeded grass species—such as wheat, rice, and corn—the plants most suitable for easy cultivation. **Table 3** shows the world distribution of 56 large-seeded grasses. These species have seeds that are 10 times larger than the average seed and account for less than 1% of the world's grasses. They are unevenly distributed, with much greater numbers occurring in the Mediterranean region. It is noteworthy that corn, domesticated in Mesoamerica, had a number of characteristics that made it particularly difficult to use as a food source and that might well account for the thousands of years that separate its cultivation from that of wheat and rice (**Figure 5**).

Table 3 World Distribution of Large-Seeded Grasses

Area	Number of Species	
West Asia, Europe, North Africa	33	Mediterranean zone (32), England (1)
East Asia	6	
sub-Saharan Africa	4	
Americas	11	North America (4), Mesoamerica (5), South America (2)
northern Australia	2	
Total	56	

Thus, the geographic distribution of suitable plant and animal species largely determined which human populations had the opportunity to switch to an agricultural lifestyle. Even today, after thousands of years and many attempts, human populations remain almost entirely dependent on the same few food crops and domestic animals

Figure 4

(a)

(b)

Figure 5
The probable ancestor of modern corn, teosinte grass **(a)** has a hard outer casing making it difficult to process as a food. Humans living in Mesoamerica bred this original grass into the useful crop plant corn **(b)**.

that launched our population boom: wheat, rice, and corn, and sheep, pigs, and cattle. Although not used as a major food animal, the horse played a critical role as a mode of transportation and a source of power.

Not surprisingly, those human populations that experienced rapid growth based on large agricultural food surpluses began to spread out across the globe. In most cases, they brought their domesticated plants and animals and their technological innovations with them. A striking example of this was seen in North America where wheat, barley, and other crops, as well as cattle and pigs, arrived from Europe and became the dominant food sources for future generations. Often this was not to the benefit of the indigenous human populations, which were often displaced, or worse, by the invaders who brought not only assorted plants and animals, but also guns, steel, and germs.

The Impact of Disease

The establishment of large sedentary human populations has a profound impact on the evolution of pathogens. Many **epidemic** diseases—those capable of spreading rapidly through large populations and causing significant mortality—can only evolve where population densities are high. For example, recent studies indicate that the measles virus dies out in any human population numbering fewer than one half million. In addition, many of the epidemic diseases that have significantly reduced local populations, such as small pox, tuberculosis, and influenza, are believed to have been diseases of domestic animals that later evolved into diseases that could infect humans (**Table 4**). Epidemic diseases arose in regions of the world where human populations were dense and where animal domestication was widespread: Europe, North Africa, Southeast Asia, India, and China. The best known example is the Black Death, or bubonic plague, caused by the bacterium *Yersinia pestis*. During the fourteenth century this epidemic spread from China to Europe resulting in millions of deaths. In turn, human populations continuously exposed to these diseases evolved some resistance.

epidemic describes any disease that spreads rapidly through a population

Table 4 Probable Origins of Epidemic Diseases

Human disease	Animal with most closely related pathogen
measles	cattle (rinderpest)
tuberculosis	cattle
smallpox	cattle (cowpox) or other livestock with related pox viruses
flu	pigs and ducks
pertussis (whooping cough)	pigs and dogs
falciparum malaria	birds (chickens and ducks?)

When European explorers reached the shores of the Americas and made contact with populations of Aboriginal peoples, they brought with them many serious diseases to which the Aboriginals had little or no resistance. In many cases, large epidemics decimated Aboriginal populations. For example, in 1837, smallpox was introduced into a Mandan village and, within a few weeks, the population plummeted from 2000 to fewer than 40 individuals.

Although on a local scale tragedy and devastation were commonplace, on a global scale the human population was steadily increasing in size. In terms of sheer numbers, the global human population is thought to have increased to an estimated 300 million by A.D. 1 and, as agricultural practices continued to spread around the world, is thought to have reached 500 million by the year 1650.

The Impact of Science and Medicine

In the 17th century, science-based knowledge and technology started to become major influencing factors in European societies. Biology, chemistry and physics began to shed new light on the natural world. During a brief period of a few hundred years, our understanding of and interaction with the natural environment would be revolutionized. The cell and germ theories in biology would provide the foundation for modern medicine. Advances in chemistry and physics would permit the harnessing and conversion of fossil fuel energy into mechanical energy. The result was the Industrial Revolution, in which power-driven manufacturing resulted in the mass production of industrial and consumer goods. Further advances in chemistry led to the production of countless new compounds, including fertilizers, which further increased and intensified food production. The steam engine and, later, the internal combustion engine would provide unrivalled sources of power for industry and lead us into the modern world.

Although many of these changes would begin to have a negative impact on the environment, the most dramatic and immediate effect on human populations was the significant increase in food supply and the reduction in the number of people required to work the land. Populations became increasingly urbanized as tractors and trains replaced teams of men and horses. In addition, death rates dropped with improved access to safe drinking water, sewage systems, and the development and widespread use of vaccines. In this same few hundred years, Earth's population would double to 1 billion and go on to double once more in the following hundred years. Science and technology allowed humans to sidestep, at least temporarily, the natural limits to growth that had existed for millions of years. By the mid-twentieth century, humans had what seemed like unlimited access to energy, water, food, and mineral resources while, at the same time, modern medicine had dramatically reduced rates of mortality—most significantly, mortality due to infectious disease. The greatest effect was on the death rate of the very young, which decreased sharply. The human population explosion was under way, with the population doubling about every 50 years (**Figure 6** and **Table 5**).

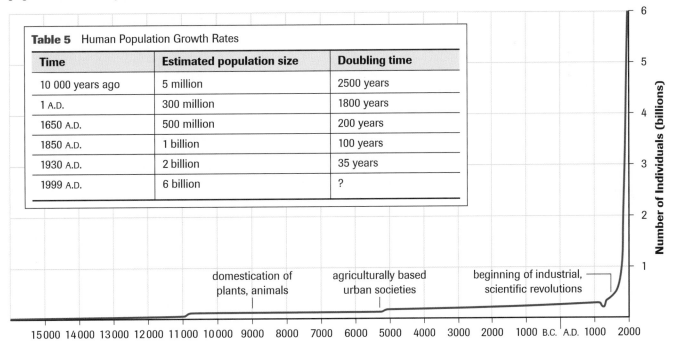

Table 5 Human Population Growth Rates

Time	Estimated population size	Doubling time
10 000 years ago	5 million	2500 years
1 A.D.	300 million	1800 years
1650 A.D.	500 million	200 years
1850 A.D.	1 billion	100 years
1930 A.D.	2 billion	35 years
1999 A.D.	6 billion	?

Figure 6

SUMMARY	*Human Populations*

- Humans have employed their tool-making skills to successfully inhabit almost every terrestrial environment on Earth.

- Approximately 11 000 years ago, humans began to cultivate food crops and domesticate livestock. Domestication first arose in those regions of the world that possessed ideally suited plant and mammal species. Many regions had few or no suitable candidates.

- Epidemic diseases evolved in large human populations. As these populations spread around the world, they carried these deadly diseases to other human populations often with devastating consequences.

- Advances in science and technology revolutionized human understanding of biology, chemistry, and physics and lead to the Industrial Revolution and rapid progress in food production, sanitation, and medicine.

- These changes allowed humans to avoid the natural limits to growth that had existed for millions of years. The result has been a dramatic decline in death rates—particularly among the very young—and a rapidly growing human population.

▶ *Section 15.1 Questions*

Understanding Concepts

1. Does the human population exhibit a uniform, clumped, or random distribution? Explain.

2. What regions of the world contain the highest population densities? Suggest possible reasons.

3. How are the locations of wheat and rice production influenced by
   (a) global climate patterns?
   (b) human population distribution?

4. How did large-scale food production influence
   (a) family size?
   (b) daily life?
   (c) technological advances?

5. How was the rise of human societies influenced by the distribution of suitable plant and animals species? Use examples of each in your answer.

Applying Inquiry Skills

6. The Maya were able to support a very large and complex society in Mesoamerica.
   (a) What was their main food crop?
   (b) What might explain the fact that, although they had a large population, they had few epidemic diseases of their own and were largely decimated by diseases only after contact with early European explorers?

7. Compare and contrast the relative abundance of humans and production of wheat and rice in North America and Africa.
   (a) Did these crops originate in either of these locations? If not, why is it possible for them to be grown where they are now?
   (b) How do you think these two regions compare in terms of present economic activity? How might this be related to food production?

Making Connections

8. Contact your local health department, or check your own immunization record, and record those diseases against which Canadian children are still routinely vaccinated. Which of these diseases were epidemics in the past?

9. Conduct research on the Internet to determine several key advances in the science disciplines of biology, chemistry, and physics that were among the most influential in human history. Account for their importance.

 www.science.nelson.com

The population of Mexico City exceeds 20 million and increases by 2000 people each and every day. How is it possible for a city to grow by 750 000 per year and what are the possible consequences of such a rapid growth rate?

The study of human populations is referred to as **demography**. These studies rely on data collection, statistical analyses, and theoretical models to interpret and predict important patterns of human population growth and distribution. As with all species, human population growth depends on birth and death rates. Over the past 300 years, since the beginning of the Industrial Revolution, human birth rates have remained relatively stable, declining from about 30 to about 25 births per 1000 individuals per year. However, death rates have dropped dramatically from 29 per 1000 per year to 13 deaths per 1000 individuals per year! The result is an average growth rate of 16 per 1000 per year, or 1.6% (**Figure 1**). Although this number seems small, it represents an increase of about 96 million people per year and an overall doubling time of 43 years. In Chapter 14, you learned that the doubling time (t_d) can be approximated by the following equation (where r is the growth rate):

$$t_d = 0.69/r$$

On a daily basis, the numbers are equally dramatic, with the population increasing by about 260 000, the difference between the 480 000 daily births and the 220 000 daily deaths. This is truly a population explosion.

As we have seen, the characteristics of human populations in different geographic regions are not uniform. **Table 1** depicts differences in growth rates in different countries of the world. Demographers and population ecologists are very interested in the implications of these growth rates on the health of both humans and the natural environment, at the local and global levels. What key factors might be responsible for the significant variation in population growth rates in different regions of the world?

demography the study of the growth rate, age structure, and other characteristics of human populations

Figure 1
Annual percent increase of human population by country, 1999

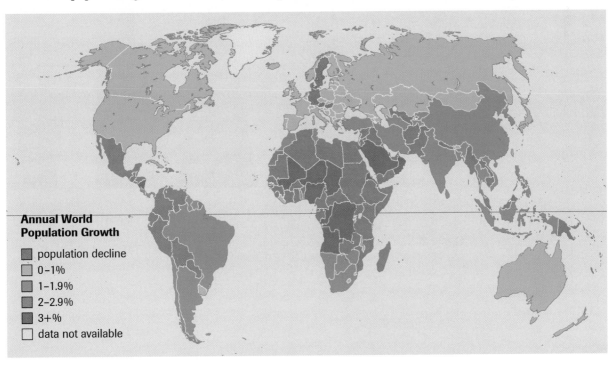

Annual World Population Growth

- population decline
- 0–1%
- 1–1.9%
- 2–2.9%
- 3+%
- data not available

Table 1 Population Data for Selected Countries

Country	Population (millions)	Annual birth rate per 1000 people	Annual death rate per 1000 people	Annual % growth rate of natural growth
Canada	31	11	8	0.3
Japan	127	9	8	0.1
Mexico	100	24	5	1.9
Philippines	77	29	6	2.3
France	59	13	9	0.4
India	1009	26	9	1.7

A major factor in the wide range of growth rates in various parts of the world is the differences in their age structures. The age structures of populations are often depicted as population pyramids. **Figure 2** illustrates the general patterns of age structures typical of populations undergoing different growth rates. Using these patterns as a guide, demographers can compare different populations. For example, the population structure of the United States appears relatively stable, as the number of individuals is evenly distributed across most age classes. In contrast, the population of Kenya comprises a very large number of young people who are likely to reproduce in the near future—long before an equivalent number of deaths occur in the older age classes (**Figure 3**). Kenya is just one example of many less industrialized countries that exhibit a **growth momentum** due to their current large populations of young people. Why might there be such a marked difference in the age structures of most industrialized countries and those in less industrialized countries?

growth momentum the inherent future growth in a population that results from a disproportionately large fraction of the population being of younger age

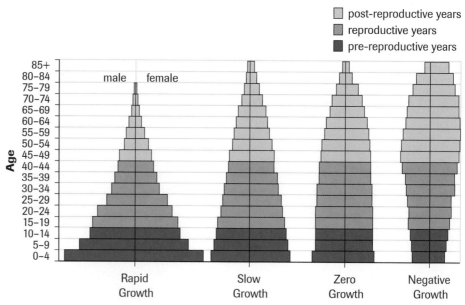

Figure 2
Age structures of populations exhibiting differing growth rates

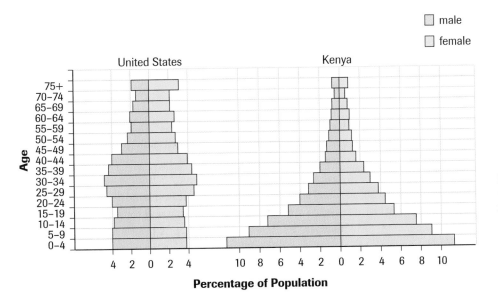

Figure 3
Population pyramids of the United States and Kenya (1990). Note the bulge in the United States age classes from 25 years to 44 years of age, representing the "baby boom" decades following the Second World War.

The Demographic Transition Model

Ultimately, the only factors that dramatically influence human population growth are birth and infant death rates. Factors that either increase or decrease death rates among older members of a population have little effect on overall growth rates if such individuals are in their post-reproductive years. Demographers have noted that, in recent decades, infant death rates have declined significantly in virtually all human populations. In some countries, birth rates have also declined. This is most evident in countries with improved living standards based on successful economies, improved health care, and better educational opportunities—particularly for women.

Warren Thompson, an American demographer, observed certain correlations between changes in population and four stages of economic development and their resulting standard of living. He formulated the **demographic transition model** (**Figure 4**) based on his observations. The model suggests that as societies shift from predominantly agricultural production to industrialization, population size increases with transitions in birth and death rates. According to this model, in a highly industrialized and economically secure country, the population growth rate slowly decreases, while a country that is still developing an industrial economy will continue to experience rapid population growth. During the "preindustrial stage" of this model, living conditions are harsh and birth and death rates are

demographic transition model
a four-stage model that describes the relationship between economic development and changes in population patterns

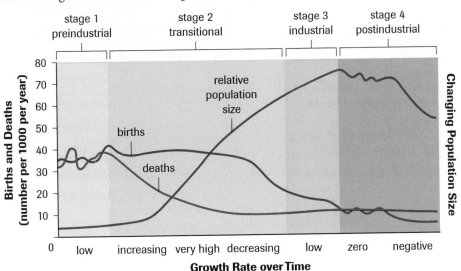

Figure 4
The demographic transition model describes changes in population in four stages of economic development.

Human Populations **707**

zero population growth the condition in which the number of individuals added to a population from births and immigration equals the number of individuals removed from a population by deaths and emigration, resulting in a constant population size

quality of life social, economic, and cultural factors that support and enhance the development of a human population; examples of such factors, which can be refined and collected as statistical indicators to be monitored for change, include housing conditions, food supply, access to education, access to a livelihood, health care, and personal security

very high, resulting in slow population growth. It is important to emphasize that the very high death rates are of the young and, therefore, occur before these individuals have had a chance to reproduce. Until about 200 years ago, most predominantly agricultural societies were in this stage. In the "transitional stage," industrialization begins to produce a decline in death rates while birth rates remain high, so that total population size increases. The "industrial stage" sees the population begin to stabilize as birth rates decline. The "postindustrial stage" features birth and death rates in balance. If countries also have balanced immigration and emigration the result is **zero population growth**.

The demographic transition model must be applied with caution, as a country is not necessarily composed of a single group of people but may contain a number of distinct populations differing in geographic and/or socioeconomic conditions. Consider the Inuit populations in northern Canada, whose birth rate is among the highest in the world (33 per 1000 individuals in the 1990s) and is higher than the average birth rate for that of Canada as a whole (11 per 1000 individuals in 2000), probably because large families are culturally important to Inuit society (**Figure 5**). The **quality of life** and preindustrial economy of the Inuit also contrast with the higher quality of life experienced by most people in southern Canada, which is considered to be in the industrial stage of the demographic transition model.

Figure 5
The demographic transition model, which considers national patterns, does not point out social and economic differences within a country.

Zero Population Growth
In November 1993, at a population conference in New Delhi, representatives of 56 national science academies urged the world to move to zero population growth during the lifetimes of their children—by about 2040. Although it is clear that Earth has a finite carrying capacity, some scientists suggest that stopping population growth may not necessarily result in lower consumption and pollution, as demonstrated by patterns in industrialized nations.

In March 2002, Statistics Canada released the results of the 2001 census. From 1996 to 2001, Canada's population grew by 4.0%, matching the slowest rate in our recorded history. And for the first time since World War II, the number of immigrants exceeded the number of births. More than half of Canada's 30 007 094 people now live in the four major urban regions of southern Ontario's Golden Horseshoe, greater Montreal, the lower mainland of British Columbia, and the Calgary–Edmonton corridor, with over 79% living in urban centres of 10 000 or more. During those five years, Ontario's population increased by 656 000.

TRY THIS activity   *Age-Structure Diagrams*

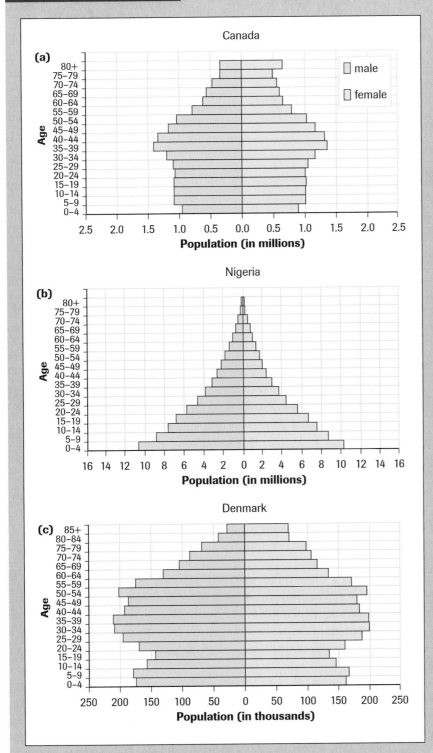

(a) Canada

(b) Nigeria

(c) Denmark

The Worldwatch Institute is preparing to launch a campaign to identify population growth patterns around the world. You have been hired as the demographer to describe the impacts of growth patterns in different areas in the world. For your presentation, you decide to present an analysis of three representative age-structure diagrams from the year 2000. Age-structure diagrams are models that demographers use to illustrate changing patterns in the proportion of people within age groups in a population (**Figure 6**).

(a) For each of the three age-structure diagrams, identify the age group present in the largest proportion.

(b) Based on the age-structure diagrams, what predictions can be made about future growth trends in each of these countries? Explain these predictions.

(c) Summarize the similarities and differences observed in each of the age-structure diagrams.

(d) Why is it important for government policymakers to have information about the proportion of their country's population in different age groups?

(e) Describe some social and economic conditions that may exist in each of the countries above, based on data from the age-structure diagrams

Figure 6
Age-structure diagrams for the year 2000 for **(a)** Canada, **(b)** Nigeria, and **(c)** Denmark

gross national product (GNP)
the total value of all goods and services produced by a country in a particular time frame, usually one year; the value can be given per capita

The **gross national product (GNP)** is often used as an index of a country's level of industrialization. For example, the 1992 per capita GNP of the United States, a highly industrialized country, was in excess of U.S.$23 000, while that of Brazil, a moderately industrialized country, was U.S.$2770. In sharp contrast, the per capita GNP of Ethiopia was only U.S.$110 (**Table 2**).

Table 2

	United States (highly industrialized)	Brazil (moderately industrialized)	Ethiopia (less industrialized)
Fertility rate[1]	2.1	3.0	6.9
Infant mortality rate[2]	8.3	66	110
Life expectancy (years)	76	57	52
Doubling time (years)	98	40	22
Per capita GNP ($U.S.)	$23 120	$2770	$110

[1] average number of children born per woman in her lifetime

[2] infant deaths per 1000 births

The additional data in **Table 2** illustrates another general relationship between industrialization and population characteristics—that of the increase in average life expectancy within industrialized countries. This increase results primarily from a decrease in mortality among infants and young children, but also from a general reduction in mortality of people of all ages. Perhaps surprisingly, even most of the poorest countries of the world have experienced recent sharp decreases in infant mortality rates. This has occurred as a direct result of science and technology and, in particular, is due to access to safer drinking water, improved sanitation, antibiotics, and vaccines. Through one of humanity's greatest achievements, the world was rid of smallpox, one of the deadliest epidemic diseases, using a global vaccination campaign by the World Health Organization that began in 1967 and lasted 13 years. Yet each year more than 10 million children die of readily preventable diseases and malnutrition.

In sharp contrast, increases in the longevity of older (post-reproductive) members of a population have only a minimal and temporary impact on growth rates. In wealthy countries such as Canada, where infant mortality is already extremely low, the focus of much medical research and health-care spending is directed at extending and improving the quality of life of an aging population. Although this relationship may seem surprising, it explains why the populations of countries with the greatest life expectancy are not characterized by high growth rates—in fact, many have among the lowest growth rates in the world.

Although increasing adult life expectancy does not significantly affect population growth rates it does have other potentially dramatic impacts. The United Nations has described the world's aging population as a "profound demographic revolution." In the last century the average human life span increased by 30 years. Currently, one in ten people are over 60; the UN estimates that by 2050 that number will have jumped to one in five worldwide. Such a shift in the population's age has major implications for pension and health care systems. Elderly people require greater medical attention and specialized home care and residential facilities. Most are not in the workforce for the last 30 years of their lives. However, in industrialized countries, this particular group of elderly people will be the wealthiest and healthiest in human history.

Better nutrition and modern medical advances, particularly medications for the treatment of high blood pressure and heart disease, are primary factors accounting for the greater numbers of people living beyond 65 years of age. Much current medical research promises to further increase human life expectancy.

Projections of Future Trends

Will human populations in the 21st century develop a sustainable world or a world of scarcity? Projections of a dire future for humanity began as early as 1798 with Thomas Malthus, the mathematician who predicted that exponential human population growth will surpass the arithmetic growth of food resources (**Figure 7**). In 1968, a biology professor at Stanford University, Paul Ehrlich, published *The Population Bomb*, in which he suggested that famine and the world's economic state would be so poor in the 1970s and 1980s that hundreds of millions of people would starve to death. Ehrlich's predictions were based on his insistence that food production and water resources were at their limits; instead, improvements in food technology and access to potable water reduced the number of deaths from famine. In the 1990 sequel to this book, *The Population Explosion*, Ehrlich warned that as Earth's population continues to grow, it will rapidly deplete the planet's resources and result in famine, global warming, acid precipitation, pollution, and other severe ecological crises. Ehrlich has persisted in his stance that humanity is living in unsustainable ways that stretch the ecological limits of the planet's ability to support human life.

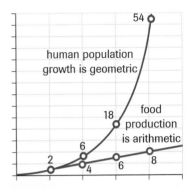

Figure 7
According to Malthus, changes in human population growth will surpass the ability to produce enough food resources.

Between 1950 and 1990, the global human population averaged an increase of 70 million people per year. The growth has now surpassed 90 million people per year. Whether the world's population will continue to increase, however, is a matter of debate. Warren Sanderson, a history and economics professor from State University of New York, predicts that the world's population will become stable, peaking at 9 billion by the year 2070 and will then start to decline.

Future changes in world population growth may indeed affect the ability to maintain sustainable societies. If population growth continues, as illustrated by **Figure 8(a)** on page 712, density-dependent factors, such as food production, land use, quality of water, and availability of nonrenewable resources, could potentially undermine Earth's carrying capacity for future generations. In many countries of the world, population growth has stabilized. But some ecologists argue that even if population growth stabilizes worldwide, as illustrated by **Figure 8(b)** on page 712, human lifestyles of consumerism are exceeding Earth's carrying capacity in ways that are unsustainable. Individual levels of consumption are increasing, especially in North America. Some observers go so far as to predict that even if the human population were to crash, as depicted in **Figure 8(c)** on page 712, humans are degrading Earth at such an incredible rate that it may still be impossible to regenerate the resources needed for the remaining population to survive.

Concerns about the world's future are rooted not only in the ability of a country to support its own populations economically but also in the long-term carrying capacity of the planet to support current and future generations, as well as the lives of other organisms. Bjorn Lomborg, a statistician from the University of Aarhus in Denmark, reminded ecologists recently that they must offer firm evidence to support their claims that increasing occupation of this planet by humans is degrading the support systems on which life relies. Lomborg is not alone in his opinion. Economist Steven Moore, representing a nonprofit public research foundation, also supports the notion that many environmental fears about population growth are not backed by sufficient scientific evidence. Ecologists argue that the human population cannot afford to wait for concrete evidence while Earth's life support systems diminish. They urge economists to support sustainability, with humanity's welfare as their goal.

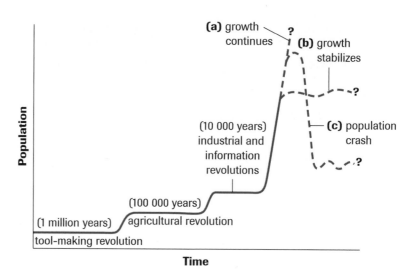

Figure 8
Advances in technological development can lead to further growth in the human population. Three dashed lines represent possible future outcomes: **(a)** growth continues, **(b)** growth stabilizes, and **(c)** population crash

TRY THIS activity   *Effects on Life Support Systems*

To deal with the consequences of continued human population growth, it is important to be able to make predictions about future growth trends. United Nations demographers cite three population projections for the year 2050 (**Figure 9**). With a group of two or three students, perform the follow tasks:

- Choose one of the three UN world population projections for 2050 and discuss factors that could lead to it.

- Regarding the population projection you chose, brainstorm a list of effects it might have on Earth's life support systems, and discuss the most realistic consequences on your list.

- Summarize your discussions and develop recommendations for a class presentation.

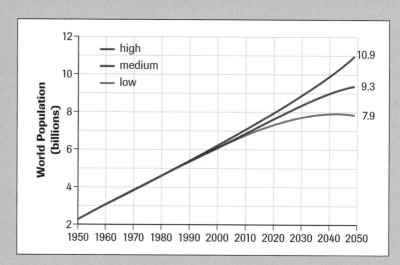

Figure 9

The implications of a rapidly growing population are profound. By the year 2020, Earth's human population is expected to exceed 8 billion and to continue to grow until it reaches a stable population size late in the century. Does Earth have the capacity to support such a large population? If not, will ecological problems arise because there are too many people in one region, or because of the increasing rates of resource consumption and environmental degradation as a result of human activity in general? The average Canadian consumes 50 times as many goods and services as the average person in India. Proportionately this means that 30 million Canadians place the same demands on Earth's ecosystems as would a population of 1.5 billion Indians. These demands include the requirement for food and water, energy and mineral resources, and the burdens placed on natural environments, such as pollution and habitat destruction. The remainder of this chapter provides an overview of some of the challenges we face as we attempt to meet the demands of a growing human population in a sustainable way.

SUMMARY *Human Demographics*

- During the past 300 years, the global human birth rate has declined slightly while the death rate has dropped dramatically. The result has been a population explosion.
- Birth rates and death rates vary significantly by region.
- The demographic transition model suggests that, as countries become more industrialized, there is a decline in population death rates, followed sometime later by a similar drop in birth rates.
- Less industrialized countries typically experience growth momentum resulting from large populations of young people.

▶ **Section 15.2** *Questions*

Understanding Concepts

1. If the human growth rate dropped to 0.56%, what would be the doubling time for the population?

2. How can age-structure diagrams be used to help predict the future trends in population growth in a population?

3. (a) What is the shape of an age-structure diagram for a population that has growth momentum?
   (b) Name a country that currently exhibits this pattern.

4. Demographers have recognized that increases in life expectancy have little long-term effect on the size and growth rate of a population. How do they account for this observation?

5. The United Nations and the World Health Organization predict that the world's human population will stop growing sometime in this century. What does this mean about the number of births per day and the number of deaths per day?

Applying Inquiry Skills

6. Present the data in **Table 1** (page 706) in a visually informative manner using one or more graphs. Be creative and effective in your presentation.

7. The population of Canada is approximately 30 million. What is its growth rate? How does this compare to the annual growth rate of the entire world's population?

8. The human population is growing most rapidly in many of the poorest countries of the world. Use the demographic transition model to explain this pattern.

Humans, like other heterotrophs, need food for survival. As populations grow, the need for increased food production is greater. Technology has had an enormous impact on our ability to grow food. The plough, invented about 5000 years ago, was the first tool that permitted humans to dramatically alter the soil surface by enabling them to till heavy but fertile soils that could not be worked by hand. Oxen and horses were the first beasts of burden used as a source of power in agriculture. Since the industrial revolution, tractors, fertilizers, and pesticides have further enhanced agricultural productivity. In 1900, one Canadian farmer could produce enough food to feed about seven people. Today, the average Canadian farmer produces enough food to feed close to 100 people (**Figure 1**)!

From 1950 to the mid-1980s, increases in global food production steadily outpaced increases in demand (based on population growth). However, during the past 15 years global food production has levelled off and actually declined on a per capita basis. **Figure 2** illustrates this trend for grain production—humans' most important food source.

Figure 1
Modern intensive and highly technological farming practices produce very large food yields.

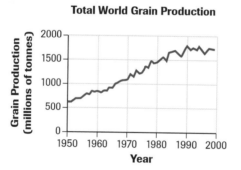
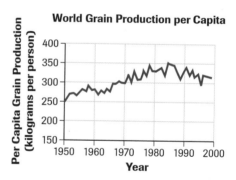

Figure 2
Trends in world grain production from 1950–2000

Even though food production is no longer outpacing population growth, we still produce enough food on a global scale to adequately feed all of the world's people. Yet, tens of millions continue to suffer from starvation and undernourishment every day. According to the United Nations, the amount of food consumed each day by the average global citizen should be no less than 9900 joules (J), the minimum critical diet. Below this level, chronic malnutrition may result. **Figure 3** shows the average food energy intake for different countries. According to the United Nations Food and Agricultural Organization (FAO) and the World Bank, almost 500 million people regularly consume less than the minimum critical diet, and over two thirds of these people live in Asia, Africa, and Latin America. However, there are two forms of malnutrition: undernourishment and overnourishment. Health statistics from industrialized regions, including North America and Europe, reveal a growth in the incidence of illness caused by overeating, such as coronary heart disease and obesity.

Well into this century much attention will be focused on the efficient production, distribution, and utilization of food resources needed to nourish a dense and expanding population. Success will only be achieved if and when malnutrition is eliminated and when those who do have access to plentiful food supplies make health-wise choices in their preparation and consumption.

DID YOU *KNOW* ?

World Hunger
Each year, more than 40 million people die from hunger and hunger-related diseases. This is equivalent to 300 fatal jumbo jet crashes a day, with almost half of the passengers being children.

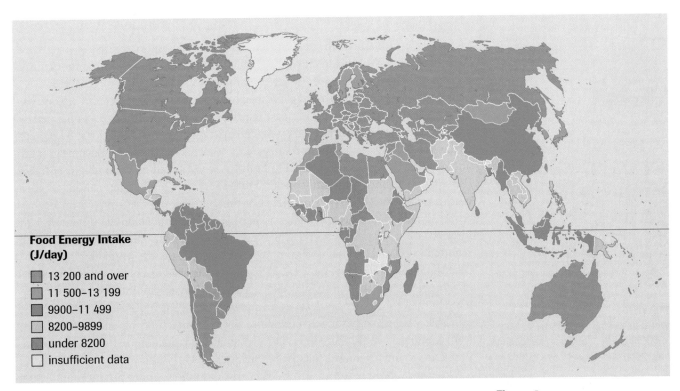

Figure 3
Global food consumption as measured by food energy intake (J/day). A daily intake of 9900 J is considered to provide adequate nutrition for the average person. People receiving less than 8200 J/day are chronically malnourished.

The causes of world hunger are complex. One of the basic problems is that some regions of the world are endowed with fertile soils and favourable climates, while others have poor soil conditions, drought, flooding, and other harsh conditions that severely limit the ability to grow food. **Figure 4** illustrates the disproportionate availability of **arable land** in various regions of the world.

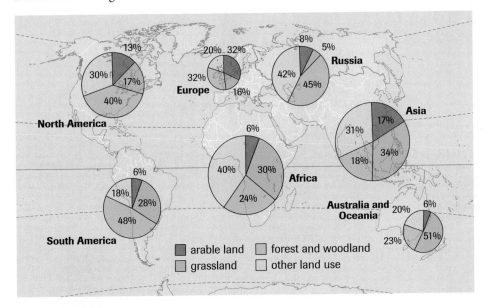

arable land land capable of being used for the purpose of cultivating crops

Figure 4
Major categories of land use. The sizes of the circles are proportional to total land area. "Other land use" includes deserts, mountainous regions, and urban areas.

Most agricultural experts agree that the principal cause of hunger is, and will continue to be, poverty, which prevents the poor from growing food or buying food, regardless of how much is available. For example, despite impressive gains in total and per capita food production, about 40% of India's population suffers from undernourishment because they do not have enough money to buy or grow the food necessary to meet their basic needs.

Food Sources

Humans depend largely on grain production, livestock, and ocean fisheries for food. According to the Worldwatch Institute, yield thresholds—the maximum production of fish, grain, and livestock—are currently estimated to be at, or close to, maximum.

Seafood has been gathered and consumed by people for centuries. Global seafood harvests, however, peaked in 1989 at an estimated 100 million tonnes (or 19 kg per person) and the Worldwatch Institute predicts that no increase in seafood catches can be expected to the year 2030. Ocean pollution, damage to aquatic habitats that are spawning grounds, and overharvesting have combined to limit the world's seafood catch (**Figure 5**). Wild fish stocks were once sufficiently plentiful to feed adjacent human populations as well as marine predators. Abundant stocks, such as those off the Atlantic coast of Canada, have attracted many fishing fleets to harvest food resources for populations far distant from the location of the fish stocks. On discovering these cod stocks 500 years ago, Europeans reportedly declared that there were enough fish to feed the entire world for all time. That prediction has a hollow ring now, as overfishing has depleted stocks beyond estimated sustainable yields and the continental shelf habitat has become so disturbed that cod populations may not recover for decades. To add to the concern, in February 2002, scientists published findings suggesting that the entire North Atlantic fishery is under serious threat from continued overfishing. They called for significant reductions in total catches to prevent a severe and prolonged collapse of the industry.

World meat production surged from nearly 44 million tonnes produced in 1950 to 211 million tonnes in 1997 (**Figure 6**). Livestock ranching, however, has led to forest degradation in Central and South America and the introduction of foreign species almost everywhere commercial ranching exists. Currently, 36% of the world's grain goes to feed livestock. If this amount were reduced by just 10%, 67 million tonnes of grain could be available to sustain 225 million people. Agriculture that provides animal protein and

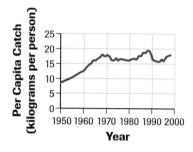

Figure 5

World fish catch per person, 1950–1997. The worldwide per capita fish catch has dropped since 1989.

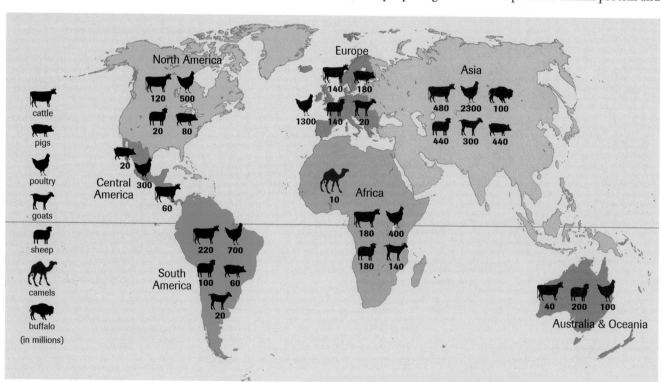

Figure 6

The world is home to more than 10 billion domestic animals used for livestock, including approximately 3 billion ruminants, 1 billion pigs, and 6 billion poultry.

products has also resulted in significant water scarcity in northern China, the Middle East, northern India, and sub-Saharan Africa. A number of the world's large rivers, including the Colorado River and China's Yellow River, are so heavily utilized for irrigation supplies that virtually no water reaches the river mouth during the summer months. Wastes produced by livestock—on a global scale, 130 times more than those produced by humans—have been implicated in the pollution of waterways, the proliferation of toxic algal blooms, and the loss of many different fish species.

Growth in grain production from 631 million tonnes in 1950 to 1780 million tonnes in 1990 is similar to the expansion in global livestock production. The Worldwatch Institute predicts, however, that growth will slow to 2149 million tonnes by 2030, an increase of only 369 million tonnes from 1990. Reasons for a slower increase in grain production are varied, the most serious of which is loss of cropland at a rate worldwide of 10.1 million hectares per year—equivalent to the size of southern Ontario.

Much of the lost agricultural land is being converted to nonagricultural uses, particularly to meet increased housing needs. The most severe loss, however, is to **soil degradation**. In regions where people rely on wood for fuel, deforestation exposes soil that may have marginal fertility for growing crops. Loss of roots of living plants that would normally retrieve nutrients from deeper layers of soil can disrupt nutrient cycles, allowing these nutrients to be leached away. To improve productivity, in the past, many farmers were encouraged to grow crops season after season, but such continuous cropping gradually strips soil of organic matter and of nutrients that cannot be replaced by fertilizers. Droughts can accelerate soil erosion; as crops wither, soil surfaces dry and are subject to wind erosion. Excessive wetness can lead to waterlogged soil, which also reduces crop productivity by limiting soil oxygen supplies needed by plant roots. In many areas where grain producers rely on irrigation, salts and mineral deposits can gradually accumulate in the soil through a process called **salinization**. A high salt content in soils disrupts the flow of nutrients into plants and restricts the water that plants can withdraw. Severe salinization of lands around the Aral Sea has resulted in the collapse of an agricultural region that was once more productive than California (**Figure 7**).

Although food production worldwide has slowed, the need for it has not. Research into improved food production continues, while related ecological issues, such as water and air qualities, are also topics of intense study.

DID YOU KNOW?

A Desert World?
The world's deserts are spreading. This desertification threatens about one third of the entire world's land surface! Human activity is the main cause. On arid lands, drought-resistant plants help maintain some soil moisture by shading and adding organic matter to the soil surface. If these plants are removed by overgrazing or excessive cutting of fuel wood, the soil is exposed, dries out, and the limited nutrients in the surface are removed by wind erosion.

soil degradation the removal of soil nutrients and loss of topsoil rich in organic matter, caused by erosion from the movement of water and wind

salinization the accumulation of excess of salts in soil that restricts the amount of water and essential nutrients plants can withdraw from the soil

Figure 7
Unsustainable use of water for vast irrigation projects near the Aral Sea drained this large water body, which could no longer rely on consistent inflows from rivers fed by melting glaciers. Evaporation of the Aral Sea has exposed deep layers of saline soils, which has resulted in an ecological disaster.

More Efficient Energy Distribution

On one of his CDs, the musician Moby states: "In a world where people are starving, it seems criminal to fatten up cows with grain that could be keeping people alive." Is this observation supported by science? Ecologists often use **energy pyramids** as models to visualize the transfer of food energy from one **trophic level** to another through a **food chain** (**Figure 8**). These models are pyramidal in shape to illustrate that energy flows through an ecosystem in an inefficient manner. Not all the energy trapped by producer organisms in the first trophic level of the pyramid is transferred to consumers in the next level, since the majority of the energy dissipates as heat. Approximately only 10% of the energy in one organism is transferred to an organism in the next trophic level. Therefore, a food chain with a greater number of organisms would have a higher loss of usable energy as it flows through the food chain. Organisms in the uppermost trophic level (at the end of long food chains) receive a small fraction of the energy available at the lowest level in the food chain.

energy pyramid a model that illustrates energy flow from producers at the beginning of food chains to consumers farther along

trophic level a level within a food chain or energy pyramid that demonstrates energy flow from producers (at the primary trophic level) to consumers (at secondary or tertiary trophic levels)

food chain a sequence of linked organisms that feed on one another, starting with a producer and continuing with consumers; energy flow decreases proportionately for each additional consumer level

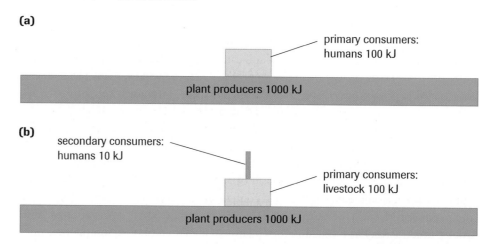

Figure 8
These energy pyramids compare the relative efficiency of energy transfer from **(a)** the primary trophic level, and from **(b)** a secondary trophic level.

▶ *SAMPLE* problem

Energy From Food Production

A food chain consists of grain, cattle, and human populations. If the amount of energy in the harvested grain is 730 kJ, how much energy is available for the human population?

Solution
At each trophic level, 10% of biological energy is transferred.

Energy transferred from the grain to the cattle = 730 kJ × 10% = 73 kJ

The cattle obtain 73 kJ of energy from the harvested grain.

Energy transferred from the cattle to the humans = 73 kJ × 10% = 7.3 kJ

The humans obtain 7.3 kJ of energy from the cattle.

Answers
1. (a) 65.7 MJ
   (b) 6.57 MJ

▶ *Practice*

Understanding Concepts

1. In a forest ecosystem, wolves and humans feed on deer, and deer feed on plants. If the plants eaten by the deer are able to trap 657 MJ of energy through the process of photosynthesis, how much energy is available for the
   (a) deer?
   (b) humans and wolves?

Energy pyramids illustrate that humans can obtain more energy from consuming producing organisms—grains, vegetables, and fruits—directly rather than from feeding them to livestock and subsequently eating livestock products. The evidence suggests that eating organisms early in food chains or energy pyramids (e.g., plants) can sustain large populations in countries with insufficient food production. For example, in 2000, China harvested 462.5 million tonnes of various grains, an amount that was sufficient to ensure that the entire population could be fed. In the past, the people of China consumed fish in addition to grains, but very few meat products. More recently, many Chinese changed their diets to include more meat products, which led to challenges in food production. The raising of livestock resulted in the clearing of agricultural land formerly used exclusively to cultivate crops for human consumption. Some of the land was converted to pasture and some grains were diverted to feed livestock. The total energy value of the livestock was less than what had been provided directly by the grains. As Chinese people continue to eat farther along food chains, more grains will have to be produced to sustain their energy needs.

In many regions of the world, livestock are used to provide food for humans when crop plants cannot be readily grown. For example, in regions where the poor soils, rugged terrain, or weather patterns make land unsuitable for cultivation, grazing herds of goats, sheep, and cattle may be able to convert unpalatable native plants into meat and milk products suitable for human consumption (**Figure 9**).

Figure 9
These grazing sheep are utilizing land unfit for growing crops and they are consuming plant species that are not suitable for human consumption.

▶ **TRY THIS** activity   *Test Your Energy Intake*

You can relate the foods you consume to the amount of land required to produce those foods.

- On a graphic organizer, record everything you eat in 48 h by the type of food (animal or plant product) and the amount that you ingest.
- Use a kilojoule counter, or record kilojoule values from food packaging, to calculate the number of kilojoules you consumed from animal products and from plant products.
- Multiply each of your results by 365 to get an average annual intake of kilojoules.
- Divide the annual average intake of kilojoules from animal products by 835 kJ per m² to determine the amount of land needed to produce the animal products you consume.
- Divide the annual average intake of kilojoules from plant products by 8350 kJ per m² to determine the amount of land needed to produce the plant products you consume.

(a) Which food products required less land?

(b) Compare your results with those of your classmates.

(c) How might you and your classmates change your diets to contribute to global sustainability?

(d) How might your results compare to people living in other parts of the world?

More Sustainable Food Production

Agricultural research scientists with the Canadian International Development Agency (CIDA) are involved in many projects around the world to support better food production. Each project is designed to address a specific agricultural problem. An integrated pest management program in Africa, for example, is aimed at saving the cassava (or manioc) crops, which provide high-energy values for more than 200 million people (**Figure 10**). Cassava had almost become extinct from the ravages of the cassava mealybug, which had no natural predator. Scientists involved in a project supported financially by CIDA proposed the introduction of an exotic species of parasitic wasp for release through aerial spraying over African lands where the cassava mealybug was prevalent. The wasps dramatically reduced the mealybug populations, allowing a return to successful cultivation of cassava crops that provide much-needed food for the human populations.

Figure 10
Cassava, or manioc, is a staple crop in much of Africa, where flour is produced from the plant's starchy tuberous roots. Many Canadians enjoy tapioca pudding, which is produced from this root starch.

The Food and Agriculture Organization (FAO) plays a major role in global food security and production. On November 3, 2001, an FAO conference adopted an International Treaty on Plant Genetic Resources for Food and Agriculture, which represents global agreement on management of the world's agricultural biodiversity. Currently, only about 30 crops provide human populations with 95% of their food energy. Seed diversity has decreased as international programs have sought to focus on the most efficient plant species. The hundreds of rice species that used to be grown in a variety of countries have been reduced to those with the best durability and highest yields. Rice is a vital grain crop with high nutritional and energy values that feeds millions of people in regions of low prosperity. Given the diminishing number of crops and varieties being grown, FAO scientists and officials have expressed concern that the genetic diversity of crops is decreasing, and that entire crops could become extinct. To save the genetic diversity of grain crops, the FAO has established its own gene and seed banks.

As a food production technology, genetic modification has been actively promoted for several years. Crops are being genetically designed to withstand extreme temperatures, to be resistant to pests and herbicides, to have improved flavour and nutritional content, and to produce bigger yields. Genetically modified organisms (GMOs) are a subject of controversy for many reasons. Recently, Bt-corn made the news when it was discovered that some flour made from a variety of Bt-corn plants had been accidentally introduced into some human food items. Bt-corn has been genetically modified with a gene from the bacterium *Bacillus thuringiensis*. The gene codes for a protein that is then produced within the plant tissues and is toxic to corn borers, thus providing resistance to the pest without the spraying of pesticides. This particular variety of corn is not to be consumed by humans—it is to be fed to livestock or processed into fuel.

Some critics wonder when GMO technologies will deliver on their promises to solve the world's food shortages. Others are concerned that large corporations that own patents to GMOs may be contributing to the loss of genetic diversity. Farmers who want to plant GMOs for increased food production must every year pay companies from which they purchase seed, since the seeds have often been changed so that mature plants will not reseed themselves. Farmers accustomed to having seeds last several seasons may not expect or be able to afford to pay yearly, especially poor farmers in countries struggling to stabilize their economies or repay huge debts to international funding agencies.

▶ *EXPLORE* an issue

Take a Stand: Can Biotechnology Increase Food Supplies?

Decision-Making Skills

- Define the Issue
- Analyze the Issue
- Research
- Defend the Position
- Identify Alternatives
- Evaluate

In 1798, Thomas Malthus hypothesized that "the power of population is infinitely greater than the power in the earth to produce food for man." Many countries do not have the financial means to import food nor have enough productive farmland—without cutting down valuable forests—to produce enough food to sustain their populations. Delegates to the Tokyo International Forum of the Conference of the World's Scientific Academies in May 2000 said, "Over the next 50 years, the worldwide demand for food is expected to triple in response to population growth, increases in per capita income, and continuing attempts to improve nutrition among the very poor. Meeting this demand will require dramatic advances in food production, distribution, access, and security." Biotechnology in agriculture is being explored as a potential solution to feeding populations in countries that are unable to meet current or future food needs. Proponents for the use of biotechnology declare that it will provide ways for farmers to produce larger harvests from existing land, eliminate diseases that devastate crops, reduce pesticide and fertilizer use, improve the nutritional content of food, and increase a crop's ability to withstand harsh environmental conditions. Opponents of agricultural biotechnology argue that it is unsafe, as it could eliminate genetic diversity and take the control of crop cultivation for basic survival from farmers and put it into the hands of corporations.

- Identify the issue presented above.

- Who might have an interest in this issue? Identify the perspectives these parties might have that reflect their interest in the issue.

- Research Canada's role in agricultural biotechnology and agricultural technologies. For example, you might want to look into the work of the Canadian International Development Agency. Your research should include a brief description of how the technology works and should focus on the scientific, societal, environmental, and technological impacts of the various biotechnologies.

 www.science.nelson.com

(a) Complete a risk-benefit analysis of your research.

(b) Write a supported opinion letter to the Food and Agriculture Organization of the United Nations on whether you agree with using agricultural biotechnology to increase food supplies.

SUMMARY   *Life Support Systems: Food and Soil*

- Although total world food production has increased dramatically during the past century, the per capita production of food has started to decline during the past decade.

- The present annual world food supply is adequate to feed the entire human population but both food production and food distribution are very uneven among different regions of the globe.

- Soil quality is threatened in many regions of the world by human activity that can lead to erosion, salinization, nutrient loss, and desertification.

- Consuming plant foods directly provides significantly more joules of food energy than feeding the same plant foods to livestock for the production of animal food products.

- Advances in biotechnology may provide methods to increase food supplies.

Understanding Concepts

1. Describe how the following major factors might influence the ability of humans to produce large quantities of food in a given region of the world:
   (a) technological innovation
   (b) soil quality
   (c) climate

2. Using a graphic organizer of your choice, summarize the challenges that food scarcity poses to global food production.

3. What do energy pyramids show about the suitability of plants versus animals for meeting human food demands?

4. How have grazing animals been used to effectively increase the land area available to feed the human population?

Applying Inquiry Skills

5. **Table 1** provides estimates of the average primary productivity of different habitats in joules per m² per year. Complete the table, comparing the theoretical value of food energy that could be supplied for human consumption if we were to feed on the plants directly or if we were to feed on livestock.

Table 1

Type of ecosystem	Primary productivity (J/m²/year)	Energy available to primary consumers (J/m²/year)	Energy available to secondary consumers (J/m²/year)
tropical rain forest	38 000		
temperate grassland	9 200		
open ocean	5 000		

Making Connections

6. Many alternatives in agricultural practices are leading to improvements in land use. Use the Internet to research one of the following farming practices and report your findings to the class:
   (a) low-tillage cultivation
   (b) drip irrigation
   (c) crop rotation

    www.science.nelson.com

7. Vegans are advocates of consuming plants only. They claim that a larger number of people could be sustained if populations switched to a vegan diet. Their opponents argue that there is not enough productive land to grow plant products to sustain that many people.
   (a) With two or three other students, discuss arguments and counterarguments for each position.
   (b) Create a PMI chart to summarize your thinking. See Appendix A4 for a description of a PMI chart.

People in Toronto survive, as residents of other cities do, by drawing largely on food, water, energy, and other resources from farms, forests, watersheds, and mines well outside the city itself. The city also produces enormous quantities of waste products, which have to be disposed of in ways that will not degrade the urban environment. In 2001, Toronto household garbage was being trucked to American landfill sites, as sites in the Greater Toronto area are almost full. Urban development around Lake Ontario has spread so much that there are few open land areas left to build in. Prime agricultural land and moraines are now being targeted for housing and commercial construction. The Oak Ridges Moraine runs for about 160 km from the Niagara Escarpment to just east of Rice Lake (**Figure 1**). Thirty percent of the moraine is still covered in forests, which provide important habitats for many animal and plant species already displaced by **urban sprawl**. Beneath the moraine is one of the most significant **aquifers** in Canada. This underground water reservoir feeds 65 rivers and streams, which are a direct source of drinking water for more than a quarter of a million people. Despite its recognized ecological importance to southern Ontario, the moraine is at risk from having to accommodate housing developments.

urban sprawl the growth of low-density development on the edges of cities and towns

aquifers porous, water-saturated layers of sand, gravel, or bedrock that can store and yield significant volumes of water; precipitation and runoff enter recharge zones and percolate gradually down to an aquifer

Figure 1
An aerial view of a small segment of the Oak Ridges moraine, source for much of the drinking water in southern and central Ontario

More than 2.5 million people reside within a 160-km radius of Toronto. The possibility of employment in the city attracts immigration from other parts of Ontario and Canada, as well as from other countries. Given its rapid population growth, Toronto is facing numerous ecological and social challenges. Like cities the world over, Toronto must handle homelessness, traffic congestion, air pollution, water and fuel supply, and a critical lack of land for building, agriculture, and waste disposal. The challenges of sustaining life support systems for urban populations seem to be increasing.

Figure 2
The people of Walkerton, Ontario, suffered from serious contamination of their drinking water by a deadly strain of *E. coli* bacteria.

Water Resources

In May 2000, a deadly strain of *E. coli* 0157:H7 bacteria invaded the water supply of the Ontario town of Walkerton (**Figure 2**), killing 7 people and infecting 2300 others. The tragedy began when animal feces infected with this strain of *E. coli* leached into the town's water supply. Subsequent failures in the town's chlorinating equipment allowed the bacteria to survive, and inadequate monitoring procedures failed to warn residents not to drink the water.

Over the past 20 years, much larger-scale livestock operations have been introduced into rural Canada. To protect water supplies, stiffer regulation of such operations and more frequent monitoring of water quality are in place to ensure that pathogens and all leachate from manure remains out of groundwater. Besides contamination from livestock wastes, water supplies can become polluted by excessive fertilization of croplands, pesticides, acid precipitation, runoff from storm water off pavement, industrial wastes, mine tailings, household wastewater, and raw sewage. Water treatment is not consistently available, even in countries with a high quality of life. The scarcity of clean water for drinking and irrigation is a critical health concern in many countries.

The Supply and Demand for Water Resources

Worldwide, approximately 70% of all diverted fresh water is used to irrigate cropland, while 20% is used by industry, and 10% is used by residents and cities. Increased consumption and unsustainable uses of water have resulted in overdrawing of supplies from rivers and groundwater. Falling water tables and aquifer depletion are affecting the ability of groundwater to meet current and future needs. Melting snow and alpine glaciers are the sources of many river systems in western Canada and Europe, but many are melting at such a fast rate that concerns are being raised about how long the glaciers might be able to continue supplying water. Researchers in Canada predict that glaciers supplying the South Saskatchewan River may last another 200 years, and researchers from Switzerland's Fribourg University are predicting that 50% to 90% of Swiss glaciers will have melted completely by the end of the 21st century. Part of the ecological crisis of the Aral Sea resulted from the slowing of glacial melt to the Darya River, so that it no longer discharged into the sea. The water cycle itself is under severe stress, as demonstrated by such human-made drainage pattern changes as the Yellow River in China running dry part of every year and the Nile River in Africa having very little volume when it reaches the Mediterranean Sea. In Ontario, low water levels have been experienced in many rural wells in Bruce and Grey counties, and in areas where governments have permitted huge withdrawals of spring water for bottling.

Currently, 1.4 billion people lack access to clean drinking water, most of them in regions with poor economic conditions. According to the United Nations, 500 million people in 31 different countries will face water shortages that threaten agriculture, industry, and the health of their people (**Figure 3**). By the year 2025, 2.8 billion people in 48 countries are expected to face severe water stress and, by 2050, the number could rise to 7 billion. Decreases in water supply have limiting effects on the production of grain and other food crops. Falling water tables have been observed in such major food-producing areas as the North China plain, the southern United States Great Plains, and much of India (**Figure 4**). In India, Pakistan, Iran, Egypt, China, and Mexico, water shortages are so severe, there is not enough water to satisfy drinking, food production, and hygienic needs. Further, water systems in many countries are often in such poor condition that precious supplies leak from pipes intended to carry water to homes or crops. Water, arguably the world's most precious resource, may also be the planet's most poorly managed resource.

⊙ **ACTIVITY 15.4.1**

The Impacts of Resource Consumption (p. 746)
In what ways are the rates of consumption correlated to quality of life, population size, and environmental degradation? In this activity, you will complete a study to assess the relationship between human populations, resource consumption rates, indicators of quality of life, and the health of the natural environment.

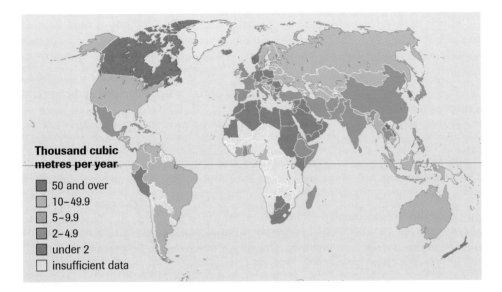

Thousand cubic metres per year
- 50 and over
- 10–49.9
- 5–9.9
- 2–4.9
- under 2
- insufficient data

Figure 3
Internal renewable water resources per capita measured in thousand cubic metres per year. Countries with under 2000 m³/person/year are considered to be chronically short of water.

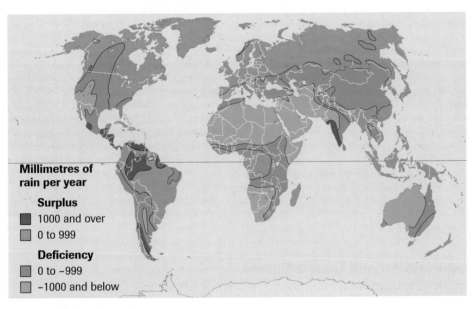

Millimetres of rain per year
Surplus
- 1000 and over
- 0 to 999
Deficiency
- 0 to –999
- –1000 and below

Figure 4
Global water surplus and deficiency measured in millimetres of rain per year. Surplus areas have enough water to support vegetation and crops without irrigation. Deficiency areas lack the minimum amount of water required to maintain native vegetation and crops.

Solving Our Water Problems

Governments are taking action to encourage their populations to use water in a more sustainable manner. Obvious actions, such as repairing leaky pipes, recycling wastewater for irrigation, and assessing fines for water waste, are a good start. Mexico City has replaced 650 000 toilets with models that consume less water, while Toronto has similar plans. Melbourne, Australia, is one of a number of communities implementing a user-pay system for water that was previously charged on the basis of property taxes. The government is expecting a large reduction in water consumption once property owners have to pay for it.

In agriculture, most farmers worldwide use a flood irrigation method to deliver water to crops. Water from groundwater wells, dammed reservoirs, or surface canals is distributed to crops by means of gravity flow through unlined ditches (**Figure 5**). Only 50% to 60% of the water reaches the plants—the rest evaporates, seeps into the ground, or is lost to runoff. Because of these inefficiencies, more water than necessary is pumped into the system to ensure crops receive sufficient amounts. Methods of reducing this form of water loss include lining irrigation ditches to reduce the amount of water that seeps into the ground, using holding ponds to store rainfall, and capturing unused irrigation water for future use. Wells can be used instead of dams to focus the irrigation to the local area. Computers may also be used to monitor the moisture level of the soil and deliver water only when the soil is too dry.

Some other farmers are abandoning flood irrigation altogether and turning to sprinkler systems (**Figure 5(b)**)in which 70% to 85% of the water reaches the crops. Water is still pumped from groundwater sources but less is needed because less is lost. Unfortunately, these systems are so expensive to install that few farmers can afford them. About 50 years ago, Israel developed an irrigation system that is 90% to 95% efficient, called drip irrigation, to cope with very dry conditions and ensure that water is not being wasted. Perforated piping is run underground to deliver water almost directly to the roots of each plant. This localized system ensures less water seepage and waste. Like the sprinkler systems, however, drip irrigation is very costly and not feasible for poorer farmers. Some high-tech solutions include the use of lasers to identify differences in surface elevations in fields, and genetic engineering to grow water-efficient crops. Soil and satellite sensors and computers are used to determine the amount of water needed so that only the necessary amount is used.

(a)

(b)

Figure 5
Countries with an adequate electrical supply are turning away from **(a)** the construction of large dams and reservoirs for irrigation, and choosing **(b)** improvements to familiar methods that draw from local groundwater.

desalinization removal of salt, usually from seawater, to produce water suitable for drinking or irrigating croplands

Salinization and Desalinization

In 2000, Environment Canada announced a policy to respond to concerns about excessive amounts of salt deposited on roadways during winter. Although it makes highways safer for traffic, the excessive salt in runoff poses risks to plants, animals, birds, lakes, sediment, and groundwater. Canada is preparing legislation to restrict the use of salt on icy highways in the winter. Partly in response to this announcement, the City of Toronto has begun installing Road Weather Information Stations that use sensors embedded in roads to monitor pavement temperature, air temperature, wind speed, and pavement moisture as well as the amount of salt present. This information goes to city staff to determine whether monitored roads should be salted and, if so, at what rate.

In response to concerns about water shortages, technological means have been found to increase the amount of water available for human use. One method involves the removal of salt from salt water, a process called **desalinization**. Desalinization can be accomplished through two methods. In distillation, salt water is heated until it evaporates, and salts are left behind as the fresh water condenses. A second method involves a process called reverse osmosis. In this method, salt water is pumped at high pressure through a thin membrane that allows water molecules through but leaves the salts behind. Desalinization is not an alternative to conserving water since a great deal of it is energy intensive, making the process very expensive. Desalinization also produces large amounts of waste salt, which cannot be put back into water or buried underground where it might recontaminate water. Despite these drawbacks, desalinization offers relief to regions with severe water shortages.

Threats to the Atmosphere

The addition of harmful substances to Earth's atmosphere has increased greatly as more countries undergo major changes. Expanding urbanization and increased industrialization in cities such as Los Angeles, Bangkok, Mexico City, Beijing, New York City, Toronto, and Prague have resulted in degraded environments and human health problems from increases in **air pollution**. Smoke and noxious emissions come from vehicles as well as industrial and hospital smokestacks, and affect the well-being of local residents and ecosystems. Distant ecosystems are negatively affected as well, through **acid deposition**, the precipitation of dilute sulfuric and nitric acids in rain, snow, gas, and dust. The source of these acids is usually air pollutants, such as sulfur dioxide, SO_2, from burning coal, and nitrogen oxides, NO_x, from internal combustion engines such as in automobiles.

Regions that are downwind from coal-burning power plants, smelters, or factories, or major urban areas with large volumes of vehicular traffic, are most affected by acid precipitation. Acid-producing chemicals are sometimes exported to other countries by prevailing winds. Acid rain is blown, for instance, into Norway, Switzerland, Austria, Sweden, the Netherlands, and Finland from more industrialized areas of Europe, such as Germany. Sulfur dioxide emissions from the Ohio Valley can be traced to regions of southeastern Canada. Similarly, acidic emissions from Canada often blow southward into the United States. Some of the worst cases of acid deposition are found in China, where coal is burned to produce 73% of energy requirements. **Figure 6** (page 728) shows the regions of the world where acid deposition is a problem and where it threatens the quality of the air. Acid deposition results in changes in soil pH that can have the following drastic effects on forests and crops:

- Essential soil nutrients, such as calcium and magnesium salts, leach from soil, reducing plant productivity and the chemical buffering activity of soil.

- Aluminum ions can be released, impeding the uptake and transport of nutrients, such as nitrogen, potassium, and phosphorus, from soil water to the roots of plants.

- Trees and plants become weakened as a result of these processes, increasing their susceptibility to other diseases.

- The dissolving of insoluble soil compounds is promoted, releasing lead, cadmium, and mercury ions that are highly toxic to plants and animals.

Acid deposition also alters the pH of many aquatic systems, reducing fish populations through acid stress or by impeding reproduction. Tiny organisms that fish require for food may suffer reproductive failure as well. In Canada, fish populations in 150 000 lakes are declining because of excess acidity from acid rain and other forms of acid deposition. In Norway and Sweden, at least 16 000 lakes are without fish due to high acidity levels. Water with low pH can also leach toxic metals, such as lead and copper, from water pipes into drinking water.

air pollution presence of one or more chemicals in sufficient quantities in the atmosphere to cause harm to humans and other forms of life

acid deposition a mixture of sulfur dioxide and nitrogen oxide pollutants that can reach Earth's surface in the form of rain, gas, or solid particles; also called acid precipitation

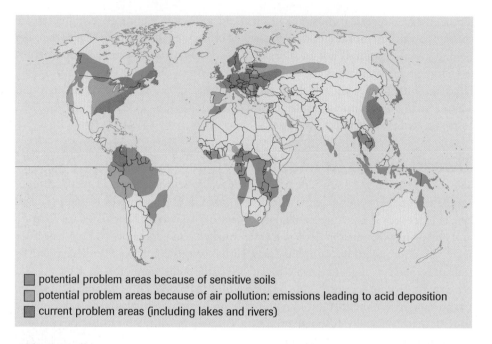

Figure 6
Regions in the world where acid deposition is already a problem, and places where a potential threat exists

potential problem areas because of sensitive soils
potential problem areas because of air pollution: emissions leading to acid deposition
current problem areas (including lakes and rivers)

Figure 7
Smoke from annual fires throughout Canada's boreal forest contributes to atmospheric pollution.

greenhouse effect a result of certain atmospheric gases, such as CO_2, water vapour, and methane, trapping heat in the atmosphere by letting visible sunlight penetrate to Earth's surface, while absorbing most wavelengths of infrared radiation that radiate from Earth's surface

Recent images from NASA's Terra Spacecraft provide striking evidence that air pollution is much more than a local problem. The Terra Spacecraft measures the amount of atmospheric carbon monoxide produced from the burning of fossil fuels. The first set of observations monitored from March to December 2000 showed air pollution generated by forest fires in the western United States, emissions from consumption of fossil fuels across much of the northern hemisphere, and clouds of carbon monoxide from forest and grasslands fires in Africa and South America, spreading as far as Australia. In Canada, forest fires that burn vast tracts of boreal forest contribute to atmospheric pollutants (**Figure 7**).

Related to concerns about air pollution is global warming, along with climate change in general, two factors related to the phenomenon known as the **greenhouse effect**. The greenhouse effect involves certain gases in Earth's atmosphere, such as water vapour, carbon dioxide, methane, and nitrous oxide that are efficient absorbers of heat. They allow the wavelengths of visible light and some ultraviolet radiation from the Sun to pass through to Earth's surface (**Figure 8**). When the light is absorbed by the surface, much of it is reradiated back out as infrared radiation (heat) and is then absorbed by these gases. Thus, the atmosphere forms a thermal heat blanket and results in the characteristic range of temperatures in Earth's atmosphere. This range of temperatures can change, however, in response to changes in the concentrations of these gases. When the concentrations increase, additional global warming is likely to occur. Such human activities as the burning of fossil fuels release billions of tonnes of carbon dioxide into the atmosphere each year, while waste sites (which release methane) and high-temperature automobile combustion (which releases nitrous oxide in addition to CO_2) are adding to atmospheric concentrations of other greenhouse gases. Predicting the precise impacts of changing concentrations of greenhouse gases is difficult and complex, but the vast majority of climate specialists agree that average air temperatures are rising and that increasing emissions of greenhouse gases, due to human activities, are a contributing cause. What remains unclear is what specific climate changes will occur in particular geographic regions of Earth and how severely such changes will affect the ecosystems they support.

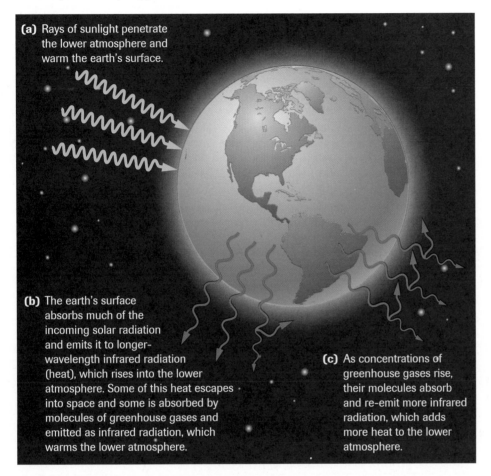

(a) Rays of sunlight penetrate the lower atmosphere and warm the earth's surface.

(b) The earth's surface absorbs much of the incoming solar radiation and emits it to longer-wavelength infrared radiation (heat), which rises into the lower atmosphere. Some of this heat escapes into space and some is absorbed by molecules of greenhouse gases and emitted as infrared radiation, which warms the lower atmosphere.

(c) As concentrations of greenhouse gases rise, their molecules absorb and re-emit more infrared radiation, which adds more heat to the lower atmosphere.

Figure 8
Greenhouse gases—such as water vapour, carbon dioxide, methane, and nitrous oxide—absorb heat radiation, preventing it from escaping back into space. These gases have the same effect as the glass that prevents infrared radiation from leaving a greenhouse.

The Intergovernmental Panel on Climate Change (IPCC) published a report in 2001 describing some effects of global warming on life support systems.

- The global average surface temperature has increased over the 20th century by about 0.6°C.
- Snow cover and ice extent have decreased.
- Global average sea level has risen and ocean heat content has increased.
- Changes have also occurred in other important aspects of climate (e.g., precipitation, cloudiness, droughts, floods, El Niño).
- Emissions of CO_2 due to fossil-fuel burning are virtually certain to be the dominant influence on the trends in atmospheric CO_2 concentration in the 21st century.

Table 1 (page 730) shows the variety of ways in which the above effects could be harmful to life.

Table 1 Possible Effects of Global Warming

Areas affected	Examples of effects
weather extremes	• prolonged heat waves and droughts • increased flooding • more intense hurricanes, typhoons, tornadoes, and violent storms
water resources	• changes in water supply • decreased water quality • increased drought • increased flooding
biodiversity	• extinction of some plant and animal species • loss of habitats • disruption of aquatic life
forests	• change in forest composition and locations • disappearance of some forests • increased fires from drying • loss of wildlife habitat and species
sea level and coastal areas	• rising sea levels • flooding of low-lying islands and coastal cities • flooding of coastal estuaries, wetlands, and coral reefs • beach erosion • disruption of coastal fisheries • contamination of coastal aquifers
agriculture	• shifts in food-growing areas • changes in crop yields • increased irrigation demands • increased pests, crop diseases, and weeds in warmer areas
human populations	• increased deaths • more environmental refugees • increased migration
human health	• increased deaths from heat and disease • disruption of food and water supplies • spread of tropical diseases to temperate areas • increased respiratory disease • increased water pollution from coastal floods

SUMMARY   *Life Support Systems: Water and Air*

- The world's vital supply of fresh water is threatened by pollution, overconsumption, and climate change.

- The shortage of fresh water can be alleviated by wiser use and by the application of new technologies, such as drip irrigation systems and desalinization.

- Acid deposition threatens both water and soil quality and has damaged many aquatic and terrestrial ecosystems.

- Human activities are a major contributing factor to the increasing concentrations of greenhouse gases in the atmosphere and the threats posed by global warming and climate change.

- Climate change resulting from increased greenhouse gas concentrations is likely to have unpredictable and potentially serious consequences for many ecosystems.

▶ *Section 15.4* **Questions**

Understanding Concepts

1. Identify and describe three factors that contribute to stresses on global water resources.

2. Compare and contrast the flood irrigation method with the drip irrigation method, describing the advantages and disadvantages of each.

3. (a) Why is desalinization of seawater not a complete solution to the water shortage problem?
   (b) Describe two desalinization methods, indicating advantages and disadvantages in each case.

4. (a) Define *acid deposition* and explain why it is not just a local problem.
   (b) Describe two sources of acid precipitation.

5. List and describe several technologies or activities that could be used to help conserve water and improve water quality, both locally and globally.

6. Describe the mechanism by which certain gases trap heat within Earth's atmosphere.

Applying Inquiry Skills

7. Compare the global water surplus and deficiency map (**Figure 4** on page 725) with the map of population distribution (**Figure 1** on page 698).
   (a) Describe any obvious patterns between these two maps.
   (b) Do people always live in places where there is an adequate supply of water? If not, provide a few examples of relatively dry places where significant populations live. How might these people obtain water for their needs?

Making Connections

8. Some ecologists recommend that countries with a high standard and quality of living should assist less fortunate countries by financing and exporting nonpolluting industrial technologies and more efficient irrigation systems. Do you agree or disagree with this scenario? Discuss your opinion, with an emphasis on social and economic issues.

9. With two or three other students, brainstorm advantages and disadvantages of selling Canadian water to the United States. Prepare an advertising campaign on behalf of a provincial government to promote or protest the sale of water to the United States.

10. Air pollution in Ontario kills 1800 people each year. Although a considerable portion of Ontario's air pollution comes from sources south of the border, claims have been made that Ontario has not taken strong action at home to reduce emissions from major sources.
    (a) Identify three major sources of air pollutant emissions in Ontario.
    (b) Research the health, economic, and societal effects of air pollution.
    (c) Describe technological and political actions taken to minimize or eliminate emissions.
    (d) Explain whether you feel confident that sufficient efforts are being made to reduce pollutants in Ontario air.

 www.science.nelson.com

Changes to the quality of life in many countries, including Canada, mean that human populations have become accustomed to an increasing number of consumer goods and technological excellence in medical care, water treatment and distribution, fuel production, and public sanitation. People in these countries have also become more aware of the need to maintain a more sustainable lifestyle (**Figure 1**). Many use public transportation, conserve water, and practise the "five Rs" of waste reduction (refuse, reduce, reuse, recover, and recycle). Actions that may help sustain global carrying capacity, however, need to address the widespread and rapid degradation of ecosystems, resource depletion, and growing levels of pollution that spread across national boundaries. In 1987, the United Nations sponsored the Commission on Environment and Development, chaired by then Prime Minister of Norway, Gro Harlem Brundtland, which published a landmark report, *Our Common Future*, that sounded an international alarm over the state of the environment and the future well-being of the human population. As expressed in the report and other international initiatives, ecological and demographic crises have many causes. Resolutions require scientific and technological expertise to identify the most effective responses to these crises; also necessary is the cooperation of different levels of government within and among countries.

What is Earth's carrying capacity for a healthy and sustainable human population? Some scientists say that Earth can support up to 40 billion people, while others argue that human population numbers have already exceeded Earth's capacity to provide clean drinking water, food, and resources at present rates of consumption. Determining the carrying capacity for humanity is a challenge for many reasons. It is difficult to apply equations to dynamic, complex ecosystems because they are continually undergoing natural change and being modified by human action.

Monitoring Carrying Capacity

Biologists have systems in place for monitoring specific components of carrying capacity, such as soil, water, or air quality. By using existing data, or establishing **baseline data** for an ecosystem not previously studied, technicians and scientists can assess changes in quality from one monitoring period to another. Biologists look for patterns and trends in ecosystem dynamics, much as demographers look for changes in demographic patterns. Specific **indicators** found in the field, or samples taken from the field, also help to identify changes to the quality of ecosystem components that support life (**Figure 2**). Water quality can be determined using such empirical indicators as smell and turbidity, as well as biological and chemical indicators. Biological indicators for water quality include organisms that may appear, disappear, or become unhealthy as the quality of water declines. Frogs and invertebrates, for example, are especially sensitive to pollutants in water and respond very rapidly to changes in quality. Such organisms are referred to as **indicator species**.

Monitoring and assessing changes to carrying capacity pose significant challenges for biologists who try to account for the intricate interrelationships among changing environmental components. Currently, several models are used by biologists to assess how humans are affecting the **thresholds**, or tolerance limits, of an ecosystem's carrying capacity. The **ecological footprint**, a concept first developed by William Rees and Mathis Wackernagel in the 1990s, is a model that allows people to visualize the impact of their consumptive activities on Earth, and helps them determine whether they are living in a sustainable manner.

Figure 1
Recycling has become an everyday responsibility for many people.

baseline data initial information about both abiotic and biotic factors of an ecosystem that can be used to monitor changes in the ecosystem

indicators chemicals, organisms, or other components of an ecosystem that can be measured or observed to identify changes in the quality of conditions in an ecosystem

indicator species a species sensitive to small changes in environmental conditions; monitoring the health of their population is used to indirectly monitor the health of the environment in which they live

thresholds the tolerance limits, which, if exceeded, will result in harmful or fatal reactions within ecosystems

ecological footprint an estimate of the amount of biologically productive land and water needed to supply the resources and assimilate the wastes produced by a human population

Figure 2
Biologists spend many hours gathering and analyzing baseline data and information about similar species in other water bodies. They then use software programs that create projections based on the data.

More than 4 million residents live in the Greater Toronto Area (GTA), an area of 7061 km². However, it has been estimated that this population uses an area equivalent to 30% of Ontario (323 192 km²) to obtain resources and assimilate wastes. We call this Toronto's ecological footprint. In general, people who live in Ontario have large ecological footprints, as they are among the highest per capita consumers of energy and producers of waste on the planet. Calculating ecological footprints for different populations can provide biologists with approximate indicators of whether a population is living within the carrying capacity of the local environment, or exceeding it.

Monitoring thresholds and assessing the potential recovery of depleted or threatened resources is also challenging. For example, quantifying the annual yield for the harvest of a species of marine fish is extraordinarily difficult. Fish are mobile and are affected by changes in climate, pollution levels, disruption of their spawning habits, and predation. Thus, estimates of population counts for a single species could prove inaccurate. Scientists can monitor the rate of depletion and recharging of an aquifer by checking, at regular intervals, how far below the ground the upper surface of the water is. It is much more difficult to quantify the volume of an underground aquifer or to determine the quality of the water deep in the aquifer. Similarly, changes in air quality can be monitored for specific locations, but determining global patterns requires statistical analyses of large amounts of data for many locations and projections based on computer-generated models.

Soil organisms, such as mycorrhizal fungi, that help transport nutrients from soil water to roots are essential to the healthy growth of many plants. Canadian researchers have investigated potential changes to these mutualistic organisms as a result of increased acidification of the soil and water. In 1991, a research study showed that the establishment of mycorrhizal fungi on the root systems of jack pine was not affected directly by a decrease in soil pH but rather by a change in the ratio of calcium and aluminum ionic concentrations caused by a lower soil pH.

ACTIVITY 15.5.1

Comparing, Examining, and Reducing Ecological Footprints (p. 748)
Do all humans have equal footprints? In this activity, you will compare ecological footprints for various countries. You will also examine closely one major aspect of the footprints of your classroom and home and subsequently suggest ways to reduce the size of those footprints.

Restoring Degraded Ecosystems

In their 1992 joint report, *Population Growth, Resource Consumption, and a Sustainable World*, the United States National Academy of Sciences and the Royal Society of London stated, "Sustainable development can be achieved, but only if irreversible degradation of the environment can be halted in time." One method being used to replenish depleted resources and increase lost biodiversity is restoration ecology. Ecosystems may be degraded to the point that they can no longer support a natural diversity of living organisms. Restoration ecology applies ecological principles to repair, renew, or reconstruct degraded ecosystems. A key strategy in restoration ecology is **bioremediation**— a process in which bacteria, fungi, and plants are used to clean up polluted sites. Approximately 1000 species of bacteria, such as those from the genus *Pseudomonas* and those designed through recombinant DNA technology, have been found to thrive in polluted environments (**Figure 3**). These bacteria break down pollutants with enzymes and convert toxic chemicals into materials that are more easily digested. As they digest the materials, the bacteria produce more enzymes, which convert the digested pollutants into far less harmful wastes. The bacteria continue this clean-up process until the toxic chemicals are gone and the bacteria die of starvation. This method is becoming widely used to remove oil spilled in marine ecosystems. Studies are also being conducted to determine the feasibility of using lichens, bacteria, and plants to remove metals from ecosystems damaged by mine tailings.

bioremediation the use of organisms, usually bacteria, to detoxify polluted environments such as oil spill sites or contaminated soils

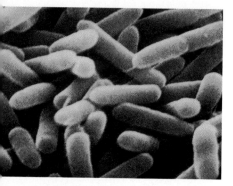

Figure 3
Pseudomonas bacteria are used to clean oil spills

Bioremediation in Action

Chlorinated organic solvents like trichloroethene (TCE), and tetrachloroethene, better known as perchloroethylene (PCE or PERC), are suspected carcinogens commonly used in paint thinners, antifreeze, dry-cleaning, and industrial processes that involve grease removal. These chlorinated hydrocarbons are volatile liquids at room temperature that seep into the ground and cling to soil when they are spilled. Cleanup of TCE- or PCE-contaminated sites was very difficult before the discovery in 1997 of certain anaerobic bacteria tentatively named *Dehalococcoides ethenogenes* Strain 195 that actually respire chlorinated hydrocarbons like TCE and PCE and convert them into nontoxic ethene gas. Enzymes within Strain 195 catalyze a series of reactions in which chlorine atoms are sequentially removed from PCE (**Figure 4**). These bacteria use PCE the same way aerobic organisms use oxygen in cellular respiration. A group of Cornell University microbiologists discovered these bacteria in the sludge from a now abandoned sewage plant in Ithaca, New York.

trichloroethene   cis-1,2-dichloroethene   vinyl chloride   ethene

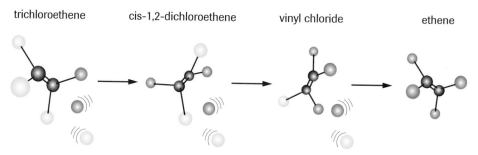

Figure 4
The sequential degradation of PCE to ethene by Strain 195

Strain 195 is not the first microorganism discovered to remove chlorine from organic compounds. Other microbes, however, usually do not remove all the unwanted atoms. Instead, they produce intermediate reaction products that can be as bad or worse than TCE. Removing one chlorine atom from PCE yields TCE, an equally toxic substance. Stripping two chlorine atoms produces cis-dichloroethene, also a suspected carcinogen. Eliminating the third yields vinyl chloride, a known carcinogen.

Only removal of all four of the chlorine atoms from PCE leaves a truly innocuous byproduct, ethene. The chlorine atoms that are produced in the process react with hydrogen gas produced by fermenting organic matter in the soil, forming hydrochloric acid, which reacts with metal ions in the ground to form harmless salts.

At some sites, chlorinated solvents are naturally (intrinsically) biodegraded by subsurface microorganisms. This process is referred to as intrinsic biodegradation or intrinsic bioremediation. If engineering steps are required to promote or enhance chlorinated solvent biodegradation by the subsurface microorganisms, then the process is referred to as enhanced or accelerated bioremediation. This strategy involves determining when intrinsic or enhanced bioremediation of chlorinated solvents should be used and designing and engineering an appropriate system that can significantly reduce capital, operations, and maintenance costs compared to other remediation technologies.

Chlorinated solvent bioremediation is an environmentally friendly, commonsense approach to site remediation because it destroys chemicals in place rather than transferring them to other locations (e.g., landfills) or to other states (e.g., gases in the atmosphere). Bioremediation is an innovative, cost-effective technology that is ideally suited to long-term treatment of these chemicals. Chlorinated solvent bioremediation (intrinsic or enhanced) can be used either on its own, or in concert with conventional remediation technologies, to destroy solvents present in soil and groundwater.

Waste: The Legacy of Consumption

The source of water and air pollutants is the disposal of waste. Household, commercial, and industrial byproducts have proliferated during the era of consumer goods, adding stress to the atmosphere, aquatic environments, and land. In Ontario, before 1920, most garbage was organic refuse, which could be disposed of by burning or by **biodegradation** in pits, backyards, and small dumpsites. Some people routinely disposed of garbage in water bodies, such as lakes and rivers. Since 1920, the proportion of organic refuse has decreased, and the proportions of paper, glass, metal, and plastics have increased exponentially. It has been estimated that 9.8 kg of garbage is generated and thrown away each week by a typical individual in present-day North America. Massive quantities of garbage are now routinely disposed of in enormous landfill sites and burned in incinerators—both of which lead to the production of toxic chemicals and the potential for air, ground, and water pollution. Our growing mountains of garbage are a legacy of the high rates of consumption by North American society.

biodegradation the natural recycling of wastes through the breakdown of organic matter by a variety of organisms, but especially bacteria and other microbes

DID YOU *KNOW* ?

Evolution Assists Restoration Biologists
Some plants have and are evolving the ability to grow on highly acidic mine tailings, which are the large deposits of silt-like material left over from the refining of metal-bearing ores. These tailings presently cover large areas of land and are too toxic to support most plant life. In addition, such tailings are a potential threat to the natural environment since erosion or a dam break could result in contamination of surface water (**Figure 5**). By encouraging the growth of plants that have evolved some resistance to such toxic chemicals, partial remediation of the sites can occur. If the tailings surface is covered by a thick growth of vegetation, the risk of serious erosion is greatly reduced.

Figure 5
A spill of more than 100 000 tonnes of mine tailings severely polluted this northern Ontario stream.

(a)

(b)

Figure 6

Increasing human populations and demands for consumer goods have increased the need for more large engineered landfills **(a)** and for well managed recycling operations **(b)** than in the past to handle the huge increases in paper, plastic, and household hazardous wastes.

The collection, recycling, and disposal of garbage has become a vast economic enterprise operated by municipalities and private industry. Most municipalities have very large tracts of land devoted to disposal, where the garbage is stored, as shown in **Figure 6(a)**. Until the 1960s, garbage was routinely dumped into swamps and marshes, which were regarded as wastelands rather than as biologically productive areas. It was thought that once a former wetland was filled with wastes, it could be covered with soil and used as a building site. Many municipalities now collect source separated waste as organic, recyclable, and solid trash, as shown in **Figure 6(b)**, and restrict the deposit of hazardous wastes into landfill sites. Concern remains about the disposal of products in breakable containers with minor amounts of toxic substances, such as nail polish. Landfill sites receive hundreds of thousands of discarded bottles containing trace amounts of hazardous materials resulting in significant accumulation. Researchers have found hazardous organic chemicals—apparently from residues of common household cleaners and from plastic containers such as shampoo bottles—leaching into groundwater from landfill sites that have never received industrial wastes. Many municipalities dispose of sewage sludge from treatment plants in landfill sites, a practice thought to enhance biodegradation. Because of concerns about the spread of germs and the leaching of toxic pollution out of garbage dump sites into adjacent water bodies, more care is being taken to reduce leachate problems with new landfills. As our waste stream has become increasingly toxic, we have responded by constructing better landfill liners to reduce contact between garbage and the natural environment. Many concerned citizens, environmentalists, and scientists strongly advocate a different approach—dramatically reducing the volume of garbage we generate and eliminating hazardous materials entirely from the waste stream.

Case Study Is Biodegradation Good for Garbage?

How much biodegradation occurs in landfill sites? Researchers from the University of Arizona began the Garbage Project to find out. Teams of archaeologists, microbiologists, engineers, and biochemists have excavated more than a dozen landfill sites in the United States. They have applied sampling and archaeological excavation techniques to record the age and content of garbage removed from gradually older levels in landfill sites. Samples are carefully stored and removed for biochemical and other analyses. The conditions below the surface of landfills—where there is no sunlight or oxygen, and there are high temperatures (40°C and higher)—ought to be ideal for colonies of anaerobic bacteria to degrade wastes. The Garbage Project has found, however, that relatively little biodegradation occurs; wastes are intact and clearly recognizable in layers of garbage deposited four to six decades ago.

Since as much as 35% to 45% of garbage deposited since the 1960s is paper, a microbiologist working with the Garbage Project researchers analyzed 28 samples of garbage from the excavation of 1 biologically active landfill site and 12 samples from a less active site. She searched specifically for evidence of anaerobic bacteria that degrade cellulose, which could be expected to degrade paper wastes. Although the microbiologist found high concentrations of many kinds of bacteria, the distribution of cellulase enzymes was thin, and there were no cellulose-degrading bacteria in any samples from either site.

Only one of the Garbage Project's excavations struck totally biodegraded materials, found at the lowest levels of the landfill site, where garbage had been deposited seven to eight decades before. The difference between this site and the others was that it had full contact with water. Formerly a wetland, it was located adjacent to a tidal river and had its supply of moisture to the lower levels renewed regularly. Without moisture, this research found, the biological activity of landfill sites occurred at extremely slow rates, if it occurred at all.

Environmental groups counter that the entire exercise of attempting to research and better manage or design landfills is misguided. They point out that the five major components in the waste stream should never reach a landfill. Organic material (1) such as household food scraps should be composted; all paper and wood fibre (2) and scrap metals (3) should be recycled and, because glass (4) is both physically and chemically inert, it can simply be crushed and either recycled or safely discarded. Plastics (5) are resistant to decay and can be difficult to recycle but, since they represent a nonrenewable resource, technological advances must be encouraged to make better and sustainable use of these materials.

> ### ▶ Case Study *Questions*
>
> **Understanding Concepts**
>
> **1.** Describe the process of biodegradation in a typical landfill site.
>
> **2.** What effects might biodegradation in deposited garbage have on the environment surrounding a landfill site?
>
> **3.** How might changes to the proportion of organic wastes being deposited into a landfill site affect biodegradation?
>
> **Making Connections**
>
> **4.** The Garbage Project recommended that, rather than encouraging biodegradation, the goal for waste disposal should be for dry and relatively inert long-term underground storage. Prepare arguments for or against biodegradation in a landfill site.
>
> **5.** How reasonable is it for humans to continue to dispose of millions of tonnes of garbage each day in landfills and incinerators around the world? Research the advantages and disadvantages of this approach.
>
> www.science.nelson.com

Limiting Wastes

One of the greatest efforts necessary to ensure sustainability is limiting toxic pollutants and waste production. Controlling, monitoring, determining sources of, and assessing the dangers posed by, trans-boundary pollution remain ongoing focuses for scientific research.

There are two basic ways to deal with global pollution. The first way is to prevent, reduce, or eliminate the production of pollutants by use of such cleaner, nonpolluting technologies as wind power, solar photovoltaic systems, and microturbines. The second is to continue cleaning up the pollutants after they have been produced. Ecologists and economists agree that prevention is far more effective and far less expensive. Governments could encourage prevention, for instance, by granting incentives such as tax write-offs and by implementing specific pollution regulations and laws. Alternative automobiles could be more widely marketed to reduce the amount of emissions from combustion of fossil fuels (**Figure 7**, page 738). Whether they are electric, fuel-cell powered, or a hybrid, these cars create little pollution directly. However, they are not pollution free—the electricity generated to charge car batteries or produce hydrogen fuel from the electrolysis of water must be supplied by some original energy source. If that energy source is wind or solar power, the process may have little net environmental impact. If the electricity comes from nuclear reactors, large electric dams, or generating stations that burn fossil fuels the environmental impact will continue to be substantial. Davis, California, has

Figure 7
Personal vehicles of the future may use hydrogen fuel cells to generate electricity. Although these cars emit no pollutants directly, they do not eliminate the need for an original energy supply for the production of the hydrogen fuel.

eco-cities urban centres that are planned to minimize their impact on the local and regional environments and to foster sustainable lifestyles

taken measurable steps toward curbing the amount of pollution from transportation. The city has closed some streets to automobiles, allowing only bicycles or in-line skates, and has built a network of bicycle paths to enhance accessibility to cyclists. As a result, bicycles now account for 40% of all transportation within the city.

A useful model can illustrate the kinds of shifts in thinking that can move societies toward greater sustainability for future generations. Most societies with high purchasing power live in what economists refer to as a high-throughput economy, illustrated in **Figure 8(a)**, meaning that they try to sustain economic growth by increasing the movement of resources into the economy. These resources are output as wastes and pollution that do not support sustainability. For greater future sustainability, high-throughput economies must be transformed into matter-recycling economies in which economic growth can continue with the recycling of resources, without depleting resources or emitting too much pollution. This is only a partial solution to excessive resource consumption. In addition, recycling processes may produce their own pollutants. A better solution to reducing consumption and limiting toxic effects is to transform high-throughput economies into low-throughput economies, as shown in **Figure 8(b)**. This is accomplished by reusing and recycling most nonrenewable resources, using potentially renewable resources at a pace that will allow natural regeneration, making efficient use of energy resources, reducing unnecessary consumption, emphasizing pollution prevention, and reducing waste production.

Some Ontario cities, such as Mississauga, have implemented policies to divert garbage. Area households can dispose of three bags of garbage per week without charge; any additional bags must have a special tag that is purchased from the city. Frequency of collection of materials for recycling has also been increased and city officials hope that, within 20 years, garbage will be reduced by 70%. There remain concerns that some residents will simply use larger bags or dump garbage to avoid paying. Many Ontario municipalities now collect organic garbage for composting, along with recyclable materials and trash. To ensure a sustainable future, however, residents and businesses need to reduce their individual waste production.

The development of **eco-cities** could contribute to the transformation of consumer societies into low-throughput societies. Ecologists and urban planners agree that populations should be concentrated in cities, and rural areas should be left for preservation of resources and biodiversity. But future urban communities need to reduce consumption, wastes, and pollution to provide consistent life support systems for their populations. Eco-cities could use solar and other alternate power sources to make more efficient use of energy and matter resources, reduce waste and pollution, and encourage biodiversity. Organic farms and gardens, and surrounding woods, could support human needs as well as provide habitats for diverse indigenous species. In model eco-cities, people would routinely walk, cycle, or use public transit. Are eco-cities just a dream? Davis, California, and Curitiba, Brazil, are already well on their way, making great progress toward enhanced sustainability.

While ecologists have been conducting research and presenting concerns about changes in environmental conditions, they have also been contributing better ways to monitor carrying capacity, restore degraded ecosystems, and manage wastes. Ecologists have also been encouraging people to change their attitudes and activities to support improved sustainability as individuals, business and industry managers, and employees. Science is also working to enhance food production while maintaining the sustainability of life support systems on which people and nonhuman populations rely.

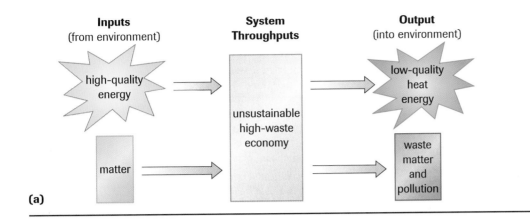

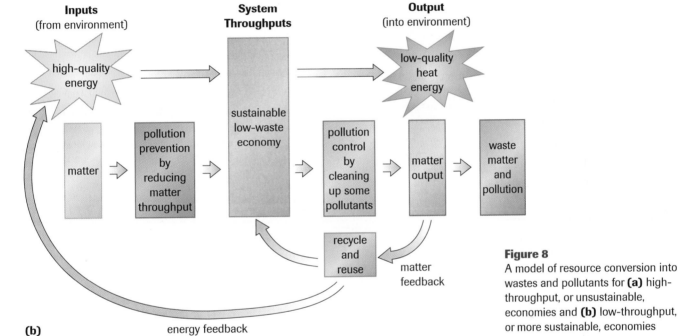

Figure 8
A model of resource conversion into wastes and pollutants for **(a)** high-throughput, or unsustainable, economies and **(b)** low-throughput, or more sustainable, economies

SUMMARY *Restoration Ecology and Waste Management*

- Owing to consumption patterns and the dynamic nature of ecosystems, it is challenging for scientists to quantify the carrying capacity of the global or local environment for human populations.
- Societies that use large amounts of consumer goods produce large quantities of waste materials that must be disposed of, reused, or recycled.
- To maintain sustainability, scientists are endeavouring to restore degraded ecosystems and improve methods of waste disposal and management.
- More sustainable approaches will require a dramatic reduction in resource consumption patterns rather than simply relying on technologies to dispose of large volumes of potentially toxic waste materials.
- Low throughput strategies provide a more sustainable approach to resource management.
- Some progress is being made in designing ecologically friendly urban centres that could reduce human impact on the environment.

Understanding Concepts

1. Explain some of the challenges biologists face in determining and predicting the carrying capacity for the expanding human population of a large city.

2. In your own words, describe bioremediation.

3. What advantages does the use of *D. ethenoagenes* Strain 195 have as a method of destroying perchloroethylene and related chlorinated organic solvents?

4. On average, each Canadian generates and throws away about 4.9 kg of household garbage each week. An equal amount is typically added to the municipal waste stream from other sources. Of the total 9.8 kg per person per week, approximately 12% is plastic that will never biodegrade.
   (a) Estimate how large a landfill site is needed to accommodate the wastes of a community of 150 000 residents for 20 years.
   (b) Estimate the quantity of garbage that will never biodegrade in this landfill site.
   (c) What additional factors would have to be taken into account for an actual landfill site in a real community?

5. Compare and contrast high-throughput, matter-recycling, and low-throughput economies. Present your comparison in a Venn diagram or other graphic organizer.

Making Connections

6. Methane digesters are a suggested technological solution to waste disposal. Methane digesters convert garbage into compost and methane gas, which can be used as a power source. Research the benefits and drawbacks of methane digesters, and write a brief position paper on whether you support methane digesters rather than new landfill sites for Ontario communities.

7. In the 1990s, the city of Toronto seriously considered a plan to dispose of 20 million to 30 million tonnes of waste in an abandoned open-pit mine in northeastern Ontario. The plan called for the shipping of the garbage some 500 km by rail to a landfill site in the 200-m deep pit. Conduct Internet and or library research to learn the details of the proposal and comment on the suitability of such a plan as a waste-management strategy for the city.

 www.science.nelson.com

8. As high product consumers, Canadians produce the most garbage per person of any society in the world. In 1996, almost 9 million tonnes of municipal solid wastes were generated in Ontario alone, 80% of which was deposited in landfills. How does this influence the ecological footprint of Canada's 30 million people relative to the populations of less industrialized countries?

Our Common Future 15.6

Humans are the only species able to study Earth and its physical, chemical, and biological nature (**Figure 1**). Although our understanding of the complexity of Earth's biosphere and many interconnected ecosystems remains in its infancy, the demands that our population is placing on these same systems are growing. The many stresses being placed on our planet's life support systems result largely from our overconsumption of Earth's mineral, water, and living resources. As the human population of 6 billion continues to follow a pattern of growth, these pressures are increasing and are producing stresses on a global level.

Larger-scale and much longer-term research initiatives and modelling are now underway to assess the causes and implications of these global and **cumulative effects**. Ecosystems are dynamic—as more stresses are applied to them, more changes and interactions are observed. In addition to the effects on food production, water availability, atmospheric cleanliness, and general waste production, there are other ecological effects, evident throughout the world, stemming from the large human population (**Table 1**).

cumulative effects the accumulated changes to natural environments caused by several interactions, factors, or changing conditions

Figure 1
Earth as seen from the Apollo 8 spacecraft (1968)

Table 1 Some Potential Ecological Effects Caused by Human Activity

Ecological effect	Description
Excessive use of fertilizers	The application of large quantities of inorganic fertilizer has enabled growing populations to produce much more food on the same land base, but may result in the release of the greenhouse gas nitrous oxide from soil, cause water pollution, and actually reduce soil fertility after extended use. Alternative cultivation practices can reduce fertilizer demand without reducing productivity.
Deforestation	Growing populations, mostly in equatorial regions, are cutting down large areas of forest for new farmland and settlements. These same forests purify the air, act as giant sponges regulating the flow of water, influence regional and local climate, provide numerous habitats for wildlife species, and support food webs, energy flow, and nutrient cycling.
Loss of biodiversity	Humans have disturbed or degraded 40%–50% of Earth's land surface by converting grasslands and forests to crop fields and urban areas, often displacing indigenous species of plants and animals. Many species have become extinct or have been threatened with extinction as a result.
Ozone depletion	A layer of ozone in the lower stratosphere shields Earth's surface from about 95% of the Sun's harmful ultraviolet radiation. Ozone depletion has resulted from the release of large quantities of CFCs and other ozone-degrading chemicals into the air. The sources of these emissions have largely been from highly industrialized populations.
Pesticides	The widespread use of pesticides has increased harvests of crops by destroying many of the insects that compete with humans. However, the use of pesticides poses risks to terrestrial and aquatic ecosystems. The risks of long-term exposure to these trace amounts of pesticide residues in food, air, and water are unknown.
Widespread pollution	Air, water, and soil pollution result from routine human activity, as well as disasters such as oil spills. Pollution problems are most severe in large urban populations where most waste is generated.

Cumulative effects as a result of rapidly growing human populations can clearly be observed in Mexico City. The city was built on the bed of a lake drained by the Spanish conquistadors during the 16th century after they had polluted it with raw sewage. As **Figure 2** shows, the spongy clay soil of the lakebed has not been able to support the weight of extensive building for the city's exploding population; even in the medieval era, large public buildings had started to sink. They have continued to subside for centuries, as the level of groundwater below the city drops. Ironically, the city has always been subjected to flooding during the rainy season from May to October. The Spanish had hoped that draining the four lakes in the Mexico City valley would alleviate the flooding.

Figure 2

Many parts of Mexico City are sinking as a result of severe depletion of the aquifer, which may not serve the city's people for much longer.

A perennial problem for Mexico City administrators has been acquiring a consistent water supply for the steadily growing population. Countless canals, tunnels, and more complex water systems have been constructed, one reversing the water flow in a river across mountains several hundred metres higher than either the river or Mexico City, which itself is more than 2000 m above sea level. High consumption of electrical resources is required to pump water up and across the mountains down to the millions of residents in Mexico City. Although there is groundwater below the city, the aquifer has had such tremendous volumes of water removed that water stored in the aquifer for thousands of years is being now tapped. Much of this water is poor quality, as it has a high mineral and salt content. It is now difficult for the aquifer to be replenished, as forests in recharge zones have been cut to make room for more housing. The loss of forests results in increased surface runoff thereby reducing the volume of water that penetrates the soil—water needed to replenish the aquifer. With a population in 2001 in excess of 22 million drawing on this water supply, predictions are that the aquifer has only decades left before it dries up.

At least 8 million people in Mexico City are without sewer facilities. Vast amounts of human waste are deposited daily into gutters and vacant lots. When dried, these wastes are picked up by the winds, which deposit a fecal dust on other parts of the city, leading to widespread salmonella and hepatitis infections. Water pollution is estimated to cause 100 000 premature deaths per year. Industrial and household wastes in the water have turned rivers leaving Mexico City into foul-smelling, polluted rivers that deliver drinking and irrigation water to several surrounding regions.

In addition to the water crises, Mexico City is subject to earthquakes and has a serious air pollution problem. The well-being of future generations will depend largely on the ability of current generations to minimize harmful effects on human life support systems and to facilitate movement toward more sustainable lifestyles and resource use.

The Brundtland Commission's report, *Our Common Future*, defined the term *sustainable development* as "development that meets the needs of the present without compromising the ability of future generations to meet their own needs" (World Commission on Environment and Development, 1987). Canadian geneticist and outspoken environmentalist David Suzuki frequently encourages scientists to consider the long-term implications of their research and to extend their studies to sustainable and integrated whole ecosystems, rather than focusing on isolated components. The adoption of global **systems thinking** seems desirable to help ensure the preservation of global carrying capacity. Fritjof Capra, a physicist and systems theorist with the International Centre for Ecological Studies in England, describes systems thinking as an understanding that "the essential properties of an organism, or living system, are properties of the whole, which none of the parts possess. They arise from the interactions and relationships among the parts. These properties are destroyed when the system is dissected, either physically or theoretically, into isolated elements." Earth is a living system; all organisms are interconnected and affected by one another. It makes sense for human populations to take a greater collective interest and responsibility for enhancing future sustainability by addressing concerns through global systems thinking. Governments develop foreign policies and trade policies. Greater cooperation between national governments and industries with global interests could support integrated approaches to sustainability for future generations.

We began the chapter with the recognition that humans are a highly intelligent and successful species. We have, indeed, been able to take full advantage of many of Earth's habitats and resources. Although our success has lead to many stresses on the various aspects of the biosphere, we do have the capacity to recognize these stresses and make changes to our behaviours if we desire. Modern science and technology can and will provide many of the tools needed to achieve a sustainable future for the human population. Alternative clean energy sources are needed, as are sustainable agricultural practices that produce more food while reducing or avoiding soil and water degradation and overconsumption. We will also need to reduce our impact on the other species with which we share this planet. The loss of biodiversity is of grave concern and cannot go unchecked. Although the tools of science will provide much-needed knowledge and understanding, all societies, governments, and economies must be prepared to deal with the human side of these problems. Humans need to take responsibility for their actions, whether on a local scale or on a global scale. The wealthiest 20% of people on Earth are presently enjoying a much higher quality of life than the remaining 80%, while imposing a far greater share of the stresses on the planet's life support systems. Although much progress is now being made, many challenges and opportunities await tomorrow's citizens, scientists, and decision makers.

systems thinking the analysis of problems that focuses on the implications for an entire system rather than on the implications for individual components or parts of the system

SUMMARY   *Our Common Future*

- The cumulative effects of human activity are placing many stresses on the planet's life support systems.

- Large populations, especially in urban centres, are often unsustainable and result in widespread negative impacts on the quality of life and the natural environment.

- Systems thinking recognizes the interactions of many components of an ecosystem rather than considering them in isolation.

- Science and technology can provide some of the tools that will be required to foster a sustainable future for the human population.

▶ *Section 15.6 Questions*

Understanding Concepts

1. Outline six major stresses being placed on the natural environment as a result of large-scale human activity.

2. (a) How has Mexico City earned its reputation as one of the most polluted cities in the world?
   (b) In what ways are urban centres near you similar to Mexico City in terms of life support systems? How are they different?

Applying Inquiry Skills

3. (a) What is meant by the expression *global systems thinking*?
   (b) Conduct library and/or Internet research to identify a local, provincial, or national policy that runs counter to global systems thinking.

 www.science.nelson.com

4. Consider the Brundtland report's definition of *sustainable development*. In relation to each of the following, what factors might you have to consider when planning for a sustainable future?
   (a) energy use in transportation
   (b) food production
   (c) urban planning

Making Connections

5. Some believe that cities, by definition, cannot be sustainable. Conduct library and/or Internet research to explore both sides of this issue and then write a brief position paper.

 www.science.nelson.com

Ecologist

Ecologists study the relationships among organisms and the relationships organisms have with their abiotic environment. They analyze the effects of factors such as population size, pollution, temperature, and climate on ecosystems. Ecologists spend time in the field conducting research, for example, monitoring animal populations, studying how plants grow and develop, and determining how to protect ecosystems and their indigenous wildlife and plants. Ecologists also spend time in laboratories using sophisticated technology to analyze samples, or in offices to advise governments and private companies on environmental management strategies. Some ecologists also teach at universities and colleges. Depending on the position, ecologists need a college diploma or university degree. Ecologists may go on to pursue a master's or doctoral degree. Ongoing education is required to be an ecologist, as ecosystems are always changing and new techniques become available to detect these changes and the responses of organisms to them.

Demographer

Demographers analyze the size, nature, and movement of human populations. Often they specialize in an area such as health, housing, education, agriculture, or economics. Demographers collect and analyze population data, prepare population forecasts, examine the effects of population changes on society, and are involved in the creation of social policy. Demographers may also study how social and economic factors affect family patterns and forecast demands for public services. A university degree with an emphasis on mathematics and statistical analysis is a minimum requirement. For anyone wanting to conduct research into demographics, higher-level university degrees are necessary.

Wildlife Biologist

Wildlife biologists, including zoologists and marine biologists, may carry out laboratory and fieldwork to determine where and how organisms live, and how they interact with their surroundings. They also study the origin, makeup, and structure of nonhuman populations, as well as their growth and the diseases that affect them. They may research evolutionary biology, cellular and molecular biology, or genetics. They try to find ways to help conserve species at risk and advise local authorities on how to manage these species. Wildlife biologists may teach at colleges and universities, write reports and scientific articles, and give talks to community groups. To become a wildlife biologist, a four-year university degree is required and often a master's or doctoral degree as well.

▸ Practice

Making Connections

1. Discuss some skills that would benefit someone who wants to become an ecologist, a demographer, or a wildlife biologist.

2. Conduct research to determine what types of technologies are used in each of the above careers and how they contribute to the effectiveness of the job.

3. Conduct an online search to find out which businesses, government branches, and other organizations employ people in these three careers.

 www.science.nelson.com

4. Suggest some other careers that might benefit from an understanding of population ecology.

The Impacts of Resource Consumption

Human populations in different parts of the world place variable demands on the natural environment. The acquisition and consumption of such resources are ultimately used in an attempt to benefit humans and increase their quality of life. Excessive resource consumption however, can also result in environmental degradation and thus potentially reduce the quality of life and jeopardize the ability of the environment to provide for future generations.

In this activity you conduct a study to assess the relationships between human populations, resource consumption rates, and indicators of quality of life. In addition, you will evaluate the impact of such consumption rates on the health of the natural environment.

Questions

What resources are in the greatest demand by human populations? In what ways are the rates of consumption correlated to quality of life, population size, and environmental degradation?

Procedure/Analysis

1. Consider what general relationships exist between pairings of the following variables: total population size, resource consumption rates, per capita consumption rates, quality of life, pollution and environmental degradation.

   (a) Record your ideas.

2. Examine the data presented in **Figure 1** and **Table 1** and **Figure 2** and **Table 2**. The number of people per doctor is one indicator of quality of life, while the production of carbon dioxide per person is an indicator of energy consumption and affluence as well as pollution and environmental impact.

Table 1 The Number of People per Doctor by Country

Country	People per doctor	Country	People per doctor
Italy	207	China	1 063
Russia	222	Mauritius	1 165
UK	300	Egypt	1 316
Norway	308	Vietnam	2 279
France	334	India	2 459
Germany	367	Nigeria	5 208
Sweden	394	Sudan	10 000
USA	421	Burma	12 528
Canada	464	Bangladesh	12 884
Australia	500	Kenya	21 970
Romania	538	Chad	30 030
Japan	608	Mozambique	36 225
Saudi Arabia	749		

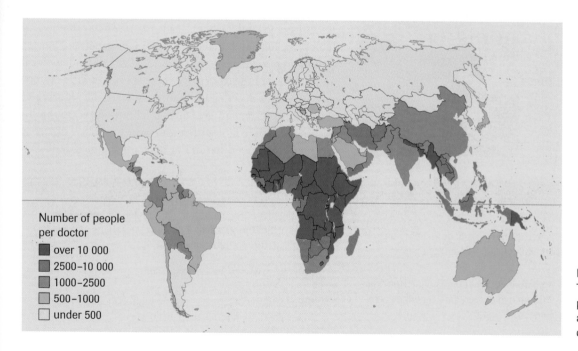

Number of people per doctor

- ■ over 10 000
- ■ 2500–10 000
- ▨ 1000–2500
- ▨ 500–1000
- □ under 500

Figure 1
The number of people per doctor in a country is a quality of life indicator

ACTIVITY 15.4.1 *continued*

Table 2 Carbon Dioxide Production by Country

Country	Tonnes per year per person	Country	Tonnes per year per person
United Arab Emirates	34	Czech Rep.	13
USA	19	Saudi Arabia	13
Singapore	18	Ukraine	12
Kazakhstan	18	North Korea	11
Trinidad	17	Kuwait	11
Australia	15	Germany	11
Canada	14	Turkmenistan	11
Norway	14	Denmark	10
Russia	14	Belgium	10
Estonia	14	Belarus	10
		UK	10

2. Select a number of specific countries and/or regions and plot the relationship between people per doctor and tonnes of carbon dioxide per person per year.

(b) How are these two variables correlated?

(c) Are there exceptions to the general trend? Offer a possible explanation.

(d) What does the carbon dioxide production per person indicate about the ecological footprint of individuals living in various parts of the world?

3. Although there is certainly no direct causal relationship between the numbers of doctors in a country and the amount of fossil fuels burned per capita, there may well be a relationship between resource consumption and higher standard of living. Working with a group of two or three other students you will investigate a variety of measures of resource consumption, environmental impact and indices of quality of life. Your teacher may allow you to select your own variables or assign them. Some examples are listed in **Table 3**.

Table 3

Measures of resource consumption	Environmental impact	Indicators of quality of life
• fertilizer and pesticide use • energy consumption • mineral extraction and use	• hazardous and solid waste production • habitat loss • urban smog	• average life expectancy • adult literacy rate • access to safe drinking water and sanitation

4. Use a variety of sources including the library and the Internet to obtain current data related to your chosen measures. Be sure to take into consideration whether the data are presented in gross values or on a per capita basis.

 www.science.nelson.com

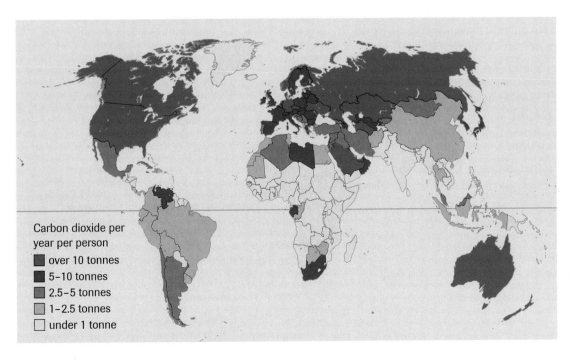

Carbon dioxide per year per person

- over 10 tonnes
- 5–10 tonnes
- 2.5–5 tonnes
- 1–2.5 tonnes
- under 1 tonne

Figure 2
The production of carbon dioxide per person is an indicator of energy consumption

5. After collecting and tabulating your data, perform similar analyses given in step 3 and look for relationships between the various measures.

(e) Present your findings in visual and tabular form.

(f) Summarize your findings, indicating which factors seem to be most closely correlated. Did you find any evidence to suggest that the overall impact of individuals on the environment is directly related to an improved quality of life?

(g) What do your findings imply about the relative importance of total population size verse the impact of per capita consumption rates?

Synthesis

(h) Choose a single resource that seemed to be strongly correlated to both environmental impacts and the quality of people's lives. Research and consider strategies or alternatives that might reduce the environmental impact of a particular resource use while maintaining or enhancing the quality of life for human populations. Present your findings to the class.

🔍 **ACTIVITY 15.5.1**

Comparing, Examining, and Reducing Ecological Footprints

Canadians now depend on resources that come from beyond the local community, often from outside Canada. Consider the 10 000 km² that surround your home. How long could the resources directly available within that area sustain the human population living within that same area? The concept of ecological footprints helps us visualize and compare ecological effects of different human lifestyles. Canadians have an estimated average ecological footprint of 7.7 ha per person. It is assessed that, if the amount of productive land on Earth were divided equally among every human alive in 2002, there would be 1.7 ha available for each human. If all people in the world had a footprint as large as each Canadian, the global population would need another four Earths to sustain itself! Many factors must be considered to assess an ecological footprint. These factors include food consumed, energy used, water required, and waste generated.

Questions

Why might various countries have significantly different ecological footprints? Do all people in the same country have the same ecological footprint? Can humans alter their ecological footprints?

Part I: Comparing Ecological Footprints

1. Study **Figure 1**, which graphically represents the ecological footprints of five countries

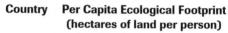

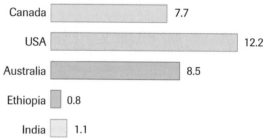

Figure 1

A country's total ecological footprint is calculated by multiplying the per capita footprint by the country's total population.

Analysis

(a) With two or three other students, list possible reasons for the differences among the per capita footprints of the five countries illustrated in **Figure 1**.

(b) Suggest why the countries' total footprints do not simply repeat the pattern of the per capita footprints.

ACTIVITY 15.5.1 *continued*

Synthesis

(c) Why are the total footprints for Canada and Australia very small compared to the United States while their per capita footprints are only a little smaller?

(d) Why is the total footprint for India very large compared to Canada's while its per capita footprint is very much smaller?

(e) Why does the United States have both the largest per capita footprint and the largest total footprint?

(f) Eric Krause, an environmental policy official with the City of Toronto, has noted "it would be better for the environment if 10% of the population reduced its ecological footprint by 90%, than 90% of the population reducing its ecological footprint by 10%." Explain this statement.

Part II: Your Classroom's Ecological Footprints

2. With two or three other students examine the following lists of items that may be found in your classroom's wastebasket and in its paper recycling receptacle.

Wastebasket
old textbook; 3 ring binder with paper; 3 pop cans; 2 apple cores; bag with ham sandwich; 4 marked tests; running shoes; 2 plastic water bottles; 6 cm piece of chalk; waxed paper; pieces of used masking tape; paper takeout coffee cup

Paper Recycling Receptacle
foil gift wrapping paper; cling wrap; 25 sheets of blank paper; crumpled muffin bag; newspaper; topographic maps; pop can; drink box; 10 copies of a handout; poster with tape attached; magazine; last term's biology notes

Analysis

(g) List the items that should not be present in the wastebasket. Record where each of these ought to be placed.

(h) List the items that should not be in the paper recycling receptacle. Record where each of these ought to be placed.

Synthesis

(i) Explain why your lists provide an indication of your classroom's ecological footprint.

(j) Suggest specific ways that your classroom could reduce its footprint in addition to merely putting these items in a better place.

(k) Write a letter to your school board recommending specific actions it could take to encourage schools to reduce their ecological footprints.

(l) Why should schools and businesses not hold contests to see which institution recycles the most?

Part III: Your Home's Ecological Footprints

3. With two or three other students examine the following list of items that may be found in your home's wastebaskets and kitchen garbage can.

Home Wastebaskets and Kitchen Garbage Can
torn T-shirt; plastic price stickers; 'exhausted' marking pen; orange peelings; shells from 5 eggs; 16 foil bags from instant soup mixes; 4 cat food tins; used paper towels; chicken bones and skin; 5 metal hangers; empty cereal box; 6 plastic yogurt tubs; 4 'dead' batteries; burnt out light bulb; broken clay flowerpot; chewed gum; wax from Gouda cheese; 12 jars and lids; carrot tops; frozen orange juice 'can'; plastic pouches from cat 'treats'

Analysis

(m) List the items that should not be present in these home waste receptacles. Record where each of these ought to be placed.

Synthesis

(n) Suggest specific ways that your family could reduce its footprint in addition to merely putting these items in a better place.

(o) If you observe that a neighbour does not put out recycling containers regularly, you might assume that the family is not recycling as much as possible. How could you prove whether you were right? If you were wrong, what might be an explanation for your observations?

(p) Explain why examining the contents of waste and recycling containers allows a good comparison of ecological prints for families, schools, communities, and businesses.

(q) Write a letter to your local municipal or regional council recommending specific actions it could take to encourage families and businesses to reduce their ecological footprints.

Key Expectations

- Explain the demographic changes that have affected populations over the past 10 000 years, including epidemics, the rise of agriculture, the Industrial Revolution, and the development of modern medicine. (15.1)

- Locate, select, analyze, and integrate information on the effects of human population growth on the environment and on quality of life, working independently or as part of a team, and using appropriate research resources. (15.1, 15.2, 15.3, 15.4, 15.5, 15.6)

- Compile, organize, and interpret data, using appropriate formats and treatments, including tables, flow charts, graphs, and diagrams. (15.1, 15.2, 15.3, 15.4, 15.5, 15.6)

- Outline the advances in medical care and technology that have contributed to an increase in life expectancy, and relate these developments to demographic issues. (15.2)

- Explain, using demographic principles, problems related to the rapid growth of human populations and the effects of that growth on future generations. (15.2, 15.3, 15.4)

- Use examples of the energy pyramid to explain production, distribution, and use of food resources. (15.3)

- Describe examples of stable food-production technologies that nourish a dense and expanding population. (15.3)

- Analyze Canadian investments in human resources and agricultural technology in a developing country. (15.3, 15.6)

- Analyze the carrying capacity of Earth in terms of the growth of populations and their consumption of resources. (15.3, 15.4, 15.5, 15.6)

- Describe general principles and strategies for achieving sustainable development. (15.3, 15.4, 15.5, 15.6)

- Identify and describe science-and-technology based careers related to populations dynamics (15.6)

Key Terms

acid deposition

air pollution

aquifers

arable land

baseline data

biodegradation

bioremediation

cumulative effects

demographic transition model

demography

desalinization

eco-cities

ecological footprint

energy pyramids

epidemic

food chain

greenhouse effect

gross national product (GNP)

growth momentum

indicators

indicator species

quality of life

salinization

soil degradation

systems thinking

thresholds

trophic level

urban sprawl

zero population growth

▶ *MAKE* a summary

Imagine you have been hired as a population ecologist by a member of Parliament who wants you to draft recommendations for a government policy. This policy, which will include recommendations from other specialists, is to be implemented nationally to create a more demographically and ecologically sustainable future. Your draft recommendations must have the following components:

- Outline the actions that have to be taken and why.

- Explain what ecological effects each action will help to alleviate, mitigate, or enhance.

- Predict potential consequences of, or backlash from, each of the actions.

In your notebook, indicate whether each statement is true or false. Rewrite a false statement to make it true.

1. Population size is the only factor that determines human carrying capacity.

2. An ecological footprint measures how much land and water an ecosystem occupies.

3. People can potentially obtain more food energy by consuming from the lower trophic levels of a food chain.

4. Drip irrigation is much more efficient than flood irrigation.

5. The public should not worry about preventing pollution because there are methods for cleaning it up.

6. High-throughput societies are more sustainable than low-throughput societies.

7. Human population growth has remained steady as a result of high fertility rates in many countries around the world.

8. Food scarcity is one of the major challenges faced by an unsustainable global economy.

9. The vast majority of wastes in Ontario biodegrade quickly in landfills.

10. Images from NASA's Terra Spacecraft shows that air pollution is no longer a problem.

In your notebook, indicate the letter of choice that best answers the question.

11. An individual with a small ecological footprint would practise which of the following activities?
    (a) have high resource consumption
    (b) eat large amounts of imported foods
    (c) leave the water running while brushing teeth
    (d) use solar-powered technologies
    (e) build a home on fertile agricultural land

12. Why is desalinization not a feasible method for increasing the supply of fresh water?
    (a) too expensive
    (b) produces a large amount of waste
    (c) consumes a great deal of energy
    (d) all of the above
    (e) none of the above

13. To increase the amount of energy available from food, which of the following strategies can people pursue?
    (a) Eat earlier in the food chain.
    (b) Eat farther along the food chain.
    (c) Eat more meat products.
    (d) Drink more water.
    (e) Take more vitamins.

14. According to the demographic transition model, what happens to population size as societies move from agricultural production to industrialization?
    (a) It increases as death rates exceed birth rates.
    (b) It decreases as death rates exceed birth rates.
    (c) It increases as birth rates exceed death rates.
    (d) It stabilizes with no changes in birth and death rates.
    (e) It stabilizes as birth and death rates are balanced.

15. Which of the following does not represent a probable result of global warming?
    (a) The global average surface temperature has increased by 0.6°C.
    (b) Snow cover and extent of ice have decreased in the north and south polar regions.
    (c) The global average sea level has risen.
    (d) Ocean heat content has decreased.
    (e) Fossil fuel burning has resulted in increased CO_2 concentrations in the atmosphere.

16. In *The Population Bomb*, what statement did Ehrlich make that turned out to be false?
    (a) Millions of people would deplete the planet of its valuable resources.
    (b) There are plenty of resources to sustain millions of people.
    (c) Famine would lead to the death of millions of people.
    (d) Human population grows exponentially while food production grows arithmetically.
    (e) The world's population would become stable.

17. Restoration ecology involves which two strategies?
    (a) bioremediation and desalinization
    (b) augmenting ecosystem processes and desalinization
    (c) bioremediation and augmenting ecosystem processes
    (d) desalinization and irrigation
    (e) bioremediation and irrigation

NEL   An interactive version of the quiz is available online.
GO ▶ www.science.nelson.com

Human Populations   **751**

Understanding Concepts

1. Explain how the rapid growth of a human population has the potential to adversely affect the local environment and the quality of life. Describe solutions currently in place to reverse or compensate for these effects.

2. Some countries with a low quality of life have argued that, to meet basic human needs and to reduce injustice and achieve equity for their populace, as outlined in the Brundtland report, they would have to develop an industrialized economy. Would this solution support sustainable development? Explain your reasoning.

3. Some people believe that Earth will never run out of resources because technology will be able to replace depleted resources or substitute materials for them. Do you agree or disagree with this line of thinking? Explain your answer, remembering to consider the effects that some technologies can have on the environment and sustainability.

4. What lifestyle choices do you make on a regular basis that contribute, either directly or indirectly, to the degradation of resources, especially the atmosphere and fresh and salt water. Are any of these lifestyle choices that you could easily change? Explain your thinking.

5. Analyze the energy pyramid in **Figure 1**. How could this food chain be adjusted to provide more food energy for a dense and expanding population? Explain your reasoning.

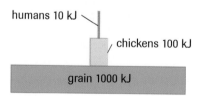

Figure 1

6. Freshwater shortages can be offset by desalinization technologies that allow salt water to be converted into drinkable, usable water. Describe benefits and drawbacks of desalinization processes.

7. Some demographers suggest that the major issue regarding the world's future population will be a global decrease rather than an ongoing increase. Comment on this statement, offering evidence that supports your response.

8. Increasing urbanization and the lack of ecological sustainability as cities grow have emerged as significant challenges. Identify some ecological effects of urbanization on the carrying capacity of the urban environment in each of the following:
   (a) your community or a large city nearby with which you are familiar;
   (b) a large city in another country.

9. How has soil degradation resulted in significant losses of agricultural land?

10. Describe the impact of increased worldwide meat production on
    (a) agriculture;
    (b) water resources;
    (c) waste management.

Applying Inquiry Skills

11. Create a time line that shows changes in lifestyles over the 10 000 years since the start of agriculture that have enabled human populations to thrive.

12. The Ontario Minister of the Environment has commissioned you and your research team to investigate the sources and ecological effects of acid rain in Ontario as outlined below:
    • the causes of acid rain in Ontario, past and present
    • the effects of acid rain on wildlife species, plant species, and agriculture
    • current technologies in place to deal with these ecological effects
    • effects of acid rain on the economy and health of Ontario

    After completing the research, prepare a presentation for the minister to take to a national meeting of environment ministers. For the presentation, design a concept map that summarizes the research completed and identifies the relationships or links between the effects of acid rain and its impact on Ontario.

13 Drinking-water treatment facilities ensure the delivery of safe, clean water to the majority of Canadians. In light of events in Walkerton, Ontario, in 2000, Canadians are questioning the ability of local water treatment plants to treat water adequately.
   (a) Investigate changes made by the government related to water treatment, testing, staffing, and reporting that contributed to events in Walkerton.

(b) Research and outline three initiatives taken by the Ontario Ministry of the Environment since the Walkerton events to improve water regulations and treatment systems.

14. Birth rate and death rate are expressed as the number of births and deaths for every 1000 people in a given year. The percent annual natural increase in a population can be determined using the following formula:

$$\% \text{ annual increase} = \frac{(\text{birth rate} - \text{death rate})}{100}$$

(a) Study **Table 1** that shows population data for selected countries. Using the equation, calculate the percentage increase for the countries listed.

Table 1 Sample Population Data for 2000 for Selected Countries[1]

Country	Birth rate	Death rate
Brazil	19	9
China	16	7
Denmark	12	11
India	26	9
Indonesia	23	6
Italy	9	10
Kenya	29	14
Mexico	24	5
Nigeria	40	14
Russia	9	14
South Africa	22	15
United Kingdom	12	10
United States	14	9

[1] Rates are per 1000 individuals

(b) Explain how birth and death rates determine the rate of natural population growth.

(c) Identify the countries you would consider to have a large annual increase in growth. Describe two factors that may account for these large numbers.

(d) Why do some countries, such as Italy and Denmark, experience almost negligible changes in annual growth?

Making Connections

15. With two or three other students, discuss the answers to the following:

(a) Do you think it would be possible, theoretically, for humans to survive in a world that consisted of only water, shelter, oxygen, and grain?

(b) Do you think humans would want to?

16. G. Tyler Miller, Jr., president of Earth Education and Research, estimated that "it would take 12.9 billion impoverished individuals living in India to have as much impact on the environment as 258 million Americans." The population of India comprises 16% of the world's population, while the population of the United States comprises only 4.7%. Suggest social and economic conditions in India that limit its population's impact on the environment. Consider, for example, why the per capita use of resources is lower in India and how the lack of industrialized technology may limit ecological degradation in that country.

17. Obtain from print or electronic sources a copy of the full version of the 2001 Canadian census survey and complete it. Discuss with two or three other students

(a) how the census provides government officials with valuable information on demographic issues;

(b) some major questions that were not asked but might have been.

 www.science.nelson.com

Extension

18. From electronic or print sources, locate an article relating to human population growth. Critically analyze the article, especially for validity, possible bias, and underlying assumptions. Based on your analysis, write a critical commentary of, or response to, this article. Answers to the following questions may help guide your critical analysis:

(a) Who is the author? From what perspective is he or she writing this article?

(b) What is the author's purpose in writing and publishing the article?

(c) Is it from a credible source? How do you know?

(d) What are the author's credentials? Is the author a member of a lobby group or an employee of a government or a company with an economic interest in the article topic?

(e) Does the author use scientific evidence to support what he or she is saying, or is the article mostly based on opinion? If it is supported by scientific evidence, is the evidence valid or authoritative? How do you know?

 www.science.nelson.com

PERFORMANCE TASK

Population Dynamics of Species at Risk

In this unit task, you will explore the population dynamics among populations of a species at risk, other organisms in its community, and humans. You will apply the skills and strategies of scientific inquiry (planning, recording, analyzing, and interpreting), culminating in an informed and creative presentation. The method of presentation should be determined in consultation with your teacher.

For the purpose of this task, it is important that you select a species at risk for which adequate population data is available. For example, population data for endangered species of large mammals and birds are often readily available while similar information for populations of small animal and plant species is often unknown. Also, note that you may focus on the status of a population in a particular country or geographic region rather than considering all populations of a species. For example, in many regions of Africa elephants are threatened while in others they are plentiful and their numbers are on the increase.

(a) **(b)**

(c) **(d)**

Figure 1
Some species at risk include the swift fox **(a)**, whooping cranes **(b)**, orca **(c)**, and the burrowing owl **(d)**.

You will be provided with a list of suggested species (**Figure 1**). After choosing one of these species, or after you have obtained teacher approval to research another species, you can begin your research assignment. The specific expectations for your research are outlined in **Table 1**. Details of the assessment criteria are also provided.

Table 1 Specific Expectations for Species at Risk Performance Task

Component	Specific expectations and suggestions
Characteristic of populations	• Research and describe the primary population characteristics of the chosen species, such as population density, distribution, and dispersion. • Describe the age structure and genetic diversity of the population. Is the population threatened with a loss in genetic diversity? • If possible, provide an estimate of the environment's carrying capacity for this species and a minimum viable size of the population. • Select an appropriate graphic organizer to summarize your findings.
Population growth	• Research and provide historical data (from past to present) on the estimated population size for the species. Include estimates of population size when the species was relatively plentiful; that is, before conditions resulted in a population decline. • Present the data in an appropriate graphical format. • Describe the pattern of growth exhibited by the species. • Extrapolate the future population size for five years assuming that the most recent patterns and trends continue. • List and describe possible limiting factors of the population that have contributed to the species' "at risk" status.
Species interactions	• Identify at least five other species with which the selected at risk species interacts. • Classify the types of interactions identified. • Do any of the species listed influence the population size and growth of the species at risk? If so, describe how. • Include illustrations, images, and diagrams.

Human influence	• How has the human population influenced the "at risk" status of the species?
	• Consider the possible role of the following:
	  - habitat destruction or encroachment
	  - water, air, and soil pollution
	  - commercial harvesting
	  - hunting
	  - climate change
	  - agricultural and industrial practices

Conservation biology	• Using the Internet and print resources, research and describe programs already underway to protect or enhance the viability of the species at risk. GO www.science.nelson.com • Formulate and propose an action plan to address the issues raised in your investigation. What actions could and should be taken to offer immediate protection for the species? What actions could and should be taken to ensure the long-term viability of the species? • You may be asked to present your plan as either a written or oral report before a classroom environmental review panel. • Document two careers closely associated with the monitoring, study, and/or conservation of this species.

Understanding Concepts

1. Define and give examples of populations that exhibit the following dispersion patterns:
   (a) clumped
   (b) random
   (c) uniform

2. Students sampled aquatic insect larvae living on a small section of river bottom measuring 2.0 m by 0.8 m. They found approximately 45 000 black fly larvae in their sample.
   (a) What was the population density of this species?
   (b) Estimate the number of black fly larvae living in a similar habitat of river bottom measuring 50 m by 10 m.

3. According to the 2001 census, the population of Canada had reached 30 007 094 people.
   (a) What is the population density of Canadians if Canada's land area is 9 976 000 km^2?
   (b) Using our population as an example, explain why the ecological density of a species is usually greater than its crude density.

4. Natality, mortality, immigration, and emigration are all terms related to any population.
   (a) Describe briefly what each term means.
   (b) Explain briefly how each process affects a population.
   (c) Which of the terms do not relate to a closed population? Explain why not.
   (d) Brainstorm several examples of closed populations that occur naturally. Contrast these with examples of closed populations produced by human intervention or other activities.

5. Calculate the absolute growth of the Ontario population between July 21, 2000 and June 30, 2001, given the following data from Statistics Canada. During that year there were:
   (i) 130 672 births
   (ii) 87 565 deaths
   (iii) 149 868 immigrants
   (iv) 32 156 emigrants

6. List five environmental resources for which there might be intraspecific competition. Provide examples of such competition.

7. Explain why small populations often experience relatively slow growth for several generations. On a sigmoidal curve, what name is given to this region of slow growth?

8. What conditions are necessary for a population to experience prolonged exponential growth?

9. Explain clearly what happens once a population reaches a dynamic equilibrium.

10. Classify each of the following as density dependent or density independent limiting factors:
    (a) fertilizer run-off resulting in the eutrophication of a freshwater body.
    (b) the impact of predators on a population of snow geese.
    (c) the spread of an infectious disease in a wild population.
    (d) affects of habitat loss on wild populations.

11. Explain, using an example for each, the differences between intraspecific and interspecific interactions.

12. Explain briefly the differences between passive and aggressive defense mechanisms.

13. What obvious defense mechanism is utilized by the sea urchin, *Strongleocentrotus franciacanus*, shown in **Figure 1**. Of what possible adaptive benefit(s) is its bright colouration?

Figure 1
Sea urchin

14. Describe and give two examples of each of the following:
    (a) commensalism
    (b) mutalism
    (c) parasitism

15. Explain why some examples of mutualism are obligatory and others are not.

16. Parasites are often classified according to where they live. What two terms are involved? Explain each.

17. Describe how each of the following factors influenced human population growth patterns on a global scale.
    (a) the distribution of native large seeded grasses
    (b) the availability of large mammal species and their ease of domestication
    (c) the scientific and industrial revolutions

18. In the past 500 years a decrease in death rates has had a far more dramatic impact on population growth than an increase in birth rates. What key technological innovations account for much of the drop in death rates?

19. Why do medical advances that extend the lives of elderly individuals not have a significant impact on population growth rates?

20. Suggest five or more reasons why high density housing developments put less strain on the environment than low density housing developments.

21. Some argue that water filtration systems which prepare drinking water must be designed to remove all types of contaminants. Others argue that contaminants must be prevented from reaching water supplies in the first place. Explain why the best approach involves both of these concepts.

22. Describe what aspects of your life would need to be assessed in order to establish your personal ecological footprint.

23. Use a triple Venn diagram to compare and contrast the three models of population growth: logistic growth, geometric growth, and exponential growth.

24. Explain how regional populations show more irregular changes in size than large global populations.

25. Discuss the observation that a small population may have greater effects on the local and the global environment than a large population. Support your answer with specific examples of human patterns of consumption.

Applying Inquiry Skills

26. Meadow voles, sometimes referred to as field mice, are extremely common small rodents that breed actively throughout the year with females becoming fertile at under two months of age. The data in **Table 1** represents the growth over time of a population of these voles living in a small grassland.

Table 1 Monthly Growth for a Population of Meadow Voles (Field Mice)

Month	Meadow vole population
December	3 920
January	5 488
February	7 683
March	10 756
April	15 058
May	21 081
June	29 513
July	41 318
August	57 845
September	80 983
October	113 376

(a) Make a graph of the data.
(b) Based on your graph, is the meadow vole population exhibiting logistic, geometric, or exponential growth? Explain.
(c) Based on your answer to (b), use the appropriate formula to estimate the size of the vole population for the next three months.

27. A fisheries biologist studied a population of cichlids in a 120-ha section of Lake Tanganyika in central Africa. She found that the fish lived only in scattered reed beds that accounted for 15% of the study area. Data were collected using the mark-recapture method and summarized in the table below.

Table 2 Mark-Recapture Analysis of Cichlids

Number of cichlids marked	185
Number of cichlids captured in second sample	208
Number of marked cichlids recaptured	13

(a) How many cichlids would you estimate to be in the study area?
(b) What is the density of the cichlid population per hectare?
(c) What is the dispersion pattern of the fish?
(d) How might the dispersion pattern affect your interpretation of their density?
(e) How might the interpretation of the mark-recapture data be affected if behavioural studies determined that this species of cichlid is highly territorial?

28. Analyze the graph of the growth curve representing the human population of Ireland before 1900 (**Figure 2**).
    (a) Label the different phases of this growth curve.
    (b) Suggest possible reasons for the decline in carrying capacity at time D.
    (c) Research the events that actually occurred in Ireland before 1900 that led to this decline in carrying capacity.

 www.science.nelson.com

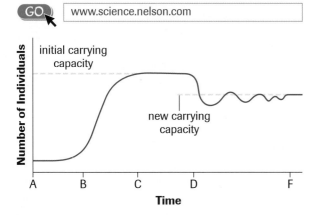

Figure 2

29. Examine the data in the table below which compares Canadian population statistics for the periods 1861–1871 and 1991–1996.

Table 3 Canadian Population Statistics

	Population at end of period (1000's)	Births (1000's)	Deaths (1000's)	Immigrants (1000's)	Emigrants (1000's)
1861–1871	3 689	1 370	760	260	410
1991–1996	29 672	1 936	1 024	1 137	229

(a) What was Canada's annual per capita birth and death rates in the period from 1861–1871?
(b) How did these values change during the next 130 years? Suggest reasons for these changes.
(c) Calculate the annual percent growth rate for these two time intervals.

Making Connections

30. Healthy, natural ecosystems play an important role in controlling the emergence and spread of infectious diseases by maintaining a balance between predators and prey, and parasites and their hosts. The deer tick, *Ixodes dammini*, feeds on white-tailed deer and occasionally bites people. Lyme disease is spread by the bite of ticks

that are carrying the bacteria *Borrelia burgdorferi*. Today, the only area in Ontario with a risk of Lyme disease is Long Point on the shore of Lake Erie. Research Lyme disease and answer the following:
    (a) Is Lyme disease a serious health concern? Justify your answer.
    (b) Is there a natural balance between predators and the deer population of Long Point?
    (c) What strategies in both wildlife management and human behaviour might reduce the risk of acquiring this disease?

 www.science.nelson.com

31. Some suggest that lakes without fish populations due to acid deposition can simply be stocked with young fish.
    (a) Who might be satisfied with this solution? Explain.
    (b) Explain why this action does not solve the problem.

32. Great gray owls, *Strix nebulosa*, are very vulnerable to human disturbance such as clearing of forests for timber or farmland. Their population dispersion and geographic range are most affected by the availability of nest sites. Other factors such as food supply, influence the carrying capacity of the local environment. These birds are top predators and feed heavily of mice and voles.
    (a) How might use of pesticides influence the health of gray owl populations?
    (b) How could forestry practices be altered to limit negative impacts on the range and dispersion patterns of these owls?
    (c) Many environmental organizations advocate taking action to ensure this and similar wildlife populations are given adequate protection. What reasons do you think these groups would offer in support of their position? Explain.

33. As water becomes scarce, the competition for water between cities and countryside intensifies. In this competition, farmers almost always lose. The demands for water use by people in cities must by balanced by the needs for water used for irrigation.
    (a) What water uses would you consider essential for city dwellers?
    (b) Outline several changes in lifestyle and/or technologies that could be used to reduce water demand in homes and industries.
    (c) How might farmers reduce their water demands? Consider alternative irrigation technologies as well as crop selection.

34. The Parakana tribe are an indigenous group of people living in the Xinga-Para region of the Amazon rain forest. They were first contacted by the outside world in 1970 and had an estimated population size of about 1000. In the years that followed they were relocated due to construction of the TransAmazonica Highway and a large hydroelectric project. They lost their traditional hunting and fishing lands as well as the sacred burial grounds of their ancestors. Today they number only 100 individuals.
    (a) What factors most likely contributed to their decline in numbers?
    (b) By international standards cultures are designated as "extinct" when the number of members is 100 or less. Do you think that this traditional culture can and will survive?
    (c) What group(s) of people likely benefited from the construction of the highway and power dam?
    (d) What ethical and moral issues arise when populations living in one region decide to utilize the resources of people living in a different part of the world?
    (e) How might such serious impacts be avoided or resolved before entire cultures are destroyed? Explain.

35. Research one CIDA funded project that invests in food-production technologies. For the project, identify the following:
    a) What specific problem(s) does the project address?
    b) What role does CIDA have in the project?
    c) Outline specific details of the project.
    d) Evaluate and discuss the potential of the project to enhance the long-term sustainability of the population and natural environment.

 www.science.nelson.com

Extension

36. Lake Victoria in central Africa was home to large populations of hundreds of different species of cichlid fishes. Although small in size they were a source of food for the local human population. In the 1960s the Nile perch was introduced into Lake Victoria in hopes of establishing a population of this much larger edible species. Research the outcome of this intentional introduction.
    (a) Describe the initial population growth rate of the perch population.
    (b) Describe the ecological relationships of the perch. What were its new sources of food? Did it have any predators?
    (c) What impacts did its introduction have on cichlid species?
    (d) Describe the long term success or failure of the introduction. Is it a valuable food source for the local human population? What were the long-term impacts on the native species inhabiting Lake Victoria?

 www.science.nelson.com

Appendixes

A1 Scientific Inquiry

Planning an Investigation

In our attempts to further our understanding of the natural world, we encounter questions, mysteries, or events that are not readily explainable. To develop explanations, we investigate using scientific inquiry. The methods used in scientific inquiry depend, to a large degree, on the purpose of the inquiry.

Controlled Experiments

A controlled experiment is an example of scientific inquiry in which an independent variable is purposefully and steadily changed to determine its effect on a second dependent variable. All other variables are controlled or kept constant. Controlled experiments are performed when the purpose of the inquiry is to create, test, or use a scientific concept.

The common components of controlled experiments are outlined below. *Note that there are normally many cycles through the steps during an actual experiment.*

Stating the Purpose

Every investigation in science has a purpose; for example,

- to develop a scientific concept (a theory, law, generalization, or definition);
- to test a scientific concept;
- to perform a chemical analysis;
- to determine a scientific constant; or
- to test an experimental design, a procedure, or a skill.

Determine which of these is the purpose of your investigation. Indicate your decision in a statement of the purpose.

Asking the Question

Your question forms the basis for your investigation: the investigation is designed to answer the question. Controlled experiments are about relationships, so the question could be about the effects on variable A when variable B is changed.

Hypothesizing/Predicting

A hypothesis is a tentative explanation. To be scientific, a hypothesis must be testable. Hypotheses can range in certainty from an educated guess to a concept that is widely accepted in the scientific community.

A prediction is based upon a hypothesis or a more established scientific explanation, such as a theory or a law. A prediction is a tentative answer to the question you are investigating. In the prediction you state what outcome you expect from your experiment.

Designing the Investigation

The design of a controlled experiment identifies how you plan to manipulate the independent variable, measure the response of the dependent variable, and control all the other variables in pursuit of an answer to your question. It is a summary of your plan for the experiment.

Gathering, Recording, and Organizing Observations

There are many ways to gather and record observations during your investigation. It is helpful to plan ahead and think about what data you will need to answer the question and how best to record them. This helps to clarify your thinking about the question posed at the beginning, the variables, the number of trials, the procedure, the materials, and your skills. It will also help you organize your evidence for easier analysis.

Analyzing the Observations

After thoroughly analyzing your observations, you may have sufficient and appropriate evidence to enable you to answer the question posed at the beginning of the investigation.

Evaluating the Evidence and the Hypothesis/Prediction

At this stage of the investigation, you evaluate the processes that you followed to plan and perform the investigation.

You will also evaluate the outcome of the investigation, which involves evaluating any prediction you made, and the hypothesis or more established concept the prediction was based on. You must identify and take into account any sources of error and uncertainty in your measurements.

Finally, compare the answer you predicted with the answer generated by analyzing the evidence. Is your hypothesis acceptable or not?

Reporting on the Investigation

In preparing your report, your objectives should be to describe your planning process and procedure clearly and in sufficient detail that the reader could repeat the experiment exactly as you performed it, and to report your observations, your analysis, and your evaluation of your experiment accurately and honestly.

Observational Studies

Often the purpose of inquiry is simply to study a natural phenomenon with the intention of gaining scientifically significant information to answer a question. Observational studies involve observing a subject or phenomenon in an unobtrusive or unstructured manner, often with no specific hypothesis. A hypothesis to describe or explain the observations may, however, be generated after repeated observations, and modified as new information is collected over time.

The stages and processes of scientific inquiry through observational studies are summarized below. *Note that there are normally many cycles through the steps during the actual study.*

Stating the Purpose

Choose a topic that interests you. Determine whether you are going to replicate or revise a previous study, or create a new one. Indicate your decision in a statement of the purpose.

Asking the Question

In planning an observational study, it is important to pose a general question about the natural world. You may or may not follow the question with the creation of a hypothesis.

Hypothesizing/Predicting

A hypothesis is a tentative explanation. In an observational study, a hypothesis can be formed after observations have been made and information gathered on a topic. A hypothesis may be created in the analysis.

Designing the Investigation

The design of an observational study describes how you will make observations relevant to the question.

Gathering, Recording, and Organizing Observations

There are many ways to gather and record observations during an investigation. During your observational study, you should quantify your observations where possible. All observations should be objective and unambiguous. Consider ways to organize your information for easier analysis.

Analyzing the Observations

After thoroughly analyzing your observations, you may have sufficient and appropriate evidence to enable you to answer the question posed at the beginning of the investigation. You may also have enough observations and information to form a hypothesis.

Evaluating the evidence and the Hypothesis

At this stage of the investigation, you will evaluate the processes used to plan and perform the investigation. Evaluating the processes includes evaluating the materials, the design, the procedure, and your skills. The results of most such investigations will suggest further studies, perhaps correlational studies or controlled experiments to explore tentative hypotheses you may have developed.

Reporting on the Investigation

In preparing your report, your objectives should be to describe your design and procedure accurately, and to report your observations accurately and honestly.

A2 Decision Making

Modern life is filled with environmental and social issues that have scientific and technological dimensions. An issue is defined as a problem that has at least two possible solutions rather than a single answer. There can be many positions, generally determined by the values that an individual or a society holds, on a single issue. Which solution is "best" is a matter of opinion; ideally, the solution that is implemented is the one that is most appropriate for society as a whole.

The common processes involved in the decision-making process are outlined below. *Note that you may go through several cycles before deciding you are ready to defend a decision.*

Defining the Issue

The first step in understanding an issue is to explain why it is an issue, describe the problems associated with the issue, and identify the individuals or groups, called stakeholders, involved in the issue. You could brainstorm the following questions to research the issue: Who? What? Where? When? Why? How? Develop background information on the issue by clarifying facts and concepts, and identifying relevant attributes, features, or characteristics of the problem.

Identifying Alternatives/Positions

Examine the issue and think of as many alternative solutions as you can. At this point it does not matter if the solutions seem unrealistic. To analyze the alternatives, you should examine the issue from a variety of perspectives. Stakeholders may bring different viewpoints to an issue and these may influence their position on the issue. Brainstorm or hypothesize how different stakeholders would feel about your alternatives. Perspectives that stakeholders may adopt while approaching an issue are listed in **Table 1**.

Researching the Issue

Formulate a research question that helps to limit, narrow, or define the issue. Then develop a plan to identify and find reliable and relevant sources of information. Outline the stages of your information search: gathering, sorting, evaluating, selecting, and integrating relevant information. You may consider using a flow chart, concept map, or other graphic organizer to outline the stages of your information search. Gather information from many sources, including newspapers, magazines, scientific journals, the Internet, and the library.

Analyzing the Issue

In this stage, you will analyze the issue in an attempt to clarify where you stand. First, you should establish criteria for evaluating your information to determine its relevance and significance. You can then evaluate your sources, determine what assumptions may have been made, and assess whether you have enough information to make your decision.

There are five steps that must be completed to effectively analyze the issue:

1. Establish criteria for determining the relevance and significance of the data you have gathered.

2. Evaluate the sources of information.

3. Identify and determine what assumptions have been made. Challenge unsupported evidence.

4. Determine any causal, sequential, or structural relationships associated with the issue.

5. Evaluate the alternative solutions, possibly by conducting a risk-benefit analysis.

Table 1 Some Possible Perspectives on an Issue

cultural	focused on customs and practices of a particular group
environmental	focused on effects on natural processes and other living things
economic	focused on the production, distribution, and consumption of wealth
educational	focused on the effects on learning
emotional	focused on feelings and emotions
aesthetic	focused on what is artistic, tasteful, beautiful
moral/ethical	focused on what is good/bad, right/wrong
legal	focused on rights and responsibilities
spiritual	focused on the effects on personal beliefs
political	focused on the aims of an identifiable group or party
scientific	focused on logic or the results of relevant inquiry
social	focused on effects on human relationships, the community
technological	focused on the use of machines and processes

Defending the Decision

After analyzing your information, you can answer your research question and take an informed position on the issue. You should be able to defend your preferred solution in an appropriate format—debate, class discussion, speech, position paper, multimedia presentation (e.g., computer slide show), video, brochure, poster, or other creative formats of your choice.

Your position on the issue must be justified using the supporting information that you have discovered in your research and tested in your analysis. You should be able to defend your position to people with different perspectives. In preparing for your defence, ask yourself the following questions:

- Do I have supporting evidence from a variety of sources?
- Can I state my position clearly?
- Do I have solid arguments (with solid evidence) supporting my position?
- Have I considered arguments against my position, and identified their faults?
- Have I analyzed the strong and weak points of each perspective?

Evaluating the Process

The final phase of decision making includes evaluating the decision the group reached, the process used to reach the decision, and the part you played in decision making. After a decision has been reached, carefully examine the thinking that led to the decision. Some questions to guide your evaluation follow:

- What was my initial perspective on the issue? How has my perspective changed since I first began to explore the issue?
- How did we make our decision? What process did we use? What steps did we follow?
- In what ways does our decision resolve the issue?
- What are the likely short- and long-term effects of our decision?
- To what extent am I satisfied with our decision?
- What reasons would I give to explain our decision?
- If we had to make this decision again, what would I do differently?

Using a Risk–Benefit Analysis Model

Risk–benefit analysis is a tool used to organize and analyze information gathered in research. A thorough analysis of the risks and benefits associated with each alternative solution can help you decide on the best alternative.

- Research as many aspects of the proposal as possible. Look at it from different perspectives.
- Collect as much evidence as you can, including reasonable projections of likely outcomes if the proposal is adopted.
- Classify every individual potential result as being either a benefit or a risk.
- Quantify the size of the potential benefit or risk (perhaps as a dollar figure, or a number of lives affected, or in severity on a scale of 1 to 5).
- Estimate the probability (percentage) of that event occurring.
- By multiplying the size of a benefit (or risk) by the probability of its happening, you can assign a significance value for each potential result.
- Total the significance values of all the potential risks and all the potential benefits, and compare the sums to help you decide whether to accept the proposed action.

Note that although you should try to be objective in your assessment, your beliefs will have an effect on the outcome—two people, even if using the same information and the same tools, could come to a different conclusion about the balance of risk and benefit for any proposed solution to an issue.

A3 Lab Reports

When carrying out investigations, it is important that scientists keep records of their plans and results, and share their findings. In order to have their investigations repeated (replicated) and accepted by the scientific community, scientists generally share their work by publishing papers in which details of their design, materials, procedure, evidence, analysis, and evaluation are given.

Lab reports are prepared after an investigation is completed. To ensure that you can accurately describe the investigation, it is important to keep thorough and accurate records of your activities as you carry out the investigation.

Investigators use a similar format in their final reports or lab books, although the headings and order may vary. Your lab book or report should reflect the type of scientific inquiry that you used in the investigation and should be based on the following headings, as appropriate.

Title

At the beginning of your report, write the number and title of your investigation. In this course the title is usually given, but if you are designing your own investigation, create a title that suggests what the investigation is about. Include the date the investigation was conducted and the names of all lab partners (if you worked as a team).

Purpose

State the purpose of the investigation. Why are you doing this investigation?

Question

This is the question that you attempted to answer in the investigation. If it is appropriate to do so, state the question in terms of independent and dependent variables.

Hypothesis/Prediction

Based on your reasoning or on a concept that you have studied, formulate an explanation for what should happen (a hypothesis). From your hypothesis you may make a prediction, a statement of what you expect to observe, before carrying out the investigation. Depending on the nature of your investigation, you may or may not have a hypothesis or a prediction.

Experimental Design

This is a brief general overview (one to three sentences) of what was done. If your investigation involved independent, dependent, and controlled variables, list them. Identify any control or control group that was used in the investigation.

Materials

This is a detailed list of all materials used, including sizes and quantities where appropriate. Be sure to include safety equipment such as eye protection, lab apron, latex gloves, and tongs, where needed. Draw a diagram to show any complicated setup of apparatus.

Procedure

Describe, in detailed, numbered steps, the procedure you followed to carry out your investigation. Include steps to clean up and dispose of waste.

Observations

This includes all qualitative and quantitative observations you made. Be as precise as possible when describing quantitative observations. Include any unexpected observations and present your information in a form that is easily understood. If you have only a few observations, this could be a list; for controlled experiments and for many observations, a table will be more appropriate.

Analysis

Interpret your observations and present the evidence in the form of tables, graphs, or illustrations, each with a title. Include any calculations, the results of which can be shown in a table. Make statements about any patterns or trends you observed. Conclude the analysis with a statement based only on the evidence you have gathered, answering the question that initiated the investigation.

Evaluation

The evaluation is your judgment about the quality of evidence obtained and about the validity of the prediction and hypothesis (if present). This section can be divided into two parts:

- Did your observations provide reliable and valid evidence to enable you to answer the question? Are you confident enough in the evidence to use it to evaluate any prediction and/or hypothesis you made?

- Was the prediction you made before the investigation supported or falsified by the evidence? Based on your evaluation of the evidence and prediction, is the hypothesis you used to make your prediction supported, or should it be rejected?

The leading questions that follow should help you through the process of evaluation.

Evaluation of the Experiment

1. Were you able to answer the question using the chosen experimental design? Are there any obvious flaws in the design? What alternative designs (better or worse) are available? As far as you know, is this design the best available in terms of controls, efficiency, and cost? How great is your confidence in the chosen design?

   You may sum up your conclusions about the design in a statement like: "The experimental design [name or describe in a few words] is judged to be adequate/inadequate because…"

2. Were the steps that you used in the laboratory correctly sequenced, and adequate to gather sufficient evidence? What improvements could be made to the procedure? What steps, if not done correctly, would have significantly affected the results?

   Sum up your conclusions about the procedure in a statement like: "The procedure is judged to be adequate/inadequate because…"

3. Which specialized skills, if any, might have the greatest effect on the experimental results? Was the evidence from repeated trials reasonably similar? Can the measurements be made more precise?

   Sum up your conclusions: "The technological skills are judged to be adequate/inadequate because…"

4. You should now be ready to sum up your evaluation of the experiment. Do you have enough confidence in your experimental results to proceed with your evaluation of the hypothesis being tested? Based on uncertainties and errors you have identified in the course of your evaluation, what would be an acceptable percent difference for this experiment (1%, 5%, or 10%)?

State your confidence level in a summary statement: "Based upon my evaluation of the experiment, I am not certain/I am moderately certain/I am very certain of my experimental results. The major sources of uncertainty or error are…"

Evaluation of the Prediction

1. Calculate the percent difference for your experiment.

$$\% \text{ difference} = \frac{|\text{experimental value}| - |\text{predicted value}|}{|\text{predicted value}|} \times 100\%$$

   How does the percent difference compare with your estimated total uncertainty (i.e. is the percent difference greater or smaller than the difference you've judged acceptable for this experiment)? Does the predicted answer clearly agree with the experimental answer in your analysis? Can the percent difference be accounted for by the sources of uncertainty listed earlier in the evaluation?

   Sum up your evaluation of the prediction: "The prediction is judged to be verified/inconclusive/falsified because…"

2. If the prediction was verified, the hypothesis behind it is supported by the experiment. If the results of the experiment were inconclusive or the prediction was falsified, then doubt is cast upon the hypothesis. How confident do you feel about any judgment you can make based on the experiment? Is there a need for a new or revised hypothesis, or to restrict, reverse, or replace the hypothesis being tested?

   Sum up your evaluation: "[The hypothesis] being tested is judged to be acceptable/unacceptable because…"

A4 Graphic Organizers

Graphic organizers such as those outlined in this section can help you to solidify your understanding of a topic, and assist you in formulating a clear, concise answer.

PMI Chart

A PMI chart is used to examine both sides of an issue. Positive aspects of a topic or issue are recorded in the P (plus) column. Negative aspects are recorded in the M (minus) column. Interesting or controversial questions are recorded in the I (interesting) column (**Table 1**).

Table 1: A PMI Chart

P	M	I

Table 2: A KWL Chart

K	W	L

KWL Chart

A KWL chart can help you identify prior knowledge and experience, decide what new information you want to learn about, and reflect on your learning. Before you begin a new concept, lesson, or unit, list what you know about a topic in the K column and what you want to know in the W column. After studying the new topic, list what you learned in the L column (**Table 2**).

Venn Diagram

A Venn diagram is used to show similarities and differences in two or more concepts. Write all similarities between the concepts in the overlapping section of the circles and all unique traits of each concept in the nonoverlapping parts of the appropriate circles (**Figure 1**).

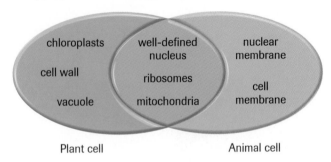

Figure 1
Venn diagram: plant and animal cells

Fishbone Diagram

A fishbone diagram is used to identify separate causes and effects. In the head of the fish, identify the effect, topic, or result. At the end of each major bone, identify the major

subtopics or categories. On the minor bones that attach to each major bone, add details about the subtopics or possible causes of each effect or result (**Figure 2**).

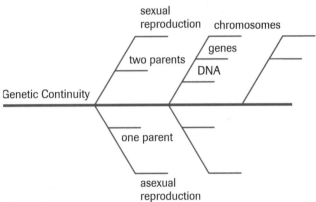

Figure 2
Fishbone diagram: genetic continuity

The Concept Map

Concept maps are used to show connections between ideas and concepts, using words or visuals. Put the central idea in the middle of a sheet of paper. Organize the ideas most closely related to each other around the centre. Draw arrows between the ideas that are related. On each arrow, write a short description of how the terms are related to each other (**Figure 3**).

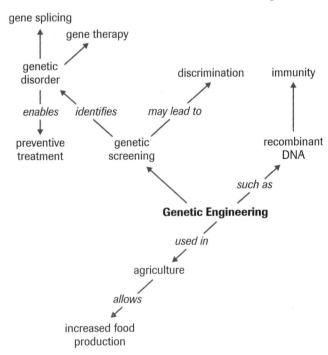

Figure 3
Concept map: genetic engineering

A5 Math Skills

Scientific Notation

It is difficult to work with very large or very small numbers when they are written in common decimal notation. Usually it is possible to accommodate such numbers by changing the SI prefix so that the number falls between 0.1 and 1000; for example, 237 000 000 mm can be expressed as 237 km and 0.000 000 895 kg can be expressed as 0.895 mg. However, this prefix change is not always possible, either because an appropriate prefix does not exist or because it is essential to use a particular unit of measurement. In these cases, the best method of dealing with very large and very small numbers is to write them using scientific notation. Scientific notation expresses a number by writing it in the form $a \times 10^n$, where $1 < |a| < 10$ and the digits in the coefficient a are all significant. **Table 1** shows situations where scientific notation would be used.

Table 1 Examples of Scientific Notation

Expression	Common decimal notation	Scientific notation
124.5 million kilometres	124 500 000 km	1.245×10^8 km
154 thousand picometres	154 000 pm	1.54×10^{-5} pm
602 sextillion /mol	602 000 000 000 000 000 000 000 /mol	6.02×10^{23}/mol

To multiply numbers in scientific notation, multiply the coefficients and add the exponents; the answer is expressed in scientific notation. Note that when writing a number in scientific notation, the coefficient should be between 1 and 10 and should be rounded to the same certainty (number of significant digits) as the measurement with the least certainty (fewest number of significant digits). Look at the following examples:

$$(4.73 \times 10^5 \text{ m})(5.82 \times 10^7 \text{ m}) = 27.5 \times 10^{12} \text{ m}^2 = 2.75 \times 10^{13} \text{ m}^2$$
$$(3.9 \times 10^4 \text{ N})(5.3 \times 10^{-3} \text{ m}) = 21 \times 10^1 \text{ N·m} = 2.1 \times 10^2 \text{ N·m}$$

On many calculators, scientific notation is entered using a special key, labelled EXP or EE. This key includes "$\times 10$" from the scientific notation; you need to enter only the exponent. For example, to enter

7.5×10^4	press	7.5 EXP 4
3.6×10^{-3}	press	3.6 EXP +/−3

Logarithms

In the exponential equation $y = a^n$, a is the base and n is the exponent. $y = a^n$ can be written as $\log_a y = n$ ($a > 0$ and $a \neq 1$) and is read as "the logarithm of y with base a is equal to n." For example, $10^2 = 100$ can be written as $\log_{10} 100 = 2$.

On many scientific calculators, the key "LOG" calculates the logarithm of a number with base 10. For example, to enter

$\log_{10} 2$	press	LOG 2

Logarithm of a product is one of the logarithm laws. The law is as follows:

$$\log_a(mn) = \log_a m + \log_a n$$

This law is useful when calculating the pH of solutions. The definition of pH is the negative logarithm of the hydronium ion concentration, $-\log_{10}[H_3O^+_{(aq)}]$, where the concentration is measured in moles per litre of solution (mol/L).

In pure water at 25°C, the $H_3O^+_{(aq)}$ is 1.0×10^{-7} mol/L.

$$\begin{aligned}
\text{pH} &= -\log_{10}[H_3O^+_{(aq)}] \\
&= -\log_{10}(1.0 \times 10^{-7}) \\
&= -((\log_{10} 1.0) + \log_{10}(10^{-7})) \\
&= -(0 + (-7)) \\
&= 7
\end{aligned}$$

Therefore, the pH of pure water is 7.

Uncertainty in Measurements

There are two types of quantities that are used in science: exact values and measurements. Exact values include defined quantities (1 m = 100 cm) and counted values (5 cars in a parking lot). Measurements, however, are not exact because there is some uncertainty or error associated with every measurement.

There are two types of measurement error. **Random error** results when an estimate is made to obtain the last significant figure for any measurement. The size of the random error is determined by the precision of the measuring instrument. For example, when measuring length, it is necessary to estimate between the marks on the measuring tape. If these marks are 1 cm apart, the random error will be greater and the precision will be less than if the marks are 1 mm apart.

Systematic error is associated with an inherent problem with the measuring system, such as the presence of an interfering substance, incorrect calibration, or room conditions. For example, if the balance is not zeroed at the beginning, all measurements will have a systematic error; if using a metre stick that has been worn slightly, all measurements will contain an error.

The precision of measurements depends upon the gradations of the measuring device. **Precision** is the place value of the last measurable digit. For example, a measurement of 12.74 cm is more precise than a measurement of 127.4 cm because the first value was measured to hundredths of a centimetre whereas the latter was measured to tenths of a centimetre.

When adding or subtracting measurements of different precision, the answer is rounded to the same precision as the least precise measurement. For example, using a calculator, add

$$11.7 \text{ cm} + 3.29 \text{ cm} + 0.542 \text{ cm} = 15.532 \text{ cm}$$

The answer must be rounded to 15.5 cm because the first measurement limits the precision to a tenth of a centimetre.

No matter how precise a measurement is, it still may not be accurate. **Accuracy** refers to how close a value is to its true value. The comparison of the two values can be expressed as a percentage difference. The percentage difference is calculated as:

$$\% \text{ difference} = \frac{|\text{experimental value} - \text{predicted value}|}{\text{predicted value}} \times 100\%$$

Figure 1 shows an analogy between precision and accuracy, and the positions of darts thrown at a dartboard.

How certain you are about a measurement depends on two factors: the precision of the instrument used and the size of the measured quantity. More precise instruments give more certain values. For example, a mass measurement of 13 g is less precise than a measurement of 12.76 g; you are more certain about the second measurement than the first. Certainty also depends on the measurement. For example, consider the measurements 0.4 cm and 15.9 cm; both have the same precision. However, if the measuring instrument is precise to ± 0.1 cm, the first measurement is 0.4 ± 0.1 cm (0.3 cm or 0.5 cm) or an error of 25%, whereas the second measurement could be 15.9 ± 0.1 cm (15.8 cm or 16.0 cm) for an error of 0.6%. For both factors—the precision of the instrument used and the value of the measured quantity—the more digits there are in a measurement, the more certain you are about the measurement.

Significant Digits

The certainty of any measurement is communicated by the number of significant digits in the measurement. In a measured or calculated value, significant digits are the digits that are certain plus one estimated (uncertain) digit. Significant digits include all digits correctly reported from a measurement.

Follow these rules to decide if a digit is significant:

1. If a decimal point is present, zeros to the left of the first non-zero digit (leading zeros) are not significant.

2. If a decimal point is not present, zeros to the right of the last non-zero digit (trailing zeros) are not significant.

3. All other digits are significant.

4. When a measurement is written in scientific notation, all digits in the coefficient are significant.

5. Counted and defined values have infinite significant digits.

Table 2 shows some examples of significant digits.

Table 2 Certainty in Significant Digits

Measurement	Number of significant digits
32.07 m	4
0.0041 g	2
5×10^5 kg	1
6400 s	2
100 people	3

(a) 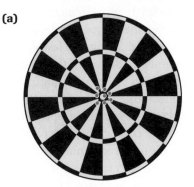 **(b)** 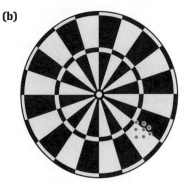 **(c)**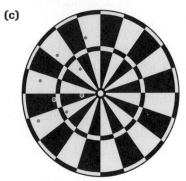

Figure 1
The positions of the darts in each of these figures are analogous to measured or calculated results in a laboratory setting. The results in **(a)** are precise and accurate, in **(b)** they are precise but not accurate, and in **(c)** they are neither precise nor accurate.

An answer obtained by multiplying and/or dividing measurements is rounded to the same number of significant digits as the measurement with the fewest number of significant digits. For example, we could use a calculator to solve the following equation:

77.8 km/h $\times$ 0.8967 h = 69.76326 km

However, the certainty of the answer is limited to three significant digits, so the answer is rounded up to 69.8 km.

Rounding Off

The following rules should be used when rounding answers to calculations.

1. When the first digit discarded is less than five, the last digit retained should not be changed.

    3.141 326 rounded to 4 digits is 3.141

2. When the first digit discarded is greater than five, or if it is a five followed by at least one digit other than zero, the last digit retained is increased by 1 unit.

    2.213 724 rounded to 4 digits is 2.214

    4.168 501 rounded to 4 digits is 4.169

3. When the first digit discarded is five followed by only zeros, the last digit retained is increased by 1 if it is odd, but not changed if it is even.

    2.35 rounded to 2 digits is 2.4

    2.45 rounded to 2 digits is 2.4

    −6.35 rounded to 2 digits is −6.4

Measuring and Estimating

Many people believe that all measurements are reliable (consistent over many trials), precise (to as many decimal places as possible), and accurate (representing the actual value). But there are many things that can go wrong when measuring.

*   There may be limitations that make the instrument or its use unreliable (inconsistent).

*   The investigator may make a mistake or fail to follow the correct techniques when reading the measurement to the available precision (number of decimal places).

*   The instrument may be faulty or inaccurate; a similar instrument may give different readings.

For example, when measuring the temperature of a liquid, it is important to keep the thermometer at the proper depth and the bulb of the thermometer away from the bottom and sides of the container. If you sit a thermometer with its bulb at the bottom of a liquid-filled container, you will be measuring

the temperature of the bottom of the container and not the temperature of the liquid. There are similar concerns with other measurements.

To be sure that you have measured correctly, you should repeat your measurements at least three times. If your measurements appear to be reliable, calculate the mean and use that value. To be more certain about the accuracy, repeat the measurements with a different instrument.

Every measurement is a best estimate of the actual value. The measuring instrument and the skill of the investigator determine the certainty and the precision of the measurement. The usual rule is to make a measurement that estimates between the smallest divisions on the scale of the instrument.

Probability

In scientific investigations, probability is a measure of the likelihood of a specific event occurring and is usually expressed as a number between 0 and 1. A probability of 0 means the event will not occur; a probability of 1 means the event will definitely occur. Probabilities may also be expressed as fractions or as percents.

There are two types of probability: theoretical probability and experimental probability. Theoretical probability is the likelihood of an event occurring based on the information known about certain conditions. This is an expectation.

$$\text{theoretical probability} = \frac{\text{number of desired outcomes}}{\text{total number of possible outcomes}}$$

Example

Black fur colour is a dominant trait in guinea pigs, while white fur colour is a recessive trait. What is the theoretical probability of a pair of heterozygous black guinea pigs (*Bb*) producing offspring with white fur (*bb*)?

Using a Punnett square (**Figure 2**), we show that if four offspring were produced, it is expected that three would have black fur and one would have white fur.

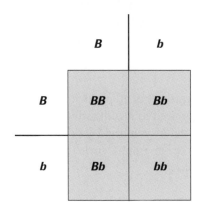

Figure 2
Punnett square showing a *Bb* × *Bb* cross

$$\text{theoretical probability} = \frac{\text{number of desired outcomes}}{\text{total number of possible outcomes}}$$

$$= \frac{\text{number of offspring with white fur}}{\text{total number of possible offspring}}$$

$$= \frac{1}{4}$$

$$= 0.25$$

$$= 25\%$$

The theoretical probability of producing white offspring is 1/4 or 25%. So if a litter had eight offspring, then you could expect two to have white fur.

Experimental probability is based on the recorded outcomes or events of an investigation. The more often an experiment is repeated or the more observations made, the closer the experimental probability will be to the theoretical probability.

$$\text{experimental probability} = \frac{\text{number of desired outcomes observed}}{\text{total number of observations}}$$

Example

Black fur colour is a dominant trait in guinea pigs, while white fur colour is a recessive trait. Two heterozygous black guinea pigs (*Bb*) were crossed. The litter contained six offspring with black fur and one with white fur. What is the experimental probability of producing offspring with white fur?

$$\text{experimental probability} = \frac{\text{number of desired outcomes observed}}{\text{total number of observations}}$$

$$= \frac{\text{number of offspring with white fur}}{\text{total number of possible offspring}}$$

$$= \frac{1}{7}$$

$$\doteq 0.14$$

$$= 14\%$$

The experimental probability of having offspring with white fur is 14%. If you performed the same analysis on a large number of litters, you would expect the experimental probability to be the same as (or very close to) the theoretical probability.

Graphs

There are many types of graphs that you can use to organize your data. You need to identify which type of graph is best for your data before you begin graphing. Three of the most useful kinds are bar graphs, circle (pie) graphs, and point-and-line graphs.

Bar Graphs

When at least one of the variables is qualitative, use a bar graph to organize your data (**Figure 3**). For example, a bar graph would be a good way to present the data collected from a study of the number of plants (quantitative) and the type of plants found (qualitative) in a local nursery. In this graph, each bar stands for a different category, in this case a type of plant.

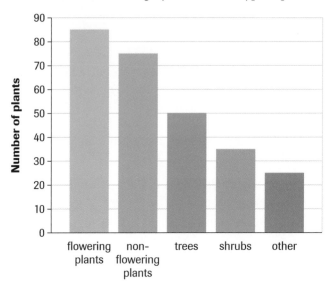

Figure 3
Bar graph

Circle Graphs

Circle graphs and bar graphs are used for similar types of data. A circle graph is used if the quantitative variable can be changed to a percentage of a total quantity (**Figure 4**). For example, if you surveyed a local nursery to determine the types of plants found and the number of each, you could make a circle graph. Each piece in the graph stands for a different category (e.g., the type of plant). The size of each piece is determined by the percentage of the total that belongs in each category (e.g., the percentage of plants of a particular type).

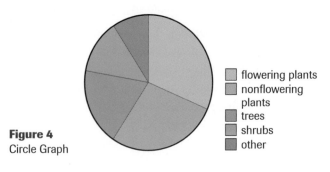

Figure 4
Circle Graph

Point-and-Line Graphs

When both variables are quantitative, use a point-and-line graph. For example, we can use the following guidelines and the data in **Table 3** to construct the point-and-line graph shown in **Figure 5**.

Table 3 Number of Brine Shrimp Eggs Hatched in Salt Solutions of Various Concentrations

Day	2% salt	4% salt	6% salt	8% salt
1	0	0	0	1
2	0	11	2	3
3	0	14	8	5
4	2	20	17	8
5	5	37	25	15
6	6	51	37	31

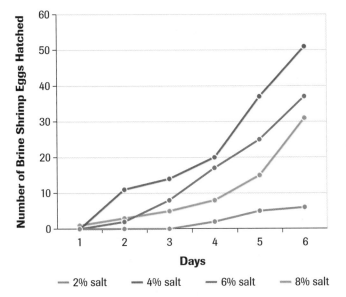

Figure 5
Point-and-line graph

1. Use graph paper and construct your graph on a grid. The horizontal edge on the bottom of this grid is the *x*-axis and the vertical edge on the left is the *y*-axis. Do not be too thrifty with graph paper—a larger graph is easier to interpret.

2. Decide which variable goes on which axis and label each axis, including the units of measurement. The independent variable is generally plotted along the *x*-axis and the dependent variable along the *y*-axis. The exception to this is when you plot a variable against time: regardless of which is the independent or dependent variable, always plot time on the *x*-axis. This convention ensures that the slope of the graph always represents a rate.

3. Title your graph. The title should be a concise description of the data contained in the graph.

4. Determine the range of values for each variable. The range is the difference between the largest and

smallest values. Graphs often include a little extra length on each axis, to make them appear less cramped.

5. Choose a scale for each axis. This will depend on how much space you have and the range of values for each axis. Each line on the grid usually increases steadily in value by a convenient number, such as 1, 2, 5, 10, or 50.

6. Plot the points. Start with the first pair of values, which may or may not be at the origin of the graph.

7. After all the points are plotted, draw a line through the points to show the relationship between the variables, if possible. Not all points may lie exactly on a line; small errors in each measurement may have occurred and moved the points away from the perfect line. Draw a line that comes closest to most of the points. This is called the line of best fit—a smooth line that passes through or between the points so that there are about the same number of points on each side of the line. The line of best fit may be straight or curved.

8. If you are plotting more than one set of data on one graph, use different colours or symbols to indicate the different sets, and include a legend (**Figure 5**).

Spreadsheets

A spreadsheet is a useful tool for creating different graph types such as column, bar, line, and pie to display data.

The following steps show how the data for the plants in a local nursery can be used to create a column (or bar) graph.

Step 1
Enter the data in the spreadsheet.

Step 2
Select the graph type that will display the data in the most appropriate form.

Step 3
Select the data from the spreadsheet that will be plotted on the graph.

Step 4
Enter the remaining graph information, such as the graph title and axes titles.

Step 5
Indicate where the graph is to be located in the spreadsheet. It can be included on the same sheet as the data or on a new sheet.

B1 Safety Conventions and Symbols

Although every effort is undertaken to make the science experience a safe one, there are inherent risks associated with some scientific investigations. These risks are generally associated with the materials and equipment used, and the disregard of safety instructions that accompany investigations. However, there may also be risks associated with the location of the investigation, whether in the science laboratory, at home, or outdoors. Most of these risks pose no more danger than one would normally experience in everyday life. With an awareness of the possible hazards, knowledge of the rules, appropriate behaviour, and a little common sense, these risks can be practically eliminated.

Remember, you share the responsibility not only for your own safety, but also for the safety of those around you. Always alert the teacher in case of an accident.

In this text, chemicals, equipment, and procedures that are hazardous are highlighted in red and are preceded by the appropriate Workplace Hazardous Materials Information System (WHMIS) symbol or by .

WHMIS Symbols and HHPS

The Workplace Hazardous Materials Information System (WHMIS) provides workers and students with complete and accurate information regarding hazardous products. All chemical products supplied to schools, businesses, and industries must contain standardized labels and be accompanied by Material Safety Data sheets (MSDS) providing detailed information about the product. Clear and standardized labelling is an important component of WHMIS (**Table 1**). These labels must be present on the product's original container or be added to other containers if the product is transferred.

The Canadian Hazardous Products Act requires manufacturers of consumer products containing chemicals to include a symbol specifying both the nature of the primary hazard and the degree of this hazard. In addition, any secondary hazards, first aid treatment, storage, and disposal must be noted. Household Hazardous Product Symbols (HHPS) are used to show the hazard and the degree of the hazard by the type of border surrounding the illustration (**Figure 1**).

	Corrosive
	This material can burn your skin and eyes. If you swallow it, it will damage your throat and stomach.
	Flammable
	This product or the gas (or vapour) from it can catch fire quickly. Keep this product away from heat, flames, and sparks.
	Explosive
	Container will explode if it is heated or if a hole is punched in it. Metal or plastic can fly out and hurt your eyes and other parts of your body.
	Poison
	If you swallow or lick this product, you could become very sick or die. Some products with this symbol on the label can hurt you even if you breathe (or inhale) them.

Danger

Warning

Caution

Figure 1
Hazardous household product symbols

Table 1 The Workplace Hazardous Materials Information System (WHMIS)

Class and type of compounds	WHMIS symbol	Risks	Precautions
Class A: *Compressed Gas* Material that is normally gaseous and kept in a pressurized container		• could explode due to pressure • could explode if heated or dropped • possible hazard from both the force of explosion and the release of contents	• ensure container is always secured • store in designated areas • do not drop or allow to fall
Class B: *Flammable and Combustible Materials* Materials that will continue to burn after being exposed to a flame or other ignition source		• may ignite spontaneously • may release flammable products if allowed to degrade or when exposed to water	• store in properly designated areas • work in well-ventilated areas • avoid heating • avoid sparks and flames • ensure that electrical sources are safe
Class C: *Oxidizing Materials* Materials that can cause other materials to burn or support combustion		• can cause skin or eye burns • increase fire and explosion hazards • may cause combustibles to explode or react violently	• store away from combustibles • wear body, hand, face, and eye protection • store in proper container that will not rust or oxidize
Class D: *Toxic Materials Immediate and Severe* Poisons and potentially fatal materials that cause immediate and severe harm		• may be fatal if ingested or inhaled • may be absorbed through the skin • small volumes have a toxic effect	• avoid breathing dust or vapours • avoid contact with skin or eyes • wear protective clothing, and face and eye protection • work in well-ventilated areas and wear breathing protection
Class D: *Toxic Materials Long Term Concealed* Materials that have a harmful effect after repeated exposures or over a long period		• may cause death or permanent injury • may cause birth defects or sterility • may cause cancer • may be sensitizers causing allergies	• wear appropriate personal protection • work in a well-ventilated area • store in appropriate designated areas • avoid direct contact • use hand, body, face, and eye protection • ensure respiratory and body protection is appropriate for the specific hazard
Class D: *Biohazardous Infectious Materials* Infectious agents or a biological toxin causing a serious disease or death		• may cause anaphylactic shock • includes viruses, yeasts, moulds, bacteria, and parasites that affect humans • includes fluids containing toxic products • includes cellular components	• special training is required to handle materials • work in designated biological areas with appropriate engineering controls • avoid forming aerosols • avoid breathing vapours • avoid contamination of people and/or area • store in special designated areas
Class E: *Corrosive Materials* Materials that react with metals and living tissue		• eye and skin irritation on exposure • severe burns/tissue damage on longer exposure • lung damage if inhaled • may cause blindness if contacts eyes • environmental damage from fumes	• wear body, hand, face, and eye protection • use breathing apparatus • ensure protective equipment is appropriate • work in a well-ventilated area • avoid all direct body contact • use appropriate storage containers and ensure proper non-venting closures
Class F: *Dangerously Reactive Materials* Materials that may have unexpected reactions		• may react with water • may be chemically unstable • may explode if exposed to shock or heat • may release toxic or flammable vapours • may vigorously polymerize • may burn unexpectedly	• handle with care, avoiding vibration, shocks, and sudden temperature changes • store in appropriate containers • ensure storage containers are sealed • store and work in designated areas

B2 Safety in the Laboratory

General Safety Rules

Safety in the laboratory is an attitude and a habit more than it is a set of rules. It is easier to prevent accidents than to deal with the consequences of an accident. Most of the following rules are common sense:

- Do not enter a laboratory unless a teacher or other supervisor is present, or you have permission to do so.
- Familiarize yourself with your school's safety regulations.
- Make your teacher aware of any allergies and other health problems you may have.
- Wear eye protection, lab aprons or coats, and gloves when appropriate.
- Wear closed shoes (not sandals) when working in the laboratory.
- Place your books and bags away from the work area. Keep your work area clear of all materials except those that you will use in the investigation.
- Do not chew gum, eat, or drink in the laboratory. Food should not be stored in refrigerators in laboratories.
- Know the location of MSDS information, exits, and all safety equipment, such as the fire blanket, fire extinguisher, and eyewash station.
- Avoid sudden or rapid motion in the laboratory that may interfere with someone carrying or working with chemicals or using sharp instruments.
- Never engage in horseplay or practical jokes in the laboratory.
- Ask for assistance when you are not sure how to do a procedural step.
- Never attempt any unauthorized experiments.
- Never work in a crowded area or alone in the laboratory.
- Always wash your hands with soap and water before and after you leave the laboratory. Definitely wash your hands before you touch any food.
- Use stands, clamps, and holders to secure any potentially dangerous or fragile equipment that could be tipped over.
- Do not taste any substance in a laboratory.
- Never smell chemicals unless specifically instructed to do so by the teacher. Do not inhale the vapours, or gas, directly from the container. Take a deep breath to fill your lungs with air, then waft or fan the vapours toward your nose.

- Clean up all spills, even water spills, immediately.
- If you are using a microscope with a mirror, never direct the mirror to sunlight. The concentrated reflected light could hurt your eyes badly.
- Do not forget safety procedures when you leave the laboratory. Accidents can also occur outdoors, at home, and at work.

Eye and Face Safety

- Always wear approved eye protection in a laboratory, no matter how simple or safe the task appears to be. Keep the safety glasses over your eyes, not on top of your head. For certain experiments, full face protection may be necessary.
- If you must wear contact lenses in the laboratory, be extra careful; whether or not you wear contact lenses, do not touch your eyes without first washing your hands. If you do wear contact lenses, make sure that your teacher is aware of it. Carry your lens case and a pair of glasses with you.
- Do not stare directly at any bright source of light (e.g., a burning magnesium ribbon, lasers, the Sun). You will not feel any pain if your retina is being damaged by intense radiation. You cannot rely on the sensation of pain to protect you.
- Never look directly into the opening of flasks or test tubes.

Handling Glassware Safely

- Never use glassware that is cracked or chipped. Give such glassware to your teacher or dispose of it as directed. Do not put the item back into circulation.
- Never pick up broken glassware with your fingers. Use a broom and dustpan.
- Do not put broken glassware into garbage containers. Dispose of glass fragments in special containers marked "Broken Glass."
- Heat glassware only if it is approved for heating. Check with your teacher before heating any glassware.
- Be very careful when cleaning glassware. There is an increased risk of breakage from dropping when the glassware is wet and slippery.

- If you need to insert glass tubing or a thermometer into a rubber stopper, get a cork borer of a suitable size. Insert the borer in the hole of the rubber stopper, starting from the small end of the stopper. Once the borer is pushed all the way through the hole, insert the tubing or thermometer through the borer. Ease the borer out of the hole, leaving the tubing or thermometer inside. To remove the tubing or thermometer from the stopper, push the borer from the small end through the stopper until it shows at the other end. Ease the tubing or thermometer out of the borer.

- Protect your hands with heavy gloves or several layers of cloth before inserting glass into rubber stoppers.

Using Sharp Instruments Safely

- Make sure your instruments are sharp. Surprisingly, one of the main causes of accidents with cutting instruments is using a dull instrument. Dull cutting instruments require more pressure than sharp instruments and are, therefore, much more likely to slip.

- Always transport a scalpel in a dissection case or box. Never carry the scalpel from one area of the laboratory to another with an exposed blade.

- Select the appropriate instrument for the task. Never use a knife when scissors would work best.

- Always cut away from yourself and others.

Fire Safety

- Immediately inform your teacher of any fires. Very small fires in a container may be extinguished by covering the container with a wet paper towel or a ceramic square to cut off the supply of air. Alternatively, sand may be used to smother small fires. A bucket of sand with a scoop should be available in the laboratory.

- If anyone's clothes or hair catch fire, tell the person to drop to the floor and roll. Then use a fire blanket to help smother the flames. Never wrap the blanket around a person on fire; the chimney effect will burn the lungs. For larger fires, immediately evacuate the area. Call the office or sound the fire alarm if close by. Do not try to extinguish larger fires. Your prime concern is to save lives. As you leave the classroom, make sure that the windows and doors are closed.

- If you use a fire extinguisher, direct the extinguisher at the base of the fire and use a sweeping motion, moving the extinguisher nozzle back and forth across the front of the fire's base. Different extinguishers are effective for different classes of fires. The fire classes are outlined below. Fire extinguishers in the laboratory are 2A10BC. They extinguish classes A, B, and C fires.

- Class A fires involve ordinary combustible materials that leave coals or ashes, such as wood, paper, or cloth. Use water or dry chemical extinguishers on class A fires.

- Class B fires involve flammable liquids such as gasoline or solvents. Carbon dioxide or dry chemical extinguishers are effective on class B fires.

- Class C fires involve live electrical equipment, such as appliances, photocopiers, computers, or laboratory electrical apparatus. Carbon dioxide or dry chemical extinguishers are recommended for class C fires. Do not use water on live electrical devices as this can result in severe electrical shock.

- Class D fires involve burning metals, such as sodium, potassium, magnesium, or aluminum. Sand, salt, or graphite can be used to put out class D fires. Do not use water on a metal fire as this can cause a violent reaction.

- Class E fires involve a radioactive substance. These require special consideration at each site.

Heat Safety

- Keep a clear workplace when performing experiments with heat.

- Make sure that heating equipment, such as the burner, hot plate, or electric heater, is secure on the bench and clamped in place when necessary.

- Do not use a laboratory burner near wooden shelves, flammable liquids, or any other item that is combustible.

- Take care that the heat developed by the heat source does not cause any material close by to get hot enough to burst into flame. Do not allow overheating if you are performing an experiment in a closed area. For example, if you are using a light source in a large cardboard box, be sure you have enough holes at the top of the box and on the sides to dissipate heat.

- Before using a laboratory burner, make sure that long hair is always tied back. Do not wear loose clothing (wide long sleeves should be tied back or rolled up).

- Always assume that hot plates and electric heaters are hot and use protective gloves when handling.

- Do not touch a light source that has been on for some time. It may be hot and cause burns.

- In a laboratory where burners or hot plates are being used, never pick up a glass object without first checking the temperature by lightly and quickly touching the item, or by placing your hand near but not touching it. Glass items that have been heated stay hot for a long time, even if they do not appear to be hot. Metal items such as ring stands and hot plates can also cause burns; take care when touching them.
- Never look down the barrel of a laboratory burner.
- Always pick up a burner by the base, never by the barrel.
- Never leave a lighted burner unattended.
- Any metal powder can be explosive. Do not put these in a flame.
- When heating a test tube over a laboratory burner, use a test-tube holder and a spurt cap. Holding the test tube at an angle, with the open end pointed away from you and others, gently move the test tube back and forth through the flame.
- To heat a beaker, put it on the hot plate and secure with a ring support attached to a utility stand. (A wire gauze under the beaker is optional.)
- Remember to include a cooling time in your experiment plan; do not put away hot equipment.

To use a burner:

- Tie back long hair and tie back or roll up wide long sleeves.
- Secure the burner to a stand using a metal clamp.
- Check that the rubber hose is properly connected to the gas valve.
- Close the air vents on the burner. Use a sparker to light the burner.
- Open the air vents just enough to get a blue flame.
- Control the size of the flame using the gas valve.

Electrical Safety

- Water or wet hands should never be used near electrical equipment such as a hotplate, a light source, or a microscope.
- Do not use the equipment if the cord is frayed or if the third pin on the plug is missing. If the teacher allows this, then make sure the equipment has a double-insulated cord.
- Do not operate electrical equipment near running water or a large container of water.
- Check the condition of electrical equipment. Do not use if wires or plugs are damaged.
- If using a light source, check that the wires of the light fixture are not frayed, and that the bulb socket is in good shape and well secured to a stand.
- Make sure that electrical cords are not placed where someone could trip over them.
- When unplugging equipment, remove the plug gently from the socket. Do not pull on the cord.

Handling Chemicals Safely

Many chemicals are hazardous to some degree. When using chemicals, operate under the following principles:

- Never underestimate the risks associated with chemicals. Assume that any unknown chemicals are hazardous.
- Use a less hazardous chemical wherever possible.
- Reduce exposure to chemicals as much as possible. Avoid direct skin contact if possible.
- Ensure that there is adequate ventilation when using chemicals.

The following guidelines do not address every possible situation but, used with common sense, are appropriate for situations in the high school laboratory.

- Obtain an MSDS for each chemical and consult the MSDS before you use the chemical.
- Know the emergency procedures for the building, the department, and the chemicals being used.
- Wear a lab coat and/or other protective clothing (e.g., apron, gloves), as well as appropriate eye protection at all times in areas where chemicals are used or stored.
- Never use the contents from a bottle that has no label or has an illegible label. Give any containers with illegible labels to your teacher. When leaving chemicals in containers, ensure that the containers are labelled. Always double-check the label, once, when you pick it up, and a second time when you are about to use it.
- Carry chemicals carefully using two hands, one around the container and one underneath.
- Always pour from the side opposite the label on a reagent bottle; your hands and the label are protected as previous drips are always on the side of the bottle opposite of the label.
- Do not let the chemicals touch your skin. Use a laboratory scoop or spatula for handling solids.

- Pour chemicals carefully (down the side of the receiving container or down a stirring rod) to ensure that they do not splash.

- Always pour volatile chemicals in a fume hood or in a well-ventilated area.

- Never pipet or start a siphon by mouth. Always use a pipet suction device (such as a bulb or a pump).

- If you spill a chemical, use a chemical spill kit to clean up.

- Return chemicals to their proper storage place according to your teacher's instructions.

- Do not return surplus chemicals to stock bottles. Dispose of excess chemicals in an appropriate manner as instructed by your teacher.

- Clean up your work area, the fume hood, and any other area where chemicals were used.

- Wash hands immediately after handling chemicals and before and after leaving the lab, even if you wore gloves. Definitely wash your hands before you touch any food.

Handling Animals, Plants, and Other Organisms Safely

- Do not perform any investigation on any animal that might cause suffering or pain, or that might pose a health hazard to you or anyone else in the school.

- Animals that live in the classroom should be treated with care and respect, and be kept in a clean, healthy environment.

- Ensure that your teacher is aware of any plant or animal allergies that you may have.

- Never bring a plant, animal, or other organism to school without receiving prior permission from the teacher.

- Keep cages and tanks clean—both for your health and the health of the organism. Most jurisdictions recommend no live mammals or birds in the laboratory. Reptiles often carry Salmonella.

- Wear gloves and wash your hands before and after feeding or handling an animal, touching materials from the animal's cage or tank, or handling bacterial cultures.

- Do not grow any microorganisms other than those that occur naturally on mouldy bread, cheese, and mildewed objects. Anaerobic bacteria should not be grown.

- Cultures should be grown at room temperature or in the range of 25°C to 32°C. Incubation at 37°C may encourage the growth of microorganisms that are capable of living in the human body.

- Bacteria from soils should not be grown because of the possibility of culturing tetanus-causing organisms.

- Spores collected from household locations, such as telephones or bathrooms, should not be cultured in the laboratory. The body can destroy small numbers of these bacteria, but may not be able to cope with large numbers.

- All surfaces and equipment used in culturing microorganisms should be washed down with a disinfectant (e.g., a solution of bleach).

- Apparatus used in microbiology should be autoclaved because liquid disinfectants and germicidal agents generally cannot guarantee complete sterilization. The oven of an ordinary kitchen stove may be used.

- Wild or sick animals should never be brought into the lab. Dead animals, wild or tame, that have died from unknown causes should also not be brought into the lab.

- Preserved specimens should be removed from the preservative with gloves or tongs, and rinsed thoroughly in running water.

- Before going on field trips, become familiar with any dangerous plants and animals that may be common in the area (e.g., stinging nettles and poisonous plants).

Waste Disposal

Waste disposal at school, at home, and at work is a societal issue. To protect the environment, federal and provincial governments have regulations to control wastes, especially chemical wastes. For example, the WHMIS program applies to controlled products that are being handled. Most laboratory waste can be washed down the drain or, if it is in solid form, placed in ordinary garbage containers. However, some waste must be treated more carefully. It is your responsibility to follow procedures and to dispose of waste in the safest possible manner according to the teacher's instructions.

Flammable Substances

Flammable liquids should not be washed down the drain. Special fire-resistant containers are used to store flammable liquid waste. Waste solids that pose a fire hazard should be stored in fireproof containers. Care must be taken not to allow flammable waste to come into contact with any sparks, flames, other ignition sources, or oxidizing materials. The method of disposal depends on the nature of the substance.

Corrosive Solutions

Solutions that are corrosive but not toxic, such as acids, bases, and oxidizing agents, should be disposed of in a container provided by the teacher, preferably kept on the teacher's desk. Do not pour corrosive solutions down the drain.

Heavy Metal Solutions

Heavy metal compounds (e.g., lead, mercury, and cadmium compounds) should not be poured down the drain. These substances are cumulative poisons and should be kept out of the environment. A special container should be kept in the laboratory for heavy metal solutions. Pour any heavy metal waste into this container. Remember that paper towels used to wipe up solutions of heavy metals, as well as filter papers with heavy metal compounds embedded in them, should be treated as solid toxic waste.

Toxic Substances

Solutions of toxic substances, such as oxalic acid, should not be poured down the drain, but should be disposed of in the same manner as heavy metal solutions, but in a separate container.

Organic Material

Remains of plants and animals can generally be disposed of in school garbage containers. Before disposal, organic material should be rinsed thoroughly to rid it of any excess preservative. Fungi and bacterial cultures should be autoclaved or treated with a fungicide or antibacterial soap before disposal.

First Aid

The following guidelines apply in case of an injury, such as a burn, cut, chemical spill, ingestion, inhalation, or splash in the eyes.

- Always inform your teacher immediately of any injury.
- Know the location of the first-aid kit, fire blanket, eye-wash station, and shower, and be familiar with the contents and operation of them.
- If the injury is a minor cut or abrasion, wash the area thoroughly. Using a compress, apply pressure to the cut to stop the bleeding. When bleeding has stopped, replace the compress with a sterile bandage. If the cut is serious, apply pressure and seek medical attention immediately.

- If the injury is the result of chemicals, drench the affected area with a continuous flow of water for 15 min. Clothing should be removed as necessary. Retrieve the Material Safety Data Sheet (MSDS) for the chemical; this sheet provides information about the first-aid requirements for the chemical.
- If you get a solution in your eye, quickly use the eye-wash or nearest running water. Continue to rinse the eye with water for at least 15 min. This is a very long time—have someone time you. Unless you have a plumbed eyewash system, you will also need assistance in refilling the eyewash container. Have another student inform your teacher of the accident. The injured eye should be examined by a doctor.
- If you have ingested or inhaled a hazardous substance, inform your teacher immediately. The MSDS provides information about the first-aid requirements for the substance. Contact the Poison Control Centre in your area.
- If the injury is from a burn, immediately immerse the affected area in cold water or run cold water gently over the burned area. This will reduce the temperature and prevent further tissue damage.
- In case of electric shock, unplug the appliance and do not touch it or the victim. Inform your teacher immediately.
- If a classmate's injury has rendered him/her unconscious, notify the teacher immediately. The teacher will perform CPR if necessary. Do not administer CPR unless under specific instructions from the teacher. You can assist by keeping the person warm and by reassuring him/her once conscious.

Appendix C REFERENCE

C1 Numerical Prefixes and Units

Throughout *Nelson Biology 12* and in this reference section, we have attempted to be consistent in the presentation and usage of units. As far as possible, *Nelson Biology 12* uses the International System (SI) of Units. However, some other units have been included because of their practical importance, wide usage, or use in specialized fields. For example, Health Canada and the medical profession continue to use millimetres of mercury (mm Hg) as the units for measurement of blood pressure, although the Metric Practice Guide indicates that this unit is not to be used with the SI.

The most recent *Canadian Metric Practice Guide* (CAN/CSA-Z234.1-89) was published in 1989 and reaffirmed in 1995 by the Canadian Standards Association.

Other data in this reference section has been taken largely from Lange's *Handbook of Chemistry*, Fifteenth Edition, McGraw-Hill, 1999.

Numerical Prefixes

Prefix	Power	Symbol
deca-	10^1	da
hecto-	10^2	h
kilo-	10^3	k*
mega-	10^6	M*
giga-	10^9	G*
tera-	10^{12}	T
peta-	10^{15}	P
exa-	10^{18}	E
deci-	10^{-1}	d
centi-	10^{-2}	c*
milli-	10^{-3}	m*
micro-	10^{-6}	m*
nano-	10^{-9}	n*
pico-	10^{-12}	p
femto-	10^{-15}	f
atto-	10^{-18}	a

* commonly used

Common Multiples

Multiple	Prefix
0.5	hemi-
1	mono-
1.5	sesqui-
2	bi-, di-
2.5	hemipenta-
3	tri-
4	tetra-
5	penta-
6	hexa-
7	hepta-
8	octa-
9	nona-
10	deca-

Some Examples of Prefix Use

0.0034 mol = 3.4×10^{-3} mol = 3.4 **milli**moles or 3.4 mmol

1530 L = 1.53×10^3 L = 1.53 **kilo**litres or 1.53 kL

SI Base Units

Quantity	Symbol	Unit name	Symbol
amount of substance	n	mole	mol
electric current	I	ampere	A
length	L, l, h, d, w	metre	m
luminous intensity	I_v	candela	cd
mass	m	kilogram	kg
temperature	T	kelvin	K
time	t	second	s

Some SI Derived Units

Quantity	Symbol	Unit	Unit Symbol	Expression in SI base unit
acceleration	$\vec{a}$	metre per second per second	m/s^2	m/s^2
area	A	square metre	m^2	m^2
density	ρ, D	kilogram per cubic metre	kg/m^3	kg/m^3
displacement	$\vec{d}$	metre	m	m
electric charge	Q, q, e	coulomb	C	$A \cdot s$
electric potential	V	volt	V	$kg \cdot m^2/(A \cdot s^3)$
electric field	E	volt per metre	V/m	$kg \cdot m/(A \cdot s^3)$
electric field intensity	E	newton per coulomb	N/C	$kg/(A \cdot s^3)$
electric resistance	R	ohm	Ω	$kg \cdot m^2/(A^2 \cdot s^3)$
energy	E, E_k, E_p	joule	J	$kg \cdot m^2/s^2$
force	F	newton	N	$kg \cdot m/s^2$
frequency	f	hertz	Hz	s^{-1}
heat	Q	joule	J	$kg \cdot m^2/s^2$
magnetic flux	Φ	weber	Wb	$kg \cdot m^2/(A \cdot s^2)$
magnetic field	B	weber per square metre Tesla	Wb/m^2 T	$kg/(A \cdot s^2)$
momentum	P, p	kilogram metre per second	$kg \cdot m/s$	$kg \cdot m/s$
period	T	second	s	s
power	P	watt	W	$kg \cdot m^2/s^3$
pressure	P p	pascal newton per square metre	Pa N/m^2	$kg/(m \cdot s^2)$
speed	v	metre per second	m/s	m/s
velocity	$\vec{v}$	metre per second	m/s	m/s
volume	V	cubic metre	m^3	m^3
wavelength	λ	metre	m	m
weight	W, w	newton	N	$kg \cdot m/s^2$
work	W	joule	J	$kg \cdot m^2/s^2$

C2 Greek and Latin Prefixes and Suffixes

Greek and Latin Prefixes

Prefix	Meaning	Prefix	Meaning	Prefix	Meaning
a-	not, without	em-	inside	micr-, micro-	small
ab-	away from	en-	in	mono-	one
abd-	led away	end-, endo-	within	morpho-	form, shape
acro-	end, tip	epi-	at, on, over	muc-, muco-	slime
adip-	fat	equi-	equal	multi-	many
aer-, aero-	air	erythro-	red	myo-	muscle
agg-	to clump	ex-, exo-	away, out	nas-	nose
agro-	land	flag-	whip	necro-	corpse
alb-	white	gamet-, gamo-	marriage, united	neo-	new
allo-	other	gastr-, gastro-	stomach	neur-, neuro-	nerve
ameb-	change	geo-	earth	noct-	night
amphi-	around, both	glyc-	sweet	odont-, odonto-	tooth
amyl-	starch	halo-	salt	oligo-	few
an-	without	haplo-	single	oo-	egg
ana-	up	hem-, hema-, hemato-	blood	orni-	bird
andro-	man	hemi-	half	oss-, osseo-, osteo-	bone
ant-, anti-	opposite	hepat-, hepa-	liver	ovi-	egg
anth-	flower	hetero-	different	pale-, paleo-	ancient
archae-, archaeo-	ancient	histo-	web	patho-	disease
archi-	primitive	holo-	whole	peri-	around
astr-, astro-	star	homeo-	same	petro-	rock
aut-, auto-	self	hydro-	water	phag-, phago-	eat
baro-	weight (pressure)	hyper-	above	pharmaco-	drug
bi-	twice	hypo-	below	phono-	sound
bio-	life	infra-	under	photo-	light
blast-, blasto-	sprout (budding)	inter-	between	pneum-	air
carcin-	cancer	intra-	inside of, within	pod-	foot
cardio-, cardia-	heart	intro-	inward	poly-	many
chlor-, chloro-	green	iso-	equal	pseud-, pseudo-	false
chrom-, chromo-	colour	lact-, lacti-, lacto-	milk	pyr-, pyro-	fire
co-	with	leuc-, leuco-	white	radio-	ray
cosmo-	order, world	lip-, lipo-	fat	ren-	kidney
cut-	skin	lymph-, lympho-	clear water	rhizo-	root
cyan-	blue	lys-, lyso-	break up	sacchar-, saccharo-	sugar
cyt-, cyto-	cell	macro-	large	sapr-, sapro-	rotten
dendr-, dendri-, dendro-	tree	mamm-	breast	soma-	body
dent-, denti-	tooth	meg-, mega-	great	spermato-	seed
derm-	skin	melan-	black	sporo-	seed
di-	two	meningo-	membrane	squam-	scale
dors-	back	mes-, meso-	middle	sub-	beneath
ec-, ecto-	outside	meta-	after, transition	super-, supra-	above

Greek and Latin Prefixes

Prefix	Meaning	Prefix	Meaning	Prefix	Meaning
sym-, syn-	with, together	ultra-	beyond	xanth-, xantho-	yellow
telo-	end	uro-	tail, urine	xer-, xero-	dry
therm-, thermo-	temperature, heat	vas-, vaso-	vessel	xyl-	wood
tox-	poison	vita-	life	zoo-	animal
trans-	across	vitro-	glass	zygo-	yoke
trich-	hair	vivi-	alive		

Greek and Latin Suffixes

Suffix	Meaning	Suffix	Meaning	Suffix	Meaning
-aceous	like	-lysis	loosening	-phyll	leaf
-blast	budding	-lyt	dissolvable	-phyte	plant
-cide	kill	-mere	share	-pod	foot
-crin	secrete	-metry	measure	-sis	a condition
-cut	skin	-mnesia	memory	-some	body
-cyte	cell	-oid	like	-stas, -stasis	halt
-emia	blood	-ol	alcohol	-stat	to stand, stabilize
-gen	born, agent	-ole	oil	-tone, -tonic	strength
-genesis	formation	-oma	tumour	-troph	nourishment
-graph, -graphy	to write	-osis	a condition	-ty	state of
-gynous	woman	-pathy	suffering	-vorous	eat
-itis	inflammation	-ped	foot	-yl	wood
-logy	the study of	-phage	eat	-zyme	ferment

C3 Data for Some Radioisotopes

Name	Symbol	Uses	Half-life
carbon-14	^{14}C	radiometric dating—effective dating range: 100 to 100 000 years	5730 years
fluorine-18	^{18}F	medical—to image tumours and localized infections (PET)	110 minutes
indium-111	^{111}In	medical—to study the brain, the colon, and sites of infection	2.8 days
iodine-125	^{125}I	medical—to evaluate the filtration rate of kidneys and to determine bone density measurements	42 days
iodine-131	^{131}I	medical—to view and treat thyroid, liver, kidney diseases, and various cancers	8.0 days
phosphorus-32	^{32}P	medical—to treat polycythemia vera (excess red blood cells)	14.3 days
potassium-40	^{40}K	radiometric dating—effective dating range: 100 000 to 4.6 billion years	1.3 billion years
strontium-89	^{89}Sr	medical—to relieve the pain of secondary cancers lodged in the bone	46.7 hours
technetium-99*	^{99}Tc	medical—to view the skeleton and heart muscle in particular; but also the brain, thyroid, lungs, liver, spleen, kidney, gall bladder, bone marrow, and salivary glands	6.02 hours
uranium-235	^{235}U	radiometric dating—effective dating range: 10 million to 4.6 billion years	713 million years

* the most commonly used isotope in medicine

C4 RNA Codons and Amino Acids

The codons in mRNA are nucleotide bases arranged in groups of three. There are 61 of these base triplets that correspond to 20 amino acids. In **Table 1**, read the first nucleotide from the first (left) column, the second from one of the middle columns, and the third from the last (right) column. For example, the triplet UGG represents tryptophan, and tyrosine is represented by both UAU and UAC. mRNA reads the arrangements of amino acids on the DNA and carries the information to the ribosomes for the synthesis of proteins.

Table 1: Messenger Ribonucleic Acid (mRNA) Codons and Their Corresponding Amino Acids

First Base	Second Base				Third Base
	U	**C**	**A**	**G**	
U	UUU phenylalanine UUC phenylalanine UUA leucine UUG leucine	UCU serine UCC serine UCA serine UCG serine	UAU tyrosine UAC tyrosine UAA STOP** UAG STOP**	UGU cysteine UGC cysteine UGA STOP** UGG tryptophan	U C A G
C	CUU leucine CUC leucine CUA leucine CUG leucine	CCU proline CCC proline CCA proline CCG proline	CAU histidine CAC histidine CAA glutamine CAG glutamine	CGU arginine CGC arginine CGA arginine CGG arginine	U C A G
A	AUU isoleucine AUC isoleucine AUA isoleucine AUG methionine*	ACU threonine ACC threonine ACA threonine ACG threonine	AAU asparagine AAC asparagine AAA lysine AAG lysine	AGU serine AGC serine AGA arginine AGG arginine	U C A G
G	GUU valine GUC valine GUA valine GUG valine	GCU alanine GCC alanine GCA alanine GCG alanine	GAU aspartate GAC aspartate GAA glutamate GAG glutamate	GGU glycine GGC glycine GGA glycine GGG glycine	U C A G

* AUG is an initiator codon and also codes for the amino acid methionine.
** UAA, UAG, and UGA are terminator codons.

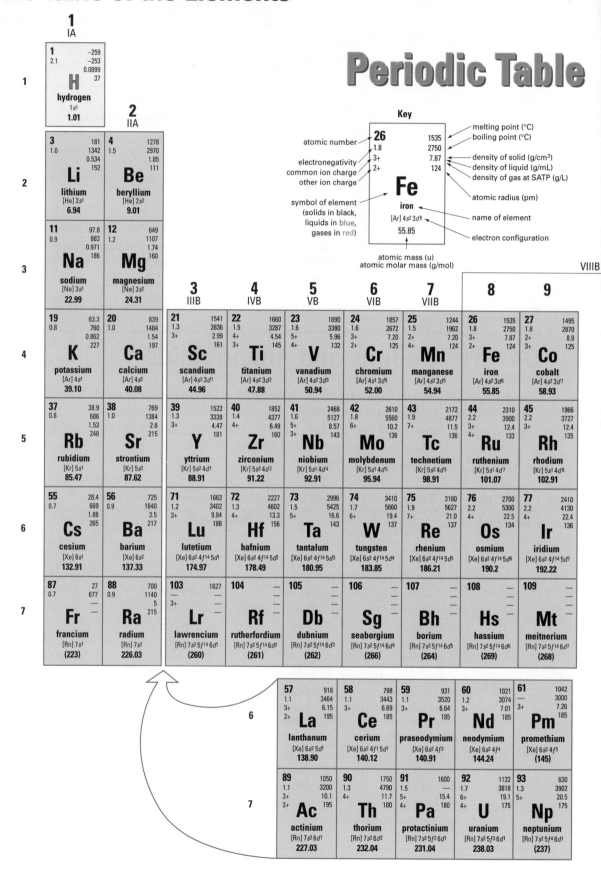

Periodic Table

Key

atomic number →	**26**	1535 ← melting point (°C)
	1.8	2750 ← boiling point (°C)
electronegativity →	3+	7.87 ← density of solid (g/cm³)
common ion charge →	2+	← density of liquid (g/mL)
other ion charge →		124 ← density of gas at SATP (g/L)
	Fe	← atomic radius (pm)
symbol of element →		
(solids in black,	**iron**	← name of element
liquids in blue,	[Ar] 4s² 3d⁶	
gases in red)		← electron configuration
	55.85	

atomic mass (u)
atomic molar mass (g/mol)

Group 1 / IA

1	−259
2.1	−253
	0.0899
	37
H	
hydrogen	
1s¹	
1.01	

Group 2 / IIA

3	181	4	1278
1.0	1342	1.5	2970
	0.534		1.85
	152		111
Li		**Be**	
lithium		beryllium	
[He] 2s¹		[He] 2s²	
6.94		**9.01**	

11	97.8	12	649
0.9	883	1.2	1107
	0.971		1.74
	186		160
Na		**Mg**	
sodium		magnesium	
[Ne] 3s¹		[Ne] 3s²	
22.99		**24.31**	

VIIIB

Group 3 / IIIB

19	63.3	20	839
0.8	760	1.0	1484
	0.862		1.54
	227		197
K		**Ca**	
potassium		calcium	
[Ar] 4s¹		[Ar] 4s²	
39.10		**40.08**	

21	1541	22	1660	23	1890	24	1857	25	1244	26	1535	27	1495
1.3	2836	1.5	3287	1.6	3380	1.6	2672	1.5	1962	1.8	2750	1.8	2870
3+	2.99	4+	4.54	5+	5.96	3+	7.20	2+	7.20	3+	7.87	2+	8.9
3+	161	3+	145	4+	132	2+	125	4+	124	2+	124	3+	125
Sc		**Ti**		**V**		**Cr**		**Mn**		**Fe**		**Co**	
scandium		titanium		vanadium		chromium		manganese		iron		cobalt	
[Ar] 4s² 3d¹		[Ar] 4s² 3d²		[Ar] 4s² 3d³		[Ar] 4s¹ 3d⁵		[Ar] 4s² 3d⁵		[Ar] 4s² 3d⁶		[Ar] 4s² 3d⁷	
44.96		**47.88**		**50.94**		**52.00**		**54.94**		**55.85**		**58.93**	

37	38.9	38	769	39	1522	40	1852	41	2468	42	2610	43	2172	44	2310	45	1966
0.8	686	1.0	1384	1.3	3338	1.4	4377	1.6	5127	1.8	5560	1.9	4877	2.2	3900	2.2	3727
	1.53		2.6	3+	4.47	4+	6.49	5+	8.57	6+	10.2	7+	11.5	3+	12.4	3+	12.4
	248		215		181		160	3+	143		136		136	4+	133		135
Rb		**Sr**		**Y**		**Zr**		**Nb**		**Mo**		**Tc**		**Ru**		**Rh**	
rubidium		strontium		yttrium		zirconium		niobium		molybdenum		technetium		ruthenium		rhodium	
[Kr] 5s¹		[Kr] 5s²		[Kr] 5s² 4d¹		[Kr] 5s² 4d²		[Kr] 5s¹ 4d⁴		[Kr] 5s¹ 4d⁵		[Kr] 5s² 4d⁵		[Kr] 5s¹ 4d⁷		[Kr] 5s¹ 4d⁸	
85.47		**87.62**		**88.91**		**91.22**		**92.91**		**95.94**		**98.91**		**101.07**		**102.91**	

55	28.4	56	725	71	1663	72	2227	73	2996	74	3410	75	3180	76	2700	77	2410
0.7	669	0.9	1640	1.2	3402	1.3	4602	1.5	5425	1.7	5660	1.9	5627	2.2	5300	2.2	4130
	1.88		3.5	3+	9.84	4+	13.3	5+	16.6	6+	19.4	7+	21.0	4+	22.5	4+	22.4
	265		217		188		156		143		137		137		134		136
Cs		**Ba**		**Lu**		**Hf**		**Ta**		**W**		**Re**		**Os**		**Ir**	
cesium		barium		lutetium		hafnium		tantalum		tungsten		rhenium		osmium		iridium	
[Xe] 6s¹		[Xe] 6s²		[Xe] 6s² 4f¹⁴ 5d¹		[Xe] 6s² 4f¹⁴ 5d²		[Xe] 6s² 4f¹⁴ 5d³		[Xe] 6s² 4f¹⁴ 5d⁴		[Xe] 6s² 4f¹⁴ 5d⁵		[Xe] 6s² 4f¹⁴ 5d⁶		[Xe] 6s² 4f¹⁴ 5d⁷	
132.91		**137.33**		**174.97**		**178.49**		**180.95**		**183.85**		**186.21**		**190.2**		**192.22**	

87	27	88	700	103	1627	104	—	105	—	106	—	107	—	108	—	109	—
0.7	677	0.9	1140	3+	—		—		—		—		—		—		—
	—		5		—		—		—		—		—		—		—
	—		215														
Fr		**Ra**		**Lr**		**Rf**		**Db**		**Sg**		**Bh**		**Hs**		**Mt**	
francium		radium		lawrencium		rutherfordium		dubnium		seaborgium		borium		hassium		meitnerium	
[Rn] 7s¹		[Rn] 7s²		[Rn] 7s² 5f¹⁴ 6d¹		[Rn] 7s² 5f¹⁴ 6d²		[Rn] 7s² 5f¹⁴ 6d³		[Rn] 7s² 5f¹⁴ 6d⁴		[Rn] 7s² 5f¹⁴ 6d⁵		[Rn] 7s² 5f¹⁴ 6d⁶		[Rn] 7s² 5f¹⁴ 6d⁷	
(223)		**226.03**		**(260)**		**(261)**		**(262)**		**(266)**		**(264)**		**(269)**		**(268)**	

Row 6 (Lanthanides)

57	918	58	798	59	931	60	1021	61	1042
1.1	3464	1.1	3443	1.1	3520	1.2	3074	—	3000
3+	6.15	3+	6.69	3+	6.64	3+	7.01		7.26
2+	195		185		185		185	3+	185
La		**Ce**		**Pr**		**Nd**		**Pm**	
lanthanum		cerium		praseodymium		neodymium		promethium	
[Xe] 6s² 5d¹		[Xe] 6s² 4f¹ 5d¹		[Xe] 6s² 4f³		[Xe] 6s² 4f⁴		[Xe] 6s² 4f⁵	
138.90		**140.12**		**140.91**		**144.24**		**(145)**	

Row 7 (Actinides)

89	1050	90	1750	91	1600	92	1132	93	630
1.1	3200	1.3	4790	1.5	—	1.7	3818	1.3	3902
3+	10.1	4+	11.7	5+	15.4	6+	19.1	5+	20.5
2+	195		180	4+	180	4+	175		175
Ac		**Th**		**Pa**		**U**		**Np**	
actinium		thorium		protactinium		uranium		neptunium	
[Rn] 7s² 6d¹		[Rn] 7s² 6d²		[Rn] 7s² 5f² 6d¹		[Rn] 7s² 5f³ 6d¹		[Rn] 7s² 5f⁴ 6d¹	
227.03		**232.04**		**231.04**		**238.03**		**(237)**	

of the Elements

Each element block lists: atomic number, melting point (°C), boiling point (°C), electronegativity, density, atomic radius (pm), oxidation states, symbol, name, electron configuration, and atomic mass.

Group 18 — VIIIA

- **2 — He**, helium, [−272 / −269 / X / 0.179 / 50], $1s^2$, 4.00

Period 2

- **5 — B**, boron, [2300 / 2.0 / 2550 / X / 2.34 / 88], [He] $2s^2 2p^1$, 10.81
- **6 — C**, carbon, [3550 / 2.5 / 4827 / X / 2.26 / 77], [He] $2s^2 2p^2$, 12.01
- **7 — N**, nitrogen, [−210 / 3.0 / −196 / 1.25 / 70], [He] $2s^2 2p^3$, 14.01
- **8 — O**, oxygen, [−218 / 3.5 / −183 / 1.43 / 66], [He] $2s^2 2p^4$, 16.00
- **9 — F**, fluorine, [−220 / 4.0 / −188 / 1.70 / 64], [He] $2s^2 2p^5$, 19.00
- **10 — Ne**, neon, [−249 / −246 / X / 0.900 / 62], [He] $2s^2 2p^6$, 20.18

Period 3

- **13 — Al**, aluminum, [660 / 1.5 / 2467 / 2.70 / 143], [Ne] $3s^2 3p^1$, 26.98
- **14 — Si**, silicon, [1410 / 1.8 / 2355 / X / 2.33 / 117], [Ne] $3s^2 3p^2$, 28.09
- **15 — P**, phosphorus, [44.1 / 2.1 / 280 / 1.82 / 110], [Ne] $3s^2 3p^3$, 30.97
- **16 — S**, sulfur, [113 / 2.5 / 445 / 2.07 / 104], [Ne] $3s^2 3p^4$, 32.06
- **17 — Cl**, chlorine, [−101 / 3.0 / −34.6 / 3.21 / 99], [Ne] $3s^2 3p^5$, 35.45
- **18 — Ar**, argon, [−189 / — / −186 / X / 1.78 / 95], [Ne] $3s^2 3p^6$, 39.95

Groups 10, 11 (IB), 12 (IIB) — Period 4

- **28 — Ni**, nickel, [1455 / 1.8 / 2730 / 2+ / 8.90 / 3+ / 124], [Ar] $4s^2 3d^8$, 58.69
- **29 — Cu**, copper, [1083 / 1.9 / 2567 / 2+ / 8.92 / 1+ / 128], [Ar] $4s^1 3d^{10}$, 63.55
- **30 — Zn**, zinc, [420 / 1.6 / 907 / 2+ / 7.14 / 133], [Ar] $4s^2 3d^{10}$, 65.38
- **31 — Ga**, gallium, [29.8 / 1.6 / 2403 / 3+ / 5.90 / 122], [Ar] $4s^2 3d^{10} 4p^1$, 69.72
- **32 — Ge**, germanium, [937 / 1.8 / 2830 / 4+ / 5.35 / 123], [Ar] $4s^2 3d^{10} 4p^2$, 72.61
- **33 — As**, arsenic, [817 / 2.0 / 613 / 5.73 / 121], [Ar] $4s^2 3d^{10} 4p^3$, 74.92
- **34 — Se**, selenium, [217 / 2.4 / 684 / 4.81 / 117], [Ar] $4s^2 3d^{10} 4p^4$, 78.96
- **35 — Br**, bromine, [−7.2 / 2.8 / 58.8 / 3.12 / 114], [Ar] $4s^2 3d^{10} 4p^5$, 79.90
- **36 — Kr**, krypton, [−157 / — / −152 / X / 3.74 / 112], [Ar] $4s^2 3d^{10} 4p^6$, 83.80

Period 5

- **46 — Pd**, palladium, [1554 / 2.2 / 2970 / 2+ / 12.0 / 4+ / 138], [Kr] $4d^{10}$, 106.42
- **47 — Ag**, silver, [962 / 1.9 / 2212 / 1+ / 10.5 / 144], [Kr] $5s^1 4d^{10}$, 107.87
- **48 — Cd**, cadmium, [321 / 1.7 / 765 / 2+ / 8.64 / 149], [Kr] $5s^2 4d^{10}$, 112.41
- **49 — In**, indium, [157 / 1.7 / 2080 / 3+ / 7.30 / 163], [Kr] $5s^2 4d^{10} 5p^1$, 114.82
- **50 — Sn**, tin, [232 / 1.8 / 2270 / 4+ / 7.31 / 2+ / 140], [Kr] $5s^2 4d^{10} 5p^2$, 118.69
- **51 — Sb**, antimony, [631 / 1.9 / 1750 / 3+ / 6.68 / 5+ / 141], [Kr] $5s^2 4d^{10} 5p^3$, 121.75
- **52 — Te**, tellurium, [450 / 2.1 / 990 / 6.2 / 137], [Kr] $5s^2 4d^{10} 5p^4$, 127.60
- **53 — I**, iodine, [114 / 2.5 / 184 / 4.93 / 133], [Kr] $5s^2 4d^{10} 5p^5$, 126.90
- **54 — Xe**, xenon, [−112 / — / −107 / X / 5.89 / 130], [Kr] $5s^2 4d^{10} 5p^6$, 131.29

Period 6

- **78 — Pt**, platinum, [1772 / 2.2 / 3827 / 4+ / 21.5 / 2+ / 138], [Xe] $6s^1 4f^{14} 5d^9$, 195.08
- **79 — Au**, gold, [1064 / 2.4 / 2808 / 3+ / 19.3 / 1+ / 144], [Xe] $6s^1 4f^{14} 5d^{10}$, 196.97
- **80 — Hg**, mercury, [−39.0 / 1.9 / 357 / 2+ / 11.85 / 1+ / 160], [Xe] $6s^2 4f^{14} 5d^{10}$, 200.59
- **81 — Tl**, thallium, [304 / 1.8 / 1457 / 1+ / 11.85 / 3+ / 170], [Xe] $6s^2 4f^{14} 5d^{10} 6p^1$, 204.38
- **82 — Pb**, lead, [328 / 1.8 / 1740 / 2+ / 11.3 / 4+ / 175], [Xe] $6s^2 4f^{14} 5d^{10} 6p^2$, 207.20
- **83 — Bi**, bismuth, [271 / 1.9 / 1560 / 3+ / 9.80 / 5+ / 155], [Xe] $6s^2 4f^{14} 5d^{10} 6p^3$, 209.98
- **84 — Po**, polonium, [254 / 2.0 / 962 / 2+ / 9.40 / 4+ / 167], [Xe] $6s^2 4f^{14} 5d^{10} 6p^4$, (209)
- **85 — At**, astatine, [302 / 2.2 / 337 / 142], [Xe] $6s^2 4f^{14} 5d^{10} 6p^5$, (210)
- **86 — Rn**, radon, [−71 / — / −61.8 / X / 9.73 / 140], [Xe] $6s^2 4f^{14} 5d^{10} 6p^6$, (222)

Period 7

- **110 — Uun**, ununnilium, [Rn] $7s^2 5f^{14} 6d^8$, (269, 271)
- **111 — Uuu**, unununium, [Rn] $7s^2 5f^{14} 6d^9$, (272)
- **112 — Uub**, ununbium, [Rn] $7s^2 5f^{14} 6d^{10}$, (277)
- **113**
- **114 — Uuq**, ununquadium, [Rn] $7s^2 5f^{14} 6d^{10} 7p^2$, (285)
- **115**
- **116 — Uuh**, ununhexium, [Rn] $7s^2 5f^{14} 6d^{10} 7p^4$, (289)
- **117**
- **118**

Lanthanides

- **62 — Sm**, samarium, [1074 / 1.2 / 1794 / 3+ / 7.52 / 2+ / 185], [Xe] $6s^2 4f^6$, 150.36
- **63 — Eu**, europium, [822 / 1527 / 3+ / 5.24 / 2+ / 185], [Xe] $6s^2 4f^7$, 151.96
- **64 — Gd**, gadolinium, [1313 / 1.1 / 3273 / 3+ / 7.90 / 180], [Xe] $6s^2 4f^7 5d^1$, 157.25
- **65 — Tb**, terbium, [1356 / 1.2 / 3230 / 3+ / 8.23 / 175], [Xe] $6s^2 4f^9$, 158.92
- **66 — Dy**, dysprosium, [1412 / 1.2 / 2567 / 3+ / 8.55 / 175], [Xe] $6s^2 4f^{10}$, 162.50
- **67 — Ho**, holmium, [1474 / 1.2 / 2700 / 3+ / 8.80 / 175], [Xe] $6s^2 4f^{11}$, 164.93
- **68 — Er**, erbium, [1529 / 1.2 / 2868 / 3+ / 9.07 / 175], [Xe] $6s^2 4f^{12}$, 167.26
- **69 — Tm**, thulium, [1545 / 1.2 / 1950 / 3+ / 9.32 / 175], [Xe] $6s^2 4f^{13}$, 168.93
- **70 — Yb**, ytterbium, [819 / 1.1 / 1196 / 3+ / 6.97 / 2+ / 175], [Xe] $6s^2 4f^{14}$, 173.04

Actinides

- **94 — Pu**, plutonium, [641 / 1.3 / 3232 / 4+ / 19.8 / 6+ / 175], [Rn] $7s^2 5f^6$, (244)
- **95 — Am**, americium, [994 / 1.3 / 2607 / 3+ / 13.7 / 4+ / 175], [Rn] $7s^2 5f^7$, (243)
- **96 — Cm**, curium, [1340 / — / 3110 / 3+ / 13.5], [Rn] $7s^2 5f^7 6d^1$, (247)
- **97 — Bk**, berkelium, [986 / — / 14 / 3+ / 4+], [Rn] $7s^2 5f^9$, (247)
- **98 — Cf**, californium, [900 / — / 3+], [Rn] $7s^2 5f^{10}$, (251)
- **99 — Es**, einsteinium, [860 / — / 3+], [Rn] $7s^2 5f^{11}$, (252)
- **100 — Fm**, fermium, [1527 / —], [Rn] $7s^2 5f^{12}$, (257)
- **101 — Md**, mendelevium, [1021 / — / 3074 / 2+ / 3+], [Rn] $7s^2 5f^{13}$, (258)
- **102 — No**, nobelium, [863 / — / 2+ / 3+], [Rn] $7s^2 5f^{14}$, (259)

C6 Plant and Animal Cells

Plant Cell

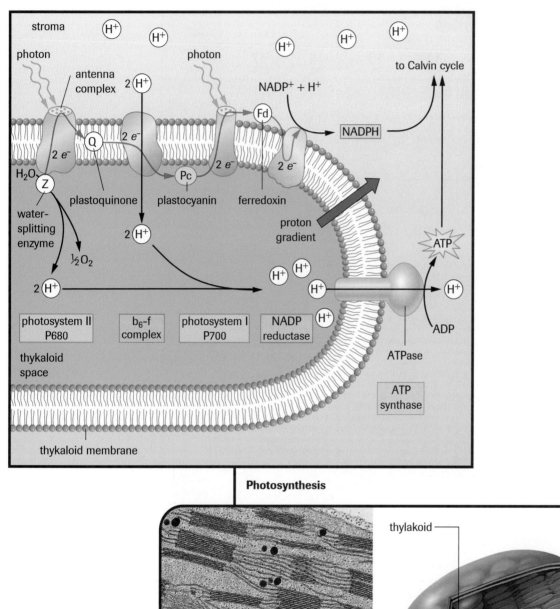

stroma

photon

antenna complex

$2 (H^+)$

NADP$^+$ + H$^+$

to Calvin cycle

photon

$2 e^-$

$2 e^-$

Fd

NADPH

Q

$2 e^-$

H$_2$O

Z

Pc

$2 e^-$

$2 e^-$

water-splitting enzyme

plastoquinone

plastocyanin

ferredoxin

proton gradient

½O$_2$

$2 (H^+)$

$2 (H^+)$

ATP

(H^+) (H^+)

$2 (H^+)$

(H^+)

(H^+)

| photosystem II P680 | b$_6$-f complex | photosystem I P700 | NADP reductase |

(H^+)

ADP

ATPase

thykaloid space

ATP synthase

thykaloid membrane

Photosynthesis

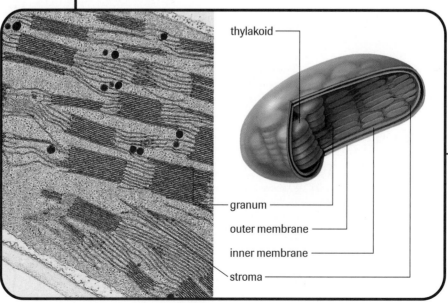

thylakoid

granum

outer membrane

inner membrane

stroma

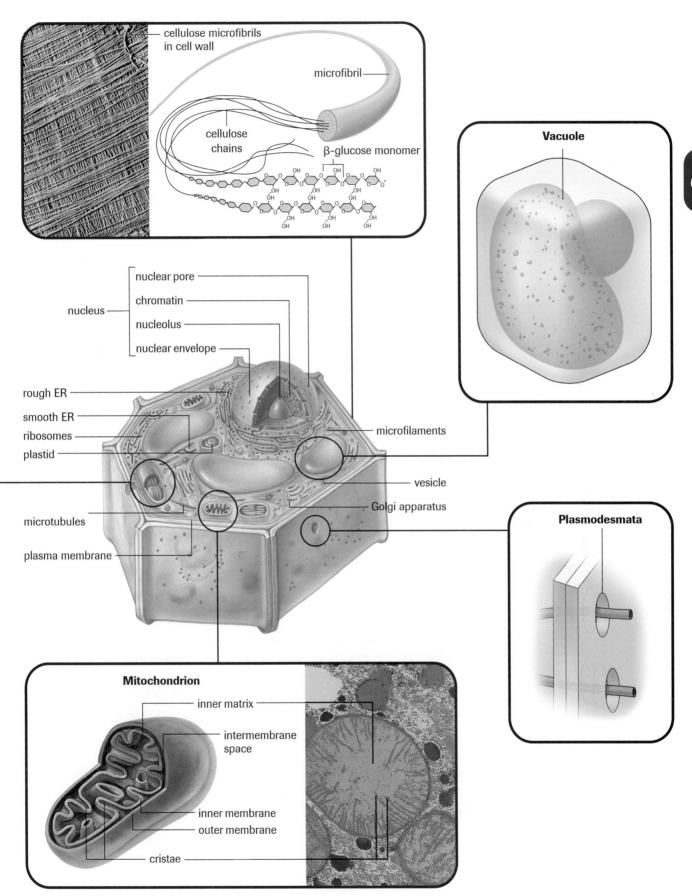

cellulose microfibrils in cell wall

microfibril

cellulose chains

β-glucose monomer

OH

Vacuole

nuclear pore

chromatin

nucleus

nucleolus

nuclear envelope

rough ER

smooth ER

ribosomes

plastid

microfilaments

vesicle

Golgi apparatus

microtubules

plasma membrane

Plasmodesmata

Mitochondrion

inner matrix

intermembrane space

inner membrane

outer membrane

cristae

Animal Cell

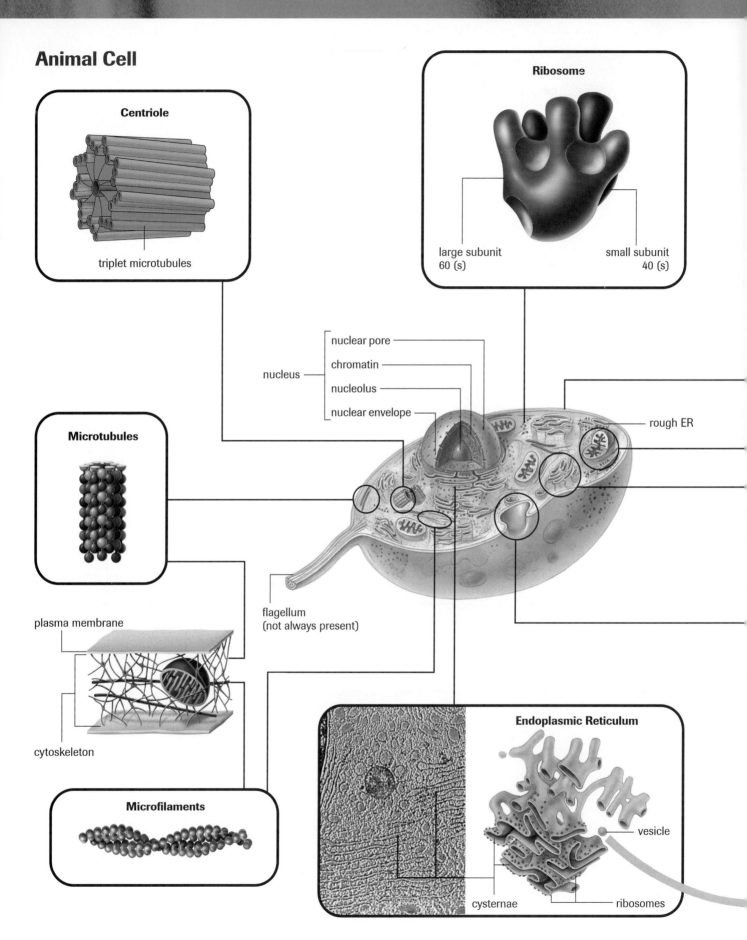

Centriole

triplet microtubules

Ribosome

large subunit 60 (s)

small subunit 40 (s)

nuclear pore

chromatin

nucleus

nucleolus

nuclear envelope

rough ER

Microtubules

plasma membrane

cytoskeleton

flagellum (not always present)

Microfilaments

Endoplasmic Reticulum

vesicle

cysternae

ribosomes

Plasma membrane

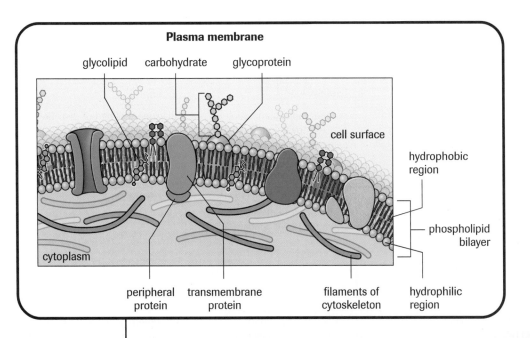

glycolipid   carbohydrate   glycoprotein

cell surface

hydrophobic region

phospholipid bilayer

cytoplasm

peripheral protein   transmembrane protein   filaments of cytoskeleton   hydrophilic region

Mitochondrion

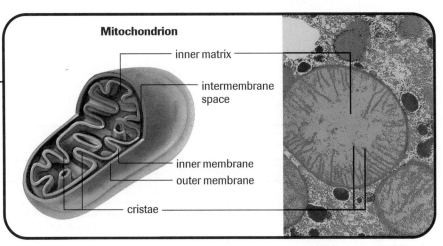

inner matrix

intermembrane space

inner membrane

outer membrane

cristae

Vacuole being formed

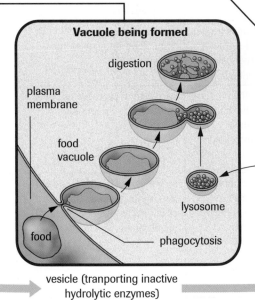

digestion

plasma membrane

food vacuole

food

lysosome

phagocytosis

Golgi Apparatus

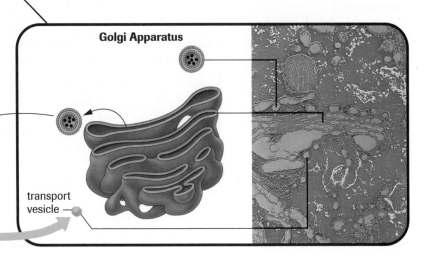

transport vesicle

vesicle (tranporting inactive hydrolytic enzymes)

Methods of Cell Membrane Transport

Passive Processes

	Vehicle	Substances	Methods	Examples
Diffusion	directly through phospholipid bilayer	hydrophobic and small hydrophilic molecules	random molecular motion causes movement down a concentration gradient	movement of carbon dioxide out of cells
Facilitated Diffusion	passive selective protein carrier	small hydrophilic molecules and ions	carrier protein binds specific molecules and moves them down their own concentration gradient	movement of glucose into cells
Osmosis	directly through phospholipid bilayer	water	water diffuses through membrane down its own concentration gradient	movement of water across a selectively permeable membrane from a hypotonic solution into a hypertonic solution until solutions on both sides of the membrane are isotonic

Active Processes

	Vehicle	Substances	Methods	Examples
Endocytosis				
Phagocytosis	membrane vesicle	solid particles	particles engulfed by pseudopod action	ingestion of bacteria by macrophages in blood
Pinocytosis	membrane vesicle	extracellular fluid	liquid engulfed by pseudopod action	pinocytosis by cells lining small blood vessels
Receptor-mediated Endocytosis	membrane vesicle	specific molecules	specific receptor proteins trigger formation of endocytic vesicle resulting in transport of certain molecules into the cell	transport of cholesterol in LDL from blood into cell of the human body
Exocytosis	secretory vesicle	substances formed within the cell	secretory vesicles in the cytoplasm fuse with the cell membrane and spill their contents into the extracellular fluid	secretion of digestive enzymes by cells of the pancreas
Active Transport	active selective protein carrier	ions and small to medium sized hydrophilic molecules	carrier proteins use cell energy in the form of ATP or electrochemical gradients to transport specific molecules across a membrane	movement of the Na^+ and K^+ ions against concentration gradients across nerve cell membranes

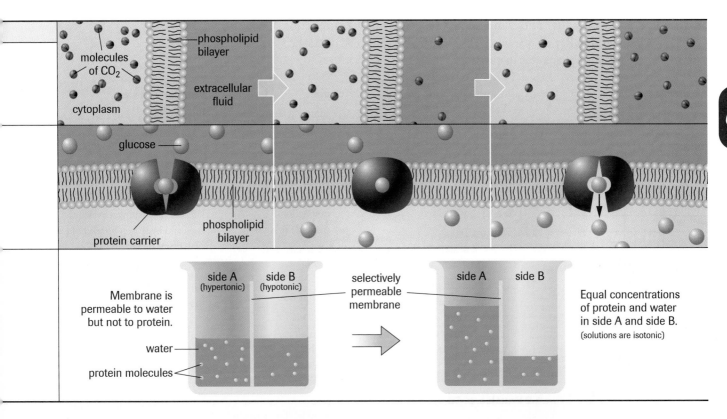

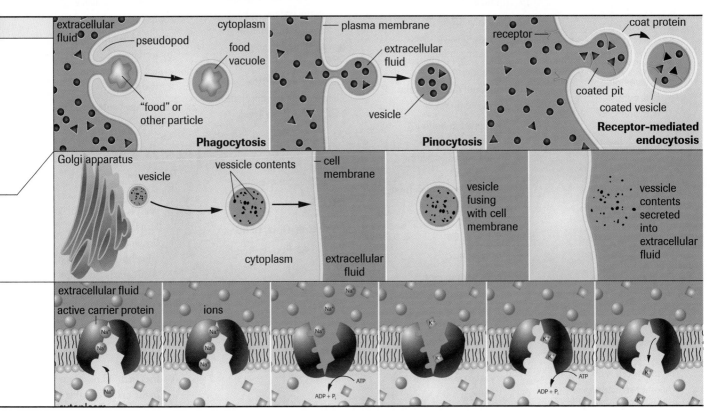

Answers

This section includes answers to all Chapter Self-Quiz questions, as well as all short answers to Practice questions and all numerical and short answers to Section, Chapter Review, and Unit Review questions. The short answers provided do not necessarily represent complete answers to questions.

1.1 Practice, p. 10

3. protons

Section 1.1, p. 23

1.

	protons	neutrons	electrons	valance
a) S	16	16	16	2–
b) Ca	20	20	20	2 +
c) N	7	7	7	3–

2. all three need two electrons to fill the outer orbital

   33

4. (a) water—hydrogen
   (b) I—London forces

5. mutual repulsion of valence electron pairs

6. CCl_4—nonpolar; NH_4—polar

9. acid—HCN, HCOOH; base—CN, HCOO

11. (a) 8.5, basic
    (b) 2.6, acidic

14. (a) hydrophobic
    (b) hydrophilic
    (c) hydrophobic

1.2 Practice, p. 27

1. NH_2 (amino), COOH (carboxyl), C-S-H (sulfhydryl), C–O (carbonyl)

1.2 Practice, p. 34

2. carbonyl hydroxyl

3. (a) and (d)

5. It forms a closed ring structure.

6. (a) store excess substances like glucose
   (b) forms cell walls and other structures

7. amylose—alpha glucose, cellulose—beta glucose

8. there are no enzymes to digest cellulose

1.2 Practice, p. 40

11. ester bond between carboxyl and hydroxl

15. (a) cutin
    (b) prevents water loss
    (c) hydrophobic

17. (a) hydrogenation adds hydrogen atoms to solidify oils
    (b) imitate butter

18. may form nonsoluble blockages in arteries

1.2 Practice, p. 50

19. amino and carboxyl

22. (a) alpha helix; beta-pleated sheets
    (b) hydrogen bonds

28. Feed patient with protein rich liquid diet to rebuild denatured proteins.

29. (a) dipped in boiling water
    (b) denatures browning enzymes
    (c) to prevent discoloured fruit

Section 1.2, p. 56

2. 1–C, 2–H, 1–O

7. (a) The number of covalent bonds that an atom can form.
   (b) C–4, N–3, O–2
   (c) only one possible bond

11. (a) 3, 2
    (b) carboxyl and hydroxyl

14. (b) COOH

(c) acidic

15. The amino acids must be eaten, they cannot be made by the body.

17. (a) disulfide bridges, ionic bonds, hydrogen bonds, hydrophobic interactions
    (b) disulfide bridges

19. A–T (2 bonds) and C–G (3 bonds)

Section 1.3, p. 68

2. (a) anabolic
   (b) catabolic
   (c) anabolic
   (d) anabolic
   (e) catabolic

9. (a) zero
   (b) there is a lot of energy in the system, but not enough free energy to allow anabolic reactions to occur

10. concentrated negative charge in the triphosphate tail of the ATP molecule

Section 1.4, p. 77

8. regenerated intact, ready to catalyze same reaction again

Chapter 1 Self-Quiz, p. 85

1. T	8. F	15. B
2. T	9. T	16. D
3. T	10. F	17. B
4. F	11. B	18. C
5. T	12. A	19. D
6. T	13. D	20. C
7. F	14. B	

Chapter 1 Review, pp. 86–87

4. (a) no enzyme
   (b) cows, sheep, rabbits contain symbiotic bacteria and protists in their digestive tracts

5. nonbranching, covalent and hydrogen bonds hold fibres together

9. estrogen, progesterone

10. hydrophobic

19. (a) kinetic
    (b) potential
    (c) potential
    (d) kinetic
    (e) potential

21. autotrophs can convert EM energy into chemical potential energy; heterotrophs cannot

27. (a) disulfide bridges
    (b) tertiary
    (c) will change proteins in the skin

28. wood, paper, cotton clothing

Section 2.1, p. 93

1. (a) glucose
   (b) water

2. (b) yeast is heterotroph; fern is autotroph; lichen is autotroph; amoeba is heterotroph; *Anabaena* is autotroph

Section 2.2, p. 115

1. ADP + phosphate, and substrate enzyme

2. (a) cytoplasm
   (b) the chemical reactions in the cytoplasm that

break glucose into two pyruvate molecules

4. (a) 2 pyruvate, 4 ATP, 2 (NADH + H⁺), 2 ADP

   (b) pyruvate and NADH

11. 36 ATP, 6 H_2O, 6 CO_2

13. NADH and $FADH_2$

14. (a) hydrogen
    (d) chemiosmosis (oxidative phosphorylation)
    (e) Peter Mitchell

15. (b) oxygen

20. (a) 2.0 m²
    (b) 7757 kJ/24h

21. (b) from the female egg cell

Section 2.3, p. 124

1. fatty acid

3. lactic acid buildup

4. CO_2

5. alcoholic fermentation— ATP, ethanol, CO_2 lactic acid fermentation— ATP and lactate

6. (a) 2
   (b) 2
   (c) none

7. yeast

9. pain

13. (a) 3 mmol/L

Chapter 2 Self-Quiz, p. 133

1. F	8. F	15. B
2. F	9. T	16. B
3. T	10. T	17. A
4. F	11. F	18. D
5. T	12. B	19. B
6. F	13. C	
7. F	14. E	

Chapter 2 Review, pp. 134–135

3. (a) Q—electron carrier
   (b) $FADH_2$—electron and proton carrier

(c) pyruvic acid—electron and proton donor

4. (a) NADH—3 ATP; $FADH_2$—2 ATP

5. (a) 36 ATP
   (b) 2 ATP

7. glycolysis in cytoplasm; pyruvate oxidation in matrix of mitochondria; Krebs cycle in matrix of mitochondria; electron transport chain along the inner membrane of the mitochondria

8. electrons and protons added

11. alcoholic fermentation

13. low oxygen levels

15. cytoplasm

16. 2

17. (a) matrix
    (b) inner membrane

18. death

19. 32%

20. A inner membrane; B cristae; C matrix; D outer membrane

30. (a) lactic acid fermentation
    (b) low oxygen and pH inhibit most microorganisms, increase shelf life of food

Section 3.1, p. 145

1. (a) lettuce—autotroph; rabbit and wolf—heterotroph

2. (a) contain chlorophyll
   (b) infolds of cell mebrane

4. (a) maximizes gas diffusion

Section 3.2, p. 154–155

1. $6 CO_2 + 12 H_2O + \text{light energy} \rightarrow C_6H_{12}O_6 + 6 H_2O + 6 O_2$

2. ATP, NADPH, and oxygen

3. (a) form of energy that travels in wave packets called photons
   (b) photons are EM wave packets of light
   (d) green has more energy and a shorter wave length

5. (a) removes oxygen and releases carbon dioxide
   (b) removes carbon dioxide and releases oxygen

8. (a) chlorophyll a and b, carotenoids, xanthophylls and anthocyanins
   (b) masked by chlorophyll
   (c) wavelengths 400 nm— 700 nm support photosynthesis

9. (a) heavy—stable, radioactive—unstable
   (c) mass spectrometer

11. (b) 480 nm (blue)— most absorbed
    (c) 520 nm (green)— least absorbed
    (d) chlorophyll a
    (e) no absorption by chlorophyll
    (f) chlorophyll a and b, and carotenoids

Section 3.3, pp. 166–167

2. (b) thylakoid
   (c) ATP, NADPH
   (d) light

3. (a) oxygen
   (b) water

4. (a) 2
   (b) b_6-f complex, Q, plastocyanin
   (c) lost as heat

6. A—Z, B—b_6-f complex, C—photosystem I, D—NADPH, E—ATP synthase

9. (a) rubisco
   (b) CO_2 and RuBP
   (c) PGA
   (d) stroma

10. (a) ATP and NADPH are absent
    (b) light energy

12. 12 CO_2, 36 ATP, 24 NADPH, 36 photons

14. blue > green

Section 3.4, p. 172

1. (b) oxygen
   (c) 20%

3. (a) A—oxaloacetate, B—CO_2, C—pyruvate, D—mesophyll cell, E—bundle-sheath cell
   (b) plasmodesmata

4. (a) C_4—wet environment, CAM—dry environment
   (b) C_4—sugar cane, corn; CAM—pineapples, cacti

6. (a) late at night
   (b) evening
   (c) reduce water loss
   (d) CO_2

Section 3.5, p. 178

1. irradiance

5. (a) A—light compensation point
   (c) B—light saturation point

Chapter 3 Self-Quiz, p. 191

1. T	7. T	13. B
2. F	8. F	14. B
3. T	9. T	15. A
4. F	10. T	16. E
5. F	11. B	17. C
6. T	12. C	18. E

Chapter 3 Review, pp. 192–193

1. Short

2. 400–700 nm

3. light reaction, forms oxygen, protons, electrons

8. inside the thylakoid, no

19. A

Unit 1 Review, pp. 196–199

1. (a) polar
   (b) nonpolar
   (c) nonpolar
5. liver and muscle cells
9. D
12. (a) amino acid
    (b) carbohydrate
    (c) polypeptide
    (d) fatty acid
14. NAD, $FADH_2$, cytochrome C and coenzyme Q
19. 380–750 nm; PAR is 400–700 nm
20. (a) thylakoid membrane
    (b) infolds of the cell membrane
22. (a) electrochemical
    (b) light—stroma donates protons to thylakoid, reverses in dark
24. (a) thylakoid; stroma
    (b) light reaction
    (c) ATP and NADPH
33. glycolysis—1.8%, pyruvate oxidation—1.8%, Krebs cycle—28.4%
34. (b) 15 mL/kg/min
    (c) 1125 mL/min; 6750 mL/min
43. (a) ATP and RuBP

Section 4.2, p. 216

3. 3′-TACGGAAT-5′
6. A is 20%, G is 30%, C is 30%
7. A is double stranded
8. 1.02 m; 300 million turns
9. 220 million nucleotides

Section 4.3, p. 223

7. 25%

Chapter 4 Self-Quiz, p. 229

1. F   7. T   13. D
2. T   8. F   14. C
3. F   9. F   15. A
4. F   10. T   16. C
5. F   11. A
6. F   12. D

Chapter 4 Review, pp. 230–231

6. TTAACGTAT
16. A is 22.5%; T is 22.5%; G is 27.5% and C is 27.5%

Section 5.1, p. 236

6. (a) B, C, A, D
   (b) 1—C–A, 2—B–C, 3—A–D

Section 5.2, p. 241

8. Only 16 combinations for 20 amino acids
9. start is AUG, stop could be UAA or UAG or UGA
10. CCUAGUCCAGGUCCGU UAAAUCGUACGGGGUU
12. 4, 254
13. (a) 15

Section 5.3, p. 249

2. 5′-AUGAUGCCAUC CAUAU-3′
3. (a) ATAATA; string of As and Ts
   (b) site of RNA polymerase binding; where DNA unwound more easily

Section 5.4, p. 254

3. Thr—UGA, UGG, UGU, UGC
   Ala—CGA, CGG, CGU, CGC
   Pro—GGA, GGG, GGU, GGC

Section 5.6, p. 263

4. UV radiation, x-rays, and certain chemicals
7. (a) yes
   (b) no
   (c) yes
   (d) yes

Chapter 5 Self-Quiz, p. 273

1. F   7. T   13. C
2. T   8. F   14. C
3. T   9. T   15. C
4. T   10. E   16. A
5. F   11. A
6. F   12. A

Chapter 5 Review, pp. 274–275

1. (a) introns; exons; variable number tandem repeats (VNTRs); telomeric; centromeric
   (b) promoters; promoter; operator; operon
   (c) transcription; RNA polymerase; mRNA; 5′; 3′; termination sequence
7. Lys-Tyr-Ser
10. (a) AUG, GGG, CCC, GUU, CGG
14. pairs provide 16 combinations for 20 amino acids
17. four, 625
18. (a) Gly-Gln-Lys-Gln-Glu

6.1 Practice, p. 281

1. (a) between CCC and GGG; between 9 and 10; between 39 and 40
   (b) 3
   (c) blunt
2. sticky
4. 18
5. GAATTC and CTTTAAAG

6.1 Practice, p. 282

10. condensation reaction, DNA and water

6.1 Practice, p. 284

11. DNA is negatively charged.
14. (a) It can insert itself into rungs of DNA and fluoresce under UV light.

Section 6.1, p. 292

15. (a) *Eco*RI—2; *Bam*HI—2; *Pst*I—3
    (b) 2400
16. (a) *Eco*RI—2, 1236 + 6660; *Bam*HI—2, 1399 + 6497; *Hind*III—3, 1071 + 429 + 6396; *Eco*RI and *Hind*III—5, 1236 + 1071 + 429 + 372 + 4788; all three—7, 1236 + 2333 + 1399 + 1056 + 1071 + 429 + 372
    (b) *Bam*HI; cuts the gene in half
17. 24

6.3 Practice, p. 300

6. TAATCGTA
20. 5′-GCTTACGAA-3′

Chapter 6 Self-Quiz, p. 319

1. T   7. F   13. B
2. F   8. T   14. B
3. T   9. T   15. D
4. T   10. F   16. C
5. F   11. A
6. T   12. E or D

Chapter 6 Review, pp. 320–321

5. 1.05×10^6
10. TC GA; *Alu*I
19. 9
20. no
22. 5′-TTCGATCCGCGA-3′

Unit 2 Review, pp. 324–327

1. protein
2. sugar-phosphate
3. semiconservative
4. template
5. nucleotides
6. codon
7. amino acid
8. RNA polymerase
9. transcription

10. translation

11. anticodon

12. restriction enzyme, recognition sites

13. hydrogen; to separate during replication and transcription

27. 15% T, 30% A, 20% C, 35% G

33. 1.7 million nm; 500 000 turns

34. A—single, B—double, C—single

35. 75%

36. 3

38. (a) (i) 896, 2758, 346;
       (ii) 1254, 1255, 1491;
       (iii) 2100, 107, 1793;
       (iv) 896, 358, 1255, 1145, 346;
       (v) 1254, 846, 107, 302, 1491;
       (vi) 896, 358, 846, 107, 302, 1145, 346

39. (a) 1750

40. (a) 5′-ATGTATCGTA CATATTCGTATTTA CATCATGATTAA-3′
    (b) strand given in question (a)
    (c) 5′-AUGUAUCGUA CAUAUUCGUAUU UACAUCAUGAUUAA -3′
    (d) start (Met)-Tyr-Arg-Thr-Tyr-Ser-Tyr-Leu-His-His-Asp-stop
    (f) fourth codon is ACA, can change to ACU, ACC, or ACG and still code for Thr

Section 7.2, p. 341

1. The erect hair traps warm air.

4. Wearing clothes, sitting in the sun/shade.

7. Sweat cannot evaporate under water.

8. A—sweating; B—evaporation; C—adjustment; D—shivering; E—adjustment

Section 7.4, p. 348

3. (a) D—kidney
   (b) E—ureter
   (c) C—renal artery
   (d) F—urinary bladder

Section 7.5, p. 352

5. (a) afferent arteriole—C, and efferent arteriole—A
   (b) increase filtration
   (c) the glomerulus is not permeable to larger molecules
   (d) glomerulus, B, or proximal tubule, D
   (e) D and G
   (f) G
   (g) all areas of the nephron, but most concentrated in G
   (h) D—the proximal tube is the area of active transport. The ATP required for active transport is provided by cellular respiration, which occurs in mitochondria

6. b, c, e, f, a, d

Section 7.6, p. 356

1. ADH is antidiuretic hormone, produced in hypothalamus, stored in pituitary.

3. hypothalamus

10. (a) proteins are not filtered
    (b) urea is secreted
    (c) glucose is reabsorbed

Section 7.7, p. 356

1. The precipitation of mineral solutes from the blood

7. rejection of the donor kidney by the recipient

8. (a) D

8. (c) glucose in the nephron creates an osmotic force, drawing in fluids

Chapter 7 Self-Quiz, p. 367

1. B   4. D   7. C
2. A   5. A   8. E
3. A   6. A

Chapter 7 Review, p. 368–369

1. ADH and aldosterone

2. Hormone 1 increases the permeability of the nephron to water, while hormone 2 increases permeability to Na$^+$.

4. concentration of urea in the blood will decrease

5. dialysis solution must contain fewer wastes than the blood for diffusion to occur

6. C only

8. Blood moves into glomerulus, but not out easily. Pressure increases filtration and urine production increases.

9. site of active transport, which requires energy from ATP

10. Salt increases osmotic pressure in blood, causing hypothalamus to shrink and release ADH.

8.1 Practice, p. 377

1. Target tissues have specific receptor sites, which bind with hormones.

4. Cyclic AMP functions as a messenger, activating enzymes in the cytoplasm to carry out their normal functions.

Sections 8.1–8.2, p. 383

2. Insulin decreases blood sugar levels. When blood

sugars are high insulin is released.

3. Glucagon increases blood sugar levels. When blood sugars are low glucagon is released.

5. The cells are provided with more glucose to increase the rate of cellular respiration.

6. High levels of ACTH stimulate the production and release of cortisol. High levels of cortisol decrease the production and release of ACTH.

Section 8.3, p. 387

3. Thyroxine increases metabolic rate. By increasing carbohydrate utilization, blood sugar levels can drop.

8. Parathyroids release PTH, which raises calcium levels. Low calcium levels can lead to rapid, uncontrolled muscle twitching, referred to as tetany.

Section 8.4, p. 392

5. Hormones that have a pronounced effect in small, localized areas of the body.

7. Growth hormone and anabolic steroids increases muscle mass, which adds strength but not endurance. Marathon runners need to increase oxygen delivery.

9. They are natural products of the human body. Genetic differences cause differing concentrations in a normal population.

Chapter 8 Self-Quiz, p. 407

1. B   4. B   7. A
2. D   5. A   8. A
3. C   6. D

Section 9.1, p. 417

5. The sensory receptor would detect touch, but the muscle would not react.

6. III, II, I

Section 9.2, p. 426

2. A large size allows electrodes to be placed inside the neuron.

Section 9.3, p. 434

3. R—medulla oblongata; controls involuntary muscles

   S—pons; relays impulses

   T—sensory cortex of the cerebrum; body sensations such as touch

   V—cerebellum; balance

Section 9.5, p. 444

4. In bright light, the light sensitive pigment in rods breaks down faster than it can be rebuilt.

Chapter 9 Self-Quiz, p. 457

1. (a) axon
   (b) action potential
   (c) autonomic nerves
   (d) acetylcholine
   (e) sensory neuron
   (f) accommodation
   (g) occipital lobe of the brain
   (h) refractory period

2. C   5. B   8. C
3. D   6. C   9. C
4. B   7. B

Section 10.1, p. 465

2. Lysozymes destroy the cell wall of bacteria.

4. Monocytes become macrophages which engulf microbes.

7. The increased numbers of white blood cells usually indicate a microbial infection.

Section 10.3, p. 475

4. The immune system mistakenly identifies some of the body's own cells as pathogens.

Section 10.4, p. 481

1. A, D, E

2. Pathogenic organisms cause disease.

6. She was a carrier of a serious disease, Because she suffered no symptoms, she was able to move freely through society, spreading the microbes.

Section 10.5, p. 490

1. Each antibody is made to fit a specific active site on the virus. Each virus has a different geometry.

3. Antibiotics are not effective against virus, only against bacteria.

Chapter 10 Self Quiz, p. 495

1. E   5. C   9. A
2. B   6. B   10. B
3. A   7. B
4. C   8. C

Chapter 10 Review, p. 496–497

2. The necessary antibodies are already present.

10. Cells that can differentiate into a number of specialized cells.

11. A disorder caused when lymphocytes treat healthy body cells as pathogens.

12. The number of suppressor T cells decrease with age.

17. Many microbes are found in fecal materials.

25. (a) Virus need living cells to be produced.
    (b) Patients are occasionally allergic to the

chicken eggs used to produce the vaccine.

Unit 3 Review, p. 500–503

2. excrete urea; maintain water balance; maintain ion balance

7. (a) steroids and protein hormones

Section 11.2, p. 516

5. Argon is a gas that escapes from molten rock.

Section 11.3, p. 518

2. A common ancestor can evolve to form new species.

4. Cuvier believed in mass extinctions, followed by replacement by migrating species. He did not believe that species changed over time.

Sections 11.4-11.5, p. 528

2. rock movement during an earthquake in South America; noted fossils of marine organisms above sea level

3. collected mockingbirds in the Galapagos; compared similarities of species on the Cape Verde Islands to the mainland of Africa

5. in the past, cormorants could fly

8. Darwin applied Malthus' principles to the theory he was formulating on the origin and evolution of species.

11. Bats have only arrived recently and have not had time to evolve into new species.

12. (a) 2.5×10^{27}
    (b) predation
    (c) absence of major predators

Chapter 11 Self-Quiz, p. 539

1. A   5. D   9. A
2. C   6. E   10. D
3. C   7. C   11. C
4. B   8. D

Chapter 11 Review, p. 540–541

1. does not change

6. (a) Earth—4.6 billion years old
   (b) bacteria—3.8 billion years old
   (c) land plants—400 million years old
   (d) mammals—200 million years old

7. at least 713 million years old

15. (a) Curie
    (b) Kelvin
    (c) Lyell
    (d) Cuvier
    (e) Buffon
    (f) Lamarck
    (g) Malthus
    (h) Darwin
    (i) Mendel
    (j) Wallace

Sections 12.2–12.3, p. 555

9. 267 beneficial mutations

Section 12.6, p. 576

1. (a) flowers that bloom at different times
   (b) animals that have different mating rituals
   (c) members of the same species that live in different habitats, e.g, fish that live in shallow water and fish that live in deep water
   (d) all wind blown pollen lands on the stigma of flowers, but only that of the same species can fertilize

Chapter 12 Self-Quiz, p. 581

1. C	5. C	9. E
2. E	6. A	10. D
3. B	7. E	11. C
4. A	8. B	

Chapter 12 Review, p. 582–583

19. $C = 0.98$, $c = 0.02$

20. 9.5%

Section 13.3, p. 605

1. speciation, adaptive radiation

Section 13.4. p, 614

1. indicates the degree of relatedness between species

6. (a) They differ at position 66.
   (b) human—(I = isoleucine) TAA, TAG, TAT; rhesus—(T = threonine) TGA, TGG, TGT, TGC
   (c) TAA → TGA; TAG → TGG; TAT → TGT

Section 13.5, pp. 621–622

3. Humans have 23 pairs of chromosomes, chimpanzees have 24; humans have bipedal motion, a greatly enlarged brain, the use of complex language, the construction and use of complex tools.

5. (a) *A. anamensis*, *A. afarensis*, *H. habilis*, *H. erectus*
   (b) *Ardipithecus*, *A. africanus*, *A. aethiopicus*, *A. robustus, A. boisei*, *H. neanderthalensis*
   (c) *A. robustus, A. boisei*

Chapter 13 Self-Quiz, p. 635

1. F	6. F	11. F
2. T	7. F	12. F
3. T	8. T	13. F
4. F	9. T	14. F
5. F	10. F	15. T

16. (a) clade
    (b) punctuated equilibrium
    (c) synapomorphy
    (d) primary abiogenesis
    (e) adaptive radiation
    (f) coevolution
    (g) convergent evolution
    (h) gradualism
    (i) divergent evolution
    (j) homonid

17. (U+V), (R+S), (R, S+T), (R, S, T, U, V), (Q), (ALL SPECIES)

18. (a) U
    (b) R, S equally
    (c) all others equally

19. Q, R, S, T

20. 2, 4, 7

21. won't resemble any one species more than another

22. 1, 3, 5

23. no reason to think that any living species appears more ancient than another

Chapter 13 Review, pp. 636–637

6. (a) convergent evolution

8. (a) yes
   (b) no
   (c) yes

10. (b) approximately 37 mya

12. (a) molecular record
    (d) 0 to 1, 2 to 3, 45 to 50
    (e) compare the amino acid differences between humans, gorillas, and chickens

Unit 4 Review, pp. 640–643

30. 2.6 bya

35. (a) X–A = 0.5, a = 0.5;
        Y–A = 0.5, a = 0.5,
        Z–A = 0.2, a = 0.8
    (b) Y

36. sickle cell allele = 12.5%

Section 14.1, p. 659

3. (a) clumped
   (b) 4095, 4.6/ha

5. (a) Table 1—population = 4550, density = 2275/ha; Table 2—population = 404, density = 404/ha

14.2 Practice, p. 668

4. no

Section 14.2, p. 670

1. 3.2

2. 87

3. 6.2%

4. (a) exponential
   (b) 336 000

5. (a) exponential,

7. (a) – 0.2%

Section 14.3, p. 675

3. (a) density dependent
   (b) density independent
   (c) density independent

5. (a) density dependent

Section 14.4, p. 688

5. (a) camouflage
   (b) chemical toxins

Chapter 14 Self-Quiz, p. 693

1. T	6. F	11. B
2. T	7. T	12. A
3. F	8. D	13. C
4. F	9. C	14. E
5. F	10. E	15. C

Chapter 14 Review, p. 694–695

2. 14/ha

9. (a) camouflage
   (b) mimicry
   (c) stinger and toxins
   (d) camouflage

12. (a) 1125
    (b) toxins

15. (a) owls
    (b) bacteria

Section 15.1, p. 704

1. clumped

6. (a) corn

Section 15.2, p. 713

1. 123 years

3. (a) wide base, narrow at top

5. equal, or births < deaths

Section 15.3, p. 722

3. plants use less energy, and transfer more energy to humans

5. tropical rainforest— 3800, 380; temperate grassland—920, 92; open ocean—500, 50

Section 15.5, p. 740

4. (a) 1.5×10^9 kg
   (b) 1.8×10^8 kg

Section 15.6, p. 744

1. excessive use of fertilizers, deforestation, loss of biodiversity, ozone depletion, pesticides, pollution

Chapter 15 Self-Quiz, p. 751

1. F	7. F	13. A
2. F	8. T	14. C
3. T	9. F	15. D
4. T	10. F	16. C
5. F	11. D	17. C
6. F	12. D	

Chapter 15 Review, pp. 752–753

14. (a) Brazil 0.1%
    China 0.09%
    Denmark 0.01%
    India 0.17%
    Indonesia 0.17%

Italy –0.01%
Kenya 0.15%
Mexico 0.19%
Nigeria 0.26%
Russia –0.05%
South Africa 0.07%
United Kingdom 0.02%
United States 0.05%

Unit 5 Review, pp. 756–759

2. (a) 28 125/m^2
   (b) 14 million

3. (a) 3/ km

5. 160 819

10. (a) density independent
    (b) density dependent
    (c) density independent
    (d) density independent

27. (a) 2960
    (b) 25/ha
    (c) clumped

29. (a) birth rate—37%; death rate—21%
    (c) 1861–1871 = 12%, 1991–1996 = 6%

36. (a) lag phase, slow population growth

A

A (acceptor) site site in the ribosome where tRNA brings in an amino acid

absolute age an estimate of the actual age of a rock or fossil specimen

absorption spectrum a graph illustrating the wavelengths of light absorbed by a pigment

accommodation adjustments made by the lens and pupil of the eye for near and distant objects

acetylcholine neurotransmitter released from vesicles in the end plates of neurons, which makes the postsynaptic membranes more permeable to Na^+ ions

acid deposition a mixture of sulfur dioxide and nitrogen oxide pollutants that can reach Earth's surface in the from of rain, gas, or solid particles; also called acid precipitation

acquired traits those changes in an individual resulting form interaction with the environment

action potential the voltage difference across a nerve cell membrane when the nerve is excited

action spectrum a graph illustrating the effectiveness with which different wavelengths of light promote photosynthesis

activation energy the difference between the energy level of the transition state and the potential energy of reacting molecules

activator a substance that binds to an allosteric site on an enzyme and stabilizes the protein conformation that keeps all the active sites available to their substrates

active immunity occurs when the body produces antibodies against an invading antigen

active site the location where the substrate binds to an enzyme

actualism the theory that the same geological processes occurring in the present also occurred in the past

adaptation any trait that increases an individual's ability to survive or reproduce compared to organisms that do not have the trait

adaptive radiation process in which divergent evolution occurs in rapid succession, or simultaneously, among a number of groups to produce three or more species or higher taxa

adenosine triphosphate (ATP) a nucleotide derivative that acts as the primary energy-transferring molecule in living organisms

adrenal cortex outer region of the adrenal gland that produces glucocorticoids and mineralocorticoids

adrenal medulla found at the core of the adrenal gland, the adrenal medulla produces epinephrine and norepinephrine

aerobic cellular respiration harvesting energy from organic compounds using oxygen

aerobic fitness a measure of the ability of the heart, lungs, and bloodstream to supply oxygen to the cells of the body during physical exercise

afferent arterioles small branches that carry blood to the glomerulus

agarose gel-forming polysaccharide found in some types of seaweed that is used to form a gel meshwork for electrophoresis

Agrobacterium tumefaciens type of bacteria that infects wounded plant cells and creates a crown gall

air pollution presence of one or more chemicals in sufficient quantities in the atmosphere to cause harm to humans and other forms of life

aldosterone hormone that increases Na^+ reabsorption from the distal tubule and collecting duct

Allee effect density-dependent phenomenon that occurs when a population cannot survive or fails to reproduce enough to offset mortality once the population density is too low; such populations usually do not survive

allele frequency the proportion of gene copies in a population of a given allele

alleles particular forms of a gene; many genes have two or more alleles

all-or-none response a nerve or muscle fibre responds completely or not at all to a stimulus

allosteric inhibitor a substance that binds to an allosteric site on an enzyme and stabilizes the inactive form of the enzyme

allosteric sites receptor sites, some distance from the active site of certain enzymes, that bind substances that may inhibit or stimulate an enzyme's activity

α helix a type of polypeptide secondary structure characterized by a tight coil that is stabilized by hydrogen bonds

altruism behaviour that decreases the fitness of an individual that is assisting or cooperating with a recipient individual whose fitness is increased; such behaviour can evolve only when the assisting or cooperating individual increases its fitness indirectly (i.e., though the fitness of its kin)

amino acid the monomer unit of a polypeptide chain that is composed of a carboxylic acid, an amino group, and a side group that differentiates it from other amino acids

amino terminus the free amino group at one end of a polypeptide

aminoacyl-tRNA a tRNA molecule with its corresponding amino acid attached to its acceptor site at the 3′ end.

anabolic reactions reactions that produce large molecules from smaller subunits

anabolic steroids substances that are designed to mimic many of the muscle-building traits of the sex hormone testosterone

analogous features structures similar in function but not in origin or anatomical structure (e.g., wings of birds and bees)

anneal the pairing of complementary strands of DNA through hydrogen bonding

antenna complex a web of chlorophyll molecules embedded in the thylakoid membrane that transfers energy to the reaction centre

anthocyanins pigments in vacuoles that give rise to the red colour in autumn leaves

anthropoid the higher primates, including all extinct and living monkeys, apes, and humans

antibodies proteins formed within the blood that inactivate or destroy antigens

anticodon group of three complementary bases on tRNA that recognizes and pairs with a codon on the mRNA

antidiuretic hormone (ADH) causes the kidneys to increase water absorption

antigens substances, usually protein in nature, that stimulate the formation of antibodies

antinociceptive transmitters signal molecules that dampen pain sensation via a neurological pathway

antiparallel describes two adjacent nucleotides running in opposite directions relative to one another

antisense oligonucleotides complementary DNA or RNA that anneals to mRNA and inhibits

aqueous humour watery liquid that protects the lens of the eye and supplies the cornea with nutrients

aquifers porous, water-saturated layers of sand, gravel, or bedrock that can store and yield significant volumes of water; precipitation and runoff enter recharge zones and percolate gradually down to an aquifer

arable land land used for the purpose of cultivating crops

archaebacteria prokaryotic organisms with greater genetic similarity to eukaryotic cells than eubacteria (includes methanogens, extreme thermophiles, and extreme halophiles)

astigmatism vision defect caused by abnormal curvature of surface of the lens or cornea

auditory canal carries sound waves to the eardrum

autoradiogram gel pattern imprinted on X-ray film by radioactive probes

axon extension of cytoplasm that carries nerve impulses away from the cell body

B

B cell lymphocyte, made and processed in the bond marrow, that produces antibodies

bacteriophage any bacteria-infecting virus

basal metabolic rate (BMR) the minimum amount of energy on which an organism can survive

baseline data initial information about both abiotic and biotic factors of an ecosystem that can be used to monitor changes in the ecosystem

basilar membrane anchors the receptor hair cells in the organ of Corti

beneficial mutation an inheritable change in a cell's DNA that results in an additional or enhanced gene product or regulatory function

β-carotene a carotenoid pigment found in photosystems that forms vitamin A in animals

β-galactosidase the enzyme responsible for the breakdown of lactose into its component sugars, glucose and galactose

β-oxidation the sequential removal of acetyl groups in the catabolism of fatty acids

β-pleated sheets polypeptide secondary structures that form between parallel stretches of a polypeptide and are stabilized by hydrogen bonds

biodegradation the natural recycling of wastes through the breakdown of organic matter by a variety of organisms, but especially bacteria and other microbes

biogeography the careful observation and analysis of the geographic distribution of organisms

bioremediation the use of organisms, usually bacteria, to detoxify polluted environments such as oil spill sites or contaminated soils

biotic potential the maximum rate a population can increase under ideal conditions

blunt ends fragment ends of a DNA molecule that are fully base paired, resulting from cleavage by a restriction enzyme

bond energy the minimum energy required to break one mole of bonds between two species of atoms; a measure of the stability of a chemical bond

bonding capacity the number of covalent bonds an atom can form with neighbouring atoms

bottleneck effect a dramatic, often temporary, reduction in population size usually resulting in significant genetic drift

Bowman's capsule cuplike structure that surrounds the glomerulus

Bt toxin poisonous substance produced by *Bacillus thuringiensis*, which acts as a natural herbicide

buffers chemical systems containing a substance that can donate H^+ ions when they are required and containing a substance that can remove H^+ ions when there are too many in a solution

C

C₃ photosynthesis photosynthesis in which a three-carbon intermediate is first formed when CO_2 is incorporated into organic acid molecules

C₄ photosynthesis a photosynthetic pathway of carbon fixation that reduces the amount of photorespiration that takes place by continually pumping CO_2 molecules (via malate) from the mesophyll cells to the bundle-sheath cells, where rubisco brings them into the C_3 Calvin cycle

calcitonin hormone produced by the thyroid gland that lowers calcium levels in the blood

Calvin cycle a cyclic set of reactions occurring in the stroma of chloroplasts that fixes the carbon of CO_2 into carbohydrate molecules and recycles coenzymes

Cambrian explosion a period, beginning about 565 million years ago and lasting about 40 million years, during which the animal kingdom underwent rapid speciation and diversification; the origin of almost all major groups of animals can be traced to this period

carbon fixation the process of incorporating CO_2 into carbohydrate molecules

carboxyl terminus the free carboxyl group at one end of a polypeptide

carrying capacity the maximum number of organisms that can be sustained by available resources over a given period of time

catabolic reactions reactions that break macromolecules into their individual subunits

cataract a condition that occurs when the lens or cornea becomes opaque, preventing light from passing through

catastrophism Cuvier's theory that numerous global catastrophes in the past had repeatedly caused the extinction of species that were then replaced by newly created forms

central nervous system (CNS) the body's coordinating centre for mechanical and chemical actions; made up of the brain and spinal cord

centromeres constricted region of chromosomes that holds two replicated chromosome strands together

cerebellum part of the hindbrain that controls limb movements, balance, and muscle tone

cerebral cortex outer lining of the cerebral hemispheres

cerebrospinal fluid cushioning fluid that circulates between the innermost and middle membranes of the brain and spinal cord; it provides a connection between neural and endocrine systems

cerebrum largest and most highly developed part of the human brain, which stores sensory information and initiates voluntary motor activities

chaperone proteins special proteins that aid a growing polypeptide to fold into tertiary structure

chemiosmosis a process for synthesizing ATP using the energy of an electrochemical gradient and the ATP synthase enzyme

chemoautotrophic describes an organism capable of synthesizing its own organic molecules with carbon dioxide as a carbon source and oxidizing an inorganic substance as an energy source

chemoautotrophs organisms that can build all the organic compounds required for life from simple inorganic materials without using light energy

chlorophyll the light-absorbing green-coloured pigment that begins the process of photosynthesis

chloroplasts membrane-bound organelles in plant and algal cells that carry out photosynthesis

cholinesterase enzyme, which breaks down acetylcholine, that is released from postsynaptic membranes in the end plates of neurons shortly after acetylcholine

choroid layer middle layer of tissue in the eye that contains blood vessels and dark pigments that absorb light to stop reflection

chromatin complex of DNA and histone proteins located in the nucleus of eukaryotes

cladistics a phylogenetic system of classification used to infer and construct cladograms based on shared derived traits

cloned a fragment of DNA that has been introduced into a foreign cell, resulting in exact copies of the original DNA fragment being made when the foreign cell replicates and divides

closed population a population in which change in size and density is determined by natality (birth rate) and mortality (death rate) alone

clumped dispersion the pattern in which individuals in a population are more concentrated in certain parts of a habitat

cochlea coiled structure of the inner ear that responds to various sound waves, and converts them into nerve impulses

coding strand the strand of DNA that is not used for transcription and is identical in sequence to mRNA, except it contains uracil instead of thymine

codon sequence of three bases in DNA or complementary mRNA that serves as a code for a particular amino acid

coenzymes organic nonprotein cofactors that are needed for some enzymes to function

coevolution process, sometimes referred to as reciprocal adaptation, in which one species evolves in response to the evolution of another species

cofactors nonprotein components, such as dissolved ions, that are needed for some enzymes to function

collecting duct tube that carries urine from nephrons to the pelvis of a kidney

commensalism a symbiotic relationship in which one organism benefits and the other organism neither benefits nor is harmed; it is categorized as a +/0 relationship

community all populations in a given ecosystem at a given time

competent cell a cell that readily takes up foreign DNA

competitive inhibitors substances that compete with the substrate for an enzyme's active site

complement proteins plasma proteins that help defend against invading microbes by tagging the microbe for phagocytosis, puncturing cell membranes, or triggering the formation of a mucous coating

complementary base pairing pairing of the nitrogenous base of one strand of DNA with the nitrogenous base of another strand; adenine (A) pairs with thymine (T), and guanine (G) pairs with cytosine (C)

condensation reaction (dehydration synthesis) a reaction that creates a covalent bond between two interacting subunits, linking them to each other

cones photoreceptors that operate in bright light to identify colour

conformation the three-dimensional shape of a protein determined by the sequence of amino acids it contains

constant-length strands mixture of strands of DNA that have been replicated and are of equal length

contractile vacuole a structure in unicellular organisms that maintains osmotic equilibrium by pumping fluid out from the cell

convergent evolution occurs when two or more species become increasingly similar in phenotype in response to similar selective pressures

copy number number of copies of a particular plasmid found in a bacterial cell

corepressor a molecule (usually the product of an operon) that binds to a repressor to activate it

cornea transparent part of the sclera that protects the eye and refracts light toward the pupil of the eye

corpus callosum nerve tract that joins the two cerebral hemispheres

corpus luteum a mass of follicle cells that forms within the ovary after ovulation; secretes estrogen and progesterone

cortex outer layer of the kidney

cortisol hormone that stimulates the conversion of amino acids to glucose by the liver

coupled transcription-translation a phenomenon in which ribosomes of bacteria start translating an mRNA molecule that is still being transcribed

crassulacean acid metabolism (CAM) a photosynthetic mechanism in which stomata open at night so that plants can take in CO_2 and incorporate it into organic acids, and close during the day to allow the organic acids to release CO_2 molecules that enter the C_3 Calvin cycle to be fixed into carbohydrates

cristae the folds of the inner mitochondrial membrane

crude density population density measured in terms of number of organisms of the same species within the total area of the entire habitat

cumulative effects the accumulated changes to natural environments cause by several interactions, factors, or changing conditions

cumulative selection the accumulation of many small evolutionary changes over long periods of time and many generations, resulting in a significant new adaptation relative to the ancestral species

cuticle the water-resistant waxy layer on the external surfaces of a leaf or stem

cyanobacteria the largest group of photosynthesizing prokaryotes

cyclic adenosine monophosphate (cyclic AMP) secondary chemical messenger that directs the synthesis of protein by ribosomes

cyclic electron flow flow of photon-energized electrons from photosystem I, though an electron transport chain that produces ATP by chemiosmosis, but no NADPH

cytokinesis division of the cytoplasm and organelles of a cell into two daughter cells

D

daughter isotope what a parent isotope changes into during radioactive decay; may be stable or may be radioactive and capable of further decay

deamination the first step in protein catabolism, involving the removal of the amino group of an amino acid as ammonia

deletion the elimination of a base pair or group of base pairs from a DNA sequence

demographic transition model a four-stage model that describes the relationship between economic development and changes in population patterns

demography the study of the growth rate, age structure, and other characteristics of human populations

dendrites projections of cytoplasm that carry impulses toward the cell body

density-dependent factor a factor that influences population regulation, having a greater impact as population density increases or decreases

density-independent factors factors that influence population regulation regardless of population density

deoxyribonucleic acid (DNA) a double-stranded polymer of nucleotides (each consisting of a deoxyribose sugar, a phosphate, and four nitrogenous bases) that carries the genetic information of an organism

deoxyribonucleoside triphophates molecules composed of a deoxyribose bonded to three phosphate groups and a nitrogenous base

deoxyribose sugar sugar molecule containing five carbons that has lost the hydroxyl group on its 2′ carbon

depolarization diffusion of sodium ions into the nerve cell resulting in charge reversal

desalinization removal of salt, usually from seawater, to produce water suitable for drinking or irrigating croplands

diabetes chronic disease that occurs when the body cannot produce any insulin or enough insulin, or is unable to use properly the insulin it does make

dideoxy analogue nucleoside triphosphate whose ribose sugar does not possess a hydroxyl group on the 2′ and the 3′ carbon

disruptive selection selection that favours two or more variations or forms of a trait that differ from the current population average

distal tubule conducts urine from the loop of Henle to the collecting duct

disulfide bridge covalent bonds between cysteine residues in a polypeptide that stabilize tertiary structure

divergent evolution occurs when two or more species evolve increasingly different traits, resulting from differing selective pressures or genetic drift

DNA fingerprinting pattern of bands on a gel, originating from RFLP analysis or PCR, that is unique to each individual

DNA gyrase the bacterial enzyme that relieves the tension produced by the unwinding of DNA during replication

DNA helicase the enzyme that unwinds double-helical DNA by disrupting hydrogen bonds

DNA ligase the enzyme that joins DNA fragments together by catalyzing the formation of a bond between the 3′ hydroxyl group and a 5′ phosphate group on the sugar-phosphate backbones

DNA polymerase I an enzyme that removes RNA primers and replaces them with the appropriate deoxyribonucleotides during DNA replication

DNA polymerase III the enzyme responsible for synthesizing complementary strands of DNA during DNA replication

DNA primers short sequences of DNA nucleotides that are complementary to the opposing 3′-to-5′ ends of the DNA target sequence that is to be replicated

dynamic equilibrium in homeostasis, the condition that remains stable within fluctuating limits; in populations, the condition in which the birth rate equals the death rate and there is no net change in population size

E

eco-cities urban centres that are planned to minimize their impact on the local and regional environments and to foster sustainable lifestyles

ecological density population density measured in terms of the number of individuals of the same species per unit area or volume actually used by the individuals

ecological footprint an estimate of the amount of biologically productive land and water needed to supply the resources and assimilate the wastes produced by a human population

ecological niche an organism's biological characteristics, including use of and interaction with abiotic and biotic resources in its environment

ectoparasites parasites that live and feed on the outside surface of the host, such as lice, ticks, and parasitic mites

efferent arterioles small branches that carry blood away from the glomerulus to a capillary net

electrochemical gradient a concentration gradient created by pumping ions into a space surrounded by a membrane that is impermeable to the ions

electromagnetic (EM) radiation a form of energy that travels at 3×10^8 m/s in wave packets called photons that include visible light

electromagnetic spectrum photons separated according to wavelength

electron transport chain (ETC) a series of membrane-associated protein complexes and cytochromes that transfer energy to an electrochemical gradient by pumping H^+ ions into an intermembrane space

electronegativity a measure of an atom's ability to attract a shared electron pair when it is participating in a covalent bond

endergonic reaction a chemical reaction in which the energy of the products is more than the energy of the reactants; chemists call it an endothermic reaction

endocrine hormones chemicals secreted by endocrine glands directly into the blood

endoparasites parasites that live and feed within the host's body

endorphins natural painkillers belonging to a group of chemicals called neuropeptides; contain between 16 and 31 amino acids

endosymbiosis relationship in which a single-celled organism lives within the cell(s) of another organism; recent findings suggest this may be very common

endosymbiotic physical and chemical contact between one species and another species living within its body, which is beneficial to at least one of the species

energy the ability to do work

Glossary

energy pyramid a model that illustrates energy flow from producers at the beginning of food chains to consumers farther along

enkephalins natural painkillers belonging to a group of chemicals called neuropeptides; contain five amino acids and are produced by the splitting of larger endorphin chains

entropy a measure of the randomness or disorder in a collection of objects or energy; symbolized by S

enzyme–substrate complex an enzyme with its substrate attached to the active site

epidemic describes any disease that spreads rapidly through a population

epidermis layer the transparent, colourless layer of cells below the cuticle of a leaf or stem

epinephrine hormone, produced in the adrenal medulla, that accelerates heart rate and body reactions during a crisis (the fight-or-flight response); also known as adrenaline

equilibrium a condition in which opposing reactions occur at equal rates

essential amino acids amino acids that the body cannot synthesize from simpler compounds; they must be obtained from the diet

ester linkage a functional group linkage formed by the condensation of a carboxyl group and a hydroxyl group

estrogen female sex hormone that activates the development of female secondary sex characteristics, including development of the breasts and body hair, and increased thickening of the endometrium

ethanol fermentation a form of fermentation occurring in yeast in which NADH passes its hydrogen atoms to acetaldehyde, generating carbon dioxide, ethanol, and NAD^+

ethidium bromide a carcinogenic, flat molecule that inserts itself among the rungs of the ladder of DNA and fluoresces under UV light

eubacteria prokaryotic organisms that are distinguished from archaebacteria and eukaryotic cells by differences in genomic, chemical, and structural traits

eukaryotic cells cells possessing a cell nucleus and other membrane-bound organelles

eustachian tube air-filled tube of the middle ear that equalizes pressure between the external and internal ear

evolution the process in which significant changes in the inheritable traits (i.e., the genetic makeup) of a species or population occur over time

excitation the absorption of energy by an atom's electron

exergonic reaction a chemical reaction in which the energy of the products is less than the energy of the reactants; chemists call it an exothermic reaction

exons segments of DNA that code for part of a specific protein

exonuclease an enzyme that cuts out nucleotides at the end of a DNA strand

exploitative competition interspecific competition that involves consumption of shared resources by individuals of different species, where consumption by one species may limit resource availability to other species

exponential growth a pattern of population growth where organisms reproduce continuously at a constant rate

extreme halophiles species of archaebacteria that live in such extremely saline environments as landlocked saline lakes

extreme thermophiles species of archaebacteria that live in very hot aquatic environments, such as hot springs and hydrothermal vents on the ocean floor

F

facultative anaerobes organisms that obtain energy by oxidizing inorganic substances with or without oxygen

farsightedness a condition that occurs when the image is focused behind the retina

fecundity the potential for a species to produce offspring in one lifetime

feedback inhibition a method of metabolic control in which a product formed later in a sequence of reactions allosterically inhibits an enzyme that catalyzes a reaction occurring earlier in the process

fermentation a process in which the hydrogen atoms of NADH are transferred to organic compounds other than an electron transport chain

filtration process by which blood or body fluids pass through a selectively permeable membrane

fission asexual reproduction typical of bacteria in which the cell divides into two daughter cells

fitness general term referring to lifetime reproductive success of an individual

5′ cap a 7-methyl guanosine added to the start of a primary transcript to protect it from digestion in the cytoplasm and to bind it to the ribosome as part of the initiation of translation

fixed manner of describing the frequency of an allele within a population when only a single allele is present for a particular gene (i.e., the allele's frequency is 100%)

flow phase phase of the menstrual cycle marked by shedding of the endometrium

fluorescence the release of energy as light as an atom's electron returns to its ground state

fluoresces glows under UV light because of the excitation of a molecule's electrons

follicles structures in the ovary that contain the egg and secrete estrogen

follicle-stimulating hormone (FSH) in females, a gonadotropin that promotes the development of the follicles in the ovary; in males, a hormone that increases sperm production

follicular phase phase marked by development of ovarian follicles before ovulation

food chain a sequence of linked organisms that feed on one another, starting with a producer and continuing with consumers; energy flow decreases proportionately for each additional consumer level

fossilization the process by which traces of past organisms become part of sedimentary rock layers or, more rarely, hardened tar pits, volcanic ash, peat bogs, or amber

fossils any preserved remains or traces of an organism or its activity; many fossils are of such hardened body parts as bone

founder effect genetic drift that results when a small number of individuals separate from their original population and find a new population

fovea centralis area at centre of retina where cones are most dense and vision is sharpest

frameshift mutation a mutation that causes the reading frame of codons to change, usually resulting in different amino acids being incorporated into the polypeptide

free energy (Gibbs free energy) energy that can do useful work

functional groups reactive clusters of atoms attached to the carbon backbone of organic molecules

fundamental niche the biological characteristics of the organism and the set of resources individuals in the population are theoretically capable of using under ideal conditions

G

ganglian collections of nerve cell bodies located outside of the central nervous system

gel electrophoresis separation of charged molecules on the basis of size by sorting through a gel meshwork

gene duplication a mutation leading to the production of an extra copy of a gene locus, usually resulting from unequal crossing over

gene flow the movement of alleles from one population to another through the movement of individuals or gametes

gene pool total of all alleles within a population

gene regulation the turning on or off of specific genes depending on the requirements of an organism

gene therapy the alteration of a genetic sequence in an organism to prevent or treat a genetic disorder

genes a sequence of nucleotides in DNA that perform a specific function, such as coding, for a particular protein

genetic drift changes to allele frequency and combination as a result of random chance; such changes are much more pronounced in small populations

genetic engineering altering the sequence of DNA molecules

genetic screening process by which an individual's DNA is scanned for genetic mutations

genome the complete set of chromosomes of an organism, containing all its genes and associated DNA

genotype frequency for a particular pair of homologues genes, the proportion of those of a particular genotype (e.g., the frequency of *AA*, *Aa*, and *aa*)

genotype the set of all alleles possessed by an individual organism (also, a specific set of alleles at one or more loci)

geometric growth a pattern of population growth where organisms reproduce at fixed intervals at a constant rate

glaucoma disorder of the eye caused by buildup of fluid in the chamber anterior to the lens

glial cells nonconducting cells important for structural support and metabolism of the nerve cells

globular proteins protein molecules composed of one or more polypeptide chains that take on a rounded, spherical shape

glomerulus high-pressure capillary bed that is the site of filtration

glucagon hormone produced by the pancreas; when blood sugar levels are low, glucagon promotes conversion of glycogen to glucose

glucocorticoids various hormones, produced by the adrenal cortex, designed to help the body meet the demands of stress

glycolysis a process for harnessing energy in which a glucose molecule is broken into two pyruvate molecules in the cytoplasm of a cell

glycosidic linkages covalent bonds holding monosaccharides to one another that are formed by condensation reactions in which the H atom comes from a hydroxyl group of one sugar and the —OH group comes from a hydroxyl group of the other

glycosyl bond a bond between a sugar and another organic molecule by way of an intervening nitrogen or oxygen atom

goiter disorder that causes an enlargement of the thyroid gland

gonadotropic hormones hormones produced by the pituitary gland that regulate the functions of the testes in males and the ovaries in females

gonadotropin-releasing hormone (GnRH) chemical messenger from the hypothalamus that stimulates secretions of FSH and LH from the pituitary

grana a stack of thylakoids

greenhouse effect a result of certain atmospheric gases, such as CO_2, water vapour, and methane, trapping heat in the atmosphere by letting visible sunlight penetrate to Earth's surface, which absorbing most wavelengths of infrared radiation that radiate from Earth's surface

gross national product (GNP) the total value of all goods and services produced by a county in a particular time frame, usually one year; the value can be given per capita

ground state the lowest possible potential energy level of an atom's electron

growth hormone (GH) hormone, produced by the pituitary gland, that stimulates growth of the body; also known as somatotropin (STH)

growth momentum the inherent future growth in a population that results from a disproportionately large fraction of the population being of younger age

guard cells photosynthetic epidermal cells of a leaf or stem that form and regulate the size of an opening called a stoma

H

habitat the place where an organism or species normally lives

half-life the time it takes for one half of the nuclei in a radioactive sample to decay

harmful mutation an inheritable change in a cell's DNA that impairs the proper operation of a gene product or regulatory function or adversely affects mitosis or meiosis

helper T cells T cells with receptors that bind to fragments of antigens

heterotrophic describes an organism unable to make its own primary supply of organic compounds (i.e., feeds on autotrophic, other heterotrophic, or organic material)

heterotrophs organisms that feed on other organisms to obtain chemical energy

histones positively charged proteins that bind to negatively charged DNA in chromosomes

homeostasis process by which a constant internal environment is maintained despite changes in the external environment

hominid humans and other extinct members of the lineage arising from the most recent common ancestor that humans share with the apes

hominoids the apes, humans, and extinct members of their clade

homologous features structures that share a common origin but may serve different functions in modern species (e.g., dolphin flippers and human hands)

homoplasies similar traits found in two or more different species, resulting from convergent evolution or from reversals, not from common descent; also called analogous features

hormones chemicals released by cells that affect cells in other parts of the body

host cell a cell that has taken up a foreign plasmid or virus and whose cellular machinery is being used to express the foreign DNA

housekeeping genes genes that are switched on all the time because they are needed for life functions vital to an organism

hybridization for atoms, a modification of the valence orbitals that changes the orientation of the valence electrons; in genetic coding, complementary base pairing between strands of nucleic acids via hydrogen bonding

hybrids the offspring of genetically dissimilar parents or breeding stock

hydrolysis reaction a catabolic reaction in which a water molecule is used to break a covalent bond holding subunits together

hydrophilic having an affinity to water; the tendency of polar and ionic substances to dissolve in water

hydrophobic having an aversion to water; the tendency of nonpolar molecules to exclude water

hyperpolarized condition in which the inside of the nerve cell membrane becomes more negative than the resting potential

hypothalamus region of the vertebrate's brain responsible for coordinating many nerve and hormone functions

I

immiscible describes liquids that form separate layers instead of dissolving

immutable unchanged and unchanging, believed (before evolutionary theory became accepted) to be characteristic of life forms

indicator species a species sensitive to small changes in environmental conditions; monitoring the health of their population is used to indirectly monitor the health of the environment in which they live

indicators chemicals, organisms, or other components of an ecosystem that can be measured or observed to identify changes in the quality of conditions in an ecosystem

induced-fit model a model of enzyme activity that describes an enzyme as a dynamic protein molecule that changes shape to better accommodate the substrate

induced mutations mutations caused by a chemical agent or radiation

inducer a molecule that binds to a repressor protein and causes a change in conformation, resulting in the repressor protein falling off the operator

inflammatory response localized nonspecific response triggered when tissue cells are injured by bacteria or physical injury, characterized by swelling, heat, redness, and pain

ingroup members of a clade having one or more synapomorphies

insertion the placement of an extra nucleotide in a DNA sequence

insulin hormone produced by the islets of Langerhans in the pancreas; insulin is secreted when blood sugar levels are high

interference competition interspecific competition that involves aggression between individuals of different species who fight over the same resource(s)

intermembrane space the fluid-filled space between the inner and outer mitochondrial membranes

intermolecular bonds chemical bonds between molecules

intramolecular forces of attraction the covalent bond that holds the atoms of a molecule together, and the ionic bond that holds ions together in a salt

interspecific competition interactions between individuals of different species for an essential common resource that is in limited supply

interstitial fluid fluid that surrounds the body cells

intraspecific competition an ecological interaction in which individuals of the same species or population compete for resources in their habitat

introns noncoding region of a gene

inversion the reversal of a segment of DNA within a chromosome

iris opaque disk of tissue surrounding the pupil that regulates the amount of light entering the eye

isomers molecules with the same chemical formula but with a different arrangement of atoms

isotope different atoms of the same element containing the same number of protons but a different number of neutrons

K

killer T cells T cells that puncture the cell membranes of cells infected with foreign invaders, killing the cells and the invaders; also destroy body cells that have become infected by a virus

kin selection the natural selection of a behaviour or trait of one individual that enhances the success of closely related individuals, thereby increasing the first individual's fitness indirectly

kinetic energy energy possessed by moving objects

Krebs cycle a cyclic series of reactions that transfers energy from organic molecules to ATP, NADH, and $FADH_2$ and removes carbon atoms as CO_2

L

***lac* operon** a cluster of genes under the control of one promoter and one operator; the genes collectively code for the enzymes and proteins required for a bacterial cell to use lactose as a nutrient

***Lac*I protein** a repressor protein that binds to the *lac* operon operator, preventing RNA polymerase from transcribing the *lac* operon genes

lactate (lactic acid) fermentation a form of fermentation occurring in animal cells in which NADH transfers its hydrogen atoms to pyruvate, regenerating NAD^+ and lactate

lactose a disaccharide that consists of the sugars glucose and galactose

lag phase the initial stage in which population growth rates are slow as a result of a small population size; characteristic of geometric, exponential, and logistic population growth

lagging strand the new strand of DNA that is synthesized in short fragments, which are later joined together

lamellae (singular: lamella) unstacked thylakoids between grana

leading strand the new strand of DNA that is synthesized continuously during DNA replication

leukocytes white blood cells that may engulf invading microbes or produce antibodies

light reactions the first set of reactions of photosynthesis in which light energy excites electrons in chlorophyll molecules, powers chemiosmotic ATP synthesis, and results in the reduction of $NADP^+$ to NADPH

light-compensation point the point on a light-response curve at which the rate of photosynthetic CO_2 uptake exactly equals the rate of respiratory CO_2 evolution

light-saturation point the light intensity at which the carbon fixation reactions reach a maximum overall rate

limiting factor any essential resource that is in short supply or unavailable

lineage the descendants of a common ancestor

LINEs repeated DNA sequences of 5000 to 7000 base pairs in length that alternate with lengths of DNA sequences found in the genomes of higher organisms

liposomes spherical arrangements of lipid molecules that form spontaneously in water

loci the locations along a DNA molecule of a particular gene (singular: locus)

log phase the stage in which population growth rates are very rapid; characteristic of geometric, exponential, and logistic growth

logistic growth a model of population growth describing growth that levels off as the size of the population approaches its carrying capacity

loop of Henle carries filtrate from the proximal tubule to the distal tubule

luteal phase phase of the menstrual cycle characterized by the formation of the corpus luteum following ovulation

luteinizing hormone (LH) in females, a gonadotropin that promotes ovulation and the formation of the corpus luteum; in males, a hormone that regulates the production of testosterone

lymphocytes specialized white blood cells that produce antibodies

lymphokine protein produced by the T cells that acts as a chemical messenger between other T cells and B cells

M

macroevolution large-scale evolutionary change significant enough to warrant the classification of groups or lineages into district genera or even higher-level taxa

macromolecules large molecules sometimes composed of a large number of repeating subunits

macroparasites larger parasites, such as tapeworms, fleas, and ticks, that are readily visible

macrophages phagocytic white blood cells found in lymph nodes or in the blood of the bone marrow, spleen, and liver

mark–recapture method sampling techniques for estimating population size and density by comparing the proportion of marked and unmarked animals captured in a given area; sometimes called capture-recapture

matrix the fluid that fills the interior space of the mitochondrion

maximum oxygen consumption, VO$_2$ max the maximum volume of oxygen, in milliliters, that the cells of the body can remove from the bloodstream in one minute per kilogram of body mass while the body experiences maximal exertion

medulla area inside of the cortex

medulla oblongata region of the hindbrain that joins the spinal cord to the cerebellum; one of the most important sites of autonomic nerve control

memory B cell cell that retains information about the geometry of an antigen or antigens

meninges protective membranes that surround the brain and spinal cord

mesophyll layers the photosynthetic cells that form the bulk of a plant leaf

messenger RNA (mRNA) the end product of transcription of a gene, mRNA is translated by ribosomes into protein

metabolic rate the amount of energy consumed by an organism in a given time

metabolism the sum of all anabolic and catabolic processes in a cell or organism

methanogens species of archaebacteria that produce methane as a waste product

methyl group CH$_3$

methylases enzymes that add a methyl group to one of the nucleotides found in a restriction endonuclease recognition site, altering its chemical composition

micelles spheres formed when phospholipids are added to water

microevolution changes in gene (allele) frequencies and phenotypic traits within populations and species; microevolution can result in the formation of new species

microfossils microscopic remains of tiny organisms or structures that have hard and resistant outer coverings

microparasites parasites, such as plasmodia or trypanosomes, that are too small to see with the naked eye

mineralocorticoids hormones of the adrenal cortex important for regulation of salt-water balance

minimum viable population size the smallest number of individuals needed for a population to continue for a given period of time

miscible describes liquids that dissolve into one another

missense mutation a mutation that results in the single substitution of one amino acid in the resulting polypeptide

mitochondria eukaryotic cell organelle in which aerobic cellular respiration occurs

mitosis division of the nucleus of a eukaryotic cell into two daughter nuclei with identical sets of chromosomes

monoclonal antibodies antibodies secreted by hybridomas; new cells created by the fusion (or cloning) of two cells

monophyletic group (clade) all the descendants of a single common ancestor

motor neurons neurons that carry impulses from the central nervous system to effectors; also known as efferent neurons

mRNA transcript mRNA that has been modified for exit out of the nucleus and into the cytoplasm

multidrug-resistant tuberculosis *Mycobacterium tuberculosis* that is resistant to both antibiotics usually used to treat it: isoniazid and rifampin

multiple-cloning site region in plasmid that has been engineered to contain recognition sites of a number of restriction endonucleases

mutagenic agents agents that can cause a mutation

mutations changes in the DNA sequence that are inherited

mutualism a symbiotic relationship in which both organisms benefit; as neither is harmed it is categorized as a +/+ relationship

myelin sheath insulated covering over the axon of a nerve cell

N

nearsightedness a condition that occurs when the image is focused in front of the retina

negative feedback process by which a mechanism is activated to restore conditions to their original state

nephrons functional units of the kidneys

neurilemma delicate membrane that surrounds the axon of some nerve cells

neurons nerve cells that conduct nerve impulses

neurotransmitters chemicals released from vesicles into synapses

neutral mutation has no immediate effect on an individual's fitness; most neutral mutations are silent or occur in non-coding DNA

neutralization reaction the reaction of an acid and a base to produce water and a salt

nicotinamide adenine dinucleotide, NAD$^+$ coenzyme used to shuttle electrons to the first component of the electron transport chain in the mitochondrial inner membrane

nitrogenous base an alkaline, cyclic molecule containing nitrogen

nanometer (nm) the equivalent of 10^{-9} m

nodes of Ranvier regularly occurring gaps between sections of myelin sheath along the axon

nomograms graphical methods for determining the value of an unknown quantity when the values of other quantities that it is mathematically related to are known

noncompetitive inhibitors substances that attach to a binding site on an enzyme other than the active site, causing a change in the enzyme's shape and a loss of affinity for its substrate

noncyclic electron flow the process in which photon-energized electrons flow from water to NADP$^+$ through electron transport chains in thylakoid membranes, producing NADPH by reduction and ATP by chemiosmosis

nonsense mutation a mutation that converts a codon for an amino acid into a termination codon

norepinephrine also known as noradrenaline, it initiates the fight-or-flight response by increasing heart rate and blood sugar

nucleomorph tiny nucleus containing genomic material found within a eukaryotic endosymbiotic structure originally believed to be derived from primitive red alga

nucleosome a complex of eight histones enveloped by coiled DNA

nucleotides molecules that consists of a five-carbon sugar (deoxyribose or ribose) with a nitrogenous base attached to their 1′ carbon and a phosphate group attached to their 5′ carbon

O

obligate aerobes organisms that obtain energy by oxidizing organic substances using oxygen

obligate anaerobes organisms that cannot live in the presence of oxygen and obtain energy by oxidizing inorganic substances

obligatory mutualism a symbiotic relationship in which neither species involved could survive without the other

Okazaki fragments short fragments of DNA that are a result of the synthesis of the lagging strand during DNA replication

olfactory lobes areas of the brain that process information about smell

oligosaccharides sugars containing several simple sugars attached to one another

open population a population in which change in number and density is determined by births, deaths, immigration, and emigration

operator regulatory sequences of DNA to which a repressor protein binds

operon a cluster of genes under the control of one promoter and one operator in prokaryotic cells; acts as a simple regulatory loop

orbitals volumes of space around the nucleus where electrons are most likely to be found

organ of Corti primary sound receptor in the cochlea

osmoreceptors specialized nerve cells in the hypothalamus that detect changes in the osmotic pressure of the blood and surrounding extracellular fluids (ECF)

ossicles tiny bones that amplify and carry sound in the middle ear

otoliths tiny stones of calcium carbonate embedded in a gelatinous coating within the saccule and utricle

outgroup the first group to have diverged from the other members of a clade being considered in a phylogenetic analysis

oval window oval-shaped hole in the vestibule of the inner ear, covered by a thin layer of tissue

ovulation release of the egg from the follicle held within the ovary

oxidation a chemical reaction in which an atom loses one or more electrons

oxidative phosphorylation mechanism forming ATP indirectly through a series of enzyme-catalyzed redox reactions involving oxygen as the final electron acceptor

oxidizing agent the substance that gains an electron in a redox reaction; the substance that causes the oxidized atom to become oxidized

oxygen debt the extra oxygen required to catabolize lactate to CO_2 and H_2O

P

P (peptide) site site in the ribosome where peptide bonds are formed between adjoining amino acids on a growing polypeptide chain

palaeontology the scientific study of fossil remains

pandemic an epidemic that spreads over a very wide geographical area, in some cases sweeping the entire globe

parasitism a symbiotic relationship in which one organism (the parasite) benefits at the expense of another organism (the host), which is often harmed but usually not killed; it is categorized as a $+/-$ relationship

parasympathetic nervous system nerve cells of the autonomic nervous system that return the body to normal resting levels after adjustments to stress

parathyroid glands four pea-sized glands in the thyroid gland that produce parathyroid hormone to regulate blood calcium and phosphate levels

parathyroid hormone (PTH) hormone produced by the parathyroid glands, which will increase calcium levels in the blood and lower the levels of phosphates

parent isotope changes into a daughter isotope as radioactive decay occurs

passive immunity occurs when antibodies are introduced into the body directly; provides only temporary protection from the agent, unlike active immunity

pathogens disease-causing organisms

peptide bonds the amide linkage that holds amino acids together in polypeptides

peripheral nervous system (PNS) all parts of the nervous system, excluding brain and spinal cord, that relay information between the central nervous system and other parts of the body

peritubular capillaries network of small blood vessels that surround the nephrons

permineralized fossil remains of organisms replaced by mineral deposits when dissolved minerals precipitate from a solution in the space occupied by the organism's remains

phagocytosis process by which a white blood cell engulfs and chemically destroys a microbe

phenotypes observable traits of an organism that result from interaction between genes and the environment

phosphate group group of four oxygen atoms surrounding a central phosphorous atom found in the backbone of DNA

phosphorylation the process of attaching a phosphate group to an organic molecule

photoautotrophs organisms that can build all the organic compounds required for life from simple inorganic materials, using light in the process

photons packets of EM radiation; also known as quanta

photophosphorylation the light-dependent formation of ATP by chemiosmosis in photosynthesis

photorespiration oxidation of RuBP by rubisco and oxygen in light to form glycolate, which subsequently releases carbon dioxide

photosynthetic efficiency the net amount of carbon dioxide uptake per unit of light energy absorbed, also called quantum yield

photosynthetically active radiation (PAR) wavelengths of light between 400 nm and 700 nm that support photosynthesis

photosystem I a photosystem embedded in the thylakoid membrane containing chlorophyll P700

photosystem II a photosystem embedded in the thylakoid membrane containing chlorophyll P680

photosystems clusters of photosynthetic pigments embedded in the thylakoid membranes of chloroplasts that absorb light energy

phylogenetic tree (cladogram) diagram depicting the proposed evolutionary relationships, or common ancestry by descent, of groups of species or populations

phylogeny the theoretical evolutionary history of a species or group

pinna outer part of the ear that acts like a funnel, taking sound from a large area and channeling it into a small canal

pituitary gland gland at the base of the brain that, together with the hypothalamus, functions as a control centre, coordinating the endocrine and nervous systems

plasmids small circular pieces of DNA that can exit and enter bacterial cells

plasmodesmata membrane-lined channels between plant cells that allow for the movement of some substances from cell to cell

pluripotent cells cells that are capable of differentiating into a number of different specialized cells, such as neurons or muscle cells

point mutations changes in single base-pairs along the DNA molecule

polarized membrane membrane charged by unequal distribution of positively charged ions inside and outside the nerve cell

poly-A polymerase enzyme responsible for adding a string of adenine bases to the end of mRNA to protect it from degradation later on

poly-A tail a string of 200 to 300 adenine base pairs at the end of an mRNA transcript

polyacrylamide artificial polymer used to form a gel meshwork for electrophoresis

polymerase chain reaction (PCR) amplification of DNA sequence by repeated cycles of strand separation and replication

polymorphism any difference in DNA sequence, coding or noncoding, that can be detected between individuals

polyploids individuals with three or more complete sets of chromosomes

pons region of the brain that acts as a relay station by sending nerve messages between the cerebellum and the medulla

population a group of individuals of the same species living within a particular area or volume

population density the number of individuals of the same species that occur per unit area or volume

population dispersion the general pattern in which individuals are distributed through a specified area

population dynamics changes in population characteristics determined by natality, mortality, immigration, and emigration

population size the number of individuals of a specific species occupying a given area/volume at a given time

porphyrin the light-absorbing portion of the chlorophyll molecule, containing a magnesium atom surrounded by a hydrocarbon ring

positive feedback process by which a small effect is amplified

postsynaptic neuron neuron that carries impulses away from the synapse

postzygotic mechanisms reproductive isolating mechanisms that prevent maturation and reproduction in offspring from interspecies reproduction

potential energy energy stored by virtue of an object's position within an attractive or repulsive force field

potential energy diagram a diagram showing the changes in potential energy that take place during a chemical reaction

predation an ecological interaction in which a predator (a member of one species) catches, kills, and consumes prey (a member of another species)

presynaptic neuron neuron that carries impulses to the synapse

prezygotic mechanisms reproductive isolating mechanisms that prevent interspecies mating and fertilization (e.g., ecological isolation, temporal isolation, and behavioural isolation)

primary abiogenesis theory that the first living things on Earth arose from nonliving material

primary electron acceptor a compound embedded in the thylakoid membrane that is reduced by an excited chlorophyll electron

primary transcript mRNA that has to be modified before exiting the nucleus in eukaryotic cells

primary structure the unique sequence of amino acids in a polypeptide chain

primase the enzyme that builds RNA primers

progesterone female sex hormone produced by the ovaries that maintain uterine lining during pregnancy

prokaryotic cells cells possessing no intracellular membrane-bound organelles or nucleus

promoter sequence of DNA that binds RNA polymerase upstream of a gene

pronociceptive transmitters signal molecules that amplify pain sensation via a nuerological pathway

prostaglandins hormones that have a pronounced effect in a small localized area

protein hormones group of hormone, composed of chains of amino acids, that includes insulin and growth hormone

proteins complex molecules composed of one or more polypeptide chains made of amino acids and folded into specific three-dimensional shapes that determine protein function

proton–motive force (PMF) a force that moves protons through an ATPase complex on account of the free energy stored in the form of an electrochemical gradient of protons across a biological membrane

proximal tubule section of the nephron joining the Bowman's capsule with the loop of Henle

pseudogenes DNA sequences that are homologous to functioning genes but are not transcribed

Q

quadrats sampling areas of any shape used for estimating population size; frames can be real or virtual

quality of life social, economic, and cultural factors that support and enhance the development of a human population; examples of such factors, which can be refined and collected as statistical indictors to be monitored for change, include housing conditions, food supply, access to education, access to a livelihood, health care, and personal security

quaternary structure two or more polypeptide subunits forming a functional protein

R

radioactive decay the release of subatomic particles from the nucleus of an atom, which results in the change of a radioactive parent isotope into a daughter isotope; when the number of protons is altered, a different element is formed

radioactive tracers radioisotopes that are used to follow chemicals through chemical reactions and trace their path as they move through the cells and bodies of organisms

radioisotopes radioactive atoms of an element that spontaneously decay into smaller atoms, subatomic particles, and energy

radiometric dating calculation of the age of rock—and of embedded fossils or other objects—through the measurement of the decay of radioactive isotopes in the rock

random dispersion the pattern in which individuals are spread throughout a habitat in an unpredictable and patternless manner

reabsorption transfer of glomerular filtrate from the nephron back into the capillaries

reaction centre a transmembrane protein complex containing chlorophyll *a* whose electrons absorb light energy and begin the process of photosynthesis

reading frame one of three possible phases in which to read the bases of a gene in groups of three

realized niche the biological characteristics of the organism and the resources individuals in a population actually use under the prevailing environmental conditions

receptor sites ports along cell membranes into which nutrients and other needed materials fit

recognition site a specific sequence within double-stranded DNA, usually palindromic and consisting of four to eight nucleotides, that a restriction endonuclease recognizes and cleaves

recombinant DNA fragment of DNA composed of sequences originating from at least two different sources

redox reaction a chemical reaction involving the transfer of one or more electrons from one atom to another; a reaction in which oxidation and reduction occur

reducing agent the substance that loses an electron in a redox reaction; the substance that causes the reduced atom to become reduced

reduction a chemical reaction in which an atom gains one or more electrons

reflex arc neural circuit through the spinal cord that provides a framework for a reflex action

refractory period recovery time required before a neuron can produce another action potential

relative age an estimate of the age of a rock or fossil specimen in relation to other specimens

release factor a protein involved in the release of a finished polypeptide chain from the ribosome

renal pelvis area where the kidney joins the ureter

replication bubble the region where two replication forks are in close proximity to each other, producing a bubble in the replicating DNA

replication fork the region where the enzymes replicating a DNA molecule are bound to untwisted, single-stranded DNA

repolarization process of restoring the original polarity of the nerve membrane

repressor protein regulatory molecules that bind to an operator site and prevent the transcription of an operon

reproductive isolating mechanisms any behavioural, structural, or biochemical traits that prevent individuals of different species from reproducing successfully together

residue an amino acid subunit of a polypeptide

resource partitioning avoidance of, or reduction in, competition for similar resources by individuals of different species occupying different nonoverlapping ecological niches

resting potential the voltage difference across a nerve cell membrane during the resting stage (usually negative)

restriction endonucleases enzymes that are able to cleave double-stranded DNA into fragments at specific sequences; also known as restriction enzymes

restriction fragment length polymorphism (RFLP) analysis a technique in which DNA regions are digested using restriction endonuclease(s) and subjected to radioactive complementary DNA probes to compare the differences in DNA fragment lengths between individuals

retina innermost layer of tissue at the back of the eye containing photoreceptors

ribosomal RNA (rRNA) a form of RNA that binds with ribosomal protein to form ribosomes

ribozymes an RNA molecule able to catalyze a chemical reaction

ribulose bisphosphate carboxylase/oxygenase (rubisco) an enzyme in the stroma of chloroplasts that catalyzes the first reaction of the Calvin cycle; also catalyzes the oxidation of RuBP to CO_2

RNA primer a sequence of 10—60 RNA bases that is annealed to a region of single-stranded DNA for the purpose of initiating DNA replication

RNA polymerase enzyme that transcribes DNA into complementary mRNA

rods photoreceptors that operate in dim light to detect light in black and white

S

salinization the accumulation of excess salts in soil that restricts the amount of water and essential nutrients plants can withdraw from the soil

Sanger dideoxy method DNA sequencing technique based on DNA replication that uses dideoxy nucleoside triphosphates

Schwann cells special type of glial cell that produces the myelin sheath

sclera outer covering of the eye that supports and protects the eye's inner layers; usually referred to as the white of the eye

secondary structure coils and folds in a polypeptide caused by hydrogen bonds between hydrogen and oxygen atoms near the peptide bonds

secretion movement of materials, such as ammonia and some drugs, from the blood back into the distal tubule

self-pollinating plants plants that habitually fertilize themselves and produce viable offspring

semicircular canals fluid-filled structures within the inner ear that provide information about dynamic equilibrium

semiconservative process of replication in which each DNA molecule is composed of one parent strand and one newly synthesized strand

sensory adaptation occurs once you have adjusted to a change in the environment; sensory receptors become less sensitive when stimulated repeatedly

sensory neurons neurons that carry impulses from sensory receptors to the central nervous system; also known as afferent neurons

sensory receptors modified ends of sensory neurons that are activated by specific stimuli

sexual dimorphism striking differences in the physical appearance of males and females not usually applied to behavioural differences between sexes

sexual selection differential reproductive success that results from variation in the ability to obtain mates; results in sexual dimorphism and mating and courtship behaviours

signal molecule a molecule that activates an activator protein or represses a repressor protein

silent mutation a mutation that does not result in a change in the amino acid coded for and, therefore, does not cause any phenotypic change

SINEs repeated DNA sequences of 300 base pairs in length that alternate with lengths of DNA sequences found in the genomes of higher organisms

single-stranded binding proteins (SSBs) a protein that keeps separated strands of DNA apart

social parasites parasites that complete their life cycle by manipulating the social behaviour of their hosts

sodium-potassium pump an active transport mechanism that moves sodium ions out of a cell and potassium ions into a cell against their concentration gradients

soil degradation the removal of soil nutrients and loss of topsoil rich in organic matter, caused by erosion from the movement of water and wind

somatropin a drug identical to human growth hormone (somatotropin) used to treat growth deficiency

Southern blotting a procedure that allows the DNA in an electrophoresis gel to be transferred to a nylon membrane while maintaining the position of the DNA band fragments

speciation the evolutionary formation of new species

species organisms that resemble one another in appearance, behaviour, chemistry, and genetic makeup, and that interbreed, or have the ability to interbreed, with each other under natural conditions to produce fertile offspring

spectroscope an instrument that separates different wavelengths into an electromagnetic spectrum

spermatogenesis process by which spermatogonia divide and differentiate into mature sperm cells

spliceosomes particles made of RNA and protein that cut introns from mRNA primary transcript and joins together the remaining coding exon regions

spontaneous generation the idea that living organisms arise from nonliving matter

spontaneous mutations mutations occurring without chemical change or radiation but as a result of errors made in DNA replication

stabilizing selection selection against individuals exhibiting variations in a trait that deviate from the current population average

start codon specific codon (AUG) that signals to the ribosome that the translation commences at that point

stationary phase the phase in which population growth rates decrease as the population size reaches the carrying capacity and stabilizes, the defining characteristic of logistic population growth

steroid hormones group of hormones, made from cholesterol, that includes male and female sex hormones and cortisol

sticky ends fragment end of a DNA molecule with short single-stranded overhangs, resulting from cleavage by a restriction enzyme

stomata openings on the surface of a leaf that allow for the exchange of gases between air spaces in the leaf interior and the atmosphere

stop codons specific codons that signal the end of translation to a ribosome

strand a single nucleotide polymer

stroma the protein-rich semiliquid material in the interior of a chloroplast

stromatolites shaped rock formations that result from the fossilization of mats of ancient prokaryotic cells and sediment

strong acids acids that ionize completely in aqueous solution

strong bases bases that ionize completely in aqueous solution

substitution the replacement of one base in a DNA sequence by another base

substrate the reactant that an enzyme acts on when it catalyzes a chemical reaction

substrate-level phosphorylation mechanism forming ATP directly in an enzyme-catalyzed reaction

summation effect produced by the accumulation of neurotransmitters from two or more neurons

supercoiling DNA folded into a higher level of coiling than is already present in nucleosomes

suppressor T cell T cell that turns off the immune system

symbiosis various interactions in which two species maintain a close, usually physical, association; includes parasitism, mutualism, and commensalism

sympathetic nervous system nerve cells of the autonomic nervous system that prepare the body for stress

synapomorphies shared traits that evolved only once and have been inherited by two or more species

synapses regions between neurons, or between neurons and effectors

systems thinking the analysis of problems that focuses on the implications for an entire system rather than on the implications for individual components or parts of the system

T

T cell lymphocyte, manufactured in the bone marrow and processed by the thymus gland, that identifies and attacks foreign substances

T4 DNA ligase an enzyme that joins together DNA blunt fragments

Taq polymerase DNA polymerase, extracted from Thermus aquaticus, that is able to withstand high temperatures

telomeres long sequences of repetitive, noncoding DNA on the end of chromosomes

template a single-stranded DNA sequence that acts as the guiding pattern for producing a complementary DNA strand

template strand the strand of DNA that the RNA polymerase uses as a guide to build complementary mRNA

terminator sequence sequence of bases at the end of a gene that signals the RNA polymerase to stop transcribing

tertiary structure supercoiling of a polypeptide that is stabilized by side-chain interactions, including covalent bonds, such as disulfide bridges

testosterone male sex hormone produced by the interstitial cells of the testes

thalamus area of the brain that coordinates and interprets sensory information and directs it to the cerebrum

theory of gradualism a theory that attributes large evolutionary changes in species to the accumulation of many small and ongoing changes and processes

theory of punctuated equilibrium a theory that attributes large evolutionary changes to relatively rapid spurts of change followed by long periods of little or no change

thermal proteinoids a polypeptide that forms spontaneously when amino acids polymerize on hot surfaces; proteinoids are able to self-assemble into small cell-like structures in cool water

thermoregulation maintenance of body temperature within a range that enables cells to function efficiently

Thermus aquaticus species of bacteria found in hot springs

threshold level maximum amount of material that can be moved across the nephron

threshold level minimum level of a stimulus required to produce a response

thresholds the tolerance limits, which, if exceeded, will result in harmful or fatal reactions within ecosystems

thylakoid lumen the fluid-filled space inside a thylakoid

thylakoid membrane the photosynthetic membrane within a chloroplast that contains light-gathering pigment molecules and electron transport chains

thylakoids a system of interconnected flattened membrane sacs forming a separate compartment within the stroma of a chloroplast

thyroid gland a two-lobed gland at the base of the neck that controls metabolic processes

thyroxine iodine-containing hormone, produced by the thyroid gland, that increases the rate of body metabolism and regulates growth

Ti plasmid plasmid, found in Agrobacterium tumefaciens bacteria, that is able to enter plant cells

transcription factors proteins that switch on genes by binding to DNA and helping the RNA polymerase to bind

transcription the process in which DNA is used as a template for the production of complementary messenger RNA molecules

transfer RNA (tRNA) a form of RNA that is responsible for delivering amino acids to the ribosomes during the process of translation

transformation introduction of foreign DNA, usually by a plasmid or virus, into a bacterial cell

transgenic animals animals that have genes from other species incorporated into their DNA

transgenic organism in which foreign DNA has been artificially incorporated into its genome

transition state in a chemical reaction, a temporary condition in which the bonds within reactants are breaking and the bonds between products are forming

transitional forms a fossil or species intermediate in form between two other species in a direct line of descent

translation the process by which a ribosome assembles amino acids in a specific sequence to synthesize a specific polypeptide coded by messenger RNA

translocation the transfer of a fragment of DNA from one site in the genome to another location

transpiration the loss of water vapour from plant tissues, primarily through stomata

transposable elements segments of DNA that are replicated as a unit from one location to another on chromosomal DNA

triacylglycerols (triglycerides) lipids containing three fatty acids attached to a single molecule of glycerol

trophic level a level within a food chain or energy pyramid that demonstrates energy flow from producers (at the primary trophic level) to consumers (at secondary or tertiary trophic levels)

***trp* operon** a cluster of genes in a prokaryotic cell under the control of one promoter and one operator; the genes govern the synthesis of the necessary enzymes required to synthesize the amino acid tryptophan

tympanic membrane thin layer of tissue, also known as the eardrum

U

uniform dispersion the pattern in which individuals are equally spaced throughout a habitat

uniformitarianism the theory that Earth's surface has always changed and continues to change through similar, uniform, and very gradual processes

upstream region of DNA adjacent to the start of a gene

urban sprawl the growth of low-density development on the edges of cities and towns

urea nitrogen waste formed from two molecules of ammonia and one molecule of carbon dioxide

ureters tubes that conduct urine from the kidneys to the bladder

urethra tube that carries urine from the bladder to the exterior of the body

uric acid waste product formed from the breakdown of nucleic acids

V

vaccines antigen-containing substances that can be swallowed or injected to provide continued immunity to specific diseases

vagus nerve major cranial nerve that is part of the parasympathetic nervous system

valence electrons electrons located in outermost *s* and *p* orbitals that determine an atom's chemical behaviour

Valence Shell Electron Pair Repulsion (VESPR) theory a method for predicting molecular shape based on the mutual repulsion of valence electron pairs

van der Waals forces intermolecular forces of attraction including London forces, dipole–dipole forces, and hydrogen bonds

variable number tandem repeats (VNTRs) repetitive sequences of DNA that vary among individuals; also known as microsatellites

variable-length strands mixture of strands of DNA that have been replicated and are of unequal length

vascular bundles a system of tubes and cells that transport water and mineral from the roots to the leaf cells and carry carbohydrates from the leaves to other parts of a plant, including the roots

vector organism—most often an insect—that carries and transmits diseases to other organisms

vectors vehicles by which DNA may be introduced into host cells

vestibule chamber found at the base of the semicircular canals that provides information about static equilibrium

vestigial features rudimentary and nonfunctioning structures that are homologous to fully functioning structures in closely related species

virulent virus a harmful microbe that is effective at overcoming the defence mechanisms of the host

W

weak acids acids that partially ionize in aqueous solution

weak bases bases that partially ionize in aqueous solution

work the transfer of energy from one body or place to another

X

xanthophylls pigments in chloroplast membranes that give rise to the yellow colour in autumn leaves

xenotransplants transplants from one species to another; the word xeno means "strange" or "foreign"

Z

Z protein a protein associated with photosystem II that splits water into hydrogen ions, oxygen atoms, and electrons

zero population growth the condition in which the number of individuals added to a population from births and immigration equals the number of individuals removed from a population by deaths and emigration, resulting in a constant population size that neither increases nor decreases

Baseline data, 732
Bases
    conjugate, 21
    defined, 20
    properties, 20
    strong, 21
    weak, 21
Basilar membrane, 447
Basilosaurus, 613
B cells, 467
    memory, 470
Beadle, George, 234–35
Beagle (ship), 519–21
Beneficial mutations, 553
Berg, Paul, 293
Berger, Hans, 418
Bernstein, Julius, 418
Best, Charles, 379–80
Beta blockers, 390
β-carotene, 152
β-galactosidase, 255
β-lactamase, 561–62
β-oxidation, 118
β-pleated sheets, 44fig., 45
Bicarbonate–carbon dioxide buffer system, 355–56
Bicarbonate ions, 22
Bile salts, 38
Biochemists, 125
Biodegradation, 735, 736–37
Bioengineering, 489–90
Bioethicists, 304
Biogeography, 522
Bioinformaticians, 304
Biological warfare, 488–89
Biology, careers in, 125, 304, 491, 627, 745
Biomass, 193
Biomaterials industry, 24
Biomedical engineering, 24
Bioremediation, 276, 734–35
Biotechnology
    agricultural applications, 306–7
    food supplies and, 721
    forensic applications, 307–8
    medical applications, 305–6
Biotic potential, 662
Birds, collected by Darwin, 522–23
Birth control pills, 397
Birth process, in humans, 336
Biston betularia, 531
Black death. *See* Bubonic plague
Blackman, F.F., 150–51
Black widow spiders, 505–6
Blood
    antifreeze in, 340–41
    calcium levels in, 385–86
    collection of, 248
    transfusions, 248
Blood-brain barrier, 427
Blood cells, 464
Bloodletting, 623
Blood pressure, and kidneys, 354–55

Blood sugar
    hormones affecting, 378–83
    stress and, 388
Blunt ends, 279
Body odour, and testosterone, 394
Bohr, Niels, 11
Bond energy, 59
Bonding capacity, 26
Bone marrow, 464
Borrelia burgdorferi, 758
Bottleneck effect, 551
Botulism, 480
Bovine somatotropin (BST), 294, 307
Bovine spongiform encephalopathy (BSE), 460
Bowden, Lori, 332
Bowman's capsule, 347, 351table
Boyer, Herbert, 293
Bradykinin, 473
Brain, 428–30
    mapping, 430–33
    nerves within, 414
    neuroimaging, 433
Brazil nuts, 605
Breast cancer, 626
Breathing movements, 435
Breeding stock, choice of, 525
Bright's disease, 357–58
Brønsted–Lowry concept, 21
Brown fat, 339
BSE (bovine spongiform encephalopathy), 460
BST (bovine somatotropin), 307
Bt toxin, 306
Bubonic plague, 476–77
Buffers, 20, 22
Buffer systems, of blood, 355–56
Buffon, Georges de, 517, 518
Burgess Shale, 595

C

Caffeine, 389, 392, 408, 498
Calcitonin, 385
Calcium, 10
    levels in blood, 385–86
Calcium chloride transformation protocol, 286, 287–88
Calories, 35
Calorimeters, 111
Calvin, Melvin, 9, 161, 163
Calvin cycle, 147, 161–64
Cambrian explosion, 595
Canadian Council on Animal Care, 658
Cancer
    breast, 626
    as genetic disease, 261
    skin, 261
Cannon, Walter, 335
Cape Verde Islands, 521
Capons, 398
Capra, Fritjof, 743

Carbohydrates, 27–28, 29–34, 117
Carbon-12, 8, 9
Carbon-13, 8
Carbon-14, 8, 9
Carbon (C), 8, 24, 25–27
Carbon dioxide
    emissions, 730
    molecules, 18
Carbon fixation, 147, 168–72
Carbonic acid–bicarbonate buffers, 22
Carbon tetrachloride, 16
Carbonyl group, 25
Carboxyl group, 25, 27, 41
Carboxyl terminus, 44
Carotenoids, 152
Carrying capacity, of ecosystems, 660, 732–33
Carson, Rachel. *Silent Spring*, 399
Castration, 393
Catabolic reactions, 28
Catabolism
    lipids, 118
    proteins, 117
Catalase, 562
Catalysts, 69
Cataracts, of eye, 443
Catastrophism, 513
Cations, 12
Cech, Thomas, 588
Cell phones, 262
Cells, 6
    blood, 464
    competent, 287
    freezing of, 340–41
    host, 287
    organelles and, 264–65
    pluripotent, 471
    stem. *See* Stem cells
Cellular proteins, 22
Cellular respiration, 65, 88–116. *See also*
    Respiration
    energy transfer in, 95–96
    photosynthesis vs., 179–83
Cellulose, 31, 32–33
Cenozoic era, 595–98
Central nervous system (CNS), 412–13, 427–34
    inhibitory impulses in, 425
Centromeres, 266
Cerebellum, 429–31
Cerebral cortex, 428, 431
Cerebral palsy, 500
Cerebrospinal fluid, 427
Cerebrum, 428
Chaperone proteins, 50
Character displacement, 680
Chargoff, Erwin, 210, 211
Chase, Martha, 208–9
Chaussy, Christian, 358
Cheese, 75–76
Chemical bonding, 11–13
Chemical messengers, 373, 412

Chemical therapy, 484–85
Chemiosmosis, 106–8, 158fig., 159–60
Chemoautotrophism, 589
Chemoautotrophs, 90
Chemotaxis, 463
Chicxulub Crater, 598
Chimpanzees, 615
Chitin, 31, 33–34
Chlamydia pneumonia, 497
Chlorella, 150, 163
Chlorophyll, 138, 139, 151, 152–53
Chloroplasts, 140–41, 143–44
    DNA in, 207
Cholera, 92, 476
Cholesterol, 38
Cholinesterase, 423–24
Choroid layer, 439
Chromatin, 266
Chromatography, 153, 163–64
Chromosomes, 266
Chymosin, 75
Cicadas, 576
Circulatory system, of animals, 18
Citrate, 112
Citrulline, 235
Clades, 606
Cladistics, 606–7, 608–10
Cladograms, 606–9
Cleaning industry, enzymes used in, 76
Cloning, 287
    of DNA, 290–91
Closed population, 662
Clostridium botulinum, 488
Clumped dispersion, 652
Cochlea, 446, 447
Coding strand, 242
Codons, 240
Coenzymes, 72
Coevolution, 605
Cofactors, 72
Cohen, Stanley, 293
Cold stress, human response to, 338–41
Cole, K.S., 418
Collagen, 46
Collecting ducts, 347, 351table
Coloration, for survival, 33
Combustion, 60
Commensalism, 685
Communities
    competition and, 677–80
    interaction within, 676–77
    symbiosis in, 677, 684–88
Community, defined, 676
Competent cells, 287
Competition between populations, 677–80
Competitive inhibitors, 73
Complementary base pairing, 213
Complementary DNA (cDNA), 232
Complement proteins, 466
Compounds, 12
Computerized tomography (CT), 433
Condensation reactions, 28

Cones, 440
Conformation, 41
Connor, Jim, 380
Conservation biologists, 627
Constant-length strands, 296
Contact lenses, 444
Continuous ambulatory peritoneal dialysis (CAPD), 358
Contractile vacuole, 343–44
Convergent evolution, 602
Cooperation, and evolution, 569
Copy numbers, 286
Coral reefs, 521
Corepressors, 257
Corn
    oxygen and, 165
    transgenic, 306
Cornea, 439, 442
    surgery on, 443–44
Corpus callosum, 428
Corpus luteum, 396
Cortex, 346
Corticotropin, 381
Cortisol, 374
Cosmic rays, 261
Cotton plants, 306
Coupled transcription–translation, 264
Covalent bonds, 12, 14
Cowpox, 483
Cranial nerves, 435
Crassulacean acid metabolism (CAM), 170–71
Creation stories, 510
Creutzfeldt-Jakob disease (CJD), 293, 460
Crick, Francis, 204, 211, 212
Cristae, 99
Crops. *See also* Agriculture
    remote sensing technology, 167
    transgenic, 306–7
    watering of, 726
Crude density, 651
Cryptomonads, 265
C_3 photosynthesis, 162
C_4 photosynthesis, 168–70
C_4 plants, 168–70
Cumulative effects, 741
Cumulative selection, 565–70
Curds, 75
Curie, Pierre, 513
Curtis, H.J., 418
Cuticle, on leaves, 141
Cutin, 39
Cuvier, Georges, Baron, 513
Cyanide, 105
Cyanobacteria, 139–40
Cyborgs, 329
Cyclic adenosine monophosphate (cyclic AMP), 52, 374–75
Cyclic electron flow, 160–61
Cystic fibrosis, 259, 261–62, 305
Cystic fibrosis transmembrane regulator (CFTR) gene, 259

Cytokinesis, 217
Cytosine (C), 52, 53, 204, 210, 213

D

Dacron, 24
Darwin, Charles, 517, 519–21, 522–27, 529–31, 573
Darwin, Erasmus, 517, 518
Daughter isotopes, 514
Da Vinci, Leonardo. See Leonardo da Vinci
Dawkins, Richard, 569
Dawson, Sir William, 517
Deamination, 117, 342
Death, 122, 434
De Buffon, Georges. *See* Buffon, Georges de
Deer ticks, 758
Defence
    lines of, 462–65
    mechanisms, 682–84
Dehydration synthesis, 28
Delbruck, Max, 218
Deletions, 259
Democritus, 441
Demographers, 745
Demographic transition model, 707–8, 710–11
Demography, 705
Denaturation, 49–50
Dendrites, 414
Density-dependent factors, 671–73
Density gradient centrifugation, 218
Density-independent factors, 674–75
Deoxyribonucleic acid (DNA), 52–54. *See also* Double helix; *and headings beginning* DNA
    building complementary strands, 220–21
    chemical composition, 210–15
    cloning of, 290–91
    complementary (cDNA), 232
    databases, 308
    directionality, 213–15
    discovery of, 206
    helical structure. *See* Double helix
    as hereditary material, 207–9
    palindromic sequences, 277, 278
    recombinant. *See* Recombinant DNA
    repair, 222
    RNA vs., 237, 238
    strand separation, 219–20
    structure, 210–15
    as transforming principle, 207–9, 227
    upstream of, 242
Deoxyribonucleoside triphosphates (dNTPs), 221, 301–2
Deoxyribose, 52
    sugars, 210, 213–14, 237, 238fig.
Depolarization, 419, 420fig., 421
Depressant drugs, 437
Desalinization, 726–27
Descartes, René, 430
Desertification, 717
Detergents, 76

Index (tab)

Haldane, J.B.S., 586
Half-life, 9, 514
Hallucinogenic drugs, 426
Hammerling, Joachim, 206
Hardy, Godfrey, 547, 548
Hardy–Weinberg principle, 547–49
Harmful mutations, 553
Hawaiian islands
    adaptive radiation and, 604
Hearing, 445–46
    damage to, 449
    equilibrium and, 448
    sound and, 446–47
Heat stress, human response to, 8, 338
Helicobacter pylori, 497
Helium, 12
Helix
    alpha. *See* α helix
    double. *See* Double helix
Helix aspersa, 576
Helmont, Jan Baptiste van, 149
Helper T cells, 245–47, 469
Hemodialysis, 358–59, 361
Hemoglobin, 18, 41, 46, 49
    amino acid sequence of, 235–36
Hemophilia, 305
Henslow, John, 519
Hereditary information, location of, 206–7
Hereditary material, DNA as, 207–9
Heroin, 436–37
Hershey, Alfred D., 208–9
Heterotrophism, 589
Heterotrophs, 90, 136
Heterozygosity, 544
Hexose sugars, 29
Highly active antiretroviral therapy
    (HAART), 247
Hill reaction, 156
*Hind*II, 280
Hippopotamuses, 613
Histamine, 473
Histones, 266, 553
Hoary bat (*Lasiurus cinerus cinerus*), 528
Homeostasis, 349
    autonomic nervous system and, 435
    control systems, 334–37
    defined, 334–35
    discovery of, 335
    feedback systems, 335–36
Hominids, 617
Hominoids, 616
Homo
    erectus, 618, 619
    ergaster, 618
    habilis, 585, 618, 619
    heidelbergensis, 618
    neanderthalensis, 515, 618, 619, 621
    rudolfensis, 622
    sapiens, 515, 618, 619, 622
Homologous features, 523–24
Homoplasies, 602
Homozygosity, 544

Honeycombs, 39
Honeycreepers, 604
Hooke, Robert, 511
Hooker, J.D., 523, 529
Hoppe-Seyler, Felix, 207
Hormones, 412
    affecting blood sugar, 378–83
    affecting metabolism, 384–87
    defined, 372
    endocrine, 372
    follicle-stimulating, 394
    gonadotropic, 394
    gonadotropin-releasing, 394
    luteinizing, 394
    metamorphosis and, 402
    nontarget, 372
    peptide, 394
    reproductive, 393–99
Horses, 524
Host cells, 287
Housekeeping genes, 255
Human DNA sequencing, 204
Human Genome Project, 204, 232, 302, 308
Human genome sequencing, 204
Human immunodeficiency virus (HIV),
    245–48, 298, 305, 467, 476, 480–81, 488
Human population(s), 696
    agriculture and, 700–702
    demographics, 705–12
    disease and, 702
    future trends, 711–12
    growth, 699–704
    growth momentum, 706
    medicine and, 703
    science and, 703
    water content within body, 16
Hummingbirds, 116, 558
Hunger, 714, 715
Hunter-gatherers, 699
Huntington's Chorea, 305
Huperzine A, 459
Hutton, James, 517
Huygens, Christiaan, 441
Hybridization, 299
    of valence orbitals, 15
Hybridomas, 487
Hybrids, 573
Hydrocarbons, 25
Hydrogen, 8, 16
Hydrogen bonds, 17, 213
Hydrogen molecules, 13
Hydrogen sulfide, 16
Hydrolysis, 75
Hydrolysis reactions, 28
Hydrolyzation, 47
Hydronium ions, 20, 21
Hydrophilia, 19
Hydrophobia, 19
Hydroxide ions, 20
Hydroxyl group, 25, 27
Hyla, 571fig., 574
Hymenoptera, 569

Hyperglycemia, 379
Hyperopia, 443
Hyperplasia, 386
Hyperpolarization, 424
Hyperthyroidism, 384
Hypertrophy, 386
Hyperventilation, 22
Hypoglycemia, 379
Hypothalamus, 338–39, 429
    stress and, 388
Hypothermia, 340

I

Ice, 340–41
Ichthyosaurus, 585
Iguanas, fever and, 623–25
Immiscibility, 18
Immune system
    malfunctions, 473–75
    memory of, 470
    response, 466–72
Immunity
    active and passive, 482–90
    induced, 482–90
Immunoglobulins, 41
Immunologists, 491
Immutability, of life forms, 510
Incus, 445, 446
Indicators, 732
Indicator species, 732
Induced-fit models, 70
Induced mutations, 261
Inducers, 256
Infections, 477
Inflammatory response, 463
Influenza, 497
Ingenhousz, Jan, 150
Ingram, Vernon, 235–36
Ingroups, 608
Inheritance
    adaptation and, 517–18
    evolution and, 525
Inhibin, 394
Inorganic chemistry, 24
Insecticides, 424
Insect pollination, 567–69
Insertion, of nucleotide, 259
Insulin, 45, 253, 255, 269, 285, 293, 357, 372,
    373, 378–79, 380
Insulin-like growth factors, 386–87
Interference competition, 677
Intergovernmental Panel on Climate Change
    (IPCC), 729
Intermembrane space, 99
Intermolecular bonds, 17
International Olympic Committee (IOC),
    391–92
Interneurons, 413
Interspecific competition, 677–80
Interstitial fluid, 351
Intramolecular forces of attraction, 13

Methionine, 41–42, 251–53, 264
Methylases, 281
Methyl group, 281
Mexico City, 742–43
Micelles, 37
Microarray technology, 232, 308
Microcystis aeruginosa, 139–40
Microevolution, 571
Microfibrils, 32–33
Microfossils, 512
Microparasites, 685
Microsatellites, 298
Microscopy, 373
Miescher, Friedrich, 204, 206, 207
Milk
    homogenized, 7
    production, 307
Miller, Freda, 471
Miller, Stanley, 586–87
Milstein, Cesar, 487
Mineralocorticoids, 381, 382fig.
Minimum viable population size, 673
Minkowski, Oscar, 373
Miscibility, 18
Missense mutation, 259
Mitchell, Peter, 106–7
Mitochondria, 88, 99–100, 207, 264
Mitosis, 217
Moa (bird), 604
Molecular biologists, 125
Molecules
    carbon dioxide, 18
    hydrogen, 13
    nitrogen, 13
    oxygen, 13, 18
    polarity, 16, 18–19
    radiolabelled, 9
    shape, 15
    stability of, 59
    water. *See* Water
Monarch butterflies, 306–7
Moniz, Antonio Egas, 432
Monoclonal antibodies, 487
Monocytes, 463
Monophyletic groups, 606
Monosaccharides, 29–30
Morin, Guy Paul, 308
Motor cortex, 428
Motor neurons, 413
mRNA transcript, 244
Mules, 573fig.
Mullis, Kary, 298
Multidrug-resistant tuberculosis, 562–63
Multiple-cloning sites, 286
Multiple sclerosis (MS), 414, 474–75
Muscle fatigue, 89
Mutagenic agents, 261
Mutations, 552–53
    causes of, 261–62
    evolution of complex features by,
      565–67
    by HIV ?, 247

    inheritance and, 531
    types of, 259–61, 553–54
Mutualism, 684
Myasthenia gravis, 424
Mycena, 60
Mycobacterium tuberculosis, 561, 562
Myelin-associated glycoprotein (MAG), 415
Myelin sheath, 414
Myopia, 443, 625–26
Myxedema, 384

N

N-acetylglucosamine, 33, 34fig.
Nanometer (nm), defined, 211
Napaeozapus insignis, 694
Natural-product research chemists, 627
Natural selection, 529–33, 565
Nearsightedness, 443
Negative feedback, 335–36
    in control of thyroid hormones, 385
    in male reproductive system, 394
Negative ions, 419, 421
Neogloboquadrina pachyderma, 516
Neon, 12
Nephridiopores, 344
Nephritis, 357–58
Nephrons, 346–47
    andiuretic hormone (ADH) and, 354
Nephrostome, 344
Nerve cells, 413–14
Nerves
    cranial, 435
    optic, 415
    reconnection of severed, 415–16
Nervous system(s), 412–13
    stress and, 388
Neural circuits, 416
Neurilemma, 414
Neuroglial cells, 413
Neuroimaging, 433
Neurons, 413, 419
Neurospora crassa, 234–35
Neurotransmitters, 423–25
Neutralization reaction, 20
Neutral mutations, 552–53
Neutrons, 8
Neutrophils, 463
Newton, Sir Isaac, 441
Niagara Falls, 517
Nicotinamide adenine dinucleotide
    (NAD+), 52, 72
Nicotinamide adenine dinucleotide
    (NAD+), 95–96
Nicotinamide adenine dinucleotide phos-
    phate (NADP+), 72
Nile perch (fish), 759
Nitrogen-14, 8
Nitrogen molecules, 13
Nitrogenous base, 210
Noble gases, 12
Nodes of Ranvier, 414

Noncompetitive inhibitors, 73
Noncyclic electron flow, 158, 159–60
Nonpolar covalent bond, 14
Nonsense mutation, 259
Noradrenaline. *See* Norepinephrine
Norepinephrine, 381, 410, 425
Nuclear medicine, 9–10
Nucleic acids, 27–28, 28, 52–54
Nuclein, 206
Nucleomorphs, 265
Nucleosomes, 266
Nucleotides, 45, 52–54, 210, 233
Nucleus, of atom, 8, 11
NutraSweet, 56
Nutriceuticals, 51

O

Obligate aerobes, 92
Obligate anaerobes, 92
Obligatory mutualism, 684
Oil spills, 276
Okazaki fragments, 221
Oleic acids, 35, 36
Olfactory lobes, 428
Oligosaccharides, 30
One gene–one polypeptide hypothesis,
    234–36
Oocyte, 395
Oogenesis, 395–96
Oparin, Alexander, 586
Open population, 662
Operators, 255
Operons, 255
    lac, 255–56, 258table
    trp, 256–57, 258table
Ophioglossum reticulatum, 554
Opiates, 436–37
Optic nerves, 415
Orangutans, 615
Orbitals, 11–12
Orb-web spiders, 570
Orchids, 573
Organ donation, 471
Organelles, and cells, 264–65
Organic chemistry, 24
Organ of Corti, 447
Organ transplants, 471
Ornithine, 235
Osmoreceptors, 353
Ossicles, 445, 446
Osteoblasts, 10
Osteoporosis, 10
Otoliths, 448
Outgroups, 608
Oval window, 445
Ovaries, 395–96
Ovulation, 395–96, 398
Owen, Richard, 522
Owls, great gray, 758
Oxidation, 66–67, 91
    β–, 118

▶ Credits

Photo Credits

Unit Opening Photos: Unit 1: Spektrum Akademischer Verlag, Heidelberg; Unit 2: Meckes/Ottawa/Photo Researchers Inc; Unit 3: courtesy Ming Chen, University of Alberta; Unit 4: ©Greg Harold/Auscape International; Unit 5: ©Nada Pecnik/ Visuals Unlimited.

Chapter 1: 5 (a) courtesy of M.M. Perry and A.B. Gilbert; (e) ©Larry Stepanowicz/ Visuals Unlimited; 7 ©Doug Gibbon Photography; 9 ©SIU/Visuals Unlimited; 10 ©Photo Researchers Inc; 24 (a) ©James Stephenson/Photo Researchers, Inc; (b) David Campione/SPL/Photo Researchers, Inc; (c) Princess Margaret Rose Orthopaedic Hospital/SPL/ Photo Researchers, Inc; (d) SIU/Visuals Unlimited; 31 ©Dr. Jeremy Burgess/SPL/Photo Researchers, Inc; 32 ©Don W. Fawcett/Visuals Unlimited; 33 ©Biophoto Associates/Science Source/Photo Researchers, Inc; 34 (b)Dave B. Fleetham/ Visuals Unlimited; (c)Phil A. Dodson/Photo Researchers Inc; 35 ©Rod Planck/Photo Researchers, Inc; 36 ©Dick Hemingway; 39 t ©Wm Ober/Visuals Unlimited; (bottom left) Emma Lee/Life File/ PhotoDisc; (bottom right) ©Scott Camazine/Photo Researchers, Inc; 40 (top) Greg Kuchik/PhotoDisc; (left) C-Squared Studios/PhotoDisc; (middle left) ©P.Motta/ Photo Researchers, Inc; (middle right) ©John Gerlach/Visuals Unlimited; (right) ©John D. Cunningham/Visuals Unlimited; 45 ©John Gerlach/Visuals Unlimited; 49 ©Andrew Syred/SPL/Photo Researchers; 51 ©Dick Hemingway; 60 ©Photo Researchers, Inc; 61 ©Dick Hemingway; 70 © Visuals Unlimited.

Chapter 2: 89 ©Mugshots/Corbisstockmarket.com; 92 (a) George J. Wilder/Visuals Unlimited; (b) M. Kalab/Visuals Unlimited; 93 (a) ©Dick Hemingway; (b) ©Gary W. Carter/Visuals Unlimited; (c) ©Visuals Unlimited; (d) ©A.M. Siegelman/Visuals Unlimited; (e) ©Sherman Thomson/Visuals Unlimited;100 ©D. Fawcett/Visuals Unlimited; 116 ©Jack Milchanowski/Visuals Unlimited; 119 (left) ©Nick Gunderson/PhotoDisc; (right) ©Craig Brewer/ PhotoDisc; 121 (Fig 5) CP Archive/AP Photo/Amy Sancetta; (Fig 6) UIC Photo Services; 125 (top left) John A. Rizzo/PhotoDisc; (top right) Kim Steele/PhotoDisc; (bottom left) Keith Brofsky/PhotoDisc; 134 ©T. Kanariki-D. Fawcett/Visuals Unlimited.

Chapter 3: 137 ©Fritz Polking/Visuals Unlimited; 138 (a)©Inga Spence/Visuals Unlimited; (b) Eric V. Graves/Visuals Unlimited; (c)M.Abbey/Visuals Unlimited; (d) ©J. Richardson/Visuals Unlimited; 139 ©Photo Researchers,Inc; 140 (Fig 4a) ©John D. Cunningham/Visuals Unlimited; (b) ©A.Siegelman/Visuals Unlimited; (Fig 5) ©Stephen/Sylvia Sharnoff; (Fig 6a) ©Visuals Unlimited; (b) ©Visuals Unlimited; 141 (a) ©Jerome Wexler/Visuals Unlimited; (b) ©Walt Anderson/Visuals Unlimited; (Fig 8) ©D. Cavagnaro/Visuals Unlimited; 142 ©David M. Phillips/Visuals Unlimited; 144 (a) ©C. Gerald Van Dyke/Visuals Unlimited; (c)George Chapman/Visuals Unlimited; 149 (Fig 5) ©CORBIS/MAGMA; © courtesy of Stanford University Archives; 153 (a) ©Visuals Unlimited; (inset) ©Jerome Wexler/Visuals Unlimited; (b) ©Visuals Unlimited; (inset) ©Visuals Unlimited; 155 Craig Brewer/PhotoDisc; 156 ©Christine Case/Visuals Unlimited; 161 ©Ted Spiegel/CORBIS/MAGMA; 163 ©Lawrence Berkeley National Library; 164 courtesy Lawrence Berkeley National Library; 165 ©RMF/Visuals Unlimited; 170 (a) ©Photo Researchers, Inc; (b) ©Visuals Unlimited; 171 (left, middle) Eyewire Collection; (right) ©Visuals Unlimited; 176 (a) ©Wally Eberhardt/Visuals Unlimited; (b) ©David Sieren/Visuals Unlimited; 177 Dick Hemingway; 186 Dave Starrett.

Chapter 4: 202 (left) Custom Medical Stock Photo; (right) Canadian Cancer Society; 203 (Fig 3,4) ©Dick Hemingway; (Fig 5)

©Jackson/Visuals Unlimited; 206 Novartis Research Foundation/ Friedrich Miescher Institute; 208 ©Oliver Meckes/E.O.S./MPI-Tubingen/Photo Researchers, Inc; 211 (Fig 4) ©Barrington Brown/ Photo Researchers, Inc; (Fig 5) Omikron/Photo Researchers, Inc; 212 By courtesy of the National Portrait Gallery, London.

Chapter 5: 233 ©M. Abbey/Visuals Unlimited; 234 ©Glenn M. Oliver/Visuals Unlimited; 236 (a) ©Visuals Unlimited; (b) ©Stanley Flegler/ Visuals Unlimited; 242 R.C. Williams, Proc. Nat. Academy of Science, 74 (1977): 2313; 245 ©Science Photo Library/Photo Researchers; 247 (a) ©Prof. Luc Montagnier, Institut Pasteur/CNRI/Science Photo Library/Photo Researchers Inc; (b & c) NIBSC/Science Photo Library/ Photo Researchers Inc; 261 (Fig 2) ©Bettmann/CORBIS/ MAGMA; (Fig 3) ©Ken Greer/Visuals Unlimited; 262 (Fig 4) Daisuke Morita/PhotoDisc; 264 ©Oscar L. Miller Jr./ Visuals Unlimited; 270 (Fig 1) ©Jack Griffith; (Fig 2) ©Dr. Oscar L. Miller/ Visuals Unlimited; (Fig 3) R.C. Williams, Proc. Nat. Academy of Science 74 (1977): 2313; 271 (Fig 4) ©Visuals Unlimited; (Fig 5) ©Visuals Unlimited.

Chapter 6: 277 ©Vanessa Vicky/Photo Researchers, Inc; (inset) ©Visuals Unlimited; 284 ©David Parker/SPL/Photo Researchers Inc; 284 ©Jack Bostrack/Visuals Unlimited; 304 (top left) Scott T. Baxter/PhotoDisc; (top right) ©William Taufic/First Light.ca; (bottom left) ©William Taufic/First Light.ca; (bottom right) Ryan McVay/PhotoDisc; 306 (Fig 2) Dept. of Plant Pathology, Cornell University; (Fig 4) ©Visuals Unlimited; 307 (Fig 3) CP Photo/Moe Doiron; Custom Medical Stock Photo; 316 ©Photo Researchers Inc; 322 (top) ©Al Foley/ Visuals Unlimited; (bottom left) ©Gordon E., Smith/Photo Researchers, Inc; (bottom right) Dick Hemingway.

Chapter 7: 331 (Fig 2) Dick Hemingway; (Fig 3) ©Photo Researchers, Inc; 333 (top) ©Mark Oleskyn; (bottom) Janice Palmer; 338 ©Richard Hermann/ Visuals Unlimited; 339 (left) CP Photo/ Andrew Vaughan; (right) Steve Mason/ PhotoDisc; 340, 341 courtesy Kenneth Storey; 343 ©Cabisco/ Visuals Unlimited; 358 ©David M. Phillips/Photo Researchers, Inc; 360 AP Photo/ho/PPI Therapeutics; 361 Brad Nelson/CMSP.

Chapter 8: 371 CP Photo/Paul Chiasson; 372 ©C.O.R.E. Digital Pictures Inc; 373 ©Dee Breger, Lamont-Doherty Earth Observatory; 379 ©Dick Hemingway; 380 (Fig 3) courtesy Banting House National Historic Site; (Fig 4) courtesy of the Clinical Islet Transplant Program; 384 © Biophoto Associates/ Science Source /Photo Researchers, Inc; 385 ©Ken Greer/ Visuals Unlimited; 386 Biophoto Associates/Science Source/Photo Researchers, Inc; 388 ©Bettmann/CORBIS/MAGMA; 391 ©CP Photo/ Peter Dejong; 393 ©G. Shih-R. Kessel/Visuals Unlimited; 395 ©David M. Phillip/Visuals Unlimited; 402 (top) ©Gilbert L. Twiest/ Visuals Unlimited; (middle) ©Gustav Verderber/ Visuals Unlimited; (bottom) ©Gilbert L. Twiest/ Visuals Unlimited.

Chapter 9: 411 CP Photo/ Kevork Djansezian; 414 G.W. Willis, MD/Visuals Unlimited; 415 CP Photo/ Clifford Skarstedt; 416 ©Will & Deni McIntyre/Photo Researchers Inc; 418 ©SIU/ Visuals Unlimited; 431 CP Photo; 432 Damasio H, Grabowski T, Frank R, Galaburda AM, Damasio AR: The Return of Phineas Gage: Clues about the Brain from a famous patient. Science, 264:1102-1105, 1994. Department of Neurology and Image Analysis Facility, University of Iowa; 433 ©Science Photo Library/Photo Researchers Inc; 440 ©Ralph C. Eagle/Photo Researchers, Inc; 444 Bill Kamin/Visuals Unlimited; 449 (Fig 6) Ryan McVay/ PhotoDisc; Fig 7a,b) ©Prof. P. Motta/A. Caggiati/ University "La Sapienza," Rome/SPL/ Photo Researchers Inc; (Fig 7 c) Robert Preston and Joseph E. Hawkins, Kresge Hearing Research Institute, University of Michigan; 459 ©Simon Fraser/SPL/Photo Researchers, Inc.

Chapter 10: 461 (Fig 1) ©John Guistina/PhotoDisc; (Fig 2) Image courtesy of Dr. R. Higgins, School of Veterinary Medicine, University of California at Davis; 462 ©Science VU/Jim Deleon; 463 ©RMF/Visuals Unlimited; 464 ©Juergen Berger/Max-Planck Institute/SPL/Photo Researchers Inc; 470 ©Dr. Andrejs Liepins/Science Photo Library/Photo Researchers, Inc.; 476 ©CAMR/A.B. Dowsett/SPL/Photo Researchers, Inc; 477 (Fig 2) ©Nicole Duplaix/CORBIS/MAGMA; (Fig 3) ©Kent Wood/Photo Researchers, Inc; (Fig 4) ©Larry Stepanowicz/ Visuals Unlimited; 479 ©Bettmann/CORBIS/MAGMA; 484 ©Bettmann/CORBIS/MAGMA; 485 ©Bettmann/CORBIS/MAGMA; 488 Visuals Unlimited; 491 (top left) ©Dick Hemingway; (top right) ©SIU/Visuals Unlimited; (bottom left) ©Melanie Carr/CMSP; (bottom right) ©Jeff Greenberg/Visuals Unlimited; 498 (top left) ©Kevin & Betty Collins/Visuals Unlimited; (top right) ©T.E. Adams/Visuals Unlimited; (bottom left) ©Mike Abbey/Visuals Unlimited; (bottom right) ©T.E. Adams/Visuals Unlimited.

Chapter 11: 507 (a) ©Gary W. Carter/Visuals Unlimited; (b) ©Charles Philip/Visuals Unlimited; (c) ©Gerald & Buff Corsi /Visuals Unlimited; (d) ©Gerald & Buff Corsi/Visuals Unlimited; (e) ©Bill Banaszewski /Visuals Unlimited; (f) ©Barbara Gerlach /Visuals Unlimited; (Fig 4) ©Cheryl A. Ertelt/Visuals Unlimited; 508 ©Jeanne White/Photo Researchers, Inc; 509 ©A.J. Copley/Visuals Unlimited; 510 MARY EVANS PICTURE LIBRARY; 511 (a) ©Dean Conger/CORBIS/ MAGMA; (b) ©National Museum of Kenya/Visuals Unlimited; (c) ©Ken Lucas/Visuals Unlimited; 512 ©Jeff J. Daly/Visuals Unlimited; 513 ©David Parker/Science Photo Library/Photo Researchers, Inc; 515 (left) ©A.J. Copley/ Visuals Unlimited; (right) ©Mark Newman/ Visuals Unlimited; 516 ©James Amos/Photo Researchers, Inc; 517 ©John Sohlden/Visuals Unlimited; 519 (Fig 1) Visuals Unlimited (Fig 2) The Natural History Museum, London; 520 ©Jeff Gage, courtesy The Florida Museum of Natural History; (b) Joe McDonald/Visuals Unlimited; 521 (a) ©Michael Maslan Historic Photographs/CORBIS/MAGMA; (b) ©Inga Spence/Visuals Unlimited; (Fig 6) NASA/Science VU; 529 ©CORBIS/MAGMA; 530 ©Joe McDonald/Visuals Unlimited; 533 (Fig 5) ©Stephen Dalton/Photo Researchers, Inc; (Fig 6) ©Joe McDonald/Visuals Unlimited; 536 ©Rob and Ann Simpson/Visuals Unlimited; 541 Korite International.

Chapter 12: 542 (a) Doug Fraser; (b) Christopher J. Crowle/Visuals Unlimited; 543 ©Will & Deni McIntyre/Photo Researchers, Inc; 544 ©Merlin D. Tuttle/Bat Conservation International; 550 ©Rob & Ann Simpson/Visuals Unlimited; 551 ©Joe McDonald/Visuals Unlimited; 555 ©Gerald 7 Buff Corsi/Visuals Unlimited; 559 courtesy Tom Smith, Center for Tropical Research, University of California; 560 (Fig 10) ©Photo Researchers, Inc; (Fig 11) ©Visuals Unlimited; 561 (Fig 12) ©Merlin Tuttle/ Photo Researchers, Inc; (Fig 13) Simon D. Pollard/Photo Researchers, Inc; 564 (Fig 15) ©Photo Researchers, Inc; 567 (Fig 2) ©Kay Coleman/Visuals Unlimited; (Fig 3) ©Kvell B. Sandved/Visuals Unlimited; (Fig 4) ©Charlie Heidecker/Visuals Unlimited; 568 (a) ©Eyewire Collection (b) ©Photo Researchers, Inc; (Fig 6 left) ©Photo Researchers, Inc; (Fig 7 right) ©Photo Researchers, Inc; 570 (Fig 9) ©Merlin D. Tuttle/Bat Conservation International/ Photo Researchers, Inc; 571 (left) ©Gary Mezaros/Visuals Unlimited; (right) ©Ray Coleman/Visuals Unlimited; ©Doug Fraser; (Fig 3) ©P.Starborn/Visuals Unlimited; (Fig 4) ©Doug Fraser; 573 ©John Eastcott/ The Imageworks; 574 (a) ©Ken Lucas/Visuals Unlimited; (b) ©David Wrobel/Visuals Unlimited; 575 Royal B.C. Museum; 582 ©Leonard Lee Rae/Visuals Unlimited.

Chapter 13: 584 ©Ken Lucas/Visuals Unlimited; 585 ©Ken Lucas/Visuals Unlimited; (inset) ©Visuals Unlimited; 586 ©William J. Weber/Visuals Unlimited; 588 ©Sidney Fox/Visuals Unlimited; 589 (a) ©J. William Schopf, UCLA; (b) ©Science VU; (c) ©Brian Rogers/Visuals Unlimited; 590 ©NASA/SPL/ Photo Researchers, Inc; 592 (a) J. William Schopf, UCLA; (b) George J. Wilder/Visuals Unlimited; (c) J. William Schopf, UCLA; (d) ©Sherman Thomson/Visuals Unlimited; 595 (left) A.J. Copely/Visuals Unlimited; (top right) ©Ken Lucas/Visuals Unlimited; (bottom) ©Ken Lucas/Visuals Unlimited; 598 NASA; 601 ©Charles O'Rear/CORBIS/MAGMA; 602 (a) ©Marty Snyderman/Visuals Unlimited; (b) Dave B. Fleetham/Visuals Unlimited; 605 ©K. B. Sandved/Visuals Unlimited; 613 ©Eda Rogers; 615 (Fig 1) ©Joe McDonald/Visuals Unlimited; (Fig 2a) ©Fritz Polking/Visuals Unlimited; (Fig 2b) ©Gerald & Buff Corsi/Visuals Unlimited; 617 (left) ©Carolina Biological/Visuals Unlimited; (right) ©National Museum of Kenya/Visuals Unlimited; 618 (a) ©Photo Researchers, Inc; (b) ©Photo Researchers, Inc; 619 (a) ©Visuals Unlimited; (b,c) ©Carolina Biological/Visuals Unlimited; 623 ©Drew Conroy/Oxford Farms; 624 ©Joe McDonald/Visuals Unlimited; 627 (top left) ©Inga Spence/Visuals Unlimited; (top right) ©Janice Palmer; (bottom) ©Mark S. Skalney/Visuals Unlimited; 636 (left) ©John D. Cunningham/Visuals Unlimited; (c) ©John W. Trager/Visuals Unlimited.

Chapter 14: 647 ©Rob & Ann Simpson/Visuals Unlimited; 649 A.B.P.L./Photo Researchers, Inc.; 650 ©Arthur Morris/Visuals Unlimited; 651 ©Gary Meszaros/Visuals Unlimited; 652 (Fig 3) ©Nada Pecnick/Visuals Unlimited; (a)© Photo Researchers, Inc; (b) ©Fritz Polking/Visuals Unlimited; (c) ©Will & Deni McIntyre/Photo Researchers, Inc; 653 (Fig 5) ©Tim Hauf/Visuals Unlimited; (Fig 6) ©Bayard H. Brattstrom/Visuals Unlimited; 657 ©Doug Fraser; 660 ©Janice Palmer; 669 (Fig 8) ©Arthur Morris/Visuals Unlimited; (Fig 9) ©Photo Researchers, Inc; 670 ©Photo Researchers, Inc; 671 ©Gary Meszaros/Visuals Unlimited; 672 (Fig 3) ©Steve McCutcheon/Visuals Unlimited; (Fig 4) ©Inga Spence/Visuals Unlimited; 673 (Fig 5,6) ©Ken Lucas/Visuals Unlimited; 674 (Fig 7) ©John Gerlach/Visuals Unlimited; (a,b) ©Janice Palmer; 676 (a) ©Photo Researchers, Inc; (b) ©Gil Lopez-Espina/Visuals Unlimited; (Fig 2) ©Photo Researchers, Inc; 679 (l-r) ©William J. Weber/Visuals Unlimited; ©Gary W. Carter/Visuals Unlimited; ©Rob & Ann Simpson/Visuals Unlimited; ©Jeffrey Howe/Visuals Unlimited; 682 ©Lincoln P. Brower/Sweet Briar College; 683 (Fig 11) ©Photo Researchers, Inc; (Fig 12a) ©Bill Beatty/Visuals Unlimited; (b) ©Rick & Nora Bowers/Visuals Unlimited; (Fig 13) ©Alex Kerstitch/Visuals Unlimited; 684 ©John Gerlach/Visuals Unlimited; 685 (Fig 15) ©Photo Researchers, Inc; (fig 16) ©John D. Cunningham/Visuals Unlimited; 687 ©D.M. Caron, BES/Visuals Unlimited; 688 (Fig 18) ©Photo Researchers, Inc; 691 ©Joe McDonald/Visuals Unlimited; 694 (a) ©Visuals Unlimited; (b) ©William J. Weber/Visuals Unlimited; Joe McDonald/Visuals Unlimited (c) ©Photo Researchers, Inc; (d) ©Joe McDonald/Visuals Unlimited.

Chapter 15: 697 ©Bill Kamin/Visuals Unlimited; (inset top) ©Charles Preitner/Visuals Unlimited; (inset) CP Picture Archive/Pierre Obendrauf; 699 (a) ©D. Cavagnaro/Visuals Unlimited; (Fig 3) ©Bettmann/CORBIS/MAGMA; 701 Eyewire Collection; 708 CP Photo/Kevin Frayer; 714 ©Kent Foster/Visuals Unlimited; 717 ©A. J. Cunningham/Visuals Unlimited; 719 Emanuele Taroni/PhotoDisc; 720 ©Visuals Unlimited; 723 ©Lou Wise; 724 CP Photo/Kevin Frayer; 726 (a) Kent Knudson/PhotoDisc; (b) Karl Weatherly/PhotoDisc; (c) D. Falconer/PhotoDisc; 728 CP Photo/Kevin Frayer; 732 ©Janice Palmer; 733 © H.W. Robinson/Visuals Unlimited; 734 ©David M. Phillips/Visuals Unlimited; 735 ©Doug Fraser; 736 ©Janice Palmer; 738 CP

Photo/Toronto Star/Dave Cooper; 741 NASA; 742 ©John D. Cunningham/Visuals Unlimited; 745 (top left) ©Janice Palmer; (top right) ©PhotoDisc; (bottom) ©Doug Fraser; 754 (top left) CP Picture Archive/ Adrian Wyld; (top right) CP Picture Archive/ Roberto Borea; (bottom left) CP Picture Archive/Larry Steagall; (bottom right) CP Picture Archive/Adrian Wyld; 756 ©Doug Fraser.

Text and Figure Credits

Chapter 10: 471 (Fig 9) Source: Canadian Institute for Health Information (2002). Canadian Organ Replacement Register e-Quarterly Report on Transplant, Waiting List and Donor Statistics: Third Quarter, July 1 to September 30, 2001. http://www.cihi.ca/facts/Q3corrstats.shtml.

Chapter 11: 526 (Fig 6) Source: *Evolutionary Analysis 2/E* by Freeman/Herron, ©2001, p. 69. Reprinted by permission of Pearson Education, Inc., Upper Saddle River, NJ; 527 (Tables 1 and 2) Source: *Evolutionary Analysis 2/E* by Freeman/Herron, ©2001, p. 55. Reprinted by per-mission of Pearson Education, Inc., Upper Saddle River, NJ; 526 (Fig 6) From E.R. Leng (1962). Results of long-term selection for chemical composition in maize and their significance in evolutionary breeding systems. Zeitschrift fur Pflanzenzuchtung 47: 67–91.

Chapter 12: 551 From *Biology: The Unity and Diversity of Life w/CD-ROM, 9th edition,* by C. Starr and R. Taggart ©2001; 556 (Fig 1) Adapted from *Biology, 5th Edition,* fig. 20.4 p. 407. Peter H. Raven and George B. Johnson, the McGraw-Hill Companies, Inc.; 558 (Fig 6) From Canadian Journal of Fisheries and Aquatic Sciences 38: 1636–1656. Fisheries and Oceans Canada, Reproduced with the per-mission of the Minister of Public Works and Government Services Canada, 2002; 559 (Fig 9) From *Biology: The Unity and Diversity of Life w/CD-ROM, 9th edition,* by C.Starr and R. Taggart ©2001. Reprinted with permission of Wadsworth, an imprint of the Wadsworth Group, a division of Thomson Learning. Fax 800-730-2215; 563 (Figure 14) Source: World Health Organization; 574 (Fig 7) Source: Northern Prairie Wildlife Research Center, 1997. http://www.npwrc.usgs.gov/narcam/ idguide/.

Chapter 13: 611 From *Phylogenetic relationships among cetartio-dactyls,* Masato Nikaido, Alejandro P. Rooney and Norihiro Okada. Proc. National Academy of Sciences, USA, Vol. 96 pp. 10261–10266, August 1999 *Evolution;* 619 (Fig 9) After Cavalli-Sforza et al. The History and Geography of Human Genes. Princeton University Press, 1994. Princeton, NJ; 620 (Fig 10) Source: *Evolutionary Analysis 2/E* by Freeman/Herron, ©2001. Reprinted by permission of Pearson Education, Inc., Upper Saddle River, NJ; 624 (Fig 3) From Vaughn et al. (1974) Reprinted with permission from *Nature* 1974, 252: 473-474. MacMillan Magazines Ltd. (Fig 4) Reprinted with permission from Kluger, Ringler and Aver 1975. M.J. Kluger, D. H. Ringler and Aver, M.R.. Fever and Survival. *Science* 188:166-168 Copyright 1975, American Association for the Advancement of Science; 632 (Fig 6) From J.J. Yunis and O. Prakash (1982). The Origin of Man: A Chromosomal Pictorial Legacy, *Science,* 19 March 1982, vol. 215: 1525–1530.

Chapter 15: 709 (Fig 6) Source: U.S. Census Bureau. International Database.